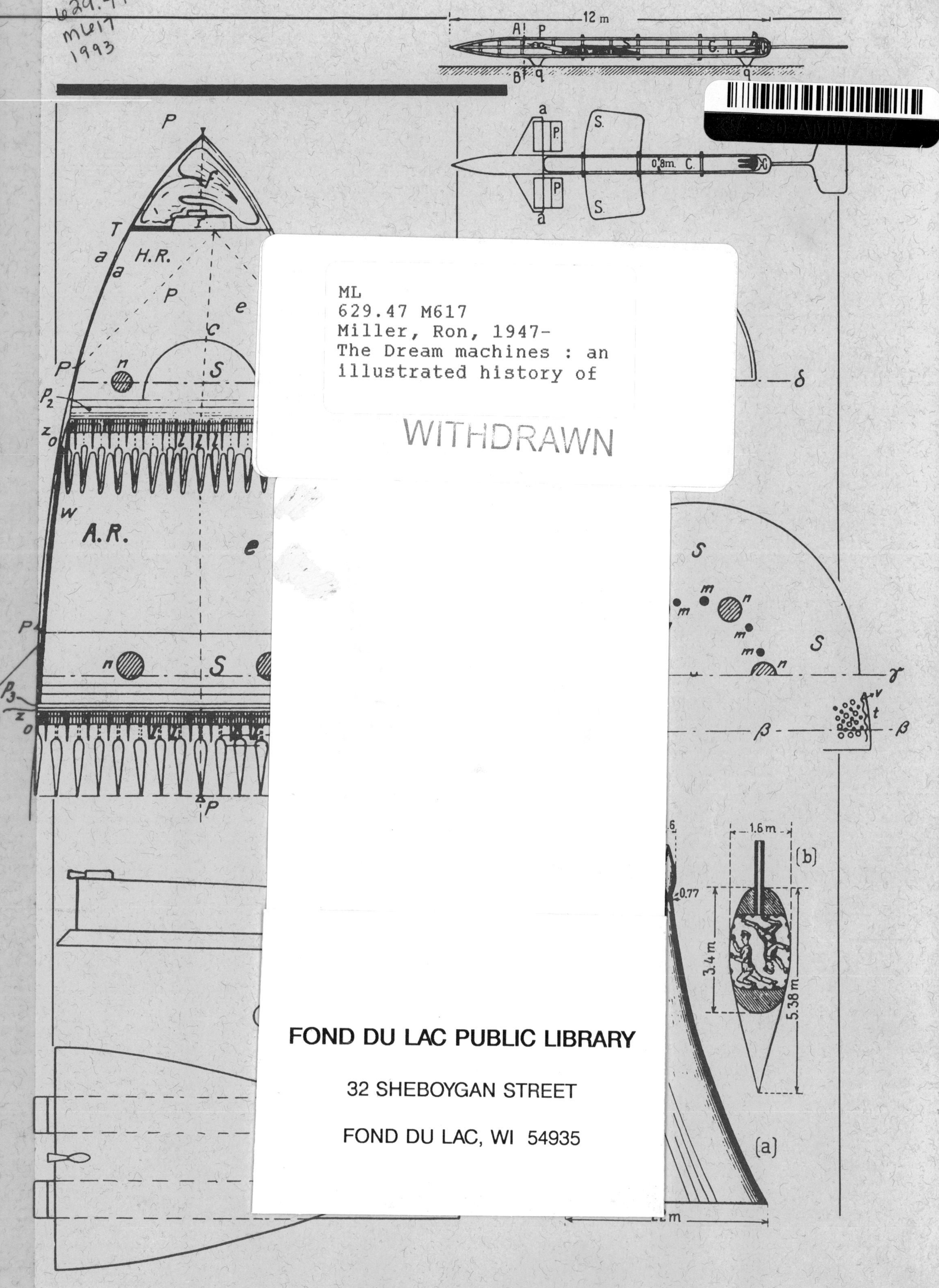
12 m
A
P
B
q
G
a
S.
P.
0.8m.
C.
S.
P
a
H.R.
T
e
c
S
n
P₂
W
A.R.
m
n
β
δ
1.6 m
(b)
0.77
3.4 m
5.38 m
(a)

THE DREAM MACHINES

An Illustrated History of the Spaceship in Art, Science and Literature

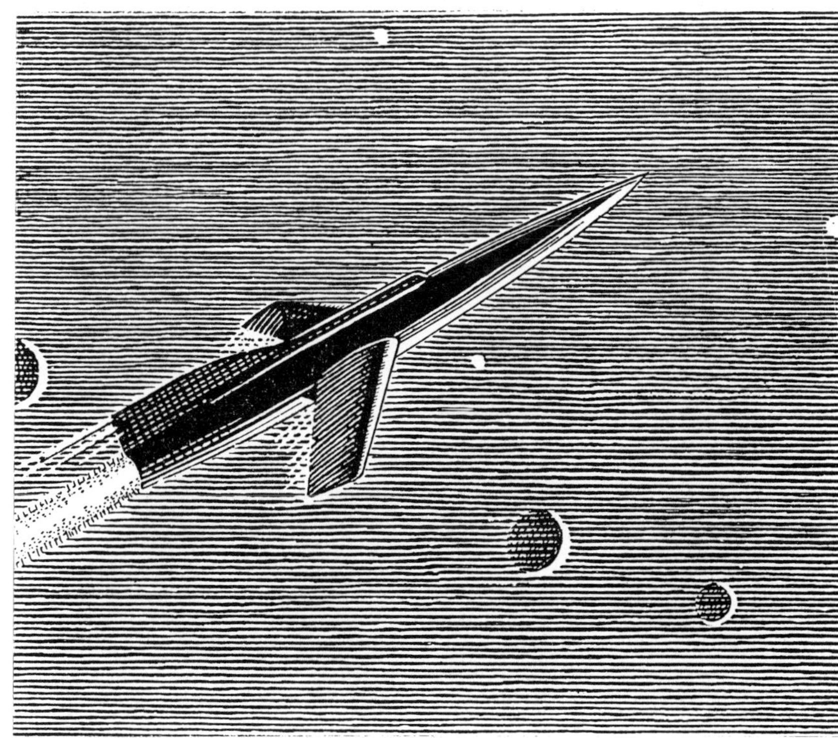

DMP 49

THE DREAM MACHINES

An Illustrated History of the Spaceship in Art, Science and Literature

By Ron Miller

Original illustrations by Ron Miller and Rick Dunning

Foreword by Arthur C. Clarke

KRIEGER PUBLISHING COMPANY
MALABAR, FLORIDA
1993

Original Edition 1993

Printed and Published by
KRIEGER PUBLISHING COMPANY
KREIGER DRIVE
MALABAR, FLORIDA 32950

Library of Congress Cataloging-In-Publication Data

Miller, Ron, 1947-
The dream machines : an illustrated history of the spaceship in art, science, and literature / by Ron Miller : original illustrations by Ron Miller and Rick Dunning : foreword by Arthur C. Clarke.
p. cm.
Includes bibliographical references and index.
ISBN 0-89464-039-9 (alk. paper)
1. Space ships--History. I. Title.
TL795.M54 1992
629.47--dc20 92-2343
CIP

10 9 8 7 6 5 4 3 2

DEDICATION

This book is dedicated to Frederick C. Durant, III, for inspiring my interest in the history of rockets and space travel, and for encouraging me to take notes—and to Pip Durant, for just being Pip.

F. R. Paul, 1930

CONTENTS

Hardy

FOREWORD

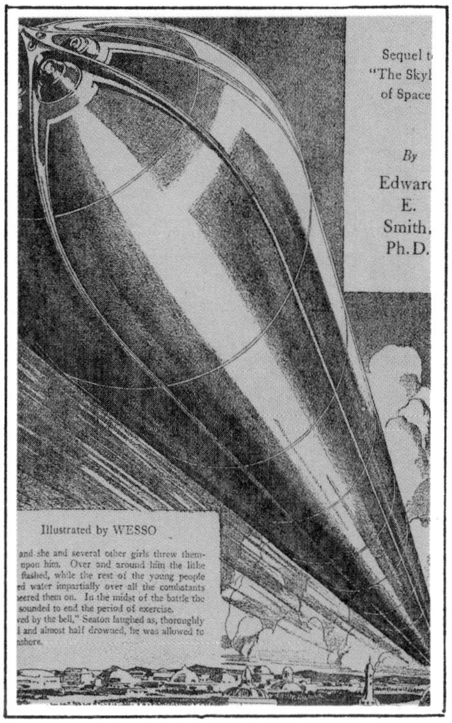

"The power had met the dream"—C. S. Lewis

I have been a designer of spaceships all my life; if my schoolboy exercise books still existed, you'd understand why my nickname was "Spaceship" some 60 years ago. . . . Of course, those schoolboy doodlings were largely fantasies inspired by the science fiction magazines which did so much to fire my imagination. My favorite of all spaceships is still the mile-long torpedo "Wesso" drew for *Skylark Three* in a 1930 *Amazing*. A close runner-up—by the same artist—is the cover of the March 1930 *Astounding Stories* (the first science fiction magazine I ever owned) showing something that looks like a cross between a submarine and a conservatory. They don't make spaceships like that anymore. . . .

It's hard to believe that a few years later I was involved in a much more serious project—the British Interplanetary Society's pre-war design for a lunar spaceship. I still recall with affection and nostalgia those early days in pubs and coffee shops with a handful of like-minded space enthusiasts of whom, alas, I must be the last survivor. Although we grossly underestimated what space travel would cost, and many of our ideas now look rather quaint, the basic design of the BIS spaceship still stands up quite well. And after WW2, of course, we were on a much more professional basis; it is in the BIS 1948/9 journals that designs were first published for the nuclear-powered rockets that may one day give us freedom of the solar system. . . .

I commend Ron Miller, one of today's leading space artists, for this labour of love in compiling the history of the spaceship. In some form or other it has been a persistent dream for many centuries. We have been lucky enough to have lived in the age when "the power has met the dream."

ARTHUR C. CLARKE, CBE

Chancellor
International Space University

RAKETEN FAHRT

VON MAX VALIER

NASM

ACKNOWLEDGMENTS

This book would have been impossible without the help and encouragement of a great many people. The first to be thanked must be my wife, Judith, whose support and patience have been both invaluable and faultless; then (in no particular order) Frederick I. Ordway, III; Frederick C. Durant, III; Frank Winter of the National Air & Space Museum; Lee Saegesser of the NASA History Office; Ted Talay; Darrell Romick; Forrest J. Ackerman; Lee Staton; Tony Hardy; David Reynolds; Rick Dunning; Bob Burns; Bob and Dennis Skotak; Randy Liebermann; Richard Hallion; Phil Edwards of the National Air & Space Museum Library; Robert Shaw; Greg Kennedy; Ernst Stuhlinger; Saunders B. Kramer; John Frassanito and Associates; Dr. Helmuth Hauck of MBB; Mark Dowman; Barbara and Mike Peck of the From Out of the Past bookstore; Harry H.-K. Lange; James Hagler of the Alabama Space & Rocket Center; Dr. R. C. Parkinson of British Aerospace; Yolanda Mendoza of General Dynamics Space Systems; Kaori Sasaki of NASDA; Benjamin Donahue of Boeing Advanced Civil Space Systems; Arthur C. Clarke; Christine Dutrey of CNES.

While I am grateful to these people for their generous help, any faults that this book may have are entirely my responsibility.

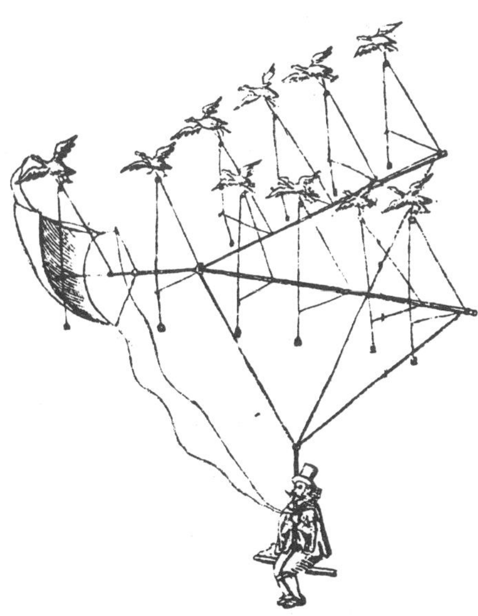

"They went through the air and space without fear and the shining stars marked their shining deeds."

—Inscription on the Astronomical Clock, York Cathedral, York, England

PREFACE

About This Book

This book does not, indeed in all honesty cannot, claim to be definitive. I'm not even sure if it is possible for it to *ever* be complete, no matter how many editions and revisions it may be fortunate enough to go through. Like some infinitely repeating fractal diagram, the closer one looks into the subject, the more detail one discovers. I hope that any reader who uncovers some error or who has additional information about any of the following entries—many of which I already know are sparsely covered—will contact me through the publisher. I would be even more interested in hearing from those who have information about speculative spacecraft that I have overlooked, of which I am certain there are plenty.

It is also important to remember what this book is *not*. It is not a history of rocketry, for one thing. Strictly speaking, it is not a history of space stations, space suits, space colonies or lunar and planetary bases, either. On the other hand, many of these things do show up here and there. In the earlier parts of this history, almost *anything* that had to do with inspiring human beings to think about leaving the earth is pertinent and worth mentioning. Thus, we find entries dealing not only with early devices intended for interplanetary travel, but speculations about life on other worlds, the development of reaction-powered vehicles and, though no one at that time ever connected it with space travel, the invention and improvement of the rocket. By the time of the late nineteenth century, the history of the rocket developed numerous branches that had very little to do with inspiring or developing space travel: military rockets, signal rockets, life-saving rockets, and so forth. At this point, this book ignores all developments except those bearing directly upon the idea of space travel.

Similar restrictions applied to concepts like space stations, satellites and space suits, for example. In the earlier years, these were rare and unusual enough ideas, and certainly very tightly linked with the idea of travelling beyond the earth. Discussions of satellites and spacesuits in the nineteenth century are, I think, perfectly germaine. However, these, too, eventually evolved into separate subjects of their own.

For similar reasons there are few movie spaceships mentioned after the 1960s. Before then, especially up to the end of the 1950s, the appearance of a spaceship in a motion picture or on television was an unusual occurance. After the 1960s they became commonplace. Moreover, for the most part they ceased being reflections of or influences upon contemporary technology.

There is a great deal of material here concerning the development of manned rocket aircraft—a subject not ordinarily considered directly connected with the history of spacecraft . . . with the possible exception of the X-15. However, since many of the spaceships now being flown . . . such as the space shuttle or Buran, or being planned . . . such as the National Aero-Space Plane, Sänger II, Hope II, HOTOL or Hermes, either take off or land, or both, like conventional aircraft, they are clear descendants of rocket-powered airplanes.

This book is not a history of astronautics or space travel, either. The reader will not find every manned space flight listed within its chronology, but rather only the first flights made in certain spacecraft, or manned flights that are of some particular or special interest.

As this history progresses, page by page, the reader will find fewer and fewer entries that have anything to do with anything other than spaceships proper. Which brings us to this question: What determines whether or not something is a spaceship? I have to admit here that, whatever objective definitions there may be, the final decisions were highly subjective. To paraphrase one science fiction author's definition of science fiction, a spaceship is that thing I'm pointing at when I say "That is a spaceship."

Throughout this book both the metric and English systems of measurement are used, and I am fully aware that this may seem at first to be both inconsistent and potentially confusing. However, since I am quoting from so many sources from so many different nations, covering nearly five centuries of scientific and cultural progress, it would have been impossible to be both consistent and true to the original materials. It would not have been proper, I felt, to have altered the original texts to bring them into line with modern usage, nor would it have been strictly accurate since the conversions would too often have only been approximate (Jules Verne, to make matters worse, often used both English and metric units within the same novel). Therefore, at the risk of appearing inconsistent, but with the end in mind of maintaining the flavor and intent of my sources I have chosen to use either the metric or English system when one or the other was used by the original author.

One more word. I apologize to what I hope will be my many female readers. The limitations of the English language are such that it is often unavoidable to use phrases such as "manned" or "one-man" or "two-man," for example. It would be, I felt, much less clumsy to express my regrets at having to use these words and phrases than to have to resort to creating ungainly compounds with "person" as the affixal. In any case, when the words "man" or "manned" is used in the text it is most often used in the generic, rather than the sexist, sense of "mankind."

I make no apologies when I am quoting the text of others, however; they are responsible for their own prejudices.

Ron Miller

Fredericksburg, Virginia

SHOOTING A BULLET OFF THE EARTH

When a bullet is fired from a horizontal gun, the curve of the path that it pursues is dependent upon the horizontal velocity of the bullet and gravity. The higher the horizontal velocity the flatter will be the curve and the further will the bullet travel before it strikes the earth. If the velocity of the bullet were 26,100 feet per second the curve of its path would be parallel to the circumference of the earth. Accordingly the bullet would pass around the earth without touching it, and return to its starting point in one hour and twenty-three minutes. It would continue to revolve about the earth as long as the velocity was maintained. At the same time it would be constantly attracted by the earth and would be forever falling away from a straight line towards the earth without ever reaching it. If the speed of the bullet were 37,000 feet per second it would fly off never to return to the earth.

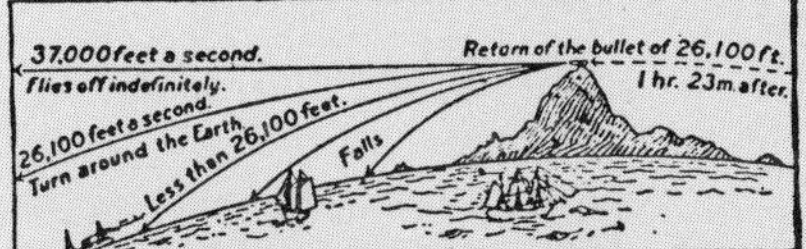

Scientific American, 1921

F. R. Paul, 1920

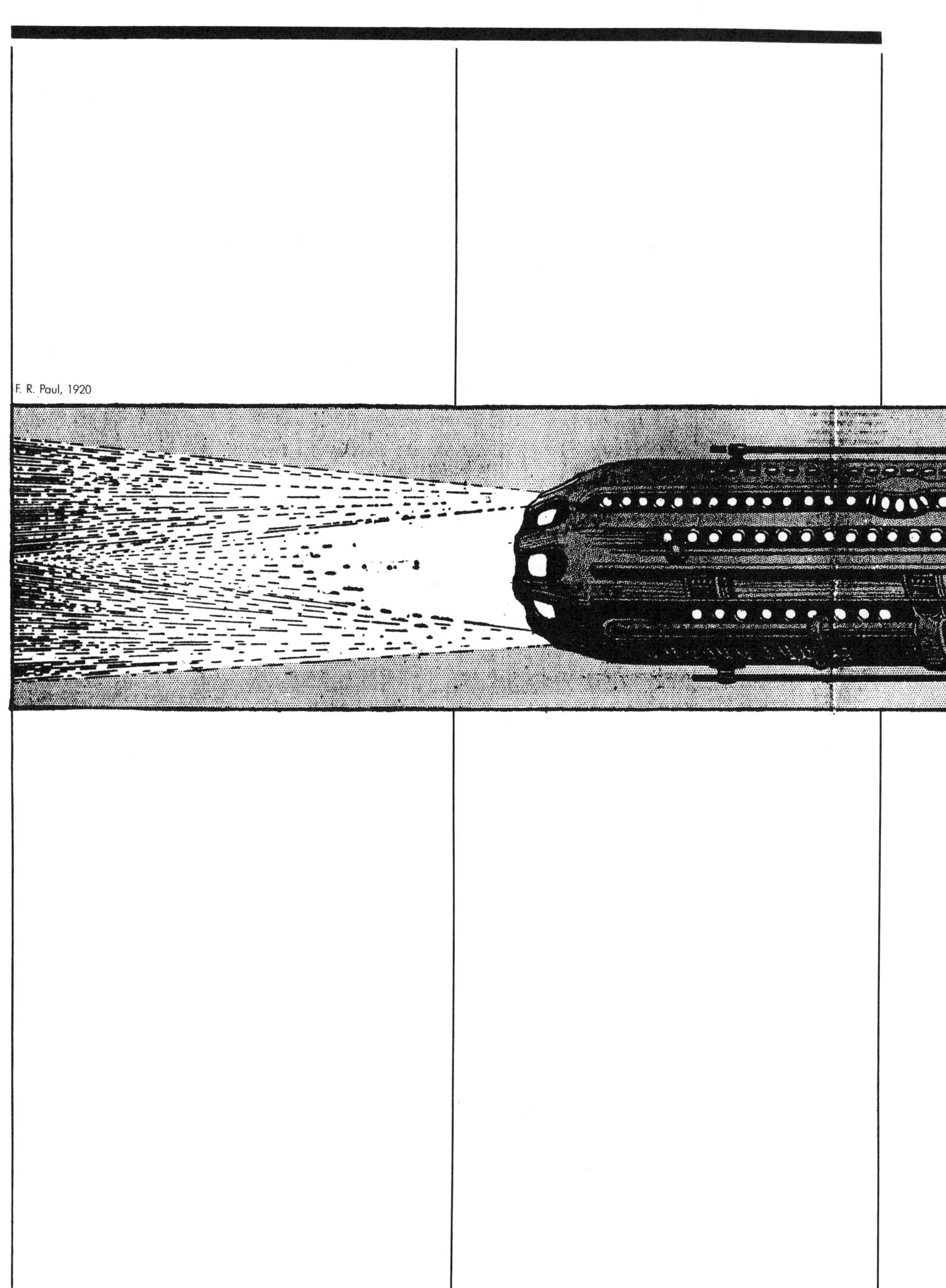

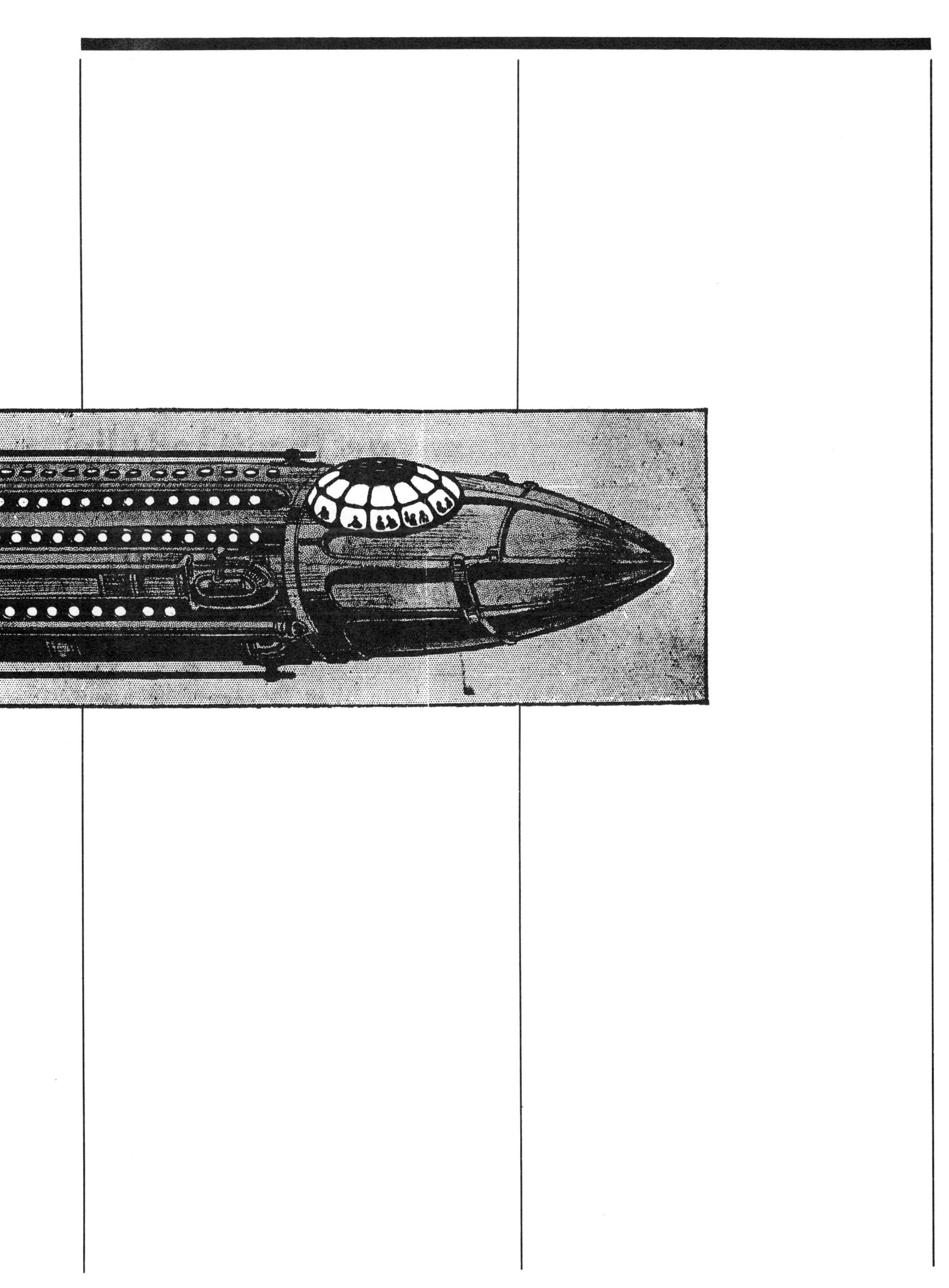

PART I

THE ARCHAEOLOGY OF THE SPACESHIP

Until the invention of the astronomical telescope by Galileo Galilei in 1610*, the heavens were thought to be no great distance from the earth, and the sun and the moon were thought to be the only material bodies with which we shared the universe. Even at that, the nature of the moon, for example, was an object of much debate: was it in fact a body like the earth? or was it something more ethereal? The stars, while some were brighter than others, were nevertheless thought to be at more or less the same distance from the earth, though just what that distance might be was a matter for discussion. The planets were merely a special class of bright stars that wandered among the other "fixed" stars; the word "planet" itself simply means nothing more than "wanderer." Otherwise there was nothing particularly unusual about them.

It was unthinkable to the ancients that those twinkling lights might be *places* that could be travelled to and only the moon served as a destination in a rare handful of fantasies. And even it was not regarded as something altogether physical, but rather a kind of ethereal Never-never Land. Some few of the early Greek philosopher-scientists speculated on the relative distances of the sun, moon and plan-

---

*Galileo probably was not the inventor of the instrument nor the first to use it—that credit most likely belongs to an unknown Dutch optician—but there appears to be little question that he was the first to turn the telescope toward the sky and note what he saw there.

ets, such as Anaximander in 600 B.C. Pythagoras and Aristotle both theorized that the moon might be spherical. But these and others were all based on quantitative measurements—little thought, if any, was given as to what the moon *was*. When it was considered, speculation knew few limits. Anaximander thought that the moon might be a kind of fiery chariot wheel and Anaxagoras suggested that it was an incandescent solid (albeit with "plains, mountains and ravines"). But by the time Plutarch was writing, the burgeoning faith of Christianity was laying the foundation for the religious fundamentalism and intolerance that eventually produced the thousand-year-long Dark Ages. During that bleak millenium the earth was clearly the center of the universe, there were no other worlds than this one and the moon was a perfect, pristine sphere since God would be incapable of creating anything less. If it showed spots, these were nothing but the reflection of our own imperfect world in the moon's mirrorlike surface. Change and decay were limited to the earth; the heavens, the handiwork and domain of God, were immutable and eternal. To question any of this was dangerous heresy.

Galileo's revelation changed all of that forever. With his first observations he immediately realized that the moon was not a pristine disk or sphere, but rather a world as imperfect as our own, with mountains, valleys, plains and hundreds of odd, circular ring mountains and craters. The planets did not have the same appearance in his telescope as did the other stars, which no matter what the magnification used still looked like points of light. Instead, the planets were revealed as tiny disks with indistinct features. They were obviously worlds like the moon and the earth. And if they were indeed worlds like our own, did not that imply other similarities? Would they not have landscapes and living inhabitants? Surely there would be animal life and perhaps civilizations? Would there be great cities and mighty kingdoms up there in the heavens? And if there were, might there not also be great treasures? These questions were far from rhetorical.

The Church forced Galileo to recant his discoveries and his interpretations of them, but the damage had already been done. When human beings looked skyward they no longer saw abstract points of light. They saw the infinite possibilities of new worlds.

At the time of Galileo's discovery of new worlds in the sky, there were new worlds being discovered right here on earth. Scarcely more than a century earlier the continents of North and South America had been discovered quite by accident, lying unsuspected and unknown on the far side of the Atlantic Ocean. Since then, John and Sebastian Cabot had explored the coasts of North America for Great Britain, while the Portugese and Spanish were laying the groundwork for a vast empire in the southern continent. Between 1519 and 1522 Magellan and Del Cano made their epic voyage around the now undoubtedly spherical earth. By the time of Galileo, hundreds of ships and thousands of explorers, colonists, soldiers, priests and adventurers had made the journey to these amazingly fertile, rich and strange new lands. Now they learned that an Italian scientist had found that not only did our own earth harbor unsuspected worlds, but that the sky was full of them, too.

How frustrating it must have been. The new worlds of the Americas, which could not even be seen and which existed for the vast majority of Europeans only in the form of traveller's tales and evocative if imaginative charts, nevertheless could be visited by anyone possessing the funds or courage. But now here were whole new earths—Venus, Mars, Jupiter, Saturn and the moon—which could be seen by anyone and even mapped; whole new planets with unimaginable continents and riches . . . yet there was no way to touch them! They were like a banana dangling just beyond the reach of a monkey.

It is little wonder that Galileo's discoveries could not be suppressed. Their publication was quickly followed by a spate of space travel stories: *Somnium, The Man in the Moone, Voyage to the Moon, A Voyage to the World of Cartesius, Iter Lunaire, John Daniel, Micromegas, A Voyage to the Moon* and countless others. There were poems, songs, stage plays and sermons, all inspired by the possibility of travelling to the new worlds in the sky. If it were not presently possible to reach them in reality, it could at least be done by proxy.

Bishop Wilkins had no personal doubts that these voyages would eventually be made. He wrote in his *Discovery of a New World* (1638), "You will say there can be no sailing thither [to the moon] . . . We have not now any Drake, or Columbus, to undertake this voyage, or any Daedalus to invent a conveyance through the air. I answer, though we have not, yet why may not succeeding times raise up some spirits as eminent for new attempts, and strange inventions, as any that were before them? . . . I do seriously, and upon good grounds affirm it possible to make a flying-chariot . . . "

Galileo's discoveries, and the discoveries of other great astronomers soon afterwards (the rings of Saturn, Saturn's great moon Titan, the dusky markings on Mars and even a new planet, Uranus), had a another profound effect on the evolution of the spaceship, in addition to inspiring the need for such a machine. Since the moon and planets were now known to be real worlds, it was no longer possible to employ them as merely metaphorical symbols. It was one thing to speak of visiting a vast mirrored disk suspended in the heavens, a disk that, so far anyone knew, had no real physical existence. Now that the moon was known to be a real place, transportation there could not be shrugged off onto some vaguely described magic. If one were to write seriously about travelling to the moon or planets, then the method of getting to them had to have at least the ring of plausibility. Even Bishop Francis Godwin with his fantastic moonbound swans was compelled to add such materialistic and realistic details as the construction of the birds' harnesses and the framework that bound them together. He even computed their top speed. Cyrano de Bergerac, although writing a burlesque, felt constrained to limit himself to pseudoscientific methods of spaceflight. Though he was striving for strictly comic effects it is important to note that none of his methods depended upon magic or the supernatural. He took a great deal of care in describing the fantastic devices he used in his attempts to travel to the sun and moon, even managing to stumble, however accidentally, upon the use of rockets.

These and many other authors of the time were discovering *verisimilitude*—the evocation of a sense of reality by the use of masses of convincing detail . . . or convincing-*sounding* detail, at least.

Still, the writers of space travel stories before the end of the 1700s were groping in the dark: there simply was no method by which a human being could leave the surface of the earth. In all the history of mankind no one had ever left the earth any farther than human muscles could push.

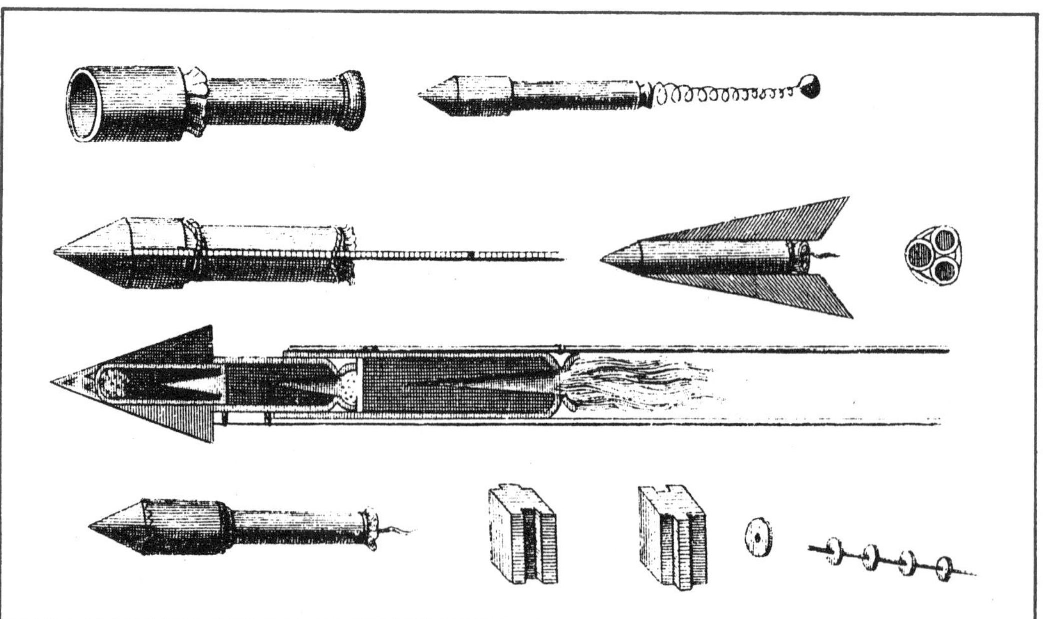

Frezier (1747)

ca. 360 B.C.

Archytas of Tarentum (428–347 B.C.) constructs a model pigeon which is made to fly by the reaction of steam or compressed air escaping from holes in its tail. Archytas is a Greek philosopher, mathematician and general who is also an intimate friend of Plato. His pigeon is made of wood and suspended by a string, balanced by a counterweight, "which preserved the equilibrium and the pigeon was propelled by the blowing of the air mysteriously encased therein," [also see 62]. A description can be found in Aulus Gellius's *Noctis Atticae* (A.D. 130–180) to the effect that " . . . a pigeon formed in wood by Archytas, was so contrived, as by a certain mechanical art and power to fly: so nicely was it balanced by weights, and put in motion by hidden and enclosed air . . . " Gellius suspected that there might be "some lamp or other fire within it which might produce such a forcible rarefaction . . . "

ca. 356 B.C.-A.D. 160

A story by Antonius Diogenes, *Of the Wonderful Things Beyond Thule*, may be one of the earliest tales of a trip to the moon. Unfortunately, this has not survived and only a mention of it remains in the *Bibliotheca* of Photius. Even the author's dates are obscure . . . he may have been a contemporary of either Alexander the Great or of Lukian [see 160], a range of over 500 years.

ca. 46–120

Plutarch, in his *Of the Face Appearing in the Orb of the Moon*, suggests that the moon might resemble a smaller edition of the earth: " . . . as this Earth on which we are has in it many great Sinuosities and Vallies, so 'tis probable that the Moon also lies open, and is cleft with many deep Caves and Ruptures, in which there is Water, or very obscure Air, to the bottom of which the Sun cannot reach or penetrate . . . "

62

Hero of Alexandria demonstrates his "aeolipile," a hollow sphere that is made to rotate by the effect of steam escaping from nozzles placed around its circumference [also see 360 B.C.].

150

Claudius Ptolemy describes the sun and planets as orbiting around the earth.

160

Lukian of Samosata writes the first novel of interplanetary travel, *Vera Historia*, which pretends to be a missing chapter from the adventures of Odysseus. In this story, the hero's ship is caught up in a whirlwind that carries the boat "three hundred furlongs" into the air. While aloft, another powerful wind strikes the sails and the ship is carried off into the sky for seven days and seven nights. On the eighth day Odysseus sees "a great country in [the air], resembling an island, bright and round and shining with a great light. "It is the moon and from this vantage point, Odysseus observes that, " . . . as night came on we began to see the many other islands hard by, some larger, some smaller, and they were like fire in color [perhaps the first hint that there are other worlds in space than the earth?]. We also saw another country below, with cities in it and rivers and seas and forests and mountains. This we inferred to be our own world."

In an earlier story, *Icaromenippus*, Lukian described a flight to the moon by more mechanical means. After determining that the moon was no less than "three thousand furlongs" away, Menippus " . . . tooke a good bigge Eagle, and a strong Vulture, and cut their wings at the first joynt . . . I cut off the right wing of the one, and the left wing of the other which was the vulture, as handsomely as I could, and buckling them about mee, fastened them to my shoulders with thongs of strong leather, and at the ends of the uttermost feathers made me loopes to put my hands through, and then began to trie what I could do, leaping upwards at the first withall, and sayling with my armes . . . " After considerable practice and numerous, ever-enlarging trial hops, Menippus finally makes the attempt to fly to heaven, for his planned interview with Zeus. Nearing the moon, he becomes understandibly weary and sets down upon that body for a rest.

850

Chinese use gunpowder to create fireworks for religious festivals.

1145

Franciscus Gratianus declares that belief in a plurality of worlds is heretical.

1232

Chinese use rockets in combat, repelling the Mongols who were besieging the town of Kai-Fung-Fu. Described as "arrows of fire," these were apparently true rockets.

1242

Roger Bacon, a Franciscan monk, develops a formula for gunpowder. He also creates distilled saltpetre (potassium nitrate), which allows for a faster rate of burning. He conceals his discovery in an anagram ["Annis Arabum 630 . . . Item pondus totum sit 30. Sed tamen sal petrae LURU VOPO VIR CAN VTRIET sulphuris; et sic facies tonitruum et coruscationem, si scias artificium"] which, when translated and decoded (by Colonel Henry W. L. Hime), reads: "Take 7 parts of saltpetre, 5 of young hazelwood, and 5 of sulfur, and you may make thunder and lightning if you understand the artifice."

1249

Arabs use rockets in combat at the siege of Damietta.

1277

Etien Tempier, the Bishop of Paris, under the authority of Pope John XXI, officially condemns the proposition that there can be only one world: God's plenitude and creative power cannot be limited.

1280

In *Liber Ignium ad Comburendos Hostes*, Marcus Graecus writes: "Note there are two compositions of fire flying in the air. This is the first. Take one part of colophonium and as much native sulfur, six parts of saltpetre. After finely powdering, dissolve in linseed oil or in laurel oil, which is better. Then put into a reed or hollow wood and light it. It flies away suddenly to whatever place you wish and burns up everything." The second method requires, " . . . one pound of native sulfur, two pounds of linden or willow charcoal, six pounds of saltpetre, which three things are very finely powdered on a marble slab. Then put as much powder as desired into a case to make flying fire or thunder."

1280

Al-Hasan al-Rammah, a Syrian military historian, publishes instructions for making gunpowder and rockets called "Chinese arrows," in his book *The Book of Fighting on Horseback and with War Engines*.

1379

Rockets are known in Western Europe. They are used in the siege of Chioggia, Italy.

1405

In Frankfurt-am-Main a kite is raised by means of a rocket employing the system of Konrad Keyser von Eichstadt (who, in that year, described making a rocket stabilized by a stick).

1420

Joanes de Fontana, an Italian military engineer, fills a sketchbook with speculative rocket projects. His book, eventually titled *Bellicorum Instrumentorum Liber* (Book of War Machines), contains many rocket designs for military use, including a rocket-powered torpedo. Although it was evidently never actually constructed, Fontana designed the first rocket-car. It was a platform with a rocket at the rear, mounted on rollers rather than wheels (to facilitate travel over rough terrain), and equipped with a battering ram at the front so that it could be used to breach walls and gates. He never intended any of his rocket vehicles to carry passengers.

ca. 1500

Wan-Hoo, a Chinese official (whose name, roughly translated, means "crazy fox"), is credited with being the first to attempt a manned rocket flight. He fitted a chair within a bamboo framework connecting a pair of large kites. Forty-seven powder rockets were attached to the kites. At a prearranged signal, 47 coolies with matches simultaneously touched off all 47 rockets. Wan-Hoo vanished in a brilliant flash and a cloud of smoke, reaching the heavens a little more suddenly than he had perhaps intended.

Most authorities consider the story of Wan-Hoo apocryphal, due to the large number of internal inconsistencies as well as an inability to discover any published reference to the tale earlier than 1909. It is most likely that the story was fabricated during the Chinoiserie period in Europe, during the 17th and 19th centuries, which was characterized by a fascination with all things Oriental. The story, however, is so charming that Wan-Hoo, fictional or not, has had a lunar crater named for him.

ca. 1529–1569

Conrad Haas, the chief of the artillery arsenal in Sibiu, Romania, contributes a chapter on rocketry to a book on pyrotechnics and ballistics. Haas suggests several innovative ideas: the two-stage rocket, the use of a warhead that would explode upon impact, and swept-back fins to give stability to the rocket's flight instead of the traditional stick. In one woodcut illustration, Haas is possibly the first person to suggest the use of the rocket for manned flight. It is a castellated, two-story cylinder with windows and a door, atop the upper portion of a large "stick."

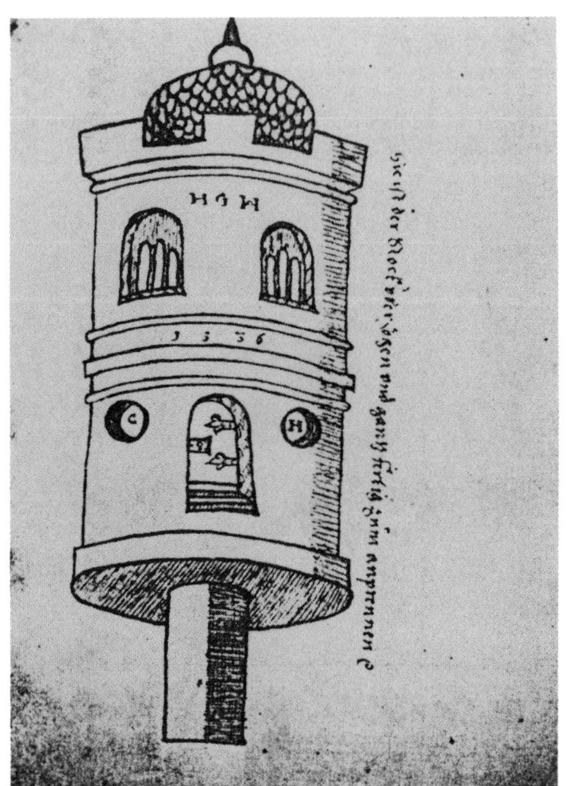

1543

Nicholas Copernicus publishes his theory of the solar system, *De Revolutionibus Orbem Coelesticum*, in which he proposes the idea that the sun is at the center of the solar system, with the earth and the other planets orbiting it. He retains the circular orbits of Ptolemy and Tycho Brahe, which require that the planets also move in epicycles to account for their retrograde movements.

1605–1615

Miguel de Cervantes, in his novel *El Ingenioso hidalgo Don Quixote de la Mancha*, describes how Don Quixote and Sancho Panza return from the aerial kingdom of

Kandy when fireworks are attached to the tail of their horse, Clavileno, which "immediately blew up with a prodigious noise, and brought Don Quixote and Sancho Panza to the ground half singed." Poor Clavileno is probably the first reaction-powered spacecraft in literature.

1609

Johannes Kepler's *De Motibus Stellae Martis* appears, in which he corrects Tycho Brahe's and Copernicus's concept of circular orbits for the planets. That the planets travel in ellipses around the sun, with the sun occupying one focus of the ellipse, becomes the first of Kepler's laws of planetary motion.

1610

Galileo Galilei's discovery of the moon as a world, and Jupiter as a planet with satellites of its own, provides the impetus for exploring worlds beyond our own, just as the discovery of the New World across the Atlantic is inspiring a wave of exploration of the earth.

The moon is even included in such terrestrial geographies as Peter Heylyn's *World Geography* (1652), indicative of its status as an equal among other such new worlds as North and South America. "The New World in the Moon," he writes, "was first of Lucian's discovering . . . But of late times, that World which he there fancyed and proposed but as a fancy only, is become a matter of more serious Debate; and some have laboured with great pains to make it probable that there is another *World* in the Moon, inhabited as this is by persons of divers Languages, Customs, Politics and Religions; and more then so, some means and ways proposed to Consideration for maintaining an entercourse and Comerce betwixt that and this . . . "

1622

Charles Sorel, in his *Comical History of Francion*, writes that "Some men have affirmed that there are many Worlds, which some have placed in the Planets, and others in the fixed Starrs; For my part, I believe there is a world in the moon." In describing a trip there—"which hath yet never entered into the thoughts of mortal man"—he supposed a "Prince as ambitious as Alexander, who shall come to conquer this world." In doing so, he "shall make great provision of Engins, either to descend or ascend . . . " And if not a new Alexander, then a new Archimedes who "shall make all manner of structures, and ladders, by the means whereof he shall enter into the Epicycle of the moon . . . " Though Sorel never described his "Engins," he was among the first to suggest that travel into space would be resolved by mechanical means.

1623

Lagari Hassan Çelebi, in a probably apocryphal story, flies from the top of the Topkapi Palace in Istanbul using a kind of rocket-powered glider. According to the account he flew 980 feet over the Bosporus with his 10-foot rocket: a large central tube surrounded by six smaller ones. His rocket, launched in honor of the birth of Sultan Murad IV's daughter, Kaya Sultan, was loaded with about 54 pounds of gunpowder "paste" (or moistened gunpowder). It is not clear whether his rocket was winged or not . . . the translation of the Turkish word is uncertain.

Before Celebi took off he told the Sultan: "Your Majesty, I leave you in this world while I am going to have a talk with the prophet Jesus. " After his brief flight he reported that, "Your Majesty, Prophet Jesus sends his greetings to you."

1634

Johannes Kepler's novel *Somnium* (The Dream) is published posthumously (he died in 1630). While his method of transporting his character to the moon harks back to the magic of Lukian and earlier writers, his description of our satellite is based upon the most accurate knowledge available at the time.

Kepler's hero, the Icelander Duracotus [sic], is a student of the great astronomer Tycho Brahe and as a result develops an intense desire to visit the moon. His mother is a witch who reveals that both the earth and the moon are inhabited by demons. They are normally prevented from passing from one world to the other because of the brilliance of the sunlight, which they cannot tolerate. However, during a lunar eclipse there is a kind of bridge of darkness when the earth's shadow falls onto the moon. During this time it is possible for the demons to make the journey from world to world. Duracotus's mother makes the necessary arrangements and the demons agree to carry her son to the moon. Now the story becomes influenced by Kepler the scientist. The demons must make some allowances for the human's need for air and his intolerance of extreme cold. Kepler was aware that Arab astronomers had calculated a height of no more than a thousand miles for the earth's atmosphere, so that most of the journey to the moon must be made in a vacuum. Duracotus is consequently given an anesthetic sleeping draught to ease the discomfort, as well as moistened sponges for him to hold against his nostrils.

As soon as Duracotus arrives at that point in space where the gravity of the moon and the earth balance each other [see 1865], he curls up into a ball because "When the attractions of the Moon and of the Earth equalize each other, it is as though neither of them exerted any attraction. Then the body itself, being the whole, attracts its minor parts, its limbs, because the body is the whole."

1638

John Wilkins publishes his *A Discourse Concerning A New World and Another Planet: The First Book, The Discovery of a New World*. The second edition of 1640 discusses the need for "flying-chariots" to take men to the moon: "I do seriously, and upon good grounds affirm it possible to make a flying-chariot; in which a man may sit, and give such motion unto it, as shall convey him through the air. And this perhaps might be made large enough to carry divers men at the same time, together with food for their *viaticum*, and commodities for traffic."

Bishop Wilkins suggests the possibility of colonizing the moon: "It is the opinion of Keplar [sic], that as soon as the art of flying is found out, some of their nation will make one of the first colonies that shall transplant into that other world . . . "

He considers the problems of space travel in such detail that this book becomes literally a bible for generations of authors of early science fiction. He believes that the influence of the earth's gravity (Its "sphere of magnetic virtue") does not extend far beyond the limits of the atmosphere, which he in turn estimates at about 20 miles. "So you see the former thesis remains probable: that if a man could but fly, or by any other means get twenty miles upwards, it were possible for him to reach unto the moon." Once beyond the pull of the earth, the space traveller will encounter no further impedance. "He might there stand as firmly as in the open air . . . And if he may *stand* there, why might he not also *go* there?" He does not believe that the "extreme coldness" nor the "extreme thinness" of the upper atmosphere will prove any great obstacle. He also thinks it possible that, once beyond the earth's gravity, the need for food and sleep will be greatly lessened. He might be speaking to today's astronauts when he writes that " . . . we cannot desire a softer bed than the air, where we may repose ourselves firmly and safely as in our chambers."

Bishop Wilkins is one of the very first to take the idea of space travel seriously, and to approach its possibility with scientific thinking. Wilkins's unbounded enthusiasm and faith in man's eventual conquest of both the sky and space is a positive influence that will reach into the future as far as the time of Jules Verne.

Francis Godwin, under the pseudonym "Domingo Gonzales" writes the novel *The Man in the Moone*. Although he has written the first interplanetary journey in English literature, Godwin's science is very much behind the time in which he writes. This is particularly true in his descriptions of the conditions of outer space.

Domingo Gonzales (who is also the hero of the story) discovers a species of unusual swans, called "ganzas," on the island of St. Helena, where he had been marooned. Finding that the birds can be trained to fly while carrying heavy weights, Gonzales constructs a light framework from which a seat is suspended. A number of the ganzas are attached to points on the frame. Since our hero's intention has been only to fly, he is shocked to learn that the ganzas are unique in that they periodically migrate to the moon. As it happens to be ganza migrating season, he is carried to the moon with the strange birds. The ganzas are powerful birds, reaching a speed of up to 175 miles an hour. After a trip of 12 days, Gonzales reaches the moon.

Godwin bravely attempts a description of the conditions of translunar space: "I must now declare to you the nature of the place in which I found myself. All the clouds were beneath my feet, or, if you please, spread out between me and the earth. As to the stars, since there was no night where I was, they always had the same appearance; not brilliant, as usual, but pale, and very nearly like the moon of a morning. But few of them were visible, and these ten times larger (as well as I could judge), than they seem to the inhabitants of the earth. The moon which wanted two days of being full, was of a terrible bigness." [Also see 1687 and 1706.]

NASM

1648

John Wilkins describes in his book *Mathematicall Magick* (part 2, *Daedalus, or Mechanicall Motions*) the "four several ways whereby this flying in the air hath been or may be attempted. Two of them by the strength of other things, and two of them by our own strength: (1) By spirits, or angels. (2) By the help of fowls. (3) By wings fastened immediately to the body. (4) By a flying chariot."

1656

A. Kircher, in *Iter Extaticum Kircherianum*, travels in a dream to the stars.

Three of Cyrano de Bergerac's space travel schemes. Left to right: rocket-powered car, bottles of dew being attracted to the sun, solar-powered "jet."

1657

Cyrano de Bergerac publishes his satirical fantasy, *Histoire Comique: Contenant les Etats et Empire de la Lune* (which had been written 10 years earlier). In this book de Bergerac describes a number of highly imaginative methods for reaching the moon. Only one of them really interests us here: the first description of a manned rocket flight in literature. De Bergerac has found himself stranded in Quebec, still trying to find a way of travelling to the moon. He constructs a flying-machine with wings and a "spring," though he gives us no clue as to how it was to work. At the next full moon, Cyrano casts himself and the machine from the top of a high cliff. He discovers too late that a miscalculation has been made, a discovery confirmed by the crash that immediately follows. While Cyrano consoles his bruised body with brandy in his room, a group of soldiers of New France carry his machine away. Setting it up in the middle of Quebec's marketplace, they fasten rockets all around it. Their peculiar motive for doing this was the hope that when they launched it, the flames and sparks along with the spring-operated wings would make everyone think that they were seeing a dragon. They had just lit the fuses when Cyrano shows up. He leaps into his machine to save it, just a split second too late. With a roar, he is launched into the air. "The fearful horror that dismayed me," writes Cyrano, "did not so thoroughly overwhelm the faculties of my soul but that I could recollect afterwards

NASM

all that happened to me at this moment. You must know then that the flame had no sooner consumed one line of rockets (for they had placed them in sixes by means of a fuse which ran along each half-dozen), when another set caught fire and then another, so that the blazing powder delayed my peril by increasing it. The rockets at length ceased through the exhaustion of material and, while I was thinking that I should leave my head on the summit of a mountain, I felt (without my having stirred) my elevation continue; and my machine, taking leave of me, fell towards the Earth."

Cyrano thus becomes perhaps the first human second stage in astronautical history. By virtue of some beef marrow he had rubbed over his body earlier (to relieve his bruises), Cyrano finds himself continuing on to the moon (it being a popular superstition at one time that the moon sucked up the marrow of animals).

In his later, unfinished, novel *Histoire des Etats et Empires du Soliel* (1652), Cyrano describes another spacecraft that enables his hero to travel to the sun. "It was a large very light box which shut very exactly. It was about six feet high and three wide in each direction. This box had holes in the bottom, and over the roof, which was also pierced, I placed a crystal vessel with similar holes made globe shaped but very large, whose neck terminated exactly at and fitted in the opening I had made in the top. The vessel was expressly made with several angles, in the shape of an icosahedron, so that as each facet was convex and concave my globe produced the effect of a burning mirror." A plank was provided inside as a seat and outside was a small sail for regulating the amount of sun shining on the crystal globe. As the sun shone upon the icosahedron its interior was heated. The air inside expands, creating a vacuum. "A furious abundance of air" rushes into the holes in the bottom of the box, rushing through and out the top, thereby carrying the car with it.

1659

Hans Jacob Christophel von Grimmelshausen publishes the first German work dealing with space travel: *Die Fliegende Wandersmann nach den Mond*. It is a retelling of a French version of Godwin's novel [see 1638].

1666

Margaret Cavendish, the Duchess of Newcastle, writes *The Blazing World*, in which her heroine, perhaps the first woman space traveller in history, makes the rounds of the moon and the other planets.

1668

Christoph Friedrich von Geisler of Berlin experiments with very large rockets, of 50 to 100 pounds, for launching bombs.

ca. 1670

Honore Fabri, a French Jesuit priest, works on the design of a large flying machine propelled by the compressed air contained in a pipe.

The anonymous *The Lunarians: Or News from the World of the Moon is intended for the "Lunaticks of this World."* The hero travels to the moon on a kite "of the height of a large sheet, fixing himself to the tail of it." The story has been attributed to Samuel Butler solely on the strength that a kite is also mentioned in his poem "Hudibras."

1677

In Thomas Shadwell's comedy, the character "Sir Nicholas Gimcrack, the great scientist" has the ambition of travelling to the other worlds in space. "I doubt not," he says, "but in a little time the art of flying will so improve, it will be as common to buy a pair of wings to fly to the Moon, as to buy a pair of wax-boots to ride to Sussex with . . . "

1680

Czar Peter the Great establishes a rocket manufactory in Moscow, the first of its kind. It was later moved to St. Petersburg. The rockets were to be used for signalling and illumination for the Russian army.

1686

Isaac Newton publishes the third law of motion in his *Principia*: "To every action there is always opposed an equal reaction . . . " This is the principle of *recoil* and is the reason rockets work. The gases resulting from the burning of the rocket's fuel stream backward from its nozzle at a high velocity, with the result that the rocket is pushed forward. The final velocity of the rocket at the time the fuel is used up or the engine is turned off depends upon the ratio of the weight of the rocket compared to the weight of the fuel. The smaller the ratio, the better is the rocket's performance (this is called a rocket's *mass-ratio*).

It has been suggested that Newton himself proposed a reaction-powered steam car. It would have been basically a boiler on wheels. The steam escaping through a narrow tube in the rear would propel the vehicle.

Also in the *Principia* Newton is the first to suggest—if unintentionally—the possibility of artificial earth satellites. Newton is able to explain why the moon does not fall to the earth, and that this same explanation would serve to keep any object forever orbiting the earth, without falling.

Bernard le Bouvier de Fontenelle writes *Entretiens sur la Pluralité des Mondes*. Once the discoveries of Galileo Galilei and Copernicus have transformed the planets from merely lights in the sky into worlds in their own right, speculation begins about the possibilities that these worlds might be inhabited.

Fontenelle speculates on what the inhabitants of each of the known planets might be like, based on what is known concerning the physical conditions on those worlds—mainly limited to guesses at what their temperatures might be. Mercurians, for example, "are so full of Fire that they are absolutely mad . . . The Day there is very short, and the Sun appears to them like a vast fiery Furnace at a little distance whose Motion is prodigiously swift and rapid; and during their Night, *Venus* and the Earth (which must appear considerably big) give light to them."

1687

Mrs. Aphra Behn adapts Godwin's story as the play *The Emperor of the Moon*. In it, a Doctor Baliardo has a passion for the moon and discusses Gonzales's trip [see 1638, 1659 and 1706]:

Doct. [Baliardo]
That wondrous Ebula, which Gonzales had?

Char. [Charmante]
The same—by Vertue of which, all weight was taken from him, and then with ease the lofty Traveller flew . . . to *Olympus* Top, from whence he had but one step to the Moon. Dizzy he grants he was.

Doct.
No wonder, Sir, Oh happy great Gonzales!

1690

Voyage du Monde de Descartes is a satirical novel by Jesuit priest and historian Gabriel Daniel, attacking the philosophy of Rene Descartes, who wrote that mind and matter had separate, independent existances. Daniel has a caricature of the great philosopher travel to the moon when he is dosed with such a powerful quantity of snuff that the resulting sneeze ("three or four Times with mighty Violence") separates his mind from his body. " . . . In an instant my soul perceived the unspeakable pleasure to scud high into the air." On the moon, he discovers the late thinker "still busily engaged in correcting the mistakes of the Almighty."

He speculates that if terrestrial animals were to be brought to the otherwise barren moon, they would flourish.

1698

Christian Huygens publishes *Cosmothereos* (or *The Celestial Worlds Discover'd*). Like Fontenelle [see 1686], Huygens speculates on the habitability of other worlds, though he is more sceptical of the possibility than his predecessor. He is, for example, certain that the moon harbors no indigenous life . . . in fact, it probably has no water and certainly no air. "I cannot imagine how any Plants or Animals, whose whole nourishment comes from liquid Bodies, can thrive in a dry, waterless, parch'd Soil." He does admit, however, that extraterrestrial lifeforms might have different constitutions than those here on the earth, and admits that "those great and noble Bodies have somewhat or other growing and living upon them, tho very different from what we see and enjoy here. Perhaps their Plants and Animals may have another sort of Nourishment there . . . "

ca. 1700

Selenographia: The Lunarian, or Newes from the World in the Moon to the Lunaticks of This World is written and published anonymously. In this satirical novel a flight is made to the moon by way of a kite—such devices being but recently introduced to England. The hero is inspired when "observing likewise with an attentive eye & serious Rumination the admirable Invention of Paper Kites, which the Politick Boyes raised to a wonderfull height by ye help of a little gale of wind, which heightened likewise his fancy to this great enterprise . . . " The erstwhile astronaut makes "himselfe a kite of ye height of a large sheet, and fixing himself to the tayle of it by the help of some trusty friends, to whom he promised Mountains of Land in his new-found World." The takeoff is made, appropriately enough, on April First, "a day alwaies esteemed prosperous for such adventures, Bestriding ye taile of his Pegasus, as Millers mount their Asses on ye Rump."

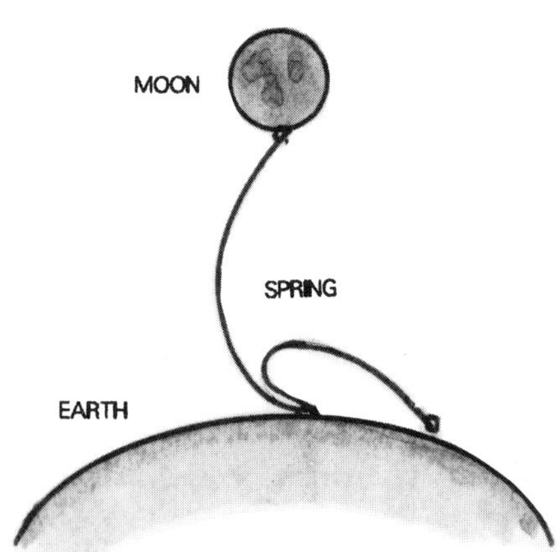

1703

Davis Russen, in his novel *Iter Lunaire; or, A Voyage to the Moon* describes what must be one of the most naïve of all space travel methods. An enormous steel spring is attached to the peak of some high mountain. "Since Springiness is a cause of forcible motion, and a Spring will, when bended and let loose, extend itself to its length; could a Spring of well tempered Steel be framed, whose Basis being fastened to the Earth and on the other end placed a Frame or Seat, wherein a Man, with other neccessaries, could abide with safety, this Spring being with Cords, Pullies, or other Engins bent and then let loose by degrees by those who manage the Pullies, the other end would reach the Moon, where the Person ascended landing, might continue there, and according to a time appointed, might again enter into his Seat, and with Pullies the Engin may again be bent, till the end touching the Earth should discharge the Passenger again in safety."

Perhaps it should be explained that Russen takes de Bergerac's work seriously! So much so, in fact, that he devotes much of his own book to rationalizing Cyrano's "science" and reading cryptic double and triple meanings into the French author's irreverent farce.

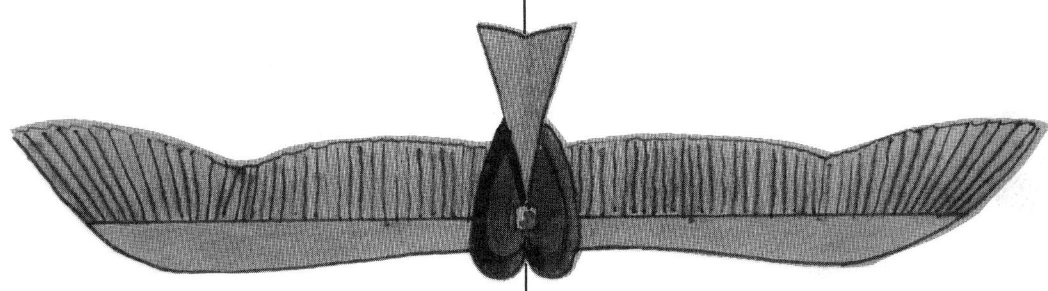

1705

Daniel Defoe describes in his book *The Consolidator* the spaceship of that name. Supposedly invented "two thousand years before the flood" by the Chinaman Mira-cho-cholasmo, it is "a certain engine formed in the shape of a Chariot, on the backs of two vast bodies with extended wings, which spread about fifty yards in breadth, composed of feathers so nicely put together that no air could pass; and as the bodies were made of Lunar earth [sic], which would bear the fire, the cavities [within the otherwise solid bodies] were filled with an ambient flame, which fed on a certain spirit [i.e., a flammable, distilled liquid] deposited in a proper quantity to last out the voyage. And this fire so ordered as to move about such springs and wheels as kept the wings in a most exact and regular motion, always ascendant; thus the person being placed in this airy Chariot drinks a certain dozing draught that throws him into a gentle slumber, and dreaming all the way, never wakes till he comes to his journey's end . . . "

Here DeFoe describes vividly not only a flying machine powered by an internal combustion engine, but the use of hibernation to ease the tedium of a long and dangerous spaceflight. Many of the physical details of the *Consolidator*, however, are symbolic and pertinent to Defoe's political satire. The number of feathers, for example, refers to the members of the House of Commons and the large, central feather to the Speaker. [Also see 1741.]

1706

Thomas d'Ufrey produces a comic opera based on Godwin's *Man in the Moon*. The elaborately staged play features the lunar world and Domingo's ganza-powered flying machine suspended from the ceiling. D'Ufrey follows up this success with adaptations of Cyrano de Bergerac's novels. [Also see 1659, 1687 and 1638.]

1707

Pier Jacopo Martello describes a spaceship in his long poem "Gli occhi di Gesu" (The Eyes of Jesus). It is a swallow-shaped boat with a beak-shaped prow and retractable tail. It flies by virtue of the "magnetic" energy generated by a pair of amber globes (amber is an excellent generator of static electricity; the static electricity machine had only just recently been invented in England, employing rotating spheres of amber; this mysteri-

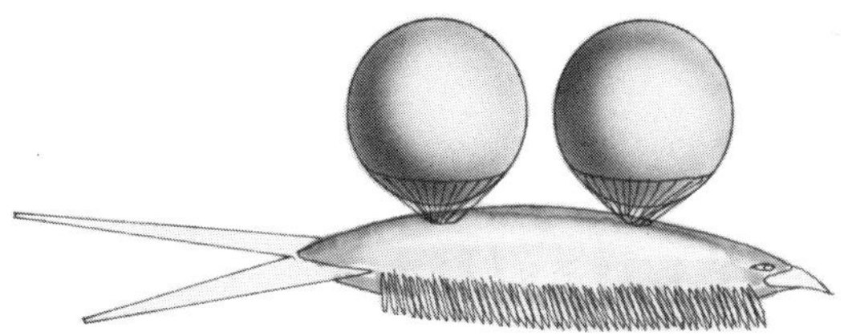

Martello

ous power gripped the popular imagination in much the same way that radium did at the beginning of our own century). It is propelled by one hundred wings, attached to oars rowed by harnessed apes dressed in yellow or blue livery.

The ship is equipped with landing gear, consisting of twenty hooked feet, which catch onto the surface of the moon. There is a crew, including steersmen and pilots who navigate using compasses and the polestar.

1711

The anonymously written book *Gongam, or the Wild Man in Space, in the Bottomless Seas and Earth* describes a flight across the world on a magic arrow, starting from the highest level of the atmosphere.

1711

In the anonymous *Satyre Menippée, de la vertu du Catholicon d'Espagne* is a supplement describing adventures on the moon without, however, any explanation about how the travellers arrived.

1724

The virtually unknown author, Labadie, has the hero of *Les aventures de Pomponius Chevalier Romain* voyage to the moon. It is a political satire that includes a catalog of the books in the lunar library.

Louis-Bertrand Castel, in his *Traité physique*, develops the theory that terrestrial objects transposed to the moon would fall back to the earth. He also speaks in passing of the possibility of human flight. However, he writes, even if one could fly, the uppermost layer of air would form an impassable barrier, since our substance has no affinity with that of the moon.

1726

May 2

An advertisement placed in the *Country Gentleman* [London] reads in full:

"The famous *Planetary Caravan*, which I spoke of before, being now entirely finish'd and render'd convenient for all such Persons, who have any Desire to visit the *Moon, Venus, Mercury*, or any other of the Planets, is remov'd from Mr. *Deard's* Toyshop in *Fleet-street*, to Mr. *Fawkes's* great Booth in the *Tennis-Court* near the *Hay-Market*, where Passengers may be accomodated with every Thing Proper for so long a Journey. This Machine sets out from thence to the *Moon* very soon (only waiting at Present to introduce *Faustina*, who is to make her Entry into the *Opera*, at the Roof of the *Theatre*, over the Heads of all the rest of the Singers.)

"Any Person who intends to go this Way, or send any of the Friends, must send their Names before the first Day of *June* next, and likewise must deposite their Earnest Money, in the Hands of the said Mr. *Fawkes*, which being one half of the Fare to the *Moon*, will come to a Hundred and Twenty Five Pounds. The *Machinist* contents himself with this moderate Price, (being only one Farthing a Mile,) purely to serve his Country and facilitate the Means of Transportation, having long observ'd, how usefull this Project has been to the Inhabitants of the Island.

"In the same Place also, may be seen the *Planetary Curricule*, which is a Vehicle prepar'd only for two Persons, being a lighter Carriage, and very fit for a Couple of Lovers, who have a Mind to spend their Honeymoon in *Venus*, and perhaps shou'd take a Fancy to come back again in Haste.

"N. B. *For the Encouragement of all such Persons, who go long Journeys,* Fawkes *is order'd to take only one Thousand Pounds for a Million of Miles, which will be a saving of 125 £. for every Million of Miles, and render this Aetherial Navigation more easy to the Adventurers."*

Although this was no doubt intended satirically, there are some interesting details, not the least of which is that when the distance to the moon is worked out, based upon the information given, it comes to exactly 240,000 miles.

NASM

1727

"Samuel Brunt" is the main character in the pseudonymously written *A Voyage to Cacklogallinia*. The novel is a Swiftian satire on the recent South Sea Bubble scandal. In it, Brunt finds himself in the strange land of Cacklogallinia, whose inhabitants are all birds. In a scheme to raise money, it is proposed to mount an expedition to the moon, to mine the gold that must be there. Brunt's objections to the possibilities of such a project being successful are waved aside. Since Brunt, unlike his hosts, cannot fly, a specially made "Palanquin" is made to carry him, powered by a crew of low-class Cacklogallinians. It is lined with down against the "extreame Coldness" expected between the worlds. The palanquin is streamlined as well, and "made sharp at each End, to cut the Air."

A century has passed since Godwin had his hero carried to the moon by birds. "Brunt" cannot, even in a satire, ignore the development of scientific knowledge in that interval. Test flights are made from the top of a mountain, to acclimate the passengers to the thin atmosphere. Even then, moistened sponges need to be used to aid breathing. Once beyond the "magnetick Power of the Earth," the bird men experience weightlessness. They are able to rest "in the Air, as on the solid Earth." Brunt later himself discovers one of its joys when "I could with as much Ease lift a Palanquin of Provisions, which did not on Earth weigh less than 500 weight, as I could on our Globe raise a Feather." This region of lessened "magnetick Power" is supposed to be about 180 miles above the surface of the earth.

The eventual journey to the moon takes a full month. On the way, Brunt makes the observation that the phases of the earth and the moon appear to be opposite of one another. He is worried about falling onto the moon from such a great height, but is reassured that since the moon is so much smaller than the earth, its attraction is proportionally less as well.

1728

Murtagh McDermot, in his novel *A Trip to the Moon*, describes this method used to *return* his hero from the moon (he had been carried there by a whirlwind): "We already know how gunpowder will raise a ball of any weight to any height. Now I design to place myself in the middle of ten wooden vessels, placed one within an other, with the outermost strongly hooped with iron, to prevent its breaking. This I will place over seven thousand barrels of powder, which I know will raise me to the top of the atmosphere." As a safety precaution he provides himself with a pair of wings, fastened to his arms. The journey is successful and the selenites help him to return to the earth by digging a gun barrel one mile deep into the lunar soil (anticipating Jules Verne's giant cannon—see 1865), and using a powder train a mile long to ignite it.

any weight to any height. Now I design to place myself in the middle of ten wooden vessels, placed one within an other, with the outermost strongly hooped with iron, to prevent its breaking. This I will place over seven thousand barrels of powder, which I know will raise me to the top of the atmosphere." As a safety precaution he provides himself with a pair of wings, fastened to his arms. The journey is successful and the selenites help him to return to the earth by digging a gun barrel one mile deep into the lunar soil (anticipating Jules Verne's giant cannon—see 1865), and using a powder train a mile long to ignite it.

ca. 1741

The anonymously written book *A New Journey to the World in the Moon* describes a spaceship very similar to the *Consolidator* [see 1705]. The first part of this little book accurately describes the conditions of outer space and the difficulties in travelling through it. " . . . a Passage thither is altogether impracticable by the Help of any *Machine* . . . for a Journey thither would require a Passage through a vast *Abyss*, or *Vacuum*, between the Atmospheres of this Globe and the *Moon's*, which could not be subsisted in by any living inhabitant of this Globe, who require a sufficient Quantity of Air to breathe in to support the Union of Animal Life . . .

"The vessel also wou'd not be out of the Attraction of the Body of the Earth, if it was possible to carry it beyond its Atmosphere, and therefore its flight must necessarily be there unsupported . . . "

The author's rather timid answer to this remarkable grasp of the problems facing future space explorers is to have his hero *dream* his way to the moon; he falls into a reverie and awakens on the lunar surface. There he meets the inhabitants of our satellite who explain that they have solved the problems of spaceflight, in spite of the objections the author has made. The Selenite's spacecraft is described in virtually the same terms as DeFoe's *Consolidator*. Its method of operation, however, is so incomprehensible as to be worth quoting here: instead of the 513 feathers of Defoes's *Consolidator*, this machine has 513 "Knotches, to receive the same number of Catches, of the Moon's Attraction"; that is, the moon's "Attraction may fix on the Superficies of its Catches, by the number of Knotches prepar'd to lay hold on the Spokes of the Chariot Wheels, whereby the Body of the Engine, or Chariot, is kept in a most regular and uniform Motion, with the Body always ascendant . . . It is, with the Persons plac'd in it, circularly drawn by a spring Screw into the Lunar World; for by the Catches of the Moon's Attraction we are loosen'd from the Gravity we otherwise owe to our own Globe, and so are transmitted, in a regular Motion, from our own more refined Atmosphere into [the moon's], without any Damages of want of Air, by the way for Support and Breathing, which are manifest Hindrances to your journeying thither after the same Manner . . . "

1742

A very borderline space travel novel is published by Baron Ludwig von Holberg: *A Journey to the World Under-Ground by Nicholas Klimius* (originally written in Latin). The title character discovers that the earth is hollow with its center occupied by the small planet Nazar. Klimius first enters into orbit around this new world (since " . . . all Bodies plac'd in Aequilibrium naturally affect a circular Motion."), eventually descending onto its inhabited surface.

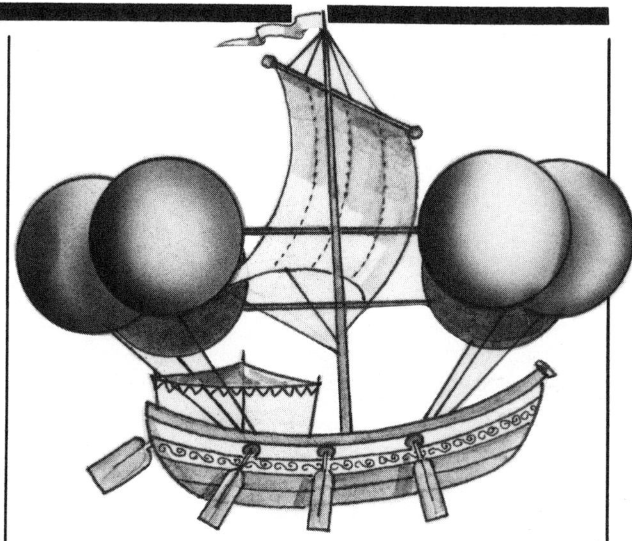

1745

Eberhard Christian Kindermann is inspired by the flying machine described by Francesco de Lana-Terzi, a Jesuit priest, in writing his *Swift Journey by Airship to the Upper World*. De Lana-Terzi's 1670 airship was a balloon (and perhaps the first theoretical description of such a device). It was to have been a gondola suspended from four spheres made of thin copper sheets. When all of the air had been evacuated from these globes, the air their volume displaced would weigh more than the globes themselves and they would rise into the atmosphere. De Lana-Terzi's only error was in ignoring atmospheric pressure which would have crushed the thin spheres like eggshells.

In Kindermann's story, five German sailors take off in a machine modeled on the Jesuit's, with the improvement of two additional copper globes, for a total of six. The gondola is made of light sandalwood and is equipped with a sail, oars and a rudder. At the stern there is an awning to protect the passengers. They stock their supplies with water and food, as well as mushrooms soaked in water. These latter are to be held to the nose to ease breathing at high altitudes. Their goal is to discover "whether it was true that on that day—July 10, 1744—the planet Mars [would appear] with a satellite for the first time since the world was in existence" . . . and thus confirming the theory of German astronomer-mathematician Leonhard Euler. The mariners circumnavigate the moon and go on to the other plan-

1745

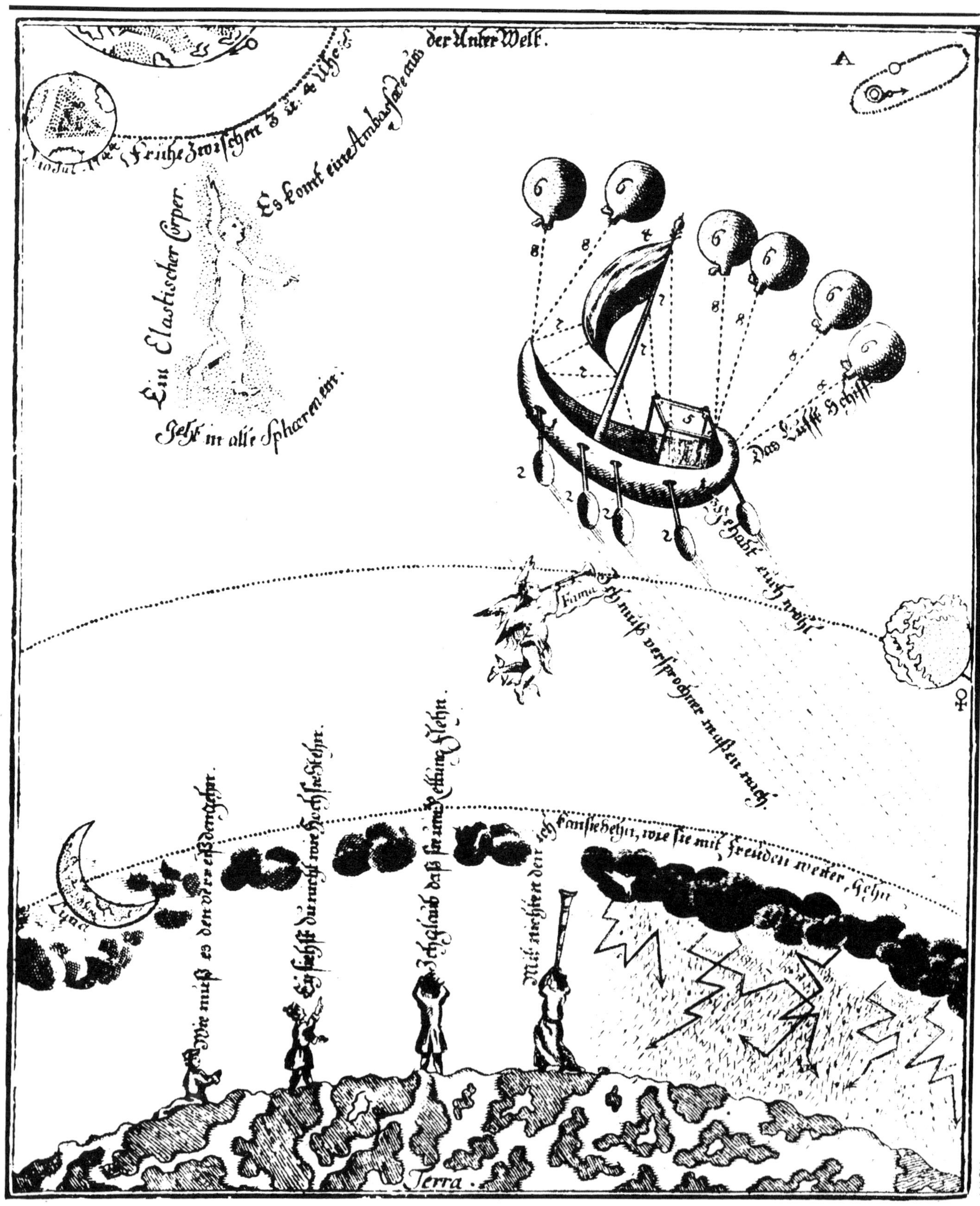
der Unter Welt.
A
Ein Elastischer Cörper
Das Lufft Schiff
Fama
Luna
Terra

NASM

ets, as their earth-bound friends watch through powerful telescopes.

Later, Kindermann suggested that flights to Jupiter could be made in order to bring back exotic plants, in the same way that "monkeys and peacocks from Asia" were being brought to Europe.

1747

Amedée Francois Frezier's *Traite de Feux d'Artifice* contains a woodcut illustration of various types of fireworks rockets including one with a remarkably modern appearance: a three-stage rocket with a delta-winged third stage.

1751

"Mr. [or Rev.] Ralph Morris" describes the flying machine *Eagle*, which is used for a voyage to the moon, in his novel *The Life and Astonishing Transactions of John Daniel*. It is a "most surprising Engine," run by the working of hand-operated pumps that operate a pair of flapping wings. The narrator, John Daniel, visits his son, Jacob, who has been working on a mysterious invention. Jacob shows him his workroom where " . . . the several pieces of which [the machine] consisted, most of which were made of iron; and though exceeding strong and tough, they were so thin, light, and taper, that I could not have imagined so great a force of iron could have been wrought into so little a weight; there were several pieces of woodwork too, and one somewhat like a pump, but all so nicely wrought, as only to preserve strength, without superfluous weight; but then, the whole being in such a number of separate pieces, it was no easy matter to conceive what sort of a figure it would compose, when each was adapted to the other; nor could I, from the best idea of its single parts, dive into several of its consequences; but this I only could observe in the general, that I never saw pieces of work better executed in my life, than the several parts, separately examined, seemed to be." Loading the machine, christened the *Eagle*, onto a cart, the father and son carry it to

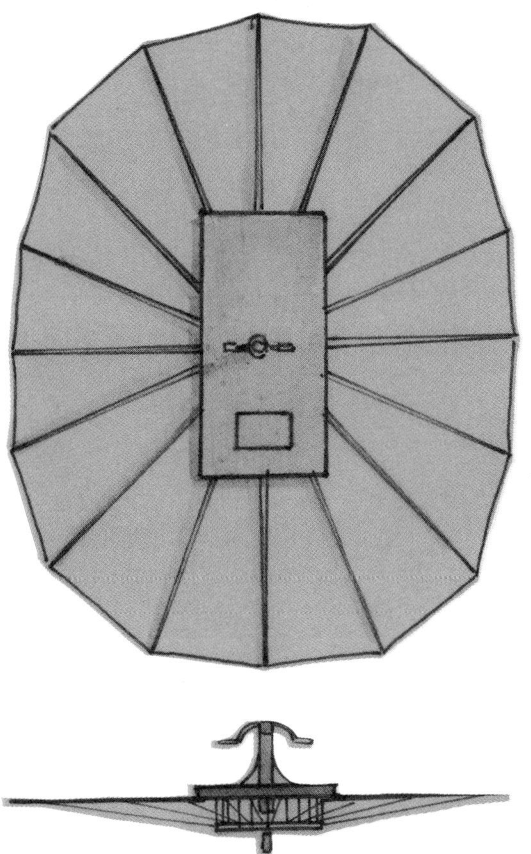

the top of a hill for a test flight.

"He [Jacob] first of all struck four poles into the earth at proper distances, measuring them with four bars, in the ends of the longest of which, on the flat sides, were four holes, into which the four points of the upright poles were to enter, at about three feet high from the ground, then letting the ends of the shorter pieces, of which there were several, all tenanted at the ends, into mortises or grooves on the inward edges of the two long pieces, he pinned them in very tight, leaving about a foot space unfilled up near one end, where he had contrived a trapdoor to lift up and shut down at pleasure: so that when the whole woodwork was framed, it looked like a stage or floor, upon which he could mount, by getting under it, and opening the trap-door.

"In the middle of this floor was a hole about four inches diameter, to let in a pipe like a pump, to the upper part of which was an handle on each side, and a pendant iron between them, which ran through the pipe be-

1751

Ordway

neath the floor; and the pipe itself was held firm in the floor by four long irons fastened to its body, and screwed down to the floor in a square figure. This was the whole form of the upper surface of the floor.

neath the floor; and the pipe itself was held firm in the floor by four long irons fastened to its body, and screwed down to the floor in a square figure. This was the whole form of the upper surface of the floor.

"Near the extremities of this floor every way, at proper distances, on the under edge, were driven in several flat and broad-headed staples, into each of which were thrust and screwed in a thin iron rib, about three inches broad next the floor, and from thence tapering to a point, at the length of about three yards, so wrought and tempered, as to be exceeding tough and elastic, with each a female screw at about three feet distant from the edge of the floor; these were all clothed with calico dipt in wax, each running into a sort of scabbard or sheath, made proper in the cloth to receive it; and being all screwed to their staples and the floor, made an horizontal superficies of calico, including the floor of about eight yards diameter, but somewhat longer than broad.

"On the under side of the floor was a circle of round iron, about five feet diameter, with several upright legs, about a foot long, equal in number to the above described ribs, and standing in the middle space between them; each of which legs entering upward through a recipient hole in the floor, was screwed tight by a nut on the upper side of the floor. Between the legs, on the interspaces of the round iron ring, just under each rib, hung balances, exactly poised upon the ring, with all their ends nearly meeting in the center, under the pipe-hole, each of which, by an iron chain fixed to it, was linked to the sucker iron of the pipe or pump, and the other end was, with a like chain, linked to an iron loop, screwed into the female screw of the rib, just placed over it; and then all the clothing was hooked upon little pegs all around the outward edge of the floor, so close as to keep the air from passing in any quantity."

"It was of almost an oval form, and each wing extended at least three yards at the sides from the floor, but at the two ends it was somewhat more; and there being a handle on each side or the other of the pump, he could make it go which way he pleased, by altering his own standing, as he told me, either on the one side or the other of the pump; for the side he stood on being the heaviest, and the other consequently mounting the highest, it would always move that way, which end was the highest."

Jacob makes some cautious tests (fearful that any over-energetic pumping of the wings might cause the *Eagle* to take off prematurely). John observes how the apparatus is operated: " . . . when the pumphandle was pressed downward, as in pumping, that in raising the sucker the pendant iron raised the end of the balances next to it, when the other extremities of the balances hooked to the several ribs necessarily descending, drew their corresponding ribs downward; and the uplifting of the handle consequently gave the ribs liberty, through their springing, to return to their horizontal position again, so that they were raised and depressed, proportionably to the motion and force of the handle, and exactly answered the use and play of wings in birds."

With the *Eagle* John and Jacob make an inadvertent voyage to the moon (they never realize where they had been until after returning to the earth).

The *Eagle* is one of the most carefully realized spacecraft in all fiction, and is a remarkable advance beyond Godwin's ganzas or Kepler's demons. It has even earned mention in a number of books on aeronautical history, such as J. E. Hodgson's *History of Aeronautics in Great Britain*. It set the stage for the carefully researched scientific verisimilitude of Poe and Verne [see 1835 and 1865].

1754

Jesuit priest Saverio Bettinell publishes his heroic-comic poem, "Il Mondo della Luna." It is a long poem, in twelve parts, that describes the inhabitants of the moon.

1758

Emmanuel Swedenborg comes to a conclusion similar to Huygens's [see 1698] concerning life on other worlds, writing: "Earths in our Solar System, which are called Planets, and Earths in the Starry Heavens . . . thousands, yea, ten thousands of Earths, all full of inhabitants."

1761

An Indian rocket corps of 1,200 men, led by Hyder Ali [see ca. 1780 and 1792] defeats the British at the battle of Panipat. Rockets were launched in salvoes of 2,000, with ranges of up to 1/2 mile.

1763

Biagio Caputi publishes his poem "Estasi e Rapimento sopra la Luna de Archerio Filo Seleno."

1764

A Trip to the Moon by "Sir Humphrey Lunatic, Bart." is published. The narrator is an admirer of the works of Cyrano de Bergerac and, during a dream, finds himself seated "in a Kind of Triumphal Car, surrounded by a great Number of human Figures . . . " These people, who are inhabitants of the moon, carry him there. The means by which they do this is explained by their leader thus: " . . . the imperceptible Method of thy Conveyance I cannot explain to thy Comprehension; let it suffice to say that some Rays of Attraction, sent down from the Mount of Observation, a Spot which from Earth appears to the Nose of the Man in the Moon, drew thee from the Place where thou lay'st asleep . . . "

1765–1766

Madame Marie-Anne Roumier publishes her novel *Voyages de Milord Ceton dans les sept planètes*. Following the death of Charles I, Milord Ceton and his sister Monime leave England and travel to seven different planets on the wings of the angel Zachiel.

ca. 1766

Filipo Morghen publishes his collection of illustrations, *Raccolta delle cose più notabili vedute dal Cavaliere Wild Scull, e dal Sigr.*

de la Hire nel lor famoso viaggio dalla Terra alla Luna (A Collection of the most notable things seen by Sir Wild Skull and Signor de la Hire on their famous voyage from the earth to the moon). The journey is made in a flying machine shaped like "a large light Box" equipped with wings. In the second edition scientist de la Hire (based on the real astronomer and mathematician, Phillipe de la Hire) is replaced by one "Giovanni Wilkins" [see 1638]. In the third edition of the book, a hydrogen balloon is attached to the car.

ca. 1770

Indian war rockets are brought to England and examined by Captain Thomas Desaguliers at the Royal Laboratory, Woolwich, England. They fail to duplicate their reported range or accuracy.

1775

Louis Guillaume de la Folie, a scientist specializing in industrial chemistry, describes a flight in an electrical spaceship in his popular science book *Le Philosophe sans Pretention ou l'Homme Rare*. Although the book is generally nonfiction, La Folie interspersed numerous little fantasies and tales with the more technical articles.

The spaceship is invented on the planet Mercury (our hero learns via a visit from a Mercurian in a dream). It is an elaborate construction of geared wheels, glass spheres, springs, wires, a metal plate rubbed with camphor and covered with gold leaf, all in a framework made of glass-covered wood. The Mercurian inventor, Scintilla, seats himself in the midst of this device. He operates the wheels, which begin to turn with great speed, and the machine lifts into the air. It flies by creating a vacuum above it, with the air pressure beneath providing the lift. A volunteer eventually makes a journey to our own planet, where he is marooned when his flying machine is destroyed in a crash landing.

NASM

La Folie's motive in writing this little tale is obvious: like Jules Verne a century later, he wants to educate his readers on a number of scientific facts by sugar-coating them with a little fiction. In the course of the story, La Folie is able to describe the conditions existing between worlds as they were known in the late eighteenth century, gravity and the planets. Like Verne, he has a character voice the objections to space travel (which, of course, are successfully refuted by the hero). The flying machine itself is patterned after the many static electricity machines being built and experimented with at this time in Europe.

ca. 1780

Rajah Hyder Ali, the Prince of Mysore, uses rockets against the British East India Company. They are iron-cased with sticks 8 to 10 feet long and weigh between 6 and 12 pounds. They have ranges up to 1 1/2 miles.

The British lion hoisted by an American rocket labled "Declaration of Independence."

PART II

THE INVENTION OF THE SPACESHIP

The invention of the lighter-than-air manned balloon was a major revolution, and revelation, in mankind's perception of his capabilities of exploring the universe. On two counts: For the first time in history a human being had gotten further away from the earth than the distance one could jump. And it was not accomplished by magic or occult means but by the use of a manmade machine, a device of science. Although by the end of the eighteenth century balloons had drifted everywhere across the landscapes of France, England and other European countries, they seldom reached an altitude of more than a few thousand feet. The moon was a great deal farther away than that, to say nothing of the planets. Nevertheless it seemed to the most enthusiastic and imaginative dreamers that if it were possible for a human being to lift himself from the surface of the earth even half a mile, then a flight to the moon was merely a matter of magnitude. This is the altered perception that is most important to realize: that by means of a manmade instrumentality, employing well understood physical principles, it was possible to leave the earth. Therefore the problem of travelling to the other worlds that shared the universe with the earth ought also be surmountable by means of science and mechanics. That is, even if the wiser heads were aware that it was unlikely that anyone would ever travel to the moon in a balloon—hot air, gas or otherwise—they were also cognizant that the idea of travelling there *somehow* was no longer a matter relegated to pure fantasy.

As the nineteenth century progressed, it began to seem to the average person and to scientists and engineers as well that there might quite literally be nothing that was beyond the abilities of science and engineering. Between 1800 and 1865, an astounded public saw the introduction of electric batteries, steam trains and steamboats, ironclad warships, photographs, gas lighting, telegraphy, high-speed rotary printing presses and color printing, electric motors, calculating machines, blast furnaces, anaesthetics, revolvers, electric lighting, typewriters, sewing machines, Bessemer converters and transatlantic telegraph cables, among literally thousands of other inventions and discoveries in technology and science. Meanwhile, engineers were building vast iron bridges, cutting canals through deserts and jungles and spanning continents with railroads.

At this same time, explorers were opening the hitherto unknown territories of Africa and the Poles. For the first time, engineers, explorers and scientists were considered public heroes; they were held in an esteem previously reserved for generals and admirals.

By the arrival of the latter half of the nineteenth century there seemed to be little that science and technology could not accomplish. In the field of space literature, Edgar Allan Poe introduced scientific verisimilitude. His novelette *Hans Pfall*, in spite of its satiric and comic overtones, was packed with realistic and well-researched details; so much so that his description of a high altitude balloon flight reads almost interchangeably with one of the stratosphere balloon flights of the 1930s or 1950s.

Poe was the first author since Kepler to take the scientific basis of a fictional story seriously and consequently was a major influence on a Frenchman who was a great admirer of Poe and his works and was an erstwhile author of scientific romances himself. If Edgar Allan Poe was the grandfather of realistic space fiction, Jules Verne was surely the father. Verne has had more positive influence on the development of astronautics than possibly any other author of fiction or nonfiction, at least until the early decades of this century; and even these latter authors—such as Hermann Oberth, Konstantin Tsiolkovsky, and others—owed their introduction to spaceflight to Jules Verne.

Whenever space travel was the subject, it was assumed that it would be accomplished by some sort of mechanical device; Jules Verne's classic *From the Earth to the Moon* (1865) is a literal paean to the engineering arts and American enterprise. However, while it seemed perfectly clear what a spaceship must be like in order to survive the ordeals of interplanetary space and, most importantly, allow its passengers to survive as well, it was far from clear what would be the most plausible method of getting the spaceship from the surface of the earth and into space. For reasons explained in the entry devoted to Verne's novel (see 1865), rockets were seldom considered. In fact, while there were numerous proposals for reaction-powered aircraft during the nineteenth century, there were less than half a dozen suggestions made that rockets might be useful in space travel. Even Verne himself, who was the first to realize that rockets would be effective in a vacuum, did not dare have his projectile launched by them.

By and large, the propulsive method of choice for the last century (and, for that matter, well into this one) was antigravity, which was, in reality, only the magic and occultism of previous centuries given a pseudoscientific guise. The important difference is that it was felt necessary to put on that guise. And no matter what the method of propulsion ultimately chosen, writers still had to deal with the known reality of conditions beyond the earth's atmosphere and on other planets. The airlessness of the interplanetary void, its extremes of cold and heat, the danger of meteorites, the problem of providing food and oxygen . . . all had to be dealt with believably, especially since the facts of astronomy were quite well known to the science-knowledgeable nineteenth century reader.Between the time of Jules Verne's two lunar novels and the flight of the first liquid-fueled rocket in 1926, most of the theoretical groundwork for spaceflight was laid, and most of the possibilities had been imagined. To mention a very few: Edward Everett Hale described the first artificial manned satellite in his novelette *The Brick Moon* (1869), in which he listed nearly every function applied to modern satellites. In 1881 Hermann Ganswindt first described his interplanetary spaceship. While never quite grasping the principles of rocket propulsion, Ganswindt did take into consideration the possible need for artificial gravity. He created this by spinning his spacecraft; he anticipated Hermann Oberth by nearly 40 years by suggesting that two spacecraft could be joined by a cable and spun around their common center. Although he made errors in detail, he was one of the first to suggest the use of rockets in spaceflight, and the drawing he commissioned to illustrate his invention is one of the few nineteenth century depictions of a manned rocket operating in space.

The year 1880 saw the appearance in *St. Nicholas* magazine of the charming short story "A Christmas Dinner With the Man in the Moon" by Washington Gladden. The giant spaceliner *Meteor* travelled to the Moon on the "great electric currents" that passed between the earth and its satellite. The iron hull of the spaceship was magnetized to take advantage of the currents. The *Meteor* was spindle-shaped and equipped with giant paddle wheels that raised it to the upper atmosphere. Because of the thinning atmosphere, Gladden equipped his passengers with respirators.

A sign of the changing times came with the publication, in 1880, of Percy Greg's two-volume novel, *Across the Zodiac*. In the story, a mysterious force called "apergy" was used to negate gravity, providing the means for a voyage through space to Mars. The spaceship *Astronaut* was a monstrous affair with 3-foot thick metal walls. The deck and keel were described as "absolutely flat, and each one hundred feet in length and fifty in breadth, the height of the vessel being about twenty feet." The apergy receptacle was placed above the generator, both located in the center of the ship, and from there "descended right through the floor a conducting bar in an antapergetic sheath, so divided that without separating it from the upper portion the lower might revolve in any direction through an angle of twenty minutes." This sheath was used to direct a "stream of repulsive force" against the Sun or any other body.

Greg's "apergy" was apparently such an appealing element that it appeared in several other novels, notably John Jacob Astor's *A Journey in Other Worlds* (1894).

It is not particularily important what happened on Mars, which was reached in a little over 40 days. What is important is that the red planet was beginning to receive the attention of writers of space fiction that it astronomically deserves. By the onset of World War I more than one hundred novels and stories had been published, all dealing with flights to Mars. All of this was a result of increasing observational knowledge about the planet itself and the development of some meticulously worked-out concepts of the origin of the Solar System, of which Mars was an especially interesting component. It was the period of the discovery of canali on Mars by Giovanni Virginio Schiaparelli (who interpreted them as naturally occuring channels or grooves) and their popularization—or perhaps sensationalization—by Percival Lowell (to whom they were artificially constructed canals), the discovery in 1877 of Mars's two small moons by Asaph Hall (which up-to-date Greg described), and of other phenomena that seemed to suggest that Mars might be much like the earth, only older. For years, Lowell excited professionals and laymen alike with his proposition that Mars was inhabited by an advanced race of intelligent beings.

Kurd Lasswitz's *Auf zwei Planeten* carried the Mars theme several notches higher in the literary scale. Published in 1897 (but not translated into English until 1971 as *Two Planets*), it took a very logical look at the supposition that since Mars was the happy abode of a higher intelligence than earth, it would not be earthmen who would first go to Mars, but rather the opposite. Thus, it was Martian space travellers who flew to the earth and set up a base on the North Pole. Why they chose such a seemingly inconvenient location is explained in the dialogue: "You must realize . . . that the Martians can only land on earth in the areas of the north or south poles. Their spaceships try, as soon as they have reached the outer border of the atmosphere, to approach in the direction of the axis of the earth. But it is dangerous for them to enter the atmosphere. Therefore, everyone agreed with the suggestion my father had made to build a station outside the atmosphere but in the direction of the axis of the earth, on which the ships would remain and from where they would descend to the earth in a different manner." The method of crossing space relied on a gravity-nullifying material, although the actual propulsion was provided by rockets. Lasswitz's novel was enormously influential on the growing interest in rocketry and spaceflight then taking place in Germany.

Capitalizing on the growing popularity of Mars, H. G. Wells wrote his acclaimed *War of the Worlds*, serialized in magazine form in 1897 and published as a book in 1898. The story dealt with the hair-raising tale of a Martian invasion of our planet. What appeared initially to be a successful conquest of the earth ended in failure as Wells had the Martians die from terrestrial diseases against which their organisms had no defense. In a retaliatory vein, American astronomer Garrett P. Serviss wrote *Edison's Conquest of Mars*, which began serialization even before the last installment of the Wells novel saw print! In this story, Serviss created the first-ever scenes of massed fleets of interplanetary spaceships. Serviss was an experienced astronomer and science writer, and while he was eventually to write far better-polished fiction, this, his first novel, has a much sounder scientific basis than even the Wells original.

In 1889 a three-volume set of novels, *Aventures extraordinaires d'un savant Russe* (The Extraordinary Adventures of a Russian Scientist) by G. Le Faure and H. de Graffigny, was published. It is a veritable catalog of imaginative spacecraft, ranging from Vernian projectiles to rockets to solar sails. The three books dealt with adventures on the moon, the inner planets, comets, asteroids and the giant outer planets. A fourth volume, *Les mondes stellaires*, was published but immediately withdrawn and destroyed with the result that only a handful of copies are reported to have survived.

By the end of the century, the spaceship as we know it today had developed.

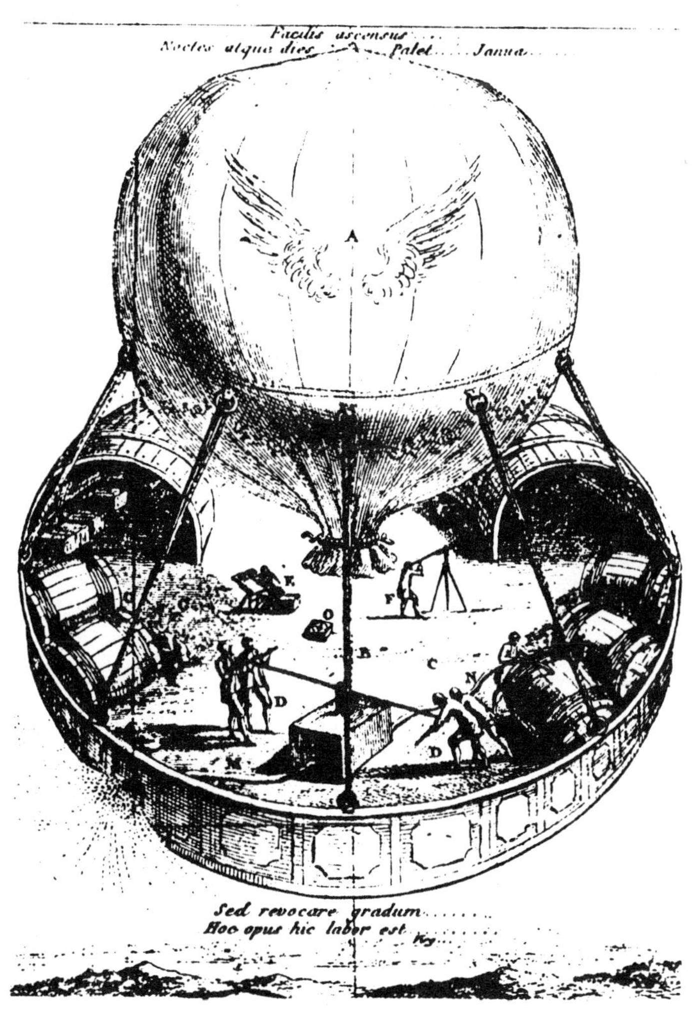

Reaction-propelled balloon from *Sur son projet de voyager avec la Sphère Aérostatique de M. de Montgolfier*, by M. de Lettre (1783).

1783

JUNE 5

The Montgolfier brothers (Joseph Michel and Jacques Etienne) make the first public experiment with the hot-air balloon they have invented. Throngs of spectators come to see the launch of the 110-foot paper bag.

AUGUST 27

The first hydrogen-filled balloon—a 10-foot globe of rubber-impregnated taffeta—is flown from the Champ de Mars in Paris. It has been invented by Jacques Alexandre Cesar Charles, the distinguished physicist. It flies to a height of more than half a mile and travels to a distance of 15 miles. Fifty thousand people witness its flight.

NOVEMBER 21

The scientist Pilatre de Rozier and the Marquis d'Arlandes become the world's first aeronauts. At 2:00 in the afternoon the elegantly decorated Montgolfier balloon rises from the garden of the Chateau de la Mouette in Paris. The adventurers rise to 300 feet and sail 5 miles in 20 minutes, landing safely in a field. For the first time in history human beings have risen above the earth farther than their muscle power can take them. It becomes immediately clear that the problem of flight to other worlds is also one that can be solved by science and technology, and that the difference between flying a few thousand feet in a balloon and a trip to the moon is only a matter of magnitude.

1784

"Vivenair" publishes a short account of a trip to Uranus (then, temporarily, called "Georgium Sidus," in honor of George III) by balloon. The story, printed in pamphlet form, is entitled A *Journey lately performed through the Air, in an Aerostatic Globe . . . To the newly discovered Planet, Georgium Sidus.*

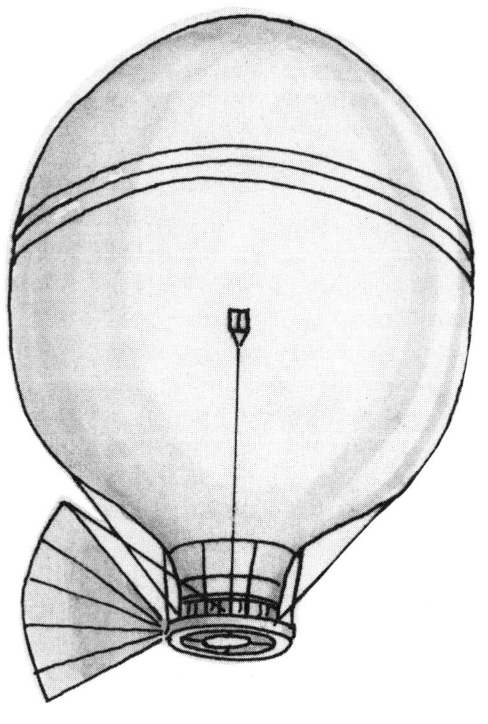

JULY 11

Two French priests, the Abbot Miollan and Father Janinet (in association with a M. Bredin), attempt the launch of a steerable, reaction-powered hot-air balloon. It is constructed with large holes in its sides, each covered with a moveable valve. The theory is that when one of the valves is opened, the hot air within will rush out through the open hole, propelling the balloon in the opposite direction. The ill-fated experiment takes place at the Luxembourg Gardens in Paris. The day is a hot one and the balloon cannot not gain enough buoyancy to lift off the ground. Ultimately, it catches fire. Its two inventors flee the grounds while the crowd of spectators tear the balloon into souvenir shreds.

In spite of their naïvete, Miollan and Janinet are among the first to attempt building a man-carrying, reaction-powered flying machine.

ca. 1784

The description of the pyrotechnics accompanying the reception of the French Ambassador by the King of Siam (recounted in *Petits de la Croix*) includes this: "There were rockets as big as one of our hogsheads, and of a proportionable length . . . The inventor of this fire-work sitting himself down on the end of one of these rockets, ordered it to be fired, and was whisked up into the air higher than any four steeples in the world could reach were they set one upon another. The rocket having spent its strength, and being ready to fall down, all luminous with the infinite number of stars that broke from it every moment, the engineer opened a sort of umbrella he had carried with him, which, when it was extended, was little less than thirty feet in diameter. This umbrella was made of feathers, and so very light, that the air supported it without any trouble . . . Insomuch that the engineer supported by this great umbrella, came to the ground, surrounded with stars, as gently as if he had wings, and could have flown with them."

The matter-of-fact style and lack of overt exaggeration leads this author to half-believe that the above describes an actual event! True or false, it is certainly one of the earliest descriptions of human flight by means of rocket propulsion.

1785

Rudolph Erich Raspe publishes his *Baron Münchausen's Narrative of his Marvellous Travels*. In one of the Baron's adventures he is carried to the moon in a sailing ship, in a reprise of the journey taken by Lukian's hero [see 160].

William Congreve, Jr., at the age of 13, writes to his father that he was "fully bent on going to the Moon in an aerial balloon . . . " and encloses a sketch of his proposed Montgolfier aerostat. [Also see 1804–1805.]

The novel entitled *An Account of Count D'Artois and his Friend's Passage to the Moon In a Flying Machine called An Air Balloon*, attributed to Daniel Moore, is published. The lunar-bound balloon is constructed and launched from France and after travelling to the moon the aeronauts meet the selenites whose "Language, Manners, Religion, etc." they are able to study.

1792

Hyder Ali's son, Tippu Sahib, enlarges his father's rocket corps from 1,200 to 5,000 men, as well as issuing larger rockets. He uses them against the British in the Third Mysore War. He defeats them several times between 1782 and the time of his death in 1799 at the battle of Seringapatam. [Also see 1761 and 1780.]

1793

"Aratus" in *A Voyage to the Moon* describes a trip to that satellite by balloon, facilitated by currents of air, where English-speaking snakes are discovered who walk upright.

1804–1805

William Congreve orders large rockets to be built to his specifications. They are constructed at the Royal Laboratory in Woolwich. Within a year his rockets are as large as 24 pounds with a range of up to 2,000 yards. His iron-cased rockets weigh 32 pounds, are 42 inches long and carry an explosive charge of 7 pounds. They are stabilized with a 15-foot stick. Congreve tests his rockets at Woolwich, increasing range by using faster-burning powders. By 1805 a rocket carrying a 6-pound lead ball could travel about 6,000 feet.

1806

Claude Ruggieri (of the family of famous fireworks artisans), in Marseilles, launches a living ram to an altitude of 600 feet, returning it safely to earth by means of a parachute. [See ca. 1830.]

1808

"Mr. Nicholas Lunatic, F.R.S.," the pseudonymous author of *A Voyage to the Moon* (in *Satyric Tales*) describes a lunar voyage by way of a hydrogen balloon steered by oars. Having "a most exalted opinion of the power of aerostatic machines," he proceeds to make "an immense bag of silk, of spherical form, more than twice the size of any balloon I had ever seen. This I strengthened by a coat of varnish, and covered it with a strong network, then filled the silk with inflammable air [hydrogen]." The car is a boat-shaped affair, with a pair of oars to guide the balloon. The narrator's motive is to "not confine my voyages to this Globe, but to endeavour to explore the unknown regions of the Moon!" The balloon almost escapes without him, but Nicholas leaps into the basket in the nick of time, and the balloon ascends "swifter than the ball from the cannon's mouth."

1811

The anonymous author of *Münchausen at Walcheren* describes the Baron's flight on a rocket, though not into space. The original edition boasts five hand-colored illustrations by Cruikshank.

1813

William Moore, an instructor of mathematics at the Royal Military Academy, publishes *Treatise on the Motion of Rockets* in which he anticipates rocket-powered spaceflight when he calculates that a 24-pound Congreve rocket, unchecked by air resistance, would reach a speed of 2,896.9895 feet per second at burnout and would therefore "never return, but continue to move forever, or fly off to an infinite distance."

1815

Edward Francesca Burney writes *Q.Q. Esq.'s Journey to the Moon*. Possibly the earliest suggestion for a "space gun" is used to launch the hero on his comic journey to the moon. [Also see 1728 and 1865.] Four large cannons are used. They are bound together, their muzzles pointed vertically. Into each muzzle is placed a heavy pole that protrudes a foot or so from the mouth of each cannon. Balanced atop these is the small, circular platform Q.Q. sits curled up upon. Burney's astronaut is also equipped with the first spacesuit: a bag-like helmet fitted with circular glass lenses and an oxygen supply. A valve in the top of the helmet allows stale air to escape. The projectile is made aerodynamic by sheathing it with a partly furled umbrella; the overall effect being that of a narrow cone with fluted sides. The four cannons are fired simultaneously, the poles

NASM

thrusting the tiny projectile aloft. Upon arriving at the moon and entering its atmosphere bottom-first the parachute opens, allowing Q.Q. to descend gently to the lunar surface.

1816

Thomas Erskine, in his novel *Armata*, describes the discovery of a planet attached to the earth at the south pole. Its surface is reached by a sailing ship. A sequel is published in the following year.

1820

The anonymous satirical short story "America in the Year 2318—a Quiz" is published in the British magazine *The Bee*, describing a journey to the moon through a tube 240,000 miles long [also see 1883].

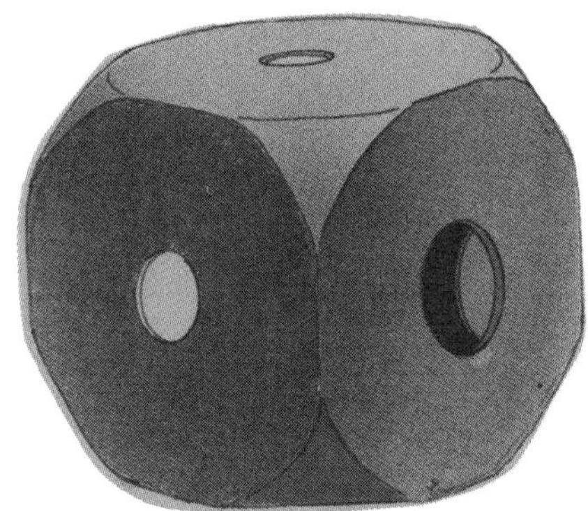

1827

George Tucker, writing as "Joseph Atterley," publishes *A Voyage to the Moon*. In it, Tucker describes what may be the first true spaceship, in that it attempts to deal with the known conditions of outer space in a realistic fashion. Lacking any available or believable motive force, Tucker has his vessel propelled by the mysterious new metal, "Lunarium." "There is a principle of repulsion as well as gravitation," writes "Atterley," and after a great deal of labor this principle is discovered to be embodied in a hitherto unknown metal. Unlike gold or lead, it has a tendency to fly *away* from the earth. Eventually it is found that while Lunarium is repelled by the earth, it is attracted to the moon.

A spaceship is constructed. It is in the form of a truncated cube (that is, a cube with its eight corners cut off) 6 feet square. It is made of copper, reinforced by strong iron bars. The strength of the cube is tested by exhausting the air inside with a vacuum pump. There is a small door in one side, just large enough to admit a body. The door is sealed with a double sliding panel, insulated with a layer of quilting. The astronauts take along a supply of air compressed in a spherical tank (partially made of Lunarium to save weight). Each of the six sides of the cube has a small circular window of thick glass set within it. The outside of the spacecraft is coated with Lunarium, lumps of which could be released by turning screws inside the vessel. In this way, its descent onto the lunar surface could be controlled. Lumps of lead are also fastened to the outer surface, to serve the same function upon the return to earth.

Tucker was a professor at the University of Virginia. He had Edgar Allan Poe as a student for one year and it has been suggested that he may have influenced his student in writing *The Unparalleled Adventures of Hans Pfaal* [see 1835].

ca. 1828

The possible earliest appearance of a well-known cartoon illustrating a man straddling a rocket-propelled flying machine is attributed to this period. Other dates ascribed to it vary from 1825 to 1841. The lithograph is captioned "Portrait of Mr. Golightly, experimenting on Mess Quick & Speed's new patent high pressure Steam Riding Rocket" and is published by C. Tilt and drawn by C. E. Madeley. The rocket is a cylinder on the upper surface of which is a saddle for the rider (with stirrups!), steam boiler and bicycle-like steering bar. The latter is connected by a pair of chains to a steering device of some kind located in the nozzle of the steam jet. On the sides of the cylinder are two labels: "Warranted not to burst!" and "Quick

NASM (2)

and Speed's Safety-Patent."

Historian Frank Winter has spent considerable effort tracing the origins of this cartoon, particularly since there is the additional mystery of the existence of an actual patent issued in 1841 (British patent #18771 of 4 January) for "Motive Power" taken out by a "Mr., Charles Golightly, of Gravel Lane, Southwark, in the County of Surrey, Gentleman." The most tantalizing feature of the patent is that it is virtually blank! Only the words "No Specification enrolled" appear. Winter has been unable to discover anything at all about the inventor (other than that Charles Golightly did once live at the address given) or even to whether Patent #8771 relates to the Golightly rocket cartoon.

The cartoon is immensely popular and will appear in one form or another on both sides of the Atlantic for many years. At least one of these variations marginally relates to spaceflight. In this ca. 1830 lithograph two figures are riding a Quick & Speed-style steam rocket, a man and a woman, illustrating the "Elopement Extraordinary, or Jack and his Lassie on a Matrimonial Excursion to the Moon, on the New Aerial Machine."

Two Golightly cartoons. Above: a German version ca. 1828.

Claud-Fortune Ruggieri, the Royal pyrotechnician of France, [see 1806] begins experiments, in Paris, of launching live animals in rockets, returning them to earth by parachute. After successfully launching a number of mice and rats, he announces plans to build a giant two-tube "combination-rocket" (not a step rocket) that will carry a full-grown ram aloft. Immediately after this announcement, Ruggieri received an offer from a "young man" who volunteered to take the place of the animal. A date for the launch was advertised, which was to take place from the Champ des Mars. Probably because the "young man" turned out to be a small boy (Wilfrid de Fonvielle—who later became a noted balloonist), the police forbade the experiment. It is not known whether or not Ruggieri ever launched the ram.

De Fonvielle himself told a substantially different version of the story (in 1869) in which he "solicited rich amateurs of extraordinary adventures to come forward with the francs necessary to enable me to repeat the experiment which Ruggieri had made upon a sheep. I declared that I was ready to be shot up in a skyrocket provided its projectile powers were carefully calculated, and that it were provided with a parachute. But it was all in vain; no capitalist presented himself."

There are substantial discrepancies in the various versions of this story. For example, de Fonvielle was born in 1826, making him far too young to have been the young man in question, and there have been several dates ascribed to the event, ranging from 1828 to 1848.

1831

In a pamphlet published by a Signor Molinari, *Scoperta della direzione del globo aerostatico*, is the description of a balloon propelled by a cluster of Congreve-type rockets.

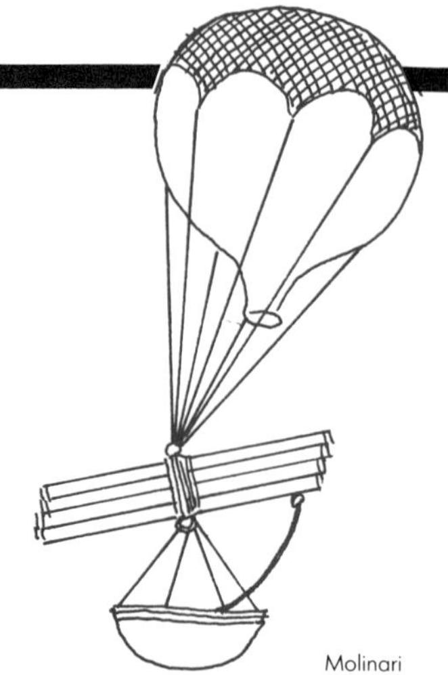
Molinari

1835

Edgar Allan Poe writes his moon-travel novelette *The Unparalleled Adventures of Hans Pfaal*. Although written as a satire it is one of the first attempts at scientific verisimilitude in space fiction. The story is crowded with specific detail which, whether it is detail that Poe made up out of whole cloth or not, contributes toward an unprecedented sense of realism.

In order to escape his creditors, Hans Pfaal decides on the daring, and original, plan of decamping to the moon. He builds a patchwork balloon out of muslin, with a wicker car "made to order." From a nameless "particular metallic substance, or semi-metal" and an equally nameless "very common acid," he creates a balloon gas 37.4 times lighter than hydrogen. The wicker basket is surrounded by an airtight rubber bag. As the balloon ascends into more and more rarified atmosphere, this will inflate into a sphere due to its internal air pressure. It is fitted with four round windows of thick glass, three around the bag's circumference and a fourth in the bottom. Also in the side of the bag is the opening for the air compressor, which supplies Pfaal with his atmosphere. A small valve at the bottom of the bag exhausts the spent air [also see Tridon, 1871].

On the outside of the bag, suspended from its bottom, is a basket with a cat and its kittens. They are Pfaal's indicators of the conditions outside his protective enclosure.

On the night prior to his departure, Pfaal inflates his monster balloon. It is the night of April 1. He attaches the wicker basket and loads it with his supplies: telescope, barometer ("with some important modifications"),

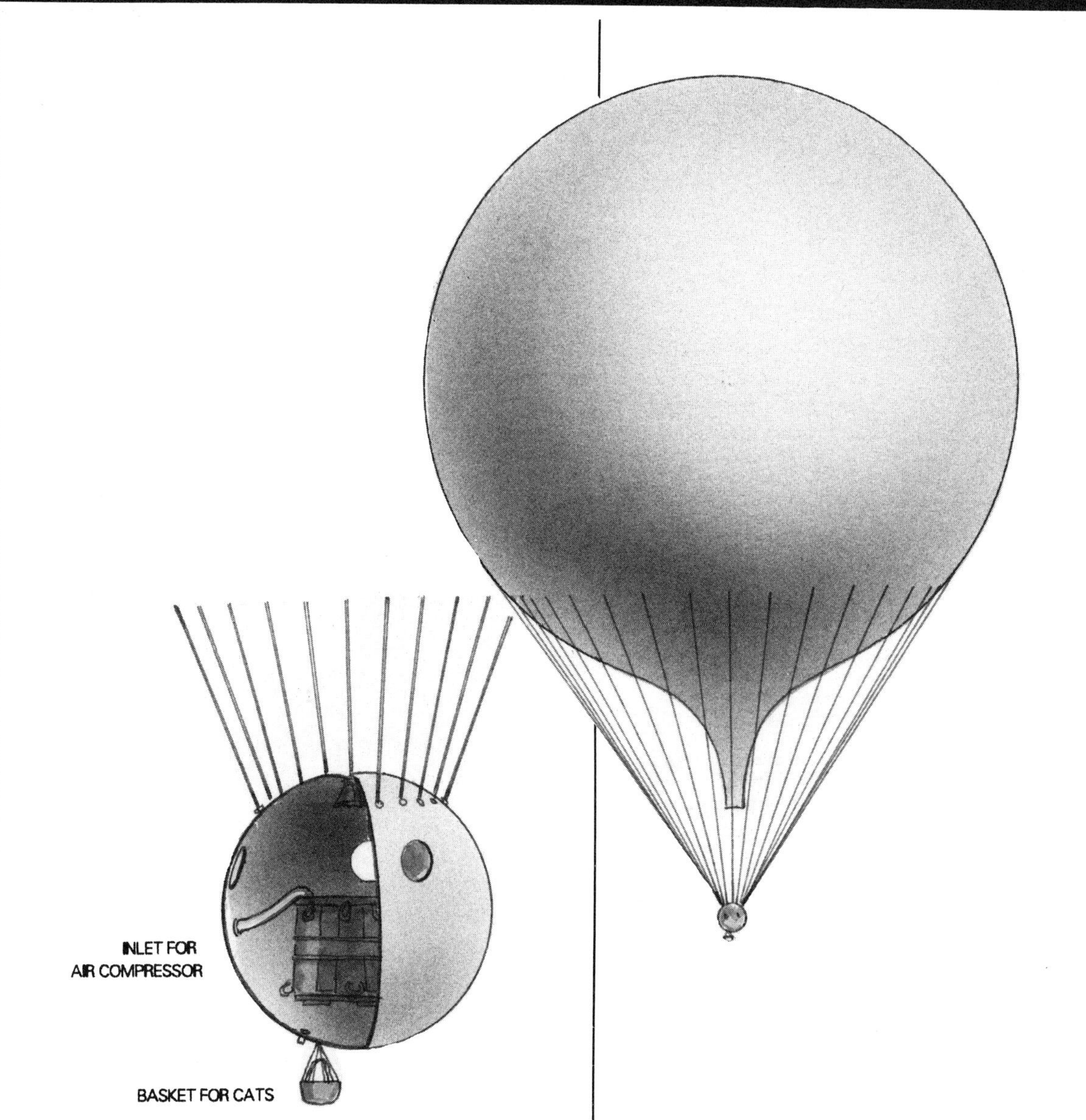

thermometer, electrometer, compass, magnetic needle, a "seconds watch," bell, speaking-trumpet, "a globe of glass, exhausted of air, and carefully closed with a stopper," the air compressor, unslaked lime [which absorbs carbon dioxide], provisions ("such as pemmican, in which much nutriment is contained in comparatively little bulk"), two pigeons and a cat.

Pfaal has his balloon inflated over a hidden pit containing several hundred pounds of gunpowder. He invites his many creditors to view his launch. At the moment of takeoff, he surreptitiously ignites the fuse leading to the powder cache. Pfaal has risen scarcely 150 feet into the air when the explosion occurs, instantly eliminating his indebtedness.

Poe accurately describes the effects of ballooning to extreme altitudes, and the appearance of the earth. These latter descriptions could have been written by the astronauts of the 1960s, for example: "By eight o'clock I had actually attained an elevation of seventeen miles above the surface of the earth . . . The view of the earth, at this period of my ascension, was beautiful indeed. To the westward, the northward, and the southward, as far as I could see, lay a boundless sheet of apparently unruffled ocean, which every moment gained a deeper and deeper tint of blue. At a vast distance to the eastward, although perfectly discernable, extended the islands of Great Britain, the entire Atlantic coasts of France and Spain, with a small portion of the northern part of the continent of Africa. Of individual edifices not a trace could be discovered, and the proudest cities of mankind had utterly faded away from the face of the earth."

At an altitude of over 8,000 miles Pfaal observes that he "found a sensible diminution in the earth's apparent diameter, besides a material alteration in its general color and appearance. The whole visible area partook in different degrees of a tint of pale yellow, and in some portions had acquired a brilliancy even painful to the eye. My view downward was also considerably impeded by the dense atmosphere in the vicinity of the surface being loaded with clouds, between whose masses I could only now and then obtain a glimpse of the earth itself."

Poe rationalizes being able to use a balloon for an interplanetary flight by assuming that the earth's atmosphere extends, albeit of extreme tenuity, nearly as far as the moon. As he explains it: "I . . . considered that, provided in my passage I found the *medium* I had imagined, and provided that it should prove to be *essentially* what we denominate atmospheric air, it could make comparatively little difference at what extreme state of rarefaction I should discover it—that is to say, in regard to my power of ascending—for the gas in the balloon would not only be itself subject to similar rarefaction . . . but, *being what it was*, would, at all events, continue specifically lighter than any compound of whatever of mere nitrogen and oxygen. Thus there was a chance—in fact, there was a strong probability—that, *at no epoch of my ascent, I should reach a point where the united weights of my immense balloon, the inconceivably rare gas within it, the car, its contents, should equal the weight of the mass of the surrounding atmosphere displaced*; and this will be readily understood as the sole condition upon which my upward flight would be arrested. But, if this point were even attained, I could dispense with ballast and other weight to the amount of nearly three hundred pounds. In the meantime, the force of gravitation would be constantly diminishing, in proportion to the squares of the distances, and so, with a velocity prodigiously accelerating, I should at length arrive in those distant regions where the force of the earth's attraction would be superseded by that of the moon."

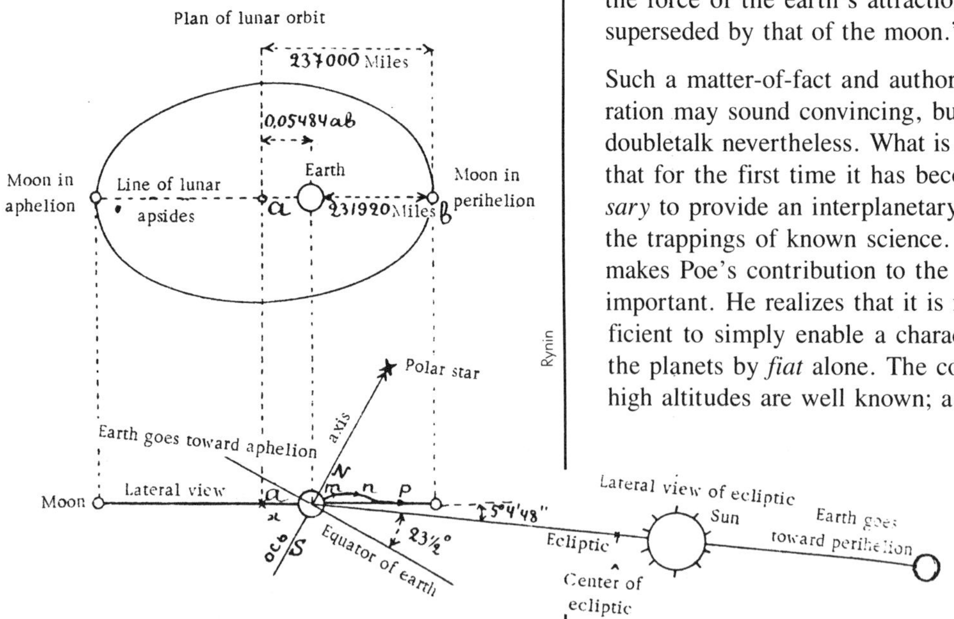

Rynin

Such a matter-of-fact and authoritative declaration may sound convincing, but it is doubletalk nevertheless. What is important is that for the first time it has become *necessary* to provide an interplanetary story with the trappings of known science. This is what makes Poe's contribution to the literature so important. He realizes that it is no longer sufficient to simply enable a character to reach the planets by *fiat* alone. The conditions at high altitudes are well known; astronomers are becoming more assured about the nature of outer space and the prevailing conditions on our own satellite. Authors wishing to write a story set in these places can no longer blithely ignore this knowledge, nor make up conditions to suit themselves.

Poe himself is entirely aware of the uniqueness of what he is doing. In a postscript to *Hans Pfaal*, Poe reminds us that the intent of earlier moon-journey stories was "always

satirical; the theme being a description of Lunarian customs as compared with ours. In none, is there any effort at *plausibility* in the details of the voyage itself. The writers seem, in each instance, to be utterly uninformed in respect to astronomy. In *Hans Pfaal* the design is original, inasmuch as regards an attempt at *verisimilitude*, in the application of scientific principles (so far as the whimsical nature of the subject would permit), to the actual passage between the earth and the moon."
Poe is a major influence on Jules Verne [see 1865].

1836

The anonymous author of *Adventures in the Moon and Other Worlds* describes a trip to the moon by a mechanical spacecraft using an unspecified means of propulsion.

1838

Boitard writes of a voyage to the sun and planets on a meteor.

Dionysus Lander, in addressing the British Association, says that "Men might as well project a voyage to the Moon as attempt to employ steam navigation across the stormy Atlantic Ocean."

1839

Rebenstein invents, in Nuremburg, a method of propelling an airplane by jets of steam or compressed CO_2.

1841

The satirical essay *The Great Steam Duck: or a Concise Description of a most useful and extraordinary invention for Aerial Navigation* is published anonymously by "A Member of the L.L.B.B." [the Louisville Literary Brass Band]. Although the steam-powered duck-shaped balloon that is described works on the principle of the ornithopter, it gains a little extra propulsion by means of the "scapepipe, passing along the bottom, [which] is conducted out of a small hole under the tail or rudder, and thus gives an additional impetus to the Aerostat, every puff." That is, the Steam Duck is at least partially reaction-powered. While this might be a little far-fetched for legitimate inclusion in this book, the Great Steam Duck does have the advantage of its charm.

1843

Emil Jir, a Russian inventor, develops a steerable balloon that uses a jet of compressed air. The reaction force will be used only to raise and lower the balloon, in order to find the most favorable winds, and not for propulsion. The source of the compressed air is to be a hand-operated compressor.

ca. 1844

Selligue suggests propelling a vessel by means of a continuous explosion of hydrogen and gaseous carbide in a metal tube at its rear.

1845

Wernher von Siemens suggests a guncotton-propelled flying machine. It is not clear, however, whether this is reaction-propelled or if the wings are made to flap by means of an internal combustion engine.

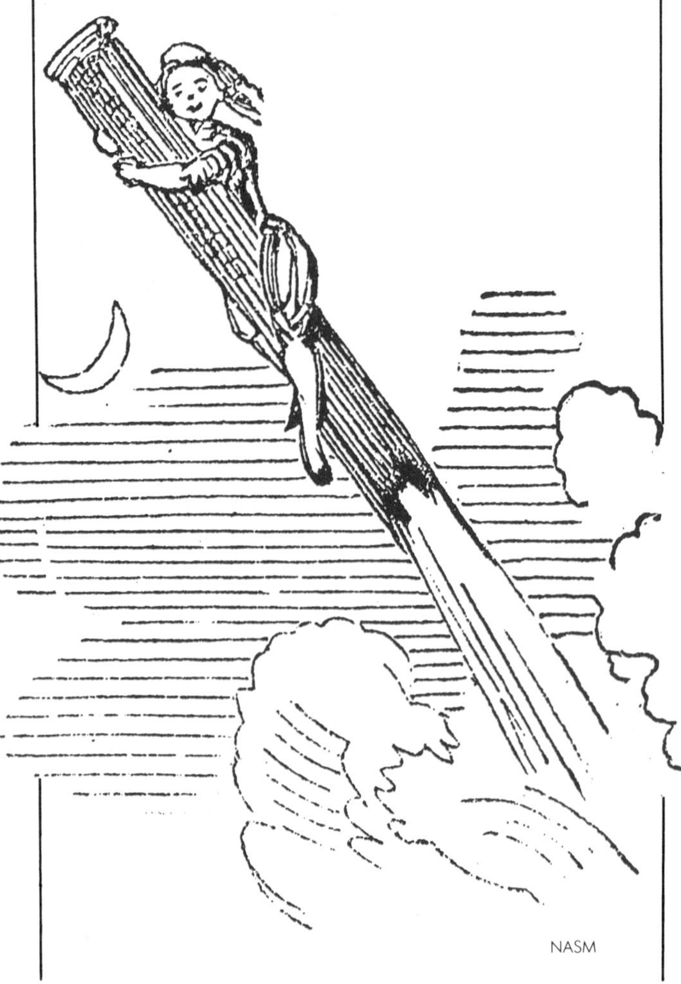
NASM

1846

C. G. Ehrenberg suggests propelling airships vertically and horizontally with rockets.

A cartoon is published in the *Illustrated London News* depicting the flight of one Joel Diavolo on a rocket. Diavolo proposed "going up on a monster rocket. Where he will come down is not yet known; but Mr. Darby [a London pyrotechnist] will endeavour to make the rocket strong enough to carry the [sic] Diavolo beyond the sphere of earth's attraction, and perhaps ultimately take up his abode there with the lunatics popularly supposed to inhabit that planet. "There is no evidence that this flight was ever actually attempted. However, it may have been inspired by the schemes of Claude Ruggieri in 1830 [see above].

1847

J. L. Riddell writes *Orin Lindsay's plan of Aerial Navigation, with a Narrative of his Explorations in the High Regions of the Atmosphere, and his Wonderful Voyage around the Moon*, in which antigravity is used, with the hero being dragged off into space by a comet.

1849

Nicholas Tretesski, a Russian engineer, in *On the methods of guiding an airship*, proposes an airship propelled and steered by a jet of steam, hot or compressed air. The airship would have jet nozzles pointing in all directions. Depending on the direction one intended to go, the appropriate nozzle would be connected to the central power source.

1851

Charles Rumball, writing as Charles Delorme, publishes *The Marvellous and Incredible Adventures of Charles Thunderbolt in the Moon*. The hero journeys to the moon

NASM

" And Travellers are out of date, I mean to cut *them* soon,
" Unless you send me some one who has travelled to the Moon."

and Jupiter by means of a steam-driven [reaction-powered?] North Pole. Although the book is intended for children, not a little care is given to solving some of the difficulties of spaceflight.

1852

Elbert Perce, in his novel *Gulliver Joi*, describes a journey to a hitherto-unknown planet, "Kailoo," in a vehicle propelled by a rocket. This is probably the first accurate, unambiguous description of the use of rocket propulsion for space travel (although Cyrano de Bergerac described a rocket-propelled vehicle in 1657, he does not get full points since the context of the story makes it clear that he did not take the method seriously, nor did he intend for his readers to do so).

The spacecraft is a hollow cylinder made of a very light substance that is nevertheless as hard as iron. It is only just large enough to contain its passenger. The rear of the cylinder is sharply pointed. The cylinder is placed in a frame to which are attached powerful steel springs. Released by a trigger, these would give the projectile its initial velocity. In the pointed end of the spacecraft is placed a strong, square steel box. This contains a newly invented powder. From one end of the box extends a small, very strong tube. When the box is heated by the "malleable flame," a kind of perpetually burning globular mass resembling molten iron, the powder inside ignites. "As long as a steady heat can be obtained enough to keep it in fusion, so long a steady blast of exceedingly powerful flame will issue from the tube of the steel box, which tube . . . extends through the aperture at the pointed end of the cylinder." The exhaust can be controlled by a stop-cock.

The spaceship is controlled by a kind of magnetic compass that automatically keeps it pointed toward the planet Kailoo. The passenger is also equipped with a powerful telescope.

The "cabin" is just large enough to allow its passenger to lie down within it. It is lined with fur and the controls are conveniently within reach.

At the time for takeoff the inventor inserts the malleable flame and "instantly a stream of fire issued from it, striking the rock with great violence . . . The old man then pulled the small trigger that confined the steel springs, and propelled by their force, and that of the flame, I shot up into the air, the long broad flame of fire streaming behind me like the blaze of a comet."

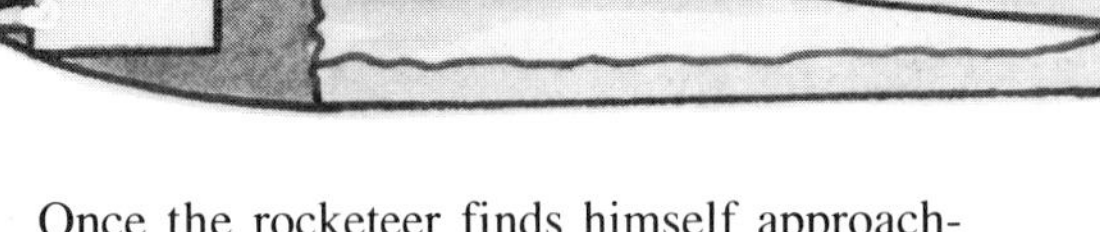

Once the rocketeer finds himself approaching Kailoo he "shut off the supply of flame that propelled" him and descends to the surface.

San Antonio Maria Claret, archbishop of Cuba, suggests using rockets to steer balloons. As Claret explains his invention: "The thrust is given by a tube placed across the balloon boat [the car or basket]. The tube must have a closed end at the front and an open one behind. It is filled with black powder so that, once ignited, a rocket-like forward thrust will be brought about . . . The amount of powder is in proportion to the speed required and in proportion to the resistance to the wind against which the balloon is to fly . . . "

One Dr. Cathelineau describes in a novel the discovery of an antigravitational substance in South America.

James Nye publishes his *Thoughts on Aerial Travelling and on the Best Means of Propelling Balloons*. He describes a 337-foot hot-air dirigible powered by 3-pound Congreve rockets fired successively every 7 seconds, using a cartridge-fed, machine-gun-like device. There is a similar device at the front end of the ship to provide control. Nye expects his airship to be capable of speeds of up to 15 mph.

1853

William Whewell contributes to the debate on the inhabitability of the planets with his book *Of the Plurality of Worlds: An Essay*. He argues against the presence of extraterrestrial life or intelligence, saying: " . . . we are brought to the conviction that God is, so far as we yet see, in an especial and peculiar manner, the Governor of the earth and of its human inhabitants; in such a way that the like government cannot be conceived to be extended to other planets, and other systems, without arbitrary and fanciful assumptions . . . "

1854

Charles Defontenay writes *Star ou de Cassiopée, histoire merveilleuse de l'un des mondes de l'espace* . . . featuring an antigravity spaceship. This might perhaps be the first interstellar voyage in fiction (although Voltaire's alien giant *Micromégas* [1752] traveled from the star Sirius).

There was an Old Man of the Hague,
whose ideas were excessively vague;
He built a balloon
to examine the moon,
That deluded Old Man of the Hague.
-- Edward Lear

1855

Miguel Estorch, a Spanish writer, describes a space gun in his novel *Lunigrafia*. Constructed in the Himalayas, it launches a 2-foot iron ball to the moon.

1856

JANUARY

Harper's magazine publishes the short story "January First AD 3000," in which a visitor from the nineteenth century is given a tour of the thirty-first. Among the illustrations by Thomas Worth Knox is one showing a huge mortar (with a bore of 3 or 4 yards) launch-

ing a pair of cannonballs linked by a chain. Each ball in one of these "bomb ferries" carries passengers since "All short distances were now traversed by bomb-carriages . . . " Also described are regular intercontinental passenger flights from Hong Kong (now capital of the world) on rocket-propelled balloons.

1857

K. I. Konstantinov experiments to determine the practicality of rocket-propelled flying machines. He concludes that contemporary rockets are "limited to one powerful impulse at the start of their action [and are] unfit for transportation of large masses during a long time [or] for considerable distances. Human force is . . . more efficient than rockets for the propulsion of aerostats."

1858

FEBRUARY 14

Pedro Maffiote, a Spanish inventor, builds a model rocket airplane. His purpose is, as he writes of it: "Having observed that the curve described by the rocket in its ascent, turning its concavity towards the direction of the wind, it occurred to me to discover the reason for this phenomenon and examine the possibility of building a machine which would regulate its own movement, making it describe a given line in space." It is constructed of bamboo and paper. Its wing is an oval disc 3 feet wide and 2 feet 3 inches deep. A single vertical stabilizer sits above the wing. Suspended beneath is an iron cylinder containing four grains of black powder. The forward end of the iron tube touches the wing, the rear end sloping down at a 30 degree angle. Its total weight is just a little over 2 ounces. It flies with a velocity of 7 1/2 to 13 feet per second.

Maffiote speculates on future improvements with these farsighted and realistic words: "Today, as the science of engineering happily moves ahead with giant strides, we can have no doubts of the possibility of building a solid, light, and larger model; but it will still be difficult to discover a chemical compound more effective than powder, having less weight while developing a greater amount of gas, and which, unlike the rocket built these days, would not be subject to unforeseen explosions . . . Today the problem of aeronavigation lies in overcoming chemistry and physics rather than in rational mechanics. We must hope that some wealthy and disinterested person will support this field of research which requires talent, perseverance, and freedom of action . . . "

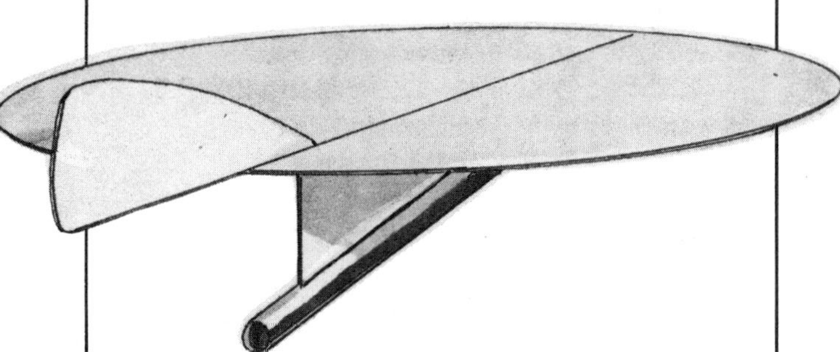

Maffiote later becomes an avid reader of Jules Verne [see 1865 and 1877].

1858–1860

Narcisco Monturiol performs experiments relative to the physiology of spaceflight by studying the effects of prolonged confinement in the submarine *Ictineo*. Monturiol studies the problems and solutions of oxygen use and regeneration, the condensation of water vapor, and temperature control as well as psychological effects of prolonged confinement in close quarters.

1860

NOVEMBER 8

Scientific American publishes a drawing similar to the Golightly cartoons of the previous decades [see ca. 1828]. The accompanying text says that "The simplest . . . of all conceivable flying machines would be a cylinder blowing out gas in the rear, and driving itself along on the principle of the rocket. Carbonic acid may be liquified, and, at a temperature of 150°, it exerts a pressure of 1,496 lbs. to the square inch. If, consequently, a cylinder were filled with this liquid, and an opening, an inch square, made in the lower end, the cylinder would be driven upward with a force equal top. 1,496 lbs., which would carry a man, with a surplus of some 1,350 lbs. for the weight of the machine." This prescient, and anonymous, author concludes that the "newly-discovered metal aluminum" would be the "proper material for flying machines."

1862

Camille Flammarion, the great science popularizer, publishes *La pluralite des mondes habites* (The Plurality of Inhabited Worlds), in which he speculates on the physiological properties of extraterrestrial creatures. [Also see 1865 and 1884.]

1863

Charles de Louvrié designs a rocket-propelled airship, fueled by hydrocarbons or vaporized petroleum oil. He calls his airship the *Aeronava* and redesigns it in 1865.

1864

The anonymous author of *The History of a Voyage to the Moon* tells the story of the journey of Crysostum Trueman to the earth's satellite after an antigravitational substance is discovered. Great care is taken in describing both the operation of the antigravity material—which is repelled by minerals, but counteracted by iron—and the spacecraft itself. Air, food, insulation and the problem of landing are all considered. The exterior of the ship is made of seasoned wood caulked with tar and the inside is covered with sheet iron. A garden is taken to provide oxygen.

1865

Alexander Dumas *Père* publishes *Voyage à la Lune*. Its spacecraft uses a substance which "is repelled by the earth."

An anonymous author publishes the novel *Voyage à la Lune* [perhaps the same as above?].

Henri de Parville publishes his novel *Un Habitant de la Planète Mars*. In the story, the fossilized remains of a Martian are discovered in a meteorite.

Camille Flammarion publishes his nonfiction work *Mondes imaginaires et mondes réels*. [Also see 1862 and 1884.]

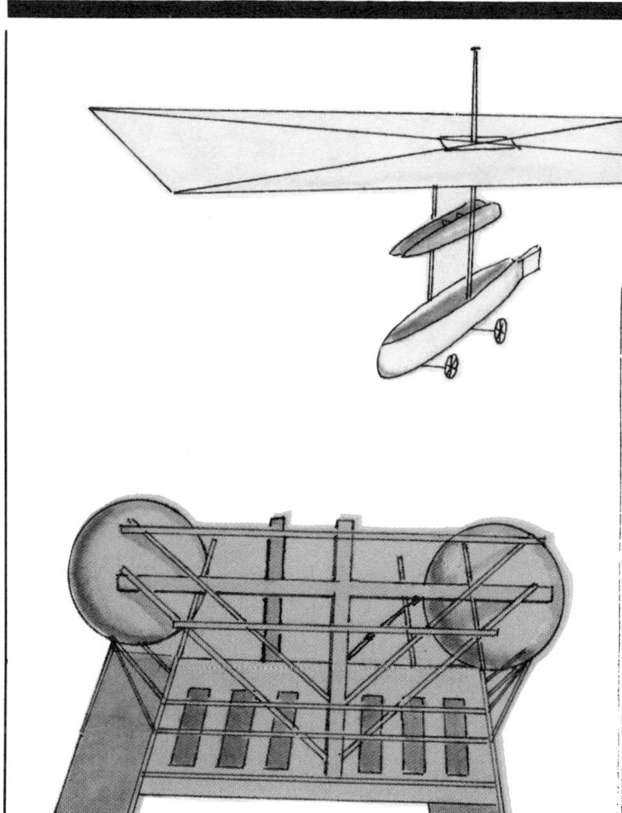

Louvrié

Trueman

Jules Verne publishes the classic space novel *From the Earth to the Moon*. It is the seminal work on spaceflight of the nineteenth century, and has had more influence on the history of astronautics than any other work of fiction. There is powerful evidence that the entire history of rocketry would have been retarded had this book not been written [see below: Tsiolkovsky, Goddard, Oberth and von Braun].

In the post-Civil War United States a club of munitions manufacturers and ex-soldiers finds itself at a loss for something to do. Its president, Impey Barbicane, makes in an announcement an extraordinary suggestion: that the means be raised to fire a cannonshell to the earth's moon.

The first sixteen chapters of the book are devoted to a detailed accounting of every stage of the operation: the mathematics of the projectile's required speed and trajectory, the time of the launch so that the projectile's path will intersect the orbit of the moon, the calculation of the size and weight of the cannonball, the size of the cannon and the amount of explosive needed, etc. In the calculations of his mathematics Verne has the help of a cousin, Henri Garcet of the Lycée Henri IV. That there are errors in the mathematics is understandable: neither Verne nor Garcet has any precedent to refer to.

Once the plan to launch a projectile to the moon is settled upon, the next decision is where to construct the giant gun. In order to intercept the moon's orbit, it is decided that the projectile must be launched within 28 degrees of the equator. The size of the gun also requires a fairly uninhabited locality. The Gun Club wishes to keep the project within the United States, which further limits the number of choices. Finally, it comes down to two: Texas or Florida. In spite of an intense rivalry between the two states for the honor, Florida wins. The Texans tell the Floridians that their state is so flat and narrow that the explosion will blow it up. "So let it blow up!" is the reply.

A site is selected at "Stony Hill," a fictitious location (in hilless Florida) not far from Tampa (the exact location is difficult to determine from the novel; Stony Hill could be in any one of three places).

To pay for the operation, an appeal for funds is made to the world at large. Bank accounts are set up in all major capitals, and official and private donations pour in. Eventually $4,000,000 is collected from within the United States and $1,446,675 from foreign nations (including a quarter of a million dollars from Russia which, says Verne, " . . . will surprise no one aware of the Russians' strong interest in science, especially their interest in astronomical research . . . "), for a total of nearly $5.5 million (equivalent to between $16 and $33 million in today's currency).

Excavation and construction of the monster gun begin. It is to be a 900 foot shaft sunk vertically into the ground. The enormous well is 60 feet wide, but lined first with masonry and stone and then with cast iron so that the final bore is 9 feet. The giant gun is christened the *Columbiad* (oddly, the projectile itself never receives a name). The Gun Club's original plans are to simply launch a solid cannonball at the moon. However, well into the construction of the gun, a sensational telegram is received:

FRANCE, PARIS

SEPTEMBER 30, 4 A.M.

TO BARBICANE, TAMPA, FLORIDA, U.S.A.

REPLACE THE SPHERICAL SHELL WITH A CYLINDO-CONICAL PROJECTILE. I SHALL BE INSIDE WHEN SHE LEAVES. ARRIVING ON STEAMER *ATLANTA*.

MICHEL ARDAN

The whole aspect of the grand project is al-

tered. Instead of merely touching the moon vicariously, a man will actually travel to that planet. Three men will go, as it turns out: Ardan, Barbicane and J. T. Nicholl. Verne based the brave Ardan closely on his friend Nadar (in turn the pseudonym of Felix Tournachon), a pioneering Parisian photographer and heavier-than-air flight promoter.

The projectile is redesigned. Instead of a spherical cannonball of immense size, it is now a bullet-shaped vehicle 12 feet tall and 9 feet wide. It is cast in one piece in the then-rare metal, aluminum, and weighs 10 tons. The interior is hollow, the walls lined with thick leather padding supported on springs. Though the walls are relatively thin, the floor is extremely thick. Entry is made into the projectile through a small manhole near the nose. There are also four portholes of thick glass, one in the nose, one in the floor and two in the walls. During the launch all of the portholes are covered with

metal plates. The air the astronauts are to breathe is provided by apparatus invented by Reiset and Regnault: oxygen produced by the heating of potassium chlorate and carbon dioxide absorbed by caustic potash. In order to prove that it is possible to survive an extended period within the projectile, Maston volunteers to stay hermetically sealed within the vehicle for a "week or so." When the time is up, Maston reappears not only perfectly healthy, but he has even grown fat!

To track the flight of the projectile and observe its landing on the moon, a telescope of unprecedented size is erected in the Rocky Mountains. Built on the summit of Colorado's Long's Peak, it has a tube 280 feet long and a mirror 16 feet in diameter. The great Hale Telescope on Mount Palomar, built in 1948, has a mirror only 4 inches wider than the one described by Verne.

The day of departure approaches, and the *Columbiad* is loaded with 400,000 pounds of guncotton. It will eventually be detonated from a distance by electricity.

The projectile is stocked with its equipment, supplies and provisions. Included are thermometers, barometers, telescopes, lunar charts, guns, pickaxes, crowbars and saws, extra clothing, seeds and shrubs, preserved meats and vegetables, brandy and water. And a pair of dogs. Lighting and heating are provided by a tank of gas. Although one is not specifically mentioned in the text, a camera is included in one of the illustrations to the original French edition.

Finally, all is ready. The launch is to take place on December 1, at 10:46:47 p.m. The imminent firing of the cannon has attracted tourists from all over the world. Hundreds of thousands have flocked to the Tampa area to witness the great event and a veritable tent city has grown up overnight. Souvenir and refreshment stands are everywhere and making fortunes.

The projectile is lowered down the shaft of the gun onto the bed of nitrocellulose. And at exactly 1 hour 13 minutes and 13 seconds before midnight, the switch is pressed. "The instantaneous result was a terrifying, incredible, unearthly detonation that could be compared to nothing already known, not to the roll of thunder, not to the eruption of a volcano. An immense geyser of flame shot from the entrails of the Earth as if from a crater. The ground heaved, and only a few of the spectators could catch a moment's glimpse of the projectile triumphantly cleaving the air through clouds of blazing vapor." The three astronauts are on their way.

From the Earth to the Moon has probably been the subject of more critical scrutiny than any other novel Verne ever wrote. It has been fashionable in the last few decades—especially since the beginnings of the Space Age—to look back sneeringly at Verne's science. One well-known science fiction author, who truly ought to have known better, referred to the speculations in *From the Earth to the Moon* as ". . . the kind of pseudo-science that gives science fiction a bad name. Everything, including Verne's elaborate calculations, is completely wrong . . . Verne was very proud of this novel and of its scientific accuracy, which is a good measure of Verne's almost total lack of knowledge about science." Yet Wernher von Braun, the father of the Saturn V moon rocket, said that ". . . the science in *From the Earth to the Moon* is nearly as accurate as the knowledge of the time permitted . . . he was read with great respect by working scientists, so carefully did he do his scientific homework." Von Braun concluded that ". . . the debt modern astronauts owe [Verne] is apparent." Why such a widespread dichotomy in opinion concerning the science in this novel? The "error" in *From the Earth to the Moon* that is inevitably singled out is the use of the giant cannon for launching the manned projectile. Rightly so: it simply would not have worked—at least as described by Verne. Going from a standing start to escape velocity (about 7 miles per second) in just 708 feet would subject the projectile's passengers to a force of nearly 30,000 g's. Barbicane, Ardan and Nicholl would have been spread into a thin coating on the floor of the projectile. This is not the worst of it: there is the air within the bore of the gun to contend with. Above the projectile is a column of air 708 feet high and

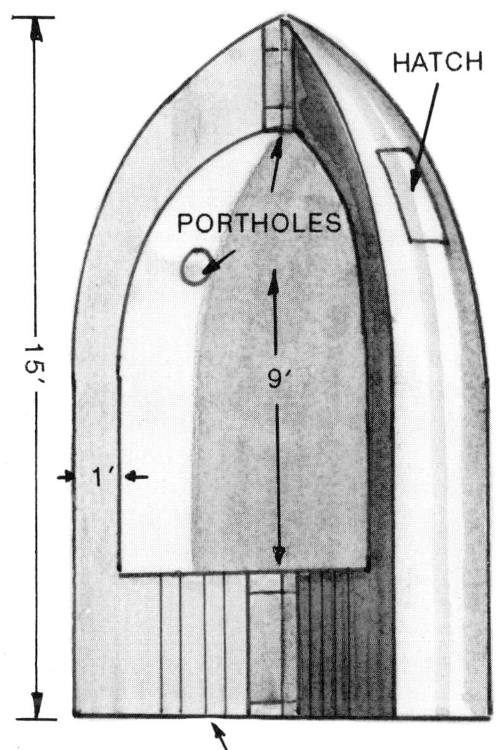

9 feet wide; there is just no way that this mass of air could get out of the way of a 10-ton projectile moving at several times the speed of sound. It would be compressed into a virtually solid body. The projectile would be trapped between the white-hot blast from below and the nearly as hot compressed air above. The projectile, instead of shooting to the moon, would instead have showered central Florida with a rain of aluminum.

But is this "mistake" indeed an error on Verne's part? Did he really not know any better? There is no direct evidence to the contrary, but there is a great deal of internal evidence within the novel that Verne, indeed, knew exactly what he was doing.

The immediate question is: if Verne was trying to write a novel showing how it might be realistically possible to reach the moon, what other methods had he at hand? With the self-imposed limitation of using only contemporary technology, his only other choice would have been the rocket. That he was aware of the possible uses of rockets in space, and that they would function in a vacuum (something few, if any, other science fiction authors, and fewer scientists, seemed to be aware of), is illustrated by his use of rockets in steering the projectile.

Why, it is usually asked by his critics, did he not use them for the entire journey?

The reason is that Verne was not only striving for accuracy, but *believability*. He was certainly aware that all of the accurate details in the world were for nought if no one believed them. And that was the problem with the use of rockets. The largest consistently successful rockets built up to the time Verne was writing were the war rockets of Britishers William Congreve and William Hale. It is Congreve's rockets that we sing of in our national anthem, " . . . by the rocket's red glare, the bombs bursting in air . . . ," when they were used against Fort McHenry in the War of 1812. The largest of these were 6.5 inches in diameter (42-pounders) with a maximum range of 3,000 yards. Congreve designed rockets with 8-inch diameters and weighing up to 1,000 pounds. These never saw service because they were then considered too heavy and impractical. Hale's rockets, developed from the 1840s, had ranges of up to 2,200 yards. These rockets may have been more accurate and more

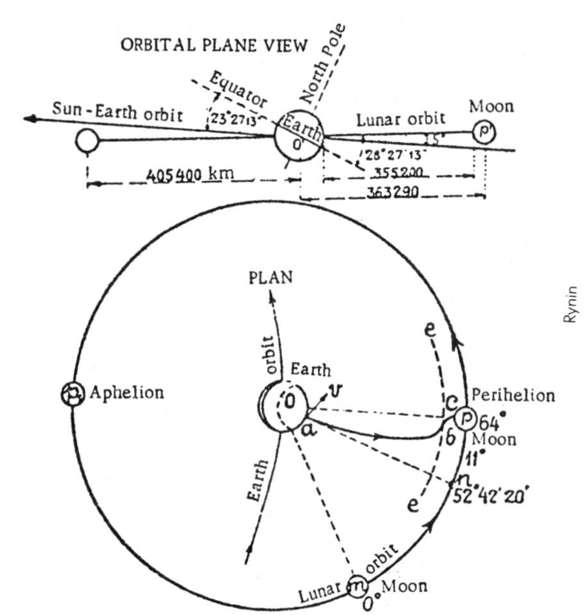

Rynin

reliable, but they were still, fundamentally, skyrockets. To have used the rocket principle in getting his characters to the moon, Verne would have had only two options: making his spacecraft a single enormous powder rocket, or making a compound rocket by combining literally tens of thousands of individual rockets (as the British Interplanetary Society did in their original moon rocket designs of 1939). Either plan would have been thoroughly ludicrous—not only realistically, but especially to readers to whom a rocket was little more than a toy. But people *did* believe in the seemingly limitless power of the cannon. As Verne was careful to point out, the decade immediately preceding his book had seen incredible advances in the art of artillery. One of Dahlgren's large guns was capable of throwing a 100-pound shell 5,000 yards, Rodman's *Columbiad* shot a projectile weighing 1,000 pounds 6 miles. There seemed little limit to the size of the gun or the shell it launched.

That Verne was aware of many of the difficulties inherent in his scheme is evidenced by the care with which he "answers" the very questions his critics raise. To this end he introduces a major character whose sole purpose is to voice these doubts, allowing Verne, through his other characters and the action of the story, to respond with believable (if imaginary) solutions. That this character criticizes the moon venture in virtually the same terms as do Verne's later critics simply shows that Verne himself was aware of the problems. It also puts into the mouth of a character the very objections which may have been forming in the minds of his readers. Some of the objections raised by this character, Captain Nicholl, are: that it would not be possible to cast a 900-foot cannon, that it would not be possible to load it (or if it were, that the weight of the projectile itself would detonate the explosives), that the *Columbiad* would burst when fired, and that the shell could not possibly rise more than 6 miles. He places a wager upon each of his exceptions and Verne, in the course of the story, has Nicholl lose each bet, one at a time.

Verne has other methods of adding verisimilitude to his story. Aware that even his most innocent reader might question the effects on the human body caused by such a violent launch, Verne has his gun work by simply *making* it work, by *fiat*. His precautions against the forces of the launch are so detailed that his average reader would simply take it for granted that Verne was aware of the dangers and had adequately prepared for them. Verne could be even more subtle than that. Before the launch of the human astronauts, Barbicane and Co. load a cat and a squirrel into a hollow shell. They are fired from a mortar several thousand feet through the air, dropping into the waters off Pensacola. When the shell is recovered, the cat leaps out unharmed (the squirrel, however, did not survive the hungry cat). "After this experiment," writes Verne, "all hesitation, all fear vanished; besides, Barbicane planned to work further to perfect the projectile and to eliminate entirely the internal effects of the firing." All hesitation and all fears of the reader vanish as well, which is Verne's real purpose for this episode.

To "eliminate entirely" the effects of the launch, Barbicane installs a shock absorber in the floor of the projectile. This consists of a layer of water 3 feet deep on top of which floats a wooden disc. The disc fits the inside of the projectile closely, like a piston. The water beneath it is separated into three layers by two thinner discs, designed to rupture at the time of the takeoff. The water is to be forced through a system of pipes, where it is ejected from near the projectile's nose (lightening the vehicle at the same time). The astronauts also plan to lie prone upon the disc on thick mattresses.

A number of space and rocket scientists have, since the publication of *From the Earth to the Moon*, amused themselves by developing variations on Verne's space gun by which it might be made to work. Max Valier, in his book *Der Vorstoss in den Weltenraum—eine technische Moglichkeit* [1924], analyzed the Gun Club's project. His conclusion, after several pages of mathematics, was that what was needed was a projectile of tungsten steel filled with lead 21.5 feet long and 3.5 feet wide. Valier's gun would, like Verne's, have a barrel 900 feet long. This one, however, would have to be situated on the equator at an altitude of 15,000 feet. The barrel would be cast in concrete with a rifled steel lining. Before the launch, the air would be evacuated from the gun and its muzzle sealed by an airtight disc. At the moment of firing, the small amount of residual air ahead of the projectile would be sufficient to blow off the cap before the projectile reached the opening. There would be no possibility of carrying human passengers.

Somewhat later, the Baron von Pirquet examined the problem. He made the required mountain 20,000 feet high, placing even more of the earth's atmosphere below the gun. Finding that the gases from the detonating explosive would not expand fast enough if the whole charge were placed in the bottom of the gun, he placed most of the charge on the bottom of the unmanned projectile itself. He also supplied supplementary firing chambers along the length of the barrel [see LeFaure and de Graffigny, 1889].

Others, such as Tsiolkovsky (who in 1895 expressed the opinion that a gun several kilometers long, placed horizontally, might work if the astronaut were immersed in liquid at the time of launch; Gen. Antonio Stefano thought a projectile 150 mm in diameter would need a gun only 800 meters long), also attacked the question of Verne's gun and there, of course, have been fictional space guns other than Verne's. There have

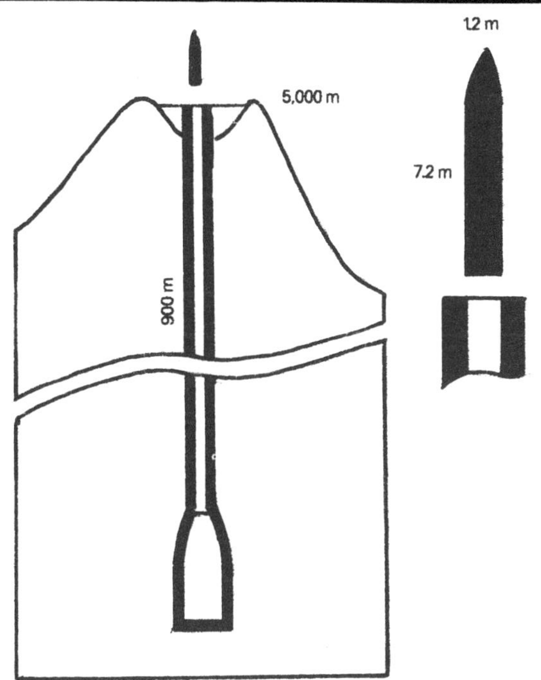

also been serious proposals for actual space guns and one for firing instrument-carrying shells has actually been used.

Verne's accurate foretelling of space travel and its problems far outweighs any errors he may have made in the telling. Among the many "firsts" Verne could claim for this novel are: the first approach to spaceflight on a mathematical basis; *From the Earth to the Moon* can be rightly considered the grandfather of astronautics. He was the first to approach the problem of space travel on realistic terms. Verne correctly understood the operation of rockets in the vacuum of space and was the first to suggest their use [although also see Perce, 1851]. He was the first author to appreciate the necessity of reaching escape velocity to leave the earth. Verne was aware of many of the special needs of space travellers and allowed for them; and he described the effects of weightlessness (though he was wrong in explaining its cause). Verne chose a site for his launch in Florida that is only 137 miles from Cape Canaveral. This was no coincidence; Verne chose this site for at least one of the same reasons NASA did. Verne was also cognizant of the many social effects a space program would initiate; he even described the commercialization of his moon launch. Verne was also one of the first to suggest the possibilities of "light . . . as [a] mechanical agent" for spaceship propulsion.

From the Earth to the Moon has been enormously influential on the history and development of rocketry and spaceflight. Tsiolkovsky, the great Russian theoretician who laid a great deal of the mathematical groundwork for modern astronautics, said, "Possibly the first seeds of the idea were sown by that great fantastic author Jules Verne—he directed my thinking." Hermann Oberth, "the father of the V2," wrote, "I always had in mind the rockets designed by Jules Verne." [Also see 1905–1906.] Already mentioned is von Braun's expression of debt. Other space pioneers Verne influenced include American rocket pioneer Robert H. Goddard; rocket engine designer Valentin Glushko (who wrote that "at the age of 13, while studying at a technical school, I read two books by Jules Verne . . . which shaped my life-long interest."); cosmonaut Yuri Gagarin, the first man into space; and astronaut Frank Borman. Astronomer Robert Richardson, in *Man and the Moon* (1961), wrote, "There can be no doubt that Jules Verne's *Trip to the Moon* with all its faults has exerted a powerful effect on human thought in preparing our minds for this greatest of all adventures."

Achille Eyraud describes a "moteur à reaction" which is used to propel a space-travelling balloon in his novel *Voyage à Vénus*. Eyraud's novel is often described as preceding Verne's in predicting the use of rockets in space travel; however, since *Voyage à Vénus* is published in the same year as *From the Earth to the Moon*, the point is a moot one [also see 1851].

Eyraud correctly explains that the reaction effect that propels his spacecraft is the same as that produced by the recoil of a gun and that it flies in exactly the same way that a skyrocket does, although it is clear that Eyraud does not understand the principles involved.

The spaceship in this novel is not propelled by a true rocket, in the sense that it operates by the reaction of gas produced by a burning fuel. Although Eyraud uses the recoil of a gun to explain the operation of his "moteur," it works by ejecting water. This is recovered in a container towed behind the spacecraft, which would, of course, negate any reactive effect: the spaceship would simply stand still.

1866

N. Sokovnin, a Russian inventor, proposes in *The Airship* a dirigible propelled by jets of compressed air.

October 13

Leslie's Weekly publishes the story "To Venus by Balloon."

1867

Nicholas Telescheff proposes a jet-propelled airplane.

Butler and Edwards, two Englishmen, design or patent (but do not build) several delta-winged reaction-powered aircraft. One is a catamaran-like double-delta design propelled by pusher propellors, which are in turn to be propelled by jets in the tips of the blades.

1869

Edward Everett Hale, in his novelette *The Brick Moon*, is the first to describe an artificial earth satellite, as well as the first to suggest the navigation satellite.

Space scientist John Nicolaides has compiled a long list of the forecasts made in this story: Hale foresaw the navigation satellite as well as satellites used for geodetics, mapping, reconnaissance, communications, sea surveillance, bioscience, meteorology, human habitation, and orbital rendezvous. It must be admitted, however, that some of these require a little definition stretching on Nicolaides's part.

In the story, which Hale writes with his tongue firmly in cheek, a group of American entrepreneurs decide to launch a satellite into earth orbit. It is to act as a navigational

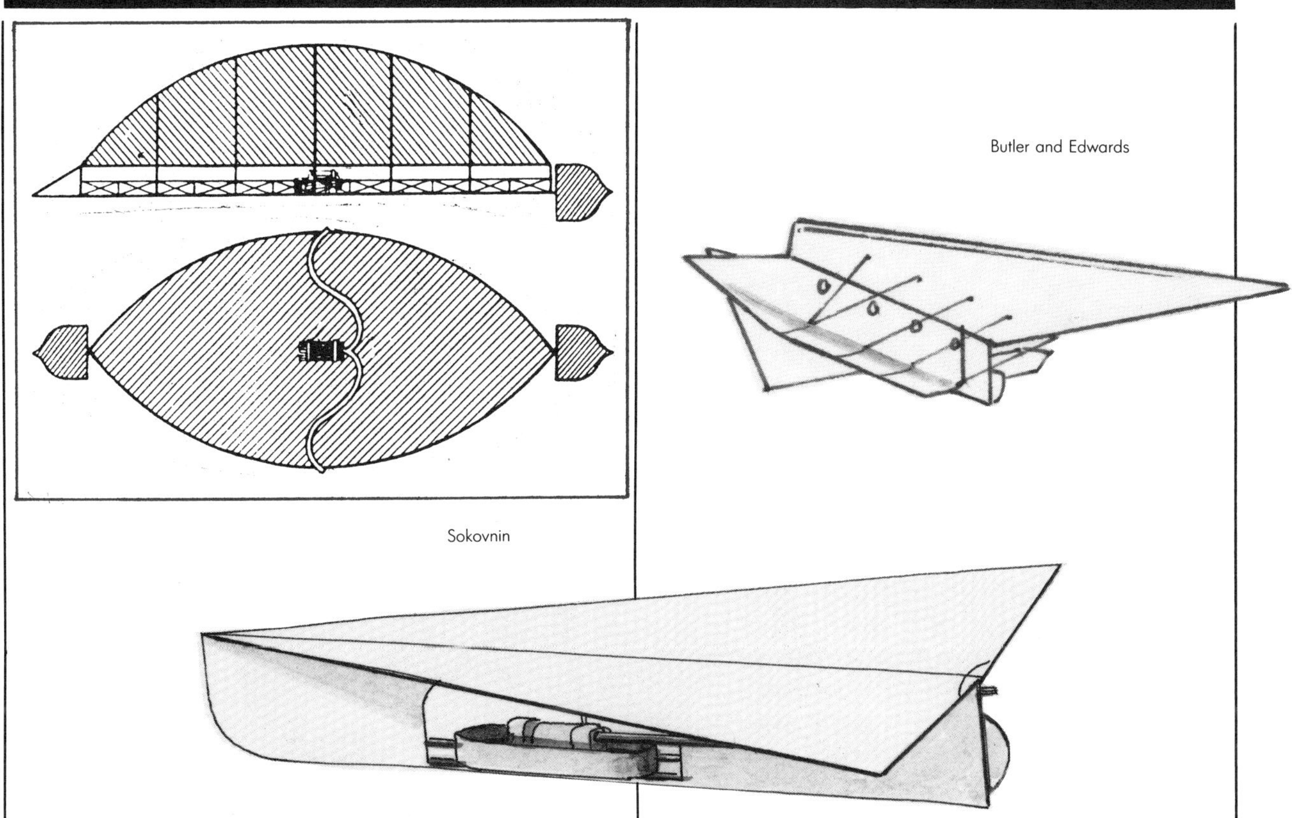

aid to mariners, its polar orbit allowing them to accurately calculate their longitude in the same way they determine their latitude (by measuring the altitude of Polaris, the north star). If the initial effort is successful, three additional satellites will be launched, at slightly different altitudes (all circa 4,000 miles) so their orbits will not intersect, which will allow one of them always to be visible from anywhere on the earth.

The Brick Moon is to be an enormous sphere of brick, 200 feet in diameter. Brick is chosen for its refractory properties, to withstand the friction of the moon's passage through the atmosphere at the time of launch. The moon is hollow, containing thirteen smaller hollow spheres of brick. The resulting internal arches give the moon great strength. To save weight even further, circular holes are left in the interior spheres where they are not in contact with one another.

The launch is to be effected by the use of gigantic flywheels. Two of these are constructed near a waterfall where a huge waterwheel powers them. These are built of wood with heavy iron rims. The flywheels are set up vertically and in the same plane, their rims almost but not quite touching. They are spun in opposite directions, so that anything dropped into the space between them will be thrown violently upwards. One of the two wheels is slightly smaller, in order to apply a slight deflection to the path of the moon. It is not desired to launch the moon vertically, but rather in a path that will put it into the proper orbit. The Brick Moon, once completed, will be allowed to slide down a ramp onto the juncture of the two spinning flywheels. Hale spends considerable time going into the details of the Brick Moon's construction, and the methods of raising the funds to build it.

After the builders have been absent from the site for some days, they return to discover the Brick Moon missing! A heavy rain has shifted the ways, the moon has slipped down onto the flywheels and has been prematurely launched into space. Worse, the workmen and their wives had taken to living within the nearly completed moon and they have been carried along with it. The Brick Moon is assumed to be lost until reports of a new asteroid begin to arrive from astronomers

around the globe. Of course it is the Brick Moon.

It is quickly found in telescopes. Oddly, it is no longer red, but green and small black specks are seen moving upon it. The specks are the workmen, who survived the launch. Communications are established (the workmen jump up and down *en masse* on the limb of the moon, in Morse code, while on the earth large letters of cloth are placed on the ground—until a better method is invented later). Some of the Brick Moon's mysteries are explained: the workers are surviving because "Our atmosphere stuck to us" (the barometric pressure is 0.3 inches "our weight"). There is regular rainfall and plant life is not only flourishing but rapidly evolving ("Write to Darwin that he is all right. We began with lichens and have come as far as palms and hemlocks."). It is orbiting at exactly 5,109 miles above the surface of the earth.

The "Bricks" (as the inhabitants of the moon are known) observe the earth from space and transmit their observations to terrestrial scientists. They watch storms below, let geographers know that the North Pole is an open sea and the Antarctic a cluster of islands and clear up some mysteries about the interior of Africa.

Although the Bricks are virtually independent of the earth, there are a number of personal and luxury items they require and it is decided to launch them to the Brick Moon. A quantity of goods are wrapped securely within several heavy woolen carpets. This parcel is placed inside a large bag filled with sand. This in turn is placed within another, larger sand-filled bag—until there are five such layers. The idea is that as the bundle passes through the atmosphere, each layer would in turn burn away by friction with the air. This is similar to the method employed in the ablative heat shields used in modern spacecraft.

The experiment is less than successful. The bag bursts somewhere on its way to its rendezvous with the Brick Moon. The message from the disappointed Bricks reads, "Nothing has come through but two croquet balls and a china horse." The remaining goods go into orbit around the Brick Moon, forever out of reach ("They had five volumes of the *Congressional Globe* whirling round like bats within a hundred feet of their heads.").

Hale's story is, as mentioned already, basically humorous, and his science is a mixture of mostly very good and a little very bad. Nevertheless, it is well received, even by the scientific community, Hale even getting a kind note from Asaph Hall, the discoverer of the twin moons of Mars. Yet, among the still very readable humor, are two of the most advanced ideas suggested during Hale's century: the artificial satellite and the navigational satellite. The Brick Moon is ultimately nearly duplicated, in form as well as function, in the 1960s by the 135-foot diameter Echo balloon satellite, and by the first navigational satellite in 1962.

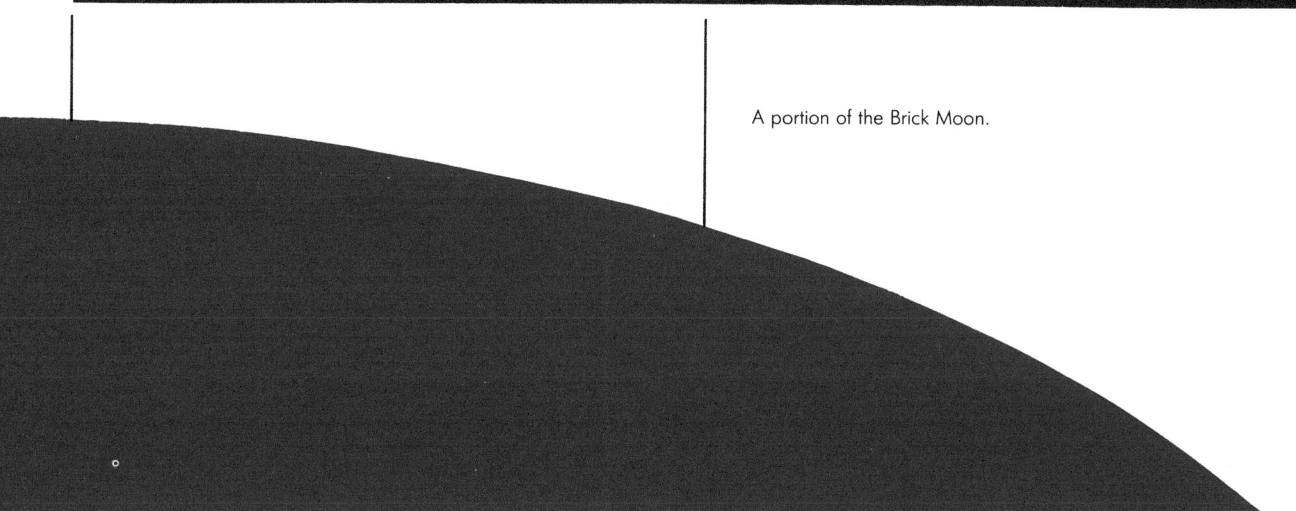

A portion of the Brick Moon.

1871

Tridon, a French inventor, designs a pressurized balloon gondola for high-altitude flights. The pilot keeps the airtight cloth capsule pressurized with a hand pump and breathes compressed oxygen [see Poe, 1835].

An alleged attempt to launch a manned skyrocket, in Delaware, ends in an explosion and the loss of an unknown rocketeer, described as an "old Delaware philosopher, ambitious of aeronautic fame" (as reported in the Transatlantic Clippings department of the London *Graphic*). The rocket ascended "at an enormous speed. But alas . . . the rocket and its parachute were seen to turn sharply in mid-air and fall, while the rash aeronaut was found close to his laboratory fearfully burned and mangled."

1872

SEPT.–OCT.

Frederico Gomez Arias, director of the Escuela de Nautica in Barcelona, presents a paper entitled *Memoires Concerning Aerodynamic Propulsion*. In Part One, he describes what he calls his "aero-dynamic motor-propulsor." Even though he makes no mention of spaceflight (despite being a reader of Flammarion and Verne), his work is important in establishing many of the characteristics of the rocket-powered aircraft. He anticipates many of the innovations of later theoreticians such as Tsiolkovsky (propellants), and Ganswindt (propellant feed systems).

In *Colleccion de Problemas* he writes: " . . . a cosmic object subjected to Earth's gravitational force, falling should not achieve when rising from the terrestrial surface a velocity greater than 11,200 m/s, that is, the same that should be imparted to this object to lose it in the infinity of space and prevent its descent. When boosted horizontally, a speed of 8000 m/s is enough to convert the object into a small satellite describing a curve of greater radius than the Earth's . . . "

Arias describes a wingless, vertical takeoff and landing (VTOL), rocket-propelled vehicle. Arias's craft has an enclosed nacelle for the pilot: a large bucket-shaped cockpit made of heavy rubber. The engine is supported on three struts above the pilot. Below the pilot's bucket are pneumatic shock absorbers for landing.

Arias's engine has a central combustion chamber with two horizontal, projecting tubes. The ends of these are turned down to provide vertical lift. The combustion chamber is fed by a revolver operated by a kind of paddlewheel in front of one of the exhaust vents. A second pair of nozzles faces horizontally. By turning a valve, the escaping gases will be diverted to these for level flight. The materials Arias suggests for his machine are iron, steel and aluminum and, for the firing chamber, a platinum-iridium alloy.

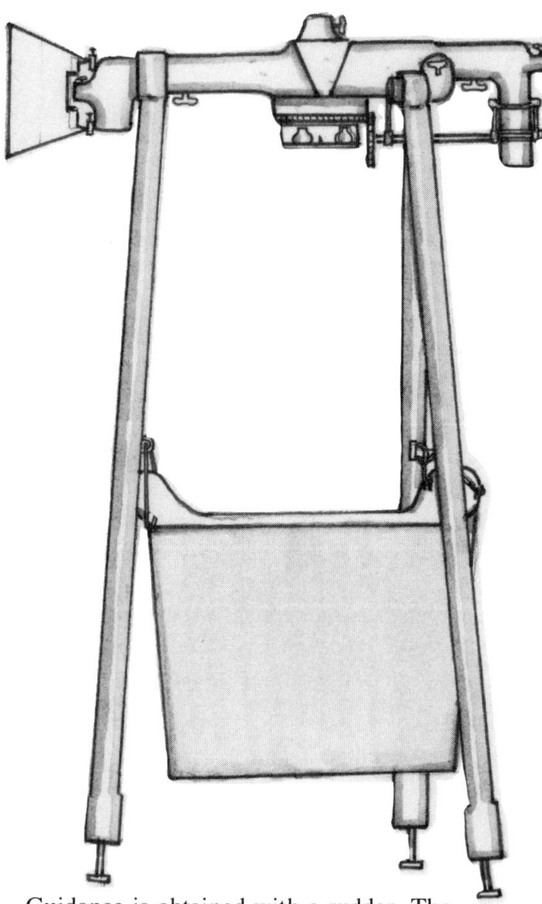

Guidance is obtained with a rudder. The rocket engine, he writes, "is based on the continuous production of gases, that, thrown into the atmosphere at a high calculated speed through a convenient section produces the necessary reactive effect for propulsion . . . " For propellants, he suggests gunpowder, guncotton (nitrocellulose) or nitroglycerine, as well as the possibility of hydrogen-oxygen. Like numerous others, Arias makes the error of believing that thrust is produced by the action of the escaping gases pushing against the air.

Arias does consider the problems that might arise if his craft were to be piloted to extreme altitudes. He arrives at several solutions to the problem of air storage and renewal, and the removal of CO_2 wastes. He suggests the use of the Rouquayrol diving apparatus (also used by Jules Verne in *20,000 Leagues Under the Sea*). In 1890 he publishes a drawing of a sealed suit intended to be worn at high altitudes. It is connected to an air compressor and a CO_2 removal system. If it is not the first serious design for a spacesuit, it is certainly the first high-altitude suit [see 1890].

1873

"Paul Aermont" publishes the juvenile novel *Narrative of the Travels and Adventures of Paul Aermont Among the Planets.*

Ivanin, a Russian general, proposes powering aircraft with existing war rockets.

1873–1876

In Moscow, Konstantin E. Tsiolkovsky speculates on the possibility of achieving "cosmic velocities" with centrifuges. [Also see 1913–1916, 1916, 1927.]

1874

Andrew Blair publishes his novel *Annals of the Twenty-Ninth Century*. A flight to the moon is described, which is made hazardous because of the many tiny satellites which clutter the space around the earth. The flight fails after travelling only 50,000 kilometers in 4 weeks. Later tries are successful and the moon is eventually colonized. In the third volume of the trilogy, trips to Venus and Jupiter are described.

1875

Arthur Penrice describes, in his novel *Skyward and Earthward*, a flight in a specially designed balloon to the moon and Mars. A unique feature of the book is a hand-colored Mercator map of Mars on which is traced the routes of the explorers.

The four-act, twenty-three-scene play "Voyage dans la Lune" is produced in France. Under the title "A Trip to the Moon" it is presented as an opera at New York's Booth Theatre (opening on March 19, 1877), with music by Jacques Offenbach. It was adapted from the Jules Verne novel [see 1865] by MM. Leterrier, Vanloo and Mortier.

In the operetta the voyage to the moon is ini-

A steam jet-propelled balloon from *Scientific American* (1872).

tiated by a cannon 20 leagues (about 50 miles) long!

1877

Jules Verne publishes his second interplanetary novel *Hector Servadac* (published in English as *Off on a Comet*). The earth, in the region of Algeria, is hit a glancing blow by a small asteroid. It carries off with it bits and pieces of dry land, water and a handful of humans. The asteroid, named Gallia by its reluctant inhabitants, follows a highly eccentric orbit that takes it as far as the orbit of Saturn. One of the Gallians is an astronomer who calculates that the asteroid will collide once again with the earth as its orbit swings it back toward the sun. Preparations are made which take the form of a large hot-air balloon. The survivors of the original collision take to the air as Gallia once again grazes the earth and are safely transferred home.

Verne's novel takes great pains to accurately describe conditions on a world much smaller than the earth. The Gallians also have to deal with the cold and darkness of deep space, to say nothing of the problem of food and water on the barren worldlet. Verne's descriptions of Jupiter and Saturn—especially a passage describing the latter's rings as they would appear to a Saturnian—are beautifully done.

1880

Percy Greg publishes his interplanetary novel *Across the Zodiac*. In it he describes the enormous spacecraft *Astronaut*. (This is certainly the first use of the word "astronaut", though it is not used in the modern sense. Greg was simply playing upon the name Jason's men were given when they sailed aboard his ship, *Argo*, in search of the Golden Fleece: the Argonauts. The word "astronautique"—in its modern sense—was

coined in 1927 by J. H. Rosny and was quickly adapted in English as "astronautics." "Astronaute" [English "astronaut"], as it is used today, seems to have been first used by Robert Esnault-Pelterie ca. 1928—also see December 26, 1927.)

The two-volume novel describes a trip to Mars in the year 1830. The *Astronaut* which the nameless hero constructs for the journey is a huge vessel 150 feet long, 50 feet wide and 20 feet high, with metal walls 3 feet thick. "my *Astronaut*," writes the narrator, "somewhat resembled the form of an antique Dutch East-Indiaman [a resemblance belied by the following description], being widest and longest in a plane equidistant from floor to ceiling, the sides and ends sloping outwards from the floor and again inwards towards the roof. The deck and keel, however, were absolutely flat . . . " There are glass windows along the sides and in the ends, as well as crystal lenses in the floor and roof that act as telescopes.

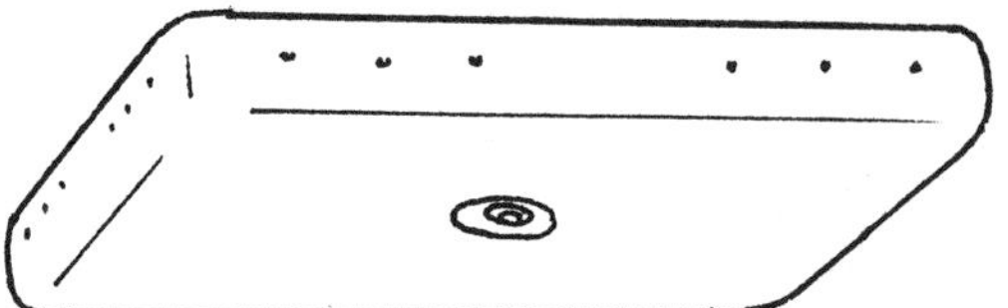

The interior is luxuriously appointed. The floor is carpeted with alternating layers of cork and cloth. "At one end I placed my couch, table, bookshelves, and other necessary furniture, with all the stores needed for my voyage, and with a further weight sufficient to preserve equilibrium. At the other I made a garden with soil three feet deep and five in width, divided into two parts so as to permit access to the windows." The purpose of the plants is to absorb waste. The narrator also hopes to cultivate them on Mars, if possible.

In the center of the *Astronaut* is the propulsion system, a mass of machinery 20 feet by 30 feet. "The larger portion of this area was, of course, taken up by the generator, above which was the receptacle of the Apergy [also see Astor, 1894]. From this descended right through the floor a conducting bar in an antapergic sheath, so divided that without separating it from the upper portion the lower might revolve in any direction through an angle of twenty minutes (20′). This, of course, was intended to direct the stream of the repulsive force against the sun . . . "

The *Astronaut* is another spaceship where its author feels forced to resort to the use of antigravity, in lieu of anything else he feels would be less believable. Unlike Tucker's Lunarium [see above], Apergy is a repulsive force only. Greg understandibly gives few details concerning its operation. We are told that it "acts through air or in a vacuum in a single straight line, without deflection, and seemingly without diminution." Instead of being a physical substance like Lunarium or H. G. Wells's Cavorite [see below], Apergy is a force which must be generated. It is carried through a bar sheathed in an "anapergic" insulation. To move the *Astronaut* it is only required to "turn the repulsion upon the resistant body (Sun or planet), and so propel the vessel in any direction I pleased."

S. S. Nezhdanovsky begins studying the principles of jet propulsion in the 1880s. In July of 1880 he suggests the possibility of jet-propelled aircraft, saying: "A jet projectile can be made with the use of an explosive; the products of its burning are discharged from an ejector type device." By the end of that year he concludes: "I suppose that we can and should build the aircraft. It will be able to carry a man in the air for five minutes at least." As a source of reactive power he considers, for example, a magazine-type gun not dissimilar to that later developed by RobertGoddard.

Between 1882 and 1884 he begins developing his idea for a jet nozzle in which the escaping burnt gases carry with them a quantity of ambient air as additional reaction mass. As fuel Nezhdanovsky considers nitroglycerine, carbon dioxide, gunpowder, compressed air, steam, and various explosive mixtures, including a combination of oxidizer and fuel.

In 1882–1884 he proposes the possibility of using liquid fuels, as well as using the fuel itself to cool the engine by surrounding it.

"Isn't it possible," he writes, "to make a flying inclined surface with a rocket to supply speed in a horizontal direction? Which is more profitable, a single rocket or a rocket with an inclined surface? To sketch a rocket with an inclined surface, isn't it the most simple flying machine?"

This idea is more or less realized by Geshvend in the design of his rocket-propelled flying machine [see 1887]. Nezhdanovsky's jet engine is a series of nested, truncated cones through which the exhaust gases of the combustion pass. Ambient air is drawn into the path of the gas and supplements the thrust.

DECEMBER

Washington Gladden publishes his short story "A Christmas Dinner With the Man in the Moon," in *St. Nicholas* magazine.

The giant space liner *Meteor* takes a group of children and their uncle on a holiday visit to the moon. The spaceship, which resembles a Winan "cigar-steamer" (The

NASM

TOMMY'S RIDE

By George Cooper

The crackers cracked; the guns went bang;
Folks shouted; and the bells they rang;
All hearts were full of joy and pride,
When Tommy took his famous ride.

It wasn't in a big balloon
That he sailed up to meet the moon;
But all the money in his pocket
He spent upon a single rocket.

He planted it against the wall,
And there it towered, slim and tall;
Then silly Tommy—such a trick!—
Must tie himself to the stick.

Whizz! went the rocket in the air;
The people stopped to wildly stare;
The dogs they barked with all their might,
But Tommy soon was out of sight.

The old man in the moon looked out
To see what it was all about;
Said he to Tommy, "Is that you?
Come in and see me,—how d'ye do?"

Away went Tommy, fast and far;
He tried to catch a pretty star;
He saw the clouds go sailing by,
Like boats of pearl along the sky.

But soon he slower went, and then—
Down, down, he fell to earth again!
Down, down:—the old man in the moon
Said, "Call again some afternoon."

Down, down; sweet faces o'er him beam;
How lucky this was all a dream!
Safe in his little crib he lay;
And it was Independence Day.

Winan ships were also the inspiration for Jules Verne's submarine *Nautilus*), is 150 feet long and shaped like a cylinder with pointed ends. Just forward of the ship's middle are a pair of huge paddle wheels. These are arranged so that they are open on the down stroke and closed when swinging up. They are used to lift the ship to the upper limits of the earth's atmosphere (about 200 miles). Inside is an elliptical salon furnished with overstuffed lounges and easy chairs, as well as small, thick plate-glass windows and electric heaters.

The *Meteor* sits atop a high wooden trestle, from which it is launched. The passengers are given respirators in order to protect them from the thinning air. These take the form of rubber caps that cover the nose and are supplied with air from containers of "condensed air"—a miraculous invention of Edison in which 25,000 cubic feet-of air is compressed into a block resembling Parian marble.

Once at the limits of the atmosphere, the *Meteor* strikes one of the "great electric currents" that pass between the earth and the moon. These travel at 20,000 mph and the ship is covered with soft iron, magnetized by on-board dynamos, to take advantage of them. Carried along by the current, the trip takes only 12 hours.

ca. 1880

Fyodor Geshvend, an engineer of Kiev, Russia, proposes a kind of winged railroad car powered by a jet of steam [also see 1887].

1881

The *Boston Daily Globe* publishes a "newspaper" supposed to have been issued by the inhabitants of a comet: "The Comet's Tale." Among the news items is one describing a flight into space by a group of terrestrials.

Hermann Ganswindt, in a lecture given in the Philharmonic Hall in Berlin, describes his interplanetary vehicle [see 1891 and 1926] for the first time.

1881–1882

Nicholai Ivanovich Kibalchich, a Russian inventor, in *Preliminary Design of a Rocket Airplane*, designs a rocket-powered man-carrying platform, intended for atmospheric flight only. The rocket engine works on the "machine-gun" principle of a large number of individual charges fired in rapid sequence. The platform has a hole in its center, above which is mounted the engine. It fires through the central hole in the platform. Once the platform is in the air, the firing chamber can be rotated for horizontal flight. Kibalchich makes an important contribution in suggesting that the platform be stabilized by mounting the engine in gimbals, so that it can be swiveled in any direction while firing [also see Peterson, 1892].

Although Y. I. Perelman refers to Kibalchich's device as a "spaceship," there is no indication that it is intended for space travel.

A Christmas Dinner With the Man in the Moon.

1882–1883

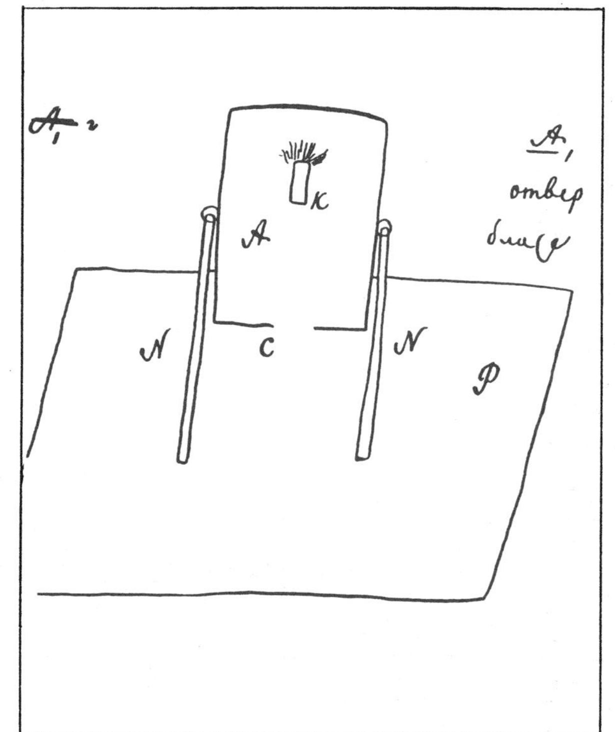

Kibalchich

1882

"Nunsowe Greene's" novel *A Thousand Years Hence* describes voyages by spacecraft to the moon, Venus, Mars, several asteroids, Vulcan and the sun.

1883

In "Free Space" K. E. Tsiolkovsky proposes for the first time the use of a reaction as a propulsive force in outer space, although this manuscript will not be published in his lifetime. Tsiolkovsky includes drawings, one of which schematically illustrates a concept for a spherical, reaction-propelled spaceship. The ejection of spherical masses of different sizes provides thrust and control, and stabilization is created by gyroscopes.

André Laurie (pseudonym of Paschal Grousset) publishes what has to be one of the most extravagantly impossible moon travel novels of all time: *Les Exiles de la Terre* (published in English as *Conquest of the Moon*).

In the novel a syndicate is established to exploit the probable mineral riches of the earth's satellite. Their only problem is the exact method by which the lunar surface is to be reached. A number of possibilities are suggested and eventually discarded: launching a projectile from an enormous cannon (in a gentle dig at Grousset's friend and occasional collaborator, Jules Verne), building a vast iron tubular "tunnel" to connect the two planets [see 1820], and so forth. Eventually, the syndicate accepts the amazing proposal of a young French engineer,

"These considerations have forced upon me the conclusion that there is only one solution of the problem . . . forcing the *moon to come down into our atmospheric zone* . . . This would annihilate the distance between us, and do away at the same time with many other difficulties . . . Our satellite would henceforward be *at our mercy* . . . We could go there either in a balloon or by means of a tubular railway. We could turn to account all her resources, and, getting hold of her riches, bring them to earth . . . unless, indeed, we might choose to settle down permanently in our lunar colony. (*Laughter and great applause.*)

"In fine, the whole question may be thus resolved: *we must not go to the moon; we must force her to come to us.*"

[Above] Laurie was possibly inspired by this concept illustrated by French science-fantasy artist Albert Robida, in which the moon is shown being drawn down to earth by magnets.

In order to accomplish this, an enormous electromagnet has to be built. A mountain in the Sudan is discovered to be composed of almost pure iron ore; it is to this mountain that the Luna Company takes its operations. The great mass of iron is found to be actually resting upon a sandy base, so it becomes possible to isolate it from the desert by fusing the sand into glass. Twenty-five huge solar reflectors perform the task of creating a layer of insulating glass 50 inches thick. These same solar engines will also provide the energy that will eventually be needed to magnetize the mountain. To accomplish the transformation of the mountain into a monster electromagnet, miles of cable are wrapped around it.

An observatory is erected at the summit. It is for all practical purposes a spaceship, so far as its construction is concerned. It is capable of being hermetically sealed and is equipped with a vast array of scientific apparatus, provisions, water, oxygen-making equipment, weapons and so forth. In due course it becomes a spaceship in fact as well as appearance.

The day of the great experiment arrives. The giant magnet is energized. The moon, as expected, begins to spiral slowly in toward the earth. At its closest approach an unforeseen accident occurs: the entire mountain is wrenched from the surface of the earth, and lands on the moon!

1884

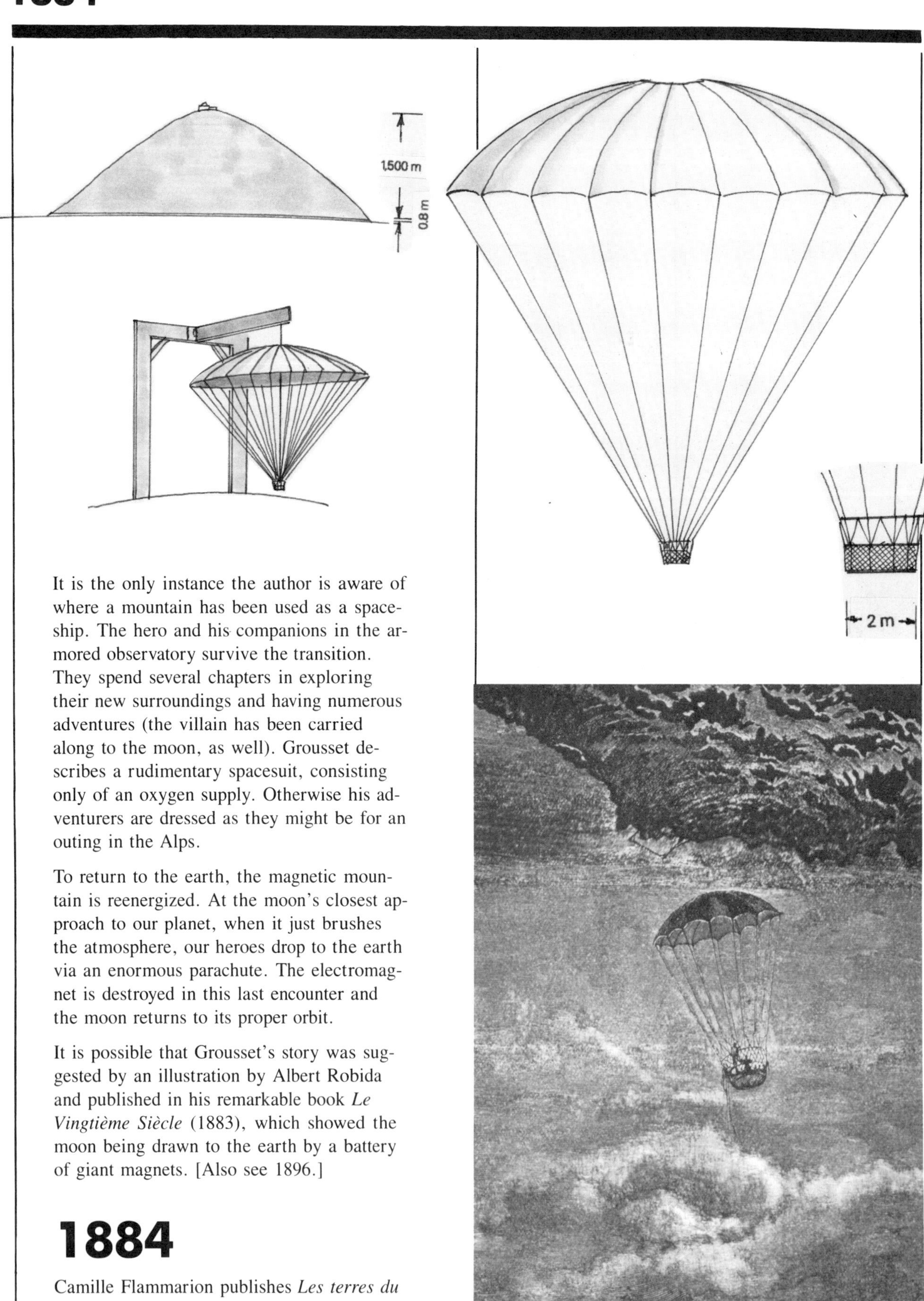

It is the only instance the author is aware of where a mountain has been used as a spaceship. The hero and his companions in the armored observatory survive the transition. They spend several chapters in exploring their new surroundings and having numerous adventures (the villain has been carried along to the moon, as well). Grousset describes a rudimentary spacesuit, consisting only of an oxygen supply. Otherwise his adventurers are dressed as they might be for an outing in the Alps.

To return to the earth, the magnetic mountain is reenergized. At the moon's closest approach to our planet, when it just brushes the atmosphere, our heroes drop to the earth via an enormous parachute. The electromagnet is destroyed in this last encounter and the moon returns to its proper orbit.

It is possible that Grousset's story was suggested by an illustration by Albert Robida and published in his remarkable book *Le Vingtième Siècle* (1883), which showed the moon being drawn to the earth by a battery of giant magnets. [Also see 1896.]

1884

Camille Flammarion publishes *Les terres du ciel; voyage astronomique sur les autre mondes* (Lands in the sky; an astronomical

voyage to other worlds). [Also see 1862 and 1865.]

Flammarion continues his theme of exploring the possibilities of life on other worlds. This book, like his others, is lavishly illustrated not only with woodcuts of actual astronomical views, but excellent imaginary extraterrestrial landscapes as well.

1885

General Russell Thayer of Philadelphia designs an aerial warship. It is a cannon equipped dirigible powered by a high speed air compressor, coupled to a "carbonic acid gas" (CO_2) engine. This is connected to a tank located beneath the passenger deck or platform. Compressed air is forced into this until the desired pressure is achieved. The rear part of the reservoir is so made that at given intervals the compressed gas is suddenly released, "producing a powerful forward thrust." Another motor Thayer devises for his aerial warship consists of a powerful turbine that forces air through a narrow nozzle that points sternward. A series of hollow, concentric, truncated cones placed over the nozzle draw in outside air through their annular openings. This increases the volume of the discharge (while decreasing its velocity). In experiments made with water, the addition of five such cones increases efficiency by fifty per cent.

Thayer's schemes are taken quite seriously (and rightly so; they are realistic designs worked out in great detail) and working models are considered by the British War Office, among others. [Also see 1887].

1886

"Pruning Knife" (Henry Francis Allen) publishes his utopian treatise *The Key of Industrial Co-operative Government*. In it he

NASM

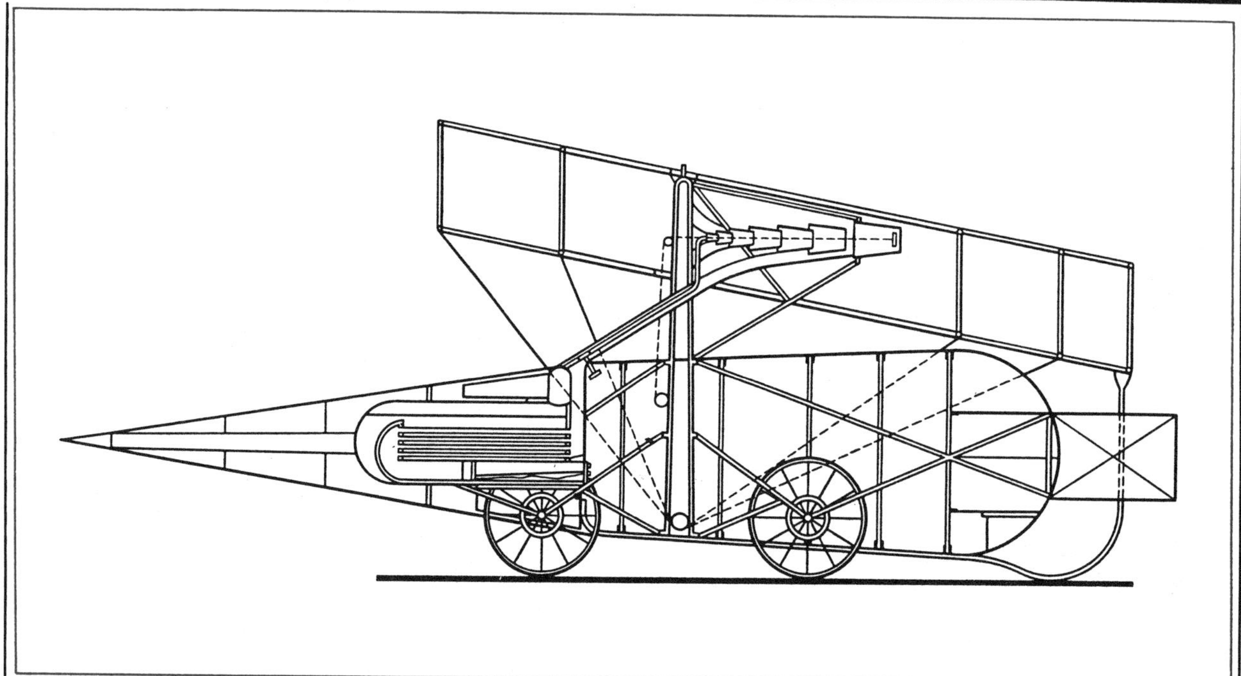

describes a flight to Venus in a winged flying machine.

In the anonymously written novel *Man Abroad* is the description of the colonization of the solar system—and the terraforming of its planets—and travel in a faster-than-light spacecraft.

Engineer Evald experiments with a small rocket-propelled airplane at the riding school of the Horse Guards in St. Petersburg.

1887

Hudor Genone, in the novel *Bellona's Husband*, describes a flight to Mars aboard an "ethereal disc" propelled by antigravity.

Fyodor Geshvend invents a reaction-powered flying machine designed to fly by the force of a jet of steam ejected from a specially designed nozzle [see 1889]. It will have a takeoff speed of 70 mph. It is equipped with a pair of oval wings, arranged biplane-fashion, the long axis of the ovals in line with the direction of flight. The wings have an area of 350 square feet, producing a lift of 2,900 pounds. A trip from Kiev to St. Petersburg, a distance of over 600 miles, will take only 6

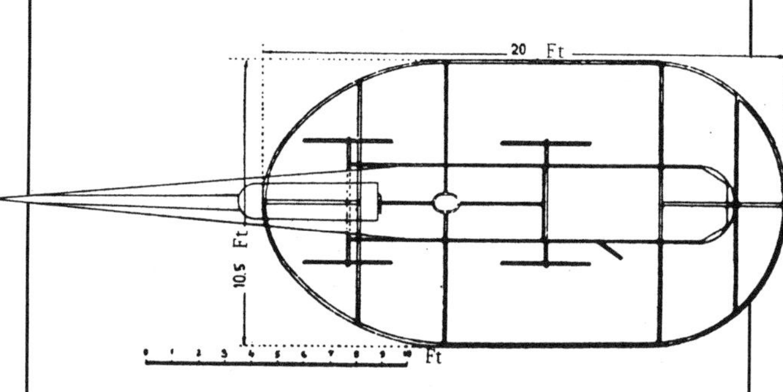

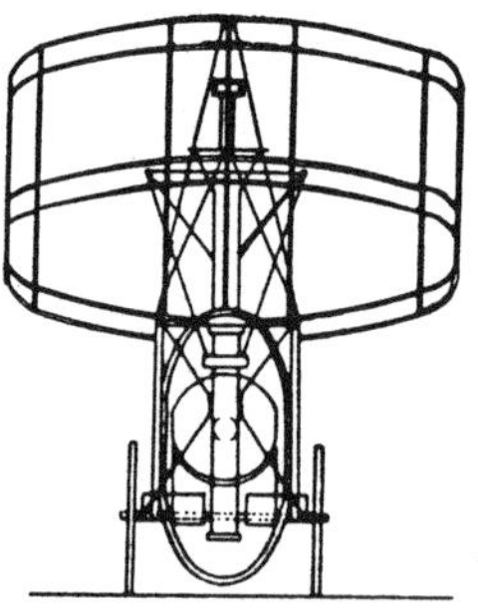

hours, with five stops of 10 minutes each. The steam plane will carry three passengers and one engineer, who steers the craft by means of a rudder.

The nozzle through which the steam escapes was made in the form of seven concentric cones, which allow air to be aspirated into the jet stream, increasing its force. [Also see 1880.]

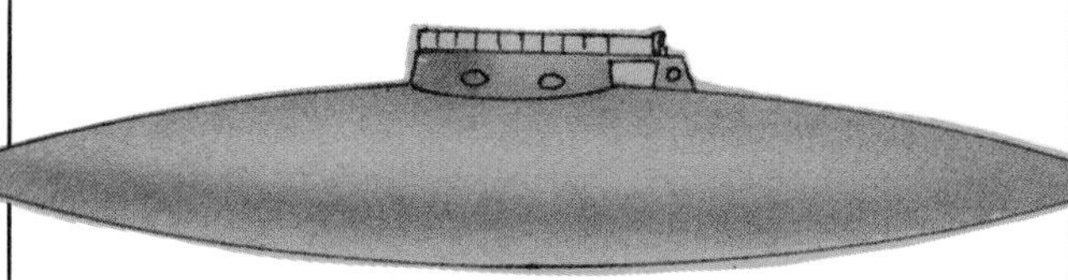

1887–1893

Rev. Wladislaw S. Lach-Szyrma publishes his series of stories *Letters From the Planets*. The spaceship described belongs to a race of Mercurians, who take the unnamed narrator on a tour of the solar system. It is powered by an undisclosed source of anti-gravity. It is spindle-shaped, with sharply pointed ends, a small cabin with two oval windows and a "directing magnet" on its top deck. Lach-Szyrma had also published *A Voice From Another World* (1874) in which the Venusian Aleriel describes his journeys in an ether ship around the solar system and which was expanded as *Aleriel* (1883). Aleriel is also the main character of *Letters from the Planets*.

1888

A French inventor is awarded a patent for a reaction-powered airship propelled by the recoil of a gun mounted in its gondola.

Mark Twain writes his satirical novelette *Captain Stormfield's Trip to Heaven* though it is not published until January 1908 (in *Harper's* magazine).

1889

G. Le Faure and Henri de Graffigny (pseudonym of Raoul Marquis) publish the four-volume novel *Aventures extraordinaires d'un savant russe* (The Extraordinary Adventures of a Russian Scientist) with an introduction by Camille Flammarion [also see 1913–1916]. The novels are a veritable catalog of imaginative spacecraft designs. Nearly a dozen different ideas are suggested for travelling to or communicating with other worlds.

1. The itinerary of the spaceship *L'Ossipoff* is described, which is launched from a cannon built within the crater of the volcano Cotopaxi [see 1937]. The explosive is the suddenly released lava, hitherto plugged by a slab of obsidian. The speed imparted to the projectile is roughly 10,000 m/sec. The five passengers protect themselves against the shock of the launch by rolling themselves up in mattresses.

The projectile is made of a nickel-magnesium alloy, weighing altogether over 590 kg. Oxygen is stored in the form of solid tablets and carbon dioxide is removed from the atmosphere with potassium hydroxide. Lights are electric and heating is provided by an alcohol burner. It is furnished in much the same Victorian parlor fashion as Verne's projectile.

2. Earlier in the novel another space gun and projectile are described, for a flight to the moon. The steel gun, 73 meters long, is sunk vertically into the ground like Verne's *Columbiad* [1865]. It is charged with over a ton of the new explosive "selenite" (several tons of which are sufficient to blow up the earth, we are told). It is made of nitrogelatin and potassium carbonate.

For the projectile to reach the necessary velocity within the length of the gun, twelve supplementary charges are placed in the walls of the gun, each holding 680 kilos of selenite. As the base of the projectile passes each pair of supplementary charges, they are fired, giving extra impetus to the projectile. The gun is placed on Colombia's Malpelo Island.

The projectile is about 3 meters tall and 2 meters in diameter. The gun is fired electrically, and the projectile eventually arrives safely on the moon. The passengers have been provided with cushions and spring shock absorbers to help them survive the landing.

3. A spacecraft used by the inhabitants of the moon is detailed. It had a boat-shaped hull, with flat bottom and open deck. The hull bisects a large, oval wing. A special mixture of substances, upon exploding, creates gases which jet from the stern, propelling the aircraft.

4. Travellers, on a fragment of Mercury that has been torn away by a comet, decide to escape from it as it passes near Mars. They fill a hydrogen balloon and when the fragment passes through the atmosphere of Phobos they make the transfer [also see Verne, 1877]. The travellers are wearing pressure suits, since the atmosphere of the comet is composed of carbon dioxide. The balloon descends toward Mars and is caught in an artificially produced jet stream in the Martian atmosphere—created by the Martians to aid the flights of their aircraft.

5. Since there is a common atmosphere linking Mars and Phobos (according to the authors), the Martians are able to travel to their moon by way of propellor-driven aircraft. A gondola is suspended beneath a large metal cylinder with a conical nose. It is about 160 meters long and 12 meters in diameter. The cylinder rotates on its long axis, driven by a powerful electric motor. A helical propellor 25 meters in diameter moves the ship at a speed of 700 km/h.

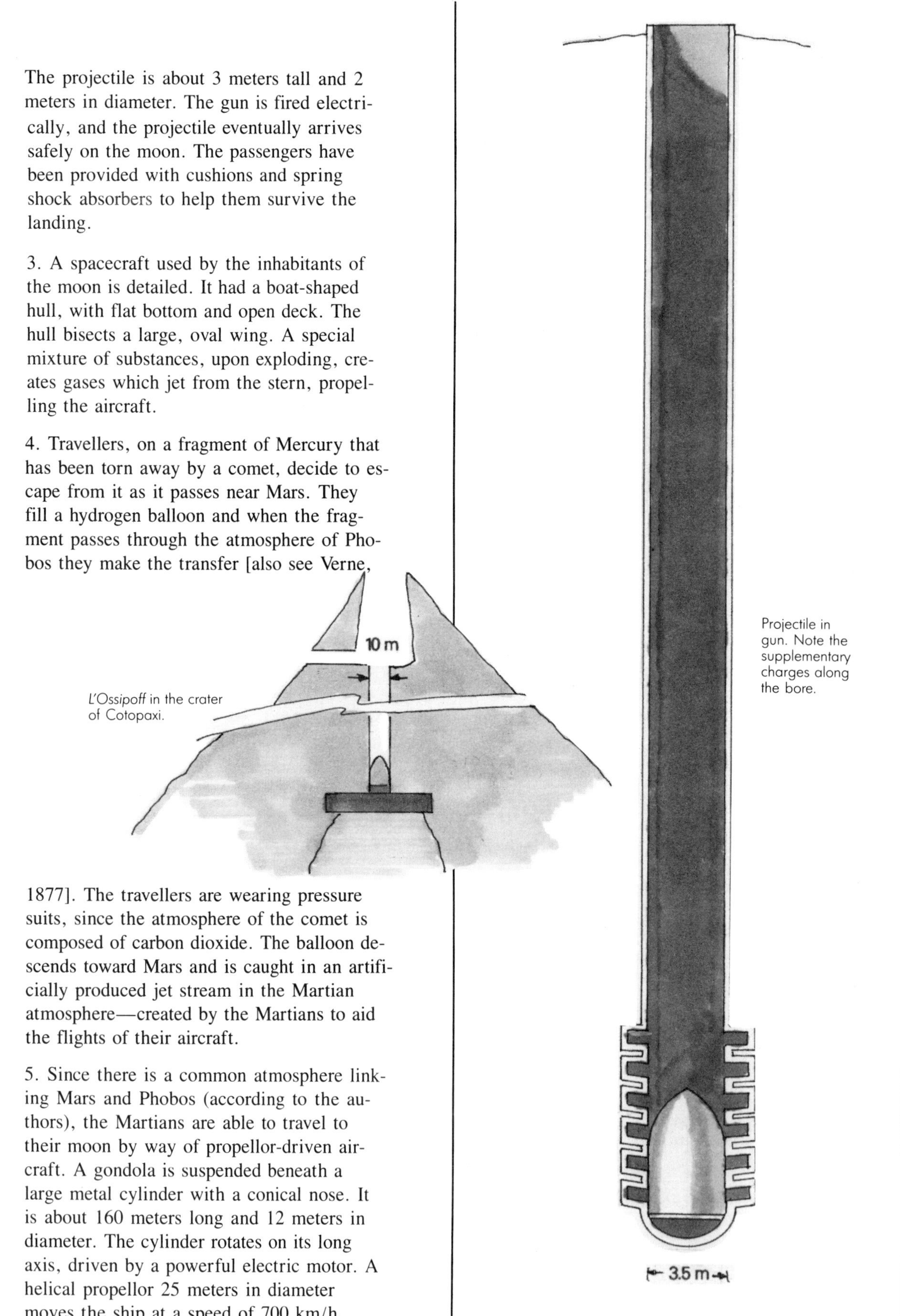

L'Ossipoff in the crater of Cotopaxi.

Projectile in gun. Note the supplementary charges along the bore.

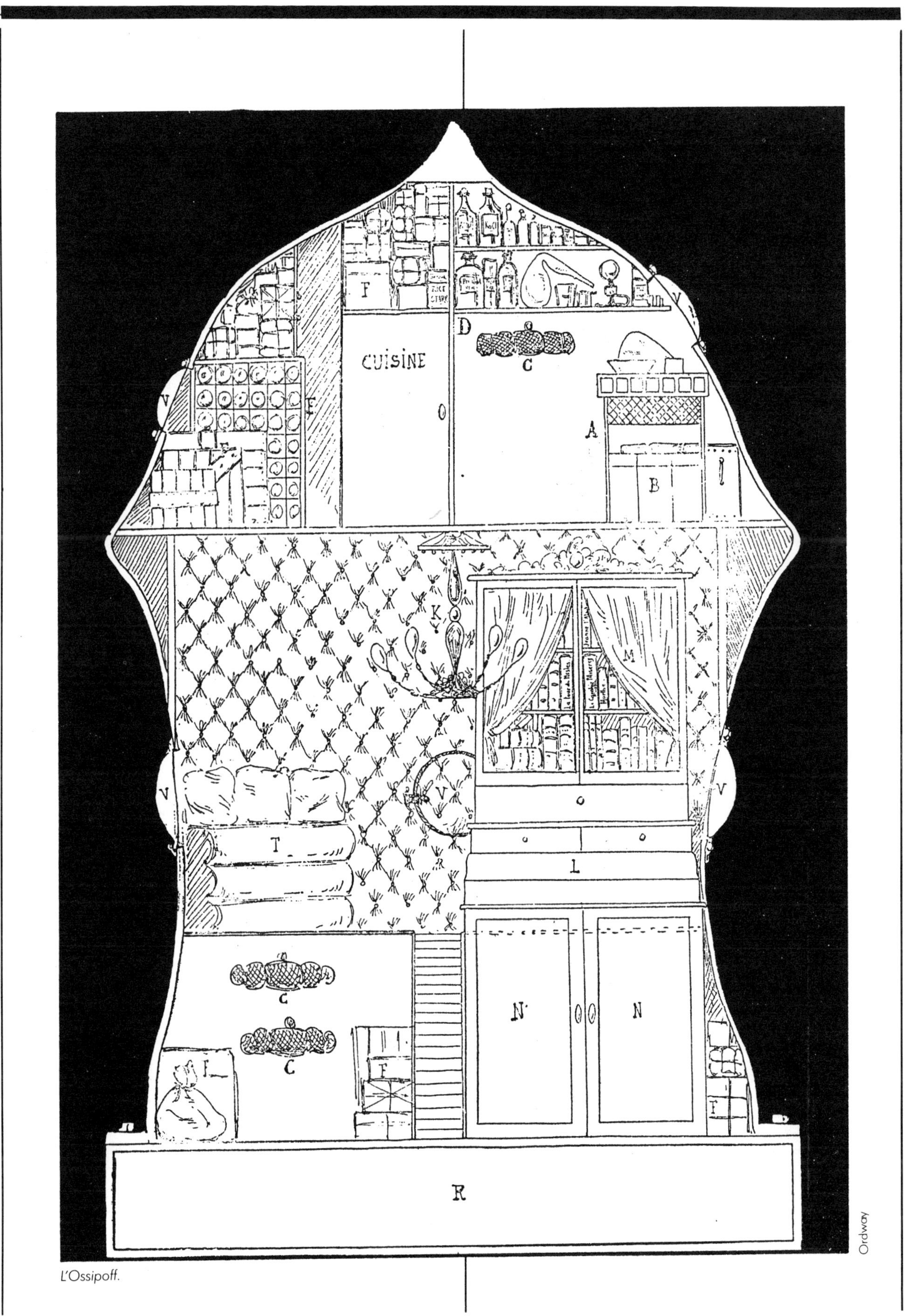

L'Ossipoff.

Ordway

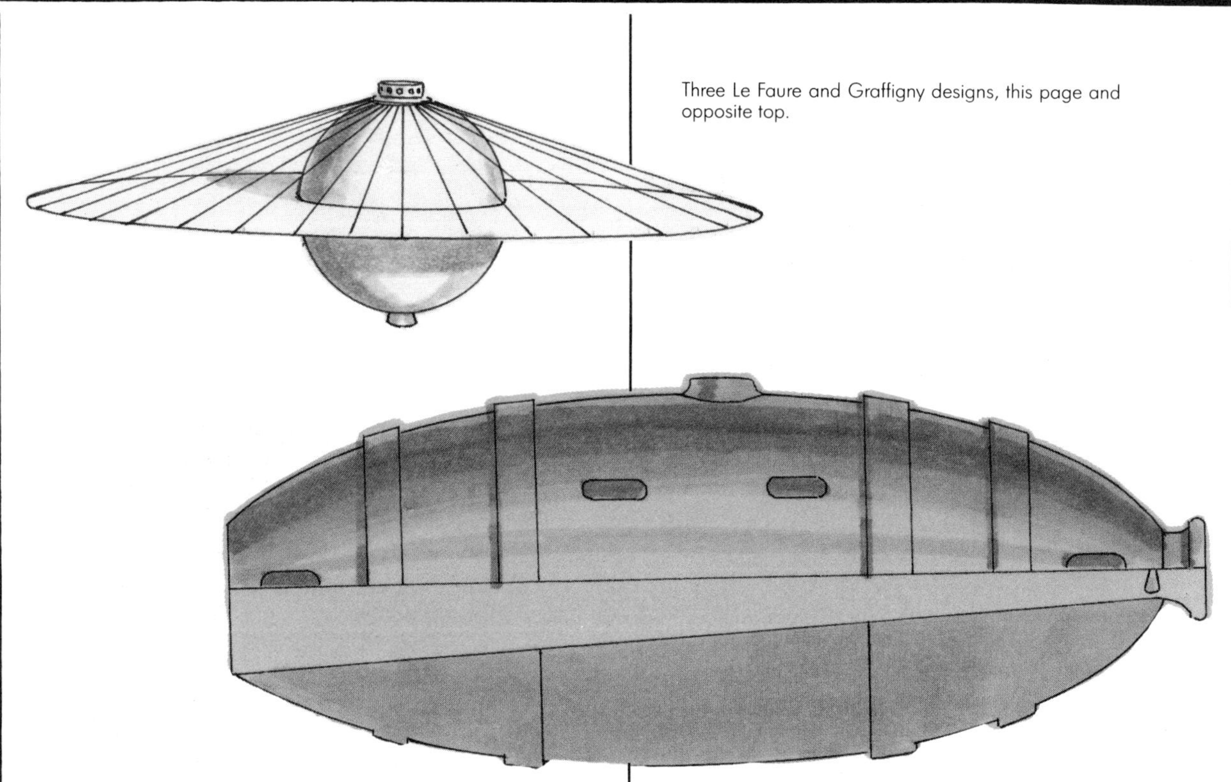

Three Le Faure and Graffigny designs, this page and opposite top.

6. A flight is made from Mars to Jupiter. The spacecraft is a barrel-shaped cylinder 7 meters long and 5 meters in diameter. Down its axis runs a hollow pipe, narrowing from 1.5 meters at one end to a small opening at the other. A propellor inside the pipe sucks interplanetary matter into the large end and expels it from the smaller end, moving the ship forward by reaction. The central 6 meters of the ship is divided into cabins: three hold the passengers, one the supplies and mechanisms and one the kitchen. A small cabin near the propellor is reserved for the pilot.

7. Communication is established between the moon and Venus. Beams of light from one planet to the other are picked up by selenium photocells. An apparatus similar to Edison's heliophone modulates the light of a powerful reflector. The beam is picked up by a similar device and turned back into sound.

8. A flight from the moon to Venus is made by the use of pressure of light from the sun. The ship itself is a hollow sphere, about 10 meters in diameter, made of selenium. In its bottom is an opening about 1 meter wide. Inside is a chamber for passengers that can rotate independently of the sphere. Surrounding the sphere is a disc of selenium nearly 30 meters across. To launch the ship, it is placed in the center of a huge reflector, 250 meters in diameter, which is made of separate, movable mirrors. When this focusses the light of the sun onto the spaceship, it is repelled away from the moon at a speed of 28,000 km/h. Nearing the atmosphere of Venus, the ship inverts. The lower part of the sphere is jettisoned and the upper half now becomes a cup-shaped gondola. Cables connecting it to the selenium ring are lengthened and the flat ring becomes a parachute for the descent. The travellers are encased in space suits, since they are now exposed to outer space. The same ship is eventually used for a trip from Venus to Mercury.

9. An artillery-shell-shaped spacecraft is made of a substance found on the moon that is attracted by the sun. Around the waist of the ship, like a flared skirt, are twenty-four vanes. These are painted black on one side while the other is coated with the new element. Each vane is movable so the ship can be steered. A trip to Mercury is made at a speed of 20 km/s.

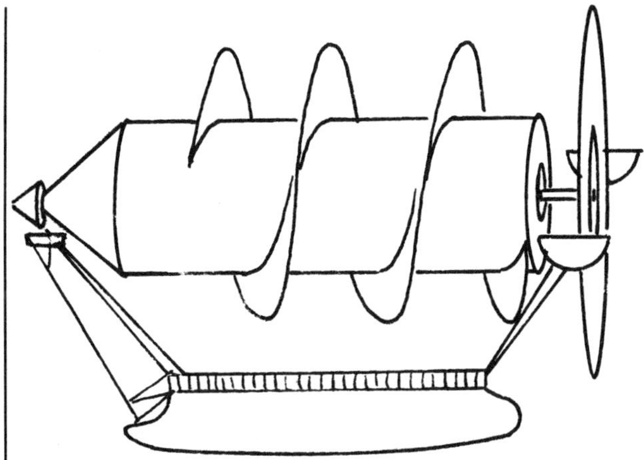

The probable source of the spaceship above was the old balloon print at the right.

Richard R. Montgomery publishes the "dime novel" adventure *Two Boys' Trip to an Unknown Planet* in which the heroes, who have ventured too high in their balloon, are rescued by extraterrestrials in a spaceship. This has a vaguely bird-shaped hull, with bat-like wings and six balloons supporting it.

Hugh McColl, in *Mr. Stranger's Sealed Packet*, describes a tiny, 20-foot, transparent spaceship, *The Shooting Star*. It is propelled to Mars by an Apergy-like force [see 1880 and 1894].

1890

C. C. Dail, in his novel *Willmoth, the Wanderer* describes the spaceship journey of a native of Jupiter to Venus and finally to the prehistoric earth.

F. Gomez Arias designs a space suit [see 1872].

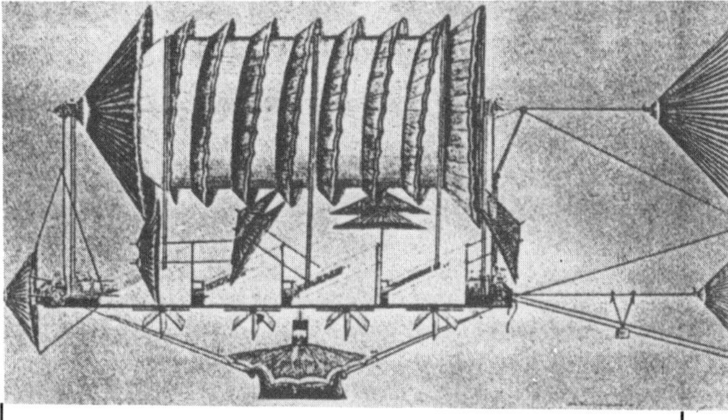

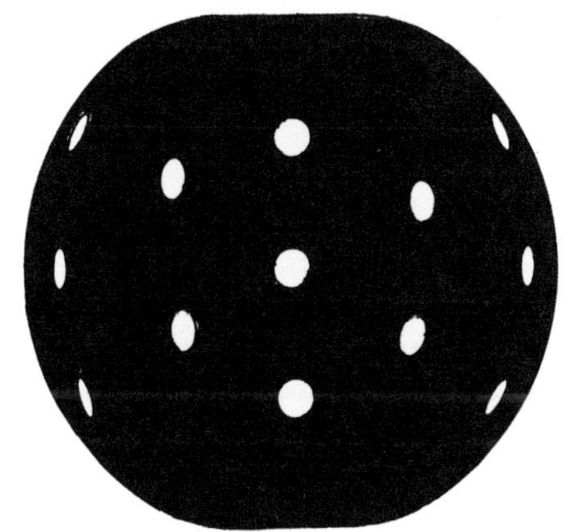

1891

Robert Cromie publishes *A Plunge Into Space* (which boasts an introduction by Jules Verne). The spaceship is the jet black *Steel Globe*, 50 feet in diameter, powered by

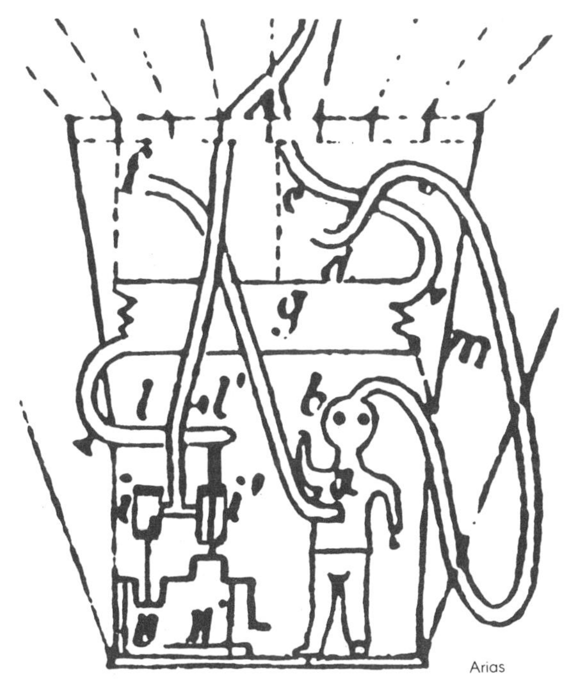

Arias

an electrically generated antigravity. "It was almost a perfect sphere, with only a certain flattening at the top and bottom . . . " Inside, "a spiral staircase wound round the interior circumference of the globe. This staircase, or rather sloping path, had one very curious feature. The handrail was duplicated, so that one could walk on the underside of the spiral as conveniently as on its upper surface . . . Again, the roof and the floor of the globe were identically fitted. Below, there were comfortable armchairs, luxurious couches, writing-tables and bookcases. Exact duplicates of these hung from above, head-downwards, so to speak.

"Across the center of the *Steel Globe* a commodious platform hung like a ship's lamp. On this a very large telescope was fixed, and the platform was literally packed with astronomical instruments."

The hull of the *Steel Globe* is pierced by numerous triple-glazed portholes and the spaceship is fully provisioned with food and water.

At the turn of a screw, the *Steel Globe* leaps into space from its secret construction site in Alaska, its destination Mars.

H. M. Bien publishes *Ben-Beor*, a novel describing a lunar civilization. The author is taken into space by a "fiery chariot." "Propelled by the power of two enormous wings it carried me upward with ease and comfort . . . Many, many years of mundane reckoning must have elapsed, when at last I came in sight of a luminous heavenly body."

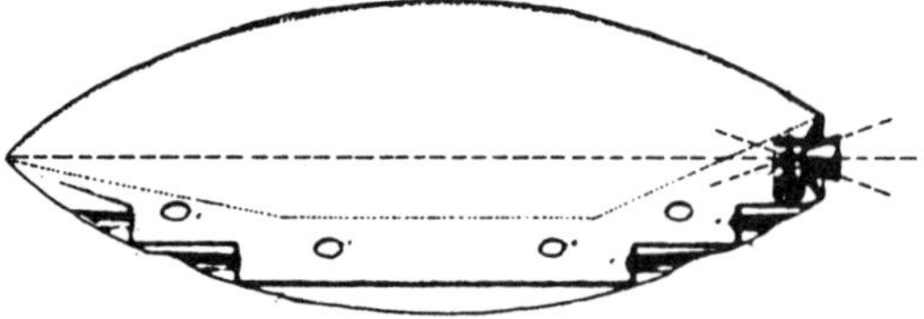

1892

Nicholas Peterson, an engineer working in Mexico City, patents a rocket-powered airship. It uses the principle of the Colt revolver, replacing the cartridges with rockets. The exhaust is mounted in a universal joint for steering (another early anticipation of the gimbal-mounted engine—see also Kibalchich, 1882).

Kenneth Folingsby, in his novel *Meda: A Tale of the Future* describes how criminals are sentenced to death by being set adrift in space. The hero, sentenced to this fate, is given navigational instruments and undergoes preflight training.

Charles Guyon publishes his novel *Voyage to the Planet Venus*. In a tip of his hat to Jules Verne, Guyon mentions that writer by name when the Venusians build an enormous cannon to launch the terrestrial explorers back to the earth.

The French chemist, Eugene Turpin, is imprisoned for supposedly divulging secrets pertaining to the national defense. It is said that while in prison he conceives of interplanetary voyages by rocket. Turpin and his fate are also the inspiration for the character of Thomas Roch in Jules Verne's novel *Face au Drapeau*. Verne makes Roch the inventor of a rocket-powered missile intended to be launched from the deck of a submarine (the first such suggestion). It is reported that Turpin is so enraged by the similarities—which Verne does not make particularly flattering—that he sues Verne for libel. (unsuccessfully).

1893

K. E. Tsiolkovsky writes the science fiction novelette *On the Moon*. There is no spacecraft; the narrator imagines himself to be on the moon during a dream, and the story mainly exists to give the author an excuse to describe lunar conditions.

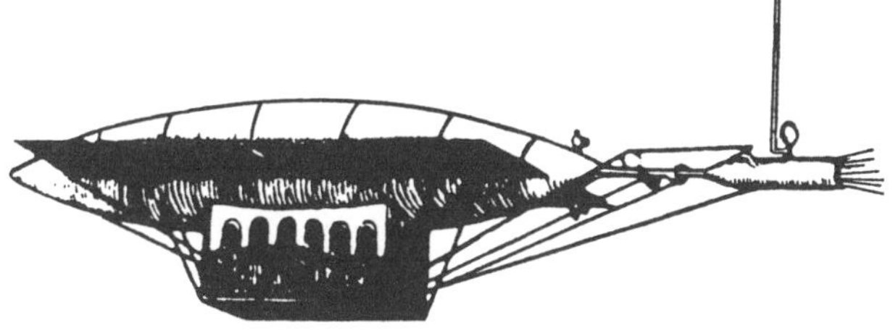

Sumter B. Battey patents a rocket-propelled airship. Like Peterson [1892], the rocket is based on a machine-gun principle.

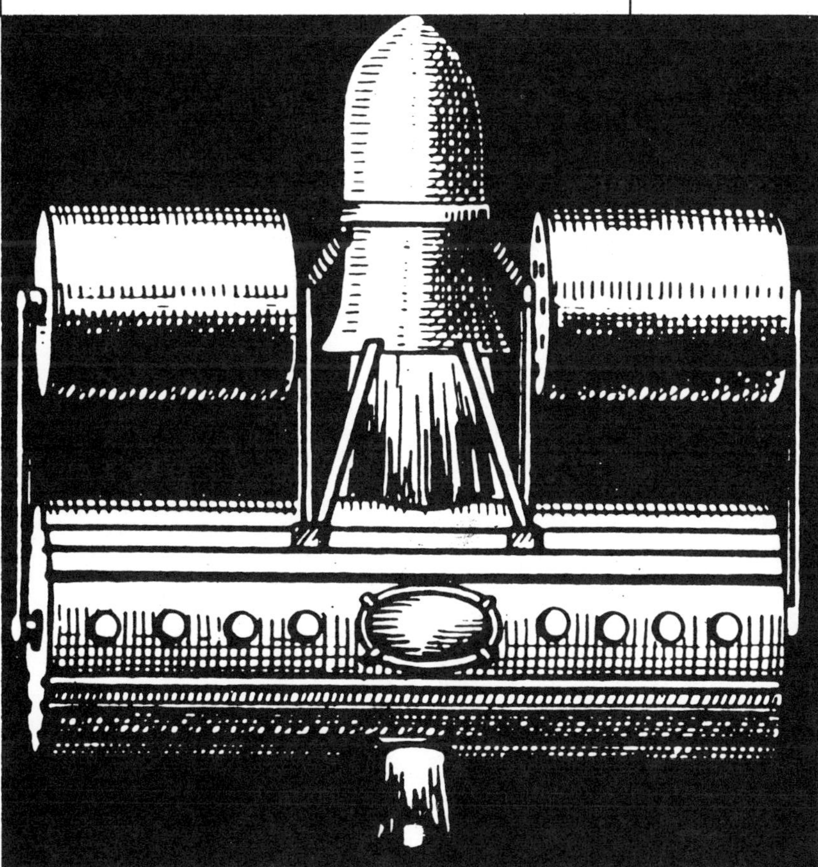

Milton W. Ramsey describes in his novel *Six Thousand Years Hence*, a cosmic catastrophe in which a nebula tears away from the earth a city of 750,000 people. The city is carried through space, along with the moon and, later, Mercury. Visits by airship are made to Jupiter and Saturn as the city passes them.

Hermann Ganswindt, a German law student and erstwhile inventor, designs a rocket-propelled spaceship. It consists of a large, cylindrical passenger "cabin" suspended beneath the engine housing and fuel receptacles. The spaceship is of the "tractor" variety: that is, the rocket motor pulls the spacecraft rather than pushes it. This motor, a thick-walled block of steel, is suspended above the passenger car, its exhaust firing through an open well in the center of the cabin. The fuel consists of heavy steel cartridges charged with a load of dynamite. Thousands are contained in the large, rotating drums that flank the firing chamber. When the cartridge is fired, one half, weighing 2 or 3 pounds, will be ejected. The other half of the cartridge will be thrown against the top of the firing chamber, transmitting its energy to the rocket. Ganswindt never properly understands the mechanics of rocket propulsion, being perpetually convinced that a rocket flies by reason of its exhaust pushing against the air behind it. In order for his spaceship to operate in the vacuum of space, he feels that it is necessary to provide something for the exhaust of his rocket to push against: hence the ejected steel shells.

The passenger cabin is equipped with spring shock absorbers to blunt the machine-gun-like progress of the rocket. The entire ship is also able to rotate on the axis centered on the central exhaust well. This provides artificial gravity for the passengers while the ship is in free fall. While rotating, centrifugal force would make the ends of the cylinder a pair of new floors, "up" now being toward the central well.

Ganswindt also suggests the possibility of connecting two similar ships by a cable and setting them spinning around their common center.

He describes the launch of one of his spaceships: It is to be carried as close to the edge of the atmosphere as possible by helicopters. Due to the ship's highly unaerodynamic form, the less atmosphere it has to pass through, the better. Only at this altitude will the rocket engine be started. The deeper penetration of interplanetary space can be made by establishing supply stations on artificial moons. He believes that it will be possible to eventually reach Alpha Centauri, though it will require an acceleration of 10 g's maintained for a very long time.

For all his errors in detail, Ganswindt is still one of the first to seriously suggest the use of rockets in spaceflight [also see 1881 and 1926].

G. LeFaure publishes the novel *Les Robinson Lunaires*. A book intended for younger readers, it features a trip to the moon by propellor-driven airship. The rigid, spindle-shaped envelope supports a hermetically sealed passenger car (which is equipped with an anchor). Its passengers are stranded on the moon when the airship breaks in two.

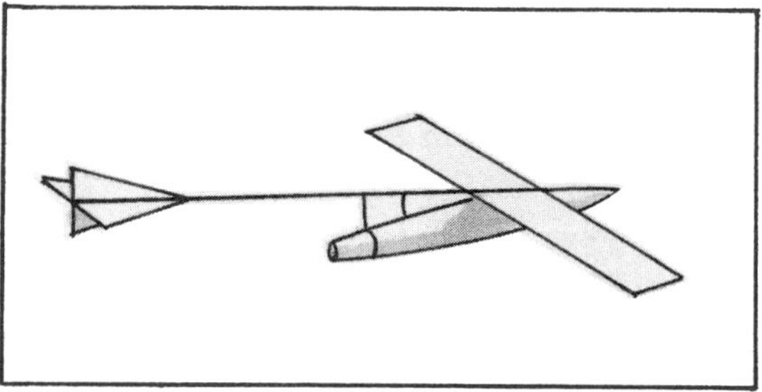

JULY 1

Alexander Graham Bell builds and flies a model rocket-powered airplane. The rocket is contained in a narrow brass tube, with tin wings and tail surfaces. The model flies 75 feet, rising 30 feet into the air. Bell performs his experiments at his home in Nova Scotia. He systematically tries to develop an ideal fuel for his rockets, testing promising compositions by mounting small rockets made of paper "quills" at the ends of a rotor.

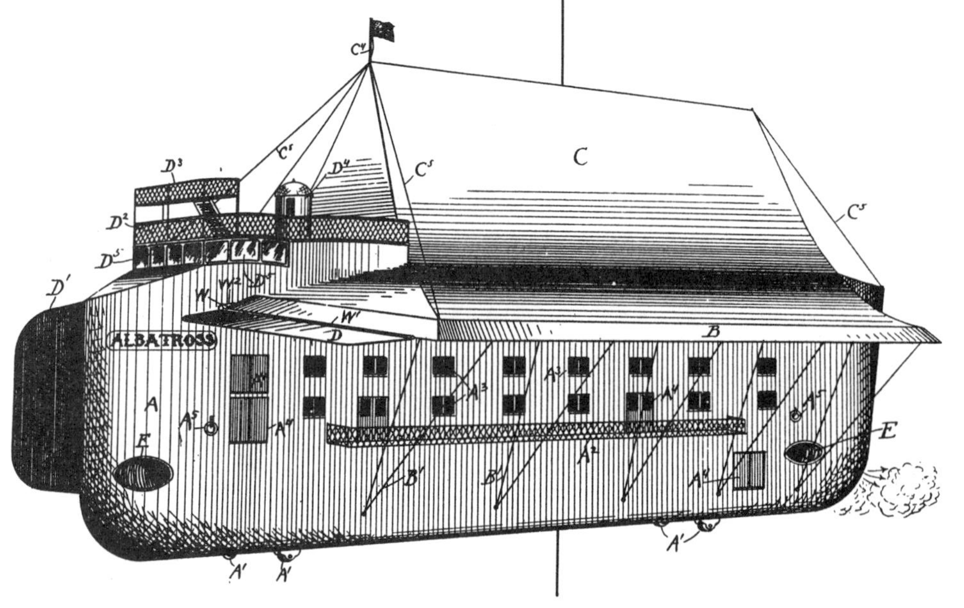

(No Model.) 3 Sheets—Sheet 3.
E. PYNCHON.
AIR SHIP.
No. 508,753. Patented Nov. 14, 1893.

Les Robinsons Lunaires.

ca. 1893–1894

Leon Lepontis of the Westinghouse Electric and Manufacturing Co. Laboratory proposes a variety of pulse-jet for propelling boats. Two combustion chambers have a common discharge nozzle. The combustion chambers are identical in that both have air intake shutters and inlets for a liquid or gaseous fuel. Only one combustion chamber is fired at a time. The escape of the burning gases from the common nozzle will have the additional effect of drawing air and fuel into the empty chamber, which is then ignited in its turn.

1894

John Jacob Astor publishes his remarkable novel *A Journey in Other Worlds*. Set in the year 2000, it describes a voyage by anti-gravity spaceship to Jupiter and Saturn. However, one of the most fascinating sections of this book—the only work of fiction by the multimillionaire real estate financier—is its description of life a century in its future.

(Another feature of this unique book is its illustrations. They are by Dan Beard, more usually associated with outdoor subjects, who later will found the Boy Scouts of America.)

The spaceship *Callisto* is powered by "Apergy," inspired by Percy Greg [see 1880]. The *Callisto* is built (after an advertisement for bids) in the form of a squat cylinder 15 feet high and 25 feet in diameter. It is constructed of "glucinium" (beryllium), with tall glass windows surrounding both floors. It is topped with a glass-domed roof 10 feet high. All of the windows are equipped with heavy shades and the walls and floors with a 6-inch layer of asbestos. The two floors have ceilings of 6.5 and 7.5 feet. The floors are open grids or grillwork. Around the base of the dome is a raingutter.

The *Callisto's* takeoff is 11 p.m. December 21, 2000, from New York City. A journey is made to Jupiter, Saturn and the trans-Neptunian planet Cassandra. During a discussion concerning cosmology and the possibilities

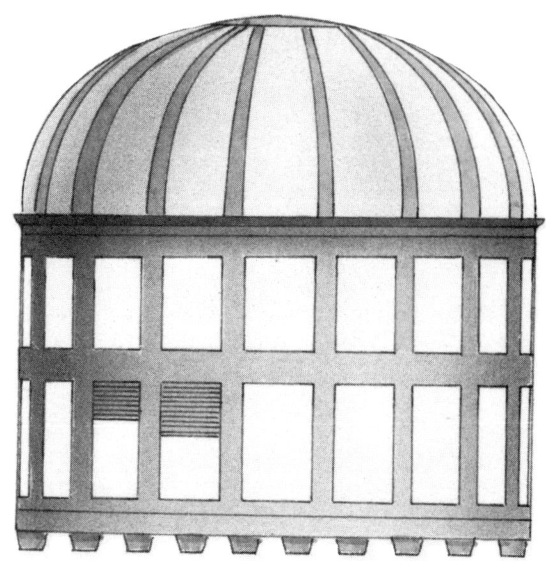

of life on other worlds, Astor describes a remarkable creature: "On [other worlds], the most highly developed species have hollow, bell-shaped tentacles, into which they inject two or more opposing gases from opposite sides of their bodies, which, in combination, produce a strong explosion. This provides them with an easy and rapid locomotion . . . " This is one of the earliest references to hypergolic rocket fuels and bell-shaped rocket engines.

Gustavus W. Pope publishes the novel *Journey to Mars*. The "Ethervolt car" is propelled by "magnetic propulsion." A sequel is published in 1895.

1894–1895

Fred T. Jane, for *Pall Mall* magazine, creates the illustrated series "Guesses at Futurity," in which appear a number of spacecraft and, in one instalment, a lunar colony. In 1897,

A Journey in Other Worlds.

Jane publishes his novel *To Venus in Five Seconds* [see 1897] which involves a matter-transmitter (possibly the first). Jane also illustrates a number of H. G. Wells's novels, but is best known for the series of reference books which bear his name, the first being *Jane's All the World's Fighting Ships.*

1895

Charles Dixon, in his novel *Fifteen Hundred Miles an Hour*, describes a flight to Mars in the year 1875, aboard an electrically powered spaceship.

K. E. Tsiolkovsky describes the launching of an earth satellite by means of a centrifuge in *Visions of Earth and Sky*, a "science fiction" novelette (actually a thinly disguised popular treatise on conditions in space, especially on the asteroids). He also describes a kind of space station: a hollow metallic sphere " . . . full of air, light and plants which regenerated the atmosphere . . . " is connected by a chain 500 meters long to a large mass, the whole sent spinning around the common center, producing an effect of one earth gravity.

Among other things Tsiolkovsky also discusses the potential use of a centrifuge in researching the effects of acceleration on human beings, experiments with shooting a shell containing a man from a cannon while he is immersed in liquid, the possibility of dropping a "laboratory" from the top of the Eiffel Tower to study weightlessness, and so on.

(A French engineer in 1891, M. Carron, seriously proposes this latter idea as a kind of "thrill ride." A spindle-shaped bomb would be dropped from the apex of the Eiffel Tower, into a 60-meter-deep pit filled with water. The slender, bullet-shaped shell would attain a terminal velocity of about 172 mph at the end of its 325-meter fall. The 10 foot diameter cabin would be carefully padded and supported on 20 inch springs. Carron thinks that the subsequent thrill would be worth a 50-franc fee.)

Charles Aikin publishes his novel *Forty Years With the Damned* in which his hero is taken to Mars aboard a spacecraft.

Pedro E. Paulet, a Peruvian engineer, claims to have designed, constructed and operated the first liquid-fueled rocket motor. Paulet will not publish an account of his experiments until nearly three decades later, in 1927. In 1902 he designs a rocket airplane [see 1907, 1927].

One "Herbert" designs a rocket-powered flying wing.

Gustavus W. Pope publishes the sequel to *A Journey to Mars* [1895]: *A Journey to Venus.* Here we learn of great sea-going Venusian vessels to which are attached fifty "ethervolt" cars by cables, enabling them to be carried through interplanetary space.

Franz Von Hoefft suggests launching a spacecraft by means of an electromagnetic cannon [also see 1937].

Mieczyslaw Wolfke does much theoretical work on the use of rockets in outer space.

DECEMBER 13

Lu Senarens publishes the dime novel story *Lost in a Comet's Tail*. Boy inventor Frank Reade invents another one of his super-aircraft (all of which are more or less blatantly patterned after Jules Verne's *Albatross*, from 1887's *Robur the Conqueror*). While flying at a high altitude it is sucked away from the earth by the tail of Verdi's Comet. The comet eventually explodes, leaving the airship suspended in space. As the earth turns beneath it (Senarens seems to be unaware of the necessity of orbital speeds) one of his passengers accidentally "falls" overboard. Instead of dropping to the earth, she merely floats nearby.

Fortunately, Reade's airship has been constructed with airtight compartments and an oxygen supply. Diving suits (which no self-respecting airship would, apparently, ever be without!) are pressed into service for space suits. In considering methods of bringing the marooned ship back to the earth, the reactive force of explosives is mentioned. But before anything can be done, another comet swings by the earth. Its effect on the airship is the opposite of the first comet and Frank Reade and his crew return safely to the earth.

ca. 1895

Reginald L. Swaby's short story "From Man to Meteor" is published in the anthology *A Forgotten Past* (Fred J. May, ed.). A balloonist ascends too high, is caught in one of the freak currents of air (which the author states flow through space) and goes into orbit around the earth, never to return.

1896

In James Cowan's novel *Daybreak* the moon, for an unexlained reason, makes an approach to the earth so close as to skim the tops off the Himalayas and scour South America, before finally landing in the Pacific Ocean.

The moon remains intact, protruding from the side of the earth like some huge wart. It proves impossible to scale its steeply curving surface until a balloon expedition is proposed. The balloon that is constructed is a remarkable precursor to the stratosphere balloons of the 1930s. Its glass-enclosed car is equipped with powerful air compressors, as well as a stockpile of provisions and gas for light and heat.

The 1,500-mile ascent is made, to a point just beyond the moon's equator. At that altitude the moon's gravity overpowers the earth's and the balloonists make a landing.

Eventually, for equally unexplained reasons, the moon detaches itself from the earth and heads for Mars, where it eventually deposits the explorers. (Could this novel have inspired Immanuel Velikovski's peculiar notion of the planets acting like ball bearings in a pinball game?)

K. E. Tsiolkovsky begins his theoretical work on rocket propulsion as applied to the problem of interplanetary flight. In the title of an article he asks, "Will the Earth someday be able to inform the inhabitants of other planets of the existence in it of intelligent beings?"

Edwin Pallander, in the novel *Across the Zodiac*, describes a tour by spaceship through the solar system.

"Professor Gray" publishes the dime novel adventure *The Cave of Death*. An airship rises too high and spends several days adrift in space before a meteor shower carries it back to earth.

John W. Hill, in his dime novel story *The Blue Star*, describes the flight in a spacecraft to a "blue star"—a mysterious planet that comes near the earth every 4 years.

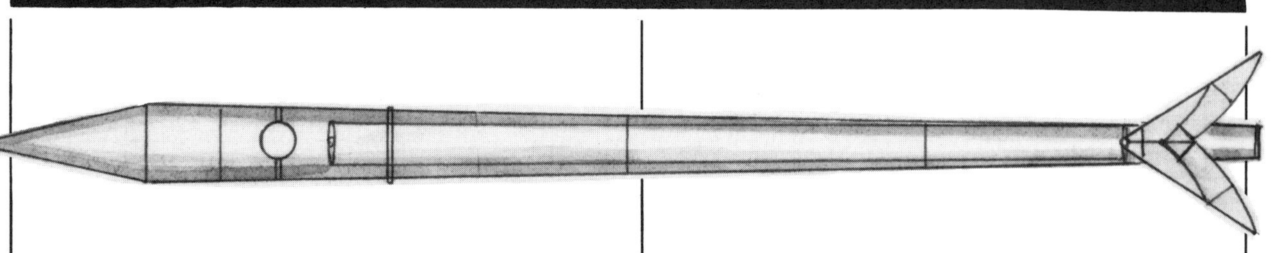

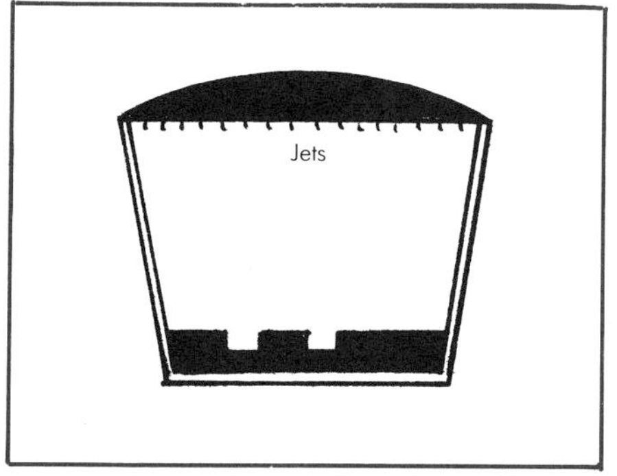

A. Fedorov proposes the design of a reaction-powered machine that would operate in space. It uses a gas such as steam, compressed air or carbon dioxide. The gas is introduced into a chamber where it will expand and escape through a nozzle. A model of a later rocket is exhibited at the 1927 Moscow interplanetary exhibition [see below].

JULY 26

An article in the *New York Journal* reports on a passenger rocket proposed by a Professor W. W. McEwen of Jackson, Michigan (a balloonist and parachutist of eighteen years' experience). It is a 60-foot aluminum cylinder that would be used for military reconnaisance. It has a domed cockpit and a rudder for steering. It will ascend to an altitude of 8,000 feet where its pilot bails out, using a personal parachute to descend to earth. A larger version of the rocket could be developed later for the "rapid transit" of "excursion parties."

McEwen claimed to have performed experiments with rocket models 7 feet 10 inches long. These took off at a speed of only 20 mph. This he regarded as an advantage since it would prevent his future passengers from encountering any sudden shock during their initial acceleration. The final passenger rocket would reach its maximum speed at an altitude of 500 feet, continuing on to 6,000 or 8,000 feet ("or more if desired"). At this altitude, the rocket was to "begin to turn in the air" and fly horizontally. For landing, McEwen's Improved Parachute (which used a lighter-than-air-gas to beneath the canopy to provide temporary lift) would automatically release and lower the rocket to the ground.

The reconnaissance rocket is under construction in Chicago. It is being made of aluminum with phosphor-bronze reinforcements. The passenger rocket would be much bigger, however: 100 feet long and 25 feet in diameter, capable of accomodating "about fifty people. "Fan-like propellors on the sides support the rocket when its fuel is exhausted. The interior arrangements include (from nose to tail) telescopes, an engineer's room, and "an apartment fited up after the manner of a small ocean yacht. Immediately below this will be kept a large water tank [for deadening the shock of takeoff—the water is jettisoned afterward]. At the bottom there will be a repository containing all the powder, which will give the rocket an impetus of a mile in forty seconds . . . " This version of the rocket would fly at six-mile altitudes and would provide a cheap answer to rapid transport. The only risk that McEwen forsaw was that of running into trees if the initial acceleration was too low.

1897

Edmund Rostand's play *Cyrano de Bergerac* repeats and somewhat embellishes Cyrano's methods of going to the moon.

MARCH-APRIL

George Parsons Lathrop publishes the story "In the Deep of Time." An "Interstellar Express car" is propelled by an antigravity metal made from helium.

1898

Fred T. Jane publishes his novel *To Venus in Five Seconds*. The journey is made by way of a variety of matter transmitter [see 1894–1895].

John Munro publishes the novel *A Trip to Venus*, his only known work of fiction. After having his hero discuss and dismiss the possible uses of the electromagnetic gun, or mass-driver, and the rocket-propelled vehicle he settles upon an undisclosed method of antigravity. Munro's description of his unchristened spaceship is vague: it somewhat resembles one of John Holland's submarines of the same period: " . . . a tubular boiler of a dumpy shape. It was built of aluminium steel [sic], able to withstand the impact of a meteorite, and the interior was lined with caoutchouc [india-rubber], which is a non-conductor of heat, as well as air-proof." The lower deck contains the driving mechanism and a small cabin. The upper deck, which had an oval plan and projects beyond the lower deck, is surrounded by an observation deck and conning tower. This deck is divided into several compartments, including a salon and smoking room. "The vessel was entered by a door in the middle, and a railed gallery or deck ran round outside." There are thick glass portholes, but electric lighting is available as well. The astronaut's air is replenished from tanks of liquid gas.

Munro's spaceship may also be the earliest description of a multistage spacecraft.

K. E. Tsiolkovsky begins a systematic study of the theory of rocket dynamics (the results will be published in 1903). In doing this he approaches the problems of motion in free space under reaction power alone, landings on the earth's surface and other questions.

Franklyn Wright and "Professor Butler" publish *The Rival Airships*, a novelette for boys. The story concerns the flight of an airship to Mars. Another story in the same series is also interplanetary: *Lost in Space*.

Kurd Lasswitz publishes the novel *Auf zwei Planeten* (*On Two Planets*).

A trio of balloonists attempting a transpolar flight discover the terrestrial base of a Martian expedition to the earth. This takes the form of a circular, artificial island located precisely at the North Pole. There is a row of columns that outlines the circumference of the island and between the columns is stretched a network of cables. This creates an electromagnetic field in the form of a narrow beam that, the explorers discover to their dismay, connects the island with a space station hovering 3,800 miles above the pole. The balloon is caught in this field—which is "diabaric," or gravitationless—and is carried directly toward the distant station. The explorers are rescued by the Martians in the nick of time and taken back to the pole, where explanations are made.

The space station is located directly above the axis of the earth because the poles are the only places where Martian spaceships can safely approach and land upon the earth (there is a smaller station above the South Pole, as well). The reason given is that the rotation of the earth (1,000 mph at the equator) makes landing anywhere else but at the relatively motionless poles impossibly dangerous. The north polar station is a huge wheel hovering one earth radius above the planet. Its position is maintained by the magnetic force beamed from the island. All of the energy required is generated by the sun. The ring is 50 meters deep and has an inner diameter of 20 meters, so that the overall diameter is 120 meters. Surrounding this ring are other thin, broad discs, like the rings of Saturn, with an overall diameter of 400 meters. These form a high-speed, frictionless flywheel that maintains the station's position perpendicular to the earth's axis. The inner ring resembles a circular hall and consisted of three floors each 15 meters in height. On

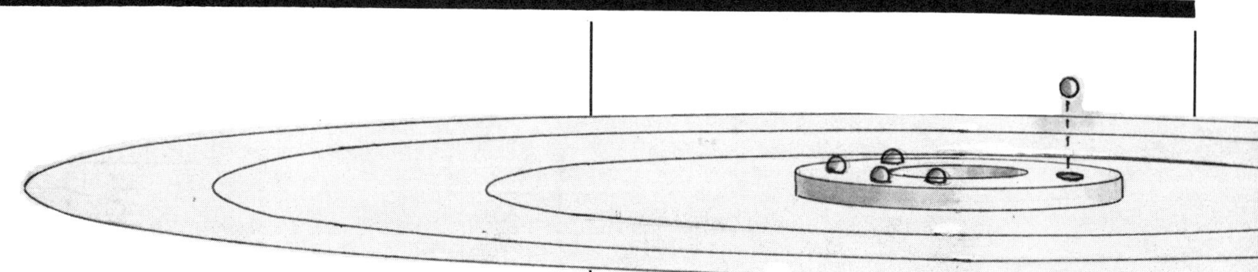

the upper surface of the station ring are depressions into which the Martian spaceships can dock. This particular station has space for eight such ships.

It is for this description that Lasswitz is often overenthusiastically credited with the idea of the space station.

The typical Martian spaceship is a perfectly featureless globe large enough to carry 60 passengers. It consists of two concentric spheres, the outer one made of the transparent diabaric material called "stellite." It is this antigravitational substance that allows the ships to penetrate space. However, in order to steer or to accelerate, the additional force of rocket propulsion is required. The rocket systems use a violent explosive called "repulsit"—successive small charges are detonated to propel the spacecraft, since any larger quantities would be too violent for the passengers to withstand.

The spacecraft is divided into three floors. The large central area is square in plan, allowing for four smaller rooms between it and the outer shell. The upper and lower areas each contain an engineer. The main area is luxuriously carpeted and lit by fluorescent lighting. There are no windows or portholes since any break in the stellite coating would cancel its antigravitational effects. There is a table in the middle of the room as well as chairs with strong, comfortable supports for the head and feet. At takeoff from the earth, each passenger stands against the upholstered wall, with his feet placed into shoe-like protrusions in the floor and his arms against armrests. A blast of compressed air launches the sphere into the ascending column of abaric energy that carries it weightless from the North Pole to the hovering space station. It is caught in a net and settled into one of the empty dimples in the central ring.

Later in the novel a kind of aerospace plane is described. It is an elongated spindle with a long, flat tail and wings that can be folded when not in use. The slender fuselage is 20 meters long and 4 meters at its widest. Its outer hull is made of the same glass-like stellite as the spacecraft. There is a space about a meter wide between the upper and lower halves of the hull. Observation windows bulge from either side of the nose, and the nose itself is transparent for the pilot. In order for the repulsit motors to be operated in the earth's atmosphere, they have been adapted to work in a more gradual, sustained release of energy—more like true rockets than machine guns. These new motors are called "repulsors," a name adopted by the German Rocket Society—the VfR—for their experimental rocket motors [see 1927].

The rocket-propelled intercontinental spaceplanes are equipped with some modern-sounding life-support equipment: oxygen masks, automatic heart-rate monitors, indicators to record the rate of breathing and pulse, etc.

Oddly, although *Auf zwei Planeten* is almost immediately translated into more than half a dozen languages, it is not available in English until 1971. Therefore, the influence of this remarkable novel is almost entirely limited to continental Europe. "I shall never forget," wrote Wernher von Braun, "how I devoured this novel with curiosity and excitement as a young man . . . From this book the reader can obtain an inkling of that richness of ideas at the twilight of the nineteenth century upon which the technological and scientific progress of the twentieth is based."

Later, in conjunction with an engineer named Sahulka, Lasswitz designs a spaceship employing the former's theories. According to Sahulka the earth is subject to a continuous rain of "ether atoms." Since the ether will pass through any solid substance there

is also an ether rain from below, in the form of the ether atoms that are passing through the globe from the antipodes. The effect of the ether upon terrestrial objects is not equal, however, since the vertical rain is considerably weakened by its passage through the bulk of the earth. However, if a shield could be made that would negate the effect of the pressure from above, the pressure from beneath would lift it. Lasswitz proposes a spherical spaceship that is controlled and guided by a movable, hemispherical shield.

Paolo Mantegazza publishes his novel *The Year 3000* in which a young couple from Rome (the capital of the United States of Europe) take a space voyage on an electric spaceship.

JAN. 12–FEB. 10

Garrett P. Serviss begins serialization of the novel *Edison's Conquest of Mars*. Serviss is a prolific author of popular science books and articles, and this is his first published work of fiction (though many more are later to follow; see 1909, for example). H. G. Wells's novel *War of the Worlds*, had been serialized in *Cosmopolitan* in 1897, and proved to be an outstanding success, the last instalment appearing in the December 1897 number. Scarcely six weeks later *Edison's Conquest of Mars* begins running in the New York *Evening Journal*, beginning with the January 12, 1898, issue. It has been written to cash in on *War of the Worlds's* popularity. Although Serviss has had no previous experience at fiction, the *Evening Journal's* editor feels that Serviss's reputation as an expert in astronomy and science will more than make up the difference. The author apparently writes the episodes as fast as they appear.

The resultant novel is as awful as it is remarkable (and gives little idea of the excellent quality of Serviss's later fiction, one hastens to add). It can boast many firsts: the description of whole fleets of spacecraft, the first ever to portray a battle fought between spacecraft in space, and one of the earliest descriptions of a spacesuit. Although style and plot may be lacking, Serviss's science is well thought out and detailed.

The story begins immediately after *War of the Worlds* ends. The earth is licking its wounds when signs appear indicating another Martian invasion may be under way. The great scientist and inventor, Thomas Alva Edison, gathers together the world's most outstanding scientists. Together they succeed in discovering the secrets of the Martian weapons. (No doubt such distinguished men as Röntgen, Kelvin and Rayleigh would have been startled to discover themselves as leading characters in a science fiction novel.) Given *carte blanche* by the nations of the world, Edison whips up an antigravity generator, to say nothing of an improved atomic disintegrator ray. Within 6 months a fleet of 100 spaceships is built and launched. The scene as they group, hovering, above the city of New York is spectacular: "The polished sides of the huge floating cars sparkled in the sunlight, and, as they slowly rose and fell, and swung this way and that, upon the tides of the air, as if held by invisible cables, the brilliant pen-

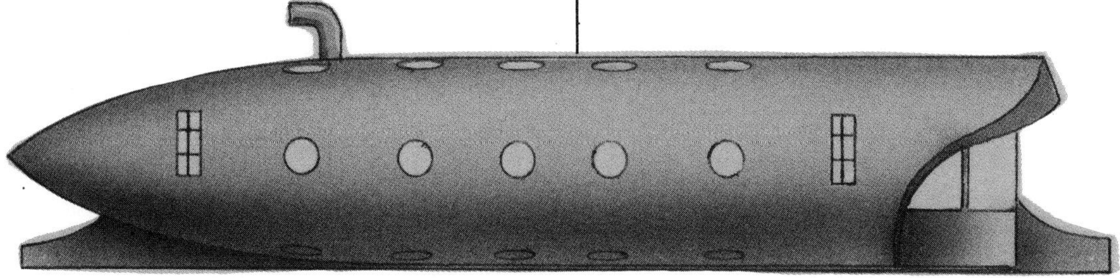

nons streaming from their peaks waved up and down like the wings of an assemblage of gigantic humming birds."

The ships are abundantly provided with windows, through which the ships's commanders can communicate via heliograph. Access to the exterior of the ships, while in space, is made possible by the use of an "air-tight dress constructed somewhat after the manner of a diver's suit, but of much lighter material . . .

"Provision had been made to meet the terrific cold which we knew would be encountered the moment we passed beyond the atmosphere . . . by a simple system of producing within the air-tight suits a temperature sufficiently elevated to counteract the effects of the frigidity without. By means of long, flexible tubes, air could be continually supplied to the wearers of the suits, and by an ingenious contrivance a store of compressed air sufficient to last for several hours was provided for each suit . . . " Communication between spacesuited astronauts is via telephone—two persons wishing to converse need only to plug into one another's helmets with a connecting wire.

The serial is illustrated with more than fifty drawings by P. Gray, showing massed fleets of spaceships and spacesuited astronauts working outside the ships.

Another spinoff of *War of the Worlds, At War with Mars*, by Weldon J. Cobb, appears as a serial in the boys's weekly, *Golden Hours* almost simultaneously with the serialization of Wells's novel.

Pierre de Sélène (a pseudonym?) publishes his novel *Un Mundo Desconocido: Dos Anos en la Luna* in which Jules Verne's cannon [see 1865] is purchased at an auction benefiting the Baltimore Gun Club. It is sold to an Englishman for $200,050 and is used for a trip to the moon.

Mary Platt Parmele describes a manned geosynchronous satellite orbiting the earth at an altitude of 400,000 miles, in her novel *Ariel*.

C. J. Cutcliffe Hyne, under the name Weatherby Chesney, publishes *The Adventures of a Solicitor,* a collection of short stories in one of which, "The Men From Mars," visitors from the red planet travel through space in specially designed inflatable suits.

AUGUST 25

One of the earliest studies by K. E. Tsiolkovsky bears this date. "The old sheet of paper," he will write, "with the final formulae of a rocket device bears that date . . . I evidently must have been working on the subject much earlier, and it was not the mere flight of a rocket that fascinated me but the exact calculations."

1899

Ellsworth Douglass, in his novel *The Pharoah's Broker*, describes a cigar-shaped antigravity spaceship that carries the hero to the planet Mars. To counter the effects of prolonged weightlessness, the spacecraft is equipped with exercise apparatus.

The anonymous *Half Hours in Air and Sky* contains the following prophetic statement: "In the infancy of physical science it was hoped that some discovery should be made that would enable us . . . to pay a visit to our

neighbor, the Moon. The only machine independent of the atmosphere, we can conceive of, would be one on the principle of the rocket. The rocket rises in the air, not from the resistance offered by the atmosphere on its fiery stream, but from internal reaction. The velocity would, indeed, be greater in a vacuum than in the atmosphere, and could we dispense with the comfort of breathing air, we might with such a machine transcend the boundaries of our globe and visit the other orbs." The unknown author of these lines shows an appreciation and understanding of the principles of rocket flight unusual even among scientists of his day.

Stanley Huntley publishes his collection of literary parodies, *Spoopendyke Sketches*. One, "A Journey to the Sun," is a spoof of Jules Verne. It describes the method "Sir Fillemup Frog" uses to travel to the sun: a pair of repeating cannons strapped to his stomach. In spite of its satirical nature, this story is one of only two or three works of pre-twentieth century fiction to use the principle of rocket propulsion in space travel.

Hermann Ganswindt self-publishes his collected works in *Das jungste Gericht*, which includes his description of his spaceship [also see 1891].

The frontispiece illustration from Mark Twain's *Captain Stormfield's Trip to Heaven* (1888).

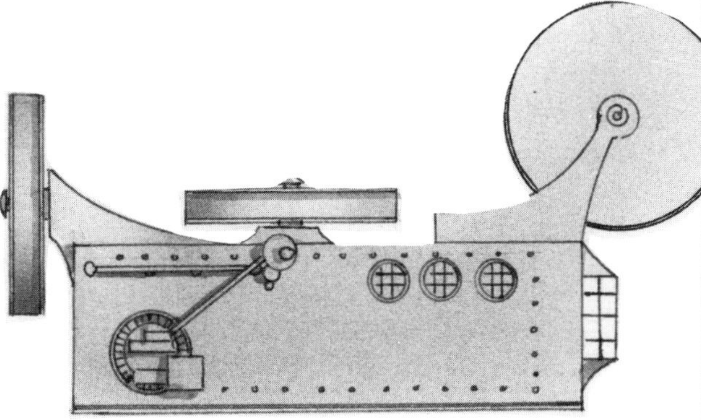

NOVEMBER

Ellsworth Douglass and Edwin Pallander publish their short story "The Wheels of Dr. Ginochio Gyves" (*Cassell's Magazine*). It is one of the more unusual spacecraft conceived to this date in that, in the strictest sense, it does not move at all! The boxcar-shaped vehicle carries several large, massive flywheels on its outside. When these are set into motion, their combined gyroscopic effect is to prevent the car from moving in any direction. It stays put while the earth in its orbit moves out from beneath it!

PART III

THE EXPERIMENTERS

The problem of spaceflight gradually evolved from the purely speculative and theoretical. A more or less developmental approach was undertaken by those interested in the possibilities of spaceflight, though this approach was in many ways dictated as much by neccessity as by intent. It was far simpler, cheaper and safer to experiment with small-scale rockets than with full-size spaceships; it was realized very early on that the exploration of space would be a fabulously expensive and difficult proposition.

In 1903 Tsiolkovsky published the first of his spacecraft designs; it employed liquid fuel and gyroscopic stabilization. In outward appearance his spaceship laid the groundwork for the modern spaceship to come.

Between 1913 and 1916, Andre Mas, Drouet and de Graffigny devised schemes for centrifugally launched spacecraft, thrown from the rims of rapidly-spinning flywheels. Arthur Train and Robert Wood described a remarkable spaceship in their 1917 novel *The Moonmaker*. *The Flying Wheel* was a 66-foot-diameter torus propelled by an atomic motor suspended in gimbals from the apex of a tripod over the center of the doughnut-shaped ship. The fuel was uranium, producing a beam of alpha-particles as it disintegrated.

The year 1923 saw the publication of Hermann Oberth's seminal *Die Rakete in den Planetenraum* (The Rocket in Planetary Space), one of the theoreticial cornerstones of modern spaceflight. In it he first proposed his "Model E," an enormous manned rocket that finally settled the outward form of the classic spaceship. It was an artillery-shell-shaped hull 35 meters tall and 10 meters in diameter that stood erect on the tips of four big fins. Oberth later elaborated upon the design in the 1929 revised edition of his book, *Wege zur Raumshiffahrt* (Ways to Spaceflight). In it he described a fictional circumlunar flight by the Model E spaceship *Luna* (on June 14, 1932). It was a three-stage rocket launched from the Indian Ocean. The pilots were ensconced in a small cabin shaped like an oblate spheroid, contained in the nose of the third stage. This was equipped with a parachute for the final descent to the earth. Oberth, with his typical meticulous care, considered every detail: how his crew were to eat in free fall, waste disposal, heating and cooling, etc.

Oberth's Model E formed the basis for the design of the spaceship *Friede*, which he provided for Fritz Lang's 1929 motion picture *Frau im Mond*, the first realistic spaceship in movie history.

At about this same time Max Valier was actively publishing his own designs for spacecraft. Although he evolved his spaceship from existing aircraft—they even took off more or less horizontally from inclined ramps—the final design was aesthetically more pleasing than Oberth's rather ultrafunctional rockets. Valier's final design was a chunky streamlined spindle with curved fins at the rear and two outrigger nacelles containing the rocket motors near the front of the craft. A similar design, but with the rockets in the rear, would have been launched from the back of an enormous rocket-powered flying wing. This version was elaborated upon in Otto Willi Gail's novel *Hans Hardt's Mondfahrt* (Rocket to the Moon) (1930).

Gail, combining the ideas of Valier and Oberth, described the spaceship *Geryon* in *Der Schuss ins All* (A Shot Into Infinity) (1925)—a three-stage rocket with folding wings (a feature of which Oberth disapproved). It reappeared in *Der Stein vom Mond* (The Stone From the Moon (1926), along with the space station *Astropol*. In this novel—which mixed space travel with the bizarre "Cosmic Ice Theory" of Horbiger then popular in Germany—the spaceship *Ikaros* made a voyage to Venus (where it remained in orbit while a small lander made the actual descent).

R. H. Romans' novel *The Moon Conquerors* (1930) described the 1945 flight of the spaceship *Astronaut*. The rocket was the result of an international competition for the best scheme for reaching space (152 of the submitted plans "were for a Goddard rocket"). It was a slender torpedo with narrow fins running its length. Romans described the rocket in some detail. It was launched horizontally, its initial velocity provided by an electromagnetic cannon (a method first proposed seriously by E. F. Northrup in *Zero to 80* [1937]). For propulsion between the earth and the moon, it used the pressure of sunlight on special black vanes.

Most of these concepts were illustrated by Frank R. Paul, who was almost solely responsible for all of the artwork in Hugo Gernsback's large stable of magazines. Trained as an architect and engineer, Paul's vaguely baroque spacecraft had an unprecedented aura of believability.

Many other fictional rockets continued to contribute to the collective and cumulative design of the spaceship. The 1922 animated film *All Aboard for the Moon* featured a streamlined rocket launched from a rooftop, carrying tourists to the moon. Miral-Vigee's 1922 novel *L'Anneau des Feu* (Ring of Fire) based its atomic-powered spaceship on the theories of Robert Esnault-Pelterie, the French aviation and space pioneer. The 1930 Hollywood musical *Just Imagine* featured the ultimate Art Deco spaceship. It became the representative prewar spaceship since it was recycled in the immensely popular Flash Gordon serials.

By this time—in the "real world"—Tsiolkovsky had published extensively and his plans included not only large manned rockets but lunar rovers and self-contained space colonies.

The Russian Nicholai Tsander designed enormous biplane spaceships that fed upon their own structure for fuel, and Franz Ulinski published his schemes for electrically propelled spacecraft. In 1925, Walter Hohmann not only designed his "powder tower" spaceship, an enormous cone-shaped rocket with an egg-shaped manned capsule at its apex, but his work on interplanetary orbits became so fundamental that these energy-saving orbits have been named for him.

Franz von Hoefft proposed an evolutionary spaceship design, employing the lifting body concept. Using standardized units spaceships could be customized for particular missions. He laid out a systematic and progressively more ambitious scenario for the exploration of outer space, employing a series of eight spacecraft, designated RH I-VIII. Hoefft's unique design resembled the blade of a shovel with a pair of slender pontoons beneath, since the larger ones were to be launched from water.

In the late 1920s Eugen Sänger began his researches into spaceflight, basing his hopes on the development of an "aerospace-plane." This eventually resulted, a decade later, in his famous "silver bird" antipodal bomber concept, the immediate precursor of today's Space Shuttle and modern aerospace planes.

By the time Goddard and the VfR flew their first liquid-fueled rockets, these dreamers and theorists—and scores of others—had not only established that space travel would ultimately take place, but had anticipated virtually every step on the road to achieving it.

During the 1920s and 1930s three highly influential organizations were formed: the Verein für Raumschiffarht (or VfR, the German Society for Spaceship Travel), the American Interplanetary Society (later the American Rocket Society) and the British Interplanetary Society (the only one of the three to exist in more or less its original form today). The first two of these three groups performed many of the earliest serious and controlled liquid fuel rocket experiments. Robert Goddard, who had made the first liquid fueled rocket flight, was working in strict secrecy, allowing little if any news of his work to be available to other researchers. The societies, however, were entirely open and freely shared the results of their experiments. Much of the development of modern rocketry, at least up until the 1960s, can be traced directly to the experiments of the VfR and ARS. Meanwhile, the BIS, prevented by law from experimenting with rockets, devoted itself to the theory and promotion of spaceflight, a service it still performs to this day. For example, in 1939 the BIS published the results of the first-ever detailed scientific and engineering study for a manned lunar rocket and lander.

With the advent of hostilities in Europe at the end of the 1930s, the work of the VfR, as well as many of its most brilliant members, was usurped by the German military machine, disappearing into well-kept secrecy, a secrecy broken only by the first V2 missiles dropping onto London. At the same time, ex-members of the ARS formed the private companies that produced wartime rockets and JATO units for the United States military and, later, the propulsion systems for the first American rocket-propelled aircraft. High-altitude balloonists were taking their "spaceships" to the limits of the Earth's atmosphere (by the mid-1930s altitudes of 12 miles or more were being reached), and rocket-powered gliders and aircraft were being flown—great advances being made in this area by the Germans and Russians.

1900

Mathematician-astronomer Simon Newcomb publishes his novel *His Wisdom the Defender* in which a newly discovered force allows the development of aircraft, one of which makes a flight into space to a distance of about 100 miles. What is of particular interest is that Newcomb is vociferous in his denunciation of the possibilities of heavier-than-air flight. In the same year in which the Wright brothers make their first successful flight, Newcomb writes: "May not our mechanicians . . . be ultimately forced to admit that aerial flight is one of that great class of problems with which man can never cope, and give up all attempts to grapple with it?"

Robert William Cole publishes *The Struggle for Empire* in which the Anglo-Saxon race has conquered the solar system. The story involves a struggle between the humans and aliens from Sirius, requiring battles among fleets of spaceships (perhaps the first interstellar conflict in literature). [Also see 1909.]

A. De Ville d'Avray publishes *Voyage dans la lune avant 1900*. In this small book of some fifty color plates, the author depicts a journey to the moon in the balloon *Intrépide*, and the adventures its passengers have once there.

JANUARY-JUNE

George Griffith (pseudonym of George Chetwynn Griffith-Jones) publishes his magazine serial, "Stories of Other Worlds" in *Pearson's Magazine* (Subsequently released as the novel *A Honeymoon in Space*). The stories tell of "the adventures of Rollo Lenox Smeaton Aubrey, Earl of Redgrave, and his bride Lilla Zaidie, daughter of the late Professor Hartley Rennick . . . " These adventures are made possible by the Professor's discovery of the "separation of the Forces of Nature into their positive and negative elements. "This includes gravity, which is divided into an attractive force and a repulsive force. He is able to construct a machine which generates either of these two forces as needed. It is, of course, only a small step to the building of the spaceship *Astronef.* All of this work is financed by Lord Redgrave who "was equally fascinated by the daring theories of the Professor, and by the mental and physical charms of Miss Zaidie."

Accompanied by the trusty old engineer, Andrew Murgatroyd, a successful trial flight is made across the Atlantic. Lord Redgrave and Zaidie are married immediately afterwards and what would be more natural than to celebrate their honeymoon in space?

The *Astronef* is spindle-shaped and about 75 feet long. A raised deck runs along two-thirds of this length; it is 20 feet wide and covered over with a cylindrical dome of glass, something in form like a greenhouse. If necessary, steel shutters could be drawn up from the teak floor to cover the toughened glass. The deck is circled by a light steel railing and two stairways lead to it, each with a hatch which can be sealed hermetically. A conning tower rises from the rear of the deck, protruding through the glass dome. In the bottom of the ship, at the aft-end behind the engine room, is another glass window, about 6 by 12 feet, set into the floor and normally covered by a pair of sliding panels. The *Astronef* is actually a sort of glass-bottomed spaceship.

The ship is fitted with an airlock in its lowest compartment, so that it may be safely exited when on airless worlds.

A pair of propellors at the stern provides propulsion when the *Astronef* is visiting a planet possessed of an atmosphere. Her top speed then is about 100 miles an hour. "The maximum power would have sufficed to hurl the vessel beyond the limits of the earth's atmosphere in a few minutes."

Within the ship is machinery to take care of every need: "warming, lighting, cooking, distillation and re-distillation of water, constant and automatic purification of the air, everything, in fact, but the regulation of the mysterious 'R. Force,' could be done with a minimum of human attention."

Their first stop is the moon, where Griffith has an opportunity to describe the couple's remarkably modern-sounding spacesuits. "These were not unlike diving dresses [made "chiefly of asbestos fibre"], save that they were much lighter. The helmets were smaller, and made of aluminium covered with asbestos. A sort of knapsack fitted onto the back, and below this was a cylinder of liquified air which, when passed through the expanding apparatus, would furnish pure air for a practically indefinite period, as the respired air passed into another portion of the upper chamber, where it was forced through a chemical solution which deprived it of its poisonous gases and made it fit to breath again. "The pressure in the helmets is regulated automatically (though the suits themselves are not pressurized, to prevent them from inflating and tearing the fabric—though the unpressurized suits would be equally fatal). For communication, the helmets are connected by a telephone wire. One of Stanley Wood's most charming illustrations for the novel is of Rollo and Zaidie standing on the moon, hand in hand, their helmets connected by a wire (carrying what sweet words?) too short to ever let them get more than a foot apart.

After a few adventures on the earth's satellite, the crew of the *Astronef* head for Mars. They land first on Phobos, using it as a convenient base for observing the Red Planet as the tiny moon orbits it.

Believing that any inhabitants of Mars would have to be naturally warlike, they begin to break out the *Astronef's* armaments. This consisted of four pneumatic guns, two forward two astern; armed with either explosive or incendiary shells; two machine guns, elephant guns, revolvers, a dozen rifles of assorted kinds and a double-barrelled shotgun (the property of her ladyship). Preparations that turned out to be well founded, since the Martians first reaction to the appearance of the *Astronef* is to attack it without warning. Lord Redgrave is shocked by this unsportsmanlike behavior. "Meanly?" he cries, "If there was anything like a code of interplanetary morals, one might call it absolutely caddish."

Being British and naturally in the right, they triumph over the Martians in a tremendous aerial battle and eventually continue on to visit Venus, the moons of Jupiter, Jupiter itself and Saturn—to say nothing of witnessing the formation of a new star and its planets! Their adventures and observations are of interest mainly in that they reflect current astronomical knowledge—now almost entirely erroneous.

1901

George C. Wallis, in his story "The Last Days of Earth," describes how the last inhabitants of our planet leave it in a spherical spaceship powered by antigravity.

H. G. Wells publishes his novel *First Men in the Moon*.

The eccentric scientist Cavor invents a material that screens gravity. In fact, no form of radiant energy can pass through it: neither light, heat, radio, X rays nor gravity. This material, which he names Cavorite is a compound of several elements, including helium (which had been discovered on earth 6 years earlier).

Cavor decides to construct a spaceship for a trip to the moon which would take advantage of Cavorite's peculiarity. He builds a large sphere of thick glass, broken only by a circular opening for the hatch. Around this is a steel framework which is fitted with numerous shutters that can be opened or closed electrically by operating switches inside the sphere. The wires pass through the glass, which is fused around them. The shutters are coated with the new compound. When the shutters are closed nothing in the way of electromagnetic energy can penetrate the

shell of Cavorite, including gravity. If one of the shutters is rolled up, however, then the spaceship is attracted to any body which might be lying in that direction.

The sphere is equipped with "solidified air," provisions, water-distilling apparatus, chemicals for removing carbon dioxide from the ship's atmosphere, and so forth.

The sphere is launched by simply closing all the shutters except those facing the overhead moon. All attraction by the earth is cut off and only the moon affects the sphere. It literally begins falling toward it. (The Russian rocket scientist Ya. I. Perelman calculates how long it will take Wells's sphere to fall from the earth to the moon: 1.5 months.)

The two astronauts discover that any time all of the shutters are closed, they are completely weightless, and float helplessly around the interior of the sphere. When one or more of the shutters are opened, however, they find themselves drifting in that direction, attracted by the gravitation of whatever object is outside.

The return trip, made only by Cavor's companion, takes only about 5 days (according to Perelman). The earth's gravity being so much greater than the moon's, the sphere "falls" much faster.

The Pan-America Exposition at Buffalo, New York, features a simulated "Trip to the Moon." It is apparently inspired by Jules Verne's novel (although artwork illustrating the trip shows a giant winged airship) and features midgets impersonating Selenites. It was designed by Frederic Thompson, who created many of the other attractions at the fair.

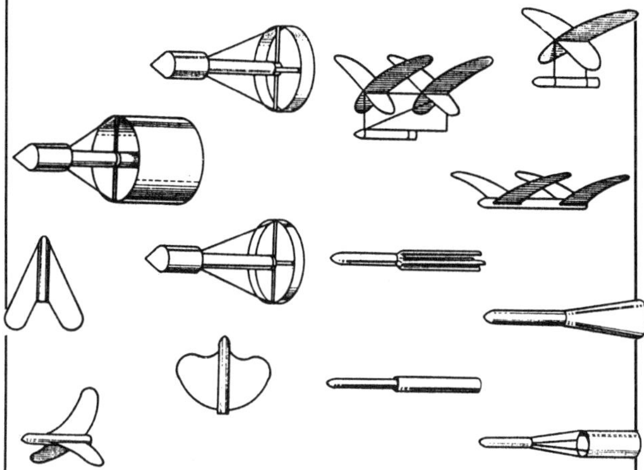

Russian research rocket models.

The illusion of actual flight is so convincing "that men have wagered in the *Luna's* cabin over the question of whether or not the boat [sic] leaves the building. The sensation of flight through the air is strong enough to bring fainting spells to some."

"The trip begins [describes a brochure] with a sight of the *Luna*, lying quietly beside her dock in the pale moonlight. You seem to be above the world some hundreds of feet. Below lies the exposition, the tower nearby . . . the *Luna* is a green and white cigar shaped thing, the size of a small lake steamer with a great cabin in the middle. Slowly she starts and gathers a long undulating motion. The exposition grounds drop . . . " Eventually the ship rises so high that everything seen from it "merges into a great globe. The globe lessens in size. It becomes a ball, then a mere speck." After passing through a violent storm, the moon is approached and the passengers are treated to a close up view of its craggy face. After landing in a crater the visitors go on to meet the Selenites and their king.

Franz A. Ulinski constructs a model of a spaceship design. It is later exhibited at the 1927 Moscow exhibition, where a label indicates that its fuel is "dust" (possibly coal or carbon dust, perhaps mixed with oil?).

MARCH 30–MAY 18

Welson J. Cobb publishes the serial "To Mars With Tesla" in the boys' magazine *Golden Hours*. It tells how the (fictional) nephew of Thomas Alva Edison, Frank Edison, teams up with the great electrical wizard Nikola Tesla in an effort to communicate with Mars. An unlikely pairing, since in real life Edison senior loathed Tesla.

ca. 1901

A Signor Edeselle of Callao, Peru and "formerlly of the United States Navy" presents a novel attraction at the Pan-American Exposition in Buffalo, New York. Described as a "human skyrocket" who had already demonstrated his feat in South America (supposedly "under Peruvian government patronage," near Callao in December of 1900 [?]), Edeselle launched himself in a rocket propelled by four tubes filled with a new fueled called "dynoascenemite." Ignited by electricity, this carried him to a height of over 15,545 feet. He safely descended by parachute over five miles from his takeoff point. Although this was alleged to have been witnessed by more than 15,000 people at the Exposition there are no substantiating accounts.

1902

A Trip to the Moon, a film by motion picture pioneer Georges Melies, is the first depiction of spaceflight in movie history. Based loosely on both Jules Verne's *From the Earth to the Moon* and H. G. Wells's *First Men in the Moon*, it has been reported that the venerable Jules Verne himself paid a visit to the sets during the filming. The film is pirated and released in the United States under the title *A Trip to Mars* [also see 1904].

Pedro E. Paulet designs a rocket-propelled aircraft. [Also see 1895, 1927]. Like his claims to precedence in constructing a working liquid-fuel rocket engine, this design for a rocket-powered aircraft has been the subject of dispute concerning its date.

The vehicle consists of a nearly spherical cabin and a more or less triangular wing capable of pivoting 90 degrees. The trailing

edge of the wing is thick and contains three banks of rockets, each wing having a total of thirty-six. For takeoff or vertical flight the wing is pivoted so that the trailing edge is facing the ground and the point of the triangle is vertical. For horizontal flight the wing pivots 90 degrees so it is parallel to the ground. A vertical stabilizer drops from the bottom of the craft like the keel of a boat. The tips of the wings are equipped with flaps for steering.

Paulet's device is apparently capable of meeting all of his criteria for "the perfect airplane," which should "(1) rise up vertically; (2) maintain itself at any point in the atmosphere; (3) fly at an altitude of more than 20,000 meters; (4) possess an exterior which would not be deformed by atmospheric agents and whose interior would be suitable for a large number of passengers and a heavy load of merchandise; and (5) descend vertically."

Samuel Pierpont Langley writes a series of letters concerning rockets to his assistant, C. M. Manly. In the first letter, dated March 9, he writes: " . . . the ultimate development of the flying machine is likely to be an affair of very small wings or no wings at all, and that it may depend for its velocity on what Mr. [Alexander Graham] Bell calls 'its momentum' in the same way that an arrow or any other missile flies. It is known that arrow derives its energy from the bow which projects it and that when this is spent the arrow will drop. We have, however, only to renew this energy . . . [and] the arrow may still be heading upward without limit, as in the case of a rocket which has no wings but goes very much better without them, renewing its energy by recoil . . . In any case it is a thing which deserves thinking over."

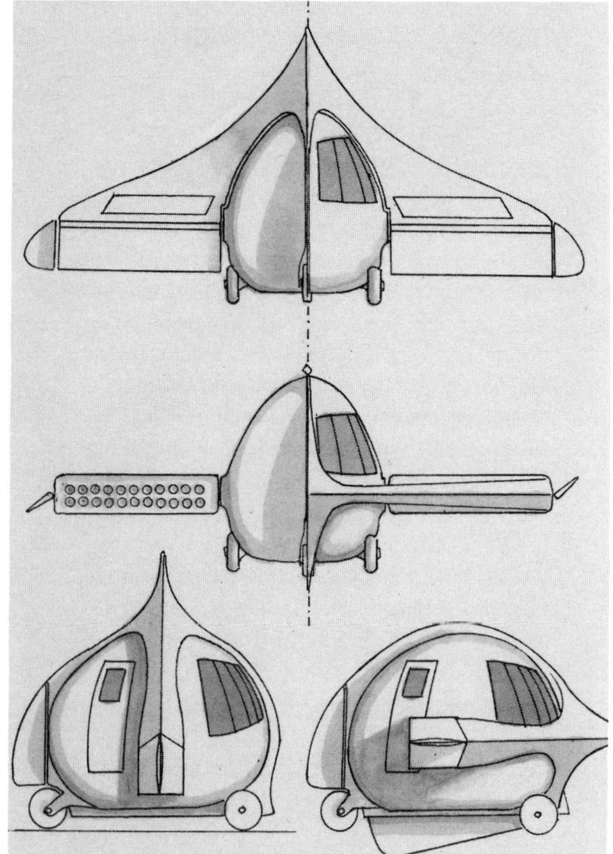

The second letter, dated September 25, contains another reference to Langley's considerations concerning wingless aircraft, without specifically mentioning rockets, as a possibility "apparently practical." Langley later writes to Professor Simon Newcomb, with a question concerning rocket propulsion, but without any indication that it applies to his earlier speculations concerning rocket aircraft.

1903

K. E. Tsiolkovsky describes a spaceship (rocket #1) in an article titled "Investigation of Outer Space by Means of Reactive Devices," in the magazine *Scientific Review* (it is further described in *The Aeronaut* [1910] and *Herald of Aeronautics* [1911]). This is his first published description of a spaceship; it is a streamlined metal hull divided into three main sections. The forward section of the ship contains the pilot and his

1903

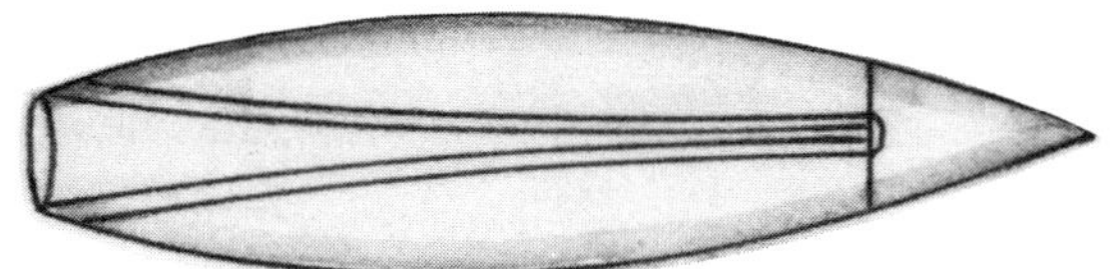

life support equipment: air purifiers, instruments, etc. The rear of the rocket is divided into two large tanks, one containing liquid hydrogen and the other liquid oxygen. The nozzle of the rocket engine passes between the two tanks (the circulation of the surrounding liquid cooling the motor). It is in the shape of a long, narrow cone, more than half the length of the rocket. The burning of the fuel takes place at the apex of the cone, the gases expand and cool as they pass toward the wide end, finally escaping at a tremendous exhaust velocity. The ship carries enough fuel for 20 minutes of burning. Movable vanes in the exhaust allow the ship to be steered. They are actuated automatically by a pair of gyroscopes. Drawings are inadvertently not published with the article, but are included with a reprint in 1911 (also reprinted in 1912 and 1914). [Also see 1903, 1914, 1916 and 1930.]

V. I. Kryzhanovskaya publishes her novel *On a Neighboring Planet*. The spaceship involved is cigar-shaped, one end terminating in a small, rotating wheel with paddles. Four thick glass windows break the surface: two in the sides and two near the nose, opposite the end with the wheel. A hatch in the top allows access to the interior. Inside, a large number of glass globes containing a spongy substance cover the circular walls. There is a cot in the middle and, near the front windows, seats with soft cushions and a steering device like the handlebar of a bicycle. In the tip of the nose is a ray projector.

The ship is lifted from the ground by a balloon. As soon as it reaches the proper height, the electrical apparatus causes the ship to shoot off. The only explanation given for the method of propulsion is that the ship is carried along on waves of universal vibration, in this case apparently following a beam of energy sent to earth from Mars.

Robert E. Hanvey publishes the novel *Myora* in which a trip to Mars is made in a flying

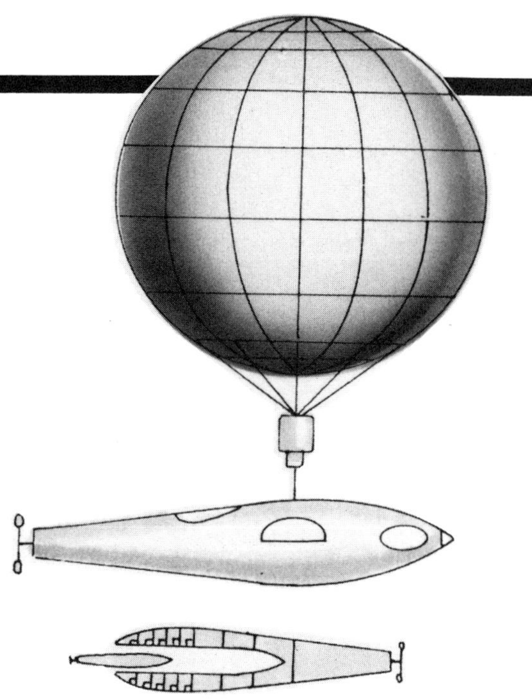

machine. Its wings are moved by engines powered by liquid oxygen.

ca. 1903

On the Silver Globe is published in Poland by J. Žulawski. It contains a description of the moon as seen by astronauts travelling in an airtight car from the crater Eratosthenes to the lunar north pole. The scenic descriptions—described more than 40 years later as being as "suggestive and dramatic as [Chesley] Bonestell's pictures"—have been checked for accuracy by astronomer friends of the author. Exploding "mines" are used to check the descent of the explorer's vehicle—launched by a space gun—onto the moon.

On the Silver Globe is followed by a sequel, *The Victor*. The first book describes the launching of two projectiles to the moon. The story takes place in the year 2005, on the centennial of Jules Verne's death. The details of the projectiles and their method of launch are left vague. The site of the takeoff is on the coast of Africa, 13 miles from the mouth of the Congo. Five passengers are in the projectile when it is fired from an enormous cast iron gun. It describes a parabolic trajectory, falling almost perpendicularly onto the moon in the region of the Sinus Medii, near the center of the hemisphere that faces the earth. The passengers survive the landing and communicate with the earth by way of radio. Unable to return to the

Two French model rocket airplanes.

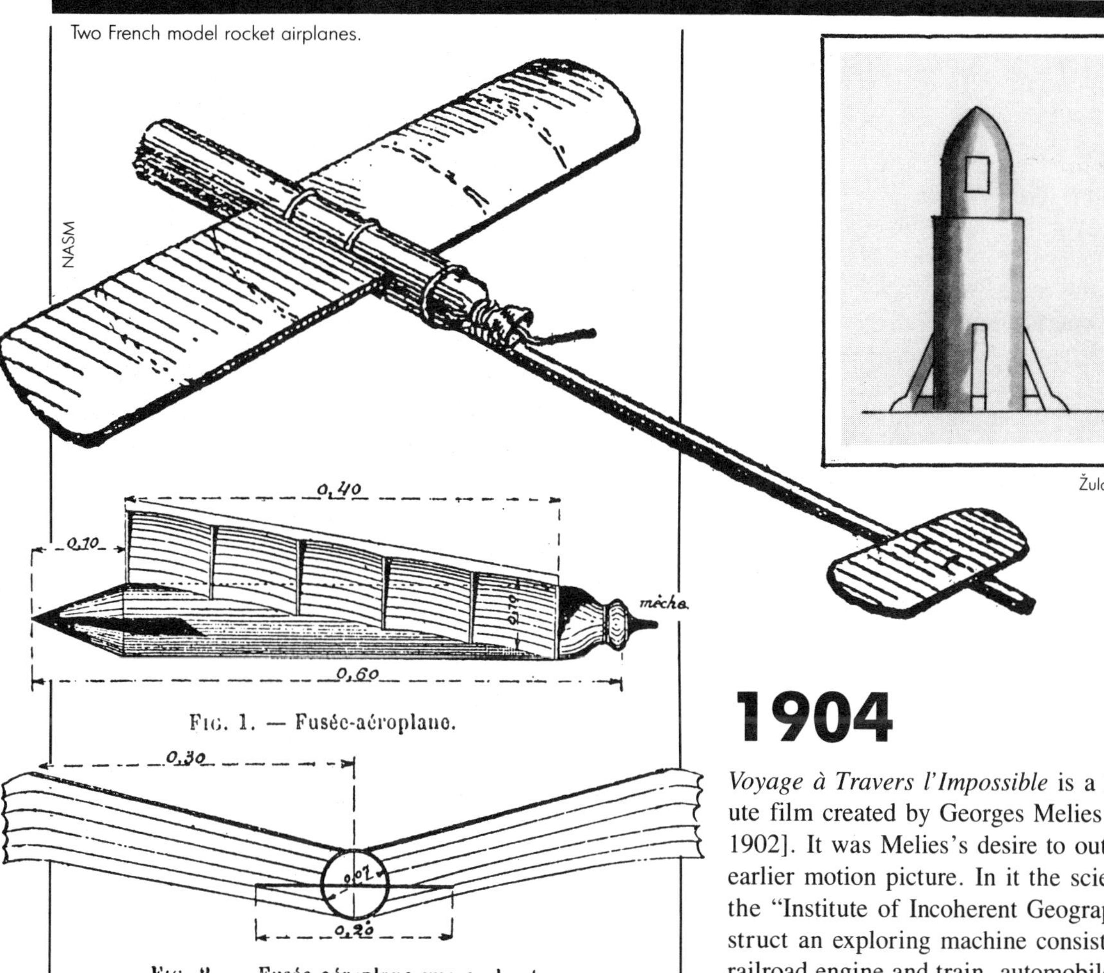

Żulawsky

earth, they convert the projectile into a hermetically sealed wheeled vehicle, powered by an electric motor.

The second expedition, consisting of two men, is less fortunate, crashing on reaching the moon, killing the astronauts.

The second novel (1921) takes place several centuries later. Another trip is made to the moon. This time the spaceship is equipped for a return trip. The ship is made in two parts: a bullet-shaped main section that fits into a sleeve with a closed bottom, like a piston in a cylinder. When the spaceship hits the moon tail-first, the central projectile compresses the air between its base and the bottom of the cylinder. The projectile locks in this position. For the return flight, all that is necessary is to press a button and the compressed air would shoot the projectile from the cylinder like a bullet from a cannon.

1904

Voyage à Travers l'Impossible is a 25-minute film created by Georges Melies [also see 1902]. It was Melies's desire to outdo his earlier motion picture. In it the scientists of the "Institute of Incoherent Geography" construct an exploring machine consisting of a railroad engine and train, automobile, airship and submarine combined. After numerous adventures, and in the process losing parts of the amazing train, the scientists leave the earth altogether, assisted by balloons attached to the roofs of the carriages. They land on the sun, leave the wrecked train and go exploring. Returning, they discover that the submarine is still intact. They launch it toward the earth, using its propellors to steer it. A parachute allows it to descend safely to the surface of the sea.

Jean Delaire publishes the novel *Around a Distant Star*. A spaceship travels to a star 2,000 light-years distant in order to use a super-telescope to observe the events on earth 2,000 years in the past: specifically, the last days of Christ. A number of alien lifeforms are encountered.

Ivan Meschersky in his *Collection of Problems in Theoretical Mechanics* develops the mathematics of rocket propulsion, predating some of Tsiolkovsky's publications.

SEPTEMBER

L. J. Beeston publishes his story "A Star Fell" in *Cassell's Magazine*. In the year 2004 a spacecraft is driven by a "new application" of electricity that allows the inventor to attain "so enormous a speed as to practically annihilate space." This prediction is tragically realized. He plans a trip to Venus but his airship is stolen by someone who does not understand the dangers of friction. The ship rises with such prodigious speed that it is overheated by atmospheric friction and "finally dissolved . . . in a streak of glowing vapor."

SEPTEMBER 17

Henri Graffigny (pseudonym of Raoul Marquis) flies a rocket-powered model airplane at Mers (Somme), France. The model is 50 cm long and 6 cm in diameter, weighing 320 gr.

1904–1905

William Wallace Cook publishes the novel *Adrift in the Unknown* as a fifteen-cent paperback. It describes a journey to the planet Mercury.

1905

The musical stage production *A Yankee Circus on Mars* is written by George Vere Hobart. A circus is sold at auction to visiting Martians who carry it back to their planet in a pair of airships.

John Mastin, in the novel *The Stolen Planet*, describes a voyage made by an antigravity spaceship, the *Regina*, through our solar system and into interstellar space. There it discovers a dead planet wandering, with the remains of a vanished civilization on it. The heroes visit Sirius, a cluster of twenty-five colored stars. After returning to the earth, an asteroid is moved from its proper orbit into one around the earth—its tidal forces causing havoc [also see 1909].

1905–1906

WINTER

Eleven-year-old Hermann Oberth is introduced to Jules Verne's *From the Earth to the Moon*, which causes him to become "fascinated with spaceflight, and even more so, because I succeeded in verifying [Verne's] magnitude of the escape velocity." He also confirms "that the time of flight was correct," two remarkable accomplishments for a young boy. "Jules Verne's idea of retarding the fall onto the moon by rockets had surprised me very much in the beginning, because there was nothing the escaping gas could push against," although a little reasoning eventually proves to him that Verne was quite correct.

1906

Percival Lowell publishes *Mars and its Canals*.

1907

Robert Esnault-Pelterie begins his researches into rockets and spaceflight. [Also see 1930.]

The Swedish astronomer Birkeland experiments with a model spaceship that is propelled *in vacuo* by hydrogen and oxygen. If this is true, then Birkeland's experiments are the very first liquid oxygen/hydrogen space-oriented experiments ever performed. He also at one time suggests the use of electromagnetic guns for launching spacecraft.

JANUARY

As a student at Worcester Tech, Robert Goddard receives rejection letters from three highly esteemed magazines. They turn down an article in which he suggests that rockets might someday be propelled into interplanetary space by means of atomic power. One editor replies: "The speculation is interesting, but the impossibility of ever doing it is so certain that it is not practically useful. You have written well and clearly, but not helpfully to science as I see it . . . "

JUNE 23

Among Fridrikh Tsander's first working notebooks is an entry on the motion of a body propelled by the reaction of ejected particles. This is the earliest indication of his interest in interplanetary travel.

OCTOBER 3

Robert H. Goddard, in his paper "On the Possibility of Navigating Interplanetary Space," points out the potential use of the pressure of sunlight as a means of propulsion. He observes that the "pressure induced by radiation [of the sun] equals 4 pounds per square mile." In the same work, Goddard speculates on the possibility of gaining energy for rocket propulsion from atomic fission.

NOVEMBER 10

A brief mention is made in Fridrikh Tsander's notebooks concerning the problems associated with the creation of a spacecraft. Some of his thoughts include "[a]ppliances for keeping the floor of the frame in horizontal position." He suggests the possibility of using gyroscopes, as used in oceangoing ships, He also concerns himself with compressors and substances for the removal of carbon dioxide and other gases, as well as the regeneration of oxygen, for which a small garden might be provided.

Cover of the dime novel *Brave and Bold Weekly*, featuring the story *At War With Mars* (1907).

The latter would also contribute toward the elimination of wastes. He considers the utilization of solar energy, and even the problem of constructing a hangar for "the building and accommodation of the spacecraft."

1907–1908

"Fenton Ash" (F. Atkins), in his story "A Son of the Stars," tells the adventures of some boys who are taken to Mars aboard a Martian spaceship. They discover the ship while its owners are attempting to repair it in the Australian desert [also see 1909 and 1910].

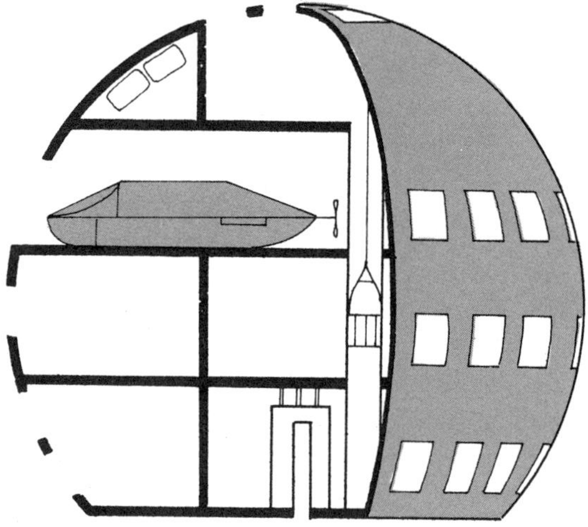

1908

A. Bogdanov describes an antimatter spaceship in his utopian novel *Red Star*. In an analogy with electricity, Bogdanov suggests the existence of matter that possesses a gravity with an "opposite sign" to that which we enjoy on the earth. This matter would be repelled rather than attracted by the earth, the sun, the moon and the other planets. This new element resembles mercury and can be stored in tanks. The spaceship that is built to capitalize upon the properties of "minus matter" is a huge sphere with a flat base. It is 55.5 feet tall and 66 feet wide and constructed primarily of aluminum and glass. The spaceship is divided into four floors. The lowest contains the engine room with its atomic engine: a narrow cylinder of osmium containing the disintegrating radioactive matter (arrangements are made for cooling). Surrounding the engine are rooms dedicated to computing, astronomy, oxygen production and water purification. The second floor contains a photographic laboratory, a gymnasium, library, lounge, etc. The third floor has cabins and a hangar for a minus-matter airship. On the top floor are the cylinders of minus matter and an astronomical observatory.

Voyage à Planète Jupiter, a film by Spanish director Segundo de Chomon, is made to compete with Melies's successful cinematic fantasies [see 1902 and 1904]. In it a trip to Jupiter is made in a dream.

F. A. Tsander experiments with regenerative greenhouses, for eventual use in spacecraft. He fertilizes his plants with "night gold" (human excrement), in order to duplicate the closed system on board a spacecraft [see 1937].

Nettie Parrish Martin publishes *A Pilgrim's Progress in Other Worlds* in which Ulysum Storries invents a flying machine with which he explores the solar system. His "Skycycle" is modeled after a bird "with a balloon attached, The body of the bird was my boat, in which I made a small cabin. I fitted up a pair of wings of steel wire and canvas; next a tail of the same material, which I worked with the electric machinery, which was inside of my boat. The head of the bird was a small room for observation, and the eyes were windows to look through; the bill held a box for tools for repairing it if necessary; the legs were grappling irons, with feet like claws; they served as ladders by which to climb into my boat. Beneath the body of the bird were two rubber-tired wheels, like bicycle wheels, which were worked by electricity when I wanted to ride on the ground . . . "

Rene Lorin, a French engineer, proposes an advanced design for a winged rocket. It would fly by the reaction caused by the com-

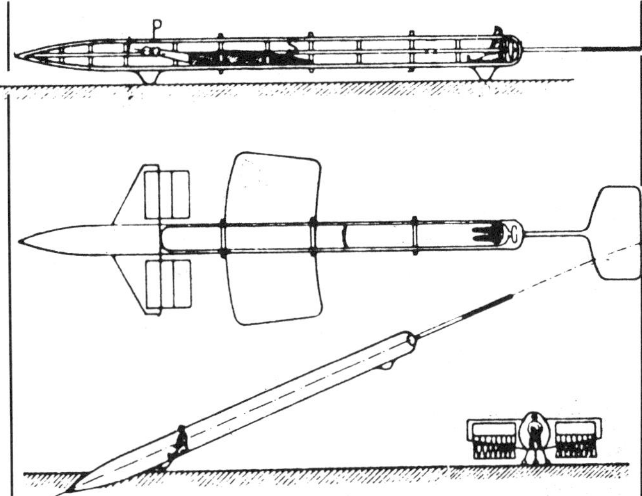

bustion of liquid fuels, the gases being ejected from a series of nozzles. The fuselage is a long, narrow cylinder with a sharply pointed nose. It has short, stumpy wings roughly midway and skids which it uses for takeoff. The rocket engines, in two banks of six, are mounted on either side of the fuselage ahead of the wings. The engines can be pivoted up and down to stabilize the plane's flight. The plane would weigh about 220 pounds. The pilot sits in the rear of the aircraft.

At takeoff the nozzles point almost straight down, gradually rotating to a horizontal position after the plane becomes airborne. The landing would be a little rough: the plane simply strikes the earth at a speed of nearly 100 feet per second, burying its sharp nose in the ground like a dart. The pilot, his seat attached to powerful elastic shock absorbers, slides down virtually the full length of the hollow cylindrical hull. [Also see 1911.]

Lorin also develops an unmanned rocket-powered aerial torpedo, guided by remote control (1910–12).

The New York Herald publishes an article by Robert C. Auld entitled "All Aboard for the Moon," in which he proposes launching a spacecraft by means of an enormous gun. He is inspired by the report that the British are planning to build "a stupendous gun or projectile machine, invented by a Mr. [A. S.] Simpson, which by the utilization of certain electrical forces can throw a projectile [of 2,000 pounds at a velocity of 30,000 feet

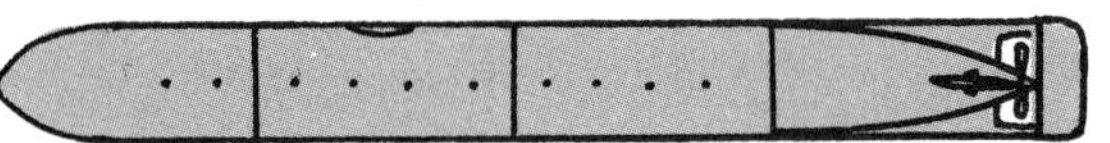

per second] an initial distance of 300 miles." Scaling up the Simpson gun, Auld finds that his proposed space gun would have to be about the size of New York's 700-foot Metropolitan Life Building (and would require an equal amount of steel to construct, at a cost of $3 million). Sections of its steel tube would be hoisted into place and welded together. The projectile would be launched by means of a series of electromagnets surrounding the cannon and energized sequentially. Central Park is suggested as the ideal place to construct the space gun!

The passengers would have every comfort, including electricity for cooking and heat. Every thirty-six hours their air would be replaced from a reserve of compressed air. Inflatable rubber cushions would ease the shock of the initial acceleration.

One Prof. Dodge speculates that upon reaching the moon the explorers could live comfortably "in thick-walled, airtight houses, and could walk out of doors in airtight divers' suits . . . Astronomers could plant their telescopes there, free from their most serious hindrance, the earth's atmosphere. Tourists of the wealthy and adventurous class would not fail to visit the satellite . . . "

FEBRUARY 8

Fridrikh Tsander returns to his spaceship designs.

JULY 26

In article titled, "De crête à crête, de ville à ville, de continent à continent" (From Hill to Hill, From City to City, From Continent to Continent), Captain F. Ferber de Rue quotes Robert Esnault-Pelterie's work. He suggests that if high altitudes are to be reached and if man wants to reach them himself, then a new method of propulsion is indicated: that of the rocket. The pilot will be enclosed and air will be manufactured for him. Such a flight will not be made, Ferber claims, in a flying machine but instead in a "dirigible projectile."

AUTUMN

Friderikh Arturovich Tsander, at the age of twenty, begins compiling his "Spaceship Notebook." It is written in German and is titled "Die Weltshiffe (Athershiffe) die den Verkehr zwischen der Sternen ermiglichen sollen. Die Bewegung im Weltenraum" (Spacecraft [Ethercraft] Will Make Possible Interstellar Travel. Movement in Space). He poses problems such as the determination of the energy required to reach any star, the work needed to move a certain mass to a particular distance from the earth, oxygen requirements, flight time to Venus and Mars, etc. Many of his notes are in a shorthand code that has not yet been deciphered. Also in 1908 he founds at his school the "Riga Student Society of Space Navigation and Technology of Flight." Tsander is greatly influenced by the novels of Jules Verne.

1909

Fridrikh Tsander expresses for the first time the idea of using the solid structural material of the spaceship itself as additional fuel or reaction mass. Between this date and 1911 he also carries out calculations on a jet engine and the necessary work required to boost a spacecraft to great altitudes. [Also see 1937.]

A. Wegener describes a reaction-powered flying machine, which is never built. It is a heavier-than-air vehicle resembling a submarine, 13 meters long and 6 meters wide, made of steel and weighing about 4,320 kg. It is claimed it could develop a speed of up to 30 meters per second.

Fenton Ash (Frank Atkins) publishes *A Trip to Mars* in which he describes the Martian spaceship *Ivenia*. It is egg-shaped and equipped with folding wings that extend for atmospheric flight. [Also see 1907–1908 and 1910.]

Robert H. Goddard begins his study of liquid-fueled rockets. He considers the possibilities of liquid oxygen and hydrogen as ideal fuels. He also suggests the concept of ion-propelled rockets.

John Mastin publishes a sequel to *The Stolen Planet* [see 1905], with one of the best of all titles: *Through the Sun in an Airship*, in which he describes a future of highly advanced science. Mars and Jupiter are inhabited. This story also features the spaceship *Regina*, whose secret of flight " . . . is the one and only scientific method of adding to and overcoming or depriving gravity . . . "

Charles Napier Richards publishes his novel *Atalanta* in which five men travel to Venus in an antigravity spaceship. The necessary force is created by the combination of a new substance plus electricity.

Arnould Galopin writes the novel *Le Docteur Oméga* in which a flight to Mars is made by three Frenchmen. The spacecraft *Cosmos* is a projectile-shaped vehicle made impervious to gravity by being built of the substance "repulsite." It is 13 meters long and 3 meters in diameter, capable of acting both as a submarine and as an automobile (with the addition of its supplementary wheels). Its interior is divided into four compartments, each lit by electric lamps powered by a generator run by an eight-cylinder 200-horsepower motor. The floors are all suspended upon universal joints in order to maintain a normal level. The windows are made of transparent repulsite. One cabin is reserved for provisions: ham, salt pork, preserves, biscuits, ale, champagne, wine and mineral water. The lowest cabin is the armory and sleeping quarters, with beds and a table.

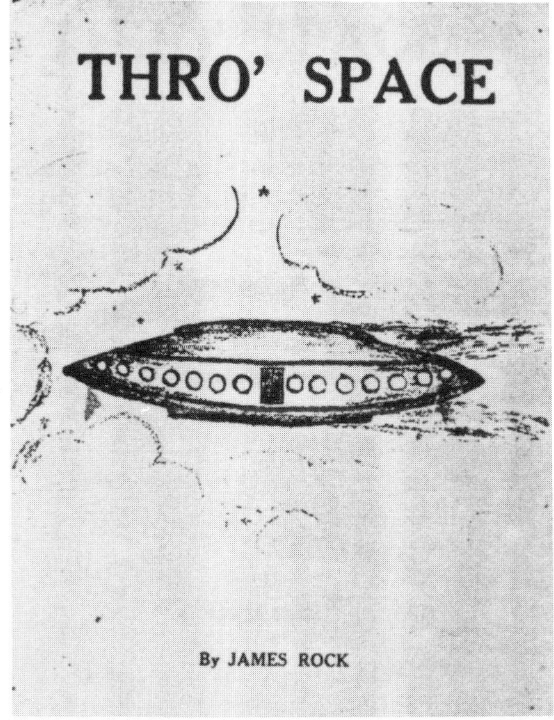

"J. Rock" (pseudonym of Clinton A. Patten?) publishes the novel *Thro' Space*, in which he describes an antigravity spaceship. Its propulsion is produced by the action of salt on a combination of helium and radium—producing "hellium." When treated with electricity, this new substance is repelled by the earth. The spacecraft is spindle-shaped, with a row of portholes down either flank. The heroes fly from the United States to Paris, and visit the moon and Venus.

Weldon J. Cobb publishes the dime novel story "Checked Through to Mars, or Adventures in Other Worlds" in the weekly *Brave and Bold*.

Voyage dans la Lune is a short fantasy film by Spanish director Segundo de Chomon [also see 1908].

Max Mechanik publishes his space travel novel *Marsiana*.

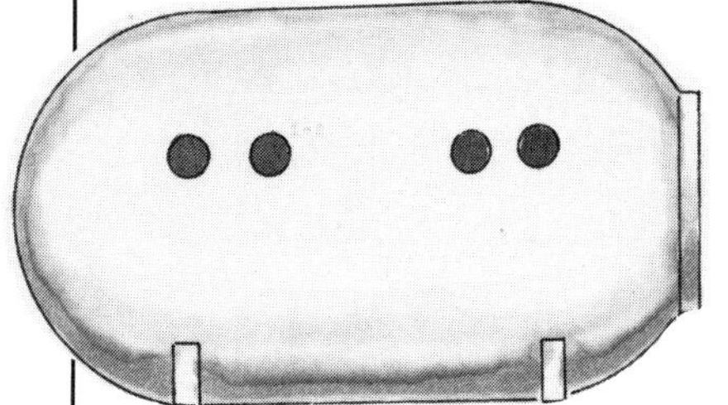
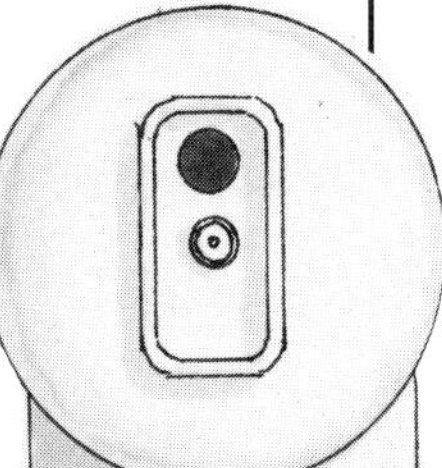

JANUARY-JUNE

Garrett P. Serviss publishes his novel *A Columbus of Space*. Serviss is author of a number of books on popular science and astronomy, as well as several science fiction novels—including *Edison's Conquest of Mars* [see 1898]. The hero of this book invents an antigravity spaceship that operates by drawing upon "inter-atomic energy," converting it into a force that can be directed against any object such as the earth. The spacecraft he builds is "round and elongated like a boiler, with bulging ends, and seemed to be made of polished steel. It total length was about eighteen feet, and its width ten feet." There is a door in one end and portholes in the sides. The interior is beautifully finished in polished wood and leather upholstery. Almost any substance can be converted by the inventor's atomic generator, including smoke and exhaled carbon dioxide.

1910

Trip to Mars is a 4-minute motion picture by Thomas Alva Edison. An inventor makes a round trip to Mars by using an antigravity powder.

N. A. Morozov, in a story written in 1891, *At the Border of the Unknown*, describes a spaceship similar to the Mars-to-Jupiter spaceship of LeFaure and deGraffigny [see 1889].

C. Antonovich describes a rocket aircraft. It consists of a platform carrying the necessary fuel and oxidizer (air). These are mixed by a device on the platform, then carried in a pair of pipes to an overhead chamber. From here the mixture is ejected through a number of small pipes and ignited by electric sparks.

"Fenton Ash" (Frank Atkins) publishes the story "Caught by a Comet." Set in 1985, an antigravity flying machine is accidentally drawn into space by the attraction of Halley's Comet. [Also see 1907–1908 and 1909.]

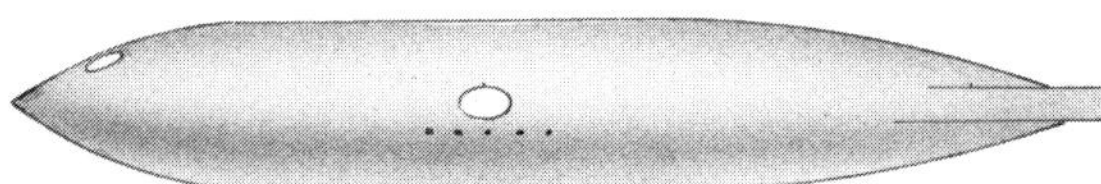

"Roy Rockwood" is the house name for the anonymous authors of a series of adventure stories for boys. In *Through Space to Mars* the spaceship *Annihilator* is described. It is a 200-foot torpedo, made of "a new metal." It is fitted with a window in the pointed nose for its pilot; a pair of oval windows, one in either side, are surrounded by airtight gun loops. The ship's walls are double with insulation. The *Annihilator* is equipped with a telescope with automatic cameras, a library, electric player piano, provisions, etc. It is powered by "etherium"—no opening is needed in the rear of the ship to provide escape for the "waves of energy" produced. The *Annihilator* also appears in a sequel, *Lost on the Moon*, where it is revealed that the power generator cannot work while the spaceship is inverted (while in interplanetary space!). The ship also apparently gains a raised pilothouse for the steersman, and is equipped with an airlock.

Donald W. Horner publishes his novel *By Aeroplane to the Sun* in which a radium-powered "aether-ship" takes a tour through the inner solar system, including stops at Venus, Mercury and Vulcan.

Henry Wallace Dowding publishes his novel *The Man From Mars*. At the end a journey to Mars is made by way of an airship.

In V. Kryshanovskaya's two novels, *Death of a Planet* and *The Legislators*, she describes a spaceship that is used to desert the earth for another planet.

The ship is a enormous oblong made of some transparent, phosphorescent substance. There is only one entrance at the end. Each of the three rooms and many small cabins possesses a window. Propulsion is effected by vibrations of the ether, which can either decompose or fuse atoms of matter. The current of ether is directed by conscious will and produced by an ether current generator consisting of two parts. The first part is a hollow metal pipe that can be lengthened or shortened. A series of keys makes it possible to control the intensity and direction of the ether current. The second part is a hollow ring that hangs from a small hook. Inside the ring are eighteen "resonators." Outside the ring are three more resonators connected by wires. From these a set of vibrating rods or needles are arranged circularly and downwards. In the middle of the ring is another, resembling a hollow drum, with two rows of circular pipes arranged like the pipes of an organ. In the very center is a rotating disc and in the lower part of the drum is small hollow sphere from which come the conductors of the force. When the motor is operating, the disc spins at an extremely high speed.

APRIL 30 TO JUNE 4

The *Chicago Ledger* serializes Will Lisenbee's story, *In the Comet's Track*, inspired by the recent appearance of Halley's Comet. The serial is illustrated by William Molt and Androwles.

The spaceship *Vesta*, powered by an unnamed source of antigravity (the inventor explains that the *Vesta* " . . . is in no sense a flying machine that goes up like a rocket and comes down like a rock. It is the result of a great discovery—one of the most tremendous of all the centuries . . . "), makes a

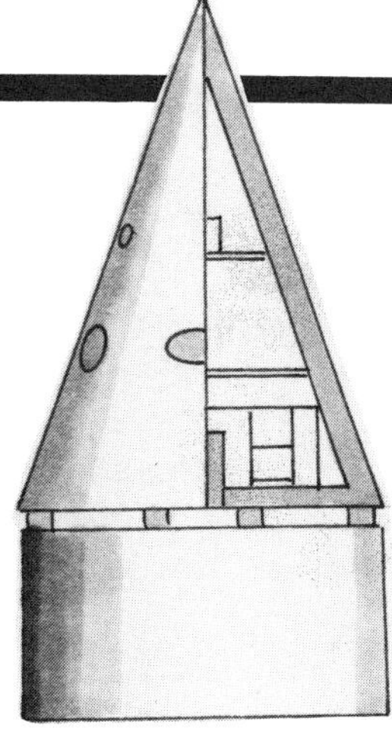

flyby of the famous comet while on its way to the moon. The *Vesta* is octagonal and shaped something like a ship's binnacle. The roof of the central "pilot room" is eighteen inches in diameter and inset with various instruments and electric lights. The cabin is well appointed with leather-covered chairs and electric radiators. There are eight rooms altogether. The spaceship was built by the Pullman Company and shipped in sections to New Mexico where it was assembled secretly. One of the characters estimates that the cost of a spaceflight to the moon at $5 million (at a rate of $0.03 a mile).

The spacecraft ascends through the earth's atmosphere at a leisurely 70 mph, increasing to 5,000 mph once it enters outer space.

The inventor explains that photographs made of the earth from space will be of great benefit to geographers. The astronauts also make a flyby of Halley's comet in order to obtain photographs of it as well as samples of its tail (which proves to be composed of dark, volcanic-looking aerolites). The *Vesta* eventually lands on the moon and the space-suited explorers leave via an airlock in order to walk on the surface.

ca. 1910

L. B. Afanasev describes an electrical spacecraft, *Galileo*, in his novel *Journey to Mars*. It is cone-shaped, about 35 feet tall, with a base about 25 feet wide, and divided into three floors. The first floor, at the base of the ship, is devoted to storage of food, water and supplies. Here also are various instruments as well as machinery for the manufacture of artificial air and carbon dioxide absorption. The second floor is a large common room, with elliptical windows, and the top floor, divided into four quadrants, accommodates the five passengers. A spiral staircase connects all of the levels. The walls of the ship are very thick and made of several partitions. Water between them helps to soften shocks.

It is not made clear how the "electrical motors" operate the *Galileo*. The ship sits on a special electrical platform at the time of takeoff. The flight to Mars requires 206 days, and after landing, the special launching apparatus must be reconstructed. This is done with the aid of the Martians. Apparently the ship is launched by means of mutual electrical repulsion—the ship and the platform having opposite charges.

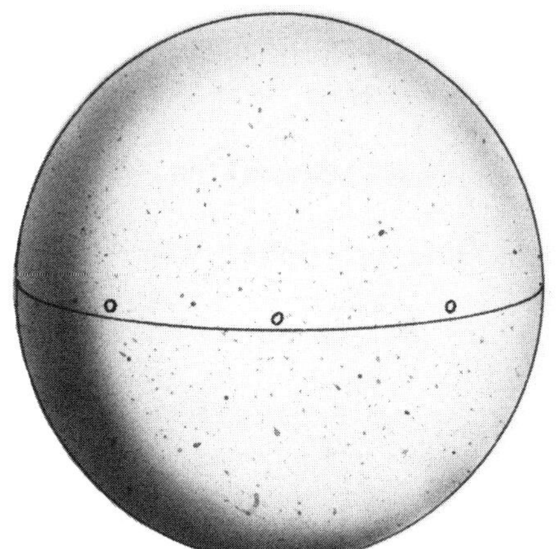

1911

F. Mader publishes his novel *Wanderwelten*—translated into English in 1932 as *Distant Worlds*. The "worldship" *Sannah* is an antigravity sphere 146 feet in diameter, coated with flint glass. It has six windows with telescopes. The central, equatorial room is 49 feet in width, length and height—that is,

1911

with 117,649 cubic feet of storage space. Supplies are stored in containers shaped like inverted pyramids. This vast storeroom is surrounded by thirty smaller rooms, each 16 × 10 feet. The uppermost is an observatory and there are workshops (including a smithy). Most of the remaining space is taken by oxygen storage.

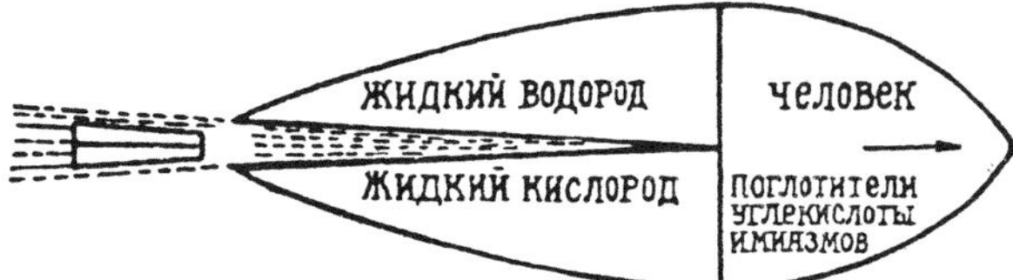

K. E. Tsiolkovsky, in *Herald of Aeronautics* magazine, publishes a summary of his 1903 paper under the title "Summary of the First Paper." This time a drawing accompanies the piece, showing the rocket to be a stubby, streamlined bullet-shape with a pointed nose and tapering tail. It is little more than a stylized schematic, however, though it indicates liquid oxygen and hydrogen tanks, space for a pilot along with carbon dioxide and "miasma" absorbers, and a narrow, conical engine fully two-thirds as long as the ship itself. In the exhaust is placed a set of steering vanes. A drawing discovered among Tsiolkovsky's manuscripts, dated 1902, shows a rocket that clearly agrees in all details with the 1903 rocket description.

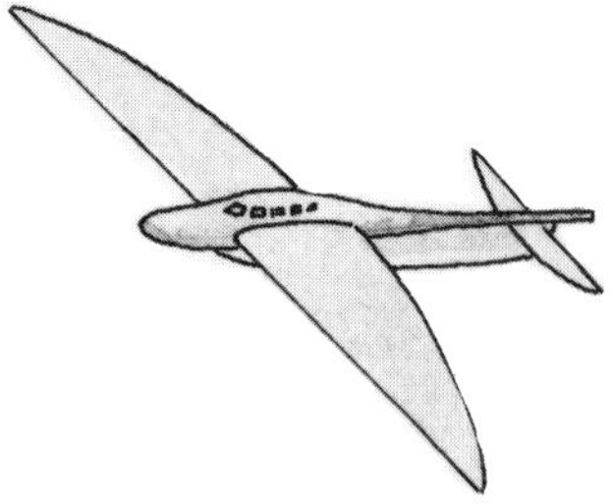

Robert Esnault-Pelterie suggests the possibility of a winged, rocket-powered manned aircraft. In outward form it resembles very much a modern sailplane or U-2 aircraft.

Rene Lorin invents a catapult-launched all-metal jet aircraft with a takeoff speed of 185 mph.

The airplane is an extremely thin, needle-like craft. Fuel, engine and payload are all contained in the front half of the fuselage. The engine is the same as that described in his 1908 plan [see above]. The pilot sits within a small compartment at the very rear of the dragonfly-like tail. This compartment is able to slide on rails.

Takeoff is initiated by an electric carriage that accelerates the plane along a 1-kilometer track. At the end of the track, the plane is moving at about 180 mph and capable of sustaining flight on its own. The pilot lands the plane by first releasing an airbrake. Once it has slowed, the jet hits at an angle, burying its nose 6 feet into the ground like an enormous dart. The pilot's compartment, sliding freely within the tubular fuselage, absorbs the landing shock by way of elastic cables.

Cartoon by Charles Rosen from *Life*, 1906: "Stranded on a Deserted Asteroid."

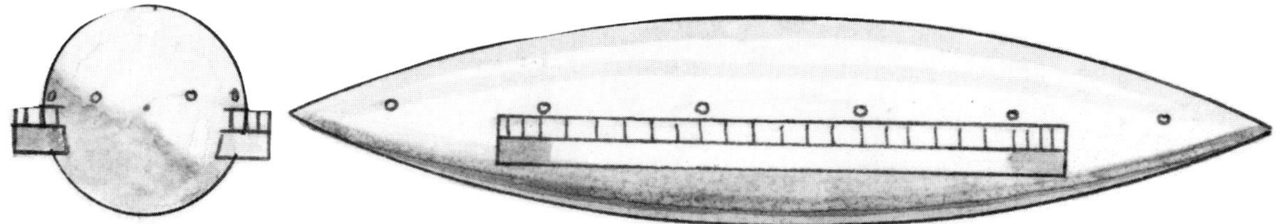

Mark Wicks, in his novel *To Mars via the Moon*, describes the spaceship *Areonal*. It is a spindle-shaped vessel "shaped somewhat like a fish," 95 feet long and 20 feet wide at its thickest and pointed at either end. It is made of "Martialium" (a combination of metals "almost as light as aluminum, yet many times harder and tougher than case-hardened steel . . . "), powered by machinery that is either "electric or magnetic." The *Areonal* is perfectly streamlined, with no protruding machinery visible. There are several windows in the sides together with a few at the top and bottom (made of "a special toughened glass from Vienna"). There are also periscopes. Along each side of the ship is an observation platform or gallery, with a railing, on to which the exterior doors opened.

The interior is divided into five compartments. The rearmost is a general living and sleeping room, with observation windows; next is a storeroom that can also be used for observation; next is a small, special-purpose room, and then a water-storage room with apparatus for air storage and production. In the nose of the vessel is the driving, lighting, and steering machinery. The controls are repeated in each of the other compartments for safety's sake.

The hull is double and provided with airlocks.

A. Gorokhov describes a jet aircraft very similar to the one proposed by Lorin [see above]. It is propelled by a jet engine burning liquid fuel (such as gasoline) mixed with atmospheric air. The exhausts are ejected from six outlets in the sides of the plane, near its midsection. Its expected top speed is in excess of 215 mph. Like Lorin's jet, Gorokhov's has extremely small wing surfaces. Both fuselage and wings are made of steel.

It is launched via a catapult, or perhaps on a carriage that is allowed to roll down a long slope, to build up speed. Like Lorin, Gorokhov allows his plane to land dart-fashion, burying its nose in the ground.

JUNE 10

Andre Bing patents in Belgium a design for a rocket engine that (according to Esnault-Pelterie) is very similar to Goddard's 1915 design. Bing's rocket is intended for instrumented exploration of the upper atmosphere. He also discusses the possibility of "travelling beyond the limits of the earth's atmosphere" using step (or "successive") rockets and nuclear energy. Exhaust gases are to be ejected in the three coordinate directions for automatic control. Bing's patent is effectively one for a spaceship, the first such ever issued.

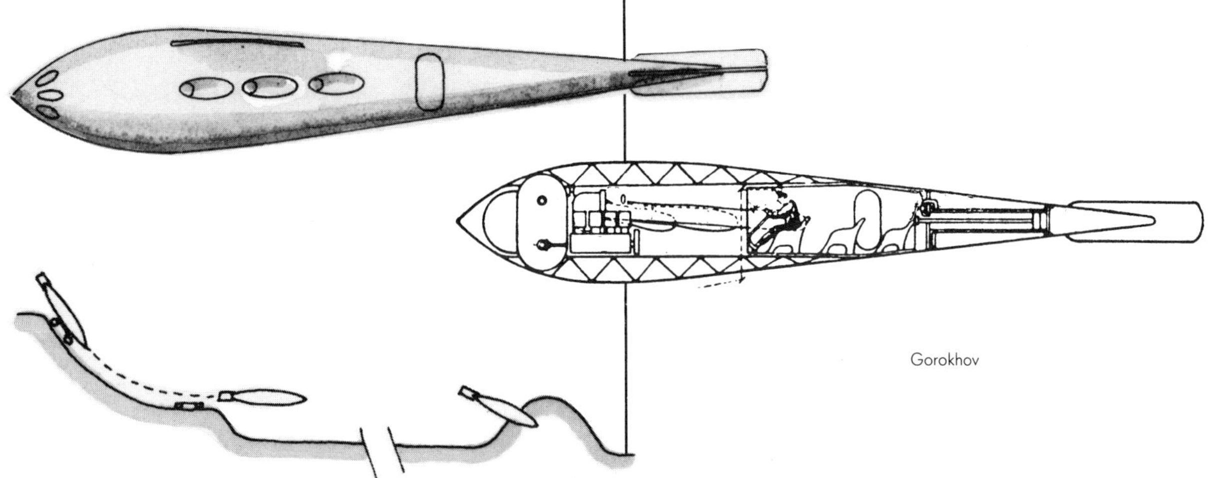

Gorokhov

FALL

Hermann Oberth performs experiments in free fall in a swimming pool at Schasburg. "While attempting to cross the pool under water, diagonally, I hit a wall which seemed almost perpendicular to me . . . From several encounters I finally recognized that this 'wall' was the bottom [of the pool]." This disorientation, he concludes, "meant that I had undergone the psychological experience of weightlessness!"

It is due, he thinks, to the numbing effects of the cold water, the excess of carbon dioxide in his system, and the lack of anything touching his body while he floated freely in the water. His normal sensory input needed for orientation becomes no longer effective. He will continue these experiments during the years of World War I [see 1915-1918].

1911–1912

K. E. Tsiolkovsky, in *Exploring Universal Space with Jet Apparatus* (Parts I and II), considers the use of radium as a source of energy for spaceflight. "If the decomposition of radium," he writes, "or other radioactive bodies, which probably includes all bodies, could be accelerated rapidly enough, then its use would probably furnish, under otherwise equal conditions, such a velocity to the jet apparatus that the arrival at the nearest [star] could be reduced to 10–40 years."

Tsiolkovsky also suggests that "Maybe in due time with electricity we can impart enormous velocities to particles ejected from the jet apparatus." [Also see 1914.]

1911–1913

Professor B. P. Veinberg carries out experiments with electromagnetic launching at the Tomsk Technological Institute. A 10 kg projectile is accelerated within an annular tube. The air has been evacuated from the torus to reduce friction. Veinberg envisions an eventual solenoid accelerator 2 miles long which could achieve speeds of 800 to 1,000 km/h. A starting "station" would hold a large num-

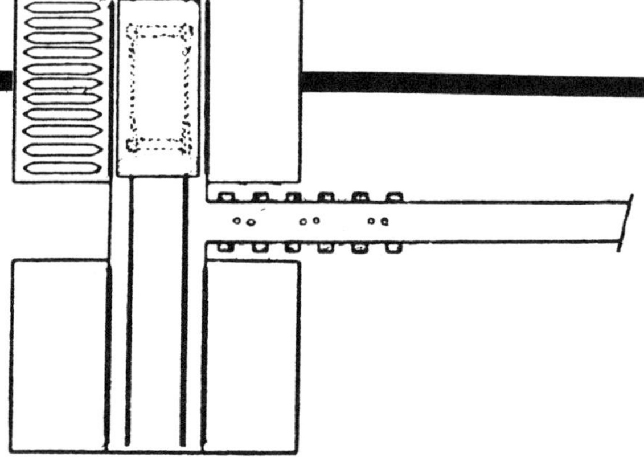

ber of passenger projectiles which would be automatically loaded into a special chamber. This would be sealed and the air exhausted. This chamber would then be moved to the breech of the solenoid gun and each projectile fired in its turn.

1912

Donald Horner's novel *Their Winged Destiny* is published. When the earth is threatened by a dark star, an escape is made by spaceship. The destination is Alpha Centauri, where the explorers discover a second earth. Also published under the title *The World's Double*, the book has a cover illustration showing helmeted astronauts climbing from the spacecraft.

Stanley Grey publishes the novel *The Helioplane* in which a trip to Mars is made. The spacecraft operates by antigravitation, created by a substance that shields it from gravity when it is cooled to absolute zero.

Auguste Piccard experiments with rocket-propelled model aircraft (using fireworks rockets). Professor Piccard will become one of the first men to reach the fringes of outer space when he balloons twice to an altitude of 10 miles, well into the stratosphere [see 1933].

Bertram Atkey, in the short story "The Strange Case of Alan Moraine," describes how an aviator is kidnaped by the inhabitants of a disc-shaped spacecraft (the first flying saucer?).

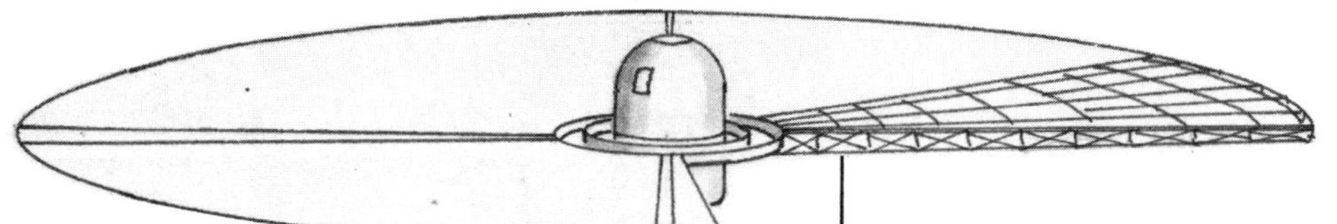

A. S. Shor, in "Aeronautics in Life," proposes space colonies with radio communication and solar-powered life support.

NOVEMBER 15

Robert Esnault-Pelterie suggests the desirability of using intra-atomic energy in spaceflight, in a report made to the Sociétè Francaise de Physique (published in an abbreviated form in the March 3, 1913 edition of *Journal de Physique*) on the possibility of flying to the moon. He estimates that a moon rocket would require 21 million foot-pounds of energy to make the trip and, adopting Bing's suggestion [see 1911], proposes to use radium for the energy.

In this report, REP (as he preferred to be known) anticipates the use of auxiliary propulsion for guidance; calculations for escape velocity; the times, velocities and durations of trips to the moon, Mars and Venus; and answers to such problems as control of the spaceship's internal temperature. In France, at least, REP is considered the founder of the science of theoretical astronautics [also see 1930].

1913

Franciszek Abden Ulinski designs an electric "cathode rocket" [see 1920].

Robert Esnault-Pelterie publishes "Consideration sur les resultants d'un allegement indefini des moteurs."

B. Krasnogorskii publishes *On the Waves of the Ether*, in which he describes a solar sailing ship. The spaceship, intended for a flight to Venus, is called *The Victor of Outer Space*. It is built in the form of a small, central gondola surrounded by an enormous circular mirror. It will fly through space by the force of the sun's radiation impinging upon

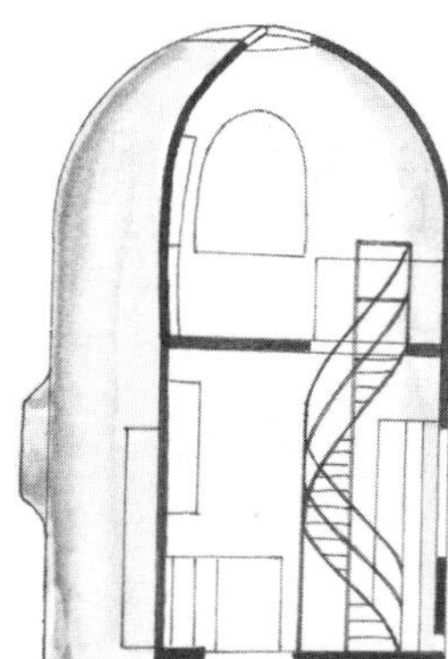

this mirror. Steering will be accomplished by changing the angle of the mirror relative to the direction of the sun.

The central gondola is shaped like a blunt bullet 4.5 meters high and 3 meters in diameter. It is divided into two floors, the lower about 2.75 meters high. It has two windows in the sides, one in the floor and an entry. Cabinets contain utensils and provisions; there are controls for maneuvering the mirror, carbon dioxide removal, instruments, a table, chairs and lamp, a stove, an oxygen outlet, and storage for liquid hydrogen and oxygen. The upper floor has a hemispherical roof, two side windows and one window in the middle of the roof. Cabinets contain instruments and supplies, and there is a table and chairs as well. A spiral staircase connects the two floors. The walls of the gondola are double, with the space between evacuated for insulation. The gondola itself is made of "maxwellium." The outside wall of the ship is silver-plated, and the inside walls padded. The windows are shaded with thick, black curtains.

Food and water for a journey of 60 days is taken: 170 kg of the former and 360 kg of the latter. For breathing, an apparatus is carried on board that not only provides oxygen for breathing, but also yields heat and light. Oxygen and hydrogen are carried in liquid form in a pair of double-walled tanks mounted within a pair of bulges on the exterior of the vessel. Pipes leading from these convey the liquid gases to another pair of tanks where the heat of the sun is allowed to warm them. The gaseous hydrogen and oxygen are drawn off as needed. They are mixed and burned to provide light and heat for cooking. The resulting water is collected in a small tank for later human consumption. Oxygen alone is released for breathing. Carbon dioxide is removed from the atmosphere by a special liquid that can be purged of the CO_2 it has absorbed by exposure to the vacuum of space, after which it can be reused. The velocity of the ship toward or away from the sun is measured by a special instrument that determines the Doppler shift of the spectrum caused by the speed of the spaceship, translating it into terms of velocity.

The total weight of the gondola is not quite 800 pounds (maxwellium being a very light metal, an alloy of aluminum, lead, vanadium and some light metals).

Surrounding the gondola is a disc, attached to it at a level with the floor of the dome. The disc is a framework of maxwellium 35 m in diameter. Its flat upper surface is covered with sheets of polished metal only 0.1 mm thick, forming the mirror. In the center of the disc is a hole 6 m in diameter. Inside is a ring, joined to the disc by a pivot so that it can rotate freely. Inside this is the gondola, attached to the ring by another pivot at right angles to the rings. The gondola is thus attached to the mirror by a universal joint, allowing it to assume any inclination desired relative to the mirror. Four screens of black silk can be deployed to cover any of the four quadrants of the mirror, to control the radiation pressure on it.

Krasnogorskii has calculated a force of about 1 pound on every square yard of the mirror. Since the ship (mirror plus gondola) weighs a total of about 1,000 pounds, this leaves an excess of radiation pressure, or propelling force, of about 100 pounds.

The initial launch of the ship from the earth is accomplished by balloons. A large cruciform structure supports the vessel at its center. From the protruding beams of the carrier rise cables that are attached to four large hydrogen balloons. These carry the spaceship to an altitude where the force of solar radiation can lift the ship off the cradle. Departure is at 6 p.m., July 28, from "Mars Field" in Leningrad.

The *Victor of Outer Space* is lifted to a height of just over 5 miles. At this point the sun's radiation lifts the spaceship. The space travellers are able to navigate the ship very much like a sailing ship on the earth's oceans. The journey to Venus has to be abandoned when the mirror is irreparably damaged by meteorites. The gondola falls back to the earth, landing safely in Russia's Lake Ladoga.

A more successful flight is described in a sequel, *Islands in the Ether Ocean* (1914).

Reid Whitley, in his short story "The Dominant Factor," describes a visit to Venus with the concentration being on the description of nonhuman forms of life.

John N. Raphael sets his novel *Up Above* in the year 1915, when a huge spaceship from an unknown planet invades the earth, first hovering in the stratosphere from where the aliens gather specimens of terrestrial life. This is an idea which will be proposed seriously by Charles Fort in his remarkable book *Lo!* [1931].

D. L. Stump describes a spaceship powered by a rocket in his novel *The Love of Meltha Laone*. The ship, launched from Pike's Peak, remains fixed in space while the earth moves out from beneath it. The ship stays in the orbit of the earth and soon another world approaches it: a planet that is normally hidden from the earth because it is on the opposite side of the sun.

Oskar Hoffman in *Mack Milford's Reisen im Universum* (Mack Milford's Voyage in Space) describes a spacecraft propelled by gravity-neutralizing magnetic waves. It travels to the moon at a speed of 5,000 km/h. Hoffman also describes a matter transmitter.

The hero breaks down his assistant and sends his atoms to the moon by radio.

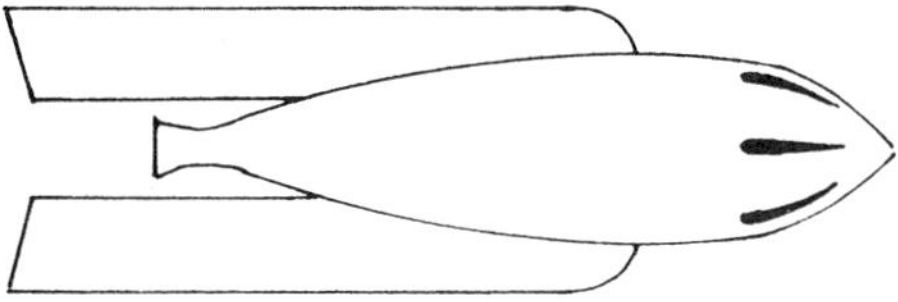

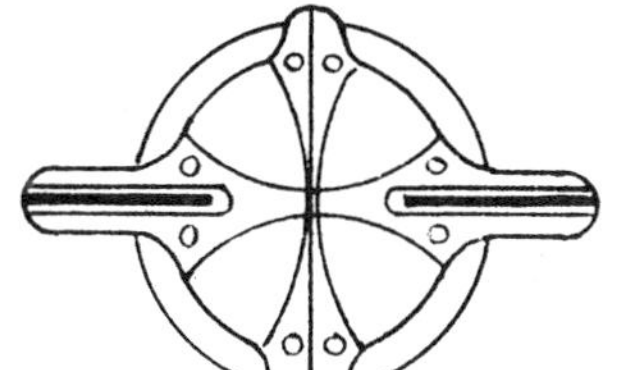

Spaceship in the barrel of the electro-magnetic gun, its fins fitted into slots in the wall of the gun.

G. A. Polevoi proposes an electromagnetically launched spaceship (it is exhibited later at the Moscow Exhibition, see 1929). A slanted tunnel is lined with solenoids which impart an initial acceleration of 1,600 m/s to the vehicle. This is a streamlined spacecraft covered with a streamlined iron "armor." The rocket's fins fit into guide slots in the sides of the tunnel. Part of launching acceleration is provided by the rocket itself, which burns one motor while still in the tunnel. At an altitude of 150 km the armor is shed and all of the rocket motors are fired. The armor descends by parachute and is returned to the launching station. [Also see 1911–1913.]

MARCH 13

F[rederick]. Rodman Law attempts to launch himself in a giant skyrocket. Law, a notorious stuntman and younger brother of famed aviatrix Ruth Law, has a rocket made especially for his daredevil stunt (supposedly for a scene in a movie about interplanetary flight, though which movie this was is unknown). It is some 10 feet long and 3 feet in diameter (though these dimensions are variously given as 44 feet in length and 24, 10 or 2 feet in diameter). The top half is a cardboard cylinder surrounding a chair in which Law will sit, in a papier-mashtub, topped by a cardboard "nosecone". Law had a "lever in front, which was meant to control the rocket's flight and keep it from whirling him off his seat" (though how this control was to have operated is unknown). The bottom half of the rocket is made of riveted sheet metal and filled with a slow-burning black powder, supplied by the Deitwiler and Street Fireworks Co. of Jersey City, New Jersey. Attached to the rocket is a "stick"—a heavy timber—nearly 20 feet long amd 4 inches square. The whole structure is propped upright by a timber scaffolding. Law plans for his rocket to launch him 3,500 feet into the air, after which he will descend by a newly-invented safety parachute (which would open automatically). The parachute, made of Japanese silk, weighed only 6.5 pounds.

The flight is to take place from a hillock in some marshland near the fireworks plant. Law hopes to land near Elisabeth, 12 miles away. Aided by an assistant, Law climbs into the nose of the rocket and fits the nosecone above his head. As 150 people and newsreel cameramen watch, Samuel L. Serpico, manager of the International Fireworks Co., lights the fuse, which "spluttered for some time. Then followed a terrific explosion. The gases . . . burst the steel shell into

many pieces. Law fell like a sack to the ground, a distance of approximately 15 feet. "Badly burned, Law is taken to a hospital, where he announces his intention of making another attempt at a rocket flight in the near future. There is no record of his ever having done so.

APRIL

Scientific American, in its monthly *Supplement*, publishes the article "Travelling Through Inter-stellar Space What Type of Motor Would You Employ?" The answer is the rocket and Law's experiment [see 1913 above] is mentioned in support [!].

1913–1916

The engineers Andre Mas and M. Drouet describe (in "La Catapulte tournante de Mas et Droute," published in *L'Avion*, 1913) a projectile launched from the rim of a spinning wheel [also see 1889, 1916 and 1927]. A wheel 80 meters in diameter will spin at a rate beginning at 30 rpm and increasing to 65 rpm. When the 2,100 kg projectile carried at its rim reaches a speed of 65,000 mps, it will separate from the wheel and shoot off into space. The wheel will be sunk into a deep furrow for half of its diameter. A powerful electric motor will begin the spin, eventually being supplanted by a steam turbine.

In 1915 H. de Graffigny proposes a very similar idea, with an enormous wheel powered by a steam turbine. A cable—subjected to 200 tons of tension—attaches the projectile to the rim of the wheel [also see]. In 1916 he comes up with another variation on the basic idea of launching a projectile by way of centrifugal force. The spacecraft is placed at the end of a counterweighted 50-meter beam which is attached to an axle at its center of gravity. The beam rotates at 44 rpm, developing a speed of about 14 km/s, sufficient to fling the projectile away from the earth. To reach such a rate of rotation a 12,000 hp motor is needed, which requires 7 hours to get up to full speed. An electrical release sends the spaceship off on a tangent.

De Graffigny also proposes a variation on these schemes, in *Le Chemin Circulaire* (1913). The vertical wheel is replaced by horizontal, circular tunnel, with a tangentially placed "launching ramp." A partial vacuum is established within the tunnel, which is 120 km in circumference. The tunnel opening is 5 × 6 m. The projectile rests upon a car which is supported by a layer of pressurized oil or water. The projectile travels around the inside of the tunnel (from which the air has been evacuated) for 55 minutes, accelerating all the while. A few seconds before launch, outside air is admitted to the tunnel.

The projectile is a cylinder 11 m tall and 4.2 m wide with a domed top. Built of a framework covered inside and out by sheet aluminum, it is divided into five floors. From bottom up: a storeroom containing supplies, food, oxygen (kept in containers within the walls), etc.; a laboratory, with a door to the outside, containing the electrical apparatus; a dining room with two windows; the staterooms and lavatory; and, under the domed roof, an observatory. Its telescope peers out through any one of three windows in the rotating dome. A "reaction motor" on the second floor exhausts through a nozzle in the base of the projectile.

The vehicle's empty weight is 1,250 kg; fully loaded it is 4,000 kg.

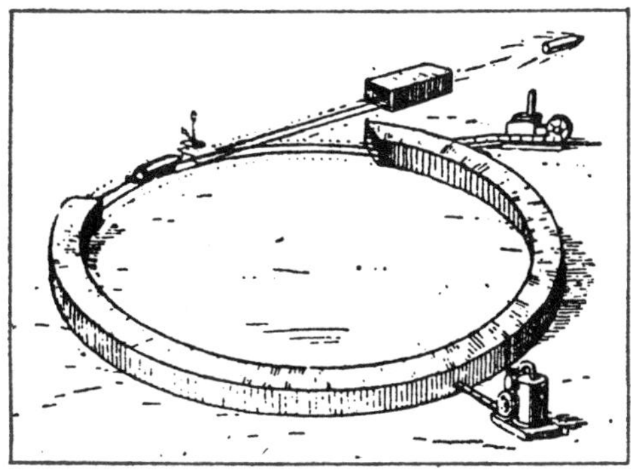

Above is an aerial view of the launching tunnel.

Middle right and left are sections and a cross section of the projectile-rocket.

In the diagram at bottom center, 2 indicates the oil bearings, 6 and 7 the linear induction accelerator.

1914

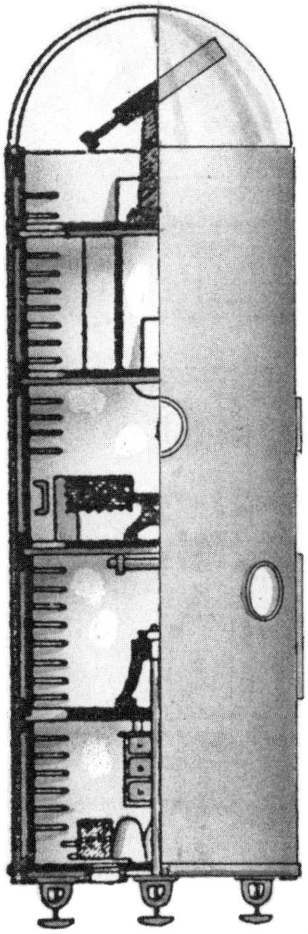

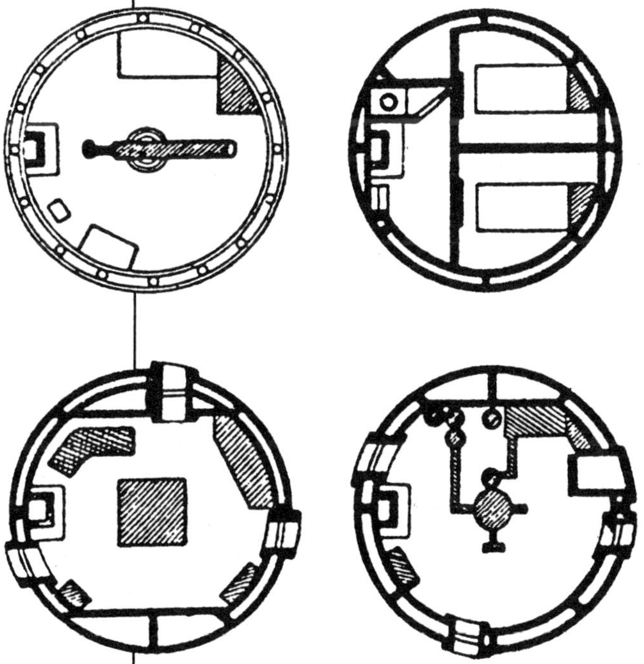

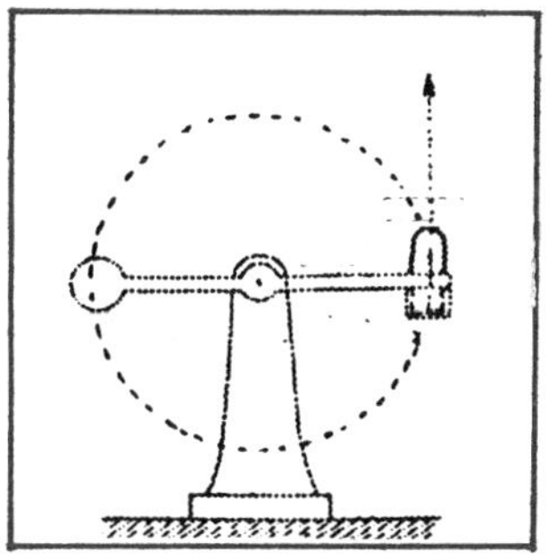

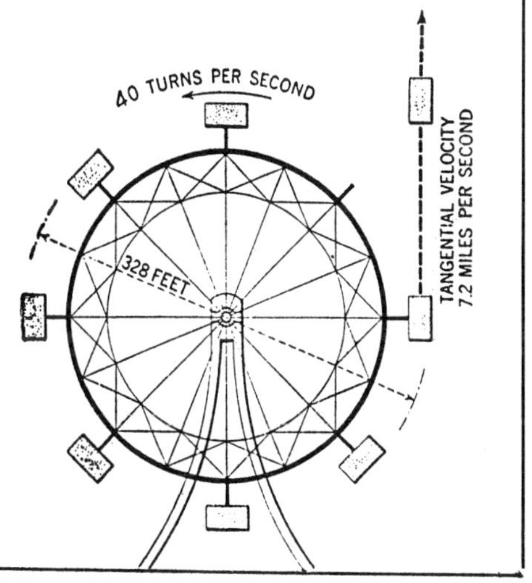

K. E. Tsiolkovsky describes a spaceship (#2) in the booklet *The Exploration of Space With Rocket-Propelled Devices* (a supplement to the publication of 1911–1912). It is similar to the rocket described in 1903 [see above] in general form and interior layout. Improvements in this rocket now include the circulation of oxygen gas, evaporated from the main tank, between the double hull to prevent the interior from overheating from air friction. The nozzle of the

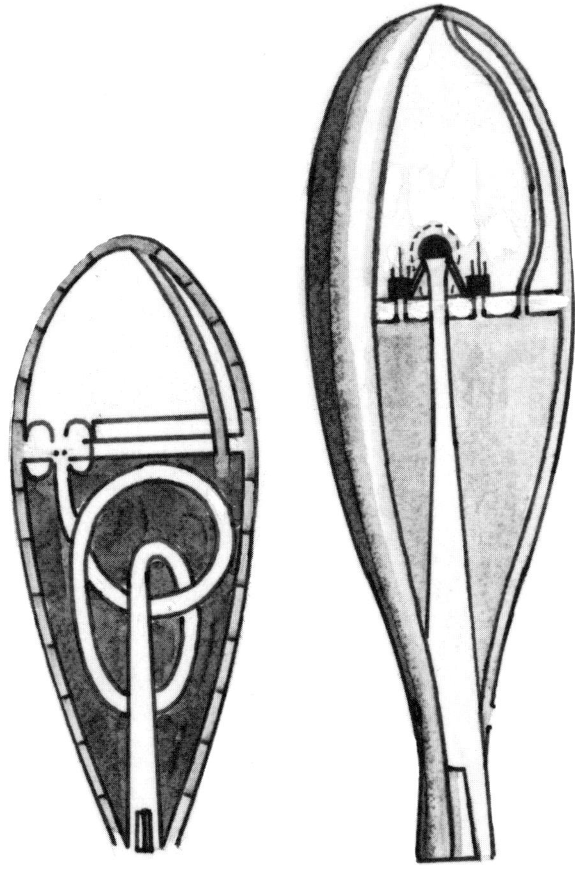

engine is now extraordinarily long and makes two complete, circular loops before exiting at the rear. The first loop is parallel to the rocket's long axis and the second loop is perpendicular to the first. The idea is that the rushing gases would act to stabilize the ship in the same way that the spinning flywheel of a gyroscope would. This is an idea that Tsiolkovsky will abandon in 1927.

Pumps for the fuel are unnecessary: the liquids are fed into the engine by the process of combustion itself, by way of a carburetor device. Steering is accomplished by the use of gyroscopes and deflection vanes in the exhaust.

The forward part of the rocket contains the cockpit and the necessary life support equipment. To protect the pilot from the expected 5 or 10 g's of acceleration at takeoff, he is provided with a water-filled compartment to lie in.

Tsiolkovsky also publishes his story "Gravity Free" this year. [See 1903, 1914, 1916 and 1930.]

Giulio Costenzi publishes a paper in *AER* magazine—the first Italian contribution to the study of spaceflight—which is, for the most part, a paraphrase of Esnault-Pelterie's 1913 study.

George Allan England, in his story "The Empire of the Air," describes the invasion of the earth by nonhuman aliens from 100,000 light years away. They eat whole planets and travel across space by means of a spacewarp—possibly the first in fiction.

R. H. Goddard takes out U.S. patents for both liquid-fueled rockets (#1,103,503) and the step-rocket (#1,103,653).

Eugen Sänger builds a model rocket-plane powered with a firework rocket. [Also see 1943.]

1915

A spaceship designed by Tsiolkovsky (#3) is described in Yakov Perelman's booklet *Interplanetary Voyages*. It is similar in general form and internal layout to the spaceships of 1903 and 1914 (see above). Pumps supply the engine with fuel and oxidizer (liquid hydrogen and liquid oxygen). Gaseous oxygen is piped from its tank to fill the space between the double hull to prevent overheating due to atmospheric friction. Excess oxygen is bled off into space.

The combustion chamber and nozzle are made of a metal with a high melting point and are lined with an even more refractory material. The outer hull, too, is refractory. Steering of the rocket is accomplished via deflecting vanes set in the exhaust.

MAY

Hugo Gernsback begins the serialization of "Baron Münchausen's New Scientific Adventures" in his magazine *The Electrical Experimenter*. The second instalment has the revivified Baron building a "space flyer" called the *Interstellar*. It is a large sphere

covered with a double netting of "Marconium." Connected to an elaborate switching apparatus on board, the netting creates an antigravitational field. The part of the netting not energized is attracted to whatever celestial body it is facing. With a Professor Flitternix, the Baron makes a journey of 102 hours to the moon. The *Interstellar* lands on a wide, treaded belt that runs around the spacecraft's equator, allowing it to roll to a stop.

1915–1918

With access to the drugs available at a hospital and military pharmacological supply station during World War I, Hermann Oberth experiments with some of the physiological effects of spaceflight. "I numbed the sense of equilibrium in my muscles and skin," he writes, "so that by floating under water with my eyes closed, and by using an airhose wound around my body, I could extend the psychological experience of weightlessness for hours. I noticed that I did not become nauseated when using these drugs." Oberth does not think that an astronaut will ever need to expose himself to prolonged periods of weightlessness, except when working outside of his spaceship. Artificial gravity could be provided easily by rotating the cabin. "In my opinion," he continues, "it is the aim of technology to provide man with conditions in space which correspond to his nature." [Also see 1911.]

On a Lark to the Planets, the spacefaring sequel to Frances T. Montgomery's *The Wonderful Electric Elephant*, describes a tour of the solar system via an electrically powered artificial elephant borne by a huge balloon. Although as a juvenile fantasy it is almost too precious to be readable, it at least can boast one of the most unusual spacecraft in literature.

1916

Fashon-Vilplet designs a giant electric gun and a model is tested. The full-size gun is intended to fire a 220-pound projectile with a muzzle velocity of nearly one mile per second.

R. H. Goddard demonstrates experimentally that rockets will not only work in a vacuum, their performance is improved.

K. E. Tsiolkovsky publishes his novel *Beyond the Planet Earth*. Although the first chapters were written in 1896, the first part is now issued in the journal *Priroda i lyudi*. The book will be available in 1920. It is as much a novel as any of Tsiolkovsky's fiction, being basically a scientific discourse with a few fictional elements to tie it together.

In the year 2017 six fabulously rich people have secreted themselves in a castle high in the Himalayan mountains [!], where they are free to indulge in their "scientific whims." There is a Frenchman, Laplace; an Englishman, Newton; an American, Franklin; an Italian, Galileo; a German, Helmhotz; and a Russian, Ivanov. [The latter is equivalent to "Jones," and I will make no further remarks concerning the author's abilities as a novelist.] The story picks up as the group is hav-

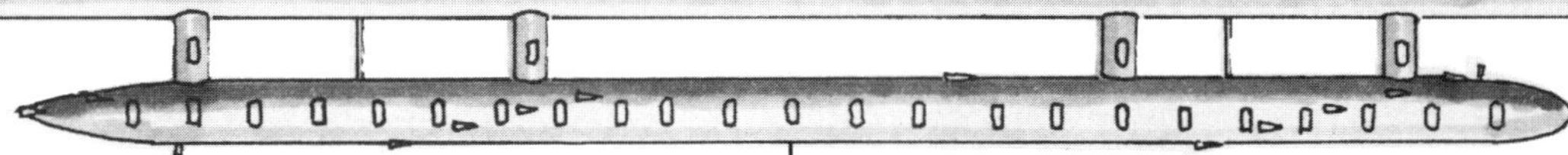

ing a discussion concerning the possibilities of spaceflight. Most of the others support the use of a giant cannon, but Ivanov vetoes this, advocating the use of a rocket. "Imagine an egg-shaped capsule," he explâins, "Inside the capsule is a pipe with an exhaust nozzle, accommodation for myself, and a stock of propellant explosives." He continues, explaining how a rocket functions. Laplace makes the apparently instantaneous mental calculation that "Using oxy-hydrogen gas it could hover in mid-air for 23 minutes and 20 seconds, if the weight of the propellant is seven times that of the vehicle and its payload."

There are several chapters of verbatim lectures given by some of the characters before they get down to the business of constructing a rocket. Franklin produces a propellant 100 times more efficient than anything known. After a month or so, a rocket vehicle is constructed (#4). It is 20 meters long, 2 meters wide and cigar-shaped. It stands vertically. Numerous portholes illuminate its interior. Three pipes run down along its walls, protruding at the lower end. It is liquid-fueled (Franklin's secret formula is not revealed, but it is evidently hypergolic). The test flight is a resounding success.

From this simple design they proceed to the "compound rocket" (#5), which is simply a great many of the simpler rockets used in conjunction. It is an "elongated, streamlined body, 100 meters long and 4 meters across and resembling a mammoth spindle." It is divided laterally into twenty sections, each of which is a complete rocket. These twenty segments are each 4 meters long. A twenty-first segment in the center section is different in that it is 20 meters long and has no propulsion of its own—it is merely a passenger compartment. The compound form of the rocket "made it possible to achieve comparatively low weight combined with tremendous thrust."

The exhaust nozzles project from the sides of the sections, forming a spiral around the circumference. Before the gases are ejected from the nozzles they are made to pass

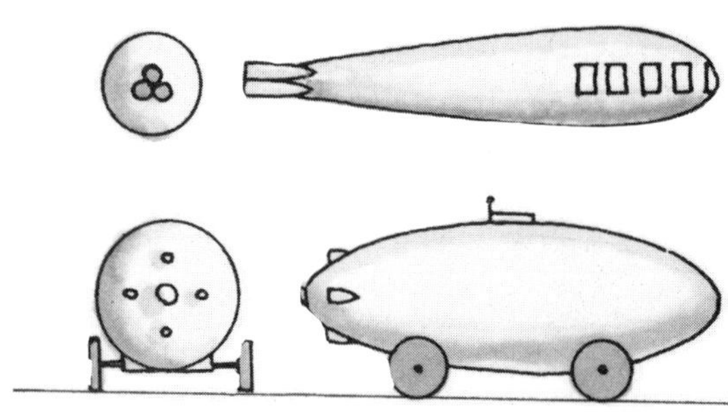

through coils that are wound at right angles to one another, thus (Tsiolkovsky supposed) providing a kind of gyroscopic stabilization [see 1914].

The walls of the rocket are made in three layers: the innermost a strong metal with quartz portholes and airtight doors, the central layer is an insulator, and the outermost a thin but very refractory metal shell. All of the necessary arrangements for the life support of the crew—including space suits—are provided for. The spacesuits are equipped with rocket backpacks (#6) that allow the cosmonauts to maneuver outside the spacecraft.

Of the rocket's available volume of 800 cubic meters, less than a third is occupied by the 240 tons of special propellants. This is sufficient for fifty flights from the earth and fifty reentries. The unfueled rocket weighs 40 tons; supplies, the greenhouse, and instruments come to 30 tons; and the passengers and "everything else" total only 10 tons. The living quarters occupy about 400 cubic meters—enough room for twenty people. Each passenger has a compartment of 32 cubic meters, of which all but 16 cubic meters is occupied by furnishings, propellant, etc. The separate compartments are connected by narrow passages. There is a 20-meter-long common room provided in the center segment.

The rocket is launched on April 10, 2017, from a ramp sloping at 25 or 30 degrees. The passengers and pilots are submerged in tanks of water, breathing through tubes, to insulate them from the effects of acceleration (a technique also advocated by Oberth, see 1929).

It should be emphasized that this compound rocket is *not* a step rocket.

The launch is successful and the rocket goes into a 1 hour and 40 minute orbit a thousand kilometers above the earth. After a time it is decided that weightlessness is an annoying novelty and the ship is started rotating on its central lateral axis, providing a sense of gravity on every compartment, with the greatest g forces in the outermost cabins. Rotation at 2 rpm provides 1/5 g.

Along one side of the rocket are a great number of windows that have been shuttered. Open, they comprise a glazed area 4 x 80 meters (one-third of the rocket's circumference). This converts a portion of the spaceship into a greenhouse. Eventually it is decided to establish the greenhouse separately from the main rocket. This takes the form of a long cylinder 2 meters wide, that has been carried in sections in the spaceship. It is made of thin cylindrical tiles of a wire-reinforced glass and is assembled into a single long pipe. A liquid soil is introduced as well as two thin tubes that run down the center of the greenhouse. One supplies gases and the other liquid fertilizers. The completed structure has a mass of 20 tons, and exposes 1,000 square meters to the sun. Waste carbon dioxide from the spaceship is piped into the greenhouse, along with other wastes, and oxygen is piped back. The greenhouse produces food as well as atmosphere.

Communication with the earth is accomplished by heliograph. Eventually thousands of similar rockets are built and launched, a swarm going into a geosynchronous orbit at 33,000 kilometers (5.5 earth radii). The first greenhouse, 1 kilometer in length and 10 meters in diameter, is occupied by and feeds 100 people. Eventually several greenhouses are joined together, in the shape of a star or some other geometric figure, and made to rotate.

Meanwhile the original spaceship has made a journey to the moon and from orbit around it sends a small, manned rocket down to the surface (#7). It is equipped with four small wheels to allow it to maneuver around the landscape.

After the moon a tour of the solar system is undertaken after which the rocket returns to the earth, splashing down into an ocean.

For all of its faults as literature, Tsiolkovsky's novel is a remarkable achievement in the history of astronautics. There are few aspects, large or small, that he overlooks: the amount of detail and thought is prodigious. He discusses space colonies, asteroid mining, space suit design, the effects of prolonged weightlessness . . . even the problem of bathing in free fall! [Also see 1903, 1914, 1916 and 1930.]

Arthur Train (a lawyer/novelist) and Robert Williams Wood (a professor at Johns Hopkins University) describe the spaceship *Flying Ring* in the novel *The Moon Maker* (a sequel to *The Man Who Rocked the Earth*; serialization of the second novel began in the October 1916 issue of *Cosmopolitan*.).

The *Flying Ring*, invented by the mysterious man known only as "Pax", is launched by its new owners in answer to a threat to the earth by an errant asteroid, 150 mile-wide Medusa (composed primarily of pitchblende), due to strike the earth.

The ship is constructed in two basic parts: the first and largest is a torus seventy-five feet in diameter. It is made of aluminum and is about fifteen feet thick. Entry is through an airlock. The interior contains a number of cabins, with windows in walls ceiling and floor, a chartroom, space for generators and twin gyroscopes. There is a telescope pointed vertically through a glass deadlight in the ceiling. The cabins are comfortably furnished, including a small kitchen. Arrangements are made for cooling the interior. A "cloakroom" contains spacesuits ("vacuum armor"). These are made of thick rubber with copper helmets and have tanks of liquid oxygen (which is allowed to slowly evaporate for breathing purposes).

Resting on the top surface of the ring are three girders, converging to a point above the middle of the torus, like a tripod. At the apex is a cylindrical, thimble-shaped chamber containing a large cylinder of uranium. Beams of the "lavender ray", a disintegrating

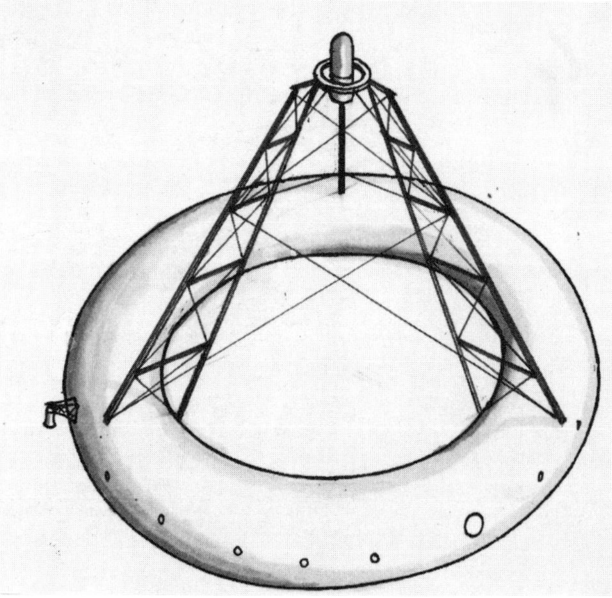

ray, are carried in pipes from the generators in the ring to the bottom surface of the uranium cylinder where its atoms "explode", creating an exhaust of superheated helium. Each cylinder is good for ten hours of flight. The chamber can be turned in any direction in a gimbal mounting, allowing the ship to be steered. This is done with an electric motor. Control is handled by a complex mechanism in the ring. The complete propulsion system is called the "tractor."

To facilitate launch, the *Flying Ring* is placed atop a wooden gantry; this allows free escape of the gasses produced by the disintegrating uranium. A wire fence more than 1,500 feet in diameter is erected around the vehicle, for safety.

At the moment of takeoff, the outlet of the uranium container glows and a ray of yellow light, helium from the disintegrating uranium, beams from it. The ground beneath the ring is blasted, the scaffolding bursting into flames, lumber and girders are thrown into the air. The *Flying Ring* no longer needs its support, however. It is now hovering in the air, supported by the helium blast from the cylinder. In ten seconds it is ninety feet in the air; a minute later it is at more than half a mile. Leaving a luminous trail behind it, the spaceship disappears into the heights. Inside the ring, the takeoff is noisy and violent; the passengers are badly affected by the acceleration. The gyroscopes, however, succeed in keeping the ship stable. At an altitude of about seventy miles all sound from outside ceases. The ship has been flying for twenty minutes at a speed of about 3.3 mps. When a speed of some eighteen mps is reached, the engine is shut down.

The authors explain that as long as the *Flying Ring* is not accelerating, there is no sensation of weight on board—the occupants are in free fall. To their credit, this is one of the earliest realizations and explanations of this phenomenon in the popular literature.

A descent to the moon is made to replace the exhausted uranium cylinder. To make a landing, the ship must be reversed, so that it approaches tail-first. The cylinder is turned to its maximum angle and fired, rotating the ring. Firing the cylinder once again in the opposite direction stops the rotation. The landing is made using the helium blast as a brake.

Eventually, the earth is saved by turning a disintegrating ray (which is mounted on the outside of the torus) on the asteroid Medusa. It is not destroyed, but its course is deflected enough that it misses the earth, eventually becoming a second small moon.

R. H. Goddard receives a grant of $5,000 from the Smithsonian Institution to develop a "method of reaching extreme altitudes."

F. A. Tsander resumes his work on spaceflight and designs a rocket plane. [See 1921 and 1924.]

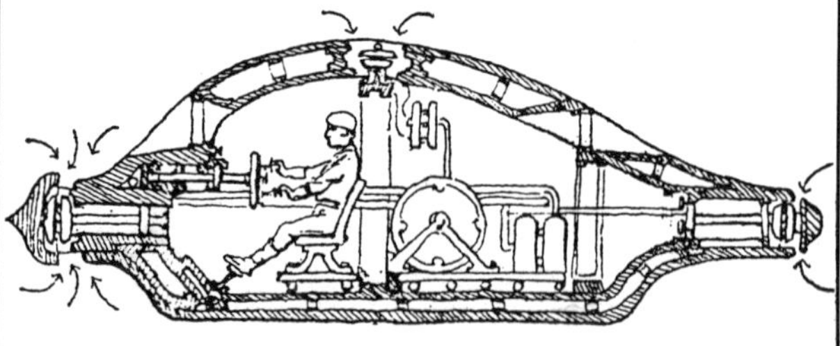

John Ames Mitchell publishes the novel *Drowsy* in which he describes an electrical spaceship as a "ten-foot fat cigar." A rather obscurely and mystically described trip to the moon is undertaken. The book is, however, distinguished by its fine illustrations by

Angus MacDonall.

Sophus Michaelis, a Danish author, publishes the novel *Himmelskibet* (Heavenship). It is made into a film with a screenplay by the author, directed by Forest Holger-Madsen. It is released in the United States ca. 1920 under the title *A Trip to Mars*. The spaceship *Excelsior* is depicted as a large, enclosed cigar-shaped craft. It is equipped with wings, a propellor and a wheeled undercarriage and apparently takes off and lands like a conventional aircraft.

The spaceship in the novel is the *Cosmopolis*, in which the kidnapped hero is taken from the trenches of World War I to Mars. The *Cosmopolis* is a globe filled with cabins, passages and so forth with a row of spheroidal glass windows covered by steel shields. There are observatories, including one devoted to solar studies that rotates to follow the sun. The weightless passengers' feet are held to the floor by suction.

The spaceship is propelled by the "radio spectrum" of Mars, that serves to both drive and guide the ship. The spectrum is generated by the *Cosmopolis* and acts like a suction, radiating a current in the direction of the planet it is headed for. To rise from the earth, the ship only has to generate an attracting force toward Mars that is greater than the earth's attraction for the ship.

1916

Victor Coissac writes in *La Conqu de l'Space* that " . . . a little vehicle, twenty times lighter than the large one, [which would be] twenty times more maneuverable and would demand twenty times less propulsive force. The travelers would be able to proceed in the following manner. They would first of all place the large vehicle in satellite orbit at a convenient distance from the planet [they wished to land on], far enough out so that the atmosphere would not significantly affect its movement and at the same time close enough so that, boarding having been achieved, they could see it with their eyeglassses. Then they would decouple from the small vehicle which would be constructed with this in mind. They would take their places inside after having provided the vehicle with everything they needed: food, furnishings, books, various instruments, etc. Next, they would decrease velocity to less than circular. The small vehicle would then slowly draw away from the large one, which the voyagers would for the time being abandon (without fear, moreover, that someone would steal it from them). It [the little vehicle] would thus descend slowly toward the ground."

Coissac also discusses atmospheric braking, multi-stage solid-fuel rockets (launched from a silo positioned atop some high mountain, such as Popocatepetl) and orbital space mirrors—the latter idea predating Oberth by eight years.

APRIL

In "Is it possible to Travel to the planets?" Henri de Graffigny proposes, once again, the use of centrifugal force for launching a spacecraft [see 1889, 1913–1916 and 1927]. The projectile is placed at the end of a beam 328 feet long. The beam is supported at its center and at the end opposite the projectile is a counterweight. The beam is made to rotate at a rate of 44 rpm, creating a speed of 8.7 mph at its ends, sufficient to launch the projectile into space. A 12,000 hp motor is required to power the rotating arm, which needs 7 hours to get up to full speed. Once in space, the projectile is steered by a rocket motor.

OCTOBER

G. Tikhov, an astronomer, submits a report at Petrograd on reaction-propelled spaceships.

1916–1925

Y. V. Kondratyuk writes his book *The Conquest of Interplanetary Space*, in which he discusses the orbital technique of landing on planetary bodies [also see 1916 and Apollo entries, below]. Kondratyuk writes that "The entire vehicle need not land; its velocity need only be reduced so that it moves uniformly in a circle as near as possible to the body on which the landing is to be made, then the machine part separates from it, carrying the amount of active agent [propellant] necessary for landing and subsequently rejoining the remainder of the vehicle."

ca. 1917

A Trip to Mars, an Italian motion picture, features a large propellor-driven spaceship similar to the one in *Heavenship* [see above]. It is possible that this film is an Italian reissue of the Danish movie.

1918

The Rocket Society of the American Academy of Sciences is formed by Dr. Matho Mietk-Liuba in Savannah, Georgia. Little is known about this apparently one-man organization. Mietk-Liuba is not listed in either the city directories or the membership of the National Academy of Science.

JANUARY 14

Robert H. Goddard writes "The Ultimate Migration," sealing it in an envelope labeled "Outline of Certain Notes on High-Altitude Research. "It will not appear in print until November 1972. In four pages of handwritten manuscript, Goddard considers the eventual abandonment of the solar system by humanity "when the sun and the earth have cooled to such an extent that life is no longer possible . . . " He suggests the possibility that an asteroid or small moon could be converted into a space ark. It would be powered by "intra-atomic energy." If this is not feasible, then hydrogen-oxygen rockets would be used, abetted by solar energy. The

PLUCK AND LUCK

COMPLETE STORIES OF ADVENTURE.

FRANK TOUSEY, PUBLISHER, 168 WEST 23D STREET, NEW YORK.

No. 976. NEW YORK, FEBRUARY 14, 1917. Price 6 Cents.

TWO BOYS' TRIP TO AN UNKNOWN PLANET.

By RICHARD R. MONTGOMERY.

AND OTHER STORIES

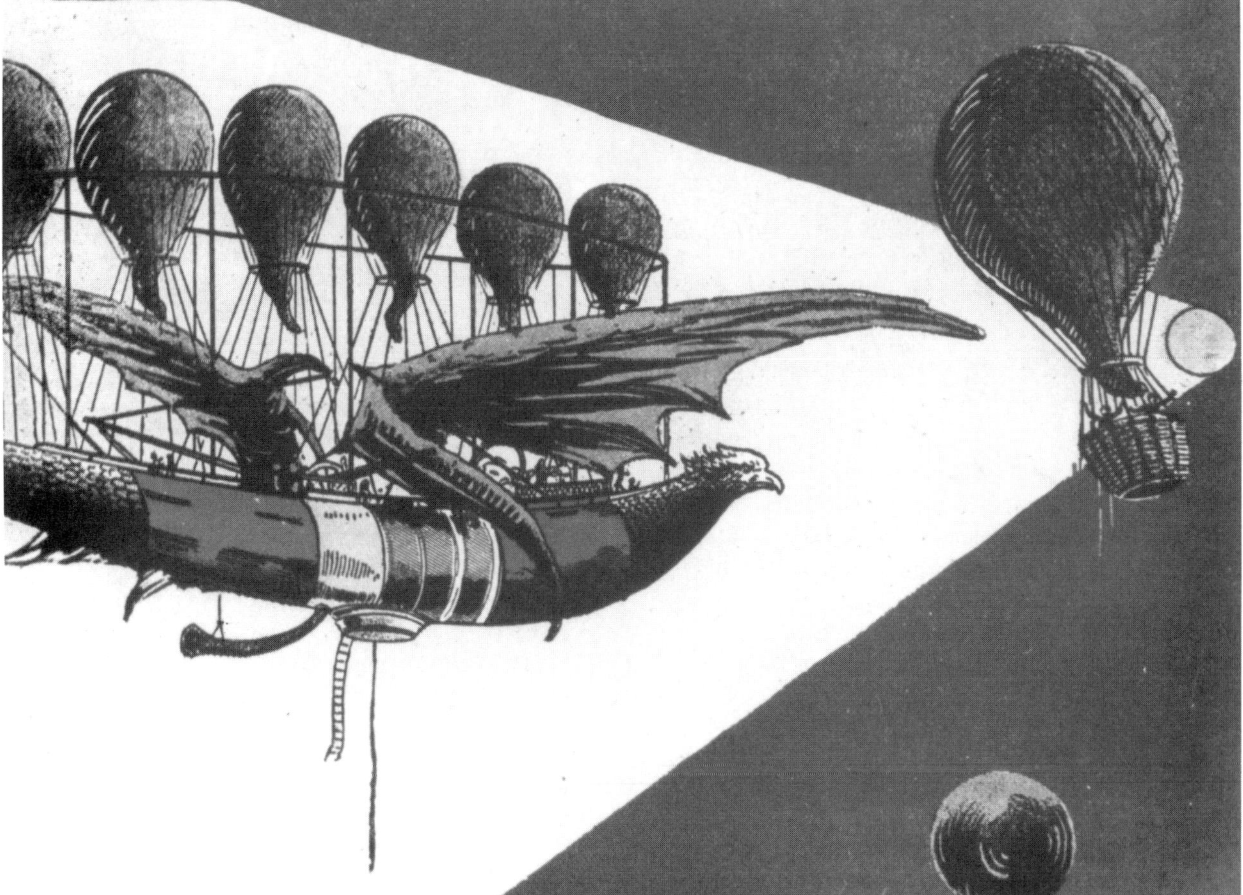

human beings on board would be placed into a state of suspended animation, only the pilot being awakened every 10,000 years or so when the nearer stars are reached, or every million years for more distant targets. An atomic alarm clock would be needed. The ultimate destination would be another star system where mankind could begin anew. For safety's sake, several expeditions would be sent out so that at least one might be successful.

1918–1919

Yuri Kondratyuk describes an apparatus for utilizing solar energy for spaceflight. Large parabolic mirrors focus heat onto the working body, creating a solar rocket engine. The thrust would be small, so these engines would work best "where significant acceleration is not required."

In the same paper, Kondratyuk also describes "reaction from material radiation." "Cathode rays," he writes, "are material particles charged and rushing along at a speed of 200,000 km/sec. Therefore they produce a corresponding reaction/recoil, which can be utilized by bringing it up to the required level." The building of such a reaction device seems to him a difficult if not improbable task, nevertheless "If successful, it promises to provide such an incredible velocity that it would be impossible to achieve even with the most powerful conventional rocket. With such velocities it may even be possible to test the theory of relativity."

Kondratyuk also describes a novel ion engine, with ions accelerated by an electromagnetic field. From one opening positively charged particles flow out at high speed, from another opening negatively charged particles flow out. Both flows meet and are neutralized. He writes, "the velocity of these particles when there is . . . a high potential difference can be made extremely high—greater than the velocity of molecules of a strongly heated gas."

1918–1920

Henri F. Melot designs, builds and operates a working jet engine. Two cylinders placed end to end have a piston inside that freely shuttles back and forth. Either end of the cylinder has spark plugs and fuel inlets. Exhaust from either end is channeled into a common "buffer chamber" to which the exhaust nozzle is attached. The result is an intermittent rocket blast. Melot's engine is never tested on an aircraft, however.

1919

The Smithsonian Institution publishes Robert Goddard's paper on "A Method of Reaching Extreme Altitudes," actually an annotated version of the research proposal he submitted in 1917. In it he suggests that it may be possible to reach the moon via rocket (using the cartridge-fed "machine-gun principle").

APRIL

Waldermar Kaempffert, editor of *Popular Science* magazine, writes the article "Hurling a Man to the Moon." He visualizes a "sky-rocket car," powered by a "radium engine." He accurately describes the conditions that must be dealt with before man can safely explore space, or leave the earth.

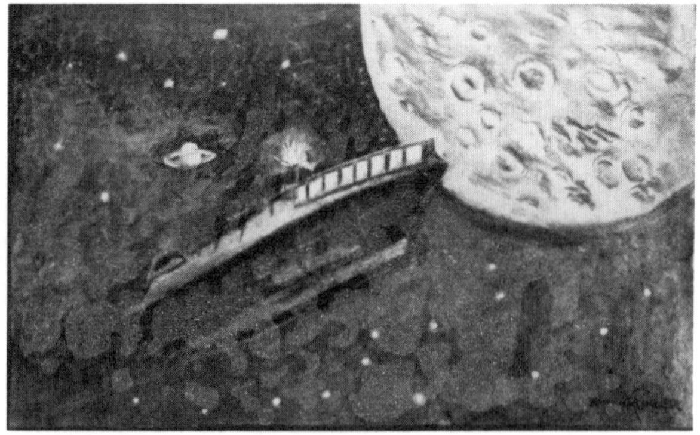

MAY-JUNE

The German *Motor* magazine publishes the short story "Der Fug Durch den Aether" (Flight Through the Ether).

JULY 1

A drawing by H. Lanos, published in *Lectures pour tous*, illustrates the takeoff of a large, rocket-propelled spaceship. It is a fat cylinder with pointed ends. The sides are broken by small, half-circular windows and the nozzles of the rockets at the rear end. It is launched vertically from a guide tower.

1920

Robert H. Goddard, in his *Report to Smithsonian Institution Concerning Further Developments of the Rocket Method*, describes his system for utilizing the energy of the sun for rocket propulsion. Thin mirrors would focus the sun's heat onto boilers, producing steam. This would in turn operate a turbogenerator. Its electricity would be used to create positive and negative ions, which would be introduced into the jet stream of the chemical (H_2 + O_2) rocket engine. The exhaust would receive additional acceleration electrostatically.

A Trip to Mars, a film released by Tower Films of New York, is the American release of *Heavenship* [see 1917].

Robert H. Goddard, in a letter sent to Charles G. Abbot, secretary of the Smithsonian Institution, describes the requirements of manned spaceflight, solar sails, etc.

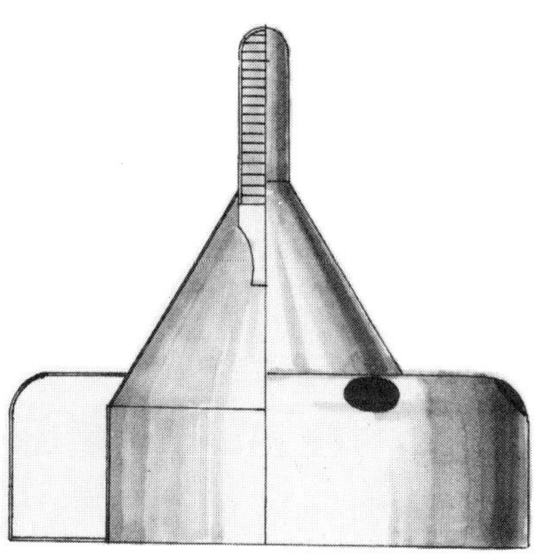

F. Ulinski describes a rocket-powered spaceship. The manned cabin is toroidal with the rocket engine held above the center at the apex of a metal cone. The rocket itself is fueled by a series of individual cartridges that are fed one at a time to the combustion chamber [see 1913 and 1920].

APRIL

In a magazine article for *Popular Science* Robert Goddard describes his scheme for hitting the moon with a flash powder charged rocket. Where Esnault-Pelterie had objected that no known explosive possesses sufficient energy to propel itself to the moon, Goddard points out that as a rocket discharges portions of its mass as it accelerates, its mass is constantly decreasing. Therefore it is not necessary for it to carry its starting weight all the way to the moon. Goddard has also, according to the article, increased his rocket's exhaust velocity from 1,000 feet per second, the best available at the time, to 7,000 feet per second. Goddard's calculations then show that to propel 1 pound of payload to the moon would require only 602 pounds of

propellant. The author points out that a visit to the moon via a manned rocket is still out of the question at the time of writing. Nevertheless, it would be possible to "reach out a long arm and tickle the moon" by sending a charge of flash powder, the explosion of which could be observed from the earth. By experiment, Goddard concludes that a minimum charge of 14 pounds is needed to make the flash visible. This would require a total amount of propellant of about 17 tons—if the weight of the flash powder and its attendant apparatus weigh about 56 pounds [see February 26, 1921].

NOVEMBER 20

Morrell, in the *London Graphic*, publishes a critique of Robert Goddard's high-altitude and moon-rocket schemes [see above]. Some of his objections include the impossibility of safely returning instruments or passengers from such great heights, what the value might be of such an experiment, how the rocket can be protected from atmospheric friction since, " . . . bodies when they speed through the air are subject to friction against the air which is sufficient to generate tremendous heat," with the result that "the rocket will generate a red heat foremost of the first hundred miles." At a speed of 6.4 mps the rocket would "vanish in an incandescent wisp of flame and smoke." An objection to the moon rocket is that the moon and the earth are moving in different directions and at high speeds, and that there are "incalculable vagaries of air currents" above 20 miles that would make steering the rocket impossible. Goddard replies to these and other objections in an article for *Scientific American* [see February 26, 1921].

1920–1924

Yuri V. Kondratyuk writes of the possibility of using high-caloric metals as fuel for spacecraft. [Also see Tsander, 1924.]

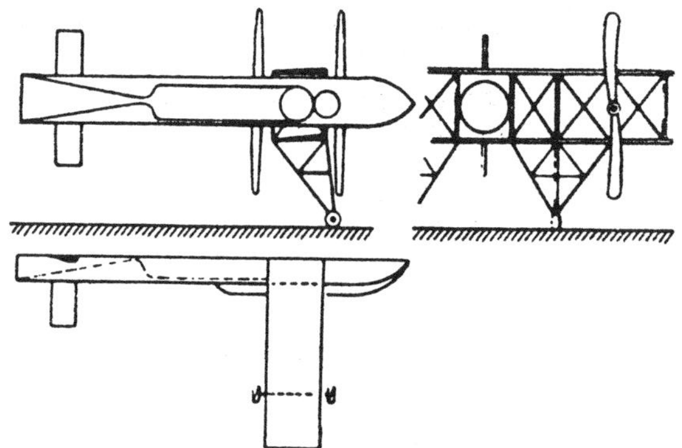

1921

F. A. Tsander suggests a composite "aerospace" vehicle that would take off with propellors, converting to rockets at high altitude. The metallic parts of the rocket no longer needed, instead of being jettisoned as in a step-rocket, would be fed into a furnace to be converted into fuel. The fuselage of the craft is a conventional rocket. The aerodynamic features—the wings, propellors, etc.—are attached to its exterior. The landing gear could be either wheels or skis for landing on water. Once the disposable airfoils have been turned into reaction mass, the rocket is left with a small set of wings of its own for its return to the earth. Tsander's design generally reflects an idea by French engineer Laurent. A model will be displayed at the 1927 spaceflight exhibition. [Also see 1924.] Tsander also devises a concept for solar sailing and a proposal for a space station.

Soviet engineers N. I. Tikhomirov and V. A. Artemyev organize in Moscow the first Soviet solid fuel rocket laboratory.

FEBRUARY 18

A. A. Andreev applies for a patent for a portable rocket backpack, that could be carried on a person's back. Fueled by liquid methane and oxygen, the rocket pack allows its wearer to fly up to 12 miles at a speed of 124 mph. The "pilot" can ascend to an altitude of 22 km in about 7 minutes, according to Andreev, an acceleration that N. A. Rynin thought might be excessive. The entire backpack weighs 110 pounds, including 17.6 pounds of fuel. The rocket pack could also be used to create an impromptu missile by attaching a mine or a bomb to it in place of a human being. [Also see 1961.]

FEBRUARY 26

Robert Goddard publishes a rebuttal (written on January 12) to his critics in *Scientific American Weekly*. Although he is mainly discussing his proposals for high-altitude sounding rockets, and the idea that an unmanned rocket could be made to hit the moon [see April 1920], he does mention that his earlier announcement inspired eighteen volunteers for the flight. To the critics [see November 20, 1920] he makes several responses: that his rockets could be safely returned from high altitudes by simple parachutes, or that perhaps a few charges could be left in the rocket to provide a braking effect; that the rocket is even more efficient in a vacuum than when operating in air; that air friction will not be a serious problem since the rocket will not reach its full velocity until much of the earth's atmosphere is behind it; and that the problem of hitting a "moving target" like the moon is only a question of the appropriate mathematics since " . . . although the speeds of the two bodies are high and different *they are known*, with great precision . . . " Any unknown air currents above 20 miles would be in air of "practically negligible density." Steering the rocket could be accomplished automatically by using "transverse impulses" controlled by photocells. Goddard concludes by saying that he believes "that the multiple charge rocket principle is correct . . . [and] . . . the experiments so far performed on the small model under test demonstrate clearly the practicability of the idea. This work is proceeding slowly because of the lack of really adequate support, although the Smithsonian Institution as much as it can on a work of this kind. But although there exists the attitude that 'everything is impossible until it is done,' there is nevertheless widespread interest being taken in the work. To the writer's mind, the whole problem is one of the most fascinating in the field of applied physics that could be imagined."

OCTOBER

Die geschichte einer Mars-expedition (The Story of a Mars Expedition), written and illustrated by engineer H. Langner, is a two-part story published in *Der Luftweg* that describes a future trip to Mars in a giant spaceship, the *Meteor VI*. The vehicle is shaped like a spindle with an extremely long tail, tapering to a needle-like point, broken only by the exhaust of one of the three rocket engines. Near the nose is a giant propellor and amidships are a pair of short, broad wings. More rocket engines project at an angle just behind the propellor. The rocket takes off vertically, aided by the propellor, which is also used for maneuvering in the atmosphere of Mars.

ca. 1921

The film *The Ship That Was Sent Off to Mars* is made by Raymond Griffith.

1922

The Sky Splitter (aka *The Stellar Express*) is a motion picture written and directed by J. A. Norling for Bray Productions. The film is

1922

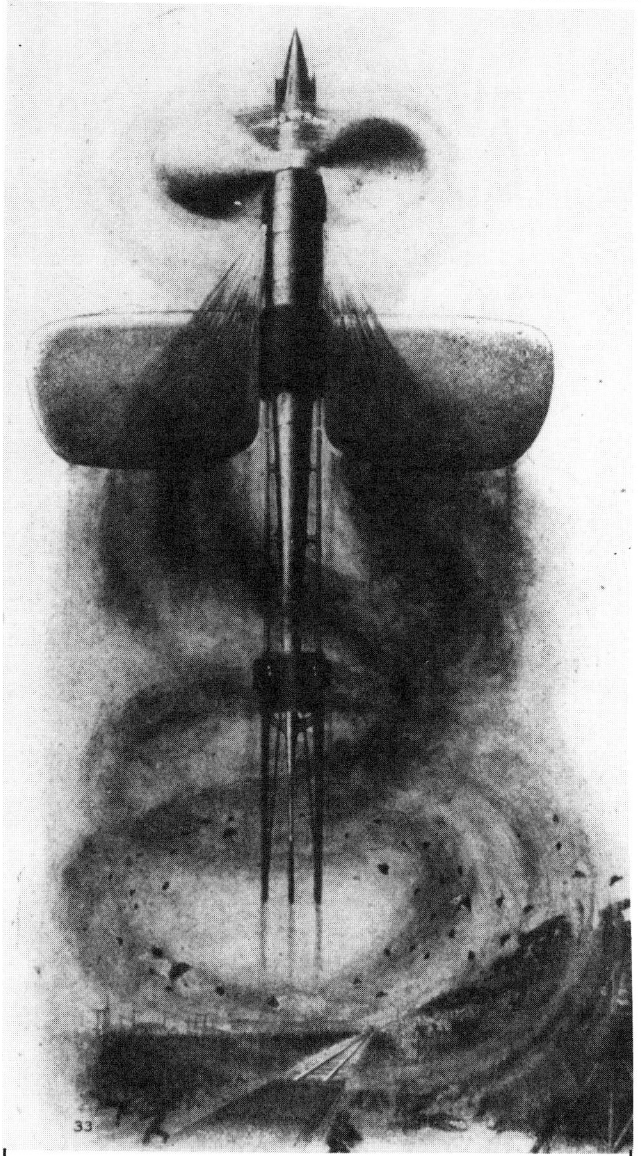

NASM

inspired by the recent discoveries by Prof. T. J. J. See concerning universal gravitation.

In the film, a Prof. Adam Cooley invents an "atomic force motor" that he first tries out on an automobile. The spaceship he eventually builds (financed by his daughter's lover) is an elongated teardrop that greatly resembles some of Max Valier's later designs. The takeoff of the spaceship (the "projectocar") is unusual: a variety of ski jump or tobaggan run is erected, the highest point being supported by an Eiffel tower-like structure. At the apex of the tower is a platform for the spaceship. By tilting the platform the spaceship is started coasting down the ramp, firing its engines when it shoots off the far end of the "jump").

Ackerman

The projectorcar, built in the film by a Professor Cooley, is equipped with telescoping wings so that it can make a gliding reentry. He discovers that his controls have no effect in airless space.

His spaceship out of control, the Professor eventually lands on a distant planet, fifty light years (or exactly 293,994,584,070,194 1/2 miles, according to Cooley's calculations) from the earth, where he builds a monster telescope that allows him to see the earth as it was half a century in the past.

The film, although presented as both a fantasy and a comedy, there is a degree of seriousness in its exposition of the concept of travelling at near-light speeds and some of the paradoxes involved.

This is perhaps the first time this theme has been explored in the cinema.

Maral-Viger, a French author, writes a novel *L'Anneau des feu* (The Ring of Fire) which describes a rocket based on the principles of Esnault-Pelterie [see 1911 and 1912].

The novel describes a trip from the earth to Mars and Saturn and back. The technical details are also based on the calculations and

ideas of Robert Esnault-Pelterie. In the novel the author suggests that 1 gram of radium in the course of 1 hour releases enough energy to raise its weight to an altitude of over 20 miles. This is equivalent to several billion horsepower. Since the energy in just one kilogram of radium is more than 5,670 times that needed for a trip to the moon, the heroes have more than they possibly need. The trip can be made at a comfortable acceleration of one-tenth of a g.

In the story the heroes discover a new element, "virium," that is 60,000 times more powerful than radium. They build a rocket propelled by the reaction of the escaping energy of virium. Three kilos is sufficient for a flight to Saturn. The disintegration of virium is hastened by directing a beam of electrons from a cathode ray tube onto it.

The ship they build—the *Fusée*—is an elongated ovoid, 45 feet tall and 12 feet wide, made of nickel steel. Half of it is painted black, lengthwise, and the other half polished. This way the rocket can be rotated to control its internal temperature. The walls of the rocket are four-layered with a near vacuum between them, also for temperature control. At the rear are four heavy legs, with shock absorbers to deaden the jolt of landing. There is a round glass hatch between the legs for entry into the ship. Three steel pipes project from the base of the ship for the exhaust of the products of the disintegration of virium.

Inside, a 3-foot diameter shaft contains a spiral staircase. The ship is divided into four floors. Above the lowest floor, which is the engine room (above which the stores of virium are kept in a lead container), are the passenger cabins. These are about 14 feet high and are equipped with windows, four in all. The top two floors, 12 feet high each, are storerooms. At the very top of the rocket, the staircase leads to a tiny observation compartment with a skylight window. There is also an airlock.

Equipment for air purification, scientific instruments, food, etc., is all detailed. Provisions are allowed at 2 pounds of food and 2 quarts of water per person per day—a total of 870 pounds of food and 625 gallons of water for four persons for four months.

The rocket is equipped with three additional jets, spaced 120 degrees apart on the sides of the ship (made, like the main jets, of molybdenum steel for maximum resistance to melting). These allow the ship to be turned while in flight.

The flight to the moon takes place at a speed of about 40 miles per second, to Mars at 500 miles per second and to Saturn at over 600 miles per second.

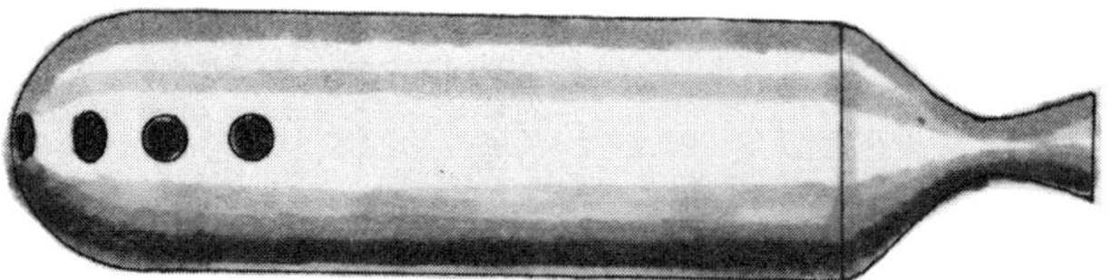

Ernest Welsh invents a spaceship propelled by detonating Melonite in compressed air. It lands by parachute. A model is exhibited in Moscow in 1927 [see below]. Welsh also invents a "Death Rocket" in 1925 as an anti-aircraft weapon.

1922–1926

There are several rumors that the Russians have a manned interplanetary rocket under construction at the Moscow airport or at the Air Force Academy in Moscow. It is reputedly of "an old-fashioned shape" [sic] and 351 feet long. Its hull is made of a fire-resistant alloy and contains a cabin with all necessary life support equipment. It is intended for a manned trip to the moon, where the landing will be effected by retro-rockets.

The rocket has been under construction for 4 years, mostly under the supervision of Italian engineers and the general direction of Tsiolkovsky and Tsander.

There are also rumors between 1925 and 1926 that a Professor V. P. Vetchinkin has designed a large manned rocket in Moscow.

1923

Alexei Tolstoy publishes his novel *Aelita*. The rocket he describes is reaction-propelled. He writes how it took off from Leningrad in 1921 for a journey to Mars. It returns to the earth in 1925, falling into Lake Michigan. The vehicle is egg-shaped, 26 feet tall and 18 feet in diameter. Around

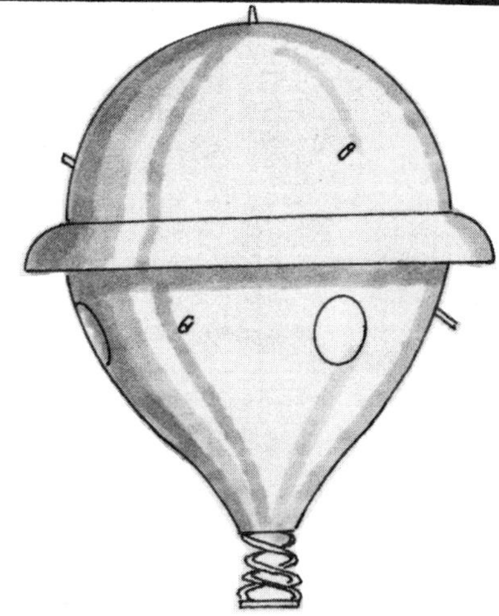

its circumference is a steel belt, flared out from the sides like a dancer's tutu. This acts as an air brake when landing. In the sides below the brake are three oval doors. The lower, narrow part of the egg terminates in a narrow neck. This ends in a spring of massive steel—a buffer for landing. The ship is built of steel, well reinforced with internal girders and ribs. Inside is a second hull, made of rubber, felt and leather.

At the lowest point of the hull are contained the engines. In each is a spark plug connected to a common magneto. Fuel is fed from a large storage chamber above the engines. It is a powder called "ultralyddite," and the ship carries enough for 100 hours of powered flight. The engines are actually machined out of the thick metal ("obin") of the ship's "neck." Narrow vertical channels have been drilled out, terminating in broad combustion chambers. By reducing or increasing the number of explosions per second, the speed of the vessel can be controlled.

Above the engine compartment is the floor of the passenger cabin. Here are the oxygen and atmosphere purification equipment, the instruments, supplies and controls. "Peepholes" in the form of narrow metal pipes fitted with lenses allow views outside the ship.

Since the ship is expected to fly at a constant acceleration, the journey to Mars is not to exceed 8 or 9 hours. It lands on Mars by reversing and landing tail-first, using its rockets and air brake to slow its descent.

The novel is made into a film in 1924, directed by Jacob Protazanov.

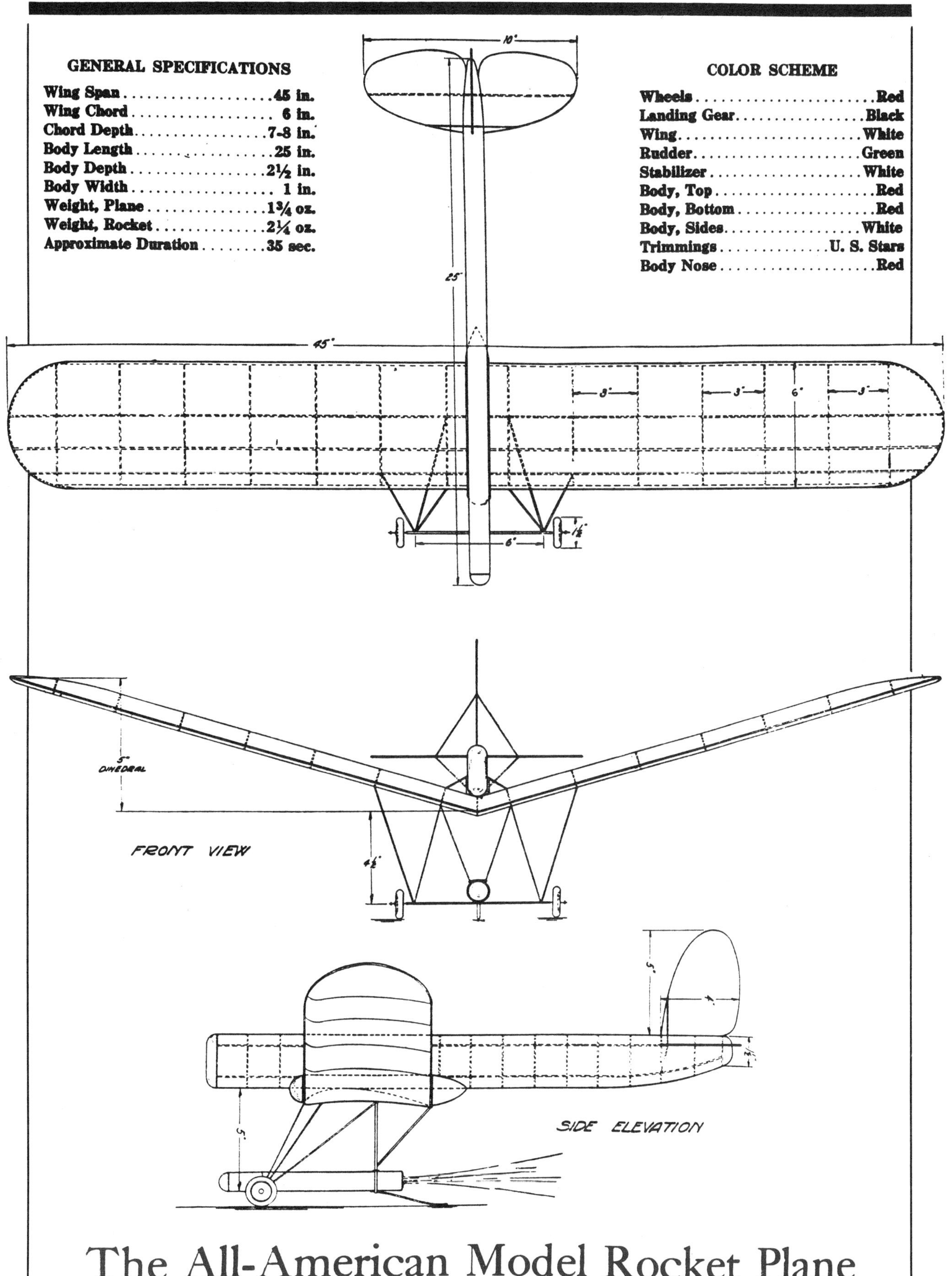

The All-American Model Rocket Plane

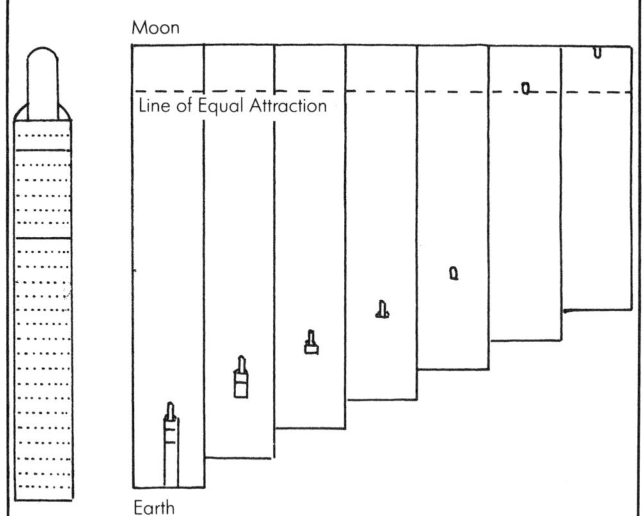

Luigi Gussalli publishes a plan for a flight to the moon in his book, *Si puo gia tentare un viaggio d'alla terra alla Luna?* He proposes a step-rocket of twenty-seven stages. Each is fuelled with 300 grams (10.5 ounces) of propellant. The stages are combined into four groups of 18, 6, 2 and 1 steps. The rocket would carry two passengers on the round trip.

To give his rocket its initial impetus, Gussalli suggests launching it with a catapult. As an engine for his spaceship, he proposes using a double-reaction turbine, similar to that designed by Melot [see 1918–1920]. He also places steering vanes in the jets of the engines.

The spaceship would consume two-thirds of its propellant in the first 6 minutes and 2 seconds of flight, jettisoning the first 18 stages as each becomes exhausted. The second group of stages, 19–24, would begin firing, lasting another 6 minutes and 2 seconds. Stages 25–26 would also burn for the same duration. The final stage, number 27, would insert the manned lunar lander into its coasting trajectory toward the moon. Two days and 30 minutes later the ship fires its retro rockets, landing on the moon exactly 48 hours, 57 minutes and 55 seconds after takeoff.

Hermann Oberth publishes *Die Rakete zu den Planetenraumen* (The Rocket in Planetary Space) a ninety-two-page book that demonstrates many of the fundamental concepts of spaceflight theory. Oberth later

greatly expands this work [see 1929]. It is a highly technical volume that tempts a number of authors to "popularize" it, such as Max Valier, Otto Willi Gail, Willy Ley, Karl von Laffert, and Felix Linke.

APRIL

G. A. Crocco, the Italian authority on aviation and aeronautics, presents a report to the Italian Academy of Sciences on the possibilities of flight beyond the atmosphere of the earth. He concludes that some sort of vehicle propelled by reaction is necessary, but does not believe that any chemical substances exist which are capable of providing the required energy. He assumes that only the controlled release of atomic energy will work and suggests two methods by which this might be done. The first method uses the energy directly, such as in the uninterrupted ejection of alpha particles, with speeds of 3 million to 6.6 million feet per second. This could be obtained by the effect of strong electrical discharges on the nuclei of atoms. In this case, it might be possible to direct the current in a single direction, as it is possible to do with a Crookes tube.

The second method assumes that most or all

of the energy released will be in the form of heat. This could be used to eject inert matter at the necessary exhaust velocities. He plans to rotate the ship to provide artificial gravity, and states that a trip to the moon would take only 4 hours, to Venus 8 days.

JULY

Science and Invention begins the serialization of Ray Cummings's "great scientific novel" *Around the Universe*. The character "Tubby," while wishing that he could know all about the stars and astronomy meets a man who introduces himself as "Professor Isaac Swift Defoe Wells-Verne." The Professor shows Tubby his space flyer and they take off for assorted adventures on the planets. The "Interplanetary Space Flyer" is the size of a small square cottage and is divided into two floors. There are tiny windows set with thick glass and a metal door. On the flat roof is an observatory dome. Inside, there is a kitchen, storeroom, a main room that serves as living room and laboratory, and upstairs, two small bedrooms.

SEPTEMBER

Clement Fezandié, in "A Car for the Moon" (one of his series of stories, under the general title of *Doctor Hackensaw's Secrets*, published in Hugo Gernsback's magazine, *Science and Invention*), describes a spaceship launched from the rim of a giant wheel. A massive flywheel, 30 feet in diameter, is suspended between a pair of enormous pylons. The wheel is spun by a powerful electric motor beneath it. The two-passenger spacecraft is attached to the surface of the wheel at the very start, but as the rotation builds up speed, the distance between the vehicle and the wheel is increased. This is done by feeding a heavy chain through the axis of the flywheel (which is supported by compressed-air bearings) and out through an opening in its rim. The end of the chain is attached to the spaceship. At the maximum radius from the wheel and at its maximum speed, the vehicle is released. [Also see 1913–1916.]

ca. 1923

A manned Mars rocket is attributed to Robert Goddard. Its passengers ride in a freely rotating sphere amidships. The ovoid spaceship is equipped with a circle of railroad-style bumpers to facilitate the landing.

1924

Max Valier publishes *Der Verstoss in den Weltenraum* (A Dash Into Space), an attempt—with variable success—to popularize Hermann Oberth's book [see 1922]. An expanded version is published as *Raketenfahrt* ("Rocketflight") in 1928. In this book he describes a spaceship that combines the features of Oberth's rockets and his own. Instead of a central nozzle, a number of smaller engines surround the ship's waist.

The U.S.S.R. Society for Interplanetary Communications is founded in Moscow. It arranges public lectures and issues the journal *Raketa*. Interest wanes, however, and the society eventually disappears.

Mark P. Madden describes in the magazine *Science and Invention* his scheme for launching a rocket equipped with a television apparatus designed by C. F. Jenkins. The bullet-shaped vehicle is propelled by firing a large number of individual cartridges, similar to the techniques Goddard has been develop-

ing. The cartridges of explosive are fed to the combustion chamber on a continuous belt. In the base of the projectile is a lens. The image it gathers is projected up the central axis of the rocket to a television apparatus of the rotating scanning disc variety. It and the radio transmitter are operated by ordinary storage batteries. The radio's aerial is a long wire that trails behind the rocket.

V. Goncharov, in his two part novel (1) *The Psychological Machine* and (2) *Interplanetary Traveller*, describes a spaceship powered by the central nervous system of the pilot.

Goncharov "explains" the phenomenon by giving these examples: how a hawk diving on its prey can fall faster than it normally would, due to its desire; a man throwing a stone can make it speed faster to its target by his desire to hit it; a person dropping a glass can catch it, though if it had fallen at a normal rate it would have hit the floor.

He describes a voyage to the moon and several other planets supposedly made in 1922. A special magnet is placed within the "psychological machine" that harnesses the waves of "universal psychological energy"—when this is lacking, it uses the energy of the pilot if it is strong enough.

In appearance, the ship is shaped like a blunt cigar or elongated football about 15 feet long and 6 feet wide and is made of an unnamed white, nonconducting metal. It absorbs the psychic energy of humans which it stores in aluminum-covered tetragonal accumulators arranged in rows down either side of the machine. From these run wires to a device called the "gramophone" because of its loudspeaker. The flight is directed by a "psychological compass," which could be someone's handkerchief. This would be inserted in a tetragonal glass case. There is a meter attached with a needle showing the direction of the person the object belongs to. Normally, the machine operates by drawing upon the mental energy of the entire globe. This it passes on to the pilot via the loudspeaker. The pilot merely concentrates his thoughts on the person who is the target and the machine moves.

A second machine is described which is a perfected version of the first. The speed of the device is equal to or faster than the speed of thought: the distance to the moon is covered in a few minutes, that to Venus in 3 hours. The most powerful of the machines can cover 2.25 billion miles in 5 minutes.

While on the moon, Goncharov's heroes discover that the inhabitants of that world employ atomic-powered, winged flying machines.

V. Vetchinkin gives a report in Moscow on interplanetary spaceships [see 1922–1926].

All Aboard for the Moon is an animated motion picture produced by the Bray Studios of New York. It is a documentary-style film featuring a radium-powered spaceship launched from the roof of a skyscraper. The rocket is torpedo-shaped and, except for a small, dome-shaped conning tower and small fins surrounding the slender exhaust nozzle, streamlined. The engine appears to be gimbel-mounted in a sort of ball-and-socket joint. A row of windows runs down either side.

The interior of the rocket is one large open area, with only a single bulkhead dividing the passenger quarters from the control room at the nose. Fuel tanks are located beneath the floor. Four cots are provided for sleeping, as well as a small table and chair, and floor-mounted swivel chairs. A pair of motorcycles are also carried on board [!].

The launcher is a V-shaped ramp lined with cylindrical rollers. A blast shield protects the building from the rocket's exhaust. The animation was by Max Fleischer (who later went on to create Betty Boop and the Popeye animated cartoons).

F. A. Tsander publishes in "Flights to Other Planets" (*Technology and Life* magazine) a design for a winged interplanetary rocket.

He believes that " . . . it will, in all probability, be possible within the next few years to fly to other planets."

The spaceship he envisions looks outwardly not unlike a conventional biplane. It has a fuselage shaped like an elongated bullet: that is, a cylinder with one pointed end. The opposite end contains the opening of the

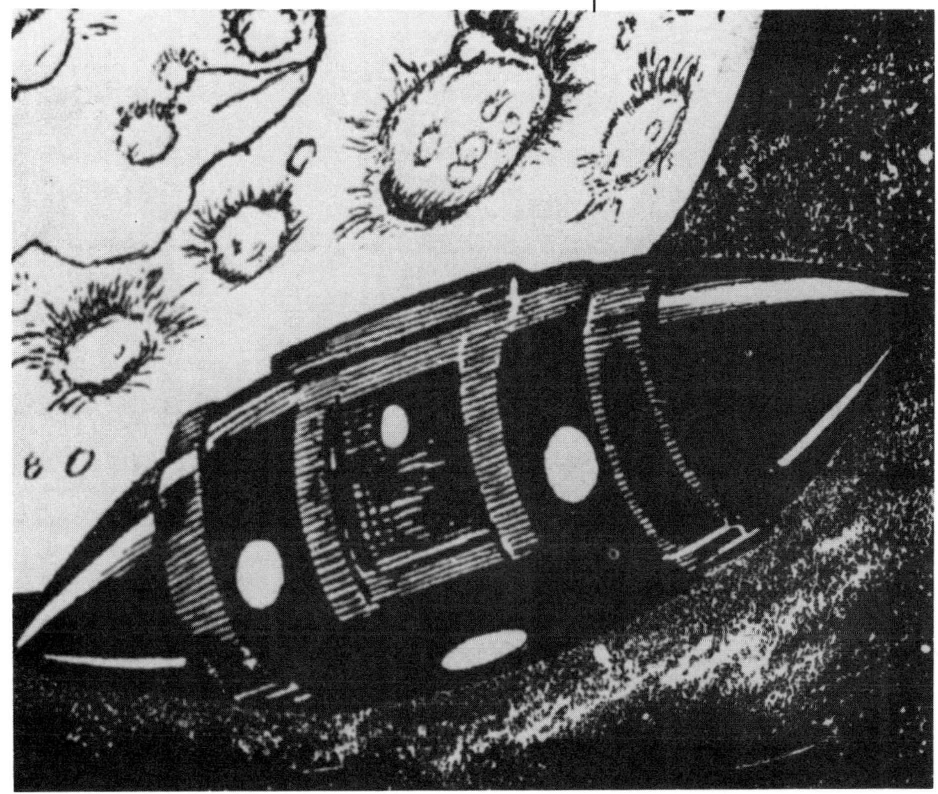

Goncharov

Ackerman

rocket engine. At about one third of the distance from the nose are placed a pair of wings, one above the fuselage and one directly below. A cruciform set of tailplanes are at the rear. The wings carry four propellors, two pushing and two pulling. They are run by an engine burning a hydrocarbon fuel and liquid oxygen. Tsander suggests that the propellors might be dispensed with and a rocket engine used for the entire flight, if possible. A large undercarriage allows the spaceship to take off like a normal aircraft. The entire ship is constructed of aluminum. Within the main fuselage is that of a smaller airplane. Its wings and tail surfaces protrude through the hull of the larger ship. It is to be used for the final descent to the earth, and is propelled by a single propellor at its nose. A model similar to this is exhibited at the 1927 Interplanetary Exhibition in Moscow [see 1927].

The spaceship is to ascend to an altitude of just over 4 miles using its propellors and wings. At this point the wings and undercarriage are pulled into the ship and consumed as fuel by burning the aluminum in pure oxygen. The metal is first melted in a special boiler before being injected into the engine. Tsander first suggested the possibility of using the structural materials of a spaceship as fuel in 1909. Beginning in 1917, he began experiments in the ignition of molten metals, obtaining the caloric values of materials like magnesium dioxide and others.

Powdered metals have often been suggested for use as rocket fuel, because of the great heat they produce when burned in oxygen—notably aluminum and magnesium. The main difficulty with their use is the solid by-product of such combustion, which is likely to eventually foul the jet of the engine. Another problem is getting the fuel fed to the engine at the proper rate.

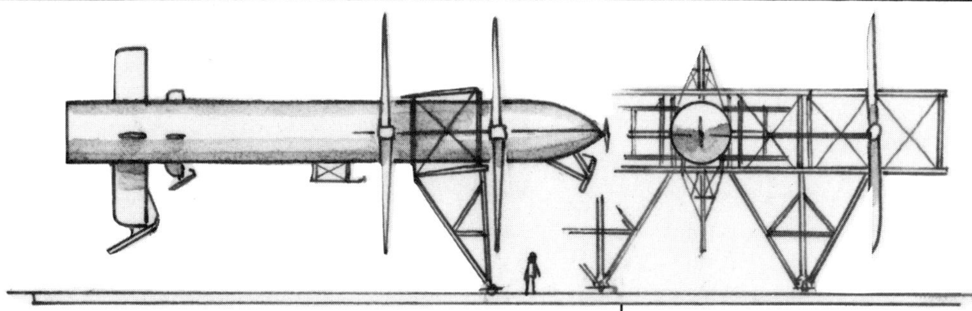

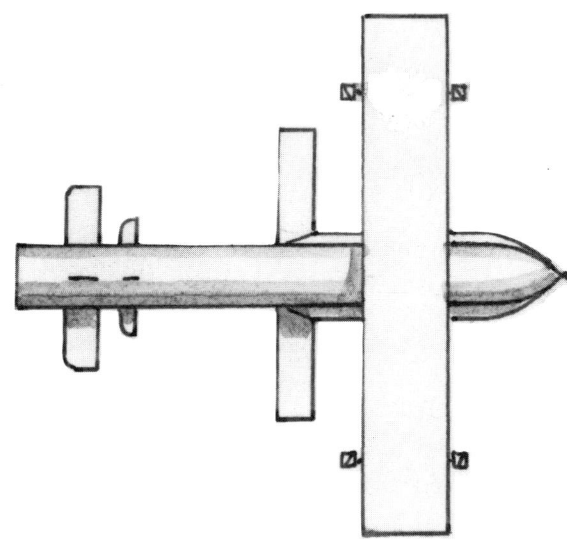

Tsander also suggests the use of solar sails for interplanetary flight. He describes thin, metal screens, rotating to maintain their rigidity. The metal screen can also be used to feed a rocket engine if necessary . While a sail has the advantage of light weight and no moving parts, it has the disadvantage of being easily damaged by meteorites. Tsander suggests making the solar sails out of iron filings. A large ring of metal would be energized, making it an electromagnet. Filings placed within the ring would be kept within its plane by the magnetic force. (The particles of iron would also have an electrostatic charge of their own so that they would maintain a constant distance from one another.) Enormous concave mirrors in earth or solar orbit would focus sunlight onto the solar sail. Tsander expects his solar sail to reach velocities far in excess of that attainable by rockets.

Tsander, in this same work, is the first to suggest the possibility of increasing the speed of a spacecraft during the flyby of a planet by using the planet's gravity. He demonstrates, for example, that by swinging close to the earth or Jupiter, a spaceship could increase its velocity by 6.3 miles per second and 15.25 miles per second, respectively. This is exactly the technique that will be used in the 1970s for the Voyager spacecraft's visits to Jupiter, Saturn and Uranus.

Turkestanov enters a flying rocket plane model in a model airplane competition held in Tiflis. The model takes off smoothly and flies well. However, the wings break off during the flight and the model crashes after flying 105 feet.

N. A. Rynin describes in his novel *In the Aerial Ocean* an "interplanetary radio ship." Its principle is based on the property of electricity or magnetism, that like charges repel and dissimilar charges attract. Considering that the earth is actually a large magnet, the inventor of the spaceship needs only to charge it with negative or positive energy in order to rise or fall. The force driving the ship's generators is cosmic rays, or radio energy beamed from the earth. To this latter end, six powerful stations are established around the globe. The appropriate ones are to beam energy to the spaceship. One of the inventor's aims is to prospect the moon for radium.

Rynin also describes the "radio ships" of the Japanese engineer Yamato. Since Yamato believes that the earth is a vast magnet whose influence extends almost indefinitely, all that would be necessary to fly into space would be to charge a spaceship with either negative or positive electricity or magnetism and the vehicle would be either attracted to or repelled from the earth. To power the ship six powerful radio stations are constructed (at Tokyo, Melbourne, Cape Town, London, Denver and Santiago). Yamato works out his scheme in some detail, including the problems of navigation in space and life support for the passengers.

K. E. Tsiolkovsky writes in his booklet *The Rocket in Space* (a revised reprint of his 1903/1911 paper) about the possibility of transmitting energy to the spaceship from the earth. A beam of electromagnetic energy of a short wavelength would be sent to the vehicle in a parallel beam. It would push the ship by its pressure, as in the solar sail concept. He also proposes the use of atmospheric braking to slow a reentering spacecraft.

"Alko" publishes in Russia some short notes on interplanetary rocket flight [same as "Alco," below?].

In N. I. Mukhanaov's novel *The Flaming Abyss* he describes spaceships of the year 2400 in which trips to the moon, Mars and the asteroids are made. Gravity is overcome by the use of the new element "nebulium." To counteract the friction of the atmosphere a device creates a cooling envelope of air around the ship. The spacecraft are shaped like elongated ellipsoids and are made of a flexible material that deforms according to stresses placed upon the hull (much like a dolphin's skin). The spaceships cruise at 100,000 km/s.

JANUARY 20

F. A. Tsander proposes forming a "Society for the Study of Space Travel."

1924

SPRING

A Group for Interplanetary Flights is organized at the Zhukovsky Academy in Moscow. In May it is reorganized as the Society for the Study of Interplanetary Flight. [Also see 1931].

APRIL

A Jet Propulsion Section is established at the Military Science Society of the Russian Air Force Academy. Its avowed aims are to bring "together . . . all persons working in the U.S.S.R. on the given question [spaceflight]; procurement of all possible information on work originating in the West; the distribution of regular information on the current situation in interplanetary navigation and, in connection with this, on publications issued; independent scientific research and, in particular, research into the military use of rockets." This section is evidently quite active: it issues a series of reports for its members, sponsors a competition for a small rocket that would have a range of 100 km, organizes a laboratory and bookshop, and forms a film group to produce motion pictures. The Jet Propulsion Section is active in the organization of the Society for Interplanetary Communications.

ca. 1924

Hans Dominick's short story "Death Rays" is set in the year 1955. In it a spacecraft is propelled by high-speed electrons. Three of these ships travel to Venus at speeds of 600 to 1,200 km/s.

Antonio de Stefano prepares calculations on a manned flight to the moon. He finds that a spaceship carrying one person and weighing 0.3 ton would require 150 tons of fuel to leave the moon and consequently would need fuel weighing 75,000 tons at takeoff from the earth. The rocket itself would weigh 120 tons at launch from the earth and 6 tons at takeoff from the moon.

A futuristic electric cannon is described by a writer publishing under the pseudonym "Alco." [Also see "Alko," above.] He claims that while working as a secret agent he discovered that the French had built a giant electric gun at Cappel in the forests of the Ardennes. It was capable of hurtling 2-ton shells over 1800 miles, with a muzzle ve-

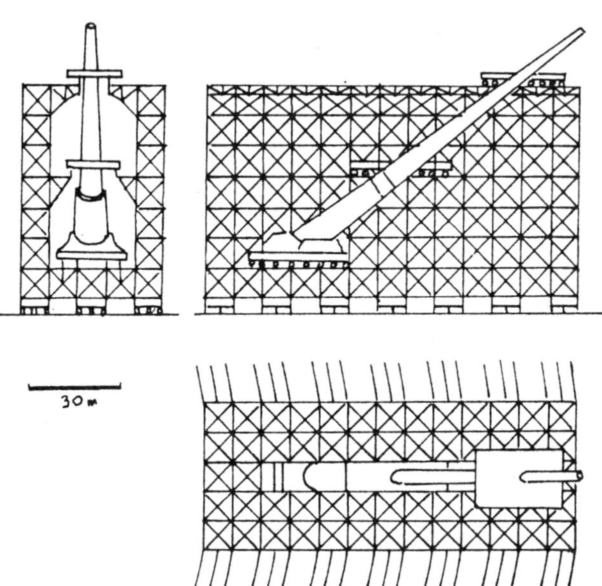

locity of 16,400 feet per second. The slender 6-foot projectile was stabilized in its flight by an internal gyroscope. The gun was 410 feet long, with a fixed elevation of 36 degrees, and was supported by an enormous gantry over 200 feet high. The power needed to energize the gun's coils was provided by ten 4,000-kW dynamos. This energy was stored in flywheels which were driven up to top speed in 11 minutes, then braked suddenly, rapidly releasing the required energy.

Most of the shell's flight was above the earth's atmosphere, its maximum altitude being over 300 miles reached in 5.5 minutes.

1925

Walter Hohmann, in *Die Erreichbarkeit der Himmelskorper* (The Attainability of the Heavenly Bodies), defines the principles of rocket orbits in space. He also describes a "spaceship." While Hohmann considers the problem of actual constructions, he does not come to any conclusions that he feels are presentable. Therefore, his "spaceships" are primarily limited to mathematical ideals. An example is his "powder tower." It is intended as an illustration of the decreasing requirement for fuel as a step rocket accelerates, rather than a realistic design for an actual spaceship. In the example shown, the rocket is intended to "burn" for 6 minutes. The tower is divided into six horizontal segments,

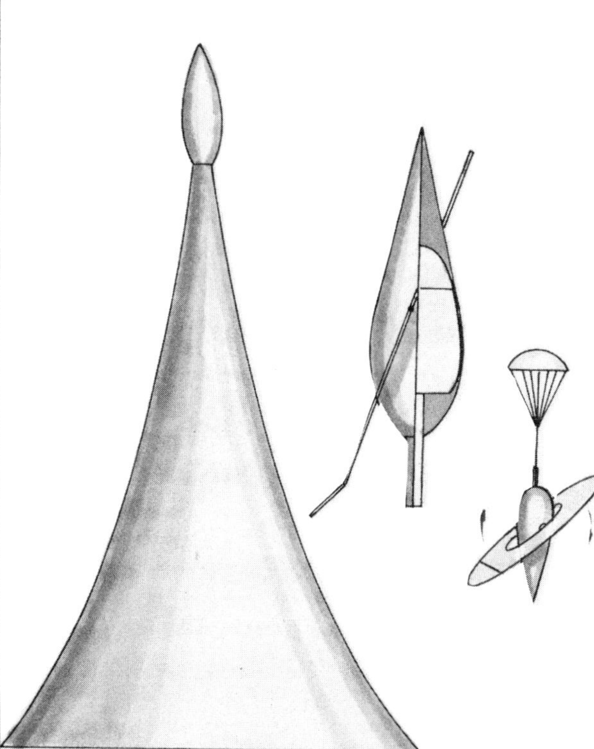

each the same thickness but different in diameter. Each of these slices represents the amount of fuel required to lift the weight above it. It is easy to see why each stage of a step rocket needs to be smaller than the one below it. As fuel is burnt, from minute to minute, there is less weight, from minute to minute, to lift. And as the rocket grows lighter, less fuel is needed.

In spite of the main rocket's speculative design, Hohmann gives some thought to its payload: a passenger-carrying capsule. This is an egg-shaped body with one sharply pointed end. It is about 17.6 feet in height and 5.25 feet in width. There is enough room for two astronauts together with enough supplies for a trip of 30 days' duration. The total weight, including the capsule's fuel, would be just over 8,800 pounds. The capsule is equipped with a braking surface for reentry, an inclined lifting surface with a rudder—with a combined surface area of just over 74 square yards—and a parachute for the final descent. In order to change the attitude of the capsule while in space, Hohmann suggests that the crew could climb in a suitable direction around the interior walls of the cabin, using handholds provided for this purpose. The spacecraft would then turn in the opposite direction. The complete spaceship is a cone-shaped tower over 105 feet tall, with the 17.6 foot capsule on top, and over 72 feet wide at the base. Its total weight at takeoff would be 2,799 tons.

Hohmann's work on interplanetary orbits is so fundamental to spaceflight that these orbits have been named for him. He first demonstrates that certain orbits are "impossible"—in that the energies required are too great to be practical. These are orbits in which the direction the spacecraft travels is opposite to that in which the planets are travelling. Seen from above the north pole, all of the planets in the solar system orbit in a counterclockwise direction. A spaceship moving from the orbit of one planet to the orbit of another opposite to the direction in which the planets are travelling would first have to overcome the earth's gravity, reduce the orbital velocity it has acquired from the earth to zero, accelerate in the opposite direction, meet the other planet on a full-speed collision course, reduce the relative speeds to zero and then acquire the new planet's orbital velocity before landing. All of these maneuvers would require a fabulous outlay of energy. Hohmann's "possible" orbits all travel in the same direction as the general rotation of the solar system. He shows that the orbit of a spaceship that begins tangent to the earth's orbit and finishes tangent to the orbit of the target planet would be the most efficient in the amount of energy required.

These *Hohmann orbits* are not the fastest paths from planet to planet; they are the ones requiring the least expenditure of fuel. This is a far more important factor in space travel than time. The less fuel a rocket requires, the smaller and less expensive it needs to be. As Hohmann's "powder tower" shows, you need fuel to lift fuel, and fuel is far bulkier and heavier than the life support systems and supplies needed for a long spaceflight. Hohmann calculates that a round trip expedition to Venus would require that the crew be gone for 762 days, or about 2 years and 1 month (146 days to reach Venus, 470 days to explore the planet while waiting for the next available "launch window," and 146 days to travel back to the earth). This would be, Hohmann says, "about the same time as a prolonged expedition to the South Pole."

In G. Arelsky's *Martian Stories*, the Martians travel to the earth in an "aerobile"—a solar-powered artificial planet with its own atmosphere. [Also see 1926.]

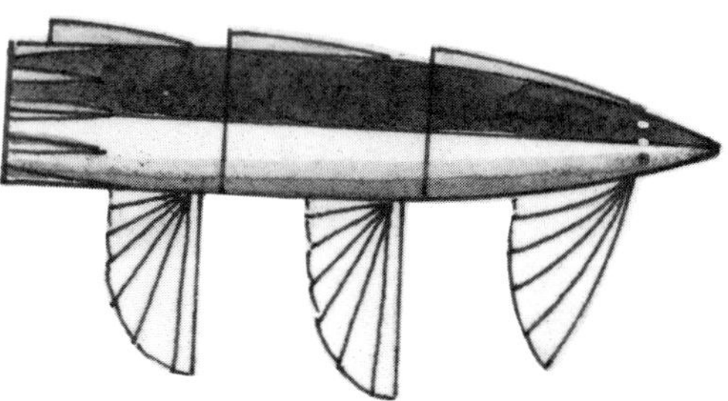

Otto Willi Gail publishes his novel *Der Schuss ins All* (*The Shot into Infinity*) in which he describes a three-stage rocket, the *Geryon*. It is a huge ship, launched horizontally on a runway. Each of its three stages is equipped with folding wings so it may be returned to earth unharmed. One half of the ship is painted black for temperature regulation. The passenger cabin is in the blunt nose and is shaped like a truncated cone. Divided into two floors, the lower contains the airlock and a central well surrounded by small cabins, a lavatory, smoking room, dining room and kitchen. Above is the control room, and above that a parachute of 120 square meters. At takeoff, the crew lie in hammocks in the control room. The first two stages are fueled with alcohol and liquid oxygen and the third stage with hydrogen. [Also see 1926, 1928 and 1929.]

Max Valier develops his step-by-step plan for the achievement of space travel. He believes that it will require a four part program:

1. Test stand experiments to further develop powder rockets, and the testing of models.
2. Ground vehicles (cars, sleds, railcars, etc.) to allow the testing of rocket propulsion.
3. A liquid propellant motor for aircraft.

Science WONDER Quarterly

FALL 1929

HUGO GERNSBACK
Editor

4. The eventual construction of stratospheric aircraft and, finally, spaceships.

This latter stage is subdivided into a number of smaller evolutionary steps as a conventional propellor-driven airplane is gradually developed into a pure rocket-propelled vehicle [see 1926–1927 and 1929].

FEBRUARY

Don Home, in an article in *Science and Invention*, "Can We Visit the Planets?" describes a space gun and a light-propelled spaceship.

The gun is a steel tube 3.5 miles tall, supported by a mast similar to those which support radio antennas. At its base is an enormous concrete "explosion chamber." The projectile would leave the muzzle with a velocity of 7 miles per second. The slim, bullet-shaped projectile contains elaborate precautions to protect its inhabitants against the terrific acceleration (estimated by Home to be in the vicinity of 50 tons). The crew of four are strapped into swivel-mounted chairs, each gyroscopically stabilized, and are suspended from the walls of the cabin by springs. The cabin itself is independent from the spaceship, able to slide up and down within it like a piston in a cylinder. Springs and hydraulic shock absorbers take up the main jolt of the launch. A 2-foot-thick disc of armor plate at the base of the projectile protects it against the initial blast. A second set of springs in the base of the cabin absorbs any remaining force. Windows line both the cabin and the outer shell, so that no matter what position the cabin takes within the projectile, the crew will be able to see out.

Home's light-propelled vehicle depends upon the eventual discovery of some method of neutralizing gravity. A powerful light beam directed from earth would move the weightless craft.

Home also describes the launch of a spaceship from the rim of a rotating wheel. A giant, vertical flywheel would have a bullet-shaped vehicle attached to its rim by a cable. At top speed, the cable would have a tension on it of 200 tons. At this point, the spaceship would be released, shooting off at a tangent to the wheel.

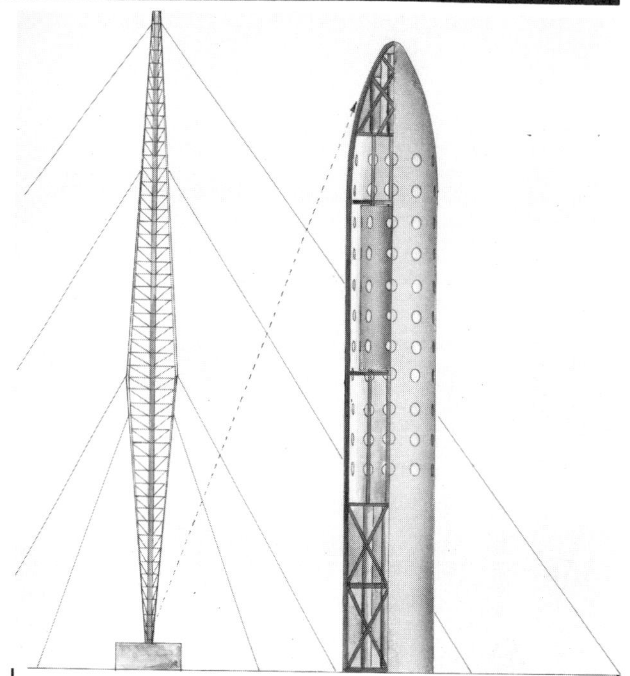

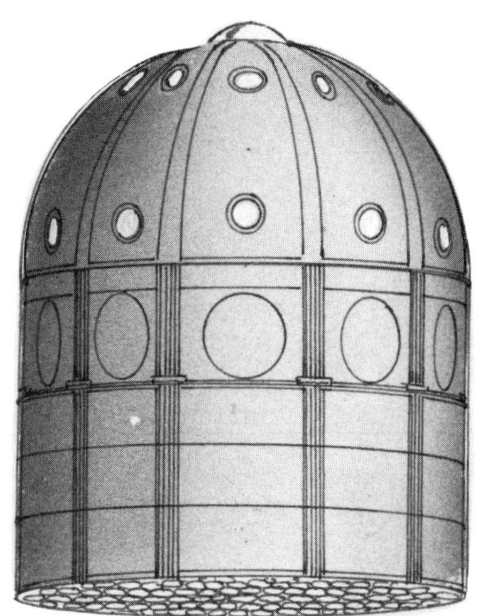

ca. 1925

J. Roberts works at Britain's Air Ministry on the problems of reaction-powered aircraft.

In Bruno Burgel's novel *Rocket to the Moon*, he describes a spaceship intended for a flight to the moon. It is a streamlined, elongated teardrop equipped with wings. It is rocket-

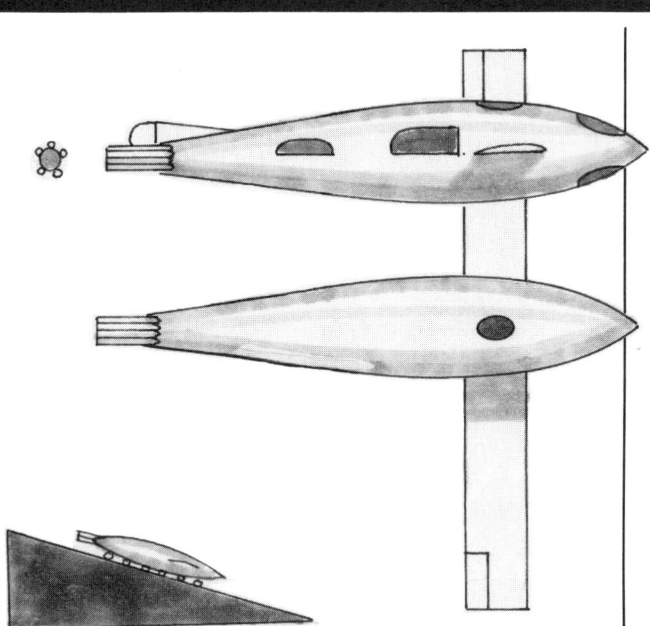

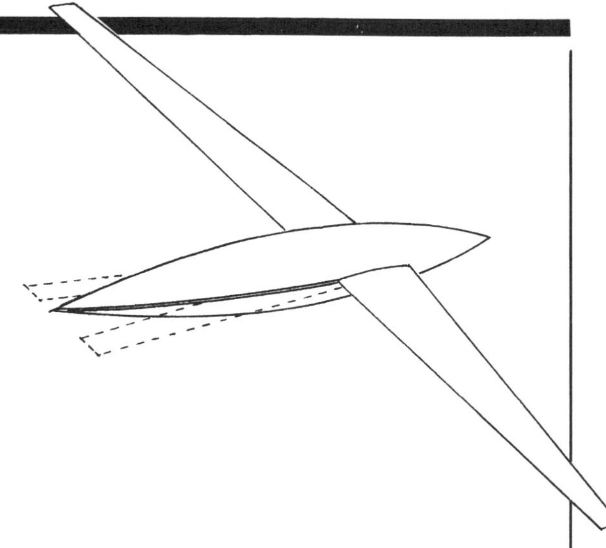

propelled, using the new explosive "uzambaranite" as fuel. Although the ship is rocket-powered, the flight to the moon is only possible because of a nebula that has invaded the solar system (in the year 3000), filling it with a fine substance capable of supporting an aircraft's wings. The spaceship in the story, the *Star of Africa*, is divided into six sections: the captain's quarters; supplies and oxygen; a mess with windows; fuel storage (with tanks of liquid helium to cool the combustion chambers and cartridges of explosives); the engine room, which includes batteries for power and the galley; and finally the five combustion chambers to which the cartridges are fed automatically. The crew consists of four persons and one passenger. The rocket itself is made of steel plate with a second, inner lining of thinner steel. A 3-inch insulating layer of wool separates the two. The wings are equipped with ailerons and there is a small rudder. Although these can steer the rocket, navigation is achieved primarily by varying the thrust of the rockets. The ship takes off by running down an inclined ramp. Although the launch is successful the ultimate fate of the *Star of Africa* is unknown.

1926

In his novel *Der Stein vom Mond* (*The Stone From the Moon*), Otto Willi Gail describes the spaceship *Icarus*. It is a squat, cylindrical space-to-space vehicle, not intended for planetary landings. It carries a "lunar rocket" for this purpose: a small, 8-meter-long winged rocket that carries a crew of three and is entered through a hatch in its nose from the main ship's control room.

A fleet of eight spaceships handles the traffic from earth to where the space station *Astropol* orbits. These are torpedo-shaped with "airplane decks" and resemble "submarines with wings." They range from 8 to 100 meters in length, smaller than the *Geryon* [see 1925]. They are launched from water and have folding wings so they can glide back to the earth.

The space station is built using the old *Geryon* (from *A Shot into Infinity*, see 1925) as its core. Made of sodium, it is shaped like a thick cylinder or bun 120 meters in diameter with a counter-rotating observatory at one axis and an airlock for spaceships at the other. It is built as a refueling station; rocket fuel is manufactured there from ice sent from the moon, using solar energy. At the end of a meter-thick tether 1,600 meters long is a "gravity cell" which rotates at 0.5 rpm, creating the effect of 1 g of gravity. The living quarters are located here. The station is in a polar orbit at 95,000 kilometers and stays above the earth's terminator. [Also see 1925, 1928 and 1929.]

Willy Ley publishes *Das Fahrt ins Weltall* (Travelling into the Universe), a popular book on rockets and space travel, written as an improvement on Valier's book [see 1924, 1928 and 1929].

1926

The Russian author S. L. Grove publishes *Journey to the Moon* in which a Tsiolkovskian spaceship is launched from Tibet. The takeoff is from an inclined ramp. The crew of two remain in communication with the earth via radio. They return to the earth by splashing down in the Caspian Sea. Most of the details of the rocket and its flight are taken bodily from Tsiolkovsky's *Beyond the Earth* [see 1916].

Illustrations by the artist S. Lodygin

In A. Platonov's *The Lunar Bomb* a spherical projectile is launched toward the moon by attaching it to the rim of a rotating disc. The plane of the disc can be set at any angle. Descent onto the moon is by a rocket device that is activated automatically—however the first flight is limited to a circumlunar expedition.

The motor of the giant disc is cooled by huge ventilators as it spins the disc at a rate of 946,000 rpm. At the right moment the "bomb" is released. Communication is maintained with the passengers of the "bomb" by radio.

T. Rockenfeller publishes "Die Fahrt zim Mond" in which he describes a spacecraft constructed according to the schemes of Franz Ulinski [see 1927]. In outward appearance it is identical to the Austrian's spherical spacecraft. It is also possible that one illustration of Rockenfeller's spaceship might have been the inspiration for the illustrator of Heuer's *Men of Other Planets* [see 1951].

G. Arelsky describes the 1930 flight of a rocket to the moon in *Gift of the Selenites*. It is launched by the "American Society of Interplanetary Communications. "It is an instrumented, unmanned rocket, 30 meters long.

Antonio de Stefano designs a gun for launching a missile into interplanetary space, and a rocket for a flight to the moon.

The Abbé Theophile Moreaux publishes *Les autres mondes sont-ils habites?* (Are the Other Worlds Inhabited?).

C. Nordmann advises against flying to the moon.

Alexander Yaroslavskii publishes the utopian novel *Argonauts of the Universe* in which he describes a flight from the earth to Mars in a radium-powered rocket. The first lap of the journey is made in the spaceship from the earth to the moon. The final leg is made by occult means.

H. Ganswindt, in two letters to N. A. Rynin, expands on his original spaceship design [see 1891]. In the first, he says that his spaceship is not to be lifted into the upper atmosphere by rockets, but rather by an airplane. It is to eventually land in a fuelless glide. In the second letter, he explains that the propellant, dynamite cartridges, is contained in the two lateral cylinders atop the spaceship, which rotate like the cylinder of a revolver.

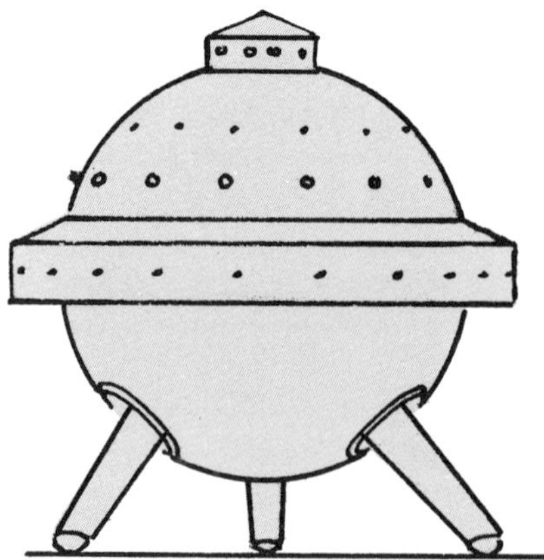

Rockenfeller

From *The Adventures of Uncle Lubin*, by W. Heath Robinson, 1925.

Each holds several hundred thousand cartridges which are fed automatically into the central steel cylinder. Part of the discharged gas would be used to heat the passenger cabin.

K. E. Tsiolkovsky publishes his new treatise on spaceflight: *Exploration of Space with Rocket-Propelled Devices*. This is a new and revised edition of his work originally published in 1903 and 1911.

He concludes that his space rocket should be launched from an inclined slope at least 300 miles long, rising 10 or 20 degrees to the horizon. The rocket is in two stages. The first stage is the "earth rocket," which never actually leaves the launching track. It is over 300 feet long and 6.5 feet wide and carries the 65-foot space rocket in its nose. The earth rocket is started first. When it reaches the end of the track it will have reached a velocity of about 2 miles per second. At this point the engine of the space rocket is ignited. The earth rocket immediately decelerates using its air brakes. Friction with the ramp has been reduced to virtually zero by using a cushion of com-

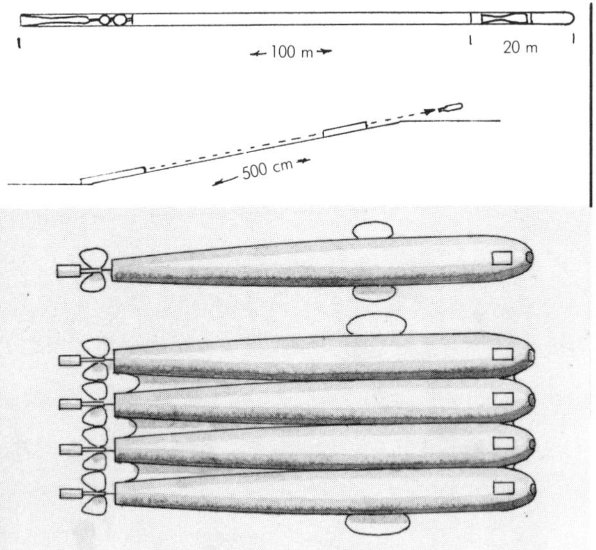

pressed air between the earth rocket and the ramp.

In this year Tsiolkovsky also designs his "cosmic rocket," a streamlined torpedo 65.5 feet long and 6.5 feet wide. At the stern are movable control surfaces set behind the opening of the rocket exhaust. For atmospheric flight, small ailerons are attached to the sides of the hull. At the blunt forward end are quartz windows coated with a substance to filter out the harmful solar radiation. Several of these rockets can be attached side by side to form "compound rockets."

Tsiolkovsky also outlines a plan for the exploration of space [also see ca. 1930]:

1. Testing with reaction-propelled aircraft, gradually increasing the altitude and decreasing the wing area.
2. Reaching high speeds and altitudes above 7.5 miles; placing the pilot in a pressurized cockpit.
3. Making the change to all-rocket power.
4. Making rocket flights with velocities up to 5 miles per second outside the atmosphere, using a rocket similar to the type described by Scherschevski [see 1927].
5. Flying farther from the earth; developing closed-cycle food and waste-disposal systems; penetrating into deep space.
6. Setting up space colonies in the asteroid belt and spreading on out through the galaxy.

The noted British astronomer, Prof. A. W. Bickerton, says that "This foolish idea of shooting at the moon is an example of the absurd length to which vicious specialization will carry scientists. To escape the earth's gravititation a projectile needs a velocity of 7 mi./sec. The thermal energy at this speed is 15,180 calories. Hence the proposition appears to be basically impossible."

NASM

MARCH 16

R. H. Goddard makes the first flight of a liquid-fueled rocket. It is the third liquid-fueled rocket built by Goddard: a fragile, spidery framework about 10 feet tall. The motor is about 2 feet long and is mounted at the forward, or upper, part of the rocket. It is supported by the frame of the fuel pipes. The fuel and oxidizer tanks are together about 2 1/2 feet long, separated from the nozzle by a gap of some 4 feet. A conical cap of asbestos shields the tanks from the blast. The whole purpose of this arrangement is to place the motor at the front, where Goddard—like many of the VfR experimenters later [see 1927]—believes the thrust should be applied for the greatest stability. The analogy is that of a wagon that is pushed versus one that is pulled, the latter being more easily kept in a straight path. This is a fallacy, as applied to rockets, that will not be discerned for some time.

The fuel tanks are pressurized by oxygen gas, initially by an outside source and in flight by the evaporation of the liquid oxygen carried by the rocket. The fuel is gasoline. Ignition is supplied by an assistant using a blowtorch attached to a long pole. The engine catches and as soon as it has burned enough fuel for the rocket's total weight to drop below 6 pounds, the contrivance takes off. It shoots with a roar to a height of 41 feet, landing 184 feet away. Its top speed is estimated to be about 60 mph.

APRIL

The first issue appears of *Amazing Stories*, the first magazine devoted wholly to the publication of science fiction. It is founded and edited by Hugo Gernsback, a native of Luxemburg, who has long been an active promoter of spaceflight. His otherwise nonfiction magazines, *Modern Electrics* and *Science and Invention*, often run stories of interplanetary adventure and exploration (such as Gernsback's own "classic," *Ralph 124C41+*). Eventually, Gernsback and his staff will play a pivotal role in the creation of the American Interplanetary Society [see 1930, below] which, according to Gernsback, "More than any other [society] traces its astronautical roots to a science fiction fatherhood."

1926–1927

Max Valier proposes the idea that a spaceship could be gradually evolved from a conventional aircraft. In doing so, he frequently enlists the services and designs of such well-known aircraft designers and builders as Hugo Junkers and Alexander Lippisch, modifying them according to his own ideas [also

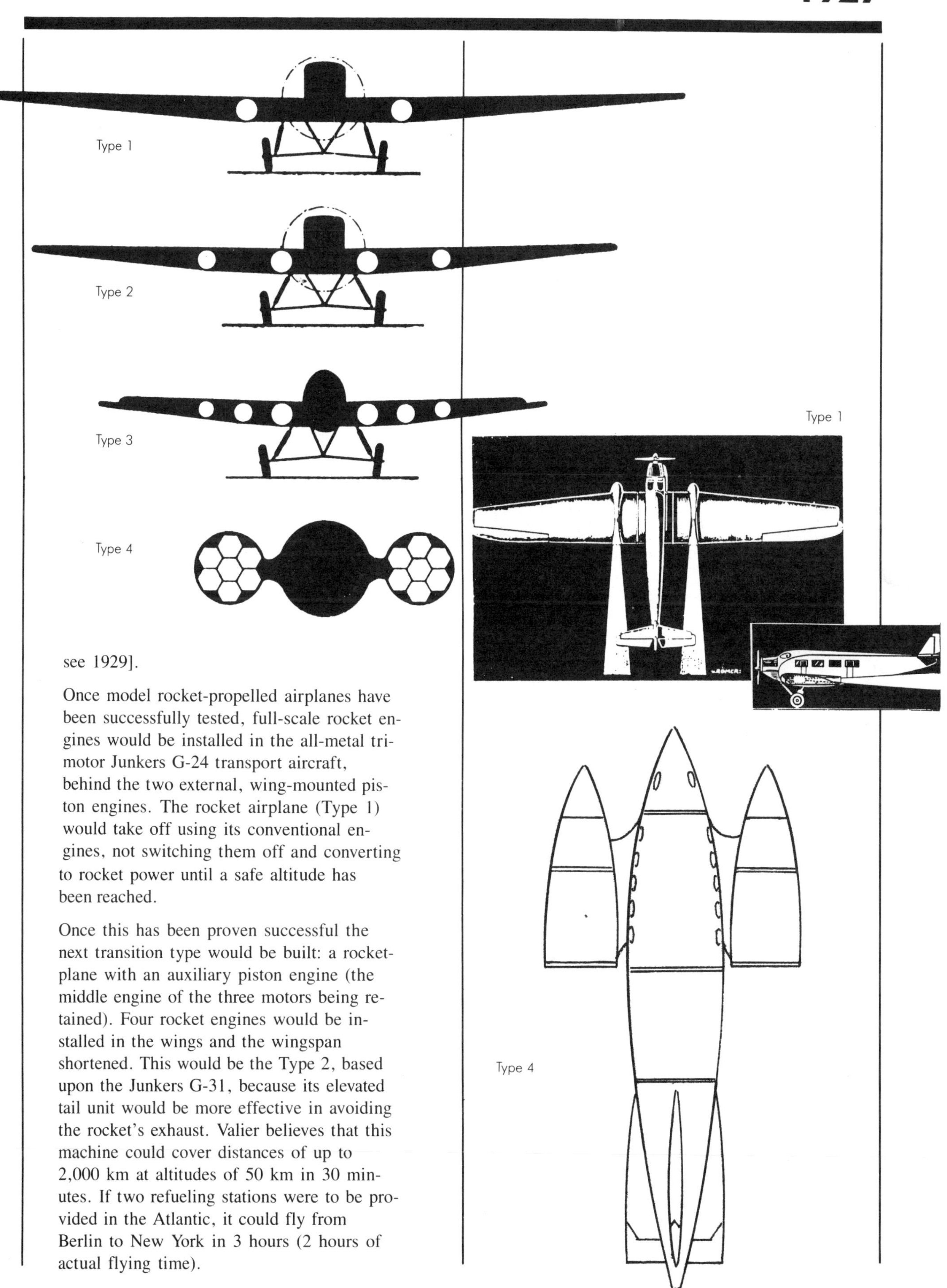

see 1929].

Once model rocket-propelled airplanes have been successfully tested, full-scale rocket engines would be installed in the all-metal trimotor Junkers G-24 transport aircraft, behind the two external, wing-mounted piston engines. The rocket airplane (Type 1) would take off using its conventional engines, not switching them off and converting to rocket power until a safe altitude has been reached.

Once this has been proven successful the next transition type would be built: a rocketplane with an auxiliary piston engine (the middle engine of the three motors being retained). Four rocket engines would be installed in the wings and the wingspan shortened. This would be the Type 2, based upon the Junkers G-31, because its elevated tail unit would be more effective in avoiding the rocket's exhaust. Valier believes that this machine could cover distances of up to 2,000 km at altitudes of 50 km in 30 minutes. If two refueling stations were to be provided in the Atlantic, it could fly from Berlin to New York in 3 hours (2 hours of actual flying time).

When enough flight experience has been gained, a pure rocket-plane would be flown. Six engines would be installed in the wings, which are shortened even further. The auxiliary piston engine is eliminated. This design, Type 3, developed in 1927, has a pressurized cabin and a spindle-shaped, streamlined fuselage. The wings are equipped with slats and landing flaps. Direct intercontinental stratosphere flights could be made with this machine.

The penultimate step is the rocket ship, also developed in 1927. It would be a wingless, finned torpedo carrying fourteen rocket engines. These would be mounted in streamlined outrigger pods, seven motors to each. Seventy-five percent of its launching weight would be fuel. The rocket would be held in a vertical position for takeoff by a launching tower. It would attain an altitude of 250 km in 5 minutes.

When challenged that no pilot would ever be foolhardy enough to fly in one of these rockets, Valier replies, ". . . in the war [World War I], I was detailed to the testing of new aircraft types more than once. Of course, one prefers to climb into an old tested machine rather than into a new one, in which one cannot tell whether the innovations will stand the test. But after all an operational sortie was no life insurance either, and in spite of this enough volunteers signed on in the air force.—I shall, of course, be a member of the first crew that flies a rocketplane."

The final objective of Valier's evolutionary scheme of development is the space ship: a design based partly on concepts invented by Hermann Oberth. Two ellipsoidal passenger compartments are contained in the nose. The bulk of the rocket is occupied by tanks of fuel, in the midst of which—at the rocket's center of gravity—are the gyroscopes. Eight large rocket engines surround the tail assembly, roughly amidship.

Also in 1927, Valier proposes two additional designs for long-range rocket aircraft: Types 4 and 5. The latter is an enormous aircraft with a twin fuselage and twenty rocket engines, all mounted in the trailing edge of the vast wing and ignited in pairs. The passengers would be carried in the thick leading edge. The transoceanic rocket plane could cover the distance from Berlin to New York in 93 minutes [see 1927].

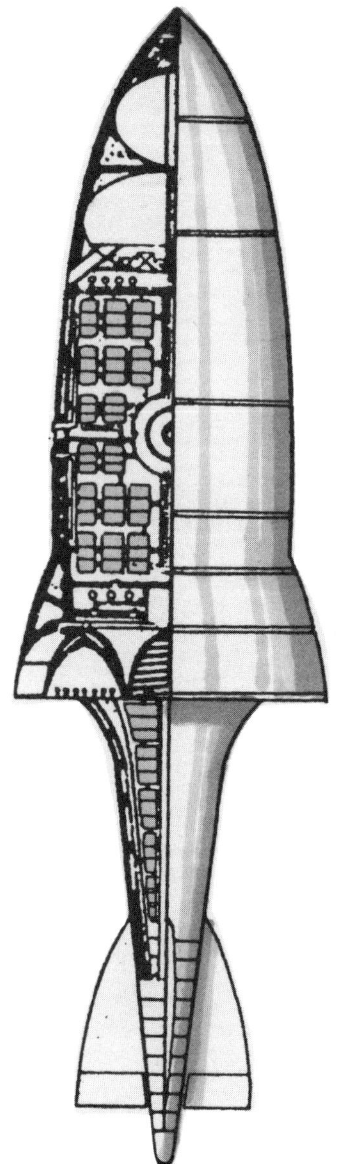

Valier also designs a rocket-propelled sport airplane in 1927 (Type 6). It is a sleek-looking high-wing, open cockpit monoplane, not dissimilar to the Curtiss "Junior," with a streamlined fuselage, twin tail and a rocket engine mounted above the wing. At the front of the rocket module is a radial piston engine (though this will ultimately be done away with).

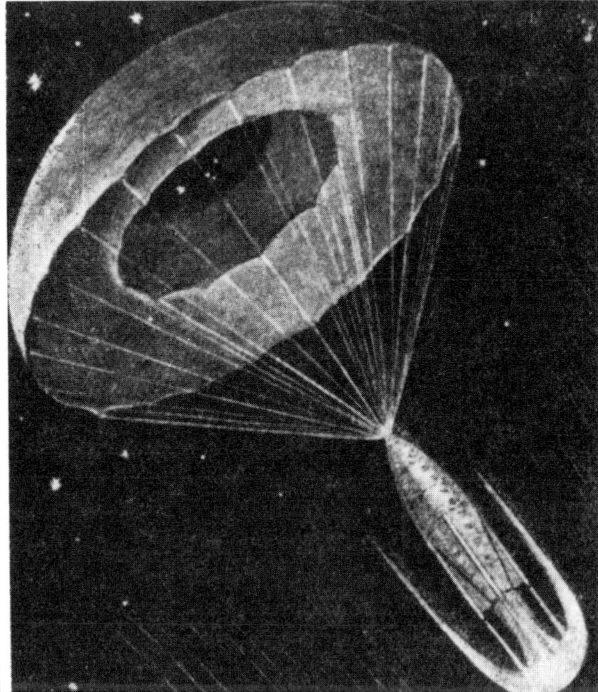

Valier rocket concepts. Left, takeoff of a Berlin-New York rocket transport (note the Tesla broadcast power antennas in the background); above, reentry of a spaceship; below, spaceship leaving the moon.

Left, sketch of a spaceship sent in a letter from Valier to Oberth.

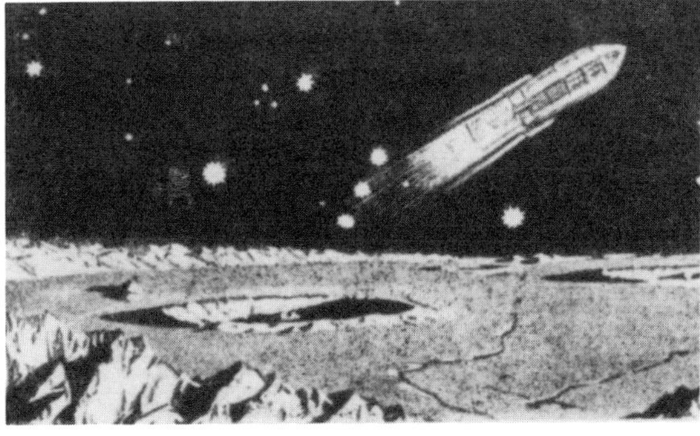

1927

Max Valier reports to the Scientific Aeronautical Association in Berlin, discussing the possibilities of spaceflight. This leads to a discussion in the aeronautical journals *Z.F.M.* and *Flugsport*.

Robert Esnault-Pelterie, in *Reflections on the Results of Reducing Engine Weight*, describes the potential of monatomic hydrogen as a rocket fuel. He goes on to discuss the merits of atomic power for space travel. "Only the forces and energy," he writes, "that appear to be contained in molecules of matter could give us the concentrated power magnitude and force that we have established [for interplanetary spacecraft]. If for a moment we assume that we have 100 kg of radium in our ship weighing 1,000 kg, and are able to release from this amount of radium all its energy in the time required by us, we will see that 100 kg of radium is more than enough for a flight to Venus and back . . . "

He again refers to atomic power in *Investigating the Upper Atmospheric Layers with Rockets and the Possibilities of Interplanetary Flight*. He deems it best that nuclear energy be used for the direct ejection of positive ions and electrons.

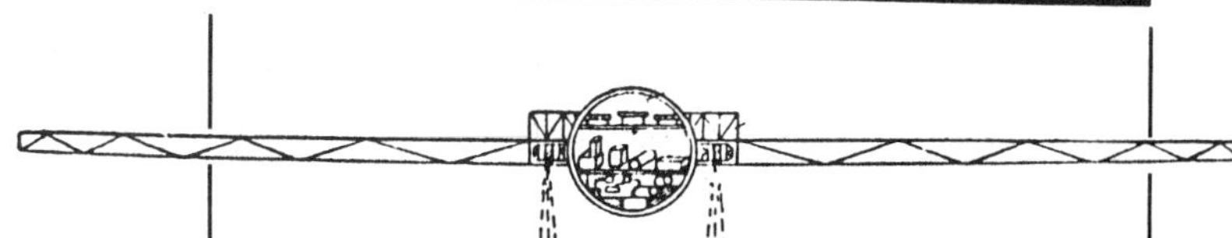

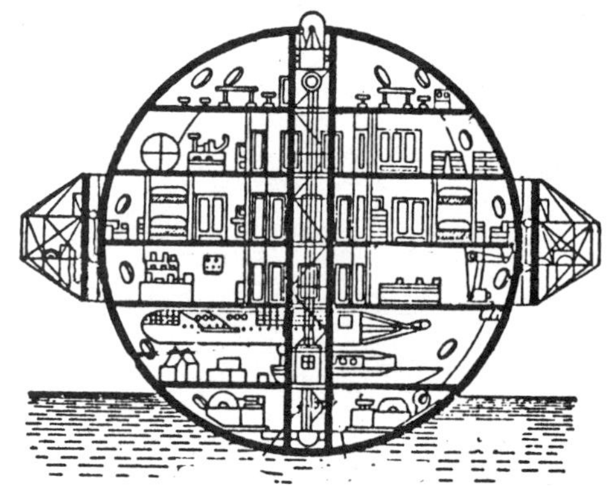

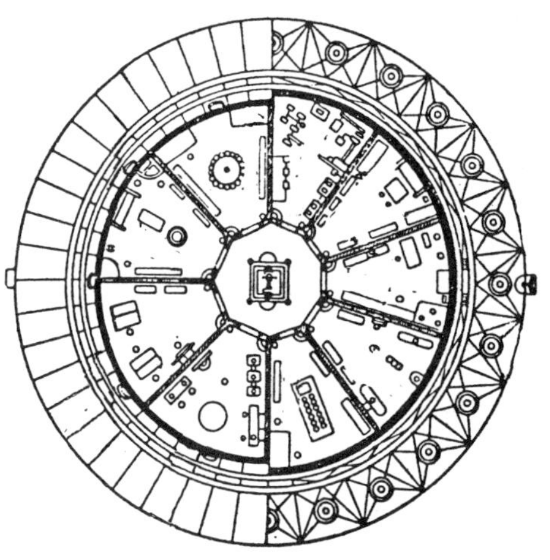

Franz A. Ulinski describes a number of spacecraft propelled by the force of ejected electrons. Basically, there are two types of ships: those that derive their power from solar energy, and those that are atomic-powered. Type one is the "interplanetary ship." Its spherical personnel cabin is surrounded by a disc of solar-activated thermopiles. The cabin is suspended on a universal joint, so that it may move independently of the disc. The sphere is surrounded at its equator by a thick belt containing the electron ejectors.

Type two is the "cosmic ship." It, too, is a sphere, though vastly larger than that of the type one ship. Also like the type one, it has an equatorial belt containing the electron ejectors. Unlike the previous ship, the "cosmic ship" is atomic-powered. The 20-meter sphere is constructed of steel, soldered with copper and reinforced with cross bars. These latter are covered with asbestos. The inner side of the sphere is covered with plywood over a layer of rubberized fabric. There are six floors connected by stairways. The ship is fully and comfortably appointed; there is even a hangar containing an aircraft. The pilot's compartment is at the "north pole" and is equipped with a speed indicator ("ethertachometer"), mass indicator, radiotelegraph, level indicator, etc. The entire compartment can be lowered to the opposite pole by elevator. Takeoff must be from water, where the 200,000 kg ship sinks to almost 8 feet.

In 1927 Ulinsky patents a new design for a solar-powered electron spaceship. The ship proper is an egg-shaped cabin with three interior floors. Surrounding its equator like a flared skirt is a disc carrying the solar thermoelectric elements. For flight within the atmosphere of a planet, the electric energy generated powers a turbocompressor with a jet nozzle. While in the vacuum of space, the energy powers the electron ejectors.

The latter requires an electrical potential of 250,000 volts. The direct current generated by the solar cells is converted to high-frequency alternating current. A large solenoid occupies about half the length of the ship and it produces a powerful magnetic field.

Several electron ejectors surround the equator of the ship. Each of these consists of a cathode brought to red heat and inserted in a solenoid. When energized, this produces an electromagnetic field. When a current of the required voltage passes between the two parts, electrons fly from the incandescent cathode to the main cathode. Between it and the anode passes a current of 250,000 volts. This drives electrons from a tungsten spiral filled with an amalgam of barium. These are ejected in great quantities and with immense acceleration.

1927

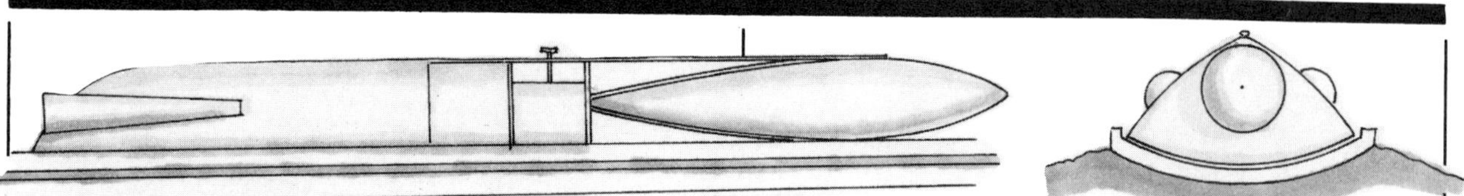

In order to propel the 3,000 kg spacecraft away from the earth, 5 grams of matter (a mercury preparation) must be converted into electrons every minute. If the ship ascends at an acceleration of 15 m/sec^2, it will in 1,800 seconds achieve enough speed to leave the earth. For this only about 15 kg of matter is required.

Ulinski makes a serious error in his reactive device for travelling within planetary atmospheres. He believes that the reactive force takes place entirely within the nozzle of the rocket and that it will be possible to recycle the expelled gases. There would be, then, no loss of matter as in the case of ordinary rockets. Unfortunately, this would only have the effect of causing the spaceship to stand still [also see Eyraud, 1865].

Ulinski's solar power units are thermoelectric, consisting of overlapping metal discs. When they are subjected to a temperature difference between them, a current of electricity begins to flow. Beyond the earth's atmosphere, it is believed, a square yard of surface could produce 1 1/2 watts or 2 horsepower of energy. Allowing for cooling by radiation, this might be reduced to 25 percent or 1/2 hp. In this case, the interplanetary electric ship would require a generating surface of 500 to 1,000 square yards. Surplus energy could be radiated away by antennas (which, in turn, could be used to receive energy beamed from the earth).

The electron ejector is the prime mover of Ulinski's ships. The basic device is a parabolic anode at whose focal point is placed a cathode. This sends a high voltage current to the anode. This in turn reflects the current in a beam. If both the anode and cathode are supplied with a source of matter, this will be projected along with the beam as a stream of particles.

A. B. Scherschevsksi describes Tsiolkovsky's 1903 and 1914 spaceships in his paper "The Space Ship." In the drawing published with this, he elaborates on Tsiolkovsky's designs. In another work, *Die Rakete* (1929), Scherschevski publishes a second drawing based on Tsiolkovsky's ideas.

The method of launching a Tsiolkovsky rocket is with a rocket-powered sled. The booster carrying the rocket ahead of it sits in a trough-shaped track, sliding frictionlessly on a cushion of compressed air.

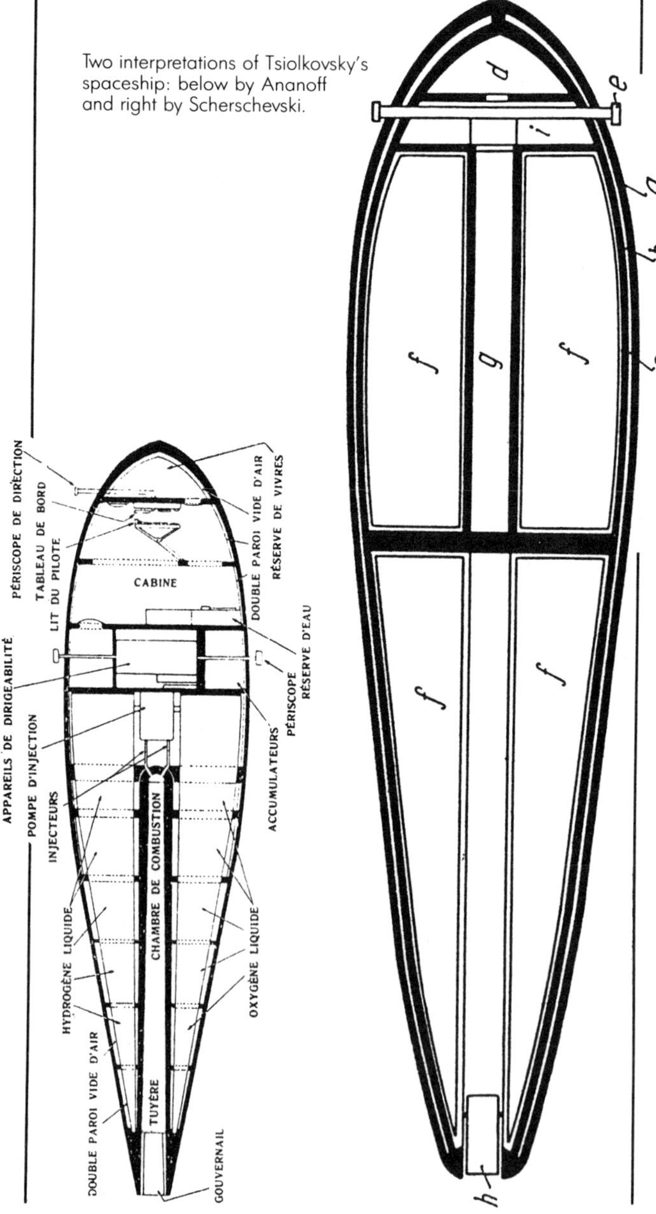

Two interpretations of Tsiolkovsky's spaceship: below by Ananoff and right by Scherschevski.

Die Rakete

Zeitschrift des Vereins für Raumschiffahrt E.V., Breslau

Jahrgang 1927.

Johannes Winkler founds the VfR journal *Die Rakete* [see July 5, below].

A. Federov displays a model of his spaceship design at the Moscow exhibition [see 1896]. He plans to achieve propulsion through "electrochemical energy resulting from the use of intra-atomic energy. "The streamlined rocket will carry six passengers. The spaceship's weight is 80,000 kg and it is 60 meters long and 8 meters wide. Two propellors are provided for takeoff, along with folding wings. The rocket engine would take over at an altitude of about 15 miles.

He hoped that his rocket would be able to take its crew to the moon for a initial stay of at least two days. If oxygen is found at the bottom of lunar craters, then future expeditions can remain longer. A base could then be established as a terminus for further exploration.

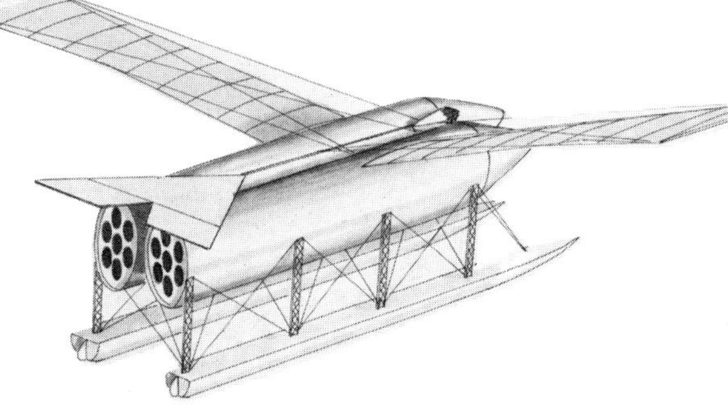

B. K. Armfeldt writes *A Leap into Emptyness*. In this story the professor of the Belorussian Academy describes the future flight of an enormous rocket to study the conditions of the upper atmosphere. The body of the ship consists of two parallel cylinders, each five or six times as long as an ocean liner. They are so large in diameter that a ship could easily steam within one. The spaceship is launched from the surface of the ocean. It is supported above the water by a virtual forest of iron braces and scaffolding attached to a pair of slender pontoons. From either side of the rocket's fuselage stretch two vast wings, canvas stretched over an iron framework. A web of control cables converge at the tiny cockpit located between the cylinders at the front of the rocket. The wings are only to support the rocket during its passage through the denser regions of the atmosphere.

The ship is fueled by a mixture of coal and gunpowder. Steering is accomplished by way of vanes placed in the exhaust of the rocket. The control cabin, carrying three passengers, is detachable and will descend separately from the main rocket by parachute. Enormous gyroscopes ensure the stability of the huge spaceship.

Its mission is to make an 1,800-mile flight at an altitude of over 300 miles. The entire duration of the flight is to be just one hour, and is to take place entirely over water, beginning near the French coast and finishing somewhere along the east coast of North America.

F. A. Radley publishes the novel *The Green Machine* in which rudimentary communication is established with Mars. Luminous signals are observed expressing the formula "2 × 2 = 4," inspiring an astronomer to travel to the red planet on the "Green Machine." This resembles nothing so much as a motorcycle. The pedals are made of a special alloy and act as brakes for the descent to Mars. Above the machine is an umbrella-like device of the same alloy, attached to the rudder. The power of the Green Machine is enormous, approaching perpetual motion. The rider must wear a spacesuit made of an elastic material.

The traveller rides to such an altitude that he enters the gravitational field of Philip's Comet, which carries him to Mars at a speed of about 1,000 miles per second. The return to the earth is made in the company of a Martian and with the aid of a giant gun. The projectile carries with it the Green Machine, which will be used for landing. The return flight takes 2 years and 10 days. Along the way the other major planets are visited.

Vivian Itin publishes the novel *The High Path* in which the element "Onteite" is used to propel a spaceship. Onteite is repelled by mass, rather than attracted to it. Graceful, dolphin-shaped airships made of a light metal have reservoirs of Onteite in their upper surfaces. The repelling force is controlled by iris-like diaphragms. "The conquerors of space" are carried to the limits of gravitation where a radioactive engine takes over.

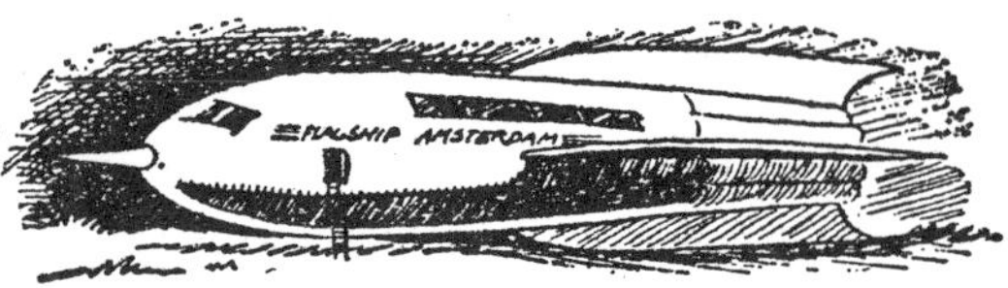

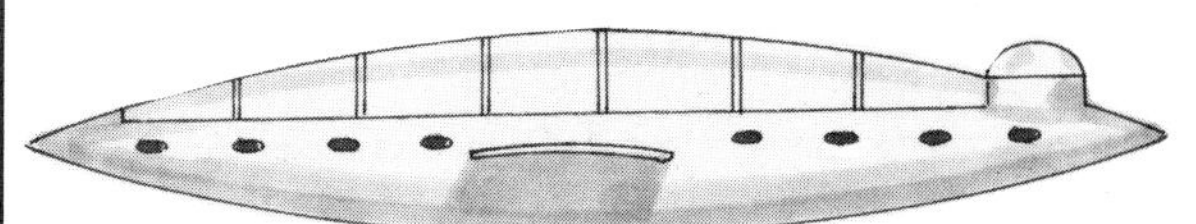

V. D. Nikolsky writes the novel *In One Thousand Years* in which a spaceship of the year 3000 is propelled by the ejection of disintegrated atoms. It is 30 meters long and shaped like a fish with a pair of short, thick wings. The upper half of the hull is transparent.

G. A. Polevoi displays a model spaceship at the Exhibition of Spaceflight in Moscow [see below]. It is to be launched from a "compressor-solenoid" tunnel at 1,600 meters per second to an altitude of 150 km. It is armor-covered to protect it against friction. This is jettisoned at 150 km. Only one of the several rocket nozzles is fired during the initial acceleration. When the armor is jettisoned, all the rockets are fired, increasing the speed to 11,000 km/s. The rocket would eventually descend via parachute.

APRIL

Henri de Graffigny, in *Une fusée Peri-Lunaire-Je sais tout*, makes a proposal for a centrifugally launched spacecraft. Instead of the rotating wheel that others have suggested [see 1889, 1913–1916 and 1916], the author of this scheme plans to build a circular, toroidal tunnel. It would be over 20 kilometers in diameter, with a railroad track laid inside. On the track runs a specially constructed truck, with skids in place of wheels. Oil under pressure is forced between the skids and the track, reducing the friction between the two by 90 percent. The truck is propelled by a linear electric motor; the fixed stator is between the rails along their full length and the moving rotor attached below the chassis of the truck. The projectile (described below) is carried on the truck. At one point the tunnel has a tangential branch line, the end of which is inclined upward.

As much of the air inside the tunnel is evacuated as is possible, and the projectile is started along the circular track. As soon as it has reached the required speed, it is switched onto the branch line. At the end of

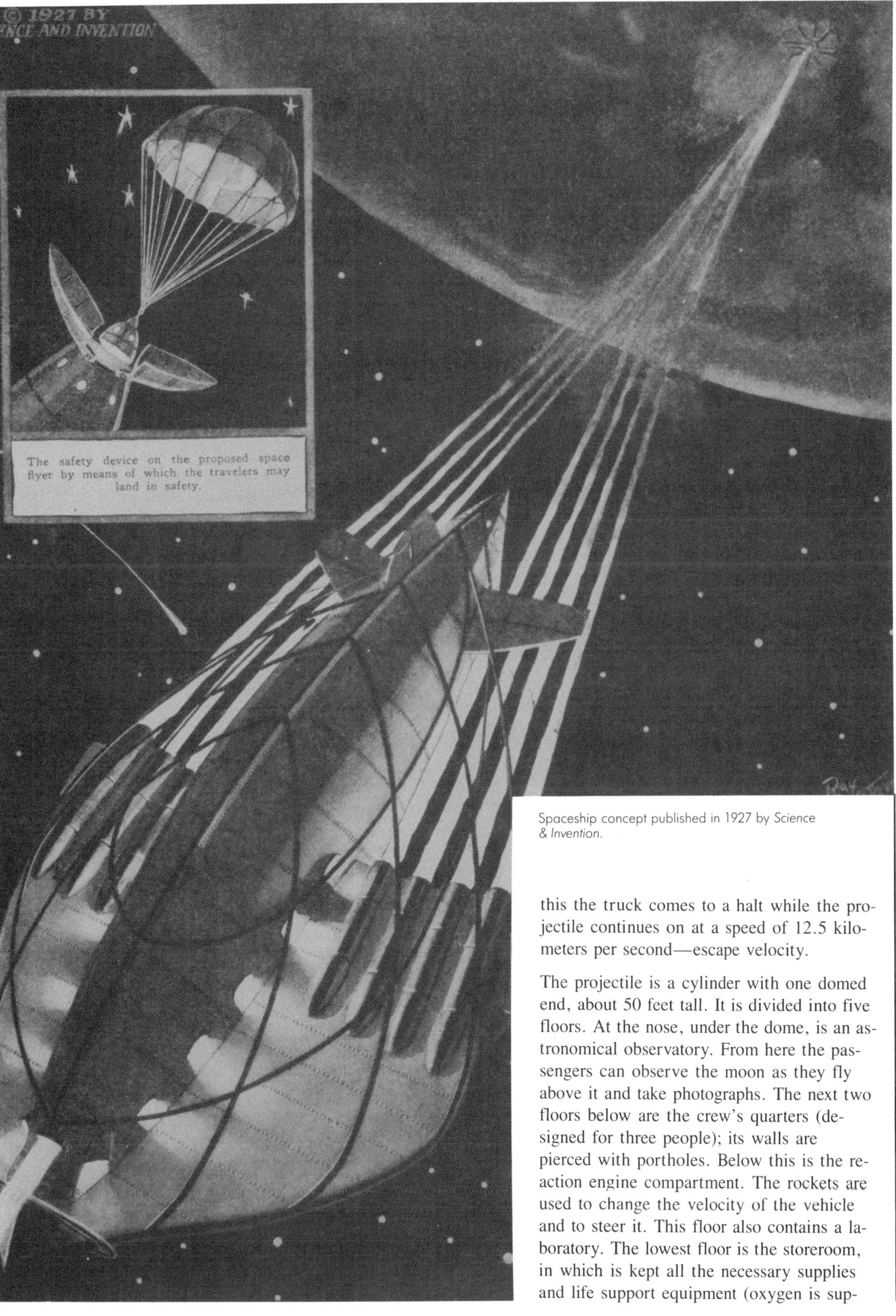

Spaceship concept published in 1927 by *Science & Invention*.

this the truck comes to a halt while the projectile continues on at a speed of 12.5 kilometers per second—escape velocity.

The projectile is a cylinder with one domed end, about 50 feet tall. It is divided into five floors. At the nose, under the dome, is an astronomical observatory. From here the passengers can observe the moon as they fly above it and take photographs. The next two floors below are the crew's quarters (designed for three people); its walls are pierced with portholes. Below this is the reaction engine compartment. The rockets are used to change the velocity of the vehicle and to steer it. This floor also contains a laboratory. The lowest floor is the storeroom, in which is kept all the necessary supplies and life support equipment (oxygen is supplied by evaporating the liquid gas). In the floor of this compartment are the outlets for the rocket motors. All five floors are connected by a ladder. The walls of the projectile are double and made of a special alloy of steel and aluminum.

The goal of the spaceflight is the moon. Eighty-three hours after launch the spaceship enters the gravitational field of the moon. By using the reaction motor, it goes into an orbit around the satellite. After examining the moon's surface by telescope, the motors are worked again and the vehicle begins the 7-day return to the earth. A safe landing is made by parachute and the firing of retrorockets.

APRIL-JUNE

A space exhibition is sponsored in Moscow by the Moscow Association of Inventors. Its full title is The First World Exhibition of Interplanetary Machines and Mechanisms. In spite of financial and bureaucratic difficulties, an extensive display of Russian and foreign inventions is arranged for public viewing. Included in the exhibition are displays relating to the work of Jules Verne and H. G. Wells, Fedorov, Kibalchich, Tsiolkovsky, Goddard, Oberth, Valier and Tsander, among others. The displays feature plans, drawings, photographs and models. The show is well attended and well received by critics and public alike.

JUNE

Max Valier publishes an article in *Discovery*: "Europe-America in Two Hours?" He discusses the development of the spaceship according to his evolutionary scheme [see 1926]. The transatlantic rocketship would start from "a very acute angle, 80 degrees, so that the thin air stratum be quickly reached. After seventeen seconds the ship is calculated to attain a speed of 400 meters per second at 3,000 meters high; after thirty-five seconds at 20,000 meters high, the rate of progress would be 800 meters per second; and after forty-five seconds at 50,000 meters over the sea and seventy kilometers horizontal distance from the starting-point, the hori-

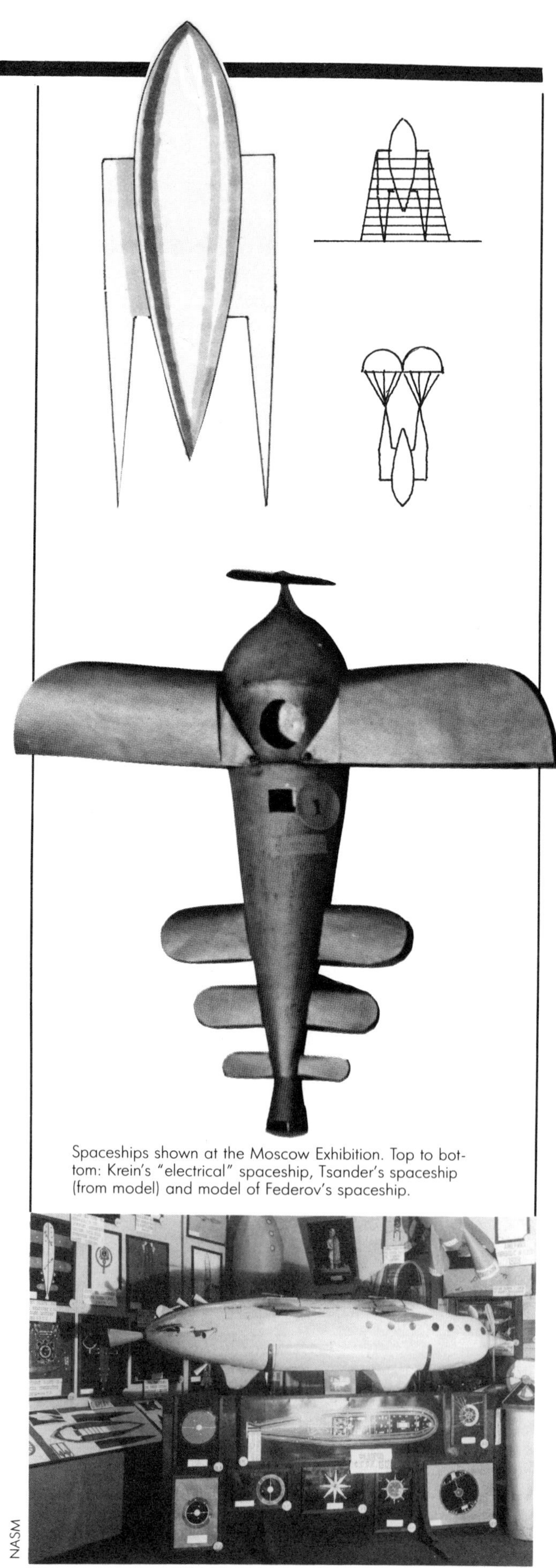

Spaceships shown at the Moscow Exhibition. Top to bottom: Krein's "electrical" spaceship, Tsander's spaceship (from model) and model of Federov's spaceship.

NASM

zontal speed would be 2,000 meters per second. At this rate New York would be reached in an hour and a half."

Eventually, Valier hopes, even greater speeds and altitudes would be reached. "As soon as a ship can be constructed giving a speed of 12,800 meters per second, it can escape, it has been calculated, from the earth; we shall be able to journey to the moon, land on its surface, and return again at will to earth; the moon may then become a colony for our world."

Charles Lindbergh had just returned to the United States after his epochal flight to Paris. *Literary Digest* comments that Lindbergh's feat will "look like the expedition of a snail across the street when Max Vallier [sic] . . . crosses the Atlantic in his 'space-ship' with rocket-propulsion in a matter of two hours."

JULY 5, 6:30 p.m.

The VfR (the *Verein für Raumschiffahrt*) is founded in Breslau (although the name of the organization is properly translated as the Society for Spaceship Travel, it is more popularly known as the German Rocket Society). There are at least ten people present at its founding in the parlor of the Golden Scepter tavern. Among those present are Johannes Winkler and Max Valier. Although one of the official founders, Willy Ley is absent from this meeting. The idea for such a club seems to have been Valier's, who had suggested it in a letter to Ley. He hopes that it will be able to raise the funds needed to finance experiments by Oberth. When Winkler applies for a charter at Breslau, the court at first refuses to accept the registration. The fact that there are just barely the legally required number of members is bad enough, but the court does not recognize the word "Raumshiffahrt," since the phrase "space travel" does not exist at this time in the German language. The court ultimately relents on the plea of the society that new inventions require new words.

The goals of the society, as expressed in their charter, are ambitious, if not grandiose: "The purpose of the union," it states, "will be that of small projects, large spacecraft can be developed which themselves can be ultimately developed by their pilots and sent to the stars." One of the Society's early slogans is "Help create the spaceship!" They hope that by careful management 200,000 marks might be raised in order to finance a spaceship.

An electromagnetic launching illustrated by the Römer brothers.

AUGUST

American Magazine reports that "Professor [William H.] Pickering is of the opinion that the only feasible method of getting to the moon is visually through the eyepiece of a good telescope."

SEPTEMBER

K. E. Tsiolkovsky publishes the booklet *The Space Rocket: Experimental Development*. In this he details the laboratory tests that need to be performed before spaceflight is possible. He publishes the layout of a static test stand for studying rocket motors (described and drawn up in more detail by Robert W. E. Lademann in *Z.F.M.* magazine, 1927).

Popular Science magazine reports on the visionary scheme of Max Valier to cross the Atlantic in little over one hour. Valier is quoted as saying that the risky flight would be preceded by experiments made with 7 to 10 foot models [see 1929].

1927

THE MOSCOW EXHIBITION OF INTER-PLANETARY MACHINES AND MECHANISMS

Above left: main entrance.
Above: Goddard corner (see p. 124 for origin of model).
Left: overall view of exhibition (left to right: Valier, Federov, Oberth and [bottom] Krein).
Facing page: Tsiolkovsky exhibit

All: NASM

Valier

1895
С. С. С.
ПРОФЕССОР Д-Р. МАТЕМАТИ
ЦИОЛКОВСКИ

OCTOBER 7

Pedro E. Paulet, a Peruvian engineer, writes a letter to a Lima newspaper, the *El Commerico*, in which he claims to have built and tested the world's first liquid-fueled rocket motor.

He alleges that while a student at the Institute of Applied Chemistry at the University of Paris, 1895–1897, he worked on an experimental rocket motor. Unfortunately, Paulet gives few technical details and Frederick I. Ordway, III, among others, is able to locate neither evidence nor witnesses that these experiments ever took place. James Wyld and George Sutton attempt to reconstruct Paulet's motor from the scant information available. Both agree on a cylindrical combustion chamber, 10 cm in diameter, with a broad, cone-shaped nozzle. Fuel (gasoline?) and oxidizer (nitrogen peroxide?) are fed into either side of the cylinder and ignited by a spark plug. Thrust is supposed to have reached 60 kg.

In 1965, the same newspaper will publish drawings of a rocket-propelled aircraft that Paulet was supposed to have designed in 1902 [see above].

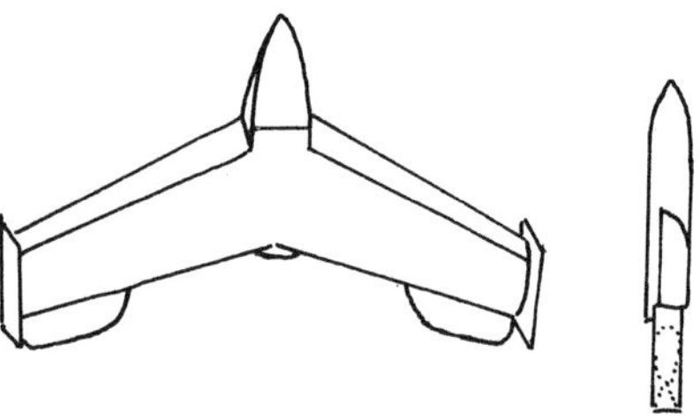

NOVEMBER 23

Experiments are performed with a rocket-propelled model plane. The tests are carried out near Breslau and the Rhône. The model is a biplane flying wing weighing just over one pound. The wing span is initially about 6 feet, but this is eventually reduced to about 5 feet. The rocket engine weighs a little over a quarter of a pound. The plane makes a 10 second flight, making a loop before gliding to a landing.

DECEMBER 15

Die Rakete, the journal of the VfR [see above], publishes an estimate of the cost of building a spaceship of one ton for a flight to the moon: 3,350,000 marks, or slightly more than the cost of a Zeppelin airship.

DECEMBER 26

J. J. Rosny *aîné* (Joseph-Henri-Honore Boex) coins the word "astronautique."

ca. 1927

A seven reel motion picture (title and origin unknown—possibly *In the World of the Planets and Stars* or *Our Heavenly Bodies*, German: UfA, written by Kornblum and Kruger) depicts a flight to the moon and Saturn in a spherical, electrically propelled spaceship. Scenes of the earth as seen from the moon and Saturn's rings seen above its clouds are extremely well done. The weightlessness of the crew on board the spaceship, after it has left the earth, is also shown.

1927–1930

Max Valier conducts his experiments with rocket-powered cars.

1927–1931

The Soviet solid fuel rocket laboratory [see 1921] is moved from Moscow to Leningrad where it becomes, in 1928, the Gas-Dynamic Laboratory (GDL) of the Revolutionary Military Council (then the Soviet ministry of defense).

On May 15, 1929 a jet/rocket department is formed in the Leningrad GDL. Here the first Soviet liquid fuel rocket motor is built and tested in 1931. This is the Experimental Reaction Motor ORM-1. Fueled by benzine and liquid oxygen it produces a thrust of 20 kg.

1928

Frau im Mond, the seminal space travel film of the prewar period, is directed by Fritz Lang. This long (over 3 hour) epic details the construction and flight of a rocket to the moon, the *Friede*, the expedition being made to discover the lunar gold that is believed to exist in abundance. The rocket has been designed by Hermann Oberth, who also acts as the film's technical advisor. He bases the movie rocket on his Model E [see 1929]. Other than the serendipitous discovery of a breathable atmosphere on the moon, *Frau im Mond* is the most authentic depiction of spaceflight to reach the screen until *Destination Moon* in 1950. When it is first shown in Britain it is described as scaring the members of the Foreign Office "witless"; eventually prints are withdrawn from distribution by the Gestapo, who also seize the 5 foot metal model of the spaceship.

International Newsreel photograph

BUILDING THE ROCKET TO CONQUER SPACE

Here we see Dr. Oberth and his aids in the workshop building the projectile that may perhaps fly to America, the moon, or anywhere its fancy may lead it.

Robert Esnault-Pelterie publishes *L'exploration par fusées de la tres haute atmosphere et la possibilité des voyages interplanetaires* (The Exploration of the Upper Atmosphere by Rockets, and the Possibility of Interplanetary Flight). This ninety-six-page treatise is derived from a paper delivered to the General Assembly of the Société Astronomique de France on June 8, 1927. [Also see 1930.]

Left, background model under construction; right, original filming model of the *Friede*.

Facing page, cutaway model of Oberth's movie spaceship; below, view of nozzles and open-ended box-like wings; bottom, close-up of top stage and cabin.

Jean Labadié describes the possibility of a flight to the moon in *La Science et la Vie*. He discusses most of the current suggestions for launching a vehicle: giant cannons, centrifugal force and so on. He concludes that only rockets have the capability and suggests the use of a streamlined step-rocket "whose ejection compartments get smaller and smaller as the combustion goes on."

He considers the question of an appropriately powerful fuel and mentions several, including liquid atomic hydrogen and nuclear energy.

Labadié's spaceship is a squat, bullet-shaped vehicle (its proportions are 1:4) divided into three main sections: an observatory equipped with a telescope, an electrical generator and combustion chamber, and at the bottom, eight nozzles.

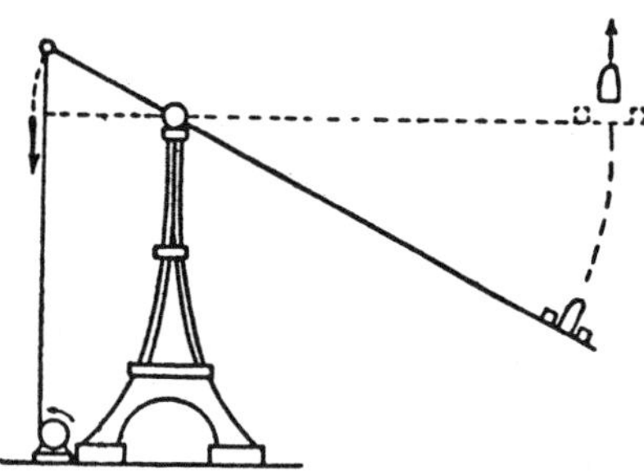

N. A. Rynin describes a theoretical "hurling machine" capable of imparting escape velocity to a projectile. A tower 300 meters high (such as the Eiffel Tower) would have a lever attached to its top. One arm of the lever would be 300 m long and the other 30 m. If the shorter arm were depressed at a speed of 1.2 km/s, the right-hand arm would raise at a speed ten times greater, or 12 km/s. Rynin rejects such a scheme as being impossible and impractical.

E. E. Smith publishes the *Skylark of Space* (written 1915–1919). It appears as a serial in *Amazing Stories* beginning in August. Richard Seaton accidentally discovers a new element which he calls "X." It has a high atomic weight, yet has no place in the periodic table. In conjunction with high-tension electrical fields it acts as a catalyst for the atomic decomposition of copper and other metals, including uranium. Seaton uses this new source of energy to construct a spaceship: the *Skylark of Space*. It is a 40-foot sphere with a 4-foot thick armored hull and several decks. There are four cabins above the main deck and four below. The *Skylark* is powered by a 400-pound bar of copper, giving it a speed approaching that of light.

Skylark Three follows in 1930. Seaton's new spaceship is 2 miles long and 1,500 feet in diameter, employing uranium rather than copper as a fuel.

This is followed by a sequel, *Skylark of Valeron*, in 1934–35. This new spaceship is intended for intergalactic travel and is 1,000 kilometers in diameter. At the center of this vast ship was an open, grassy area with replicas of the hero's terrestrial homes, including an artificial sun, breezes, streams, insects and all.

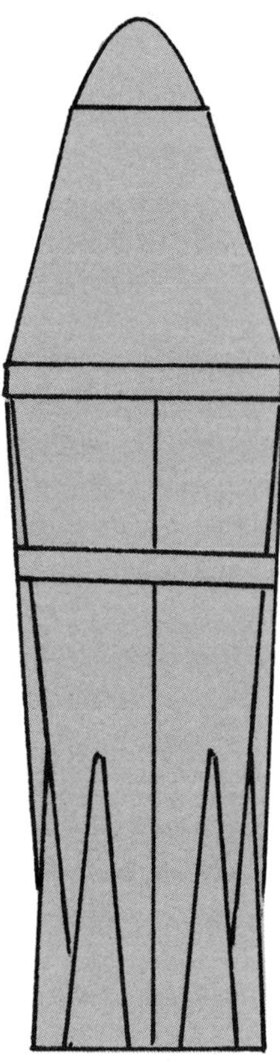

Robert Condit builds a rocket for a flight to Venus. His spaceship is a 24-foot-long bullet made of angle iron and sailcloth. It is constructed with the aid of brothers Harry B. and Sterling Uhler of Baltimore—where the launch is to take place. The rocket is fueled with 50 gallons of gasoline and has eight steel pipes for engines. The several layers of sailcloth that cover the rocket are impregnated with varnish making an airtight shell "as brittle as glass." The nose section unscrews to allow the rocket's single passenger ingress. Inside is a large tank of oxygen, a supply of concentrated food tablets and water in 1.5-inch pipes that line the interior to save space. There are also a "couple [of] flashlights and a first aid kit, and that was it." There are two glass portholes, though there is no way to steer the rocket. They plan to hit Venus by taking very careful aim at takeoff. In the nose is a 25-foot silk parachute that the pilot could push out in order that the rocket could make a safe descent.

Inside is an air compressor run by a gasoline motor. This sprays vaporized fuel into the steel tubes. A spark plug in each, attached to a battery in the ship, will keep the gasoline burning.

All aboard for Venus. Professor Robert Condit, chemist and scientist, devised this machine, in which he expects to shoot himself to the planet Venus. (P. & A. Photo.)

Motive power will be supplied by a central explosive chamber and polarized magnetic controls are expected to guide the rocket to Venus after it leaves the gravitation influence of the earth. (P. & A. Photo.)

NASM

It is estimated (by Condit, described by Harry as "a mathematical wizard") that if the ship could get off the ground, travelling at 25,000 mph, it would "pull out of the earth's gravity about 40 miles up and coast right on over to Venus."

The rocket takes 8 months to build. It is fueled and set up on a sidewalk on Morling Avenue. Handsome, athletic-looking Condit crawls inside and starts the engines. It is his plan to only take the rocket up a quarter of a mile or so "until he got the feel of her." In the words of Harry Uhler, "I never saw so much fire in my life."

The Venus rocket never leaves the ground, though it does succeed in drawing a substantial crowd. The erstwhile astronauts decide that what they need is a booster and, as they figure that would cost them at least $10,000, they give up the entire project. "Our wives were against the whole deal, anyway," laments Harry. Condit returns shortly thereafter, loads the rocket onto a truck and heads for Florida—where he disappears.

(Apparently Condit, described as a "Miami chemist," had built an earlier rocket in his home state. His goal had still been Venus, and the rocket was to have been guided there by "polarized magnetic controls.")

Otto Willi Gail publishes *Mit Raketenkraft ins Weltenall* (With Rocketpower into Space). It is a popular account of the state of rocketry and astronautics in Germany. [Also see 1926, 1928, 1929.]

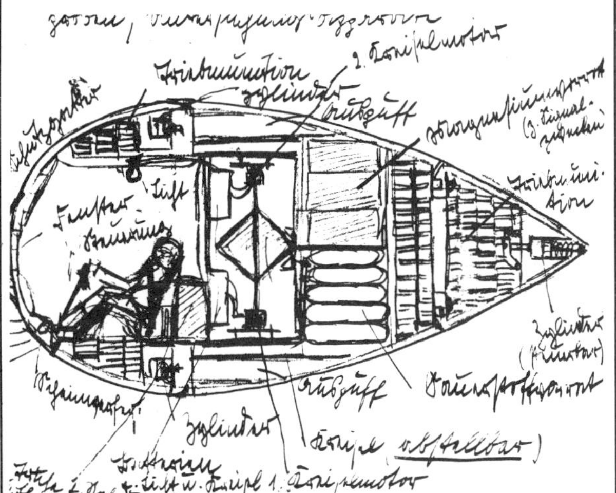

Ordway

Sixteen-year-old Wernher von Braun designs his own "rocket wagon," inspired by the experiments of Fritz von Opel.

Max Valier proposes (in *Welt am Sonntag*, July 1) his Type 8 rocket-propelled aircraft, an all-metal streamlined monoplane. It is one of the last stages in his evolutionary develop-

1928

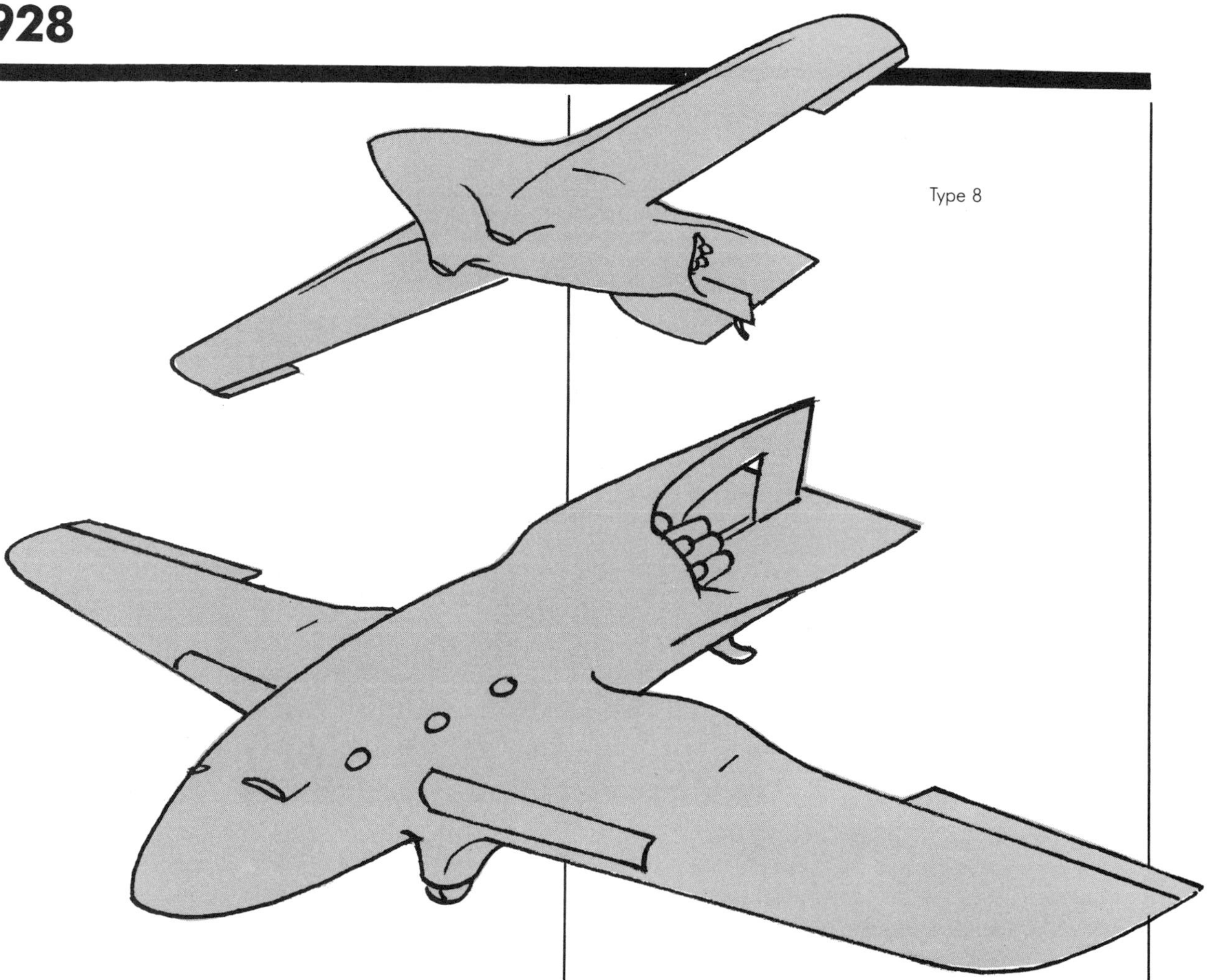

Type 8

ment of an all-rocket-propelled vehicle. Although the Type 8 is equipped with four piston-driven propellors, in a tandem pusher-puller arrangement, they are retracted and covered with sliding doors when flying at altitudes where the rocket engines are more efficient. The rocket engines are mounted in the tail of the machine, flanked by four stabilizing fins. The craft carries two pilots as well as passengers in a sound-proofed, pressurized cabin with a kitchen.

Type 9 is a flying wing all-rocket transport, not dissimilar to the Junkers G-38 or the "Storch" designed by Lippisch. Passenger compartments are to be located in the thick wing. This design is featured on the cover of his book *Raketenfarht* (1928).

Valier also discusses a Lippisch-inspired rocket-powered parabolic flying wing.

Aviation pioneer Augustus Post publishes his design for a reaction-powered aircraft. It would be "fueled" by liquid air, heated and allowed to expand into a gas. Nozzles in the tail of the craft would provide propulsion,

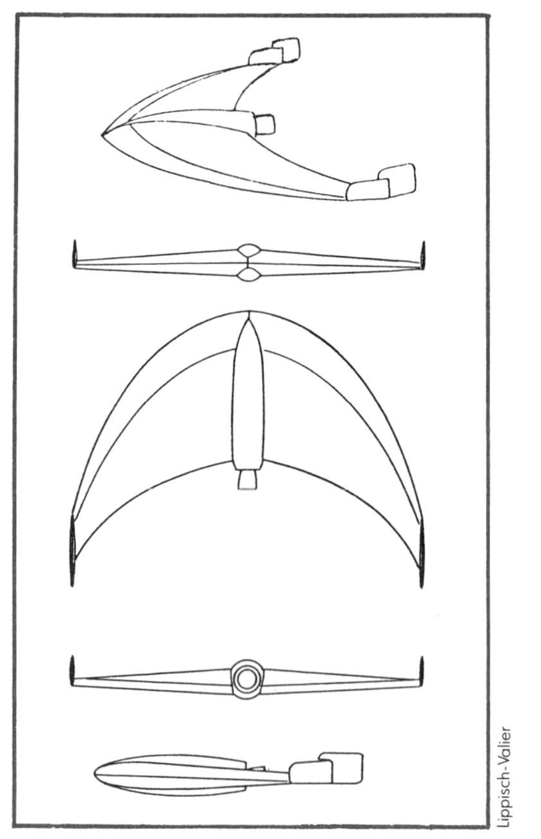

Lippisch-Valier

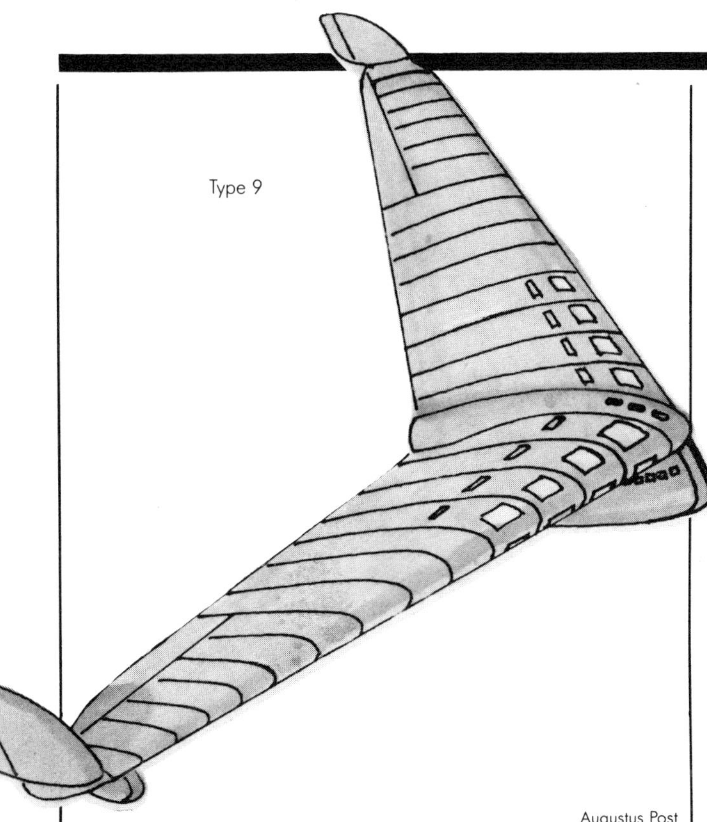

Augustus Post

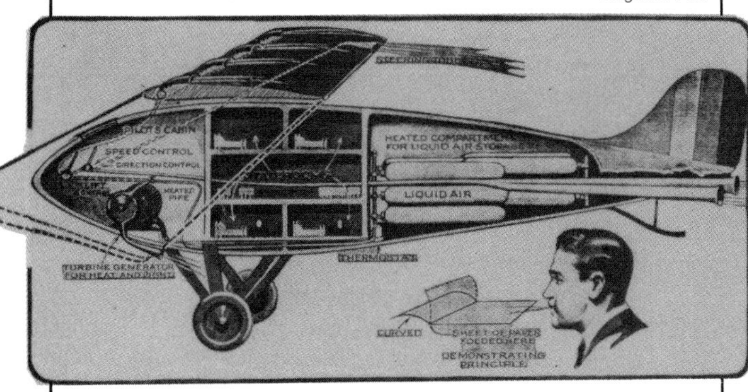

while a series of smaller outlets on the upper part of the wing's leading edge would provide additional lift. Wingtip nozzles would aid in steering. Post expects that there would eventually be stratospheric passenger flights in such aircraft, with pressurized compartments, flying at speeds of up to 500 mph (or even 1,000 mph, according to Canadian Royal Air Force Major General J. H. McBrien).

WINTER

Earl L. Bell publishes the short story "The Moon of Doom" in *Amazing Stories Quarterly*. The earth's moon has approached so near to the earth that our planet's atmosphere is being drawn off. Hermetically sealed aircraft (called "atoliners" and "atoplanes") are able to make the journey be-

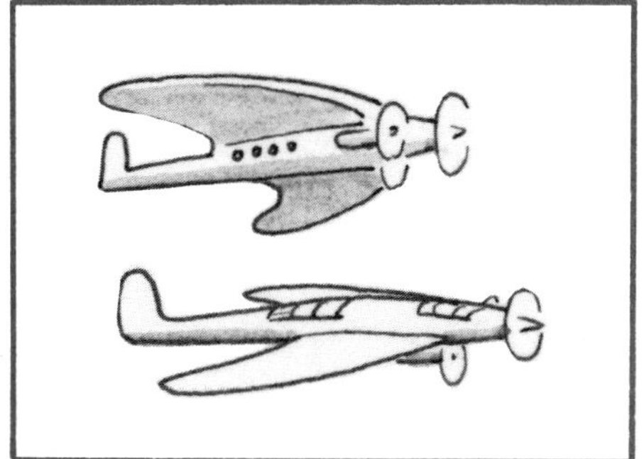

tween the two worlds by flying in the thin air bridge that connects them.

SPRING

Willy Ley publishes *Die Moglichkeiten der Weltraumfahrt* (The Possibilities of Space Travel). It is an anthology of contributions by members of the Verein für Raumshiffahrt [see 1927]. Among these are Ley, Hermann Oberth ("Basic Problems in Space Travel" and "Stations in Space"), Franz von Hoefft ("Fuels for Spaceships"), Walter Hohmann ("Routes, Travel Times and Landing Sites"), Karl Debus ("Spaceflight and the 'Habitability Fantasy' in Literature from the Renaissance to the Present Day"), Guido von Pirquet ("Impossible Ways to the Realization of Spaceflight") and Friedrich Sander.

Nikolai Alekseevich Rynin begins publication of his nine-volume history of astronautics, in Leningrad, completed in 1931 [see below].

SUMMER

Eugen Sänger submits his doctoral thesis to the Technical High School in Vienna. In it he investigates the problems of high-altitude rocket plane flights. He writes that his study compares " . . . the different ways of advancing into space; it calculates the most economical and safest method (aerospace transporter—space station—space ship) and supplies a complete theory of this method . . . The conquest of space with minimum energy display will proceed according to the following principles:

1. Transport to the altitude of a space station by means of the special aerospace-

1928

plane; further advances into space will use modified spacecraft.

2. Ascent of the aerospace-plane according to the principle of minimum energy expenditure.

3. Descent of the craft as a glider, without energy expenditure."

However, Sänger is told by a teacher that if he tries to get his doctoral degree in spaceflight, he would most likely be "an old man with a long beard before you have succeeded in obtaining your doctorate." So Sänger submits a more conventional paper for his degree.

Two journals discuss the potential use of rocket propulsion in conjunction with vehicles and spaceflight. *The Journal of the American Society of Mechanical Engineers* comes to the conclusion that ordinary powder rockets would be enormously inefficient, giving only 1/10th the energy of a like amount of gasoline or alcohol. Any sort of rocket motor would be inefficient operating at low speeds. It would only be when 1,000 mph aircraft are required that the reaction motor would be needed. For "inter-stellar [sic] navigation" propulsion would mainly be required to escape the earth's gravitation and hence "the light, comparitively short-lived rocket is our obvious method of attack."

The *Hamburger Nachrichten* reports on the rocket automobiles being tested by Fritz von Opel and Max Valier. The car is described as "a low, lightly constructed racing car without a motor. The rear end is a steel box or case, with 12 round openings from which project the blast pipes of Congreve [sic] rockets, each having a diameter of nine centimeters. The ignition wires run from the blast pipes to an automatic switch operated by the pilot by means of a foot pedal. Supporting stays are attached invertedly to the car behind the front wheels and have for the purpose the pressing of the car to the track during the run." The first test is made by the experienced racing driver Volhart. The car attains a speed of 60 mph in 8 seconds, an acceleration "equal to that of the powerful airplane catapult."

Scientific American of September 1928 observes that as impressive as this is no new principles of rocket design are involved.

FEBRUARY 1

Robert Esnault-Pelterie and Andre-Louis Hirsch establish the 5,000 franc REP-Hirsch International Astronautics Prize—sponsored by the Sociétè Astronomique de France—for the most original scientific work toward the advancement of space travel. The work has to be of a scientific nature, dealing with any of these subjects: astronomy and ballistics; physics (atomic theory, interplanetary communication, energy storage, etc.); chemistry (life support systems, preparation and storage of monatomic hydrogen, etc.); mechanics (construction of spaceships, control and guidance, etc.); metallurgy (superlight alloys, etc.); and physiology. The first recipient of the award is Hermann Oberth for a manuscript he submits. The prize money enables him to publish it as his first book [see 1929]. "It is reassuring, "Oberth writes in

DIE MÖGLICHKEIT DER WELTRAUM-FAHRT
GUNDERMANN
ALLGEMEIN-VERSTÄNDLICHE BEITRÄGE ZUM RAUMSCHIFFAHRTS-PROBLEM
HERAUSGEGEBEN VON WILLY LEY

his book, "to see that science and progress suffice to overcome national prejudices. I can think of no better way to thank the Société Astronomique de France than to pledge myself to work on behalf of science and progress and to judge people only on their personal merits." Surprisingly, this paragraph is allowed to remain when the book is later reprinted in Nazi Germany.

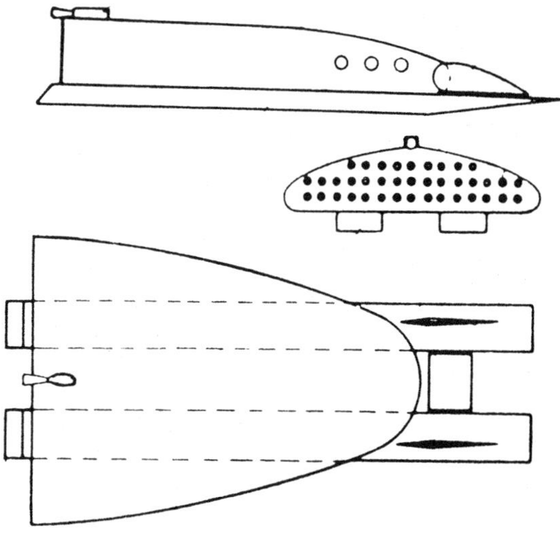

FEBRUARY 9

Franz von Hoefft develops staged, winged and lifting body spacecraft. Using standardized units, spaceships could be customized for particular missions.

Hoefft proposes a systematic and progressively more ambitious scenario for the exploration of outer space. He submits his plan in a report delivered in Breslau. In it he plans a series of eight spacecraft, designated RH I-VIII (**R**akete-**H**oefft).

RH-I is a recording rocket. It is 4 feet long with a diameter of 7.8 inches and fueled by alcohol and liquid oxygen. It would be carried to an altitude of 6.2 miles by balloon. At this point it would be released automatically, continuing on to an altitude of 62 miles. It would be automatically stabilized by a gyroscope and it would descend to earth via parachute.

RH-II is similar to the RH-I, but fueled with gunpowder.

RH-III is a two-stage rocket weighing 3 tons. The second stage carries a payload of 11 to 22 pounds of flashpowder. It is planned to impact this on the moon, observing the flash from the earth (see the similar plans made by Robert Goddard). In addition, the rocket would circle the moon, taking pictures, eventually returning to the earth.

RH-IV is similar to the RH-III and is intended to operate as a transcontinental mail rocket.

These last two are also intended to get their initial lift by way of a balloon, launching from an altitude of 3.72 miles, or they are to be launched from the top of a lofty mountain.

The RH-V is to be launched from a body of water. The rocket is a thin, flat shape—something like a shovel blade or shoe heel—with a slightly curved upper surface and a flat bottom. It is nearly 40 feet long, 26.24 feet wide and 5 feet thick, weighing 30 tons. It carries a crew of four. It would take off from a position of near-submersion, only the front end of the rocket protruding from the water at a low angle. Landing would be onto a water surface as well. After reentry into the atmosphere at a speed of about 7.75 mps, the rocket would turn perpendicular to its flight path so that the air would brake it. When the speed dropped to subsonic, the pilot would turn the ship's nose down, gliding on to a landing on the water, at 114 fps.

The RH-V is also the final stage of the RH-VI, VII and VIII rockets. In this form, flights to the moon and planets would be possible. The RH-VII is 106 feet long, 70.8 feet wide and 13.12 feet thick, weighing 600 tons at takeoff. Its first stage would be piloted and returned to the earth to land. A third stage is added in the form of the RH-VIII, whose takeoff weight is 12,000 tons.

JUNE 11

Friedrich Stamer is the first man to fly in a rocket-powered aircraft, the *Ente* (Duck). The experiment is carried out on the initiative of Max Valier, F. Sander and F. Opel. The *Ente* is one of the new canard-type gliders, that is, the tail surfaces are ahead of the

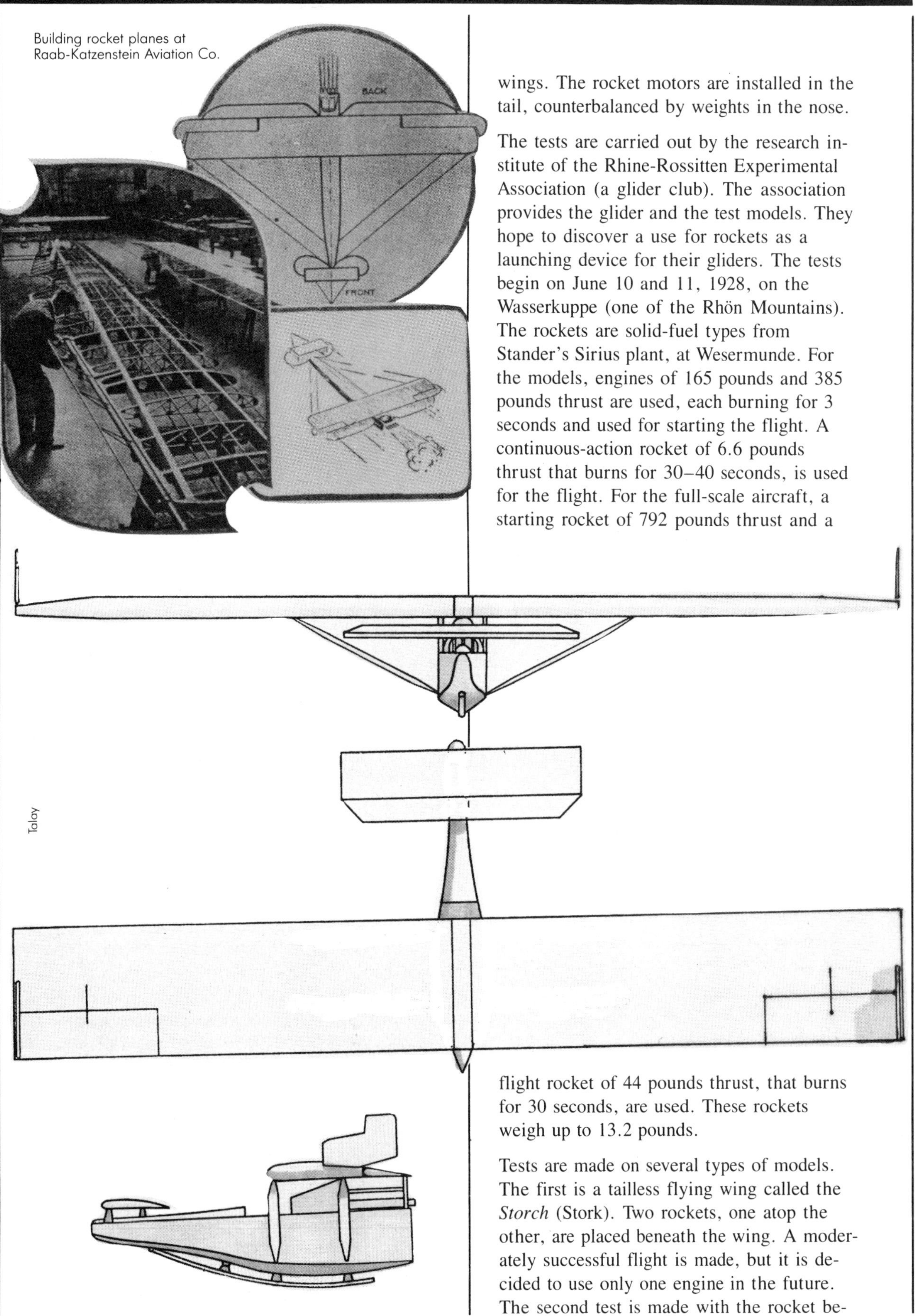

Building rocket planes at Raab-Katzenstein Aviation Co.

Talay

wings. The rocket motors are installed in the tail, counterbalanced by weights in the nose.

The tests are carried out by the research institute of the Rhine-Rossitten Experimental Association (a glider club). The association provides the glider and the test models. They hope to discover a use for rockets as a launching device for their gliders. The tests begin on June 10 and 11, 1928, on the Wasserkuppe (one of the Rhön Mountains). The rockets are solid-fuel types from Stander's Sirius plant, at Wesermunde. For the models, engines of 165 pounds and 385 pounds thrust are used, each burning for 3 seconds and used for starting the flight. A continuous-action rocket of 6.6 pounds thrust that burns for 30–40 seconds, is used for the flight. For the full-scale aircraft, a starting rocket of 792 pounds thrust and a flight rocket of 44 pounds thrust, that burns for 30 seconds, are used. These rockets weigh up to 13.2 pounds.

Tests are made on several types of models. The first is a tailless flying wing called the *Storch* (Stork). Two rockets, one atop the other, are placed beneath the wing. A moderately successful flight is made, but it is decided to use only one engine in the future. The second test is made with the rocket between the wings, but the model turns out to be unstable. The third test is with a redesigned plane. This is first tested for stability by launching it with a rubber band. It is then launched with an 11-pound thrust rocket, with a rubber band start. It flies successfully, making a smooth landing. The engine is upped to 385 pounds of thrust. It takes off "as if shot out of a gun." The test is a success, the model gliding in for a smooth landing.

These tests spur the experimenters to go on to a full-scale manned aircraft. It is fitted with two rockets, ignited by the pilot electrically. The first flights are made with two rockets of 26.4 and 33 pounds thrust, respectively. A rubber rope is used to launch the plane. The first test is unsuccessful, the plane not rising from the ground. The rockets are increased to 33 and 44 pounds thrust. The plane takes off easily with the aid of the giant rubber band. However, the 44-pound engine does not fire and the plane flies at an angle, landing after a flight of only 656 feet. The third trial uses two 44-pound rockets. The plane takes off smoothly under the power of the rubber rope and one rocket. After flying about 650 feet, the pilot turns 45 degrees to starboard, flies another thousand feet and makes a second 45 degree turn to starboard. He fires the second rocket, the first having burned out. This propels the *Ente* another 1,640 feet. Stamer turns 30 degrees to starboard and lands after 650 feet, when the second rocket has burned out. The total length of the flight is from 4,200–4,920 feet, lasting from 40–80 seconds. Stamer finds the takeoff imperceptible and the thrust uniform; flying the glider is pleasant without the vibration and torque of an engine.

The *Ente* is destroyed in its second trial. One of the rockets explodes, setting fire to the plane. Stamer lands safely, but the glider burns completely.

Scientific American (September) observes that "On the whole we are inclined to think that the rocket as applied to the airplane might be a means of securing stupendous speeds for a short interval of time, rather than a method of very speedy sustained flight."

JULY

Max Valier publishes an article in *Discovery*: "Can We Fly to the Stars?"

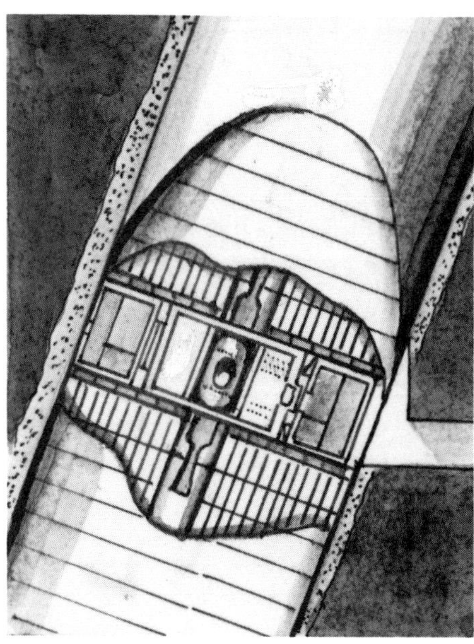

AUGUST

James R. Randolph, writing in "Can We Go to Mars?" for *Scientific American*, says that such a journey would require a rocket having several thousand cylindrical chambers, each with a separate nozzle. These would be fired consecutively from bottom to top. As each layer of motors is exhausted it would be jettisoned. There would be a pair of liquid fuel retrorockets for fine maneuvering. The passenger cabin would be rotated to provide gravity and a gyroscope would provide stability. Just above and below the cabin would be solid fuel rockets designated for slowing the rocket when reaching Mars, for leaving Mars orbit and for slowing down upon its return to the earth. The firing of all the rockets, especially during the launch, would be handled automatically by a switchboard. The complete ship, which would be launched from an inclined tunnel, would weigh as much as an ocean liner.

The rocket is not designed to land on either the earth or Mars. Instead, it would go into orbit around the latter for one year. Descent would be made via a small, collapsible air-

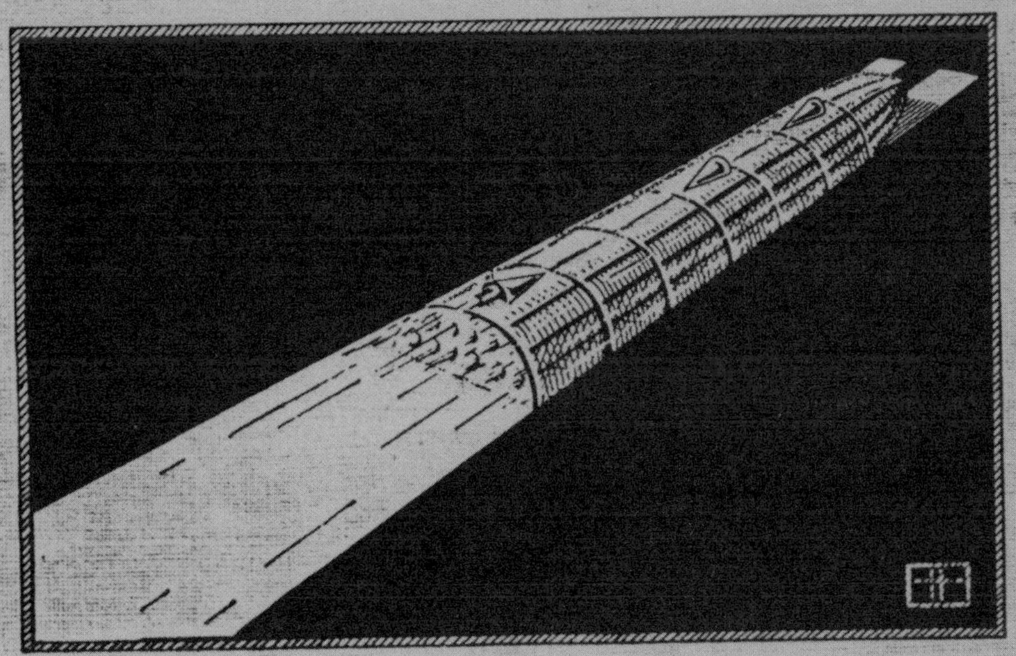

Original cover of A. B. Scherschevsky's book, *Die Rakete* (1928).

tight glider, with folding wings. Return to the earth would be made the same way.

ca. 1928

There is an unconfirmed report that the French Ministry of War is secretly building a rocket-propelled aircraft. Its speed is thought to be in the neighborhood of 370 mph.

B. Lobach-Zhuchenko describes a Tsiolkovsky-style spaceship in his book *Air Communication and Flights Over Seas and Oceans*. In this book, he writes of the flight of a future spaceship, the *Tsiolkovsky 20*. "The long, silvery, cigar-shape of the airship *Tsiolkovsky 20* appeared on the horizon carrying 500 passengers at a speed of 500 km/hr [310 mph] . . . This is its design: it embodies the rocket principle in the form of several reaction engines. The direction of the reaction can be changed by turning the engines, and this changes the direction of the flight. The reaction generally has an inclined direction and thus provides both lift and forward motion; it creates lift only if its direction is vertical.

"The rocket engines are located at several points: at the stern, on the sides, and at the bottom of the airship; two engines are even fitted at the bow in order to permit motion astern during maneuvering . . .

"It is superfluous to add that the appointments of the airship are the last word in engineering: loudspeakers, radio-telephone, television, etc., all for the convenience of the passengers."

Aleksander Borissovitch Scherschevksy describes a flying-wing rocket glider. Shershevsky is a destitute Russian aviation student who has been supporting himself in Berlin by writing articles for aviation magazines. It is through these that Hermann Oberth becomes acquainted with him, asking the young man to join in helping him to build the UfA rocket [see *Frau im Mond*, 1928]. He is, in Oberth's words, "The lazi-

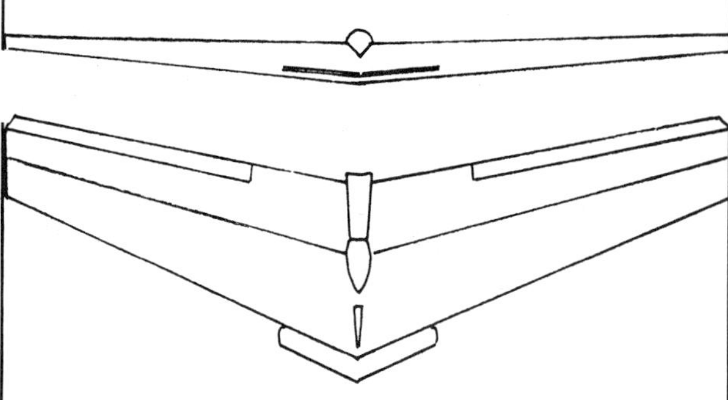

est man I had ever met."

Jean Labadié describes a spaceship in his article "From the Earth to the Moon via Rocket" in *Science et Vie* (reprinted in *Science and Invention*). It is a squat, bullet-shaped vehicle (its proportions are 1:4) divided into three main sections: an observatory equipped with a telescope, an electrical generator and combustion chamber, and at the bottom, eight nozzles.

1928–1932

Nikolai A. Rynin begins issuing his *Interplanetary Communications*, an encyclopedia in nine volumes. They contain virtually everything that is known or speculated at the time concerning spaceflight, rocketry and astronautics, including history legends and fiction. The nine volumes are titled: I. *Dreams, Legends, and Early Fantasies*; II. *Spacecraft in Science Fiction*; III. *Radiant Energy: Science Fiction and Scientific Projects*; IV. *Rockets*; V. *Theory of Rocket Propulsion*; VI. *Superaviation and Superartillery*; VII. *K. E. Tsiolkovsky: Life, Writings and Rockets*; VIII. *Theory of Spaceflight*; IX. *Astronavigation: Theory, Annals, Bibliography and Index*.

Rynin is dean of the Department of Air Communications at the Leningrad Institute of Railway Engineers, where he has established a Section of Interplanetary Travel. Members are instructors, engineers and students.

It is unfortunate that these volumes are not readily available outside the Soviet Union (an English translation will not appear until 1971, as part of NASA's Technical Translation series). In compact, logically laid-out format Rynin presents in either condensed form or verbatim the most important work

Rocket-powered cars built in 1929 by Daniel and Floyd Hungerford of Elmira, N.Y. They are propelled by gasoline/oxygen motors.

of virtually every important theorist who has worked on the problem of space travel up to this time. In addition, Rynin devotes much space to the history of rocketry and space travel in fact, legend and literature.

Volume I is an overview of the concepts of flight and space travel in early myths, folklore and legend.

Volume II, *Spacecraft in Science Fiction*, considers in some detail the works of Jules Verne and others. The book is divided according to the types of propulsion proposed: giant guns, rockets, antigravity, etc.

Volume III deals with proposed projects—real and imaginary—to communicate across space with extraterrestrials. In the last chapter of this volume, Rynin states that interplanetary travel will become a necessity within two or three centuries, demonstrating mankind's increasing numbers and the earth's decreasing resources as the motivating factors.

Invention of the rocket car according to *Judge* (March 15, 1930)

Volume IV, *Rockets*, begins with an explanation of the principle of reaction and continues with a history of the development of the rocket up to the time of the European experimenters, such as Winkler and others. The remainder of this volume deals with the mathematical theories relating to rocket propulsion developed by various authors, the final chapter being devoted to the subject of the rocket in interplanetary space. The schemes of Valier, Oberth, Ganswindt and others are described in detail.

Volume V, *The Theory of Rocket Propulsion*, was originally a paper presented to the *Proceedings of the Leningrad Institute of Road Communication Engineers* in 1929. Its first two sections introduce the concept of rocket-propelled vehicles and the following sections deal with the various formulas and equations relating to rocket motion, air resistance (which Rynin considers only in the context of leaving the earth, rather than as problem relating to reentry), fuels (including coal, gunpowder, dynamite, but also considering metals and monatomic hydrogen), the physiological effects of acceleration, etc. The second half of the volume outlines the mathematical theories of Tsiolkovsky, Hohmann, Esnault-Pelterie, Oberth and Goddard.

Volume VI has little of interest concerning either rocketry or manned rockets, though the rocket airplane is discussed.

Volume VII is a biography of K. E. Tsiolkovsky—including a brief "Autobiography" by Tsiolkovsky himself—as well as a complete bibliography of his works.

Volume VIII discusses in great detail the theories pertaining to spaceflight developed by Robert Esnault-Pelterie, Robert Goddard, Oberth and several others, usually quoting at great length from the original works.

Volume IX contains an index and bibliography.

1929

Willy Ley publishes an eighty-three-page booklet, *Die Fahrt ins Weltall* (Flight into Space), explaining in popular terms the work of Oberth, Hohmann and others. It is an expansion of his 1926 edition.

Jakov I. Perelman publishes *Interplanetary Travels* (in Russian), a bestseller that sells 150,000 copies—more than all English-language books on the subject combined, prior to World War II. The book is an expansion of a report originally prepared in 1913.

V. P. Glushko suggests the electrothermal rocket engine. The spaceship would use solar energy to generate electrical power by way of solar thermal elements placed on a disc surrounding the craft. When large amounts of electrical energy are fed into a wire, the metal explodes and a gas is instantaneously formed, flying off at a high speed. Glushko's engine capitalizes on this effect. Glushko is granted a patent in 1930 for an electrothermal rocket engine.

Hermann Oberth publishes his seminal work *Wege zur Raumschiffahrt* (Ways to Spaceflight). This book, an expanded and revised edition of his unexpectedly popular *Die Rakete zu den Planetenraum* [see 1923], is quite possibly the most influential single work on spaceflight ever published. Oberth covers in amazing detail virtually every aspect of the problem, from engineering to ballistics to space medicine.

The manned spacecraft he proposes is the "Model E," a blunt, bullet-shaped rocket 35 m tall and 10 m in diameter. It is fitted with the four large fins equipped with steering vanes that characterize most of Oberth's designs at this time. [Also see *Frau im Mond*, 1928.] It is a two-stage rocket. The first stage is fueled with alcohol, the second stage with hydrogen. Liquid oxygen is the oxidizer for both. (In a footnote, Oberth claims that with improved efficiency the spaceship's dimensions could be reduced to 17 m by 7 m.)

The nose of the second stage contains a small passenger compartment. At the maximum altitude, this separates from the main rocket, remaining connected only by electrical wires. After separation from the second stage, the two concave halves of the nose act as mirrors to focus sunlight onto the cabin, which is made of aluminum 1.5 to 2.5 cm thick. Oberth's drawings of his rocket show the cabin as being extremely small; Rynin calculates that it must be no less than 2 m high—making the entire rocket 110 m tall. The crew observes through periscopes and lie in couches for takeoff. The compartment is also equipped with leather loops for use in free fall. The crew can exit the rocket through an airlock tunnel. The return to earth is by parachute.

The launch takes place at sea. The rocket is towed into position while its fuel tanks are empty—their rigidity is maintained by filling them with compressed air. When they are filled with fuel the rocket's tail sinks, bringing it to its vertical launching position. The outside of the rocket is covered with paper to keep frost from forming on the skin.

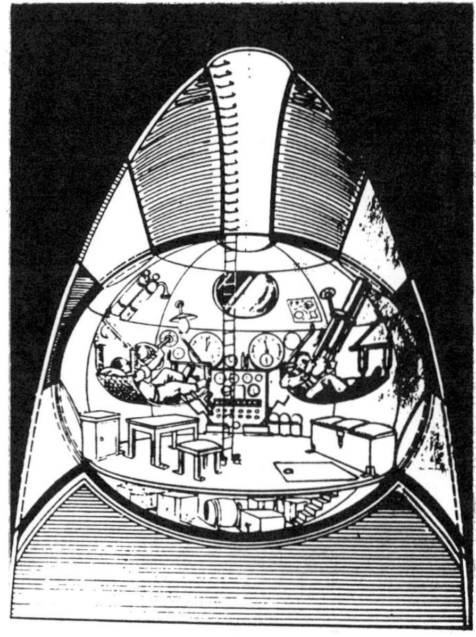

Oberth describes the fictional flight of a Model E-type spaceship, this time a three-stage version. The spaceship *Luna* is launched June 14, 1932, at 11:25:30 a.m., for a trip around the moon. The rocket has already successfully made an unmanned test flight to 4,200 km and a manned ascent to 5,000 km. Its crew of two has also gone through testing, including a centrifuge. The launch takes place from the Indian Ocean.

The pilots lie prone on their hammocks for the takeoff. After 1 minute the first stage is jettisoned, along with the tip covering the cabin. After 2 minutes, the second stage is dropped. The third stage is shut down after firing for 2 more minutes.

Once the elements of the spacecraft are in their inflight configuration a large mirror is deployed, held in place by three steel cables that can also be used to change its attitude relative to the cabin. A small telescope in the cabin acts as the eyepiece, with an effective power of 100,000x.

The cabin is a squat cylinder with slightly rounded ends, about 2 meters in height. Half of its shiny surface is covered with black paper, so that the internal temperature can be easily controlled. Oberth describes the cabin as being like "an aquarium for earth-dwellers." It has many thick quartz windows, protected with reflective plates. There is a

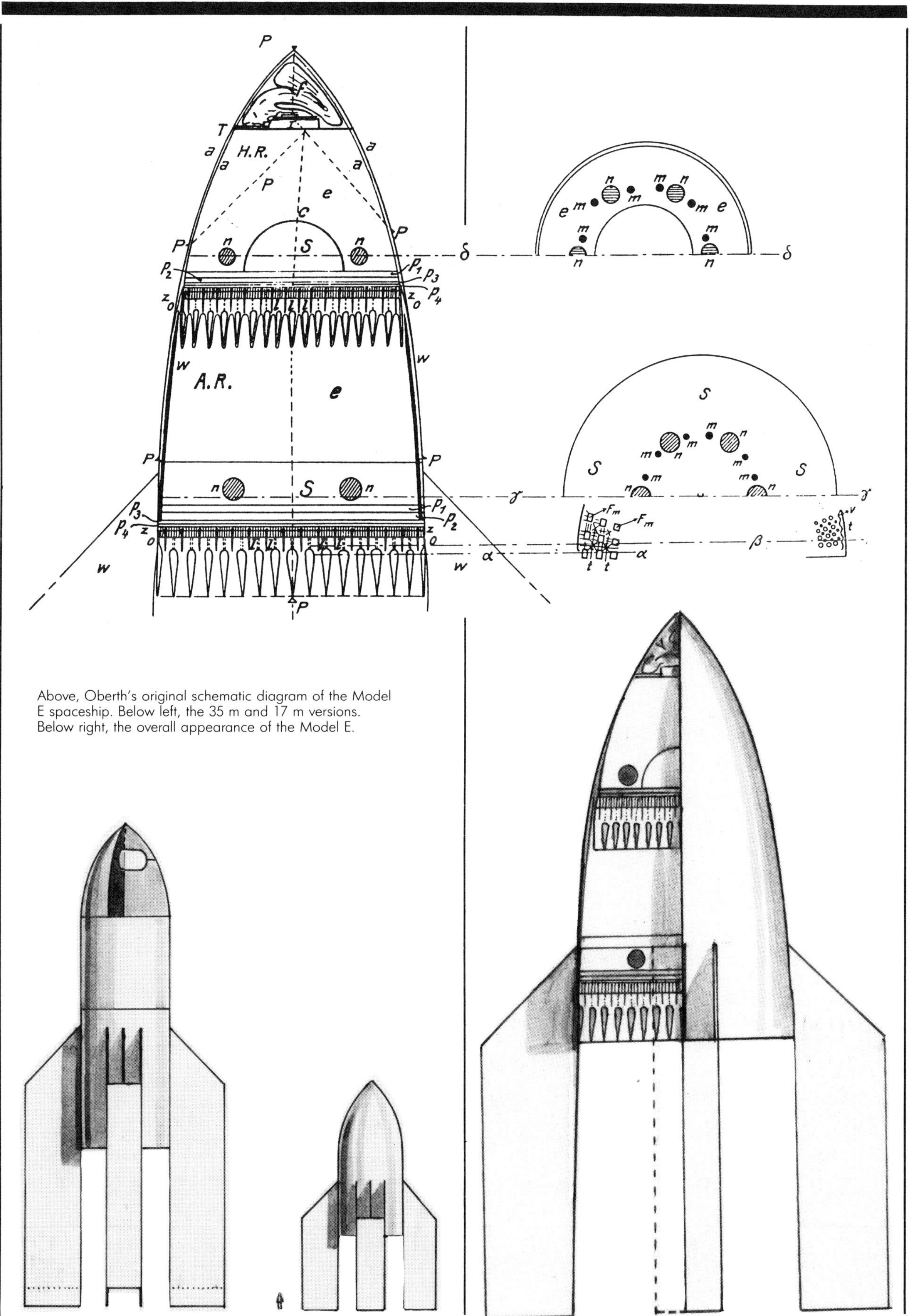

Above, Oberth's original schematic diagram of the Model E spaceship. Below left, the 35 m and 17 m versions. Below right, the overall appearance of the Model E.

miniature airlock for waste disposal (though Felix Linke criticizes this, saying that the ejected wastes would pose a hazard for future spaceflights; Oberth, in one of his few lapses, disagrees).

It is not possible to do full justice to Oberth's *magnum opus* in the space available. There is quite literally no aspect of spaceflight he does not consider to some degree, and most often in great detail—from crew testing to the experiments they would perform to the design of their instruments. It is probably safe to say that no other work of nonfiction will so influence modern astronautics as does this volume.

Konstantin Tsiolkovsky publishes the *Aims of Astronautics* in which he describes space colonies. "So far," he writes, "we cannot even dream of landing on large heavenly bodies . . . Even landings on a smaller body like our Moon is something that belongs to the very remote future. What we can realistically discuss is going to some of the minor bodies and moons, for instance, asteroids (10 to 400 kilometers in diameter)." He goes on to discuss that man need not have a planet in order to maintain a presence in space. If a container of oxygen has a mass of only 120 kg—just twice that of a man—"constructing a home for a human being is a mere trifle. There would be no harm in spending ten times as much."

"How," he continues, "is such a dwelling to be constructed? It is cylindrical, closed at each end with half-spherical surfaces . . . To make the thickness of its walls a practical proposition, the dwelling is built for several thousands or hundreds of persons . . . A third of the surface turned towards the Sun consists of latticed window panes . . . The dwelling would have the appearance of a tube of indeterminate length."

He proposes a diameter of perhaps 2 or 3 meters and a length of 3 kilometers (though it could be much longer). It would be divided into 300 compartments, each sufficient for a family and a garden capable of supplying its needs.

Tsiolkovsky proposes an alternative space station if gravity proves to be a necessity. This would take the form of a large cone, its base—closed by a transparent, spherical surface—turned toward the sun. The cone would rotate on its longitudinal axis and its inside surface would be covered with soil. To maximize use of the solar energy the colony receives, its internal area should be about four times that of the window, at the distance of the earth from the sun., To obtain this area the generatrix (that is, the length of a side) of the cone should be twice the diameter of the base. Typically, Tsiolkovsky considers even the most homely of details in designing his space habitats, never forgetting the smallest necessities for both the physical and psychological well-being of his space dwellers.

Tsiolkovsky also describes his "space rocket trains" in the book of that title. These are compound rockets, similar in some respects to that described in 1929 [see above]. The space train consists of five triple compound rockets. Each individual rocket is about 100 feet long and 10 feet wide. The complete compound rocket is therefore about 30 feet wide. The space train is pulled by the lead rocket. The exhausts are at an angle to the axis of the train, in order not to hit the following rockets. As each leading rocket's fuel is used up, it is jettisoned and the following rocket takes over. The final step in the train is the space rocket proper. Each rocket weighs 4.5 tons empty and carries an additional 4.5 tons of equipment and machinery, and 27 tons of fuel. There are four or more

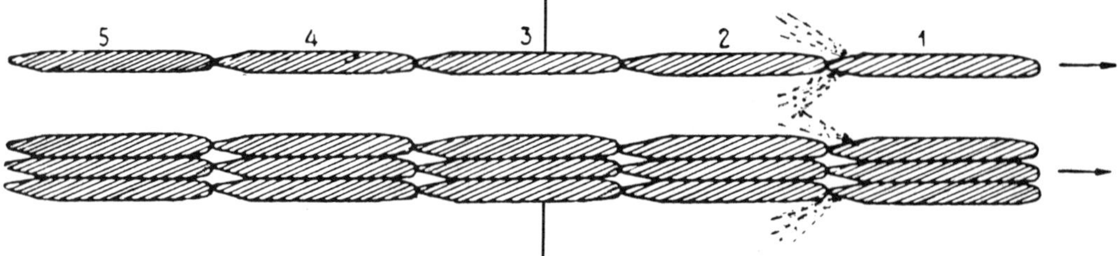

stern rocket nozzles. The entire rocket train takes off after a horizontal run of several hundred kilometers.

Tsiolkovsky calculates propulsion times, flight paths and takeoff angles, as well as fuel requirements and velocities for various combinations of components.

N. A. Rynin describes a pulse-jet engine similar to the one that will propel the V-1 "buzz bomb" of World War II.

Dr. von Dallwitz-Wegner, in "Uber Raketenpropellor und die Unmoglichheit der Weltraumshiffahrt mittels Raketenshiffen" (The Rocket Propellor and the Impossibility of Spaceflight) in *Autotechnik*, states, "The propellant does not even contain enough energy to lift its own weight beyond the earth's field of gravitation. How could it be able to take along the weight of the rocket, too?"

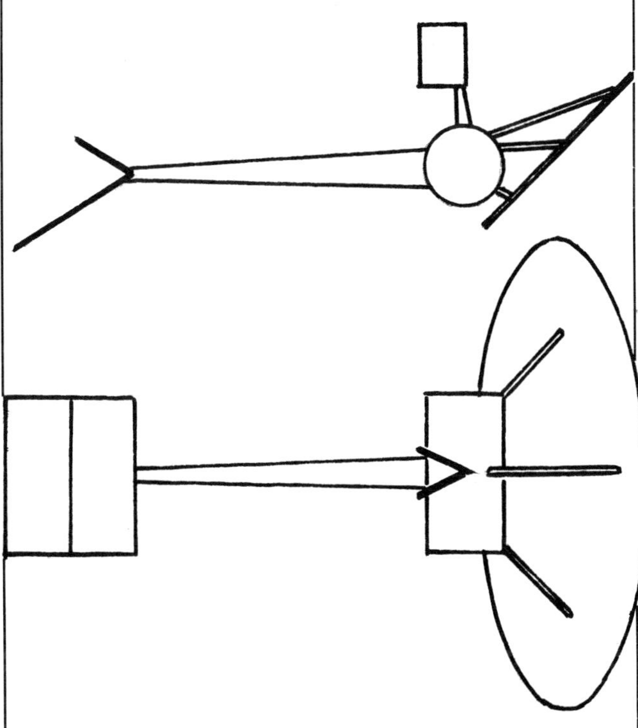

Yuri Kondratyuk publishes *The Conquest of Interplanetary Space*. In it he considers such problems of spaceflight as reentry vehicles. One such vehicle proposed by him is a cylindrical spaceship (with its longitudinal axis perpendicular to the line of flight) with a large, oval lifting surface/drag brake attached. A slender boom holds a V-shaped tail. The spacecraft would first reduce its speed by using the earth's atmosphere as a braking medium. The vehicle would then land as a glider. Kondratyuk also suggests that a spacecraft could plunge directly into the atmosphere (as the later Vostok and Mercury capsules will do).

In order to reduce the amount of propellants required for a planetary landing mission, Kondratyuk says that ."..the entire vehicle need not land, its velocity need only be reduced so that it moves uniformly in a circle as near as possible to the body on which the landing is to be made. Then the inactive part separates from it, carrying the amount of active agent [fuel] necessary for landing the inactive part and for subsequently rejoining the remainder of the vehicle." This is the method later employed by the Apollo lunar landing mission.

Jim Sorgi of Kent, Ohio, plans an ambitious program to fly to the moon in a rocket of his own construction. Sorgi, the proprietor of a small fireworks factory near Hudson, Ohio, is building his "human rocket" in a series of progressive steps, beginning with one of 50 pounds and continuing with rockets of 100, 150, 200 and 250 pounds, until he achieves one capable of carrying a man. Sorgi plans to eventually construct a rocket 90 feet long, containing 10 tons of powder, that can carry two persons to the moon or Mars. In case of accident, he plans to carry parachutes.

The passenger cabin of Sorgi's spaceship is to be slung beneath the main rocket like the gondola of an airship. Sorgi has declared that he intends to devote every last ounce of energy and every last dollar to this undertaking.

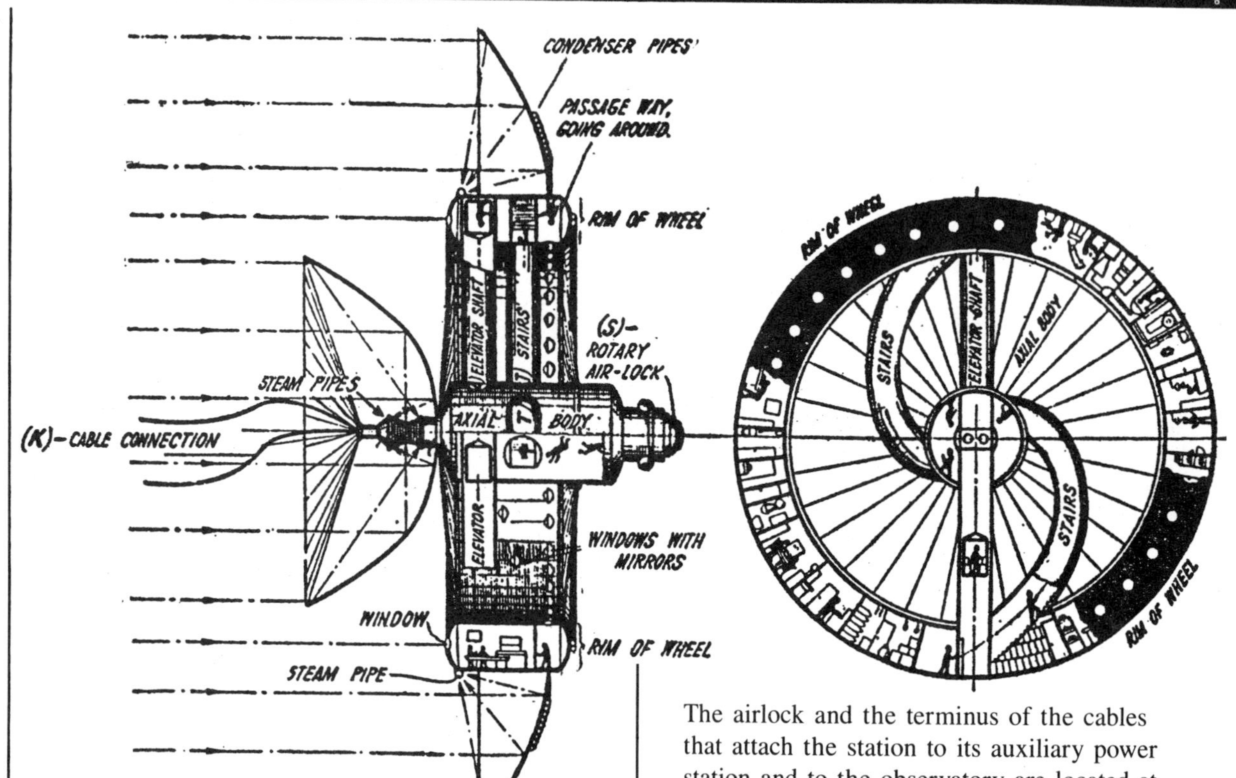

"Hermann Noordung" (pseudonym of Captain Herman Potočnik—1892–1929) describes an elaborate space station concept in his *Problems of Space Flying*. [Also see January, 1949.] Although Potočnik was Slovenian by birth his book was originally published in Leipzig in German. It was not published in his native language until 1986. Potočnik graduated in engineering in 1918 from the Technical University of Vienna. Tuberculosis prevented him from working; instead, he devoted his time to writing his book. He died in poverty soon after its publication.

His station, or "rotary house", is a toroidal structure not too dissimilar to that proposed more than twenty years later by Wernher von Braun [see 1952–1954]. The rim of the wheel contains the living and working quarters, all connected by a continuous corridor. These rooms include a dining room, cabins, laboratories, a photographic darkroom, a laundry, lavatories, etc. There is even hot and cold running water. The artificial gravity provided by the station's rotation would be the equivalent of 1 g. To accomplish this, the thirty-meter-diameter wheel rotates at a rate of 8 rpm.

The airlock and the terminus of the cables that attach the station to its auxiliary power station and to the observatory are located at one end of the cylindrical axis. In order to facilitate entering and leaving the rotating station the airlock can be spun to match the rotation or despun, as need dictates. Potočnik worked out the details of the rotating airlock with a great deal of care.

The axis is connected to the rim by both a pair of elevators and a pair of staircases. The former operate within radial tubes, while the latter are curved. Their logarithmic spiral shape allowing the astronauts to ascend or descend while maintaining their perpendicular.

One end of the axis is pointed permanently toward the sun. This end is fitted with a large parabolic mirror. Around the perimeter of the station is another mirror as well. These generate steam for electrical power.

Separate from the rotary house is the "engine house". This is an additional solar power generator in the form of a large mirror. The engine house also contains the radio station and air purification systems.

Potočnik describes a trip to his space station, orbiting at a distance of 36,000 kms, in a booster-launched space ship equipped with folded wings for its return to the earth as a glider.

Max Valier describes a transatlantic passenger rocket flight in *Die Umschau*.

The 26 minute flight begins from the Tempelhof airfield in Berlin. The rocket outwardly resembles a normal commercial aircraft. The passenger cabins are contained in the thick, stumpy wings; the rocket nozzles circle the fuselage between them. At the tail, a battery of retrorockets face forward. A pair of propellors are used to carry the vehicle to an altitude where the alcohol-liquid oxygen rockets can take over.

The cabin is hermetically sealed before takeoff. The interior is lighted by electricity and the walls and ceiling are padded. There are handstraps conveniently placed for use during weightlessness. The passengers occupy anatomically-shaped, semi-reclining couches. A restraining net is stretched over them during acceleration and deceleration.

Takeoff is made using conventional propellors; the rockets are ignited after about three minutes. After a brief period of acceleration there are 20 minutes of unpowered coasting during which the passengers are weightless. At the 24 minute mark the retrorockets are fired for two minutes, reducing the speed of the spacecraft by 35 meters per second.

The landing is made at Lakehurst, New Jersey. A report of the flight is sent back to Europe via "an Oberth-Goddard postal rocket."

JANUARY

An anonymously written Sunday-supplement article on the progress and future of rocket flight discusses some of the problems and promises of the use of rockets in warfare. The author correctly concludes that "dogfights" between high-speed manned stratosphere rockets would be, at best, impractical ("Probably the worst they could do would be to insult each other by radio . . . "). However, the author continues: "At present Russia cannot fight with the United States because her aeroplanes are unable to reach us, her navy is negligible and her monster army has no means of landing on American soil. But rockets would bring the two nations within range of each other, much to Russia's advantage because in that vast land there is so little worth destroying . . . American rockets that fell short would land in neutral European countries while the Bolsheviks would have the whole Atlantic to experiment in."

Cartoon comparing America's interest in moon rockets to a perceived imperialism.

JANUARY 7

"Buck Rogers" makes his debut appearance in American newspapers. Written by Philip Nowlan, inspired by his stories *Armageddon 2419 AD* and *The Airlords of Han*, both published in *Amazing Stories*, and illustrated by Dick Calkins ("Lt., Air Corps"), it is an immediate success.

Although Buck Rogers's name has entered our language as a synonym for space travel, spaceships will not make an appearance until 1930, after nearly 400 daily strips have been published. Although rocket-powered aircraft are introduced in strip 90 (1929), the notion of space travel seems not to have entered anyone's mind until a pair of earthwomen are kidnapped by the Tigermen from Mars. Buck decides that it is now necessary to build a spaceship, so that he'll be able to "show these Martians who's who in this solar system!"

The first spaceship to appear in Buck Rogers (discounting the Martian's antigravity sphere) is the *Satellite*. To help with its design, Buck enlists the aid of a noted scientist, one "R. H. Stoddard, the rocket expert, whose ancestor, with Max Valer [sic], had built the first rocket motors five hundred years before." Buck, "Stoddard" and author Philip Nowlan demonstrate a knowledge of the principles of rocket flight that eluded many real-life scientists. Although the *Satellite* is designed to get its initial lift from earth via antigravity, it is to be propelled by rockets because, as Buck explains, "rocket motors work better in vacuum than air. We could steer with them and attain terrific speed in outer space; then 'coast' along with no gravity or air friction to slow us down." Considering that *The New York Times* at about this same time is berating Robert Goddard for believing that rockets would work in space where they had no air to "push against," the readers of "Buck Rogers" are getting a better education about the workings of rockets than are *The Times's* subscribers.

The spaceship that Stoddard and Buck finally construct appears, from the diagram published, to be about 60 feet tall and shaped like a fat cigar. It is a tractor

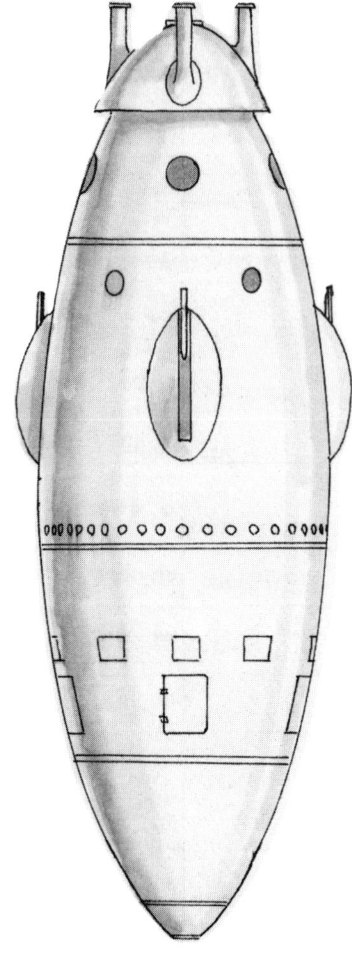

rocket—a battery of rocket nozzles surrounding its nose pull the spaceship rather than push it. Also at the nose are four retrorocket tubes and, between them, the lens of a telescope. Four egg-shaped gun turrets are spaced evenly around the ship's waist. The *Satellite* is streamlined, without fins or wings. At takeoff it is held balanced upright on its blunt stern bumper by props.

The effects of acceleration are vividly described, as is weightlessness. This is accurately explained by Professor Stoddard: "You see, the *acceleration* of the ship held us to the floor. When you cut off the motors we lost our weight." Even the effect of free fall on liquids is portrayed.

In the 388th strip, an extravehicular activity (EVA) is shown, though the space suits that Buck and his girl-friend Wilma Deering use are not actually the first to be seen in the series. The year before, an emergency abandonment of a high-altitude aircraft was shown. In order to safely bail out at an altitude of 20 miles, the crew had to don "air suits."

From EUROPE to NEW YORK

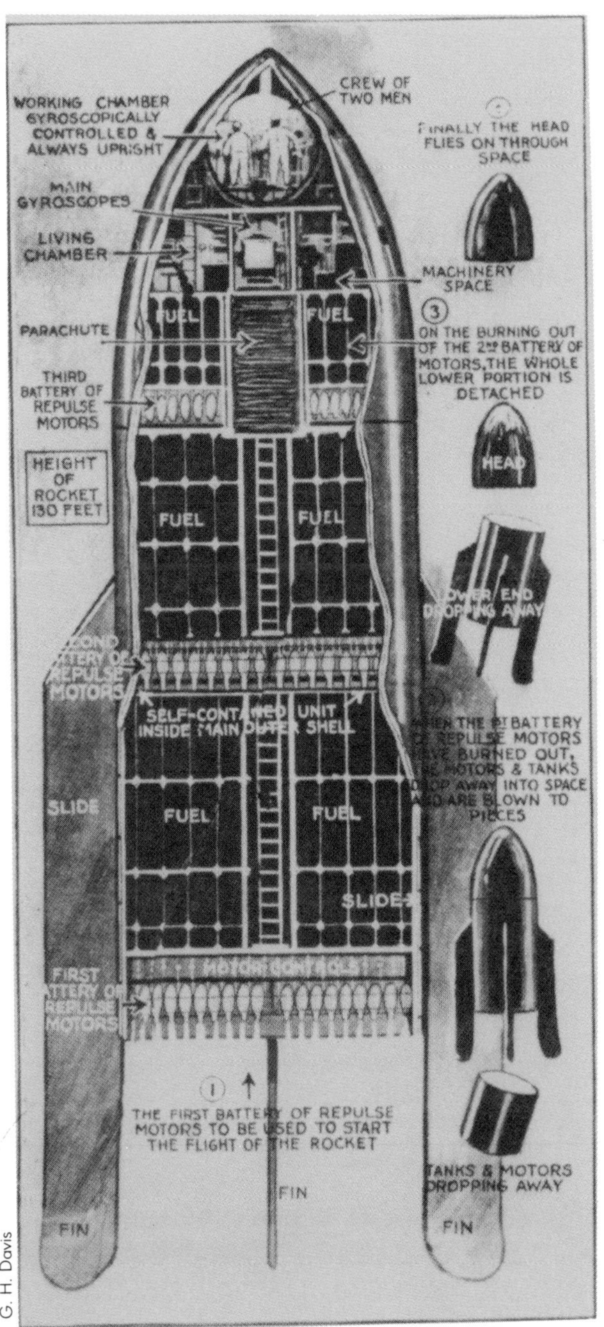

G. H. Davis

Unfortunately, the relatively high standard set early on for scientific accuracy is not maintained. By the 1940s the strip will have degenerated into outright fantasy. The believable spacecraft become art nouveau monstrosities that would do credit to Dr. Seuss. However, the Sunday color page, illustrated by Rick Yager, is much more conservative in its spaceship design. Its torpedo-shaped, slim-winged craft presages the *Terra V* of television's *Space Patrol* [see 1950–1956].

JANUARY 18

Max Valier describes his plans for a 420 mph "etherplane" in *American Weekly*. This is to be the end product of his Junkers G23 conversion [also see 1927]. The article was originally published in German in 1927.

APRIL 10

Friedrich Stamer, best known as being the manufacturer and sponsor of the solid-fuel-powered rocket cars and aircraft, allegedly builds and launches a liquid-fuel rocket, which, if true, would be the first liquid fuel rocket flown in Europe [however, see Winkler, 1931].

MAY

Max Valier announces his hopes to cross the English Channel in a rocket plane later this summer, crossing from Dover to Calais. He proposes building an aircraft along the lines of a hydroplane, with lifting wings, which would be his Type 7 [also see 1927]. Valier has commissioned the von Roemer brothers, a pair of artists who have illustrated most of his rocket designs, to work out the details of the Type 7. Later, he turns over the designs to the Espenlaub brothers, who operate an aircraft manufactory in Dusseldorf. Since funds are lacking, rocket braking tests (with the plane safely tethered on the ground) are made with the existing tow plane (rechristened the Espenlaub-Valier-Rak 3). Ordinary powder rockets are used for these tests.

Valier plans for his finished rocket plane, "built from the ground up," to be 11 m long, with a wingspan of 6.5 m. It is to be a high-wing design with the wings distinctly swept back. The tests are done to determine if the wing is strong enough to absorb the rocket's thrust, since Valier plans to install the engines directly into the wing itself, and to see how far back the exhaust flames reach, so

1929

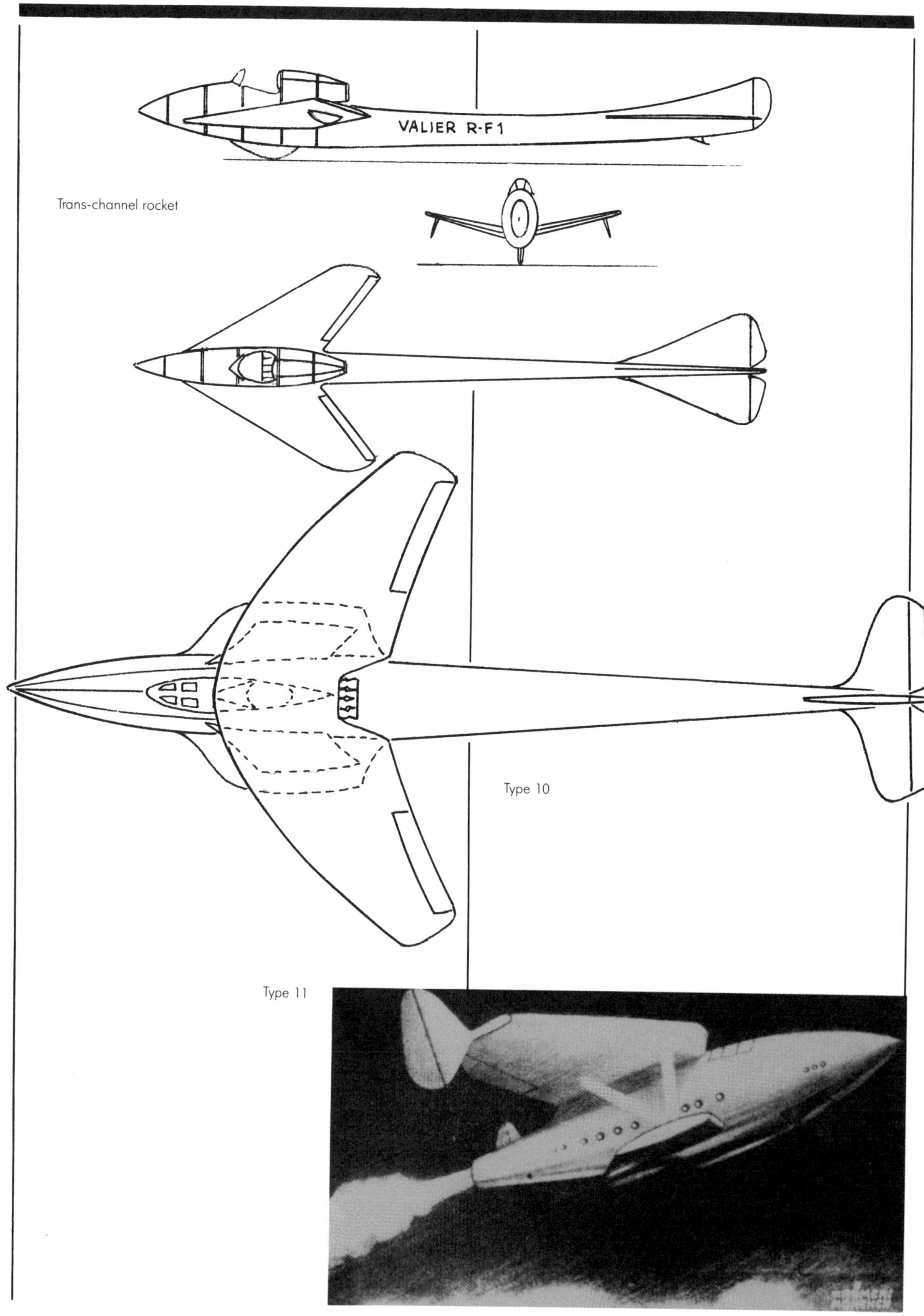

Trans-channel rocket

Type 10

Type 11

that the tail assembly can be properly designed. Although Valier runs out of money, Gottlob Espenlaub continues the tests on his own and makes his first rocket flight on October 22, 1929.

Valier plans for his channel-spanning flight to be made in a machine, the Espenlaub-constructed RF-1, that resembles an arrowhead or harpoon, its length about three times its wingspan. The long body serves as a tank for the liquid propellant. It would be able to rise without need of its wings, which would serve mostly as an aid in landing. The stubby, sharply swept-back wings project from the sides of the slender fuselage, on either side of the open cockpit. The rocket motor is mounted just behind the cockpit.

To travel the 20 miles, the first part of the flight would nearly follow the same trajectory as that taken by a shell fired from a gun. This would be done with the rocket motor firing at full thrust. By the time the level portion of the flight is reached a high speed would be attained—probably over 200 meters per second (approximately 350 mph). Here Valier would reduce his thrust to half. At "the top of the flight" the engine would be brought up to full throttle once again, using up the last of the fuel. The wings would be extended and the plane would glide to a landing. At a top speed of 400 mph, Valier expects the entire crossing to take only 3 or 4 minutes.

Later in 1929, Valier designs two new rocket seaplanes: Types 10 and 11. In his article, "Berlin to New York in One Hour," Valier compares his proposed Type 10 rocket plane with the Dornier DO-X airliner. If the giant airplane were equipped with rocket engines, its carrying capacity of 50 tons could be increased to 80 tons, while at the same time reducing the empty weight of the aircraft from 23 tons to 18. Its speed would be increased from 254 km/h to 360, or about 100 meters per second.

Valier also proposes that his rocket ship fly well into the stratosphere. At an altitude of 50 km it could attain a speed of 2 km/s, or 6,710 miles per hour. The ship would land after making a long gliding approach of

A. Gottlob Espenlaub at Dusseldorf, October 1930; top speed: 90 kmh. The 220 kg glider has a wingspan of 12 m.
B. October 2, 1929
C. Gottlob and Hans (r) Espenlaub, October 1929. This glider makes its first flight on October 23.
[Also see pages 192–193.]

about 1,900 km, using the atmosphere to brake its speed. The glide would last about 20 minutes. Takeoff acceleration would be relatively gentle, with "no greater increase in speed per second than is possible in a powerful motor car." Of the total weight of his 80-ton rocket, 34 tons would be fuel needed for takeoff, 24 tons the fuel for the horizontal flight of 4,900 km, and 2 tons of cargo, leaving an empty weight of 20 tons. Valier's intercontinental rocket is a slim-hulled aircraft with short, swept-back wings. These are just above the fuselage, with eight liquid-fueled rocket motors between wings and body. It is intended to take off and land from water and is equipped with stabilizing floats on either side of the fuselage. It is similar in overall appearance to the RF-1 rocket plane that Espenlaub was to have constructed in 1928. On February 9, 1929, Valier attempts, unsuccessfully, to launch his "Arrow" model rocket (the Rak FL-1) at Starnberg. It is a rough model of the Type 10, over 15 feet long.

The flight would not be a terribly interesting one for the passengers, according to Valier. There would be no period of weightlessness and there would be little visible from the windows: "Of the region over which one rises, almost nothing will be seen; because of the vapor and the very light cloud formations so prevalent in the higher strata. During the speedy flight over the ocean, at the highest altitudes, almost nothing will be seen of the ocean, or, for that matter, of the earth. The richest reward of the passengers will be in the sight of the black sky, while the sun (as in total eclipse) will appear surrounded by glowing red protuberances and the silvery corona."

Type 11, a stratosphere rocket aircraft, is a tailless monoplane with vertical wingtip fins and a spindle-shaped body. Like the Type 10, it is a seaplane with pontoon stabilizers on either side of the fuselage. The rocket motors are in the rear of the hull.

In the story "The Moon Strollers" (*Amazing Stories*), by J. Rogers Ullrich, a character says, "Now that the Goddard rocket has at last made a fair hit on the moon . . . "

A social club of amateur astronomers (" . . . mostly engineers or manufacturers from New York or Boston . . . ") develop and build a space suit. The Smithsonian Institution finances the construction of a rocket that will allow the suits to be tried out.

JULY

According to an article in the premier issue of *Air Wonder Stories*, an Evansville, Indiana high school instructor has designed a rocket for a trip to Mars. The vehicle resembles a "radio loop aerial" and draws its energy from space itself (though power could be transmitted from the earth)—eventually reaching the speed of light itself. The rocket, which takes off and lands vertically, has its motor mounted at the nose, where it can be pivoted for control.

AUGUST

The Junkers aircraft company uses rockets to launch a W33 Bremen-type airplane from the surface of the River Elbe. The rockets are electrically ignited. Junkers expects that it will soon be able to start a plane with a load of 11,000 pounds using only six rockets.

SEPTEMBER 30

Fritz von Opel flies a rocket-powered glider at Rabstock, near Frankfurt, Germany. It is equipped with nineteen solid-fuel motors of 50 pounds thrust apiece, and takes off under its own power. It reaches a top speed of 95 mph and is airborne 75 seconds, making a flight of 5,000 feet.. This is Opel's last experiment in rocket flight.

NOVEMBER

GIRD, the "Group for Investigation of Reaction Motion" is established in the U.S.S.R.

Max Valier writes an article for *Airways* magazine titled "Flying Beyond the Earth," in which he describes future spaceflight by means of rockets. He expresses his hopes that he will not only make the preliminary

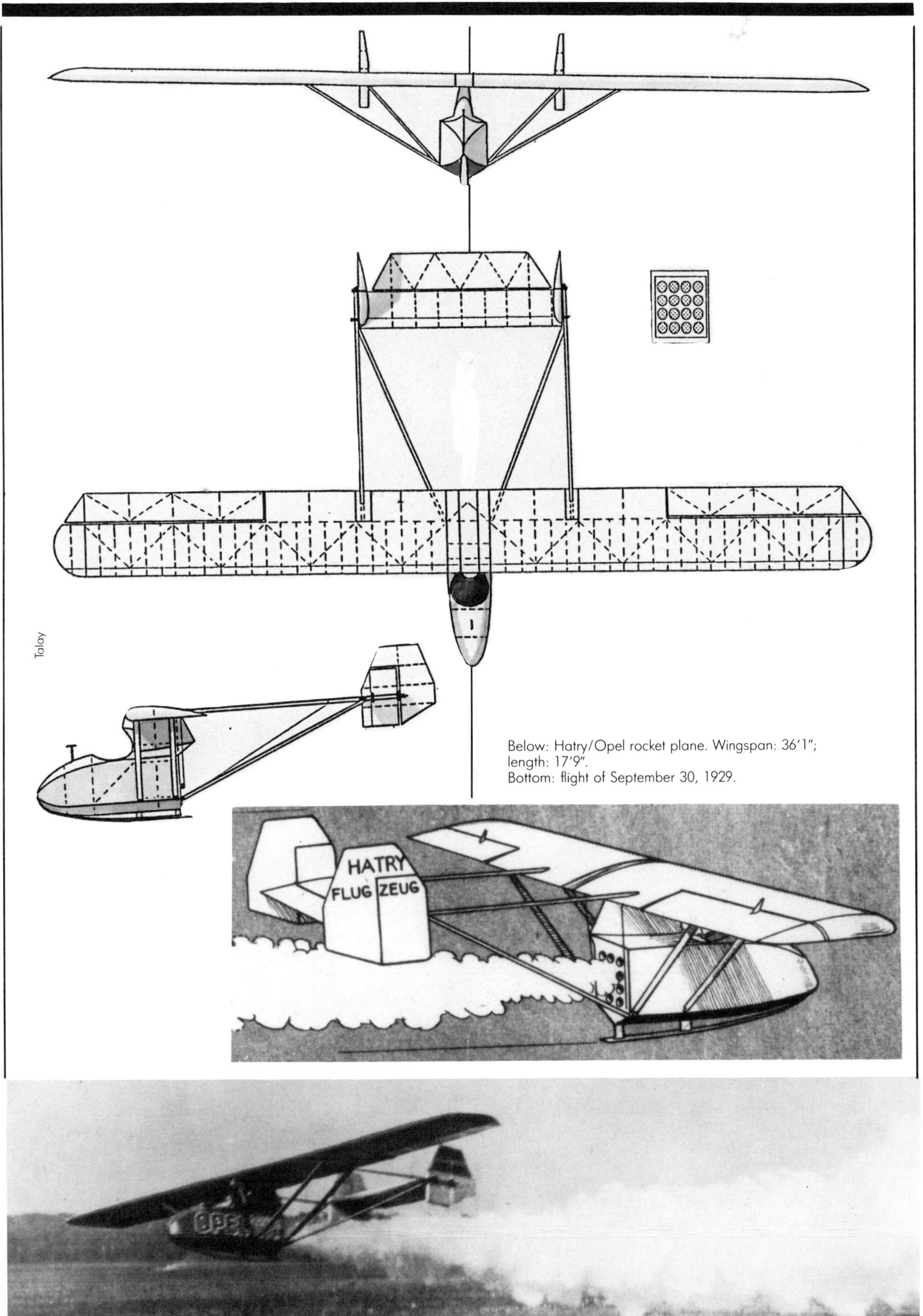

Below: Hatry/Opel rocket plane. Wingspan: 36'1"; length: 17'9".
Bottom: flight of September 30, 1929.

trial flights of such a spacecraft, but will remain at altitudes of 150 to 200 miles "for some minutes."

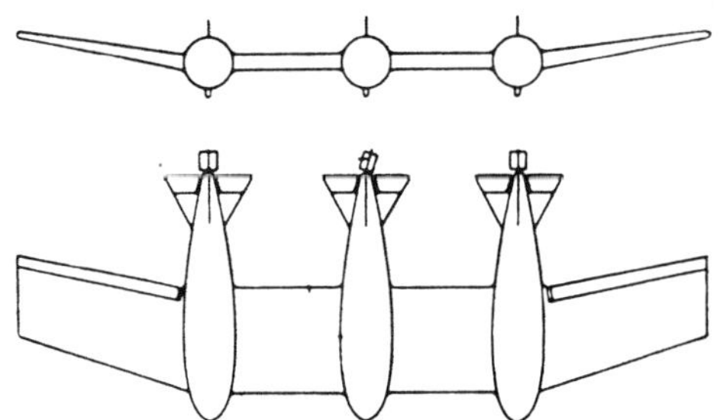

ca. 1929

An anonymous rocket-propelled aircraft is described by Rynin: it is a trimaran design with three parallel fuselages connected by a common wing. A rocket engine is located in the tail of each of the torpedo-shaped hulls.

Louis Blériot writes: "The flying-machine of the future will be a sort of shell that will easily surpass the limits of speed obtainable with present types of aircraft."

1930

William Olaf Stapledon publishes his epic "novel" *Last and First Men*. The book is, in reality, a fictional history of the human race from the present time to the year 2 billion. This remarkable book is far too complex and detailed to give any adequate resumyhere. However, this novel probably contains the least optimistic date for man's first spaceflight: roughly a quarter of a billion years in the future! The development of immense rockets is made necessary by the imminent crash of the moon into the earth. The first such "space ark" is an enormous cigar 3,000 feet long and made of "metals whose artificial atoms were incomparably more rigid than anything hitherto known. "Batteries of rockets around the hull steer the "ether ship." Within is room for a hundred passengers and supplies for 3 years. Air is created from protons and electrons and stored "under pressure comparable to that in the interior of a star." Heating is atomic. There is a field of artificial gravity generated electromagnetically. A trial circumnavigation of the moon is made, but the ship is destroyed when it collides with a meteoroid. Eventually hundreds of additional flights are made, with varying degrees of success. It is planned that the human race will ultimately migrate to Venus but two things must be done first: Venus must be made less hostile and a new form of human life must be created that can adapt to the still-harsh Venerian conditions.

One billion years in the future it becomes necessary for mankind to migrate to Neptune, since it has been discovered that the sun is about to expand into a red giant, consuming the inner planets. Once again, man is redesigned to suit his new world. Finally, 2 billion years hence, mankind faces its end. The gulf of interstellar space is uncrossable. In lieu of escape, an "artificial human dust" is cast upon the solar wind, in the hope that someday it will seed another world with human life.

Artist-astronomer Lucien Rudaux creates a "trip to the moon" for the Paris Exposition. A large, moving diorama painted on canvas gives visitors the illusion of exploring the lunar surface. Rudaux is a pioneer in the field of space art whose books, notably the classic *Sur les Autre Monds* (1937), are still influential.

Yuri V. Kondratyuk in *The Conquest of Interplanetary Space*—based on 10 years of study—describes the technique of landing on a planet using an orbiter and landing module. He also proposes the use of lifting bodies in conjunction with atmospheric braking. [Also see Houbolt, 1961.]

Kondratyuk also suggests using ozone as an oxidizer. He began working on the problem of interplanetary flight during World War I when there were very few published studies of serious work [see Tsiolkovsky and Esnault-Pelterie, although Kondratyuk claims that he was unaware of their work]. Like his contemporaries, Tsiolkovsky and Oberth, Kondratyuk develops his theories and conclu-

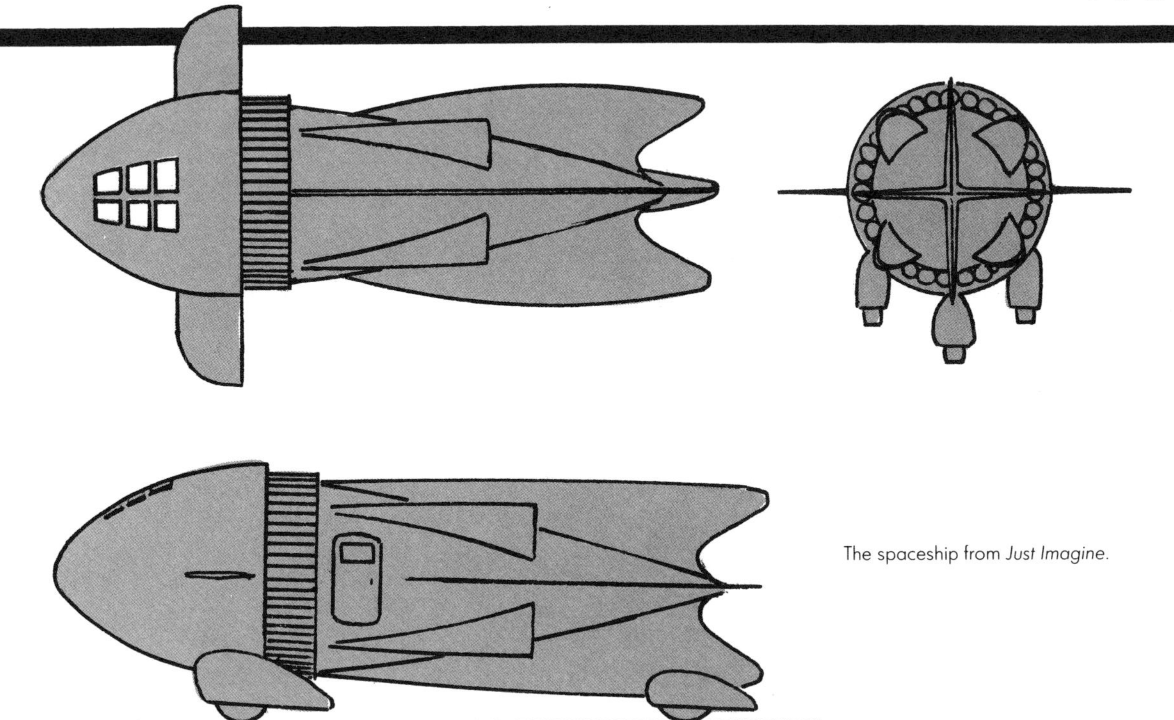

The spaceship from *Just Imagine.*

sions as a series of questions and problems, each building on the other. Since he is working independently it is inevitable that much of his work necessarily duplicates that of the others.

Once he is convinced that the principle of the rocket is the only practical means of travelling through space, he begins refining the problems facing its implementation. Ultimately, he develops a systematic approach to the exploration of space, similar to that of Tsiolkovsky [see 1930]:

1. Performance tests of the equipment in the atmosphere.
2. Flight in near-earth space.
3. Flight to the moon without landing.
4. Landing on the moon.

For the latter he proposes sending a spacecraft into lunar orbit from which a special landing module would make the actual descent to the surface. This would serve to conserve fuel since "it is possible not to land the whole rocket, but only to reduce its velocity to such a degree that it would revolve uniformly around and as near as possible to the body on which the landing must be made." In a second version of the manuscript he goes on to explain that while the main ship remains in orbit "the landing should be made with such part of the rocket as is required to land on the planet and to return back and join the rocket."

Among other concepts Kondratyuk considers are proposals for jettisoning unnecessary portions of the rocket's mass, electric propulsion, nuclear engines, the utilization of solar energy and space stations.

Just Imagine, a motion picture directed by David Butler, is Hollywood's predictable answer to Germany's *Metropolis* and *Frau im Mond* [see 1929]. In a musical comedy set in 1980, J-21 leaves his girlfriend LN-18 to travel to Mars in the company of a comic-relief character played by vaudevillian El Brendel. The spaceship that is used is the archetype of all art deco era spacecraft. It is certainly too good to waste and makes several additional screen appearances—slightly revised—as Dr. Zarkoff's invention in the Flash Gordon serials [see 1936]. The ship is bullet-shaped, with four narrow, tapering fins and a pair of retractable wings near the blunt nose. The rocket is of the "tractor" type, with a ring of rocket nozzles surrounding the rear of the forward pilot's compartment. It takes off and lands horizontally.

Dr. Theodore Wolff of Berlin writes, " . . . It can be readily calculated that the extreme limit attainable with fuels at present available is about four hundred kilometers above the surface of the earth; and at this height any body is still fully subject to the attraction of the earth, without having the slightest possibility of rising higher . . . "

The Earl of Birkenhead, in his *The World in 2030*, writes: "By 2030 the first preparations for the first attempt to reach Mars may perhaps be under consideration. The hardy individuals who form the personnel of the expedition will be sent forth in a machine propelled by a rocket; and equipped with a number of light masts which can be quickly extended, like fishing rods, from its nose. The purpose of these will be to break the impact with which, granted all possible skill and luck, the projectile would strike the surface of the planet.

"The great problem which such an expedition will face, however, is the possiblity of missing Mars altogether . . . Such a fate, indeed, may well overtake the first half-dozen expeditions which set out from the Earth to reach Mars. But, one day, a few men may arrive alive on the surface of our nearest neighbour in space. It seems unlikely that they can long hope to survive there, far less that they will be able to return to their home on the Earth. The most for which they can hope, will be to send back across the ether a few messages of information concerning Martian conditions; to transmit the results of a dozen accurate scientific observations before they perish. I should not myself be a volunteer member of that party.

"The fruit of their messages, and of their death, will be new expeditions, better equipped, better prepared to withstand the physical difficulties of life on another planet, and bearing with them in their flying machines the materials to erect another smaller machine on the surface of Mars."

A British inventor suggests a novel adaptation of rocket flight to intercontinental travel. Instead of being propelled through the air, the rocket will rise into the air, remain stationary for a length of time while the earth revolves beneath it, until the rocket is approximately above its destination, then it will descend. The altitude at which the rocket must remain suspended is so high that the passengers will have to be provided with oxygen. How the rocket is to remain motionless above the planet is a problem that has yet to "be straightened out."

Robert Esnault-Pelterie writes *L'Astronautique*, his principal contribution to the study of spaceflight. A seminal work, it covers all aspects of rocket propulsion and its use in spaceflight. Chapters are devoted to theory, the uses of rockets in high altitude research, high speed travel, trips to the moon and interplanetary flight.

His use of the word "astronautique" is its first use in a publication [also see 1927].

L'Astronautique contains the first three of his major works (*Consideration sur les resultats d'un allegement indefini des moteurs* [1913], *L'exploration par fusées de la tres haut atmosphere et la possibilité des voyages interplanetaires* [1928] and *Astronautik und Relativitatstheorie* [1928]) along with several new studies. Although he considers most of the technical aspects of space travel, including the use of atomic energy, he makes no specific suggestions for a spaceship design.

In a lecture to the French Institute, REP, as Esnault-Pelterie preferred to be known, predicts the possibility of rocket flight around the world in 1 hour and 26 minutes, and trips from Paris to New York in 24 minutes. He knows that a space-bound rocket needs to "start off slowly, so that while we are passing through the atmosphere our speed is increasing as rapidly as possible." He calculates that the top velocity would be reached after "eight minutes of travel, when we are 1,200 miles above the earth." He suggests a maximum acceleration of 3 g's for manned rockets, with the forces distributed evenly over a prone body. In another lecture, sponsored by the American Interplanetary Society (delivered, however, by G. Edward Pendray, due to REP's alleged illness), he discusses the possibilities of interplanetary travel. He predicts that it would take place after 25 years [i.e., ca., 1955] and would require large sums of money—about $2 million (!).

At another time Robert Esnault-Pelterie predicts that the first trip to the moon could be made within perhaps 15 years, circa 1945. Once again he estimates the cost at $2 million, which would include a "cheap" rocket with a range of 100 miles, a transatlantic rocket and finally the moon ship. The rocket would be a cigar-shaped vehicle propelled by chemical fuels or, more likely, by atomic energy. The journey would take about 50 hours. The rocket would land on the moon tail first, using its rockets as brakes.

V. V. Stratonoff, a Russian astronomer, predicts the future colonization of Venus.

Konstantin Tsiolkovsky, in his booklet *The Reaction-Driven Airplane*, describes an aircraft propelled by the escape of gas from a specially adapted conventional engine. In this case, the cylinders are replaced by conical nozzles through which the combustion products are expelled. The engine is air-breathing. Tsiolkovsky calculates that such an aircraft could reach speeds of up to 2,200 mph at altitudes of about 23 miles.

It is claimed that L. Ron Hubbard (the science fiction author who was later to found the lucrative pseudosciences of Dianetics and Scientology) has developed and tested a rocket motor superior to and less complicated than the later German V2 [see 1942].

The French Academy of Sciences answers the question "Will man ever travel to the moon?" with a cautious "possibly. "While the Academy will not commit itself as a body to a positive stance, the majority of its members hold the view of "Why not? "The discussion is inspired by Robert Esnault-Pelterie's book *Astronautique*. The author states that he expects a flight to the moon to be possible within 15 years. This venture would have to be backed to the amount of $2 million.

FEBRUARY

Harl Vincent publishes his story "The Explorers of Callisto" in *Amazing Stories*. It describes a flight to the moon by the spaceship *Meteor*. It is a rebuilt conventional aircraft 40 feet long with standard landing gear and tail, but without struts and guywires. It is equipped with "rocket-firing cylinders." It has in addition a 15-cylinder 600-horsepower radial engine that is used to get the ship through most of the earth's atmosphere. When the inventor is asked by his friends, "When do we go?" he casually replies, "Can you make it the day after tomorrow?"

APRIL

In a letter to *Amazing Stories* L. Partridge writes: "My own experiments have been with the idea of changing ionic or atomic motion into mass motion in an electromagnet . . . A hollow cylinder carrying in it its own power, would theoretically be capable, so excited, to continue its acceleration to the speed of 'motion *per se*'. Actually the limit even in space would be the crystallizing point [sic]. And as its crew would partake of the same acceleration no external inertia would act on them."

APRIL 4

The American Interplanetary Society is founded by eleven men and one woman, meeting in the apartment of G. Edward Pendray and his wife Leatrice (who writes professionally under the name Lee Gregory). Almost all of them are connected in some way with Hugo Gernsback's magazine *Science Wonder Stories*. David Lasser, the organization's guiding light and first president, is the managing editor of the magazine. Charles P. Mason is associate editor; Fletcher Pratt, Nathan Schachner and Lawrence Manning are all prolific writers of science fiction. The remaining members, Clyde

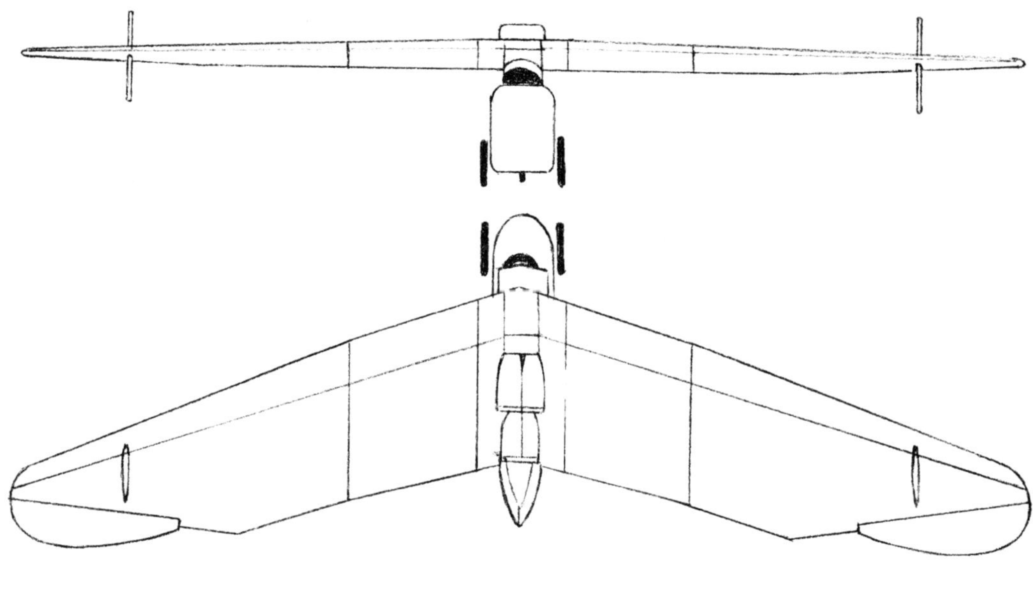

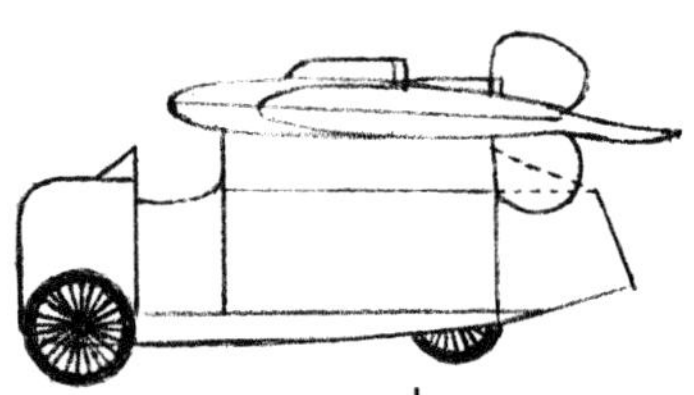

Espenlaub rocket glider.

Fitch, Dr. William Lemkin, and C. W. Van Devander, are all either occasional writers of science fiction or are fans. Gernsback himself joins the society, but apparently doesn't attend its meetings, somewhat surprisingly.

Like the early VfR [see 1927], the American Interplanetary Society is optimistic about its future. "We believed generally," writes Pendray, "that a few public meetings and some newspaper declarations are all that would be necessary to bring forth adequate public support for the space-flight program."

APRIL 28

In his article, "To Inventors of Reaction-propelled Machines," Konstantin Tsiolkovsky gives advice to children concerning some reaction-propelled boats, model airplanes, and "an ordinary rocket equipped with a chamber and manned by toy travellers, for the sake of effect."

He goes on to mention that 10 years earlier he was approached to have his novel *Beyond the Earth* adapted for the motion picture screen. It proved to be too complicated a project and the idea was dropped. However, the Mosfilm studios has recently "firmly decided" to produce the film as *Cosmic Journey*, under the direction of V. N. Zhuravlyov. Tsiolkovsky reportedly produces a number of spaceship designs—upwards of thirty—for the film, which is eventually released in 1935.

Tsiolkovsky also explains his fascination with science fiction: "Science fiction stories on interplanetary trips carry new ideas to the masses. All who work on these lines, work well, arousing interest, promoting the working of the brain, and training up people who are attracted by the idea of grand projects, and who will work in the field in the future."

APRIL-MAY

Gotfried Espenlaub successfully tests his rocket powered glider at the Lohausen Dusseldorf flying field. His airplane is tailless with sweptback wings. Each of the rockets, provided by Sander, burns for 6 seconds. Espenlaub attains a top speed of 56 mph.

After these trials, the rocket glider will be transported to Wesermunde where tests will continue under Dr. Sanders's direction. Before this occurs, however, Espenlaub crashes, badly damaging the aircraft and slightly injuring himself.

Espenlaub began building his glider the year before at Essen. It is a single-seater monoplane with a dead weight of only 100 kilograms. The pilot's seat is at the front of the aircraft where the motor would normally be, to maintain balance. The rockets are mounted behind the pilot and above the wing. For its first flights the glider is towed to an altitude of 20 meters before igniting the rockets. The first flight is for a distance of 2 kilometers.

The glider is normally launched with a rubber rope and a single rocket of 150 pounds thrust. When an altitude of 30 feet is reached, a second rocket is fired. It is while firing the third rocket that Espenlaub pushes the stick too far forward, crashing the plane. Even though Espenlaub's brother was killed recently during a similar test, Gotfried plans to continue his flights.

JUNE

The first issue of the *Bulletin* of the American Interplanetary Society appears—a simple mimeographed publication of only a few pages.

JUNE 22

John Q. Stewart, associate professor of astronomical physics at Princeton University, describes a flight to the moon for the New York *Times*. (One of the preliminaries to such a venture, he suggests, might be studying the effects of weightlessness on living organisms. This might be done by dropping guinea pigs down a shaft a quarter of a mile high, catching them unharmed at the bottom.) He proposes a sort of giant clus ter of perhaps a dozen or more cannons cum space cruiser, weighing 70,000 tons. The ship would be launched by firing 28,000 tons of powdered lead shot (in a desert area) at a rate of 2 1/2 tons per second. The 110-foot spherical ship would accelerate at a rate of

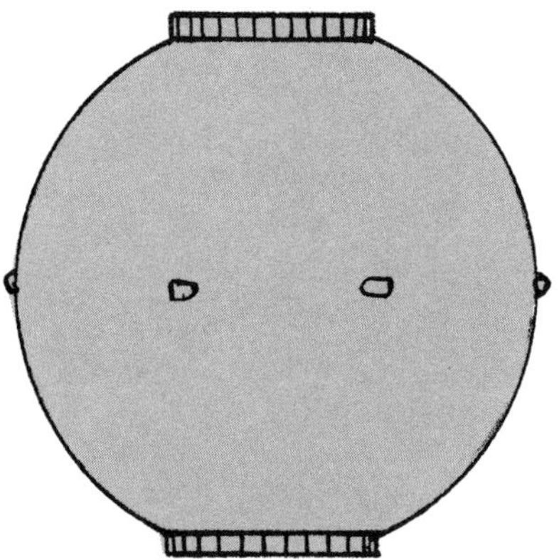

1/16 g. The spaceship would be consuming as much energy as produced by all of the public utility companies in the United States in 1 week. It would have to take off half an hour before noon, 3 days before the new moon. Two hours and 29 minutes after takeoff, the ship would be 13,200 miles from the earth traveling 190 miles per minute. The ship would then coast the remainder of the distance to the moon, arriving there 70 hours after takeoff, circling the moon before landing. Cannon facing forward would be used to brake the rocket and other cannon facing different directions would steer it. It would carry a crew of sixty with twelve scientist-passengers. Communication with the earth would be accomplished by light beam if the Heaviside layer prevented radio contact. In any case, Stewart believes that the flight and landing would be televised worldwide. Stewart predicts that such a flight as he describes could take place by the year 2050.

JULY

A letter from C. P. Mason, secretary of the American Interplanetary Society [see above], is published in *Science Wonder Stories*. He mentions that meetings are held twice monthly at the American Museum of Natural History in New York. Its members now include Robert Goddard, Clyde Fisher, Sir Hubert Wilkins, and Hugo Gernsback. It is presently assembling a library of space-flight-related materials "which will be the most comprehensive in America." The society also plans to issue an annual report and a monthly bulletin, as well as to fund research. Its present officers are: David Lasser, president; G. Edward Pendray, vice-president; and Fletcher Pratt, librarian.

AUGUST

Letters from readers in *Wonder Stories* include one from John Convoy who writes: "I am quite confident that it would be many years before it [a space ship] left the ground. The people who would hold it back would be the lawyers!" He continues, citing the difficulties in disposing of rocket stages safely and the danger they would present to inhabited areas. In the same issue C. P. Mason as secretary of the AIS [see above] reports on the society's growth. It is beginning two new projects: assembling all material pertaining to spaceflight and publishing a bulletin. It is holding meetings at the Hayden Planetarium in New York (later host to a series of historical symposia on spaceflight, see 1952–1954). Membership is $10 a year or $3 quarterly.

OCTOBER 8

Philip Barr proposes avoiding the stresses of high g-forces at takeoff by using a constant acceleration of only 22 miles per second (7,920 mph) and slowly increasing this to 500 mps after 24 hours. He makes no mention of fuel consumption.

OCTOBER 19

Gottlob Espenlaub flies a rocket-powered glider near Dusseldorf. It is propelled by Sander solid rockets [also see 1928] and develops a top speed of 55.9 mph. It is a high-wing monoplane with a cylindrical fuselage. The cockpit is immediately below the huge wing and the rocket motors are mounted in the center of the wing above the cockpit. There is some problem with the rockets' exhaust burning the leading edge of the vertical stabilizer.

FALL

An advertisement in *Wonder Stories Quarterly* states: "We will reach the moon in 1950 . . . say competent observers of scientific developments in rocket-travelling. Within twenty years the first interplanetary explorer will alight slowly on the moon's surface, using powerful liquid fuel rockets to propel and control his spaceship. As in Lindbergh's transatlantic flight, the world will cheer his intrepid adventure. As in aviation history regular communication will be rapidly established. Wall St. concerns will hire men to work the mineral deposits in the new world. Advertisements for miners and clerks to live on the lunar planet will appear in the 'Help Wanted' advertisements of the daily papers . . . "

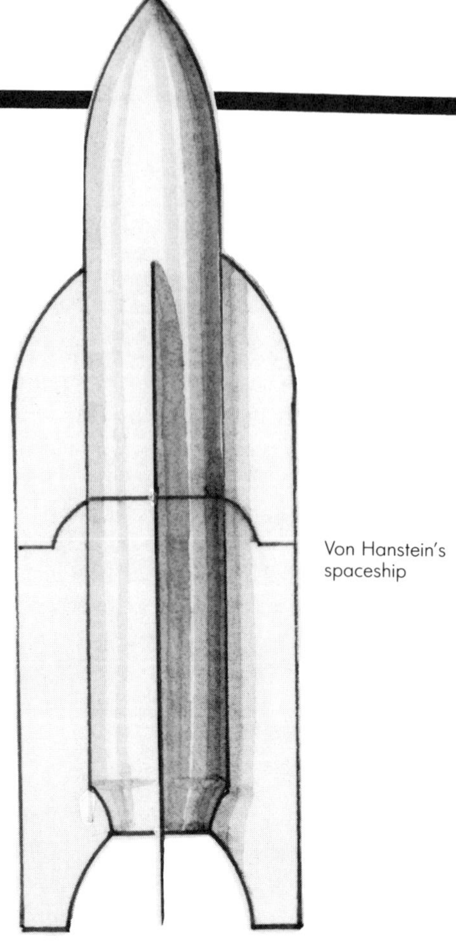

Von Hanstein's spaceship

The short story "Between Earth and Moon," translated from the German of Otfrid von Hanstein, details an extremely realistic and accurate spaceflight made by a rocket of the Oberth Model E type [see 1929]. It is a two-stage rocket 35 meters tall and 6 meters in diameter, with a beryllium hull painted black. Both stages are fueled by hydrogen and oxygen. The cabin is 5 m wide and 2 m deep, with deep, padded benches on springs. It is to be launched from an artificial floating island called "New Atlantis," half a mile in diameter [also see Space Van, 1986]. The spaceship is accidentally launched with three men on board, a German, an American and a Japanese. On the way to the moon there is a description of a space walk using a rocket-powered maneuvering unit. There is an explosion after the spaceship leaves the moon, leaving only the cabin and some attached wreckage. The astronauts manage to rendezvous with a rescue rocket.

The launch of von Hanstein's spaceship by Frank R. Paul.

WINTER

R. H. Romans, in his story "The Moon Conquerors" (*Wonder Stories Quarterly*), describes the spaceship *Astronaut*. In the story it is the winner of an international competition for the best spaceship scheme (atomic power and antigravity are dismissed . . . while 152 of the submitted plans "are for a Goddard rocket"). The finished rocket is fueled with slow-burning powder and charges of nitroglycerine and TNT (it is decided that hydrogen as a fuel would be too tricky to handle!). The *Astronaut* is equipped with

1930

The takeoff of the *Astronaut* by Frank R. Paul.

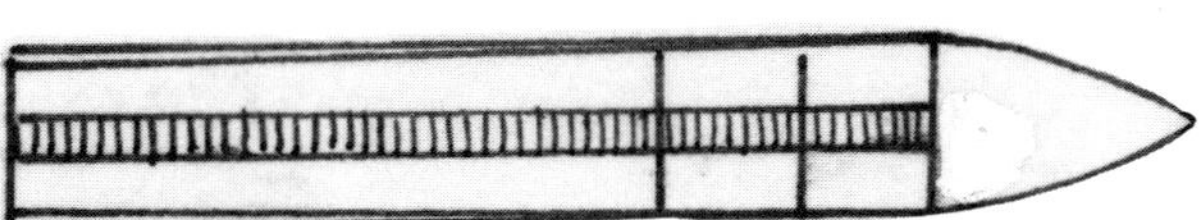

steering rockets and retrorockets. It is about the size of a railroad coach, and has a slender bullet shape with narrow fins. For visibility it is painted in bright colors: a red fuselage, yellow fins, and blue rocket exhausts, ports and periscopes. It would certainly be noticeable! Inside there are two conventional Moth airplanes for the moon landing, space for four passengers (one of whom is a woman), a padded room for the takeoff, a kitchenette, and "double portholes" for waste disposal. The walls are three airtight layers of steel with a vacuum between the layers. Periscopes replace the need for windows. The launch date decided upon is June 27, 1945, at 9:30 p.m.

Romans gives the initial launching boost to his spaceship by first using a locomotive to push the flatcar-mounted rocket to a speed of 50 or 60 mph. It then enters the first coil of an electromagnetic cannon. The rocket travels a half mile horizontally before the gun curves up the side of a mountain [also see 1937 and 1951]. The rocket's engines are started when the ship leaves the far end of the mass driver.

For propulsion between the earth and the moon, the pressure of sunlight against special folding black vanes is used.

ca. 1930

L. Korneyev, engineer for the Stratosphere Committee of the Russian All-Union Scientific Aviation Engineering and Technical Society, announces a new rocket that is expected to reach an altitude of 40 miles. Its high performance is credited to the use of a pump rather than gas pressure to feed fuel into the engine. Further improvements are expected to increase the rocket's range to 60 miles. It descends by parachute.

An experimental rocket built by Charles Bushnell of Aberdeen, Washington, explodes after a 100-foot flight. The 11-foot rocket is designed to carry a human being on a 1,000-foot flight, using 130 separate powder charges (costing the inventor $100). Crestfallen at his failure, Bushnell is undecided whether to rebuild his rocket.

W. A. Conrad, assistant professor in mathematics at the U.S. Naval Academy, estimates the cost of building a moon rocket at $100,000,000 . . . equivalent to 2 battleships.

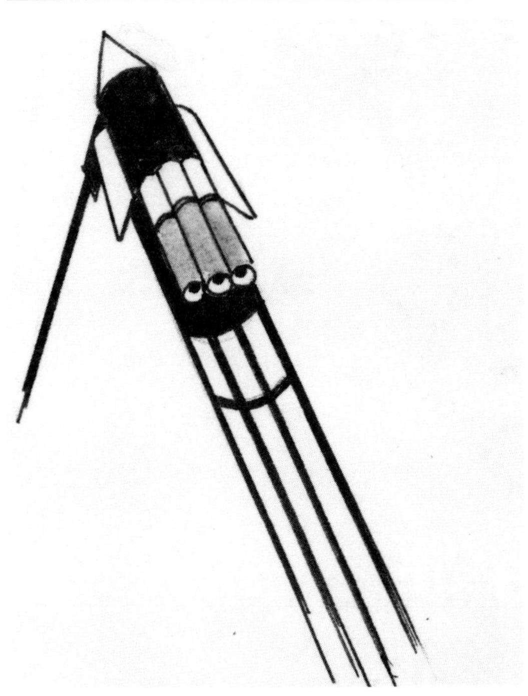

A rooster and hen are the passengers, along with 189 letters, aboard the solid-fuel rocket *David Ezra* when it is fired across the Damodar River in India, a distance of only a few hundred yards. The inventor of the rocket, his sixty-fifth, is British Interplanetary Society member Stephen S. Smith, with financing by the Maharajah of Sikkim. The rocket is equipped with wings and tailfin, allowing it to glide after its fuel is exhausted.

The chickens arrive safely.

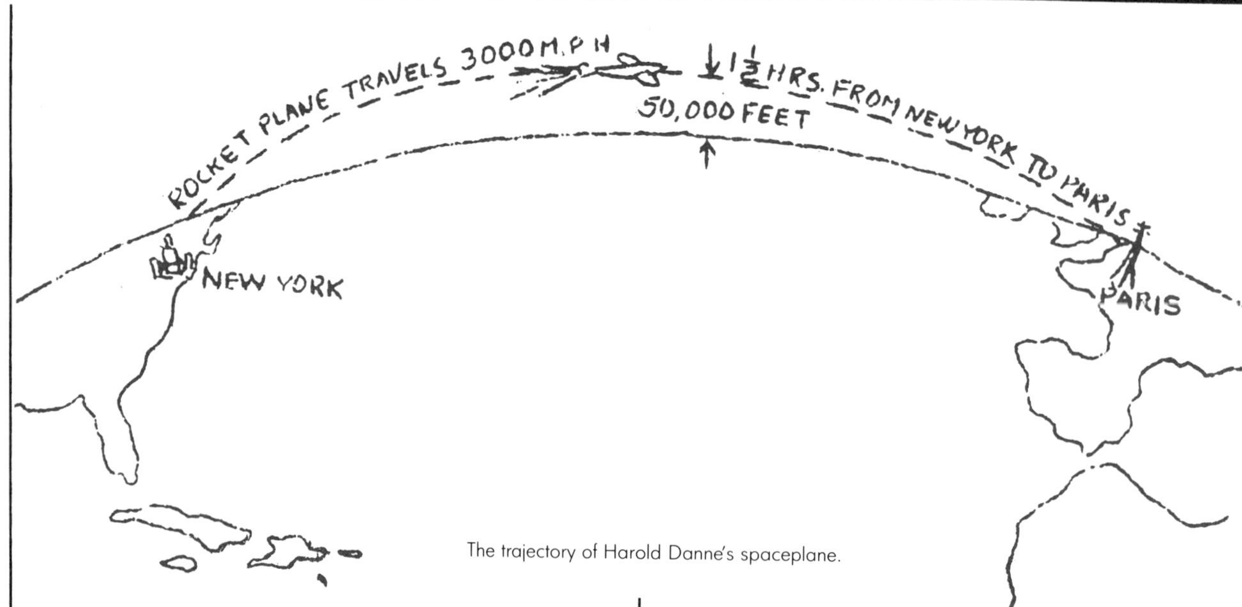

The trajectory of Harold Danne's spaceplane.

Konstantin Tsiolkovsky publishes his fourteen-point plan for the conquest of space [others have invented similar timelines, see above]:

1. *Develop a winged rocket plane*. The pilot will be protected from the wind by a "transparent substance." After spending all of its fuel during its flight, the rocket plane will land as a glider. A non-airtight cabin will limit altitudes to about 5 km. This is sufficient since the aim of this first craft is to develop pilot proficiency.
2. *Develop a rocket plane with shorter wings*. As the wings are shortened, thrust is increased.
3. *Develop a rocket plane for flights of up to 12 km*. The cabin is now airtight and oxygen apparatus must be carried. For safer landings, the rocket may alight upon water.
4. *Develop a wingless rocket*. It steered by means of vanes in the exhaust.
5. *Develop a rocket with speeds of up to 8 km per second*. This rocket will finally leave the earth's atmosphere. Like the others, it will glide to an unpowered landing.
6. *Begin spaceflight*. Longer and longer durations in space will be accomplished.
7. *Develop regenerative life support systems*. Using squash plants [such as pumpkins—see July 23, 1945] will produce food as well as oxygen.
8. *Develop a space suit*.
9. *Breed plants to aid #7*. Man will now be independent of the earth.
10. *Develop space stations and colonies*.
11. *Develop solar energy for propulsion, as well as a source of energy*.
12. *Build asteroid colonies and colonies on other small bodies*.
13. *Develop asteroid colonies further*.
14. *"Human society and its individual members become perfect."*

1930–1939

Experiments are made with winged rockets in the U.S.S.R. by the winged rocket teams of the Group for the Study of Jet Propulsion (GIRD) [see 1932] and of the Jet Propulsion Research Institute.

1931

Harold A. Danne presents a report to a meeting of the American Interplanetary Society on the possibility of transatlantic rocket planes.

He proposes an amphibian rocketship 170 feet long and 30 feet in diameter and "as perfectly streamlined as we can make it. "It will have a hermetically-sealed cabin, folding wings and retractable landing gear. The rocket would leave New York City from Floyd Bennett Field and land in France at Le Bourget Field, a distance of about 3,000 miles. The flight would last about one hour with another half hour spent climbing to an altitude of 50,000 feet (beginning at an angle of 10 degrees or less, increasing to 45 degrees as speed increases) and descending. Its (probably telescoping) wings would fold into the fuselage at speeds over 200 mph. Wingless, the spacecraft would "rely on the surface of the fuselage to keep . . . aloft and enable [it] to ride the air rushing by at such enormous speeds."

Lawrence Manning discusses methods of boosting spacebound rockets. Although he considers a three-step rocket capable of reaching space as being just "barely possible" with present technology, more powerful fuels would be needed to make spaceflight a practical fact. However, "if we can add a little extra starting push to present[ly available] power, perhaps we might not need to await new discoveries."

The first suggestion Manning makes is to start the rocket from as high an altitude as possible; a mountainous site near a railroad would be ideal. A track is laid down the incline, with a huge wheeled cradle carrying the spaceship. Gasoline engines would start the wheels of the cradle rolling and as speed increases, rocket engines attached to the cradle would take over. When full speed is attained, the spaceship's engines would start. A gentle upward curve and a wide-set wheels would keep the ship from jumping the track.

Manning did not think that even this would be enough to add the 1 mps he was looking for. He suggested the addition of sequentially-fired electromagnets so that the cradle would "be pulled along in the same way as the rotor of an electric motor."

Manning also considered the construction of a huge cannon, 5,000 feet long. The projectile/spaceship would be gradually accelerated (not by an explosion) at a rate of 100 fps to a speed of 1,000 fps, which Manning considered far too low. He suggested that an acceleration of 50 g's could achieve a velocity of 1 mps. This would require a gun barrel 3 miles long, something he believed to be not outside of probability.

Manning went on to discuss the possibility of centrifugal-launchers. A 100-foot spaceship would require a wheel (or arm) at least 500 feet in height. This would revolve at the desired speed of 1 mps. However, this increase every pound of an occupant's weight to 1 1/2 tons! However, this obstacle might be overcome by making the fulcrum longer. A wheel six miles high would only need to revolve 3 times a minute. The speed at its circumference would be 1 mps, but the centrifugal force on the passengers would be only 50 g's, a level Manning thought tolerable.

Although a six-mile wheel would be impractical to build, perhaps one the height of the Empire State Building (or about 1,200 feet) might not. It need not be a true wheel, but rather two spokes, at the end of one of which would be the 5,000-ton spaceship. This would be balanced on the opposite spoke by a counterweight.

Manning provided this table:

Diameter	Speed	RPM
1,200 feet	1,000 fps	16
3,500 feet	1,750 fps	10
4,800 feet	2,000 fps	8
3.5 miles	4,000 fps	4
6 miles	1 mps	3

Manning described what such a launch might be like for the spaceship's passengers. The pilot would be lying on a couch of springs and mattresses. At the peak acceleration, he would weigh 7,500 pounds "and his powerful chest muscles can scarcely lift his ribs enough to obtain breath, "Manning understates. Realizing this, he admits that a forced breathing apparatus might be required (such as an iron lung). A mechanical device would release the ship at the precisely correct moment (since, as Manning points out, a delay of even a microsecond could launch the ship directly into the ground).

David Lasser publishes, privately, *The Conquest of Space*. It is necessary for him to finance the book himself since no publisher is interested in such a fantastic subject. It is the first nonfiction book in the English language on the subject of astronautics. The latter half of the book is devoted to vividly describing a fictional flight into space by way of a manned rocket. Christened the *Terra*, the rocket has been built in the Swiss Alps by an International Interplanetary Commission [also see 1931]. The 150-foot, 10,000-ton rocket cost more than $100 million. The spaceship is bullet-shaped, its tail honeycombed with the rocket exhaust tubes, with three pairs of wings. Inside the spaceship is an airlock, and a small, rectangular metal room that serves as living quarters for the crew of eighteen. Hammocks (padded and laced with felt straps) hanging from the walls by strong steel springs, enclosed bookcases and lockers, and two portholes (of a "creamy, translucent glass" to filter out the harmful rays of the sun) occupy the space. There is also a control room and an observation room. The former contains six hammocks, in two rows of three, on the walls. The walls are lined with portholes and instruments. A door from here leads to the galley and washrooms, and another to the observation tower near the nose. In the nose itself are the parachutes for the return to earth. Everywhere are handholds for use during free fall. The ship's hull is double-walled, with a vacuum between, for insulation.

The *Terra* has been constructed inside a vast hanger, from which the launching track leads. The tracks rise along a gentle slope for a quarter of a mile. As the time for the launch nears, the rocket is wheeled out of the hanger on its long cradle. The *Terra* is a three-step rocket. The first step accelerates the spaceship along the track, reaching a speed of 3.5 miles per second. The undercarriage is jettisoned when the end of the track

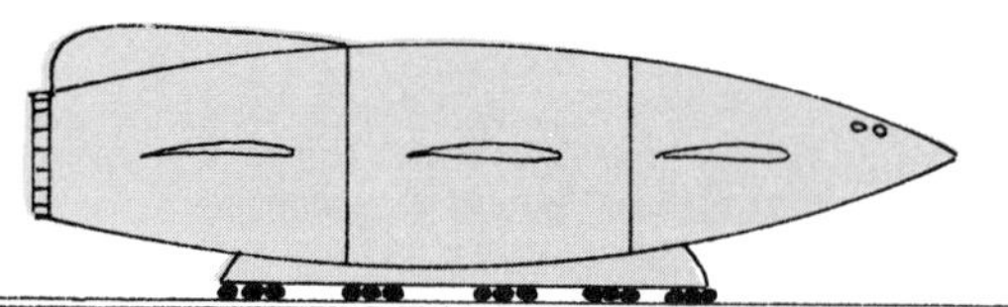

is reached. When the fuel in the first stage is exhausted, it is released and an explosive charge destroys it. The second stage takes over and it too is destroyed after it has been jettisoned. The first stage is fuelled by a mixture of alcohol and liquid oxygen, the second and third stages by liquid hydrogen and liquid oxygen.

The effects of both takeoff and weightlessness are accurately and vividly described, as is a visit to the outside of the rocket in a spacesuit. A flight around the moon is accomplished and the *Terra* returns to the earth. The rocket descends by parachute, landing in the Atlantic Ocean.

Lasser turns from this fictional account, spending most of the remainder of his book to explaining the genuine difficulties facing actual manned spaceflight—as opposed to the idealized flight of the *Terra*. He turns to fiction once again in the final chapter. Here he describes what life might be like in the year 1950, after rocket flight has become commonplace. All of the world's centers of population have been connected by passenger rocket lines. He describes what a trip from New York to Europe might be like. The rocket port is on Long Island. "From gleaming metal-sheathed hangers, the rocket planes are wheeled out into their places to receive passengers and freight. Side by side rest these great gleaming birds. 'LONDON-NEW YORK' is emblazoned on one. On another 'PARIS-NEW YORK' and on a third 'BERLIN-NEW YORK.' " These can span the ocean in an hour. Commuters from Massachusetts, Delaware and Pennsylvania transfer to autogiros for the trip into New York. These are equipped with Goddard rocket-propellor systems [see June 9, 1931].

Perhaps by 1950, Lasser hopes, Oberth's space station will be under construction in earth orbit, providing a stepping-stone to the planets.

Cattaneo's "RR" rocket, built by Piero Magni Aviation, June 1931.

Ettore Cattaneo of Milan flies a rocket-powered glider in Italy. He is the third man to fly a rocket-propelled aircraft [see 1928, 1929]. His rocket plane is a high-wing monoplane with an enclosed fuselage and twin vertical stabilizers.

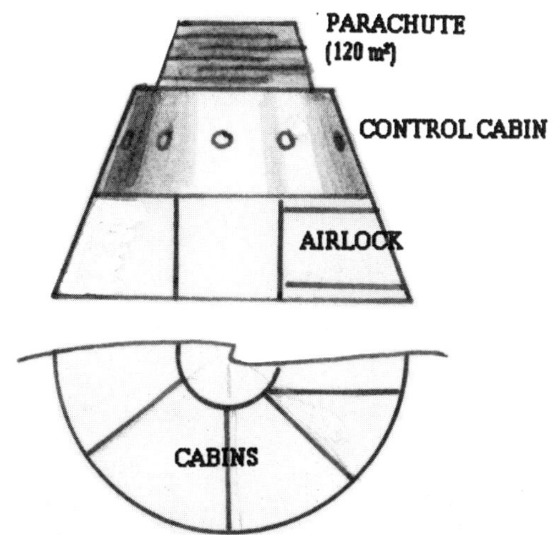

Otto Willi Gail publishes *By Rocket to the Moon* (*Hans Hardt's Mond Farht*), a juvenile novel by the popular German science fiction author. In it he describes a trip to the moon in the spaceship *Wieland* (named for a legendary blacksmith), based on the principles of Max Valier. The moonship proper is a fat torpedo with guidance fins at its tail. Portholes pierce the hull at the nose and five exhaust nozzles are clustered at the rear. It is built of steel, aluminum and beryllium (a relatively new metal now popular with science fiction authors). One of the small windows at the nose acts as a door, giving entrance into the small airlock. Beyond this is the main cabin, shaped like a truncated cone. Below is another room of the same size, divided by a partition. Half is used for a kitchen and half for sleeping quarters. There are no ladders connecting the cabins, since most of the flight will be made in weightless conditions. Instead there are light rope ladders and numerous handholds. The control room has a full and elaborate complement of instruments. In the very nose of the *Wieland* are three parachutes, each of about 400 square feet, which surround a small, spherical escape capsule. The remainder of the rocket is occupied by tanks of fuel: oxy-hydrogen gas.

While the spaceship is under construction, the crew undergoes training, including time in a high-speed centrifuge.

The *Wieland* is carried on the back of an enormous winged booster. This is propelled by a large number of rockets mounted in its broad wings. The entire assembly sits on a steel launch track 40 feet wide. It runs up a natural slope for 1 1/4 miles. At the time of takeoff its crew lie in hammocks. The booster takes off down the inclined track, rising into the air at the end, carrying the spaceship with it. At a speed of 2 1/2 miles per second, the booster disconnects and the *Wieland* fires its own engines. Gail vividly describes the effects on the crew of the rocket's acceleration. Once insertion into the translunar trajectory is accomplished, the engines are throttled back. They are not turned off, however; a constant 10 feet per second acceleration is maintained to provide a small amount of artificial gravity within the ship.

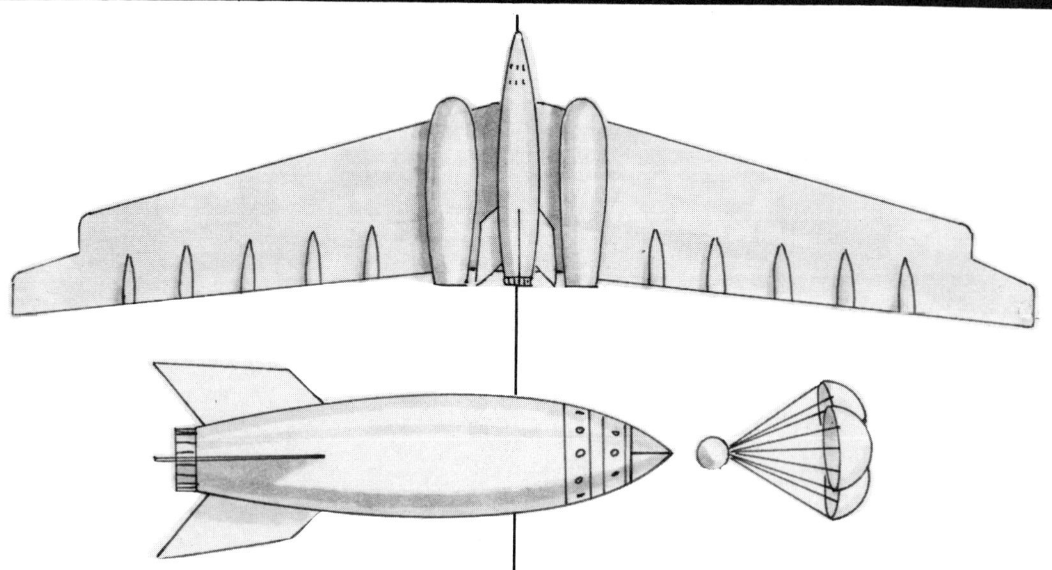

When at one point the engines are shut down the author accurately describes the many effects of weightlessness (with the glaring exception of allowing one of his characters to smoke a pipe!). It is necessary to exit the ship to construct a large reflecting telescope, which is accomplished by the spacesuited astronauts. They can maneuver themselves by firing shots from revolvers.

Landing on the moon is accomplished by firing the engines as retrorockets, against the direction of the ship's travel. Completely out of their oxyhydrogen fuel, the resourceful astronauts find a way to produce it from the materials they find on the moon.

K. E. Tsiolkovsky publishes *From Airplane to Starplane*.

Neil R. Jones publishes the novelette *The Jameson Satellite*, which describes the first earth satellite launch in 1958 (to place its deceased builder's body in orbit). The spacecraft is radium-powered.

Robert H. Wilson publishes the short story "Out Around Rigel." He describes a trip to the star Rigel in the spaceship *Comet*, which is a 30-foot cylinder with pointed ends, 15 feet in diameter. In the point at either end is the lens of a powerful telescope. Four "fins" at each end apply the working (unspecified) power. The ship is built of "helio-beryllium"

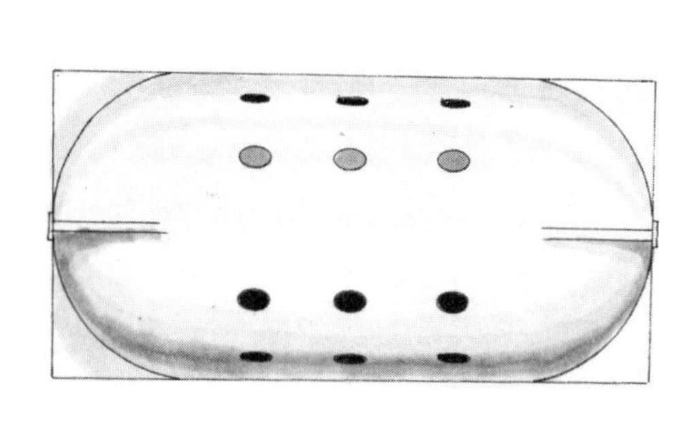

and contains a single large cabin, 10 feet high.

Eugen Sänger begins development of his rocket plane in Vienna. He does not mention his project for a semi-ballistic rocket plane publicly until the beginning of February 1933. The plane described is of fairly conventional configuration (though it resembles the later *Silver Bird* antipodal bomber—see below), propelled by a gasoline–liquid oxygen rocket motor. It would reach a velocity of 6,200 mph (Mach 10) and fly at altitudes between 37 and 43 miles. It is smoothly streamlined, "pointed in front and blunt at the back-end to give room for the exhaust. The profile of the wings has to be as thin as possible, with sharp leading edges. The wing span can then be kept short because of the negligible resistance of the wing-edges."

Eleven publishers reject Sänger's manuscript in 1932. It is finally published by the same company that produced Oberth's,

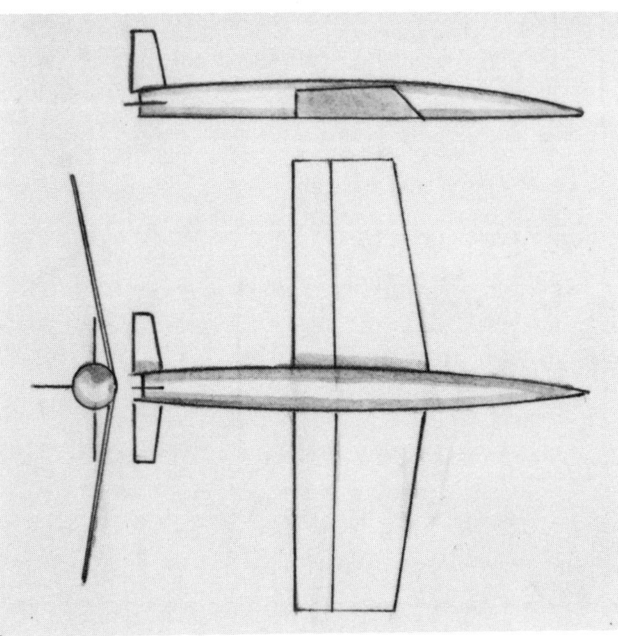

Hohmann's and Valier's books—but only after Sänger first contributes an exorbitant fee. It takes him 4 years to pay off the debt of printing what is now regarded as one of the founding classics of modern astronautics. [Also see 1943.]

N. A. Rynin and A. A. Likchachev publish *Effect of Acceleration on Living Organisms*. An early work on space medicine, it discusses experiments made with a simple centrifuge.

The *Bulletin* of the American Interplanetary Society discusses the possibilities of using cannons and giant flywheels [see 1913–1916] as boosters for rocket spacecraft. Another suggestion is to place the rocket on a kind of railroad flatcar. This would be powered and capable of reaching 1,000 mph on its downhill run. The car and rocket would travel down a sort of "ski jump" constructed on a mountainside. The track would gradually curve upwards. At the end of the jump the rocket would start its own engines, taking off without the flatcar. A similar idea is shown at the end of the film *When Worlds Collide* [1951].

Czechoslovakian author Jos. Hais Týnecký publishes his charming children's novel. *Na Měsic a Ještě Dál* (To the Moon and Beyond), in which a company of insects build a bug-shaped spaceship and travel throughout the solar system. The book's combination of whimsy and good science is beautifully illustrated by O. Stáfl.

JANUARY 23

An International Commission for Astronautics is proposed by David Lasser of the newly formed American Interplanetary Society [see 1930]. It is in the form of a letter signed by Lasser as president of the society and Robert Esnault-Pelterie on behalf of the French Committee for Astronautics and sent to Hermann Oberth, then president of the VfR [see 1927]. It suggests that the national groups of experimenters unite in a single federation.

MARCH

David Lasser, in an article published in *Scientific American* ("The Future of the Rocket"), reports on the current state of the art. He discusses some basic rocket theory and explains some of the ideas of Oberth and Goddard. He describes some of the possibilities of manned space travel. A step rocket would be necessary, of three or more stages. The first and largest would carry the rocket until a speed of 2 miles per second is reached. This stage would then be detached and would return to earth via parachute. This is repeated for succeeding stages until the final, manned stage has achieved a speed of 7 miles per second.

Lasser suggests, if no suitably powerful fuels are developed that would enable a rocket to carry enough fuel to the moon for the return trip, that perhaps a dozen smaller rockets might be launched on one-way trips to the moon. Each would carry a small quantity of surplus fuel. When the main rocket arrives it would gather together this fuel for its return trip. The small rockets would either remain on the moon or be salvaged.

MARCH 14

Johannes Winkler launches the first liquid-fuel rocket to be flown in Europe. The small—about 2-foot—HW-1 flies to an altitude of 295 feet, landing 656 feet from its launcher. It is fueled with methane, using liquid oxygen as the oxidizer and nitrogen to provide a pressure-feed. Winkler has been president of the VfR, though is now inactive in the organization since it has moved its activities to Berlin. Some of the VfR's members are distraught that Winkler, working on his own, has achieved what they have so far failed to do.

SPRING

Junkers, the German aircraft company, announces plans for a 500 mph stratosphere plane.

APRIL 11

Max Valier makes a test run in a liquid-fueled rocket car. Valier is killed soon thereafter, on May 17, when a test engine explodes in his laboratory.

MAY 14

Willy Ley and Klaus Reidel successfully launch a liquid-fuel rocket, called the *Repulsor* (a name they wish all their rockets to be known by, to distinguish them from common fireworks rockets—the word is taken from Lasswitz [see 1898], after the power system he described in his novel).

G. Edward Pendray writes about this launch in the *Bulletin* of the ARS, saying: " . . . it is clear that for the first time we have, in the one-stick Repulsor, a rocket that could be made in a large size, able to carry scientific instruments into the stratosphere or into space, or able to transmit mail, freight and even passenger from one point to another on the earth."

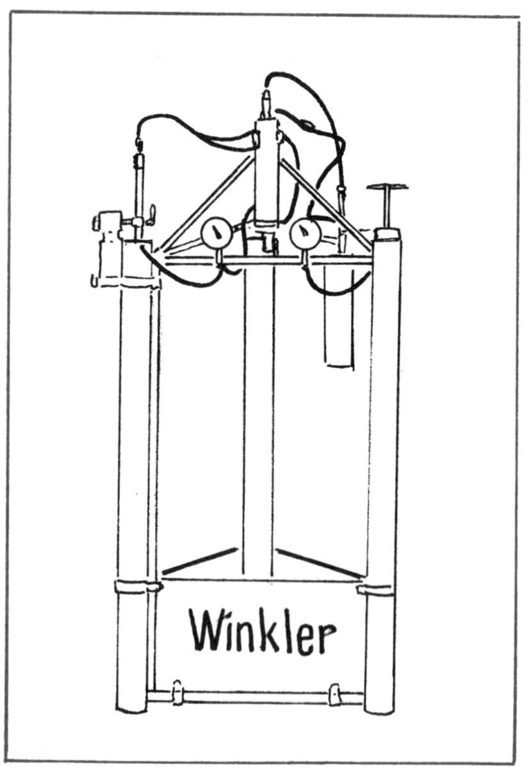

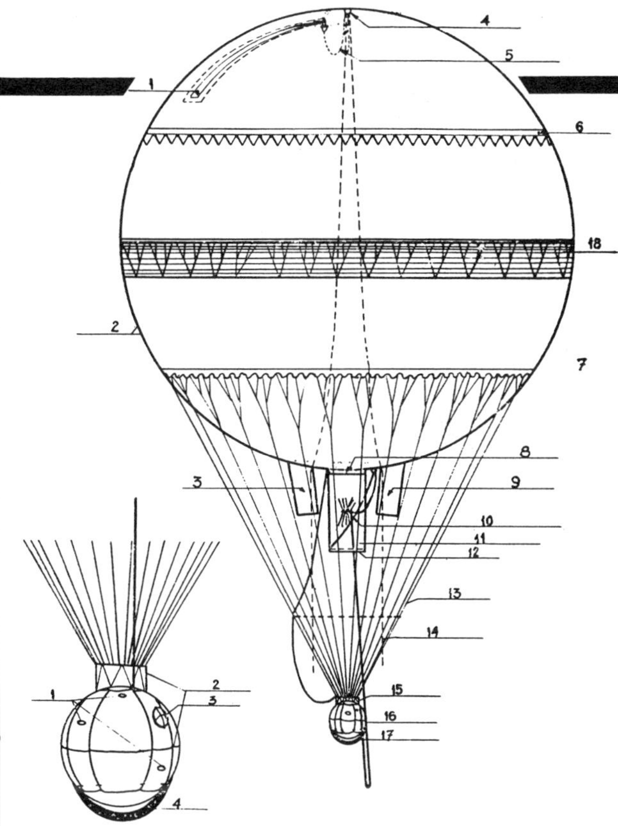

MAY 27

Auguste Piccard ascends to an altitude of 9.81 miles (51,775 feet) in a balloon. He has flown farther from the earth than any man before him. Professor Piccard's balloon, sponsored by the Belgian Fonds National de Recherché Scientifique (created by King Albert I), rises well into the stratosphere, leaving 90 percent of the earth's atmosphere below it. "The sky is beautiful up there," writes Piccard, "almost black. It is a bluish purple—a deep violet shade—ten times darker than on earth, but it is still not quite dark enough to see the stars." Professor Piccard's spaceship is an aluminum sphere 7 feet in diameter, with walls 0.138 inch thick. It is pierced by two manholes and eight tiny windows. Its interior is just large enough to hold its two passengers (the professor takes along his assistant, Paul Kipfer) and their boards of instruments. It is equipped with tanks of liquid oxygen for replenishing the air, but they are counting on the Draeger apparatus to continuously produce oxygen for breathing, and to cleanse the atmosphere of carbon dioxide. Professor Piccard credits the research done on long term confinement of crews in submarines for inspiring his own life-supporting apparatus [also see Monturiol, 1858–1860].

The gondola is attached to a monster balloon 99 feet in diameter, weighing about 1,600 pounds. The empty sphere weighs about 300 pounds; with passengers and equipment it weighs 850 pounds. The initial attempt at reaching the stratosphere is more adventurous than scientifically productive. An air leak threatens the scientists with asphyxiation, a broken cable prevents them from descending on their own, and a nonfunctioning device intended to rotate the gondola keeps its black-painted hemisphere turned toward the sun and interior temperatures climb to 104 degrees. [Also see August 18, 1932.]

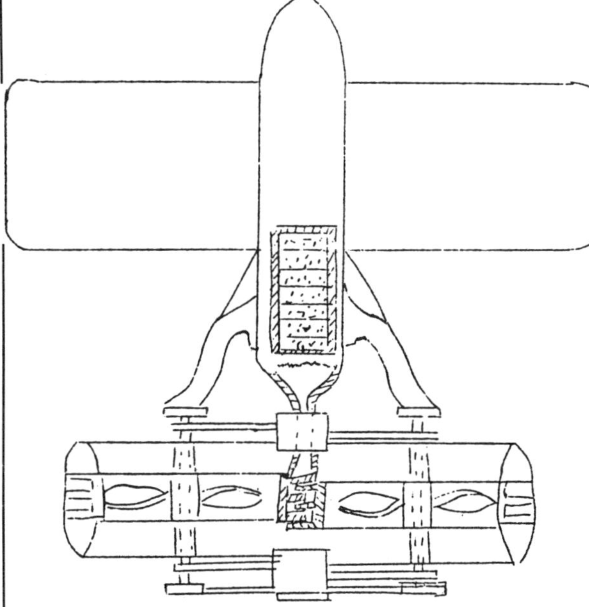

JUNE 9

Robert H. Goddard patents a rocket-propelled airplane. The exhaust gases from a rocket engine mounted in the tail of an airplane impinge on the interlocking blades of a pair of turbines. The two turbines are mounted on either side of the tail and operate a pair of conventional propellors via a pair of long drive shafts. At altitudes where propellors are no longer practicable, the turbines are disengaged. This stops the propellors and the plane flies on rocket power alone. At low altitudes, Goddard points out, most of the heat energy of the rocket fuel is wasted in ejecting the exhaust gases at uselessly high speeds. His rocket turbine would recapture some of this energy. This rocket turbine plane is capable of flying beyond the earth's atmosphere.

JUNE

A letter from Nathan Schachner, current secretary of the American Interplanetary Society, brings readers of *Wonder Stories* up to date on AIS activities during its first year. At the end of the coming year it is planned

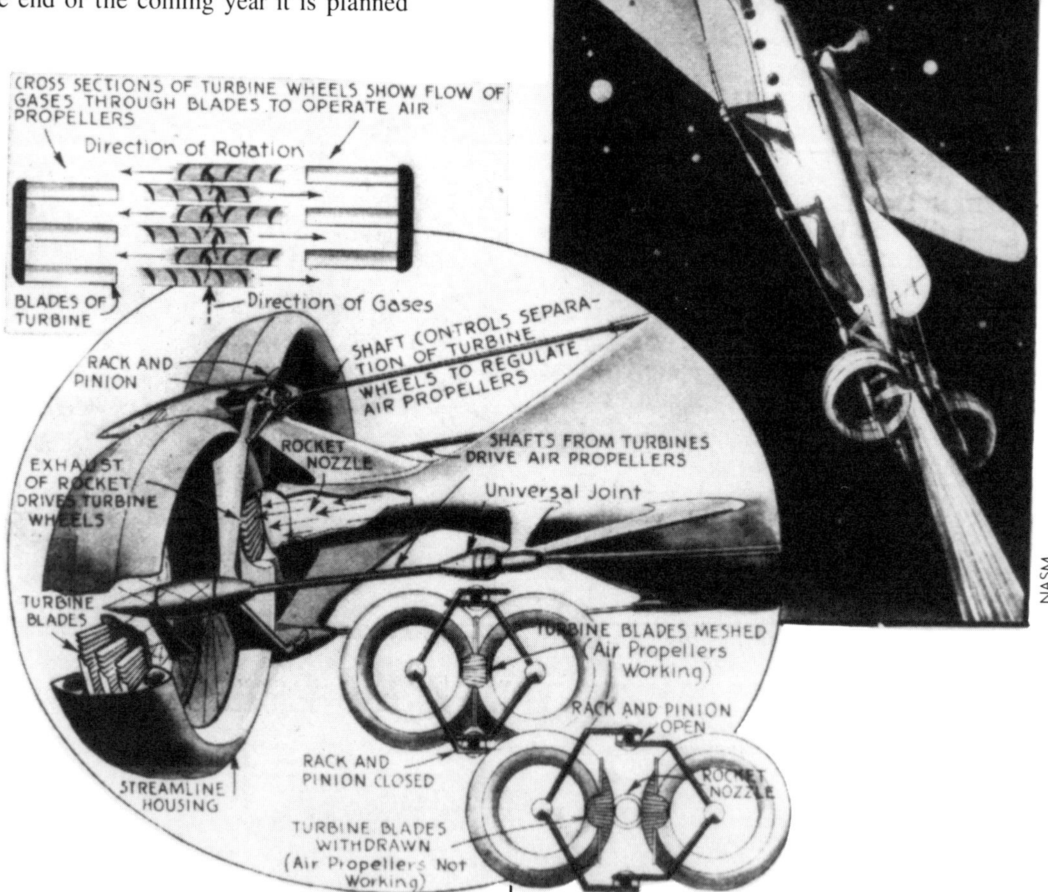

to publish a report on its researches "which will be the first extensive survey on the rocket in English." It also plans to increase the size and scope of the *Bulletin*. There are plans to launch an actual rocket before the end of the year. Undoubtedly under Lasser's urging, there is also a plan to implement an International Interplanetary Commission. The AIS now boasts members in thirty-six states as well as Canada, Mexico, France and Russia (possibly N. A. Rynin?).

JULY

Max Valier has a story published in *Wonder Stories*: "A Daring Trip to Mars" in which we learn that a woman is making the spaceflight because "Somebody must do the housekeeping . . . "

After a stop on the moon to refuel (with water that has been dissociated into hydrogen and oxygen) the journey to Mars is interrupted by an encounter with a comet. The landing on Mars is abandoned and only a fly-

Art by "Mr. Jex."

by is made before returning to the earth. The rocket, reentering the atmosphere tail first, deploys a series of air brakes to aid the retrorockets in slowing down the spacecraft. This is a long cable with a series of slightly conical metal disks spaced along it. Once the ship has slowed enough, the control cabin is ejected and makes the final descent by parachute.

OCTOBER

GIRD (Group for the Study of Reactive Motion), the Russian jet propulsion division of OSOAVIAKHIM, is founded by Friedrikh Tsander, Sergei Korolev and others. The unofficial motto of the group is "On to Mars!"

NOVEMBER

Nature magazine publishes David Lasser's article "By Rocket to the Planets. "After describing how a rocket operates and the experiments and theories of Robert Goddard and others, Lasser [also see above] tells his readers what a future trip into space aboard a rocket might be like. "At the start we will be lying in a metal-walled cabin of the rocket ship, in comfortable bunks . . . " in order to best resist the effects of acceleration. "A great roar sounds behind us," as the rocket takes off. For ten minutes the passengers feel as though their body weight has doubled. Soon the rockets are shut off, and, travelling at 7 miles a second, the spaceship coasts on into space. "We look out of a window at the side of the cot and see with amazement that our earth, fifteen hundred miles below, has shrunk appreciably in size."

Lasser describes the pleasures and problems of weightlessness; he believes that the latter, however, will seem to be nothing compared to the strange psychological effects of space travel.

The article is accompanied by 3 striking lithographic crayon drawings credited to "Mr. Jex" (Garnet W. Jex, the magazine's art director).

Hugo Gernsback describes a flight from "Berlin to New York in Less Than One Hour" in his magazine *Everyday Science and Mechanics*. The rocket transport "could make the 3,960-mile flight in about 20 minutes; although with the necessary acceleration and deceleration, the actual elapsed time would be somewhat less than 1 hour.

"The rocket transport would rise about 628 miles above the surface of the Earth. Then in descending, from about 150 miles up, near New York, decelerating rockets would be blasted in front of the machine . . . to slow it down as it re-entered the atmosphere." The rocket has retractable wings,

needed for the final glide to a landing.

NOVEMBER 30

Reinhold Tiling successfully launches a 60-inch rocket from Wangerooge Island, Germany. According to one report it ascends to an altitude of 6 miles, though this is probably an exaggeration, at which point its wings unfold and the rocket glides to a landing 5 miles from the takeoff point.

The enthusiastic Tiling declares that, "Eventually a man-sized space ship will fly the Atlantic Ocean within a few hours."

1931–1932

On the initiative of the Society for the Study of Interplanetary Flight [see 1924] two new groups are formed: the Central Group for the Study of Reactive Motion, or CGIRD, in Moscow, under the chairmanship of F. A. Tsander; and the Leningrad Group for the Study of Reactive Motion, or LenGIRD, under the chairmanship of N. A. Rynin.

In 1932 the Moscow GIRD develops the liquid fuel motor designed by Tsander, the OR-1, which, fueled by benzine and liquid oxygen, develops a thrust of 5 kg.

The trajectory (below) and the landing (left) of Gernsback's rocket transport.

Harry Bull and his spaceship model, built for an exhibition, 1931.

Also in 1932, the Soviet government creates a rocket design and experimental development establishment under the leadership of Sergei Pavlovitch Korolyov.

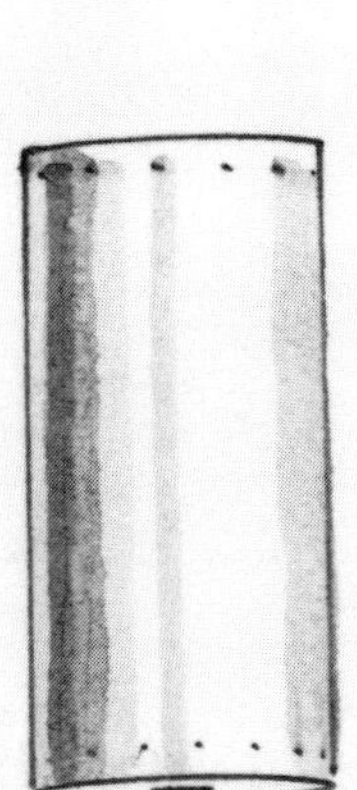

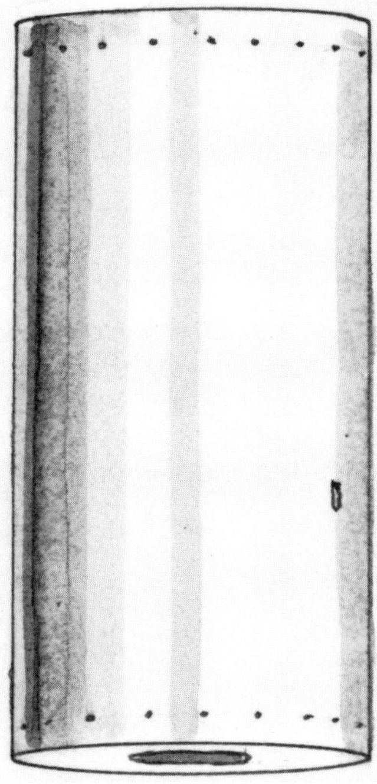

1932

Philip Wylie and Edwin Balmer publish their classic novel *When Worlds Collide*. When the earth is threatened with destruction by collision with a rogue planet, plans are made to construct spacecraft to carry a small number of human beings to safety. The gas giant that will actually collide with the earth is accompanied by an earth-like satellite. After the collision, this new planet will go into a highly elongated orbit around the sun. It is this world that is the destination of the arks.

Two rockets are ultimately built by the Americans (several other countries create their own arks). The *Noah's Ark* is a fat cylinder 135 feet tall and 62 feet in diameter. It is not streamlined. Its outer hull is a special alloy 18 inches thick, plated with a chrome-like finish. Inside is a smaller shell, the space between the two filled with an insulating material (books!). The ship is powered by atomic rockets. Later, a second, much larger rocket is built as a companion vehicle, though its exact size is unstated. Inside the arks, which are entered through airlocks, is a central spiral staircase and a taut cable that runs along the ship's axis. Floors divide the cylinder every 8 feet. The top two sections are filled with machinery and instruments. Through them run the thrust-beams against which the atomic rockets would exert their thrust. Around the upper and lower circumferences are twelve smaller, directional rockets for steering. In the stern are the engine rooms. Above them are the stockrooms for the cattle and other livestock. Above these are storerooms that continue up to the center of the rocket. The middle floors are reserved for the human cargo of 500 people, and are heavily padded. There are water taps, but few other comforts since the flight is only expected to take 90 hours.

Above the passenger quarters are more storerooms and then the forward engine rooms. The chief scientist explains the layout of the engine room: "The breeches of the main tubes are concealed behind a wall which is reenforced by the thrust-beams. These are the ones which are to break our fall; but you can see here the breeches of the smaller surrounding tubes. They are not unlike cannon, and they work on the same principle. Acting at right angles to our line of flight, they can turn the ship and revolve it end for end, in fact, like a thrown firecracker, if we should turn on jets on the opposite sides and opposite ends."

The atomic power is provided by the disintegration of beryllium into protons and nuclei. There are twenty exhaust nozzles in the base and an equal number in the top of the cylinder. [Also see 1951.]

GIRD establishes a winged rocket team under the supervision of Sergei Korolev. Its formation is inspired by the writings of F. A. Tsander and K. E. Tsiolkovsky on the use of wings on rocket vehicles. By 1933 liquid-fuel motors available at GIRD have thrusts from 66 to 110 pounds. An aircraft powered by one of these engines, to operate at its optimum values, would need to weigh in the vicinity of 88 to 132 pounds or less. Thus, the group turns first to pilotless models. In 1932–1933, GIRD mounts an OR-2 engine, alcohol and liquid oxygen fueled and with a thrust of 110 pounds, in a flying wing glider, the BICh-11, designed by B. I. Cheranovskiy. The rocket glider is then designated the RP-1 (RP = "raketoplan"). This work is overseen by Korolev. A number of gliding trials are made, but work is stopped

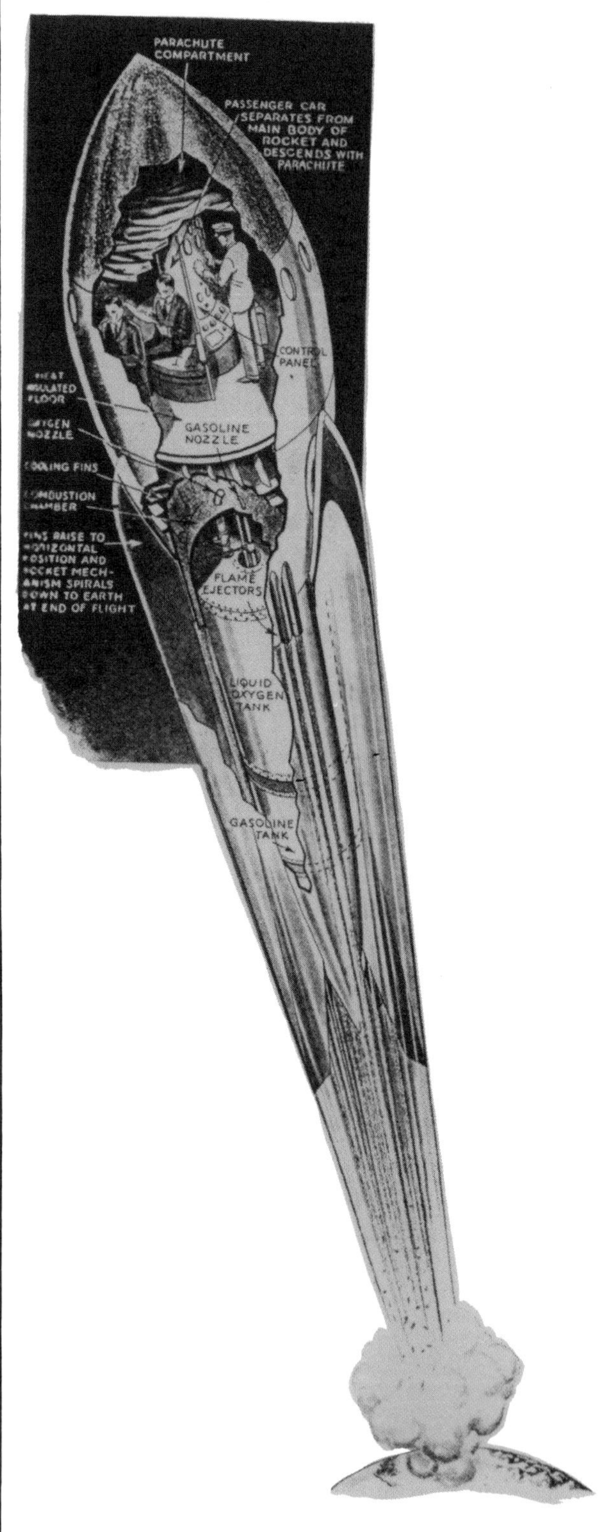

Drawing published in *Popular Science* (August 1931).

(when GIRD and GDL—the Gas Dynamics Laboratory—merge) before a powered flight can be made, and the worn-out RP-1 is retired. [Also see 1940.]

Lester D. Woodford, while an engineering student at Ohio State University, develops a liquid-fueled rocket engine. It is suggested that this engine is meant to be used for aircraft propulsion. An article in *Popular Science* magazine (September 1932) reports that Woodford "feels the perfected ship could fly to the moon and back. The model carries an automatically operated parachute for safe landing" and that Woodford is carrying out experiments with the model "on an island in a Canadian lake." Although Woodford becomes a professional engineer of some standing, assistant editor of the *Ohio State Engineer*, president of the Ohio State Aeronautical Society, and a licensed pilot—he has apparently, and mysteriously, disappeared. Rocketry historian Frank Winter has been unable to discover any trace of his whereabouts.

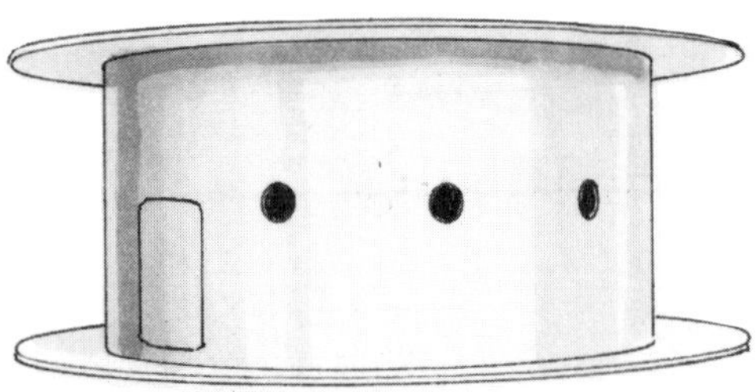

Jack Williamson publishes his short story "The Moon Era." He describes an unnamed spaceship built for a flight to the moon. It is propelled by an electrical antigravity device using a high-tension current. The ship is a squat cylinder of chrome-plated steel 8 feet tall and 16 feet wide. At top and bottom are two large disks, 20 feet in diameter, made of copper. The walls of the cylinder are 4 inches thick, and lined with a soft white fiber. There are windows and a small 4-foot door.

K. E. Tsiolkovsky publishes *The Semireactive Stratoplane*.

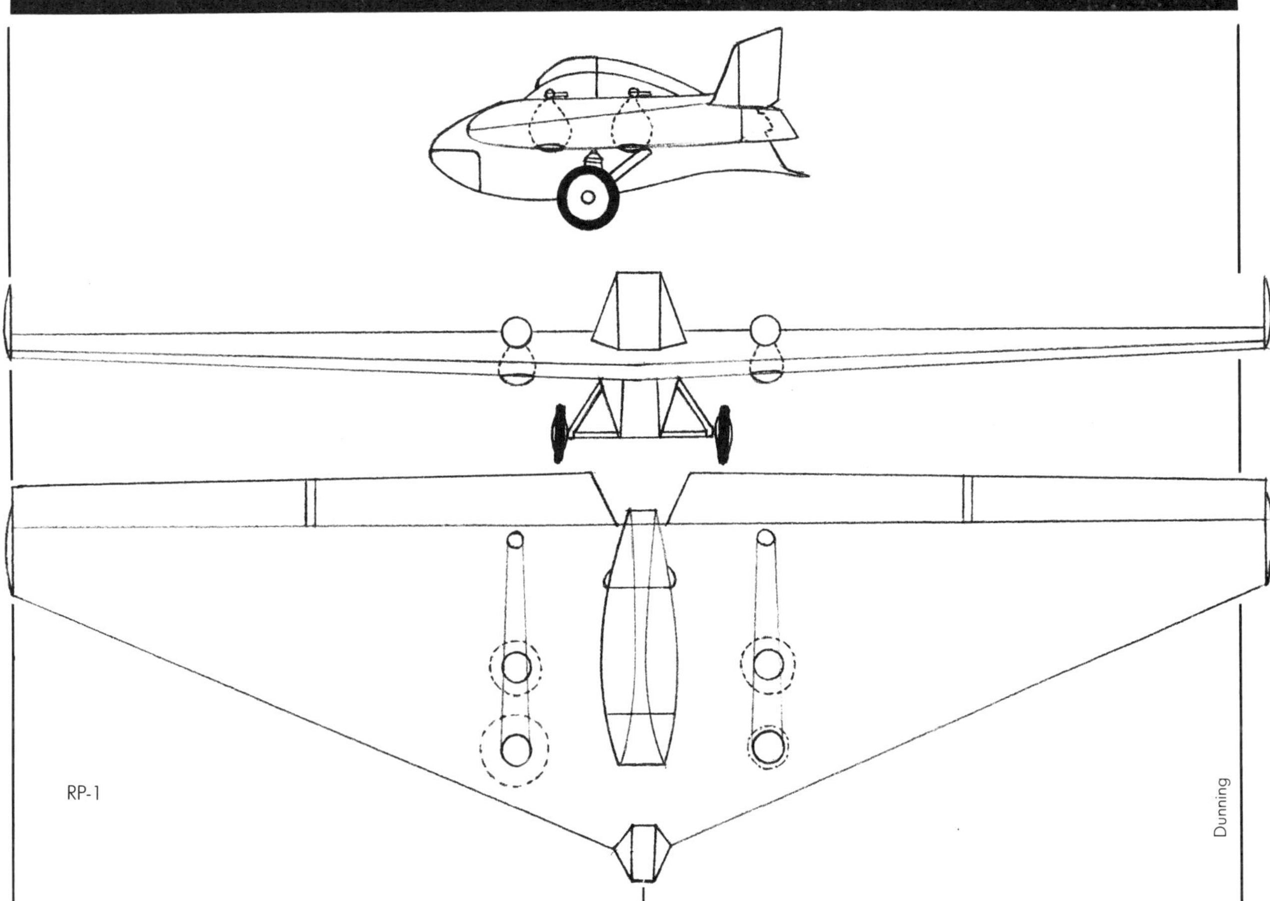

RP-1

Vladimir Mandl (one of the founders of the field of space law) describes a high-altitude rocket in his book *The Problem of Interplanetary Transport*. It is three nested or concentric cylinders, the payload being carried in the nose of the innermost rocket. The rocket nozzles are in the form of slots around the bases of the overhanging heads. Either solid or liquid fuels would be used. The intention is to use the rocket for high-altitude research.

Hearing rumors that a rocket capable of travelling to the moon has been built and is on display at Yale, astronomer O. J. Schuster comments that it does not seem to him to be an impossibility, given the unlikely event that a sufficiently powerful source of energy be discovered. In speculating about what a successful spaceship would be like, he proposes that it would be torpedo-shaped and would carry three passengers. These would be forced to lay in hammocks to protect themselves against the acceleration of takeoff. Schuster fears that weightlessness may have unfortunate effects on the physical and mental well-being of his crew, however. Schuster does not believe that the difficulties facing the first spaceflights will be easily overcome and that "it is not at all likely that any person now living will make a trip to the moon and return safely to earth."

FEBRUARY

A letter from Britisher John Beynon Harris in *Wonder Stories* hopes that David Lasser's *Conquest of Space* (1930) will be published in England: "I hear from a friend that it is causing a sensation." The editor, in his reply, mentions the *New York Herald-Tribune* review in which Lewis Gannett remarked that Lasser's book "sounds crazy."

MARCH

An explanation of the rocket principle is published in *Wonder Stories*. One of the accompanying diagrams illustrates a liquid-fuel rocket. Until the advent of World War II almost all factual information being made available to the public about rocketry and spaceflight will appear in science fiction magazines.

1932

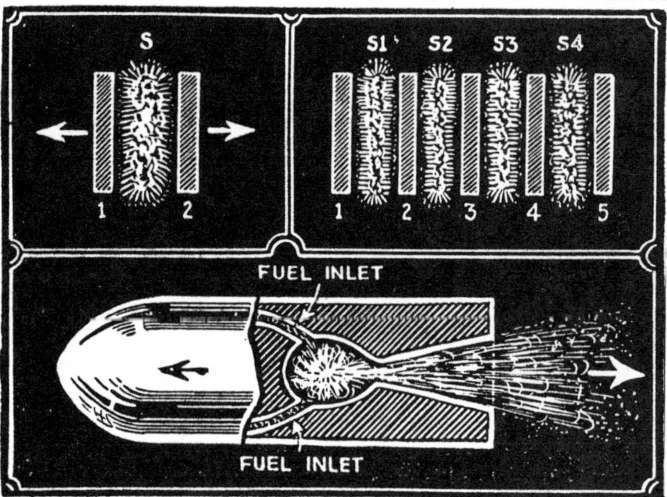

MAY

Popular Mechanics magazine predicts that "Rocket-propelled airships of the future may resemble present-day Zeppelins except that they will be equipped with stubby wings that telescope into the sides of the craft when not needed." The same article reports that German experimenters plan to launch their planned spaceships from giant catapults, using their wings to ascend. These would be withdrawn into the body of the spacecraft once it has reached the stratosphere. At an altitude of 600 miles the speed of the ship would reach about 250 miles per hour, and a journey from Paris to Chicago might only take 15 minutes. The crew of such a spaceship would be strapped into their own airtight compartment and a spectacularly surrealistic illustration by C. P. Maltman shows just such a cockpit.

AUGUST 18

Auguste Piccard makes a second trip into the earth's stratosphere via balloon [also see May 27, 1931]. This flight is far more successful than the first, from a scientific point of view. The professor and his assistant, Max Cosyns, also break their previous altitude record, reaching a point 10.07 miles above the surface of the earth.

In deciding to use a balloon for reaching high altitudes, Professor Piccard considers the potential of the rocket in these words: "What kind of craft should we use? Three possibilities offered themselves: balloon, airplane and rocket. None of these 3 had ever risen ten miles. The rocket will do so one of these days. Eventually it will go far higher, even . . . " In his book *Entre terre et ciel* (1950) Professor Piccard again discusses, in some detail, the use of rockets in manned space travel, including the possibility of atomic propulsion.

NOVEMBER 1

Wernher von Braun is entered on the German Army payroll as a civilian employee.

NASM

Spaceship published in Dutch magazine, April 1, 1931.

"It is, perhaps, apropos," he will write, ". . . that at that time none of us thought of the havoc which rockets would eventually wreak as weapons of war." He had said earlier that, "I am sure Reinickendorf is utterly inadequate even to commence that vast experimental program which must be the precursor of success. It seemed that the funds and facilities of the Army are the only practical approach to space travel."

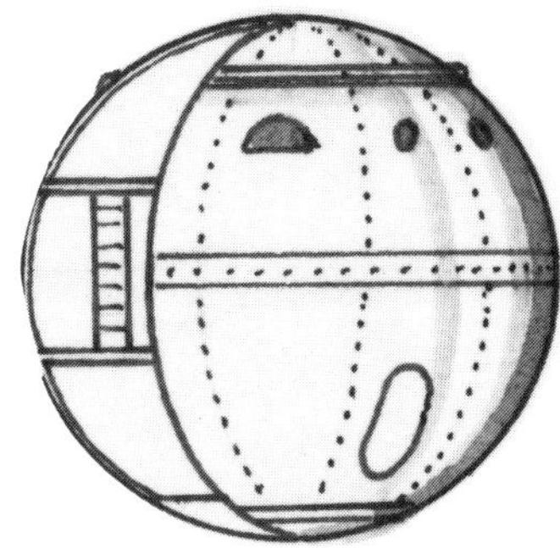

1933

C. H. Claudy, in his novel *The Mystery Men of Mars*, describes a spherical, antigravity spaceship, the *Wanderer*. It is an oblate spheroid 18 feet at its widest, made of steel and aluminum. The interior is divided into three floors. The upper is the pilot house and is provided with a circle of portholes. The middle room is equipped with three beds, a stove, table, chairs, bookcase, lavatory, etc. The lower section is for stores. A small airlock is provided for the ejection of waste. The antigravity effect causes the *Wanderer* to "fall up." Six gyroscopes are used to control the sphere's movement. Power is provided by 200 6-volt storage cells under the main floor. Internal gravity is maintained through constant acceleration; the passengers are weightless as long as the speed is constant. The passengers have to wear spacesuits while in flight in order to keep warm. They are described as looking like "Michelin men."

Frank K. Kelly publishes his short story "Into the Meteorite Orbit." He describes transcontinental passenger rockets, a space rescue and an antigravity spaceship, the

Anton Warren. It is long and silvery, its lines broken only by the exhaust jets of the "Graviton force-stream."

Edmund Hamilton writes his story "What's It Like Out There?" It is not published for nearly 20 years; most editors to whom it is submitted consider its realistic, unromantic portrayal of spaceflight too grim.

Rockets are used to boost the Russian TB-1 bomber during takeoff. Two solid-fuel rockets are placed on either side of the fuselage, above and below the wing.

Rudolf Nebel receives a loan of DM15,000 ($4,000) from the Bank of Magdeburg to build a man-carrying rocket. The "Magdeburg Project" or "Pilot Rocket" is gladly turned over in its entirety to Nebel by the VfR, who want no part in such a questionable scheme. Not the least of the strange as-

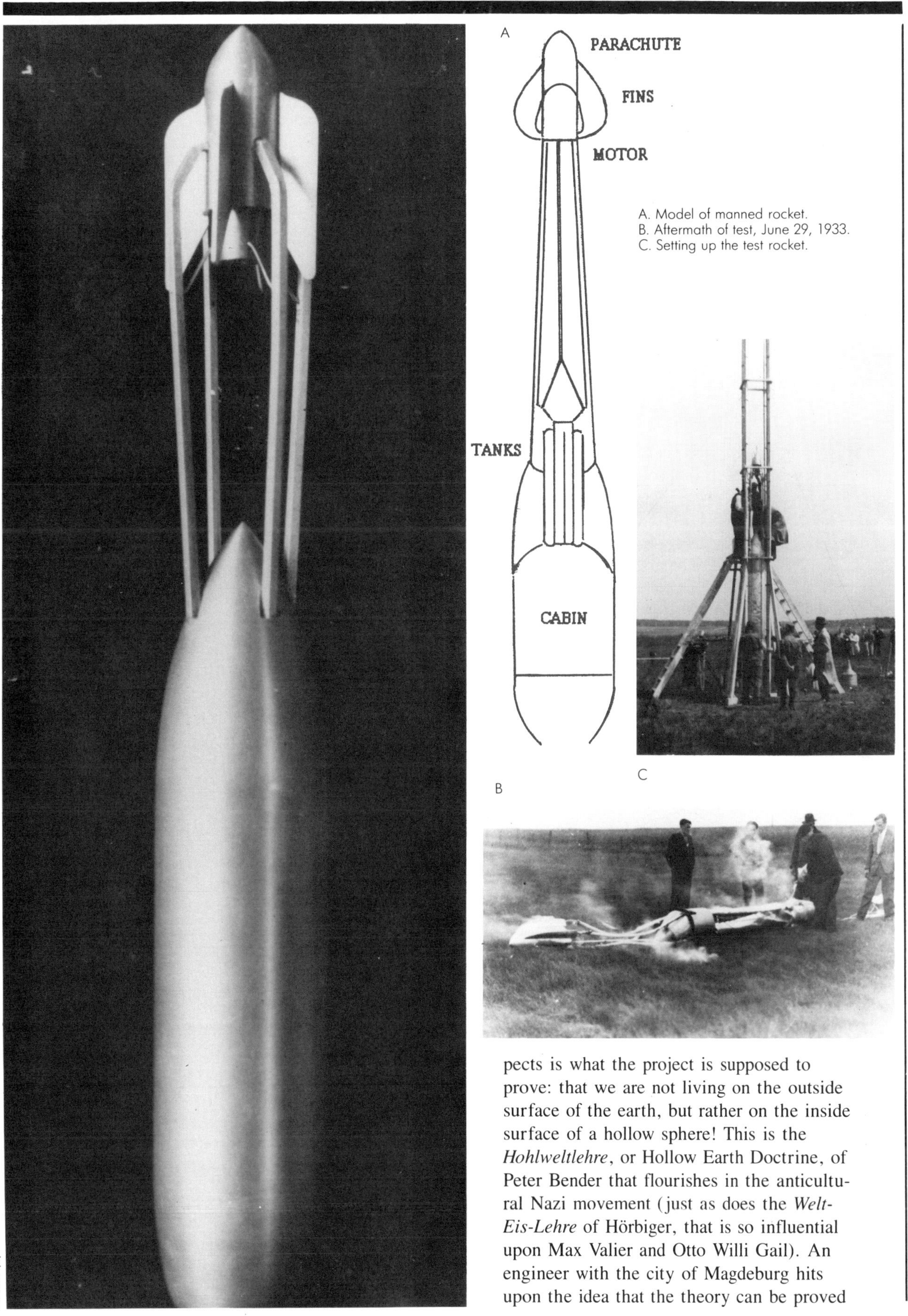

A. Model of manned rocket.
B. Aftermath of test, June 29, 1933.
C. Setting up the test rocket.

pects is what the project is supposed to prove: that we are not living on the outside surface of the earth, but rather on the inside surface of a hollow sphere! This is the *Hohlweltlehre*, or Hollow Earth Doctrine, of Peter Bender that flourishes in the anticultural Nazi movement (just as does the *Welt-Eis-Lehre* of Hörbiger, that is so influential upon Max Valier and Otto Willi Gail). An engineer with the city of Magdeburg hits upon the idea that the theory can be proved if a rocket launched vertically lands in the antipodes. To their credit, the city officials do not accept the hollow earth arguments, but instead suggest making the rocket a man-carrying one. Nebel agrees that he can build such a rocket and have it ready to launch by 11 June 1933.

The rocket is to be 25 feet tall and powered by a motor producing 1,300 pounds of thrust. The passenger cabin and fuel tanks are a single, squat, unit shaped like an artillery shell, with the motor and parachute occupying a smaller shell above the larger one. This latter is separated from the pilot's compartment by a pair of booms that also hold the fuel lines. The motor unit has a set of small vanes attached for stabilization. The planned altitude is about 1 km. At this point, the passenger will bail out, using his own parachute, while the rocket descends on the larger parachute.

A small test rocket, 15 feet tall with a motor of 440 pounds thrust, is to be built and launched first.

A test stand is built and motors are tested. Under pressure from the city to launch something, a 30 foot launching rack is erected in a cow pasture and a flight of the test rocket is planned. Out of a series of attempts the greatest distance the rocket travels is the end of the rack before sliding back down again. Eventually a more or less successful launch is made, the rocket landing 1,000 feet from the rack after a horizontal flight.

Nebel writes that "The first manned flight rocket [sic] will introduce a development which has rapid transport over the earth as its final goal. I admit that many people still laugh at such plans. However, these people also laughed and called the plans of the Mad Baron from the Bodence to be Utopia."

Darwin Lyon, supposedly a fellow of Columbia University, as part of a hoax to raise money, announces that he has successfully launched mice and birds to altitudes of 1 mile in rockets of his own design, recovering them by parachute. His work presumably has taken place in the Libyan desert. He also claims to have under construction a two-stage liquid-fueled rocket capable of reaching an altitude of between 70 and 90 miles, which he hopes to have ready in December for a January or February launch. Lyon is always able to avoid actual contact with either the public or rocket experts.

Auguste Piccard, the record-setting Belgian stratosphere balloonist [see 1931], foresees the balloon's limitations in high-altitude research. He believes that the rocket will be the natural successor, though he thinks that interplanetary rockets are still some distance in the future, when more efficient and powerful fuels are available—perhaps even atomic energy.

OCTOBER

Eugen Sänger submits a proposal for his rocket bomber to the Austrian Ministry of Defense. In his paper, "Raketenflugtechnik," although he publishes a sketch of what such a rocket bomber might look like, Sänger purposefully does not go into definite structural details. He predicts that a " . . . 5,000-kilometer flight will take place in about 5,000 seconds or at an average cruising speed of 1,000 meters/second or 3,600 kilometers/hour."

MAY 14

The American Interplanetary Society launches its first liquid-fueled rocket. It reaches an altitude of approximately 250 feet.

SEPTEMBER

G. Prokofiev, E. Bernbaum and K. Godonov take the 880,000 cubic foot *Stratostat U.S.S.R.* (the largest stratosphere balloon to date) to an altitude of 58,700 feet (11.1 miles).

OCTOBER 13

The British Interplanetary Society is founded by P. E. Cleator. Earlier in the month "three or four interested people" had met at Cleator's home at which time it was decided that there was a need for a British Interplane-

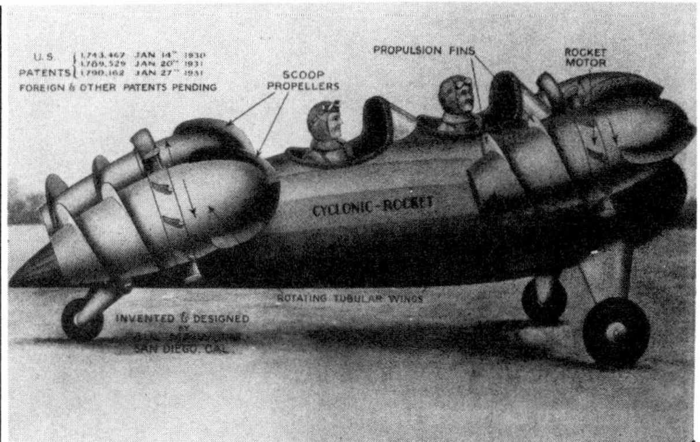

(Left) *CCCP* stratosphere balloon gondola, 1933.
(Above) Cyclonic-Rocket invented in 1931 by Paul Maiwurm.

tary Society and that the first meeting should be held on Friday the 13th. That meeting takes place in a suite of offices in Dale Street, Liverpool, and it is unanimously agreed that a journal be published, a task given over to Cleator.

"The ultimate aim of the Society," writes Cleator, "of course, is the conquest of space and thence interplanetary travel. That such a programme entails the solving of a formidable array of problems there can be no denying. It is only over the solution or otherwise of these problems that opinion becomes sharply divided.

"The skeptics, who at the moment present an overwhelming majority, rarely claim to know anything about the subject which they so readily condemn. Their reasoning is almost invariably based on the fallacy that because the idea of interplanetary travel is revolutionary, and until recently, unheard of—that because to them, it savours of the impossible—therefore it *is* impossible."

OCTOBER 29

This is the date on which a purported manned rocket flight is made to an altitude of 6 miles. The report is made in the London *Sunday Referee* that a German named Otto Fischer has flown in a 24-foot steel rocket from Rügen. Otto is supposedly the brother of the rocket's designer, Bruno Fischer. According to the story, secrecy is kept because of a fatal accident the previous year, combined with the fact that the flight is being made under the auspices of the German War Ministry. The rocket has been constructed in the town of Barmbeck, near Hamburg, and transported to Rügen. The launch takes place at 6 o'clock on a Sunday morning. Bruno and three officials take cover in a trench 200 yards away. There is an explosion, a blinding flash, and the rocket vanishes from its steel gantry. A few minutes later, it reappears, floating from a parachute attached to its nose. The rocket is equipped with steel fins that can be manipulated by the pilot to control his landing. After landing, an unharmed Otto emerges. The flight has taken only 10 minutes and 26 seconds.

This "event" is recounted in detail by Willy Ley, who demonstrates that the entire thing has been a hoax [perhaps inspired by rumors of the Magdeburg Project? see above].

NOVEMBER

The stratosphere balloon *Century of Progress* is flown by T. Settle and C. Fordney to an altitude of 61,221 feet (11.6 miles).

DECEMBER 7

In an article titled "Circulation astronautique," in *Exportateur Francais*, M.

Drouet proposes—"with all the seriousness in the world" according to Ananoff—one of the strangest of space guns. The gun would be a metallic tube erected so that one end is at sea level and the other atop one of the highest peaks of the Himalayas. If the upper end of the tube is at least 8.8 kilometers high then there will be a difference in air pressure between the bottom of the barrel and its top of about 750 grams per square centimeter. It is this pressure differential that Drouet expects to launch his projectile. The tube will be closed and the air evacuated from it. If his projectile has a bottom surface area of from 2 to 5 square meters, then the air beneath it—when the terminal end is opened—will exert a force of 30 to 37 tons, pushing the vehicle aloft with a "powerful and steady" force.

Drouet does not feel that this will be sufficient to launch the projectile as far as the tube's muzzle. He recommends that "massive quantities" of gunpowder be ignited to raise the pressure successively to 1, 5, and 10 kg/cm^2.

1933–1936

Spanish investigator Manuel Bada examines the possibilities of rocket power for stratospheric and orbital flight.

1934

Alexandre Ananoff publishes *Le Problème des Voyages interplanetaires*. A Russian emigré living in Paris, Ananoff is a prolific popularizer of spaceflight, especially of the work of Robert Esnault-Pelterie. [Also see 1935 and 1950.]

Edgar Rice Burroughs publishes *The Pirates of Venus*, in which the hero accidentally makes a flight to Venus. His intention is to make a trip to Mars but, in spite of a year's worth of calculations, the gravitational effect of the earth's moon is overlooked. The giant, torpedo-shaped rocket is launched from a horizontal track. This makes a very slight drop for the first three-quarters of a mile after which it rises at an angle of 2 1/2 degrees for the remaining quarter mile. At the end of this run, the rocket will have attained a speed of 7 miles per second, rising to 10 miles per second while passing through the atmosphere. The rocket sits upon heavily greased rollers while on the track, which has been built on some flat land near Guadalupe, Mexico.

NASM

Display at 1932 Chicago Toy Fair. Built by Mandel Bros.

The 60-ton spaceship is equipped with a cabin outfitted with controls, a berth, bookshelves, table and chair. Ahead of the cabin are oxygen and water tanks. Behind the cabin are a small galley and a storeroom containing dehydrated and canned foods sufficient for a year. Next is a battery room with a dynamo and gasoline engine. The remainder of the ship is occupied by the "rockets and the intricate mechanical device by which they are fed to the firing chambers by means of the controls in the cabin." Apparently Burroughs has in mind a system similar to Goddard's.

The cabin and the various compartments are in a torpedo-shaped hull contained within a considerably longer one. Between the two are systems of hydraulic shock absorbers.

A series of compartments run along the full length of the outer hull, each containing a parachute. These would slow the vehicle down to a speed at which it would be safe for the pilot to bail out, making the final descent onto Mars (or, as it turns out, Venus), via personal parachute.

The American Interplanetary Society changes its name to the more comfortable and less sensational American Rocket Society. There is some pressure on the British Interplanetary Society to make a similar change to a name "less imaginative," though it has, to this day and to its credit, resisted doing so.

Jack Williamson publishes his short story "Born of the Sun." The earth, it is discovered, is actually the egg of a mammoth star-beast. When the day of its hatching arrives, a small number of people are able to escape the destruction of the planet on the spaceship *Planet*: a 500-foot steel sphere propelled by antigravity.

A Russian stratosphere balloon reaches the altitude of 72,000 feet (13.63 miles). Unfortunately, the gondola breaks free during the descent and the three aeronauts are killed.

Jean Piccard takes off in a stratosphere balloon from Dearborn, Michigan, and reaches an altitude of 57,000 feet (10.8 miles). His copilot, wife Jeannette, is the first woman to enter the stratosphere.

William Swan flies a rocket-propelled glider at Atlantic City, New Jersey. The first American to fly a rocket-propelled aircraft, he reaches an altitude of 200 feet in the *Steel Pier Rocket Plane*, which is powered by twelve solid-fuel rockets. The rocket plane is a converted high-wing monoplane glider. The pilot sits below the wing in an unenclosed seat.

JANUARY

The first issue of the *Journal* of the British Interplanetary Society appears. It features a cover design chosen in a prize contest. The 1 guinea award is captured by P. E. Cleator himself, the editor.

APRIL

John Beynon Harris publishes "The Moon Devils" in *Wonder Stories*. In it he proposes "acceleration compensators" to protect astronauts during takeoff.

ROCKET AIRPLANE

FIRST AMERICAN FLIGHT

From Greenwood Lake, N. Y. to Hewitt, N. J.
By the Well Known Scientist

MR. WILLY LEY

IMPORTANT ANNOUNCEMENT

The exact date of the
FIRST AMERICAN ROCKET AIRPLANE FLIGHT
will soon be broadcast on the radio by Captain Tim Healy
on the Stamp Club Hour (WJZ 7:15 P. M.)
Listen for the announcement

Popular Science magazine publishes Donald W. Clark's plans for a scale model "Earth-to-Mars Rocket Plane." This may be the first such model spaceship. "Interplanetary rocket planes," writes Clark, "although actually far in the future, are now familiar to everyone because of the constant reference to them in radio and comic-strip adventure stories." Clark's design is not unreasonable for the period, reflecting accurately the current trend in speculative spacecraft design without obviously copying any particular one. Clark's

NASM (2)

A New Stunt for Model Makers . . . Building a Miniature

Earth-to-Mars Rocket Plane

By Donald W. Clark

INTERPLANETARY rocket planes, although actually far in the future, are now familiar to everyone because of constant reference to them in radio and comic-strip adventure stories. No one knows what these strange aircraft will look like, but there is no reason why the model maker should not delve into the future a bit on his own account.

The rocket-plane design illustrated is, of course, purely imaginary. It would be controlled by the tail fins, by the wings, which could be tilted together or separately, and by two small movable rocket tubes at the front end.

This model is made up of twenty-two simple parts, glued and pinned together. The body can be carved to shape with a knife, if no lathe is available, and smoothed with sandpaper. Saw two slots in the rear end to take the tail fins, and three long slots at the front end, as shown, to take the stabilizing fins, which should be glued on. Pin the wings to these and glue them to the body. Fasten the propulsion tubes with 1/16 in. diameter wire pins and glue. Outline the door with the point of a knife. Make cardboard templates for cutting metal parts.

Give the complete ship a coat of flat white paint. Next paint the body, propulsion tubes, wheels, shock absorber, and balancing rockets red; and the wings, stabilizing fins, tail fins, windows, tires, skid, and rear ends of tubes, black.

The model will look well if mounted on the suggested launching incline. Better still, a decorative background can be made from wall board painted blue or half a sheet of dark blue show-card board. Paint some light blue dots on it and one yellow circle to represent the moon, and, if desired, another circle in an upper corner to stand for Mars. In front of this background, place the ship as shown.

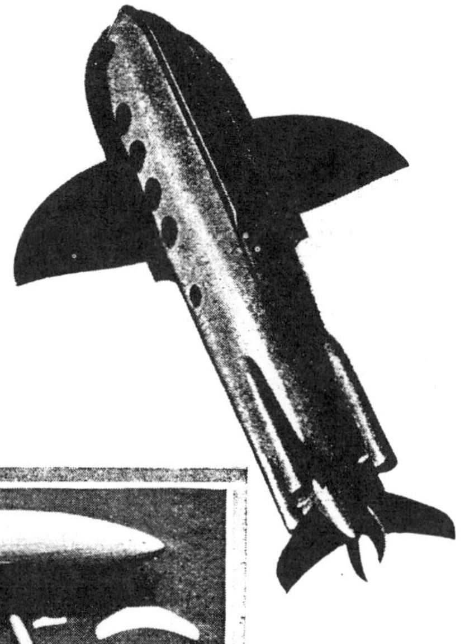

Finished rocket plane model and a photograph of the parts before assembly. If desired, the model can be lettered "Starocket—Earth-Mars Line"

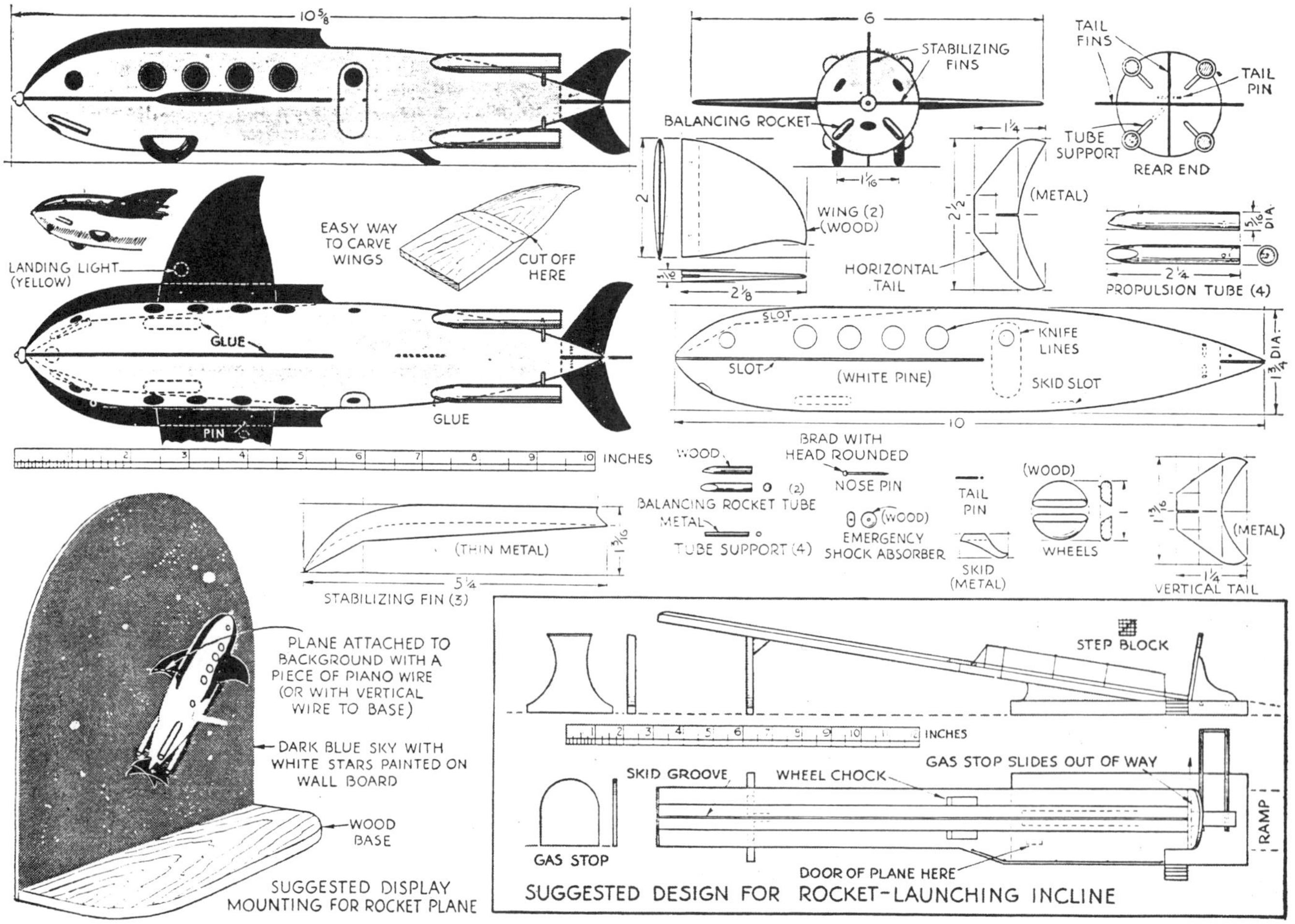

SUGGESTED DISPLAY MOUNTING FOR ROCKET PLANE

SUGGESTED DESIGN FOR ROCKET-LAUNCHING INCLINE

Side, top, and front views; the rear end; details of the body, wings, tail units, tubes, and other parts; and two methods of mounting the model

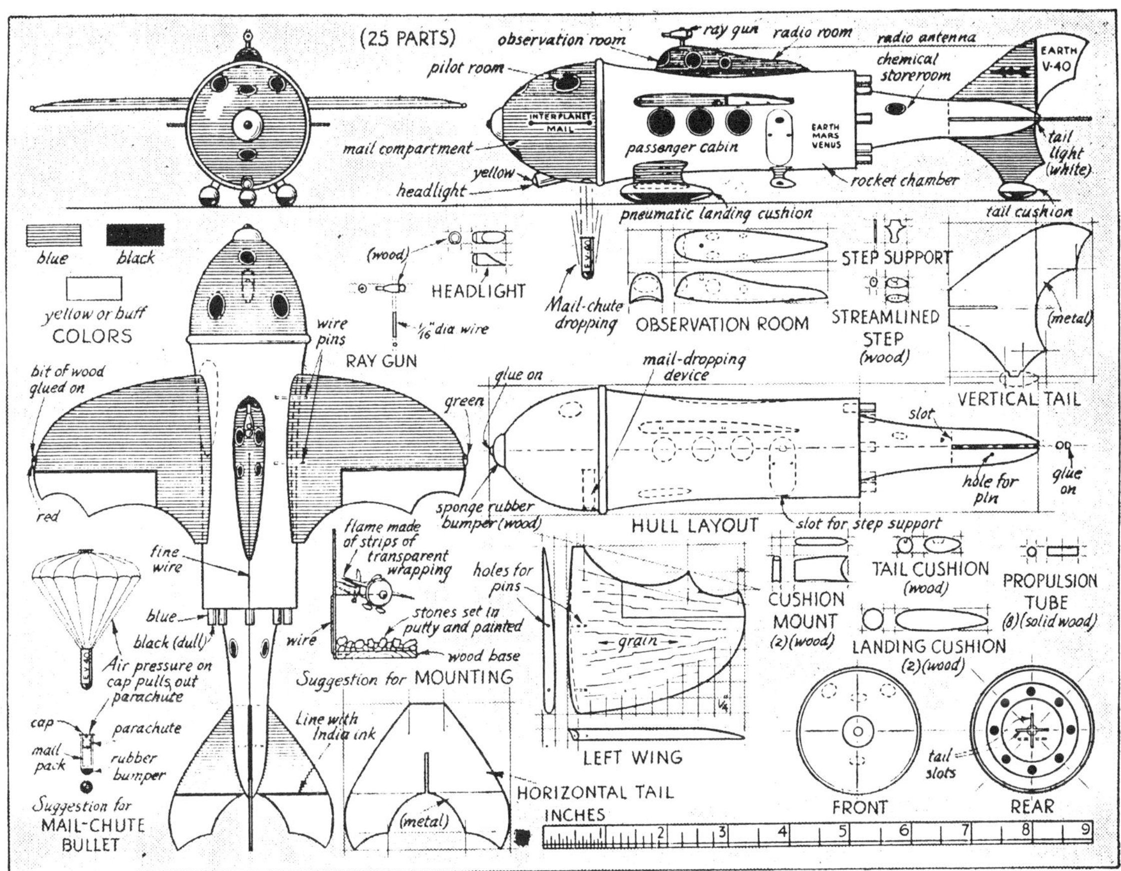

plans are reproduced here, along with a second design published in September 1935 (because of "the intense interest shown by readers . . . ").

JUNE

Eugen Sänger performs calculations on the possibility of using metals as rocket fuels [also see Tsander,].

JULY 12

Constantin Paul van Lent designs a two-stage moon rocket. The first stage is powered by four engines, each with four nozzles. The fuel and oxygen tanks form two concentric shells within the outer hull. The central axis of this stage is occupied by a massive helical spring and four hydraulic shock absorbers. The second manned stage is free to move down almost the full length of the hollow first stage.

The manned stage is a bullet-shaped vehicle divided horizontally into three sections. The lowermost contains the fuel tanks and four rocket motors. These are spaced around the central "landing pick," a device favored by Lent for planetary landings [see 1952]. The idea is that the descending vehicle would land tail first, imbedding the pick into the lunar soil. The middle section of the ship is the control cabin and the upper part, accessible by a short ladder, is an observation compartment (replaced in a later version by a parachute). The passenger ship is protected at liftoff by a jettisonable carapace.

OCTOBER 1

Henry Norris Russell of Princeton University and Mount Wilson Observatory writes an article that is printed in *Scientific American* in December as "Could a Manned Rocket Reach Mars?". Russell examines the problem of shooting a projectile to Mars via a giant cannon and, not surprisingly, arrives at the conclusion that it would not be practical. Most of the objections he raises, however, deal with the problems of accurate aiming and timing ("if our gun sends off its shell with a speed as much as 18 inches per

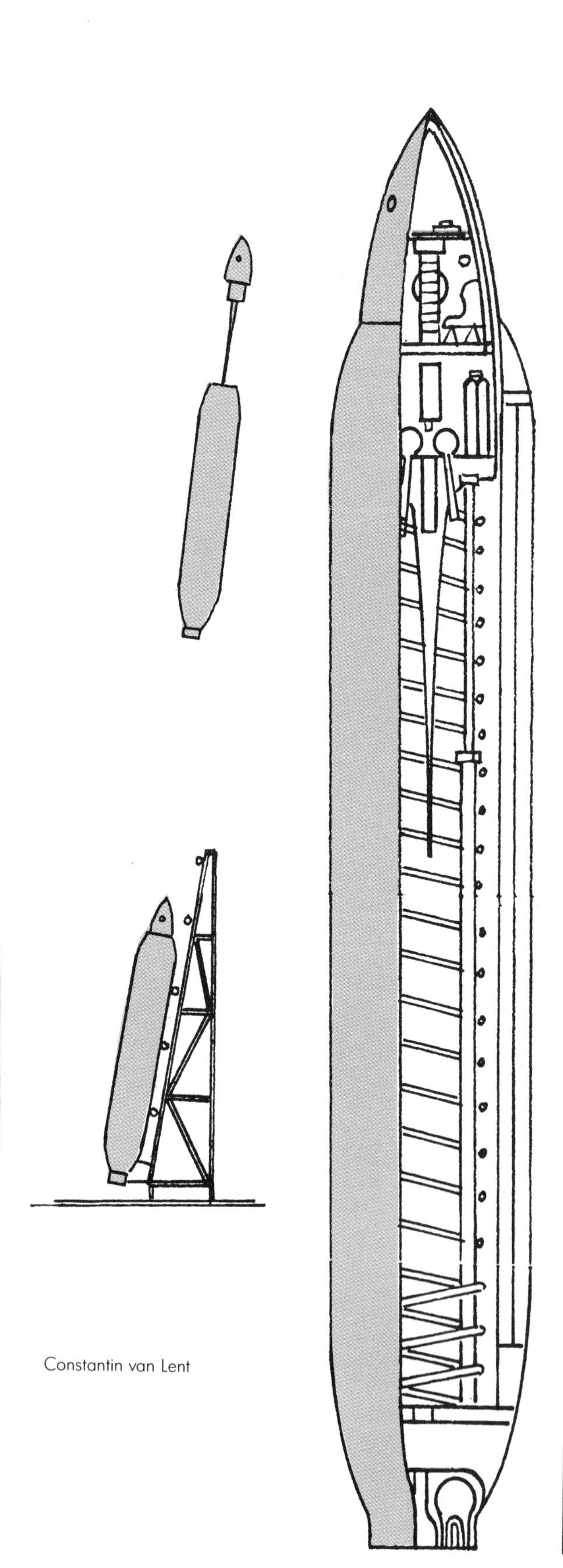

Constantin van Lent

second too great it will miss the planet altogether"). In his last paragraph, Russell dismisses rockets by saying that "Rocket ships might imaginably do better, though their success seems very remote."

1935

Alexandre Ananoff publishes *La Navigation interplanetaire*, based on a pair of lectures delivered at the Société Astronomique de France. It surveys contemporary knowledge of space travel.

Peter van Dresser, in his short story "Outbound to Jupiter," published in *American Boy* magazine, describes some of the dangers and difficulties of attempting to create artificial gravity with a system similar to that proposed by Oberth [1929]. The spaceship *President Fiske* ("She is none of your cold-blast electrostatic rockets, but an old hot-chamber, atomic-hydrogen driven warhorse.") divides into two sections once on its trajectory from the earth to Jupiter. At takeoff it is a 200-foot streamlined projectile with stubby wings. Once in space it goes through a maneuver called "splitting ship." The bow and stern sections separate while remaining attached by a strong cable. The two sections are set spinning around their common center by opposing retrorockets. As centripetal acceleration increases, the cable is gradually lengthened until 800 feet separate the two sections of the original ship. At a rate of rotation of 0.5 rpm an artificial gravity of 0.5 g is produced. The action of the story centers around the dilemma faced when the cable breaks.

In "Consider the Heavens," Forest Ray Moulton writes: "There is not the slightest possibility of such a [moon] journey . . . There is not in sight any source of energy that would be a fair start toward that which would be necessary to get us beyond the gravitational pull of the earth. There is no theory that could guide us through interplanetary space to another world even if we could control our departure from the earth; there is no

Christmas spaceship train ride at Macy's, 1934.

means of carrying the large amount of oxygen, water and food that would be necessary for such a long journey, and there is no known way of easing our . . . ship down on the surface of another world, if we could get there."

Charles G. Philip publishes *Stratosphere and Rocket Flight*, a "popular handbook on spaceflight of the future." Though inaccurate in some details, it is still the first book on the subject published in England.

Stanley Weinbaum publishes his short story "The Red Peri." He describes the spaceship of the pirate Red Peri, a red-headed female of striking beauty. Her ship, *The Red Peri*, is a tetrahedron of tubular girders, about 100 feet on a side. At the apex of the tetrahedron is located the atomic engine. Of particular interest in this story is the author's accurate description of the effects of exposing an unprotected human body to the vacuum of space. Unlike previous writers—and many writers afterward—Weinbaum does not believe a human being would freeze instantly or explode from internal pressure. In the story Red Peri and the hero must make a dash across the surface of ". . . the airless Plutonian plain, and into a temperature of ten degrees above absolute zero!"

"Instantly he is in hell. The breath rushed out of his lungs in a faint expansion mist that dissipated at once, the blood pounded in his aching ear drums, his eyes seemed to bulge, and a thin stream of blood squirted darkly from his nose. His whole body felt terribly, painfully bloated as he passed from a pressure of twelve pounds per square inch to one of nearly zero . . . " They cross a thousand feet of vacuum successfully and with

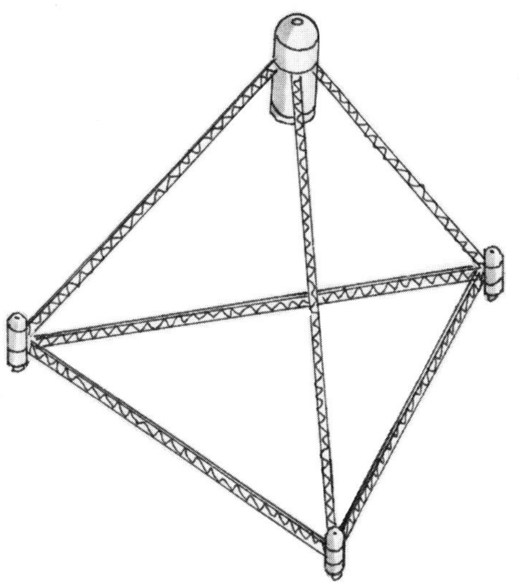

no permanent ill effects.

A rocket motor is installed in a Heinkel-112 aircraft.

Mosfilm produces the motion picture *Kosmetchesky Reis* (*The Space Ship* or *Cosmic Journey*), based on K. E. Tsiolkovsky's book *Beyond Earth* [1916]. Tsiolkovsky is the film's technical advisor and reportedly provides thirty spaceship designs.

In spite of official opposition, a scientist builds a spaceship and, with his female assistant, flies to the moon. The spacecraft bear names such as *Stalin* and *Voroshilov*. This is the last Soviet space film until *Niebo Zowiet* (1959).

November

Explorer II, the National Geographic Society-U.S. Army Air Corps stratosphere balloon, makes its first flight. Captains Orvil A. Anderson and Albert W. Stevens reach an altitude of 72,395 feet (13.9 miles). It is a record for manned balloon flights held until the 1950s [see 1957]. *National Geographic* publishes a special supplement for its readers: a foldout reproduction of "The first photograph ever made showing the division between the troposphere and the stratosphere and also the actual curvature of the earth . . . "

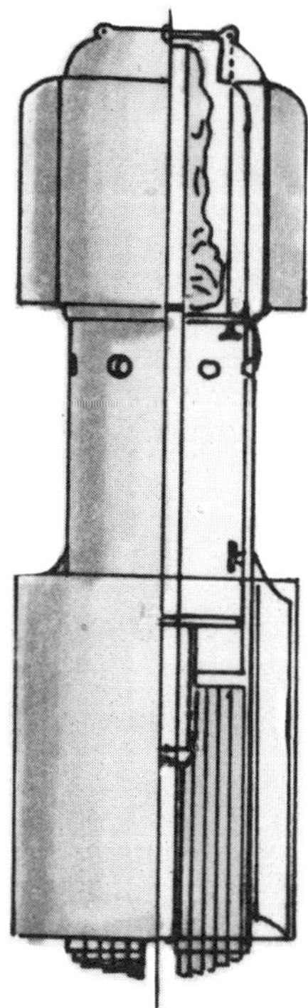

ca. 1935

A "Wyoming inventor" makes one of the first suggestions for a balloon-launched rocket. A solid-fuel rocket is to be released at an altitude of 11 miles, reaching a peak altitude of 43 miles. It would be manned and would descend to earth by way of a parachute. The balloon would be manned, too; it would carry a spherical, high-altitude gondola.

An anonymous inventor proposes a manned rocket that is launched vertically. It descends by way of unpowered rotor blades that unfold from its nose at the peak of its ascent. A number of experimenters use a similar method for successfully recovering their rockets.

1936

P. E. Cleator publishes *Rockets Through Space*, the first serious book on astronautics published in Britain. He makes several detailed proposals for spaceships. The first is a "transatlantic passenger-carrying rocketship"—a hermetically sealed streamlined "teardrop" that would take off from a metal runway inclined at a 60 degree angle. Its initial impulse could be obtained by being pushed by a high-speed locomotive, the pull of a counterweight, or by a compressed air cannon. It would reach an altitude of 600 miles over the midatlantic, and a speed of 10,000 mph. As it begins its descent, wings would unfold so it could glide to a landing. It could travel from New York to Paris, Berlin or London in less than half an hour (15 minutes if it carries cargo only). A manned rocket capable of achieving orbit would be a huge affair weighing 5,120 tons. This would be necessary simply for *getting* into space. If the astronauts wish to return, a fourth stage would be required to boost the original three, bringing the total weight up to 40,960 tons at a cost (in 1936) of £20 million. The manned stage would carry a crew of four and 60 tons of fuel. In discussing the various types of fuels available, Cleator suggests hydrogen (possibly monatomic) and oxygen, triatomic hydrogen ("hyzone") and ozone, and atomic power. The booster stages would be either destroyed by explosives or saved by means of parachutes.

NASM

If it proves necessary to provide the astronauts with gravity, the spacecraft could be split into two parts, connected by a cable. These then could be swung about the center of gravity, an idea originally suggested by Ganswindt [see 1891] and Oberth [see 1929].

Cleator also discusses the need for spacesuits "not unlike a diver's dress in appearance, the garment will completely cover the wearer. Its equipment will include a plentiful supply of air, a means of temperature control, and—who can tell?—perhaps a miniature radio receiving set. It has even been suggested that each suit be fitted with a tiny rocket motor, thus enabling the wearer to propel through space."

The distinguished journal *Nature* embarks upon what Cleator good-naturedly calls an "Unholy War" against his book. Whenever possible, it seems, the journal mentions *Rockets Through Space* when attacking the idea of space travel. The original review of the book says that "Mr. Cleator thinks it a pity that the Air Ministry evinced not the slightest interest in his ideas; provided that an equal indifference is shown by the other Ministries elsewhere, we all ought to be profoundly thankful." Later, in discussing Robert Goddard's work, *Nature* refers to Cleator's "somewhat premature book on the possibilities of using rockets for interplanetary travel . . . It is good to hear that such experiments are being carried out, and the sober objectivity of Dr. Goddard's work presents a sharp contrast to the unscientific imagination exhibited by those who seek to direct attention to the advent of interplanetary travel . . . " More than a year later, the journal refers unfavorably again to Cleator's book, in reviewing *Zero to Eighty* [see below], and in reviewing the film *Things to Come* [see 1937]: "Word has just been received that the much-heralded attempt to reach and land upon the moon has, by some mischance, been successful. We need hardly remind our readers that in the unlikely event of the foolhardy adventurers making a safe return to earth, it will be thanks neither to Mr. P. E. Cleator nor to his somewhat premature book on the possibilities of using rockets for interplanetary travel."

The French comic book *Zig et Puce au XXIeme Siecle* by Alain Saint-Ogan describes an "interplanetary projectile," invented by "William Waterproof," intended for a trip to Venus. It is a well-thought-out design employing rockets for takeoff, landing and steering. It is 60 meters tall and equipped with four telescoping landing legs. There are several concentric shells or compartments inside the heavily armored outer hull. Innermost is a spherical passenger cabin that is kept always in the same orientation through a system of gimbals. This is in turn mounted in a hydraulic cylinder that is intended to absorb the shock of takeoff. Surrounding this is space for storage, air reservoirs, machinery, etc.

Sergei Korolev designs the RP-218 rocket glider. It is a two-seat aircraft with a cluster

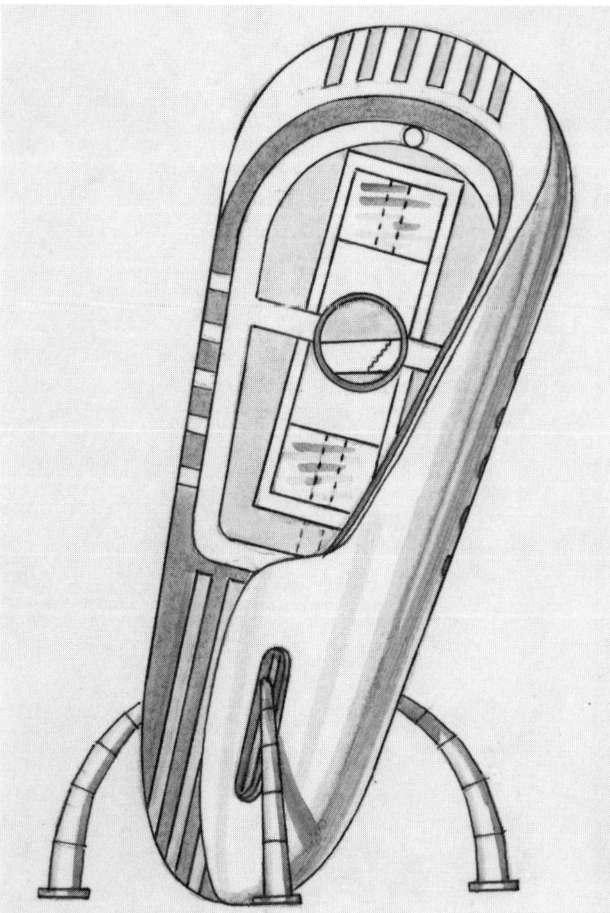
Sant-Ogan

of three nitric acid-kerosene engines with a total thrust of 900 kg. It has a takeoff weight of 1,600 kg and a wing area of 7.2 m^2. It can climb at a speed of 850 km/h with a ceiling of about 10 km from a ground takeoff and 30 km when released from a TB-3 aircraft at an altitude of 8 km.

Flash Gordon, a Universal serial in thirteen episodes, is released, starring Buster Crabbe. The rocket invented by Dr. Zarkoff, which carries Flash, the Doctor and Dale Arden to Mongo, is the spaceship from *Just Imagine* [see 1930], recycled and slightly altered.

Maurice Poirier, of Burbank, California, designs and builds a rocket that he claims will reach an altitude of 200 miles, propelled by a combination of "secret gases." He has also designed a rocket-propelled airplane that he claims is being seriously investigated by the U.S. Navy.

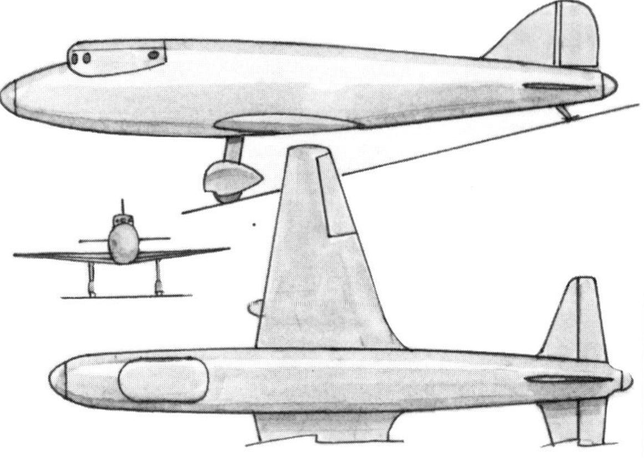
RP-218

FEBRUARY

Professor Alexander Klemin, an aeronautical expert at New York University, tells members of the Electrochemical Society that the next war would bring rocket propelled aircraft into use. "There are possibilities," he says, "of constructing an aerial weapon for destroying the enemy at a 200-mile range without risking the life of a single pilot." Stunt flying might be the first use of such rocket ships, with "[s]till later . . . a system of mail-carrying rockets across the Atlantic. Later still, we can conceive of passenger-carrying by the rocket—though there would be few volunteers for occupancy in such a craft today."

MAY 16

An anonymous article/book review entitled "Hopping off to a Dance on the Moon," in *Literary Digest* summarizes the ideas in P. E. Cleator's *Rockets Through Space* [see above]. The author spends much of his space detailing the efforts of amateur "rocketeers," such as those of the American Rocket Society, to arouse public interest and support in rocket development. In fact, the greatest obstacle facing progress in rocketry is not technical but financial: "Some believe that, with $100,000 a year for 5 years, they could shoot unmanned rockets across the ocean . . . the moon is somewhat more remote, for Mr. Cleator estimates that for every twenty tons transported to that planet, 4,380 tons of fuel would be required. And the cost of such a rocket, he fears, would be

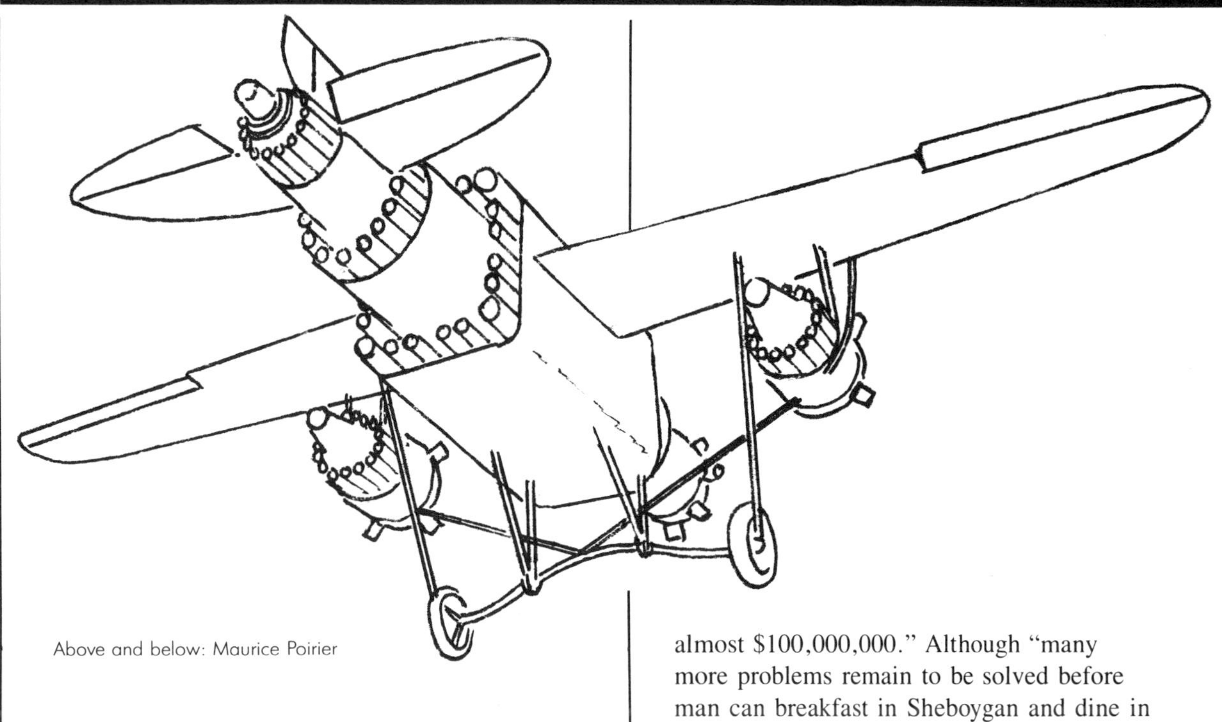

Above and below: Maurice Poirier

NASM

almost $100,000,000." Although "many more problems remain to be solved before man can breakfast in Sheboygan and dine in Shanghai" the author reminds those "skeptics who have seen these amazing 'flying needles' soar in 2-mile arcs [and] scoff that they will never be able to stay aloft long enough to be practical [that] their fathers hurled the same taunt at the Wright brothers when the first 'flying machine' plopped to earth after soaring considerably less than two miles."

AUGUST

In response to P. E. Cleator's *Rockets Through Space* [see above], F. Orlin Tremaine, editor of *Astounding*, writes in his editorial "Blazing New Trails," "Perhaps we dream—but we do so logically, and science follows in the footsteps of our dreams."

1937

The British Interplanetary Society begins studies on the design of a practical spacecraft capable of a manned journey to the moon (in this same year a prize is won for the design of spaceship launching equipment by D. H. M. Jack). The idea behind initiating the study is to determine the practicality of such a scheme in terms of contemporary technology, or a reasonable extrapolation thereof [similar to the *Collier's* symposia of 1952]. The research is undertaken by a special Technical Committee (formed in the autumn of 1936) under the direction of J. Happian Edwards. Others included H. Bramhill (draftsman), Arthur C. Clarke (astronomer), A. V. Cleaver (aircraft engineer), M. K. Hanson (mathematician), Arthur Hanser (chemist), S. Klemantski (biologist), H. E. Ross (electrical engineer), and Ralph A. Smith (turbine engineer and artist). Much credit is given by Ross himself to Edwards and Smith—the former developed the cellular rocket concept and the latter worked out its engineering.

Charles Lindbergh writes to the president of Clark University that " . . . a child born in [1896] would have been old enough to remember the discussions of his parents over the flights of Wilbur and Orville Wright. He may have been bombed during the World War, have transacted business by air mail a few years after the Armistice and have celebrated his fortieth birthday by crossing the Pacific Ocean on a 20-ton flying boat. The first half of his lifetime would have spanned the transition from the 26-pound aerodrome of Langley to the transoceanic air liner.

"Such a man . . . today could witness rocket ascents which would make him wonder whether the second half of his life would see a similar development of wingless flight.

"And if conservatives . . . should point out the problems of carrying capacity, cost and control, his vision might well recall the articles he has read on the impracticability of the airplane . . . "

"In an unguarded moment he might prophesy that we will eventually travel at speeds governed only by the acceleration which the human body can stand, and that in rocketing between America and Europa we wil accelerate half-way across the ocean and decelerate during the other half. Or, he might even point his rocket toward another planet and, without regard to fuel supply, landing facilities, or Professor Goddard, lose himself in interstellar space."

NASM

A Heinkel-112 fighter is powered by a 2,240-pound thrust reaction motor, designed by Wernher von Braun. It is fueled by liquid oxygen and liquid hydrogen. The plane is flown by Erich Warsitz, a lieutenant in the Luftwaffe. The first trials are not successful, including the loss by explosion of two of Ernst Heinkel's donated fighters. Eventually a flight is made. The takeoff is by propellor. Once the plane is aloft and flying level at about 190 mph, Warsitz ignites the rocket engine in the tail of the plane. Immediately it accelerates to over 250 mph. Unfortunately, the plane is forced to make a landing with its landing gear retracted and it is badly damaged. Nevertheless, Warsitz is the first man to pilot a liquid-fuel propelled rocket plane. Restored, the He-112 makes several more trial flights similar to the first. In the summer of 1937, Warsitz makes a flight in which he takes off by rocket power alone—the first time this has ever been accomplished.

HOW TO BUILD ROCKET SHIPS

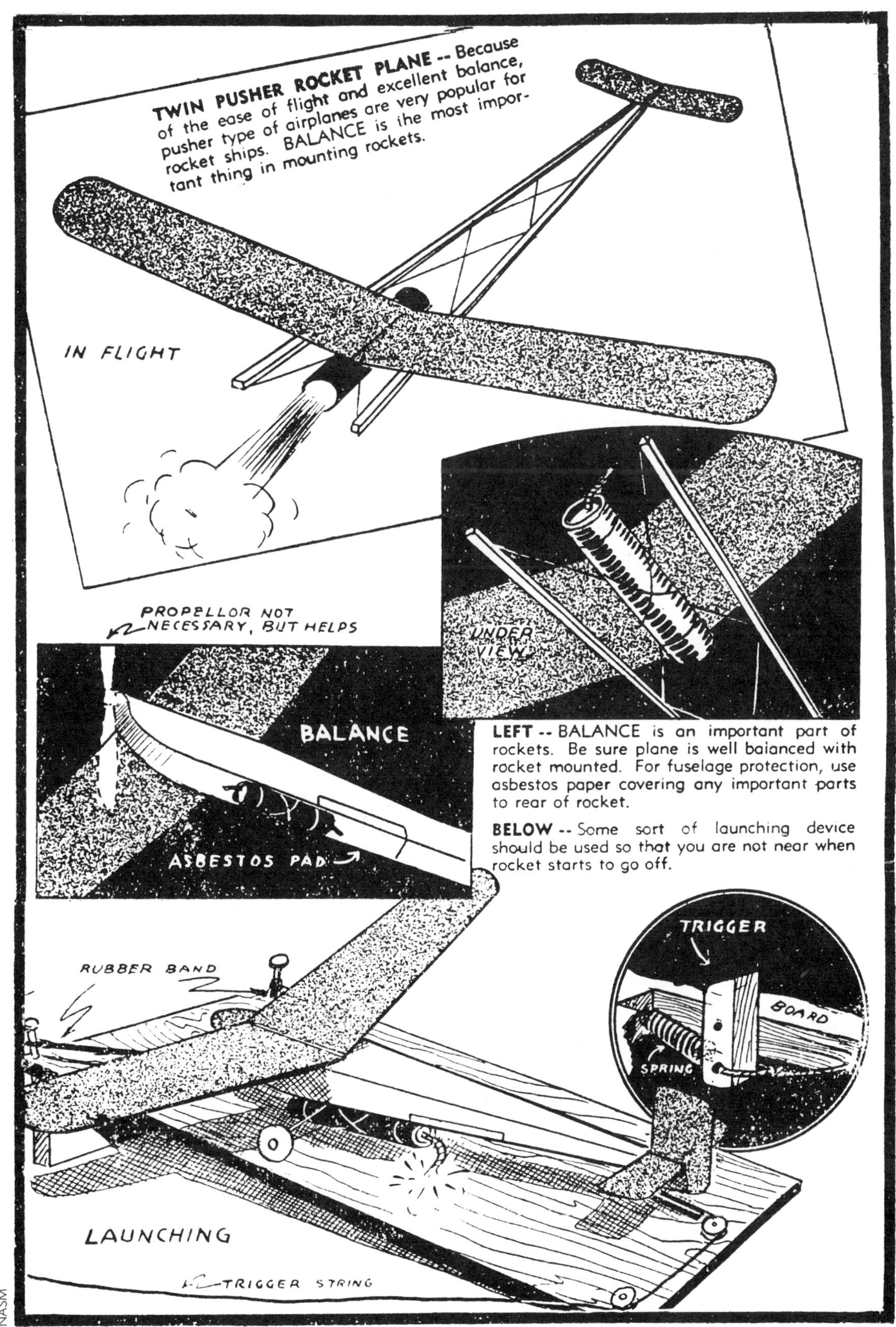

LEFT -- BALANCE is an important part of rockets. Be sure plane is well balanced with rocket mounted. For fuselage protection, use asbestos paper covering any important parts to rear of rocket.

BELOW -- Some sort of launching device should be used so that you are not near when rocket starts to go off.

1937

R. M. de Nizerolles begins his almost interminable pulp adventure serial, *Les Aventuriers du Ciel*, in which the solar system is thoroughly explored by the passengers of the spaceship *Bolide* (Meteor).

Ary J. Sternfeld publishes *Introduction to Cosmonautics*, the final chapter of which discusses the problem of interstellar flight. Sternfeld won the REP-Hirsch award in 1934 for his work in astronautics. Born in Poland and educated in Poland and in France, he moved to Russia in 1935 to work at the Institute of Scientific Research in Moscow. His works are eventually published in thirty-one languages in thirty-five different countries; two post-war books on space travel sold 600,000 copies in the Soviet Union alone.

William Dixon Bell describes in his novel *The Moon Colony* possibly one of the first suggestions for *terraforming* another world. In the story a villain is attempting to make a low-lying region of the moon habitable by launching a large number of cylinders at it by way of a (presumably) electric gun. The cylinders are filled with water ice and fertilizers. Once they are on their way, the projectiles are accelerated and steered by liquid-fueled rockets, fired by clockwork. At the start of the story he has already succeeded in creating a habitable pocket of air with a small lake.

The story also mentions 1,000-foot passenger rockets, propelled by "Goddard liquid rockets [sic]," burning carbon and liquid oxygen.

Edward Fitch Northrup publishes *Zero to Eighty* in which he develops both theoretically and experimentally the idea of the modern "mass driver," or electromagnetic gun. The 71-year-old engineer privately publishes this eccentric book, which contains descriptions, drawings and photographs of his working models, as well as a long technical appendix. What makes his book so odd is that Northrup has chosen to write it as the "autobiography" of the fictional "Akkad Pseudoman." Since Pseudoman is supposed to have written the book in the year 2000, at the age of 80, the title is explained, at least.

All of Northrup's real-life experiments with the mass driver—or electric gun, as he calls it—are ascribed to Pseudoman. By using this literary device, Northrup not only describes the machines and experiments he actually has carried out and patented, but is able to extend them in his imagination to their logical future conclusion: a manned spaceflight to the moon.

Northrup begins by describing the earliest experiments he worked on, which were actually carried out, for the most part, by his assistant, Theodore Kennedy. A photograph of one of these experiments shows a 4-inch diameter aluminum cylinder propped against the electric gun used to launch it. Even though the gun barrel is scarcely longer than the 28-inch projectile, it is able to accelerate it with a thrust of 700 pounds. In order to measure his guns' performance and acceleration without loss or damage to his projectiles, Northrup simply places two of his guns muzzle to muzzle; the projectile accelerated by one is braked by the other, using exactly the same amount of energy.

"Pseudoman" then proceeds to go on beyond the experiments actually performed by his real-life alter ego. He plans to build an electrically launched vehicle that will travel to and circle the moon. Initially, to test the idea's feasibility, he launches animals using a scaled-down version of his proposed projectile and gun. The first of these vehicles is

A. Moon rocket.
B. Launcher and projectile.
C. Two horizontal gun coils, muzzle-to-muzzle; one for accelerating, one for decelerating.
D. Electromagnetic gun launching pliers.
E. Model mail carrier.

B

E

D

A

C

over 15 feet long and a foot in diameter. It carries scientific instruments in addition to a cat and rabbit. It is gyroscopically stabilized and returns to the earth by parachute. The scale gun is 3,000 feet long, sunk vertically into an abandoned mine shaft. Needless to say, like Verne's test flight, the experiment is successful.

Work is immediately begun in a full-scale manned vehicle. It is given emergency priority when it is learned that the Russians are also planning a moon flight of their own! Pseudoman sets a launch date in late 1960.

The mass driver is not intended to provide all of the velocity necessary for the trip to the moon and back. The vehicle itself is a rocket. In a sense it is a three stage rocket in which the upper two are rocket-powered and the first stage is the mass driver, a stage that can be reused indefinitely. The spaceship has two stages and is about 28 feet long overall and about 3.25 feet wide. The upper manned stage is 17 feet long, provided with steering rockets, a life support system, air conditioning, photographic equipment, sound recorders, etc. A parachute is contained in the nose for the final descent to the earth. Both stages are propelled by liquid-fueled engines, modelled on those designed in 1916 and later by Robert Goddard. They burn liquid oxygen and gasoline, the fuel for the second stage being contained within the walls of the rocket. Even though the gun is only providing a part of the needed velocity, the astronauts still have to protect themselves against an initial thrust of some 32 g's! Of course, as Pseudoman points out, the acceleration can be lessened simply by making the gun longer—it is only a question of cost. The two passengers encase themselves in strong canvas suits which are attached by numerous cords to the inside wall of the rocket. He hopes that this arrangement will help them to withstand the shock of launching.

The monster cannon that is finally built is 124 miles long. Since it is impossible to build it pointed vertically, most of its length lies horizontally, rising gently up the slopes of a high mountain. Finally, at the summit, the muzzle points skyward. The gun's giant magnetic coils require hollow-cored, water-cooled wires. Pseudoman explains the neccessity of using the gun, as opposed to using a rocket for the entire trip. There are no fuels in use at his time, he says, that would permit a realistic mass-ratio. "The moon will never be reached by a human being in a car propelled by rockets only, unless the dream of some scientists is realized whereby we will be able to release and control the almost limitless energy stored in atoms."

The complete round trip to the moon is expected to take a total of 126 hours. Upon returning to the earth, the rocket will approach tail-first, using all five of its engines to slow it down by 10 percent of its original speed. The ship will enter the atmosphere obliquely, using the friction of the air to gradually slow it further. While still in the upper atmosphere, the parachute will be released. The rocket will still be travelling nearly parallel to the earth's surface, but the retarding effect of the parachute will start it curving downward. At about 5 miles altitude, the parachute will begin to act in a normal manner, eventually landing the ship gently.

The trip is a success. Northrup/Pseudoman accurately describes weightlessness and other effects of space travel. For some reason, the trip is timed when the moon is nearly full as seen from the earth, so that the two astronauts can see almost nothing of the far side of the moon when they circle around it. However, they do catch a glimpse of the Russian rocket! [Also see Nebel, 1969.]

F. A. Tsander describes the final design of his spaceship [also see 1931]. Outwardly, it resembles a very large biplane aircraft with two enormous propellors. The body is shaped like a long projectile. The pilot operates the spacecraft from within an enclosed cockpit. It takes off and climbs as an ordinary airplane. A special high-pressure engine using liquid oxygen enables the craft to climb to an altitude of 25 to 30 kilometers. At this point a velocity of 350–450 kilometers per second has been reached (the exhaust of the internal combustion engine may

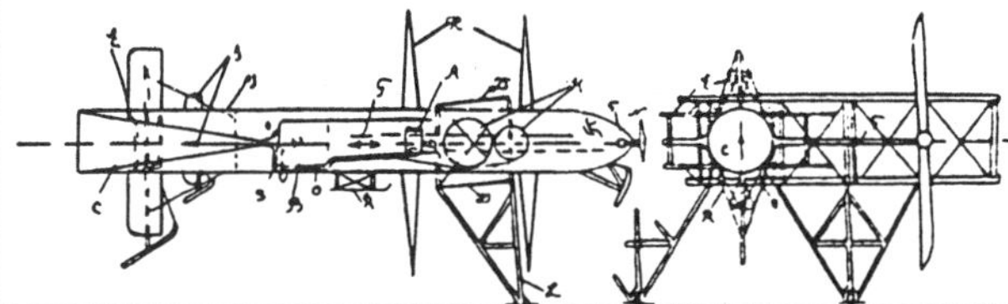

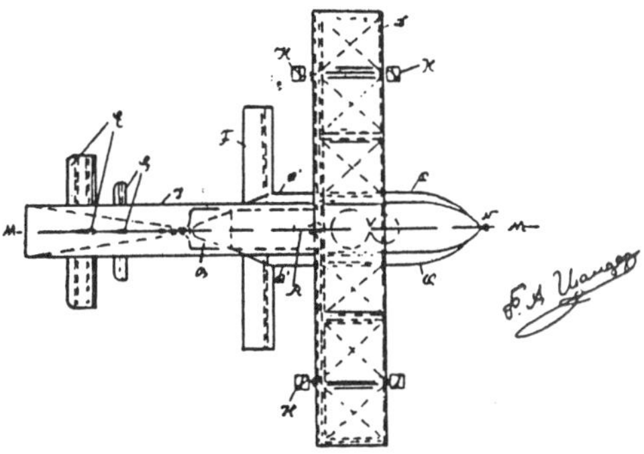

be used to augment the thrust of the propellors). The airplane engine is stopped and the rocket is ignited. The fuel is the now-unneeded parts of the rocket's airframe: wings, control surfaces, engine, etc. Ultimately the spaceship is reduced to a much smaller rocket, which has been contained within the body of the larger aircraft. This is equipped with its own small wings and tail surfaces (which can be seen protruding through the fuselage in the drawing). Along with a rocket engine it has a special chamber for melting the unused parts of the larger rocket. The molten metal is atomized and combined with a spray of liquid oxygen (which has first been used to cool the combustion chamber). Eventually even the machinery used for pulling in the wings and tail is consumed by the rocket.

Tsander suggests the possibility that, once in interplanetary space, the spaceship might unfurl large, low-weight mirrors by which it might be propelled by the force of sunlight.

The small, winged spaceship can be used for a gliding reentry into the earth's atmosphere, or into the atmosphere of another planet. Tsander feels that the use of airplane-type takeoffs and landings would better assure the safety of the passengers.

Up to 90 percent of the rocket's original weight might be used as propellant. Some of the metals Tsander considers are aluminum, magnesium, beryllium, and lithium. When combined with oxygen beryllium, for example, could generate a theoretical exhaust velocity of 6,750 meters per second as opposed to the 5,170 mps produced by hydrogen.

Loebell exhibits a model of his spaceship at the Paris Exhibition. It is an elongated teardrop with four extremely long, trailing wings or fins.

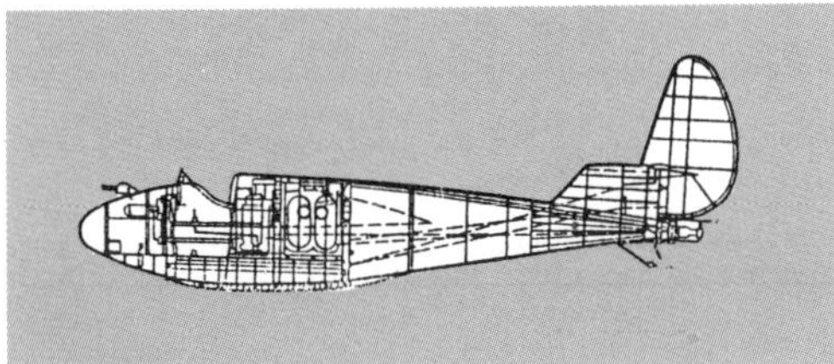

Sergei Korolev attempts to power an SK-9 glider with an ORM-65 rocket engine. This project evolves into the RP-318. Its takeoff weight is 660 kg, with 75 kg of propellant. It has a wing area of 22 m^2. It is originally propelled by an ORM-65 engine (1938–1939), which is later modified to become the RDA-1-150, with a thrust variable between 50 and 150 kg. For safety's sake the fuel supply system and the engine are tested over a period lasting from 1937 to 1940. The first flight is made in 1940 [see below].

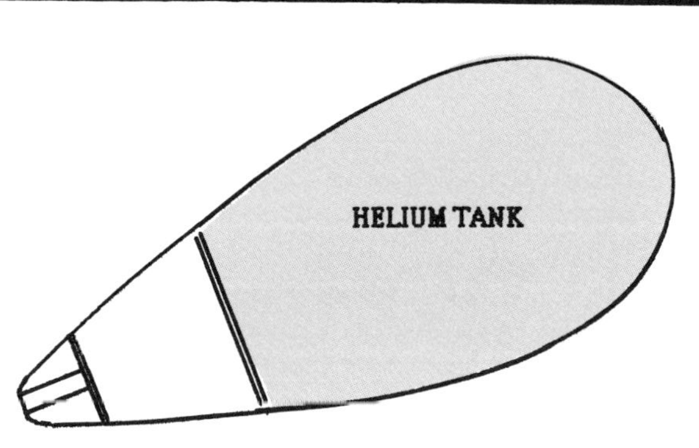

Alvin E. Moore proposes a novel spaceship design in which two-thirds of the bulbous duralumin hull is occupied by a helium reservoir "to float the craft when it returns to the earth's atmosphere."

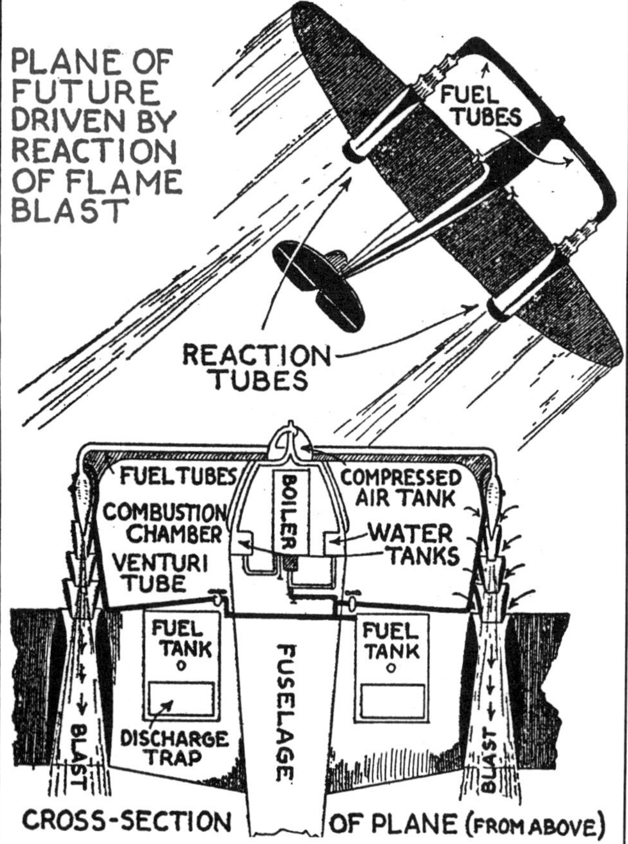

Everyday Science & Mechanics, October 1935.

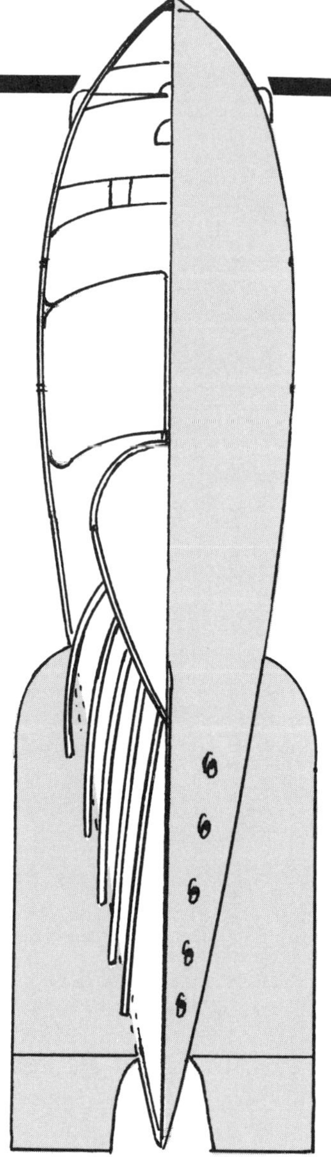

Charles G. Philip publishes a design for a spaceship that appears to be an adaptation of one of Tsiolkovsky's later designs. Twenty long "exhaust stacks" branch out from the single large combustion chamber, five each exiting between the rocket's four fins.

L. Buehrlen performs centrifuge experiments in which humans withstand g-forces with peaks of up to 17 g's, while in a supine position.

JANUARY

J. Néel publishes his plan for a future spaceport in *La Science et La Vie*. It would consist of ten departure ramps of reinforced concrete, each 627 m long and radiating from a central point. His spacecraft would attain a velocity of 1,100 km/h within 4 sec-

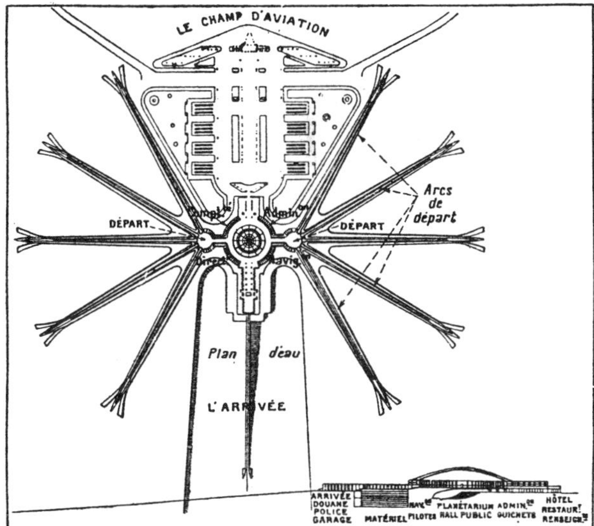

onds. If the range of the rockets is limited to 4,000 km, there would be a need for only fourteen spaceports world-wide. His rockets would take off with the aid of enormous catapults and would land by plunging into a large artificial lake adjacent to the spaceport. Néel provides the spaceport with a conventional airfield, and facilities for customs, hotels and restaurants, a planetarium, and the usual amenities of a large airport.

NOVEMBER

In a letter to *Astounding* M. Erland asks, "What about the stratosphere plane? Why can't these planes be equipped with simple rocket motors as well as an engine capable of 300 mph and use the earth's equatorial speed (1000 mph) to break loose? (From the 30,000-foot altitude.)

"Why can't a stratosphere balloon be built to support a passenger rocket? Then we could use its ten-mile height as a leg of our journey and its hydrogen gas to give the rocket an extra boost."

DEC. 1937–APR. 1938

Ground tests of the Russian rocket glider, RP-318-1, begin. The ORM 65-1 and ORM 65-2 engines are used (they are eventually transfered to the KR-212). The RP-318-1 has a wing span of 17 meters, is 7.4 meters long and weighs 700 kg. [Also see 1940.]

1937–1938

The DFS-194, a rocket-powered aircraft, is designed by Alexander Lippisch. Developed at the Research Institute for Sailplanes, it eventually evolves into the Messerschmidt Me-163 [see 1941].

1938

C. S. Lewis describes a spaceship in his classic theological science fiction novel *Out of the Silent Planet*. It is "roughly spherical," and made of two hollow, concentric globes. The inner globe contains the stores, the space between the inner and outer spheres creates the living space, which is divided into rooms. Steel-shuttered skylights in the ceiling provide light.

Flash Gordon's Trip to Mars, the second Universal Flash Gordon serial, is released in fifteen episodes, starring Buster Crabbe.

APRIL

Popular Science magazine publishes a design for a kind of combination planetarium show and thrill ride. A mockup of a spaceship with large, circular windows is wheeled through doors in the side of the planetarium dome. It is tipped upward and while "chemical vapors" are illuminated by colored lights (to imitate flames) motion pictures are displayed on the dome "to give the effect of speeding through space on a whirlwind tour of the universe."

OCTOBER

In an article for *Astounding*, Peter van Dresser answers the title question, "Why Rockets Don't Fly" with the simple answer: lack of money.

DECEMBER

In an official retort to Orson Welles's *War of the Worlds* radio broadcast (Hallowe'en,

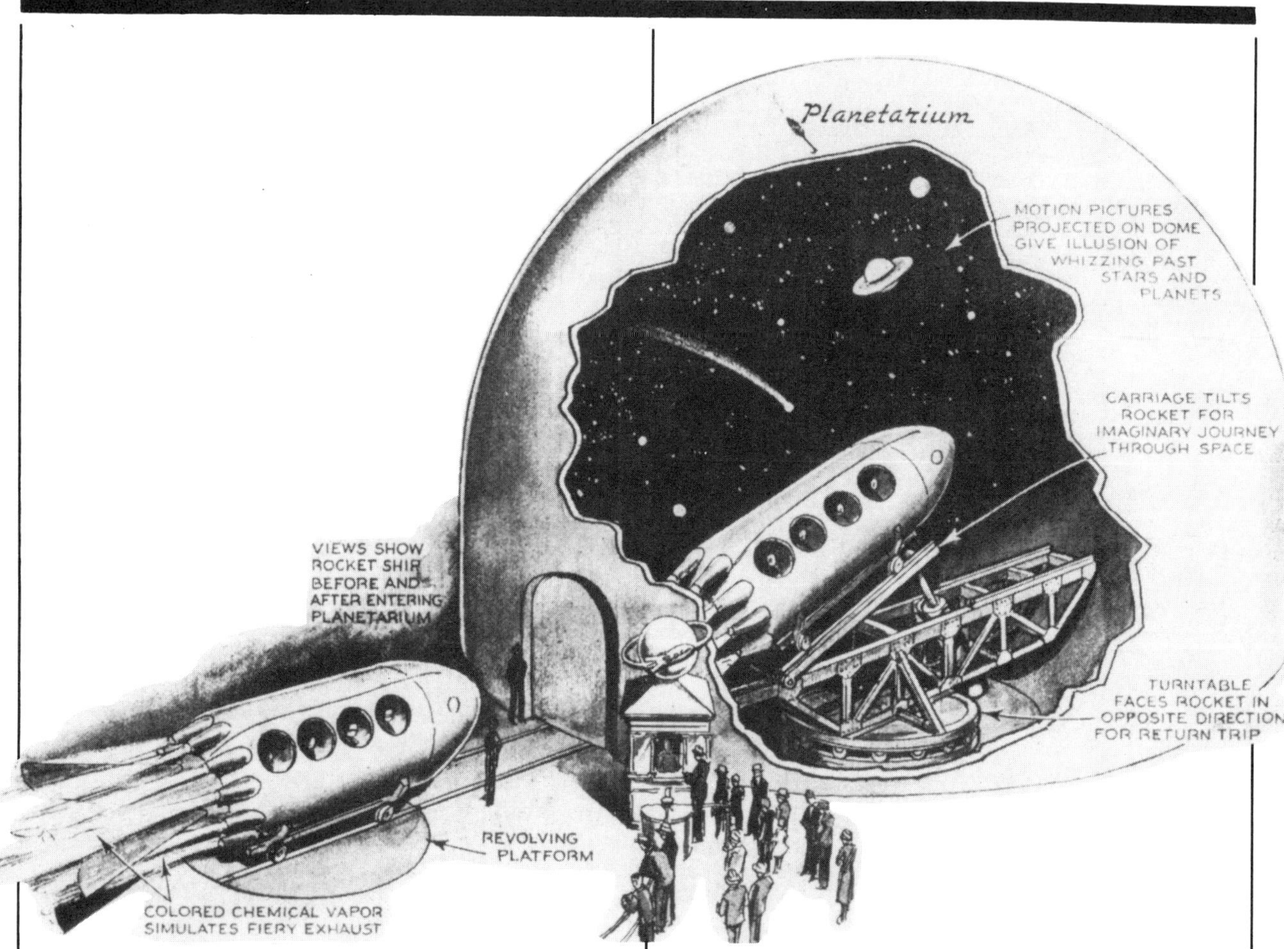

Popular Science, April 1938.

1938), Science Services says (in the article " 'Monsters' to Mars," *Scientific Monthly*) that interplanetary travel is impossible, and that to reach escape velocity would require more fuel than a spaceship could carry, even if leaving Mars.

The back-cover illustration on an issue of *Amazing Stories* depicts the "Space Ship of 2038." "Great engineering skill," reads the accompanying text, "and ingenuity will be necessary to produce a ship capable of flying to other worlds. The space ship shown here is based on theoretical extensions of known fact." The rocket shown features a "pilot and robot control room," compartments for the navigator, freight and storage, a lifeboat and its launching tube, staterooms, dining room, theater and lounge, and dining room. Artificial gravity is provided on the "gravity deck" by spinning the central part of the ship on a huge bearing/elevator shaft. Fuel is oxy-hydrogen ignited by "detonator caps" in the "major explosion chamber."

ca. 1938

Peter Vacca of Buffalo, New York, builds a rocket-powered car at cost of $16,000, the *Mars Express*. Made of aluminum, it has both a conventional V8 engine with supercharger and rocket motors (of an unspecified type). The car could reach a speed of 115 mph with its gasoline engine alone. There is no report of the results of this rocket experiment.

Rocket Flight to the Moon, a film also known as *Spaceship Number 1 Starts (Weltraumshiff I Startet)*, is made in Bavaria by the Bavaria Film-Künst. Since the Germans cancel most of their science fiction films once the war begins, this film is only a short remnant, edited from scenes already filmed. It eventually forms a portion of the German-made documentary *History of Rocket Development*. At the end of the film, *Spaceship 1* is shown leaving its hangar and

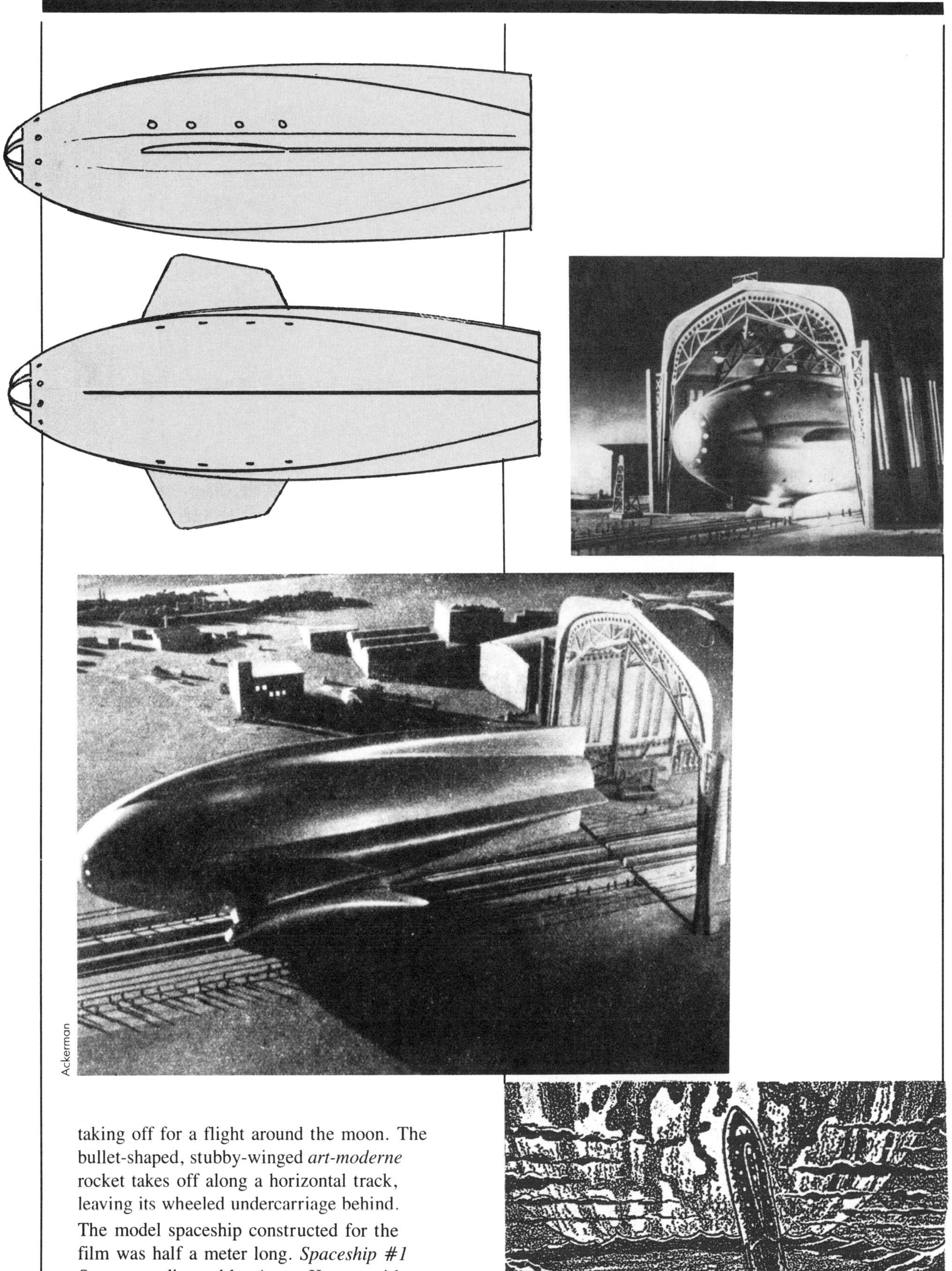

taking off for a flight around the moon. The bullet-shaped, stubby-winged *art-moderne* rocket takes off along a horizontal track, leaving its wheeled undercarriage behind. The model spaceship constructed for the film was half a meter long. *Spaceship #1 Starts* was directed by Anton Kutter, with technical advice from engineer Fritz Beck.

1938

A proposed ride to Venus that is to be constructed for the New York World's Fair at a cost of $200,000.

Frontispiece for Professor A. M. Low's novel *Lost in the Stratosphere*, which is, sadly, juvenile in more than one sense of the word, especially since the author is chairman of the BIS.

1938–1939

Eugen Sänger and Irene Bredt work on their antipodal bomber. In October, 1938 a steel model of Sänger's *Silver Bird* [see 1931, 1933 and 1943] is constructed for testing. The antipodal plane now has the form of a plano-convex hull: its underside is perfectly flat. On the basis of this research, he applies for a patent on the proposed half-ogival fuselage and wedge-shaped wing profile. The *Silver Bird* is given the far less elegant nickname of the "flat iron," by Sänger's assistants, because of its curved upper side and flat bottom. The rocket plane is to be accelerated along a horizontal launching rail by a rocket driven booster sled.

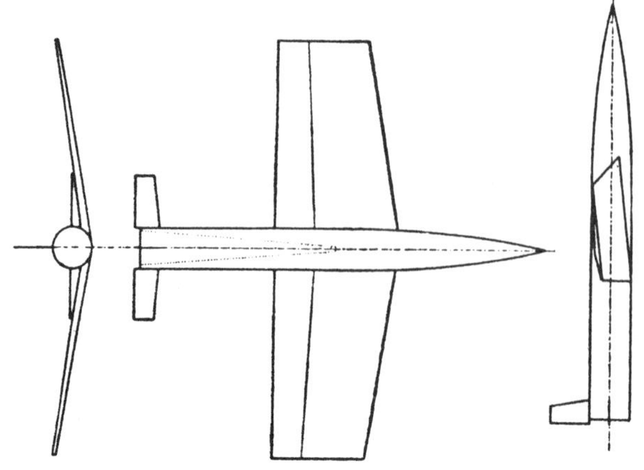

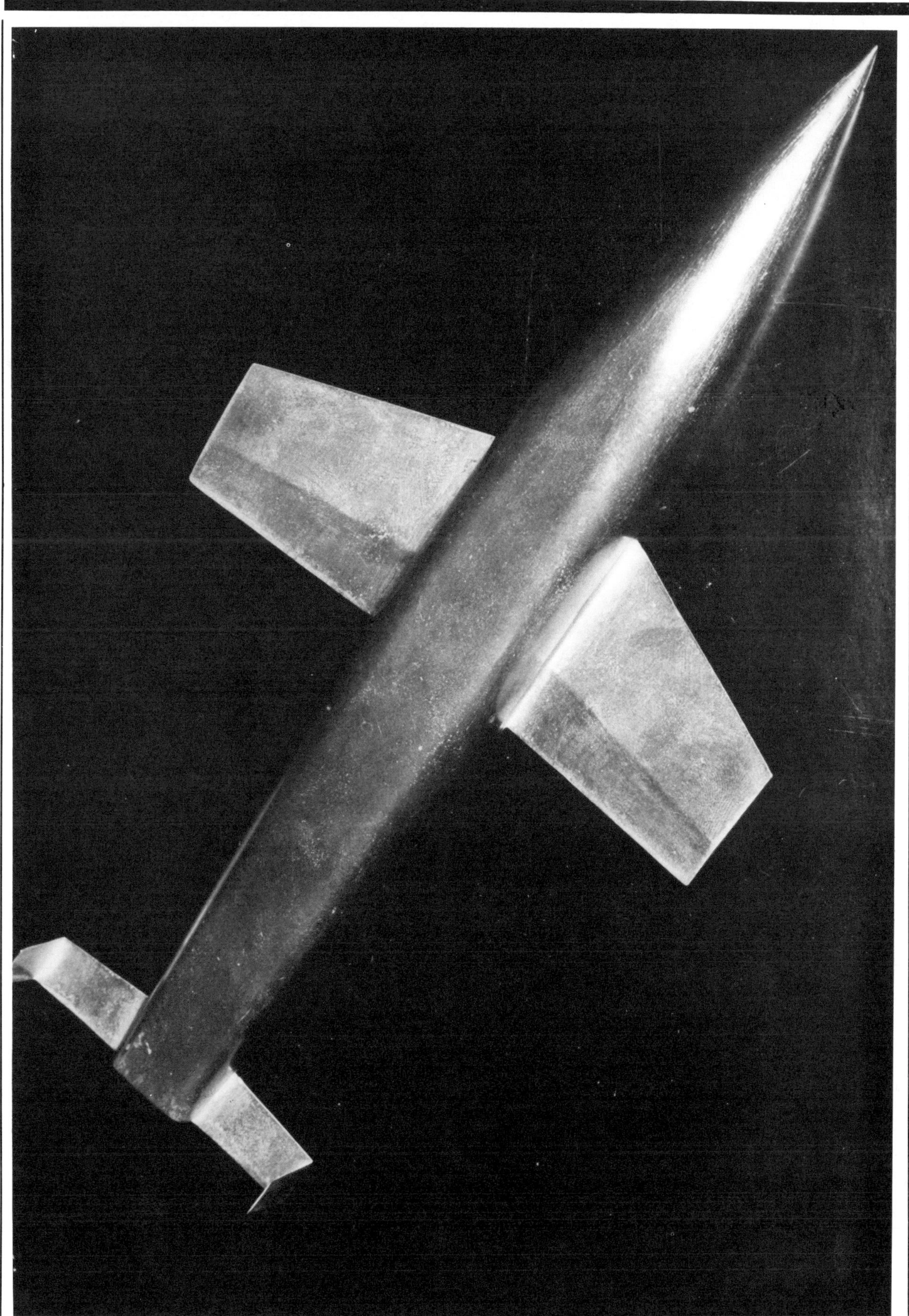

NASM

PART IV

THE WORLD WAR

The spaceship was invented during the Second World War—some say inadvertently, others say intentionally, though it does not matter very much since the end result was the same. As with all wars, research into new technologies was multiplied and the potential of the rocket as a weapon certainly was not overlooked. During this time were built the first high-speed aircraft that were designed specifically for rocket propulsion as well as the first large-scale liquid fuel rockets. The outstanding example of the latter was, of course, the infamous A-4 . . . better known as the V2. From it the great rockets that took part in the first tentative exploration of space are directly descended. It is not a little ironic that American astronautics owes more to the developments of German rocketry than to the experiments of its own Robert Goddard (whose paranoid secrecy effectively cancelled any direct influence he may have had on the development of the spaceship . . . other than as a kind of semimythical inspiration).

With the reality of the V2 confirmed, and the practicality of manned rocket aircraft demonstrated, space travel almost instantly was transported from the realm of the wholly speculative, from the realm of some distant century, into the immediately foreseeable future.

NASM

1939

At the New York World's Fair an "electric space gun" of 2039, designed by Raymond Loewy, is demonstrated as part of a miniature city of the future exhibited in the Chrysler Motors Building. The 6-minute exhibition is shown under the auspices of the American Interplanetary Society. The spectators, in rows of seats surrounding the model, after watching a motion picture history of transportation, see a spaceship—the Internplanetary Rapid Transit, or IRT—picked up by a huge magnet and loaded into a mortar-like launcher, which is turned toward the "sky. "This is accompanied by the resounding activity of bustling taxis, groaning cranes "and the other incidents to a takeoff for Mars. "The launch of the rocket (every 15 minutes) is simulated with strobe lights and steam. The spaceship is "then seen winging its way into the clouds, growing smaller and smaller until it has vanished in the night sky. This last illusion will be accomplished by the use of a refinement of the motion picture and a mechanism resembling the iris diaphragm of a camera. As far as can be learned, Mr. Loewy's model includes no depiction of how our courageous descendants will return from their jaunt into the ether."

At least one science fiction fan criticizes the AIS for condoning the depiction of a space gun for launching a manned spacecraft. "That's the layman's conception of rocket travel for people," writes Ralph Newman in *Fantastic Adventures*, "shot from a gun. From zero to 3,600 m.p.h. in a split second. "Do you suppose the model is planned just to give people a show of bang, flash, smoke? Just sensational, it seemed to me.

"The theme of the fair is The World of Tomorrow and all the other exhibits of the futuristic theme seemed plausible and practical, but this human cannonball stuff in American Interplanetary Society Bldg [sic]. . . . "

The magazine's editor replies that the exhibit should have been "graced with the advice of a competent rocket authority."

Also at the fair are two other space-related exhibits. One is the Theatre of Time and Space, which takes its visitors on a trillion-mile trip through the universe. The journey takes place in a domed theater 44 feet high capable of holding 350 people. The curves of the dome blend into the floor to give the illusion of a limitless expanse.

From a sunrise skyline of New York City, the audience is carried, with the sound of rushing wind, to a position high above the city. Continuing on into space, the audience witnesses a solar eclipse and flybys of the moon, Venus, Halley's Comet, Mars, Saturn and on into the Milky Way Galaxy.

The production is created under the supervision of Wayne M. Faunce, vice-director of the American Museum of Natural History from an idea conceived by Alan R. Ferguson. Clyde Fisher of the Hayden Planetarium is the scientific adviser, with special effects and photography by Fred Waller.

The second exhibit is a planned simulated "Flash Gordon" trip to Venus, with "an abundance of thrills guaranteed. "It is to be built at a cost of $200,000 and will depict an "epic-making trip in a gigantic rocket ship, travelling through space at the unheard-of speed of 2,000 miles per hour."

Victor F. Bolkhovitinov begins working on a rocket-propelled airplane [also see 1940–1941].

JANUARY-JULY

The British Interplanetary Society concludes its 2-year engineering study of a manned moon landing. Its spaceship is a radically new design, completely different than anything previously suggested.

Calculations show that a minimum of 1,000 tons of fuel will be required to transport a spacecraft weighing 1 ton to the moon and back. Operating according to the idea that anything that is going to be jettisoned should be jettisoned as soon as possible, the designers hit upon the cellular spaceship concept. The spaceship will be composed of many hundreds of small rockets, each complete with its own fuel and motor. They are attached in such a way that as soon as they finish firing, they drop off. Owing to the large number of small units, thrust and direction can be controlled by the rate at which fresh rockets are fired. Since each rocket is small, they can be burned until their fuel is exhausted; therefore solid-fuel motors can be used. Motors using liquid fuel will still be used where fine control is required, and steam jets are to be used for steering.

The lunar lander is to be a gumdrop-shaped vehicle bearing a strong resemblance to the Apollo Lunar Excursion Module [see 1969]. It is 11 feet tall and 13.5 feet in diameter.

The remainder of the 105-foot ship consists of the solid fuel rocket clusters. There are a total of 2,490 individual units grouped into six stages. A survey is made of 80 to 120 different potential fuel combinations. Of the 1,000,000 kg weight of the complete ship, 900,000 kg is fuel.

Dr. Arthur Janser undertakes the problem of what materials to use to construct the spaceship and chooses synthetic plastics for the most part. The outer hull is to be made of glasslike fused aluminum oxide; the inner hull required for the crew compartment is made of several layers of linen fabric, stretched over a light frame, and bonded with a compound of rubber and resin. The remainder of the cabin interior uses mainly balsa wood that has been treated with a hard resin. Windows and optical instruments are made of a transparent acrylic plastic. All of the wiring is coated with a clear polystyrene, and any tubing is made of polyvinylchloride (PVC). Liquids are contained in ethyl cellulose, protected by a coating of vinyl acetylene resin.

Each stage is a hexagonal honeycomb cluster of individual rockets. The largest rockets are 15 feet long and nearly 15 inches in diameter. There are 168 of these in each of the first five steps. The sixth stage has 450 medium-sized rockets and two tiers each of 600 small units. The units are all attached together in such a way that as all the rockets in a stage burn out they will drop away. The rockets themselves are bonded asbestos cloth inside metal tubes.

In the corners of the compartment between

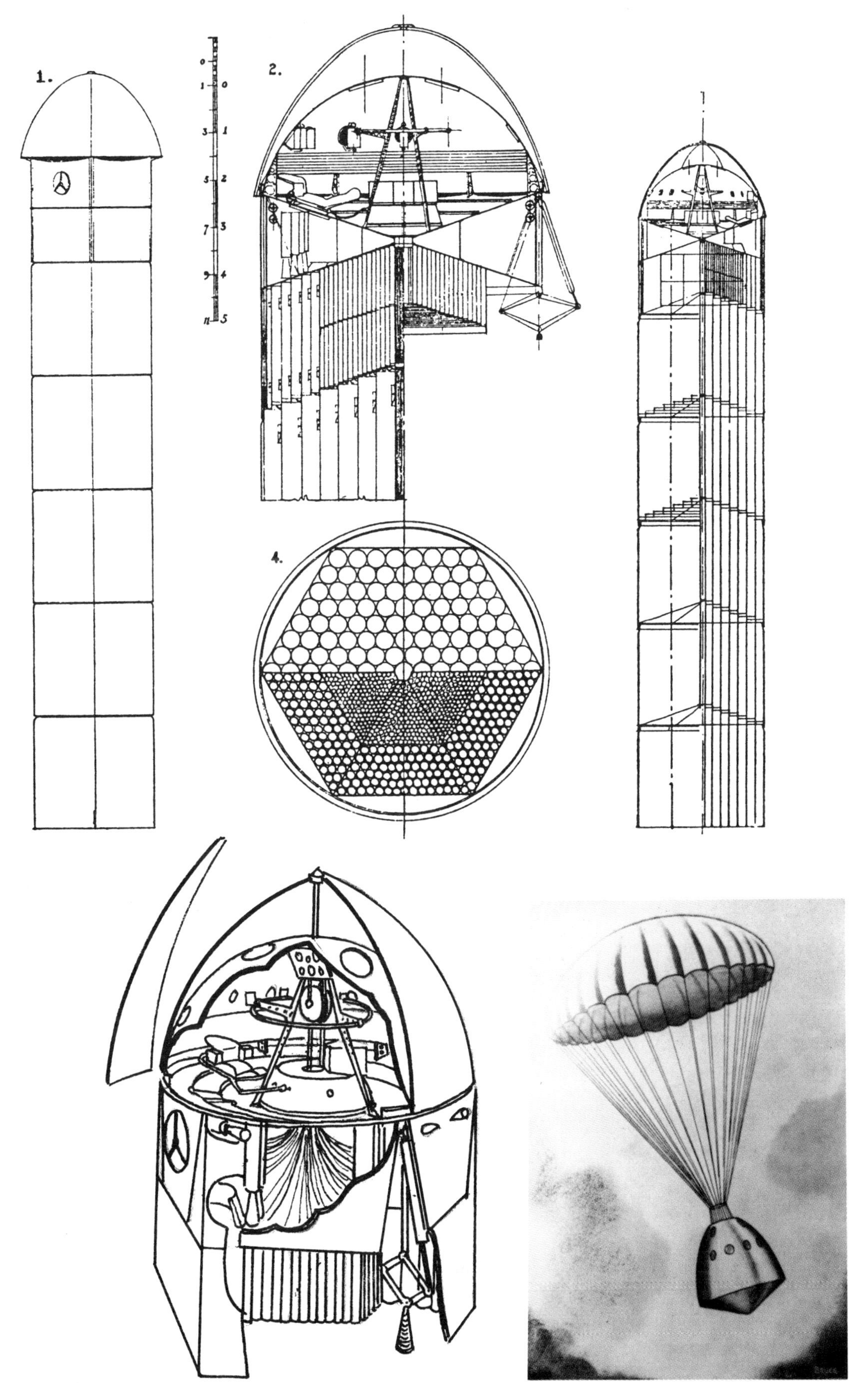
1.
2.
4.

British Interplanetary Society
Lunar Space Vessel
Schematic Detail
(Final, 1939 version)

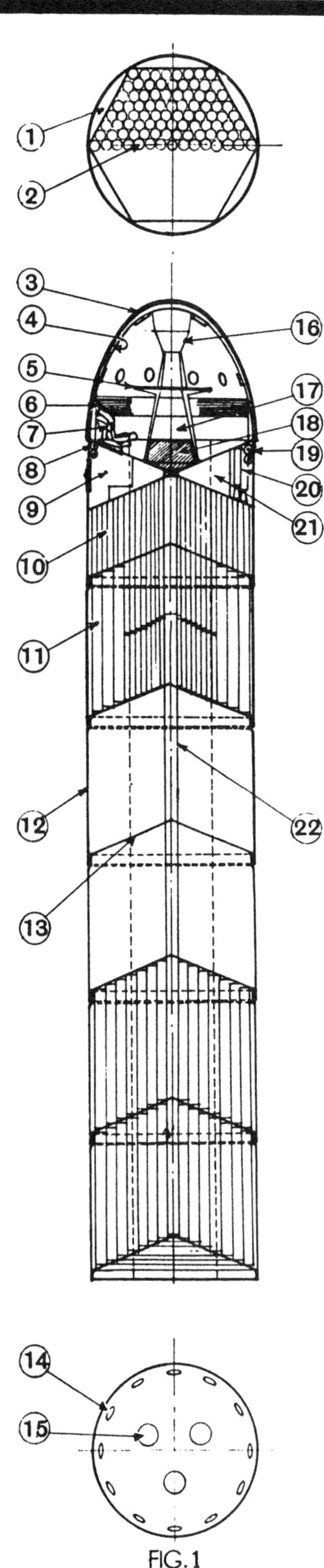

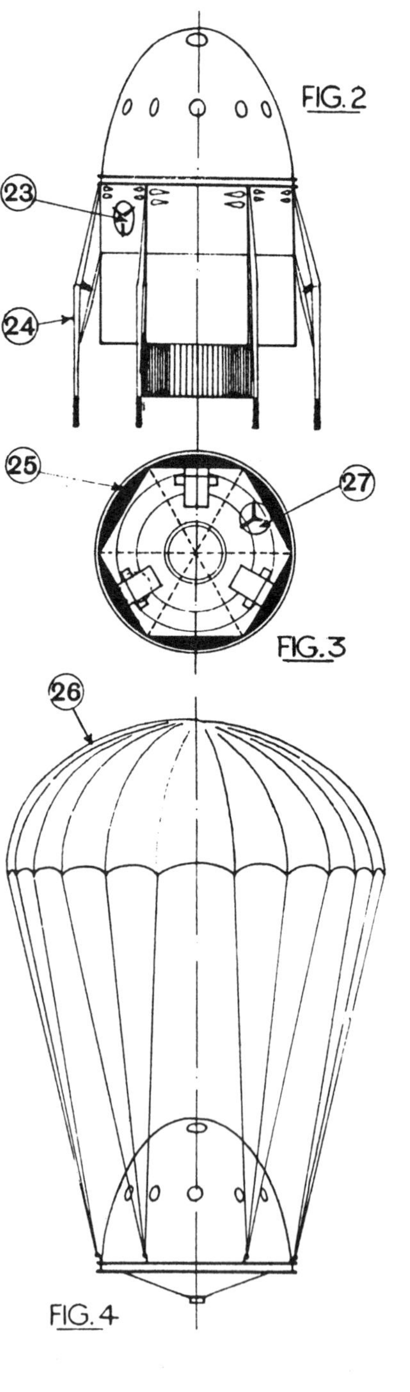

the sixth stage and the cabin are hydrogen peroxide rockets for steering and fine adjustment of velocity. Under the cabin are six liquid propellant rockets intended to both begin and eventually retard the spinning of the cabin (to provide artificial gravity).

A ceramic dome protects the plastic shell of the cabin during takeoff. The cabin contains three form-fitting couches (made of a phosphor-bronze/horsehair fabric, impregnated with rubber), a catwalk around the perimeter (parallel to the rotation axis), four windows looking forward, twelve around the circumference and six in the floor, in addition to "coelostats" to provide a steady view of the heavens while the cabin rotates.

It is obviously greatly feared that lack of gravity will have a deleterious effect on the crew of a long-term spaceflight. However, the members of the BIS committee seem not to realize the strange and no doubt disabling physiological effects of rotating such a small cabin. When standing, with their feet on the

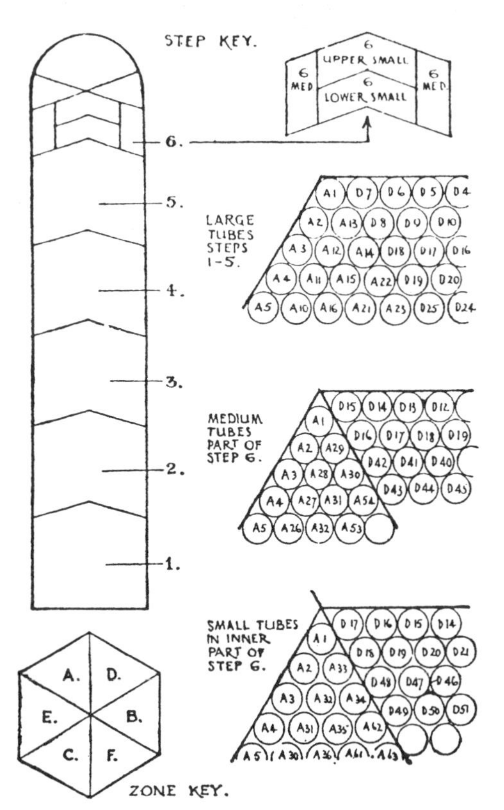

"floor"—once the outside wall—the heads of the crew will be within a foot or two of the center of rotation. Sitting, standing and moving would probably introduce serious disorientation, at best.

Other provisions for the crew's well-being are better thought out. Food for 20 days will be carried, as well as air and water—obtained by the catalysis of 500 pounds of concentrated hydrogen peroxide, with some oxygen in liquid form for emergencies and for use in the space suits. Cocoa and coffee are the main beverages. A repair kit and a medicine chest will be included—with "a little alcohol which might be raided to celebrate the lunar landing." To save weight, there will be only one cup, plate, knife, fork and spoon to be shared among the three astronauts. All power for cooking and lighting will be provided by storage batteries. Clothing is made of a fabric with a high silk content, including a tight-fitting elastic garment for the purpose of controlling blood pressure. Four space suits are taken—with one being an emergency reserve. Instruments include almanacs, mathematical tables, balsa-encased pencils and rice paper, rangefinder, telescopes, sextant, chronometer, motion picture and still cameras, etc. A deck of playing cards provides the only entertainment.

The launch of the rocket will take place from the deck of a floating platform. Ideally, the location will be a high-altitude lake not far from the equator—Lake Titicaca is one favored suggestion, as is Lake Victoria. The rocket will be inside a partly submerged caisson. High-pressure steam will give the vehicle its initial boost, with 126 of the first-stage rockets firing immediately afterwards. The rockets will be fired in rings, starting from the outermost and progressing toward the center.

For the lunar landing, special shock-absorbing legs will extend from the base of the crew module. For the eventual return to the earth, a parachute provides the final descent.

The moonship will be updated after the war [see 1947] to take advantage of the new technologies then available.

FEBRUARY 13

Abel Hermant (in *Le Temps*) determines that the gender of the word "astronef" (spaceship) is feminine . . . contradicting, it seems, all of those who have seen the rocket as a phallic symbol.

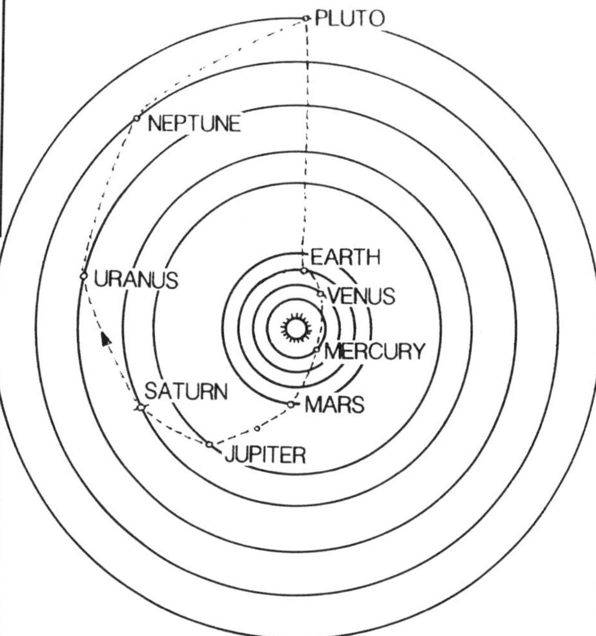

APRIL

Eando Binder publishes his novelette *The Jules Verne Express* in the magazine *Thrilling Wonder Stories*. It is the story of a "grand tour" of the solar system in which a flyby is made of every planet and the asteroid Ceres by a manned spaceship, the *Jules Verne Express*, in 9 days, 23 hours and 4 minutes. In the story, possibly the first to suggest the "grand tour" idea later used by NASA for its interplanetary space probes, the author attempts to depict some of the grimness—physical and psychological—of prolonged spaceflight. The original concept is inspired by Howard Hughes's recent round-the-world flight; the editors of the magazine suggest the Jules Verne connection. Binder admits that the velocities described in the story—up to one-tenth of the speed of light—are "a bit excessive," less from the standpoint of fuel than of human endurance. He still has to invent a "miraculous science" that eliminates the physiological effects of 60 g's acceleration.

MAY

Dr. J. W. Campbell tells the Royal Society of Canada that a rocket as massive as Mt. Robson will be required for a trip to the moon (Mt. Robson, in British Columbia, is 12,972 feet high). Campbell "pierced the legend of the rocket to the moon" by pointing out that 1 million tons of propellant will be required for every pound carried on the round trip. Assuming a spacecraft of 500 tons, a rocket the size of the mountain will be needed.

JUNE 20–JULY 3

Test flights of the first all-rocket plane take place. The Heinkel-176 uses liquid fuel and is piloted by Erich Warsitz. It is test-flown at Peenemünde, using a Helmut Walter 1,100-pound thrust engine. The plane has a pressurized cockpit that is detachable and equipped with a parachute. Its first actual flight, on June 30, lasted slightly less than a minute, but it is the flight of the first pure rocket-powered aircraft in history.

An early impression of the Heinkel-176.

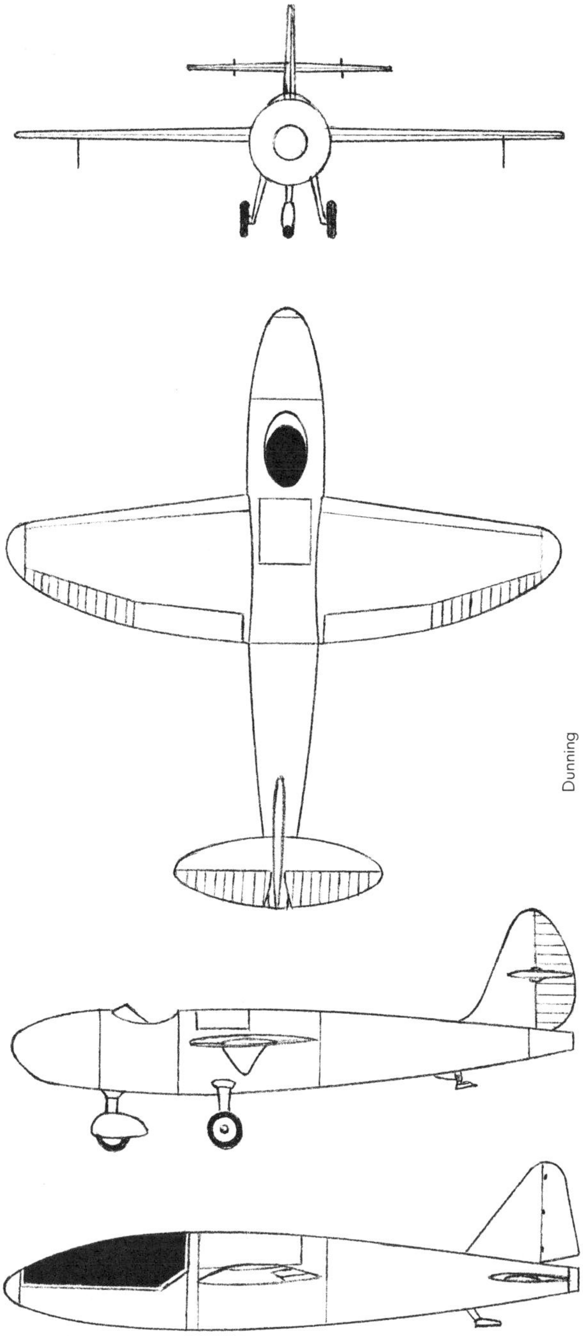

Dunning

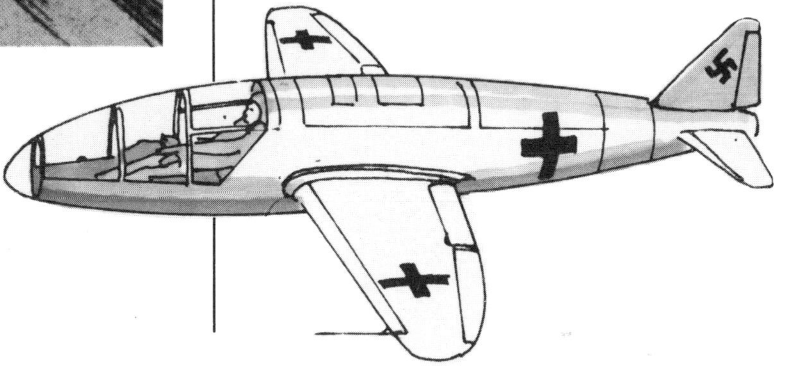

JULY 6

Wernher von Braun sends the Reich Air Ministry plans for a rocket fighter. It is a single-man liquid fuel rocket plane, launched vertically (something not attempted until the ill-fated *Natter*, see 1945, below). Its propellant is nitric acid and Visol. Launched from a simple installation, a pair of triangular scaffolds that fold up from the back of a truck and its trailer, von Braun's fighter is designed to reach a considerable altitude in a very short time.

JULY

Isaac Asimov publishes his short story "Trends" in *Astounding*. It is probably the first suggestion that the achievement of space travel might have to confront serious opposition—in this case by religious fundamentalists who want to prevent the launch (in 1973) of the first spaceship, the *Prometheus*. After the disastrous takeoff, the *New Prometheus* must be constructed in secrecy. The 20 million-strong fundamentalist group has managed to get a bill passed that sets up a government bureau that passes upon the "legality" of all scientific research. Only a secret flight to the moon in 1978 finally convinces both the public and the government that space travel must become an accepted fact.

NOVEMBER

In the short story "Pioneer-1957" (*Fantastic Adventures*), Henry Gade (pseudonym of Ray Palmer) sets an unusually early, and prescient, date for the first spaceflight.

DECEMBER

In a story published in *Amazing*, Nelson Bond describes how the new World War has been going on for 3 years, with no end in sight. Both German and American scientists are developing spaceships in order to escape from the war-torn earth. They are conducting their researches in secret bases (the German one is located at the site of the old Raketenflugplatz!). Unable to continue individually, the American and German scientists eventually trade secrets and the two spaceships, the *Goddard* and the *Oberth* take off.

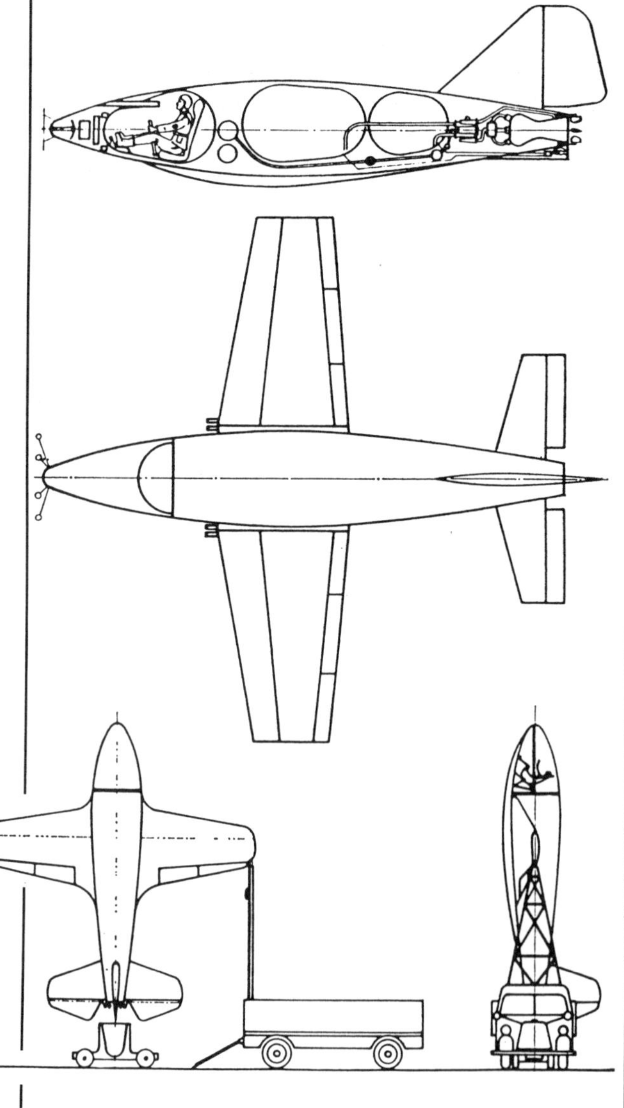

von Braun's rocket fighter.

ca. 1939

A congress of Soviet scientists predicts the use of passenger-carrying winged stratosphere rockets capable of rising to altitudes of up to than 30 miles.

1940

Flash Gordon Conquers the Universe, a Universal serial in twelve episodes, is released, starring Buster Crabbe. It is the third and last in the series.

GIRD resumes research on manned rocket planes [see 1933]. A two-man rocket plane, the RP-218, is constructed by S. P. Korolev and his team. It is powered by a cluster of three nitric acid-kerosene engines producing a combined thrust of nearly 2,000 pounds. Its takeoff weight is 3,520 pounds and it had a climbing speed of 527 mph. During the first stages of development, an ORM-65 nitric acid-kerosene motor is tried. It is fitted to an SK-9 glider and the combination is designated the RP-318. The engine develops 108.5 pounds of thrust; the plane's takeoff weight is 409 pounds. During 1938–1939, the engine is modified so that it produces between 110 and 330 pounds of thrust; it is redesignated the RDA-1-150. The new engine has been static-tested for 3 years, from 1937 to 1940 [see RP-318-1, 1940].

FEBRUARY 28

A Soviet P-5 military airplane tows the RP-318-1 rocket-propelled glider into the air. Designed by Sergei Korolev and piloted by V. P. Fyordorov it is released at an altitude of 2,600 meters and its engine ignited. Its speed increases from 80 to 140 km/h in just 5 to 6 seconds and in 110 seconds it had increased its altitude by some 300 meters.

It is a modified SK-9 2-seater airframe fitted with an ORM-65 engine, has a wingspan of 55.75 feet and is 24 feet long. A metal jacket around the engine protects the wood and fabric aircraft from the flame.

After the subsequent successful flights of the RP-318-1 Korolev remarked that "the putting of Tsiolkovky's ideas into practice has become a possibility."

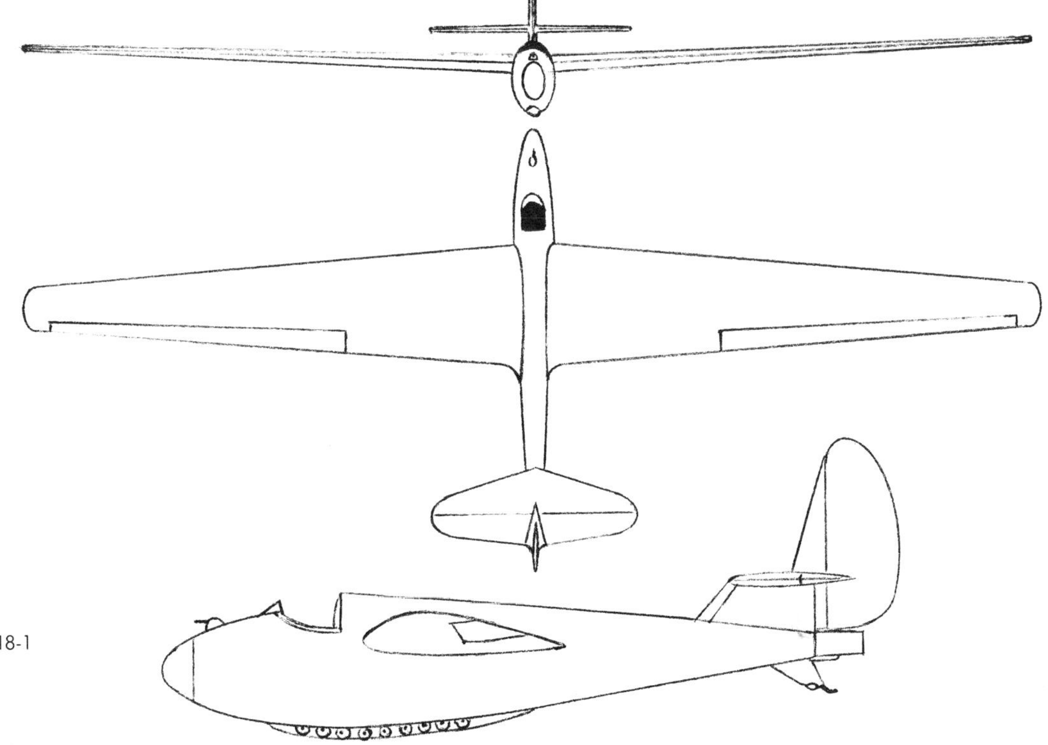

RP-318-1

Dunning

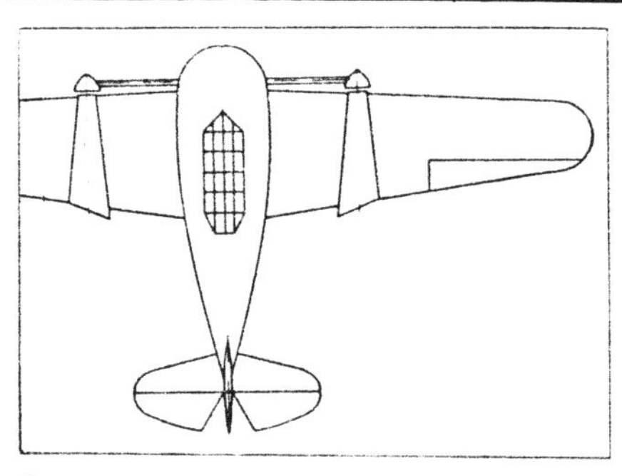
Eames

MARCH

James P. Eames proposes, in an article for *Popular Aviation*, a liquid-fueled rocket transport of the future, as well as rocket-assisted aircraft takeoff.

JUNE

Gliding flights are made in the DFS-194. This is a prototype of the Messerschmidt 163 *Komet* [see 1941].

JULY 29

The earliest calculations are made at Peenemünde for the two-stage rocket later designated the A9/A10. When the United States enters the war this project immediately leaves the speculative stage. [Also see 1941–1942.]

The A9/A10 consists of a large booster, the A10, 80 feet tall and 12 feet in diameter at its widest point. Its stabilizing fins have a span of 36 feet. It is designed to boost the A9 for 50 seconds. The A9 is similar to the winged A4b [see 1944], with thrust increased to 67,200 pounds. It will reach a maximum altitude of 210 miles and a velocity of 6,600 mph. The winged second stage will have a range of 3,000 miles, carrying a payload of 1 ton. The combined weight of the 2 rockets is about 85 tons.

To develop the powerful A10 motor—400,000 pounds of thrust is needed—it is estimated that 3 years' work will be required.

A manned version of the A9 is also proposed [see 1941–1942].

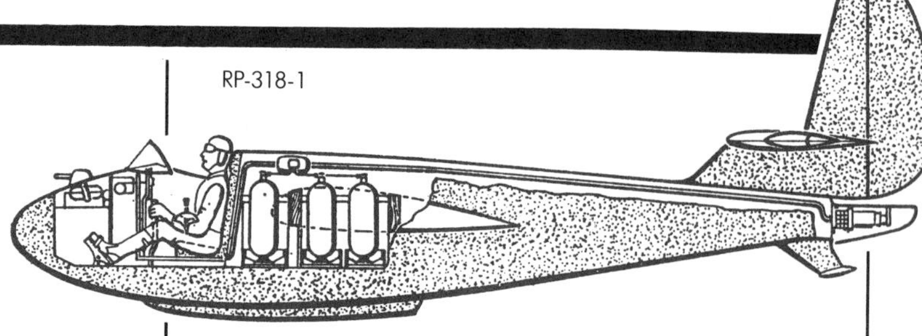
RP-318-1

AUGUST 9

R. H. Goddard launches the first rocket equipped with fuel pumps.

1940–1941

V. F. Bolkhovitinov presents a series of lectures on the aerodynamic and structural design of a rocket interceptor, perhaps the first of their kind in the U.S. S. R. He was discussing the airplane which he was then working on [also see 1939 and 1942]. The prototype of the rocketplane was ready in 1941.

1941

Robert A. Heinlein publishes his short novel *Universe* in which he describes one of the earliest depictions in fiction of the generation starship.

The idea behind the generation starship is inspired by the need to circumvent the limiting speed of light when interstellar distances are involved. Journeys with durations of centuries could be made if the spacecraft are constructed large enough to contain self-sufficient colonies of human beings. Whole generations of people will be born, live out their lives and die within the confines of the starship. It will be the distant descendants of the original crew who will eventually arrive at the destination star. A number of science fiction stories and novels have explored the problems that such a concept might generate. In Heinlein's, something goes awry and the descendants of the original crew of the starship *Vanguard* forget that they are living inside a vast cylinder, 5 miles long and 2,000 feet in diameter. To them, the interior of the starship *is* the universe, literally. Their ancestors, the Mission and the increasingly mysterious science and technology of the starship all eventually take on the aspects of a religion.

The French science fiction film *Croisières Sidérales* explores some of the paradoxes involved in near-light-speed space travel . . . one of the first times the subject has been used in cinema. [Also see *The Sky Splitter*, 1922.]

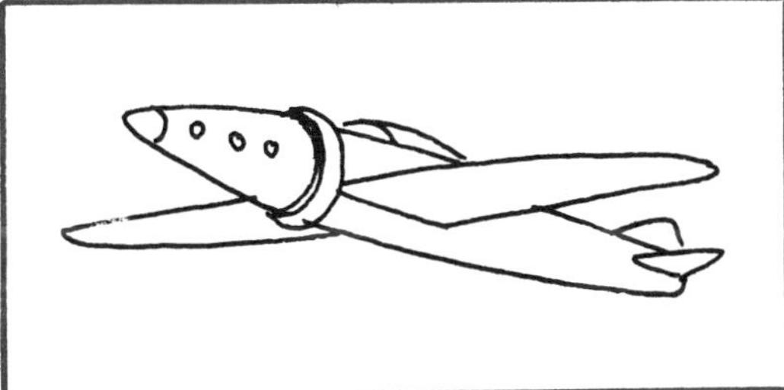

Krafft Ehricke designs a rocket-propelled airplane for the aircraft firm Campini.

Dr. Dinsmore Alter, director of the Griffith Observatory, predicts that the first manned flight to the moon will take place sometime within the next 100 years—"if not sooner." He believes that the moonflight will depend upon the development of atomic energy, especially the new element, U_{235}. He estimates that the cost of a moon rocket would be about $100 million.

AUGUST

Homer Boushey is one of the first Americans to fly a rocket-powered aircraft [also see Swan, 1934]. The plane is a commercially manufactured Ercoupe. Under the auspices of the California Institute of Technology and Aerojet-General founder Theodore von Karman, the experiment is performed for the Army Air Force. The plane is fitted with six solid-fuel rockets. On the first flight, the plane takes off under its own power, the rockets only serving as a booster, or "JATO" unit. Later, Boushey makes several flights with the plane's engine turned off, taking off and flying by rocket power alone.

"It shoots the rocket ship into the stratosphere. Guess where I got the idea, Mr. Ley"

Guy Gifford, Amazing, March 1942.

AUGUST 13

Test pilot Heini Dittmar flies the first rocket-propelled fighter, the Messerschmidt Me-163V1 *Komet*.

FALL

Captain Future ("Wizard of Science") magazine publishes an explanation and diagram of Captain Future's spaceship, the *Comet*, the fastest ship in space. The Captain Future novels are written by science fiction veteran Edmund Hamilton.

The teardrop-shaped hull has triple-sealed walls made of a secret alloy. It is powered by a battery of nine cyclotrons that convert powdered mineral fuel into raw energy. This flows through valves into the various rocket tubes that propel and steer the *Comet*. Steering is directed by a control called the "space stick," which is analogous to the control stick of a conventional airplane. Beneath the pilot's left foot is the "brake-blast pedal," which directs the atomic blast into the bow retrorockets, automatically cutting out all

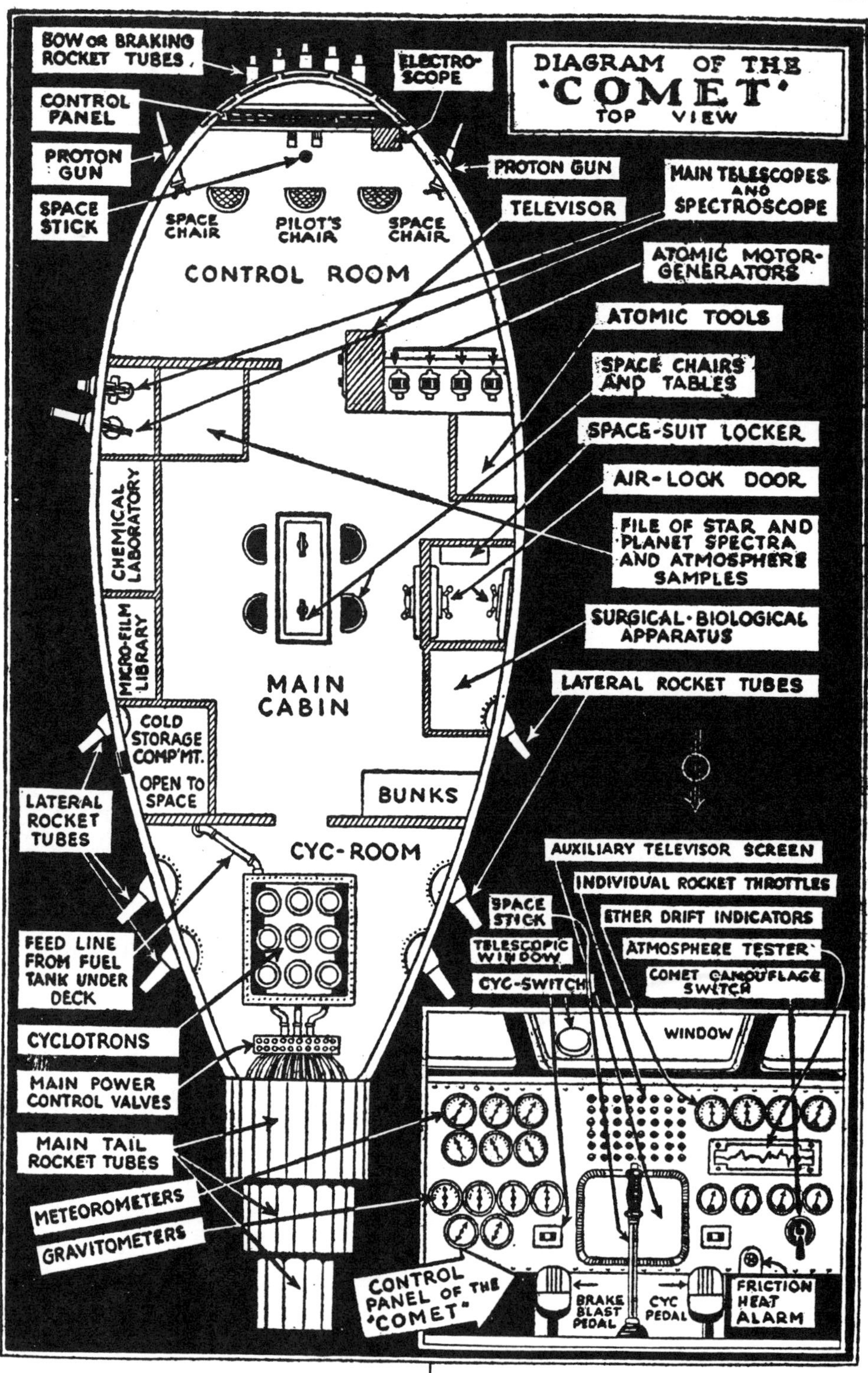

other tubes. Under the right foot is the "cyc-pedal," which controls the amount of atomic energy fed to the rocket tubes. "To make a quick stop, you simply jam both brake-blast and cyc-pedal to the floor . . . "

Ahead of the pilot's seat is a broad window. Beneath is the intrument panel with its standard meteorometers, gravitometers, ether-drift indicators, main cyc-switch, auxiliary televisor screen and microphone. In addition, the *Comet* possesses an atmosphere-tester, a comet-camouflage switch (which releases a cloud of luminous ions from the rocket tubes), and an electroscope for detecting the ion trails of other spacecraft.

Two space-chairs flank the pilot's seat and

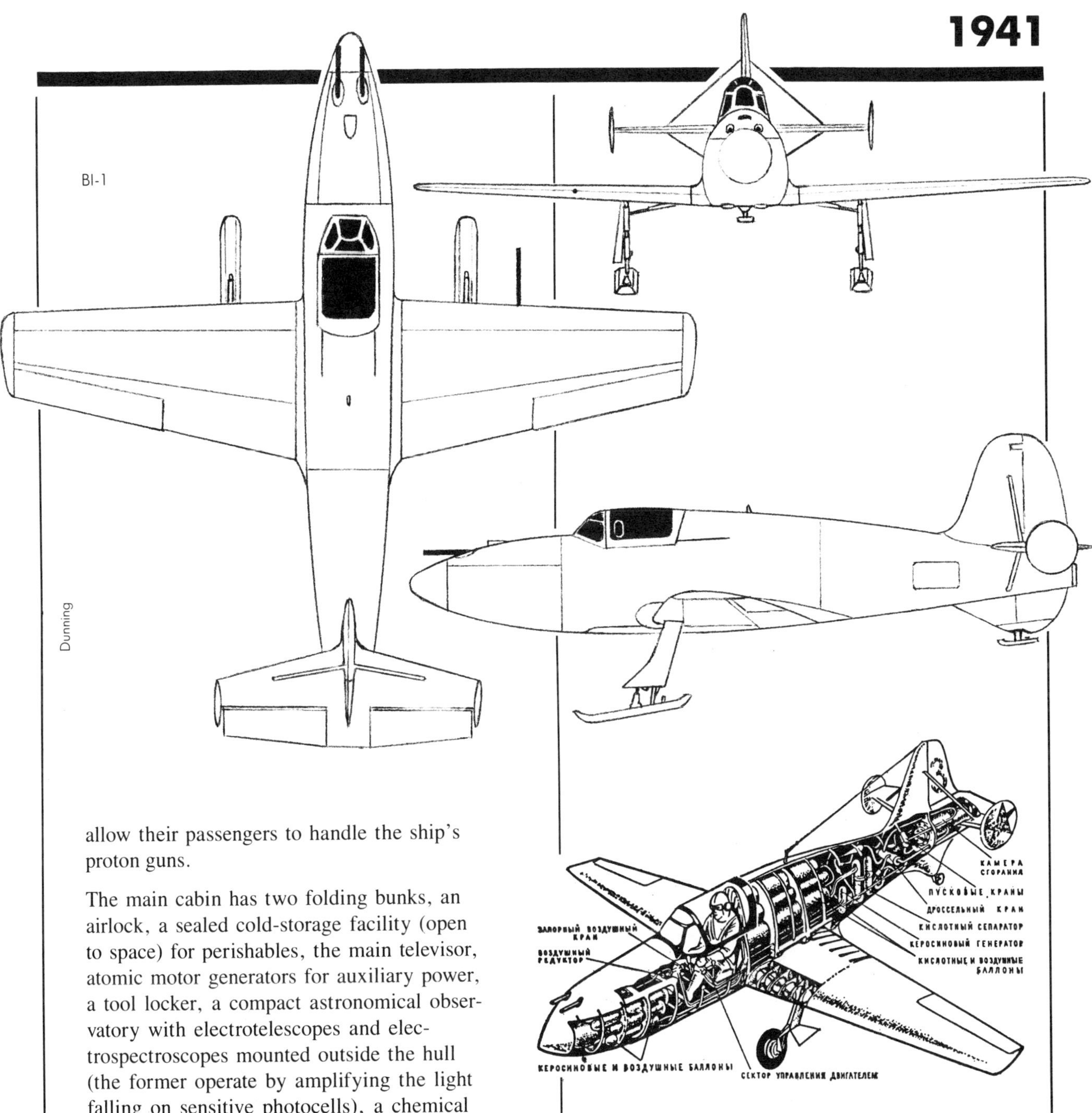

allow their passengers to handle the ship's proton guns.

The main cabin has two folding bunks, an airlock, a sealed cold-storage facility (open to space) for perishables, the main televisor, atomic motor generators for auxiliary power, a tool locker, a compact astronomical observatory with electrotelescopes and electrospectroscopes mounted outside the hull (the former operate by amplifying the light falling on sensitive photocells), a chemical laboratory, microfilm reference library, botanical collection, and so forth.

SEPTEMBER 10

The BI-1, a Russian rocket fighter is designed by A. YaBereznak and A. M. Isayev. It is initially tested as a glider, first towed aloft by a twin-engined bomber. The plane is propelled by a powerful nitric acid-kerosene engine (the RNII's D-1-A-1100) producing more than 2,200 pounds of thrust. During the war, in 1942, the fighter is flown by pilot G. Y. Bakhchivandzhi. He is killed on May 15, 1942, when the interceptor flies into the ground.

OCTOBER 2

In a Messerschmidt Me-163V1 *Komet*, after being towed to an altitude of 13,000 feet, Heini Dittmar makes the first powered flight of this rocket plane. He reaches a speed of 623.85 mph, the first aircraft to pass the speed of 1,000 km/h, at which point he experiences violent turbulence: he has encroached upon the "sound barrier" at Mach 0.84.

In another test flight on this day, the Me-163 reaches a top speed of 624 mph. It is piloted by Hanna Reitsch.

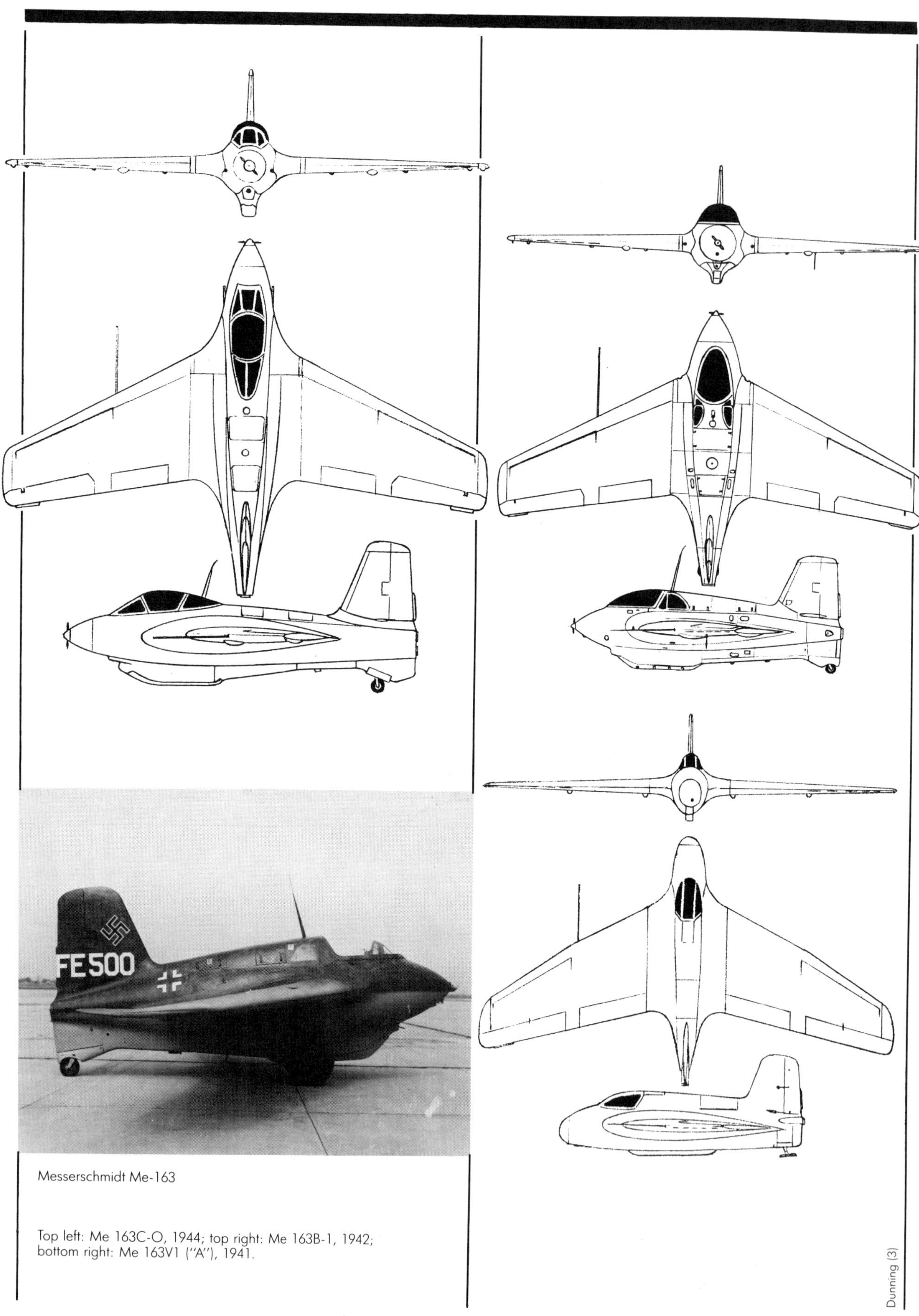

Messerschmidt Me-163

Top left: Me 163C-O, 1944; top right: Me 163B-1, 1942; bottom right: Me 163V1 ("A"), 1941.

Dunning (3)

DECEMBER 18

Reaction Motors, Inc. is founded by several former members of the American Rocket Society [see October 14, 1947].

1941–1942

The A9/A10 is proposed, a design for a two-stage intercontinental missile, with a second stage similar to the V2 [see below and July 29, 1940]. There are also plans for an eventual manned version. This is to be equipped with a pressurized cockpit and landing gear. Flaps are added to the tailfins for control, as well as broad swept wings with a 30-foot wingspan. The fourth tailfin, on the underside, is replaced by a pod containing a jet engine. It is the same diameter as the V2 but about 5 1/2 feet taller. A piloted version of the A9 would have resulted in the first manned spaceflights. [Also see A4b, 1944.]

According to *A Summary of German Guided Missiles*, by N. Harlan and G. McConnell, "A few A9s are built but it is believed that none are ever test fired. Although requiring different internal construction to provide for seventy-five square feet of wing area, it is similar in appearance to A4-b." In fact, the rocket Harlan and McConnell refer to *is* probably the A4-b.

The DFS 228 is a rocket-assisted glider used by the Russians as a reconnaisance aircraft.

Wind tunnel testing of winged A9.

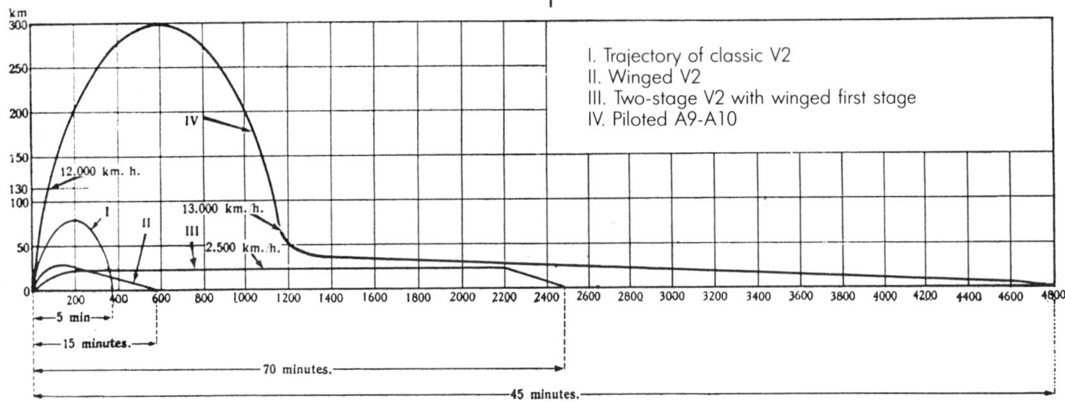

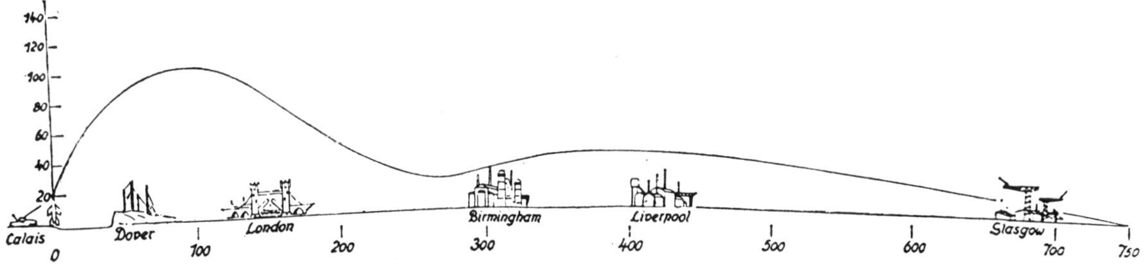

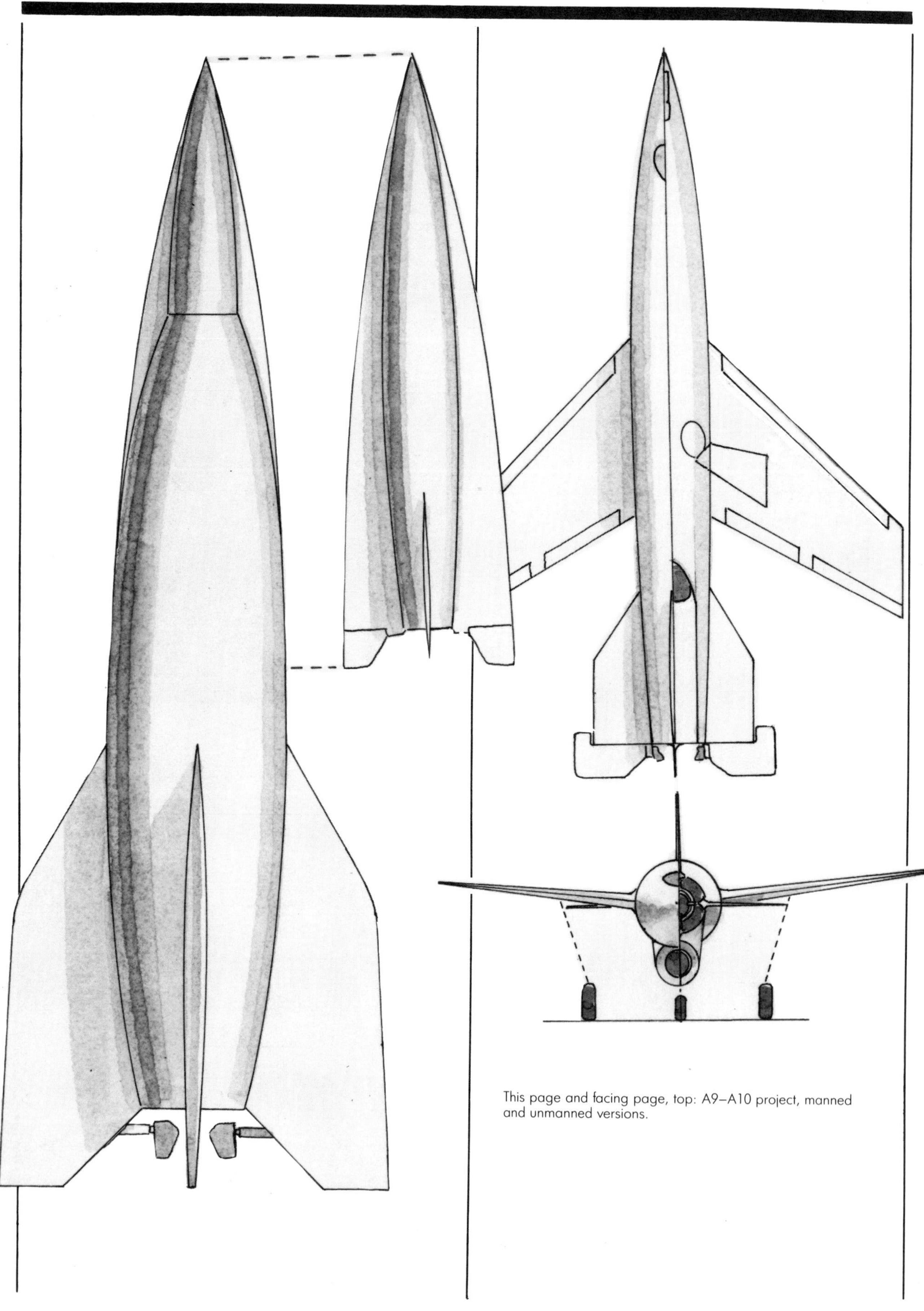

This page and facing page, top: A9–A10 project, manned and unmanned versions.

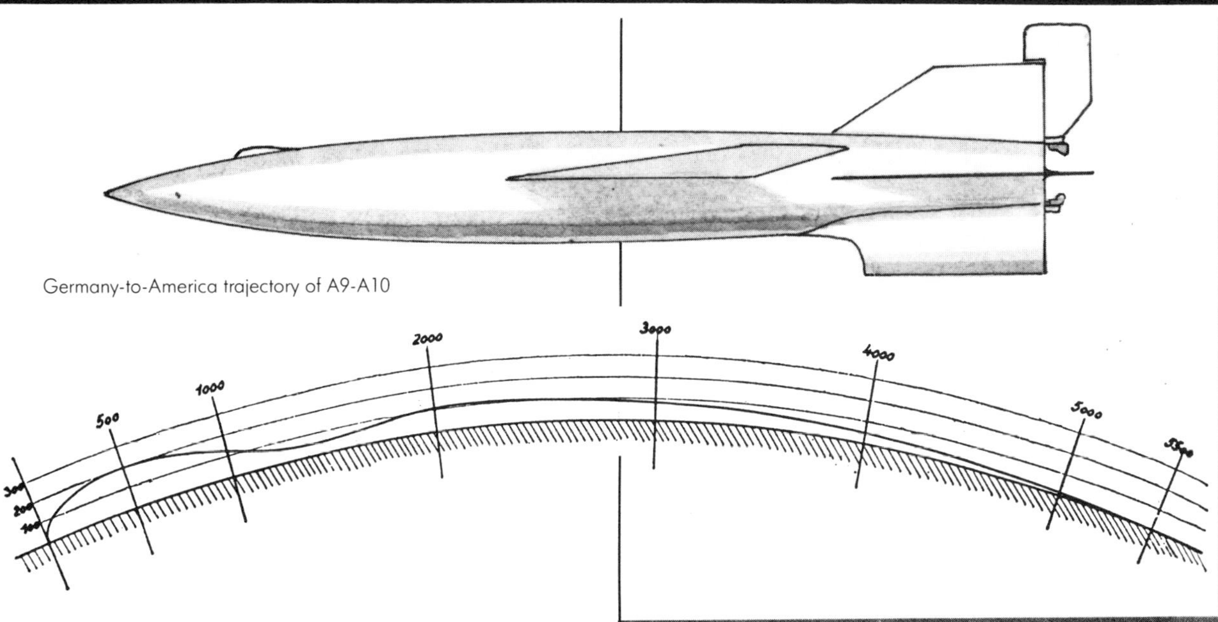

Germany-to-America trajectory of A9-A10

Artist F. Gordon-Crosby publishes his concept of a future spaceship in *Autocar* and the next year in *Flight*. The former magazine captions the art with the comment that "The possibility of the whole idea lies in the use of the rocket principle as a method of propulsion . . . " The latter adds that "Technicians appear to be agreed that theoretically it should be possible, provided certain experimental research is carried out, to fly to the moon in a rocket-propelled aircraft.

1942

MARCH 19

Aerojet Corporation is founded as an offshoot of GALCIT.

MAY 15

At Sverdlovsk, the first Soviet rocket-propelled fighter takes off. Designed by V. F. Bolkhovintinov [see 1939 and 1940–1941] and piloted by G. Ya. Bakhchivandji the BI-1 [B=Bolkhovitinov, I=Istrebitel=aircraft] flies for about 15 or 20 minutes. The rocketplane crashes, however, on its third flight, killing the pilot.

The BI-1 used the same rocket motor as Korolyov's RP-318-1 [see 1940]. Designed by L. S. Dushkin the RDA-1-150 could vary its thrust from 350 to 1,400 kg.

SEPTEMBER

Northrop produces a feasibility study for a rocket-powered interceptor, which is followed by an Army contract. This is toward the development of the XP-79 all-magnesium rocket fighter-interceptor. Its speed is designed to be in excess of 500 mph, powered by an Aerojet rocket motor with 2,000 pounds of thrust. Three glider mockups are

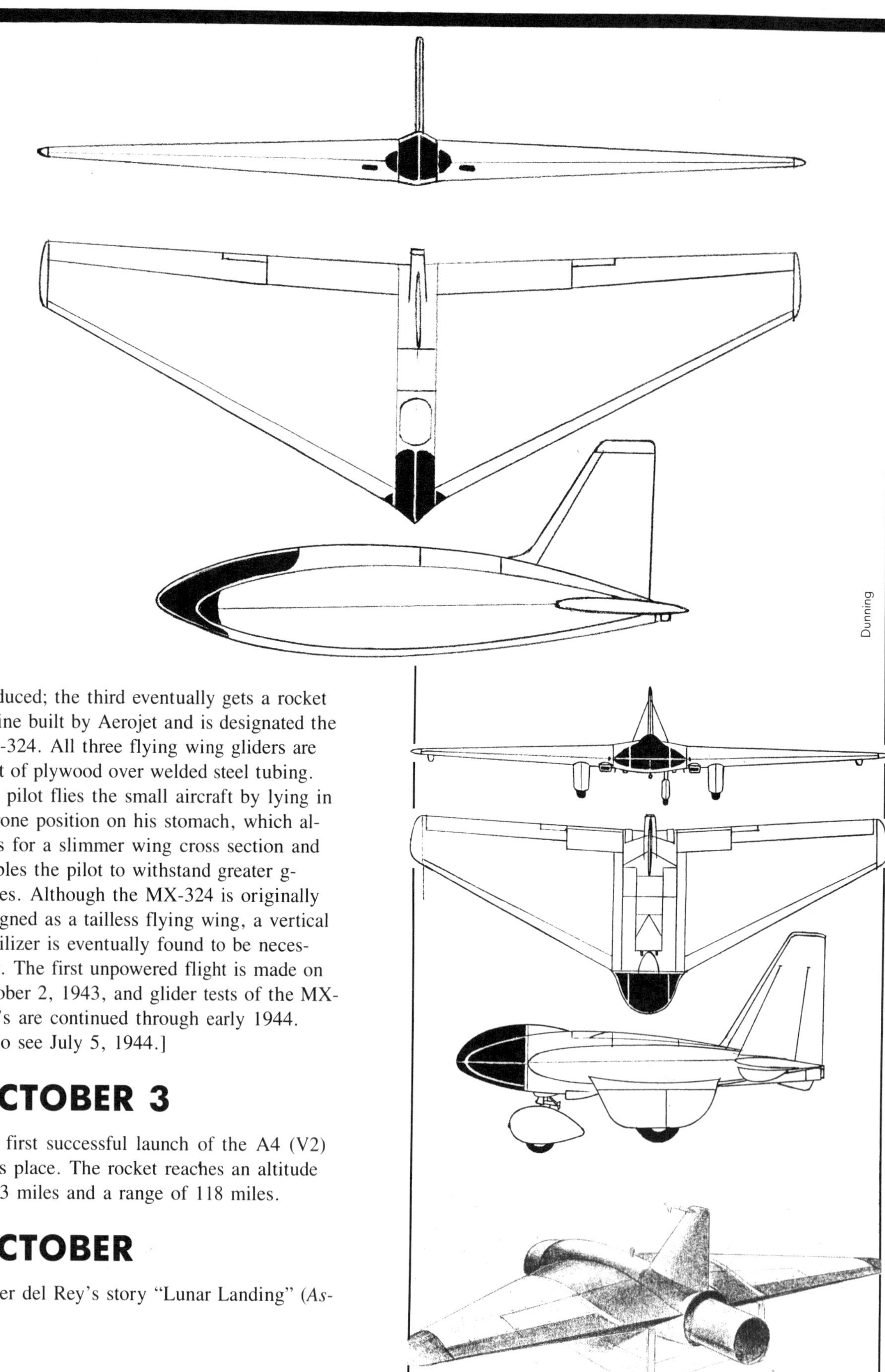

produced; the third eventually gets a rocket engine built by Aerojet and is designated the MX-324. All three flying wing gliders are built of plywood over welded steel tubing. The pilot flies the small aircraft by lying in a prone position on his stomach, which allows for a slimmer wing cross section and enables the pilot to withstand greater g-forces. Although the MX-324 is originally designed as a tailless flying wing, a vertical stabilizer is eventually found to be necessary. The first unpowered flight is made on October 2, 1943, and glider tests of the MX-324's are continued through early 1944. [Also see July 5, 1944.]

OCTOBER 3

The first successful launch of the A4 (V2) takes place. The rocket reaches an altitude of 53 miles and a range of 118 miles.

OCTOBER

Lester del Rey's story "Lunar Landing" (*As-*

tounding) proposes the use of midgets as astronauts, to save weight and space.

NOVEMBER

An "Airship of Io" is described by Henry Gade and illustrated by Frank R. Paul in *Amazing*. It is a streamlined, twin-engined, atomic-powered air and space craft 200 feet long. The U-235-fueled engines are sus-

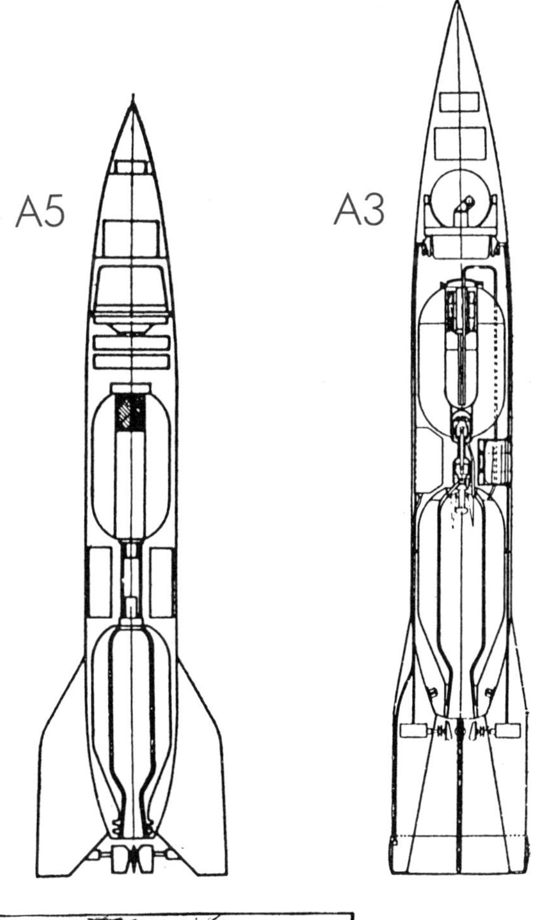

Left: The A5 and A3 are predecessors of the A4 (V2).

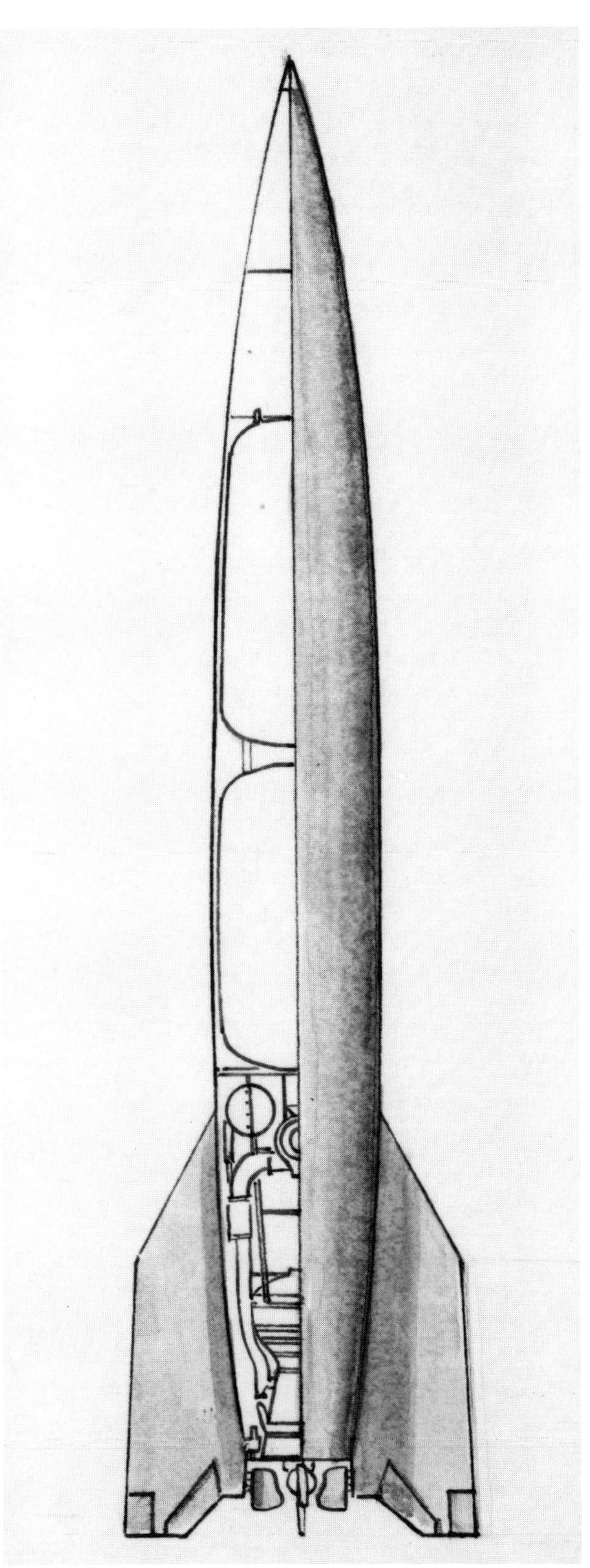

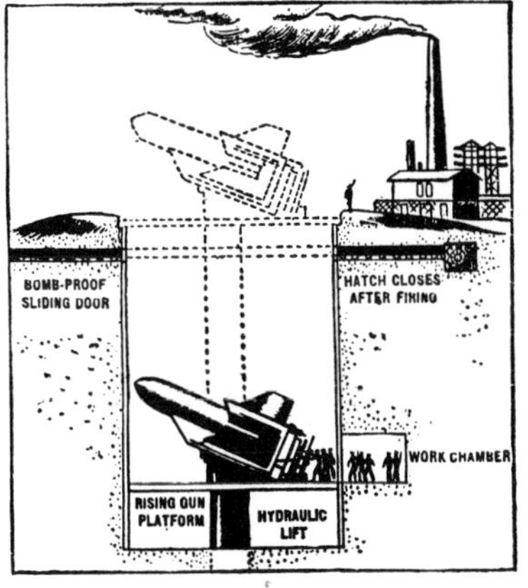

There is much Allied speculation on the true nature of the Nazi secret rocket weapon.

pended below the wings, where they provide thrust by breaking down water into its constituent atoms. The engines can be pivoted to provide reverse thrust for landing.

1943

The PE-2-R is a Russian rocket plane designed by Sergei Korolev. An RD-1 engine is installed in the tail of the aircraft.

A "ram rocket" engine is designed by Walter Lippisch.

AUGUST

Eugen Sänger and Irene Bredt, in their report *Uber einen Raketenantrieb fur Fernbomber* (A Rocket Drive for Long-Range Bombers—completed in 1941) describe the final form of the *Silver Bird* [see 1933, 1938–1939]. It is an earth-orbiting, single-stage rocket plane with a launch weight of 100 tons (including 90 tons of payload and fuel). Its liquid fuel engine will produce 100 tons of thrust. It will reach altitudes of up to 186 miles carrying a payload of 4 tons into orbit. It is able to reach such extremely long ranges by a technique Sänger has developed called "skip flying." The spacecraft will literally be skipping off the denser layers of the atmosphere like a flat stone across a surface of water. In this way he is able to achieve ranges several times those obtainable by mere aerodynamic descent.

"The rocket bomber," Sänger writes, "will differ from present day propellor-driven aircraft in the following essential points: in place of the propellor propulsion from the fuselage front it has the rocket propulsion in the fuselage stern; the fuselage is in the shape of a bullet with tapered hind part, the wings have a thin wedge-shaped profile with sharp leading and trailing edges and high wing loading at the start of the flight; the cabin is constructed as an airtight stratosphere chamber."

The space plane is similar to the one described in 1938 [above]. It is 91 feet long, 11.8 feet at its widest point, 7.5 feet wide at the engine opening. Its wingspan is 50 feet. It has a single horizontal tail plane, with a vertical stabilizer at either end. The stubby wings are knife-edged and have a triangular cross section. It has retractable tricycle landing gear, a pair beneath the wing roots and one at the nose. A single pilot sits in a pressurized cabin. Visibility is poor, only side-view slits and a periscope-like arrangement let the pilot see out. A detachable window cover can be jettisoned during landing for a forward view. In the middle of the plane is the bomb bay, able to hold bombs of up to 30 tons.

It will be brought up to its takeoff speed of 1,640 ft/sec by means of a catapult. This is a rocket-driven sled on a horizontal track. The sled's rocket is liquid-fueled.

The rocket plane is designed to skip across the upper layers of the atmosphere, dropping its bomb at any one of the low points, and then continuing on around the world back to its launch site.

ca. 1943

Ernst Stuhlinger recalls watching a V2 launch with Hermann Oberth: He had never seen Oberth before, but recognized him from photographs. Oberth was staring at a distant point in the sky, "but not at all in the direction in which the big rocket had just disappeared." After a long silence, Stuhlinger said, "It must certainly be a most gratifying experience for you, Professor Oberth, to see how beautifully your early dreams and concepts of large rockets have now come to life." The professor, however, did not answer

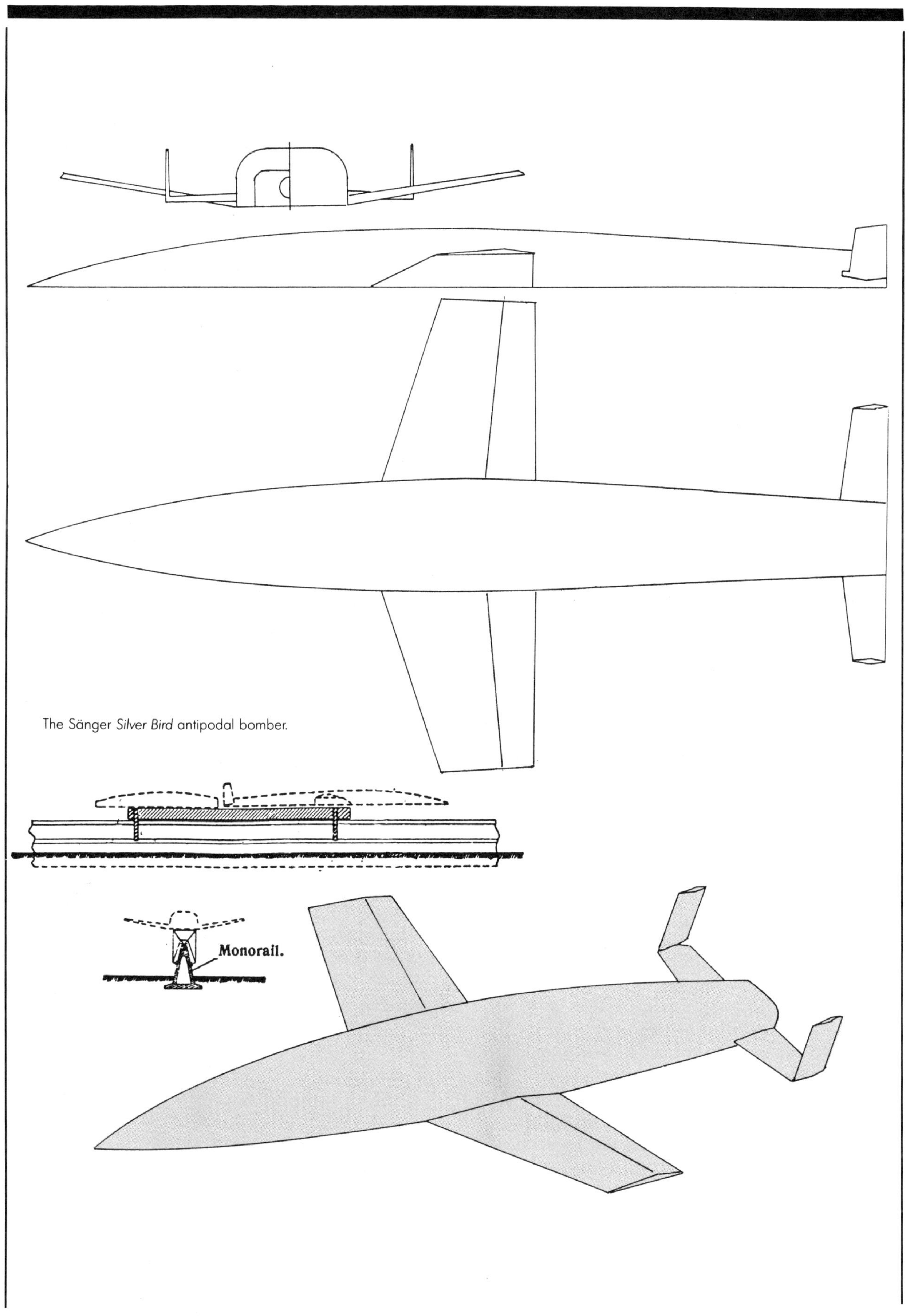

The Sänger *Silver Bird* antipodal bomber.

ROCKET AIRCRAFT OF WORLD WAR II

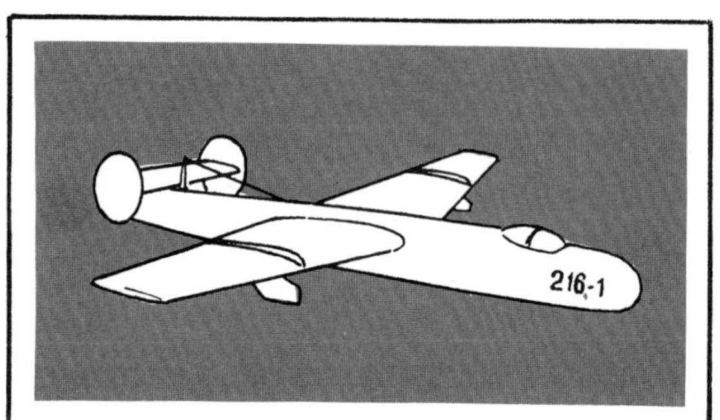

Soviet 216-1

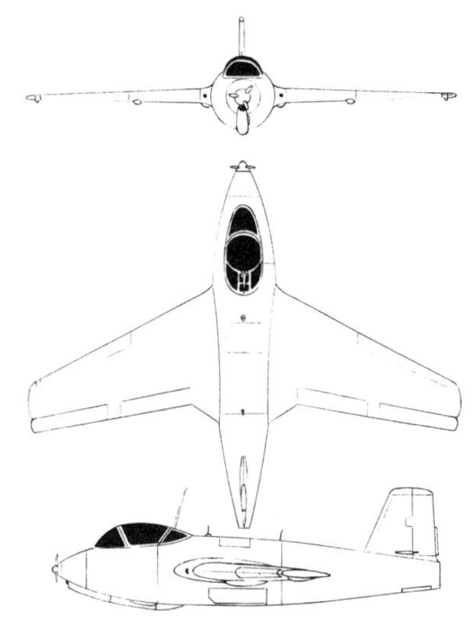

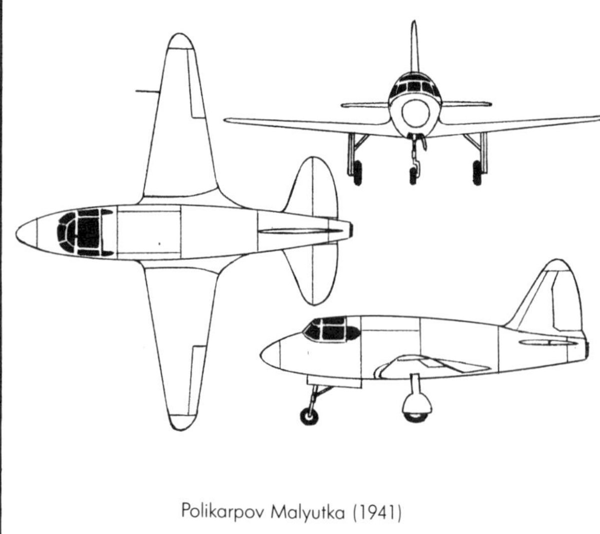

Polikarpov Malyutka (1941)

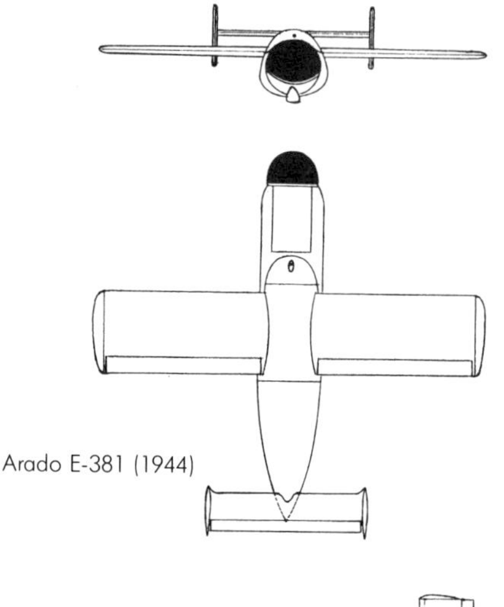

Arado E-381 (1944)

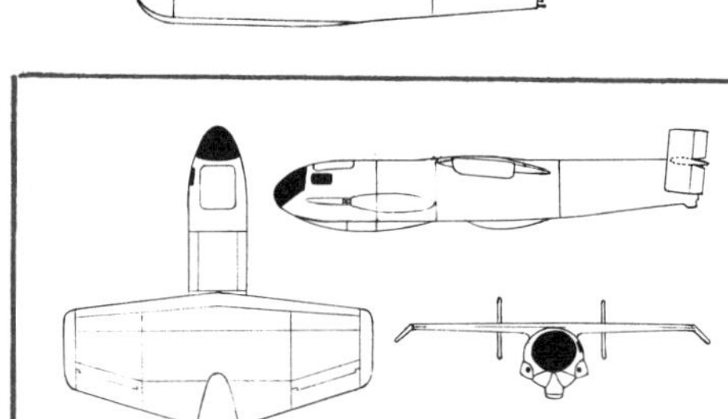

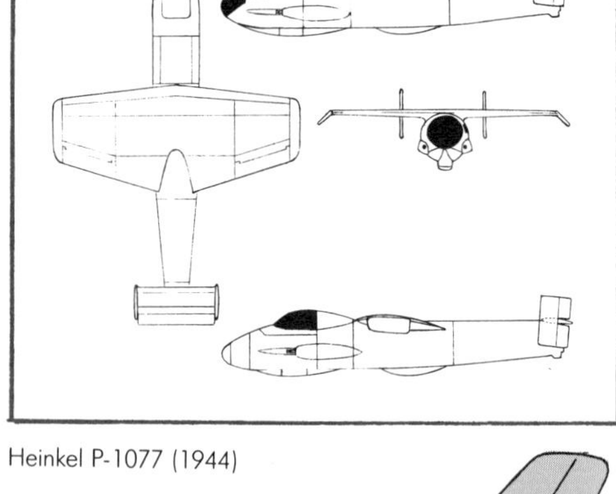

Heinkel P-1077 (1944)

Early speculation on appearance of Japanese Baka.

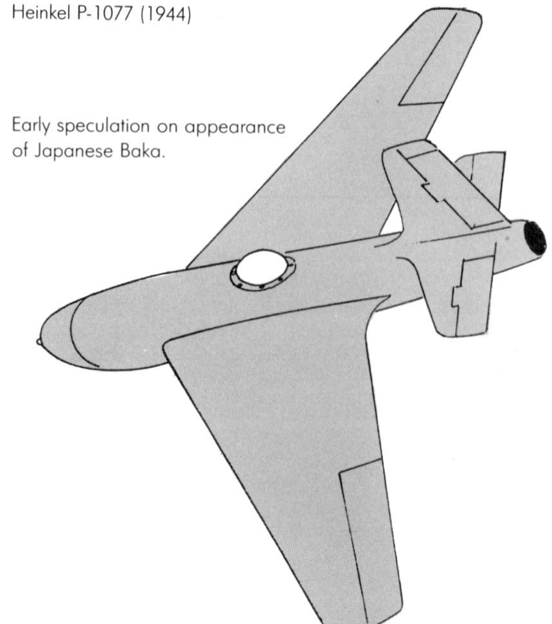

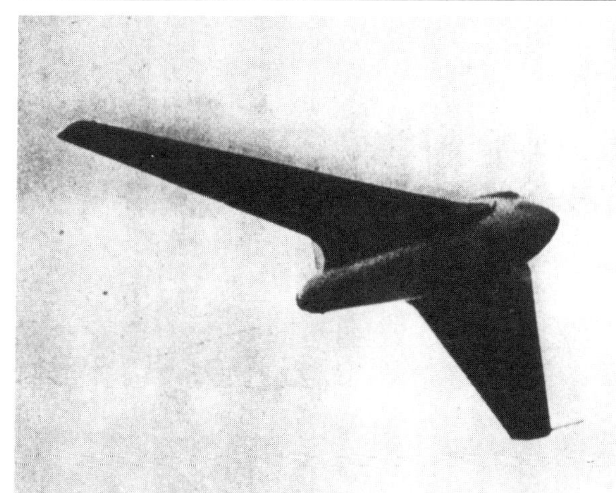

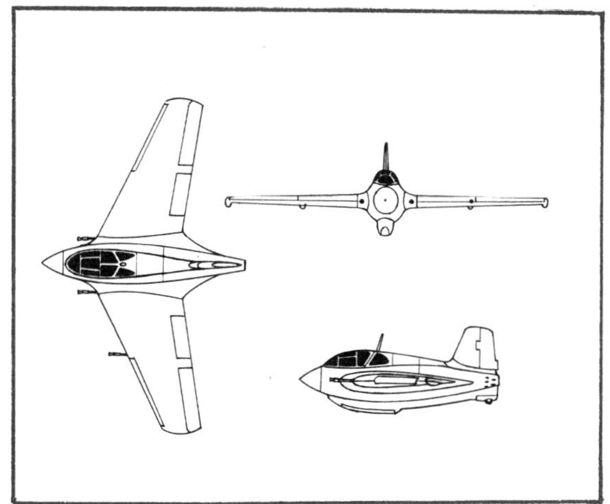

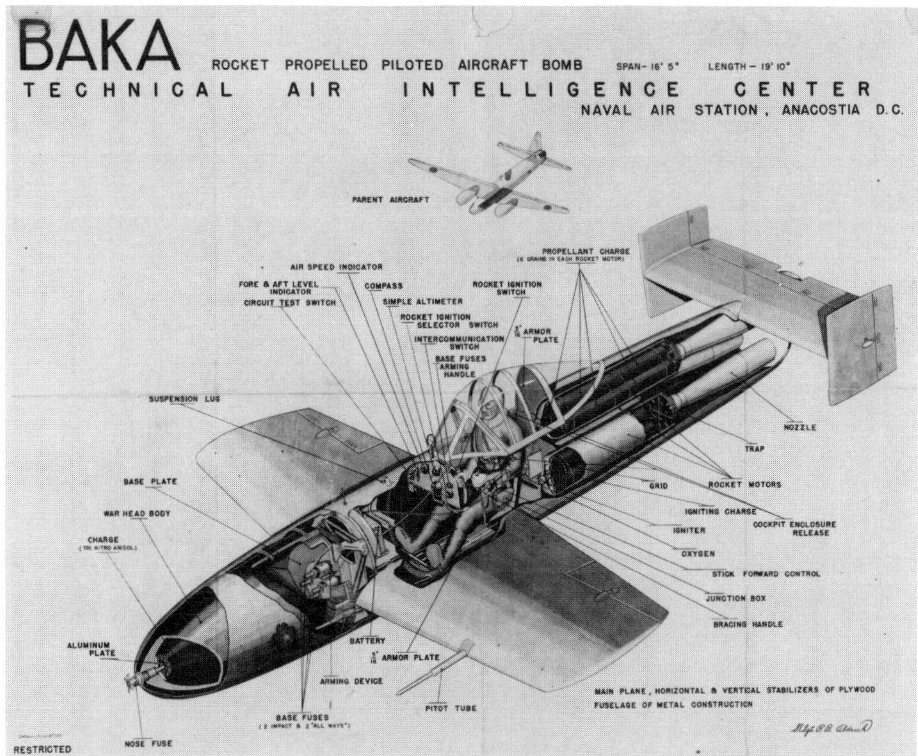

Above left:
DeHaviland DH-100 *Swallow*

Above right:
Japanese J8MI *Shusui*

Left: Baka

Junkers EF-127 (1944)

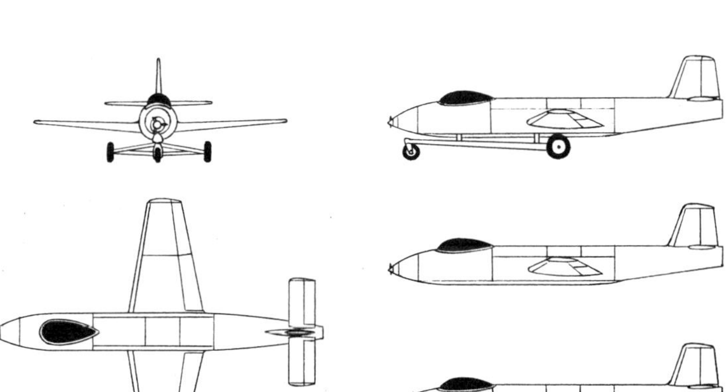

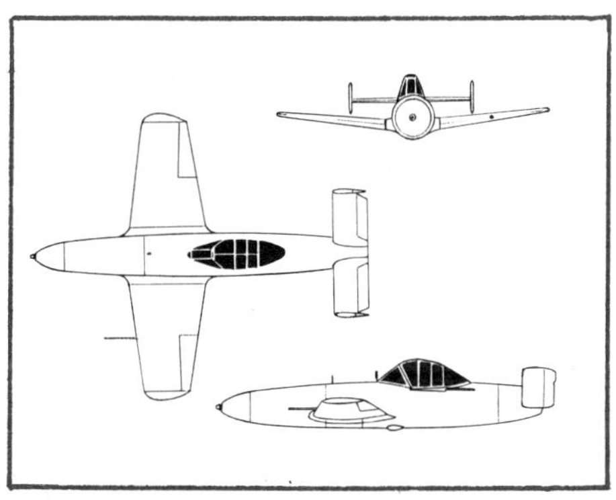

Yokosuka MXY-7 Ohka (Baka)

and Stuhlinger feared that he might have inadvertently said something offensive, or even stupid.

Finally, Oberth turned his head until he is looking far out in the opposite direction. "I have," he said carefully and slowly, "the greatest admiration for the engineers and technicians who built this rocket. But beyond that, it does not mean much. We have known before that a rocket will work within and beyond the atmosphere. This rocket is only the first little step toward a much greater project: the exploration of outer space. Out there, there are still so many things which we do not know and which are perhaps far beyond our imagination. There exploration is what really counts. We must not forget this goal in the enthusiasm that a mere technical success may give us."

1944

Willy Ley publishes the first edition of his now-classic *Rockets: The Future of Travel Beyond the Stratosphere*. It is an encyclopedic and idiosyncratic history of rocketry and space travel, and the state of the art of astronautics at the time. Ley draws heavily upon his first-hand experiences with the German Rocket Society. The book will go through several increasingly expanded revisions until its final version in 1968 (as *Rockets, Missiles and Men in Space*).

Ley proposes, in this book, a three-step moon rocket. It is a squat, artillery-shell-shaped vehicle, with the stages nested one within the other. The complete rocket is about one-third the height of the Empire State Building, and has a mass-ratio of 134:1. The returning ship slows for reentry by grazing the earth's atmosphere, much like Sänger's skip-bomber [see 1931]. A revised version of Ley's moon flight as a circumlunar mission is illustrated by Chesley Bonestell in a September 1945 *Mechanix Illustrated* article [see below].

The A4b is a winged version of the V2. It is intended to be eventually launched as the upper stage of the A9/A10 2-stage rocket.

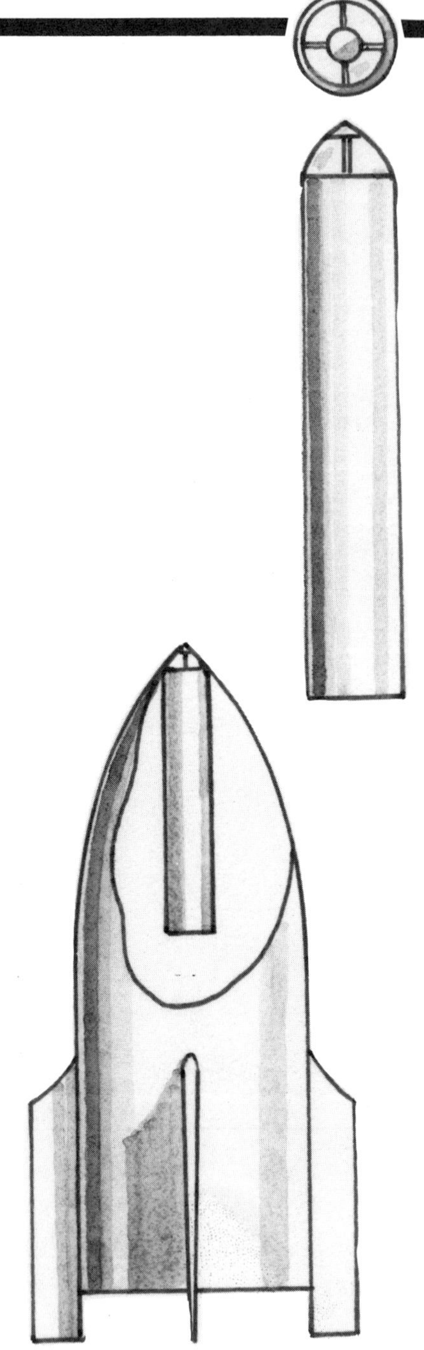

Two test flights are made at Peenemünde and drawings are made of a piloted version [see above].

The U.S.S.R. has the RD-1 engine in production. Flight trials are made between 1944 and 1945 in Pe-2, La-7, Yak-3 and Su-6 aircraft.

Tests are begun on developing the rocket-powered fighter *Natter* (Viper) [also see 1938 and 1945].

Advertisement, November 1941 (*Life*).

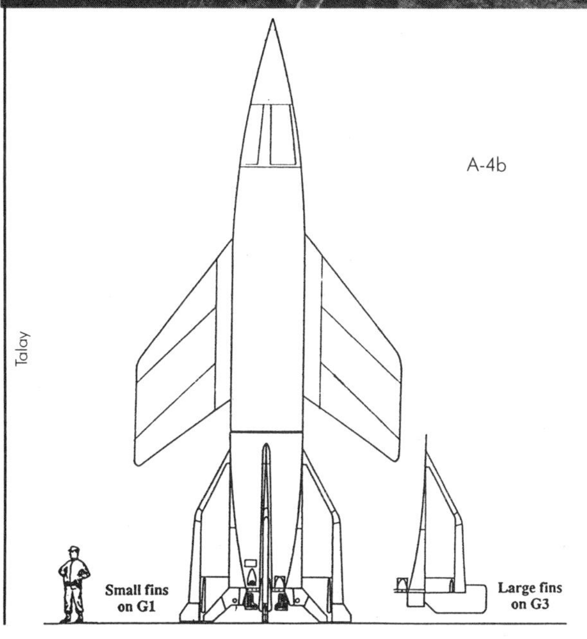

Talay

JANUARY 17

Life magazine publishes an article explaining rockets to its readers. "For a year now stories have been widely circulated, mainly by Nazi propaganda, about secret and powerful German rocket weapons." The purpose of the article is "[to] debunk a few of these stories, and to reveal the limitations as well as the possibilities of rockets in both war and peace . . . " The artwork, by B. G. Seielstad, explains how the rocket principle works and the differences between solid-fuel and liquid-fuel rockets, the various sorts of military rockets then being used by both the Allies and the Axis, and an explanation of why rocket weapons are "inaccurate, inefficient."

A reconstruction of the German "secret weapon" ("such a weapon is possible and may exist") shows it being launched at an angle from a concrete tube embedded in a hillside. "Swiss reports say that it is 45 feet long and weighs 12 tons," (the V2 is 46 feet long and weighs 14 tons). The article concludes with a large illustration of a "future passenger transport rocket."

JULY 5

The MX-324, a rocket-powered flying wing built by Northrop, makes its first powered flight [see September 1942]. It is propelled by a 200-pound thrust engine built by Aerojet, the XCAL-200, weighing 427 pounds. It is fueled by monoethylaniline and red fuming nitric acid. Ground tests of the installed motor begin on June 20 and finish in a series of taxi tests on Harper Dry Lake.

The rocket is not powerful enough to enable the little rocket plane to take off under its own power and it must be towed aloft to an altitude of 8,000 feet by a P-38 fighter. After the rocket plane is released, test pilot Harry Crosby flies for approximately 4 minutes. It is the first liquid-fuel rocket-powered aircraft to fly in the United States.

Since it proves difficult to produce the rocket motor that the proposed operational fighter version requires, the XP-79 is equipped with jet engines and redesignated the XP-79B.

DECEMBER

Bell Aircraft is commissioned by the U.S. Army Air Force to design a nuclear-powered rocket plane for postwar research use.

An editorial published in the *New York Herald-Tribune* scoffs at the aims of R. L. Farnsworth's American Rocket Society, referring to "spatial covered wagons from Illinois." The same downbeat editorial predicts that "the Nazis will get there first anyhow."

DECEMBER 12

Prof. A. M. Low, president of the British Interplanetary Society, announces that the BIS is planning to fly a passenger rocket to the moon. Whether the multimillion-dollar cigar-shaped rocket will land or not has not been decided, nor has the starting date been selected. "People laughed at us before the war," explained Low, "but look what the Germans are doing with rockets now."

1945

Ettore Ricci writes *Il Segreto della Propulsione a Reazione* (The Secret of Reaction Propulsion). The book contains an illustration of a stratospheric rocket designed by Ricci in 1942.

In *Rocket Research History and Handbook* Constantin Paul van Lent publishes a drawing of a single-stage manned rocket. It is a blunt-nosed torpedo-shape with slim fins.

BIS moon lander, ca. 1945.

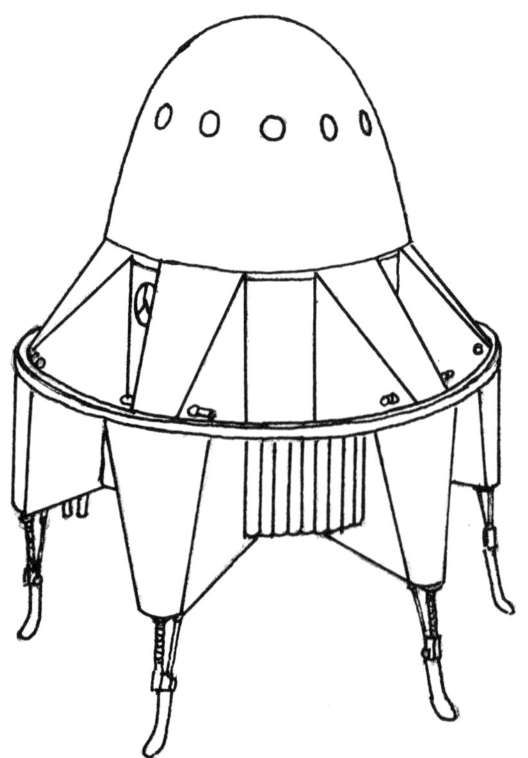

FUTURE PASSENGER TRANSPORT ROCKET

Life January 1945

The egg-shaped cabin is detachable from the main rocket (by means of manually operated clamps). It will remain attached to the booster by a long cable unwound from a wheel. By spinning the two units around their common axis, artificial gravity will be created within the passenger cabin.

G. Edward Pendray publishes *The Coming Age of Rocket Power*, a postwar survey of the current state of rocketry. Pendray is one of the founders of the American Interplanetary (later Rocket) Society and an early rocket experimenter.

Herbert S. Zim publishes *Rockets and Jets*, one of the few serious American studies of the potential for rockets in interplanetary flight published up to this time. "That rockets will reach their crowning glory in interplanetary travel is understandable," Zim

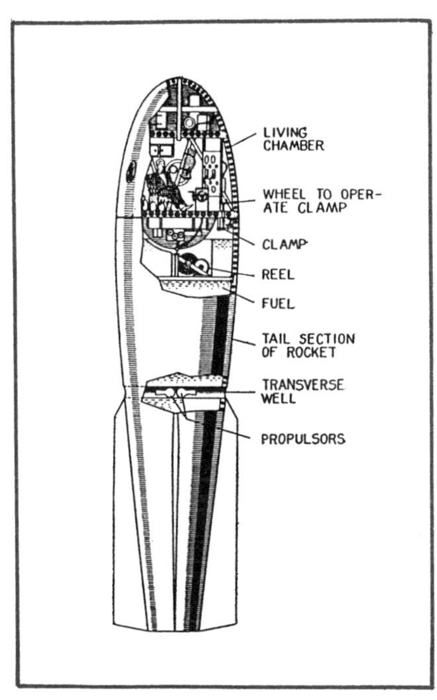

Constantin van Lent

writes, "but as yet we have no means of telling how much our dreams of this kind of travel merely reflect our desires to escape the humdrum of everyday existence."

He describes a three-stage manned rocket using gasoline and liquid oxygen as fuel. The overall weight of the rocket will be 8,051 tons, of which 6,199 tons are fuel and 100 tons is payload (50 tons of passengers and equipment and 50 tons of fuselage, etc.). Each stage will have a mass ratio of 2.72:1. The first stage will weigh 6,571 tons with 5,091 tons of fuel; the second stage will weigh 1,208 tons with 936 tons of fuel; and the third stage will weigh 272 tons with 50 tons of fuel. Zim reminds us, however, that "Much remains to be done. We have barely made a liquid-fuel rocket that works."

Jacques Martial and Robert C. Scull design rocket-powered aircraft. One is a flying wing with extremely swept-back wings and vertical tail surfaces only. It has a combination of turbojet and rocket engines and is expected to fly at 1,500 mph in the stratosphere. Another design is similar, though with a longer fuselage and stabilizers at the wingtips. It is a true rocket and is intended for long distance trajectory flights.

Burton H. Johnson of the American Rocket Society compiles a booklet of letters from executives of various aircraft companies, scientific foundations and industries, all giving their opinions of the future of rocket propulsion and spaceflight. One "visionary" thinks that a trip to the moon will be made by the year 2050. Others either refuse to respond or respond by saying "we would not be seriously interested." The head of one automobile manufacturing corporation answers, via his secretary, that he believes "this development [spaceflight] to be quite a ways off." One science foundation replies that while rocketry will be useful for weather forecasting, it regrets that it has gotten "off to a false start in crackpot interplanetary schemes."

P. H. Parkhurst proposes a spacecraft fueled by hydrogen and chlorine. The gases will be ignited by focussing the ultraviolet portion of sunlight into the combustion chamber. The rear end of the rocket will be circled by a glassed-in chamber that will admit sunlight to the engine.

A. V. Madge, a member of the Royal Astronomical Society, does not believe that interplanetary travel will ever become possible, primarily due to the vast distances involved. A rocket travelling at 1,000 mph will require more than 4 years to reach Mars. To this objection he adds the difficulties of celestial navigation. "Just how," he asks, "could a rocket be properly aimed when directed toward an object that, together with our sun, is moving at an assumed speed in a general direction rather than a specific one, from an object (Earth) whose motion is also helical?"

Madge also thinks that meteors would pose a substantial problem for erstwhile astronauts.

W. G. A. Perring tells the Royal Aeronautical Society that winged rockets equipped with 100-ton boosters will be able to travel from New York to London in less than an hour.

Referring to the pessimistic Dr. J. W. Campbell, who predicted that a Mars-bound spaceship would have to be 5 miles in diameter and as massive as Mt. Everest [also see May 1939], some experts still believe that even a modest moonrocket would cost at least $100 million and weigh 40,000 tons.

Vannevar Bush, director of the Office of Scientific Research and Development, says that "There has been a great deal said about a 3,000 miles high-angle rocket. In my opinion such a thing is impossible for many years . . . I say technically that I don't think anyone in the world knows how to do such a thing . . . it will not be done for a very long period of time to come."

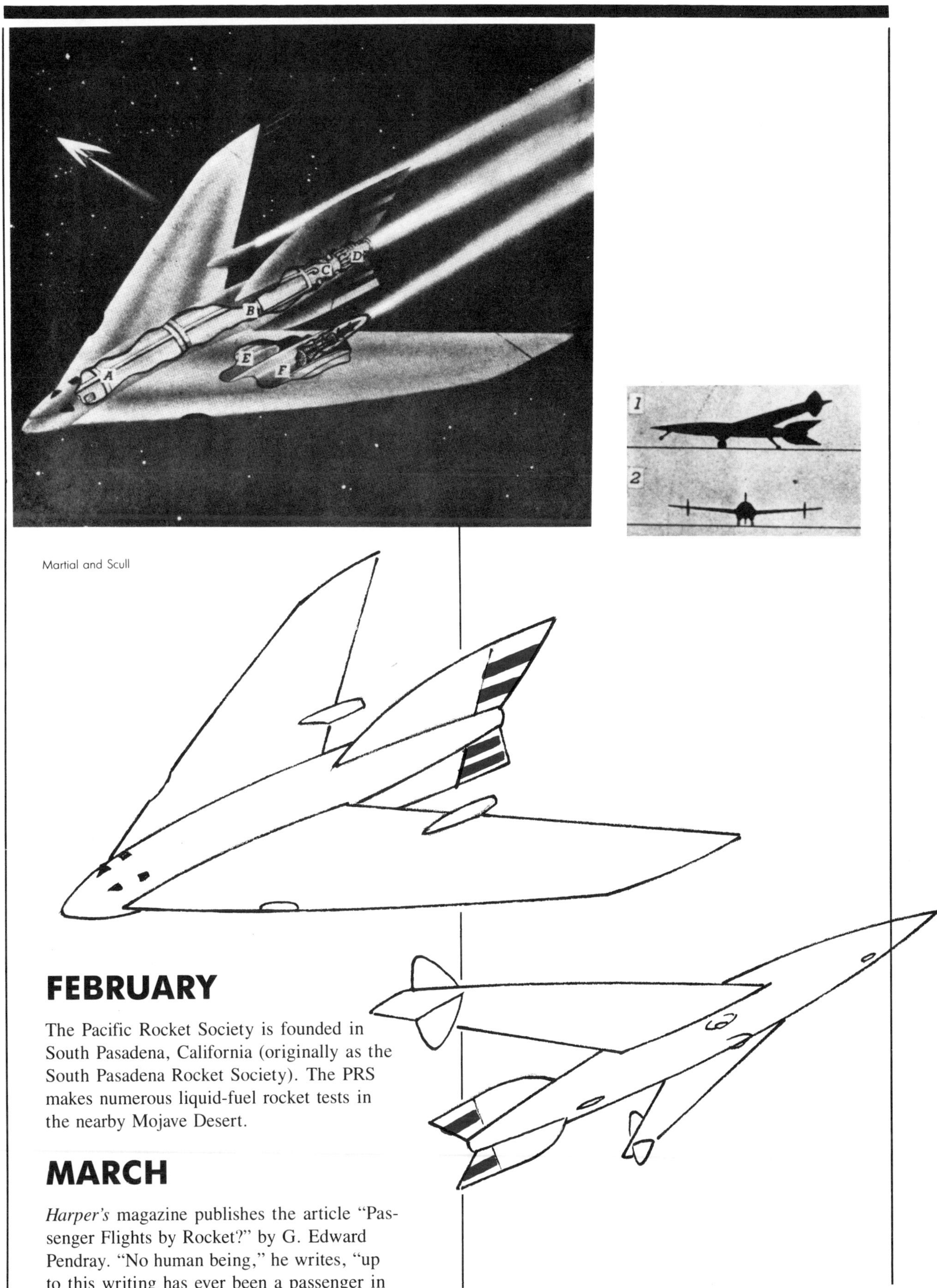

Martial and Scull

FEBRUARY

The Pacific Rocket Society is founded in South Pasadena, California (originally as the South Pasadena Rocket Society). The PRS makes numerous liquid-fuel rocket tests in the nearby Mojave Desert.

MARCH

Harper's magazine publishes the article "Passenger Flights by Rocket?" by G. Edward Pendray. "No human being," he writes, "up to this writing has ever been a passenger in a true rocket . . . The probabilities are that passengers will not travel in rockets until after these projectiles have been fully developed for carrying mail and express. Some daring venturer may then undertake to ride a large mail rocket." The minimum requirements for such an experiment, Pendray suggests, will be an enclosed cell supplied with continuously purified air at sea level pressure, a spring-mounted hammock or cot, and special shock-absorbers for the landing. There will be no windows, to save weight. There will be no manual control over the rocket's flight path since the pilot's "reflexes will be slow and erratic." The first rocket passengers "will spend a cramped and terrifying few minutes far above the earth. They will have nothing whatever to say about the course of their flight or the ending of it. And very likely they will be glad enough when it is over."

However, if rocket flight is ever to be commercialized, a few more concessions will have to be made to human comfort and safety. Pendray discusses a hypothetical rocket passenger liner of modest proportions: 9 tons overall weight, of which 6 tons are fuel and a little less than 1 ton of passengers . . . perhaps, once provisions have been made to creature comforts, baggage, safety devices and a few little luxuries (such as "padding [and] drinking water"), there will be room for only four passengers and a pilot.

A New York to Pittsburgh rocket-plane flight will cost—one way—about $75 in fuel per passenger. Adding to this items such as insurance, administrative overhead and maintenance, the actual cost of a ticket will likely be in the range of $300 to $400. At the time Pendray is writing railroad fare for the same journey is $20.77 and air fare is $25.01.

To make his passenger-rocket company more financially appealing, Pendray replaces the "minimum" size 9-ton rocket with an 18-ton vehicle carrying nine passengers, or a 36-ton rocket with twenty. Pendray foresees rivalry in the form of other reaction-propelled aircraft: "duct-engine gliders carrying passengers at 400 to 500 miles per hour,, powered perhaps by athodyds [ram-jets]," or stratosphere planes powered by turbo-jet engines at speeds of 600 to 700 miles per hour, or even "huge turbo-jet stratoplanes flying at altitudes of 10 to 12 miles, boosted . . . by auxiliary rocket motors permitting them to go as fast as 1,500 miles per hour." His rocket planes will have all of these beat in the matter of speed: his long-distance rockets will regularly achieve velocities of 5,000 miles per hour.

Pendray's passengers will be gliding far above the earth's atmosphere, "perhaps even into the aurora zone." They will be weightless, requiring seat straps and liquified foods served in squeeze tubes.

In spite of these and other discomforts, Pendray hopes that the reader's "children or . . . grandchildren may be able to take off from Paris for New York, or from Los Angeles for Honolulu, and fly faster than the sun, watching the day grow younger as they move."

MARCH 1

The Ba 349A (M-23), the *Natter*, a German manned rocket interceptor and the first to employ a vertical launch, crashes on its first and last test flight, killing the pilot. The engines used are the HWK 109-509 (in the Ba-349 A) and the HWK 109-509D (in the Ba-349 B), the same engines used in the Me-163. The plane is designed to be literally disposable. After its launch (boosted by two or more powder rockets) from the vertical track that supports it, the plane is designed to climb rapidly and attack the enemy with the battery of twenty-four high-velocity rockets it contains in its nose. The pilot then dives away. Releasing a drag parachute from the tail pulls the plane into two sections, separating the nose from the rear. The pilot then slides from his seat and bails out. A timer ejects the rocket motor unit from the tail section, and it too parachutes safely to the ground, for eventual reuse. The *Natter* simply comes too late in the war to do any good, and its development is too hasty and pressured to succeed.

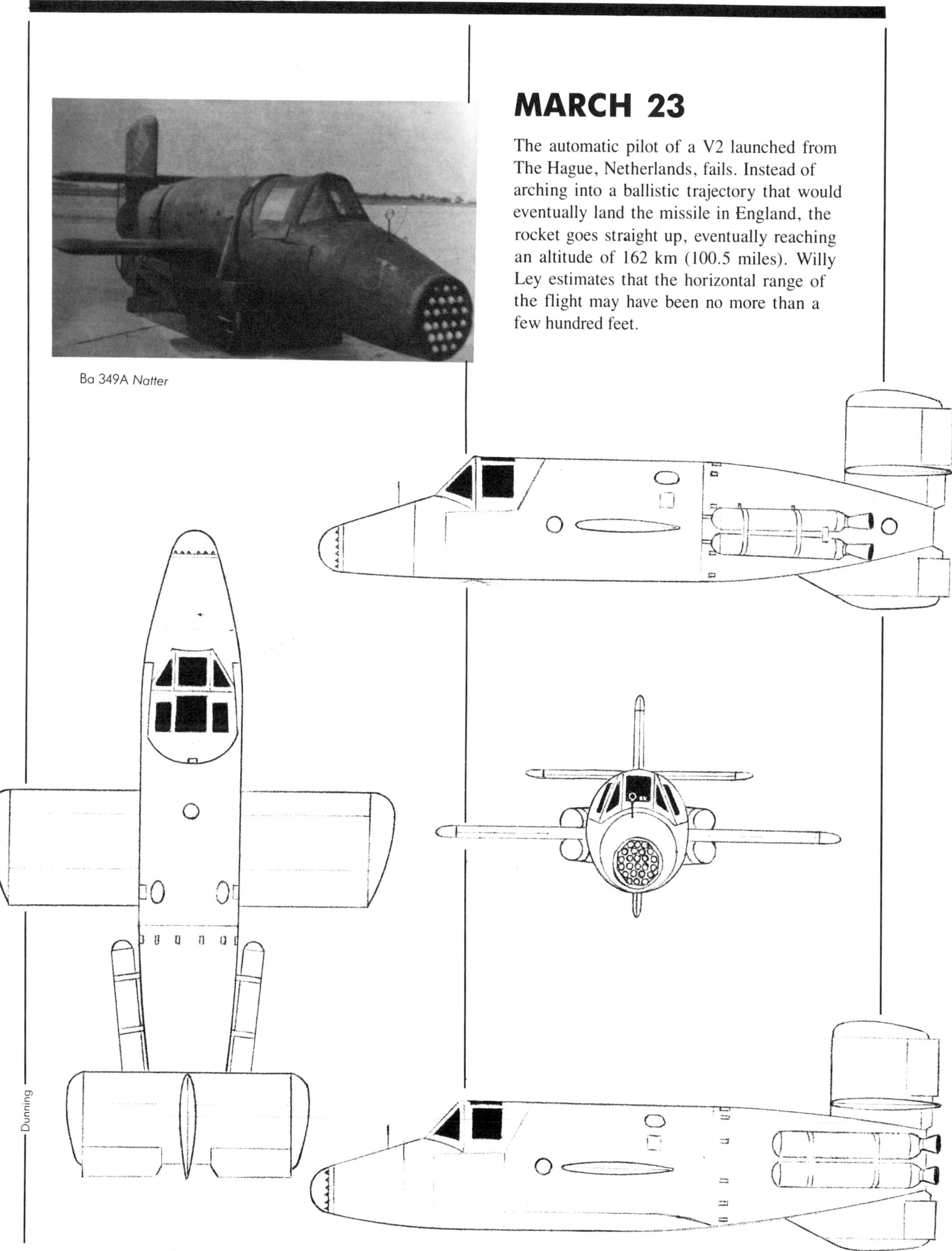

Ba 349A *Natter*

MARCH 23

The automatic pilot of a V2 launched from The Hague, Netherlands, fails. Instead of arching into a ballistic trajectory that would eventually land the missile in England, the rocket goes straight up, eventually reaching an altitude of 162 km (100.5 miles). Willy Ley estimates that the horizontal range of the flight may have been no more than a few hundred feet.

JULY 23

Life magazine reports on the "German space mirror." The Nazis, "the Army reported, hoped to use such a mirror to burn an enemy city to ashes or to boil part of an ocean." The mirror is based on the original idea Hermann Oberth published nearly 20 years earlier [see 1929], although the *Life* reporter is generous enough to point out that Oberth "had planned to use the space station not as a weapon but as a refueling point for rockets starting off on journeys into space." The finished mirror is shown in orbit in a painting by James Lewicki and its construction is illustrated by B. G. Seielstad. The mirror would be built by "ferrying prefabricated sections into space one by one." Thirty-foot holes in the circular mirror, which is to be made of sodium, accept incoming rockets. Inside the docking hole are airlocks for the passage of crewmembers and supplies. Pumpkin plants, grown under fluorescent lighting, generate oxygen, while electricity is provided by dynamos powered by solar-generated steam. The attitude of the mirror is controlled by small rockets. The whole assembly orbits at 5,100 miles.

When U.S. scientists first learn of the proposed German weapon, it is generally ridiculed. B. J. Spence, a professor of physics at Northwestern University calls it a "Jules Verne dream."

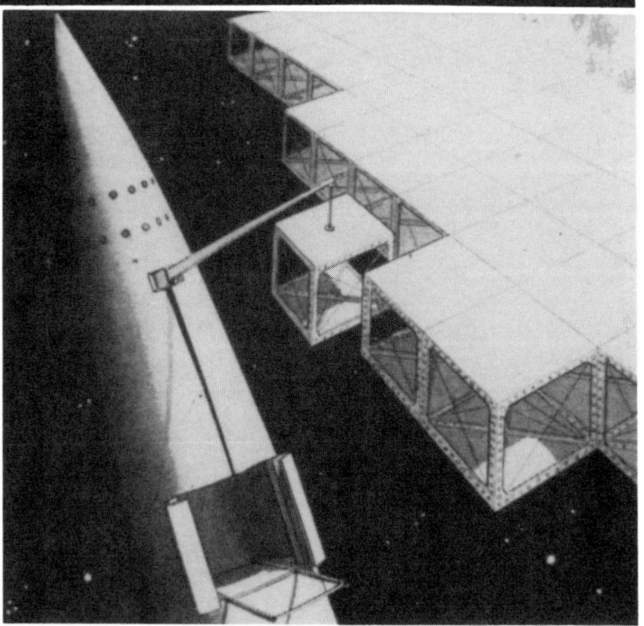

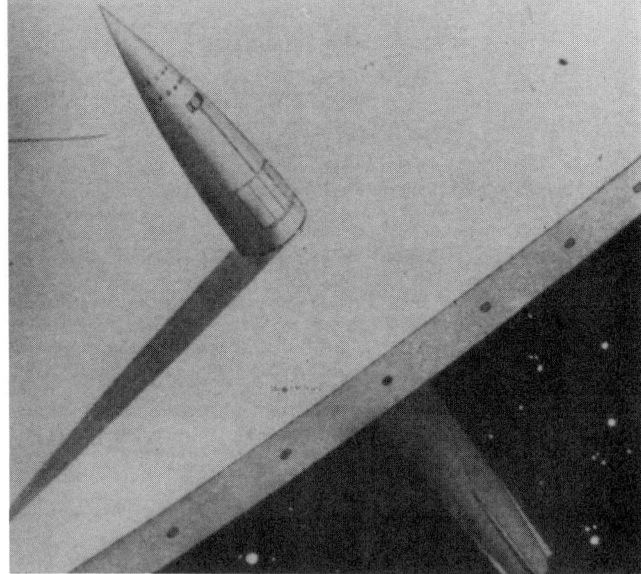

OCTOBER

Science fiction author Jerry Shelton, in "Eyewitness Report" (*Astounding*), describes a V1 bombardment and what he thinks might be a V2 in flight: "It is nerve-wracking to be constantly saying to yourself, 'How about this? Here I am running around in a sweat trying to dodge something all us science-fiction guys have been writing and dreaming of—it's actually a reality—and I'm on the wrong end, and I can't write and tell Campbell [editor of *Astounding*] that somebody is really on the way to space!' "

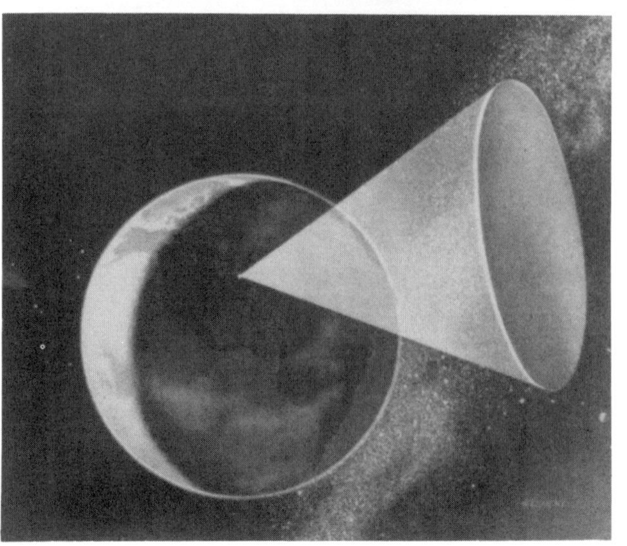

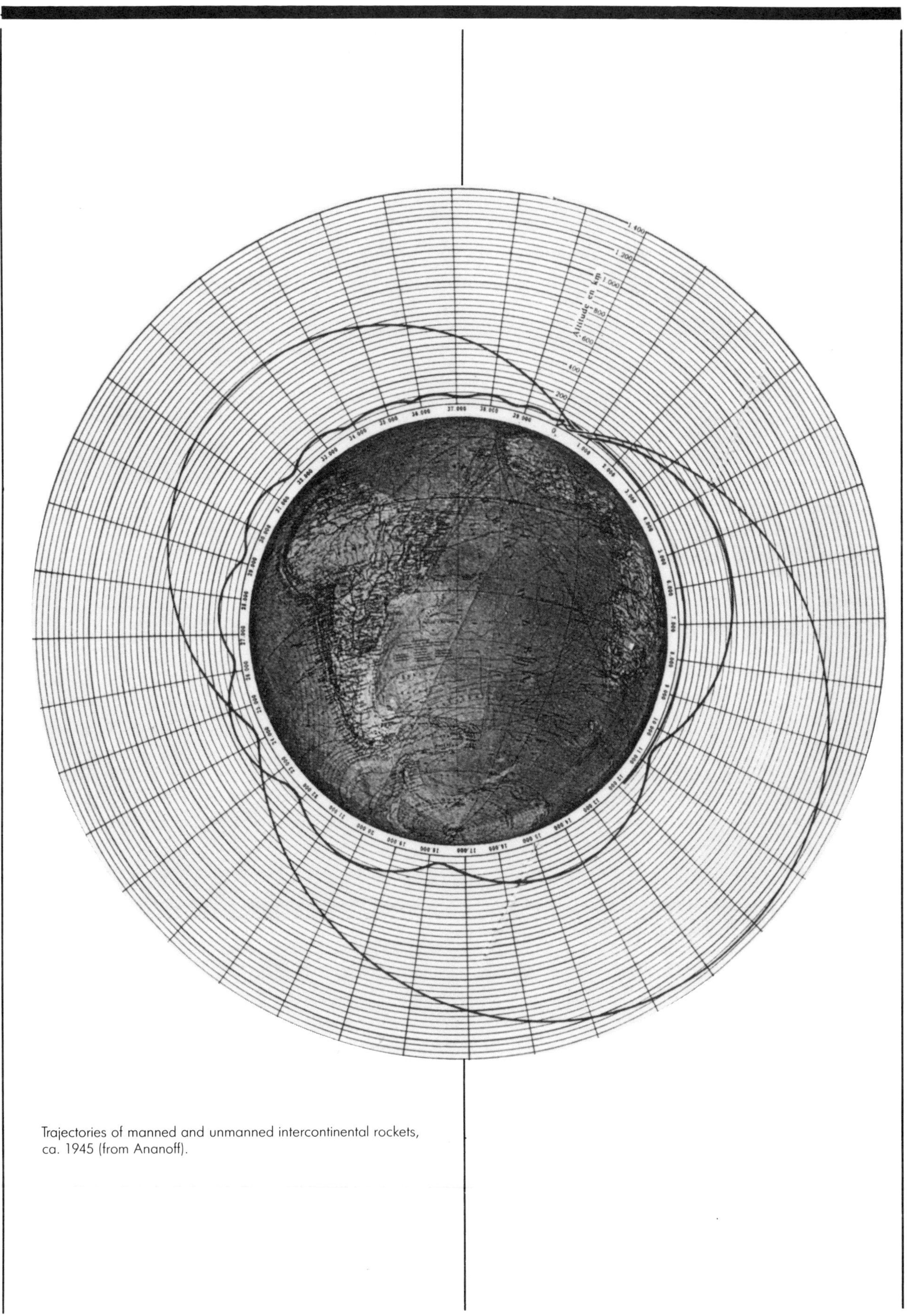

Trajectories of manned and unmanned intercontinental rockets, ca. 1945 (from Ananoff).

PART V

The Golden Age of the Spaceship

The golden age of the spaceship occurs with seeming paradoxicality during a period when there were no manned spaceflights. It would, on the face of it, seem more logical to call the present period the golden age, when scores of men and women have accumulated thousands of hours in space, when nearly two hundred spaceflights have been made, during a time when men have walked on the moon and occupied space stations and space shuttles are making regular trips into orbit. Yet it is just for these reasons that the decades following 1961 and the first manned spaceflight are not rightly the golden age of spaceflight. An appropriate analogy is the choice of period for the golden age of aviation: the late 1920s and 1930s, rather than the present. Today, millions of people have flown in aircraft, regularly cross the Atlantic and Pacific oceans, travel faster than sound and own private aircraft. Anyone with twenty-five or thirty dollars can go to a local airfield and fly in an airplane for an hour. The cumulative result is that mechanical flight is now something taken entirely for granted. Most people think no more of taking an airplane from one place to another than they once thought of taking a train or an automobile. It is this very commonplaceness that prevents the present era from being a golden age of aviation. The name belongs more rightly to a period when aviation was something unusual, unique and exciting; when a transatlantic or transcontinental flight was headline news, when families would plan to spend a Sunday afternoon at a local airport to watch the planes land, when pilots were as idolized as film and rock stars are today. Thirty years ago anyone would have immediately recognized the names of Yuri Gagarin, Wernher von Braun or Alan Shepard and could have told you the reasons for their fame. Who today can even recall the names of the original seven Mercury astronauts? or name the crewmembers of the latest space shuttle flight?

For post-World War II society, the rocket was the symbol of the fabulous world of the future that had been promised during five long, bleak years of war. It appeared everywhere and in every possible—even if unlikely—context. And where a recognizable rocket did not appear, its stylized form was an ubiquitous decorative motif.

It is both difficult to imagine and to underestimate the influence of the V2. It was larger by far than any other rocket ever built and it flew faster, higher and farther. And, not the least of its important attributes, it looked great. It took hold of the public imagination, becoming synonymous with *spaceship*, because it *was* a spaceship.

The popular science writer and expatriate German rocket experimenter Willy Ley said so himself. He pointed out that the ton of explosives the V2 carried could easily be replaced by a pilot and his life-support equipment. This was an idea that had also occured to the Nazis, who had drawn up tentative plans to boost a manned version of the V2 to New York via a super-rocket called the A-10. After the war the British Interplanetary Society took a scheme similar to Ley's to the British Ministry of Supply. They proposed using a modified V2 to launch a manned capsule into a ballistic, suborbital flight. After reaching a maximum altitude of about 190 miles, the capsule would separate from the rocket and return to the Earth by parachute. Like the lunar lander designed by the BIS in 1939, the "Megaroc" bore a remarkable resemblance to things to come, notably Alan Shepherd's Mercury capsule which inaugurated America's manned entry into space.

The V2 played roles in the first science fiction movies to appear after the war. George Pal's *Destination Moon* (1950) begins with the launch of one of the missiles, and the futuristic spaceship, the *Luna*, used for the manned trip to the moon is clearly a linear descendant. It was designed by art director Ernst Fegté. *Destination Moon* was beaten to the theaters by the low-budget *Rocketship XM*, made to cash in on the bigger movie's publicity. Its low budget forced its producers to use stock footage of the launch of a V2 for the launch of the movie rocket—even though it bore little resemblance to the model (itself lifted from a *Life* magazine article) used in the remainder of the film!

The 1951 space opera *Flight to Mars* also owed its spaceship to *Destination Moon*, but in a roundabout way. The *Mars 1* was originally designed by Chesley Bonestell as the *Luna*, but the design was rejected. The producers of *Flight to Mars*, one of whom had worked on *Destination Moon*, frugally adopted the unused spaceship for their own movie.

The *Mars 1* might be the quintessential 1950s spaceship, but the producer of *Destination Moon* had other, more realistic designs yet to come. There was the Bonestell-designed Space Ark from George Pal's *When Worlds Collide* with its spectacular if impractical launch ramp—easily the most impressive rocket launch in film history! But the peak was reached with the *Collier's* magazine-inspired spacecraft in *The Conquest of Space*. Taking the flying-wing Mars glider from the book version of the 1954 Wernher von Braun article in *Collier's* "Can We Get to Mars?" Bonestell and art director J. MacMillan Johnson adapted and even managed to improve the original design. In spite of any silliness of plotting—of which there is a great deal—the film is like watching the *Collier's* magazines and their illustrations by Bonestell, Fred Freeman and Rolf Klep brought to life. The graceful toroidal space station, a delta-winged ferry rocket, space taxis, and the elegant Mars Glider are all brought to the screen in color and extraordinary realism.

By the middle of the decade the Golden Age of space travel was in full swing, spurred primarily by the appearance of the *Collier's* magazine serial, a series of programs on space travel produced by Walt Disney and aired on his popular "Disneyland" television series, and the general postwar euphoria about the future that space travel represented. However, this was in a way self-defeating as well as self-fulfilling. The imaginative spaceships gradually gave way to the real-life ones that they helped to inspire.

1946

Harry Harper, air correspondent for the *London Daily Mail*, publishes *Dawn of the Space Age*, which is primarily a popularization of the work of the British Interplanetary Society (BIS) . . . in particular the 1939 moon rocket [see above]. The book is approved by the Technical Advisory Committee of the BIS.

The U.S. Navy designs a High Altitude Test Vehicle (HATV), an unmanned single-stage research rocket. It would weigh 101,400 pounds, loaded, with a payload of 1,000 pounds. Liquid oxygen and liquid hydrogen would power 8 28,400-pound thrust engines that surround a single 73,000-pound thrust engine. The 86-foot-long, 16-foot-diameter rocket would reach orbit at an altitude of 150 miles.

Douglas Aircraft Co. designs (with Project RAND, Northrop and North American Aviation), under a U.S. Army contract, an "experimental world-circling spaceship. "Its chief designer is F. H. Clauser. The rocket is in four stages, named (from first to fourth) "grandma" (comprising half the total length of the rocket), "mother," "daughter" and "baby." "Baby" is a winged rocket which will orbit the earth in 1 1/2 hours. Its total length is from 60 to 70 feet, with a diameter of 12 to 14 feet. It is fueled by liquid oxygen and hydrogen or liquid oxygen and alcohol.

The manned spaceship is considered to be similar to contemporary large aircraft of 50,000 pounds gross and over, so it is thought feasible to use airplane-type construction—primarily reinforced sheet metal. Due to the expected high temperatures, high strength stainless steel is to be used. Motor loads for each stage are carried to the outer skin via truncated cones extending from the base of the motor to the skin.

Although the rocket is initially designed as an unmanned satellite carrier, a manned version is also developed. Acceleration is not expected to exceed 6.5 g, most likely no more

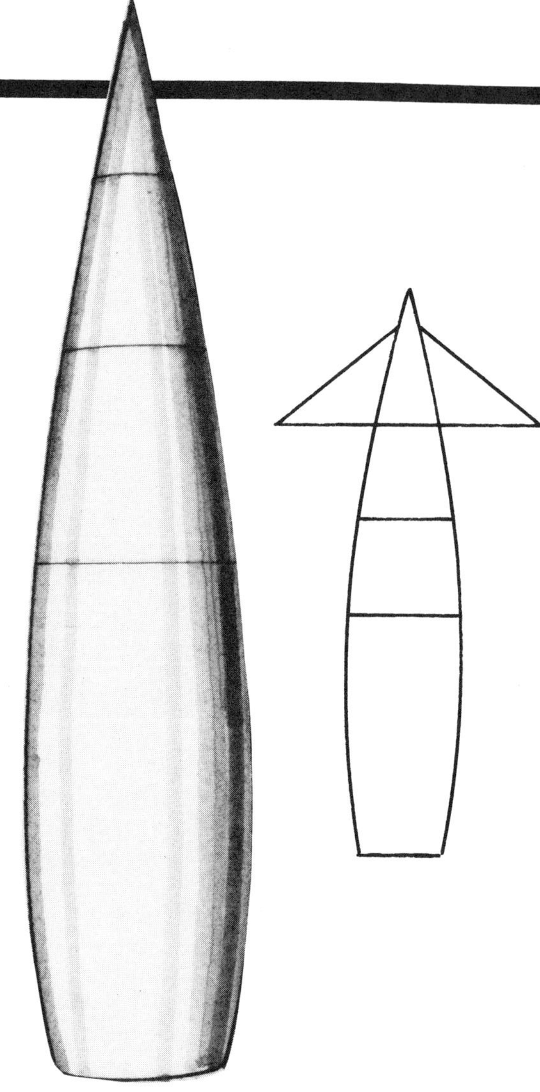

that 4 g. After reentry, the manned stage glides back to earth, landing like a normal aircraft.

The 120 R, an aircraft in which the Russians have installed RD-IX3 engines, is constructed under the direction of S. A. Lavochkina.

Luigi Gussalli writes *I Viaggi Interplanetari per Mezzo delle Radiazioni Solari* in which he proposes the use of solar sails as a method of interplanetary travel.

Megaroc is a British Interplanetary Society study in which the V2 is redesigned as a man-carrying rocket. The study group consists of R. A. Smith and H. E. Ross. The idea is to produce the minimum hardware necessary to put a man into space. It is basically a finless V2, 57 feet 6 inches tall and 7 feet 2 inches at its widest point, with a detachable capsule that will be returned to earth via parachute after a ballistic, suborbital flight. (A year earlier Willy Ley also suggested that the V2 could carry a man into space. "Take off the bomb," he wrote, "and substitute an observer, wearing a light diving suit and having a nice set of instruments around him, making a total of, say, 300 pounds. This gives you another nineteen hundred pounds of fuel. Do that and fire the V2 vertically, it is not apt to have a maximum acceleration surpassing 3 or 4 g's. It will ascend beyond two hundred miles—it will just touch empty space!")

Changes to the basic V2 include slightly increased fuel tanks and elimination of the exterior fins (in favor of graphite vanes in the

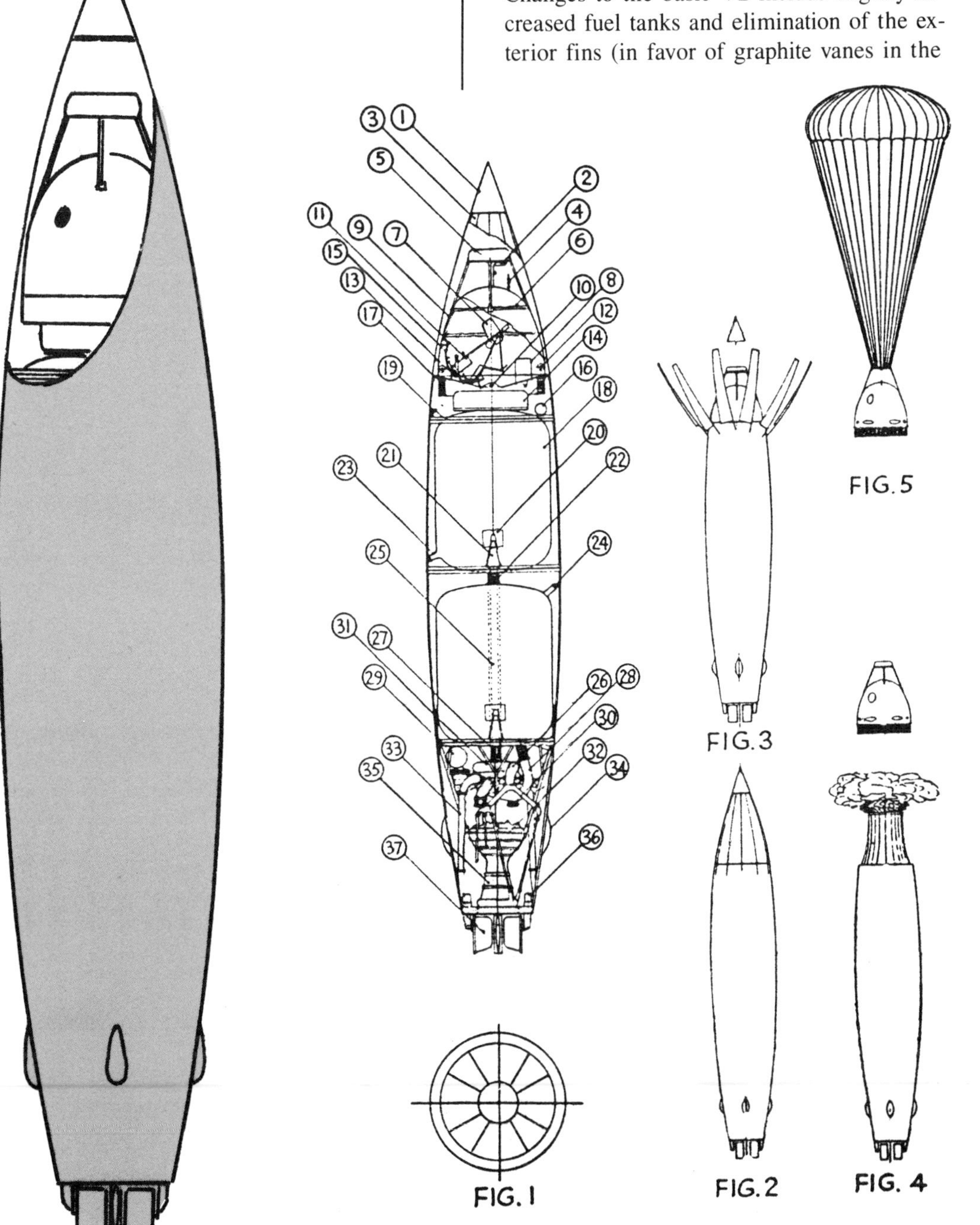

exhaust for steering). To keep the rocket upright at takeoff, it has to be launched from a scaffolding tower. The modified engine provides 60,000 pounds of thrust. After 110 seconds of powered flight, at an acceleration of 2 g, the rocket reaches an altitude of 150,000 feet. At the end of the powered flight, the streamlined fairing covering the nose is jettisoned, exposing the gumdrop-shaped passenger capsule. Its light metal-alloy shell is fitted with shutter-equipped ports and a special cradle for the pilot. This capsule is equipped with hydrogen peroxide-fuelled attitude control jets; it is also fitted with small "cold" rocket motors for orientation and for creating a spin (it is thought that it might be necessary to create an artificial gravity for the pilot). After separation from the main rocket, it continues to coast on upwards, finally reaching a maximum altitude of approximately 190–225 miles. The pilot will spend about 228 seconds in freefall. His heartbeat and other data will be telemetered back to a ground station.

As the capsule begins its descent, the parachute carried atop will be deployed, eventually landing the capsule and its passenger safely. Like the later Mercury program [see 1961], the Megaroc capsule makes a water landing and is even provided with an impact-absorbing "crumple skirt," as the Mercury capsules eventually will be.

The scheme is presented to Britain's Ministry of Supply on December 23, where it is rejected.

J. Ackeret demonstrates mathematically that for a rocket to accelerate to a speed approaching that of light, its exhaust must also be close to light speed.

Dr. Samuel Herrick of the Department of Astronomy, University of California, announces the commencement of a course in celestial navigation. "The prerequisites for my course," he writes, "on Rocket navigation are trigonometry, plane analytic geometry, differential and integral calculus. Allied courses include the determination of orbits, and numerical analysis . . . "

Twenty-four-year-old Albert Ducrocq, a student of the Ecole Polytechnique in Paris, devises an atomic-powered "astronef" weighing 3,000 tons that is intended for a trip to Mars. He hopes "that the tricolor may one day be planted on the soil of Mars."

Jerome S. Meyer, in article published in *Readers Scope*, proves that interplanetary travel is impossible, mostly because of the deleterious physiological effects of weightlessness.

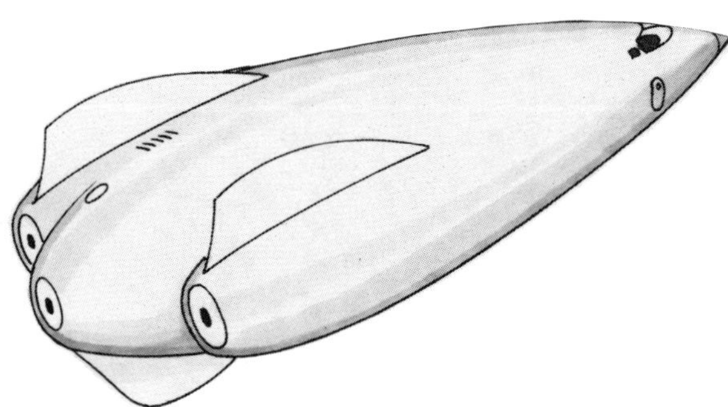

Donald S. Ritchie proposes a manned spacecraft whose three engines are fueled by red fuming nitric acid and commercial aniline. He also proposes a large spacecraft destined for a trip to Mars. It will be powered by five V2-type engines producing a combined thrust of 2 million pounds. The spaceship will weigh 1,875,000 pounds, of which 1.5 million pounds will be fuel (oxygen and hydrogen). The ship will require 287 seconds to accelerate to its final velocity of 10,228 mph. The Mars flight will take 200 days. After using the Martian atmosphere to decelerate, the rocket will settle to the ground, supported by a "tripod" of thrusters.

Robert Lee Moore (who also suggests that the United States import captured V2 rockets, replace the warheads with instruments, and fire them "straight up" to an altitude of "some 200 miles"—but fears that "army inertia and governmental red tape will prevent such an experiment") proposes two different atomic-powered manned spacecraft. A reactor weighing only a few hundred pounds would be required. Left to itself, it could produce a "jet" of high-energy neutrons. Although the mass ejected every second would be small, he thinks the jet's velocity would more than make up for this. The major difficulty with this scheme is that the reactor would quickly burn up. Water could be used to cool it, and in turn the vaporized water would be ejected as additional reaction mass. The second type of ship Moore proposes uses liquid hydrogen and liquid oxygen as coolant-reaction mass, and adds a combustion chamber to the reactor.

H. Emerson Canney, Jr., of the U.S. Army Air Corps, discusses the requirements for a manned moon rocket. He proposes a propulsion system in which two engines alternatively fire in a rapid series of discrete explosions ("thousands . . . per second"). His spaceship is a beryllium-magnesium torpedo equipped with smaller rockets for hovering and landing and short airfoils running much of the length of the fuselage (three vanes 120 degrees apart). It is launched vertically (unless vertical takeoffs are too much of a strain on the passengers in which case it may take off from a ramp inclined at 45 degrees) and lands horizontally on retractable landing gear.

R. L. Farnsworth suggests that most of the cost of developing a moon rocket could be absorbed by commercial sponsorship (he estimates the development cost at $350,000). A year earlier F. Claussen suggested that the sale of photographs of the moon's "dark side," taken by an unmanned rocket, would produce enough revenue to finance a later manned rocket to the moon. He proposed that a stock company be formed to undertake just such a project. [Also see Heinlein, 1950.]

The United States's first instructor in rocket navigation, Dr. Samuel Herrick, takes his desk at the Los Angeles campus of the University of California.

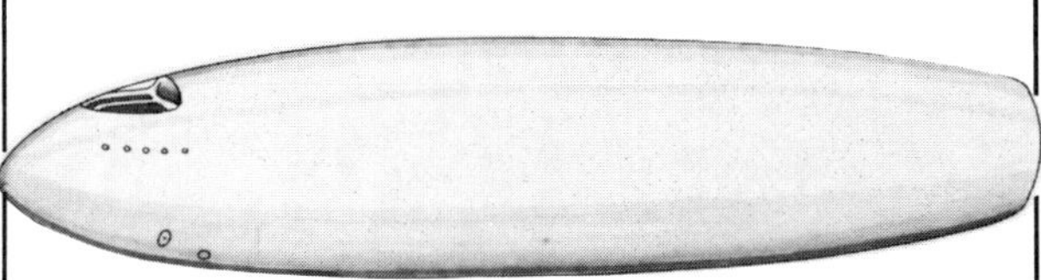

R. L. Farnsworth describes a "liquid fuel monster" for a flight from the moon to Mars. The giant rocket, designed primarily by Donald S. Ritchie, would weigh 1.875 million pounds at takeoff. The flight to Mars would take 200 days.

Colonel James G. Bain, chief of the guided missiles branch of the Army Ordnance Department of Research and Development, believes that a rocket from earth could be sent from the earth in about ten years. "If you want to put a chunk of iron about the size of your fist on the moon," said Bain, "that can be done in a relatively short time. Maybe in about ten years. If you want to land something bigger, it will take much longer."

MARCH 4

Artist Chesley Bonestell's picture journey to the moon is published in *Life* magazine. Eleven meticulous color paintings by the man who is to eventually create some of the most persuasive and well-known images of spaceflight and the universe [also see 1949, 1952–1954] take the reader on an imaginary flight to the moon. These are the most accurate and realistic portrayals of spaceflight created up to this time. Bonestell imagines his flight taking place by means of a relatively small rocket plane: a winged spaceship with three rocket engines. Oddly and picturesquely enough, in a painting designed to illustrate weightlessness Bonestell shows a pocketwatch, pitcher and wine glass floating freely, the contents of the latter drifting in spheres around the cabin!

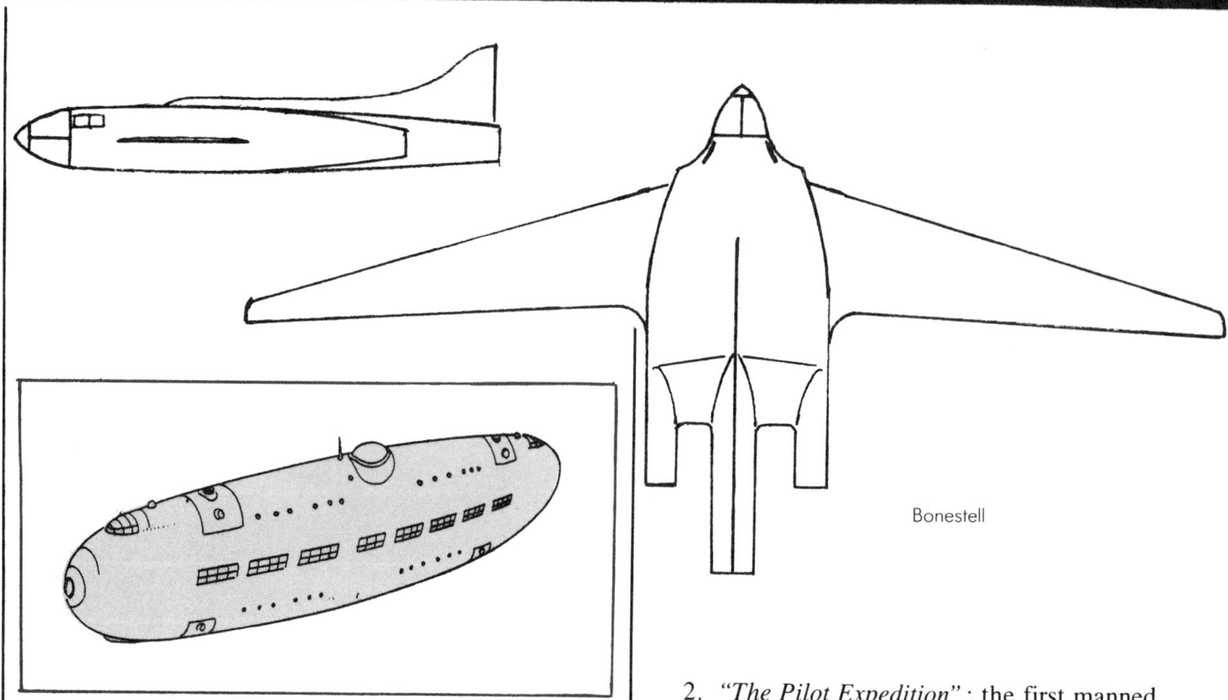
Bonestell

JUNE 23

In an article for *This Week* magazine, "139,000 Miles An Hour!", Major Alexander de Seversky discusses the possibility of an atomic-powered spaceship for a flight to the moon.

JULY 30

Major P. C. Calhoun, head of the Army Air Force guided missile branch, states that he expects to travel to the moon and back within his lifetime. Missile experts of the AAF say that they will be able to send a rocket to the moon within 18 months.

SEPTEMBER 7

In an article published in *Collier's* ("Next Stop the Moon"), G. Edward Pendray discusses the prospects of a moon landing in the near future.

Pendray proposes that the first landing and eventually colonization of the moon take place in four major stages:

1. *"The Target Shots"*: unmanned, instrument-carrying rockets that will land on the moon and telemeter back to the earth information about lunar conditions.
2. *"The Pilot Expedition"*: the first manned space mission, perhaps with a crew of five, whose goal will be to spend a full lunar day and night on the moon—28 earth days.
3. *"The Moonhead Expedition"*: this will be the first small group assigned to establish a settlement. It might consist of ten men, supplied by cargo rockets (either automatic or with small crews).
4. *"Full Colonization"*: the final phase. "A few especially courageous women may join their men in this phase . . . "

The author expects this program to take several years, or even several decades, to carry out. The pilot rocket may have to wait for the development of atomic power.

Even if chemical-fueled, the best expectations will still require 500 pounds of fuel for every pound carried to the moon. Pendray makes a conservative estimate that 7,600 pounds of payload will be required (crew of five = 900 pounds, food = 500 pounds, clothing and personal items = 200 pounds, water = 1,500 pounds, equipment and tools = 1,000 pounds, space suits = 3,500 pounds). The rocket needed to carry all of this might "be as tall as the 1,046-foot Chrysler Building in New York." Although the fully fueled moonship might weigh as much as a battleship, the lander will weigh only 40 or 50 tons.

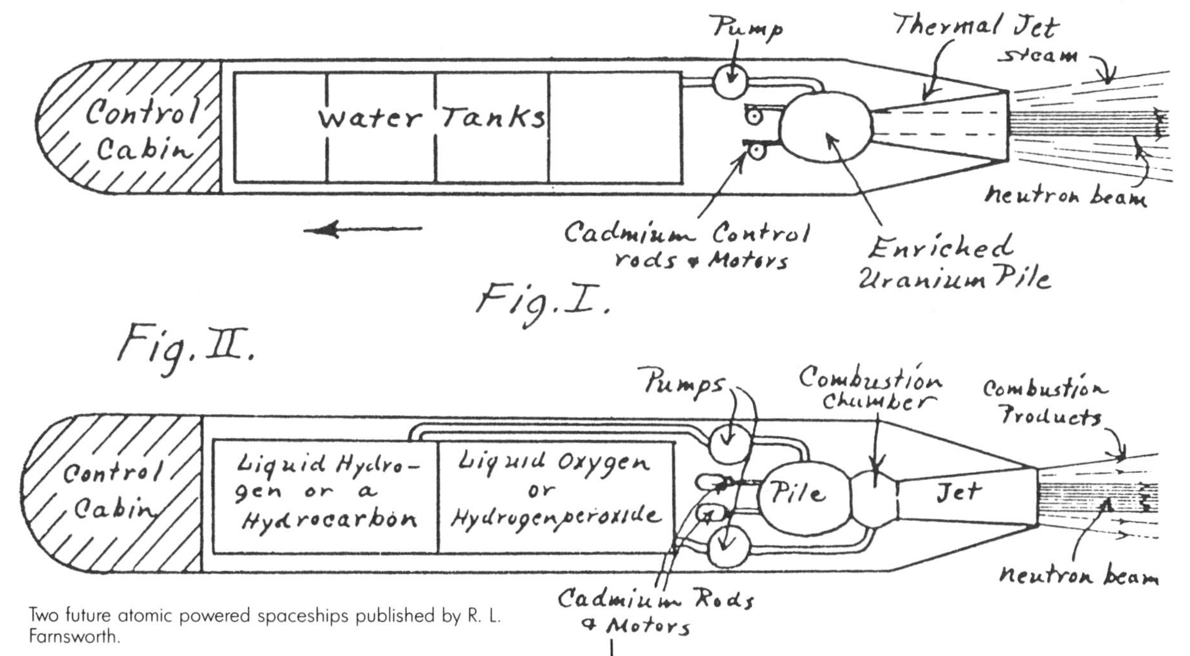

Two future atomic powered spaceships published by R. L. Farnsworth.

DECEMBER

The current issue of the *American Journal of Physics* publishes 77-year-old Henry A. Erikson's article "A Journey to the Moon and Back." Erickson, emeritus professor of physics at the University of Minnesota, states that his proposed rocket would require an exhaust speed of three miles a second, something which he considered "at present extremely improbable." Prof. Erickson's spaceship, a 135-ton aluminum monoplane, might be atomic powered, employing water, mercury or lead as its reaction mass. Approximately 5.6 tons of fuel would be required. The rocket would carry two astronauts in a spacious cockpit with plastic windows.

The rocket would be taken aloft by a carrier airplane, launching the spacecraft at an altitude of 55 miles. The rocket would enter a circular orbit around the earth 100 miles above the equator. The ship would then leave earth orbit and make a circumlunar flight, approaching as close as 10 miles to the lunar surface. The astronauts would release an engraved bronze cube to commemorate their flight.

1947

The British Interplanetary Society (BIS) develops a design for an atomic-powered moonship based on a study by Val Cleaver and Les Shepherd. It is basically the 1939 BIS moonrocket updated [see 1948 and 1949].

The landing craft is virtually unchanged from its 1937 version [see above], with its conversion to liquid fuel being the most obvious. In place of the column of nearly 2,500 solid-fuel rockets is a single atomic-powered stage. It is a squat cylinder roughly 40 feet tall and 25 feet in diameter. The complete rocket weighs some 700 tons at launch. It is entirely unstreamlined, its attitude jets being four ordinary liquid-fueled rockets attached to the sides of the cylinder. The moonship is covered with a streamlined carapace to an altitude of about 560 miles, when the sheathing is jettisoned.

Upon reaching the moon, the landing craft detaches itself from the atomic booster and makes the descent to the lunar surface. When the exploration is completed, the upper section of the lander takes off, using the lower portion as a launch base. When the manned capsule returns to the earth, it is slowed by aerodynamic braking, making the final descent via parachute.

1947

Alfred Africano calculates that a moon rocket carrying a V2-sized payload would have to weigh 5.5 million tons (using V2 technology). With liquid hydrogen and oxygen as propellant, Martin Summerfield of the California Institute of Technology has speculated upon a 72-foot five stage moon rocket weighing only 25 tons. However, this would still be an unmanned trip. A piloted rocket would require the power of an atomic engine. Five pounds of uranium would be sufficient for the round trip.

BIS moon rocket

0
5
10
15
Scale: in feet
Couch position for radial "G"
Motor position for return take-off
Torque jets
Maximum retraction
Line of fairing
Stowed position
Section at base of cabin shell looking aft

James R. Randolph writes in the *Army Ordinance Association Journal* that "In World War III we cannot limit [military] occupation to the earth alone. We must extend it out into space as far as rocket can and to our neighbor worlds in space. "Randolph extends the rather unlikely notion that an attack from Mars would afford "vastly greater opportunities for secrecy and surprise than a sudden assault from earth against an enemy country."

F. H. Heinemann, a chief engineer for Douglas Aircraft, believes that while long-range rocket missiles are possible in the near future, everyday rocket travel is still remote. Admittedly, Heinemann was reserving his remarks for high-speed atmospheric flights. He cites such difficulties as compressibility at near-Mach 1 speeds, lack of suitable power plants, lack of air pressure over 100,000

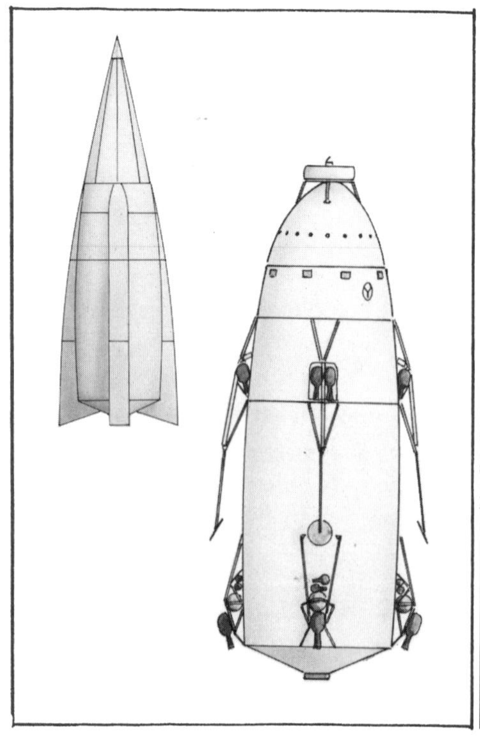

Hardy

feet, high temperatures due to friction, and the lack of strength of known materials.

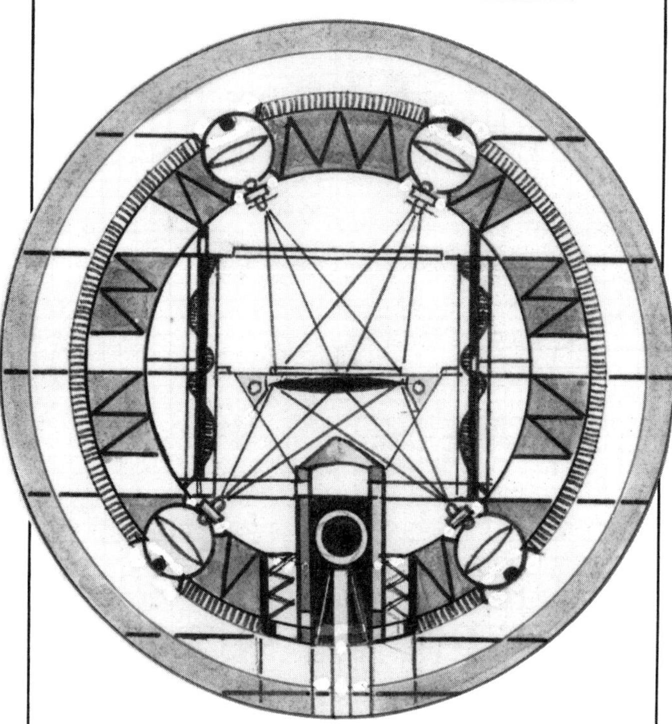

Warnett Kennedy displays a model of an atomic-powered spaceship that he has designed, so large as to be dubbed by the inventor a "man-made planet. "It is one of the features of the Designs of the Future section of the Britain Can Make It exhibition. Kennedy's vast sphere is described as "an intricate construction, with rocket tubes and telescopes, surrounded by an outer transparent sphere. This outer shell, which may represent a surface sensitive to cosmic forces, can be likened to the invisible layers surrounding our own planet . . . " The spaceship has a central control chamber where observers can study views of the earth, moon or planets via a giant camera obscura.

A six-part newspaper series is distributed by the NEA (Newspapers Enterprise Association). It describes "A Trip to the Moon and Back" and is based on an article by Henry A. Erickson, physics professor at the University of Minnesota, in the *American Journal of Physics*. The spaceship resembles a monoplane made of a sheet aluminum alloy. It carries 5.6 tons of fuel and exhaust mass and is

steered by jets in its stern. Its launch takes place from a carrier plane at 55 miles altitude and at a speed of 4.9 miles per second. It will continue on to orbit at 100 miles.

JANUARY 14

On this date the U.S. Navy succeeds in sending a rocket to Venus, according to an "article" in *Astounding Science Fiction* (June 1949) by Philip Latham (astronomer Robert Richardson). "The Aphrodite Project," supposedly an abstract of a 320-page government report, fools not a few people into thinking that the event actually has taken place! Editor John W. Campbell receives so many inquiries he has to have a reply specially printed.

The unmanned Venus rocket is supposedly launched on August 21, 1946 (at 1841 MST) and contacts Venus 146 days later on January 14, 1947.

MARCH 14

A copy of the Sänger-Bredt report [see 1943] has been captured by Soviet forces and taken to the Kremlin. G. M. Malenkov, Stalin's number two man, had said, "This V2 is not what we want . . . We must work on the development of long-range rockets. The importance of Sänger's project must be seen in the fact that it can fly very long distances. And we certainly cannot wait until the American imperialists add Sänger's rocket-plane to their B-29 and Atom Bomb."

V. F. Bolkhovitinov and G. A. Tokady dis-

A Trip To The Moon And Back

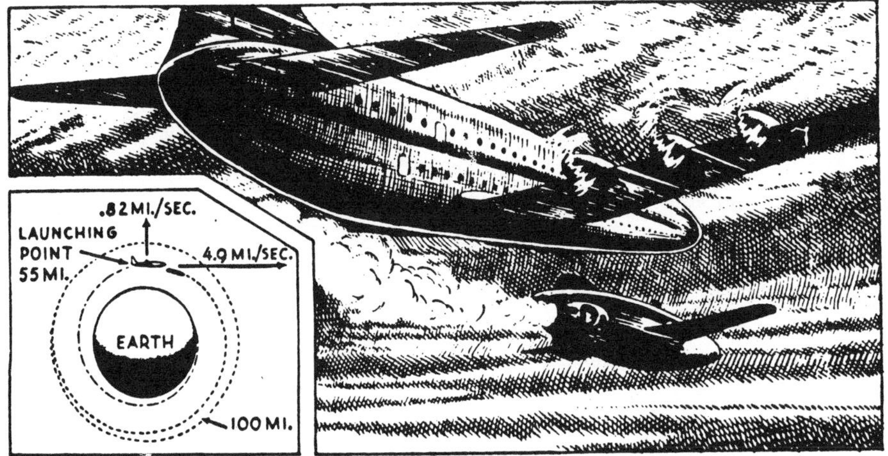

Two panels from NEA newspaper series illustrating Erickson's rocket plane.

No. 1: The Rocket Ship

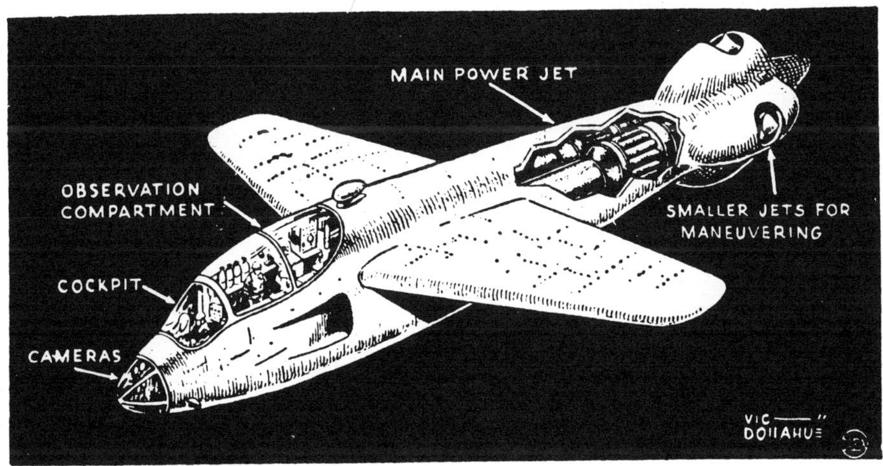

cuss the merits and shortcomings of the Sänger scheme. For the former, it seems to be no more than an enlarged version of his BI-1 [see 1942]. Why not, he asked, put something like the BI-1 on top of a multi-stage rocket? "Besides," he continued, "Should man ever succeed in putting a satellite round the earth, he will have to find a method of bringing it down, and here again wings may prove to be good."

A meeting is held at the Kremlin on this date. Among those in attendance are: G. M. Malenkov; M. A. Voznesensky, head of the State Planning Commission; D. F. Ustinov, second secretary of the CPSU; Air Marshall K. A. Vershinin; Colonel G. A. Tokady; T. F. Kutzevalov; Aleksandr Yakovlev and Arlem Mikoyan, both aircraft designers; and M. V. Khrunichev, minister of aircraft production. Tokady criticizes the scheme, otherwise it is received with enthusiasm and the report is forwarded to Stalin. Special Commission Number 2 is immediately established, which includes the still-doubting Tokady. The commission goes to Berlin to see if further information can be found there, but by August has failed to uncover any new data.

The Antipodal Bomber report is turned over in October to the Ni-88 Scientific Research Institute where German scientists are set to work on the problem. Their conclusions, like Tokady's in the beginning, criticize the bomber's optimistic engine specifications, need for high structural strength, and requirements for heat shielding. They deem the project unfeasible.

MAY

Captain B. A. Northrop writes an article, "Fortress in the Sky," for *Air Trails* magazine.

Like many others of the time, Northrop worries about the potential use of the moon as an enemy military base (or, conversely, waxes enthusiastic about its use as an American military base). Northrop believes that the existing V2, with only a few changes, could easily serve as a moon-to-earth offensive weapon.

Northrop suggests that spaceflight would be essentially naval in character, pointing out that the submarine is, in essence, a spaceship. He makes the suggestion that an existing submarine design could be adapted for space travel. The author feared, however, that an acceleration of 6 mps would prove fatal at worst and debilitating at best. There was also the fear that at extremely high acceleration the atmosphere inside the ship would tend to pool at the lowest level, perhaps even separating into it component gasses. With the use of the almost unlimited power available with atomic motors constant speed of only 500 mph could be maintained. A trip to the moon would only require about 20 days at this speed.

Northrop's spherical spaceship would be insulated with lead and asbestos and would be covered with white fabric. Steering would be accomplished with four small jets placed 90 degrees apart.

JULY

Chesley Bonestell and Willy Ley, in an article for *Mechanix Illustrated*, describe a flight to the moon and Mars by way of an atomic rocket. The rocket uses its reactor to heat hydrogen, expelling it at 65,000 feet per second as its reaction mass (as suggested by Rear Admiral W. S. Parsons). As the exhaust will be dangerously radioactive, the rocket is not launched from the surface of the earth. It is instead carried aloft on the back of a large jet-propelled aircraft. The six-engined jet will take the spaceship to an altitude of 50,000 feet and a speed of from 450 to 500 miles an hour. Here the two craft separate, the jet plane immediately swerving or diving to avoid the blast from the atomic engine. Although nothing is being actually burned in the engine, its exhaust appears as a flame when the hot hydrogen meets the oxygen in the atmosphere.

The rocket will climb almost vertically until it reaches an altitude of about 80 miles. Here it will begin flattening its path until it is almost horizontal. It then makes a half circle around the earth, taking off at a tangent toward the point the moon will be occupying 90 hours later.

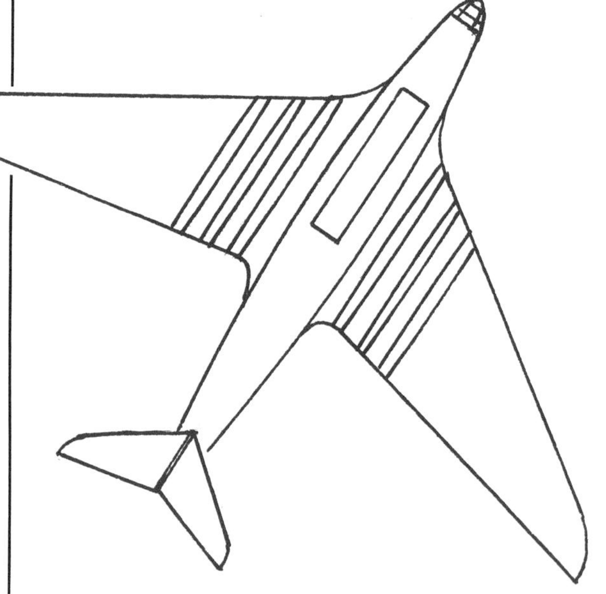

The landing is made by reversing the rocket and landing on its tail fins, using the blast of its engine to brake its fall. If water is found on the moon, the hydrogen tanks can be refilled and the journey can be continued on to Mars. Otherwise, the ship will return to the earth.

Whatever its goal finally is, the spaceship will ultimately come home. It first reduces its speed by skipping off the atmosphere like a stone on a pond, each time losing a little of its velocity. This will require four or five "skips." Ley supposes that by the time the

Mars expedition returns, this maneuver will be well-known and tested by pilots returning from the moon. Once in the atmosphere, a series of parachute-like airbrakes are released. The first are destroyed by the ship's speed, but eventually they slow it down. The final descent is made by firing the rocket engine (this time using a conventional engine, burning liquid oxygen and hydrogen).

Ley predicts that the first unmanned rocket will strike the moon by 1967 and that the first manned circumlunar flight will take place in 1982.

AUGUST 5

Look magazine publishes "Rocket Trip to the Moon," an article based on the recent book by Willy Ley, *Rockets and Space Travel*, and illustrated by Rolf Klep [also see *Collier's*, 1952–1954]. In it he describes an enormous, winged, V2-inspired spaceship one-third the height of the Empire State Building. One half of the ship will be painted white, the other black, for thermal control. The rocket will take off vertically, probably from Mt. Kenya, to take advantage of the altitude and proximity to the equator.

The spaceship will make a refueling stop at a space station (at either one of two: one orbiting at 475 miles and the other at 22,300 miles, in a geosynchronous orbit). The main station will be a nonrotating sphere surrounded by a disk-shaped "working platform." Around the perimeter of the latter will be mooring sockets for incoming spaceships. Since the space station is a weightless environment, its crew is provided a nearby pair of "rest chambers": squat, cone-shaped pods with their apexes attached by a mile-long cable. They spin around their common center, creating a sense of gravity in the pods. The space station crewmembers take periodic rest periods here.

The spaceship continues on to the moon, where it lands tail-first, supported by four extensible jacks built into its tailfins [the moonship and its landing system bear a probably noncoincidental resemblance to the famous painting by Chesley Bonestell that appears a year later—July 1948—on the cover of *Astounding* and later on the dust jacket of *Conquest of Space*]. When the ship returns to the earth it uses its wings to skip-glide back into the atmosphere, eventually landing by means of a large parachute.

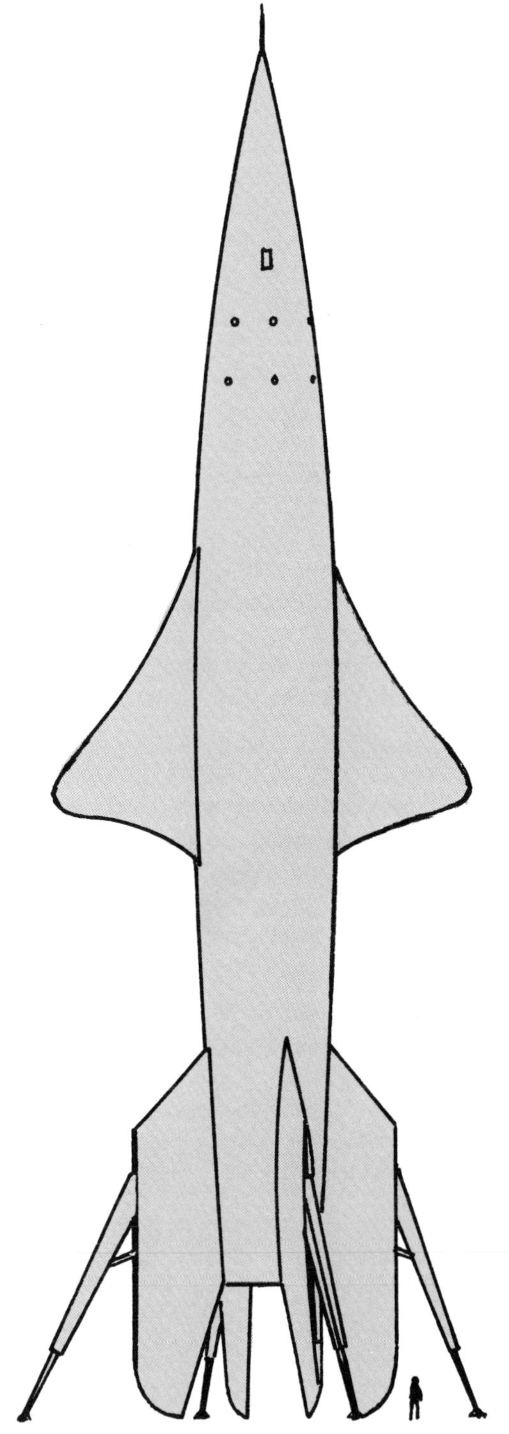

SEPTEMBER

The Russian rocket plane I-270 makes its first flight, using the RD-2M3V engine with a thrust variable from 400 to 1,500 kg.

Astronomer Robert S. Richardson writes an article for *Air Trails* magazine, "New Paths to New Planets," in which he discusses the possibility of manned flight to other worlds and what might be found there.

He suggests that a manned rocket be launched relatively "gently" at a slow speed. At around 200 miles the speed would be increased gradually to at least 6.77 mps in order to reach the moon. For a trip to Mars a speed of 7 mps would get the spacecraft there in 259 days.

OCTOBER

A. V. Cleaver describes his Interplanetary Project for the BIS. He suggests that sometime in the future an international agency for space exploration will be established, which may be the only way in which space travel can be realistically financed, particularly since Cleaver believes that the answer to practical spaceflight lies in the development of nuclear rockets. The "IPA" will probably set up an evolutionary approach such as this:

1. High-altitude research rockets, some of which may carry a crew;
2. manned circumlunar flights;
3. lunar landing; and
4. interplanetary flights, perhaps using the moon as a refueling stop.

The program will "employ thousands of workers in all grades and cost hundreds of millions of pounds."

The initial manned spaceship is envisioned by L. R. Shepherd as a structure "weighing some thousands of pounds" of which 80 percent or 90 percent will be reaction mass (hydrogen or deuterium) that will be superheated by the nuclear reactor. A large part of the ship's weight will necessarily be radiation shielding.

R. A. Smith thinks that the IPA rocket will be a step rocket more in the realm of about 1,000 tons and that the IPA project proper will start with the lunar landing. The initial development phase could take anywhere from 10 to 50 years.

Arthur C. Clarke believes that the projected spaceship might be developed for as little as $100 million. The year 1960 may see the establishment of a permanent research organization to work on the problem of spaceship design.

OCTOBER 14

The Bell X-1 (originally XS-1 #1) carries Chuck Yeager on the first manned flight faster than the speed of sound. On this date the bulletlike (literally—see below) orange aircraft reaches a speed of Mach 1.06 (700 mph) at an altitude of 43,000 feet above the Mojave Desert near Muroc Dry Lake in California.

The XS-1 was born as part of a program to develop transonic and supersonic manned aircraft. This was a cooperative program between NACA (the National Advisory Committee for Aeronautics) and the U.S. Army Air Forces. In 1945, Bell Aircraft Corporation won a contract to develop three research aircraft under project designation MX-653. The Army supplied the designation XS-1, signifying "Experimental Sonic One."

Bell will build three of the rocket-powered XS-1 aircraft. The first of these is christened *Glamorous Glennis* in honor of Yeager's wife. It is built of high-strength aluminum, with steel propellant tanks. Fuel is fed to the engine by direct nitrogen pressurization. The 6,000-pound-thrust XLR-11 engine is supplied by Reaction Motors, Inc., the pioneering rocket engine company founded by three original members of the American Rocket Society [see 1930]. It burns liquid oxygen and diluted ethyl alcohol.

The sleek little (30.6-foot) orange craft is patterned after the lines of a .50 caliber bullet. In addition to the pressurized cockpit are crammed the two propellant tanks, twelve spheres of nitrogen for fuel and cockpit pressurization, retractable landing gear, the wing structure and the engine. The XS-1 is originally designed to take off from a runway under its own power. This plan is changed, however, and all of the X-1 aircraft are carried aloft by either a B-29 or a B-50 Superfortress (although the *Glamorous Glennis*, piloted by Captain Yeager, did make a successful ground takeoff on January 5, 1949).

The top speed reached by *Glamorous Glennis* is Mach 1.45 (approximately 957 mph). Its greatest altitude is reached by Major Frank Everest: 71,902 feet. In all, the little rocket-plane makes seventy-eight successful flights, finally retiring in mid-1950. The *Glamorous Glennis* now resides in the Smithsonian Institution's National Air and Space Museum in Washington, D.C.

There are several other aircraft in the X-1 series, almost all of them successful. The XS-1 #2 (later modified as the X-1E) and the XS-1 #3 (also known as the X-1-3 *Queenie*) is lost in a ground explosion in 1951. An additional three aircraft are built and flown: the X-1D (1951), which explodes while still attached to its B-50 carrier, killing pilot Frank Everest; the X-1A (1953), which sets numerous speed and altitude records; the X-18 (1954), which acts as a test-bed for the reaction controls that are later applied to the X-15 and Mercury spacecraft; and the X-1E (1955), a modified X-1.

ca. 1947

The U.S. Army issues a recruiting poster featuring a moon rocket with the slogan: "Into the World of Tomorrow with the U.S. Army."

The Navy follows suit with a poster illustrating a rocket leaving the earth, with the moon and Saturn in the background. Its slogan reads: "The Sky *Was* the Limit. Now—a Career in Electronics."

1948

Val Cleaver and Les Shepherd, in the September issue of the *Journal* of the BIS, begin a four-part investigation of atomic rockets. These represent the first serious study of the application of atomic energy to spaceflight.

H. E. Ross and K. W. Gatland develop the orbital rocket technique. In this, small rockets carry fuel into orbit for a larger rocket. This makes the transit to the destination world, where a smaller rocket makes the actual landing [see 1949 and 1951].

Prof. H. L. Johnson of Ohio State University believes that only the lack of a rocket strong enough prevents the first flight of a spaceship, which he thinks would be most likely propelled by liquid hydrogen and liquid oxygen. Supported by the Air Force, Johnson has been experimenting with hydrogen-fueled rocket motors.

1948

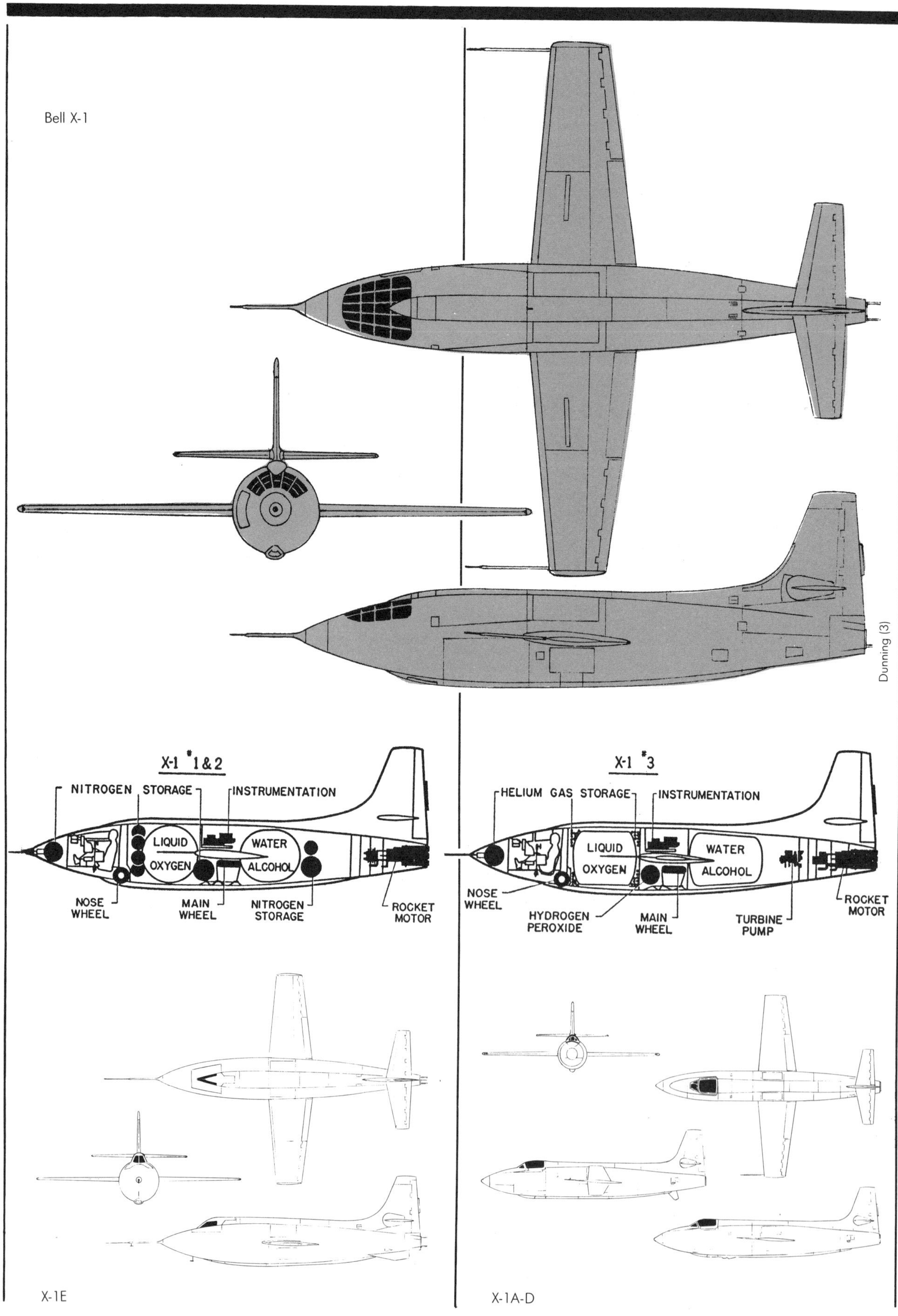
Bell X-1
Dunning (3)
X-1 #1&2
NITROGEN STORAGE
INSTRUMENTATION
LIQUID OXYGEN
WATER ALCOHOL
NOSE WHEEL
MAIN WHEEL
NITROGEN STORAGE
ROCKET MOTOR
X-1 #3
HELIUM GAS STORAGE
INSTRUMENTATION
LIQUID OXYGEN
WATER ALCOHOL
NOSE WHEEL
HYDROGEN PEROXIDE
MAIN WHEEL
TURBINE PUMP
ROCKET MOTOR
X-1E
X-1A-D

An editorial in the London *Daily Mirror* states that "Our candid opinion is that all talk of going to the Moon, and all talk of signals from the Moon is balderdash—in fact just moonshine."

Dr. George Gamow states that he believes that rocket flight to the moon is definitely "a possibility" but that the "only method within reason" would be a long spiral flight by an atomic rocket through space that might take years. Once in orbit around the earth, the spaceship would gradually increase the radius of its orbit until it finally intersected that of the moon.

JANUARY

In "The Expendable Tank Step Rocket" (*Aeronautics*) K. W. Gatland gives a brief description of a five-step atomic spaceship. The fuel tanks, containing liquid hydrogen, will be jettisoned once empty by centrifugal force [also see 1949].

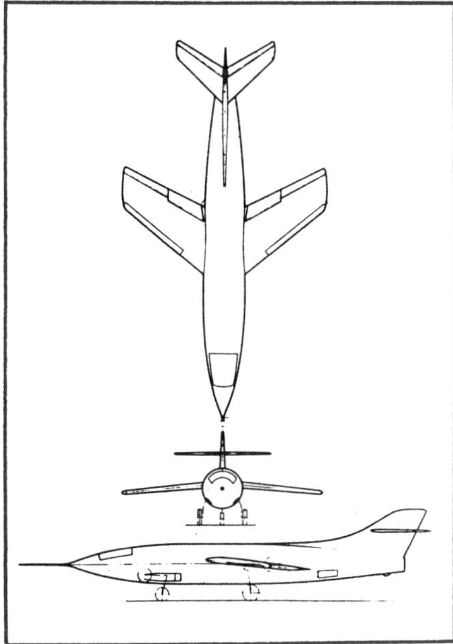

FEBRUARY

The rocket-powered D-558-II "Skyrocket" research airplane makes its first flight [see 1953].

SPRING

In "Design for Doomsday" (*Planet Stories*), Bryce Walton writes: "Either a space-flight is safe monotony, or quick death!"

AUGUST

Science Digest magazine reports that "Landing on and moving around the moon offers so many serious problems for human beings that it might take scientists another 200 years to lick them."

F. H. Clauser in an article published in the *Society of Automotive Engineers Journal* predicts that spaceflight may be only 10 or 20 years in the future.

NOVEMBER

H. E. Ross presents to the BIS his design for an "orbital base," a space station closely based upon the 1928 proposal of Hermann Noordung [see above].

Designed in collaboration with R. A. Smith, Ross's station consists of three main parts: the "bowl," which is the 200-foot mirror of the solar power generator, the "bun," which is the manned section, and the "arm," which carries the communications antennas. The bowl will be able to produce 1,000 kW of available electricity. Behind the mirror is the "bun," a toroid housing the living quarters, laboratories, etc. It is divided into two concentric levels. The station rotates at a rate of 7 rpm, producing 1 g in the outer gallery and 0.43 g in the inner gallery. Six electrically powered gyroscopes located at the hub maintain the station's attitude.

The torus is equivalent to a single story building 450 feet long and 16 feet wide. In addition to living quarters and laboratories are a mess room, kitchen, surgery, recreation room, library, two bathrooms, etc., for the crew of twenty-four. Ross and Smith have given considerable thought to the details of provisioning the station.

The antenna boom does not normally rotate with the station. At the near end is a cylindrical airlock chamber (also containing a

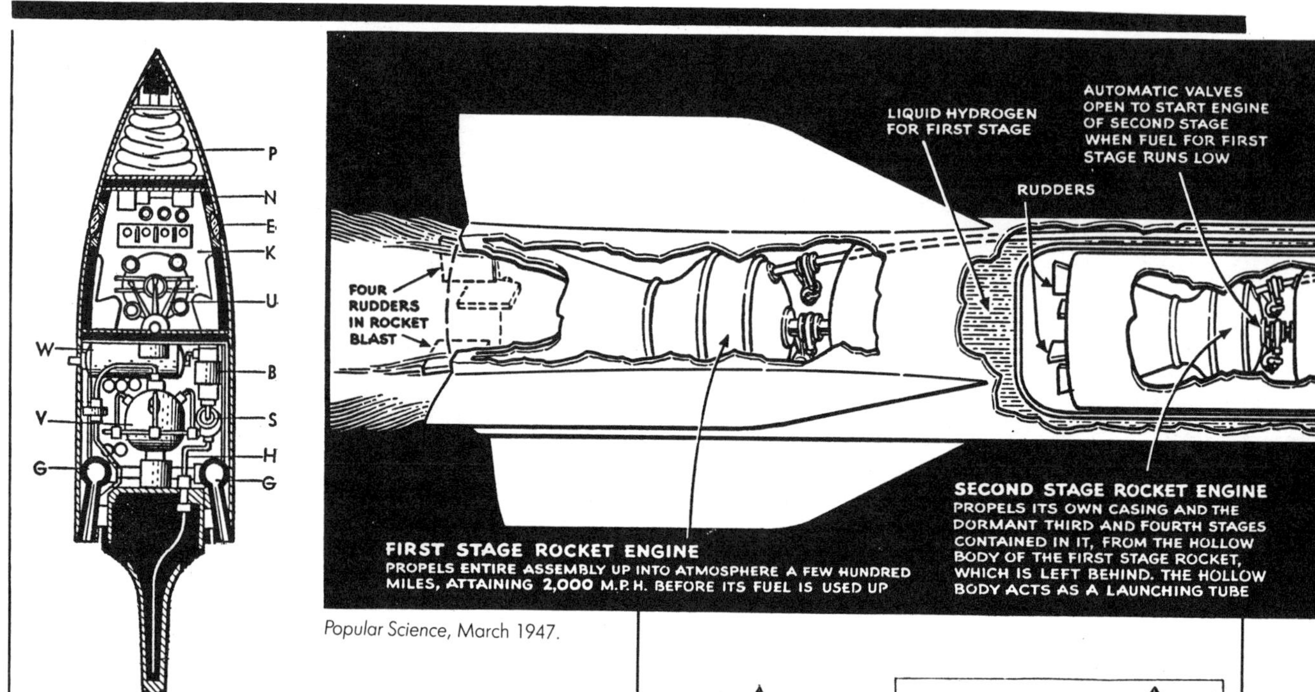

Popular Science, March 1947.

Left: Cabin of Lent manned mail rocket. Landing is made tail-first on "landing pick."

Ananoff's nuclear spaceship.

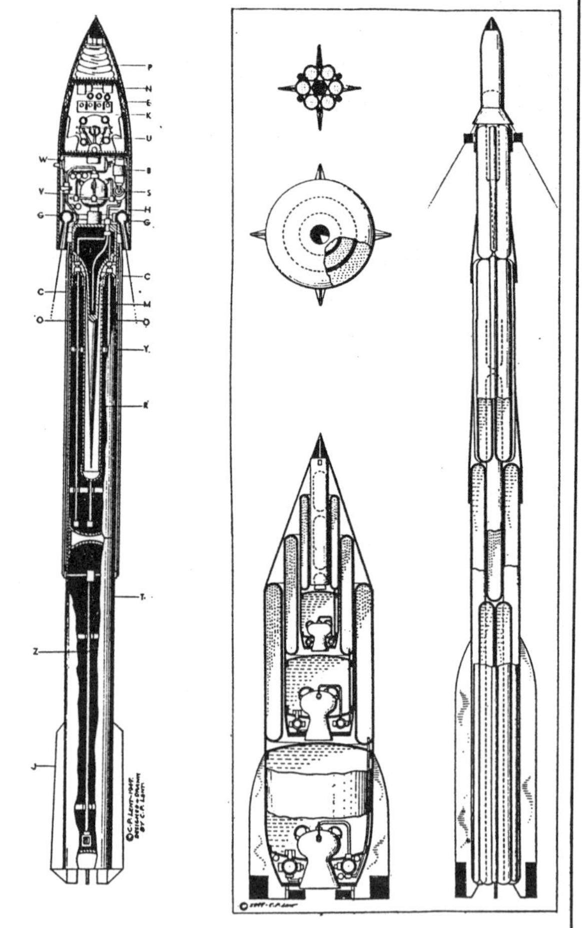

Spaceship proposals by Constantin van Lent, including a lander equipped with his landing pick device (left and right).

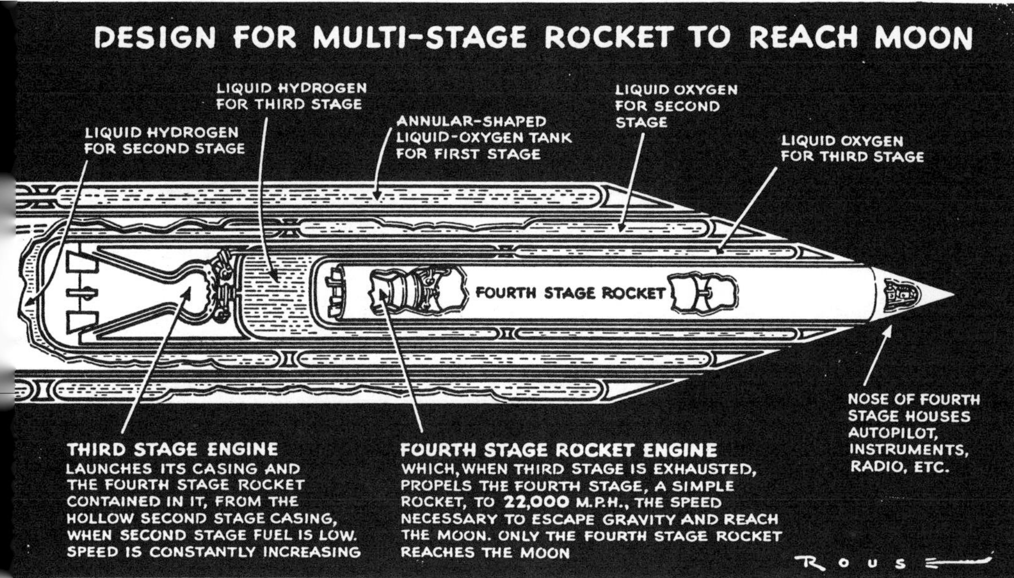

Ananoff's nuclear-electric spaceship (below left), advertisement for model rocket car (below).

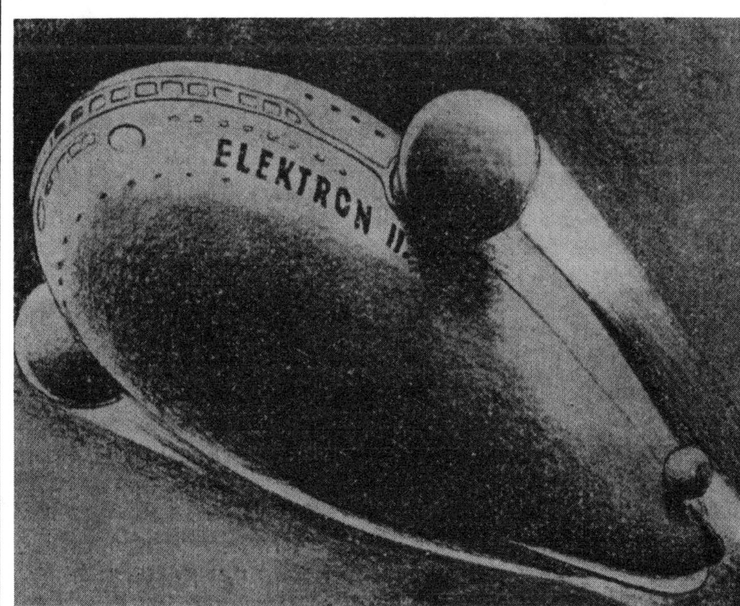

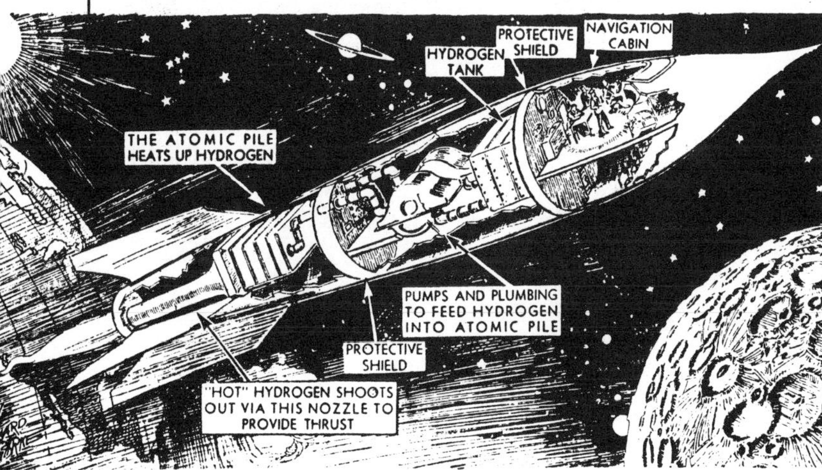

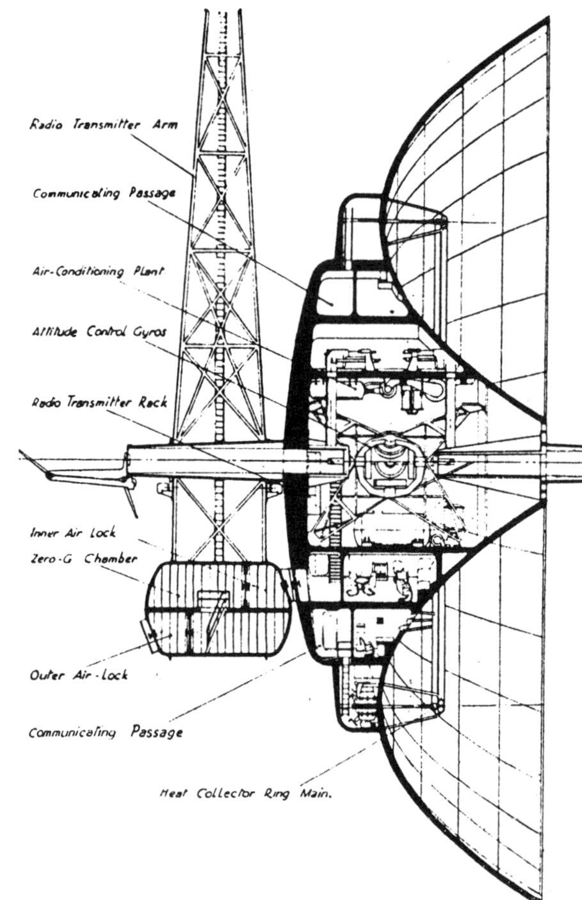

zero-g laboratory). This provides entrance and egress to and from the station. When in use, an astronaut enters the relatively stationary airlock, which is then spun to match the station's rotation. Then aligning the airlock hatch with one of several on the torus, the astronaut can make the transfer. In exiting the station, the process is reversed.

Through the axis of the mirror protrudes the tube of the "strobostelescope," a development of the coelostat invented for the 1939 BIS moonship. This provides a nonrotating view of the universe from the rotating station. There will are two of these, pointing in opposite directions along the axis of rotation.

The station is to be constructed by lifting materials into orbit by means of 442 to 600-ton cellular-step spaceships [see 1939], each carrying 30 to 40 tons of cargo. A similar process will be used for a manned flight to the moon, also described in the same article. For this, 3 442-ton ships, each carrying one pilot, will be launched simultaneously from the earth. They will rendezvous at an altitude of 500 miles. One of the three will be refueled from the other two. One of the ships will be abandoned and another will retain the fuel not used by the third. All three astronauts will transfer to the third ship, which now weighs 65.2 tons. They then depart for the moon. Establishing a 500-mile orbit above the moon, the fuel tanks (weighing 3.9 tons) are detached and left in orbit while the ship makes the descent to the lunar surface. Upon landing the ship now weighs 10 tons.

After leaving the moon, the ship rejoins the orbiting fuel tanks and transfers the remaining fuel. Returning to earth orbit, the ship makes a rendezvous with the waiting second ship, and the crew transfers to it for the final descent.

1948–1952

Wernher von Braun develops his "Mars Project" study [see 1949, 1951–1952 and 1952].

1948–1955

The XP-92, an experimental rocket-plane, is proposed by Convair using sixteen D50/CL engines. It becomes a pure jet, however, when eventually flown.

1949

Heinz Gartmann states that German atomic and rocket scientists want to go to the moon. These are the members of the newly formed Interplanetary Society for Space Research of which Dr. Gartmann is president. They hope that "sooner or later German science will be leading again in the field of rocket science." One of the scientists says, "I am of the firm conviction that before the end of my days I shall have viewed the moon from the back."

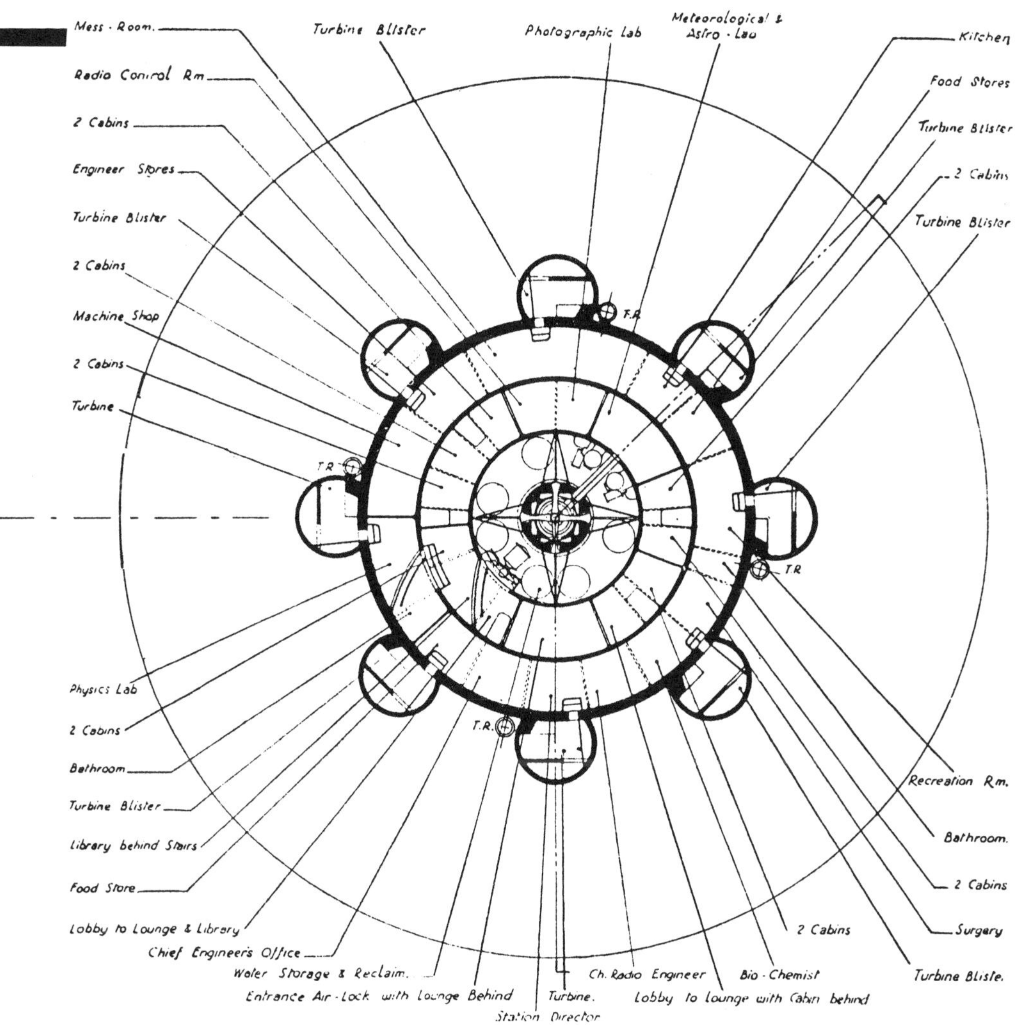

Willy Ley predicts that an artificial satellite will be orbiting the earth within the next 10 to 15 years.

Curtiss Wright discloses that the U.S. government has been secretly working on a 10,000 mph "spaceship", as well as plans for a 3,000–4,000 mph rocket plane. When reporters inquired Air Force experts for more details, they received answers that were far from expert. They were told, for example, that the 10,000-mph vehicle would either simply "float" in space with the earth whirling beneath it or it could turn on power and keep pace with the globe (and only a single pound of fuel would be needed every year for this manuever). The spaceship would have to be a huge affair, at least several hundred feet long.

A survey of scientists conducted by G. Edward Pendray for the Guggenheim Foundation discovered that "over one-fourth of them said they expected definite progress within ten years toward the development of rockets for interplanetary flight."

Tsien Hsueh-Sen, Goddard professor at the California Institute of Technology, proposes a boost-glide hypersonic aircraft with a 3,107-mile range. The transcontinental passenger carrier will be 78.9 feet long, 16.5 feet in diameter, with a wingspan of 18.9 feet, and will weigh about 48 tons, of which 36 tons will consist of hydrogen and fluorine propellant. All of the fuel will be consumed within the first 60 seconds of flight, at which time the rocket will be ascending at a speed of 9,140 mph and at an altitude of 100 miles. It will take off nearly vertically. The ten-passenger craft will travel in an arc to an altitude of 300 miles—6 minutes after takeoff—before descending back into the

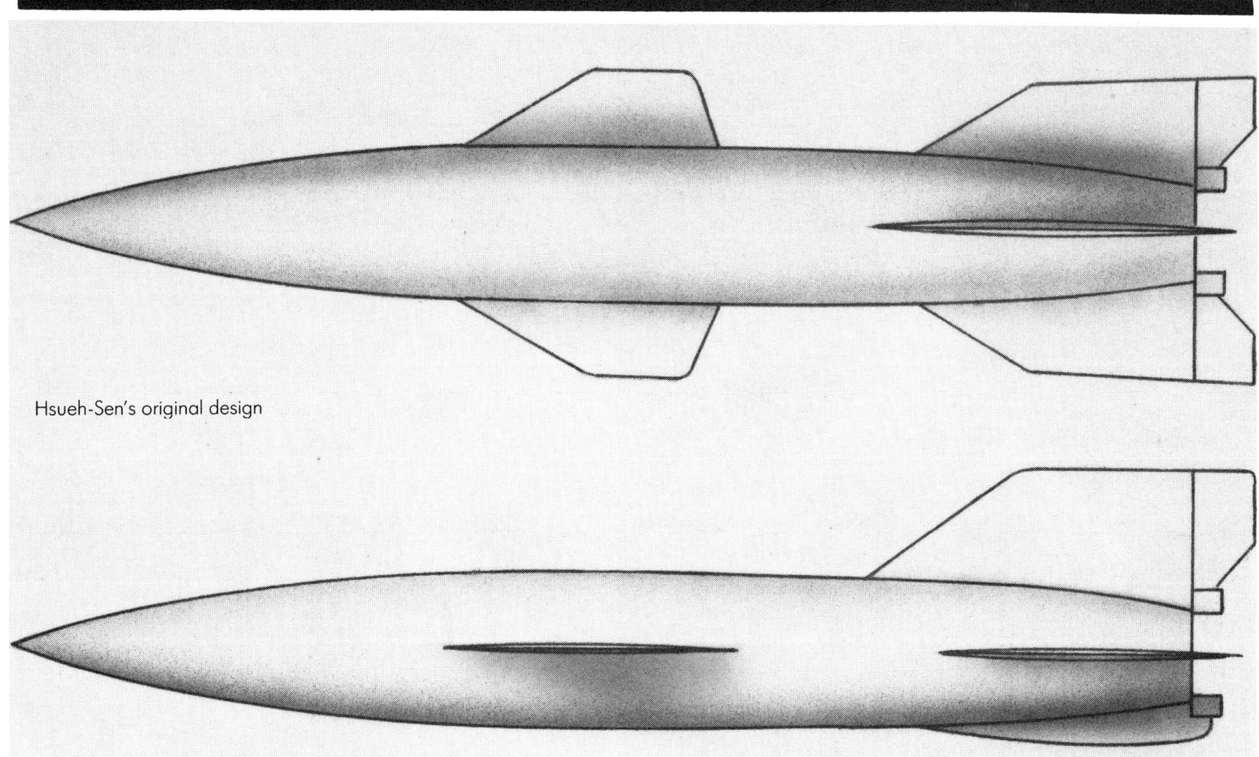
Hsueh-Sen's original design

stratosphere. It will reenter the atmosphere 1,200 miles from its takeoff point.

Leaving New York at 12 noon, the stubby-winged rocket will reenter the atmosphere near Des Moines, Iowa, gliding from an altitude of 27 miles the remaining distance to the west coast. As it nears its destination a 5,000-pound thrust turbojet will permit a controlled landing in San Francisco at 150 mph. Time: 9:45 a.m., 2 1/4 hours ahead of the sun and only 45 minutes after leaving New York.

Chesley Bonstell vividly illustrates the stages of such a flight in a pictorial article for *Pic* magazine, "Coast to Coast in 40 Minutes."

In 1959, G. Harry Stine (in *Earth Satellites*) thinks that this rocket could be adapted for high altitude research and astronaut training. Climbing vertically, it would reach a peak altitude of 55 miles.

The Conquest of Space by Willy Ley and Chesley Bonestell is published. This highly influential book on the future of space exploration goes through four printings before it is 3 months old. A manned moonship is described, employing atomic power, with a

Bonestell, *Pic magazine*

mass-ratio of 3.3:1. Takeoff is from a mountaintop near the equator. The rocket is V2-like in silhouette, with the addition of wings to aid in its return to the earth. It leaves the earth's atmosphere within 3 minutes of takeoff, the engines burning for a total of 8 minutes to reach escape velocity (putting the crew through no more than 4 g). The ship is equipped with an automatic system to guide it during the initial acceleration. At the end of this period, the engine shuts

Staton

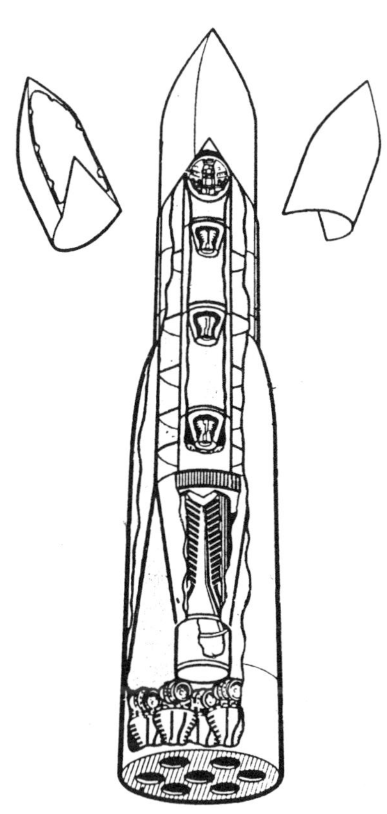

down and the crew is weightless for the 4 days it will take to reach the moon. Strategically placed loops of nylon cord help the crewmembers to move about and to stay put where they want to. At about 215,000 miles from the earth, the ship is reversed so that its tail is now pointed toward the moon. A braking maneuver of about 150 seconds is required. The landing is simply a takeoff in reverse. The rocket lands on its tail, supported by its fins and the extendable props that fold out from them.

Kunesch, Dixon, Gatland and Shepherd develop the first serious engineering concept for an atomic rocket. An atomic booster launches a three-stage manned rocket. Their study is published in the *Journal* of the BIS.

The rocket is a wingless, five-stage manned vehicle. The first stage is a chemical booster, which will help to eliminate danger of radiation at takeoff. It consists of seven 450-pound-thrust engines using liquid oxygen and liquid hydrogen. These propellants are in expendable, semicylindrical fuel tanks that shroud the second stage, which is contained within them. The nose of the rocket, too, is a pair of expendable fuel tanks for the booster. After these all drop away, with the booster itself, the atomic rocket takes over. It is a 40-ton reactor whose honeycomb structure is contained within the combustion chamber itself. The working fluid is liquid hydrogen or ammonia. With the latter, the engine develops 1,000 tons of thrust. A 20-ton shield lies between the atomic engine and the remaining three stages. The reactor is fed from semiannular expendable tanks that are jettisoned as they empty. Above the atomic rocket is a three-stage chemical rocket weighing 60 tons. Its final payload is a spherical pressurized crew compartment weighing 1.4 tons.

Curtiss-Wright reveals that government scientists are working on a "space ship" that will "flash through the stratosphere at nearly 10,000 miles an hour. "A published chart shows that a 10,000 mph "satellite" is being designed, as well as a 3,000–4,000 mph rocket-plane. Secretary of Defense Forrestal says that the "earth satellite vehicle pro-

1949

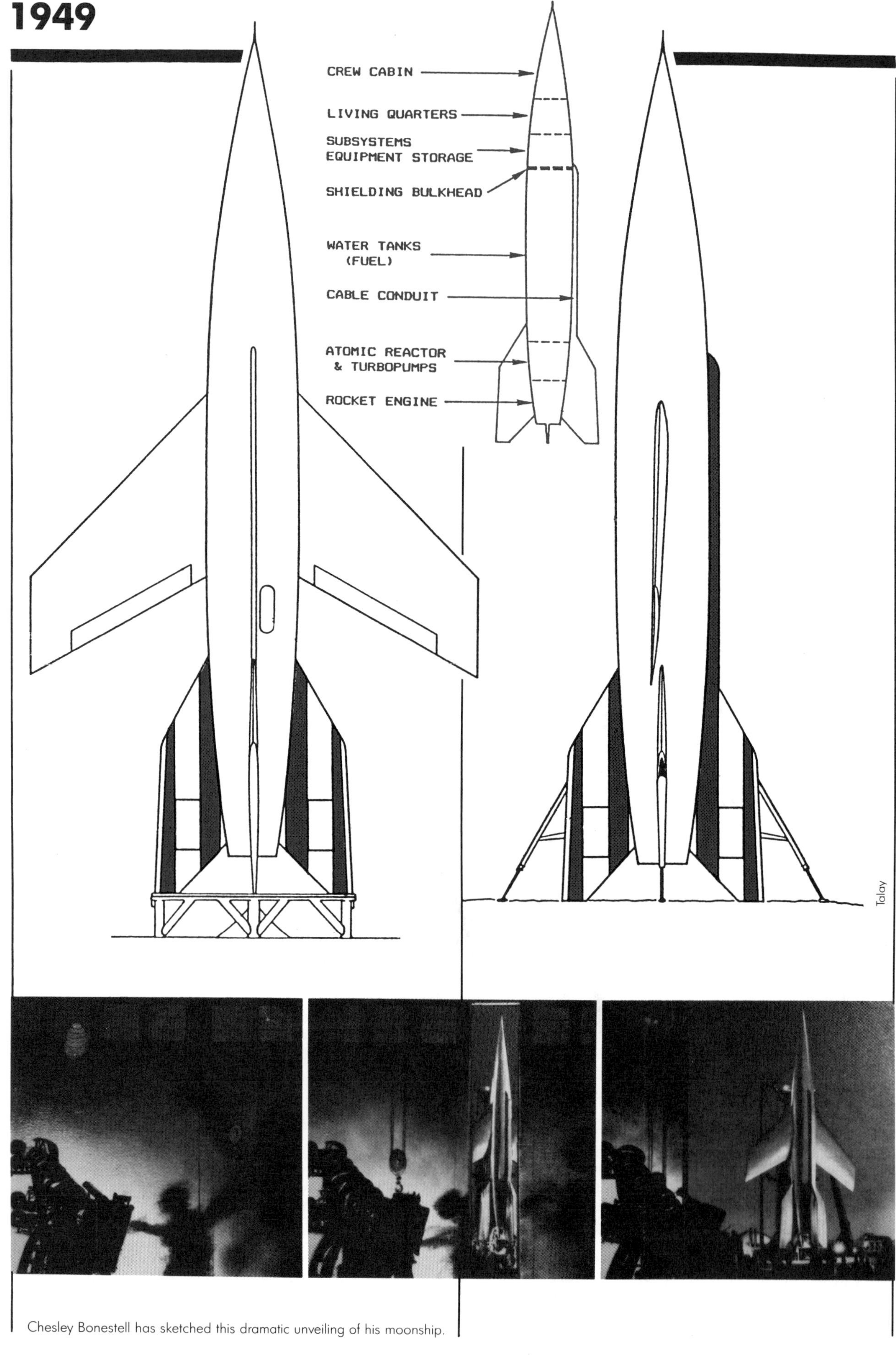

Chesley Bonestell has sketched this dramatic unveiling of his moonship.

gram" has been put under the central guided missiles project and that all three armed services are working on research programs. They are asking for $2 million with which to build a proving ground.

Ordway

Wernher von Braun publishes *Das Marsprojekt*. [Also see *Collier's*, 1952–1954 and von Braun, 1952.]

The Canadian Rocket Society plans a moon rocket for 1960. Its basic design is based on that of Captain C. Evans-Fox. The giant ship will be 200 feet long, 50 feet in diameter and weigh 1,000 tons. Estimates of its cost and "volunteer assistance" come to $2 million. It is propelled by twenty motors, each having a thrust of 100 tons. It has twin hulls and a large refrigeration plant to maintain the proper interior temperature. The crewmembers will be equipped with spacesuits they will wear from takeoff until the return to earth. The rocket is to be launched from a massive concrete blast shield that encloses the bottom third of the ship. It will land tail-first on the moon, supported by large, telescoping tripod legs.

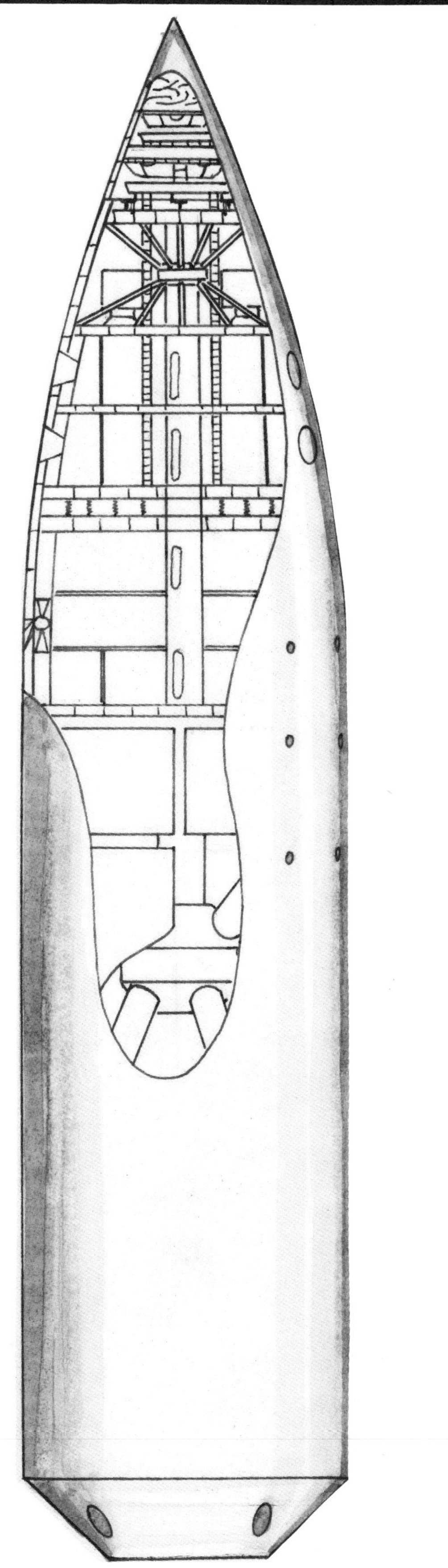

A

B

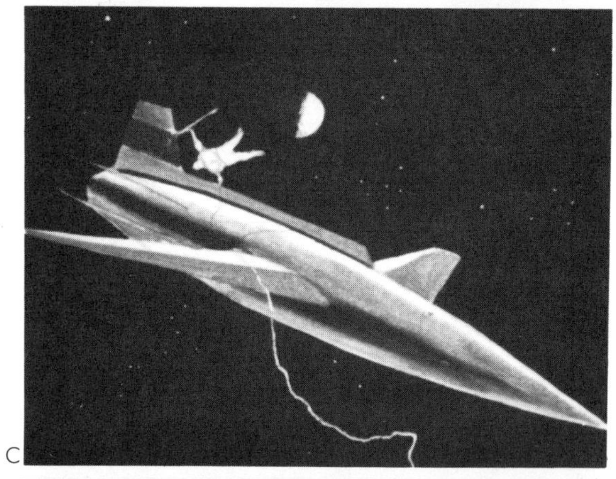

C

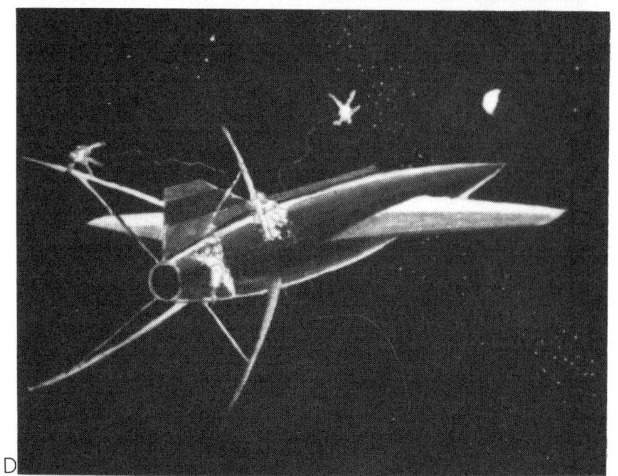

D

E

F

Storyboard by Chesley Bonestell for a filmstrip illustrating a trip to the moon (ca. 1949). A. Prelaunch, B. Takeoff, C. In transit, D. Deploying landing struts, E. Touchdown on earth, F. End of the journey.

The moonship would have to contend with the danger from meteors, the Canadian Rocket Society admits; "there would have to be some judicious steering to prevent disaster," they say.

R. L. Farnsworth of the United States Rocket Society dismisses the Canadians by saying that they are "just a bunch of amateurs who have got wrapped up in their own dreams," (although he later denies having said this, blaming it on misquotes by the media).

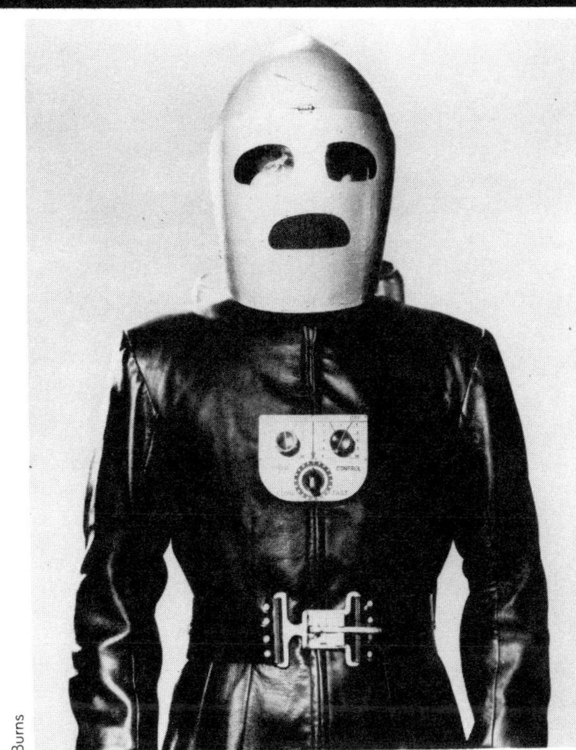

King of the Rocket Men, a twelve-chapter Republic serial starring Tristram Coffin, is released. It introduces the now-famous "flying suit": a bullet-shaped helmet, a leather jacket and a twin-rocket backpack (with chest-mounted controls that read: "On, Off, Up, Down, Slow and Fast"). The rocket is propelled by nuclear-powered "sonic propulsion." It is followed by *Radar Men from the Moon* and *Zombies of the Stratosphere* (as well as a television series, "Commando Cody, Sky Marshal of the Universe") [see 1952 and 1953].

Robert Heinlein publishes his short story "The Man Who Sold the Moon." In this piece Heinlein is one of the first authors since Jules Verne to consider the problem of financing a lunar expedition. The hero, D. D. Harriman, obsessed with going to the moon (but ultimately too old and ill to himself take the journey that he finally makes possible for others), circumvents a reluctant government and private industry, who can see no immediate purpose or profit in it, by raising the money himself though innumerable ingenious schemes. Not the least of these is the kind of licensed merchandising that is so familiar today: special commemorative postage stamps, movie, television and publication rights, and so forth.

JANUARY 17

Life magazine publishes "Rocket to the Moon," illustrated with nine spectacular paintings by Noel Sickles. It describes the flight of a 200-foot two-stage rocket. Its five-man crew is crowded into a 10-foot cabin. After takeoff, where it rises vertically "to get through the atmosphere quickly," the rocket at 300 miles "turns to fly parallel to the earth's surface and pick up sufficient speed [from the earth's rotation] to head out into space."

Almost all of the article's "scientific" details—including the erroneous ones—will be picked up and used verbatim in the 1950 motion picture *Rocketship XM* [see below]. Even the design of the movie's spaceship is copied from Sickle's artwork!

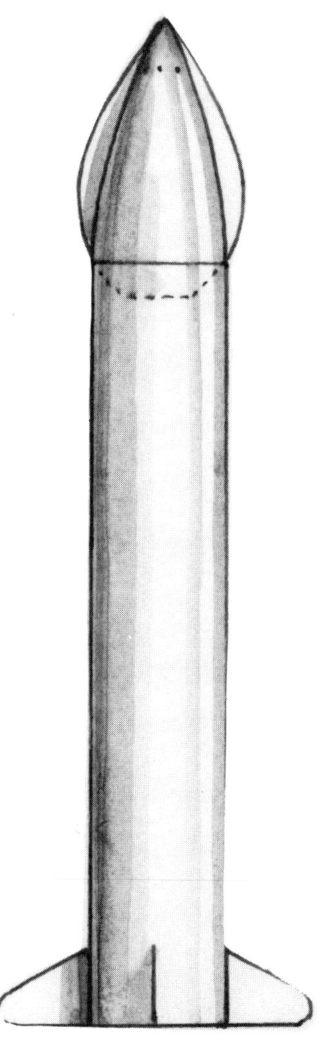

King of the Rocketmen.

Burns

MARCH

K. W. Gatland describes in the *Journal* of the British Interplanetary Society a rocket intended for a lunar landing. It will be a composite atomic-chemical vehicle. The overall rocket will be divided into three sections, from bottom to top: the "parent" stage, containing the reactor and its shielding and, above that, the working fluid; the "composite" rocket, composed of the pressurized crew compartment (similar to that of the 1939 BIS moonship)—equipped with four chemically fueled motors and four extendable landing legs, and above this a large propellant tank; and above this unit a cluster of expendable tanks.

The passenger compartment will be 22 feet in diameter (even though Gatland provides for artificial gravity by having his passenger compartment rotate, he does not think it necessary to have a particularly large radius—certainly not the 30-foot radius E. A. Pecker of the Pacific Rocket Society thought necessary for his 70-foot × 10-foot disk-shaped spaceship).

The parent craft will remain in orbit while the composite craft makes the descent to the moon (or other body). The parent craft, in fact, will never return to the surface of any planet and will therefore be available for future flights. The parent craft will be either boosted into earth orbit by chemical rockets or assembled piecemeal in orbit.

May

J. Himpan and R. Reichel, in the *American Journal of Physics*, discuss the problem of "Can We Fly to the Moon?" They conclude that while an unmanned rocket is just barely possible—though excessively costly—a manned rocket will be not be. They base their study of an instrument-carrying rocket on one propelled by a mixture of fuel oil and tetranitromethane. They dismiss the use of hydrogen and oxygen as being impractical, with the result that their mass-ratio is low. The 50-ton, three-stage rocket they describe carries a payload of only 10 kg. They demonstrate that a chemically fueled circumlunar rocket will require initial masses of fantastic proportions. However, as L. R. Shepherd of the British Interplanetary Society points out, Himpan and Reichel do not consider the use of techniques such as orbital refueling.

"The devotees of the idea of interplanetary flight," Shepherd concludes, "need not . . . be dismayed by the pronouncements of these two authors."

Ackerman

JUNE 27

Television's first children's space show, *Captain Video*, premiers. The Captain (originally played by Richard Coogan) fights the enemies of the solar system on a $25-a-week special effects budget. A year later, the program is being aired 6 days a week—with the same effects budget. Astronomer Robert S. Richardson acts as the show's technical advisor. In 1950 Al Hodge replaces Coogan in the spaceship *Galaxy*. The show finally goes off the air in 1957, after degenerating into a format for rerunning old cartoons and serials.

JULY

The Technical Committee of the British Interplanetary Society publishes the designs for a nuclear-powered manned circumlunar rocket. The design is based upon the assumption that a nuclear rocket will be developed that could produce an exhaust velocity of 10 km/s.

1949

The overall height of the rocket will be 58 m, with a base diameter of 10 m. The first booster stage will be a chemical rocket employing two expendable tanks on the design of K. Gatland, et al. [see 1953]. This stage will have seven hydrogen/oxygen-fueled motors of 450 tons thrust each and will boost the rocket to a velocity of 1.75 km/sec. The atomic rocket will then boost the ship on to its orbital velocity. From this point a three-stage chemically fueled rocket will continue on to the moon. The first stage of this rocket would have one engine of 200 tons thrust, the second stage one engine of 75 tons thrust and the final stage an engine of 25 tons thrust. The overall length of the cylindrical "crew rocket" will be 19.5 m and its diameter will be 3.5 m. It will carry a spherical, pressurized chamber weighing 1.4 tons, including the crew, instruments, provisions, etc.

CHRISTMAS

Hugo Gernsback describes a speculative space drive based upon what he called "mass-ejection propulsion." The ship's atomic generator will transform energy into mass, which will in turn propel the rocket. The small pellets created by the generator will be about the size of "buckshot," yet each will weigh several hundred tons "but only at the moment of ejection [sic]." The pellets will only be solid for 1/1000th of a second, after which time their mass will disintegrate into "nonradioactive radiations."

ca. 1949

Walter Dornberger and Krafft Ehricke design a two-stage passenger-carrying rocket (the studies are sponsored by the U.S. Air Force at Wright-Patterson Air Force Base). It is a civilian version of their later BOMI rocket bomber [see 1951]. The winged, manned stages are parallel rather than stacked, a concept that eventually influences the design of the Space Shuttle. Both orbital and suborbital versions are designed. Both stages are winged and the "spaceplane" is to ride "piggyback" fashion on the larger mothership. The first stage has five engines and the

smaller spaceplane three, all eight of which are firing at takeoff. At 130 seconds after launch, the two separate, the mothership returning to base for an aircraft-type landing. The second stage continues to climb to an altitude of 40 miles at a top speed of over 7,400 mph, crossing the United States in 70 minutes, for a total flight time of 1 hour 15 minutes (at the expense of 57,000 gallons of propellants). Since the effects of prolonged weightlessness on humans is unknown, the engines will continue to function throughout the flight, providing a constant 1/4 g. Twenty passengers could be carried on a transcontinental rocket for about twice the fare currently charged by conventional airlines.

In 1955 Dornberger and Ehricke (the latter now with Consolidated Vultee Aircraft) elaborate on the possibilities of their rocket airliner. Two different classes of space planes are proposed: one for ranges of 2,000 to 3,600 miles and another, similar in design but slightly larger, for ranges of 4,500 to 5,200 miles.

The smaller vehicle has a gross weight of 600,000 pounds for the booster and 110,000 pounds for the spaceplane, the latter carrying a payload of 8,800 pounds. The booster has five engines (with a combined thrust of 750,000 pounds) and the spaceplane three (with a combined thrust of 150,000 pounds). The larger model has gross weights of 770,000 pounds and 138,600 pounds. The spaceplane will carry the same payload as the smaller edition. The booster's six engines will have a combined thrust of 730,000 pounds and the spaceplane's three engines will produce 170,000 pounds. From nose to tail the interior of the spaceplane is divided this way: pilot's cabin, fuel tanks (gasoline), the passenger cabin, liquid oxygen tanks and the rocket motors.

The rockets will be assembled on large railroad flatcars, built in the form of "splash plates," in pits located in hangers around the periphery of the spaceport. Locomotives will haul the assembled vehicles from the assembly pits along the bottom of long "canyons," which converge into a central "crater"—a circular depression from which the launchings take place. Dornberger and Ehricke describe a rocket flight from San Francisco to Sidney. The passenger enters the rocket through a hatch at a point near the nose, descending in an elevator to one of a number of cylindrical plastic "bowls" flanking the shaft (near the rocket's center of gravity). These are designed to rotate to horizontal with respect to the spaceplane when it levels off at high altitude. Boarding takes place at the point where one of the canyons enters the launch pit.

The rocket is towed into the center of the launch pit. The walls of the pit protect the lifting rocket from stray gusts of wind. The spaceplane and booster are powered by their combined eight engines, producing 760,000 pounds of thrust. The acceleration increases to 3 g at 130 seconds after takeoff and the combined spacecraft begins to level off. The engines are then shut off and the planes coast momentarily. At this point the planes separate. The smaller rocket takes off along a set of rails set in the back of the larger plane's fuselage. Until now both vehicles have been under the command of the passenger rocket's pilot. The acceleration in the passenger rocket increases to 3.5 g. Although the acceleration then decreases, the passengers never experience zero g, since the spaceplane never follows a ballistic flight path. At 136 seconds after takeoff the plane is at an altitude of 140,000 feet, and flying at a speed of 7,300 mph. An unfortunate feature of the spaceplane's design is that its huge wings prevent the passengers from seeing the earth below them. The spaceplane glides to a landing at Honolulu where its passengers—after a half hour break—change to one of the larger rockets for the continuation of their journey to Sidney, where they land 1 1/2 hours later. Total flight time: 3 1/2 hours.

1949–1950

In the U.S.S.R. dogs are being launched up to 60 miles high in recoverable nosecones of Pobeda rockets.

1949–1952

The "Bumper" project is carried out, in which a WAC-Corporal sounding rocket is placed in the nose of a captured German V2 to create a two-stage rocket. Six of these compound rockets are launched from White Sands, New Mexico. The last two of the series are launched from the new site at Cape Canaveral, Florida.

1950

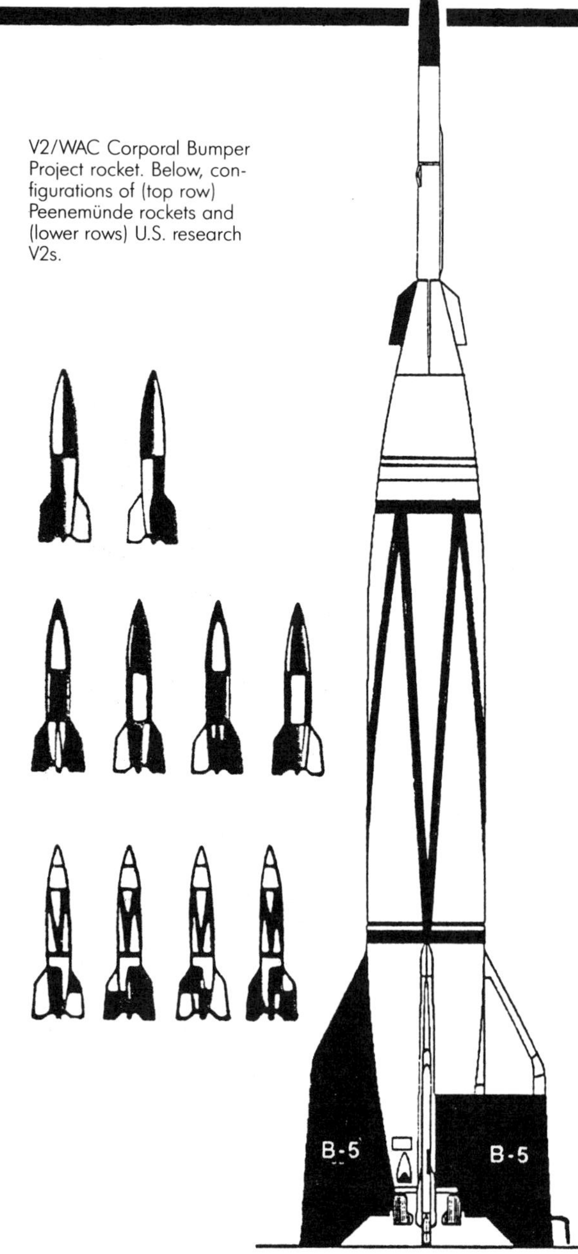

V2/WAC Corporal Bumper Project rocket. Below, configurations of (top row) Peenemünde rockets and (lower rows) U.S. research V2s.

The WAC (**W**ithout **A**ttitude **C**ontrol)-Corporal is a sounding rocket designed by Frank Malina. It has a red fuming nitric acid-aniline-burning engine that generates 1,500 pounds of thrust for 45 seconds.

Five unsuccessful launches are made until the fifth, on February 24, 1949, at 3:14 p.m. After a flight of 1 minute, the V2 reaches an altitude of 20 miles and a speed of about 1 mile per second. At this point, the 661.5 pound WAC-Corporal rocket that the V2 has been carrying in its nose like a spear takes over. It adds its own velocity to that of the V2. After 40 seconds, it, too, has expended all of its fuel. The empty V2 continues to coast upward for another 60 miles, eventually dropping to earth 5 minutes after launch. When it hits the desert, 28 miles from its pad, the WAC-Corporal is still climbing. The little rocket does not stop until it finally reaches the record altitude of 244 miles. For all practical purposes, it has entered empty space, the first manmade object ever to do so.

1950

Arthur C. Clarke publishes *Interplanetary Flight* while still chairman of the British Interplanetary Society. It is an excellent semi-technical introduction to the subject.

Darrell Romick begins a 10-year study at Goodyear developing a complex space program [see 1960].

Lovell Lawrence, president of Reaction Motors, Inc., states that he believes that within 10 years rocket-propelled airliners will be crossing the United States in 90 minutes [see 1949].

OFFICIAL HAYDEN PLANETARIUM APPLICATION

Interplanetary Tour Reservation

* * *

You are one of the first to request space tour reservation.
Your name and address will be kept on file in the HAYDEN PLANETARIUM archives.
Please list the information requested below and mail this form to Interplanetary Tour Reservations, Hayden Planetarium, New York 24, N. Y.

NAME .... (LAST) (FIRST) .... AGE........
ADDRESS........................CITY............STATE............

Check tour desired: ☐ Moon ☐ Mars ☐ Jupiter ☐ Saturn

The Hayden Planetarium in New York opens a "travel bureau" at which visitors can make reservations for interplanetary flights from the "Central Park Spaceport," beginning in 1975. One woman who registers for a flight approaches Lloyd's of London for an £18,000 insurance policy to cover her trip to Mars.

Burns

NASM

Rocketship X-M is a motion picture made quickly in order to cash in on the advance publicity for the more carefully made *Destination Moon* [see below]. Nevertheless, for all of its cheapness and inaccuracies, it is still effective, well-photographed and acted, moody and has a surprisingly downbeat, serious ending. Its cockpit set is later recycled for use in *Flight to Mars* [see below].

The spacecraft depicted in the movie, as well as most of the "scientific" explanations concerning its flight, is taken from the *Life* magazine article of January 17, 1949 [see above].

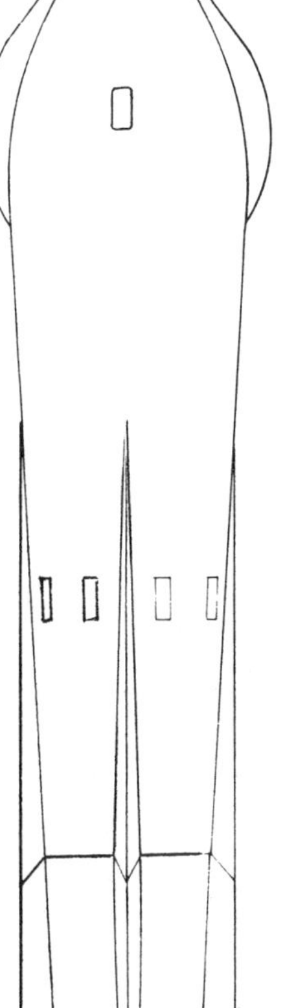

Above, preproduction painting of launch site; upper right, preflight briefing; below, original filming model of spaceship.

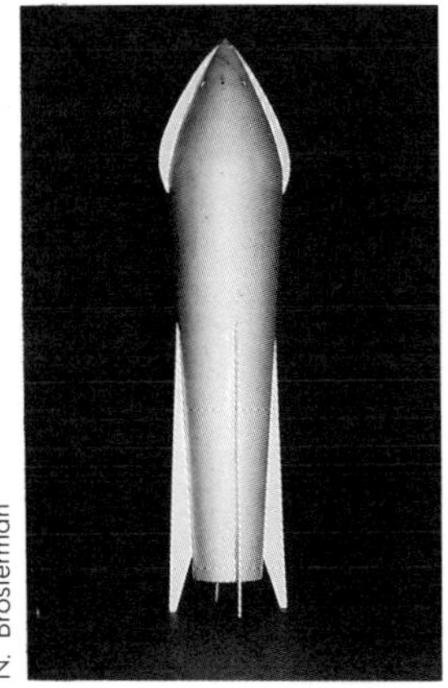
N. Brosterman

Destination Moon, a feature film produced by George Pal, illustrates in documentary-like detail a manned flight to the moon. It is the first motion picture to deal realistically with the subject of spaceflight since the silent *Frau im Mond* [see 1928].

The screenplay is written by Alford (Rip) van Ronkel and Robert A. Heinlein and is almost unrecognizably based upon the latter's novel *Rocketship Galileo*. Heinlein also acts as the film's technical advisor. It is to the credit of producer Pal and director Irving Pichel that the intention is from the very beginning to make a motion picture about spaceflight that is as accurate as possible—this against all Hollywood science fiction tradition. The decision also creates innumerable problems. The state of the art of special effects is stretched to its limit to recreate free fall, extravehicular maneuvers, moonwalks and so forth. At no time do the producers take the easy way out by saying "no one will know the difference." The film is filled with accurate and often prescient details easily overlooked or taken for granted today.

The spaceship *Luna*, designed by art director Ernst Fegté, is 150 feet tall with a loaded weight of 250 tons. Of this, 200 tons is "fuel": ordinary water to provide reaction mass for the atomic motor, 40 tons is the spaceship itself, and 10 tons for the four passengers, their accommodations, equipment, supplies, etc. Artist Bonestell [see also pages 1949, 1952–1954] is also responsible for all of the film's astronomical art: views of the earth and moon from space, and the breathtaking panoramas of the lunar surface surrounding the spaceship *Luna*.

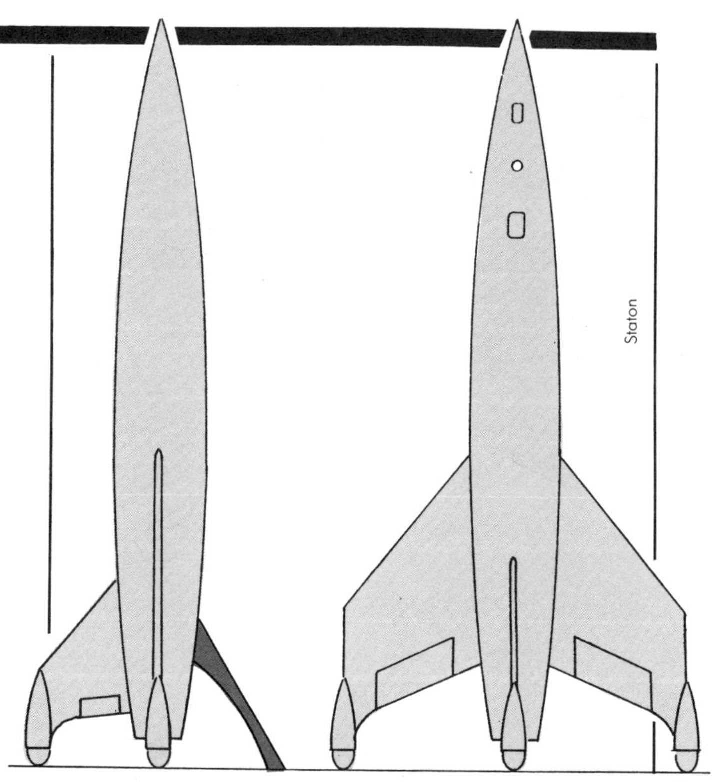
Staton

Takeoff is at 3:50 a.m. on June 20 from Lucerne Valley in California's Mojave Desert. The engine blasts for nearly 4 minutes, at the end of which time the rocket is at an altitude of 807 miles. Forty-six hours later, the *Luna* reaches her namesake (the spaceship's flight is calculated by astronomer Dr. Robert S. Richardson of Mt. Wilson and Palomar). The ship is reversed by its gyroscopes, so that it approaches the moon tail-first. In the film, the landing is difficult, using too much of the ship's fuel supply. This provides the suspenseful climax of the film, where the crew must try to lighten the rocket in order to be able to take off. It is, admittedly, an artificial dilemma: had the *Luna* been a step-rocket the problem would not have arisen; as it is, most of the excess weight the film astronauts are burdened with involves the empty and now useless fuel tanks.

Nevertheless, they solve their problem and are successful in leaving the moon. Reaching the earth, the rocket orbits three times to decelerate, eventually reentering the atmosphere. Here, for the first time, the *Luna*'s huge wings come into play, the rocket using them to glide into the lower atmosphere. Instead of landing like an aircraft, however, the *Luna* makes its final descent by way of three parachutes contained in its nose.

Burns

A special showing of the film on June 20 at the Hayden Planetarium in New York is presented to select audience of 200. An original print of the motion picture that is specially treated will be preserved at the planetarium to "Let future generations see what a pre-

Burns

space-travel age had predicted." (It's presumed that this print is now lost.) Producer Pal credits the planetarium for assistance with the film's authenticity.

According to *Oil Power*, the house magazine of the Socony-Vacuum Oil Co., a trip to the moon will be possible in 25 years at a cost of $1 billion.

R. A. Smith designs a speculative, giant ion-powered passenger spaceship.

The great Belgian cartoonist, Hergé (Georges Rémy) publishes *Objectif Lune*

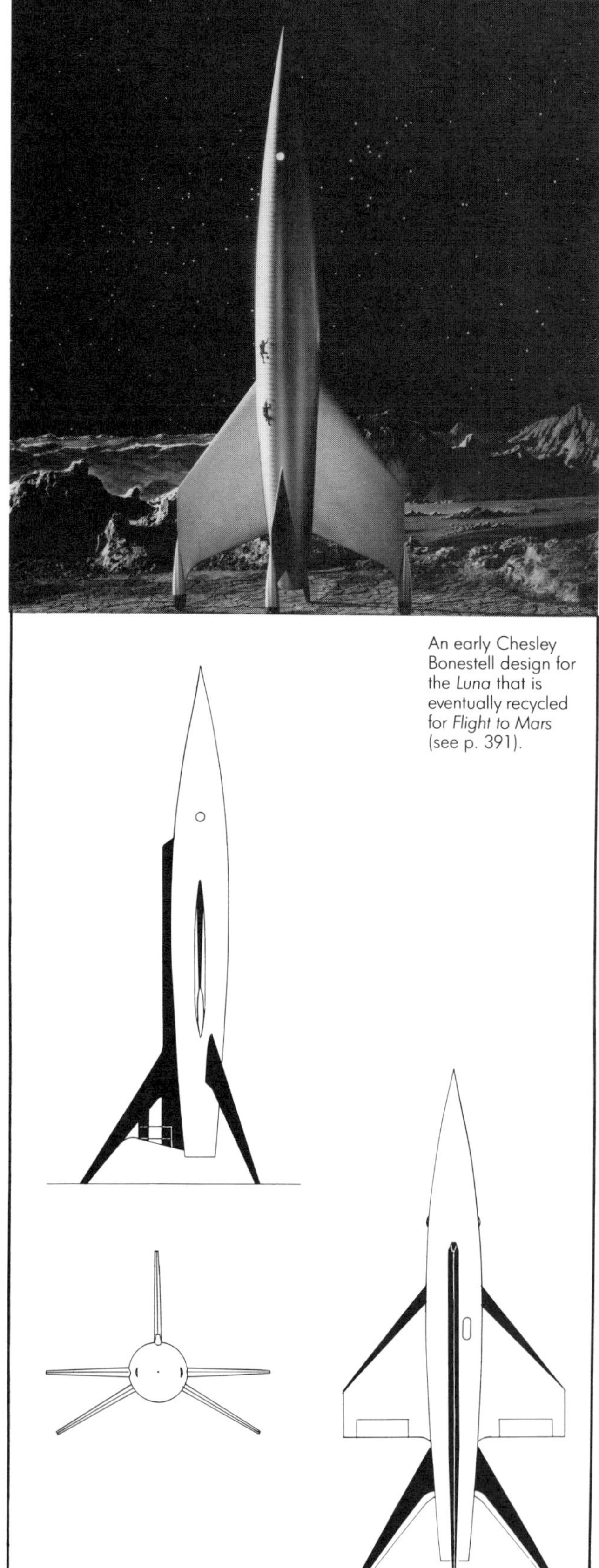

Burns

Staton

An early Chesley Bonestell design for the *Luna* that is eventually recycled for *Flight to Mars* (see p. 391).

and *On a Marché sur la Lune*, the latest entries in the series of now-classic adventures of the boy reporter, Tintin. (The first title is originally published in the magazine *Tintin* from March 30 to October 22, 1950, and the second story from October 29, 1952 to December 30, 1953. They are later gathered into book form in 1953 and 1954, respectively.) These two books are both remarkable and outstanding in their realism and attention to authenticity and scientific accuracy. One result—in addition to the memorable characters, superb storytelling and humor—has been that these books have not only never gone out of print, but have been translated into a score of different languages and published all over the world. They have inspired an interest in space in two generations of Europeans. For example, both pioneering French spationautes, Patrick Baudry and Jean-Loup Chrétien, admit to never having lost their admiration for Tintin's adventures in space.

Hergé creates his lunar adventure as a reaction against what he feels is a "vulgarization" of the subject in comic books. He believes that he can successfully mix information with adventure and humor, much as Jules Verne has done.

Hergé's main inspirations and sources of information are Auguste Piccard's *Entre Terre et Ciel* [see 1943], Pierre Rousseau's *Notre amie la Lune*, Bernard Heuvelman's *Homme parmi les etoiles* and Alexander Ananoff's *Astronautique* (1949). One illustration from the latter, in particular, of the control room of a future spaceship, is duplicated in Hergé's book. Piccard, on the other hand, is himself the model for the eccentric scientist Professor Tournesol, the inventor of the fictional spaceship. Hergé's rocket, which remains unnamed throughout, bears a resemblance to several early European rocket designs, such as the Soviet KPD-1 of 1934, for example. It is an enormous rocket, nearly 200 feet tall, and propelled by both chemical and atomic motors. The former will be used for takeoffs and landings and the latter reserved solely for use in space.

In 1988 Tintin enthusiasts Gérard Guegan and Patrick Vandevoorde, of the Mini-

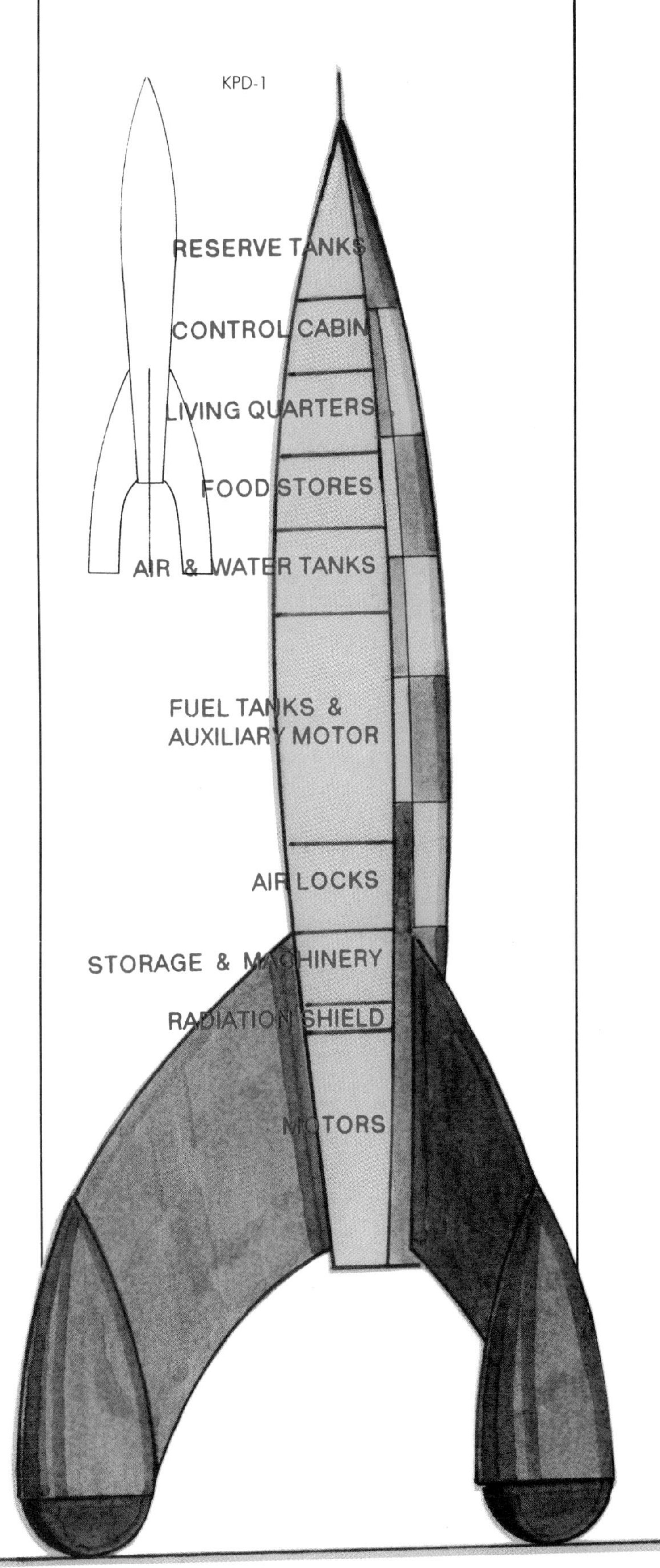

Rocket Launch Club at Kourou, Guiana (CLAMFUK), decide that an appropriate way to celebrate the coming triple anniversary of Tintin's space books, Apollo 11 and the Guiana Space Center is to launch a scale model replica of Professor Tournesol's spaceship. From August to November 1988, tests are carried out with 39.4-cm prototypes, one-quarter the size of the finished model. One of these makes six successful flights to an altitude of about 170 meters. The finished rocket is designated the RG1 and is 157.9 cm tall. It carries a gyroscope, parachute and instruments—to say nothing of miniature figures of Hergé's characters. The model is successfully launched on July 21, 1989, and reaches an altitude of some 2,000 meters.

Arthur C. Clarke publishes his novel *Prelude to Space* (written in 1947). It describes the design, building and launch of a two-stage manned orbital rocket. It is somewhat similar to Sänger's proposals [see above]. The rocket is launched on a horizontal track, 5 miles in length, in the Australian desert. The first stage is a manned flying wing, powered by an atomic ramjet. The second stage, a wingless atomic rocket, is carried atop the booster piggyback fashion.

The winged first stage, *Beta*, will be accelerated to the operating speed of its methane-fueled atomic ramjet by way of an electric linear motor. It carries the small *Alpha* into earth orbit. *Beta* remains in orbit while the smaller *Alpha* makes the round trip journey to the moon. The two ships—together named *Prometheus*—will then rejoin for the descent to the earth's surface.

In a review of the novel A. V. Cleaver (to whom the book is dedicated) criticizes the use of a nuclear ramjet. He demonstrates that at an altitude of 100 miles and a speed of 10,000 mph the ramjet will require a frontal area several million times greater than a rocket producing the same thrust.

JANUARY 7

K. W. Gatland, A. E. Dixon and A. M. Kunesch present their paper "Initial Objectives in Astronautics" to a meeting of the BIS. They advocate the use of "expendable construction" in which the forward portion of a rocket is broken down into separate tanks which are jettisoned progressively as propellant is consumed. Each of these stages consists of a pair of half-cylindrical tanks. They are held together by explosive ties. The separate stages have streamlined conical noses fitting into cone shapes recessed into the bottoms of the tanks above them.

The paper proposes a manned circumlunar rocket built on this principle. It will be a

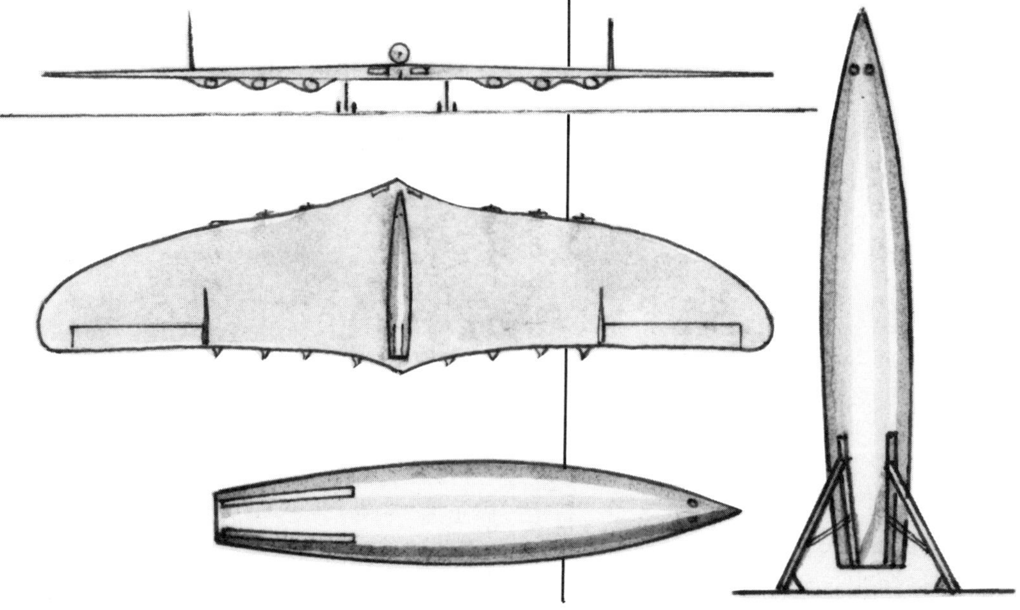

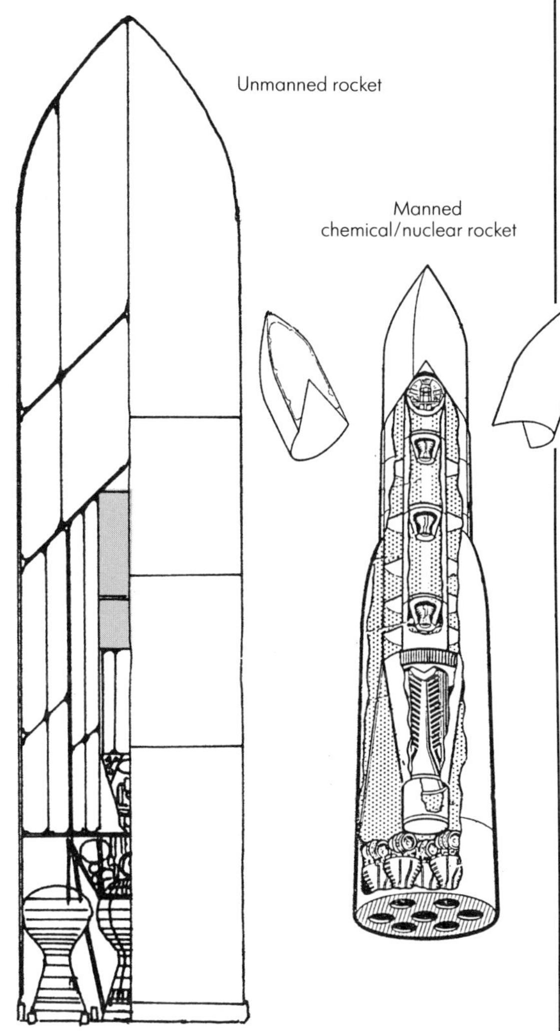

chemical expendable-stage rocket boosting a nuclear rocket that in turn boosts a three-stage chemical moon rocket. Scale models of this spacecraft were exhibited in 1949.

The first stage chemical booster has seven motors of 450 metric tons each, fueled by liquid hydrogen and liquid oxygen. This latter is carried in expendable annular and cylindrical tanks. This booster lifts the main nuclear-propulsion stage above the denser part of the earth's atmosphere. The first tanks to be jettisoned are the two nose tanks and two annular side tanks, followed by the booster itself at 75,000 feet. The atomic rocket is now exposed and it begins firing, using its ammonia propellant, which is housed in similar expendable tanks which are progressively jettisoned. The nuclear rocket has a thrust of 1,100 metric tons. Once its fuel is exhausted, at about 700 miles above the earth, it is dropped and allowed to remain in orbit. The final conventional three-step manned rocket continues on to the moon.

The rocket is launched from a location with an elevation of 13,000 feet.

The overall length of the complete rocket is 190 feet, with a diameter of 32.8 feet. It weighs 1,220 tons. The manned "payload rocket" is 64 feet long, 11.5 feet in diameter, and weighs 59 tons, of which 40.3 tons is fuel. The spherical pressurized crew cabin weighs, fully equipped, 1.4 metric tons.

The increased efficiency of fuels and motors ultimately renders such a complex system unnecessary.

MARCH

Coronet magazine publishes a minor effort by space artist Chesley Bonestell, "Mr. Smith Goes to Venus." In a series of twenty-one paintings, Bonestell tells the story of the Smith family's vacation on Venus in the year 2500. The spaceship *Diana* is an enormous rocket not dissimilar in design to other spaceships Bonestell has created during this period [see 1951]. It is launched from a mountainside ramp at a 45 degree angle. It is divided into three decks: two for cabins for its 600 passengers and a third for recreation. The *Diana* lands on Venus as a conventional aircraft, touching down on a runway, its speed checked by a pair of drogue chutes. The return flight to the earth takes only 117 hours.

NOVEMBER

Arthur C. Clarke suggests using an electric gun, or "mass-driver," to launch material mined on the moon to the earth.

ca. 1950

James A. Van Allen, in response to the number of volunteers offering to ride into space atop a V2, declares that it would be extremely immoral to accept any volunteer's offer to ride a V2 at present.

1950

An editorial in the *Saturday Evening Post* states: "In this new age of rockets, a trip to the Moon is fast becoming not only possible, but desirable."

In a review of the Willy Ley-Chesley Bonestell book, *The Conquest of Space*, astronomer Fred Hoyle writes for the London *Times*: "One has to be an optimist to believe that a trip to the Moon and back will be safely accomplished within the next hundred years. But the odds are that it will be accomplished one day."

The British Interplanetary Society proposes a three-stage manned rocket with a winged return ship. It is 182 feet tall at takeoff, with a launch weight of more than 1,000 tons. The first stage is 70 feet tall, with a single engine; the second stage is 62 feet tall; the winged, dart-like third stage is 50.25 feet. The first two stages are finned and equipped with parachutes to facilitate recovery for eventual reuse. The manned stage uses the "skip-glide" technique to lose orbital speed, using drogue parachutes for the final drop below sonic velocities. It eventually lands like an ordinary aircraft. As a safety precaution, the pilot and copilot are encased in pressurized spherical cabins that can be detached from the main body of the spaceplane, descending to the earth on separate parachutes.

Although there are a number of innovations, the rocket still shows its descent from the V2 through its aerodynamic shape, large fins, and graphite steering vanes in the rocket exhausts.

The British Interplanetary Society designs a dumbbell-shaped atomic rocket for deep space use only. Two large spheres are connected by a tubular midsection. The larger sphere contains the atomic motor and propellants, the smaller sphere the personnel. This design is featured by Arthur C. Clarke in his novel *The Sands of Mars* [see 1952].

Russian scientist M. K. Tikhonravov envisions a 1,000-ton spaceship that will carry two men on a circumlunar flight. If it can depart from a space station, it will only need to weigh 100 tons.

Other Russian scientists of about the same time have designed a moon rocket 60 meters

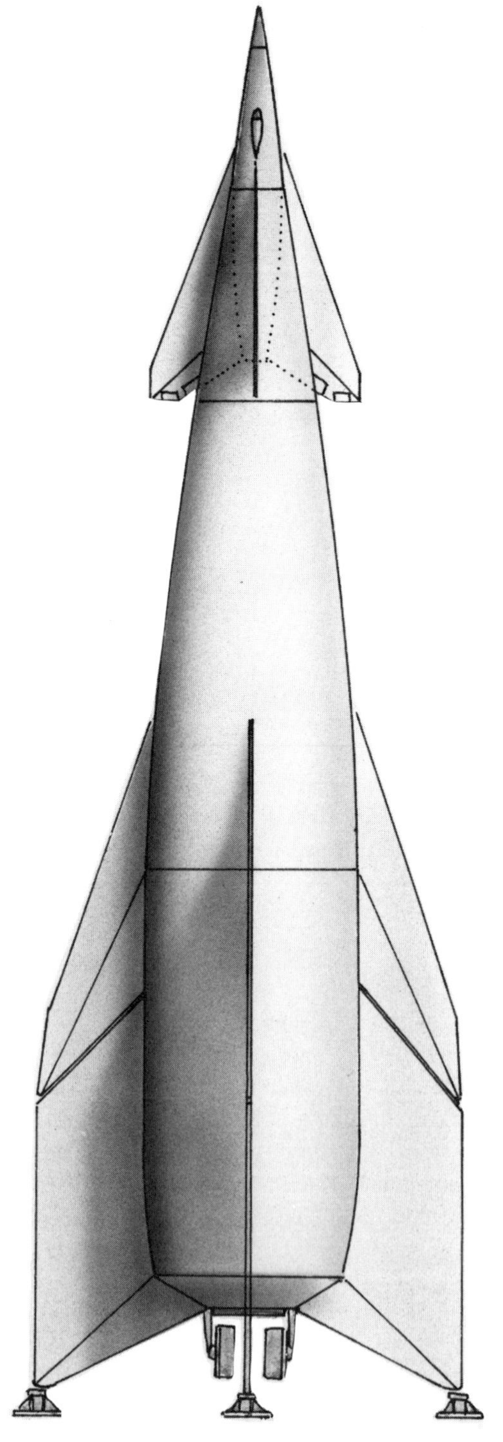

long, 15 meters in diameter and weighing 1,000 tons. It has 20 motors producing a combined 350 million horsepower.

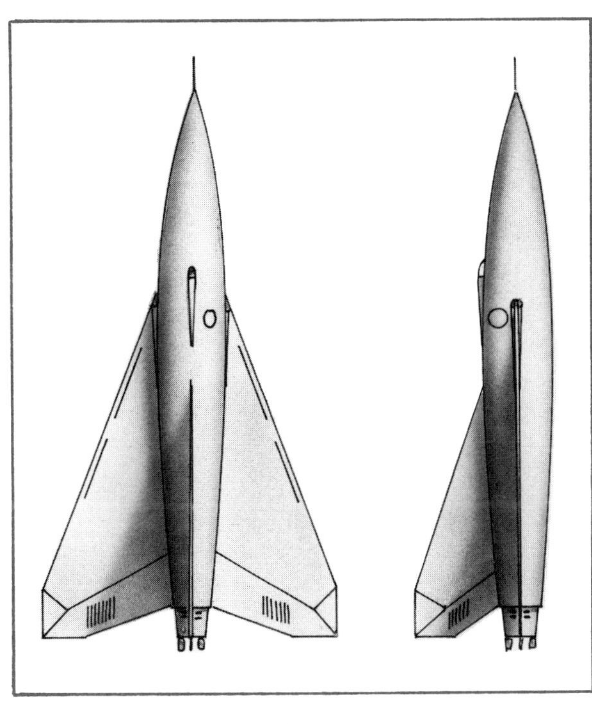

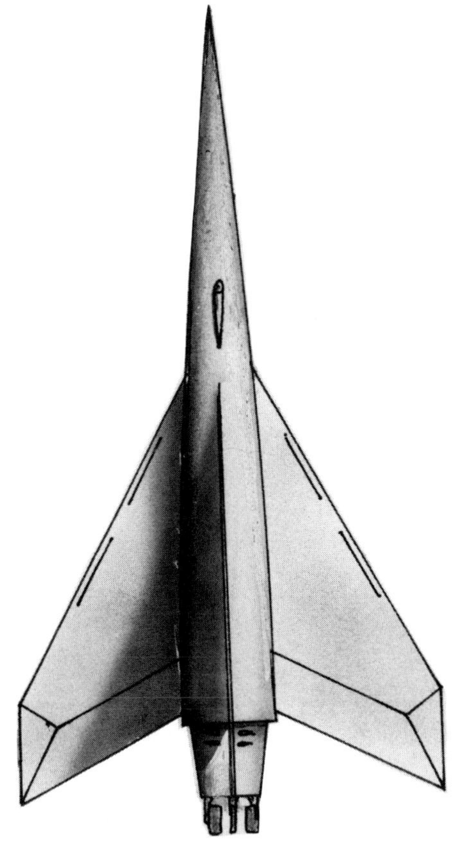

1950–1955

The early television space opera, *Tom Corbett*, debuts on the NBC-TV network. It is based on the Robert A. Heinlein novel *Space Cadet*. The series is set in the year 2352 and details the adventures of Rocketeer Tom Corbett (played by Frankie Thomas), Roger Manning the radar expert and Astro the Venusian navigator. The programs, which

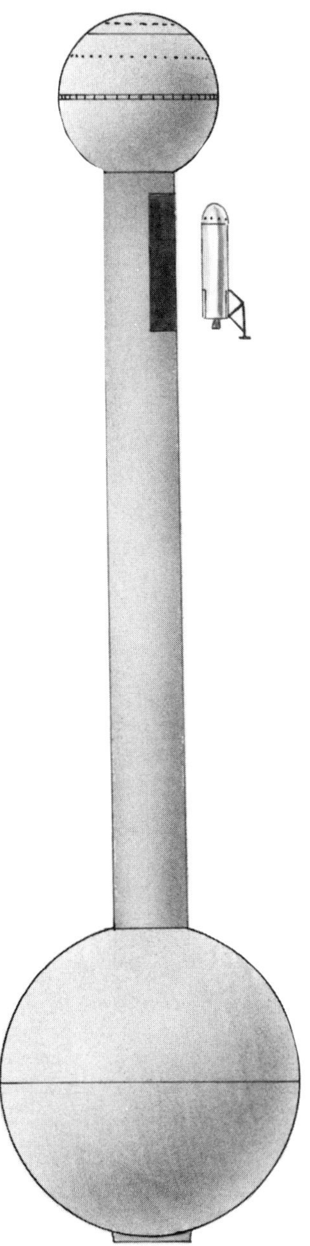

1950–1951

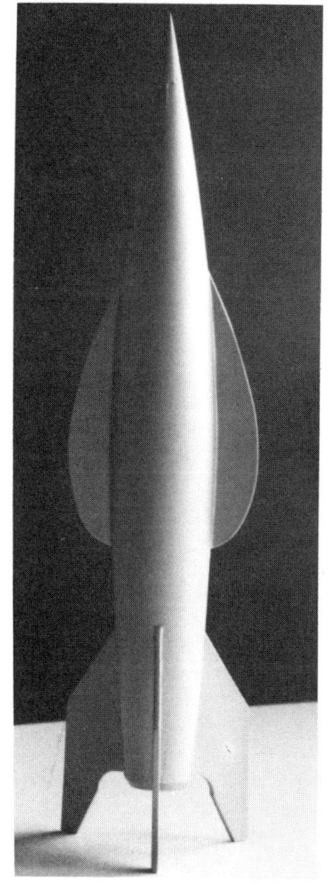
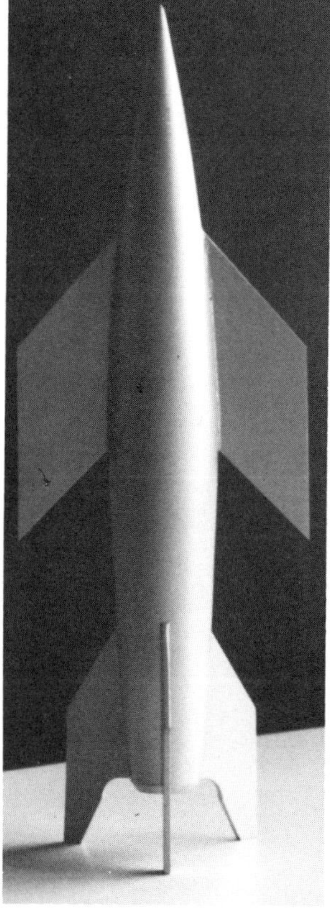

Burns

initially ran only 15 minutes, are lent authenticity by the technical advice of rocket expert Willy Ley. Perhaps not surprisingly, the show's spaceship, the *Polaris*, closely resembles the German V2. Considering that the program is broadcast live, its special effects are outstanding—many of them advancing the current state of the art of television technology.

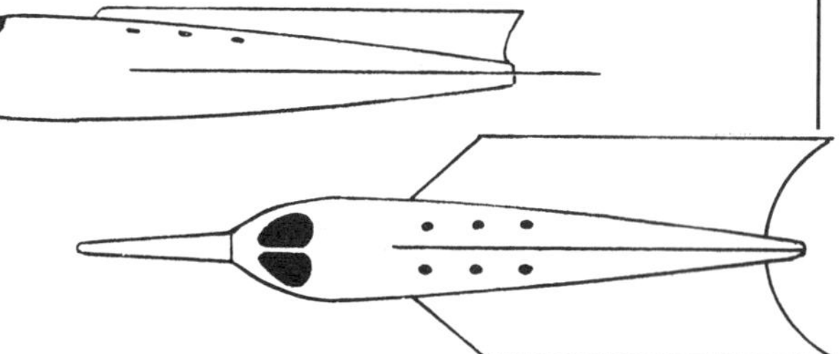

1950–1956

Space Patrol premiers, a television space opera that enjoys tremendous success. During its heyday it is broadcast 3 times a week. The stories are set in the 30th century and involves the adventures of Commander Buzz Corey (Ed Kemmer) of the Space Patrol. He and his companions cruise the known universe in their spaceship *Terra V*. Programs like *Space Patrol, Tom Corbett* and the others of this time, appearing as they do at the very dawn of the space age, serve to excite a whole generation of youngsters who will grow up with the exploration of space.

As an advertising promotion for the show's sponsor, a "full-scale" mockup of a futuristic rocket is taken on a countrywide tour. The 35-foot "Purina rocket" is a Flash-Gordon-like design and is nearly as large as a house trailer. The interior is outfitted as a clubhouse and features bunks, table, benches, telephone, kitchen and its own power supply. It costs the sponsor $35,000. In a 1953 "Name Planet X" contest, the rocket is won by 10-year-old Ricky Walker of Washington, Illinois (for some unexplained reason Ralston Purina refuses to divulge the winning name and swears Ricky to "undying secrecy"). The rocket now belongs to Robert Walker of Ghent, New York.

1951

Wernher von Braun contributes the paper "The Importance of Satellite Vehicles in Interplanetary Flight" to the second International Congress on Astronautics, where it is read by Frederick C. Durant, III. The paper outlines von Braun's idea that "Once the technique of establishing satellite rockets in stable orbits around the Earth is perfected, these can be used to refuel other rockets. Interplanetary flight will then become possible, even using present chemical propellants."

Flying Disc Man From Mars is a Republic serial in twelve episodes, starring Walter Reed. The Martians' rocket-propelled "flying disc" (which resembles more than a little Northrop's MX-324 [see 1944]) is one of the screen's first depictions of a flying saucer.

Captain Video is a Columbia serial in fifteen episodes, starring Judd Holdren. It is a spin-off of the popular television series [see above].

Leslie Greener's novel *Moon Ahead* is published. It is one of the first authentic novels about space intended for children and written in English. The spaceship *Shining Rock* is constructed in the Australian desert by the British Empire-American Moon Society . . . or BEAMS. It is a gigantic, slender cigar made of a beryllium-aluminum alloy, 110 feet long and 12 feet in diameter. It is built in England and assembled in the desert, within one of the ancient meteor craters known as the "Devil's Rings." Great secrecy is required to protect the project from interference from a cartel of financiers and industrialists who want to commercially exploit lunar minerals.

The science in the book is excellent and, which does not necessarily or even usually follow, the story is strong as well. The book is accompanied by superb illustrations by the renowned children's book designer William Pène Du Bois.

Flight to Mars is a motion picture depicting the first manned expedition to Mars, where an underground Martian civilization is discovered. The spaceship *Mars One* is based on a preliminary, rejected concept by Chesley Bonestell for the *Luna* of *Destination Moon* [see 1950]. Like most of Bonestell's spaceships, it is basically a winged V2. In keeping with the second-hand production values of *Flight to Mars* the spaceship cockpit is the one used in 1950's *Rocketship XM* and the spacesuits the Martians need to wear on the surface of their planet [!] are the ones created for *Destination Moon*.

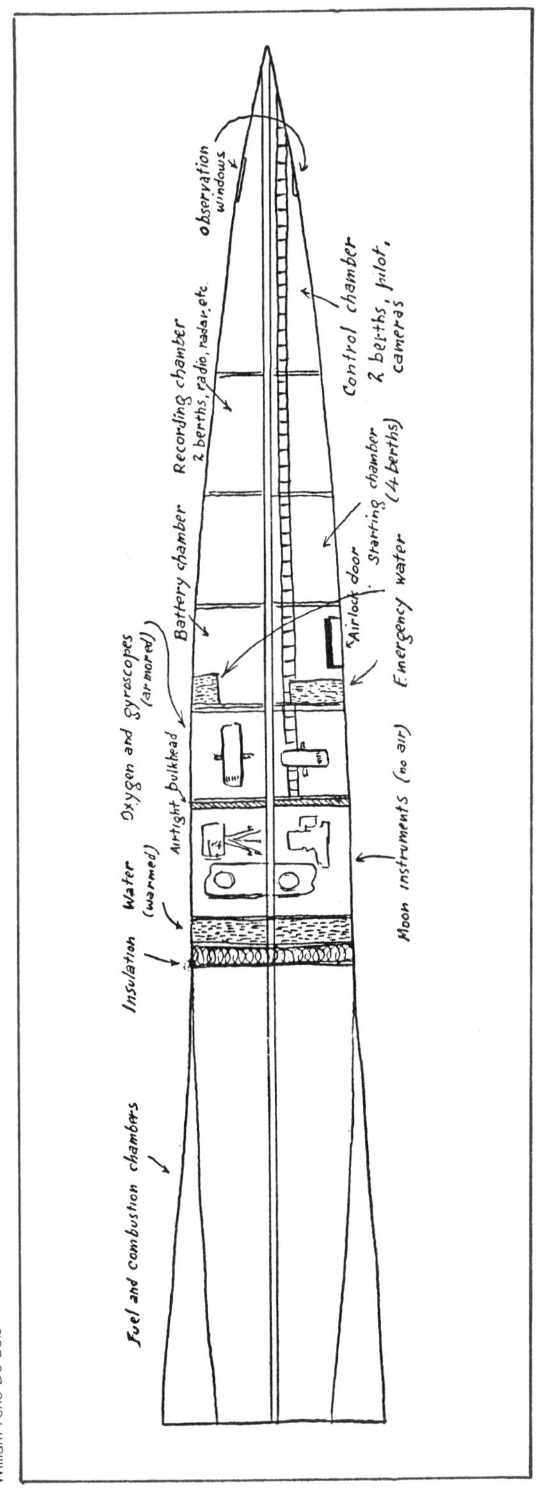

William Pène Du Bois

The *Mars One* also makes appearances in a number of subsequent low-budget science fiction films, such as *It—the Terror From Beyond Space* and *Missile to the Moon*, among others.

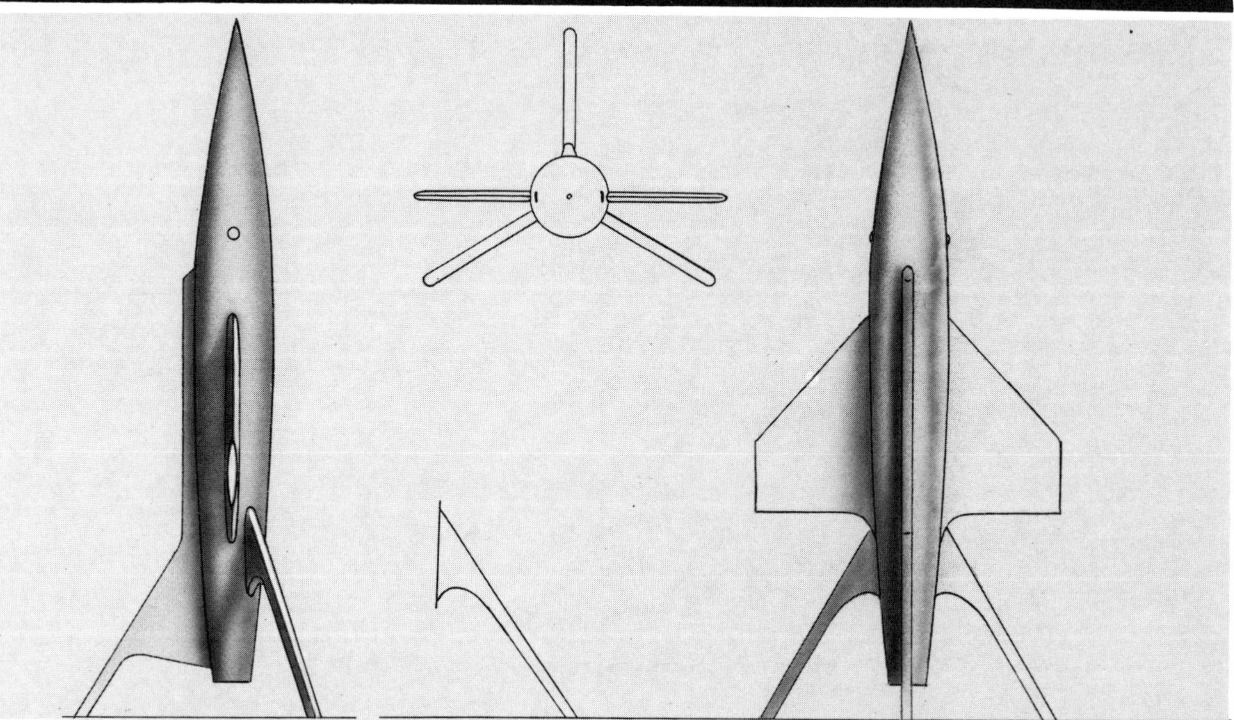

Kenneth Heuer, in his book *Men of Other Planets*, suggests that the first spaceships will be spherical rather than cigar-shaped. First for strength, second because the sphere can hold the maximum amount of volume in respect to its area; this second reason also will require less material for construction, therefore less fuel will be required. It will be atomic-powered, using water as the reaction mass. [Also see 1950.]

At a meeting of the BIS, Dr. S. F. Singer (designer, in 1953, of the MOUSE satellite) suggests that the cost of building an orbital rocket might equal that of fifty modern bomber aircraft.

When Worlds Collide is a motion picture adaptation by George Pal of the Edwin Bulmer-Phillip Wylie novel [see 1934] directed by Rudolph Mate. As in the best-selling novel, the earth is threatened with imminent destruction. A pair of rogue planets drifting through interstellar space are about to pass through the solar system (changed in the film version to a rogue star and its lone planet). Astronomers discover that their trajectory intersects the earth's orbit—unfortunately at the very place that the earth will be occupying it. Plans are immediately made to build a giant "space ark" to save as many

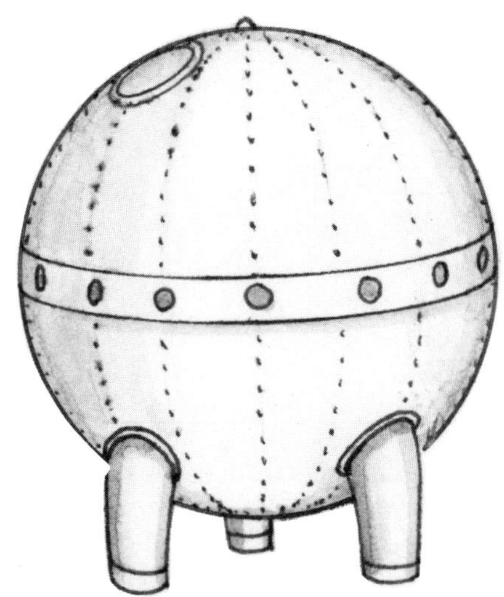

people as possible. The astronomers have discovered that only one of the two interlopers will collide with the earth, the other will go into a permanent orbit around the sun, more or less replacing our planet. This new world is the destination of the *Space Ark*. The film concerns itself primarily with the trials and tribulations of the Ark's construction before the end of the world. The *Space Ark* as depicted in the motion picture is a classic Chesley Bonestell design: sleek, needle-nosed, swept-winged. It is 400 feet long and 75 feet in diameter. Instead of taking off vertically, the *Space Ark* is mounted on a rocket-propelled undercarriage that in turn runs along a twin-railed track. Beginning horizon-

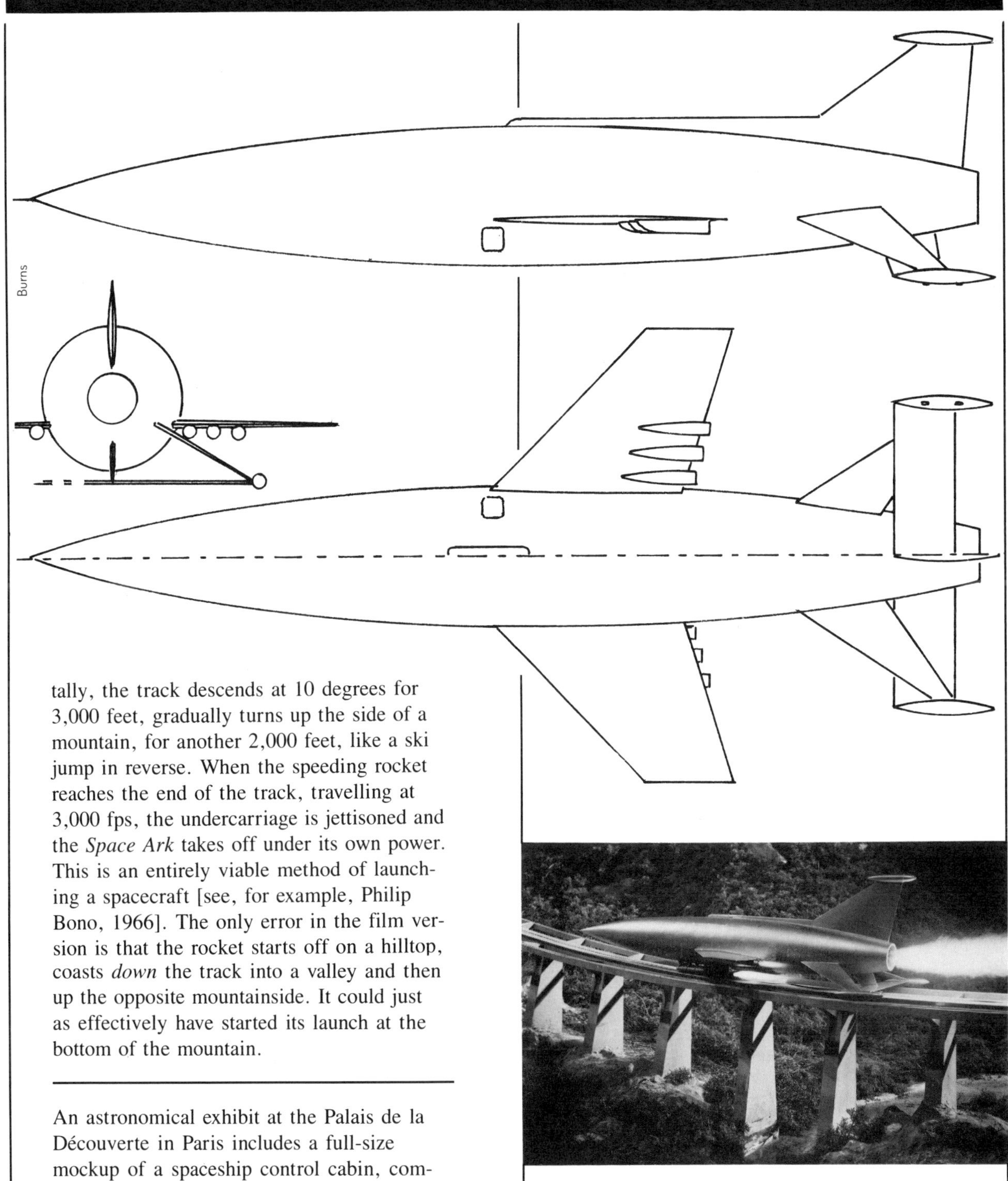

tally, the track descends at 10 degrees for 3,000 feet, gradually turns up the side of a mountain, for another 2,000 feet, like a ski jump in reverse. When the speeding rocket reaches the end of the track, travelling at 3,000 fps, the undercarriage is jettisoned and the *Space Ark* takes off under its own power. This is an entirely viable method of launching a spacecraft [see, for example, Philip Bono, 1966]. The only error in the film version is that the rocket starts off on a hilltop, coasts *down* the track into a valley and then up the opposite mountainside. It could just as effectively have started its launch at the bottom of the mountain.

An astronomical exhibit at the Palais de la Découverte in Paris includes a full-size mockup of a spaceship control cabin, complete with pilot and a crew member emerging from a hatch in the floor.

William Bergen, vice-president and chief engineer at Glenn Martin, states that he expects to see a man-carrying rocket launched within 10 to 15 years.

JANUARY 6

Sir Harry Mason Garner, chief scientist of the British Ministry of Supply, tells youngsters at the Schoolboys' Exhibition in London that he is convinced that many of them will see the first successful flights to the moon, Mars and Venus.

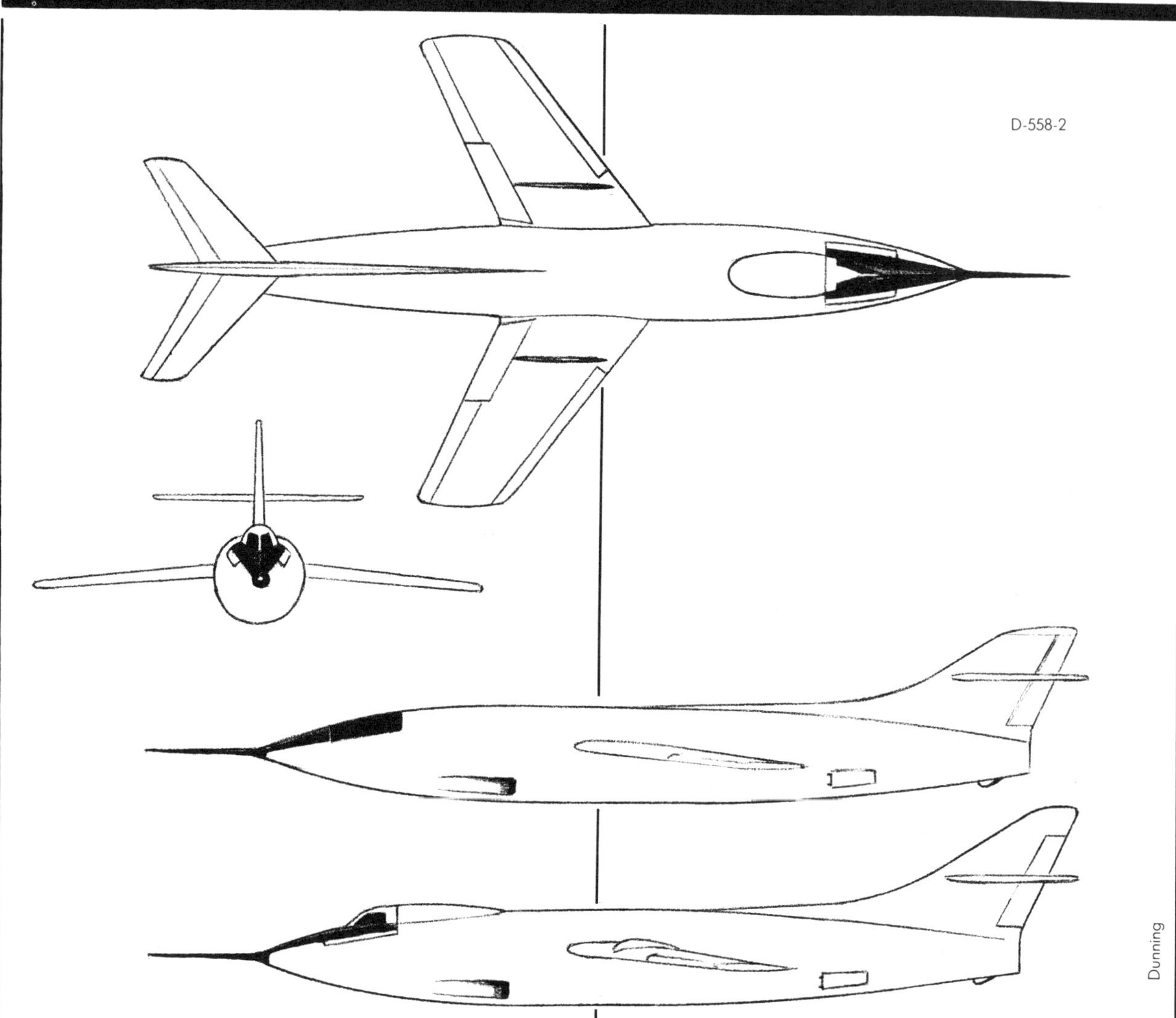

JANUARY 6

The trio of papers collectively titled "Orbital Rockets" is presented by K. W. Gatland, A. E. Dixon and A. M. Kunesch.

JANUARY 26

The first flight occurs of the rocket-powered version of the Douglas D-558-2 #2. Pilot William Bridgeman takes the air-launched rocket from 32,000 feet to 41,000 feet. He reaches a speed of Mach 1.28 on a level run. After seven more test flights, including an unofficial world's altitude record of 79,494 feet, the aircraft is turned over to NACA [see November 20, 1953].

FEBRUARY

Arthur V. St. Germain, a senior test engineer for the Fairchild guided missile division of the U.S. Navy's experimental station at Point Mugu, California, publishes a design for a spaceship. It is five-staged, 325 feet long and weighs 180 tons at takeoff. It carries a crew of two and is intended for a round trip to the moon.

MAY

Russell Saunders's (Carl A. Wiley) article "Clipper Ships of Space" in *Astounding* reintroduces the idea of the solar sail [also see 1889, 1913]. Saunders concludes that his "light-jammer" will be a magnesium hemisphere 80 kilometers in diameter and 0.15 micron thick. The rim of the sail will con-

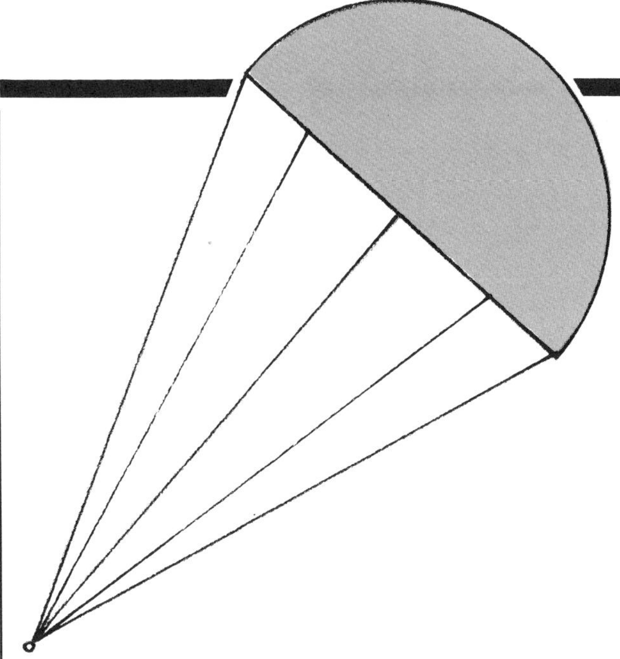

tain a wire to distribute the stress, with rigging wires going back to the ship proper, making the complete light-jammer look something like a parachute.

Saunders considers the problems of construction and navigation, as well. He sends a carbon copy of this article before publication to Willy Ley, who criticizes the idea of solar sailing, concluding that "the utilization of solar light pressure just wouldn't pay."

JULY

The British newspaper *Daily Express* and the Royal Aero Club organize an exhibition at Hendon Aerodrome at which the British Interplanetary Society is invited to provide an astronautical feature. Among other exhibits are a model of a man-carrying rocket [see 1948] as well as one of a moon rocket cutaway loaned by the Brighton and Hove Society of Model Engineers.

JULY–OCTOBER

R. L. Farnsworth suggests that the moons of Mars, Phobos in particular, may be artificial satellites—space stations erected by ancient Martians.

SEPTEMBER 3–8

Among the papers presented at the second International Congress on Astronautics in London is a summary of Wernher von Braun's "Mars Project" [see 1952]. T. Nonweiler discusses the problem of descending from orbit using aerodynamic braking. He concludes that a spacecraft in the form of a delta-shaped flying wing would have the ideal characteristics required for atmospheric braking. Nonweiler's spacecraft will weigh 20 tons and have a volume of 500 m^3—all of the payload to be contained within the wing itself, which would be just over 2 meters at its thickest point. K. W. Gatland, A. M. Kunesch and A. E. Dixon describe their minimum satellite vehicles and R. A. Smith discusses the problem of establishing contact between orbiting vehicles. E. V. Sawyer, of the Pacific Rocket Society, considers the difficulties of landing spacecraft. He favors the use of parabrakes; the most favorable and economical technique involves braking by the use of retrorockets, aerodynamic drag by flaps and parabrakes to a velocity of about Mach 0.5 and a stern-first descent with a parachute cluster at 40–60 fps, a short rocket burst for final retardation and touchdown in wide-tread shock absorbers.

SEPTEMBER 7

R. A. Smith discusses orbital rendezvous techniques in a paper read before the Second International Congress on Astronautics: "Establishing Contact Between Orbiting Vehicles."

The linking up of spacecraft, perhaps in the course of building a space station, will require spacewalking astronauts equipped with manuvering units, such as gas-jet pistols. While rendezvous could be accomplished with computers, actual docking will require human control.

OCTOBER 12

The Hayden Planetarium in New York holds a symposium on space travel [also see October 12, 1952 and 1952–1954]. Columbus Day is chosen as being a particularly significant date.

Speakers at the symposium include Dr. Albert E. Parr, director of the American Museum of Natural History, Robert E. Coles, chairman of the planetarium, Willy Ley, coördinator of the symposium, Robert P.

Haviland, Oscar Schachter, Dr. Fred L. Whipple and Dr. Heinz Haber.

ca. 1951

Rudolf Nebel [see 1933], Albert Pullenberg and Karl Poggensee think that the first manned rocket flights should be accomplished by use of man-carrying single-stage chemical rockets. Then, in 20 years, the construction of multistage, atomic-powered "true" spaceships will become feasible.

Sir Frederick Handley-Page predicts that interplanetary travel will take place within 50 years.

1951–1957

The BOMI (**BO**mber **MI**ssile) is designed by Walter Dornberger and Krafft Ehricke at Bell Aircraft [also see 1949]. It is the immediate predecessor to the later X-20 Dyna-Soar project. BOMI is a two-stage boost-glide vehicle launched to an altitude of 100,000 feet or higher and to speeds of Mach 4. Its range is up to 4,000 miles. Another version has a 6,000 mile range, while another's upper stage has the capability of returning to its launch site. The suborbital versions of BOMI are based partly upon the work of Eugen Sänger [see 1943].

BOMI's first stage is 100 to 150 feet long with a wingspan of 60 feet. The suborbital BOMI has a gross liftoff weight of 600,000 to 800,000 pounds for both stages. The five-engine version of the two-man first stage has a thrust of 770,000 pounds.

The design of the one (or two) man second stage depends a great deal upon the particular mission it is to perform. It ranges from 50 to 60 feet in length with a wingspan of 30 to 40 feet. It has three engines with a combined thrust of 110,000 to 139,000 pounds.

The engines of both stages fire during takeoff.

Payload for the military version is about 4,000 pounds.

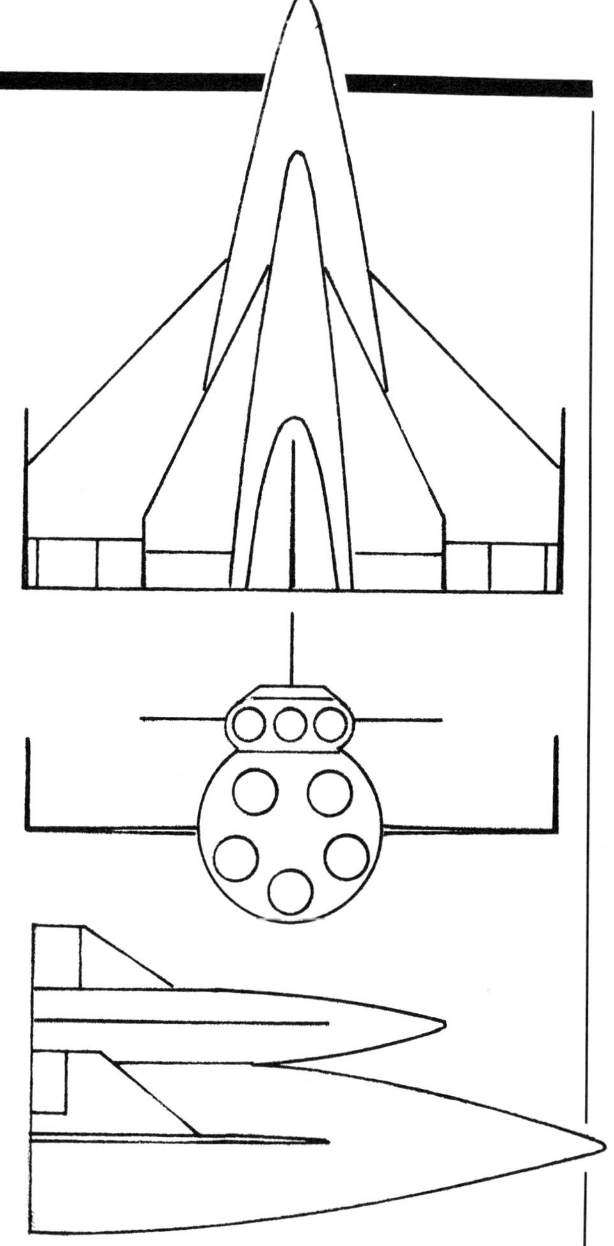

BOMI would be based in England, Spain, Africa and Canada, to allow it access to targets in the western Soviet Union.

The orbital version requires an additional 5,000 feet per second to BOMI's final velocity. To accommodate the additional fuel, the lower stage is increased to 120 to 144 feet in length and the upper to 62 to 75 feet. The second stage now has four or five engines. The launch weight is now 940,000 pounds and liftoff thrust is 1.1 million pounds.

An even further improvement is suggested by Ehricke. The orbital BOMI can be fueled by liquid hydrogen and liquid oxygen, instead of the storable fuels of the earlier versions. This doubles the length of the first stage and

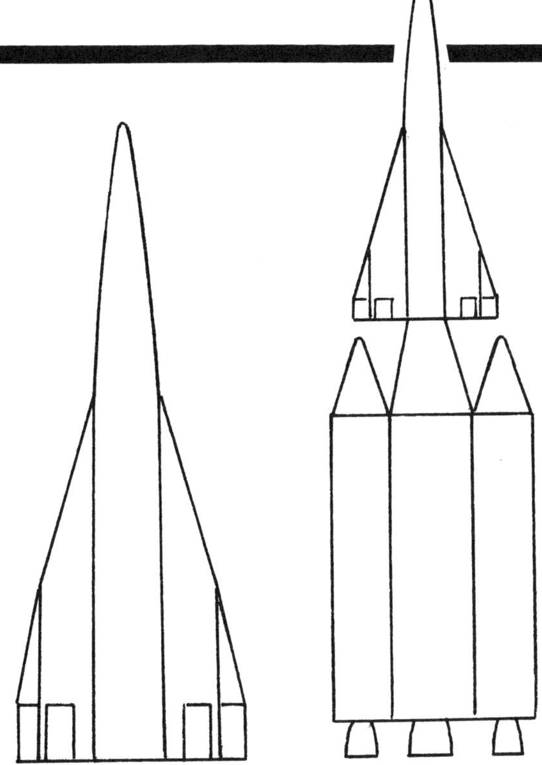

increases the size of the second stage by 50 percent.

It is hoped that, if the project is given immediate approval, that construction and flight testing could begin in 1959–1960, with tests continuing until 1965. The two stages will be, at first, test-flown separately. A reusable orbital transport could be operational by 1975.

The project is presented to the Wright Air Development Center on April 17, 1952.

Partly because of the disappointingly short range (of the suborbital version), the BOMI project is recast for reconnaisance missions, which in 1954 results in a three-stage vehicle (designated MX-2276 by the Air Force) with a liftoff weight of about 400 tons, designed to reach altitudes of 259,000 feet and a speed of 4,000 mph. Its range would be 10,600 miles.

Bell follows up the original BOMI proposal with variants such as the two-stage Mach 15 reconaissance vehicle called System 118P, Brass Bell, Robo (Rocket Bomber) and Hywards.

Eventually, in 1957, the Air Force consolidates all of these into a single research program called Dyna-Soar.

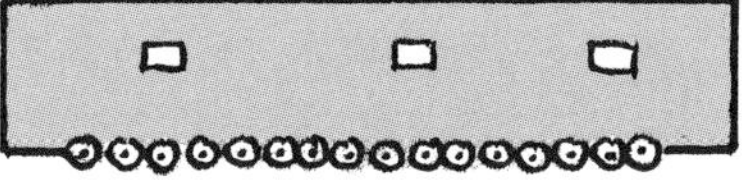

Actual size

1952

James Blish publishes his novelette *Surface Tension* which describes one of science fiction's most unusual "spaceships."

In a distant future, human beings are genetically engineered to be able to survive on alien worlds. A nearly microscopic form of human life is developed to survive within the tidal pools of a planet covered almost entirely by water. The tiny humans have been able to "domesticate" various single-celled organisms, such as paramecium. The upper surface of the tidal pool, with its tough membrane, the water's natural surface tension, is as much an irresistible lure for the story's hero as outer space is to an inhabitant of the earth. He eventually is able to devise a kind of "spacecraft" that will allow him to break through the surface tension and penetrate the outer world. The vehicle is a tiny, water filled, two-inch wooden cylinder. Diatoms along the under hull propel the ship, "their jelly treads turning against broad endless belts of crude leather." Wooden gears step up the power, transmitting it to the ship's sixteen axles.

Constantin Paul van Lent, in his book *Rockets, Jets and the Atom*, describes several proposed manned spacecraft. Lent is an engineer trained in Germany. While there he was a member of the VfR [see 1927]. After coming to the United States he became an early member, in 1930, of the American Interplanetary Society. When the society decided to produce a technical book on rocketry, Lent undertook its compilation. He ultimately was forced to publish the book himself, at his own expense. This book, *Rocket Research*, was, as Lent claims, if not the first at least one of the earliest technical books on rocketry published in the United States. This effort becomes the foundation of Lent's Pen-Ink Publishing Company through which he will issue many other books about space travel and rockets, as well as the uniquely entertaining magazine *Rocket-Jet Flying*.

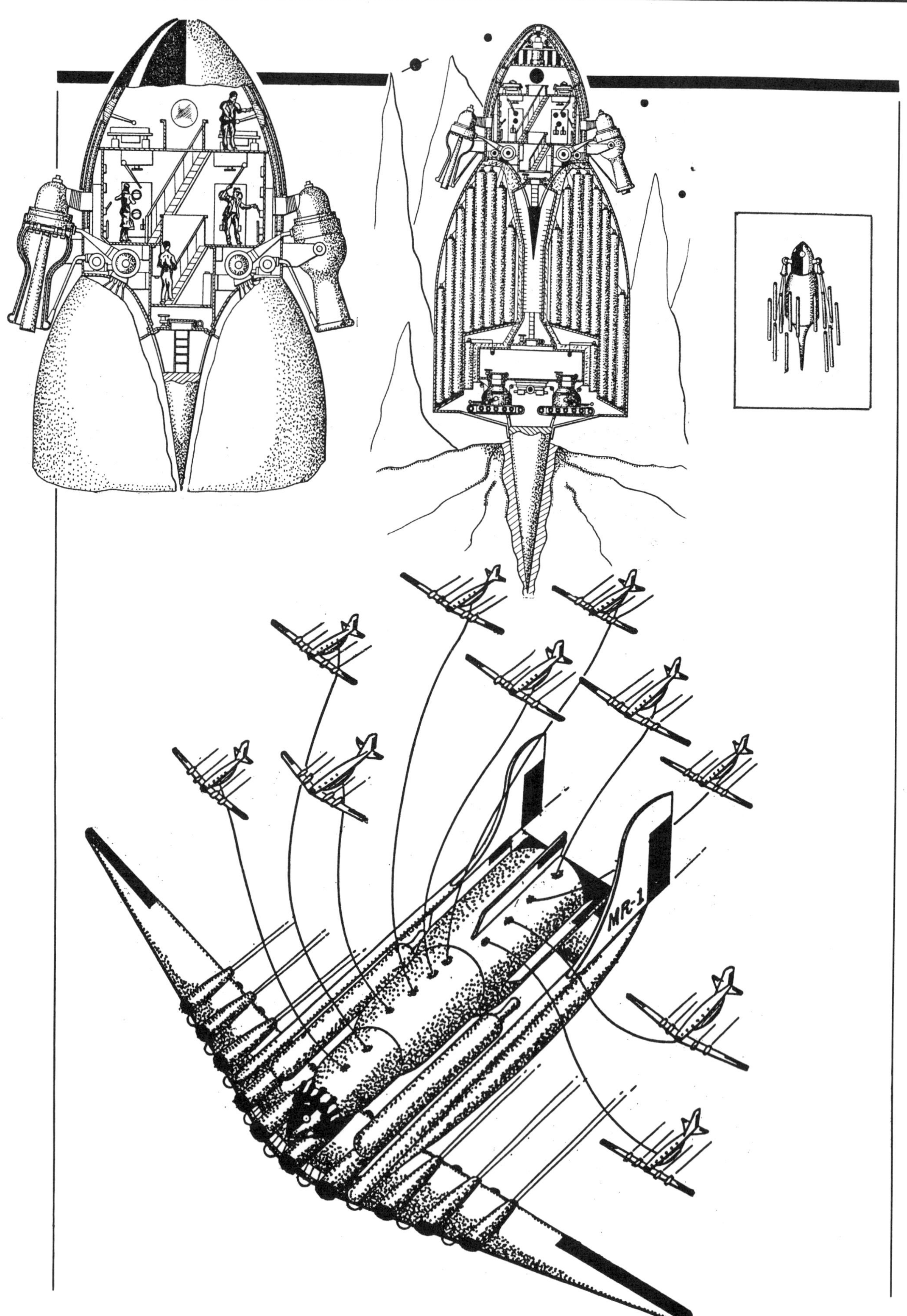
MR-1

In order of their appearance are: a three-stage rocket with the top, manned stage employing the "landing pick" Lent favors for lunar landings [also see 1934]; two four-stage rockets, one in which each stage is nested within another, and a second design which is to be composed of clusters of cylindrical fuel tanks; the final proposal requires a rocket nearly as tall as the Empire State Building and weighing 10,000 tons. It is to be used for a manned "loonar" [sic] mission. The giant four-stage rocket will be lifted bodily from the surface of the earth by a huge jet-powered flying wing carrier. At an altitude of 50 miles, the main ship will be fueled by a fleet of smaller, winged tanker rockets. The third stage will land on the moon using the author's ubiquitous "landing pick." The lower perimeter of the fourth stage will be surrounded by a cluster of rocket motors for the takeoff from the moon. Each stage of the moonship appears to have been composed of clusters of large solid-fuel rockets that will be jettisoned when spent.

The large, bullet-shaped manned stage, equipped with its own landing pick for the eventual return to the earth, is divided into three sections: a lower airlock, a control room and an observation-bunk room. To his credit, Lent indicates in one of his drawings that one of the astronauts will be a woman.

Krafft Ehricke, at the annual convention of the American Rocket Society, submits a proposal for manned earth satellites.

Radar Men From the Moon is a Republic serial in twelve episodes starring George Wallace (and a very young Leonard Nimoy). The spacecraft featured are handsome designs brought to life very realistically by the special effects of Theodore and Howard Lydecker. Also featured is the rocket-propelled hero, with his bullet-shaped helmet and rocket backpack (with controls labeled "Up, Down, Fast and Slow"). [Also see *King of the Rocket Men*, 1949.]

Burns (2)

Zombies of the Stratosphere is a Republic serial in twelve episodes starring Judd Holdren. It is a sequel to *Radar Men From the Moon* and *King of the Rocket Men*.

Kenneth Gatland publishes the first edition of *Development of the Guided Missile*. In it he describes the "close orbit earth satellite rocket" designed jointly with Kunesch and Dixon. Gatland also suggests that an atomic-powered lunar or planetary mission be an entirely orbital affair. A large spacecraft will be assembled in orbit 500 miles above the earth. A ground-launched nuclear rocket will be adapted to its new role as a booster unit for the new, larger spaceship. The passenger sphere is separated from the nuclear unit by three cylindrical sections, placed end to end, made from empty fuel tanks. The passenger sphere is attached to a conventional chemical rocket, together forming the landing craft. Upon arriving at its destination, the larger spacecraft will remain in orbit while the smaller landing craft makes the round trip to the surface.

1952

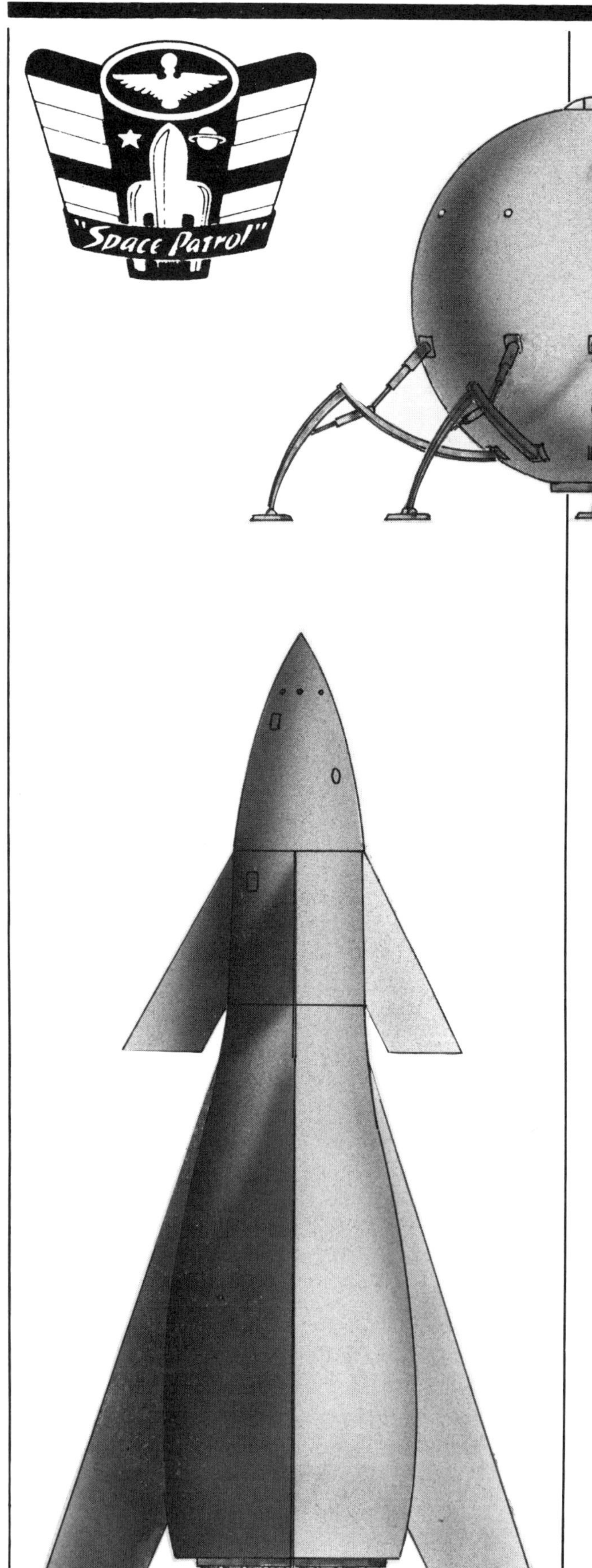

Fletcher Pratt and Jack Coggins publish the classic children's nonfiction book, *By Spaceship to the Moon*. (Pratt is one of the original founders of the American Rocket Society.)

In the book, a sequel to the successful *Rockets, Jets, Guided Missiles and Spaceships* of the previous year, Pratt outlines a moon landing scenario employing a spacecraft constructed and launched from earth orbit. The first step is the construction of a manned space station. The design elaborated upon is a prism, not unlike the later American Skylab space station in appearance. The space station, in addition to its duties as a scientific research facility, acts as a base for the assembly of the moonship. This is a small, spherical rocket equipped with ten folding legs. The large number of legs will allow the ship to adjust to any irregular terrain, though when deployed they give the ship a distinctly spider-like appearance.

The German spaceflight journal *Weltraumfahrt* publishes Wernher von Braun's study *Das Marsprojekt* as a special issue [also see 1953]. Written during his sojourn at Fort Bliss, Texas, and at White Sands, New Mexico, and based on calculations performed in 1948, it is the first expression of one of von Braun's most "grandiose" space scenarios, that eventually sees its ultimate fruition in the *Collier's* symposia [see 1952–1954]. "I believe," he writes, "it is time to explode once and for all the theory of the solitary space rocket and its little band of bold inter-

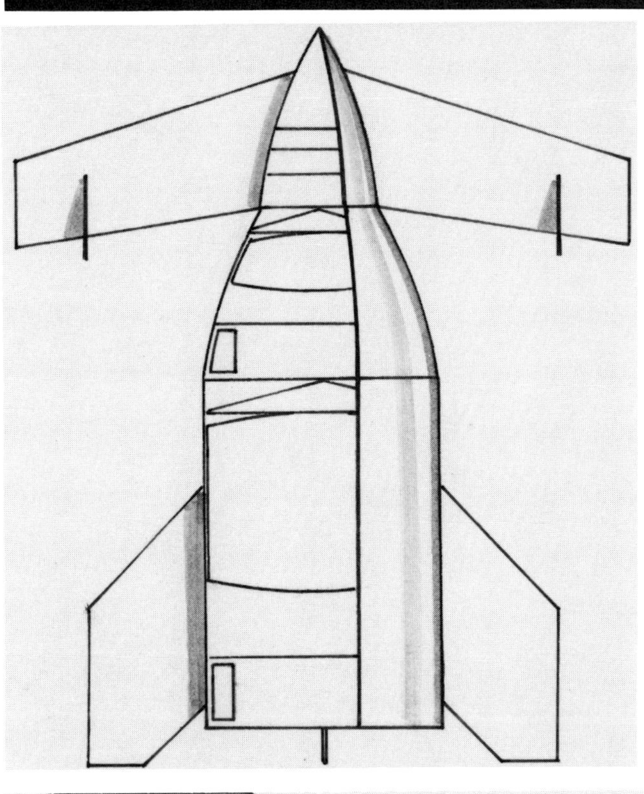

sembled in a 2-hour earth orbit from materials and supplies carried by three-stage ferry rockets. After arrival at Mars, where the fleet goes into orbit around the planet, three special winged "landing boats" will make the final descent. Two of these, after shedding their wings, will return to orbit where their crews will be transferred to the remaining seven spaceships for the return trip to earth. Thus, three spacecraft are developed: the three-stage ferry rocket, the earth-orbit to Mars-orbit spacecraft, and the winged "landing boats."

The ferry rockets are 60 meters high, with a base diameter of 20 meters, all three stages fueled by hydrazine and nitric acid. The first stage is 29 meters long, the second 14 meters long and tapering in width from 20 to 9.8 meters, and the third stage is 15 meters long with a wing span of 52 meters. It

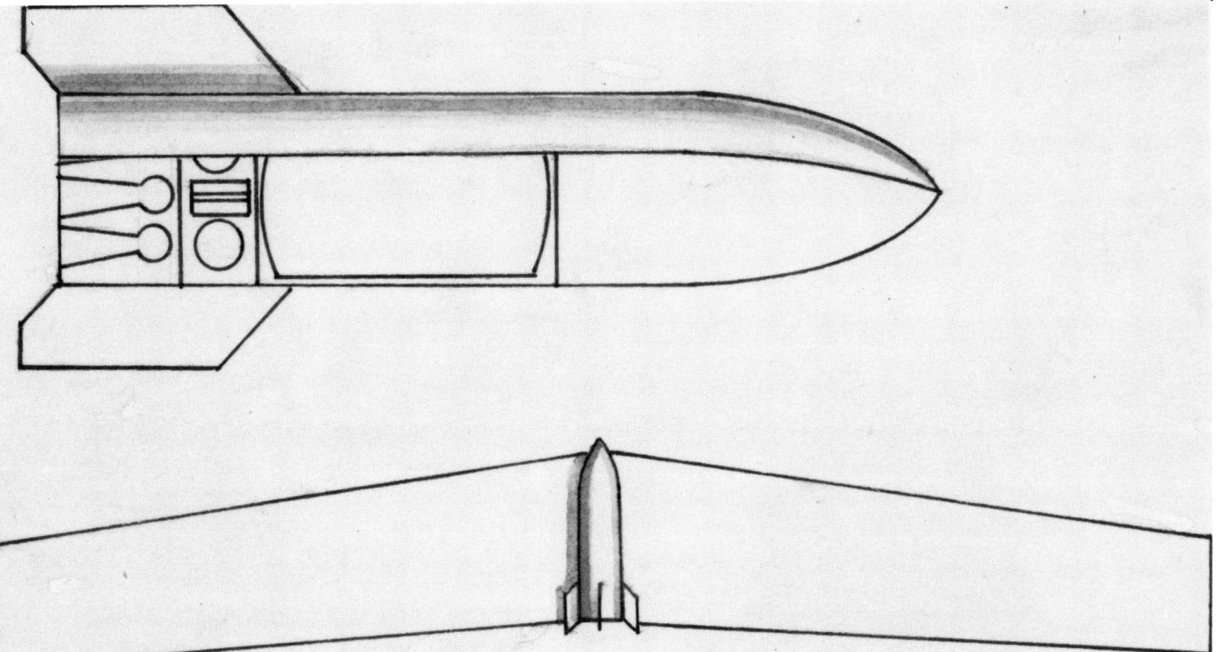

planetary adventurers . . . No one with even the most primitive knowledge of the subject can possibly believe that any dozen or so men could build and operate a functional space ship, or, for that matter, survive interstellar isolation for the required period and return to the their home planet."

Von Braun's plan is to have a flotilla of ten spacecraft manned by no fewer than seventy astronauts. Each of the ten ships are to be as-

resembles a "stubby artillery shell" with enormous glider-like wings. The third stage can carry a payload of 25 tons into earth orbit. The first-stage engines provide 12,800 tons of thrust. The fully fueled rocket weighs 6,400 tons, with 4,800 tons of fuel.

The interplanetary ships are of two types: cargo and passenger rockets. There are to be three of the former and seven of the latter. The passenger rockets serve almost exclusively the purpose of human transport. The cargo rockets carry the landing craft, two of which will make the round trip to the surface of Mars and back to orbit. All seven craft are almost identical in appearance with the main exception that the cargo rockets do not need to carry fuel for the return trip to earth.

The passenger rockets are 41 meters long with a maximum diameter of 29 meters while the cargo rockets are 64 meters long (the additional length due to the landing craft they carry). All are made of a thin-walled nylon-reinforced plastic. The passenger rockets carry 3,662.5 tons of fuel and the cargo rockets carry 3,306 tons.

The fuel is carried in eighteen tanks: four 10.1 meter spheres (for earth departure), four cylinders 3.68 × 10.1 meters (for adaption to circum-Martian orbit), four cylinders 2.4 × 10.1 meters (for departure from Mars orbit), two cylinders 2.66 × 10.1 meters (for adaption to earth orbit) and four reserve tanks 1.4 × 10.1 meters.

The "landing boats" consist of torpedo-like hulls 22 meters long with enormous glider-like wings spanning 153 meters to allow the craft to fly in the thin air of Mars (which von Braun expects to be 1/12 the density of the earth's). One of these 185-ton gliders will land on skis on the snow and ice at a Martian pole. Abandoning their ship, the crew will make the trek to the equator where they will construct a runway for the remaining wheeled gliders to land upon. This crew will join the other two in their ships for the eventual return to orbit.

Once the wings are removed and the rockets put into an upright position, the two landing craft will be ready to leave Mars. At this point these little rockets will weigh only 26.8 tons, with 5 tons of payload.

Arthur C. Clarke publishes his novels *The Sands of Mars* [see 1950] and *Islands in the Sky*. In the latter he describes the space shuttle *Sirius*. It is a passenger-cargo ferry from earth to an orbiting space station. Its launch site is Port Goddard on a manmade plateau high in the New Guinea mountains. It is a

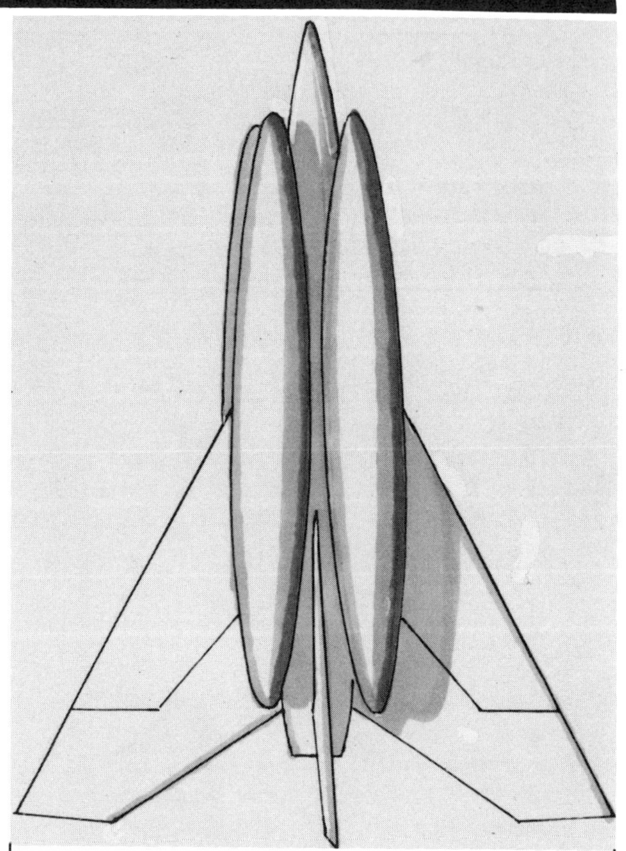

torpedo-shaped rocket that sits on four triangular wings for launch. Between the wings are four external fuel tanks, streamlined, nearly as long as the main rocket, and partially supported by the wings. These provide fuel for the launch and are jettisoned when empty.

Pat and Mike, two rhesus monkeys, are flown to 200,000 feet in an Aerobee III sounding rocket and survive. They are the first primates to enter the upper atmosphere.

Arthur C. Clarke describes the spacecraft needed for the orbital lunar landing technique. In addition to expendable freighter rockets ("Type C") and manned, winged ferry rockets ("Type D") is an enormous nuclear booster carrying a lunar lander ("Type B"). This orbital space vehicle ("Type A") is composed of five separate units: the nuclear booster stage, which is ground launched into orbit; three cylindrical tank sections recycled from the freighters, one of which acts as a habitation module and is connected directly with the last unit, the landing rocket. The latter is comprised of a sphere, the upper portion of which contains a pres-

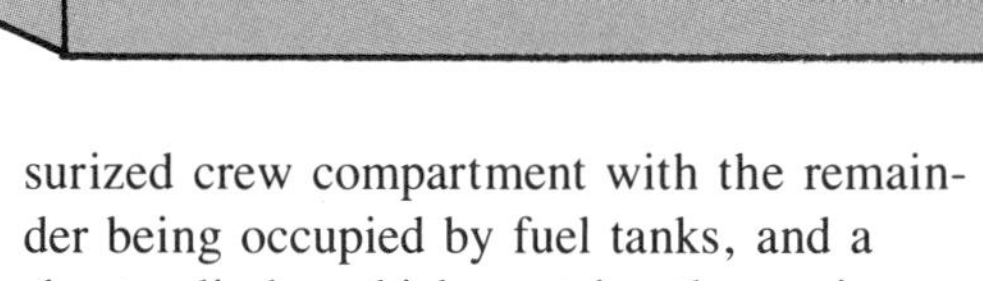

surized crew compartment with the remainder being occupied by fuel tanks, and a short cylinder which contains the engines and which also houses the folding landing legs.

The four types of rocket involved are: Type A, the ground launched nuclear booster; Type B, the landing rocket which is assembled in orbit; Type C, the unmanned cargo rockets; and Type D, the manned, winged ferry rockets. Types C and D have takeoff masses of between 200 and 300 tons.

A lucid, accurate and succinct explanation of such concepts as escape velocity, exhaust velocity and mass ratio (especially as applied to a rocket making a round trip from the earth to another planet) is published by artist Al Williamson and an uncredited writer (probably Al Feldstein). The surprising thing is that this appeared in the EC comic book *Weird Fantasy*.

Fantastic Adventures, June 1952.

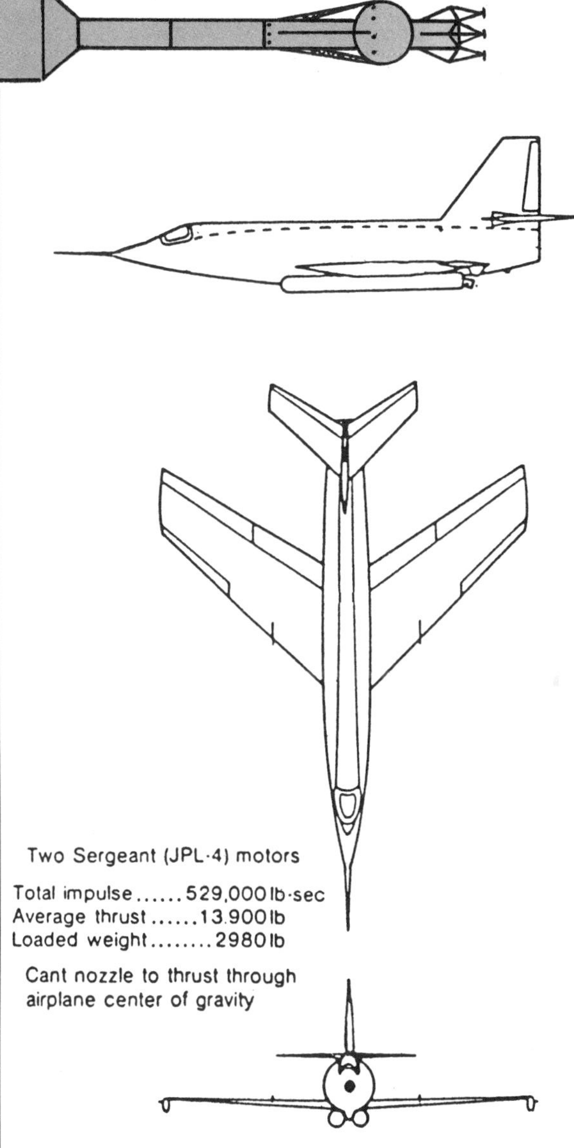

David G. Stone, head of NACA's pilotless aircraft research division, proposes that a large supersonic airplane could launch at Mach 3 a smaller manned second stage that would then accelerate to hypersonic speeds. He suggests that the second stage be a modified Bell X-2 equipped with reaction controls and two droppable solid fuel boosters. The boosted X-2 would be able to attain Mach 4.5 and orbital altitudes.

Stone's plan is evaluated by a NACA study group that ultimately rejects it on the grounds that the X-2 is too small to use as a hypersonic research aircraft. An entirely new vehicle is required [see X-15, below].

Virgil Finlay, 1953.

Dr. Hans Seebohm declares at the Third International Congress on Astronautics that the science of astronautics is an international matter and that its problems will not be solved by individuals or individual nations. Another German scientist agrees, adding that " . . . if all the money spent for armament in this world could be used for the research and development of space travel, the first trip might be made in about ten years."

JAN. 1–JUNE 26

The "Tom Corbett" radio program premieres. It is based on the popular television series [see 1950–1955, above].

JANUARY 30

Robert J. Woods of Bell Aircraft proposes to NACA's Committee on Aerodynamics that a small study group be established to look into the problems of spaceflight. On June 24 NACA adopts a resolution that (1) it increase its research on the problems of manned and unmanned flight at altitudes of between 12 and 50 miles and at speeds between Mach 4 and 10, and (2) that it devote a modest effort to determining the problems inherent to manned and unmanned flight from 50 miles to infinity and at speeds from Mach 10 to escape velocity. The NACA Executive Committee approves a similar resolution on July 14. The following month Langley Aeronautical Laboratory is authorized to set up a preliminary study group.

MARCH 3

Wernher von Braun presents a paper at a symposium held at the Professional Colleges of the University of Illinois in Chicago, in which he discusses an early version of the space station design that eventually sees its final version in the *Collier's* magazine series [see 1952–1954]. In "Multi-Stage Rockets and Artificial Satellites" von Braun describes his space station as " . . . an oddly shaped configuration floating in the pitch black of empty space. This is my concept of a space station assembled in satellite orbit." It is an inflated plastic torus 200 feet in diameter.

APRIL 8

N. R. Nicoll reads a paper to the British Interplanetary Society in Birmingham on the design of the cabin and life support system of a manned spacecraft. His design, based upon a payload limit of 3 tons, will support a crew of three for 15 days.

Nicoll's "life compartment" is not dissimilar to that of the BIS 1939 moonship [see above]. The overall shape is that of an 18-foot parabolic dome with saucer-shaped base. Three 3-inch tubular reinforcing struts of a light alloy run from apex to base. A "catwalk" of corrugated metal circles the inside wall, parallel to the central axis, and another catwalk at right angles runs around the lower wall of the dome. There are airlocks for ingress and egress as well as small bulls-eye ports made of thick plastic. These are not intended as windows, but are rather lenses designed to diffuse natural sunlight for interior illumination.

The walls of the dome are made of thin aluminum sheets (no more than 1 mm thick) separated 2 cm by small spacers. Both sides of the exterior skin and the outer side of interior one are polished. This reduces heat transfer from the sunlit exterior to 1 percent. The interior wall will be painted with a pale-green self-sealing compound.

Carbon dioxide is removed by bubbling the air through a solution of lithium oxide. Replacement oxygen will be carried in liquid form, dispensed by means similar to that used in Stevens and Anderson's stratosphere balloon flight. A small refrigeration plant will remove excess water vapor and will also remove excess heat from the compartment.

As do many others at the time, Nicoll fears the effects of prolonged weightlessness. He provides for artificial gravity by having his spacecraft rotate at 3.3 rpm, which will duplicate earth's gravity. Small, tangential hydrogen peroxide rocket motors will start the ship's rotation. The catwalk parallel to the ship's long axis then becomes the floor. The chairs (of alloy tubing covered with a foam rubber cushion on a phosphor-bronze wire mesh) need to be able to move through at least 90 degrees on curved rails.

Instrumentation is simple, at least according to Nicoll's expectations, who "assume[d] that the control panel will contain a few dials." In addition there might be a large cathode ray tube to display views from television cameras, an intercom, microphone, speaker, "and a few switches."

To provide a nonrotating view of the outside from the rotating cabin, Nicoll suggests the same coelostat used in the original BIS moonship [see 1939]. The nose of the ship is used for storage, as is the central portion of the base.

There are two airlocks, just below the saucer-shaped base, accessible through a small, circular hatch. Each lock is 3 feet × 4 feet × 7 feet, sufficient to accommodate all three astronauts if necessary. Toilet facilities will be chemical for such a short journey and can be located in one of the airlocks. One pound of food per man per day is allocated.

The total weight of the life compartment, including structure, food and oxygen, equipment and other items, comes to only 2,038 pounds, leaving nearly 2 tons out of the 3 tons originally allocated for scientific apparatus, crew, power supplies, and so forth.

APRIL 16

The North-Western Branch of the British Interplanetary Society organizes an exhibition in Manchester at the Adult Education Institute.

Among other models and displays is one illustrating the takeoff and landing of a moon rocket. A 1/300 scale spaceship is shown realistically taking off on a carefully illuminated and decorated stage, and later shown landing on the lunar surface.

MAY 19

James H. Wyld, in *Aviation Week*, favors a "Model T space ship" over the more grandiose proposals of Wernher von Braun. This will be a three-step, 12.7-ton rocket capable of carrying two men into orbit for about 2 days. It will only be slightly larger than a V2 and use gasoline and liquid oxygen. The rocket will weigh only 1/2 ton after jettisoning its empty tanks.

JUNE

A letter from Joseph Stamp in *Fantastic Mysteries* invites its readers to send him letters giving him their guesses as to what year in which spaceflight will become a reality. Stamp's guess is not before 2250 (which the editor thinks is far too pessimistic).

JUNE 24

The X-15 program is initiated when a NACA meeting resolves that it should develop an aircraft capable of exploring flight characteristics at high speeds and high altitudes [see 1954].

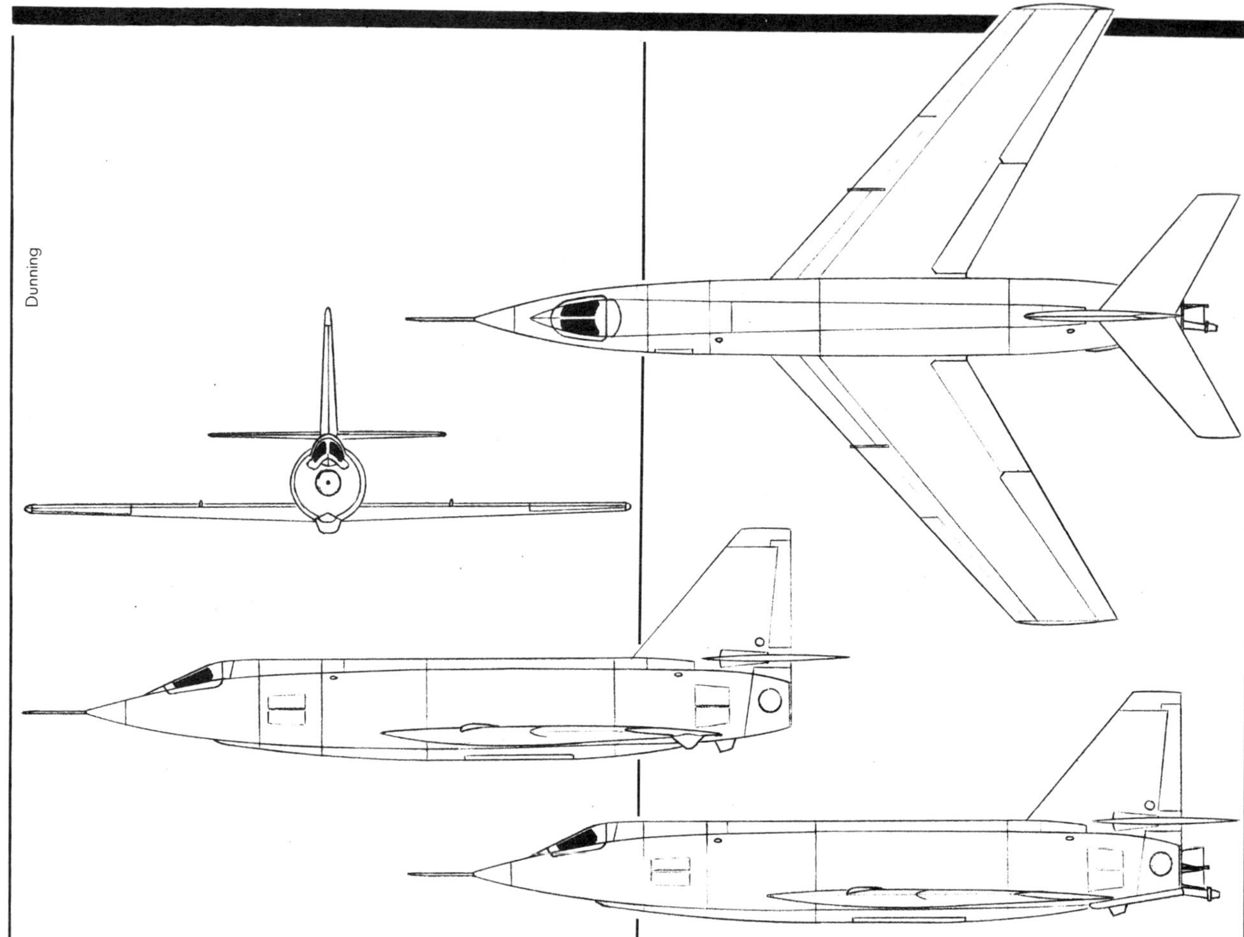

JUNE 27

The Bell X-2 rocket-plane makes its first flight, piloted by Jean Zeigler. It is an unpowered gliding flight [see 1955].

JULY

L. R. Shepherd of the BIS proposes a "Noah's Ark" starship: a million-ton atomic-powered space colony, shaped like an oblate spheroid.

AUTUMN

A. V. Cleaver states in the *News-Chronicle* "World of Tomorrow" series that the first flights to Mars, Venus and the moon will be made by A.D. 2002.

OCTOBER 13

Columbus Day is again chosen as the appropriate occasion to hold a symposium on space travel at the Hayden Planetarium in New York, the second such meeting to be organized. Featured speakers are: Robert R. Coles, chairman of the planetarium, Willy Ley, Dr. Fred Whipple, George O. Smith, Dr. Fritz Haber, Milton Rosen, and Dr. Wernher von Braun.

This symposium is the immediate predecessor to the *Collier's* magazine series of symposia [see 1952–1954].

NOVEMBER 6–9

Wernher von Braun attends the Symposia on the Physics and Medicine of the Upper Atmosphere, held at the Air University School of Aviation Medicine at Randolph Field, San Antonio, Texas. Von Braun's contribution is a paper entitled "The Return of a Winged Rocket Vehicle from a Satellite Orbit of the Earth." While at the symposium, von Braun and Fred Whipple meet Cornelius Ryan, who is representing *Collier's* magazine. This meeting ultimately generates the *Collier's* symposia and the ensuing magazine series and books [see 1952–1954.]

DECEMBER

The house organ of the Bristol Aeroplane Co. publishes a mock A.D. 2051 number. It advertises such products of the firm as the Astrason I, the Astrofreighter, 1075 Astroliner, and suggests that the "British Interplanetary Spaceways Corporation" might be hoping to reduce its last-year deficit of £1.8 billion by at least £8 billion, even though "appalled" at the prospect of having to operate the Astrazon, with its mere 1,102 passengers, nonstop to Mars.

DECEMBER 31

Two spaceship cabin mockups are exhibited at the annual Schoolboys' Own Exhibition in London. The rival boys' magazines, *Eagle* and *Lion*, display competing models. The former, a representation of comic hero Dan Dare's rocket, takes its passengers on a simulated trip to the moon via a rear-projected motion picture. The latter's version is larger, though it offers nothing more than some unconvincing sound effects. Nevertheless, both exhibits attract long lines of prospective astronauts.

ca. 1952

It is reported that no less than 112 American firms are producing toys with an "interplanetary" theme.

A series of half-hour television programs appears under the sponsorship of the Johns Hopkins Science Review, "How Man Will Conquer Space Soon." The series is broadcast by CBS. They use models, diagrams, charts and films to present the various concepts of spaceflight to the lay viewer. Wernher von Braun and Willy Ley make appearances on the series.

Actress Yvonne de Carlo states that she will want to marry the first man to fly to the moon, "because he could take me some place I've never been before."

The Midlands Branch of the British Interplanetary Society produces a design study for a spaceship: "Project Caliban."

The Liars' Club of Burlington, Wisconsin, awards a free trip to the moon to the tellers of the year's best tall stories.

1952–1954

The popular magazine, *Collier's*, creates a symposium of space experts who, in a series of illustrated articles, outline one of the first comprehensive scenarios for the exploration of space. Only the British Interplanetary Society has been earlier in considering so thoroughly the many necessary steps [see 1939]. However, the schemes of the BIS have evolved over a period of some 15 years and have not, up to this time, been collectively published.

Collier's is one of the four top-circulation general magazines that have flourished during the 1940s and 1950s and is rivalled only by *Life, Look* and *The Saturday Evening Post*. Concern about the potential military use of space leads its editors to investigate the feasibility of space travel in the near future. The series of illustrated articles that this concern finally spawns had its beginning as a symposium on space travel held at New York's Hayden Planetarium [see October, 1951]. *Collier's* editor Gordon Manning was so impressed that he decided to sponsor a symposium of his own. He assembles the team of Wernher von Braun, then technical director of the Army Ordinance Guided Missile Development Group; Fred L. Whipple, chairman of astronomy at Harvard University; Joseph Kaplan, professor of physics at UCLA; Heinz Haber, of the U.S. Air Force Department of Space Medicine; and Willy Ley, an authority on space travel and rocketry who will serve as general advisor (and who is also the only professional writer among the group). The symposium is placed under the general direction of *Collier's* associate editor Cornelius Ryan.

To translate the hardware concepts of von Braun and Ley into visual form, Ryan commissions artists Chesley Bonestell [see 1949], Fred Freeman [see 1958] and Rolf Klep.

The results are published in 22 March 1952 ("Man Will Conquer Space *Soon*"), 18 and 25 October 1952 ("Man on the Moon"), 28 February, 4 and 7 March 1953 ("Man's Survival in Space"), 27 June 1953 ("The Baby Space Station"), and 30 April 1954 ("Mars").

According to the symposium, the United States could have an artificial satellite in orbit by 1963, a fifty-man expedition to the moon by 1964 and a manned mission to Mars soon afterwards. The first two parts of the program are expected to cost a mere $4 billion.

Both *Collier's* and its team of experts are entirely serious. The technology exists to carry out their plans, however grandiose they may seem. As the magazine scrupulously points out: "Speculations regarding the future technical developments have been carefully avoided," or, as von Braun explains, "While the resulting designs may be a far cry from what Mars ships some thirty or forty years from now will actually look like, this approach will serve a worthwhile purpose. If we can show how a Mars ship could conceivably be built on the basis of what we know now, we can safely deduce that actual designs of the future can only be superior. Only by stubborn adherence to the engineering solutions based exclusively on scientific knowledge available today, and by strict avoidance of any speculations concerning future discoveries, can we bring proof that this fabulous venture is fundamentally feasible."

[From today's vantage much of the Collier's space program seems an unsophisticated, brute-force approach to space travel. However, even if the symposium members suspected some of today's technological advancements, their point was not to show that space travel was a possibility of the future, even the very near future, but that it was possible in *1952*.]

The first step proposed is the launching of a 10-foot cone-shaped "baby satellite" carrying three rhesus monkeys. It is to orbit at an altitude of 200 miles for 60 days. It is eventually allowed to reenter the atmosphere, where it burns up (after the monkeys are given a merciful dose of lethal gas).

Following the successful launching of the unmanned satellite, the go-ahead is given to begin the full-scale manned space program. It will take 10 years and cost $4 billion.

The orbital spaceship is a monster rocket, 265 feet tall—as tall as a twenty-four-story

building. It is 65 feet in diameter at its base. The fat, bottle-shaped rocket weighs 7,000 tons, as much as a light naval cruiser. By comparison, the Apollo program's Saturn V will be 363 feet tall and weigh 3,211.5 tons. The space shuttle, which the *Collier's* ship most resembles in both form and function, is 184 feet tall and weighs (at takeoff) 2,250 tons.

The first stage is powered by fifty-one rocket engines, developing a total thrust of 14,000 tons. They burn 5,250 tons of nitric acid and hydrazine. The second stage's thirty-four engines produce a combined thrust of 1,750 tons. The third, manned, stage has five engines with a total thrust of 220 tons. This stage is in reality a winged aircraft with a crew of ten (some of whom will be women) and capable of carrying a useful payload of up to 36 tons (compare with the present-day shuttle's 94 ton capacity).

The ferry rocket, as it is called, is a development and refinement of the ferry rocket proposed in Wernher von Braun's *Mars Project* [see 1952].

Symposium chairman von Braun has designed this ship to ferry into orbit the material needed for the construction of the 250-foot diameter space station that is a necessary part of the scenario. No individual component of the station is to exceed the maximum payload of one ferry rocket.

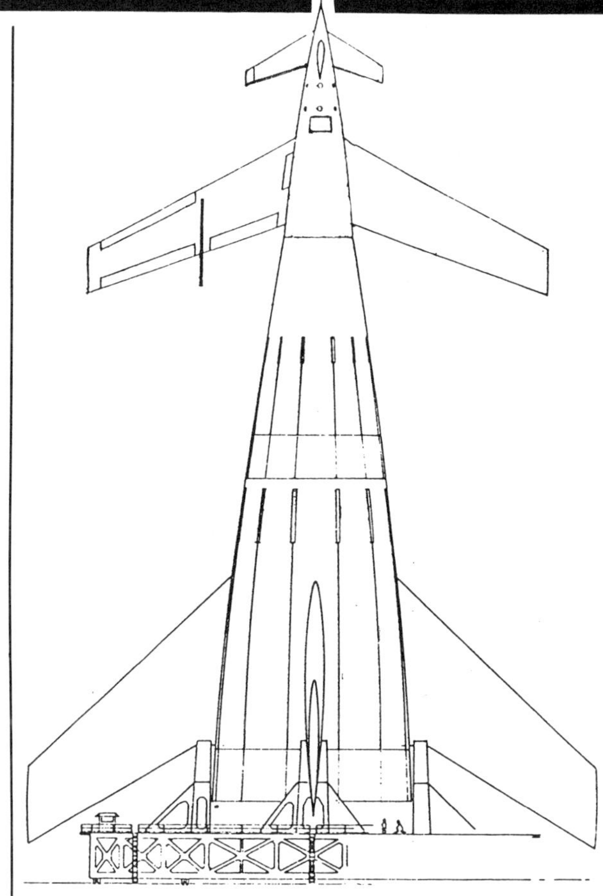

The massive launch site is to be constructed at a location where the downrange trajectory of the spacecraft can be over water. Two locations are proposed: Johnston Island in the Pacific Ocean and the Air Force Proving Grounds at Cocoa, Florida. The latter is a

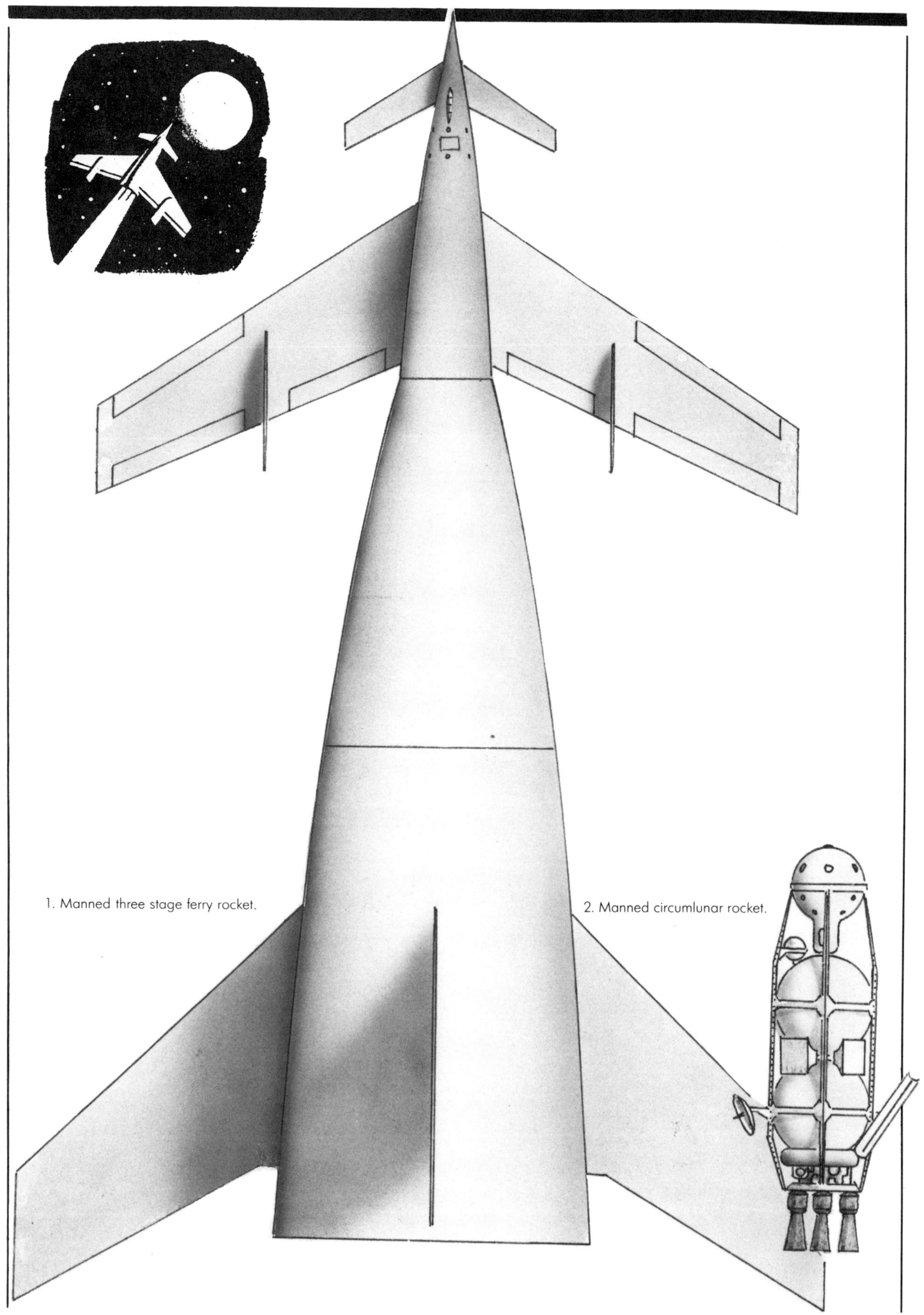

1. Manned three stage ferry rocket.

2. Manned circumlunar rocket.

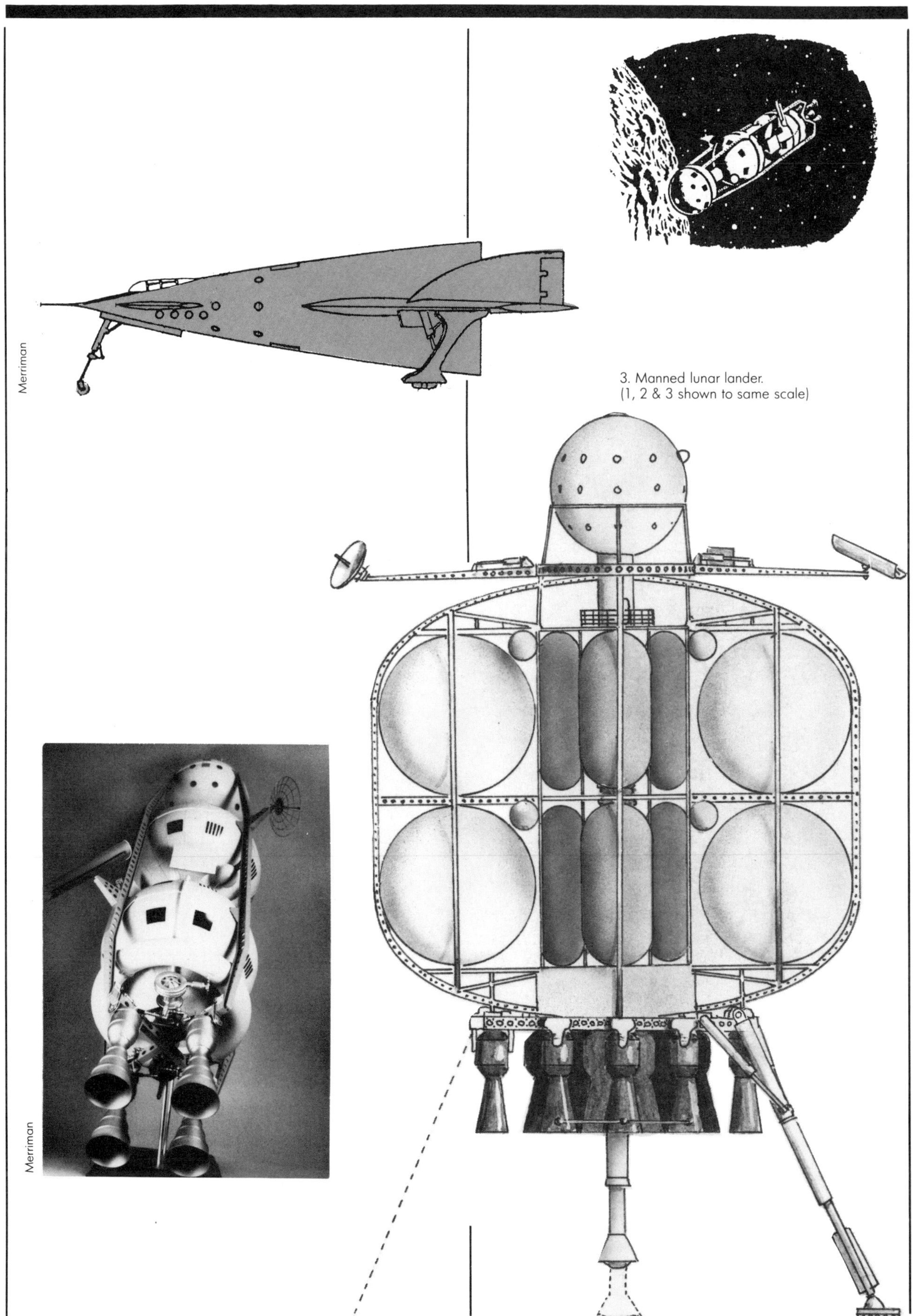

3. Manned lunar lander.
(1, 2 & 3 shown to same scale)

prescient choice: it is the location of what eventually will evolve into the Kennedy Space Center. The first and second stages of the ferry rocket are designed to be recovered at sea. The rate of launch will depend upon the recovery and turnaround time of the booster stages. It is hoped that ten or twelve ferries will be in operation during the space station construction period. At the height of the work there will be as many as one launch every 4 hours! Once the first and second stages have been recovered, they will be reassembled in a vertical assembly building. The complete rockets are then wheeled to their launch sites on giant crawlers—a sequence of events familiar to anyone who has observed the operations involved in a present-day shuttle launch [also anticipated by Fritz Lang and Hermann Oberth in the film *Frau im Mond*—see 1929].

The 200-foot wheel-shaped station will accommodate several hundred crewmembers and is scheduled for 1963 (revised to 1967 in the later book version [*Across the Space Frontier*, 1953]).

Once the station has been built and put into service, only one supply or personnel launch every third day will be necessary.

Then can begin the second phase of the master plan: the expedition to the moon. Since the magazine's design team is self-limited to the technology available in the 1950s, the authors rightly point out that there is no realistic hope of reaching the moon using only a single rocket launched from the earth. That would require, according to von Braun, a rocket taller than the Empire State Building—and ten times the weight of the *Queen Mary*! The Collier's experts propose instead a two-step operation, with the first being the construction of the moonships in earth orbit. They then travel from the earth to the moon, and back into earth orbit.

To construct the enormous lunar spaceships, an ambitious "spacelift" is to be undertaken and 360 ferry rocket flights made to carry the required materials into space. Three rockets are launched every 48 hours for nearly 8 months. In addition, 346 flights are needed to haul the necessary fuel into orbit.

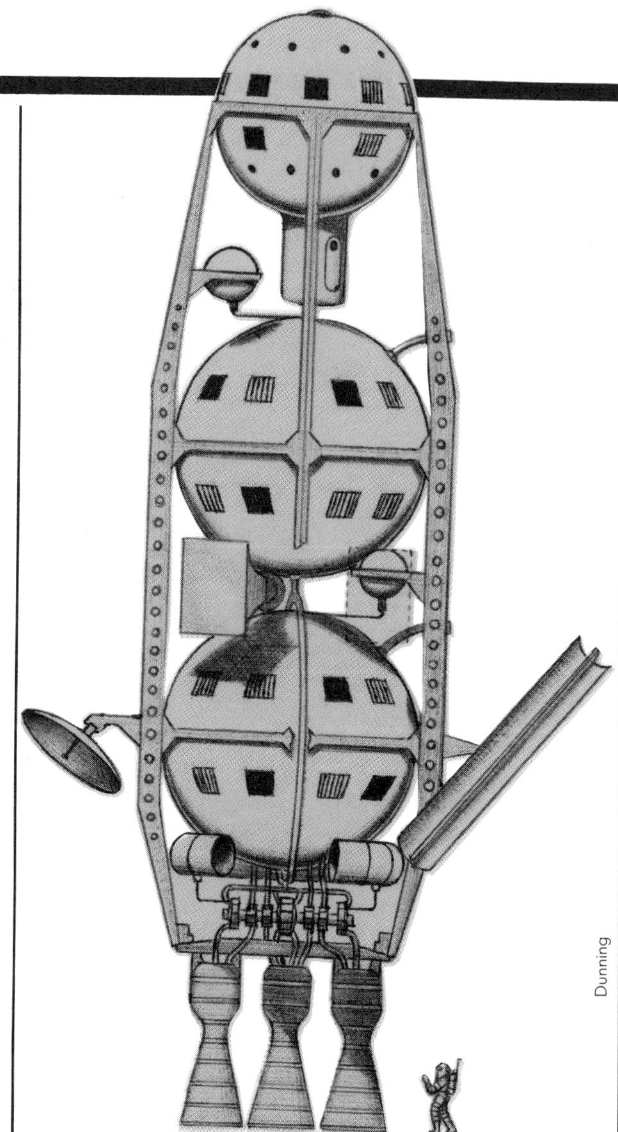

Dunning

Meanwhile, a manned trip is to be made around the moon. The vehicle is made of cannibalized fuel tanks and engines from a ferry rocket's third stage. The flight is one of reconnaisance only: the spacecraft is incapable of landing.

The three moonships that are to be assembled near the space station are huge bulks. Each is 9 feet taller than the Statue of Liberty and consists mainly of clusters of fuel tanks. These tanks are made of a flexible plastic. The fuel, pressurized to 1 pound per square inch, rigidly inflates them. More than half the fuel is carried in the four big spherical tanks—3,136 tons of propellants. All vital parts of the ships are covered with panels of 1/10 inch duralumin, separated from the inner surface by 1-inch spacers, as a shield against micrometeorites (an idea suggested by symposium member Fred Whipple). The sheets covering the spherical tanks are 7 feet square. The shielding weighs altogether 10 tons and will stop all meteors up to 1/100 inch in size.

The spherical tanks carry the fuel needed for the departure from earth orbit and will be discarded when empty. The next set of four large tanks contains fuel for the landing on the moon. They will be jettisoned on the lunar surface. The remaining smaller cylindrical tanks are for the return trip, as well as containing an emergency reserve.

One of the three ships carries no fuel for the return flight. The space these tanks would have occupied is taken by a single cylindrical cargo "silo" 75 feet tall and 36 feet in diameter. This contains all the necessary equipment, supplies and provisions for the 6-week sojourn on the moon. The cargo cylinder itself is designed to be split lengthwise into two Quonset-type buildings (an operation criticized by members of the BIS: the hut would be stronger left as a cylinder).

Construction of the moonships is made easier by the use of color-coded components. The ships themselves are painted a heat-reflecting white. Thermal control is provided by louvered temperature regulators. Each ship has a solar-powered electrical generator. A 450 square foot mirror focusses sunlight onto a mercury boiler, which in turn operates a turbine-powered generator producing 35 kW of energy. The solar collector is mounted at the end of a long boom attached to a circular track that surrounds the base of the personnel spheres. It can be moved along this track so that it can follow the sun regardless of the ship's attitude. Opposite the solar mirror is the boom carrying the 15-foot radio dish. The ships carry 100-watt transmitters operating at 3,000 megacycles per second. Von Braun hopes that this will be sufficiently powerful for signals to be detected back on the earth, though he thinks the astronauts might have to resort to Morse code to do so (by comparison, the Voyager spacecraft sent television signals from the vicinity of Saturn—almost a billion miles from the earth—using only 25 watts!).

Each 30-foot personnel sphere has five floors. From top to bottom: the control cabin, the navigation deck, living and dining quarters, the storeroom and engineering deck, and the life support equipment and machinery. Immediately below the last deck is the airlock. Fifty crewmembers are divided among the three ships. There is an overall expedition leader; fifteen crewmen, including one captain per ship, one of whom is the fleet commander; eight electronics and communications experts; six mechanical engineers; one astronomer and one surveyor; three photographers; physicists; a mineralogical team and a geophysical team.

At the opposite end of the ship are thirty rocket engines, developing a combined thrust of 470 tons. Twelve are on hinged mountings, so they can be used for steering.

The journey to the moon takes 5 days. The landing takes place in the Sinus Roris, a "bay" at the north of the mountainous region surrounding the Mare Imbrium.

The moonships are equipped with radar-actuated autopilots for the final descent to the lunar surface. The four landing legs, which are kept folded against the body of the ship during the flight, are deployed. They have hydraulically operated adjusters which will automatically compensate for uneven terrain. A central, telescoping leg extends from the midst of the engine cluster. This supports the mass of the ship, the corner legs simply keeping the ship upright. After landing, the four big tanks are discarded. The ship's weight is now reduced from its original 4,370 tons to 333 tons.

The tractors carried in the cargo silo will carry explorers as far as 250 miles from the landing site, to explore the 24 mile-wide crater Harpalus [which also figured in *Destination Moon*—see 1950] and the region surrounding it.

When the 6 weeks of exploration is completed, the scientists set up automatic telemetering stations, which will continue to radio data back to the earth. All fifty members of the expedition then reboard the two passenger ships. The cargo ship has served its purpose. It has been almost completely dismantled and will remain on the moon. After a relatively gentle 3.5 g takeoff, the crew settles in for the 5-day return trip to the space station.

1952

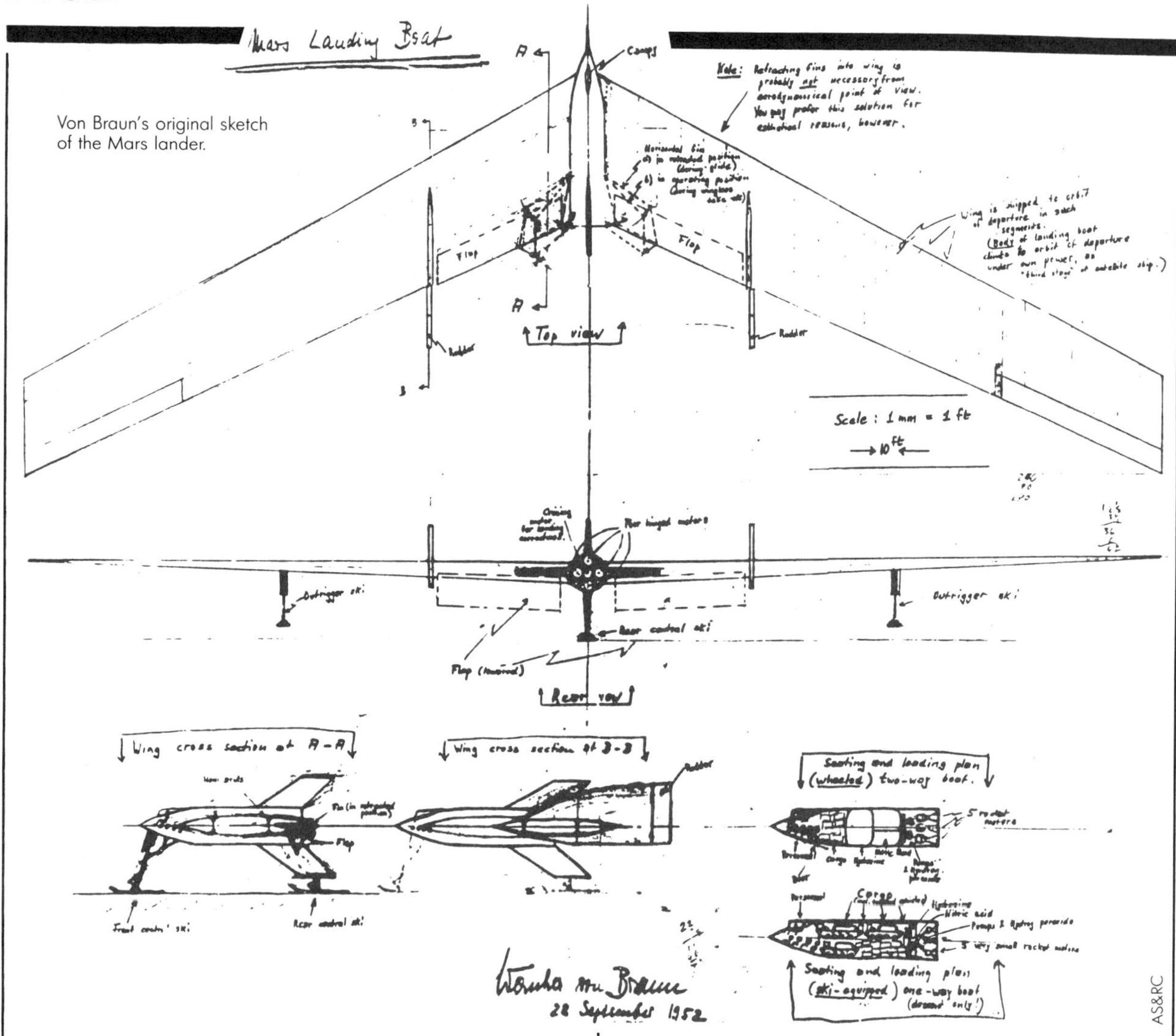

Von Braun's original sketch of the Mars lander.

AS&RC

The members of the symposium never attempt to set a date for the third and final part of their plan for the conquest of space: the manned mission to Mars.

The original *Collier's* version of the Mars expedition describes a massive assault on the planet and is based upon the calculations von Braun had prepared for his book *Das Marsprojekt* [see 1949, 1952]. It calls for a fleet of no less than ten enormous spacecraft crewed by seventy astronauts. The building of these in earth orbit will require the launch of 950 ferry rockets. The cost is estimated at "ten times" that of the post-war Berlin airlift, or about $2.24 billion. However, by the time the *Collier's* plan is put into book form (as *The Exploration of Mars*, 1956), the scheme has been radically altered. The new plan, says von Braun, "may be considered a revision of [the *Das Marsprojekt*] study." The new scenario saves 90 percent of the fuel originally called for "due to a superior over-all plan."

This revised plan requires only two ships, otherwise it is similar to the original magazine version. These are a passenger vehicle that will make the round trip to Mars and back to the earth and a winged glider that will make the actual descent to the Martian surface. Both will be constructed in orbit from materials lifted by a fleet of second-generation ferries. These are the immediate descendants of the manned ferries used in the construction of the space station and the moon ships. The new ferry is a much smaller rocket, only about 180 feet tall and 38.4 feet at its base. Unlike the earlier project, most of the ferrying flights will be made by unmanned cargo rockets. Of the 400 launches needed, 335 will be made by unmanned rockets. The very first flight will be manned, however. Its radio operator will

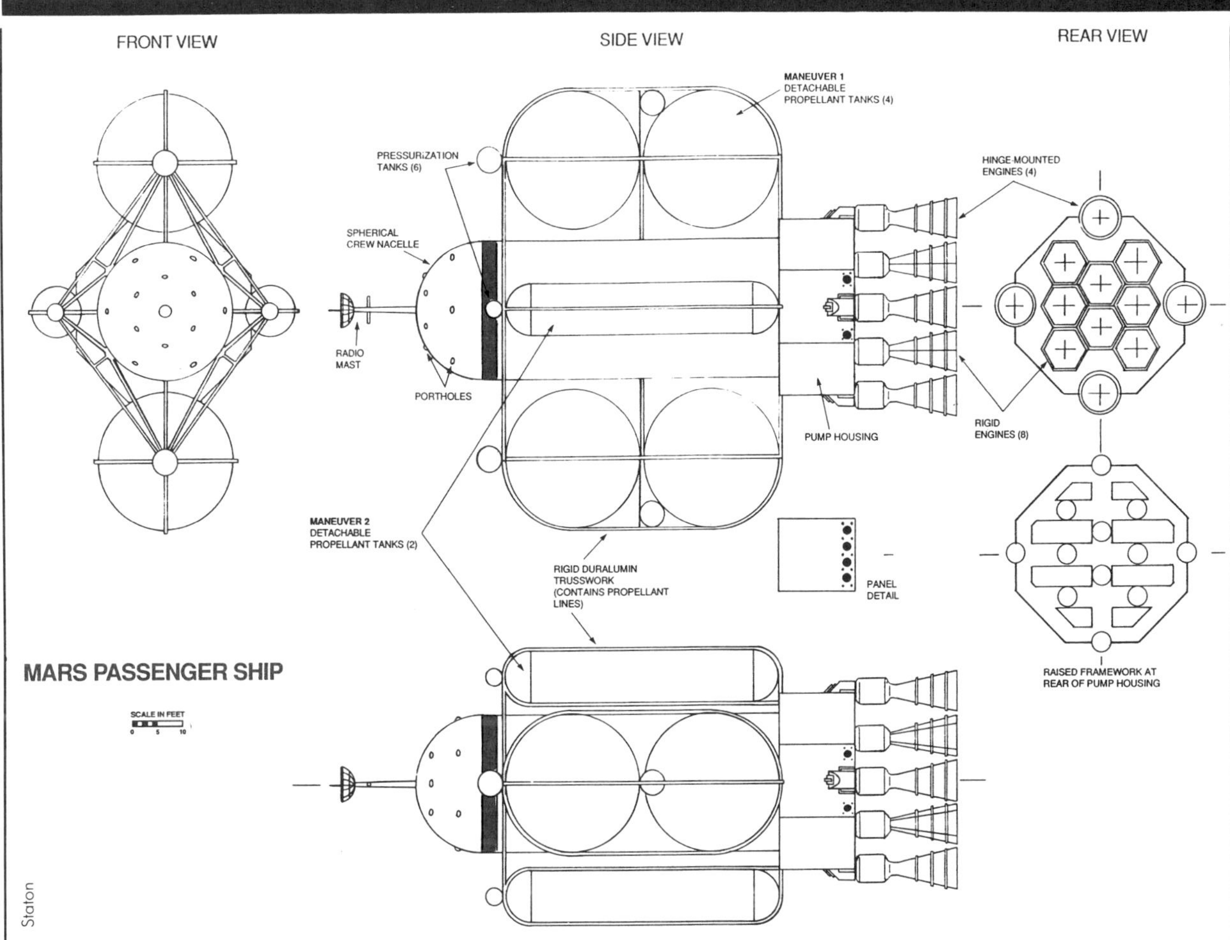

guide the subsequent cargo carriers into the proper orbit, 1,075 miles above the earth. This manned ship will remain in orbit during the entire period of materials delivery. Its crew will be rotated on a weekly basis. At a launch rate of two ships every 24 hours, the operation will be completed in about 7 weeks (von Braun does not explain why the space station couldn't assume the function of this manned orbiter).

Both of the Mars ships weigh 1,870 tons and both have a cluster of twelve rocket engines, fueled by 1,370 tons of nitric acid and hydrazine—von Braun's propellants of preference. These will develop for each ship a total thrust of 396 tons. The passenger ship closely resembles the cargo ship of the moon venture. A 26-foot personnel sphere sits ahead of a cylindrical hull that contains the 180 tons of fuel needed for the return flight from Mars orbit. The cylinder is surrounded by six additional fuel tanks: four huge spheres needed for the departure from earth orbit, and two smaller cylindrical tanks for the braking maneuver upon arriving at Mars. The passenger sphere is divided into three decks: an upper one for the controls—with an astrodome for the navigator, and two lower decks for living quarters. At the "bottom" of the sphere is an airlock that extends into the cylindrical hull. The decks are connected by a central fireman's pole passing through open wells in the floors. The sphere is constructed of fiberglass covered with a duralumin meteor shield. Air pressure is maintained at 8 psi, and the atmosphere is a mixture of oxygen (40 percent) and helium.

Twelve astronauts are to make the trip to Mars; only four need to be on board the glider at any one time during the earth-Mars leg.

(A third version of the Mars glider appears in the film *Conquest of Space* [1955] as redesigned by production designer J. MacMillan Johnson.)

In spite of its enormous size—the huge wings have a span of 450 feet—the landing

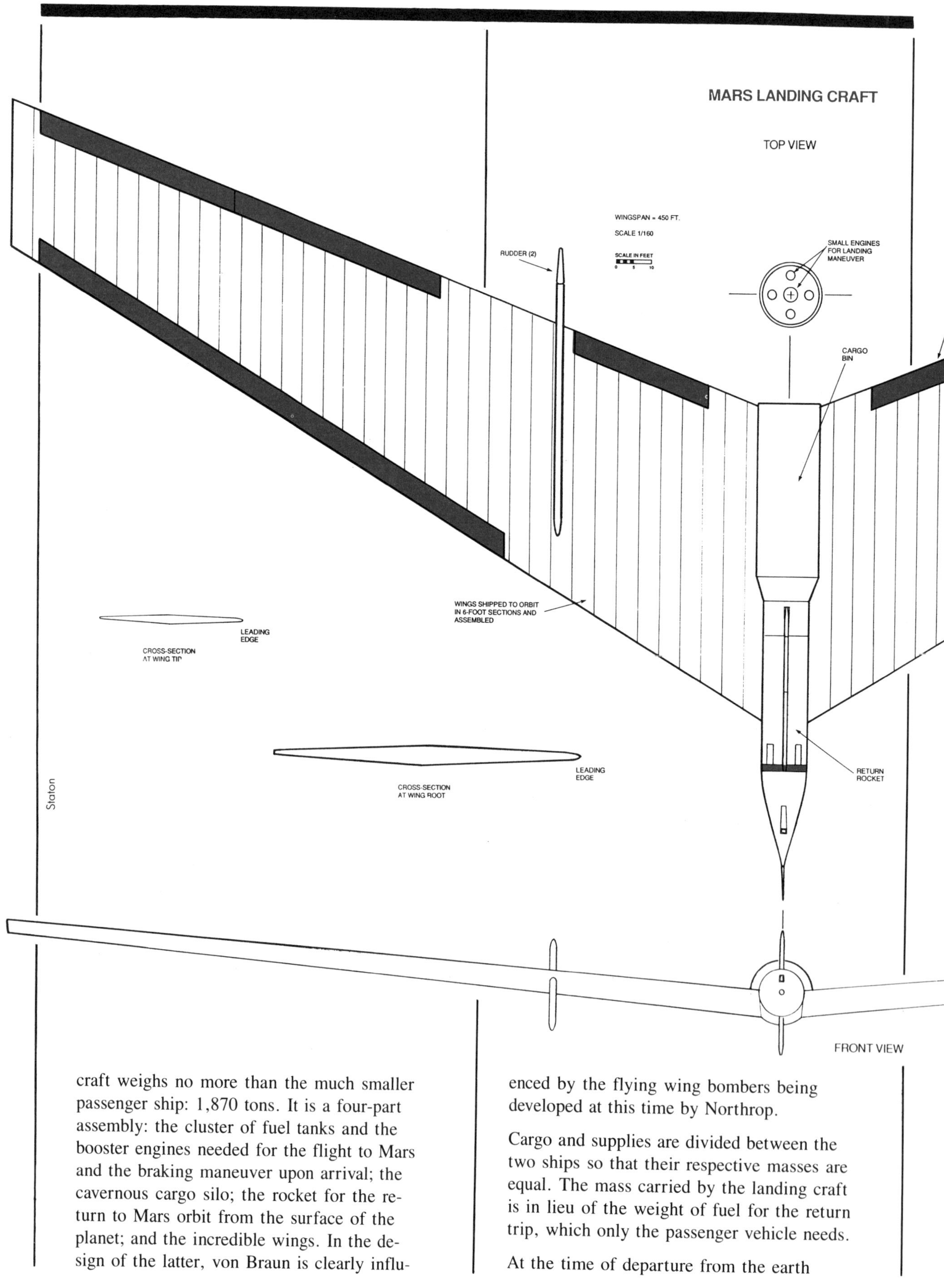

craft weighs no more than the much smaller passenger ship: 1,870 tons. It is a four-part assembly: the cluster of fuel tanks and the booster engines needed for the flight to Mars and the braking maneuver upon arrival; the cavernous cargo silo; the rocket for the return to Mars orbit from the surface of the planet; and the incredible wings. In the design of the latter, von Braun is clearly influenced by the flying wing bombers being developed at this time by Northrop.

Cargo and supplies are divided between the two ships so that their respective masses are equal. The mass carried by the landing craft is in lieu of the weight of fuel for the return trip, which only the passenger vehicle needs.

At the time of departure from the earth

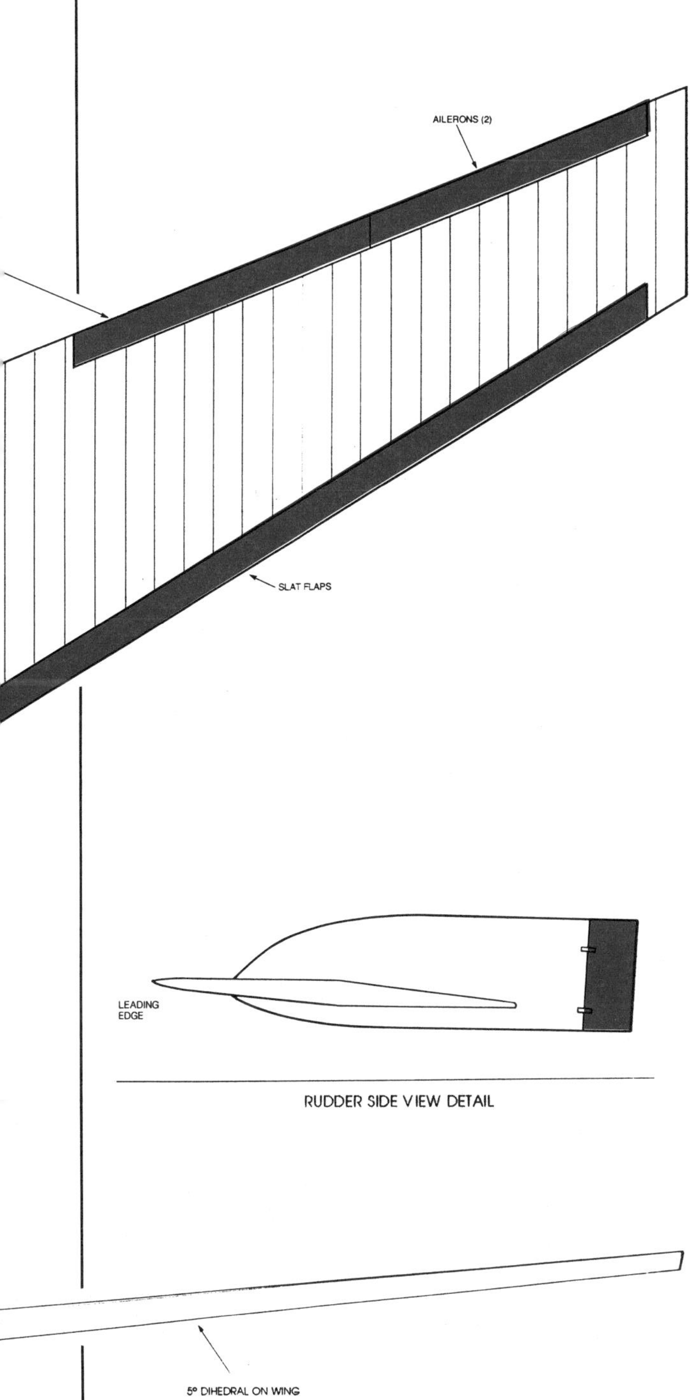

1,370 tons of fuel are burned by each ship in the 948 seconds required to escape earth's gravity. Little more need be done for the next 250 days.

Nearing Mars, each ship jettisons six of its twelve engines. Using flywheels, they are rotated until they point tail-first toward Mars. The remaining engines are fired. The ships slow down, entering an orbit 620 miles above the surface. The glider now weighs only 177 tons. Almost 1,700 tons of fuel have been consumed. The landing craft is prepared for the descent. Its remaining engines are jettisoned, as are the last four fuel tanks. Supplies are transferred to the passenger ship crew who will remain with it in orbit. Nine astronauts man the glider, the remaining three prepare themselves for their lonely, 400-day vigil in orbit. Their job is to record data sent up from the explorers on the surface. They will also use the on-board telescope to observe and photograph Mars from orbit.

Along with the nine crewmembers, the glider carries 18.7 tons of food, water and other provisions. The tractors, living quarters, scientific apparatus and other equipment account for another 35.2 tons.

The glider is slowed by firing retrorockets and begins to descend along a shallow glide path. At a few hundred feet above the surface it deploys its specialized landing gear. The main fuselage is equipped with two sets of skid-flanked caterpillar treads, one set at the nose, the other below the cargo silo. The spaceship touches down at 120 mph.

(In the magazine version of the expedition, the first ship down lands at one of the Martian poles. Its crew makes the thousand-odd-mile journey via tractor to a point further south where they construct more formal runways for the remaining gliders.)

The first order of business, once the crew has made the 18-foot drop from the lander, is to detach the return rocket from the now useless wings and cargo bay. The rocket is ready then to be raised to an upright position. Winches, wire rope and the hydrogen peroxide/fuel oil-propelled tractors are used to hoist the relatively small rocket to its vertical takeoff position. The ship is only 65 feet tall, once it is resting, V2-like, on its tailfins. In fact, it is only 18 feet taller than the V2. The rocket only needs to carry the nine astronauts and their 5 1/2 tons of specimens as far as the waiting passenger ship. All other equipment, unused supplies and materials, including the tractors, will be abandoned upon departure.

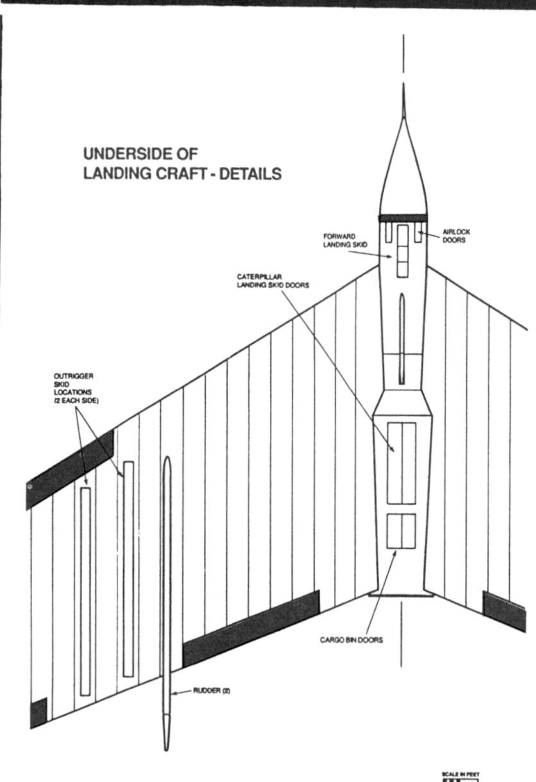

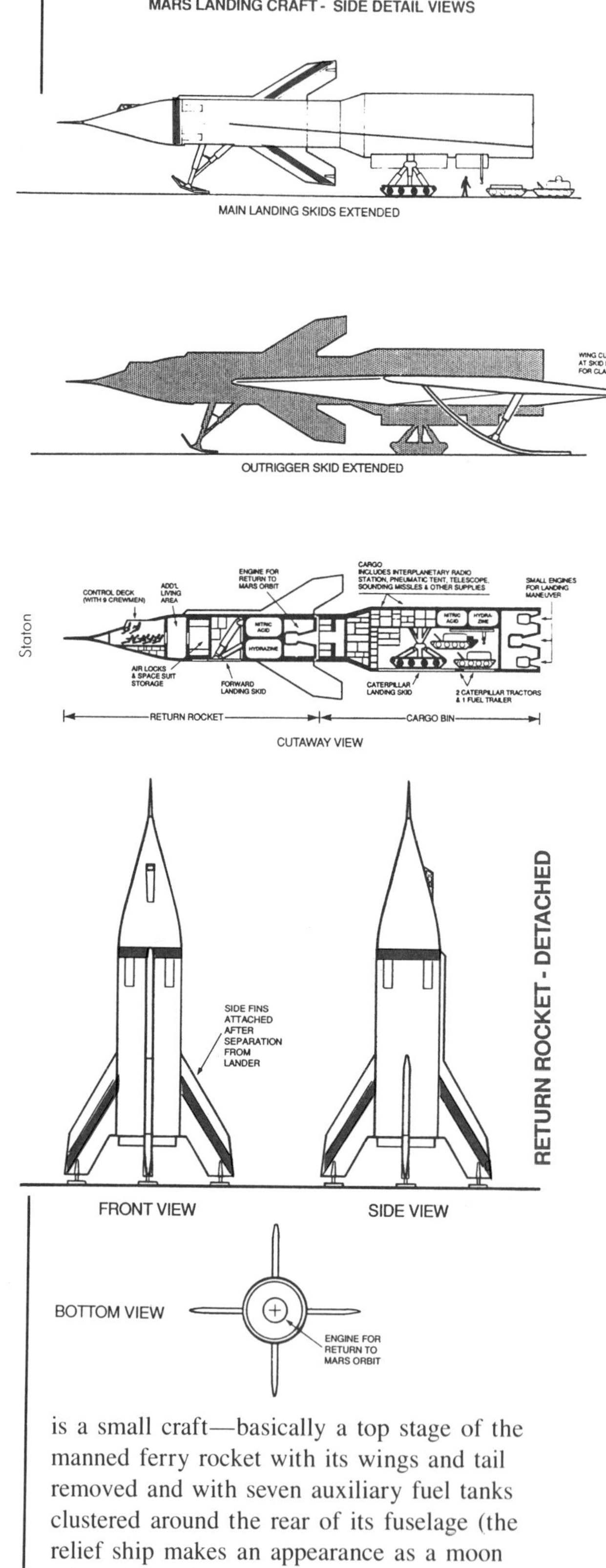

The explorers will live in a 20-foot inflatable "tent" for the year they will remain on the planet. Von Braun and his colleagues have allowed for every conceivable human need—from washing machines to pencil sharpeners.

A little more than 62 tons of fuel will be used to return the explorers to Mars orbit. By the time the two ships rejoin, the landing craft will have been reduced from its original 1,870 tons to a mere 13.8 tons. The crew and their specimens are transferred to the waiting passenger ship. The last two propellant tanks are jettisoned; only the fuel contained within the hull is required: just 180.4 tons, of which almost 130 tons are used in leaving Mars.

After 268 days, the astronauts arrive home. Approaching tail-first, four of the engines are fired (the last two rigid-mounted motors have been discarded). The ship settles into an orbit that is still 56,000 miles above the earth. The Mars ship has been reduced to 2 percent of its original mass.

The astronauts rendezvous with a relief ship, which picks them up for the penultimate leg of their return to the earth. The relief ship is a small craft—basically a top stage of the manned ferry rocket with its wings and tail removed and with seven auxiliary fuel tanks clustered around the rear of its fuselage (the relief ship makes an appearance as a moon orbiter in Walt Disney's *Man in Space* television program—see 1955). At the 1,075-mile orbit of the space station, the Mars expedition transfers to a waiting ferry for a trium-

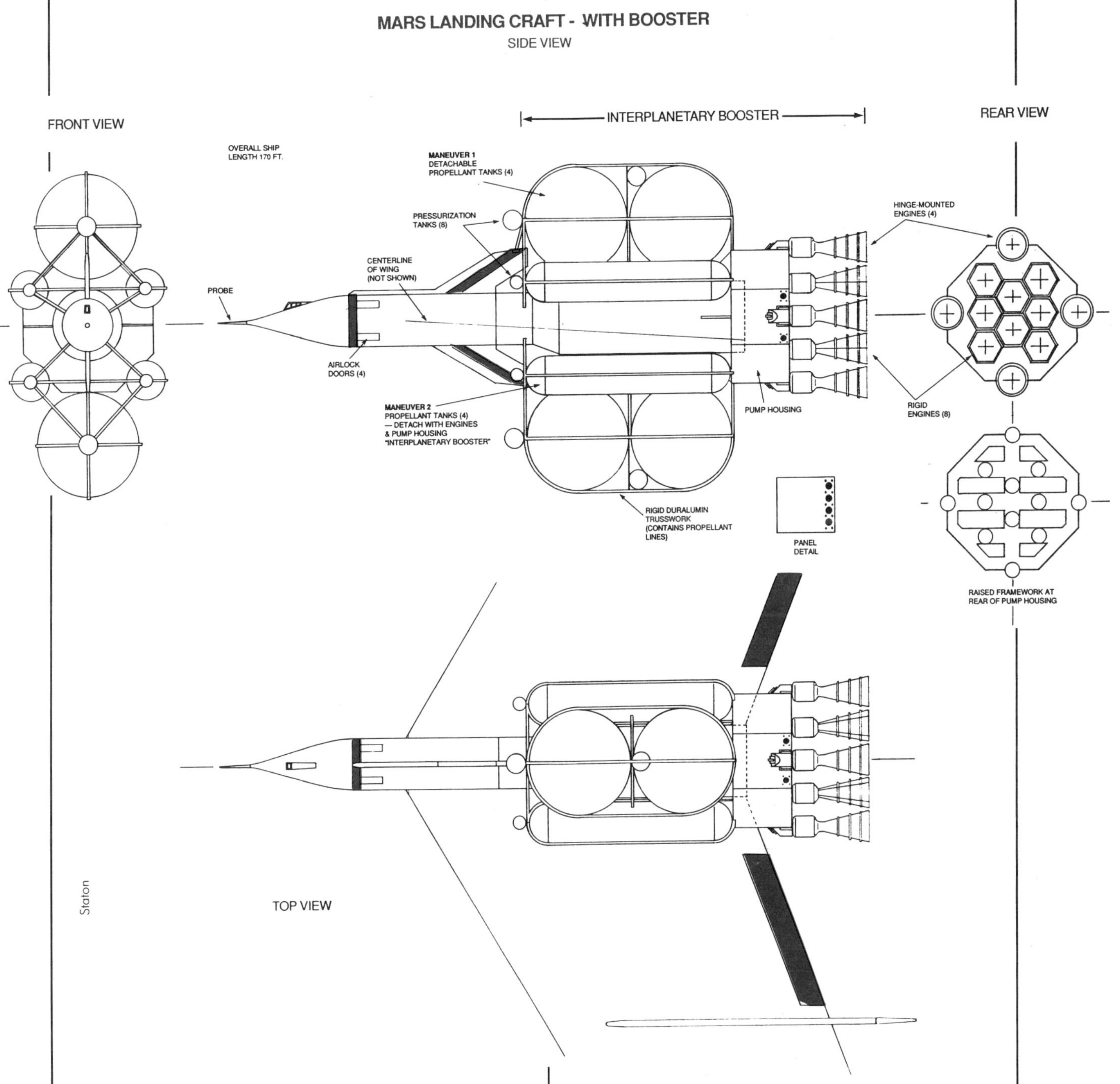

MARS LANDING CRAFT - WITH BOOSTER
SIDE VIEW

FRONT VIEW

INTERPLANETARY BOOSTER

REAR VIEW

OVERALL SHIP LENGTH 170 FT.

MANEUVER 1 DETACHABLE PROPELLANT TANKS (4)

PRESSURIZATION TANKS (8)

CENTERLINE OF WING (NOT SHOWN)

PROBE

AIRLOCK DOORS (4)

MANEUVER 2 PROPELLANT TANKS (4) — DETACH WITH ENGINES & PUMP HOUSING "INTERPLANETARY BOOSTER"

HINGE-MOUNTED ENGINES (4)

PUMP HOUSING

RIGID ENGINES (8)

RIGID DURALUMIN TRUSSWORK (CONTAINS PROPELLANT LINES)

PANEL DETAIL

RAISED FRAMEWORK AT REAR OF PUMP HOUSING

TOP VIEW

Staton

phant descent to the waiting earth.

The *Collier's* scenarios prompt mixed reactions from the readers, both civilian and official. Wernher von Braun's colleagues make the most thoughtful critiques, for the most past singling out either the grandiose nature of the project [see Leonard, 1953] or its overt military overtones. Others choose to comment upon the series' over-optimism and lack of allowance for testing and experimentation.

The magazine's letter columns are filled with generally enthusiastic responses while, on the other hand, *Time* devotes its 8 December 1952 cover story to a less-than-salutary synopsis of von Braun's ideas. The magazine admonishes an "oversold public . . . [which] happily mixing fact and fiction, apparently believes that spaceflight is just around the corner." Fritz Haber (brother of Heinz) believes that the whole idea of space suits has to be abandoned; Hubertus Strughold does not believe that man will be able to function in free fall. One of von Braun's most bitter critics says, unkindly and unfairly, "Look at this von Braun! He is the man who lost the war for Hitler . . . von Braun has always wanted to be the Columbus of space. He is thinking of spaceflight when he sold the V2 to Hitler. He says so himself. He is still thinking of spaceflight, not weapons . . . "

The series is collected in three books: *Across the Space Frontier* (1953), *Conquest of the Moon* (1953) and *The Exploration of Mars* (1956), all now highly prized collectibles, as are the magazines themselves.

In 1969, after the success of the Apollo 8 circumnavigation of the moon, Cornelius Ryan writes to Wernher von Braun, reminding him of the *Collier's* articles: "My mind went back to 15 years ago and the days when newspapers and magazines are calling me 'Blast Off Ryan' (and you even more derogatory names) because of the *Collier's* space series. I guess a little ground is broken then. I know it is for me and so I can tell you how I felt when I saw *your rocket* and *your* design streaking skywards culminating for you what surely must have been a lifetime of hoping and finally fulfillment." Dr. von Braun replies: "It is good to learn that you are keeping up with things during the Apollo 8 flight. It has indeed been a few years since the days of the articles in *Collier's*. Don't forget that you yourself had something riding on that flight. None of us have ever forgotten the impetus to space that you helped to provide 'back when.' And who knows? We may have to call on you again. I'm delighted to know that you're still a good space man . . . " In a postscript, von Braun adds: "You wouldn't believe how many people still remember 'our' *Collier's* series. Time and again complete strangers tell me that those articles really opened their eyes to these new possibilities 'out there.' You surely deserve a lot of credit for the success of Apollo 8, and this ain't no empty flattery!!"

1953

Jonathan N. Leonard, science editor for *Time*, publishes *Flight into Space* [see von Braun, below].

Spaceways stars Howard Duff in a film adaptation of a Charles Eric Maine play. He must travel into space in order to prove that he hasn't murdered his wife and hidden her body on board a satellite. The spacecraft shown in the form of models before the launch are duplicates of the von Braun ferry rockets designed for his *Collier's* series [see above]. However, stock newsreel footage of V2s are used for the actual launch scenes, and even a brief clip from *Rocketship XM* [see 1950] is used.

Eugen Sänger invents the concept of the antimatter rocket. When matter and antimatter are combined in equal masses, they are mutually annihilated in a burst of pure energy. It might be the ideal solution to interstellar travel if the problems of producing and storing antimatter can be discovered.

Project Moonbase is a film based on a screenplay by Robert A. Heinlein. It is pieced together from the initial episodes of an unaired television series "Ring Around the Moon." There are some interesting spacecraft designed for the movie by Jacques Fresco and some very good special effects, but they are ruined by poor sets, costumes and an unintentionally laughable, overly melodramatic script (which, nevertheless, features much better science than most other films of the period and at least a plot that develops from the situation and environment).

The story repeats the plot device used in *Destination Moon* where an unidentified foreign power tries to sabotage the American space effort; in this case the destruction of a military space station is the goal. The saboteur's efforts misfire, resulting in a cir-

Skotak

cumlunar flight making an emergency landing on the moon.

The script has the first manned spaceflight take place in 1966 and the first (unintentional) moon landing take place in 1970. Both, interestingly, are made by a woman, a Colonel Briteis (which the character insists is pronounced "Bry-ties," but is, of course, pronounced "Bright Eyes" by every male character). She has not been selected for these missions because of her competence, but rather, for the orbital flight, because she weighs only 90 pounds, and for the moon mission because of the fame that she accrued from her first spaceflight.

The spacecraft feature single-stage-to-orbit manned ferry rockets (the *Mexico* and the *Canada*) and a lunar lander, the *Magellan*, patterned after the recently published *Collier's* vehicles.

Murray Leinster (pseudonym of Will F. Jenkins) publishes his novels *Space Tug* and *Space Platform*. The latter describes the trials and tribulations preceding the launch of

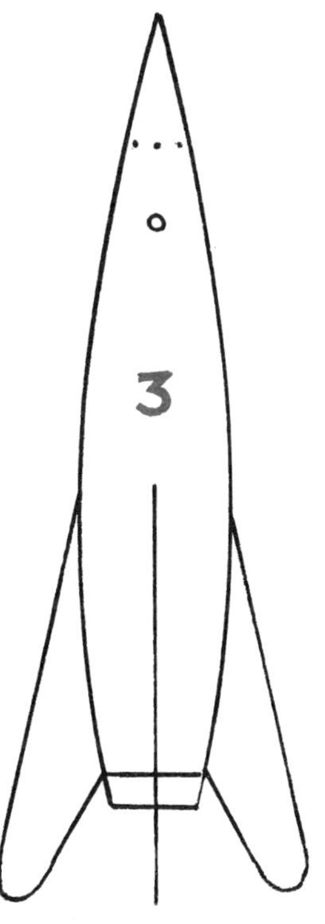

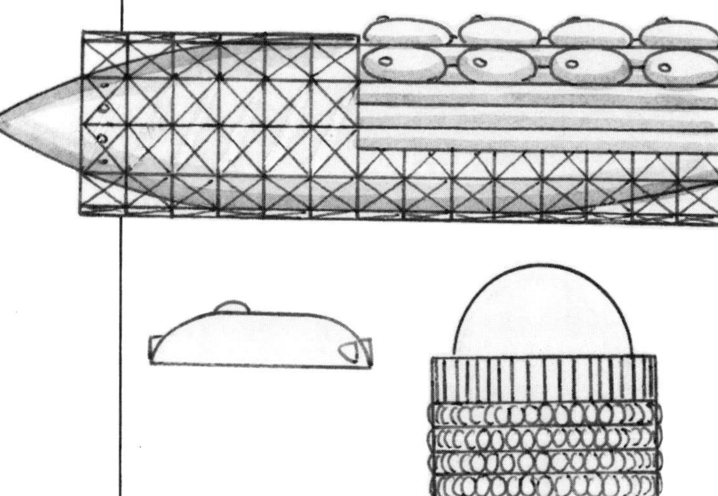

Alien spaceship in *The Man From Planet X.*

the world's first space station (like most others of the period, intended to be an orbital nuclear weapon carrier). The method Leinster chooses for launching his huge, spherical station is cumbersome and grandiose at the same time. Rather than opt for orbital assembly, he proposes constructing the complete station on the ground and launching it bodily into orbit.

The lower half of the space platform is surrounded by literally hundreds of "pushpots," cumbersome JATO-equipped jet aircraft, each individually piloted. Their combined thrust—JATO units and jets acting as the first two stages—lifts the station from the earth, taking it to a speed and altitude where a cluster of 40-foot solid-fuel motors takes over. Takeoff is initially horizontal. Once in orbit, the platform is intended to operate in zero-g conditions.

The design of the solid-fuel motors is unusual in that the entire engine, casing and all, is intended to be consumed as the rocket burns. Ideally, when the fuel is exhausted, so is the motor itself and there will be nothing left.

A hypothetical manned moon lander is also described. One of the characters, a midget, outlines some of the advantages to using people like himself to pilot spacecraft [also see del Rey, 1942]. A standard moon rocket, carrying a pilot of normal size, would require a cabin 7 feet high and 10 across. With 1,200 pounds of life support, the cabin would weigh altogether 1 1/2 tons. Fully fueled (with "old-style" fuels) the complete moonship would weigh 1200 tons—for a one-way trip. With the "new fuels," a one-way trip would require a 600-ton ship. Using a 45-pound midget pilot, however, a moon rocket would only have to weigh 60 tons.

Space Platform's sequel *Space Tug* describes the operation of the space station once it has been successfully placed in orbit. It is serviced by manned ferry rockets, which are 80 feet long and 20 feet wide, with stubby fins, carrying a crew of four. For takeoff, they are placed in a latticework cage of steel scaffolding. Attached to this is a cluster of 40-foot solid fuel boosters. Around these are attached a number of the manned "pushpots." Like the pushpots used for launching the space platform, their built-in JATO units use a beryllium-fluorine fuel that gives them an acceleration of 10 g's. Cumulatively, they impart an acceleration of 6 g's to the ferry rocket.

A moonship is ultimately constructed on earth for launch to the space platform. Looking like "something a child might have put together out of building blocks," it is assembled from welded-together cells and is 120 feet long and 60 feet wide.

Wernher von Braun takes some umbrage at critics who are accusing him of concentrating on impossibly grandiose schemes. (Such as comments like this one from J. N. Leonard's *Flight Into Space* [see above]: "Von Braun, who is capable, with an effort, of making small plans as well as big ones . . . ") He answers them in a paper delivered at the fourth International Astronautical Congress in Zurich (read by Frederick C. Durant, III) in which he explains that, "No one with any experience whatsoever in the development of large rockets . . . could possibly seriously maintain that we are in a position today to leap from the status quo to manned rocketships without a series of intermediate steps."

Von Braun's *The Mars Project* is published in English by the University of Illinois Press [see 1952]. It is reissued in 1991.

Hubert M. Drake and L. Robert Carman, under NACA auspices, design the "phase IV research plane," an X-15-like aircraft to be launched by a giant winged booster, the smaller plane riding piggyback. The booster will be a Mach 3, five-engined aircraft weighing 100,000 pounds. Top speed is expected to be Mach 18. NACA thinks the project sounds too "futuristic."

H. E. Ross redesigns his classic BIS space station [see 1948]. In place of the large solar mirror, power is now generated by means of a circular array of solar cells, the same diameter as the original mirror. Other changes in his modernization include the addition of a radio telescope on a boom balancing that carrying the communications antennas. A tubular cradle for a "tender rocket" is provided; this is a small, manned intraorbital vehicle.

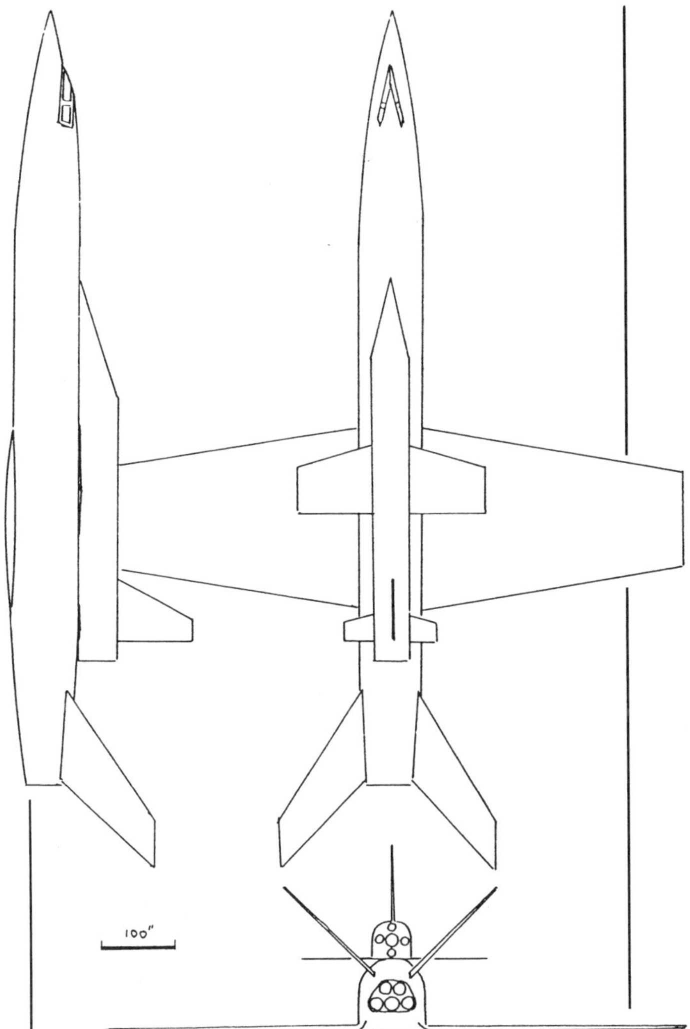

The "serial" *The Lost Planet* is cobbled together from episodes of the television series "Commando Cody, Sky Marshal of the Universe." It features a spacecraft constructed by the villain: the *Cosmojet* [also see 1949].

The U.S. Air Force's Scientific Advisory Board's Aircraft Panel proposes that the Air Force, Navy and NACA develop an advanced Mach 5 to 7 hypersonic research aircraft to study the problems of spaceflight and reentry. This coincides with a Navy/Douglas Aircraft study for an aircraft powered by a 50,000-pound-thrust Reaction Motors rocket engine, which will be capable of speeds of up to 5,000 knots and altitudes of 700,000 feet. At the peak of its trajectory its pilot will be weightless for up to 7 minutes. This aircraft is designated the D-558-3.

Kenneth Gatland in his book *Space Travel* describes an elaborate space exploration program, derived from earlier studies for the BIS [for example, see 1949 and 1951].

The "Interplanetary Project" will be located on a remote island somewhere in the Pacific Ocean. A likely candidate will be the Christmas Island group. An alternative might be Brazil or the coast of East Africa in Kenya. Far beyond the financial capabilities of almost any individual nation on earth, even the United States and Soviet Union, the project will most likely, and ideally, be sponsored by the United Nations.

The program will be incremental, much like that proposed by Wernher von Braun [see 1952–1954]. It will begin with the launching of progressively more sophisticated and larger unmanned satellite vehicles, the last of the series employing the expendable-tank construction that Gatland developed (in which the rocket's propellants are contained in tanks shaped like concentric half-cylinders that are jettisoned as they empty). A three-step orbital rocket capable of carrying 350 pounds into a 500-mile orbit will be no taller than a V2 and will weigh only 76 tons. From these will evolve a ground-to-orbit tanker that will carry 4 foot 2 inch by 17 foot propellant "capsules" into orbit, or 5 tons of payload materials. Such cargo will go toward building a space station or an orbit-to-orbit interplanetary spaceship.

An earth return vehicle will require extensive modification of the first and second steps (using conventional rather than expendable design) of the freighter rocket as well as the addition of a high speed delta-winged glider. The glider will weigh 5 to 6 tons and carry a payload of half a ton.

The freighter rocket can be converted to a passenger rocket by making the third stage conventional, rather than using the expendable construction technique. This will be necessary because the manned third stage will require wings and room must be made for them. The first stage will have fins added for stability as well. The third stage will have delta wings, retrorockets and a combination of drogue parachutes and airbrakes for the return to earth. It will require that the second stage not be of the expendable-tank design favored by Gatland et al., since this would necessitate the third stage being contained within the second. The launch weight of the winged glider rocket will be in the neighborhood of 740 tons.

When considering a site for the launching of the manned and freighter rockets, the authors consider that since "the 12 1/2 ton V2 rocket took several hours to prepare for launching, the problems of fuelling and servicing a 500 ton, three-step, freighter rocket can be appreciated. The operation will probably take at least 24 hours to complete . . . " They finally suggest these areas as being ideal: the Christmas Island group, Kenya or Brazil.

Once a space station has been built, it can be used as a base of operations for the construction of an interorbital ship. The first such to be built will be for a manned circumlunar voyage. Fifty-three freighter launches will be required and two manned launches.

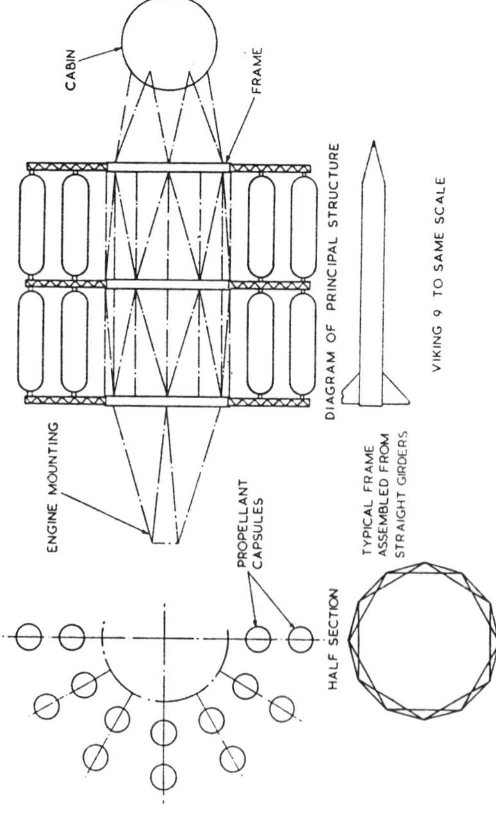

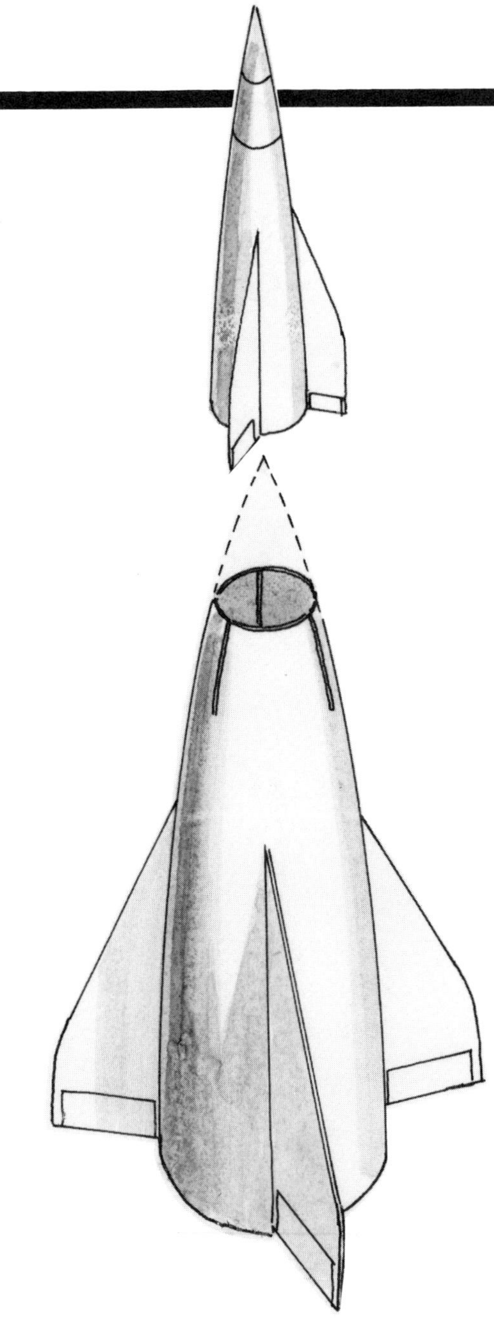

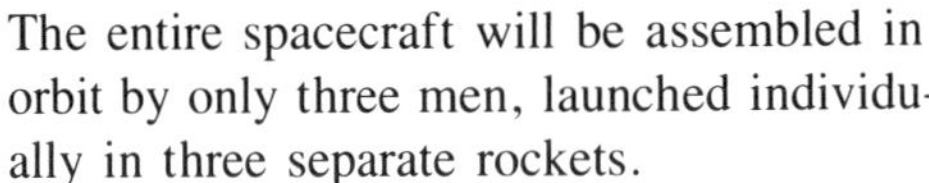

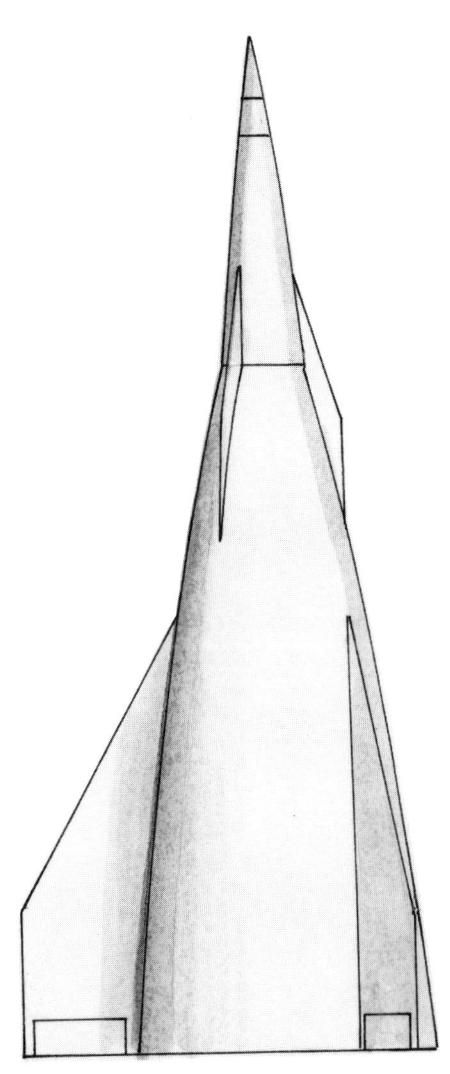

The entire spacecraft will be assembled in orbit by only three men, launched individually in three separate rockets.

Allowing one to two weeks for such an expedition, a vessel of approximately 260 tons is planned. It will consist of a cluster of forty-eight 180-kg "capsules" of fuel, each containing 5 tons of propellant. Each cylindrical capsule will be made of double-skinned aluminum and will be 1.3 m in diameter and 5.2 m long. These are arranged in two concentric rings and attached to a framework structure. A spherical crew cabin 4.6 m in diameter is mounted at one end by similar trusses and the rocket motors (providing 25 tons of thrust) at the opposite end. The support structure weighs 5 tons. The passenger sphere, itself weighing 4.5 tons, will carry 1.9 tons of crew and supplies (sufficient for 28 days).

Thirty-four of the tanks will be jettisoned when leaving earth orbit and another four when entering lunar orbit.

The ship will accelerate until reaching a speed just 200 mph short of escape velocity. This will take 34 minutes. The engines then shut down and the crewmembers leave the ship in spacesuits to detach and jettison the thirty-four empty fuel tanks. The spaceship then continues on into a lunar orbit, when four more tanks are dropped. The ship coasts on around the moon and returns to the earth.

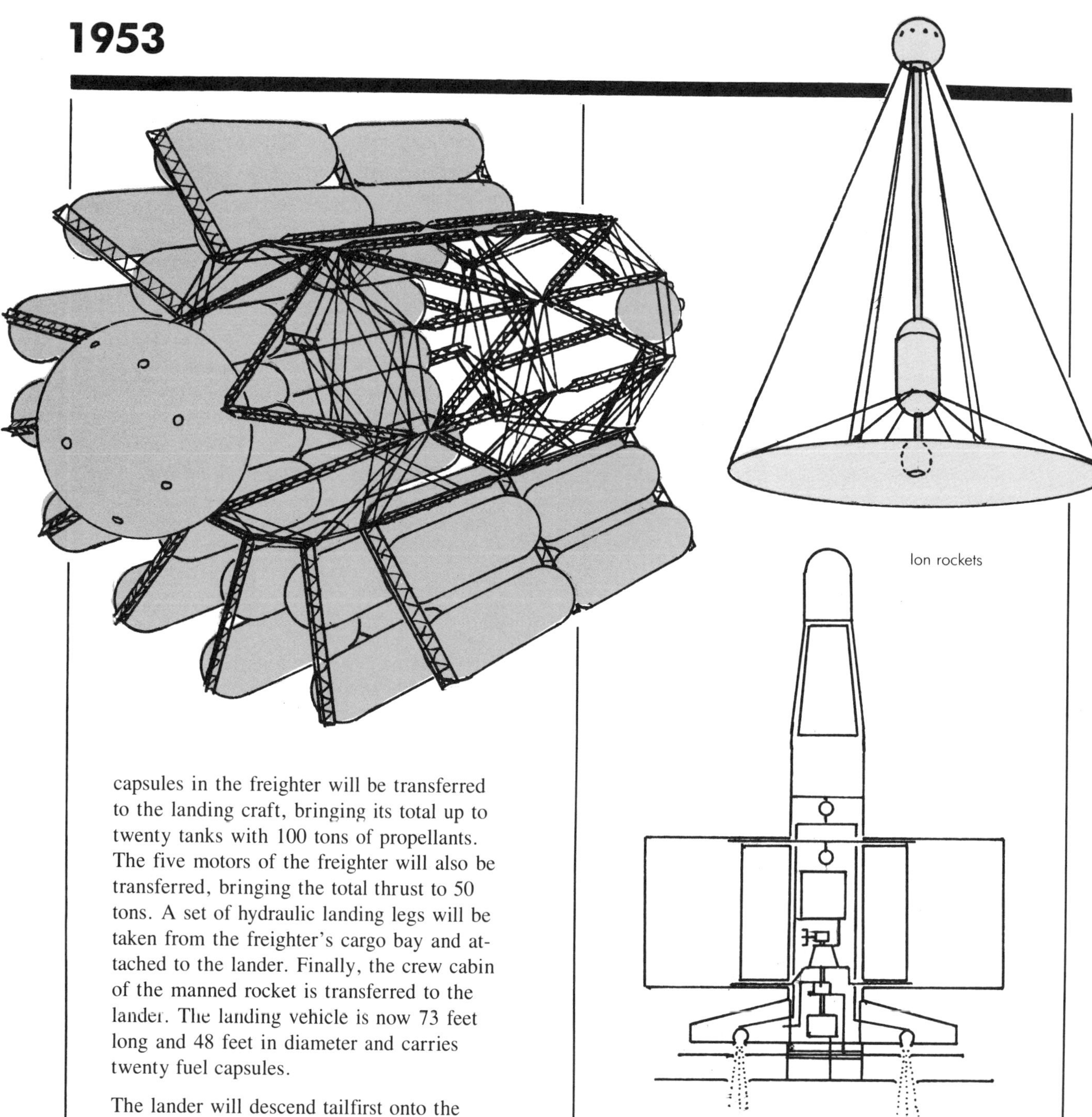

capsules in the freighter will be transferred to the landing craft, bringing its total up to twenty tanks with 100 tons of propellants. The five motors of the freighter will also be transferred, bringing the total thrust to 50 tons. A set of hydraulic landing legs will be taken from the freighter's cargo bay and attached to the lander. Finally, the crew cabin of the manned rocket is transferred to the lander. The landing vehicle is now 73 feet long and 48 feet in diameter and carries twenty fuel capsules.

The lander will descend tailfirst onto the lunar surface. Then, before takeoff the twelve tanks emptied during the landing maneuvers will be jettisoned. Once back in lunar orbit, the cabin will be replaced onto the Type A ship and the vehicle returned to the earth. The lander will be left in lunar orbit for future use. Similar techniques will be used for the exploration of Mars and Venus.

A lunar landing will require three types of spacecraft similar to the earlier reconnaissance ship in overall design but differing in detail: Type A, a manned rocket; Type B, a freighter rocket; and Type C, the landing rocket. The three astronauts will leave earth orbit in the manned vehicle and will control the two unmanned ships, the freighter and the lander. Upon arrival in circumlunar orbit, the crew will dismantle the nose section of the lander, leaving the tail section for the landing.

Gatland also describes a nuclear ion-propelled manned spaceship. The complete ship has a mass of 600 metric tons, including 300 metric tons of propellants. Another ion rocket has a spherical cabin at the opposite end of a boom carrying an enormous disc of beta-emitting materal (waste from nuclear power plants).

On the fiftieth anniversary of powered flight Lt. General James H. Doolittle commented for *Planes* magazine on what he foresaw for the coming fifty years: "I believe that if I have erred it is in the direction of conservatism . . . While many of the technical problems involved in the production of a successful earth satellite have not yet been solved, expenditure of sufficient effort and money should lead to the required answers. It is entirely probable that an earth satellite will be built within the next 50 years, and possible that attempts will be made to send missiles through space as far as the moon."

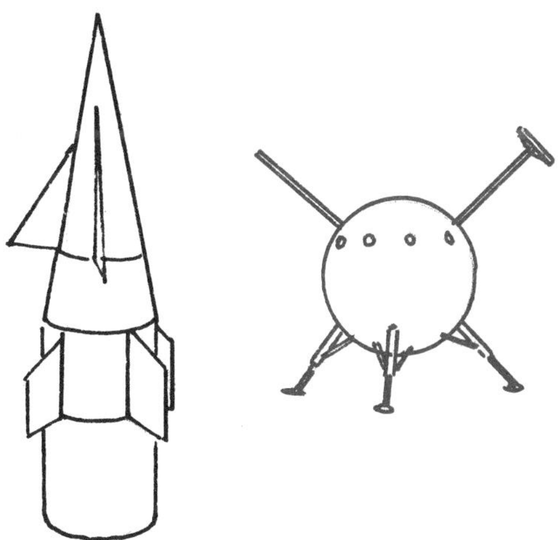

An interplanetary show in Dusseldorf displays models of a four-stage manned ferry rocket on its launch pad, a spherical moon lander and the interior of a spaceship cabin.

T. R. Nonweiler discusses the techniques of returning a glider from orbit using rocket braking, drogue parachutes, dive brakes together or in some combination.

The motion picture *Abbott and Costello Go to Mars* is released, though in spite of the title the two comedians end up on Venus instead where they meet its all-female population (entrants in the current Miss Universe competition). The spaceship is a winged V2 derivative clearly inspired by the paintings of Chesley Bonestell [see 1949]. It also makes an appearance as a background prop in *It Came From Outer Space*.

APRIL

Frank R. Paul and Alex Schomburg design a "1000-year" space ark. It is a large, hollowed-out asteroid, similar to the one suggested by Dandridge Cole in 1963 [see below; also see *Universe*, by Robert A. Heinlein (1941) and *The World, the Flesh and the Devil*, by J. D. Bernal (1929)].

MAY

Arthur C. Clarke visits Mrs. Walton C. John of Washington, D.C. (who believes she is the only American woman member of the British Interplanetary Society), to whom he predicts that man will reach the moon by 1978. Mrs. John disagrees. "Look at predictions in the past. Twenty-five years often have turned out to be five."

AUGUST 21

The D-558-2 sets an unofficial altitude record of 15 miles. The pilot is Marion E. Carl.

OCTOBER 3

A. V. Cleaver presents his paper "A Programme for Achieving Interplanetary Flight" at a meeting of the British Interplanetary Society. In it he updates and expands upon the themes he discussed in 1948 ("The Interplanetary Project").

He proposes a three-phase plan: I. the establishment of unmanned earth satellites; II. the launch of manned rockets with perhaps a circumnavigation of the moon; and III. the arrival of true interplanetary flight, with manned landings on the moon and other planets. He considers not only the technical problems, but the problems of finance and motivation as well.

Cleaver argues against those who have held that spaceflight will be too expensive for mankind *ever* to undertake by quoting some other extraordinarily expensive "projects." Compared to the £1.5 trillion spent on WW II, the £1 billion spent on aviation since 1903 (at a current rate of some £4 billion a year), or the £1 billion a year spent on atomic research by the United States, the £1 billion total cost of space exploration is fairly modest, even spent at a rate of £100 million a year.

Cleaver thinks that the three-phase program will be an international affair best undertaken by the IAF, or perhaps a similar body created especially for the project. The unmanned rocket developed during Phase I will be a relatively small 100-ton three-step vehicle carrying a payload of only a few hundred pounds to a 300 km orbit. It could be launched by 1965 at a total cost of about £50 million.

Phase II will concentrate on the construction of a few small rockets, with crews of only two or three astronauts (as opposed to the von Braunian 7,000-ton manned rockets). Cleaver's three-step manned vehicle will weigh only about 500 tons and take its crew into a 500 km orbit. The first flights of a small piloted rocket satellite will take place in 1975. Total cost of development: £250 million.

One of the practical functions of the manned satellites might be the establishment of orbiting communications stations.

Development of planetary flights will take place in the 1980s or 1990s with a lunar landing made in 2000.

Dunning

November 20

The D-558-2 *Skyrocket*, piloted by A. Scott Crossfield, becomes the first aircraft to fly at twice the speed of sound (Mach 2, or 1,291 mph). The sleek, white rocket is designed and built by Douglas Aircraft. It is powered by a Reaction Motors four-chambered rocket engine virtually identical with the one used by the Bell X-1 [see above].

The *Skyrocket* is the product of a series of D-558 research aircraft which began on a Navy initiative in 1945. The first three were powered by single General Electric turbojet engines and known as the *Skystreak* series (D-558-1). The last three, the *Skyrocket* series (D-558-2), are equipped with both a tur bojet engine and a rocket engine. Although both planes could take off from the ground, the similarity does not go much further. The D-558-1 is limited to Mach 1, while the much more powerful D-558-2 easily exceeds that speed. The straight wing and tail surfaces of the *Skystreak* are replaced by rakishly sweptback airfoils on the *Skyrockets*. Where the D-558-1's were painted glossy red, the rocket-planes are gleaming white.

The *Skyrocket* is of mixed aluminum and magnesium construction. It has a jettisonable nose section which acts as an escape capsule for the pilot. The rocket-plane is 42 feet long with a wingspan of 25 feet.

Just as with the Bell X-1, safety considerations eventually cause the *Skyrocket* to be modified for air launching. It is converted to an all-rocket version (the D-558-2 #2 and #3) and carried aloft slung below the bomb bay of a specially adapted B-29 Superfortress. By being able to carry additional rocket fuel, in the space given up by the jet engine, the D-558-2 is able to exceed Mach 2. The last in the series are flown as mixed jet and rocket.

The first *Skyrocket* was flown in 1948. The first all-rocket version was flown in 1950, immediately reaching a speed of Mach 1.88 and setting an unofficial altitude record of 79,494 feet. All three rocket-planes will be retired in 1956. The D-558-2 #2 is currently on display at the National Air and Space Museum in Washington, D.C.

NASA

December 12

The Bell X-1A sets an unofficial world speed record of 1,650 mph at an altitude of 70,000 feet. The pilot is Chuck Yaeger.

1954

S. A. Kosberg of the U.S.S.R. works on developing a monopropellant rocket engine for aircraft (the fuel planned is isopropylnitrate).

The Russians describe a possible manned moon mission.

Jack Coggins publishes a space station design. It is a fat disc, with an observatory dome at its center. An atomic power plant is contained in a sphere attached to the station by a boom. It is intended that the crew be weightless.

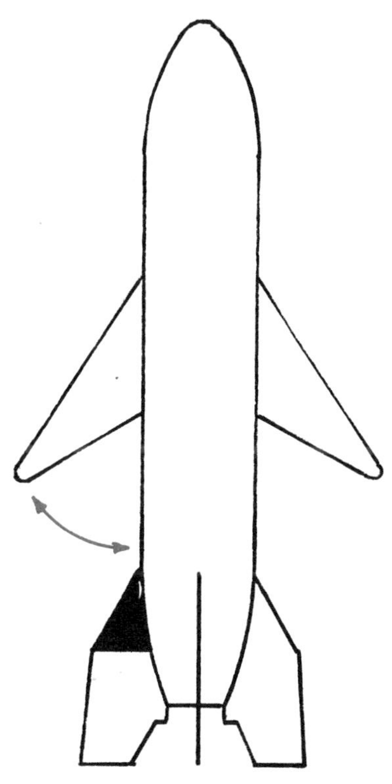

Riders to the Stars is a film written by Curt Siodmak and directed by star Richard Carlson. A mission into space is ordered to discover why cosmic rays destroy steel but leave iron meteorites unharmed. Unless the secret is learned, high-altitude spaceflight is impossible. Three one-man rockets are launched, each with a "scoop" in its nose. The idea is to snatch a virgin meteorite while it is still in space, before its entry into the earth's atmosphere has destroyed its mysterious protective coating. The film is marred by terrible special effects (the launch scene uses stock footage of two V2's and an Aerobee [!]—which just marginally resemble the rockets shown later in space), but is interesting in devoting most of its time to the selection and training processes of the astronauts.

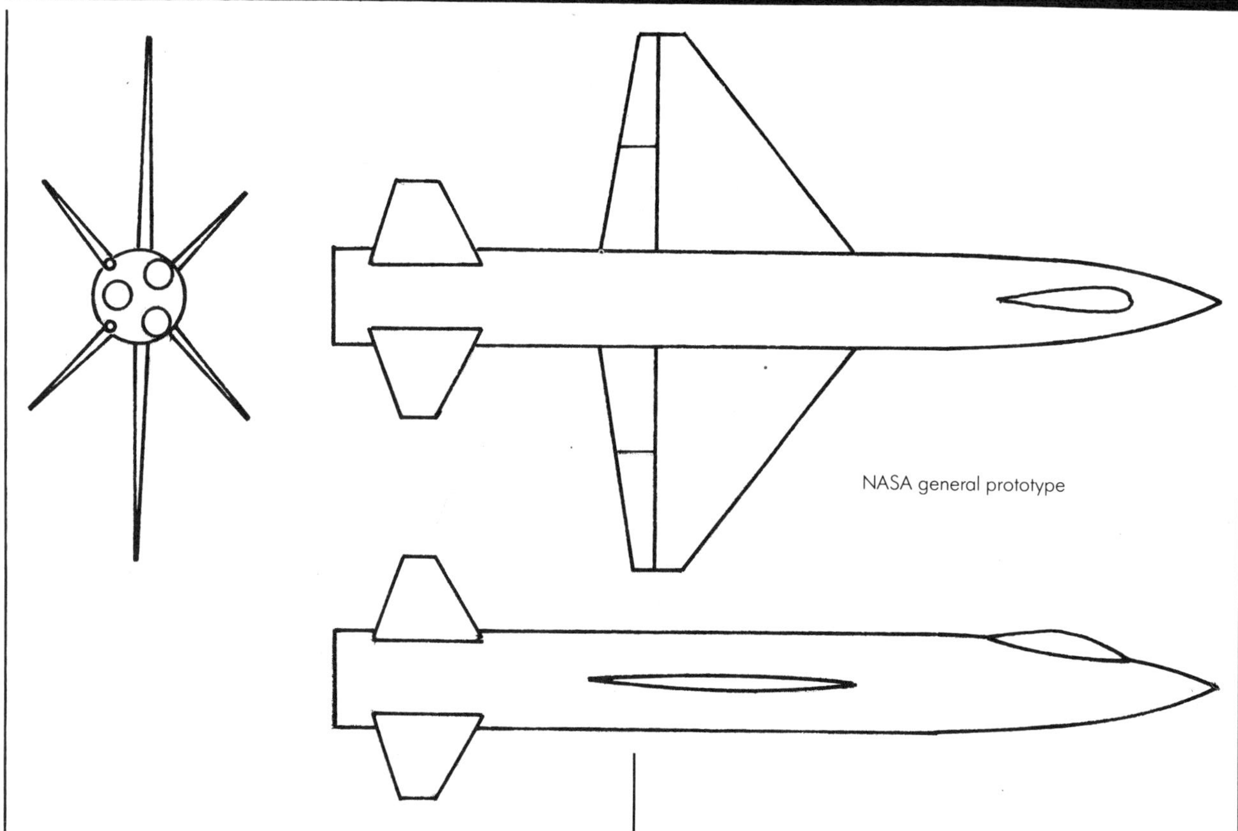

NASA general prototype

The development of the X-15 research program begins. It is a joint program sponsored by NACA (the National Advisory Committee for Aeronautics, which becomes NASA during the course of the X-15 program), the U.S. Air Force, the U.S. Navy and private industry. The purpose of the aircraft is to study the problems of atmospheric and spaceflight at speeds of Mach 6.6 and higher, and at altitudes of 12 to 50 miles. North American Aviation (NAA) is the prime contractor.

The basic design criteria have been set by the U.S.A.F. as Project 1226. It calls for a 47.5-foot aircraft with cylindrical fuselage 5 feet in diameter, double-delta wings and four stabilizing fins placed at 45 degrees to the wings (the lower fins folding for landing). The aircraft is powered by three Hermes A-1 rocket motors. Bell, Douglas, North American and Republic all respond to the call for bids on the project. Bell submits its Model D-171, an aircraft not dissimilar to its already successful series of rocket-propelled X-planes. Model D-171 is 44 feet 8 inches long, with a wingspan of 25 feet 8 inches. Douglas's Model 684 is 46 feet 9 inches long, with a wingspan of 19 feet 6 inches—resembling not a little the winning NAA design. Republic's Model AP-76 is 53 feet 4 inches long, with a wingspan of 27 feet 6 inches and a sleek fuselage unbroken by a cockpit blister.

The aircraft that is developed is the link between atmospheric flight and spaceflight. It has wings and control surfaces so it can fly within the atmosphere, but it is also equipped with wingtip and nose thrusters so that it can maneuver in the near-vacuum at the fringe of outer space. The X-15 is so much like a true spaceship that its pilot is required to wear a full-pressure space suit when flying it (and its U.S.A.F. pilots are awarded astronaut's wings). The maximum altitude reached by an X-15 is over 67 miles—not exceeded until the flights of the Mercury spacecraft. It is seriously considered at one time (by Thiokol) to make the X-15 a full-fledged spacecraft by mounting it atop a cluster of four Titan boosters or on a Navajo rocket (in the latter case, the X-15 pilot bailing out after reentry), and launching it into earth orbit. Other abandoned plans for the rocket-plane include turning it into a manned satellite booster, or replacing its stubby wings with narrow delta wings with upturned wingtip stabilizers.

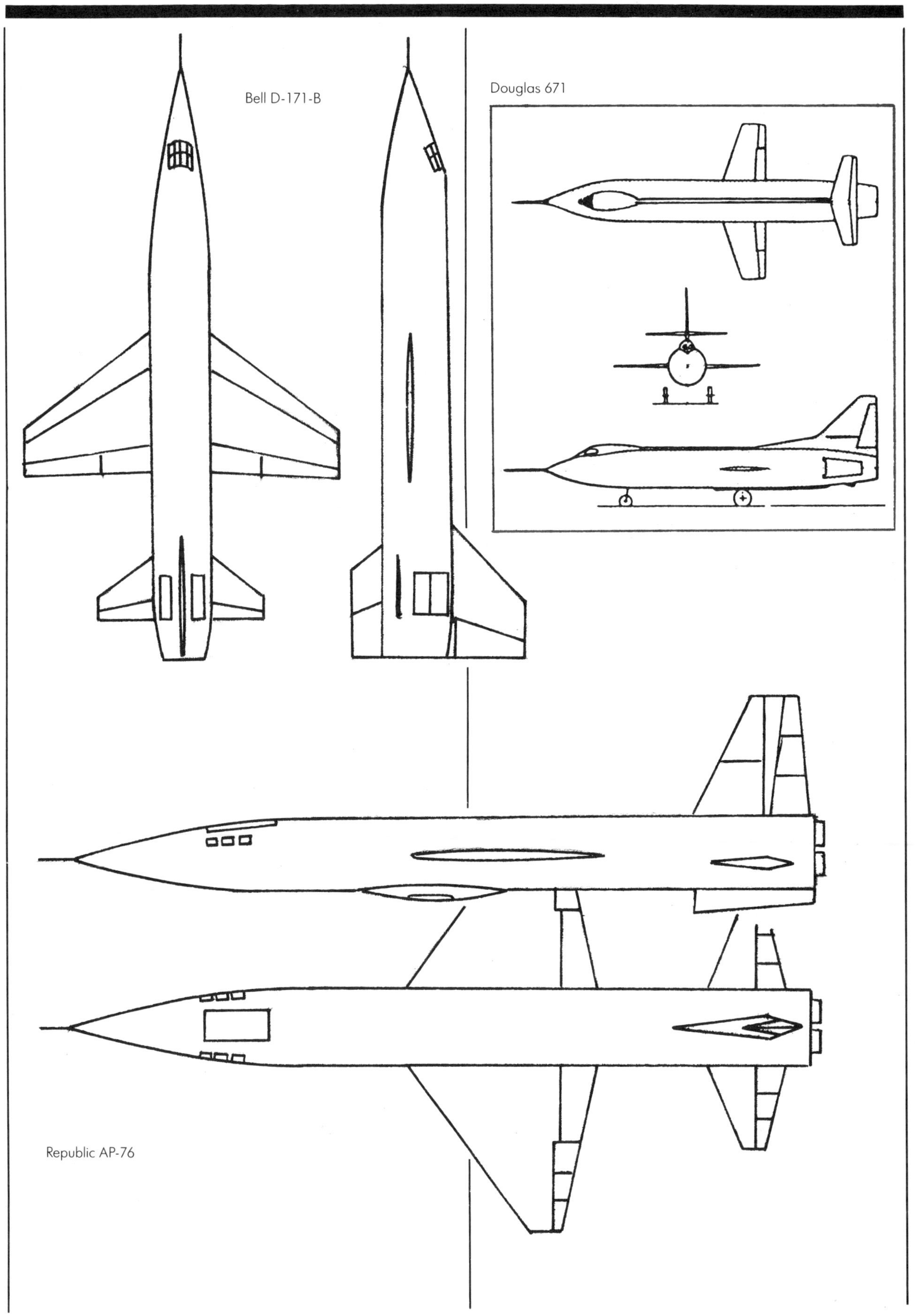
Bell D-171-B
Douglas 671
Republic AP-76

1954

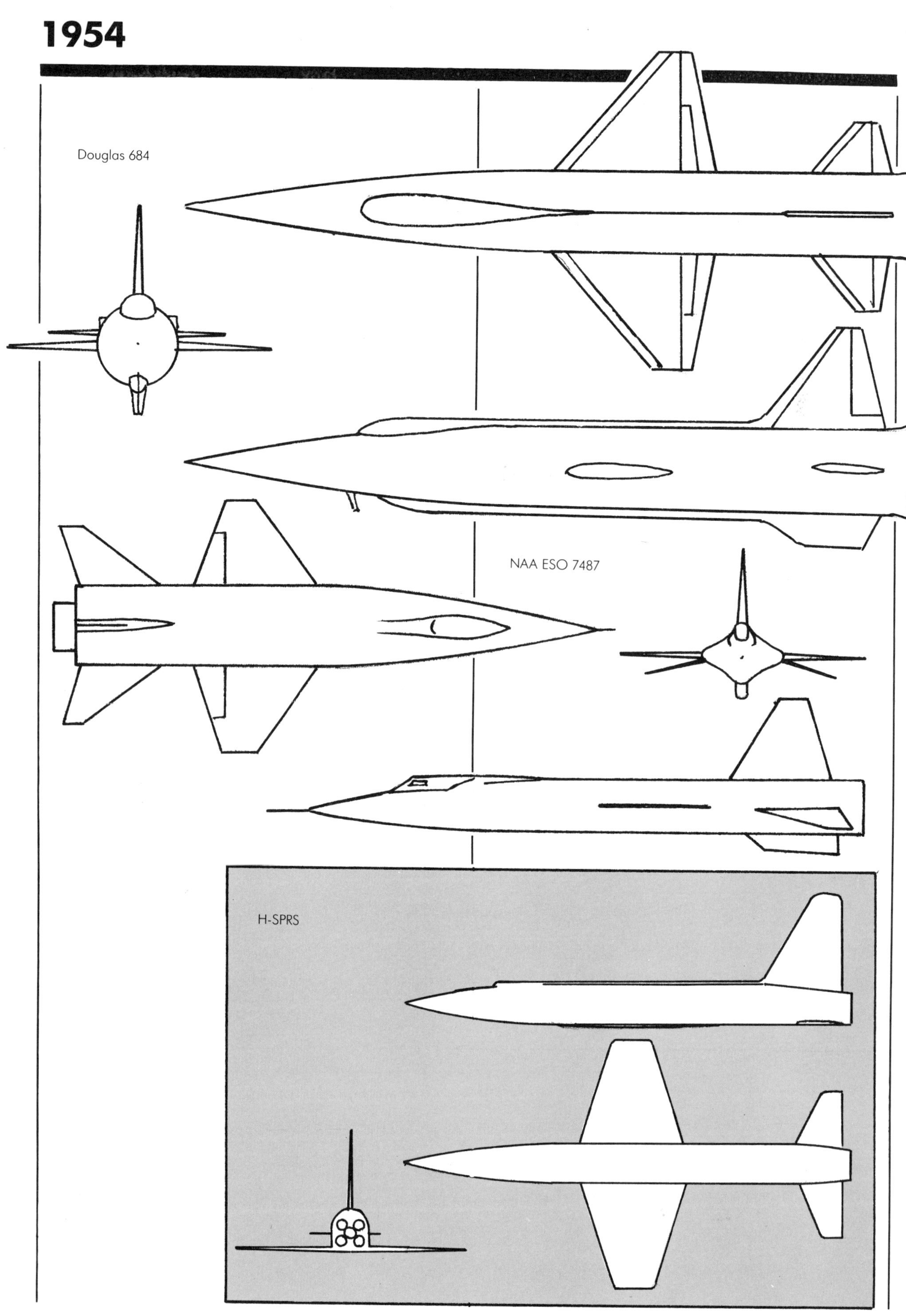

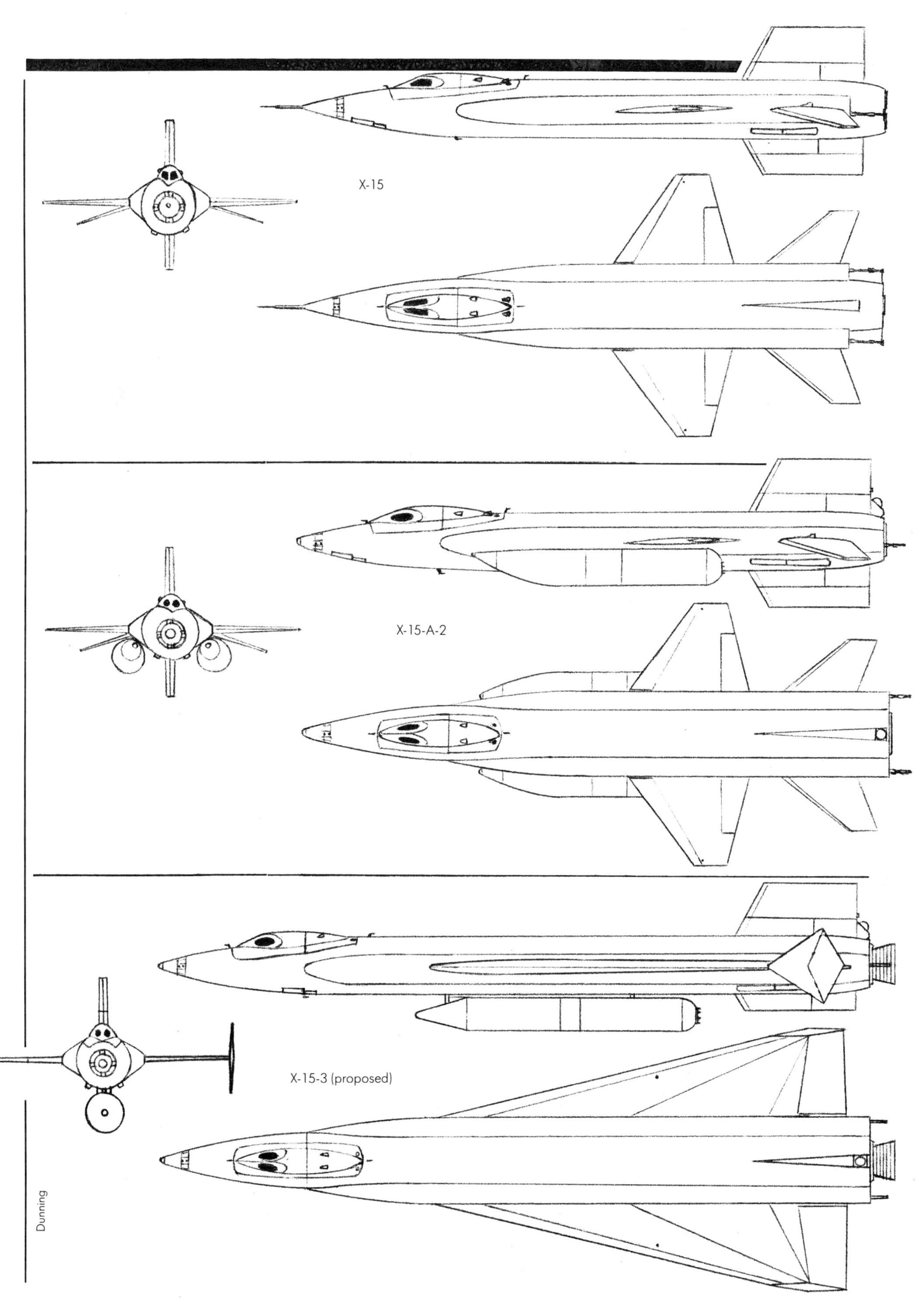
X-15
X-15-A-2
X-15-3 (proposed)
Dunning

NASA

Three X-15 aircraft are built. One is destroyed, but the remaining two have a long and successful service life, accumulating a total of 199 flights. The X-15 is capable of speeds in excess of six times the speed of sound. At such speeds, air friction heats its nickel-alloy skin to a temperature of 1,200 degrees F. The X-15 is just over 50 feet long with a wingspan of about 22 feet 4 inches. It is powered by a 57,000-pound thrust rocket engine built by Thiokol (née Reaction Motors).

The X-15 is also suggested as a recoverable booster system for the Blue Scout rocket. It would be able to place the smaller rocket into orbit at a cost saving per launch of $250,000. The primary modification to the X-15 would be the addition of a 500-lb retractable launch pylon on the underside of its fuselage and realignment of its engine. In the idea as proposed by the Ford Company's Aeronutronic Division, the B-52 carrier aircraft would climb to about 45,000 feet. Upon release from the B-52, the X-15 will climb to around 156,000 feet, launch the Blue Scout rocket, and coast on to 300,000 feet before making its reentry maneuver. The system could put a 50-pound payload into an 800-mile orbit.

Krafft Ehricke presents in the *Proceedings of the International Astronautical Congress* a design for a dumbbell-shaped space station.

The Air Research and Development Command of the U.S. Air Force sponsors several studies on boost glide systems. These are finally combined into a single plan, accepted by the U.S.A.F. in 1957. In November of that year the first directives will be issued for the X-20 Dyna-Soar program. In June 1958 two competing teams will be selected to prepare studies: Boeing on one hand and Martin Marietta and Bell Aircraft on the other. Boeing will receive the system contract in 1959, and Martin the associate booster contract.

Guy Waller expresses concern that garbage disposed from spacecraft might prove an eventual hazard to spaceflight: "Fifty years from now, if you want to steer a rocket to the moon, you'll need no gadgets, only one piece of advice: Follow the empty cans."

JANUARY

A writer to the British magazine *Practical Mechanics* baldly states that "Interplanetary travel is impossible!" This remark is based on the "fact" that "once outside the Earth's gravitational field a spaceship will be irresistibly attracted to the Sun and will proceed in that direction at ever-increasing speed!"

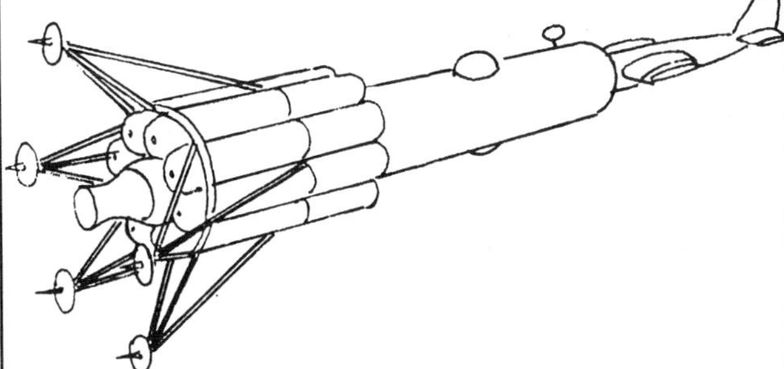

JUNE

An illustration accompanying an article published in the Soviet magazine *News* depicts a Russian spaceship designed by Kirill Stanyukovich. A cluster of cylindrical tanks—two clusters of eight tanks each—surround a single engine. Five landing legs are attached to the tanks, indicating, perhaps, that the ship is intended for a lunar landing. A cylindrical personnel cabin is mounted ahead of the fuel tanks. Inserted nose-first into the front end of the cabin is a conventional jet aircraft. It has extendable wings and is intended for the cosmonaut's return to the earth.

SUMMER

X1-A pilot Arthur Murray reaches an altitude of 90,000 feet.

JULY

The Bell X-2 reaches the speed of Mach 2.9.

AUGUST 5

The Bell X-2 makes its first glide flight.

SEPTEMBER 20

The *Japan News* reports that a local astronautical society will start registering the names of people interested in going to the moon.

NOV. 30–DEC. 3

Darrell Romick, R. E. Knight and J. M. Van Pelt present their paper "A Preliminary Design Study of a Three-Stage Satellite Ferry Rocket Vehicle with Piloted Recoverable Stages" to the ninth annual meeting of the American Rocket Society. [See 1956 for a full description.]

This is the debut of the METEOR recoverable shuttle concept, which had its origins in a study begun in late 1949.

DECEMBER

Admiral Lord Mountevans states: "For my part I should like to see some of the ingenuity and scientific knowledge now given to what I believe to be the never-never land problem of interplanetary spaceflight directed towards defence and, in particular, naval defence problems." He goes on to say, "We [the British] are a sea-going not a moon-going people."

ca. 1954

Edmund Sawyer of the Pacific Rocket Society develops techniques for planetary landings. A reentering rocket will lose speed by firing retrorockets. Once it has entered the atmosphere, drag brakes and eventually parachutes will slow it further. Parachutes and retrorockets will allow the rocket to settle down gently on shock absorbers.

T. Nonweiler of the BIS, in a paper read to the Second International Congress on Aeronautics, describes a delta-winged reentry vehicle.

A moon rocket is proposed by Dr. U. Stan: a long cylinder carrying a number of atomic bombs. These are ejected one at a time from the rear of the rocket, where they are detonated. The rocket is provided with a broad, flaring rear, just beyond which the explosion occurs. Dr. Stan estimates that only 1 or 2 dozen such "kicks" will be sufficient to

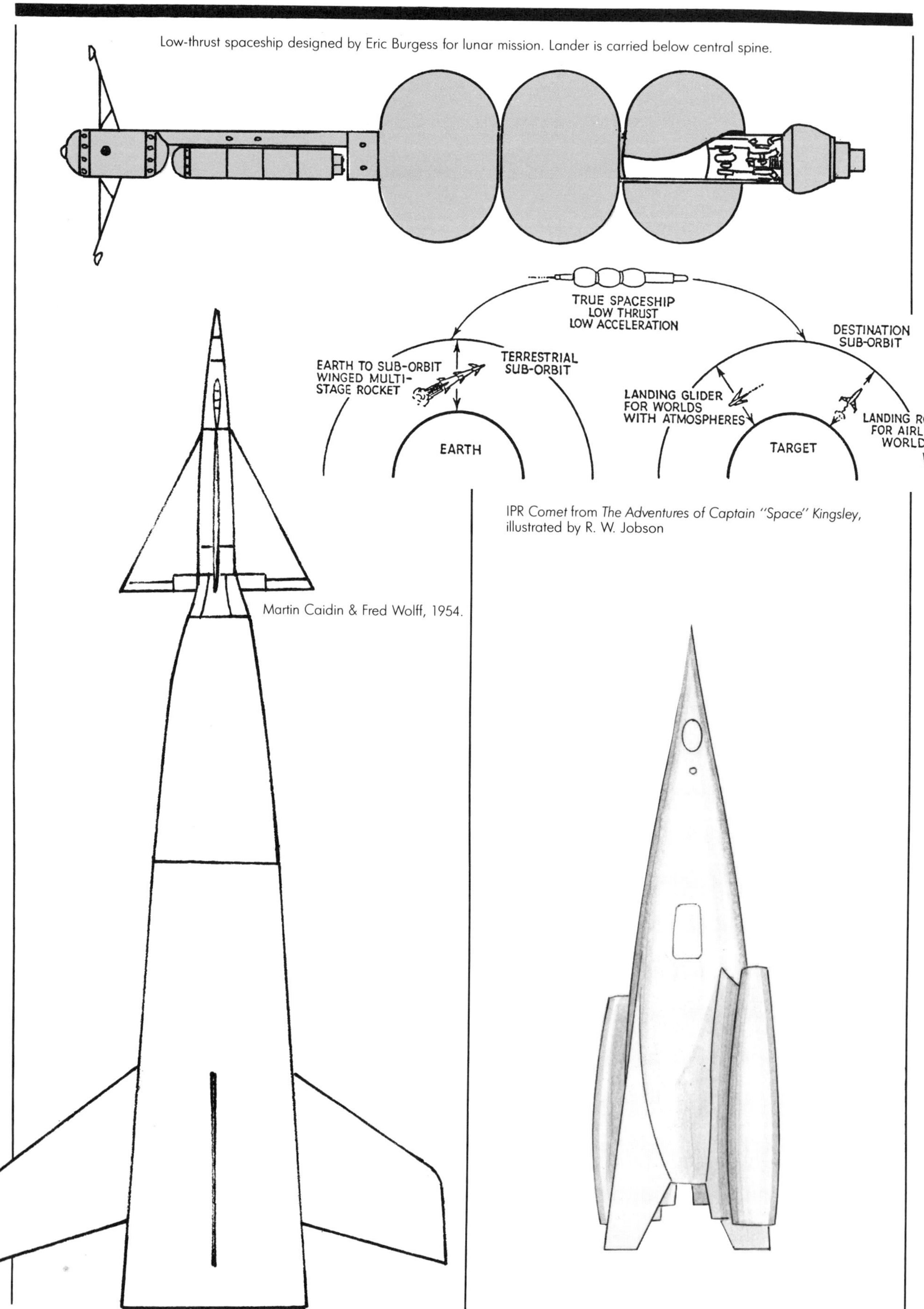

Low-thrust spaceship designed by Eric Burgess for lunar mission. Lander is carried below central spine.

Martin Caidin & Fred Wolff, 1954.

IPR Comet from *The Adventures of Captain "Space" Kingsley*, illustrated by R. W. Jobson carry the ship into space. [Also see Orion and Cole, 1959.]

1954–1955

Constantin Paul van Lent suggests launching spacecraft via an enormous gun. A vertical shaft will be bored into the summit of a 15,000-foot mountain. At the bottom of the shaft will be a spherical chamber. The neck where the shaft enters the chamber will be plugged by a thick cylinder of concrete, above which rests the rocket. Two sets of three lugs on the sides of the rocket fit into slots in the sides of the shaft. To launch the rocket an atomic bomb is detonated in the center of the spherical space. Simultaneously, a second atomic bomb is detonated on the peak of the mountain, creating, according to Lent, a momentary vacuum through which his rocket can pass.

Lent also thinks that a complete space station could be launched from the earth this way. The station will be constructed of reinforced concrete. It will be shaped like a thick disk surrounding a sphere (which will contain the station's attitude-maintaining gyroscope). This will be placed directly over the nuclear bomb-charged gun (which will be 2 to 500 feet deep and 25 feet in diameter). The bomb will be supported atop a 50-foot tower at the bottom of the gun. The remainder of the spherical chamber is filled with powdered steel and aluminum. This will turn into gas when the bomb is detonated.

One of Lent's justifications for his idea is "the planets are in eons past launched from the mother Sun by means of atom explosions."

This grandiose conception will be later simplified by Lent in 1957 to employ conventional explosives to launch a 21,000-pound rocket to the moon.

1955

Project Rover is a joint project of the U.S. Atomic Energy Commission and the U.S. Air Force to develop nuclear rocket propulsion.

The "Moonliner", designed with the advice of Wernher von Braun and Willy Ley, is erected in the Tomorrowland section of the Disneyland amusement park. Ostensibly operated by TWA [see *2001*, 1968] as an excursion liner to the moon, this graceful rocket stands in one-third scale mockup in Disneyland. Its full size is to be 80 feet tall with tripod landing legs that fold against the sides of the rocket while in flight. It carries 102 passengers and crewmembers. An atomic engine (using hydrogen as the reaction mass) propels it at a top speed of 172,000 mph. Apparently its initial boost is by way of a nitric acid/hydrazine chemical rocket. The assumed date of its maiden voyage is 1986, in time for Halley's Comet.

In the park ride, visitors are led into a circular chamber in an adjoining building. Motion picture screens are disguised as "viewports," one above the audience and one in the floor. At the moment of takeoff, the theater seats vibrate and the viewscreens show the spaceport rapidly disappearing beneath the flaming exhaust, while the sky above grows darker. As the earth recedes into the black, starry sky in the floor screen and the moon grows larger overhead, meteors shoot past and a closeup view of Halley's Comet is provided.

[It is ironic that the Moonliner was dismantled before its predicted "flight" took place; however, there are now plans to recreate this classic spaceship in a renovated Tomorrowland.]

NASM

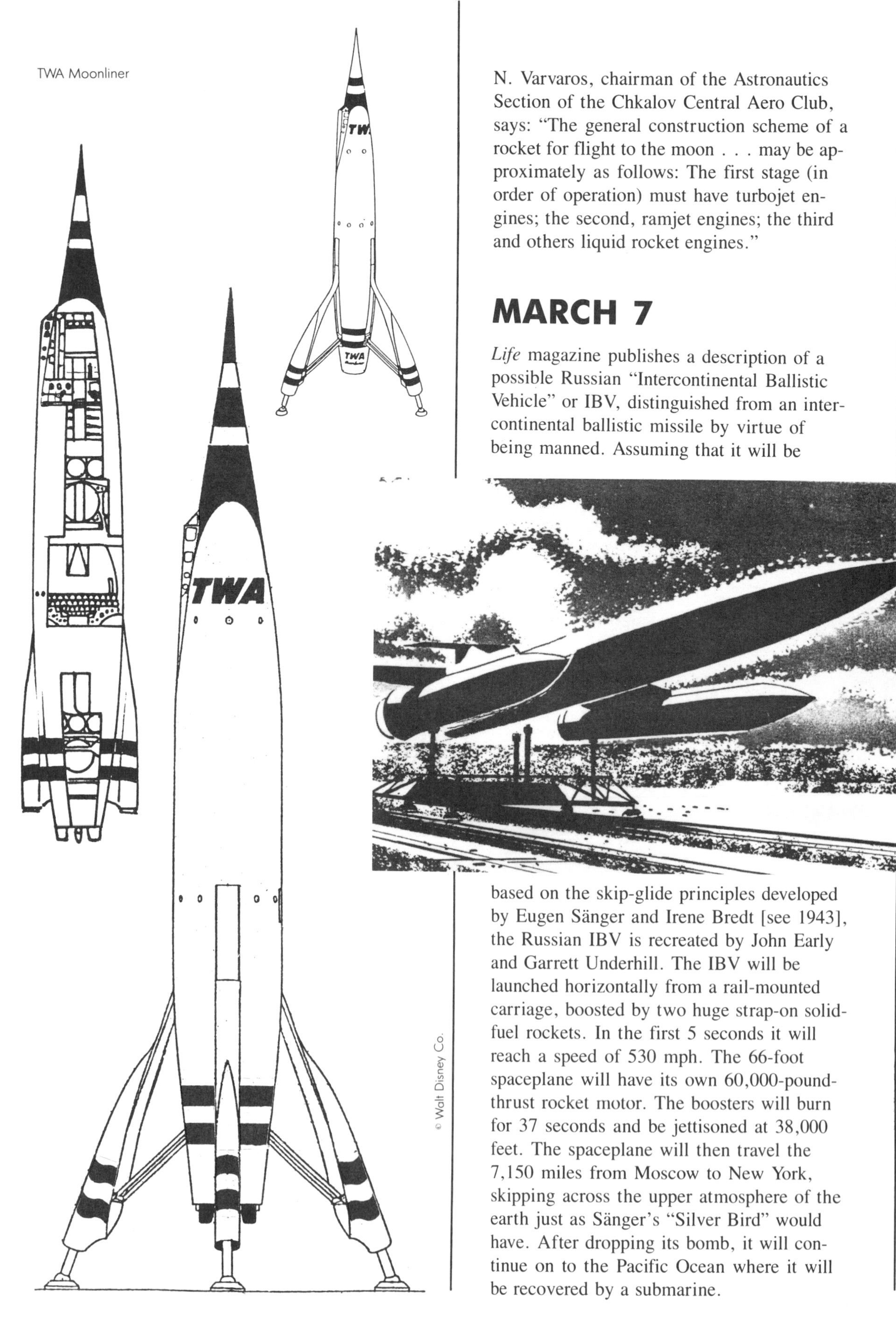

TWA Moonliner

© Walt Disney Co.

N. Varvaros, chairman of the Astronautics Section of the Chkalov Central Aero Club, says: "The general construction scheme of a rocket for flight to the moon . . . may be approximately as follows: The first stage (in order of operation) must have turbojet engines; the second, ramjet engines; the third and others liquid rocket engines."

MARCH 7

Life magazine publishes a description of a possible Russian "Intercontinental Ballistic Vehicle" or IBV, distinguished from an intercontinental ballistic missile by virtue of being manned. Assuming that it will be based on the skip-glide principles developed by Eugen Sänger and Irene Bredt [see 1943], the Russian IBV is recreated by John Early and Garrett Underhill. The IBV will be launched horizontally from a rail-mounted carriage, boosted by two huge strap-on solid-fuel rockets. In the first 5 seconds it will reach a speed of 530 mph. The 66-foot spaceplane will have its own 60,000-pound-thrust rocket motor. The boosters will burn for 37 seconds and be jettisoned at 38,000 feet. The spaceplane will then travel the 7,150 miles from Moscow to New York, skipping across the upper atmosphere of the earth just as Sänger's "Silver Bird" would have. After dropping its bomb, it will continue on to the Pacific Ocean where it will be recovered by a submarine.

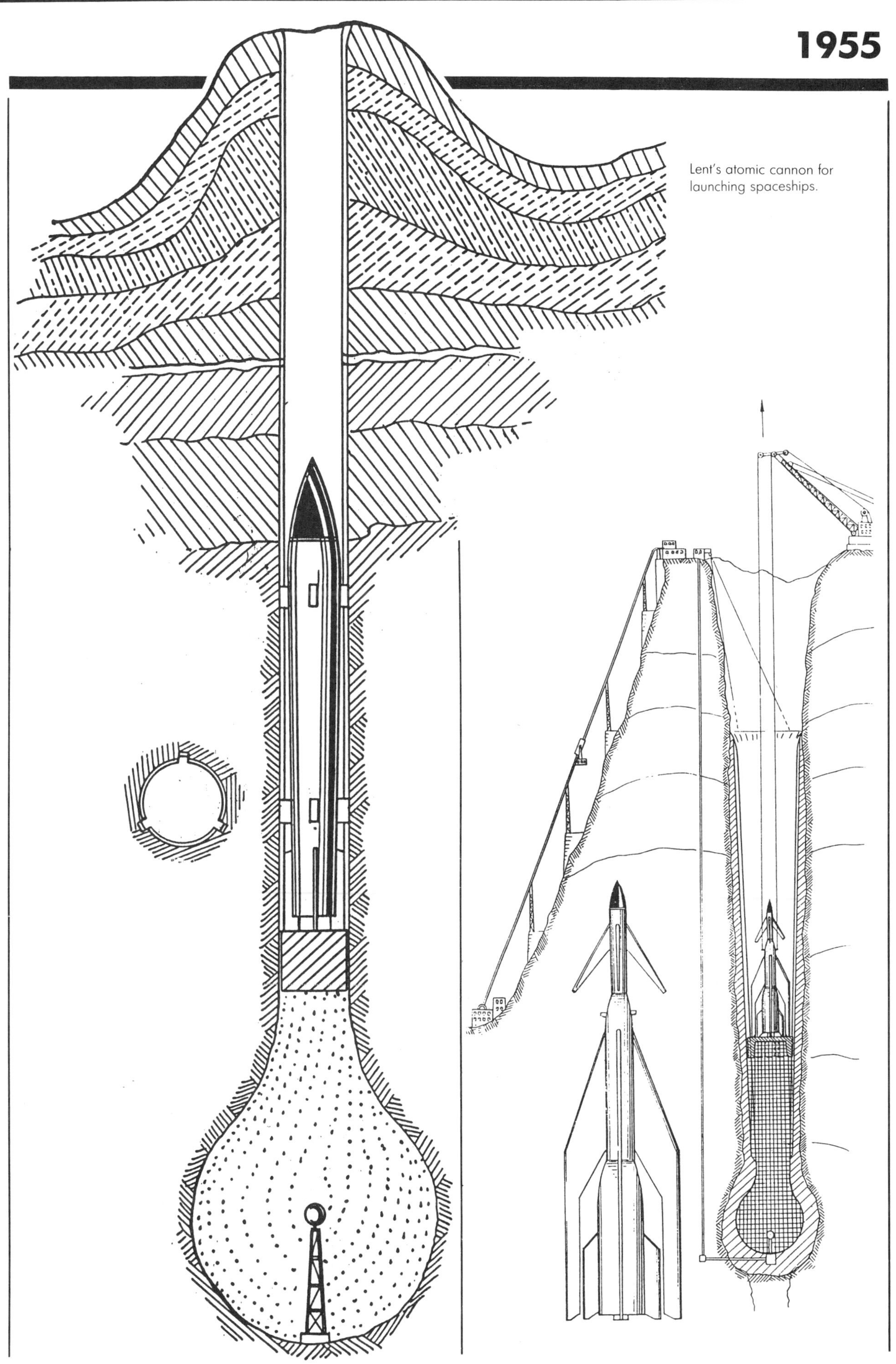

Lent's atomic cannon for launching spaceships.

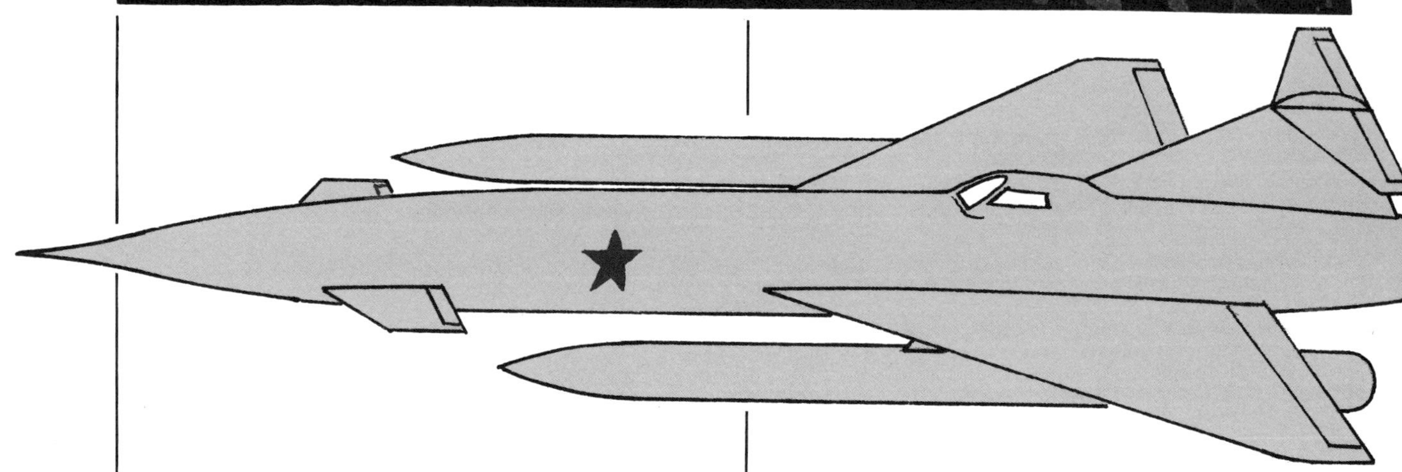

The IBV will weigh about 3 tons empty and carry a single pilot in a pressurized cabin. For fuel it will carry a kerosene-gasoline mixture and hydrogen peroxide as the oxidizer.

The article reports that engineer Grigori Tokady (who later will defect to England) quotes Stalin as saying: "Sänger planes and their construction should be our immediate objective . . . Tokaev, we wish you to exploit Sänger's ideas in every way . . . " Tokady is eventually given the task of arranging the abortive attempt to kidnap Sänger to the Soviet Union [also see April 14, 1947].

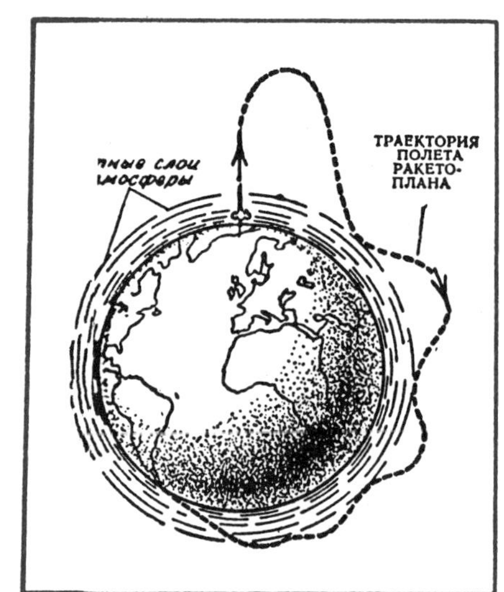

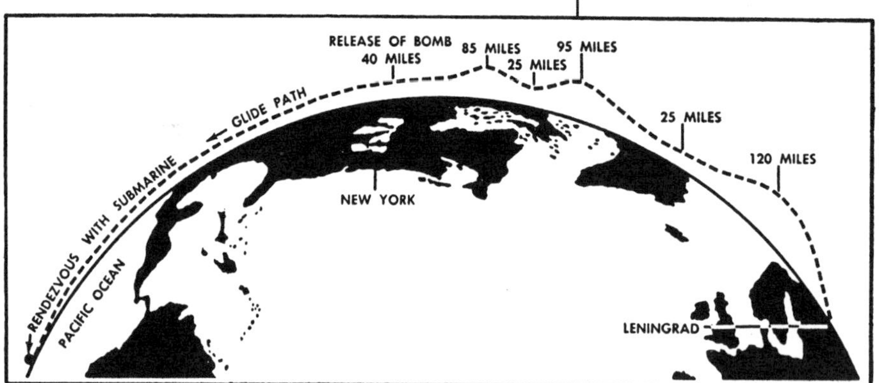

MARCH 9

The "Man in Space" segment of the *Disneyland* television series is broadcast. It is the first of three presentations on the history, techniques and future of space exploration.

The television series is inspired by the recent *Collier's* magazine articles [see 1952–1954]. Disney calls upon many of the same experts, notably Wernher von Braun and Willy Ley, as well as astronomer E. C. Slipher, electric-propulsion authority Ernst Stuhlinger and space medicine expert Heinz Haber. All of these men eventually appear in the series themselves.

The first episode, "Man in Space," is a 48-

minute film narrated by animator and director Ward Kimball. It covers the history of rocketry, from the Chinese to Project Bumper [see 1949–1952, above]. The problems facing human beings living in space are next discussed by Haber, accompanied by a delightfully humorous cartoon segment. Finally, von Braun describes the future four-stage orbital rocket he envisions. "I believe," he tells the audience, "a practical passenger rocket can be built and tested within ten years." The spaceship shown is a smaller, slightly updated version of the *Collier's* ferry (most of the updating being cosmetic restyling by the Disney staff artists). The first stage has twenty-nine rocket engines, burning 1,060 tons of fuel. The second stage has eight engines, using 155 tons of fuel, the third stage has one engine, and the manned fourth stage, carrying a crew of ten, has a single engine and 13 tons of fuel. Just as in the magazine series, the initial testing and

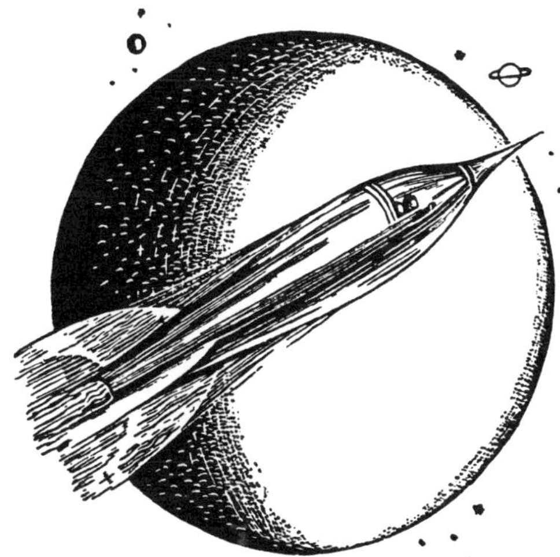

Illustration by Robert Henneberger from *The Wonderful Flight to the Mushroom Planet.*

Spaceship and manned booster designed by Morris Scott Dollins.

Dollens

training of ship and crew are described.

The program ends with a dramatization of a manned orbital rocket launch. The ship takes off from an unnamed Pacific island, attains orbit and returns to the earth glider-fashion, landing like a normal aircraft.

The second film, "Man and the Moon," features a live-action segment dramatizing a manned trip around the moon. The 53-foot spacecraft employed is the relief ship von Braun described in *The Exploration of Mars* [see above]. The program opens with a humorous animated history of lunar legends and superstitions. This is followed by the live-action segment. For this, a full-scale spaceship cockpit has been constructed. To provide authenticity, pilot's seats have been obtained from Boeing and Douglas, along with helmet prototypes.

The expedition is manned by four astronauts. For them, von Braun has developed his "bottle suit" spacesuit concept. This is in reality a minispaceship, allowing the astronaut to work in a shirtsleeve environment. Maneuvered with gyroscopes and two small rocket engines, it is equipped with seven remote handling arms operated by the pilot.

A premier showing is held for the members of the American Rocket Society, who receive it enthusiastically.

The third and final program, "Mars and Beyond," is broadcast in 1957. It follows the format of the preceding two films in that an animated, humorously presented description of the history of Mars observation and the possibilities of life on that planet is combined with a realistic, documentary-style presentation of a manned flight there. For this, spacecraft designed by Ernst Stuhlinger are depicted [also see 1957]. These are atomic-powered, ion-driven ships. The manned section of each ship is a torus, rotated to provide artificial gravity. From the perimeter of the torus, a vast disk spreads 500 feet in diameter to carry the cooling radiators for the atomic reactor. From the center of the crew torus, perpendicular to the plane of the disk, a slender column over 200 feet long descends, making the entire spacecraft look very much like an opened Japanese parasol. At the far end of the column is the reactor. Attached to the column is the landing craft.

The reactor at the "handle" of the umbrella heats silicon oil into steam. This passed up the central pipe to a turbine at the axis of the passenger torus. This in turn powers a generator producing electric current. This energizes a platinum grid through which cesium vapor is passed. The atoms of cesium are ionized. Since their charge is the same as the grid's, they are repelled. The recoil of the escaping ionized cesium creates about 110 pounds of thrust. The used silicon oil steam passes from the turbine to the "umbrella" radiator where it is condensed for reuse.

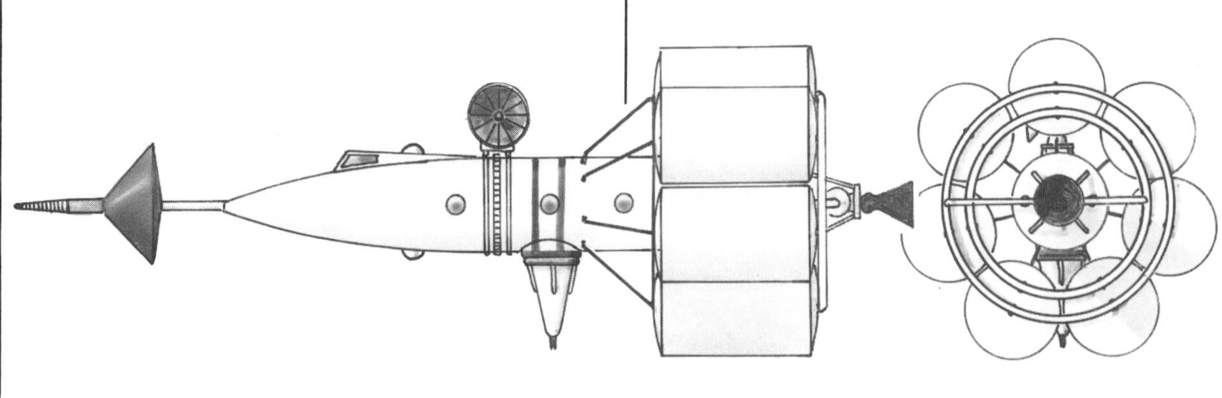

VON BRAUN MARS EXPEDITION RELIEF SHIP - 1956

Each of the Mars ships weighs 730 tons, of which 365 tons is propellants. They each carry a payload of 150 tons, including a crew of twenty. Although the thrust of each ship is only 110 pounds, this can be maintained continuously for as long as desired. Eventually, any velocity desired can be attained.

The Mars fleet orbits the earth in an ever-widening spiral, as speed builds up. After 107 days they will have circled the earth 377 times. They break free from the earth and begin the long glide to Mars. Arriving at the red planet, landing parties board the streamlined spacecraft attached to the ion-thruster on the central pipe. These ships are ordinary chemical rockets.

Now begins the descent from the 600-mile orbit. A parachute helps slow the fall of each "landing boat." When the rocket is only a few miles above the Martian surface, its engine begins firing. It eventually lands on four outrigger legs.

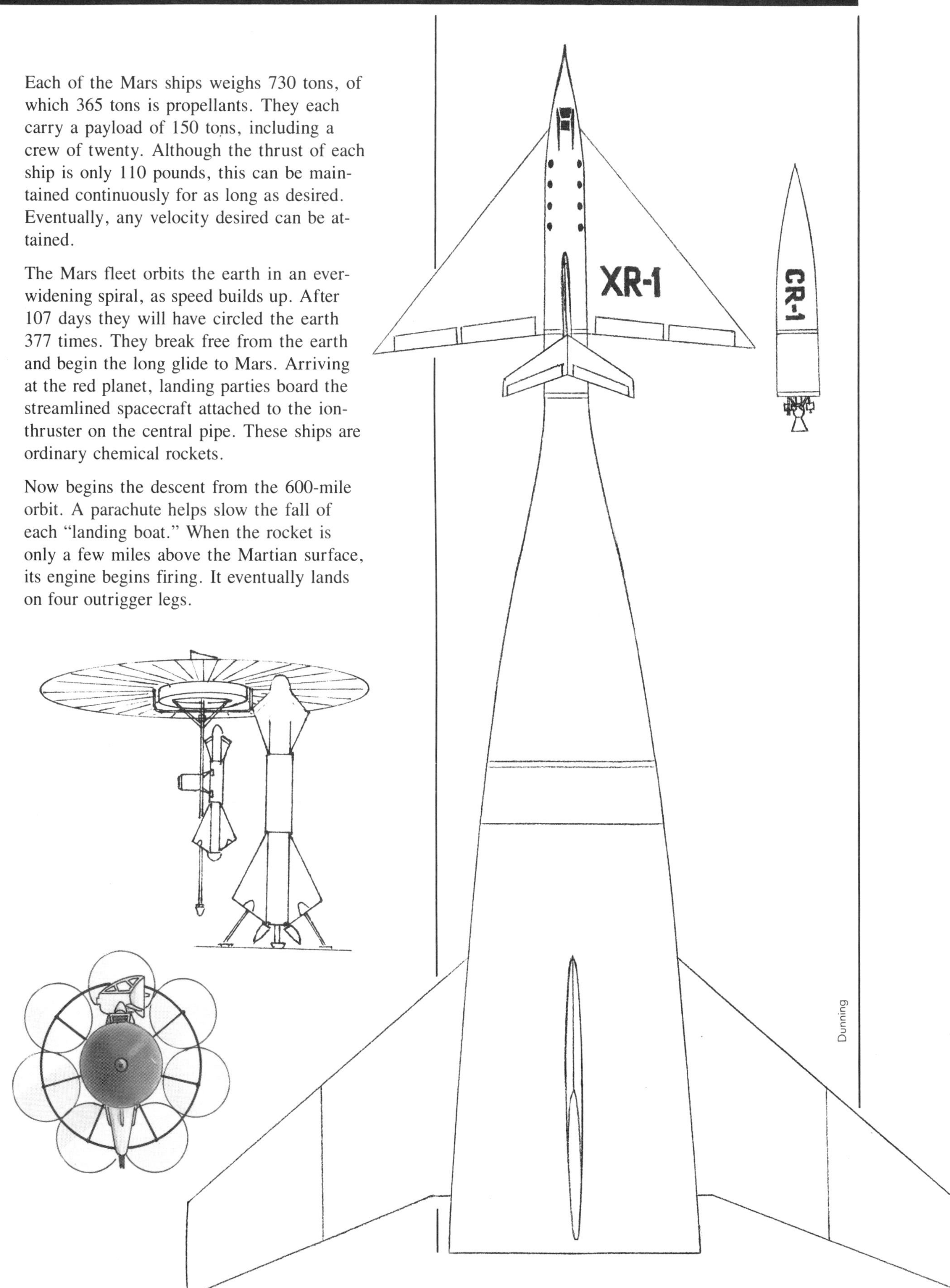

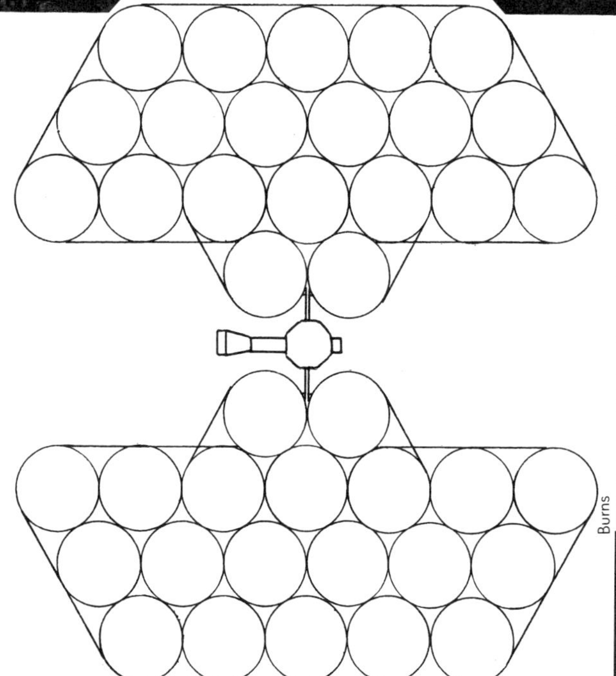

MARCH 21

Ernst Stuhlinger makes public his plans for a "butterfly spaceship" for a flight to Mars. At a symposium in New York, Dr. Stuhlinger describes how his solar-powered ship will operate. Two "wings" will each be made of twenty 150-foot circular reflectors. These will be made of aluminum foil only one-thousandth of an inch thick. Each reflector will concentrate the heat of the sun onto a boiler. The vapor from these will be used to drive a turbo-generator, creating electricity: 8,000 kilowatts. This will be used to melt cesium, vaporizing it. The ionized cesium will then be ejected from the motor, propelling the ship [also see Disney, above]. The thrust created will be only about 22 pounds, but it could be maintained over a very long period, eventually accelerating the craft to great speeds.

The ship's wingspan will be about 1,500 feet and it will weigh between 250 and 300 tons. It will carry a payload of 50 tons and ten men. The manned gondola will be carried between the vast wings. The trip to Mars will be leisurely, requiring between a year and a half to two years.

MAY

The Conquest of Space is a motion picture produced by George Pal [also see *Destination Moon*, 1950 and *When Worlds Collide*, 1951] and directed by Byron Haskin. Although the title is taken from the 1949 Chesley Bonestell and Willy Ley book [see 1949], the film is based on the later book by Bonestell and von Braun, *The Exploration of Mars*. The story concerns an expedition to Mars using the hardware made well known in the *Collier's* magazine series and in the books based upon it. Pal had originally wanted to do a much more ambitious film in which several other planets of the solar system are visited. However, studio budget constraints prevented this. The studio also insisted on the inclusion of several extremely silly sequences that jar with the film's generally sober, realistic and scientifically accurate depiction of spaceflight. Nevertheless, the movie is well worth seeing today if for no other reason than that it is like watching the *Collier's* series come to life. The portrait of Mars presented is remarkably like that of the Mars revealed by Mariner and Viking probes several decades later. Ironically, it is just these features for which the film is taken to task by several scientific reviewers! [For details about the film's spacecraft, see *Collier's* 1952–1954.]

NOVEMBER 14–18

Darrell Romick presents the paper "Preliminary Engineering Study of a Satellite Station Concept Affording Immediate Service with Simultaneous Steady Evolution and Growth" at the twenty-fifth anniversary meeting of the American Rocket Society in Chicago. He introduces his Meteor space station concept [also see 1956].

NOVEMBER 18

The Bell X-1 makes its first powered flight, piloted by Frank Everest. It reaches Mach 0.95 [see 1956].

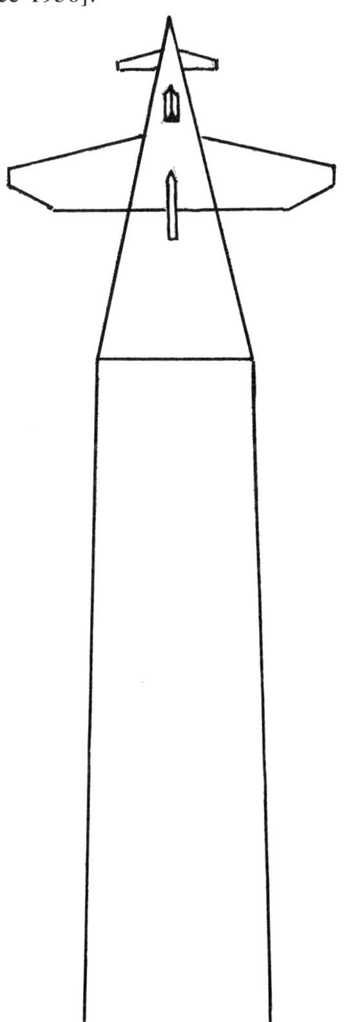

ca. 1955

In *Dawn of the Space Age*, Robert Truax discusses some of the possibilities for the future of manned spaceflight. One of these is a manned rocket for a trip around the moon and back to the earth. The suggested rocket will be a two-stage affair launching a third, winged manned stage that is approximately equivalent to a B-36 bomber in weight. Eight unmanned "cargo" versions of this stage will each deliver 10,000 pounds of fuel into earth orbit. The ninth rocket will use this cache to refuel for the trip to the moon. The manned stage will be equipped with wings for the eventual return to the earth, where it will land at the relatively sedate speed of about 65 mph.

Another possibility Truax considers will be the use of a "deep space" vehicle for the round trip to the moon. The three "hardy" astronauts will transfer to it from their winged orbital rocket, make the trip to the moon and back, and then use their waiting winged rocket for the return to the earth.

1955–1957

The U.S.S.R. launches several dogs into suborbital flights.

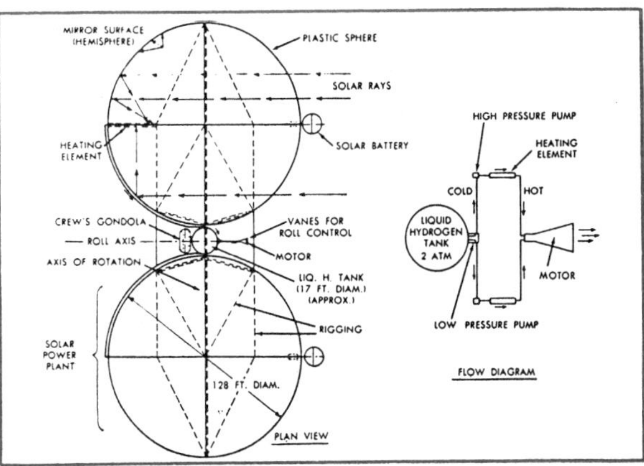

1956

Krafft Ehricke publishes his paper "The Solar-Powered Space Ship."

Satellite in the Sky is a British film based on the possibility of testing nuclear weapons in space (in this case, a "tritonium" bomb). The enormous, atomic-powered spaceship pictured takes off from an underground hangar/assembly building along an inclined ramp, similar to the one that appeared in *When Worlds Collide* [see 1951]. One of the more startling features of the spaceship *Stardust* is that it has to allow its engines to burn continuously for 45 minutes before takeoff to allow them to "warm up"!

Satellite in the Sky.

Forbidden Planet is a motion picture directed by Fred McLeod and written by Cyril Hume (from a story by Irving Block—also see *Rocketship XM*, 1950). Released by MGM, it is the first big-budget science fiction movie produced by a major Hollywood studio. Boasting outstanding special effects and a complex plot (it is a science fiction retelling of Shakespeare's *The Tempest*), it is a landmark motion picture. It is also the seed from which *Star Trek* [see 1967] will grow—in fact, *Forbidden Planet* almost seems like a pilot film for the popular television series.

The spaceship in the film is the United Planets Cruiser C-57-D. It is one of the first interstellar spacecraft to appear in films, certainly the first to be built and launched by terrestrials. The film may also be unique in being the first to portray a flying saucer from the *earth*. *Forbidden Planet* opens with the words "In the final decade of the 21st century, men and women in chemically fueled rocket ships landed on the moon. By 2200 A.D., with the perfecting of atomic propulsion, they had reached the outer limits of our solar system. Almost at once there followed the discovery of quanto-gravitetic hyper drive through which the speed of light is first attained and later greatly surpassed. And so at last mankind, now banded together in a single federation, began the conquest and colonization of deep space . . . "

Aurelio C. Robotti, in a paper presented at the Seventh International Astronautical Congress, suggests that an F-102A jet fighter can be adapted into the first stage of a three-stage satellite booster. He proposes that if a Bumper-type rocket be carried to an altitude

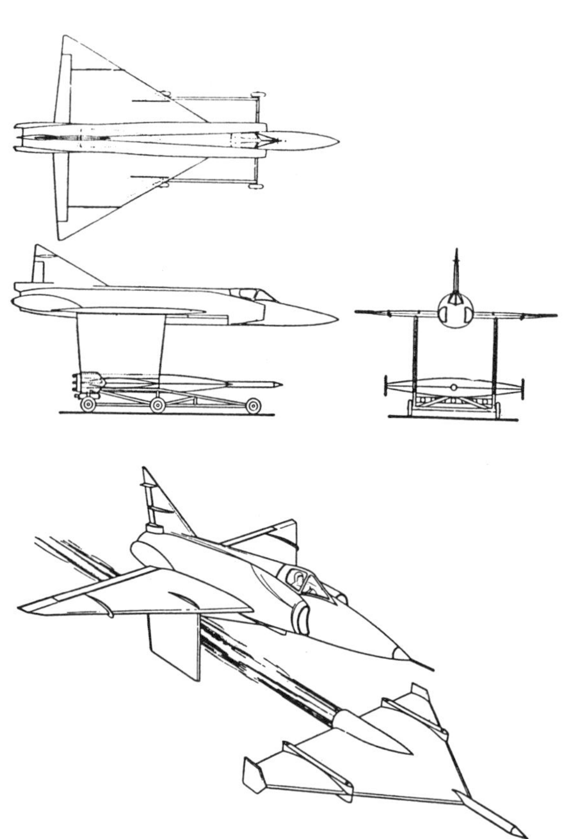

of 42,000 feet and launched with an initial velocity of 450 ft/s, that the maximum speed of the final stage WAC Corporal will be 8,350 ft/s instead of the 6,800 ft/s achieved by the V2 booster alone. The Corporal can reach an altitude of 367 miles instead of its previous record of 242 miles . . . an improvement of 50 percent.

Since the F-102A cannot practically take off with a burden of 12 tons—the weight of the V2 and WAC Corporal Bumper combination—Robotti reconfigures the V2, giving it a lifting delta-wing shape with a base and height of about 12 feet. The Bumper will then have a total weight of 27,000 pounds and a wing load of 60 pounds per square foot. The WAC Corporal will be carried in the nose of the delta. The rocket is carried beneath the F-102A by means of a pair of large struts attached to the underside of the plane's wings. The combined aircraft and rocket take off from a wheeled dolly (perhaps assisted by JATO units). The fuel tanks of the Bumper are empty at takeoff.

At a suitable altitude a tanker plane fuels

the Bumper by pumping into its tanks 1,200 gallons of fuel and 1,400 gallons of oxydizer. At 42,000 feet the pilot ignites the first-stage rocket engine and launches the satellite vehicle.

Dr. I. M. Levitt, director of the Fels Planetarium of Philadelphia, and H. H. Koelle, a German expert in rocketry, each prepares a timetable for the future of space exploration. Levitt's read:

1957–1958—Minimum satellites with or without instruments.

1960—Satellites with televison transmitters.

1963–1964—Satellites carrying animals for long periods of time.

1968—Satellites carrying men for days or weeks.

1978–1980—Complete space station.

2000—Travel to moon and planets.

Koelle's read:

1956–1960—Unmanned satellites.

1961–1965—Satellites carrying cargo and passengers.

1966–1970—Experimental space station.

1971–1977—Full-fledged space station.

1975–1985—Expedition to Mars.

JANUARY

Willy Ley publishes a timetable for the immediate future of space travel:

1957: The first unmanned artificial satellite is placed in a temporary 200-mile orbit.

1958: The first permanent unmanned satellite is placed in a 500-mile orbit.

1959–1960: Special-purpose satellites for communications, weather observation, etc., are launched as well as the first satellites carrying animals.

1965: The first piloted rocket makes a dozen orbits at a low altitude.

NASM

1967–1968: While several additional manned spacecraft are being built, construction is begun on the components for a manned space station.

1969–1970: Piloted rockets carry the space station modules into space. "Actual construction of the space station . . . will take a surprisingly short time." Twelve flights will be needed to carry the components into orbit and another six to eight to provision the station.

1970–1972: Drone rockets will take off from the space station to circumnavigate the moon and Venus (no Mars missions are planned since "the problems of Mars which still puzzle us can be solved by the space telescope which circles the earth near the space station").

1973: The first manned circumnavigation of the moon occurs.

1975: The first manned lunar landing takes place. Three landers carrying a large number of astronauts will be required.

1977: The first manned expedition is sent to a world beyond the moon: the minor planet Eros.

1980–1985: Manned expeditions travel to Mars and Venus.

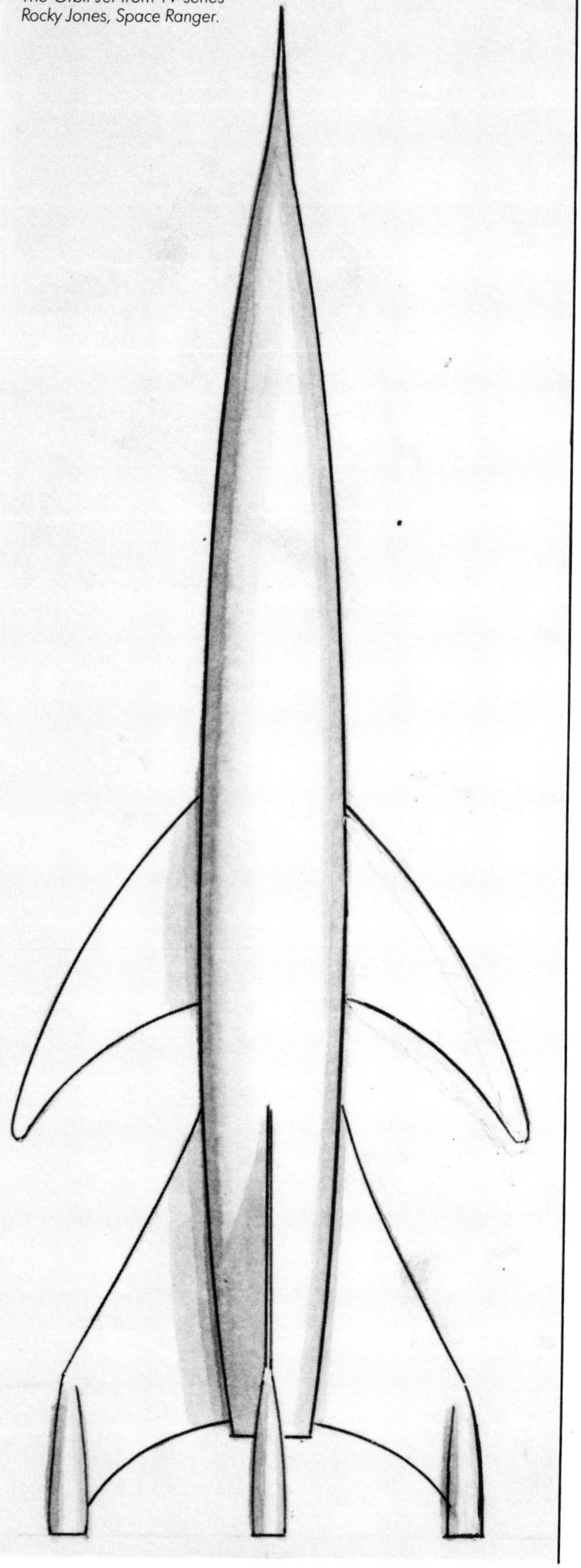

1956

The *Orbit Jet* from TV series *Rocky Jones, Space Ranger.*

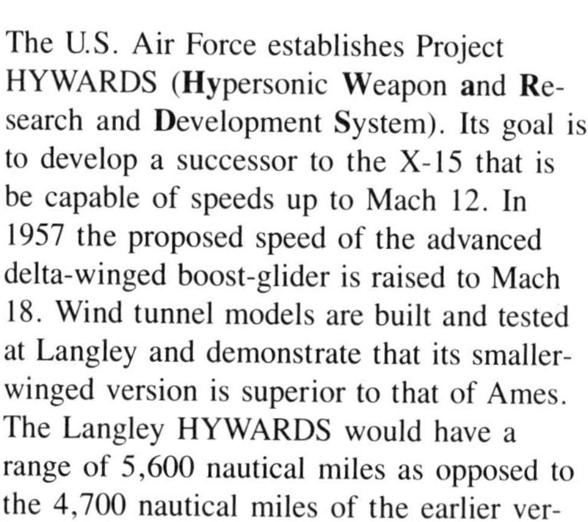

The U.S. Air Force establishes Project HYWARDS (**Hy**personic **W**eapon **a**nd **R**esearch and **D**evelopment **S**ystem). Its goal is to develop a successor to the X-15 that is be capable of speeds up to Mach 12. In 1957 the proposed speed of the advanced delta-winged boost-glider is raised to Mach 18. Wind tunnel models are built and tested at Langley and demonstrate that its smaller-winged version is superior to that of Ames. The Langley HYWARDS would have a range of 5,600 nautical miles as opposed to the 4,700 nautical miles of the earlier version.

Unfortunately, there is no launch vehicle capable of handling the weight of HYWARDS. Any weight much beyond that of the basic Mercury capsule requires an extra stage and there are no reliable stages yet developed for the Atlas. As James Hansen points out in his study of HYWARDS, but for this single impediment the U.S. might have had for its first manned satellite a winged, landable shuttle-like spacecraft.

MARCH

The U.S. Air Force starts Program 7969: "Manned Ballistic Rocket Research System."

It is intended to investigate the recovery of manned spacecraft from orbit, not to produce any hardware. Eleven proposals are submitted from the aerospace industry.

APRIL 25

The Bell X-2 makes its first supersonic flight, piloted by Frank Everest. The history of the X-2 goes back to swept-wing research being conducted as early as 1944 and to proposals for a swept-wing version of the X-1 (Bell Design-37). The U.S.A.F. contract for the development of a supersonic swept-wing research aircraft was awarded to Bell in 1945 (officially received in 1947) and called for two XS-2 aircraft.

The aircraft were completed in 1950 and gliding test flights began the following year. The X-2 is 45 feet 5 inches long, with a wingspan of 32 feet 3 inches. It is powered by a Curtiss-Wright XLR25-CW-3 throttleable two-chamber rocket engine. It can produce a thrust varying from 2,500 pounds to 15,000 pounds, and is fueled by alcohol/water and liquid oxygen. In all, twenty X-2 flights will be made.

JULY 23

The Bell X-2 sets an unofficial speed record of Mach 2.87 (1,900 mph). The aircraft is piloted by Frank Everest.

SEPTEMBER

Darrell Romick reports to the Seventh International Astronautical Congress (Rome) on the results of the Goodyear space station and ferry rocket program, begun in late 1949. The "recoverable booster system" of 1950 is named METEOR as an acronym for Manned Earth-satellite Terminal with Earth Orbital Rocket service vehicles. The original concept, roughed out by Romick in early 1950, with a diameter of only 30 feet, was much smaller than the final design [see below]. A triple ring of solid-fuel rockets around the base of the first stage was to have provided an added boost, but this was soon abandoned in favor of full recoverability. The nesting stages that are a distinctive feature of the METEOR spacecraft are described as looking like "a fish swallowing a fish swallowing a fish." The "alligator jaws" and delta wings have been part of the METEOR design from the beginning. The nearly finalized concept was presented in a paper delivered in 1954 [see above].

Wernher von Braun is one of the paper's American Rocket Society reviewers and is sufficiently intrigued to prepare a separate paper of commentary. The METEOR concept is highly publicized by the ARS and is featured in numerous magazines, including *Time* and *Newsweek*, and on the Dave Garroway *Today* television show.

In the METEOR concept, the final stage of a three-stage ferry rocket is used as the basic building module for the construction of a huge space station. The ferry's three winged stages are completely reusable. Each is manned and returns to earth, landing like a conventional aircraft. They are unpowered gliders at this time, however. After landing each is converted to a powered airplane by attaching a number of turbojet engine pods to the undersides of the wings [see Ley, 1957], adding tail fairings and other parts. The booster can then be flown back to the original launch site, where the engines, etc., are removed and shipped back to the landing area. The third stage goes into orbit. It reenters via aerodynamic braking, landing at the launch site. [Also see 1957 and 1960.]

The overall ferry rocket is 285 feet tall, 42 feet wide at its greatest diameter, while the top stage is only 9 feet wide. The first stage weighs 2,250 tons and carries 6,750 tons of

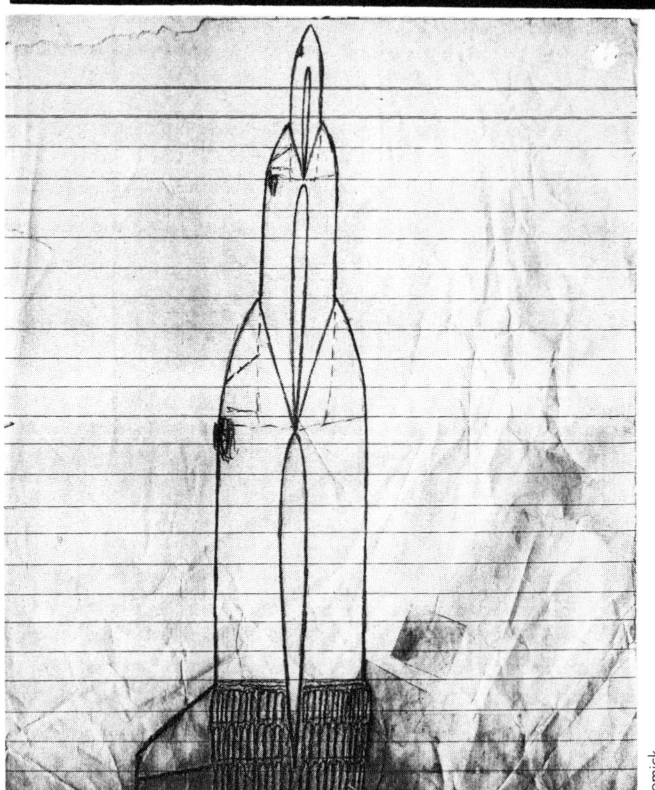

Romick

Romick's original sketch showing cluster of solid fuel boosters attached to base of first stage.

propellants, its sixty-three engines generating from 16,000 to 19,170 tons of thrust. The second stage weighs 280 tons and carries 970 tons of propellant. Its nine engines produce 22,000 tons of thrust. The final stage weighs 53 tons, carries 87 tons of fuel and has six engines producing 233 tons of thrust.

Each of the third stage rockets becomes a unit in the building of the space station. The schedule envisioned is this:

Day One. The first two ferry rockets are joined nose to nose. Fuel tanks have been cleaned so that the rockets are empty for their entire length. These can be used for temporary living space. Three unused ships stand by in case of emergency.

Day Ten. Alternate nosing together and butting of ships is progressing; by now at least ten have been joined into a long, narrow cylinder. The gaps that the ships' noses have made are being skinned over.

Week Four. Expansion begins as supply rockets bring precut and predrilled materials into orbit. A series of 75-foot diameter circular frames are erected along the length of the tube.

Week Ten. There are now sixteen ferry rocket bodies in the 1,000-foot long tube. The 75-foot diameter rings are being skinned over to make three cylindrical, airtight cells 300 feet long and 75 feet in diameter. At one end of this structure a wheel is being built, 500 feet in diameter.

Week Twelve. The wheel is almost completely enclosed, making a solid disk containing some 4 million cubic feet of living volume. It begins rotating to provide artificial gravity.

Month Six. Final enlargement is under way. The stationary cylindrical section is increased from 75 feet to 1,000 feet in diameter. The gravity wheel is increased to 1,500 feet in diameter, almost a mile in circumference.

Year Two. By now forty-nine ferry rockets have been joined to form the core of the stationary section. The vast cylinder is now 3,000 feet long and the enlarged wheel is complete.

Year Three and a Half. The hollow interior of the cylinder is divided into three airtight sections. Airlocks and ferry rocket landing bays are at the end opposite the living wheel. More than 20,000 people live and work on board. The wheel is made up of twenty-two concentric tubular rings, divided into eighty-two floors. "The primary purpose of the wheel," writes Romick, "is to provide comfortable, satisfying, convenient living conditions. Therefore the furnishings and arrangement of apartment and hotel quarters, offices, stores, etc., should resemble insofar as is practical conventional items. There should also be provided such community items as a gymnasium, stores, theaters, auditoriums, and churches.

"New arrivals, when brought to the wheel and down in the wheel elevator, should step out into an area adjoining a hotel lobby and reception center with a general appearance and arrangement similar to those on the earth. Because of this variety of arrangements some sections, particularly the lower floors, should have two or three floors per tube, whereas the rest, including all residential apartments areas, should have four floors

Romick

(Below and far right) the complete METEOR spacecraft.
(Below right) the second and third stages.

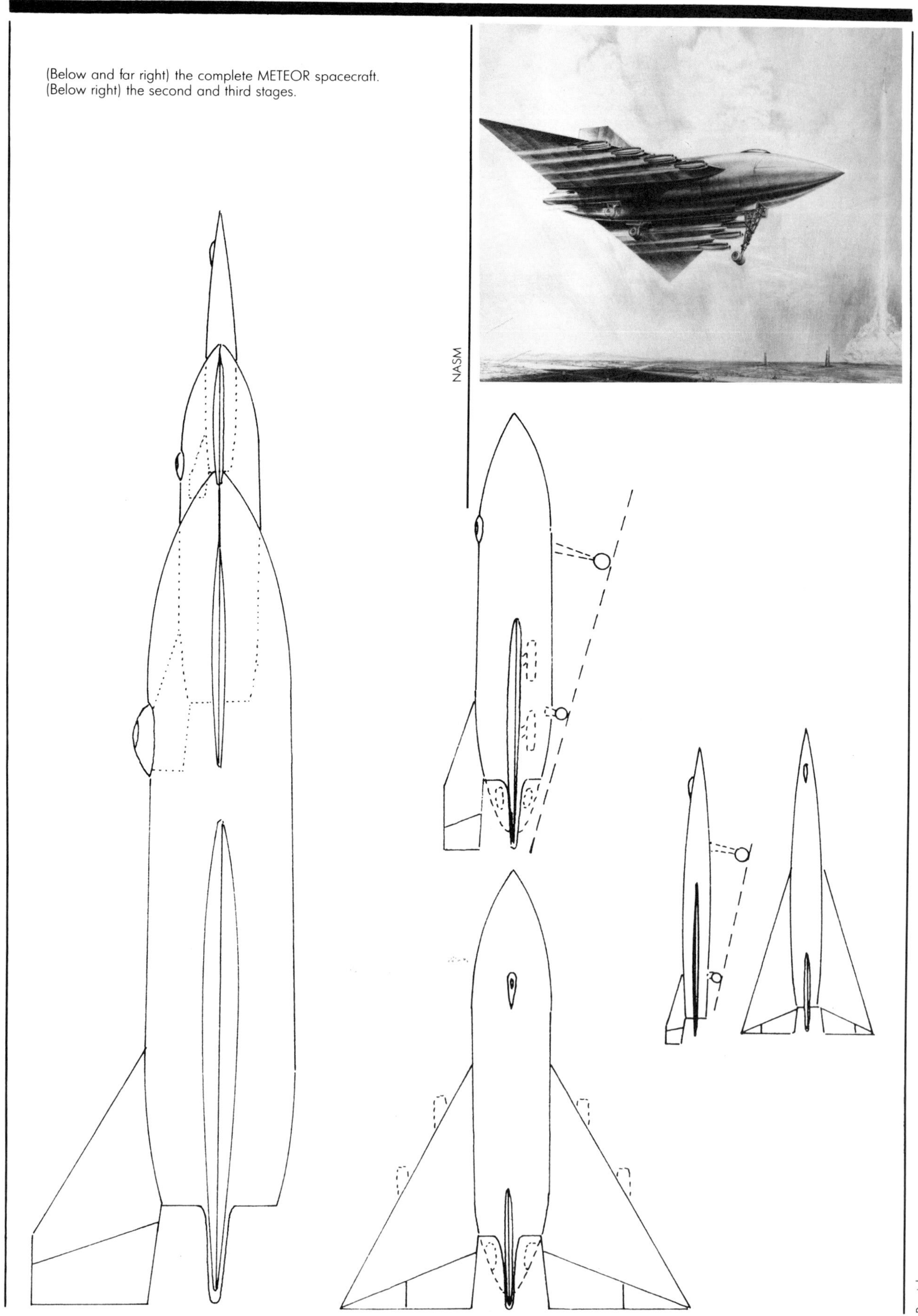

NASM

Romick

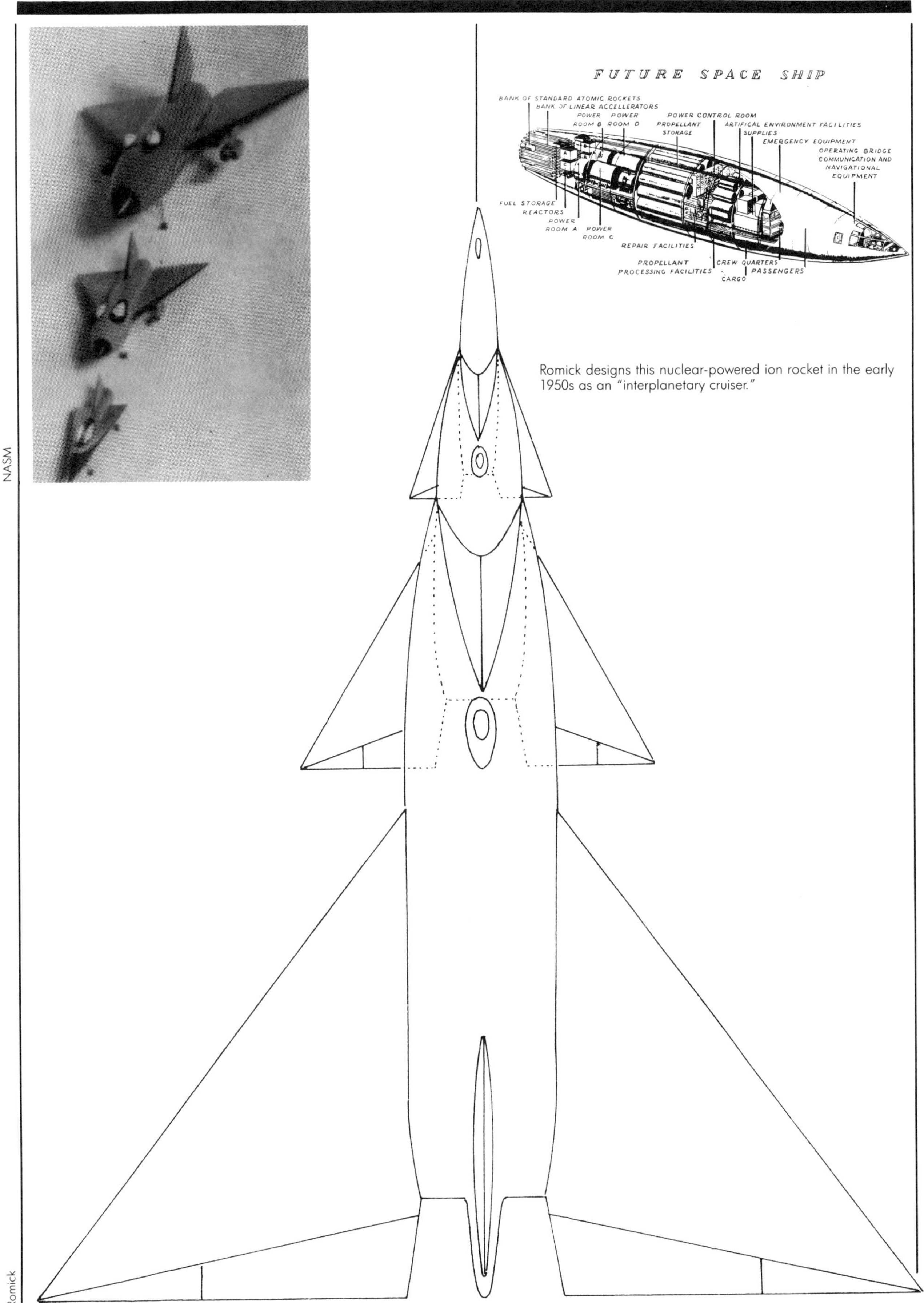

NASM

Romick

Romick designs this nuclear-powered ion rocket in the early 1950s as an "interplanetary cruiser."

per tube, giving a ceiling height of approximately eight feet . . . "

The stationary cylinder will contain zero-gravity laboratories and the life support equipment for the entire station: air conditioning, lighting, power, communication, etc. Transfer cars running along the central core carry personnel from one end of the station to the other. The docking area provides pressurized berths for ferry rockets, servicing and maintenance areas, and even space for the construction of spacecraft, which can be done in the pressurized portion of the cylinder. Large ships can be constructed in subassemblies, final erection taking place outside.

Power generation is solar. Venetian-blind-type windows in the sides of the cylinder allow sunlight to enter. Other windows regulate radiant heat intake and outgo. Observatory domes protrude from the skin of the cylinder.

Romick has calculated the ultimate passenger and freight costs of his completed scenario. Passengers travelling to the station will be charged $40,000. Those travelling from the station will be charged $15,000, while a round trip from the earth is $50,000. Mail going up costs $0.75 an ounce; express, $20.00 a pound. Freight will cost its shippers $8.00 a pound or $16,000 a ton to send it to the station.

Starting from the year 1955, Romick hopes to see the following schedule take place:

From 1955 to 1958: the development of rocket aircraft capable of altitudes of 20 to 100 miles.

From 1960 to 1969: the engineering, design, fabrication (continuing) and testing of the ferry rocket.

From 1969 on: operational ferry rockets.

From 1962 to 1971: engineering, design and fabrication of the space station and its components.

From 1971 on: erection in orbit of the station.

From 1973 on: the development and operation of lunar and planetary vehicles.

A small version of METEOR, the METEOR, Jr., is developed in 1957, and the Aero-METEOR in 1960 [see below].

SEPTEMBER 7

The Bell X-2 and pilot Iven Kincheloe set an unofficial world altitude record of 126,000 feet. The plane is equipped with a Curtiss-Wright engine producing 15,000 pounds of thrust.

SEPTEMBER 27

The Bell X-2 sets a new record when it reaches a speed of Mach 3.2 (2,100 mph). Its pilot, Milburn G. Apt, is killed when the plane goes out of control and crashes.

1957

The Army Ballistic Missile Agency proposes a million-pound booster created by clustering smaller rockets.

Ernst Stuhlinger proposes a manned mission to Mars using electrically propelled spacecraft. This scheme inspires the "Mars and Beyond" segment of the Disneyland "Man in Space" series of television programs [see above].

The payload of each umbrella-shaped ship is to be 150 tons and its electrical engine will produce an acceleration of 10^{-4} g. The payload consists of the crew and its life support, instruments, and a landing craft.

Power is to be supplied by a nuclear reactor suspended 150 feet away at the end of the "handle." A thick layer of beryllium and a sheet of boron provide additional shielding. The cooling system employs sodium-potassium, which transfers the heat to the silicon oil working fluid. The oil steam drives a turbine which is coupled to a generator. Leaving the turbine, the steam passes into the large, circular radiation cooler where it condenses and is pumped back into the heat exchanger.

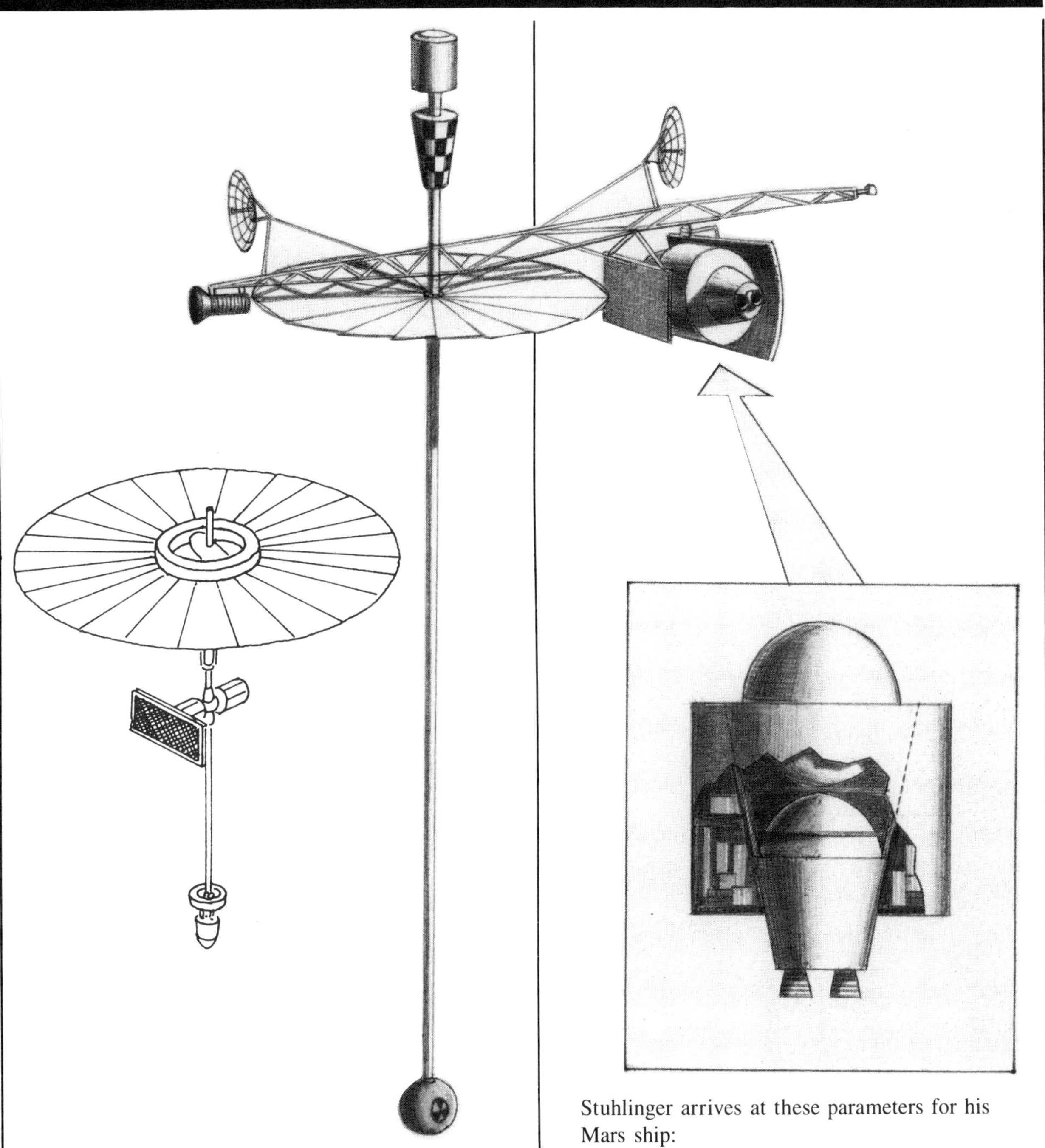

The propellant is cesium, which can be ionized easily. A temperature of 200° C produces a sufficient vapor pressure of the alkaline element. The vapor enters an ionization chamber containing hot platinum grids and the ions are extracted by an electric field. This field accelerates the ions in the thrust chamber to a velocity of about 50 miles per second. To produce and maintain a constant flow of particles from the system, the ions must be electrically neutralized soon after leaving the thrust chamber.

Stuhlinger arrives at these parameters for his Mars ship:

Total initial mass: 730 tons

Propellant mass: 365 tons

Acclerating voltage: 4,880 volts

Total electrical power: 23 megawatts

Exhaust velocity: 50 mps

Total thrust: 110 pounds

The travel time of this ship to Mars is a little over a year; the return trip takes a little less. The ratio of initial weight to the payload is less than 5:1.

The Mars ship resembles a parasol, with the living quarters at one end of the longitudinal axis and the reactor at the other. The operation of the turbine and generator causes the whole ship to rotate, a desirable side effect on two counts: it causes the condensed fluid in the radiator to flow to the outer rim, and it also creates a slight artificial gravity in the annular living quarters. The thrust chamber and its propellant tanks are mounted in such a way that the thrust force passes through the ship's center of gravity. The landing craft is mounted to the thrust chamber unit. The propulsion system operates during the entire trip except for a few short periods of powerlessness which are needed for corrective maneuvers. There will be at least ten ships in the expedition.

The trajectory of each ship will not follow an elliptical path, but rather segments of spirals. After the first 2 hours of acceleration the ship will not have moved more than 20 miles farther away from its starting point near the space station. After 100 days and 376 revolutions around the earth it will have increased this distance to 100,000 miles. By the 115th day it will have moved beyond the orbit of the moon and 12 hours past the 124th day it will have escaped the bonds of the earth's gravity and moved into an orbit around the sun. On the 195th day the thrust unit will be turned 180 degrees and the ship will begin to decelerate. On the 276th day the thrust will again accelerate the ship, carrying it in to the orbit of Mars, where it will arrive on the 347th day. On the 402nd day the ship will have spiralled down to an altitude of 600 miles above Mars, where it will go into orbit. The descent to the surface is made via an ordinary chemical rocket. The crew will not leave Mars for another 472 days. The trip back to the earth will be similar to the outgoing journey.

Dr. Alfred J. Eggers, Jr., assistant director for research and development analysis and planning at NASA's Ames Research Center originates the idea of wingless lifting body spacecraft [see 1963].

F. A. Smith proposes a manned, orbiting space telescope. It is to be a large cassegrain reflector, with a focal length of 27.5 m, attached to a pressurized observation sphere. A projecting cylindrical airlock pro-

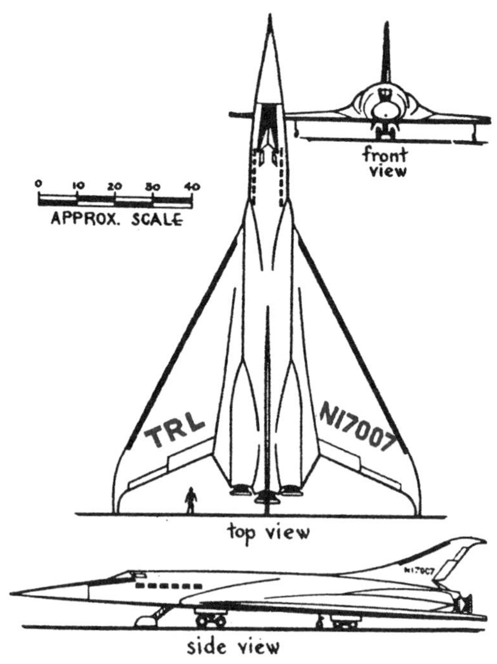

Evolutionary spacecraft designed by G. Harry Stine (ca. 1957) to fill in between existing spacecraft and Darrell Romick's METEOR project [see 1956]. Facing page, lower left and top: the *Griffon*, a long-range manned research rocket with maximum altitude of 80 miles; lower center: the *Aeolus*, a 3,000-mile transcontinental rocket transport; lower right: the two-man *Nomad*, prototype manned orbiter which is a modified *Griffon* atop a booster.

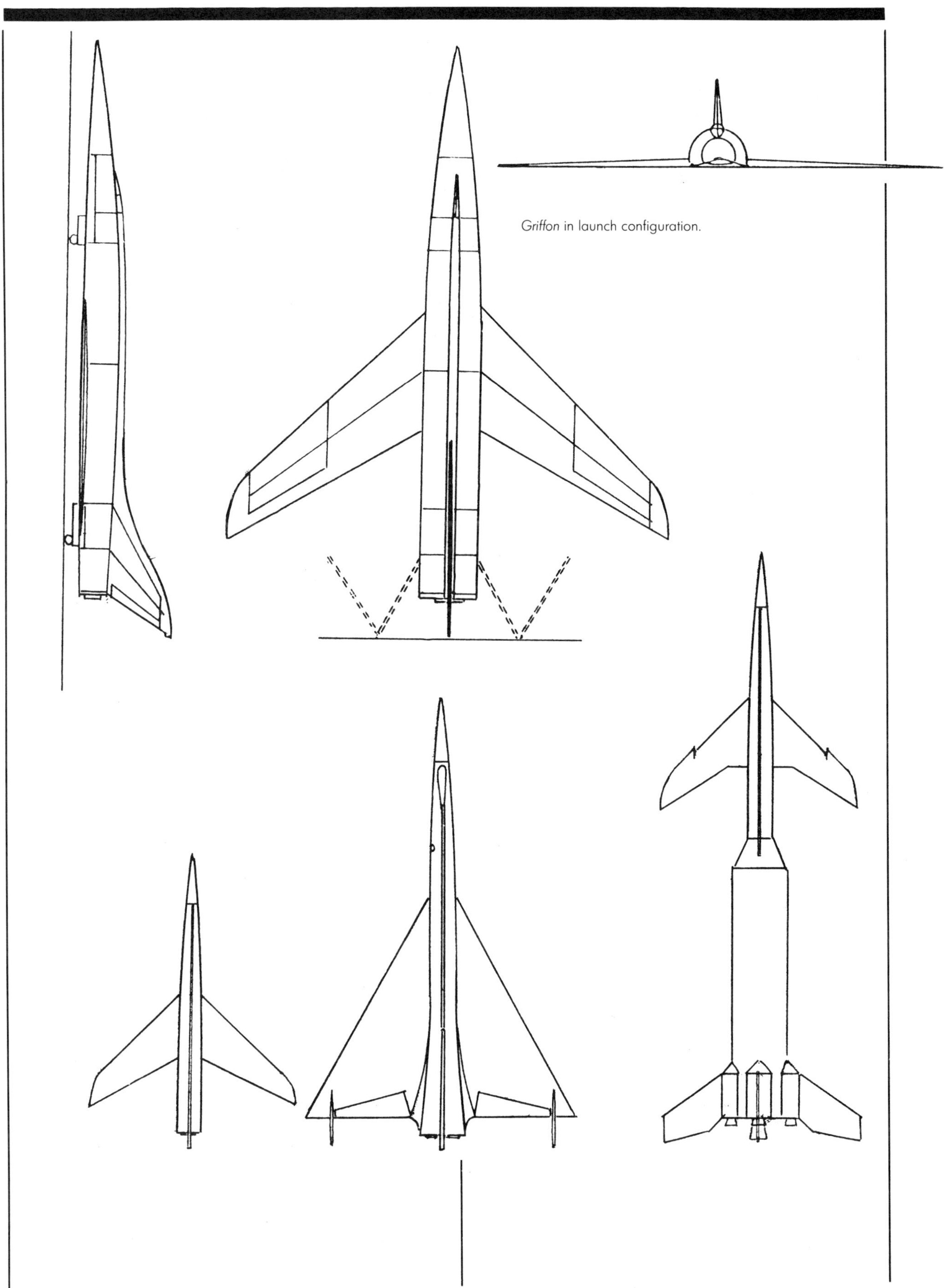

Griffon in launch configuration.

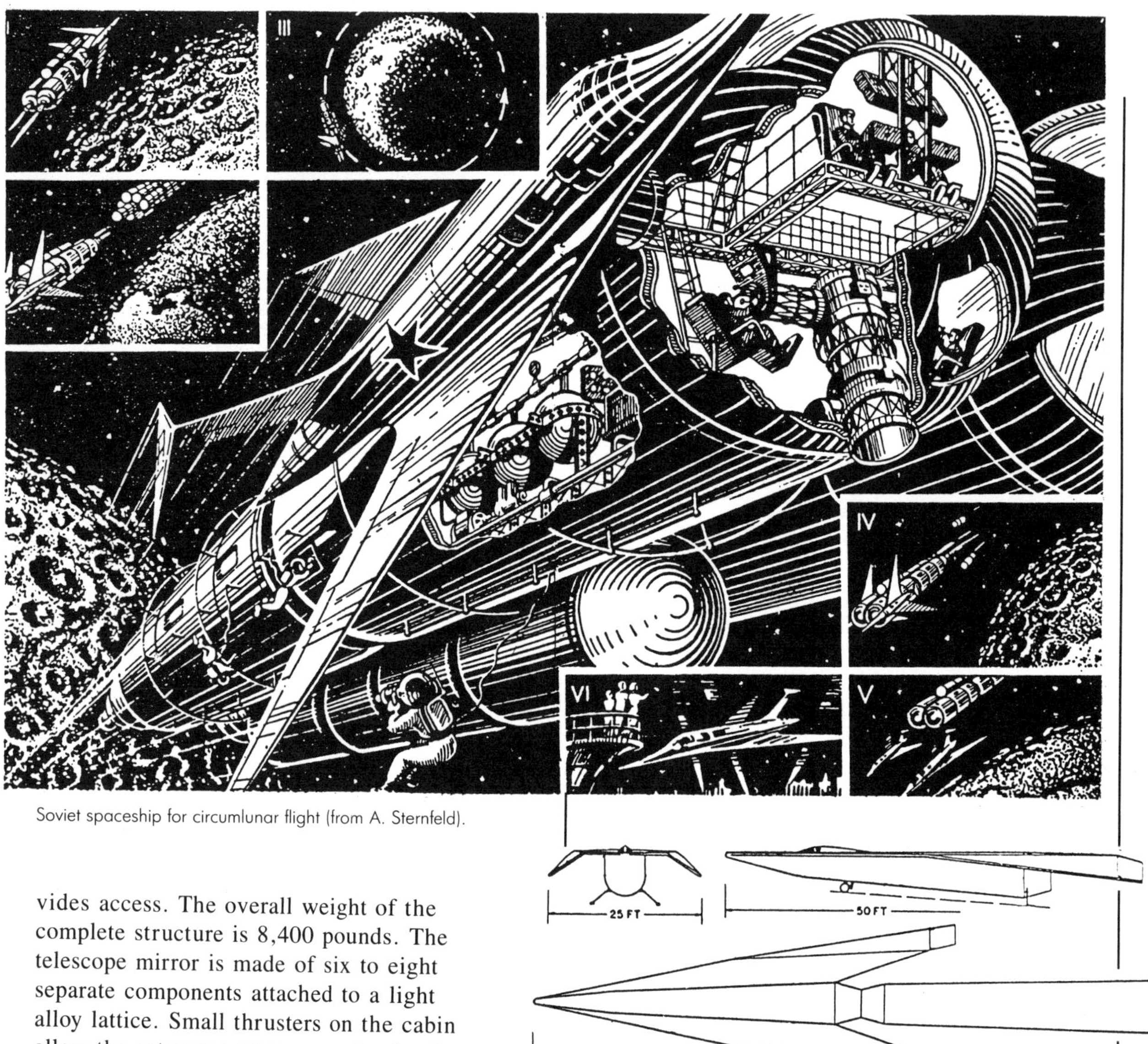

Soviet spaceship for circumlunar flight (from A. Sternfeld).

vides access. The overall weight of the complete structure is 8,400 pounds. The telescope mirror is made of six to eight separate components attached to a light alloy lattice. Small thrusters on the cabin allow the astronaut-astronomer to aim the telescope.

NACA's Ames Aeronautical Laboratory develops a "beyond X-15" Mach 10 piloted "technology demonstrator." It is to be air-launched by a B-36 for trials up to Mach 6; for higher speeds it is to be launched vertically as the second stage of a two-stage rocket. Separation from the 50-foot booster will occur at 100,000 feet and Mach 6. Takeoff will be from MacDill AFB in Tampa, Florida, and the landing will take place at Edwards AFB in California.

The technology demonstrator is to be a dart-shaped high-wing aircraft, 50 feet long and with a wingspan of 25 feet.

Eric Burgess in his book *Satellites and Spaceflight* states his belief that although "the knowledge of how to reach the Moon is now available . . . it might take 20 to 25 years to develop a suitable vehicle. It can be shown that it is not possible to construct one gargantuan rocket in which can be packed sufficient propellants to carry a crew to the Moon and bring it back again safely."

Burgess proposes the use of the orbital refueling technique often advocated by the British Interplanetary Society [see 1958, for example]. Some twenty tankers of between 600 and 1,000 tons each will be required to carry sufficient fuel for a moonship to leave earth orbit, journey to the moon, land, take off and return to earth orbit. In 1951 Kenneth Gatland had demonstrated that fifty tankers carrying fuel payloads of five tons each would allow a spacecraft assembled in a 500-mile orbit to circumnavigate the moon. H. P. Wilkins, however, does not believe that a manned circumnavigation of the moon ever need be attempted because "(a) nothing important can be gained by visual inspection of the hidden part of the surface; (b) details of the surface will already have been obtained by means of probe rockets . . . ; and (c) a manned lunar satellite could be established for the same amount of effort." From such a lunar orbiter both manned and unmanned probes could be sent to the surface.

Wernher von Braun proposes an antisatellite/antisatellite system consisting of a manned spacecraft—an orbital glider—that guides an unmanned missile carrier into orbit and directs the firing of the missiles it carries.

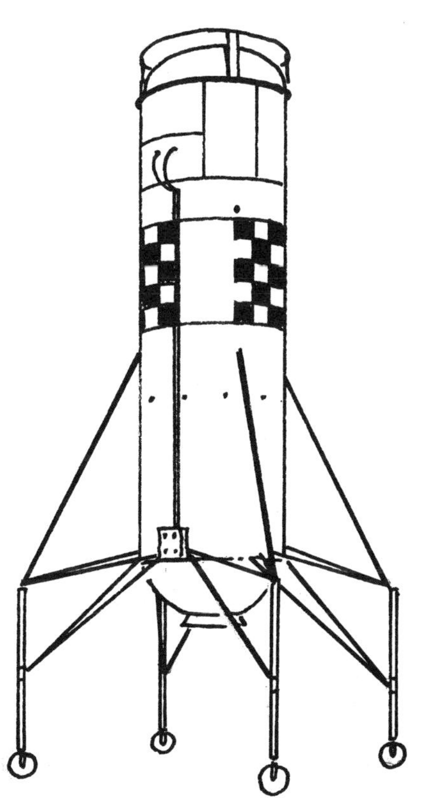

The French aviation agency, SNECMA, develops the "Flying Atar" VTOL test vehicle, which makes more than 200 test takeoffs and landings.

The original version, P1, is unmanned. Version P2 is piloted, with P3 featuring a full cockpit with rotating seat. Although developed with the end product of a VTOL annular-winged jet in mind, it is an ancestor of the "flying bedstead" which will be used to train Apollo astronauts for lunar landings.

Convair and Bell are reported to have received study contracts for manned spaceplane rocket bombers, inspired by fear that the Russians have been developing similar spacecraft. The Russians are believed to be working on a rocket-plane, the T-4A, inspired by Sänger's antipodal bomber [see 1943]. It is thought to be between 60 and 70 feet long, with three liquid-fueled T-4 rocket engines (each with a thrust of 820,000 lbs). Two smaller expendable liquid fuel boosters are carried under the wings and jettisoned during the climb to 120 miles. Takeoff is along a track or rail with a captive booster. The spacecraft is delta-winged with a T-tail stabilizer. An American version is at least two or three years in the future, according to Erik Bergaust.

David Dietz, Scripps-Howard science editor, suggests this schedule for manned exploration of the moon:

1. A rocket will be developed capable of carrying a crew and supplies to an altitude of 1,000 miles.

2. A space station will be built that can be occupied for months at a time.

3. A lunar lander will be built at the space station and launched from orbit.

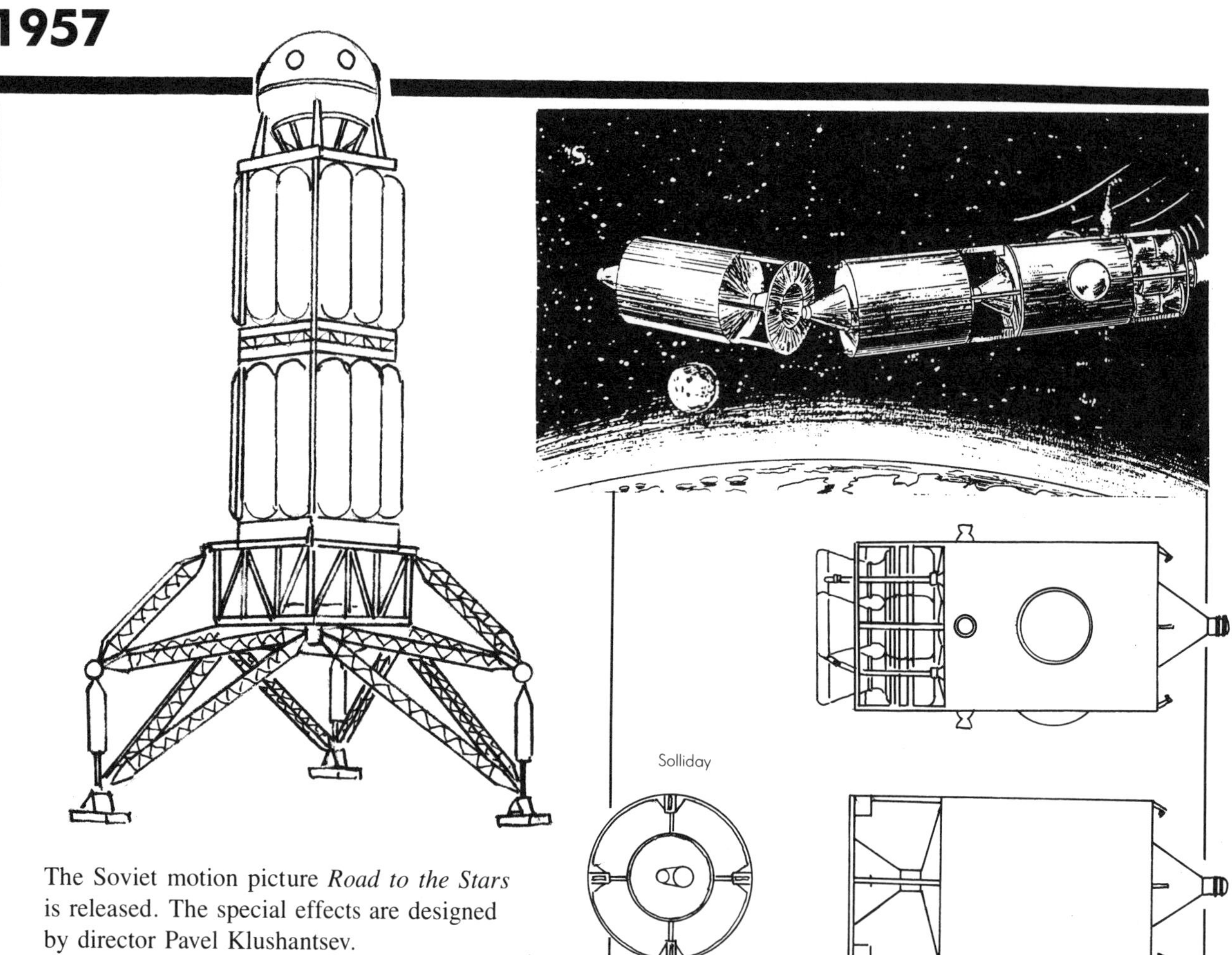

The Soviet motion picture *Road to the Stars* is released. The special effects are designed by director Pavel Klushantsev.

Horace Solliday proposes a modular concept for interorbital spacecraft. His idea is that a spaceship could be brought up from earth in sections that would merely lock together. All of the simple, cylindrical units are virtually identical with the exception of a power unit that contains the rocket motors. Other units would house fuel tanks, crew modules or instrumentation. A spacecraft of any size and for any number of purposes could be built up from these basic units.

SUMMER

An atomic bomb is buried at the bottom of a 500-foot vertical shaft in the Nevada desert. The top of the shaft is capped with a steel plate four inches thick and weighing hundreds of pounds. When the bomb is detonated, high-speed cameras record the takeoff of the "manhole cover." Subsequent calculations show that the plate was launched with a velocity six times that of the earth's escape velocity. The cover was never found.

OCTOBER

After Sputnik—the Moon is an animated film released by the Soviet Union shortly after the successful orbiting of Sputnik. Produced under the direction of Yuri Khlebtsevich, chairman of a technical committee working on rocket guidance, the film depicts an unmanned flight to the moon sometime in the early 1960s.

A three-stage rocket is carried into the upper atmosphere on the back of a winged, ten-engined carrier rocket. The smaller rocket is mounted between the carrier's twin fuselage. Once in earth orbit, the third stage of the moon rocket makes a rendezvous with an unmanned fuel tanker. Once it has refuelled, it continues on to the moon.

The moon rocket lands tail-first on four legs, then gently tips over, the nose unfolding to release a small robot tank. Guided by radio from the earth, it roams over the lunar surface televising the scenery and making automatic instrument readings.

Harald J. von Beckh, M.D., designs an anti-g escape capsule for spacecraft. The capsule is a pressurized oblate spheroid. While within the spacecraft, it is pivoted so that no matter what direction the pilot is being accelerated, he is always in a prone position relative to the spacecraft. As an ejectable escape capsule, it is equipped with parachute, dye markers, emergency radio beacon, etc.

Dr. Leonid Sedov, at the Eighth Annual Congress of the IAF, discusses some of Russia's plans for space exploration in the near future. He predicts that the Soviets will land a man on the moon within a few years, followed by unmanned moon landings (to celebrate the fortieth anniversary of the Communist Revolution).

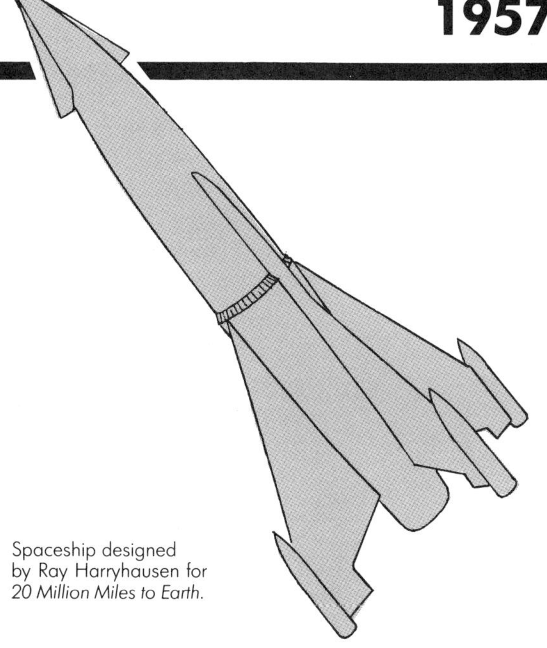

Spaceship designed by Ray Harryhausen for *20 Million Miles to Earth.*

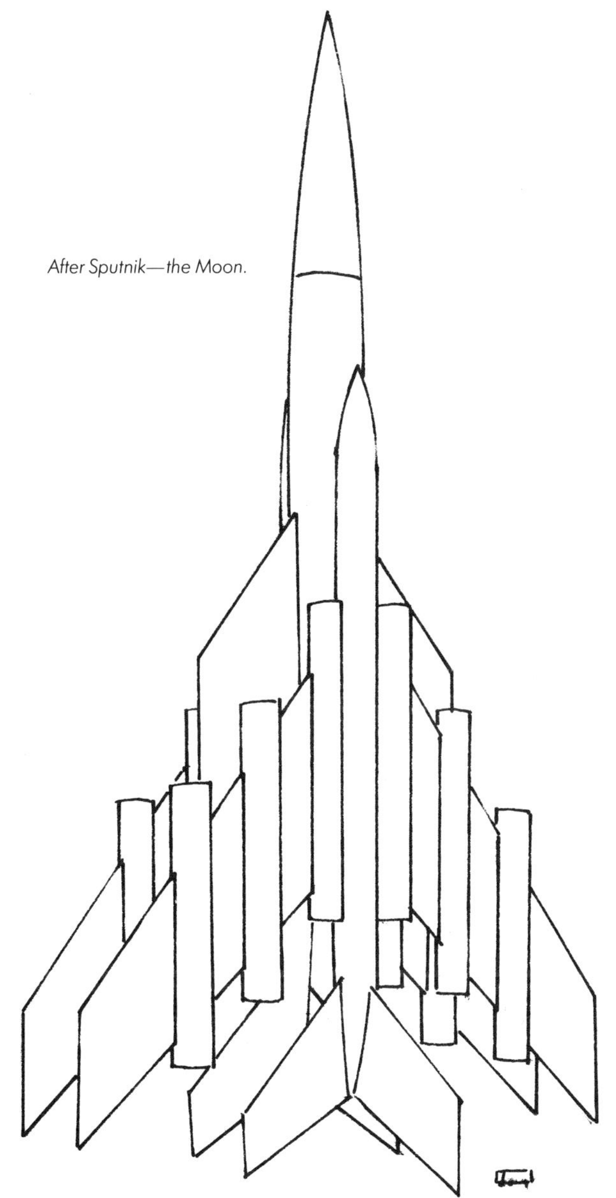

After Sputnik—the Moon.

OCTOBER 4

Sputnik, the world's first artificial earth satellite, is placed into orbit by a Russian R7-ICBM, launched from Baikonur cosmodrome. Caught virtually unaware, in spite of the predictions and warnings of experts (even to the correct launch date: the anniversary of Tsiolkovsky's birth; only the weather kept the Russians from launching on the exact day) the United States makes the goal of successful achievements in space a major priority.

OCTOBER 7–12

Darrell Romick, R. E. Knight and S. Black present their paper "METEOR Junior—A Preliminary Design Investigation of a Minimum- Sized Ferry Rocket Vehicle of the METEOR Concept." METEOR Jr. is a scaled-down version of the larger METEOR spacecraft [see 1956]. This 500-ton vehicle could carry one ton of payload to orbit, in addition to a crew of two to four (depending upon the specific mission).

The assembled rocket is to be 131 feet tall. The third stage orbiter is 39 feet long, with a wingspan of 17.5 feet. As with the original METEOR concept, all three stages are manned and recoverable.

There is also a plan for using the METEOR Jr. for a moon landing. Two additional booster stages would be attached to an existing METEOR Jr. third stage, making a three-stage vehicle. These would be used for earth departure, lunar landing and lunar departure. The winged third stage could make a return trip directly back to the earth's surface. The METEOR Jr. moonship could carry three men and 400 pounds of equipment and supplies to the moon.

OCTOBER 14

The Rocket and Satellite Research Panel with the American Rocket Society proposes a national space flight program under the aegis of a National Space Establishment, a

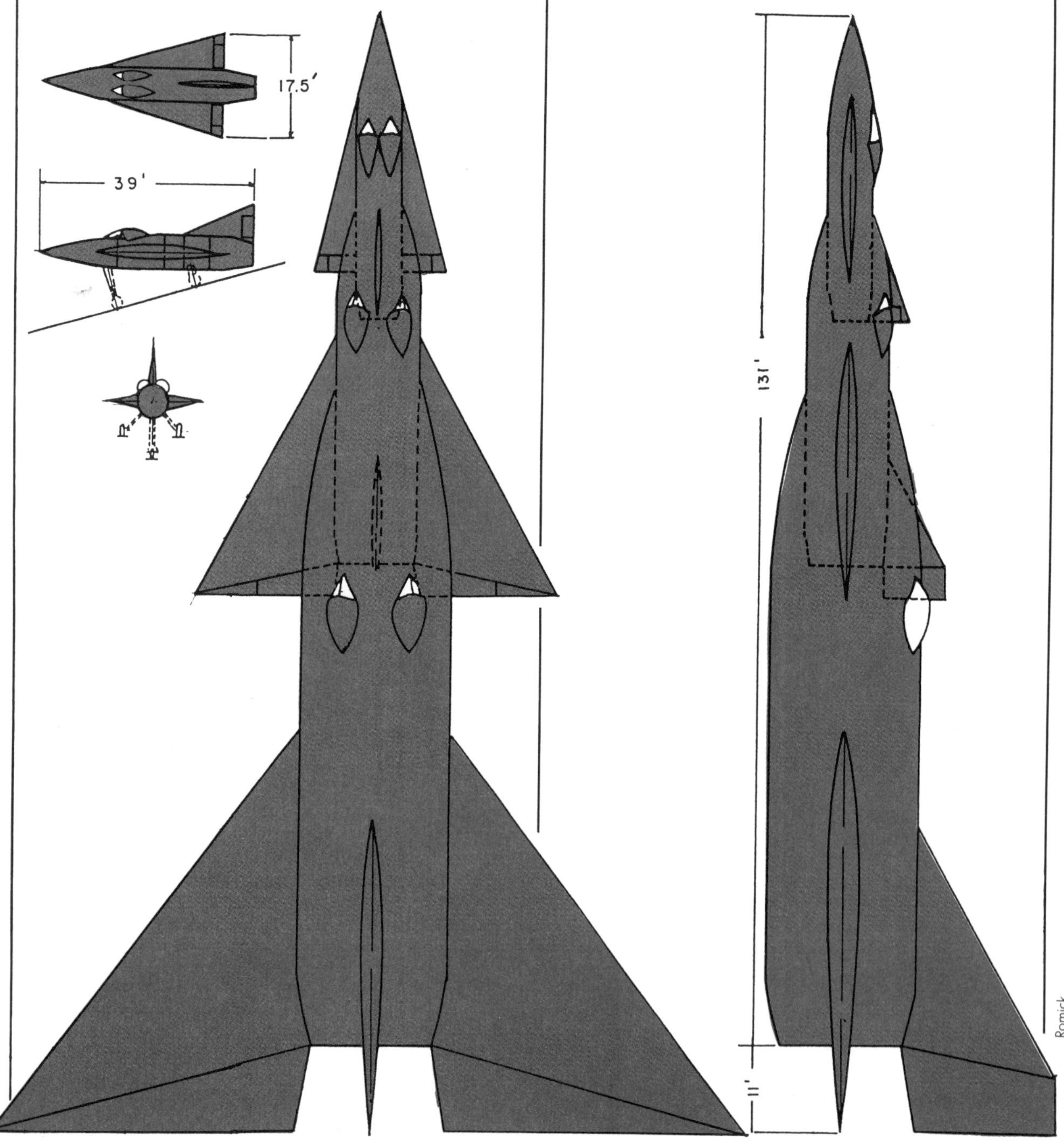

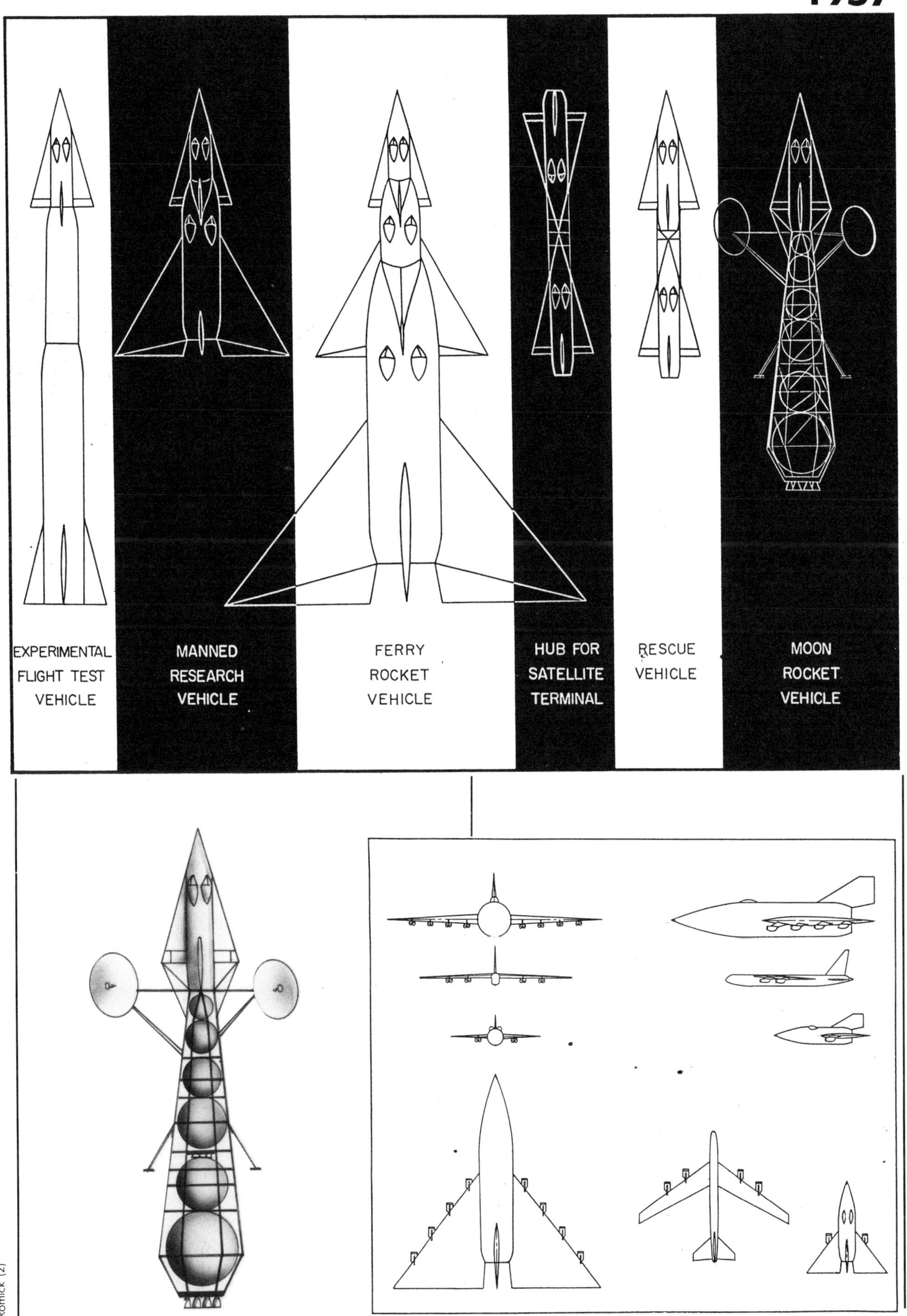
EXPERIMENTAL FLIGHT TEST VEHICLE
MANNED RESEARCH VEHICLE
FERRY ROCKET VEHICLE
HUB FOR SATELLITE TERMINAL
RESCUE VEHICLE
MOON ROCKET VEHICLE

Romick (2)

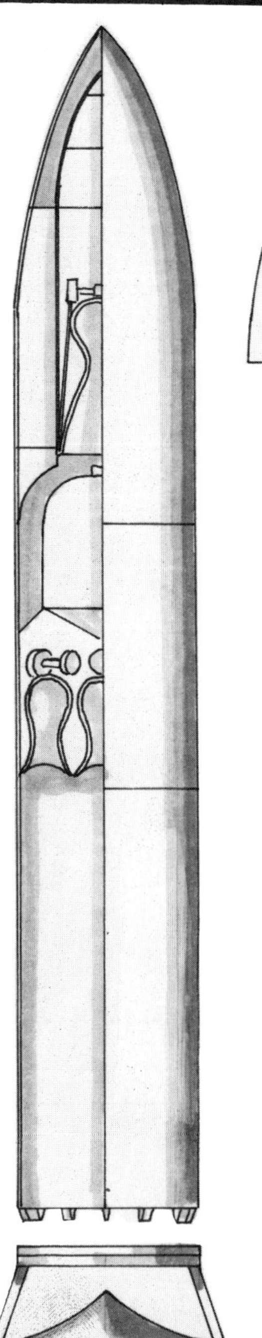

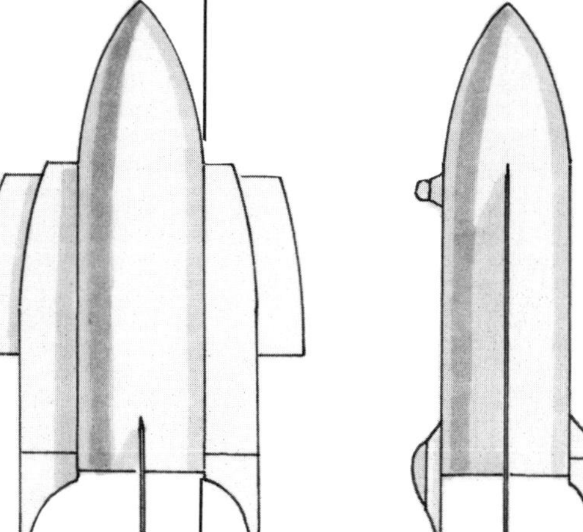

Manned spaceship designed by Hermann Oberth, featuring a reentry vehicle with telescoping wings and airplane-type landing gear.

nonmilitary organization. This group expects to achieve an unmanned hard landing on the moon by 1959, an unmanned lunar orbiter and soft lander by 1960, a manned circumnavigation in 1965 and a manned lunar landing in 1968 with a permanent base by 1968.

NOVEMBER

Sputnik 2 carries the dog Laika into orbit.

NOVEMBER 18

Life magazine publishes a design for a manned spacecraft, illustrated by Ray Pioch. It is a small vehicle with swept wings and a single vertical stabilizer. Its blunt, ceramic nose contains coolant jets that protect its inner surface from the heat of reentry. The nose itself is porous and "sweats" away the outer surface heat. The pilot occupies a pressurized cabin just below the root of the vertical fin. Guidance is by wingtip rockets. Four externally mounted retrorockets slow the craft for reentry. It enters the atmosphere from its 500-mile orbit by a series of skip-glides [see Sänger, 1943], eventually gliding to a landing.

DECEMBER 1

The Soviet newspaper *Sovietskaya Aviatsia* publishes the trajectory of a hypothetical rocket-plane. After achieving a speed of 12 to 15 kph the pilot switches off the engines and skip-glides along the top of the atmosphere to his destination [see Sänger, 1943].

DECEMBER 21

The U.S. Air Force Research and Development Command issues a directive for the development of Dyna-Soar's first phase, at this time envisioned as a delta-winged boost-glider with a single pilot. More than a dozen aerospace contractors respond, including Bell, whose work on BOMI for the Air Force initially inspired the Dyna-Soar proj-

ect. Ironically, the contract goes to Boeing and Bell is left as only a subcontractor.

CHRISTMAS

Hugo Gernsback discusses the future need for the elimination of space debris. He suggests that the empty hulks and shells of boosters and old satellites will present a hazard to manned spaceflight.

Gernsback also describes an antigravitational spaceship—a "Gravitor"—powered by atomic fusion. Instead of rocket nozzles at the spacecraft's rear there is an array of sharply pointed rods. "High-tension **cosmotronics** generates a powerful **cosmothrust** (cosmic energy thrust i.e. reaction) which operates in and above the Earth's atmosphere, thus pushing the Gravitor forward. Below the ship there is a similar pointed-rod assembly which lifts the ship."

1957–1960

Boeing proposes a large delta-winged spaceplane for space station support. By 1958 it has evolved into a form not dissimilar to an enlarged version of the Dyna-Soar (and, in fact, this proposal is one of Dyna-Soar's several predecessors). It is suggested in 1960 that it could be air-launched from beneath a modified B-70 Valkyrie bomber [also see 1960]. The spaceplane is then boosted into orbit by an expendable rocket stage.

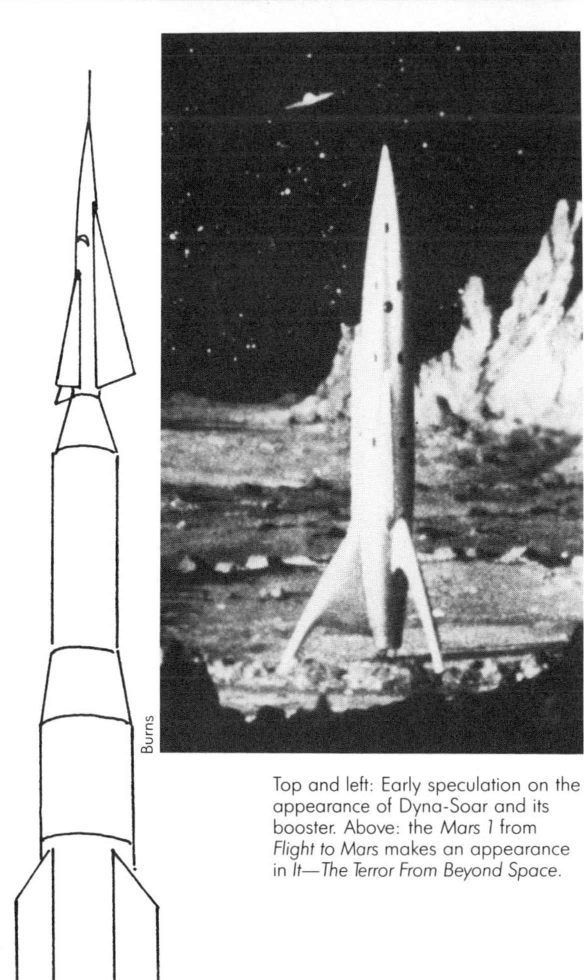

Top and left: Early speculation on the appearance of Dyna-Soar and its booster. Above: the *Mars 1* from *Flight to Mars* makes an appearance in *It—The Terror From Beyond Space.*

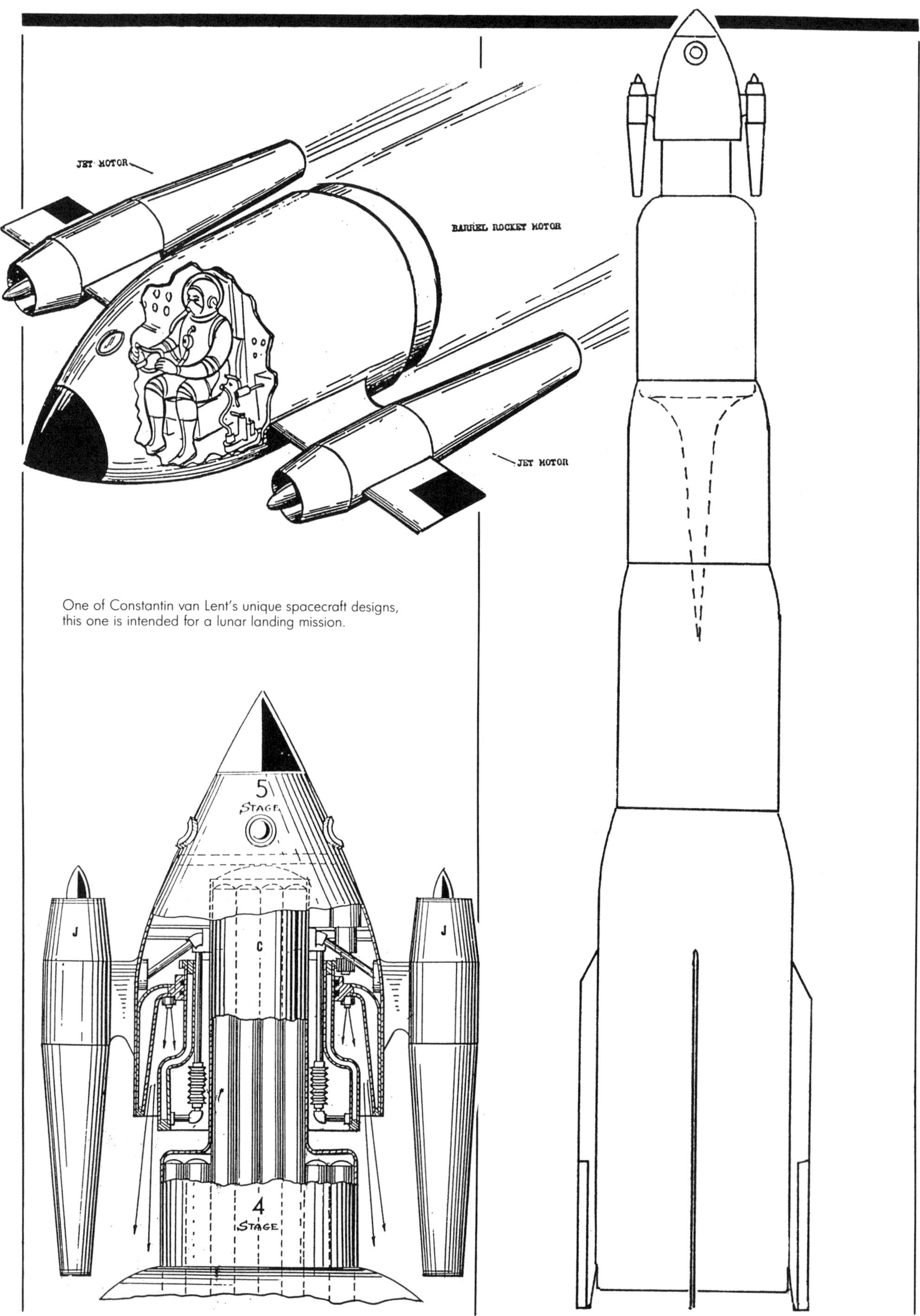

One of Constantin van Lent's unique spacecraft designs, this one is intended for a lunar landing mission.

C. P. Lent

1957

An enlarged version of this design is a central part of a manned Mars mission envisioned by Boeing engineers. It is to be launched by four large parallel boosters. The central booster is used to insert the spacecraft into a Martian trajectory. Upon arrival at Mars, a manned module and return booster remain in orbit while the glider descends to the surface. Once exploration is complete, the glider is jacked up on its landing legs to the necessary takeoff angle. The aft-section of the glider forms the takeoff platform for the nose, which is launched as an independent spacecraft. After rendezvousing with the orbiting module the composite ship returns to the earth. There it separates from the final-stage engines and uses its aerodynamic surfaces and controls to reenter the atmosphere and land.

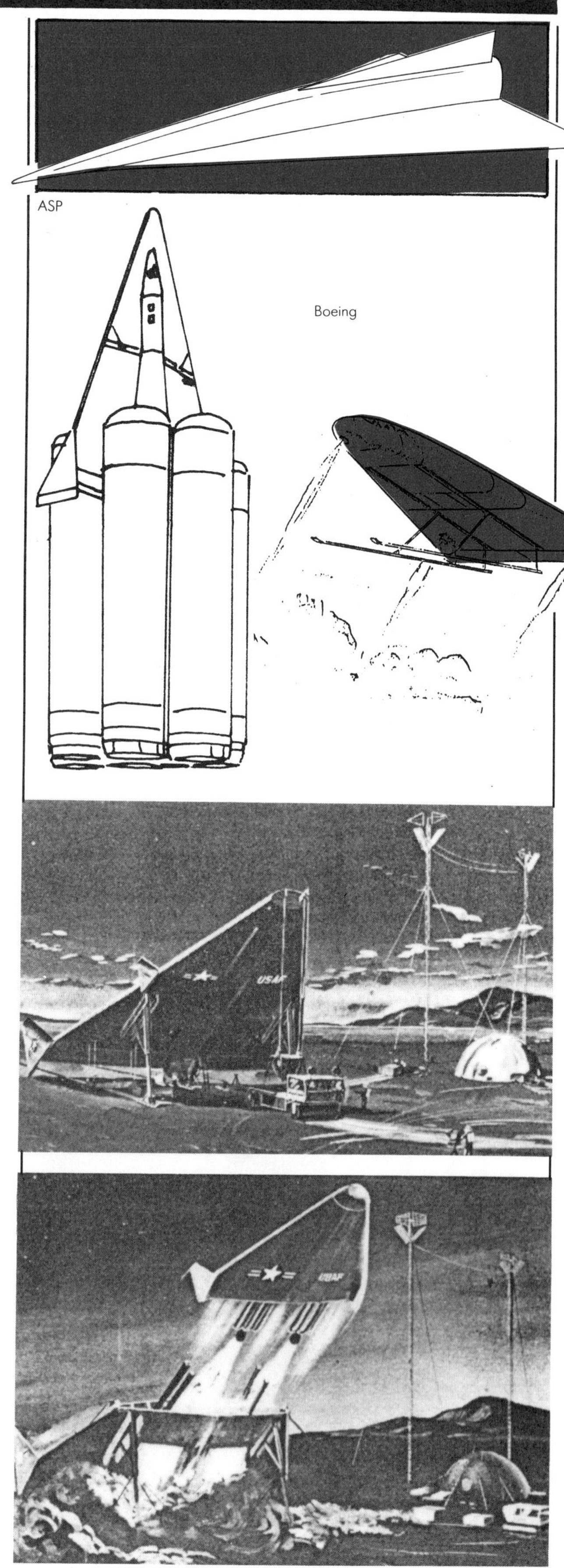

ASP

Boeing

1957–1964

The U.S. Air Force undertakes the Aerospace Plane (ASP) project as an outgrowth of its recoverable space booster studies. By 1959 this has evolved into ROLS, or the Recoverable Orbital Launch System. ROLS is a single-stage booster capable of taking off horizontally on rocket power from a B-52 runway and entering a 300-mile orbit. It employs a complex hybrid rocket-ramjet propulsion system capable of producing its own liquid oxygen by collecting, compressing and liquifying atmospheric oxygen. This combines with the liquid hydrogen it carries in its tanks and feeds to the engines.

The Air Force hopes to have the ASP operational by 1970. In attempt to avoid some of the rapidly developing complexities of the system, some designers suggest a Mach 6 aerial refueling maneuver (which would have been, according to historian John V. Becker, a "catastrophic experiment").

In 1961 the Martin Company suggests a configuration called the *Astroplane* that utilizes a nuclear-liquid rocket engine. The aircraft is to be about the same size as a B-70 and capable of using existing runways. Its shape is that of an elongated lifting body with extendable flex wings for landing and takeoff. It scoops up nitrogen from the atmosphere,

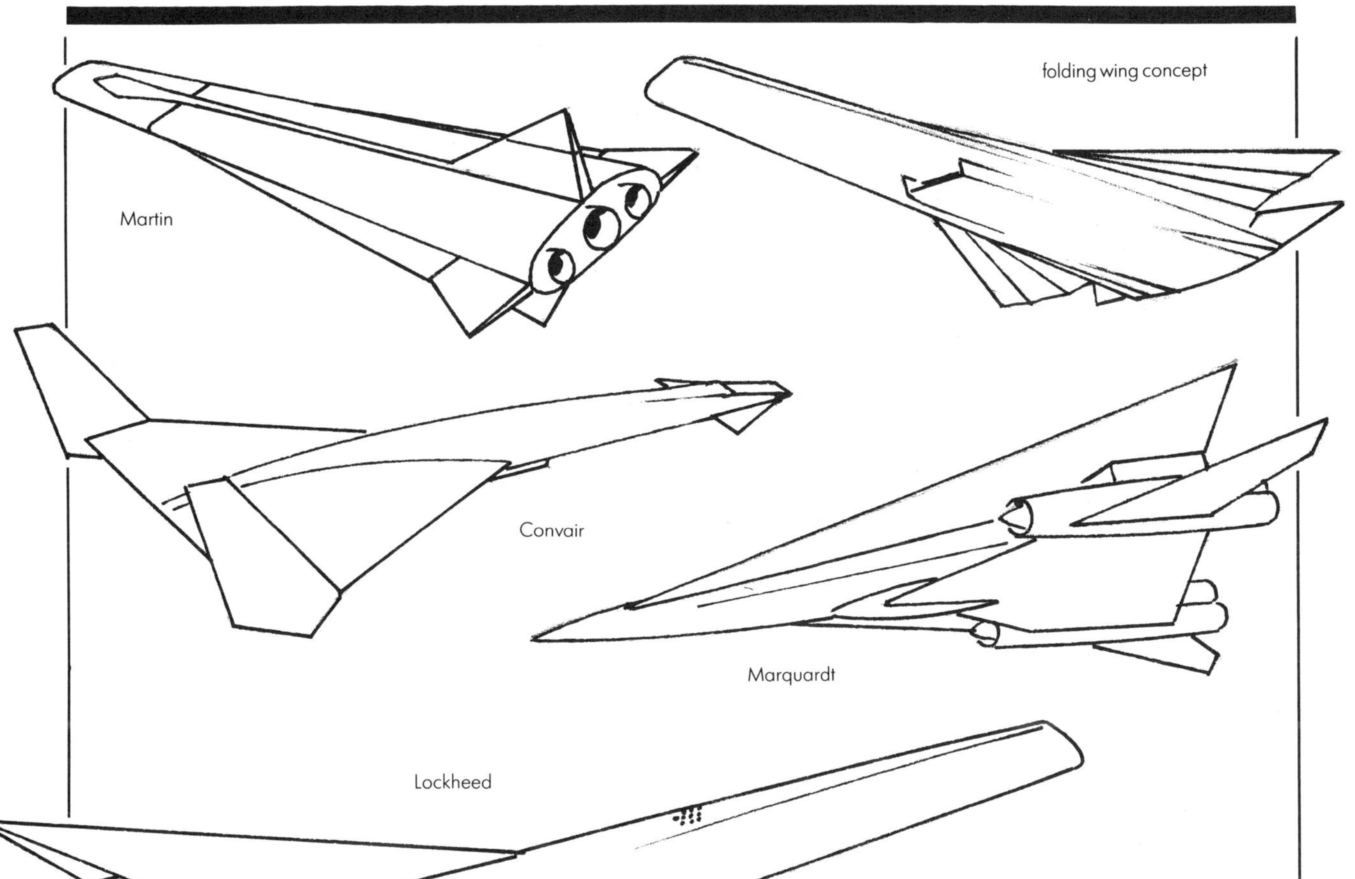

liquifying it, at altitudes as low as 320,000 feet. The magnetohydrodynamic (MHD) engine accelerates the nitrogen by electromagnetic forces through a nozzle.

The project is changed in 1962 from a single-stage design to a two-stage configuration. This allows a payload bay of about 10′ × 25′ × 40′. General Dynamics, Marquardt, Martin, Lockheed, Douglas, Republic, Boeing, Goodyear, and North American Aviation all undertake Aerospace Plane design studies. General Dynamics, Douglas and NAA receive $500,000 contracts for detailed development and Martin builds a full-scale Aerospace Plane wing-fuselage structure as part of a contract from the Flight Dynamics Laboratory. The cost of bringing ASP to full operational status by 1970 is estimated at $3 to $5 billion.

However, in 1963, the Scientific Advisory Board grows disenchanted with the project, stating that the Air Force has failed to fully identify the Aerospace Plane program, and that the Air Force's expectations, as vague as the SAB feels they are, exceed available technology. The SAB feels that " . . . the so-called Aerospace Plane program has had such an erratic history, has involved so many clearly infeasible factors, and has been subjected to so much ridicule that from now on this name should be dropped." Funding is subsequently cut from the 1964 fiscal year budget.

1958

R. W. Bussard publishes "Some Boundary Conditions for the Use of Nuclear Energy in Rocket Propulsion."

Frank Tinsley, a noted astronautical illustrator, designs a large number of speculative spacecraft.

1. Manned Satellite. This is a three-stage rocket superficially resembling von Braun's

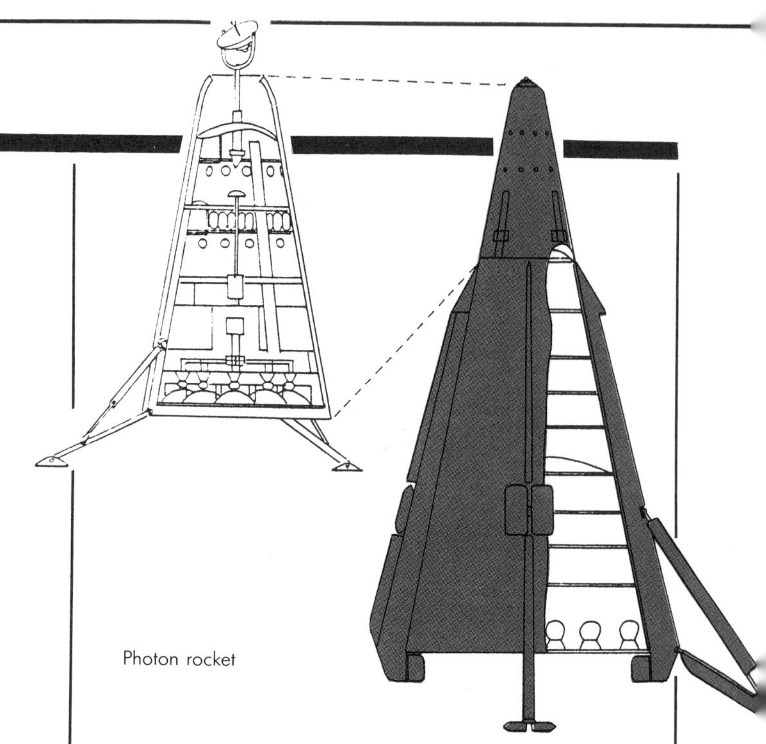
Photon rocket

ferry rocket from the *Collier's* magazine series [see above]. The first two stages are liquid fueled with multiple engines. The manned stage is canard-winged, like von Braun's. However, Tinsley's is possessed of two separate propulsion systems. One is the chemical rocket segment that provides the final impetus into orbit. Above this is a conventional, air-breathing jet engine. During takeoff, the wings are swept back to their full 40 degree slant. For reentry and landing, they swing out to a nearly straight 10 degrees.

The pressurized crew compartment is located near the rear of the rocket, between the larger wings. It is divided into half by the wing retraction cases. Between the cabin and the instrumented nose cone are the alcohol and liquid oxygen tanks. After reentry and just prior to landing, the booster rocket segment is jettisoned and the fuel tanks are dropped, leaving an open structure of girders between the pilot compartment and the nose. The two crewmembers then parachute to earth and the scientific records are parachuted down as well. Only the pilot remains in the rocket. A tricycle landing gear is lowered and the ship lands "as slowly and easily as a light, twin-engined transport."

2. *The Space Scout.* Based on specifications developed by G. Harry Stine, this spacecraft is intended to act as a kind of orbital interceptor. It is a four-stage rocket, the first three being clusters of solid propellant engines. The first stage is made up of nine solid fuel rockets, the second stage seven and the third stage four. The fourth, manned stage is liquid-fueled. It is teardrop-shaped and can fly either nose or tail-first. It is propelled by a rocket engine while in space and by a triad of jet engines while in the atmosphere. These latter are mounted in the crotches of the three V-shaped wings. Spars protruding from the air intakes of the engines act as landing legs (with expanding "shoes"). The rocket engine is mounted in the bulbous tail. This blunt end also acts as a heat shield during reentry. The manned portion of the vehicle is divided into two levels. The uppermost is the control cabin, outfitted with a reclining seat for the pilot. Below is the two-man crew compartment, with its twin reclining beds for the takeoff.

The Space Scouts are intended to remain in orbit for 3-day tours of duty. They are armed with three air-to-air missiles mounted on the wingtips.

3. *Interplanetary Photon Rocket.* A cone-shaped chemical booster lifts the photon rocket into space, where the hydrogen fusion-powered photon drive takes over. It accelerates to 98 percent of the speed of light, reaching Mars in little over an hour. The photon rocket is a narrow cone, its base containing the "neon-type reflectors [which] project reaction concentrated photons in the form of a starlight power beam." The rocket is divided into several levels. The uppermost is occupied by the communications gear, including a retractable dish antenna that can be extended beyond the nose of the ship. Below are supplies and storage. Next is the bridge deck. Below that are the crew's quarters, separated from the bridge deck by the containers of breathing oxygen. The lower third of the rocket is taken up by the hydrogen fusion pile, its water supply and the paraphernalia of the photon drive.

4. *Interstellar Generation Ship.* This is a vast, hollow sphere, over 16 miles in diameter, with an artificial sun suspended at its center, designed to be built and launched from a lunar crater. The rim of the 8-mile crater is equipped with roller bearings so the huge starship can rotate. The surface of the crater is lined with an inflatable material to cushion the outside of the sphere as it rests

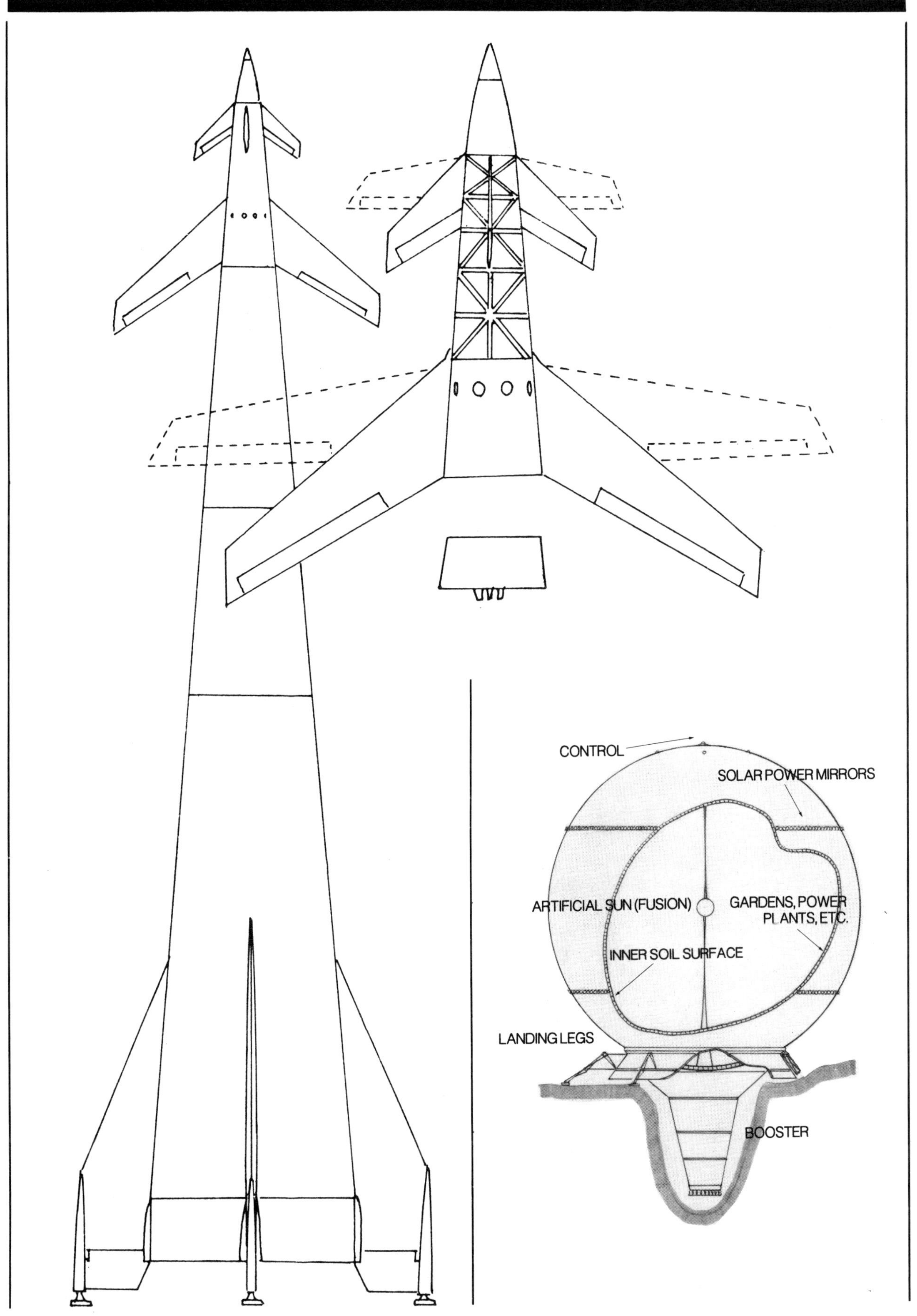
CONTROL
SOLAR POWER MIRRORS
ARTIFICIAL SUN (FUSION)
GARDENS, POWER PLANTS, ETC.
INNER SOIL SURFACE
LANDING LEGS
BOOSTER

on it. The sphere's rotation is started by firing a belt of rockets placed along its equator. When an artificial gravity has been established within it by the centrifugal force, soil is distributed on the inner surface. Landscapes are then created and houses built. Colonists begin living within the starship. A cone-shaped "stalk" attached to the bottom of the sphere, and extending into a pit excavated within the crater, is the rocket booster unit. It will drive the starship away from the solar system, ultimately being jettisoned.

On the outer surface of the ship are observatory domes and solar power generators. Between the armored outer shell and the inner living shell are chambers containing hydroponic gardens, power plants, and so on. The enormous starship is equipped with legs for landing on other worlds [!].

Tinsley also has designed a number of space station concepts. One is a disk-shaped space farm. Concentric tubes arranged on a flat, circular deck contain hydroponic farms. Collecting and drying equipment, storage areas and quarters for the small maintenance crew occupy the shallow, bowl-shaped space beneath. In the center of the disk is a solar power generator. Beneath the solar power mirror is the dome of the bridge. Algae is the main "crop," grown in a nutrient soup. It is electrostatically precipitated, the sludge drawn off by pumps and sun-dried to a flour-like powder. Once every 24 hours, a ferry rocket, driven by hydrogen-fueled atomic engines, shuttles the flour to the earth.

Another concept is a giant wheel, revolving on its axis to provide artificial gravity. It is to be constructed in orbit by cannibalizing the ferry rockets. These come apart into the sections that form the station's rim and spokes. A fixed central axis contains an astronomical observatory and docking ports. Stairways and elevators connect the hub with the rim. The flat, wide rim of the station has a central corridor off which the staterooms and laboratories open. A hydroponic farm occupies one edge of the wheel. This is on the side of the station that continuously faces the sun. On the shady side is a promenade with large windows. The circular space between the rim and the hub, divided into quadrants by the four spokes, is filled with acres of solar cells.

Lunar lander from *Space Handbook*.

The British study a rounded delta orbiter similar to the Dyna-Soar.

The RAND Corporation publishes the *Space Handbook*, the staff report of the Select Committee on Astronautics and Space Exploration.

John B. Montgomery, general manager of the Aircraft Gas Turbine Division of General Electric, proposes a turbojet-powered vertical booster for satellite rockets. Nine air-breathing engines will be placed in a ring surrounding the rocket and supported by three wings. At the tip of each wing will be another jet engine, for a total of twelve in all. After boosting the rocket, the jet unit will return to its launch site, landing vertically.

H. E. Ross updates the Ross/BIS space station. The solar mirror is replaced by a disk-shaped mat of photoelectric cells. The single telecommunications boom is now a double mast, one end carrying television cameras and radio and television antennas. The other mast carries a large radio telescope dish. The central turret, in addition to the "strobo-telescope" carried over from the original design, now has a docking berth attached for the small interorbital tenders. [Also see 1948].

A. V. Cleaver publishes his book *Project Satellite*. In the chapter "Interplanetary Flight" he describes the "orbital refueling technique," which he developed as a method of reducing the load lifted by the final stage of a rocket at any given time. [Also see 1949 and 1951.] He makes an apt analogy with the in-flight refueling of aircraft. For example, the final stage of a rocket could exhaust all of its fuel in attaining orbit. Subsequent tanker rockets could arrive in orbit with just enough fuel remaining to replenish the original rocket's empty tanks and to allow a safe return to earth themselves. (The ferry rockets, which carry payload and fuel, will be designated "Type A," while the main vehicle will be "Type B.") No one tanker need carry enough fuel to refill the first rocket's tanks. Once the first rocket is refueled it is free to continue its interplanetary mission, to the moon or Mars, for example. Once in orbit around its destination, the Type B rocket will transfer its payload to a "tender" or Type C rocket, which will perform the actual landing on the other planet.

On the return voyage, the Type C lander will rejoin the Type B mother ship, return to the earth, and then transfer crew and cargo to a waiting Type A for the final descent.

The Type A rockets will be chemically fueled and employ wings on all stages for reentry and recovery.

The Type B rocket will be assembled in orbit and will never land on the surface of a planet. It could be nuclear-powered, either of the thermodynamic type or with an ion engine.

The Type C landing craft will be a chemical rocket. Whether it possesses wings or not will depend upon the conditions prevalent on the moon or planet being visited.

Project Adam is a manned ballistic rocket launch proposed by the Army Ballistic Missile Agency, and backed by members of the Army, Navy and Air Force. The program is initiated in late 1957 when officials from Holloman Air Force Base, Randolph Air Force Base and the Navy at Pensacola are asked by ABMA to participate in the design,

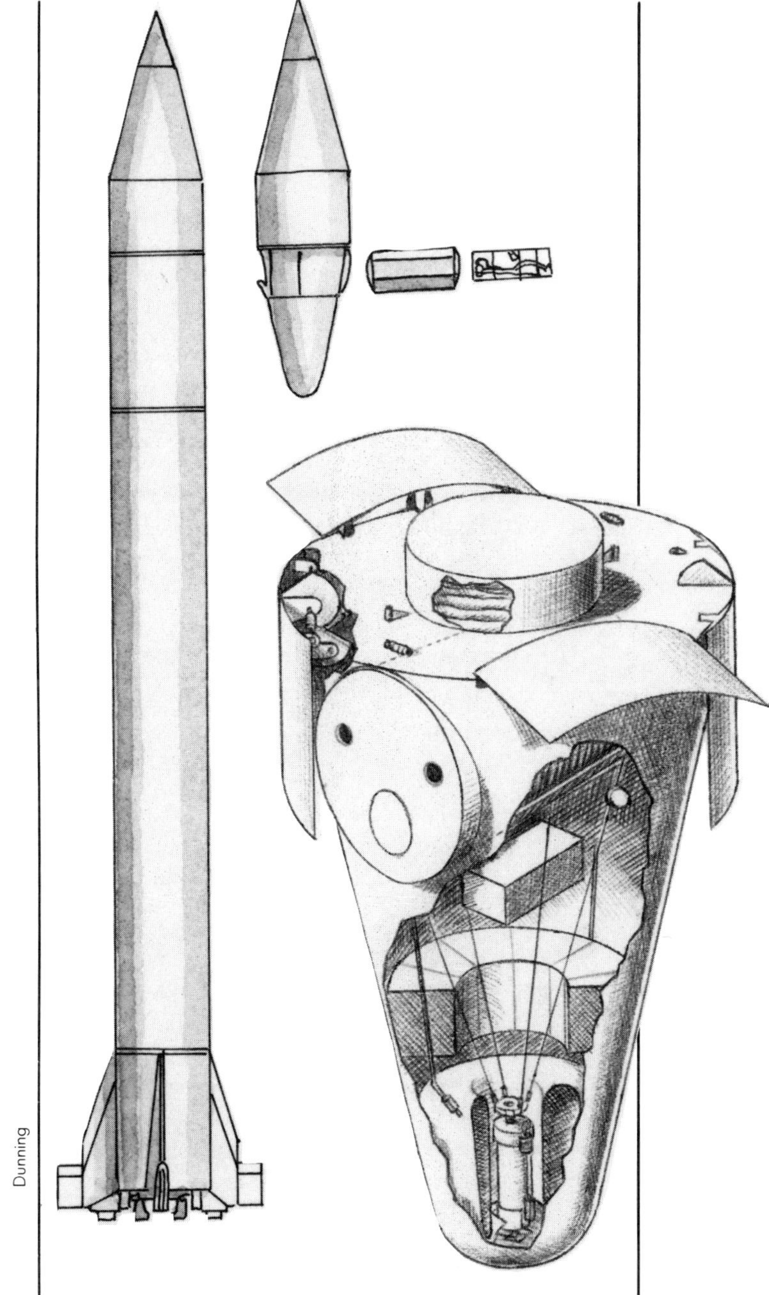
Dunning

development and planning of the Adam project.

A manned capsule is planned to be launched by a modified Redstone rocket. It would make a 150-mile high ballistic trajectory, finally landing 150 miles downrange. The justification for the project is to "improve the mobility and striking power of U.S. Army forces through large scale transportation by troop-carrying missiles." Another reason for the flight will be to analyze the effects of weightlessness on human beings. The Adam pilot will experience up to 6 minutes of freefall. The hope is to make the first flight by 1959—with the first man launched within 18 months of receipt of authority and funds.

The Redstone booster, in the Jupiter-C configuration, would carry a double cone-shaped payload. The nose cone is a protective shell covering the downward-pointing manned cone. The nose cone is jettisoned after passing the highest point of the trajectory. The "recovery body" contains instrumentation and a cylindrical capsule. Just large enough to carry one man laying prone, the capsule is made in three layers: an outer shell, a layer of ecco foam and a sponge interior. A boxlike cage surrounds the astronaut. A life support system is provided and the pilot faces a television camera and a battery of physiological test panels. The role of the astronaut is entirely passive.

Four Adam flights will cost an estimated $10–12 million.

The original plans include Lt. Col. D. G. Simons [see Man High, 1960] as designer of the capsule as well as its first pilot (though he denied this).

After separating from the nose cone, four drag vanes are to be extended to slow the fall of the capsule. A parachute is eventually released, setting the capsule gently onto the surface of the ocean. The capsule is equipped with automatic radio beacons that begin to transmit as soon as the vehicle hits the water.

An emergency ejection system is provided for the pilot. Since the manned cylinder is inserted into the capsule at right angles to the line of flight, an explosive bolt could shoot it out of the side of the rocket. A small parachute would slow it, and the cylinder would land in a cushioning basin of water nearby.

The project is blocked by the Air Force in order to boost its own X-15 plans, although a proposed orbital flight of the spaceplane is still more than 2 years in the future, as opposed to the single year required for the first Adam launch.

The XSL-01 is a spaceship designed by Ellwyn E. Angle, an engineer for the Ramo-

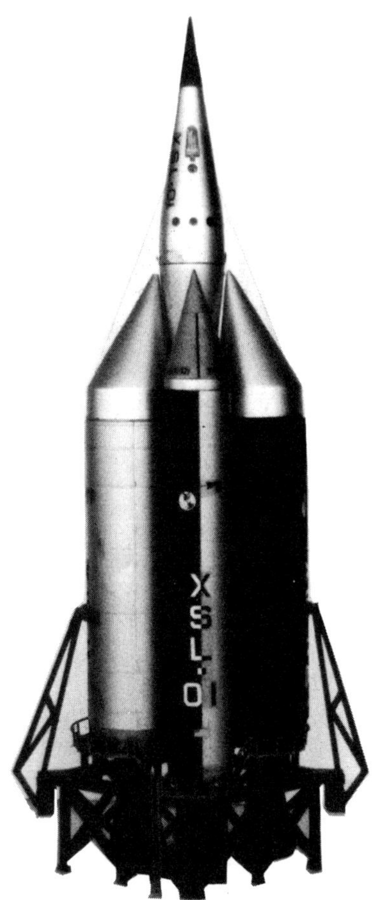

Woodridge Space Technology Laboratory, which Revell produces as a plastic model assembly kit.

The XSL-01 is a three-stage rocket (XSL = Experimental Space Laboratory). The first and second stages are two pairs of parallel boosters, with an overall length of 111 feet, the engines of which burn liquid fluorine and hydrazine. The two first stage engines produce a total of 1,800,000 pounds of thrust and the two second stage engines produce a total of 500,000 pounds of thrust. The winged, manned third stage is fueled with liquid hydrogen, which acts as the working fluid for an atomic engine producing 20,000 pounds of thrust.

The manned stage is a delta-winged spacecraft manned by a crew of three. Four cone-shaped modules surrounding its base contain landing gear, radar equipment and a solar power generator. Among the scientific equipment aboard the spacecraft is a 12-inch Newtonian reflector.

The flight plan to the moon and back includes the following: at −8 hours, the rocket is rolled to its launch site and the gantry is brought into position. At −7.5 hours the main element of the atomic engine is loaded. At −1.5 hours, the liquid hydrogen tanks are filled and at −1 hour the hydrazine tanks are loaded. The liquid fluorine is loaded at −35 minutes. At −10 minutes the crew enters the third stage. At −8 minutes the gantry is removed. At −10 seconds the four main engines are started at low thrust. At −1 second the engines are brought to full thrust. Takeoff is at zero seconds. The first stage cuts off after +205 seconds, at an altitude of 110 miles (its boosters are later recovered for reuse). At +250 seconds the atomic pile is activated. The second stage completes its burn at +350 seconds and 430 miles altitude (it, too, is eventually recovered). The atomic engine begins thrusting at +355 seconds; it cuts out at +1,380 seconds (23 minutes) at an altitude of 1,300 miles. Velocity is 31,600 fps (6 mps). At +360 seconds the doors of the radar and solar collectors are opened. At +24 minutes the crew is in freefall. At +4 2/3 days, the ship is within 24,000 miles of the moon and is turned end for end in preparation for landing. It lands on the moon, braking by the thrust of the atomic engine, in the crater Plato at +5 1/8 days. Exploration consumes 2 days and at +7 1/2 days the rocket takes off from the moon. At +12 days the rocket reaches the outer fringes of the earth's atmosphere. A series of braking ellipses are begun, each orbit slightly smaller than the preceding one. This requires 2 days. The conical storage modules are jettisoned at +14 days. The spacecraft prepares for an airplane-type descent and landing, which occurs at +14 days 4 hours.

The XSL-01 model kit is an extraordinarily well-thought-out and detailed assembly. See 1959 for a companion kit designed by the same engineer.

The XSL-01 bears a close resemblance to a moonship credited to Douglas, which has a return vehicle surrounded by six booster rockets. The conical noses of these formed part of the landing structure, as in the XSL-01. It is possible that Angle worked at Douglas after Ramo-Woodridge.

NASA's Space Task Group proposes a design for a minimum manned space laboratory to be launched by a modified Atlas Vega booster. The laboratory will consist of a Mercury capsule expanded to occupy two astronauts atop the cylindrical laboratory. After achieving orbit, the capsule will be rotated until it is parallel with the side of the laboratory module, where their access hatches are mated. The laboratory also would have an additional collapsible airlock.

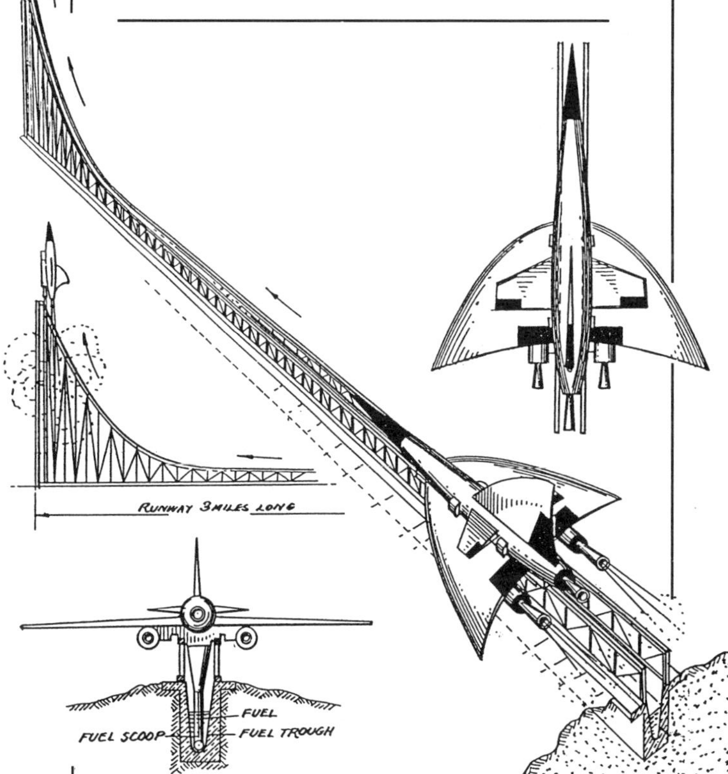

Constantin Paul van Lent proposes launching rockets using a captive air-breathing booster sled. They will travel down a 3 mile horizontal runway consisting of a double rail track. The very end of the runway curves up abruptly so that the rocket launches vertically.

The unique feature of Lent's sled is that it contains no fuel itself. The fuel instead is in a trough beneath the booster and is scooped into the sled as it passes down the track.

The rocket carried by the sled is equipped with both rocket engines and jets. It will use the latter at takeoff, switching to the rockets later.

1958

Skotak

Three-stage manned orbital rockets are featured in the film *War of the Satellites*. They are designed by Jack Rabin and Irving Block, who are also responsible for the 1950s film *Rocketship XM*. The film is produced by Roger Corman to take advantage of the publicity surrounding the successful Soviet space launches, creating the picture in just 8 days.

Donald Cox and Michael Stoiko, with illustrator N. Stanilla, publish their book *Spacepower*. Rather than discuss the technical aspects of spacetravel, the authors analyse the social, military and legal aspects. They make a detailed proposal for the control of space by a United Nations force.

They offer several timetables for the exploration of space. Summarizing them, we have:

1958—Beginning air-launched research aircraft program.

1960—Unmanned lunar probe.

1963—First large-thrust booster.

1964—First propellant tanker.

1966—First manned space vehicle (two astronauts).

1967—First recoverable booster.

1968—Two-man spacecraft launched into a minimum time orbit.

1970—Hypersonic rocket vehicle.

1972—Manned lunar circumnavigation, landing and settlement.

—Three-man spacecraft in limited time orbit.

1973—Two-man orbital capsule.

1975—two-man permanent space station.

1977—Lunar landing fleet assembled.

1980—Large-scale Lunar colonization.

—Ion-propelled vehicle developed.

1985—Unmanned Mars probe.

—Commercial rocket transport.

The first lunar orbital mission—Project Moonmen—will send two passengers around the moon. This will be followed by additional two-man orbital missions and later a three-man flight. If made under the auspices of the United Nations, the spacecraft will carry an international crew.

The lunar landing will be made with three spacecraft. Two of these will be two-passenger vehicles acting as observers and cargo carriers. The third will be a three-passenger spacecraft carrying a small "rocket personnel car" on its side. This will be a small non-streamlined vehicle carrying one or two passengers to the lunar surface while its mother ships and the observer ships remain in orbit.

FEBRUARY 10

A paper entitled "A Program of Expansion of NACA Research in Space Flight Technology," prepared by the NACA engineering staff at Lewis Aeronautical Laboratory, suggests a $241 million increase in NACA's budget to "provide basic research in support of the development of manned satellites and the travel of man to the moon and nearby planets."

MARCH

Richard L. Barwin proposes a solar sail concept in the journal *Jet Propulsion*.

APRIL 25

The Air Force Ballistic Missile Division proposes a plan for the manned exploration of space. The goal is to "achieve an early capability to land a man on the moon and return him safely to earth."

The program would take place in four stages: 1. "Man-in-Space Soonest"; 2. "Man-in-Space Sophisticated"; 3. "Lunar Reconnaisance" by unmanned probes and satellites; and 4. "Manned Lunar Landing and Return." It is expected that all four stages could be completed by 1965 at a cost of $1.5 billion.

JULY

NACA proposes "A National Integrated Missile and Space Vehicle Development Program" in which several future manned spaceflights are listed. Highlights of the program include:

June 1959—First powered flight of the X-15.

July 1960—First wingless manned orbital return flight.

August 1961—First winged orbital return flight.

November 1962—Four-man experimental space station.

July 1964—Manned lunar circumnavigation and return.

September 1964—Twenty-man space station.

July 1965—Final assembly of first 1000-ton lunar landing vehicle (with emergency manned lunar landng capability).

August 1966—Final assembly of second 1000-ton lunar lander and first expedition to the moon.

1972—Large scientific lunar expedition.

1973/1974—Establishment of permanent lunar base.

1977—First manned expedition to another planet.

1980—Second manned expedition to another planet.

OCTOBER 1

The National Aeronautics and Space Administration is officially established (and NACA abolished). It is charged with the U.S. civilian space program. President Dwight D. Eisenhower signed the NASA Act on July 29, 1958.

OCTOBER 28

NASA's Stever Committee submits its report on the future of the civilian space program. Among its other suggestions are the continuance of maintaining performance capabilities of man in space as a preliminary to more sophisticated space exploration; the development of lifting body reentry vehicles; and the development of heavy-lift boosters.

DECEMBER 17

NASA announces that its manned satellite program will be called "Project Mercury." Not by coincidence this is also the anniversary of the Wright's successful flight at Kitty Hawk, North Carolina, in 1903.

The program is assigned two broad missions: to investigate man's ability to survive and perform in the space environment, and to develop the basic space technology and hardware for future manned spaceflight programs.

ca. 1958

The National Advisory Committee on Aeronautics (NACA) designs three proposals for satellites capable of manned reentry. One is a slim dart with four narrow delta wings. Its nose is blunt, as are the leading edges of the wings. The pilot's compartment is at the very rear, just ahead of the rocket engines. It will land like a conventional aircraft. The second spacecraft is dubbed the "motor boat." It is a blunt-nosed vehicle with a flat upper surface and curved underside. The pilot sits in the conventional position, under a domed canopy with small circular portholes. It will land like a normal aircraft (or could make a water landing, justifying its nickname). The third is a small hemisphere. Its outer surface covered with a heat shield material. Retrorockets bring it down from orbit and small fins stabilize it for a nose-first reentry. It is recovered by parachute.

The Ferri sled is a reentry vehicle designed by the Brooklyn Polytechnic Institute. It is a blunt-nosed cylinder whose outer shell swings open to form a glider with hollow wings.

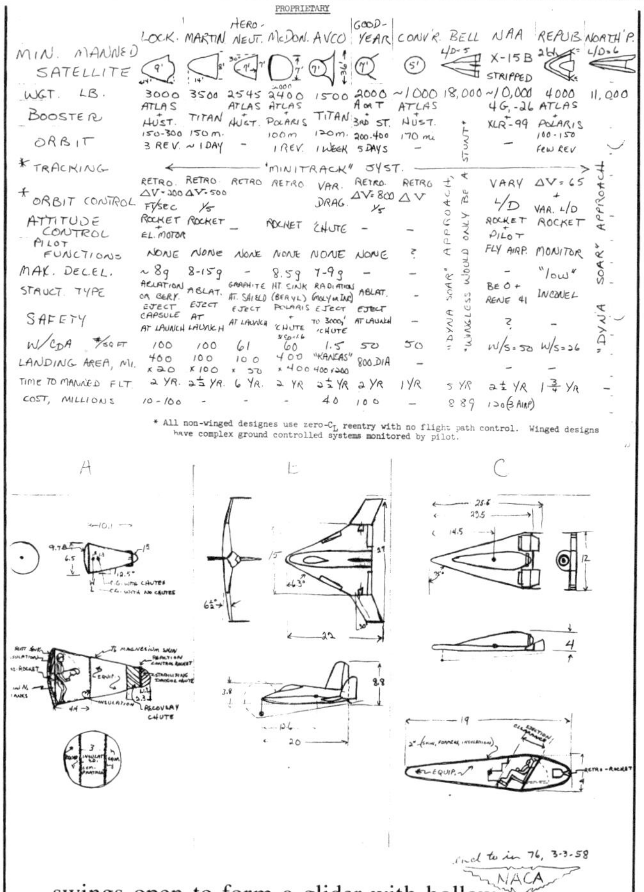

A drag brake is developed by Avco Research Laboratory. It is a spherical cabin attached to a large loop, rather like the jewel on a ring. The craft's motion through the air upon reentry creates an electric current over the surface of the loop. Inside, a magnetic field is artificially generated. The generated current "collides" with the magnetic field, slowing down the ship.

Krafft Ehricke develops a number of advanced spacecraft designs for Convair. The first is a wingless, bullet-shaped satellite booster, 126 feet tall. It is a three-stage rocket capable of placing its third stage, with a payload of 11,000 pounds, into an orbit 600 miles above the earth. The cylindrical fuel tank of this stage eventually will become the basic building block of Ehricke's space fleet. He predicts that this rocket will fly within 10 years. By comparison, the Atlas intercontinental ballistic missile, which Ehricke has recently developed, is only about 80 feet tall, weighing but 80 tons.

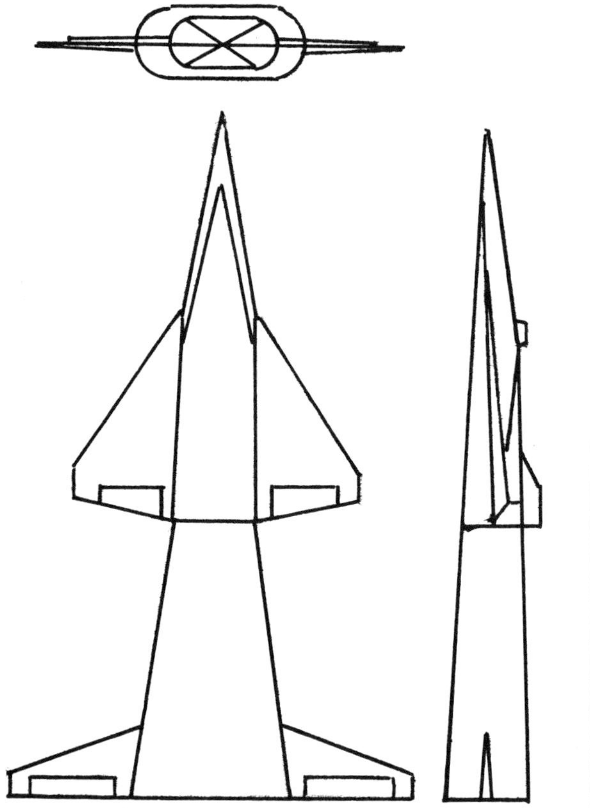

NASM (2)

Type II 2,000 lb. supply ship

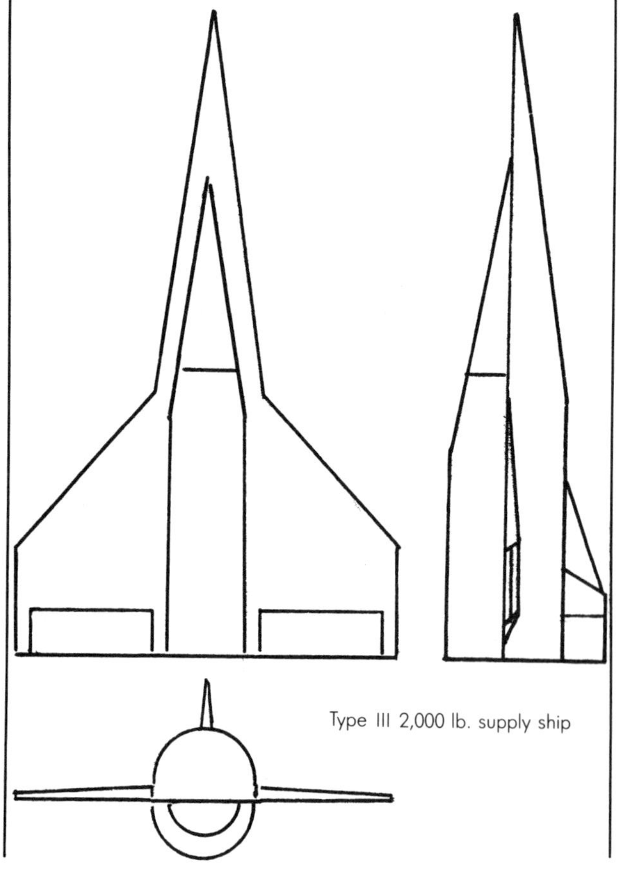

Type III 2,000 lb. supply ship

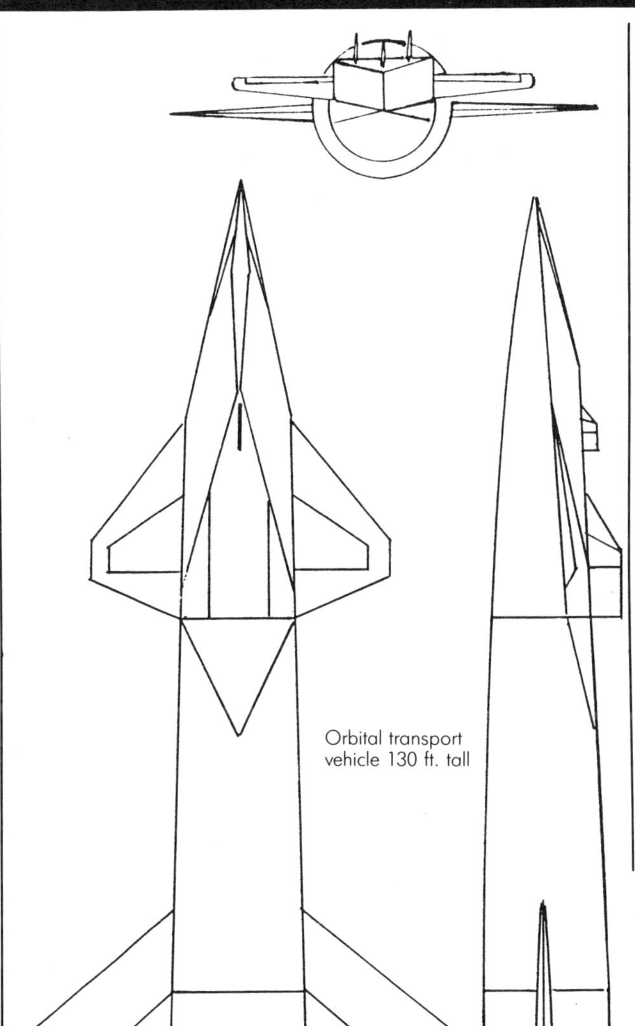
Orbital transport vehicle 130 ft. tall

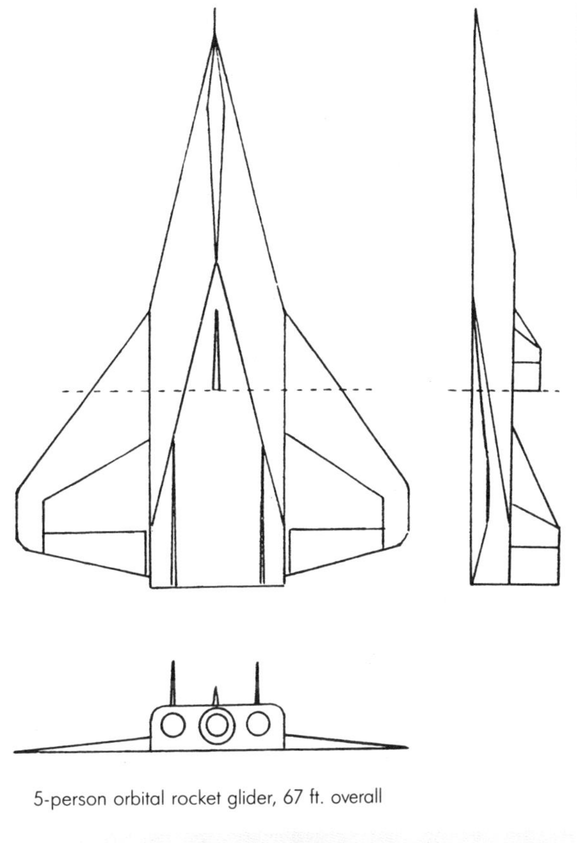
5-person orbital rocket glider, 67 ft. overall

Immediately following will be a manned "satellite tender," a three-stage, 130-foot rocket carrying a pilot and four passengers. Its manned stage is a hypersonic glider carried piggyback on the second stage. The crews of these ships will assemble the next generation of spacecraft from the orbiting fuel tanks of the satellite rockets. The first built will be a manned space station. Ehricke expects this to take place within 10 years. At either end of a 400-foot long tube will be a cluster of four parallel cylinders: the living and working quarters for the crew of four who tend the station. At the center of the tube are the station's power supply and communications. The station rotates on its short axis to provide artificial gravity.

Within 5 years after the construction of the space station, a manned flight around the moon will take place. The spacecraft used for this venture will, like the space station, be built from the standard modules of satel-

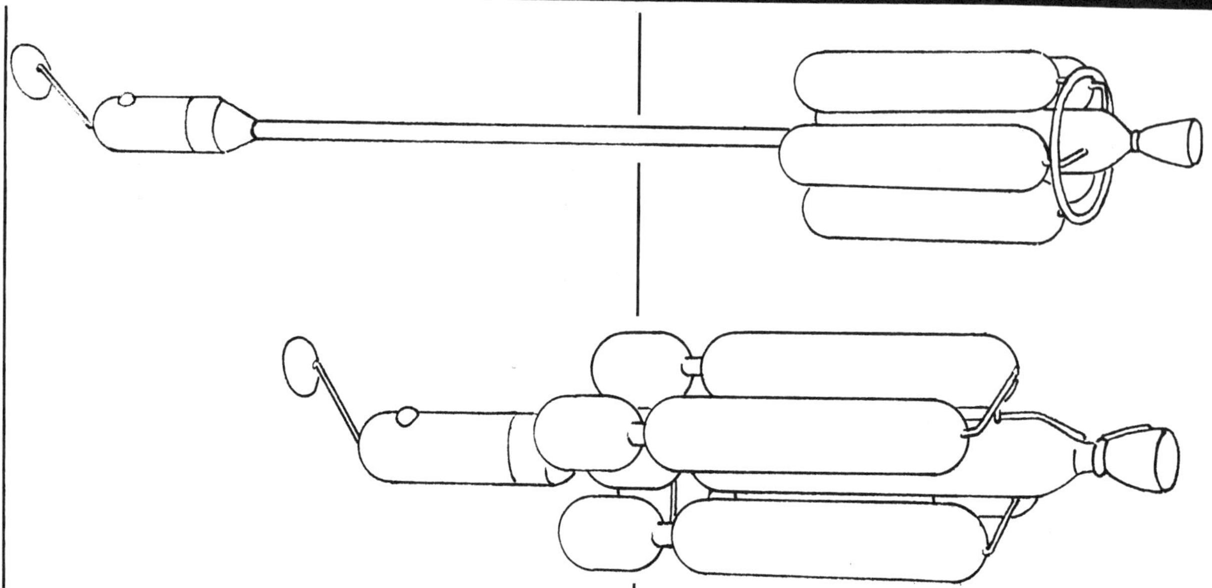

lite rocket fuel tanks. Four pairs of fuel and oxidizer tanks will be clustered at the rear of the rocket. The main cluster of tanks will be jettisoned once they are empty. At one end is a single rocket engine. At the other is a cylindrical crew capsule. The ship is not intended to land on the moon, but merely to circle it and return to earth orbit.

Next in Ehricke's scenario are manned, round trip flights to the moon. These will take place within 5 years of the moon orbiter.

For eventual manned interplanetary flights, Ehricke envisions an atomic-powered ship. It is a slim, 300-foot spacecraft—the 250-foot tubular central section being required to separate the crew from the atomic engine. Its fuel tanks are several times larger than those of the previous ship. Quarters for its crew of eight are located in a cylinder mounted at right angles to the shaft connecting it to the propulsion module. The cylinder can be made to rotate, providing the crew with artificial gravity. Once in orbit around a planet—Mars, for example—the astronauts will travel in shielded cable cars to the vehicle's center of gravity. From there they will launch themselves away from the spaceship in small, chemically fueled "satelloid" rockets. One of these will be left in Mars orbit when the time of departure comes, to act as an unmanned, instrumented satellite.

With the exception of the nuclear rockets, the advantage to Krafft Ehricke's plan is that it will make manned spaceflight a reality within 20 years "without any major technological breakthrough."

Martin develops a moonflight scenario based upon a giant booster not unlike the original Saturn (which at this time is little more than a cluster of Redstone rockets, with a Titan second stage and a Centaur third stage). The Martin rocket could launch a 9,000-pound payload around the moon. The booster is a four-stage rocket. The first stage, B-1, is a cluster of four 160,000-pound A-1 rockets; the second stage, B-2, is a single A-1; the third stage, B-3, is a 40,000-pound A-2; and the fourth stage, B-4, a 10,000-pound A-3. This configuration could launch a Mercury-type capsule into a circumlunar orbit directly from the earth.

An alternative plan requires assembling boosters for the capsule in orbit as well as refueling in orbit. A much smaller rocket is then required (about the same size as the later Atlas). This would be a three-stage vehicle carrying payloads of 5,000 pounds into orbit. It would consist of one stage each of the A-1, A-2 and A-3. [Also see 1960.]

The "Man in Space Soonest" (MISS) program is initiated, after 2 years of study, intending to put a man into orbit by October 1960. The Thor missile is to be used as the booster, plus a second stage. MISS envisions four steps of space exploration, the fourth being a manned lunar landing.

The scenario of the voyage requires the booster to use three of its stages to start the rocket on its path to the moon. The third stage remains attached. Upon arrival at the moon, the rocket is reversed and the third stage engines are used to brake for a landing, 61 hours after takeoff. Four landing legs extend from the third stage, as well as a central spike that extends down from the middle of the third stage's four engines. This supports the weight of the ship; the four outrigger legs are only used to balance it upright. The fourth stage is paralleled within the third stage and provides the rocket with the thrust needed to escape the moon. This stage is jettisoned as soon as its burn is completed. Then, 60 hours after leaving the moon and 3 hours before reentry, the ship makes a corrective maneuver. This brings the ship down into the earth's atmosphere

1958–1959

Wernher von Braun publishes the serial story "First Men to the Moon," in the newspaper Sunday supplement *This Week*. It details, in fictional form, a manned flight from the earth to the moon and back. The series and the book made from it in 1960 are profusely illustrated by Fred Freeman, veteran of the *Collier's* magazine space symposium [see above]. The half-dozen years separating the *Collier's* series and the *This Week* serial have brought an enormous leap in space technology, which is reflected in von Braun's "novel."

Two astronauts are launched from the earth in an enormous five-stage rocket. It is a slightly tapering cylinder with a delta-winged manned vehicle on top. The manned rocket resembles a conventional aircraft and carries only enough fuel for maneuvering within the earth's atmosphere prior to landing. Fully half of its interior space is occupied by the pressurized cockpit, which includes a rotating astrodome for navigation. The nose is a blunt ceramic heat shield. The rocket has standard tricycle landing gear.

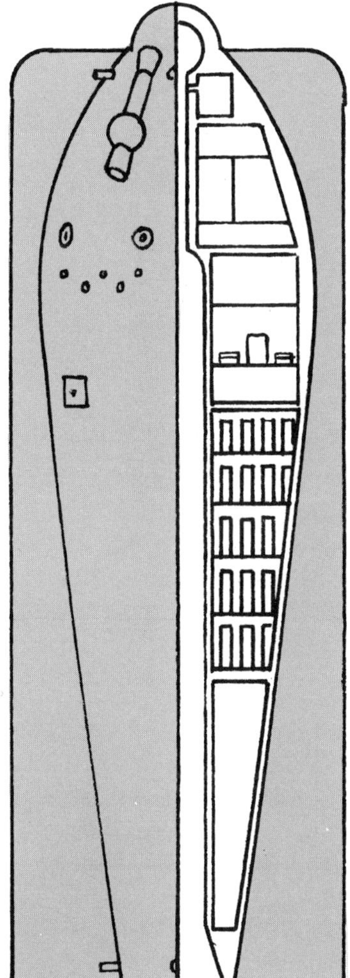

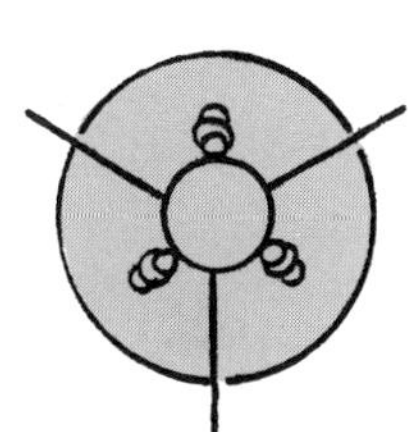

Electron gun-propelled spaceship designed by Wolfgang Schroeder. The electron beam travels down the long tunnel that passes through the length of the ship. Turbojets mounted near its nose lift it during its horizontal takeoff from a water surface.

for its long glide home. The rocket glides into the upper atmosphere inverted in order to keep aerodynamic forces from causing it to rise back into space. Downward lift is added to gravity to compensate for the 2 g's generated by the centrifugal force created by the ship's circular path at a constant altitude. Once the ship has slowed to 4 mps, centrifugal force is reduced to 1 g and the ship can be barrel-rolled into an upright position. The rocket finally makes a conventional airplane-type landing at a specially constructed landing strip in New Guinea (one of three: the others are in the Galapagos and the Cameroons).

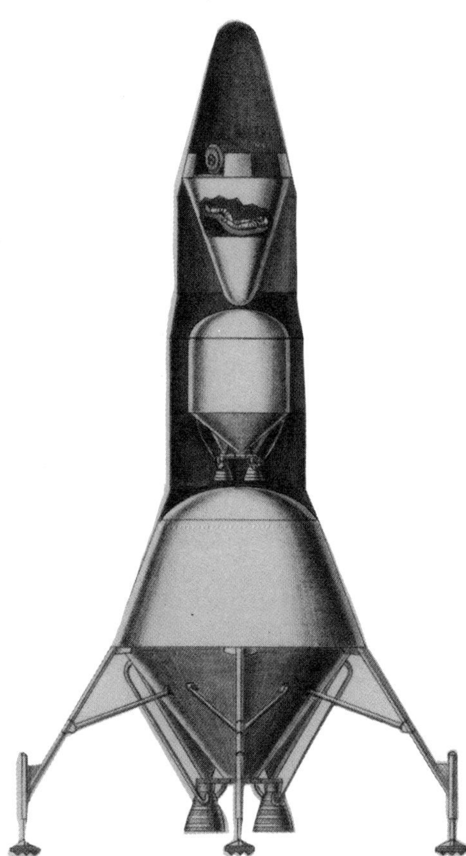

1959

Project Horizon is a U.S. Army proposal for a manned lunar base. It will put men on the moon by 1965, with a permanent base by 1966. The Army believes that anything short of being first on the moon "will be catastrophic."

A task force is created on March 20, working under the direction of Maj. Gen. John B. Medaris (Army Ordinance Missile Command) and with the full cooperation of Wernher von Braun and his team. The Phase I report is completed on June 8. It proposes a manned landing in 1965, followed by an operational base the next year. The project will cost $667 million a year from 1960 to 1968.

Later the report is recast to eliminate any overt Army/military connections, emphasizing the scientific and peaceful aspects.

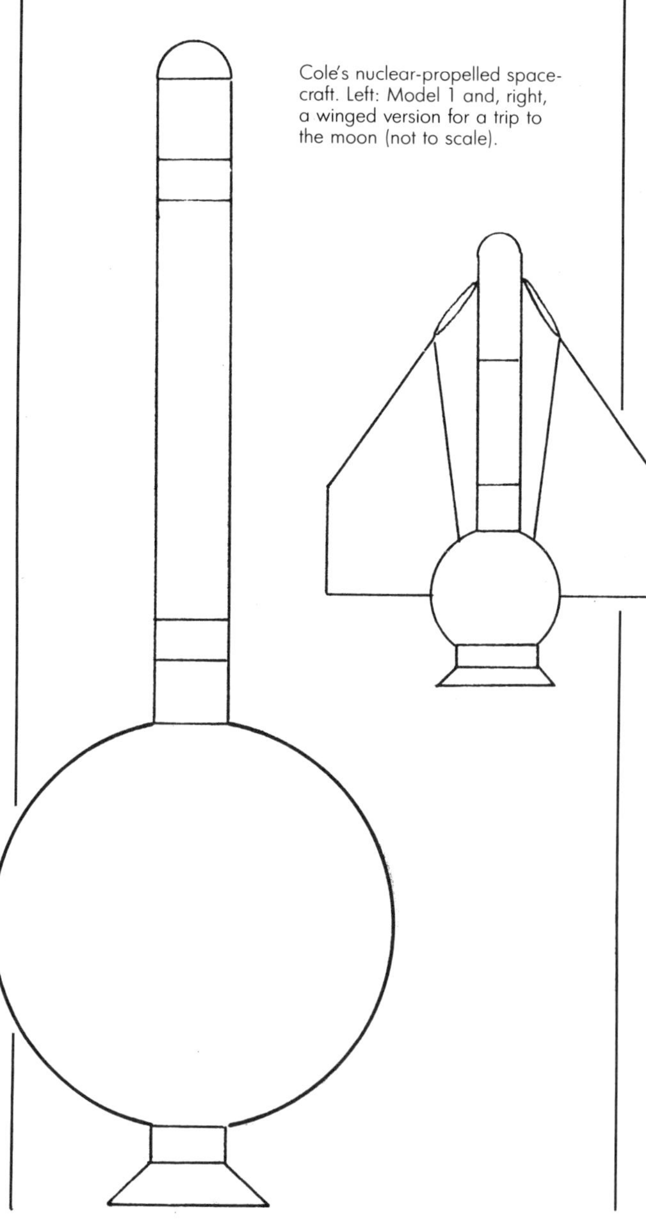

Cole's nuclear-propelled spacecraft. Left: Model 1 and, right, a winged version for a trip to the moon (not to scale).

Dandridge Cole delivers a paper at the 4th National Annual Meeting of the American Astronautical Society on "The Feasibility of Propelling Vehicles by Contained Nuclear Explosions." He began work on the idea of nuclear pulse rockets four years earlier. His Model I nuclear pulse rocket will detonate small nuclear bombs inside a spherical "thrust chamber." He believes that the detonation of just one bomb per minute will allow time for the chamber to reduce its heat by radiant cooling. Needless to say, Cole's pulse rockets are intended solely for use outside the earth's atmosphere, probably as shuttles between this planet and the moon. Although he thinks that the contained explosion will result in less atmospheric contamination ("The most obvious method of getting Model I into orbit will be to fly it out on its own power using a higher pulse frequency . . . and simply accepting the very small resulting atmospheric contamination on one such flight."), it still might be best to launch the spaceship atop a conventional booster, similar to the proposed Nova. The development of "clean bombs," he suggests, might be required for large-scale operation of nuclear pulse jets in the atmosphere.

Using 0.01 kiloton "energy capsules," or miniature nuclear bombs, acting against a 4-inch-thick steel plate (weighing 2,080,000 lb) 46 feet away will result in performance matching a conventional Mach 3 ramjet with a thrust of 4,800,000 pounds. The pressures generated within the chamber will be no more than 1,000 psi—comparable to conventional rocket motors. Using 740 pounds of water per pulse to absorb some of the shock will also have the effect of reducing some of the performance.

The free-space nuclear pulse rocket consists of two major components: the 130-foot diameter thrust chamber and a narrow cylinder 185 feet long and about 20 feet wide. The crew cabin is at the far end of the cylinder, a distance "equivalent to being 1.85 miles from a 20 kiloton burst." Between the cabin and the thrust chamber are the payload bay, a shock absorber, the large energy capsule and expellant storage area, another shock absorber and the capsule gun that fires the minibombs into the center of the chamber.

Cole suggests elsewhere an even larger version of the above spacecraft. It is an aerospacecraft almost identical in form, with the addition of large delta wings and air scoops. It could make a flight to the moon directly from the earth. It is enormous, with a thrust chamber nearly twice the size of his other spaceship's—240 feet. It draws air into the thrust chamber for reaction mass until it reaches a speed of about 15,000 fps. Water then replaces the air and the ship continues on into space as a pure rocket, carrying a payload of 23,000 tons to the moon. The gross weight of the rocket is 73,000 tons (of which over 37,000 tons is water for expellant mass). The thrust chamber weighs between 9,000 and 10,000 tons and contains the explosions of 1 kiloton capsules fired once every second. A thrust of 100,000 tons can be achieved. Cole calculates that a one-way passenger fare to the moon—allowing a 300 pound baggage allowance—might be as low as $6,000 (or only $0.025 per mile).

Cole concludes that in free space a nuclear pulse rocket will at least be able to match the performance of a conventional chemical rocket. The costs of propellant per pound of payload will be very low. The rocket itself may cost as little as $5 per pound to construct and the major operating costs will compare favorably with 1960 intercontinental air travel fares. He thinks that his nuclear pulse rockets might be feasible by 1970 or 1980. (Although it seems shocking to late twentieth-century readers, Cole insisted that launching his nuclear pulse vehicle from the earth's surface would "be no more harmful than the increase in radiation experienced by living in mile-high Denver . . . ").

The Orion Project is the invention of physicists Theodore Taylor and Freeman Dyson. Taylor believes that nuclear power is the key to space travel. He wants a ship that can be driven directly by the detonation of nuclear bombs, literally blown forward by the explosion of one bomb per second behind a shock-absorber-equipped pusher plate. With a supply of 300,000 bombs, *Orion* could accelerate to 3 percent of the speed of light, reaching Alpha Centauri in just 130 years (less time will be required if braking is accomplished by means of an interstellar ramjet).

A test vehicle is flown successfully in October. It carries five cartridges of conventional plastic high-explosive. These are ejected from the rear, through the center of the circular pusher plate, by compressed nitrogen. Each explosive charge is attached to the vehicle by a 3-foot cord and detonated by a microswitch.

The *Orion* spacecraft is designed to carry a large number of small nuclear bombs. These are to be ejected sequentially from the rear of the rocket, exploding some distance behind it. The expanding cloud of high-velocity, high-density plasma will strike the large, disk-shaped pusher plate at the rear of the spaceship (through an opening in the center of which the bombs are released). Powerful shock absorbers will cushion the instrumented or manned portion of the rocket.

Project Orion will be abandoned in 1963 when the Limited Nuclear Test Ban Treaty is signed, thus outlawing atmospheric tests of the Orion propulsion system.

A variation on the *Orion* system is studied by Pete Mohr of the Lawrence Livermore Laboratory. It has the same enormous explosion chamber, which in Mohr's version is surrounded by cylindrical habitation modules, linked end to end in a torus. *Orion* is eventually updated in a study undertaken in 1991 by Los Alamos National Laboratory. In their revised version, the pusher plate is replaced by an umbrella-like shroud that is connected to the spacecraft by a long, bungee-like tether. The explosions take place between the shroud and the spacecraft, towing it along like a sail. Since the pulsing action would somewhat resemble the motion of a jellyfish, the project is named *Medusa*.

Willy Ley designs a passenger rocket and an orbital shuttle for the Monogram plastic model kit company.

After the success of the von Braun-designed spacecraft model kits produced by the Strombecker Company [see Disney, above], several other model manufacturers are bringing out spacecraft models designed by other leading experts in astronautics. Monogram engages Willy Ley to develop four speculative spacecraft for their line. One is an unmanned television satellite, the other three are manned vehicles.

The first is a two-stage passenger rocket for transcontinental flights. Ley designs it to carry twenty-four passengers and a crew of three distances of 4,000 miles or more in one hour. The larger, manned first stage carries the smaller passenger rocket piggyback fashion. It resembles an ordinary aircraft in many ways. Although it takes off vertically, the first stage lands like a conventional plane. The second stage also lands like a conventional aircraft on reaching its destination.

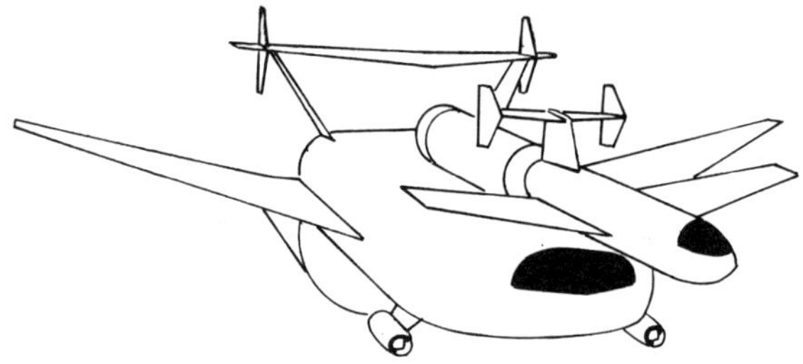

The flight scenario is like this: Ten minutes before takeoff, the passengers board the second stage. The door is at the lower end of the vertical passenger cabin and the passengers have to climb a short ladder to reach their seats. They are able to watch the countdown procedure on a screen in the cabin. At takeoff, the four large main engines fire, lifting the 1,000-ton spacecraft. Acceleration never exceeds 2.5 g's. The first stage separates about 4 minutes into the flight.

The crew of each stage consists of a pilot and co-pilot in the first stage, and a pilot, co-pilot and stewardess in the second.

At the time of separation, at about 30 miles altitude, the ship is tilted only 5 degrees from the vertical. It is 150 miles from its launch site. The first stage begins its descent. By putting the nose of the ship down and using the airbrakes, the pilot loses speed and spirals down to a landing 350 to 400 miles from the takeoff site.

After a thorough inspection, turbojet engines

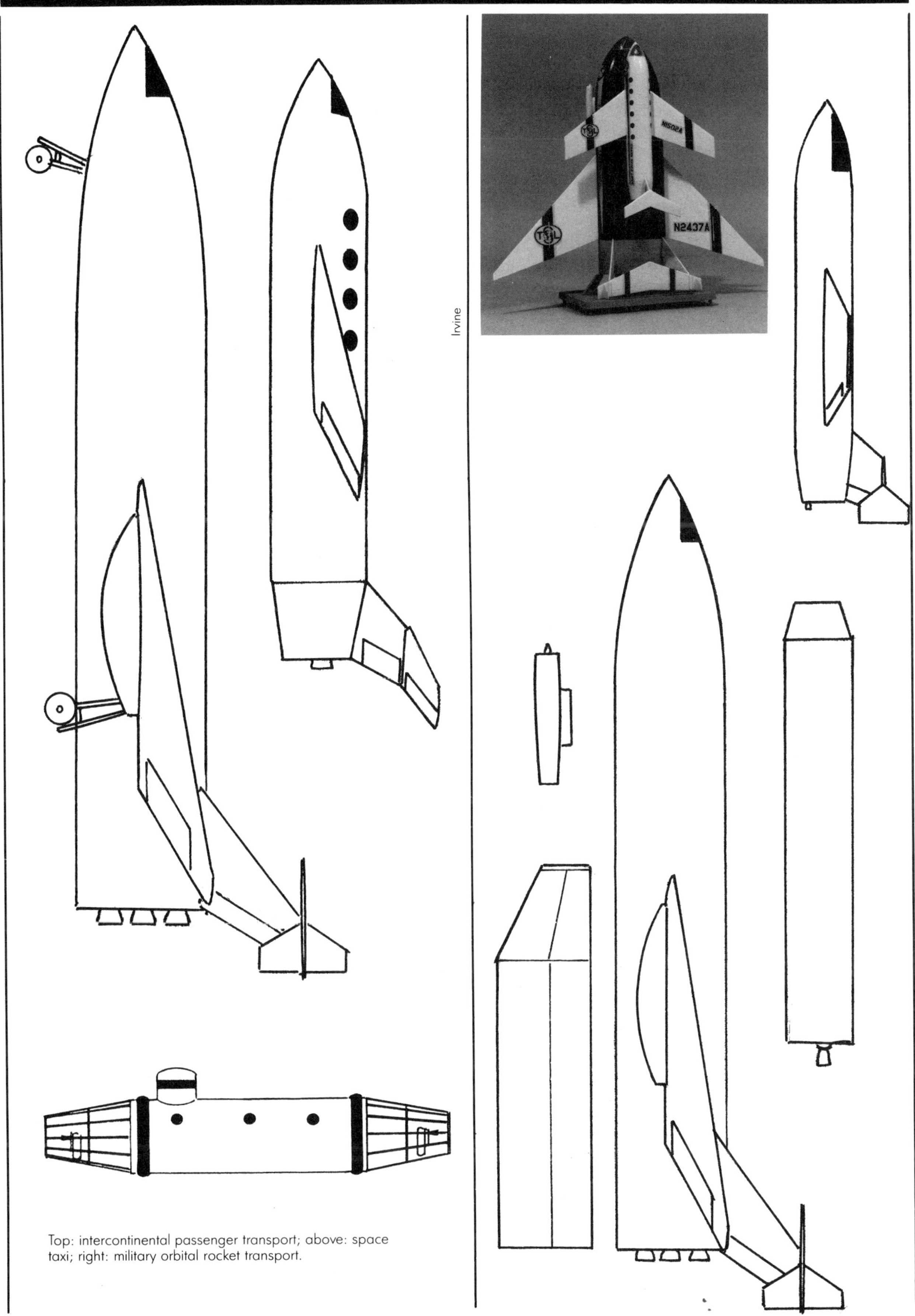

Top: intercontinental passenger transport; above: space taxi; right: military orbital rocket transport.

are attached to the fuselage and a special fuel tank in the front section of the spacecraft is filled with jet fuel. The first stage then flies back to its home base as an ordinary jet plane.

For the 4,700-mile flight from New York to Paris the pilot of the second stage cuts the engine at 3.8 miles per second, at an altitude of 70 miles. The ship continues to coast in a ballistic path, reaching the highest point, 500 miles altitude, 2,300 miles from New York. Forty minutes have passed. The passengers and crew are weightless from the time the engines cut out. If this is a problem, the rocket is equipped with a small, 100-pound thrust "comfort engine." This can burn throughout the duration of the flight, providing the passengers with a small amount of "weight."

For the next 20 minutes there is a slight sensation of negative g forces as the rocket slows down reentering the atmosphere. It lands in Paris like a conventional airliner.

The second spacecraft is almost identical to the first. The first stage is the same manned rocket used with the transatlantic passenger ship. To carry the additional load of the orbital rocket, 300 extra tons, it is fitted with a pair of expendable external fuel tanks. The piggyback rocket in this case is a two-stage affair: a small, winged manned rocket atop a tubular booster.

The flight scenario is carefully outlined: Takeoff is vertical for the first 20 seconds of the flight. The pilot of the first stage then tilts the combination rocket toward the east. After 4 minutes, the rocket is climbing at an angle of about 20 degrees. The first stage is now nearly out of fuel. It has reached a speed of 2 mps at an altitude of 40 miles (200 miles downrange). The first stage separates from the second and third stages. The first stage then returns to earth as it did in the first version. The second stage booster increases the rocket's speed to over 4 mps. When it has burned out, at about 70 miles altitude, it is jettisoned (it is not recovered). The rocket is now 400 miles downrange. The pilot lets the third stage coast to an altitude of 200 miles before igniting its engine and building up to orbital velocity. To return from orbit, the pilot of the three to five man rocket fires the engine in the direction in which the rocket is travelling. This slows the rocket down, and it reenters the atmosphere, eventually landing like an ordinary spacecraft.

Both of these designs bear a striking resemblance to actual NASA space shuttle preliminary designs of almost a decade later.

The third manned Willy Ley spacecraft is a space taxi for use between cargo rockets and a space station (it is unfortunate that this series of models did not include a Willy Ley-designed space station). The taxi is simply a 22-foot hollow cylinder with a door in one side. A pressurized conning tower allows a pilot to operate the taxi in a shirtsleeve environment (the cargo area could be pressurized, too, if necessary). Small rocket motor units at either end are sufficient to allow the taxi to maneuver. They burn liquid oxygen and acetylene and produce only 20 pounds of thrust each. Both motors can be swiveled for steering. A cage shaped like a truncated cone is attached to either end of the taxi, in which spacesuited astronauts can ride and also steer the taxi. Inflatable gaskets surround either end of the taxi's body. When it enters an airlock, these can be inflated to provide an airtight seal. The cage can then be swung aside or removed to provide access to the cargo bay.

A space station is designed by Ellwyn E. Angle for Revell as a companion to his XSL-01 spaceship [see 1958]. It is made to the same scale so that the various parts are interchangeable. The space station's cargo shuttle, for example, can replace the XSL-01's moon lander.

Helmut Hoeppner and B. Spencer Isbell propose an interstellar rocket for their Project Star, a manned flight to Alpha Centauri. The *Astra—001* is to be a photon-propelled ship. It will take off from earth orbit, where it has been assembled by lifting its component parts from earth by conventional chemical rockets. The photon propulsion unit is attached which will accelerate the ship at a constant 2 g's for half a year. This eventu-

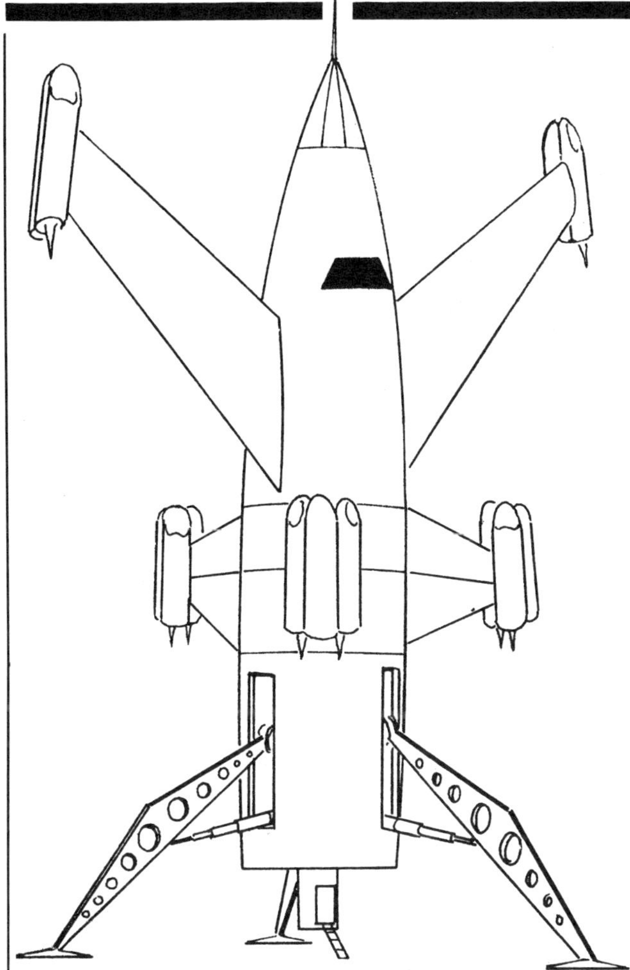

ally raises its velocity to near that of the speed of light. It then coasts at this speed for 4 years and 5 weeks. Deceleration then begins, using the photon unit as a retro-rocket. This phase lasts for another 6-month period. The photon propulsion unit will be left in orbit around any planet that is found while the main ship descends to the surface. The landing and takeoff vehicle is a winged, streamlined ship, equipped with combination turbo-ram engines in addition to conventional rockets. The jet engines are mounted on the wingtips, where access tunnels allow them to be serviced from within the ship. The rockets are be used to initiate the descent, and the jets take over once the atmosphere becomes dense enough. After a long spiral descent, the rocket lands vertically, settling onto three extendable legs. These latter are attached to a launching platform that is left behind when the ship departs. An elevator shaft extends through the long axis of the rocket, its cage descending from the bottom of the ship. A small booster section lifts the rocket back into orbit when exploration is complete. There the photon drive is reattached and the journey home is made.

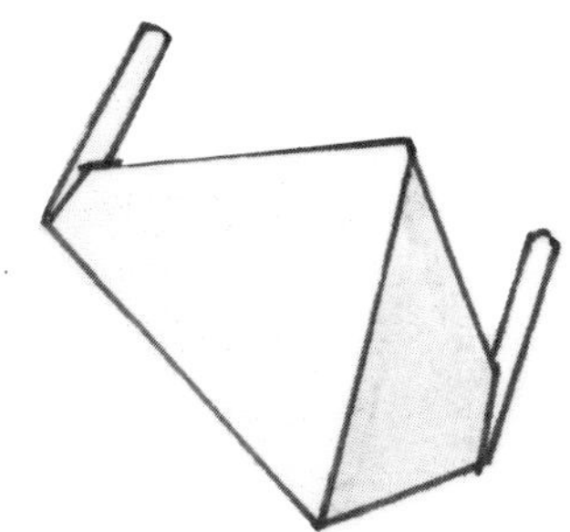

Armstrong-Whitworth in Britain designs a "pyramid" reentry vehicle. It is more or less tetrahedral in shape, though the sides are not equilateral triangles. At two corners are a pair of large stabilizer fins. Within the vehicle is a cylindrical crew compartment.

Kraft, Inc. sponsors a contest in which the first prize is a "Life-Size Aerojet Training Space Ship!" The giveaway is inspired by a new marshmallow-making process called "jet-puffing." A number of aerospace firms are solicited to participate; some decline, while others compete for the chance to share in the publicity. Kraft settles upon Aerojet-General, then manufacturing the propulsion systems for the Titan, Polaris, Bomarc, Aerobee, Vanguard and other missiles and rockets.

The mockup "trainer" Aerojet develops and builds for Kraft is 29 feet long and weighs 4,000 lbs. The interior compartment is 12 feet long and 7 feet in diameter (and is guaranteed suitable for children as young as 6 years of age). There is room inside for four crewmembers, each with a seat and a set of controls, and with a specific task to perform during the mission. The cabin is insulated and air-conditioned for both comfort and safety. There are also four space suits provided with helmets. A television set and a rear-projection motion picture system simulate a realistic spaceflight. Most of the controls and instruments have real-life counterparts in the Mercury capsules.

The finished "trainer" is mounted on a trailer equipped with hydraulics that provide

Cutaway of Kraft/Aerojet Training Spaceship

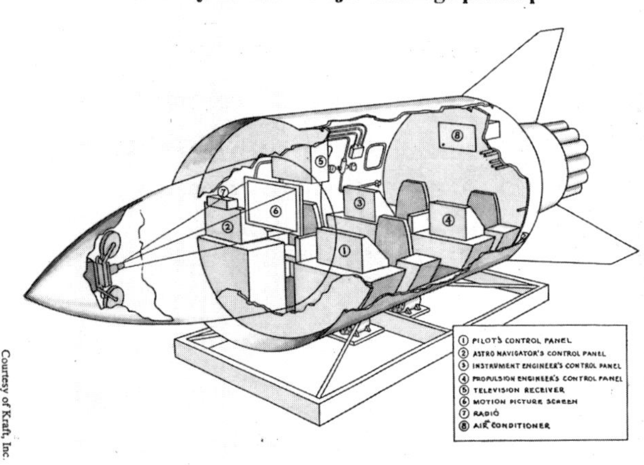

Courtesy of Kraft, Inc.

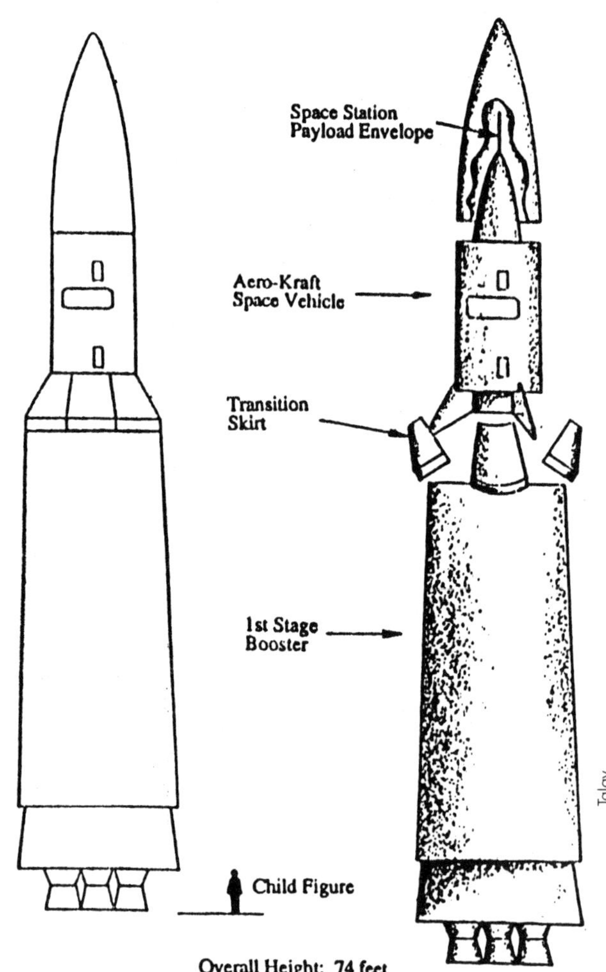

appropriate motions for the spaceship.

The mission that the lucky children who win the trainer are to perform has been realistically planned by Aerojet.

The trainer is supposedly a mockup of the second stage of a hypothetical 74 foot two-stage rocket. Aerojet bases the characteristics of the booster on the Titan missile. This first stage drops off at 300 miles, the second stage continuing on to a space station orbiting at 4,000 miles above the earth. After refueling at the station, the spaceship is free to continue on to the moon or any other destination. For an aerodynamic reentry back into the earth's atmosphere, a winged heat shield is placed over the nose of the spacecraft.

The spaceship is won by a young girl in St. Louis, Missouri, who donates it to her local elementary school. The spaceship is removed from its trailer and mounted on a concrete pad. No one seems to know its fate. [Also see 1950–1956.]

In *The Next Ten Years in Space, 1959–1969*, a publication issued by the House Select Committee on Astronautics and Space Exploration, Wernher von Braun considered that "manned flight around the Moon [would be] possible within the next 8 to 10 years, and a 2-way flight to the Moon, including landing, a few years thereafter. "Von Braun also concluded that it was "unlikely that either Soviet or American technology will be far enough advanced in the next 10 years to permit man's reaching the planets, although instrumented probes to the nearer planets (Mars or Venus) are a certainty." [add to Collier's Mars section]

Von Braun pessimistically concluded that a flight to Mars might not take place for "a century or more."

Man will land on the moon in 2 or 3 years, according to Soviet Prof. G. A. Tikhov (the so-called "Grand Old Man" of Russian astronomy).

Kurt Stehling, head of the Propulsion Group for Project Vanguard, proposes a scenario for the first manned flights to the moon and Mars, which he expected to take place around 1970.

Stehling's 600-ton, 200-foot Mars rocket would launch its 3-man crew from a site on

1. Attitude jets, 2. Solid retrorockets, 3. Nuclear jets, 4. Landing struts, 5. H_2O_2 jet. Not shown: Ion ports.

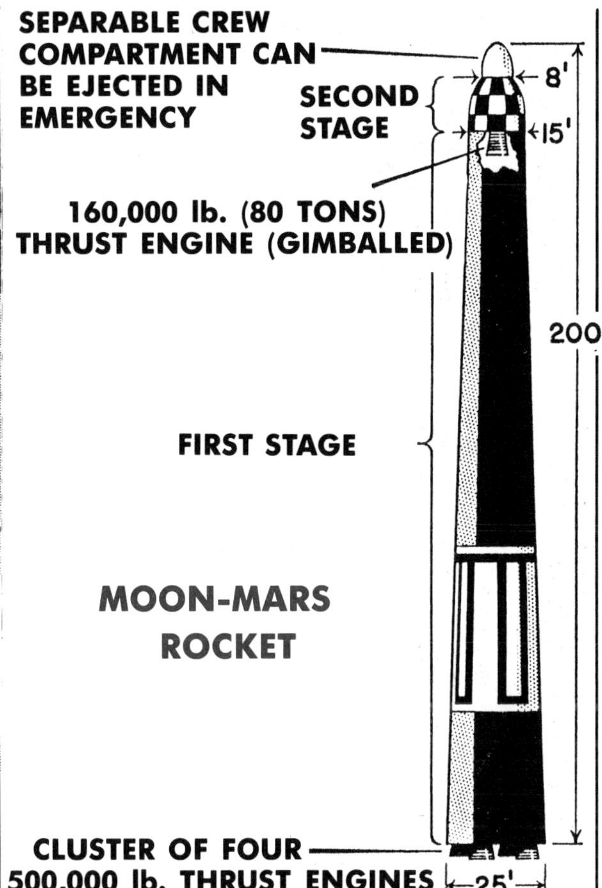

the Pacific Missile Range. The rocket's first stage would be a recoverable booster whose four engines produced a thrust of about 1,000 tons. The second stage, which would carry the crew to the moon and beyond, would have chemical (a single 250-ton-thrust engine), nuclear-thermal and nuclear-electric propulsion. The nuclear engine uses hydrogen as its working fluid for the translunar leg of the journey. Forty hours after takeoff the crew lands at the manned lunar base (established in 1966). The 23-foot second stage has an 8-foot separable crew compartment.

The 80-ton rocket lands with the aid of four 75,000-pound-thrust solid fuel rocket motors. The lunar base refuels the spacecraft with lunar-generated hydrogen.

The refueled rocket takes off for its 250-day flight to Mars using its nuclear engine in combination with solid fuel boosters. Once its has entered a trajectory for Mars, the spacecraft's cesium-"fueled" nuclear-electric propulsion system is activated.

Entry into the Martian atmosphere is facilitated by air brakes and metallic mesh parachutes.

Constantin Paul van Lent publishes a proposal for a manned satellite vehicle. Launched by a three- to five-stage booster it can carry a single astronaut in a small, cylindrical capsule equipped with its own propulsion system.

Another manned spacecraft design describes launching a 10 to 12-foot egg-shaped capsule atop a three-stage booster. The capsule, which has a diameter of 4 feet and weighs 4,000 pounds, contains a "space suit" that completely encloses the astronaut in a kind of peanut-shaped shell. A parachute attached to the "stomach" of the suit allows the astronaut to abandon the capsule after reentry.

Lent also proposes a three-stage lunar lander carrying two astronauts. The third stage is surrounded by solid fuel strap-on rockets for the lunar landing and takeoff. The lander itself sets down on four elongated fins reminiscent of Oberth's earlier designs [see 1928]. The lander is also equipped with a single

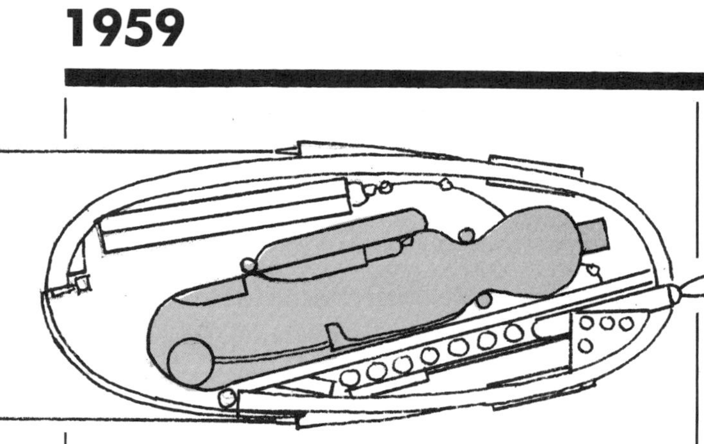

"mobile space suit"—an upright cylindrical cabin, just large enough for a single operator, with caterpillar treads. The manned capsule eventually returns to earth via parachute.

The Soviet film *Niebo Zowiet* (also known as *The Sky Calls*) is released. (It is eventually reedited and released in the United States as *Battle Beyond the Sun* [1963].) The first major Russian space movie since 1935's *Kosmitchesky Reis*, it celebrates the recent Sputnik triumphs while propagandizing the Soviet space program. A Mars mission is abandoned by the Russians in order to rescue a misfired American attempt to beat the Russians to the red planet. The ideological propaganda is eliminated in the Americanized version, which replaces Russia and the United States with the more innocuous "North Hemis" and "South Hemis."

The spacecraft are fairly typical for the period (not unlike those in *Project Moonbase* [1953], for example). The special effects range from excellent to adequate, but the film also demonstrates that Soviet science fiction films are just as susceptible to bad science as American films.

M. W. Rosen and F. C. Schwenk, at the tenth Congress of the International Astronautical Federation, argue for a direct "bold" flight to the moon using the giant Nova booster. Although the orbital rendezvous method uses smaller spacecraft, it also requires the perfection of rendezvous techniques.

The Nova mission is a 2 1/2-day flight to the moon. The first three stages of the booster are required to launch the spacecraft toward the moon, at 36,000 fps. The fourth stage lowers the spaceship onto the lunar surface, the fifth stage propels the vehicle back toward the earth, and the sixth and final stage slows the ship into an orbit around the earth.

SPACE SHIP

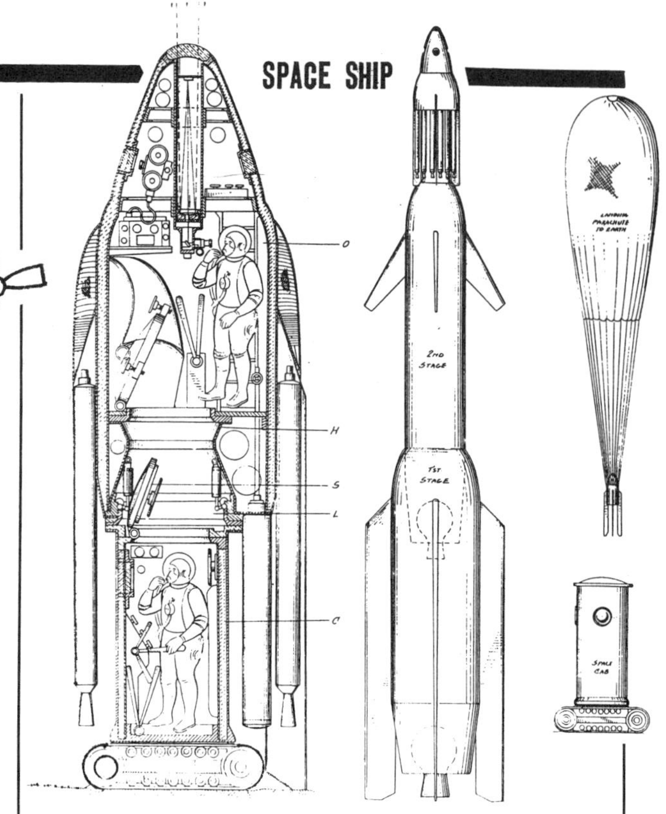

Constantin van Lent design.

The complete rocket stands 230 feet tall, with a first stage diameter of 48 feet. The 8,000 pound payload consists of equipment, earth return capsule, parachutes, and a crew of two or three astronauts. Six 1.5 million-pound-thrust engines power the first stage, a cluster of seven 16-foot-diameter tanks. The second stage is a cluster of four 16-foot-diameter tanks, as is the third stage. The latter's four engines produce a thrust of 600,000 pounds.

The fourth stage, the lander, has about one minute of hovering time available. Its four landing legs span about 40 feet for stability. The fifth stage sits atop a cylinder that pierces the tankage of the landing stage.

The manned capsule is an enlarged version of the Mercury spacecraft. It is a truncated cone 12 feet in diameter at the base and 14 feet tall. Inside are two levels, the lower with contoured couches for the crew, controls, communications and a folding airlock. The upper level contains food, the power supply, exploration gear and so forth.

Nova booster and, below right, Nova ca. 1961.

Dyna-Soar separating from its booster.

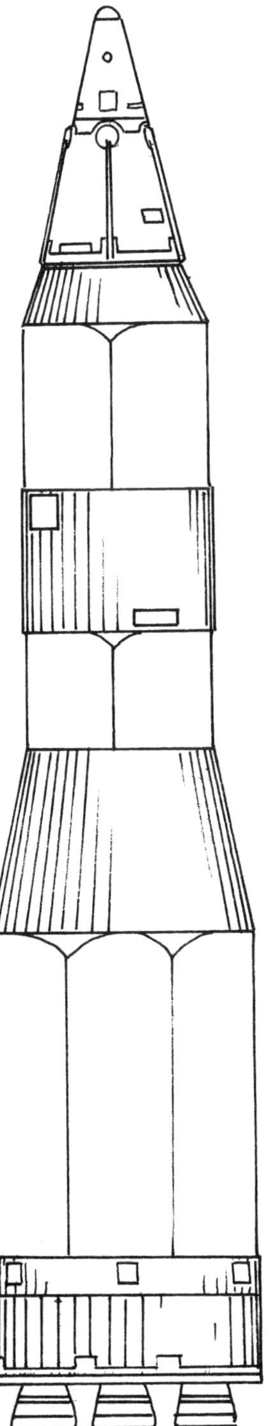

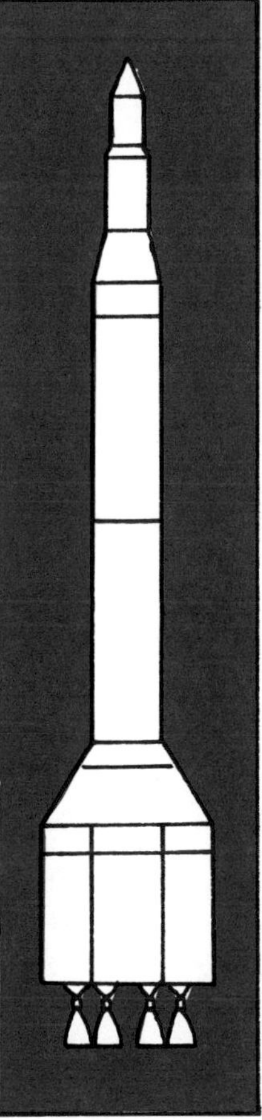

The U.S. Air Force decides to go ahead with its Dyna-Soar "space bomber" program. Boeing will develop the boost-glide vehicle and be responsible along with Wright Air Development Division for the integration of vehicle subsystems, vehicle and booster, and assembly and testing. Martin will develop the boosters, expected to be slightly modified Titans in the early part of the program. Later, the boosters may be clusters of four Titans generating 1.2 million pounds of thrust. An alternative will be expanding the Titan first stage from its 10 foot diameter to 12 feet.

The program has a $53 million budget through FY 1960. However, the initial development is expected to cost a total of at least $1 billion.

Dyna-Soar I will be a boost-glide vehicle with speeds of up to 14,000 mph and the capability of conducting reconaissance and bombing missions anywhere on the earth, as well as enemy satellite and spacecraft interception. Dyna-Soar II will be capable of deep space missions with the power to jump from orbit to orbit.

1959

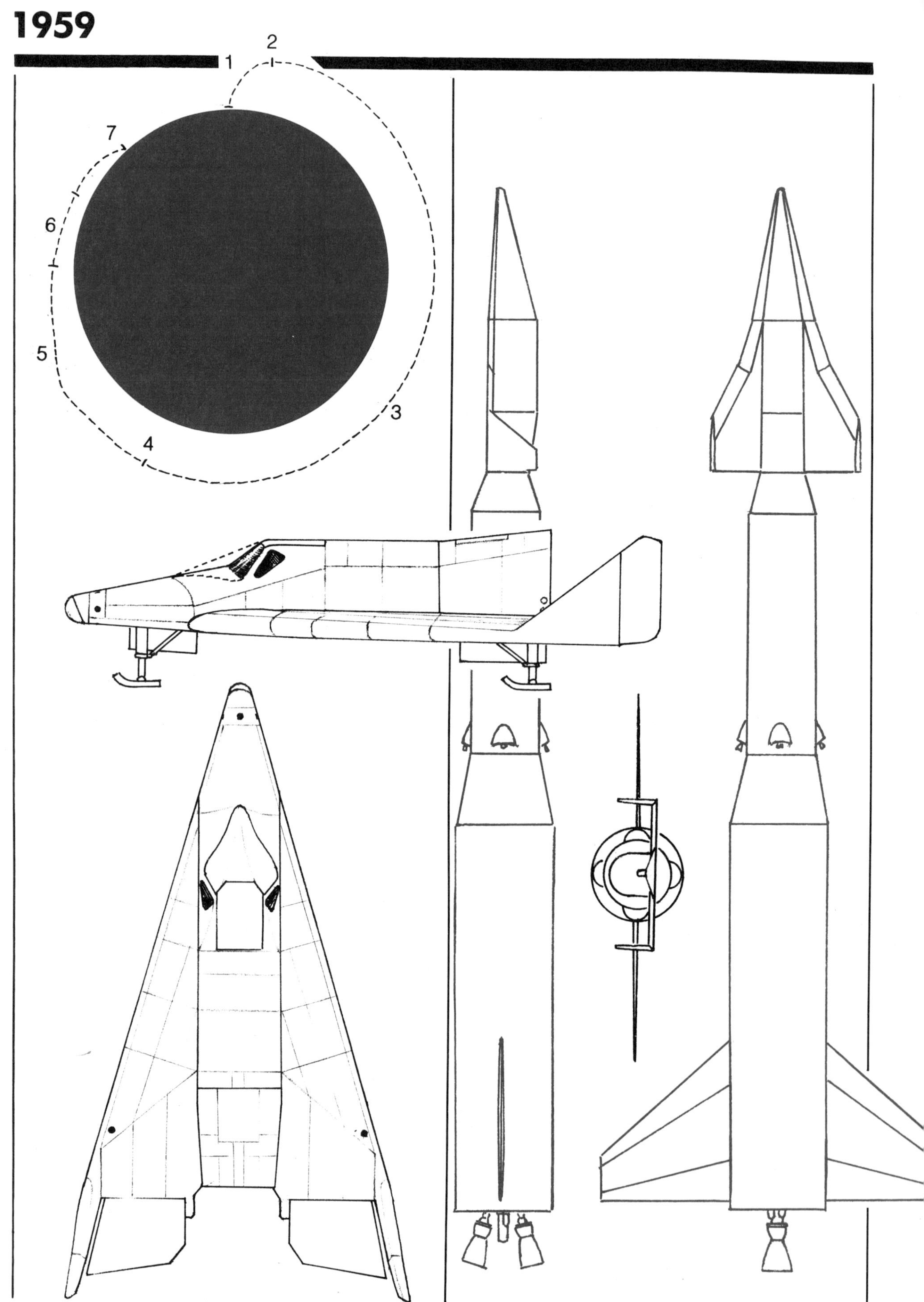

NASM

NASM

Top: Illustration from *The Chicago Ledger* for April 30, 1910, showing the spaceship *Vesta* above the earth.
Right: Two illustrations by O. Stáfl for the Czechoslovakian children's book *Na Měsíc* (1931).
Above: The cover of the sheet music for *A Trip to the Moon*, by Clifford V. Baker (ca. 1904).

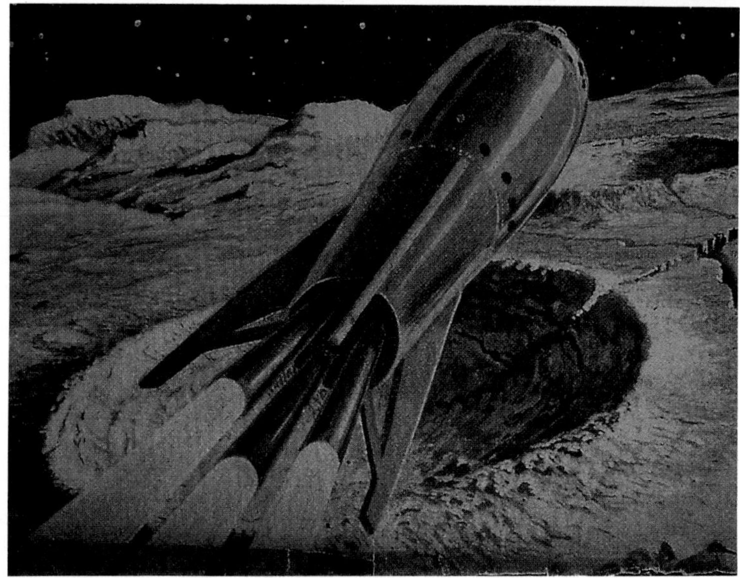

Upper left: A cover of *Science and Invention* illustrating the spaceship that appeared in the motion picture *The Stellar Express*.

Upper right: A lunar-bound spaceship created by the pioneer French space artist Lucian Rudaux, illustrating an article that appeared in *American Weekly*, a Sunday newspaper supplement.

Left: *Popular Mechanics* celebrates the premier of the seminal German space film by featuring the spaceship *Friede* on its March 1930 cover.

Above: This Frank R. Paul cover for *Thrilling Wonder Stories* illustrates the airplane-cum-spaceship from *The Flight of the Mercury*.

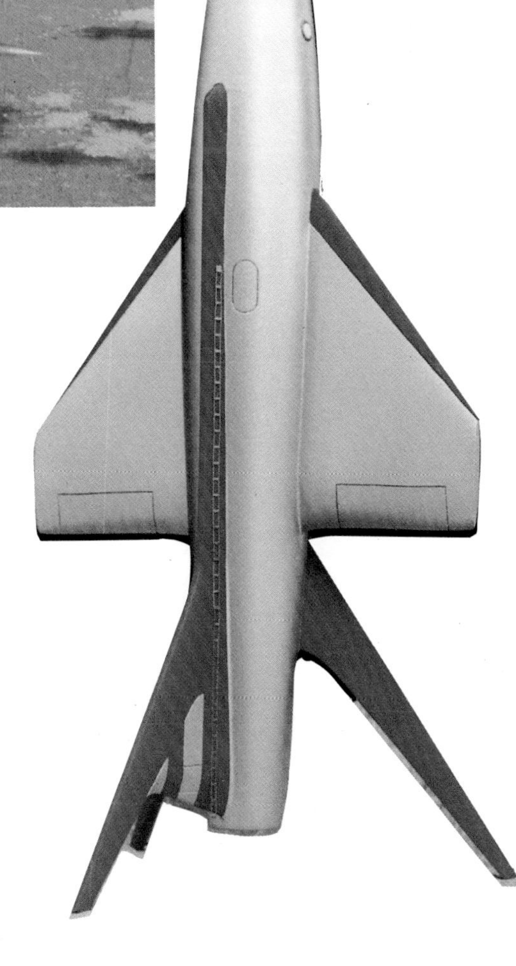

Staton

Upper left: Chesley Bonestell illustrates the transcontinental rocket transport proposed by Tsien Hsueh-Sen. Here the spaceplane is shown 500 miles above central Iowa.
Left: An archetypal *art moderne* spaceship is shown on this August 1940 *Astounding* cover by Howard Browne.

Above: Chesley Bonestell designed this beautiful rocket early in the production of George Pal's film *Destination Moon* (1950). Though it was not used in this movie, it did appear in many others, such as *Flight to Mars* and *It—The Terror From Beyond Space*.

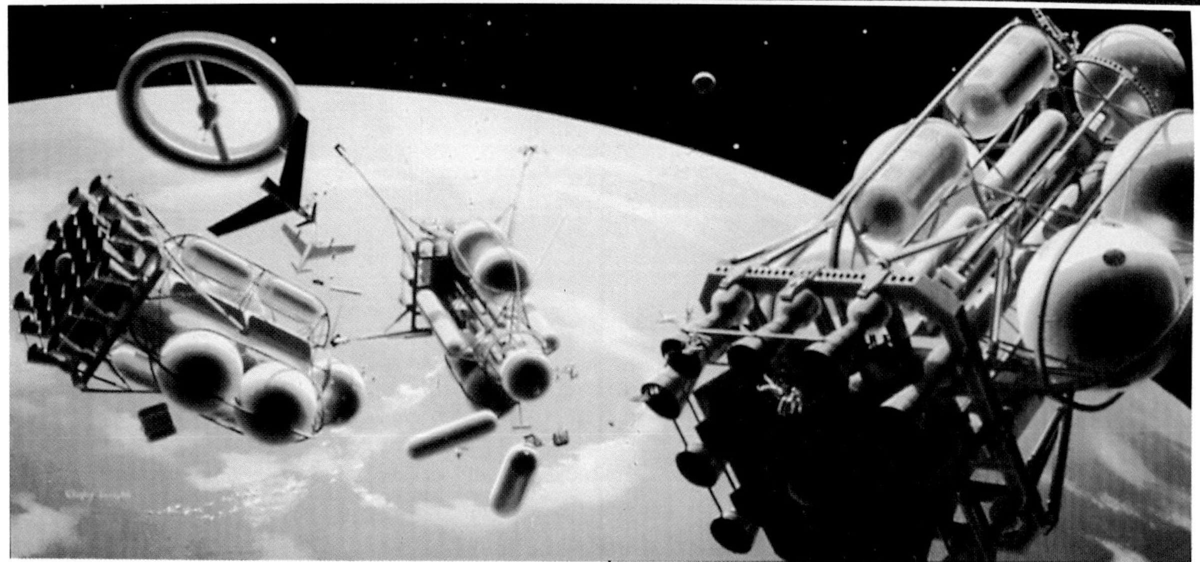

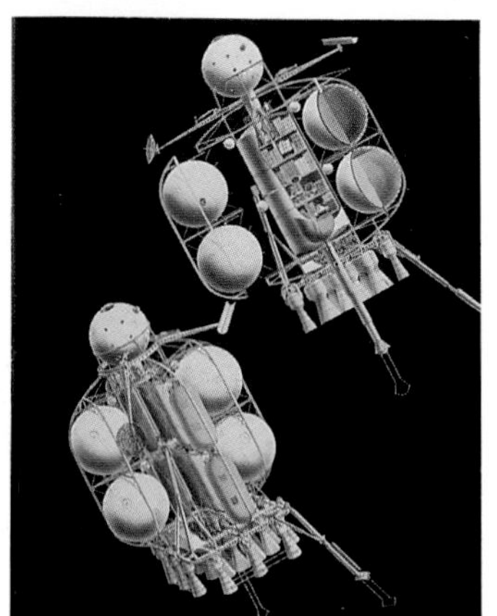

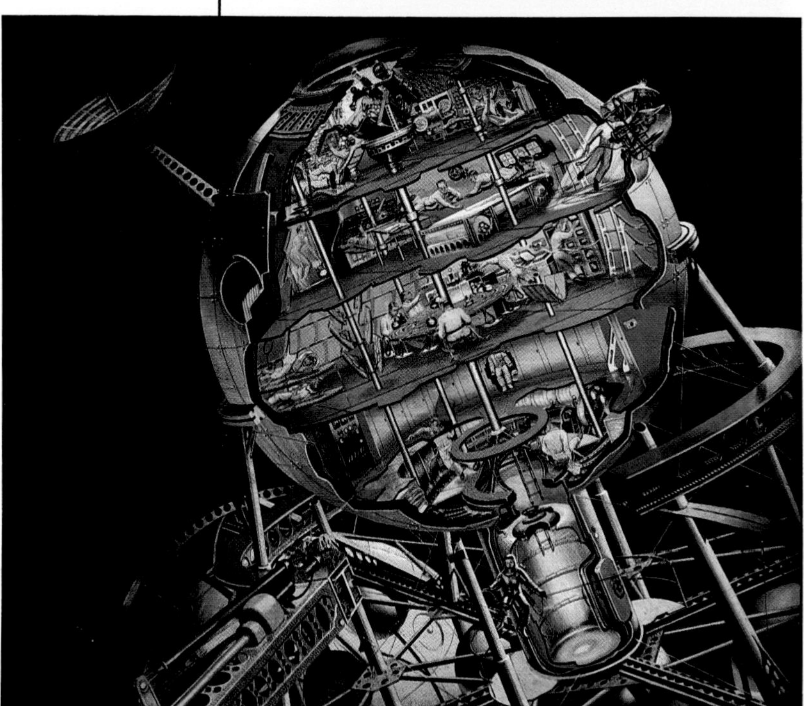

Freeman/Ordway

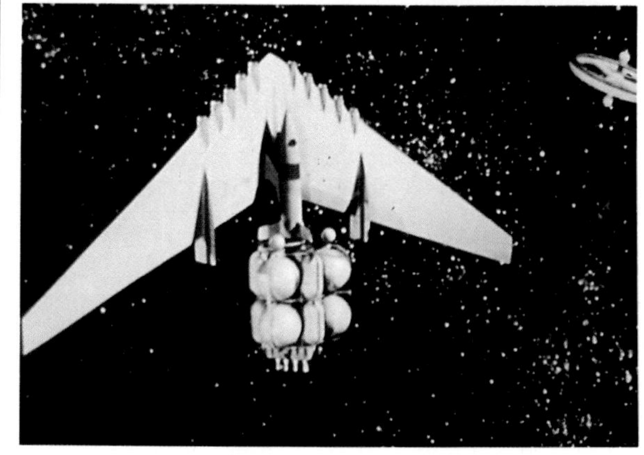

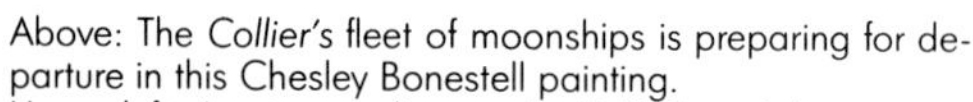

Above: The *Collier's* fleet of moonships is preparing for departure in this Chesley Bonestell painting.
Upper left: A cutaway diagram by Rolf Klep of the cargo and passenger moonships.
Lower left: The upper stage of the manned ferry rocket is separating from its second stage booster in this painting by Chesley Bonestell.
Above right: A cutaway by Fred Freeman of a moonship's personnel sphere, showing its interior details.
Right: The Mars ship and space station as they were depicted in the George Pal film *The Conquest of Space*.

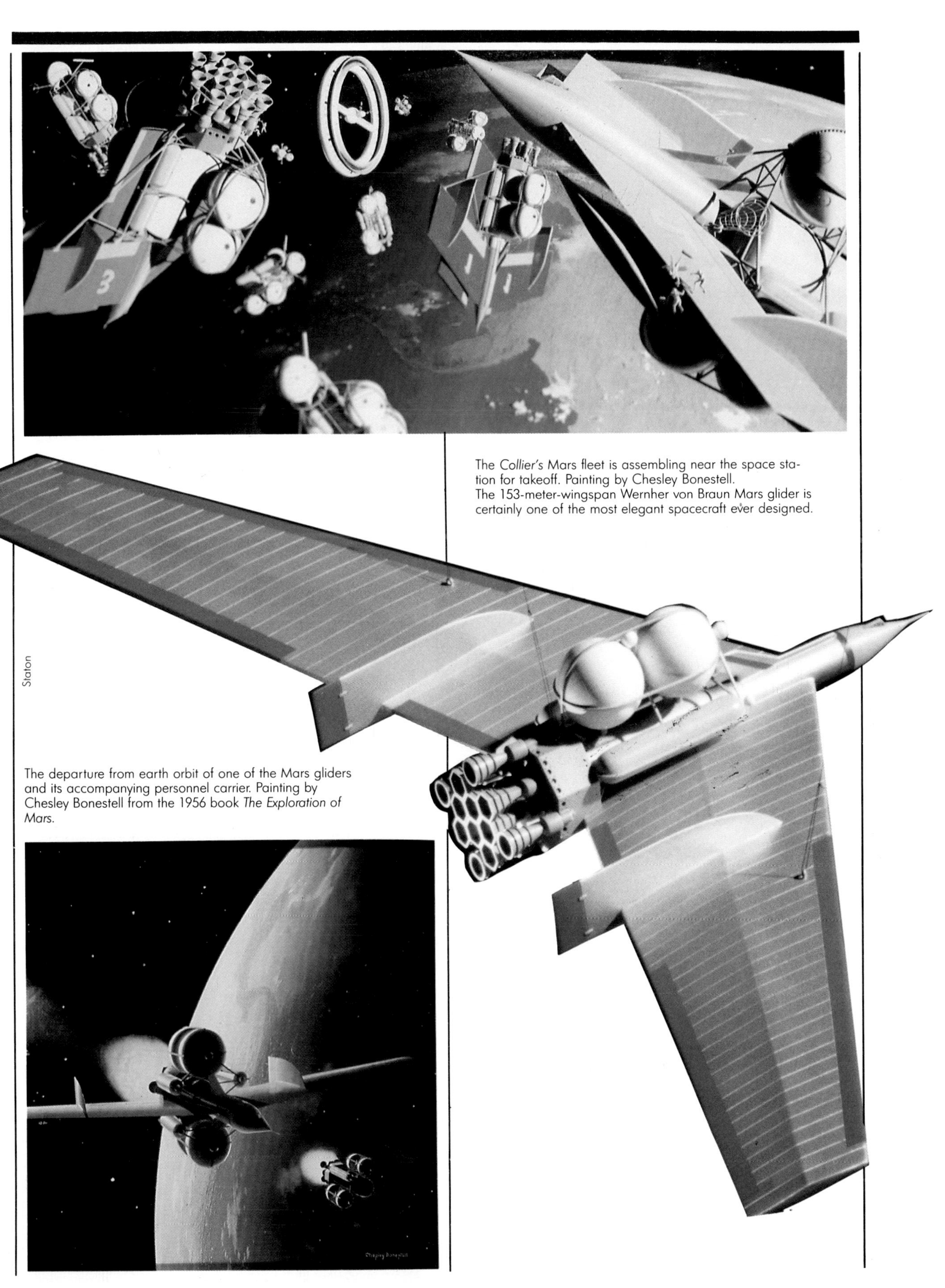

The *Collier's* Mars fleet is assembling near the space station for takeoff. Painting by Chesley Bonestell.
The 153-meter-wingspan Wernher von Braun Mars glider is certainly one of the most elegant spacecraft ever designed.

The departure from earth orbit of one of the Mars gliders and its accompanying personnel carrier. Painting by Chesley Bonestell from the 1956 book *The Exploration of Mars*.

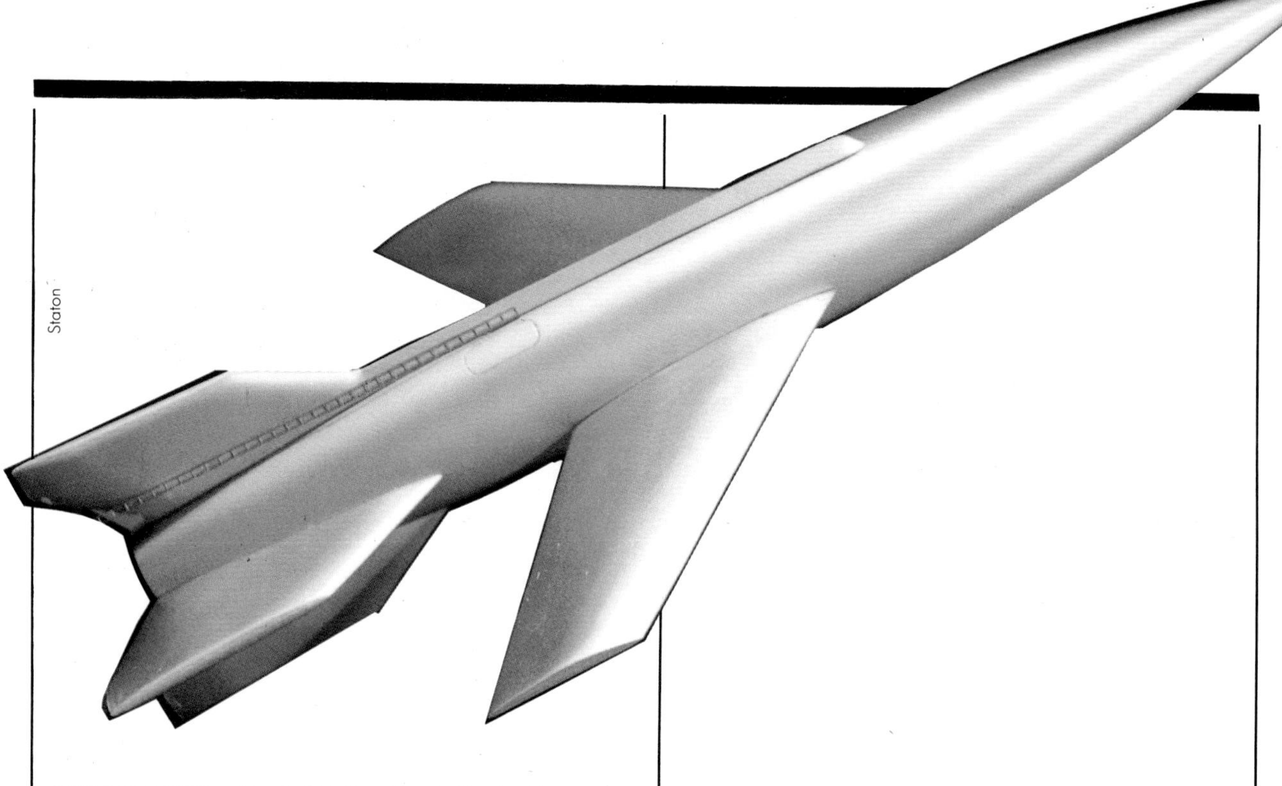

Staton

Staton

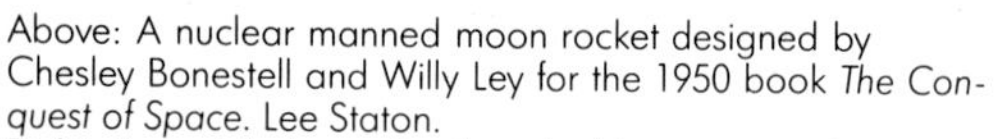

Above: A nuclear manned moon rocket designed by Chesley Bonestell and Willy Ley for the 1950 book *The Conquest of Space*. Lee Staton.
Right: A model of a rocket launched by a manned winged booster was created by Morris Scott Dollens for his unfinished space film *Dream of the Stars*.
Above: The three-stage manned ferry rocket designed in the early 1950s by R. A. Smith of the British Interplanetary Society. Above right: The rollout of the giant manned spaceplane by artist Jack Coggins, ca. 1954.

Coggins

Dollens

H. Sänger

Coggins

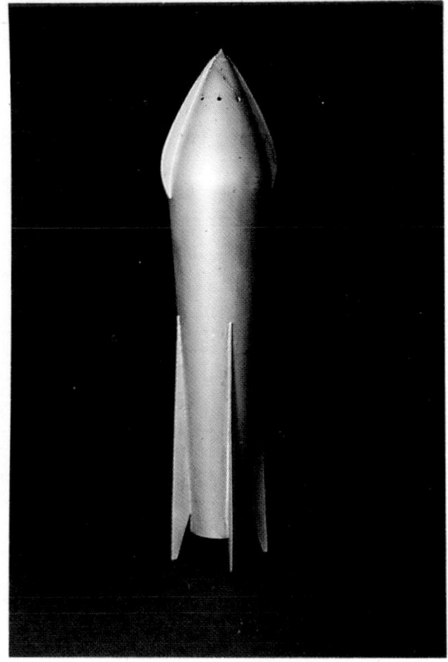

N. Brosterman

Top: Eugen Sänger created this stainless steel model of his *Silver Bird* spaceplane.
Above left: A 1935 *Amazing Stories* cover by Leo Morey.
Above right: The takeoff of a giant three-stage manned orbital rocket. Painting by Jack Coggins for the book *By Spaceship to the Moon.*
Far right: The original filming model of *Rocketship XM* (1950).

NASA

Irvine

Above left: The M2 lifting body test vehicle. NASA.
Center left: The 1967 Sänger 1 space transporter proposal.
Bottom left: The X24B lifting body test vehicle. NASA.
Top: The spaceship *Anastasia* from the long-running British science fiction comic strip, *Dan Dare*.

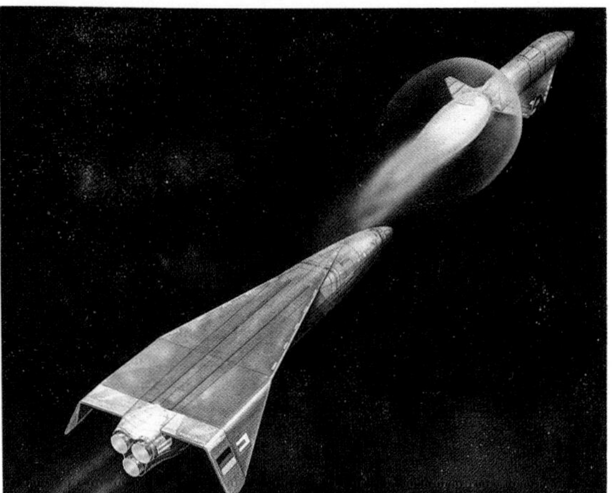

Facing page: The city-sized space station proposed by Darrell Romick as part of his METEOR project.

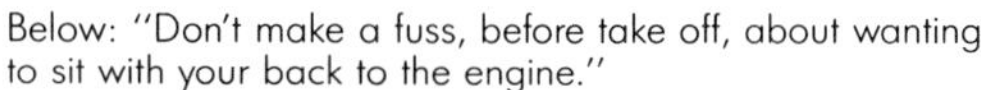

Below: "Don't make a fuss, before take off, about wanting to sit with your back to the engine."

NASA

Punch April 11, 1956

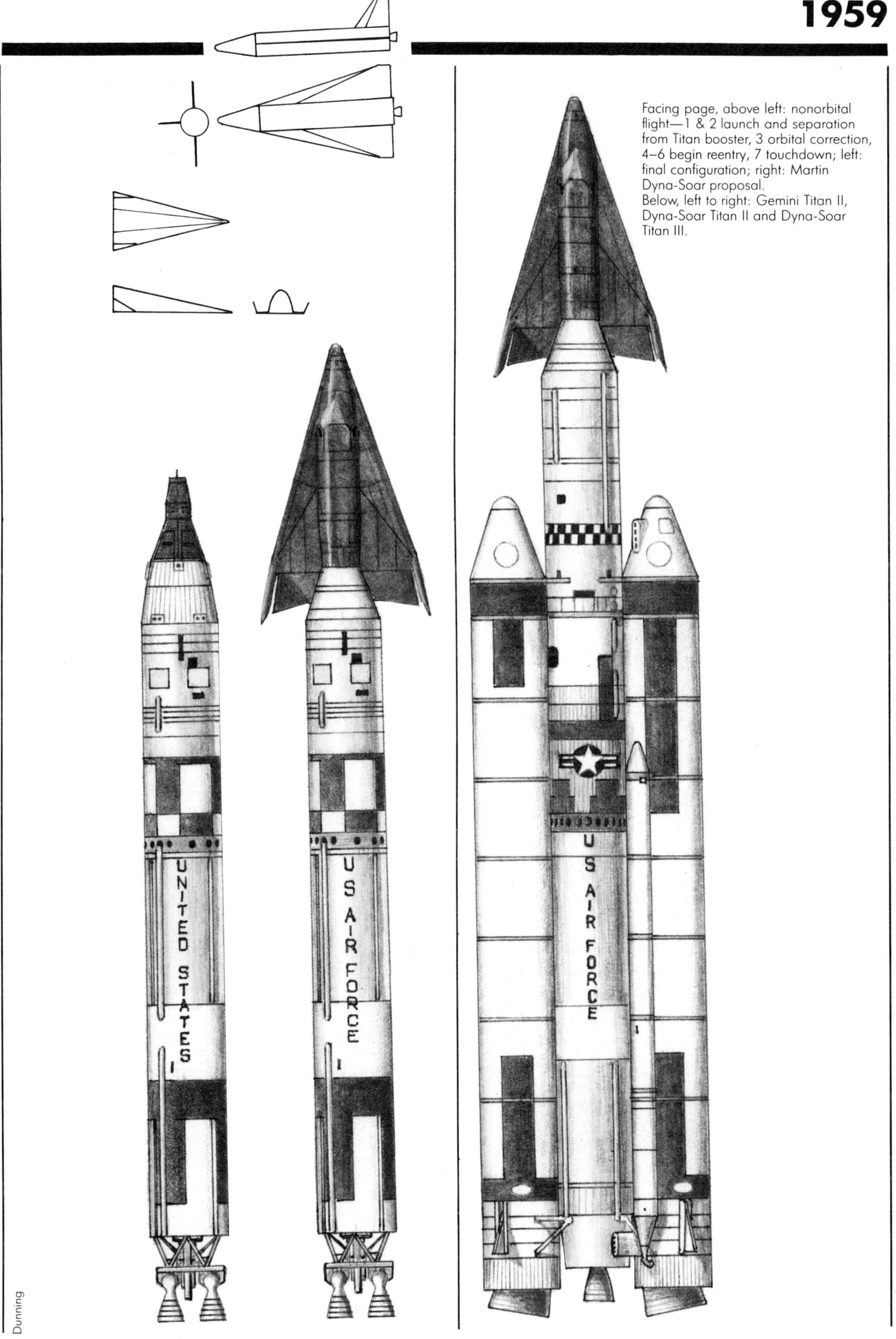

Facing page, above left: nonorbital flight—1 & 2 launch and separation from Titan booster, 3 orbital correction, 4–6 begin reentry, 7 touchdown; left: final configuration; right: Martin Dyna-Soar proposal.
Below, left to right: Gemini Titan II, Dyna-Soar Titan II and Dyna-Soar Titan III.

Dunning

1959

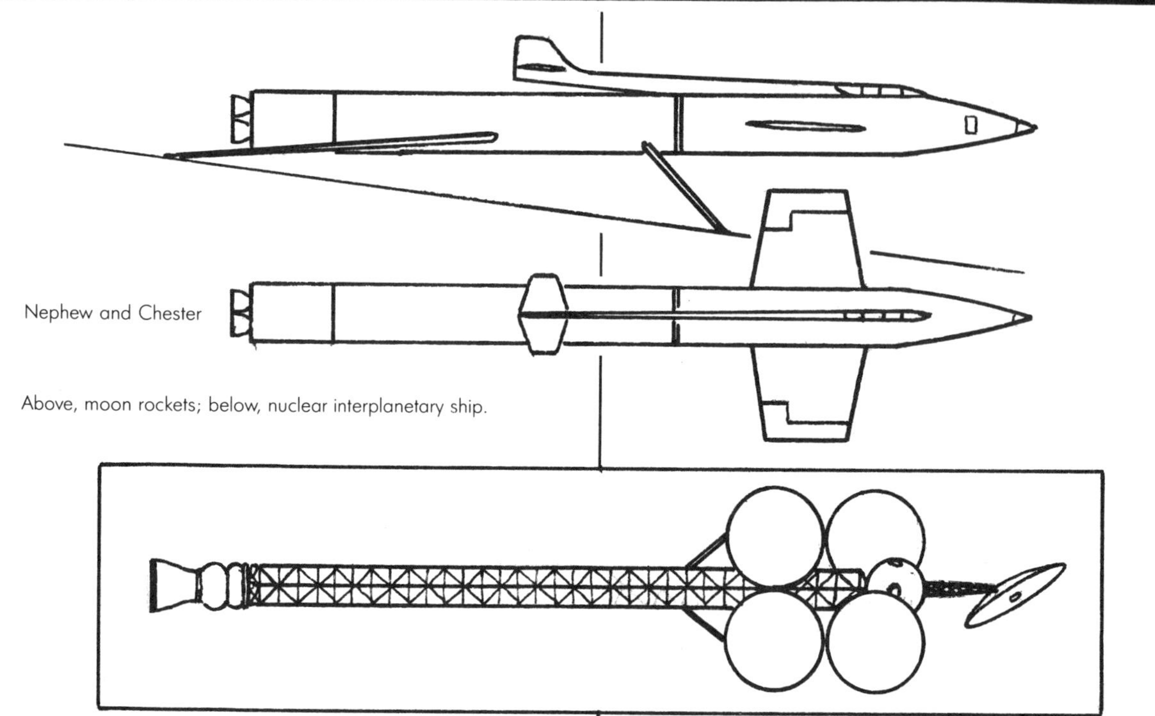
Nephew and Chester

Above, moon rockets; below, nuclear interplanetary ship.

In addition to the title subject of their superb children's nonfiction book *Moon Base*, authors William Nephew and Michael Chester describe some original spacecraft.

Second generation manned lunar landers take off from earth using ramjet boosters. Once above the earth's atmosphere a nuclear stage propels the rocket to a landing on the moon. To land, the spaceships settle onto the surface horizontally, supported by four long landing legs, making them look a great deal like waterstriders. For takeoff, the hydraulically operated legs raise the ships to an upright position. Since the nuclear booster is jettisoned before reentry into the earth's atmosphere, its no longer needed aerodynamic covering is removed for use in constructing the lunar base. Reentry is made in a winged glider.

Once the base is established unmanned nuclear-powered planetary probes can be launched from the surface of the moon, followed by manned interplanetary spacecraft. These latter consist of a personnel sphere surrounded by fuel tanks at one end of a long boom. At the opposite end is the nuclear rocket.

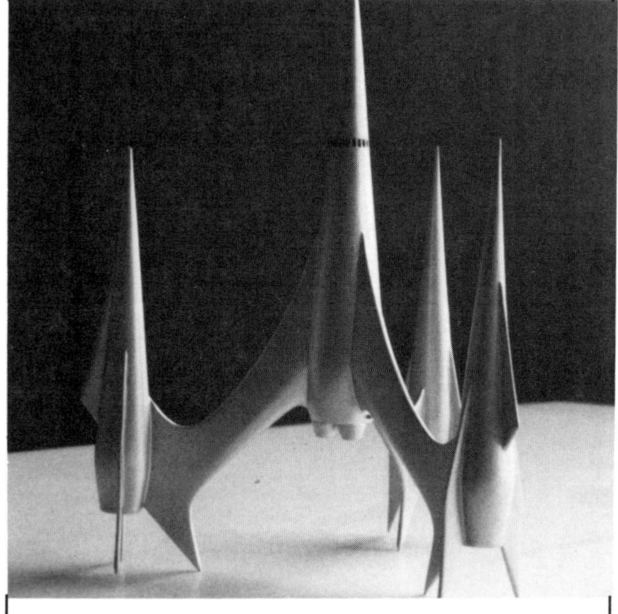

The motion picture *The Silent Star* (*Der Schweigende Stern*—also released in English as *The First Spaceship on Venus* in 1963) features what may be one of the most strikingly designed spaceships ever filmed, the *Cosmostrator* (*Kosmokrator* in the original), created by designers Anatol Rabzinowicz and Alfred Hirschmeier. This otherwise undistinguished film about a trip to Venus (based on a Stanislaw Lem novel) is an East German-Polish coproduction.

JANUARY 2

Wernher von Braun predicts that a manned circumlunar flight will take place within 8 to 10 years with a manned landing following a few years later. He makes these remarks in the staff report of the House Select Committee on Astronautics and Space Exploration. NASA administrator T. Keith Glennan, his deputy Hugh L. Dryden, Abe Silverstein, John P. Hagen and Homer Newell also foresee manned circumlunar flight within the decade, while Roy K. Knutson, chairman of the NAA Corporate Space Committee predicts a manned lunar landing by the early 1970s.

JANUARY 12

The Jet Propulsion Laboratory (JPL) proposes a schedule for future lunar and planetary exploration that includes a manned circumlunar flight and return in August 1964 and a manned circum-Mars flight and return on January 20, 1965. The former would require a gross payload of 2,300 kg and the latter a gross payload of 4,500 kg.

Above left: The USAF's Project Manhigh's capsule carries David G. Simons to a record 102,000 feet in 1957 and, above, the U.S. Navy's Stratolab carries a telescope to more than 80,000 feet.

1959

FEBRUARY 20

NASA Deputy Administrator Hugh L. Dryden and DeMarquis D. Wyatt testify before the Senate Committee on Aeronautical and Space Sciences that the long-range goals of NASA include a temporary manned space station, permanent manned orbiting laboratories, manned circumlunar flights, manned lunar landings and interplanetary flight.

APRIL 9

The astronauts for Project Mercury are selected. They are Malcolm Scott Carpenter, Leroy Gordon Cooper, John Herschel Glenn, Virgil Ivan Grissom, Walter Marty Schirra, Jr., Alan Bartlett Shepard, Jr., and Donald Kent Slayton.

JUNE 25–26

A meeting of the Research Steering Committee on Manned Space Flight considers intermediate steps toward a manned lunar landing.

Lewis Research Center says that for a direct flight to the moon and back a 10,000-pound spacecraft will require a 10 to 11 million pound booster (including the weight of the lander). The lander will descend onto the moon from a 500,000-foot altitude, dropping in free fall until applying retrorockets at the last moment.

An earth orbit rendezvous mission will require nine Saturn launches boosting nine Centaur stages into earth orbit. These will be assembled into an earth-escape booster. Three more Centaurs will be added for the lunar orbit and landing and two more for the return trip and payload. The entire operation will require 2 to 3 weeks.

The Army Ballistic Missile Agency reports on its lunar mission studies, which include refueling in earth orbit or assembling separately landed modules on the moon for the return flight (lunar surface rendezvous). ABMA proposes using either a Saturn C-2 with a 2 million-pound-thrust first stage, a 1 million-pound-thrust second stage and a 200,000-pound-thrust third stage or a new vehicle six times larger than the C-2.

Ames Research Center discusses the potential for manned flights beyond the Mercury series, including manned satellites, circumlunar flights (or a manned satellite in a highly elliptical orbit whose 250,000-mile apogee will take it close to the moon), and the use of the Nova booster for a manned lunar landing. This latter would allow two men to carry out a 1 week to 1 month expedition.

MAY 28

Able and Baker, two rhesus monkeys, are flown to an altitude of 300 miles and 1,700 miles downrange from Cape Canaveral.

AUGUST

A team of eight scientists, headed by Armstrong Whitworth's H. R. Watson and Hawker Siddeley Group's Dr. W. F. Hilton, has produced a "pyramidal" spacecraft design, based on a configuration proposed by consultant R. F. Nonwieler.

Two of the wedge-shaped spacecraft would be launched back-to-back atop a 100-foot three-stage rocket. One would contain a pressurized two-man cylinder while the other would carry additional fuel and act as an "image faring" for a stable takeoff. It would be jettisoned with the first stage.

The orbiter would be 25 feet long, 18 feet wide, and weigh about 2 tons. The 41-foot first stage of the booster would weigh 133.5 tons. Launched at a 40-degree angle it would carry the spacecraft to an altitude of 40 miles, where the 22-ton second stage would fire, taking the orbiter to 75 miles. At 80 miles the third stage would fire, carrying the spacecraft to its final 700 × 80-mile orbit.

After reentry, the vehicle would make an aircraft-type landing at only 80 mph. The entire flight, including four orbits, would take only 6.75 hours. An alternative would be to jettison the crew cylinder, returning it to earth via parachute.

CAPSULE
"HEAT SINK"
ELEVON

C. Geary

AUGUST 24

Republic Aviation publishes an advertisement of which this "Time-Table for Space Conquest" (devised by its vice president for research and development, Alexander Kartvell) is a part:

1963: Instrumented soft planetary landing

1968: Space station for staging to moon and planets

1970–1975: Moon base

AUGUST 31

A House Committee Staff Report states that lunar flights will originate from earth-orbiting space stations. The problems of orbital rendezvous will have to be balanced with those of a direct ascent.

NOVEMBER 2

The Space Task Group discusses advanced manned spacecraft. These include a three-man circumlunar vehicle (which can also be used as a space laboratory, the cabin of a lunar lander and a deep space probe).

DECEMBER 18

NASA publishes its "Ten Year Plan" for future space exploration. Manned spaceflight objectives listed include:

1960—First suborbital flight by an astronaut.

1961—First orbital manned spaceflight (Project Mercury—a projection influenced by the NASA Goett Committee).

1965–1967—Program leading to manned circumlunar flight and permanent near-earth space station.

Beyond 1970—Manned lunar landing and return.

1959–1960

CBS broadcasts the television series *Men Into Space*, featuring spacecraft and moonscapes designed by Chesley Bonestell [see 1949, 1950 and 1952–1954]. A great deal of effort goes into making the program as authentic as possible. Early in the series' development the aid of the Department of Defense is solicited and granted, but only on the condition that the DoD be accorded script approval. The DoD's interest, however, is mostly in maintaining technical accuracy. Capt. M. C. Spaulding of the U.S. Air Force Ballistic Missile Division is placed in charge of the show's technical advisors, which include personnel from the USAF Air Research and Development Command, the School of Aviation Medicine, and the Office of the Surgeon General. Location filming takes place at the Space Medicine Center at Randolph Field, Texas; Wright-Patterson Air Force Base; Cape Canaveral; Edwards Air Force Base and the Navy's testing area at Point Mugu, California.

The spacecraft used in the program include a three-stage manned ferry rocket with a winged third stage, resembling similar rockets featured in the *Collier's* series [see 1952–1954] and the Disney Tomorrowland "Man in Space" television series [see 1955]. A moon lander consists of the delta-winged third stage of the ferry rocket placed atop a *Collier's*-like cluster of cylindrical fuel tanks. After landing on four landing legs, the empty tanks are discarded, leaving the one cylindrical tank beneath the personnel rocket for the takeoff from the moon.

Many of the paintings that Bonestell creates for the television show later appear in the Robert Richardson book *Man and the Moon.*

1960

Ernst Stuhlinger proposes launch vehicles and a lander for a lunar mission, based on a NASA-initiated study at the Army Ballistic Missile Agency in 1959. This results in two different manned landers, one a modified Mercury capsule, and a manned lunar circumnav-

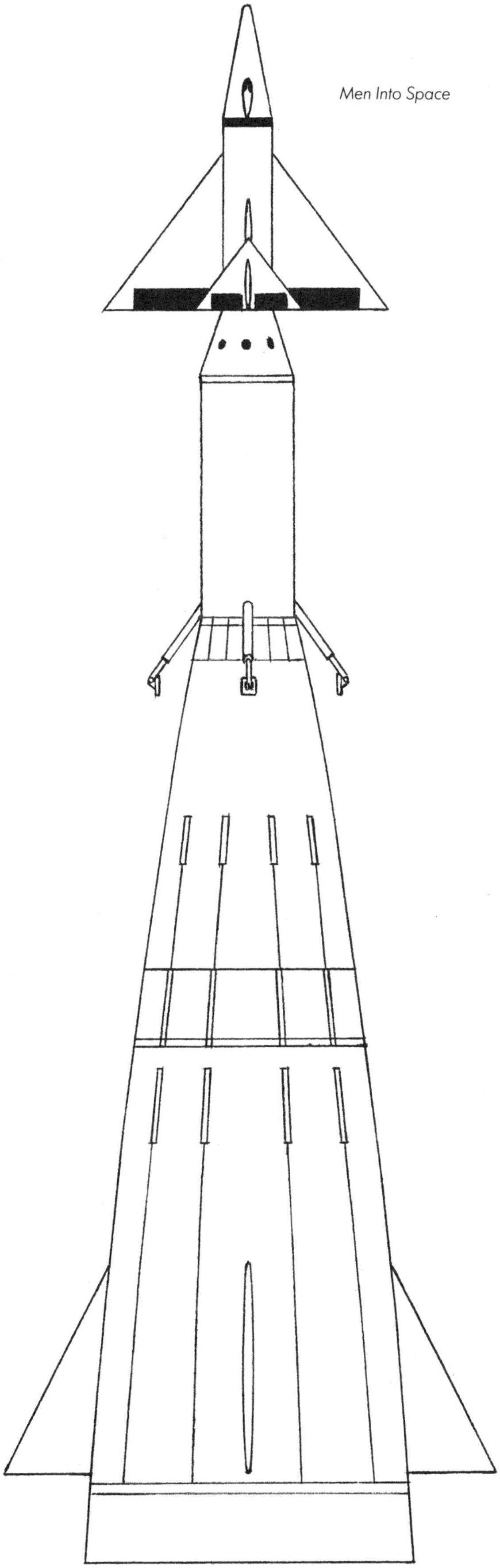

Men Into Space

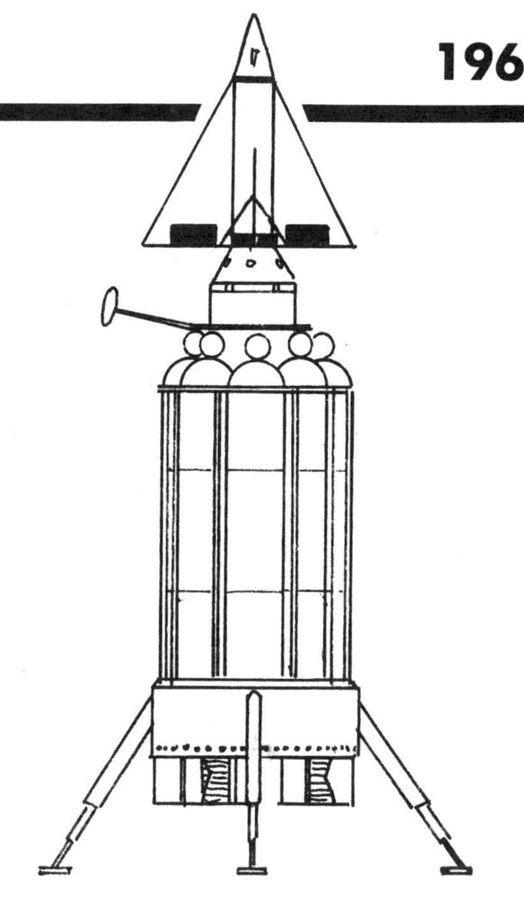

Left: launch vehicle and manned orbiter; above: lunar lander; below, variation published in Chesley Bonestell's book *Rocket to the Moon* (1961).

Dunning

igator. Launch vehicles of the Atlas and Saturn class, particularly the latter, would be capable of supporting the unmanned lunar landing program, while the manned program would require multiple Saturn launchings. Stuhlinger believes that if several Saturn vehicles were to carry out a refueling maneuver in low earth orbit, than one Saturn-boosted spacecraft could take off from orbit, land two men on the moon, and return them to the surface of the earth. Manned flights of this kind, Stuhlinger thinks, will probably not be feasible until the 1970–1975 period.

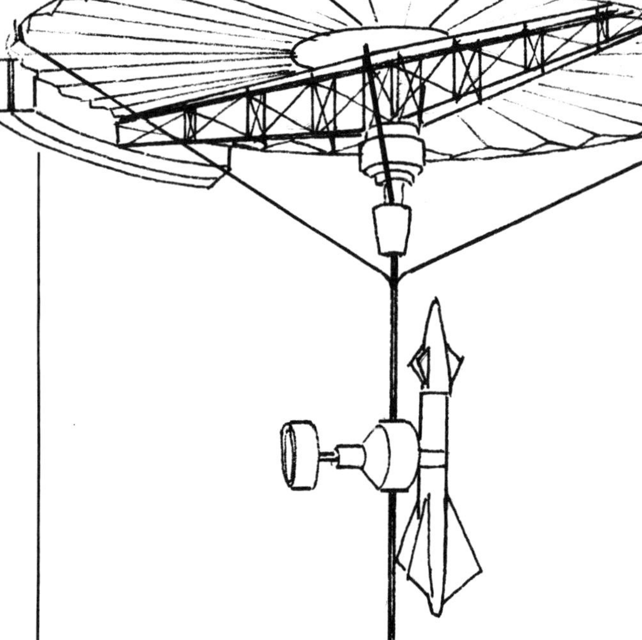

Project Outpost is developed by Krafft Ehricke at the Convair Astronautics Division of General Dynamics. It is a manned space station based on the Atlas missile. It consists of a prefabricated, fully equipped laboratory placed into orbit by a single launch of an Atlas-Centaur rocket (now under development, it is expected to be operational in 1961). The scheme capitalizes on the fact that the Atlas, relieved of its warhead and associated equipment, is the only U.S. rocket capable of orbiting the entire primary structure of a space station in a single launch. Lacking only a crew, the space station is to be placed into orbit still attached to the Centaur stage (that part containing the empty fuel tank, rocket engines and nuclear power source). One of Outpost's goals is to train astronauts for deep-space missions, "At least 500 hours of training *in orbit* will be required to qualify a space crewman for simple flights to the moon and back, plus about 4,000 hours of mixed orbital and lunar training before a man can safely embark on a planetary mission."

Outpost can house up to four men at a time in an orbit 400 miles above the earth. In addition to its function as a trainer for long space missions, it also serves as an earth observatory and research laboratory. The astronauts arrive in a "personnel glider." They separate the laboratory and propellant tank from the reactor by telescoping rails, increasing the station's length by about 50 feet. The complete station is about 106 feet long, weighing approximately 15,000 pounds. Artificial gravity (0.1 to 0.15 g) is then generated by rotating the structure end over end about 2.5 times per minute. The crew compartment, an inflated insulated capsule, is divided into four floors; from the "lowest" up: the restroom and lavatory, recreation room, sleeping quarters and a control room. The opposite end of the station is reserved for storage of heavy equipment. This tends to pull the center of gravity rearward, lengthening the distance between it and the living quarters. This, in turn, diminishes the undesirable Coriolis force side effects of the station's spin.

In the "aft" end are housed the water supply, emergency batteries, control rocket fuel, tools, and so forth. Access to the unpressurized portion of the hull is through a simple manhole near the hub. Access to the pressurized compartment is through an airlock above the control room. A yearly cargo ship carrying up to 8,000 pounds of supplies and equipment will service the station. A total of thirteen to twenty launchings a year will be required to maintain the station. Manned rendezvous will employ small lifting-body craft, launched two at a time atop Atlas boosters.

Ehricke claims that the space station could be in orbit within 5 years from the date Convair is permitted to go ahead with the project. The administration's space advisory council has already turned down the Ehricke-Convair proposals fourteen times.

Peter A. E. Stewart of the Astronautics Section, Advanced Projects Group, Hawker-Siddeley Aviation, proposes converting the discontinued Blue Streak missile into a manned spacecraft. This is to be done in three steps:

Step #1 will orbit a Mercury-type capsule. A small hydrogen-oxygen booster will give the capsule its final lift into orbit.

Step #2 is an improved second stage able to send an unmanned 450-pound soft-lander to the moon.

Step #3 is an uprated Blue Streak from step #2 placed on top of a giant booster as powerful as a Saturn. It has four broad wings containing three engines in each trailing edge, eight Blue Streak motors and four Stentor motors. The wings allow the booster to be recovered, unlike the Saturn. This configuration is able to send an advanced, unmanned soft-lander to the moon.

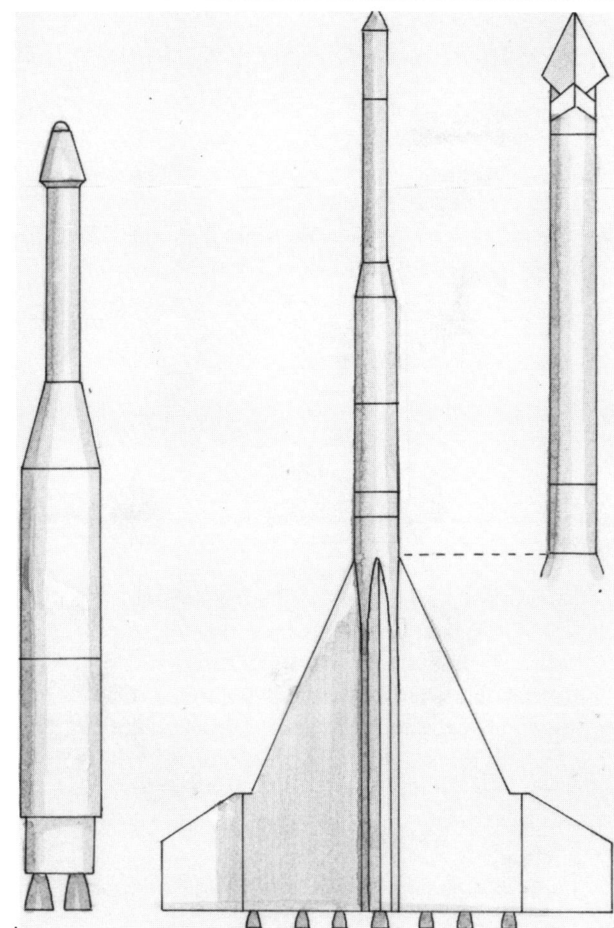

By substituting a nuclear second stage for the Blue Streak, a 5- to 7-ton manned lunar lander could be launched. Stewart expects a manned lunar expedition to take place in 1970, after a 10-year development program, landing in the Mare Imbrium at the crater Piazzi Smyth.

R. W. Bussard publishes *Galactic Matter and Interstellar Flight*, in which he describes the interstellar ramjet. The first version, RAIR, or ram-augmented interstellar rocket, scoops up interstellar hydrogen as it speeds along, using its kinetic energy to augment the onboard fuel supply. Its main thrust might be supplied by a "conventional" fusion engine, for example. This drive is also needed initially to bring the ship up to the speeds required for the ramjet to operate. It is exactly the interstellar equivalent of the World War II buzzbomb [see above]. A scoop created by a magnetic field collects interstellar gas, compressing it and ejecting it from the rear of the rocket to add to the overall thrust. Energy is transferred between the nuclear rocket and the ramjet to maximize fuel efficiency. Velocities up to 50 percent of the speed of light may be reached.

The next stage of this concept is the pure Interstellar Ramjet, often called the Bussard Ramjet, after its inventor. Like the RAIR, it scoops up interstellar hydrogen with a magnetic scoop. But instead of using the gas simply as a mass to accelerate, it will use the hydrogen as a nuclear fuel. It is potentially the most powerful starship devised; since its fuel source is entirely external, its speed and range are virtually unlimited. One version suggested requires a hydrogen fusion booster the size of the BIS Daedalus starship [see below] to bring the 100,000-ton 7,000-foot long ship up to operating speed: 2 percent of the speed of light. Constantly accelerating, it is able to cross the galaxy in just 31 years—while 100,000 years has passed on the earth! The electronic mouth of its magnetic scoop is the size of Jupiter. At the present, the interstellar ramjet poses problems beyond the knowledge of modern technology and science.

The Manned Orbiting Laboratory is announced by President Lyndon B. Johnson. It is a U.S. Air Force project, and is expected to be manned by Air Force astronauts within 3 1/2 years. The program consists of one or two unmanned launchings, followed by five manned flights in which pairs of astronauts remain on board the station for periods of up to 30 days. The MOL is 10 feet in diameter and about 41 feet long. Atop is a two-man Gemini capsule, making the total length 54 feet. It is launched by a Titan III-C booster. Once in orbit, the astronauts enter the pressurized laboratory compartment through a hatch in the capsule heat shield (or through an external inflatable tunnel). The MOL is to be designed and built by Douglas Aircraft and McDonnell Aircraft will supply the Gemini capsules.

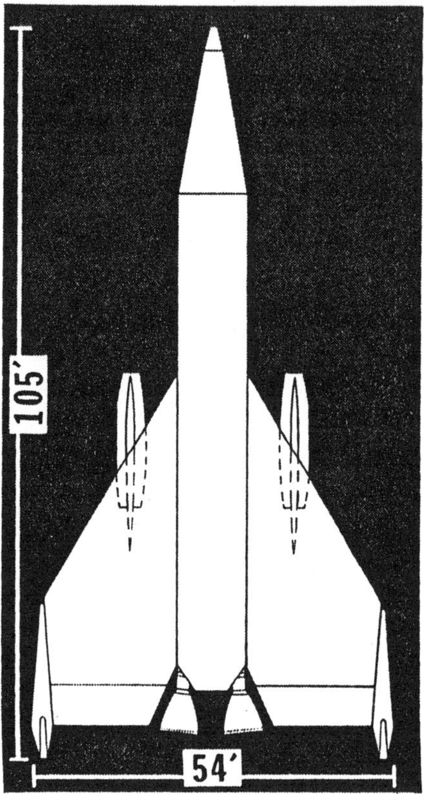

Manned booster (ca. 1960) capable being flown or towed at sea back to its launch site.

A mockup of the MOL is launched on November 3, 1966.

Betty Skelton, condescendingly if affectionately dubbed "Number 7 1/2" by the Mercury astronauts, discusses the problems of women in space with NASA, at the invitation of *Look* magazine. Skelton is an accomplished pilot of 33 at this time. During a 4-month period, she interviews NASA officials and space scientists, and undergoes much of the same training the Mercury astronauts have to endure. Brig. General Don Flickinger, assistant for bioastronautics of the Air Research and Development Command, believes that women might eventually be seriously considered, but not until three-man spacecraft are being used. Skelton fears that by the time any American woman gets the chance to go into space, she will be too old to go herself.

Nevertheless, she passes all of the training and tests with flying colors. The conclusion of *Look* is that women might eventually

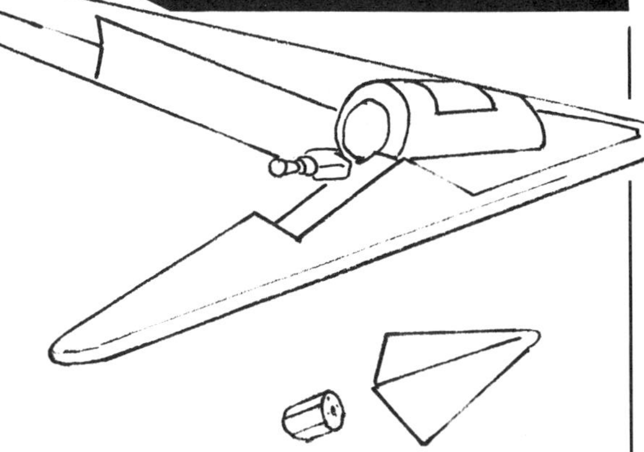

prove superior to men in space, that they have "more brains and stamina per pound than men," and require less support equipment. The future female astronaut, they predict, will be a "flat-chested lightweight under thirty-five years of age and married. Though not an outstanding athlete, she will have extraordinarily precise coordination. She is a pilot. Her interests will tend toward swimming and skiing, rather than a more muscular sport like wrestling . . . Her personality will both soothe and stimulate others on her space team . . . Her first chance in space may be as the scientist-wife of a pilot-engineer. Her specialties will range from astronomy to zoology . . . "

J. R. Camp publishes a suggestion for a solar-powered rocket engine. It is a cone-shaped metal vessel with the apex shaped into the form of a rocket motor. The convex base of the cone is made of glass or some other transparent material. Sunlight from a large reflector is reflected through it, coming to a focus at the apex, within the combustion chamber of the motor. A liquid propellant is introduced which is vaporized at a high temperature by the focussed sunlight and expelled through the exhaust nozzle. It is planned to be a low-thrust device, similar to the ion engine.

Lockheed and Hughes propose a "space ferry" to shuttle between the earth and an orbiting spacc station. It can carry up to four astronauts—a pilot and three passengers—and about 14,000 pounds of cargo. A typical rendezvous mission is accomplished in two to ten orbits over a period of 3 to 12 hours. The space ferry is launched by way of a three-stage booster (gross weight 1 million pounds). Upon arrival in orbit, the arrowhead-shaped vehicle unfolds, scissors-like, into a V-shaped flying wing configuration. A small rocket engine unit is contained in the crotch of the V. Pilot, passengers and cargo are in a central, barrel-shaped cabin. The entire design involves maximum use of the human pilot, says Lockheed, "This approach leads to a less complicated and more reliable design . . . thereby giving greater assurance of

safe return of the craft to earth."

Krafft Ehricke proposes manned missions to Mars and Venus. He assumes the years 1970–1973 as the target dates for the first manned flights to these planets. By this time experience will have been gained from manned lunar operations (1965-1970), nuclear heat-exchanger engines will have been developed (1959–1965) and operated in space (1965–1970). Long-term missions will have been conducted in space stations (1963-1970 [see above]).

The Mars mission ship is designed in two versions: for four men and for eight men. The basic difference is in the size of the

1960

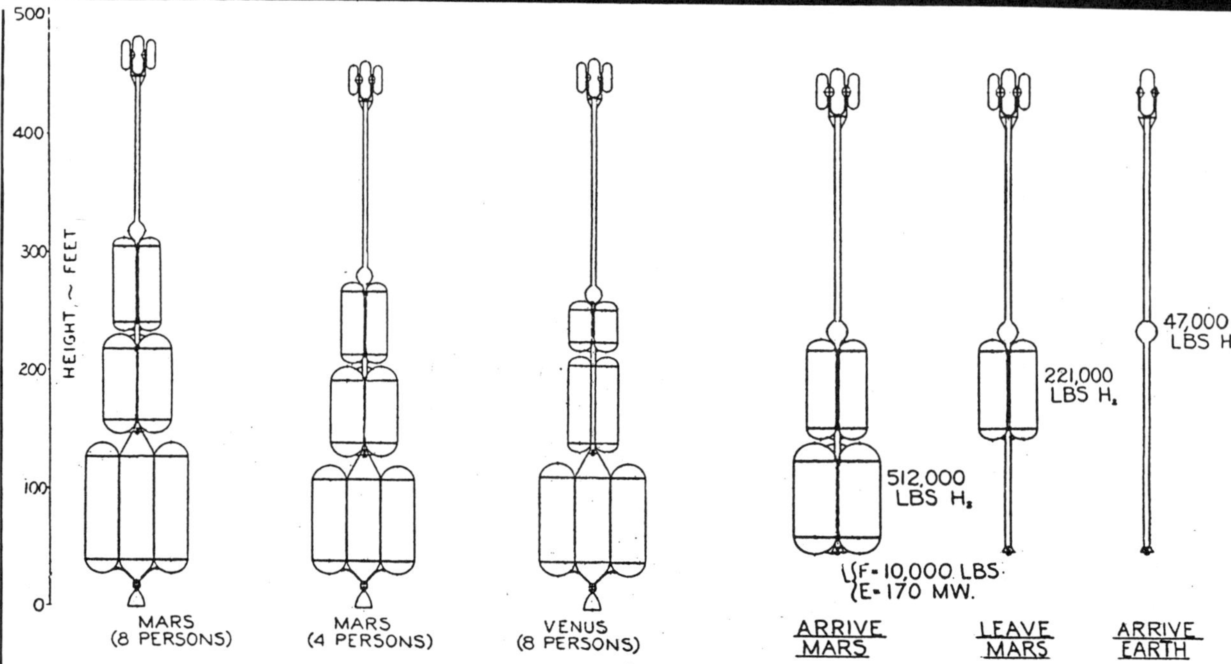

fuel tanks. The eight-man spaceship is approximately 475 feet long. The first 320 feet is occupied by three clusters of cylindrical fuel tanks. Each cluster is jettisoned when empty. The first and largest cluster is used in escaping from the earth. The second set is used for the braking maneuver upon arriving at Mars: 512,000 pounds of liquid hydrogen. The smallest cluster contains the 221,000 pounds of hydrogen needed to leave Mars orbit. All that remains of the huge spaceship once it returns to the earth is the slender central shaft, 370 feet long, the passenger capsule and a small, spherical tank of fuel weighing 47,000 pounds, needed for the final maneuver.

Dandridge Cole of Martin Aircraft and Robert Granville propose a scheme to fly a single astronaut around the moon by 1964 (1962 in an earlier version). There are five test flights and three lunar flights in the program, all to be accomplished, for the most part, from existing hardware. The authors consider their plan a natural extension of the Mercury program and their "spaceship" is a modified Mercury capsule. The major addition is the need for orbital refueling.

The Mercury astronauts themselves are ideal choices for the pilots. Their 1- to 2-hour orbital flights, scheduled for 1961, are to be increased to 24 hours. By 1962 there will have been flights lasting several days. Then, by simply adding propulsion, the nearly circular Mercury orbits can be extended into a long ellipse that will circle the moon. The total trip will take about 80 hours.

Granville and Cole suggest two ways in which the mission can be accomplished. (1) A Mercury capsule is parked in earth orbit, placed there by the currently in-development Atlas booster. Then subsequent Atlas-Centaur flights carry into a rendezvous orbit the extra propulsion stages, with their fuel. These are attached to the modified capsule, producing the necessary thrust to take it to the moon for a single orbit. (2) Use of the upcoming Saturn booster launches a Mercury-type capsule into a circumlunar orbit directly from the earth. If the Saturn is to be used, the projected date of the project will have to be moved ahead to 1964–1965.

When the Cole-Granville plans were first developed in 1958, at Martin, it was necessary to "invent" their own booster rockets. The rocket needed for the first plan had three stages, weighing 160,000, 40,000 and 10,000 pounds respectively, carrying a payload of 5,000 lb into orbit. The booster for the second plan was a four-stage rocket. The first stage was a cluster of four rockets weighing 640,000 pounds. This vehicle would have placed a payload of 9,000 pounds around the moon directly from the earth's surface.

This new scenario is more or less an updating of the Martin proposal of 1958 [see above]. As noted, the earlier version was based upon a minimum capsule weight of 9,000 pounds. Granville and Cole now believe that a capsule weight of only 1 1/2 tons will be sufficient, based upon the experience gained from the Mercury program.

Kenneth Gatland presents a paper at the Stockholm Congress of the International Astronautical Federation, entitled "A Conceptual Design for a Manned Mars Vehicle." It concludes that a manned Mars landing is possible using chemically propelled rockets (burning liquid oxygen and hydrogen). The parallel-staged concept, using plug-nozzle engines, requires 2 years and 9 months for the round trip. Minimum-energy Hohmann orbits are used that necessitate 259 days journeying for each leg and a 479-day wait on Mars. Nuclear propulsion will be required to bring travel time down to more reasonable limits. One objective is a 450-day round trip, with a crew of six to eight, and a launch date of early April 1986—the next favorable low-energy opportunity will not occur until 2001.

I. M. Levitt, director of the Fels Planetarium, proposed a ten-year timetable for the future exploration of space [also see 1956].

1960—Unmanned reconnaisance of the moon; first interplanetary probes and communication satellites.

1961—Photographs of the lunar surface will be (physically) returned to the earth.

1962—Scientists learn how much radiation shielding manned spacecraft will require; instrumented probes sent to Venus and Mars; manned reconnaisance of the moon (orbiting for up to one month); unmanned lunar landing and sample return.

1963—Small permanent space station; unmanned solar probe.

1964—Unmanned space telescope established; first ion rockets tested in space.

1965—Major space station orbited; ion rockets in use.

1966—Manned landing on the moon using vehicle assembled in earth orbit with a small ferry making descent to surface. Once the ferry has made the round trip to the surface, it is abandoned.

1967—Permanent lunar settlement; ion rockets power deep space probes.

1968—Lunar landing continue and lunar base expanded.

1969—Manned deep space probes launched to Mars and Venus.

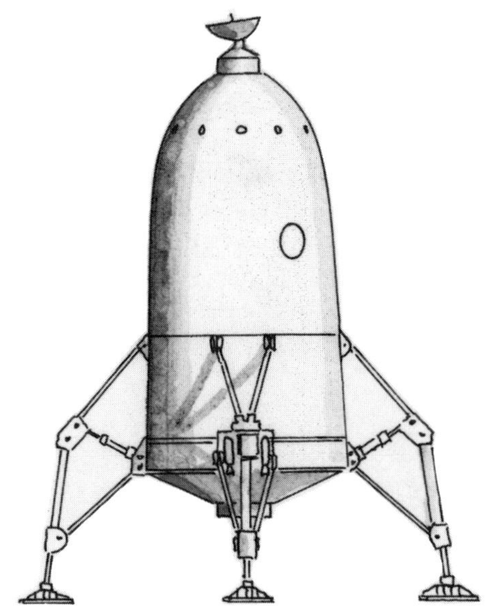

W. N. Neat suggests an updating of the old BIS moonship [see 1939]. Although he makes no specific design recommendations for a booster, he does supply a revision of the BIS manned lander, taking advantage of postwar technology and materials, especially developments in radar and radio.

Dandridge Cole designs a series of giant spacecraft. The first, *Arcturus*, is a modified Nova booster, using only two F-1 engines. Since *Arcturus* could be operational by 1965 or 1966, this design is more practical than waiting until 1967 for the six-engined Nova. Cole expects the *Arcturus* to have sufficient performance to accomplish a lunar landing

1960

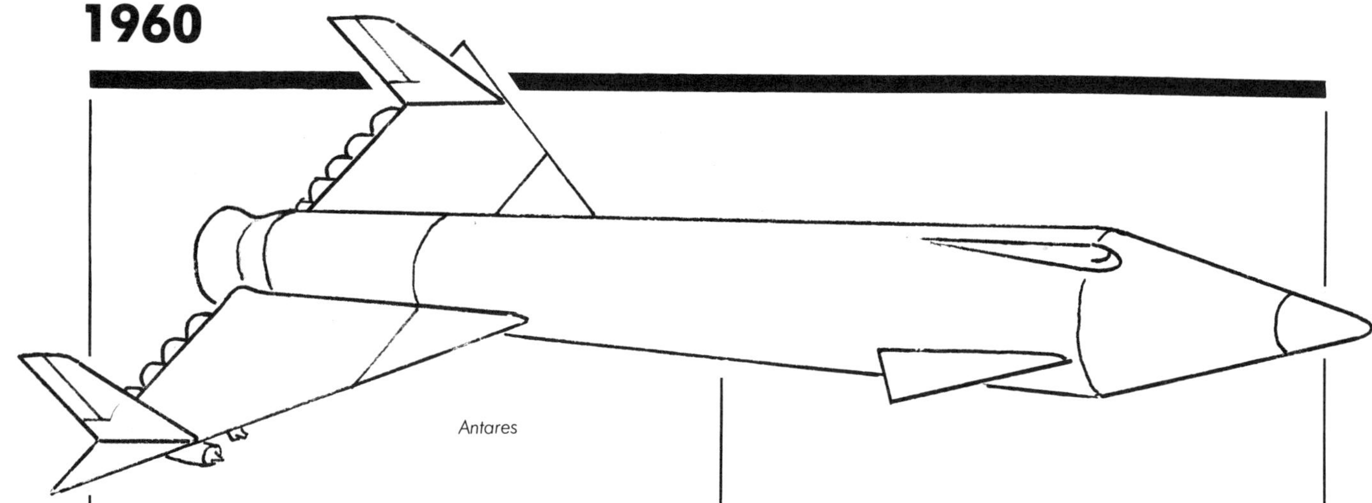
Antares

by 1966—earlier than the Nova scenario and with fewer development costs. The first stage consists of two F-1 engines and seven Titan first-stage tank assemblies, producing 2 million pounds of thrust. The second stage consists of three modified Titan tanks and a single F-1 engine, producing 500,000 pounds of thrust.

Arcturus II adds a third stage: a single Titan with 200,000 pounds of thrust. Improved propellants will increase the over-all performance. Weighing 2.4 million pounds, the *Arcturus II* has more than twice the payload capacity of Saturn and only slightly less than Nova. To give *Arcturus II* more performance, an air-breathing booster could be developed.

The next step is the *Antares*, a nuclear-powered spaceship. Because of the danger of using nuclear-powered rockets within the atmosphere, Cole proposes using an earth-captive booster. This is a turbojet-powered sled on an inclined, 15-mile-long track, which can accelerate the *Antares* to a speed of 2,000 fps. At this point the *Antares's* own ramjets take over, boosting the ship to a speed of approximately 10,000 fps. Finally, at a safe altitude, the nuclear engines ignite, carrying the rocket on into orbit. The *Antares* is winged for return to earth as a conventional aircraft. It has a gross weight of 10 to 20 million pounds and a payload capacity, to orbit, of between 2.5 and 5 million pounds.

The third spaceship is the mammoth *Aldebaran*, scheduled for the 1975–1980 time period. It, too, is atomic-powered, but by a system beyond the "crude" one used by the *Antares*. Cole does not specify which of several possible systems might be used. Cargo could be carried into space at a cost of only $10 per pound. The ship is exceedingly large—50,000 to 80,000 tons—and requires a water takeoff and landing. A single-stage winged vehicle, it could carry a payload of 30,000 tons into orbit or 22,500 tons to a soft landing on the moon.

Boeing presents its Project Parsec (initiated in 1958). The name is an acronym for "Program for Astronomical Research and Scientific Experiments Concerning Space." It describes an unmanned series of interplanetary probes and vehicles designed to pave the way for human exploration.

The project is divided into eight steps or "missions." The first is an earth satellite observatory. This is planned in both unmanned and manned versions and is capable of both terrestrial and astronomical observations. The observatory also provides facilities for low gravity and cryogenic research, as well as acting as a communications relay.

The second mission establishes a lunar colony.

Mission three develops the "counter-moon" vehicle: a satellite in the same orbit as the moon, but maintained on the opposite side of the earth. In conjunction with lunar-based observatories, it can create a stereoscopic view of the universe, with a 480,000-mile base. Ion engines hold it in position; they are used as well on the Trojan point satellites planned as part of mission six.

Mission four sees unmanned planetary probes launched into the solar system. Included is a circum-Martian vehicle that returns to the earth after its mission is accomplished. The sun is the subject of the

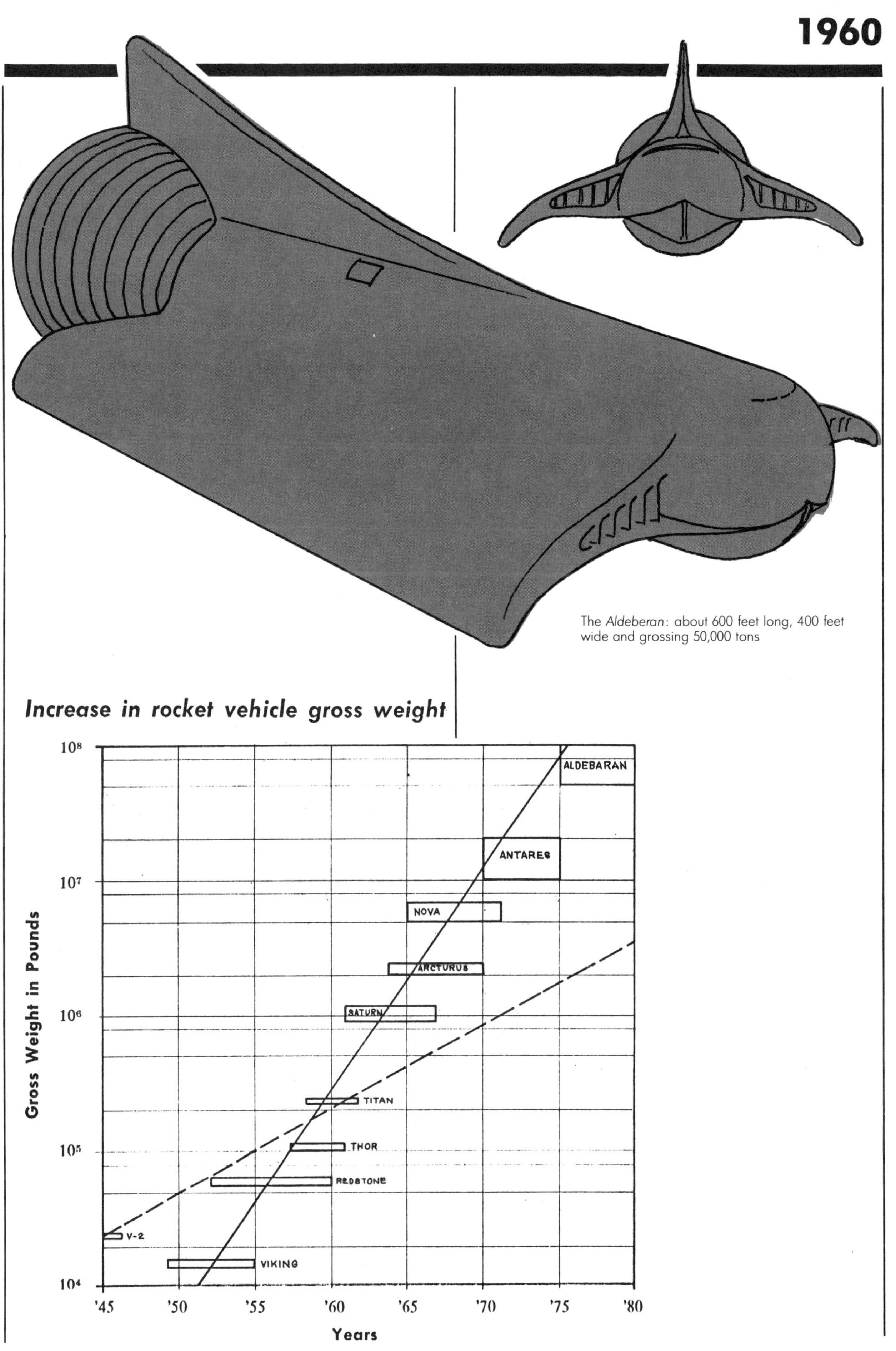

The *Aldeberan*: about 600 feet long, 400 feet wide and grossing 50,000 tons fifth mission, which places probes in close solar orbits.

The seventh mission involves manned exploration above and below the ecliptic.

Mission eight sees the exploration of the planets by manned spacecraft, including colonization.

One basic type of spacecraft can be adapted for missions six, seven and eight—the manned modules required can be modified from the earth satellite and counter-moon vehicles. This "general-purpose" capsule is a 40-foot sphere assembled in orbit. It is launched by a low-thrust nuclear propulsion system, perhaps a plasmajet, with its reactor and liquid hydrogen tanks at the end of a long boom. There are no facilities for artificial gravity. Instead, special exercises make up for the lack of gravity. One of these spacecraft can be constructed by a team of "space steeplejacks" in about 6 months. Until the personnel sphere is completed, the crew will live in an inflated, double-walled plastic balloon.

Also at about this same time, and independent of the Parsec project, Boeing engineers propose a manned Mars landing, using a specially designed spacecraft. The manned spacecraft is a large, triangular lifting-body boosted into space by seven parallel stages. In addition to the lifting body is a cylindrical module which contains living quarters, laboratories and the booster engines for the eventual return to the earth. This module is left in orbit around Mars while the lifting body lander makes the descent to the surface. Once the exploration period is completed, the nose of the lander is raised to the appropriate angle by jacks. The aft part acts as a launcher for the nose section, which returns to the orbiting module, to which it reattaches. Once the spaceship has returned to earth orbit, the module is abandoned and the lifting body lands as a glider.

In response to a call for designs for the projected Apollo spacecraft, General Electric submits a three-part spacecraft. The first section is a propulsion module 4.6 m long and 3 m in diameter. An interstage flange flared

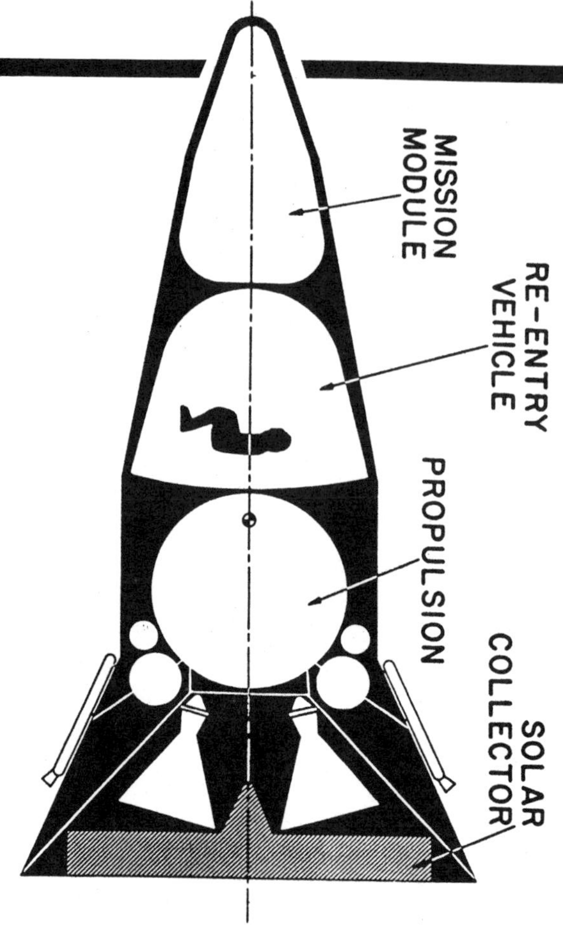

D-2 CONFIGURATION

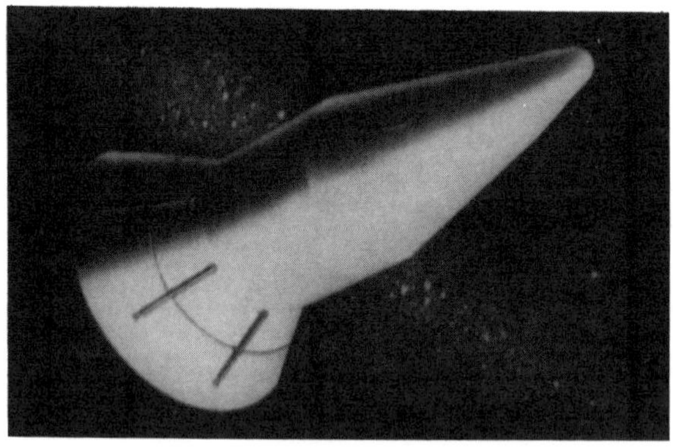

to 5.4 m attaches it to the Saturn booster. Second is a descent module about 2.5 m long and 2.8 m in diameter. Third is the pear-shaped mission module, 3.1 m long with a maximum diameter of 1.7 m. The spacecraft is intended for a circumlunar flight. Upon returning to the earth, the descent module makes a parachute landing into water.

General Electric develops a series of "life rafts" for emergency evacuation of manned spacecraft. The project, under the direction of Harold Bloom, proposes one-man and three-man versions in both short-term and

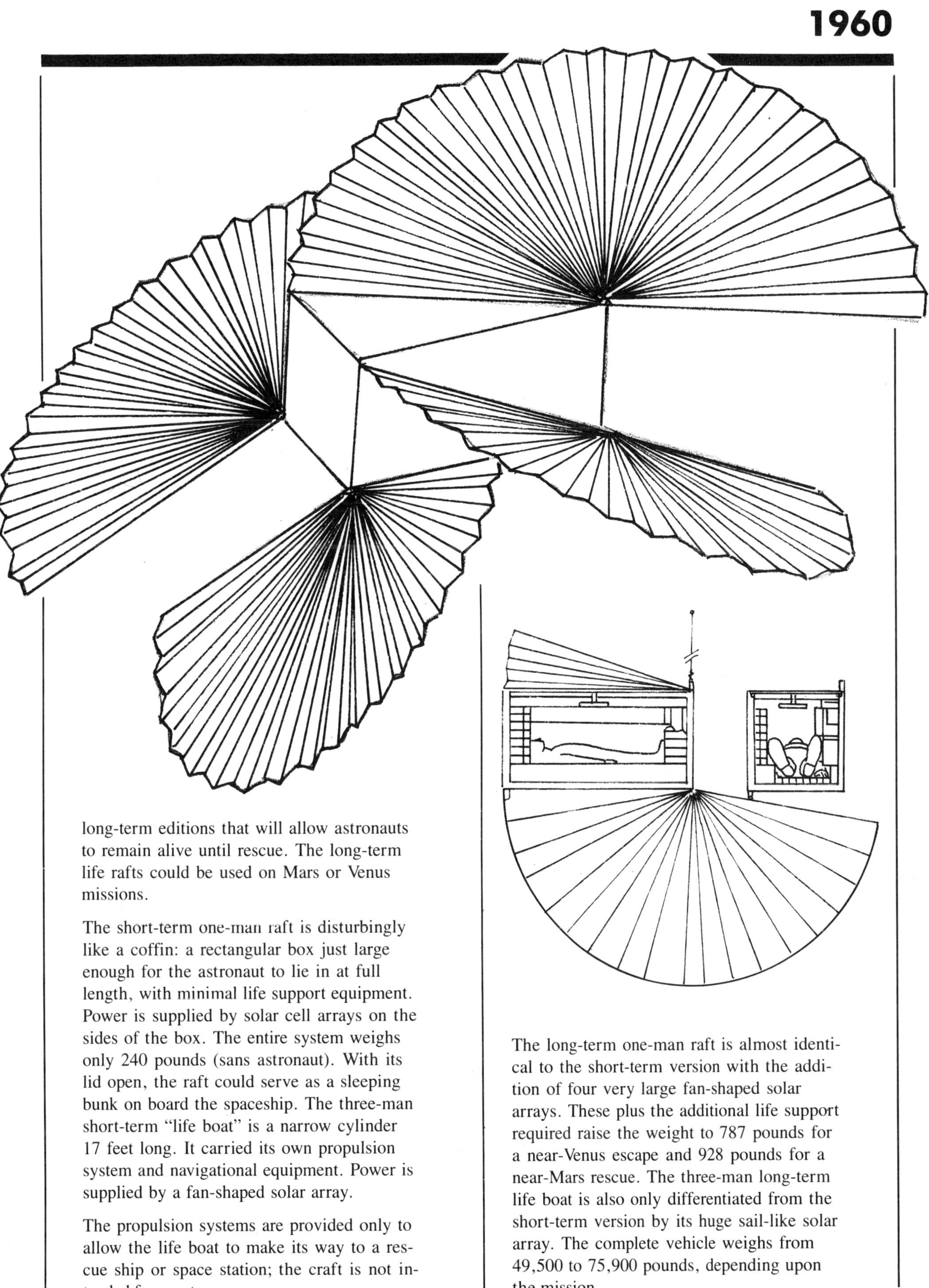

long-term editions that will allow astronauts to remain alive until rescue. The long-term life rafts could be used on Mars or Venus missions.

The short-term one-man raft is disturbingly like a coffin: a rectangular box just large enough for the astronaut to lie in at full length, with minimal life support equipment. Power is supplied by solar cell arrays on the sides of the box. The entire system weighs only 240 pounds (sans astronaut). With its lid open, the raft could serve as a sleeping bunk on board the spaceship. The three-man short-term "life boat" is a narrow cylinder 17 feet long. It carried its own propulsion system and navigational equipment. Power is supplied by a fan-shaped solar array.

The propulsion systems are provided only to allow the life boat to make its way to a rescue ship or space station; the craft is not intended for reentry.

The long-term one-man raft is almost identical to the short-term version with the addition of four very large fan-shaped solar arrays. These plus the additional life support required raise the weight to 787 pounds for a near-Venus escape and 928 pounds for a near-Mars rescue. The three-man long-term life boat is also only differentiated from the short-term version by its huge sail-like solar array. The complete vehicle weighs from 49,500 to 75,900 pounds, depending upon the mission.

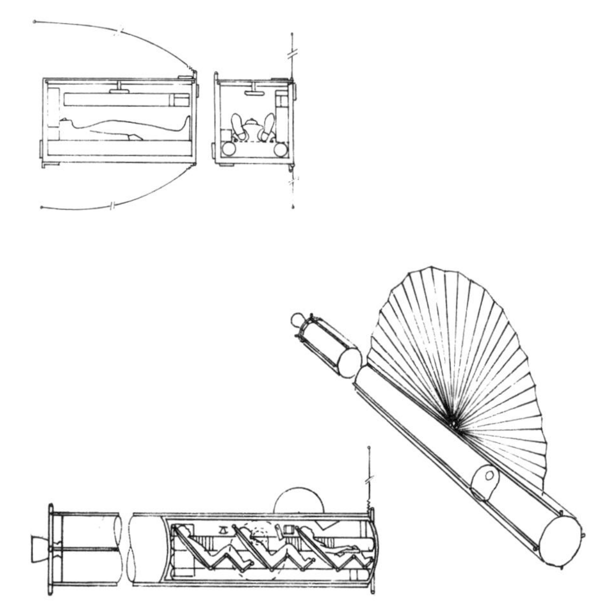

The rescue of an astronaut somewhere between the earth and the moon is expected to take only 4 days, while the rescue of an astronaut or astronauts might take from 5 months for a Venus mission to 8 or 9 months for a Mars mission.

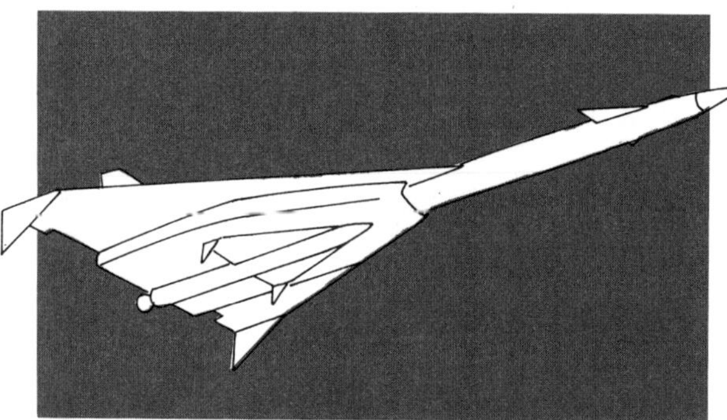

The Air Force considers using the B-70A Valkyrie bomber as a Mach 3+ launch platform for spacecraft weighing up to 15,000 pounds (the B-70 RBSS, Recoverable Booster Space System). This could result in a savings of $2.63 billion over conventional staging systems over the next 15 years. The B-70 could also be used to launch manned spacecraft such as the proposed Dyna-Soar or other shuttle-type vehicles.

NASA's Lewis Research Center develops a concept for a nuclear-electric Mars mission spacecraft. Looking something like a javelin

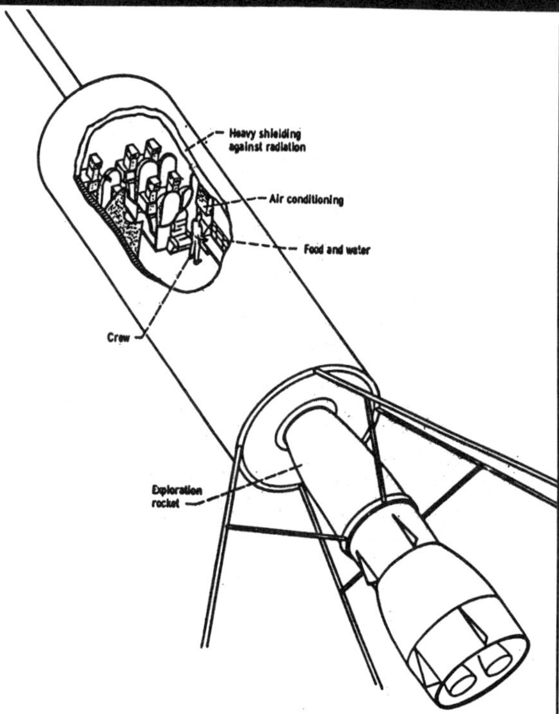

or weather vane, the slender, 400-foot spaceship has a nuclear reactor at its nose which runs electric generators located immediately behind it. Between the generator housing and the rear of the ship are a pair of winglike radiators. At the end of the vehicle opposite the reactor are fuel tanks, crew's quarters, landing module and the electric thrusters. The complete ship weighs 1/2 million pounds. The fragile spacecraft is assembled in orbit, the components launched by a pair of Saturn V boosters.

It is reported that the Soviet Union is already flight-testing its own version of the American Dyna-Soar. The "antipodal semiballistic missile"—the T-4A—is only in the advanced testing and evaluation stages and is not yet operational. The T-4A is presumed to be manned. The design is a spinoff of Eugen Sänger's antipodal bomber research, which has long fascinated the Russians [also see 1949].

The range of the T-4A is 9,936 miles. It levels off at an altitude of 186.3 miles at a sustained velocity of 11,178 mph for 4,968 miles. Its maximum speed is 13,910 mph. It carries a payload of 1 ton. It is launched from a catapult, propelled along a steel track by a booster with 300,000 to 360,000 pounds of thrust. The first stage is comprised of three kerosene-fueled rockets and

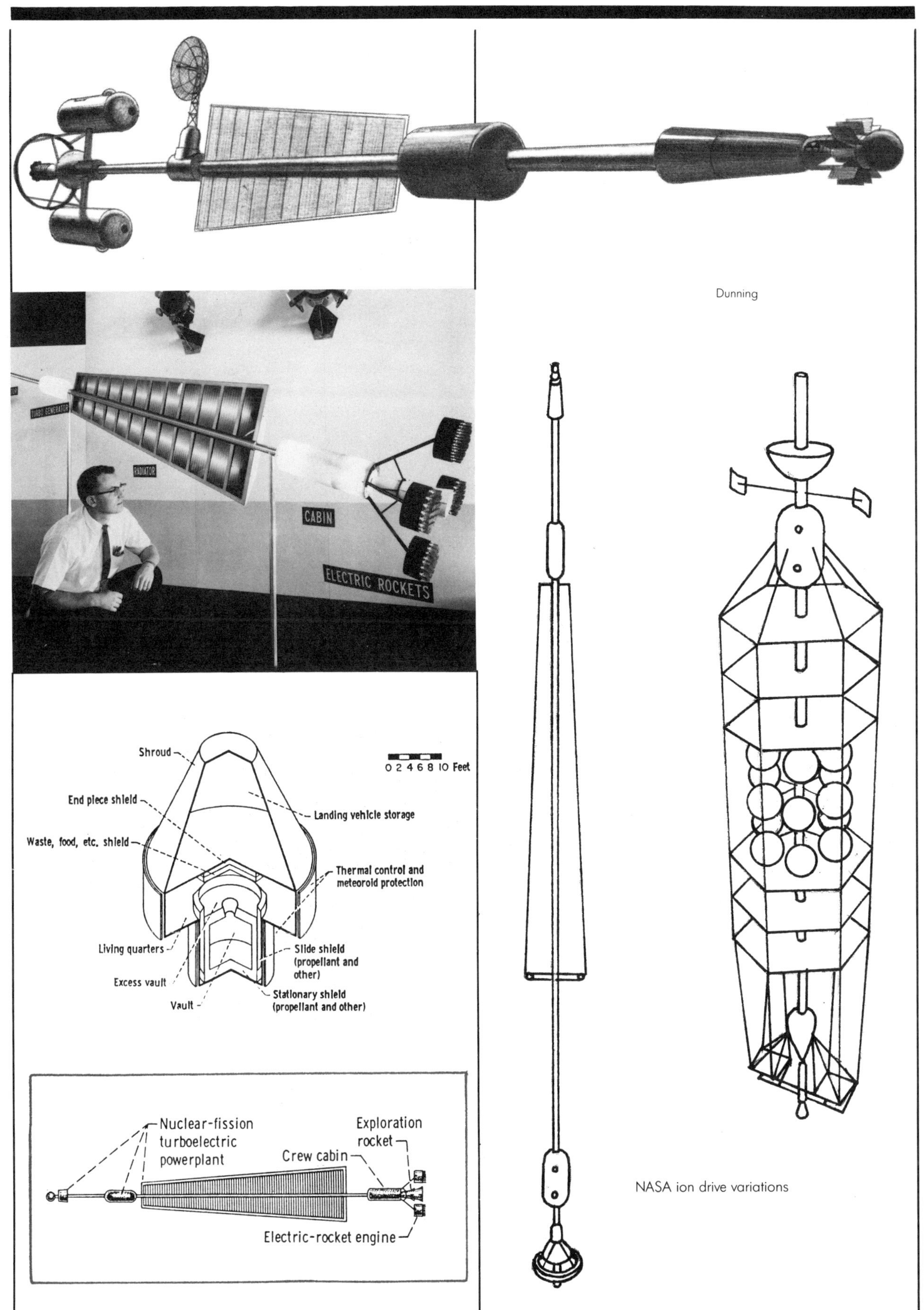

Dunning

NASA ion drive variations

1960

two solid-fuel boosters.

The T-4A is 121.02 feet long with a wing span of 65.6 feet. The main stage is 60.68 feet long and 6.88 feet in diameter.

Astrocommuter is a spacecraft designed by Saunders Kramer of Lockheed for ferrying personnel to and from his proposed space station. The delta-winged reentry vehicle would be 46 feet long (35 feet of which is the fuselage) with a wingspan of 28.3 feet. It is equipped with a pair of tubojet engines to allow it up to 48 minutes of cruising over a range of 356 statute miles. It would carry seven passengers, including the crew.

The *Astrocommuter*'s booster will be an early version of the von Braun-designed Saturn, with eight first-stage rocket engines.

In addition to the manned shuttle, the booster would also carry an *Astrotug*, an 18,000-pound space station service vehicle consisting of a manned capsule (normally two but up to eight if neccessary) and eight remote-controlled manipulator arms. Unlike the *Astrocommuter*, the *Astrotug* will not return to the earth. Instead, it remains as part of the space station's permanent retinue. An airlock allows it to rendezvous with both the station and the aft-end airlock of the *Astrocommuter*.

Kramer expects that his shuttle would cost approximately $135 million to develop. If the program could begin this year then the first flight might take place in early 1966.

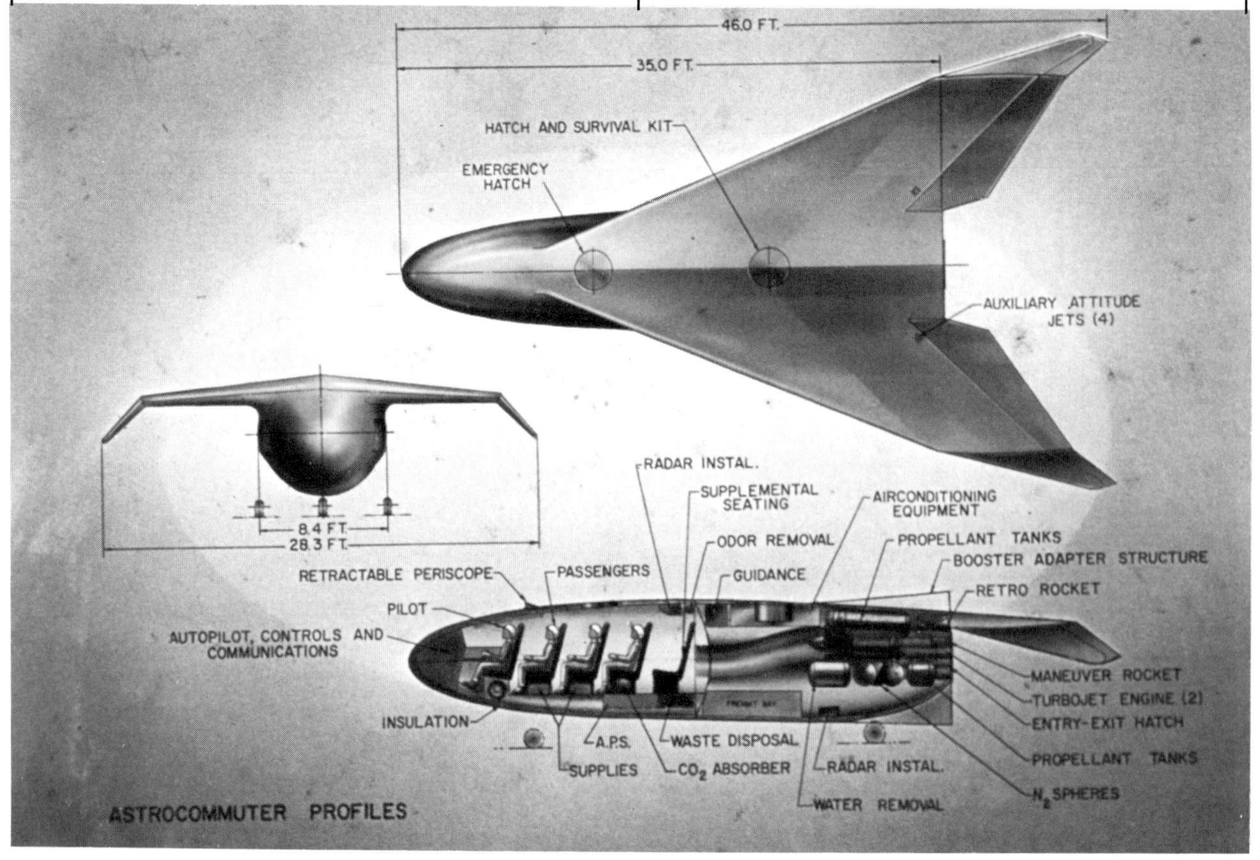

Kramer

The motion picture *12 to the Moon* features the presciently named spaceship *Lunar Eagle 1*. It carries an international crew of twelve men and women to the earth's satellite.

JANUARY

Chance-Vought completes an independent study of the manned lunar landing mission dubbed MALLAR (**Ma**nned **L**unar **L**anding **a**nd **R**eturn). It employs earth and lunar orbit rendezvous technique for a two-man 2-week mission. MALLAR requires a 6,600 pound entry vehicle, a 9,000 pound mission module and a lander weighing 27,000 pounds.

Abe Silverstein, director of the Office of Space Flight Programs, suggests the name "Apollo" for the post-Mercury program.

The U.S. Air Force accuses the U.S. Navy of pirating its Dyna-Soar space plane (by bleeding information from the Boeing and Martin companies) to build its own competing "manned manueuverable space system." The Navy's MMSS uses a cluster of Polaris missiles as its booster. The Air Force contends that the Navy's space program is being developed at Air Force expense.

JANUARY 25–27

Theodore Cotter presents a paper on solar sailing at the 28th annual meeting of the IAS. He considers two types of solar sails: a centrifugally stressed spinning disk and a cantilevered disk. He concludes that there are no major technical obstacles to an early trial.

JANUARY 28

NASA presents its 10-year plan for 1960–1970, which includes:

1960: First manned suborbital flight

1961: First orbital flight in the Mercury series

1965–1967: First launchings in a program that will lead to manned circumlunar flights and a permanent near-earth space station

Beyond 1970: First manned lunar landings

FEBRUARY

The Army Ballistic Missile Agency submits to NASA a manned lunar landing study called "A Lunar Exploration Program Based Upon Saturn-Boosted Systems."

The Space Task Group details some of the parameters required for manned space flight and manned spacecraft. These include a spacecraft capable of a circumlunar flight and earth orbit missions for evaluation and training. The earth return vehicle should also be compatible with space stations and Saturn C-1 and C-2 boosters. A spacecraft carrying a crew of several astronauts (at least three) should not weigh more than 15,000 pounds and should have a mission capability of 14 days. It should have abort capabilities up to orbital velocities and the ability to land on either land or water (in up to 10 to 12-foot waves).

The spacecraft should provide a "shirtsleeve" environment for its crew during long missions.

MARCH 23

Bell Aircraft announces its design for a hypersonic passenger transport that it expects to have operational by the 1980–1990 period. It is designed by Dr. Walter Dornberger who proposes a large lifting-body shuttle (derived from the Dyna-Soar) carried aloft on the back of a more or less conventional delta-winged aircraft with six air-breathing engines. These engines will operate in three different cycles of propulsion: normal turbojets to 50,000 feet, a transitional phase, then full ramjet propulsion. This will take the booster to 120,000 feet and a speed of 5,200 mph. At this point the crew of the shuttle will fire the rocket engines, launching their vehicle from the guide rails on the back of the carrier. The rockets will take the shuttle up to an altitude of about 40 miles and a speed of some 15,000 mph. It will make a

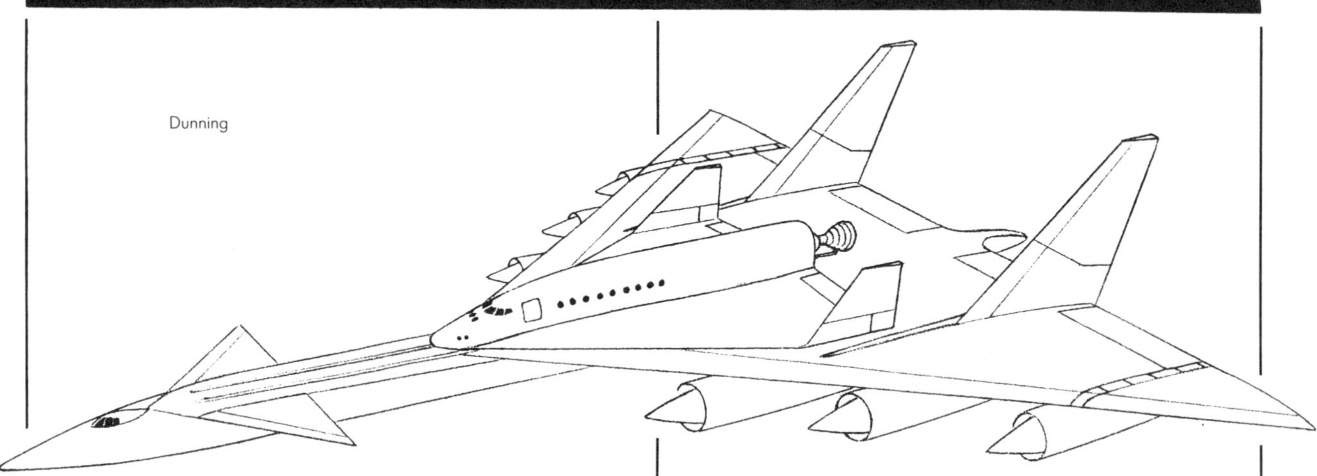

gliding reentry and auxiliary jet engines will allow the pilot to maneuver when approaching his landing, which could take place at a conventional airport.

The aerospacecraft will function as a hypersonic transport capable of reaching virtually any spot on earth in about one hour and could be a reality, according to Leston Faneuff, chairman of the board at Bell (who describes the vehicle in a lecture on the peacetime uses of space delivered at the University of California), by 1985.

APRIL

The Air Force releases funds for the first phase of its Dyna-Soar spaceplane program. After several years of trying to determine whether to go with a ballistic shape or a glider, the glider configuration wins. Under its contract Boeing will build eleven gliders, three for ground tests and four for unmanned flights. The first glider flights will be unpowered air drops from a B-52. The first manned powered flight is expected to take place in the mid-1960s.

North American Aviation revives its proposal for producing an advanced version of the X-15 that could be placed in orbit. The first step in the new program is a Mach 10 version to be air-launched from a B-52. G. R. Cramer and H. A. Barton, engineers from Thiokol, which produces the X-15's engines, urge the use of the X-15 as a safe, reliable orbiter. The XLR-99 rocket motor could be scaled up for an orbital mission by adding droppable fuel tanks, using a more energetic propellant combination (suggested are hydrazine-pentaborane, nitrogen tetroxide-hydrazine, LOX-hydrogen and LOX-hydrazine), and using a large booster rocket. Two years earlier, NAA had proposed an orbital X-15 based on the G38 Navaho booster and the S4 Atlas sustainer.

APRIL 4

In a recruiting advertisement RCA presciently advises its future systems engineers to "take a giant step."

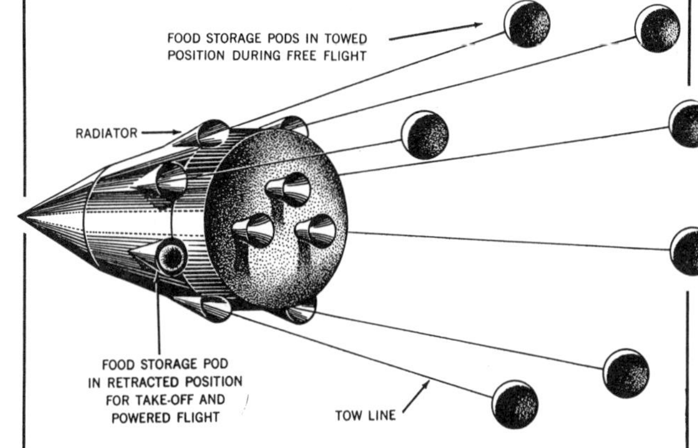

MAY 9

General Electric proposes a rather bizarre plan (FROST) for storing food reserves on spaceships. The result of a $50,000 contract for studying the problem is to store the astronauts' food in a number of spherical pods that are towed behind the spaceship while it is in orbital flight. During takeoff and pow-

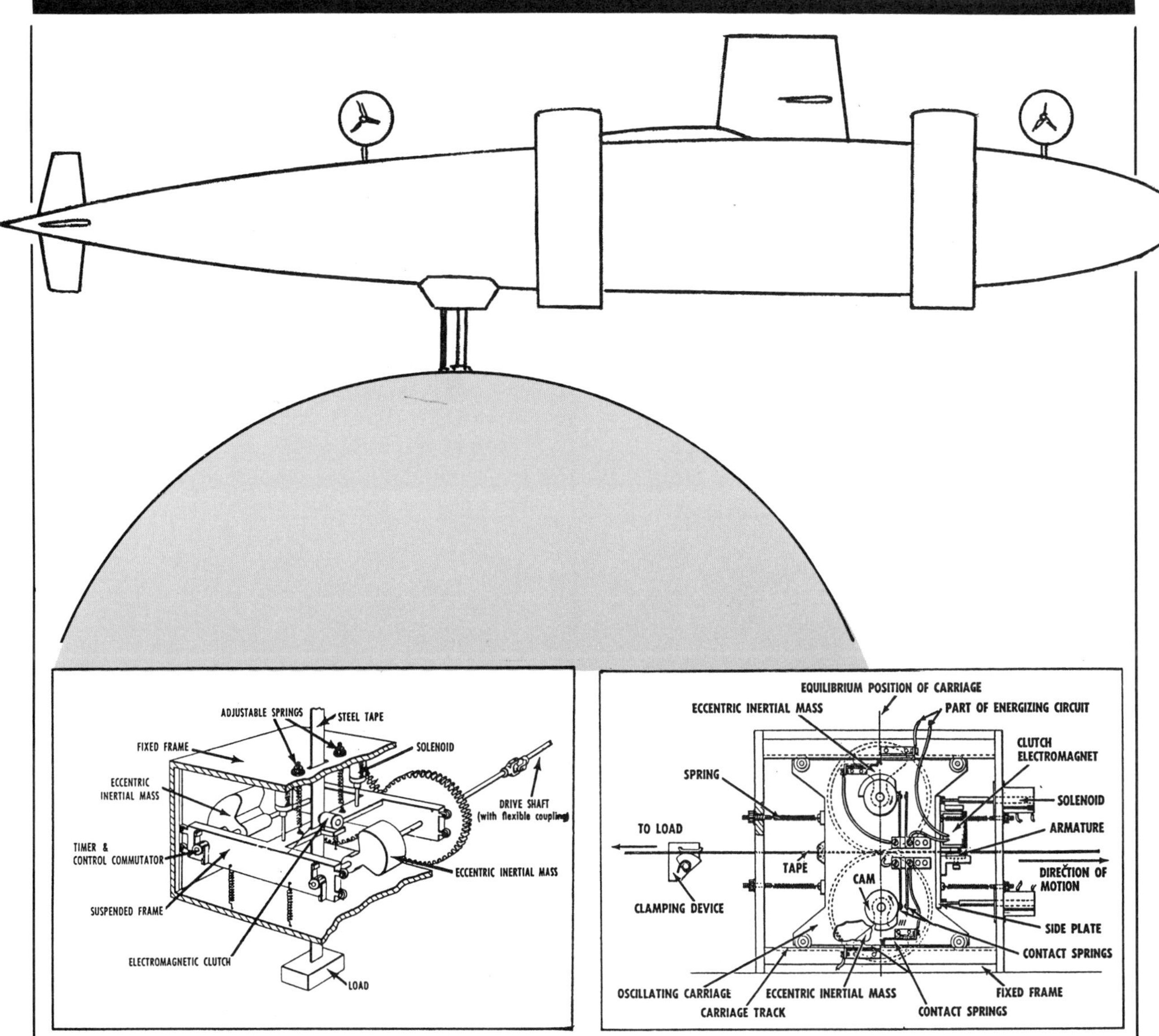

ered maneuvers, the pods are retracted into conical radiator shields.

JUNE

John W. Campbell, in an *Astounding Science Fiction* editorial espousing the wonders of a reactionless space drive invented in 1956 [see below], proposes using it to propel a converted atomic submarine into space.

He chooses a submarine of the Skate class since "The modern nuclear submarine is, in fact, a fully competent space-vehicle . . . lacking only the Dean Drive." With the device, however, the submarine will lift off the earth at a constant 1 g acceleration, which can be maintained nonstop for months, if necessary. There will be, as a consequence, no sense of free fall for the crew. Gravity will appear to be earth-normal. "In flight, the ship will simply lift out of the sea, rise vertically, maintaining a constant 1,000 cm/sec/sec drive. Halfway to Mars, it will loop its course, and decelerate the rest of the way at the same rate."

In adapting the submarine to space duty, "There is one factor that has to be taken into account, however; the exhaust steam from the turbine has to be recondensed and returned to the boiler. In the sea, sea water is used to cool the condenser; in space no cooling water is available." A huge baglike balloon will be attached to the spaceship, silvered on one side, painted black on the other, and of whatever diameter needed to operate properly (unless it is elastic and self-adjusting). This will act as the condenser for the exhaust steam.

"The tough part is the first hundred miles up from Earth; there air resistance will prevent use of the balloon condenser." Campbell suggests that the ship carry along spare water in the form of ice. By the time it melts, the ship will be above the atmosphere.

"Under the acceleration conditions described above, a ship can make the trip from Earth to Mars, when Mars is closest, in less than three days . . . It would have been nice if, in reponse to Sputnik I, the United States had been able to release full photographic evidence of Mars Base I."

The Dean Drive has been invented by Washington, D.C., businessman, Norman L. Dean, as a hobby. He builds several working models, none of which are able to lift themselves (ones that do are claimed to have been destroyed in the process of testing). The Dean Drive, in converting rotary motion into a unidirectional motion, generates a one-way force, without any reaction. That is, in the case of Newton's law that for every action there is an equal but opposite reaction, for the Dean Drive there is an action *without* any reaction, equal or otherwise. Dean's data reveals that, neglecting losses due to friction, a 150-horsepower motor could develop 6,000 pounds of thrust. In the Drive a pair of counter-rotating masses generate a nonreactive force. When John Campbell, editor of *Astounding* magazine, examines the device he admits that "I do not understand Mr. Dean's theory very clearly; my personal impression is that he doesn't understand the thing in a theoretical sense, himself."

Two counter-rotating masses (about 1/2 pound apiece) spin on shafts in a light frame. The complete model weighs about 3 pounds. Normally, with such device, a powerful oscillation is produced. However, Dean has changed the center of rotation of the masses as they spin. This point has no mass, so no energy is required to move it. "In the rotation of those counter-rotating masses, there is a particular phase-angle such that the horizontal vectors are canceled, and the vertical vector is upward, and exactly equal to the weight of the two masses. At that instant, the light framework can be moved upward *without exerting any force on the masses*." In the demonstration model Campbell sees, a small solenoid moves the frame carrying the rotating masses at the proper instant. The result of forcing the masses "to rotate about two different centers of rotation simultaneously" is "rectified centrifugal force."

Dean maintains that during operation of the counter-rotating eccentrics the heart of the system is the intricate phasing relationship which must exist. The rigid connection between paired shafts and the counter-rotation of masses produces a cancellation of forces and reactions engendered in all directions except in the direction of the desired oscillation. This is always parallel to a plane perpendicular to the axes of rotation of the two masses. The result of the cancellation is an oscillation produced by the resultant forces which represent the sum of the components of all forces acting in the direction of a plane at right angles to the shaft axes. Thus, claims Dean, such a freely suspended oscillating system is not subjected to any other reaction or force. The use of six properly phased pairs could produce an almost continuous thrust.

Western Gear Corp. runs tests on the Dean device and concludes that it can't work, although computer simulations contradict this. Alfred Africano recalls that several years earlier he had enjoyed a "ride" on a similar device developed on Long Island by Assen Jordanoff. The 500-pound man-carrying vehicle attained a speed of 1/3 mph. Twenty years prior to Dean, S. J. Byrne devised, on paper at least, an antigravity method based on displaced inertial masses, what he called the "planetary rotor." He does not believe that his invention dupicates Dean's, but rather compliments it and offers to join forces with the Washington inventor.

Ultimately, the Air Force Office of Scientific Research turns the device over to engineer Jacob Rabinow to assess. Rabinow concludes that the system "does not have any unusual properties" and that it "can not produce a unidirectional impulse." He does not even consider it an efficient vibrator or impact machine. The demonstration machine only gives the illusion of generating a force without an equal and opposite reaction by making use of the static friction of the load against the floor, similar to the way in which a person on roller skates can move across a floor by swinging his body to and fro. In the absence of static friction it does not perform as claimed.

Dean, of course, insists that Rabinow's tests were improperly performed (as a number of Dean's supporters agree). "We are going to have to live with the Drive," he writes, "whether we want to or not."

It is interesting to note that E. Mach, in his book *Die Mechanik in ihrer Entwickling* (first German edition: 1883, published in the United States as *The Science of Mechanics* in 1902, with numerous reprints up to the 1960s), described and illustrated a machine very similar to Dean's.

JULY

The House Space Committee demands that the United States raise its sights and set the goal for a manned lunar landing before 1970. The current NASA schedule has a moon landing scheduled for "after 1970"—presumably about 1972.

In a report issued on January 2, 1959, the original Select House Space Committee quotes twenty experts as saying the United States could put a man on the moon by 1968.

JULY 29

NASA publishes its goals for the next decade. These include a circumlunar flight and a manned space laboratory before 1970, and a lunar landing in the early 1970s followed by a manned space station and a planetary landing.

AUGUST 8

In an article published in *Missiles and Rockets* Robert Truax predicts that the planets will be colonized within 50 years. Within that period there will be permanent colonies on the moon, Mercury, Venus, Mars and perhaps the moons of Jupiter and Saturn. "Interplanetary space will be filled with ships making routine, scheduled trips between the planets."

Most of this advancement will be due to decreased costs and increased efficiency—especially the development of reusable boosters and spacecraft.

AUGUST 15–20

Darrell Romick presents his paper "The AERO-METEOR: A Preliminary Design Study of a Horizontal Take-off Version of the METEOR Concept" at the XIth International Astronautical Congress in Stockholm.

AERO-METEOR is basically the otherwise unchanged upper two stages of the METEOR spacecraft [see 1956] placed on the back of a large, winged booster. This is a modified version of the original first stage, with larger, heavier wings, landing gear and body structure. Instead of the nose structure for supporting the upper stages, it is equipped with a longitudinal launching rail. It also outfitted with more powerful air-breathing engines. A final rocket boost phase precedes the separation of the booster from the second stage. The operation of the second and third stages from then on remains practically unchanged from the earlier concept.

The second and third stages are carried on top of the booster in the horizontal takeoff position, supported by a central rail and steadying mounts where the junction of the second stage wing and the trailing edge of the vertical stabilizer mates with the leading edge of the vertical stabilizer of the first stage carrier. The booster resembles somewhat the contemporary B-58 bomber. The first stage requires a three-man crew.

Romick proposes two variations on the basic AERO-METEOR: one using hydrogen as fuel and one using JP-type jet fuels. The gross weight of the entire vehicle ranges from 500,000 pounds (for the all-hydrogen version with maximum use of the rocket propulsion system) to 750,000 or 800,000 pounds for the JP-fueled version.

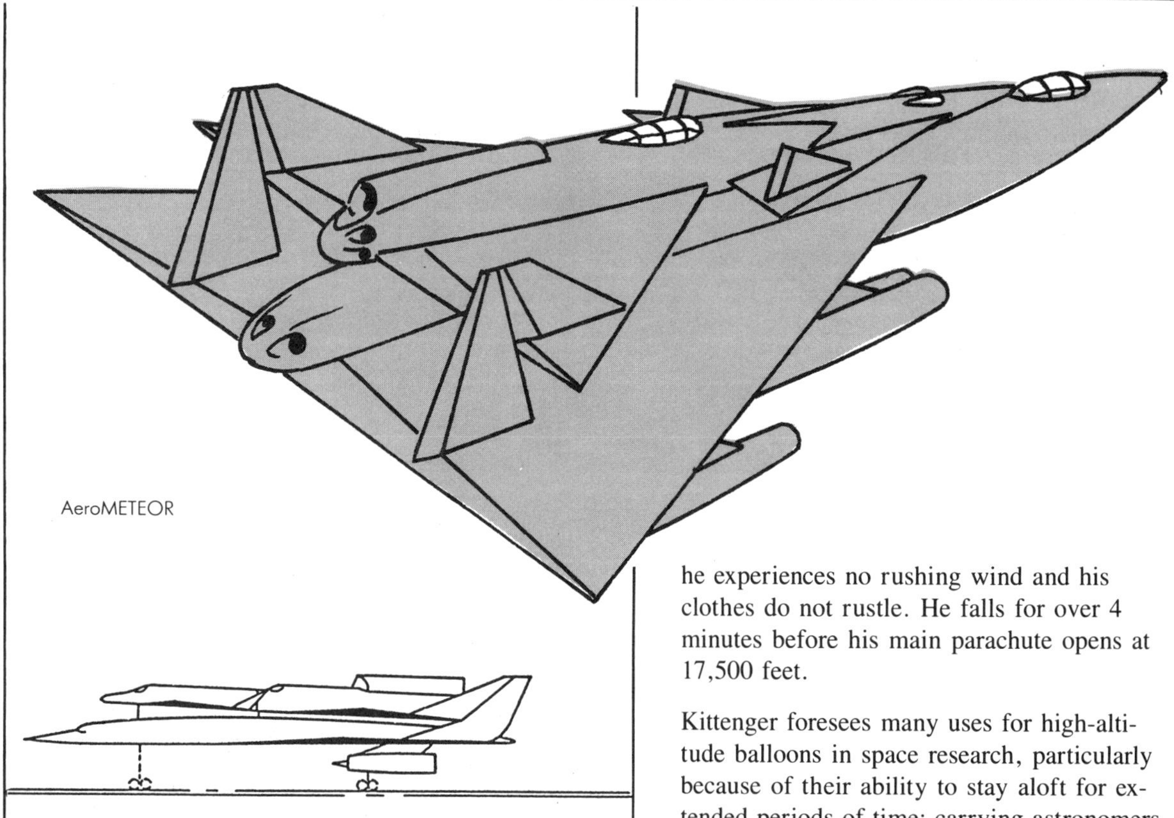

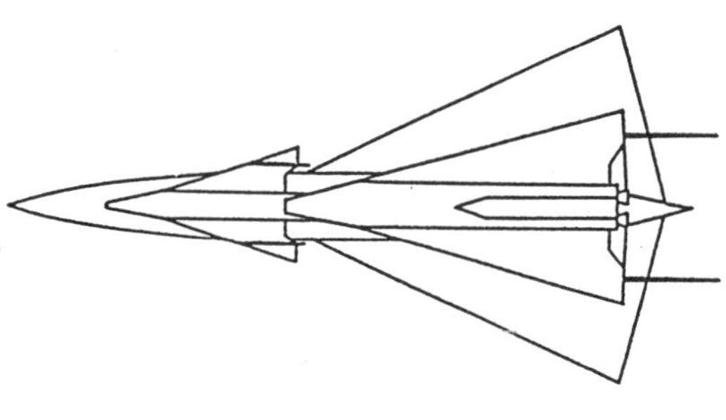

AUGUST 16

U.S. Air Force Captain Joseph Kittenger, Jr. makes a parachute jump from the open gondola of a stratosphere balloon at an altitude of 102,800 feet—some 18.5 miles. The takeoff is from Tularosa, New Mexico. The balloon is 360 feet tall and weighs 1,069 pounds; its payload is 1,250 pounds, of which 158 pounds is Kittenger.

He takes off at 5:29 a.m. and ascends at a rate of 1,200 feet per minute. When he reaches his maximum altitude, Kittenger is, for all physiological purposes, in space and needs to wear a full pressure suit and helmet. When he bails out, the surrounding atmosphere is so near to being a vacuum that he experiences no rushing wind and his clothes do not rustle. He falls for over 4 minutes before his main parachute opens at 17,500 feet.

Kittenger foresees many uses for high-altitude balloons in space research, particularly because of their ability to stay aloft for extended periods of time: carrying astronomers and their telescopes, testing life-support systems, and the training of astronauts. [Also see Project Man High, August 16, 1957.]

OCTOBER 21

The Space Task Group's Flight Systems Division assigns constraints to its design study for the proposed Apollo spacecraft: it should have a Mercury-type configuration (with a lift-to-drag ratio of 0.35), solid propellant systems for onboard propulsion, both parachutes and rotors should be considered for touchdown, and so forth.

OCTOBER 25

NASA selects three contractors to prepare feasiblity studies for the Project Apollo spacecraft: the Convair/Astronautics Division of General Dynamics, General Electric, and the Martin Company.

General Electric concentrates on conical shapes (a 9 degree blunted cone, elliptical cone, and half-cone); Martin studies the M1 and M2 lifting bodies (and the M1-1, a design halfway between the two), Mercury capsules with control flaps, as well as Dyna-Soar shapes; and Convair (subcontracting its

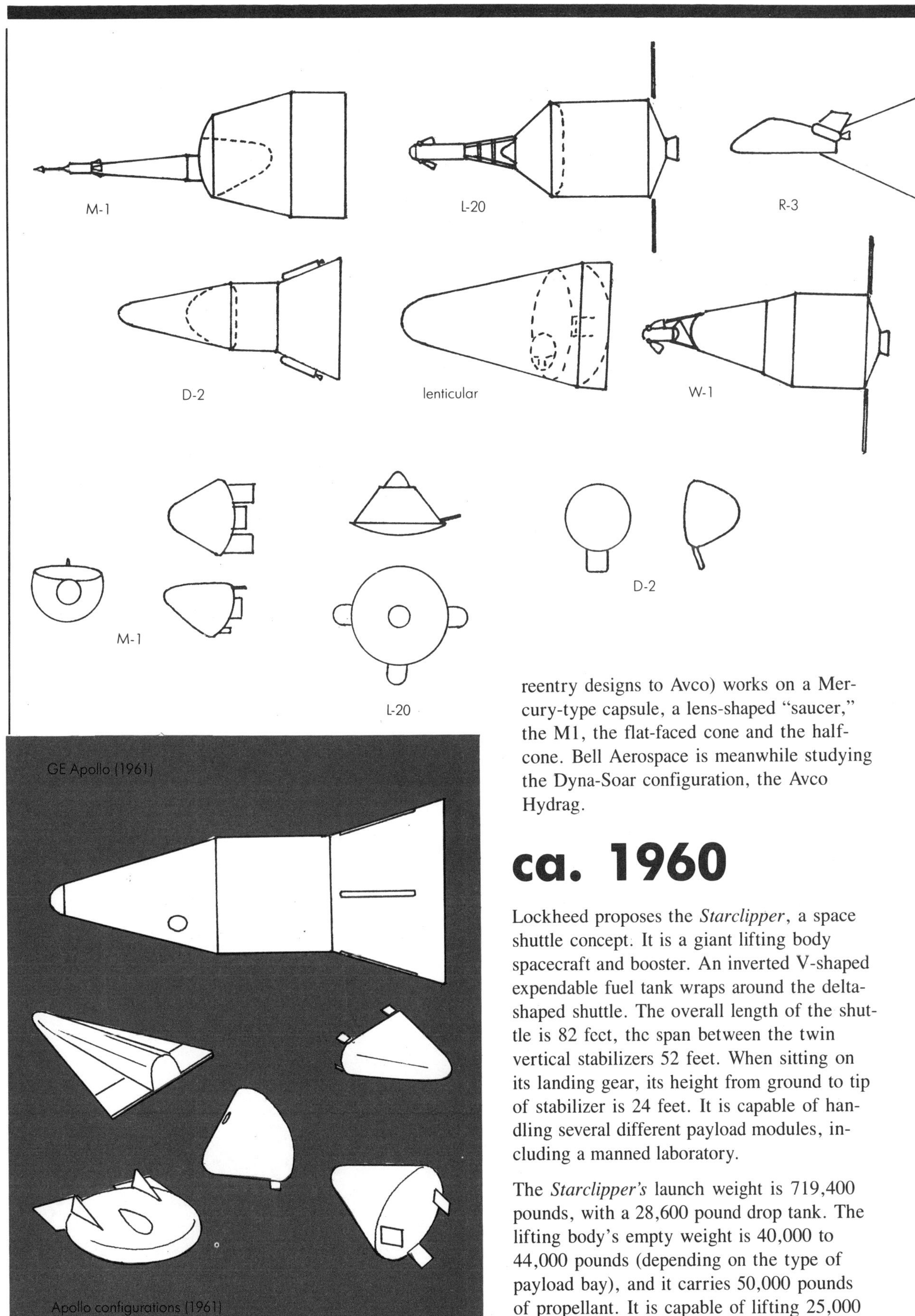

reentry designs to Avco) works on a Mercury-type capsule, a lens-shaped "saucer," the M1, the flat-faced cone and the half-cone. Bell Aerospace is meanwhile studying the Dyna-Soar configuration, the Avco Hydrag.

ca. 1960

Lockheed proposes the *Starclipper*, a space shuttle concept. It is a giant lifting body spacecraft and booster. An inverted V-shaped expendable fuel tank wraps around the delta-shaped shuttle. The overall length of the shuttle is 82 fcct, thc span between the twin vertical stabilizers 52 feet. When sitting on its landing gear, its height from ground to tip of stabilizer is 24 feet. It is capable of handling several different payload modules, including a manned laboratory.

The *Starclipper's* launch weight is 719,400 pounds, with a 28,600 pound drop tank. The lifting body's empty weight is 40,000 to 44,000 pounds (depending on the type of payload bay), and it carries 50,000 pounds of propellant. It is capable of lifting 25,000 to 50,000 pounds of payload into orbit, even-

1960

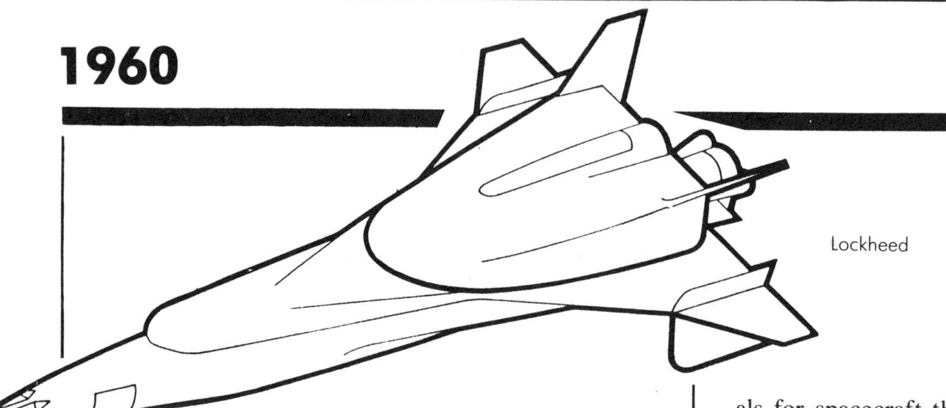

Lockheed

tually at a cost of just $5.00 per pound.

The Martin "Astrorocket" is a two-stage back-to-back shuttle. Both stages are winged and manned. The first stage is powered by underwing turbojets and a large plug-nozzle rocket engine. The total launch weight is about 2.5 million pounds. Stage separation occurs at about 40 miles and at a speed of 8,800 fps. The first stage then returns to the earth, making a conventional landing, aided by its turbojets. The system allows a crew of three to remain in orbit for up to 2 weeks.

The Astrorocket is one of numerous proposals for spacecraft that combine an X-20 Dyna-Soar-like orbiter with a booster resembling one of Ames Research Center's winged reentry vehicles from the 1950s.

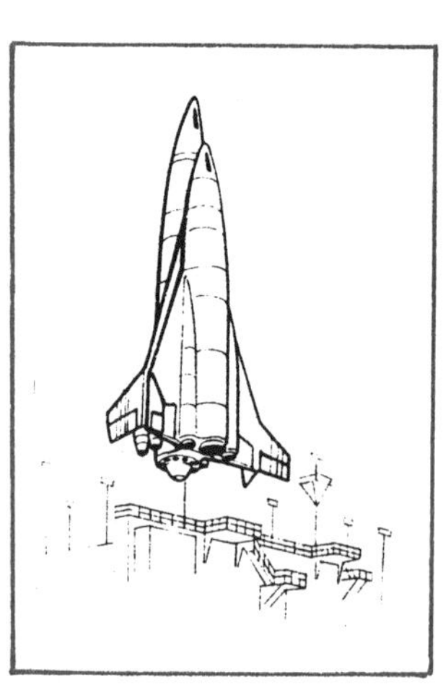

HELIOS is an atomic moonship designed by Krafft Ehricke for Convair. The name is an acronym for Hetero-powered, Earth-Launched Inter-Orbital Spacecraft. The 200-foot spacecraft could make a direct flight to the moon as early as 1970, carrying a payload of 22,000 pounds. The ship is to be launched from earth atop a large, winged recoverable booster. The booster has conventional rocket engines fueled by gasoline and liquid oxygen, developing a combined thrust of 2.7 million pounds. This delta-winged glider (with a wingspan of 90 feet) is piloted and landed like a conventional aircraft.

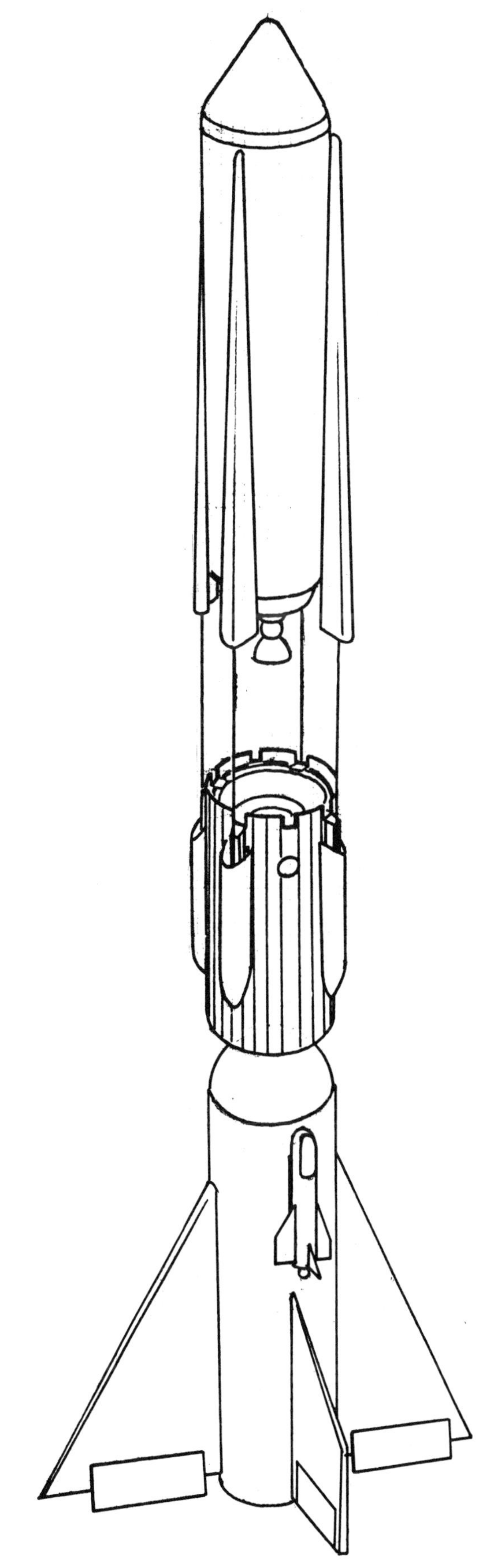

The pilot occupies a small piggyback jet, which acts as an escape capsule. Once in earth orbit, the main ship separates into two components: a cylindrical manned capsule and the atomic engine unit. They are connected by four 1,000-foot cables, the atomic rocket towing the manned capsule 1,000 feet behind. When landing on the moon, the cylindrical capsule is lowered gently by the hovering main rocket, which then lands automatically at a safe distance, on four extendable legs. The takeoff from the moon is simply the landing procedure in reverse. Once the crewmembers return to earth orbit they will transfer to reentry gliders.

Instead of a 22,000-pound payload landing on the moon (or 15,000 pounds with sufficient reserve for the return to earth), HELIOS is capable of putting 30,000 pounds in orbit around Mars.

The Revell model company is approached by Convair about making an educational model kit of the spaceship. The company responds enthusiastically, so much so, in fact, that many of its solutions for design details are eventually incorporated into the official Convair study.

The U.S. Air Force estimates that an engineering study for its proposed spaceplane will cost $20 million and should establish its feasibility before the end of 1961. An operational spaceplane is expected to be flying by 1966–1968, if priorities are high enough. Some critics of the spaceplane express doubts that a human hand could control a rocket-powered plane more powerful than a dozen X-15s, though a RAND study contradicts this fear.

The USAF spaceplane will carry only liquid hydrogen as its fuel, producing its own liquid oxygen as it passes through the atmosphere. It will take off on a conventional runway with turbojet engines, aided by rocket boosters. At the 20-mile altitude limit for its turbojets, a secondary ramjet will take over. This will drive the craft to an altitude of over 50 miles and to a speed of Mach 18. Once in the upper stratosphere, the pilot will open a large scoop that will gather quanti-

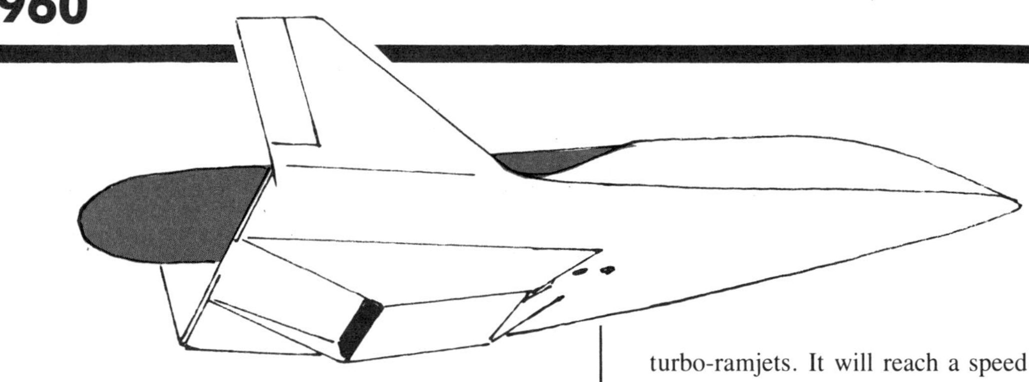

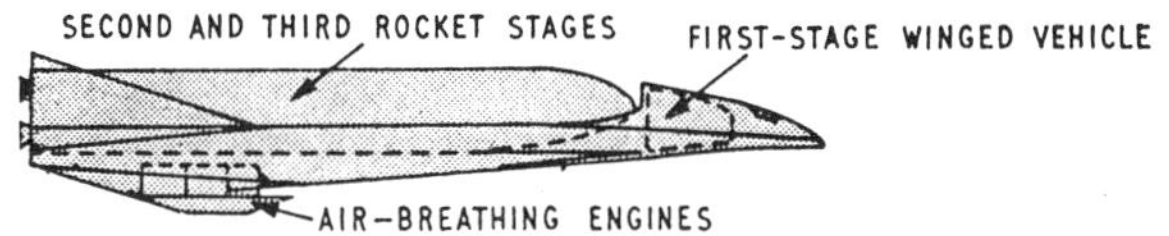

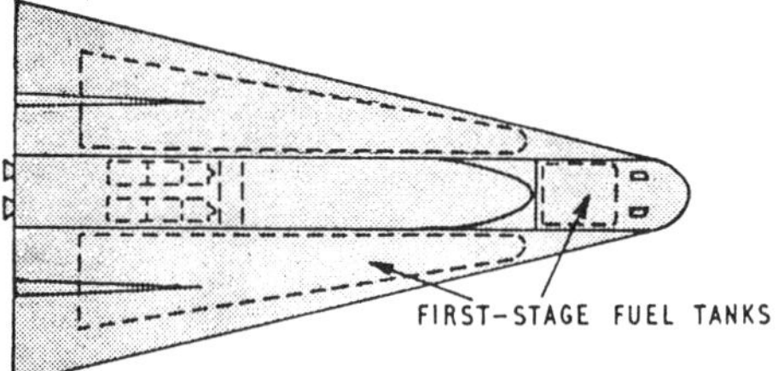

ties of the thin air, cycle it through a cryogenic unit and convert it into liquid oxygen. The great cold of the liquid hydrogen in the fuel tanks, −423 degrees F, will be used to liquefy the incoming air through standard heat-transfer techniques.

The spaceplane will weigh 500,000 pounds at takeoff. The additional liquid oxygen it takes on will increase its weight to 1 million pounds. With this additional propellant, the spaceplane could continue to thrust on into orbit.

STL's General J. H. Doolittle, addressing the Mercury astronauts, predicts that "I envision . . . certainly in the lifetime of you people, a Mach 15 transport . . . that will start in the Earth's atmosphere, go outside of it, and reenter, so that your trip from New York to Paris will take you thirty minutes."

The Royal Aircraft Establishment studies a spaceplane concept in which a large manned aircraft will be powered by a battery of turbo-ramjets. It will reach a speed of Mach 7 at an altitude of 20 miles. A small rocket-powered vehicle will then be launched, increasing its speed to Mach 12 to 14. The RAE regards the vehicle as an advanced type of reconnaisance-bomber. With an added rocket stage the vehicle will be able to achieve orbit.

R. J. Lane develops an aerospace plane for Bristol Siddeley Engines. It consists of a delta-winged, air-breathing booster that takes off like a conventional aircraft. The propulsion system is mounted in a "box" beneath the fuselage. Upon reaching a speed of about 12,000 fps, at an altitude over 110,000 feet, the two-stage piggyback rocket is launched.

"Obviously," admits Lane, "the air-breathing stage is a complex and expensive vehicle . . . " For speeds above Mach 7 he assumes that ramjets with supersonic combustion are used. An air-collection cycle is employed. At takeoff the oxygen tanks of the rocket stages are empty. Air scooped up through an auxiliary intake on the booster is compressed and cooled by a heat-exchanger using the liquid hydrogen fuel. The resulting liquid oxygen is then used to fill the rocket's oxidizer tanks.

Turbojets are used for acceleration from zero to Mach 2. From Mach 2 to Mach 7, ramjets will take over. Beyond this, up to Mach 12, ramjets with supersonic combustion are used.

In a series of project studies, based on launching a payload into a 300-mile orbit, it is found that a combination of turbo-rocket and supersonic ramjet, using hydrogen as fuel, allows a payload of about 12 percent of the takeoff weight. The air-breathing booster can have a launch weight of from 5 to 250 tons, capable of placing from 1,500 pounds to 100,000 pounds into a 300-mile orbit.

An editorial in *Life* magazine is confident that "It is taken for granted that humans will at least circle the moon in the next decade and man a space platform. A vehicle of very thin metal could be constructed for 'solar sailing' in space . . . "

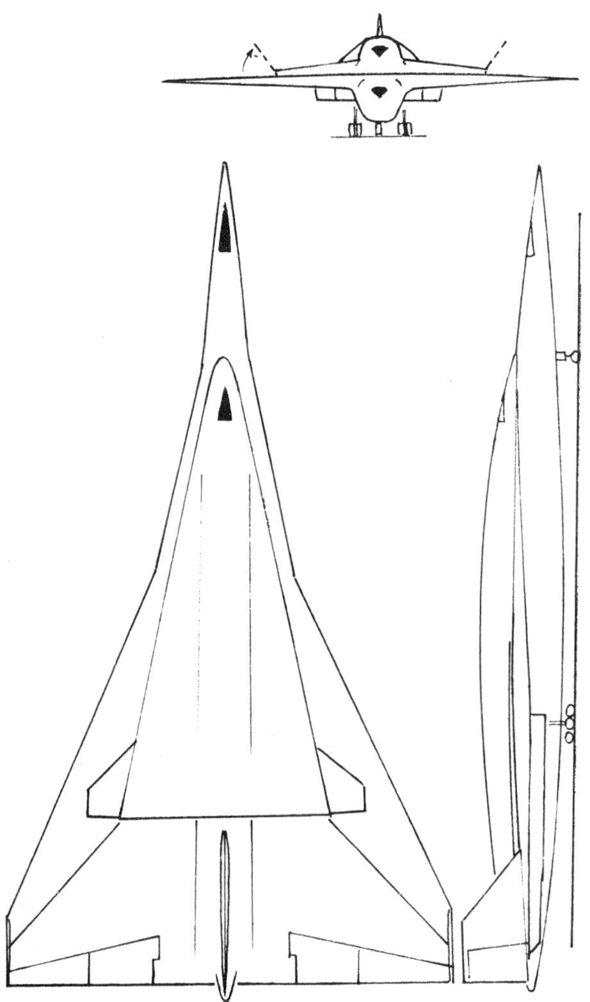

Darrell Romick designs a two-stage, horizontal takeoff and landing (HTOL) spaceplane. Both stages are delta-winged rockets. The booster stage is 282 feet long with a wingspan of 170 feet. It has a fuselage diameter of 15 feet and weighs, with the orbiter, 1,600,000 pounds. The orbiter alone weighs 831,000 pounds and has a cargo bay 8 feet in diameter and 15 feet long. It has a wingspan of 85 feet. Both booster and orbiter are to be fueled by liquid hydrogen and oxygen.

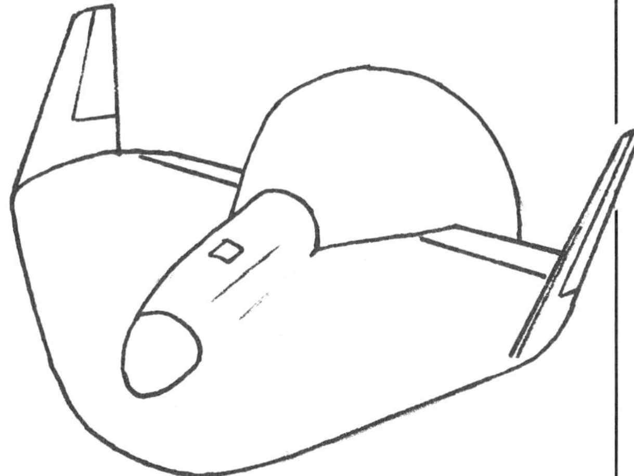

Convair studies an advanced earth-to-orbit transport that consists of a jet aircraft carrying a two-stage rocket beneath its fuselage, which is to be launched at an altitude of 60,000 feet. The rocket has a nuclear-powered second stage that will be activated once the vehicle is outside the earth's atmosphere.

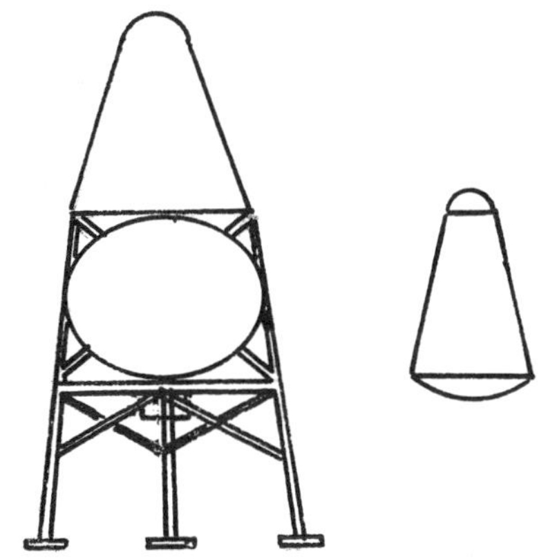

According to the Soviet magazine, *Nauka Teknika* (Technical Science), the Chelovek V Lunicke I, II, and III are proposed Russian two-man lunar orbiters projected for 1963–1965. They will be followed by the Chelovek V Lunicke IV, V, and VI one-man lunar landers, projected for 1967–1969.

Giraffes are being used as spaceflight test animals by the Russians. Its long neck has suggested the animal's use in acceleration tests; the giraffe's heart is naturally very strong because it must send blood all the way up the length of the neck to the brain. "It is as if," explains A. Klenov, "the giraffe . . . [is] all the time flying in a rocket with acceleration."

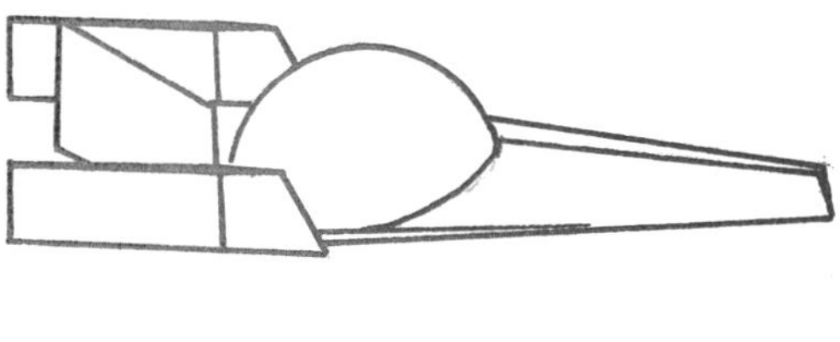

Proposed British reentry vehicles.

1961

Dr. R. A. Ibison of the Science Center of St. Petersburg, Florida creates Project Marsflight, a 24-hour "flight" inside a simulated spaceship. Intended for local high school students, the two "astronauts" making each roundtrip mission work in 4-hour shifts, manipulating 25 controls which in turn operate 214 lights, switches and meters. Other students, acting as "ground control", monitor the activities of the spaceship crew as well as provide them with mock emergencies. The St. Petersburg project is similar to many other simulated spaceship flights made by students around the United States during this period.

John Houbolt, an engineer, develops the lunar module concept. At a meeting early in this year, Dr. Houbolt (inspired by a short paper by William H. Michael written in 1960) proposes a new scheme for reaching the moon called Lunar Orbit Rendezvous (LOR). Until now, most space scientists, including Wernher von Braun, have advocated the Earth Orbit Rendezvous technique or the Direct Ascent method, which requires the development of the new Nova booster. The former requires two Saturn rockets being launched into earth orbit, one carrying fuel, the other the spacecraft. The two then rendezvous and with the extra fuel the spacecraft continues on to the moon. The alternate method advocated at the time is the Direct Ascent. This is manifested in the Nova moon project. Either is fabulously expensive.

In Dr. Houbolt's system, the Apollo spacecraft is launched into lunar orbit. A small descent module detaches itself from the main spacecraft, which remains in orbit, and makes the round trip to the surface of the moon. Not only will the Lunar Orbit Rendezvous save money but it will greatly simplify development, testing, launch and flight operations. Surprisingly, this concept meets with opposition. Even von Braun initially thinks that the idea is "no good." It takes Houbolt nearly a year to get his scenario accepted. During this time, Houbolt is reminded of Yuri Kondratyuk, the Russian scientist who conceived the LOR 50 years earlier —he too had been ignored by his government, dying in obscurity in 1952.

The details of the LOR evolve through many forms and refinements. For example, there are two possibilities for the mission: a large rocket could launch the Apollo spacecraft directly to lunar orbit, or a smaller rocket could be launched that would rendezvous with a refuelling tanker.

The lunar landing module also goes through many design changes. One of Houbolt's first concepts has the "Lunar Schooner" (as he calls it) be a bare bones flying machine. It calls for an open cockpit and room for only two astronauts. The entire vehicle weighs only 3,500 pounds. An early model put together by Grumman Aircraft Engineering Corp. is made of balsa wood and looks a great deal like a yo-yo sitting on a hockey puck. It has five landing legs made of bent paper clips. This evolves to another five-legged version, this time with a spherical crew compartment, capsule-shaped fuel tanks and two docking hatches (at one point it is expected to dock into the side of the command module).

Houbolt's first scenario envisions a five-part spacecraft. A booster, resembling the later Apollo Service Module (see 1969—Houbolt's weighs 52,300 pounds while the Apollo Service Module weighs 54,074 pounds), launches the spacecraft toward the moon from earth orbit. This is jettisoned on the way. A separate pair of fuel tanks and engine is used to place the spacecraft into lunar orbit. Houbolt allows for a pair of his little lunar landers, weighing together 10,500 pounds. After returning to lunar orbit from the surface, the astronauts then transfer to the orbiting reentry vehicle. A small engine unit then propels this from lunar orbit back toward the earth where the unit is jettisoned before reentry.

Robert C. Truax designs the "Skyliner" passenger rocket for Aerojet-General. Unlike most other spaceship schemes Truax's has no fixed launch site. Instead, the rocket is assembled in a drydock like a ship. It is then towed to the launch site by seagoing tugs. A cylindrical launcher attached to the rear of the rocket is flooded, raising the ship into a vertical position. It requires no other launch structures. [Also see Oberth—1928.] Truax expects that rockets in excess of 100 million pounds weight could easily be handled.

His "Skyliner" is a small passenger rocket launched by a cluster of parallel boosters having a total of six engines producing 6 million pounds of thrust. The three largest parallel tanks return to earth via parachute and retrorocket for recovery at sea. The passenger rocket continues on into orbit, attached to two smaller external tanks which are jettisoned when empty.

R. P. Haviland of the General Electric Company outlines a proposal for a zero g manned satellite rocket. It is similar in concept, function and design to the U.S. Air Force MOL (Manned Orbital Laboratory). The final stage of the three-stage booster goes into orbit with the manned capsule. The empty propellant tanks can then be used for living space (a concept the author credits to Krafft Ehricke). The walls of the third stage are thick to act as meteor shield. If the tank being used is the empty oxygen tank, then any leftover oxygen can be used by the astronauts. If the tank carried hydrogen, then excess hydrogen can be used to produce water. If these are indeed the propellants used, then the fuel tank is four times the size of the oxygen tank. The former becomes living space while the oxygen tank

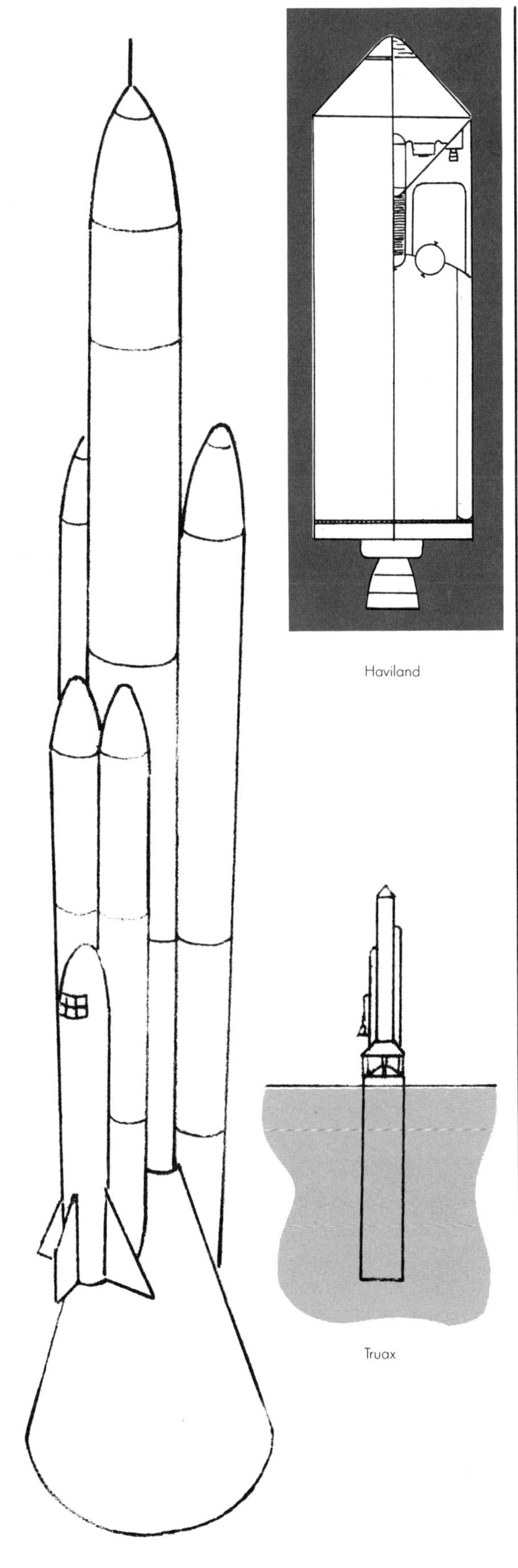

Haviland

Truax

can be used to store the gas needed for breathing and producing water. The life support oxygen is carried in narrow, cylindrical tanks surrounding the main hydrogen tank. Nitrogen for pressurizing the cabin can be carried in its own container. The hydrogen tank is about 8 feet in diameter and approximately 20 feet long. The overall diameter of the third stage, allowing for the oxygen tanks, is about 10 feet. Space between the top of the living tank and the reentry capsule is taken up by the transfer tunnel, equipment tanks, nitrogen storage and so forth.

Haviland considers a number of shapes for the reentry capsule, finally settling for a squat cone with a convex base, very similar to the Apollo command module [see 1969]. For reentry, a jettisonable blunt cone-shaped heat shield is attached to the base of the manned capsule. Surrounding the joint of the two are a series of retrorockets. The capsule and laboratory are manned by three astronauts.

Republic Aviation designs a manned boost-glide lifting body spacecraft, piloted by two astronauts.

As the X-15 program nears its original goals, the U.S. Air Force proposes a new area of use for the aerospacecraft: as the booster stage for the Blue Scout solid-fuel research rocket, which could be used to launch small satellites.

The X-15 is wing-launched from a B-52 as usual. However, the X-15 has the Blue Scout rocket mounted beneath its fuselage. The X-15 is released at 50,000 feet. Under its own power it reaches 156,000 feet (about 30 miles) where it then launches the smaller rocket, which could then place 150 pounds into a 300-mile orbit.

Douglas Missile and Space Systems proposes several manned nuclear spacecraft. Primary among them is a multiple-stage rocket which may have been designed by Ellwyn A. Angle [see 1958]. Its booster is a cluster of six parallel tanks which contain propellant

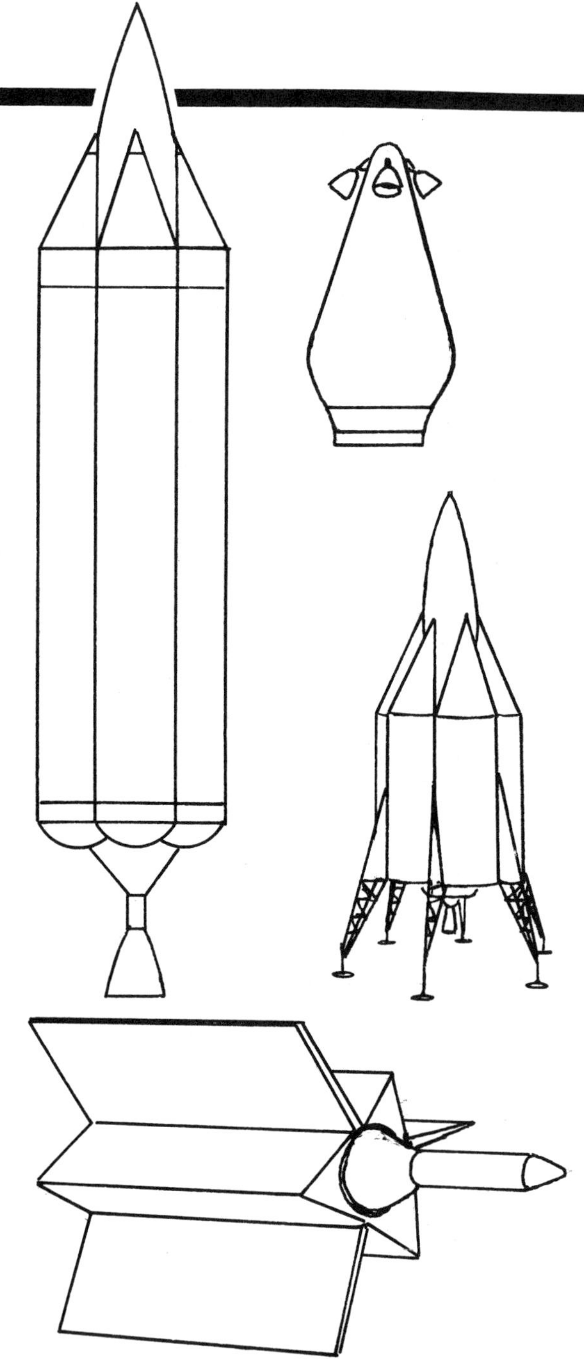

for the single nuclear engine. The manned stage is capable of landing on the moon along with the upper conical tips of the parallel tanks [also see 1958]. A larger version is capable of landing on the moons of Jupiter. Another nuclear ship is a bulbous cone with four nuclear engines clustered at its tip. The crew occupies a toroidal compartment at the base of the cone. Intended for a lunar landing, the spaceship sets down on the lower surface of the manned torus. A third configuration has a star-shaped cross section of radiators and propellant tanks and is intended for year-long missions to Mars or Venus.

Marquardt

The Titan II is proposed as the launch vehicle for a scaled-up Mercury capsule.

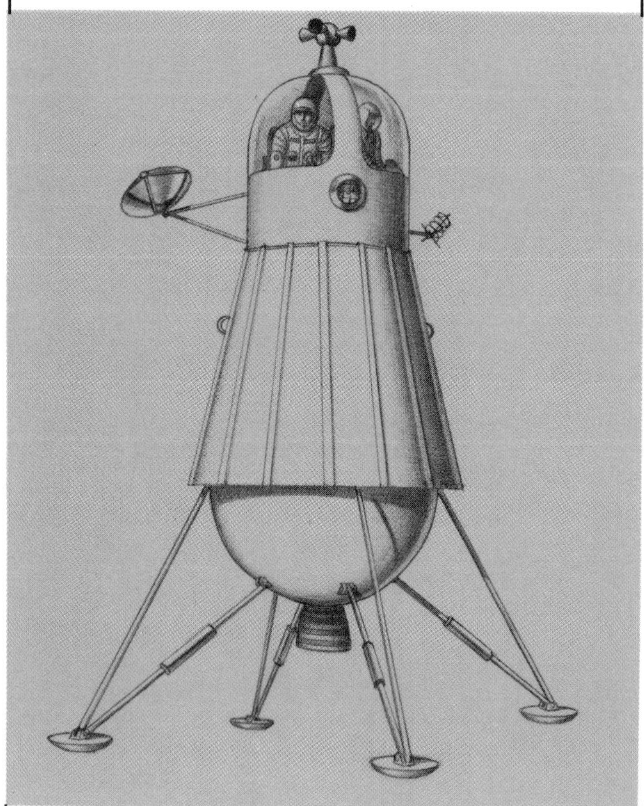

Dunning

Langley Research Center develops the MALLIR Project (**M**anned **L**unar **L**anding **I**nvolving **R**endezvous). It requires a Saturn C-1 as its booster. Using a combination of earth orbit and lunar orbit rendezvous substantially reduces the number of launches required. The initial launch places into orbit the 11,000 pound command module, the 11,000 pound lunar lander and the propulsion unit needed for the lunar landing. The next launch carries a booster for leaving earth orbit. This is jettisoned after injecting the spacecraft into its lunar trajectory. Once achieving a lunar orbit the lander separates from the command module (leaving one of the three astronauts behind) and descends to the surface. After leaving the moon the lander rendezvous with the command module which then boosts into a return trajectory, leaving behind the lander, eventually reentering the earth's atmosphere after jettisoning the propulsion unit.

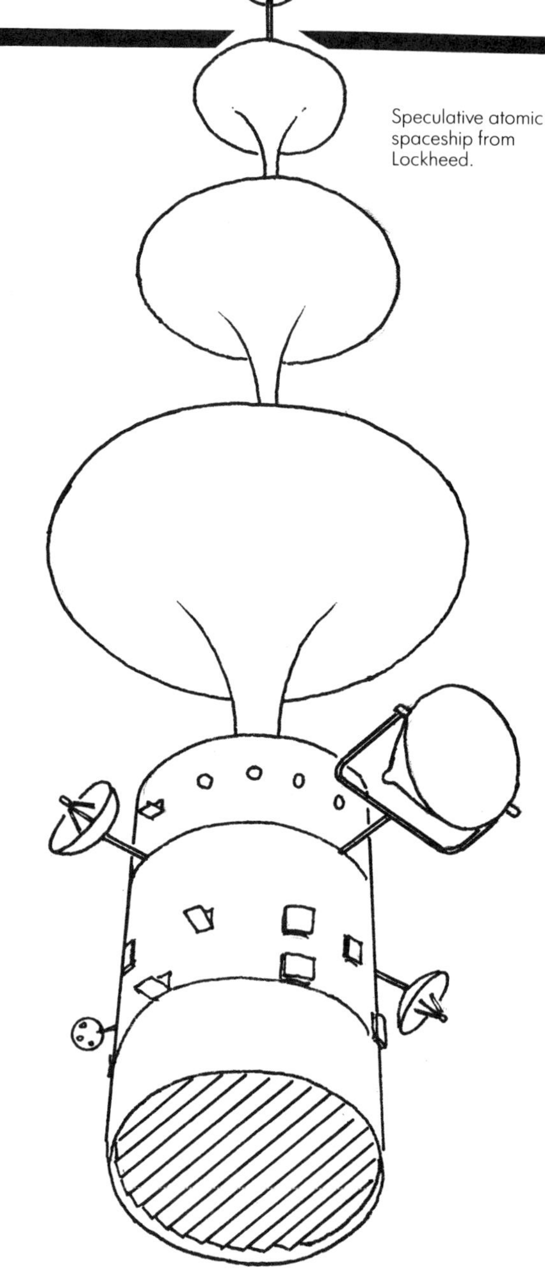

Speculative atomic spaceship from Lockheed.

Constantin Paul van Lent begins the serialization of his novel *Expedition to the Moon* (in his privately published magazine *Rocket-Jet Flying*); it was orginally written in 1941.

In his inimitable style, Lent describes the flight of the spaceship *Cosmos*. It is one of his most elaborately, carefully and consistently worked-out spacecraft designs. It is a five-stage affair fifty stories tall, each stage resembling a fat artillery shell. The first three stages are empty of fuel for reasons that are explained below and carry only oxydizers. The top two stages are only partially fueled. Fuel for the return trip to earth is sent to the moon beforehand in umanned landers. This is in the form of 50 by 2 foot

cylindrical solid fuel strap-on boosters that add 6 million pounds of thrust to the fourth stage.

The *Cosmos* is given its initial lift from the surface of the earth by being carried aloft on the back of a huge jet aircraft. This latter is for all essential purposes an enormous flying wing with a twin boom tail. The spacecraft is carried between the booms. As the carrier jet circles the earth twice at the equator, at 700 mph, twenty mid-air refueling operations are carried out, transferring rocket fuel from aerial tankers to the spaceship. At the end of the fuel transfer, the carrier increases its speed to 800 mph and launches the *Cosmos*. The first three stages are used to boost the rocket toward the moon.

The fourth stage of the rocket is filled with cylindrical fuel tanks. It has six engines, each 5 feet in diameter and producing 1/2 million pounds of thrust. Also attached to the rear of the fourth stage, between the engines, are the three caterpillar-tracked individual combination spacesuit/lunar rovers ("mobile space suits"). At the center of the base of the rocket is the huge landing pick. The landed rocket is supported by three extendable legs.

The fifth stage is the crew cabin and earth return vehicle. It is 20 feet in diameter and 60 feet long. A vertical tunnel and a spiral ladder 50 feet long connect the cabin with the airlock at the base of the fourth stage. Attached to the sides of the fifth stage are a pair of jet engines, for use during the return to earth. (In 1956 Lent described a version of the fifth stage reduced in size for use by a single astronaut. Both this stage and the booster would use a new type of rocket motor devised by Lent, called the "barrel motor.")

The spaceship lands on the moon using three of its six motors to brake its fall, settling on its landing pick, and using the three hydraulic legs to steady it in an upright position. (In later chapters, Lent adds an enormous inflatable "cushion" at the base of the lander, filled with mercury vapor, that acts as a kind of float to prevent the rocket from sinking into the deep lunar dust he expects to find on the moon.

In 1964 Lent will elaborated upon the landing pick device, first proposed by him in 1941. It is ideal, he thinks, for landing on the moon since it provides an immediate, single-point support regardless of the type of terrain. Upon landing, the sharp point imbeds itself deeply into the soil. It will be a distinguishing feature of virtually every rocket and spacecraft he designs over a period of more than a quarter century. He believes this design is vindicated since several research rockets have landed by a combination of parachute and spike nose. More recently, the landing pick has been suggested for proposed comet probes.

JANUARY 9

The first meeting of the Manned Lunar Landing Task Group determines that the immediate objectives of the NASA program should include: the exploration of the solar system, the development of astronautical technology, a manned lunar landing and return to the earth, limited lunar exploration, and a scientific lunar base. In order to accomplish the landing, it is decided that a payload of 60,000 to 80,000 pounds is needed.

FEBRUARY 7

The Low Committee of the Manned Lunar Landing Task Group prepares its final conclusion: that a manned lunar landing can be made by means of either the direct ascent method or the earth orbit rendezvous method. The former requires the giant Nova booster, and the latter, multiple Saturn C-2 launches. Either way, the program will cost about $7 billion through 1968.

JANUARY 31

Ham, a chimpanzee, makes a suborbital flight aboard a Mercury-Redstone 2. The 1-ton spacecraft reaches an altitude of 155 miles.

APRIL

The Redstone booster is approved for the Mercury program after a successful March 24 flight. The rocket carries an unmanned capsule 311 miles downrange into the Atlantic Ocean.

APRIL 12

Yuri Gagarin makes the first manned spaceflight in Vostok 1. He makes a single orbit of the earth in 89.1 minutes (though he may have fallen slightly short of a full orbit due to a technicality). He is launched from Tyuratam Cosmodrome and the entire flight, from launch to landing, takes 118 minutes.

The 4,725-kg Vostok spacecraft is composed of two sections: a spherical reentry module and an 8 foot 5 inch diameter service module shaped like two truncated cones placed base to base. This contains retrorockets as well as several communications antennas. The 7 foot 6 inch reentry sphere is covered with an ablative material. Inside are three small viewing ports, TV and film cameras, radio, life support and controls. Gagarin rides in a seat that is ejected on rails out of a circular hatch when the capsule descends to an altitude of 23,000 feet (Gagarin separates from the seat at 12,000 feet and descends by parachute to his pickup point in the Saratov region). Between the two modules is a ring of spherical oxygen and nitrogen tanks (the cabin is pressurized using an earth-normal atmosphere). The service module has batteries capable of 10 days operation.

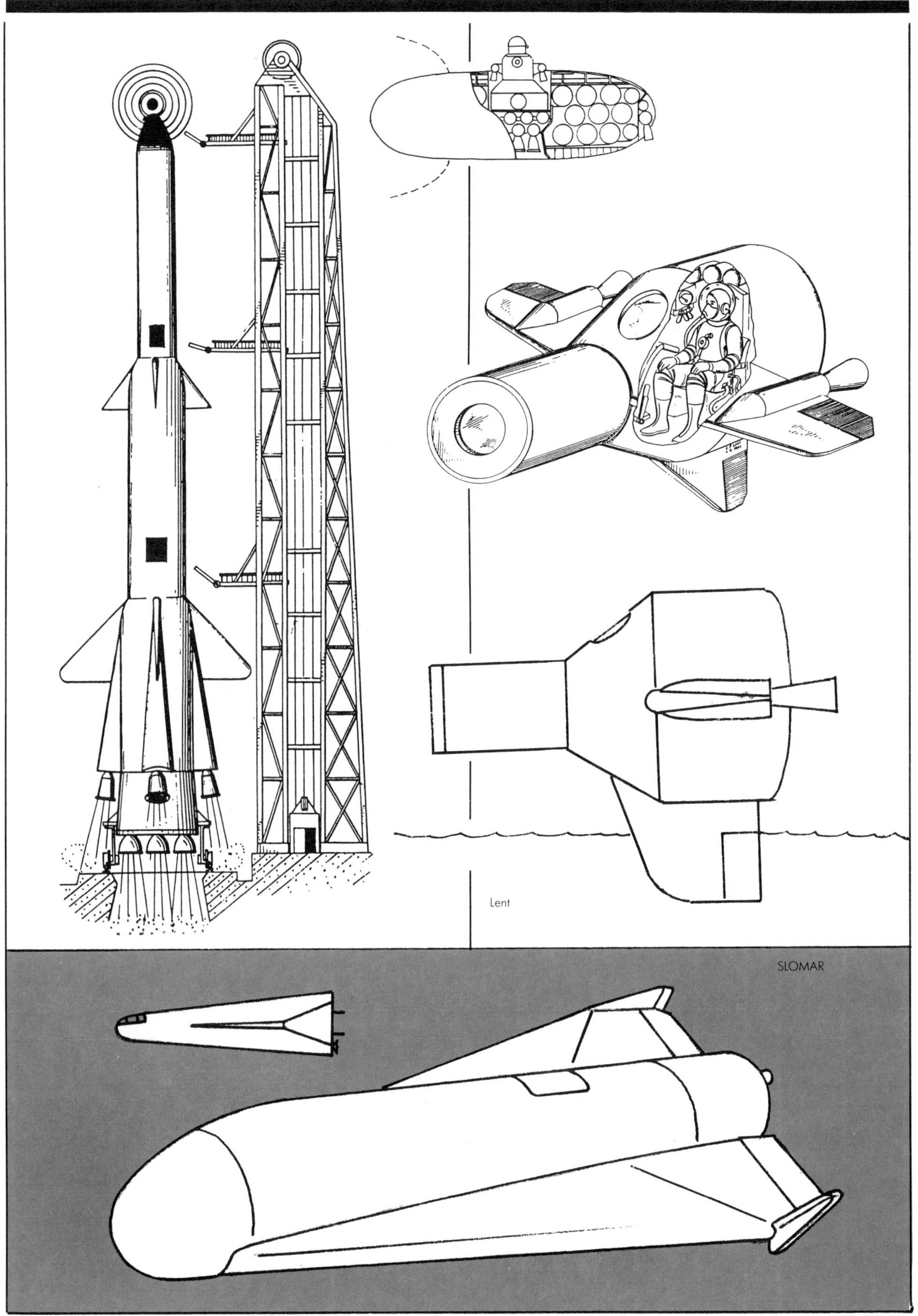
Lent
SLOMAR

The launch vehicle is the A-1 Vostok rocket. Six Vostok missions will be flown, from Gagarin's to Tereskhova's on 16 June 1963 [see below].

At the time of Gagarin's flight, no one in the West knows what his spacecraft looks like and speculation runs wild. None of the art or photographs released by the Soviets realistically depicts the Vostok's actual appearance. *Missiles and Rockets* magazine has to base its speculations on designs appearing on Soviet commemorative postage stamps, leading it to announce that the "Soviet Manned Spacecraft is Winged." "Evidence mounted this week that the manned Soviet Cosmic Ship *Vostok* is an early prototype of a winged military spacecraft . . . which used glide techniques to reenter the atmosphere and land in a preselected field . . . The [drawings on the stamps] depict the *Vostok* as a glider similar in concept to the Air Force Dyna-Soar."

APRIL 20

The Bell rocket pack makes its first free flight, "piloted" by Harold Graham [also see Andropov, 1921].

MAY

President John F. Kennedy recommends the development of nuclear powered rockets. This eventually leads to the development of the NERVA engine [see below].

The Air Force's Space Systems Division undertakes a secret study of the requirements for a manned lunar landing, using a three-man M2-type lifting body for the eventual return to the earth. The project is named LUNEX.

MAY 5

Alan Shepard, in the Mercury 3 (*Freedom 7*) spacecraft, is launched from Complex 5 at Cape Canaveral on the first U.S. manned spaceflight. His suborbital mission lasts only 15 minutes 22 seconds and takes him 116.5 miles above the earth and 302 miles down-

Russian secrecy forces the West to speculate on the appearance of the Vostok rocket and Gagarin's spacecraft.

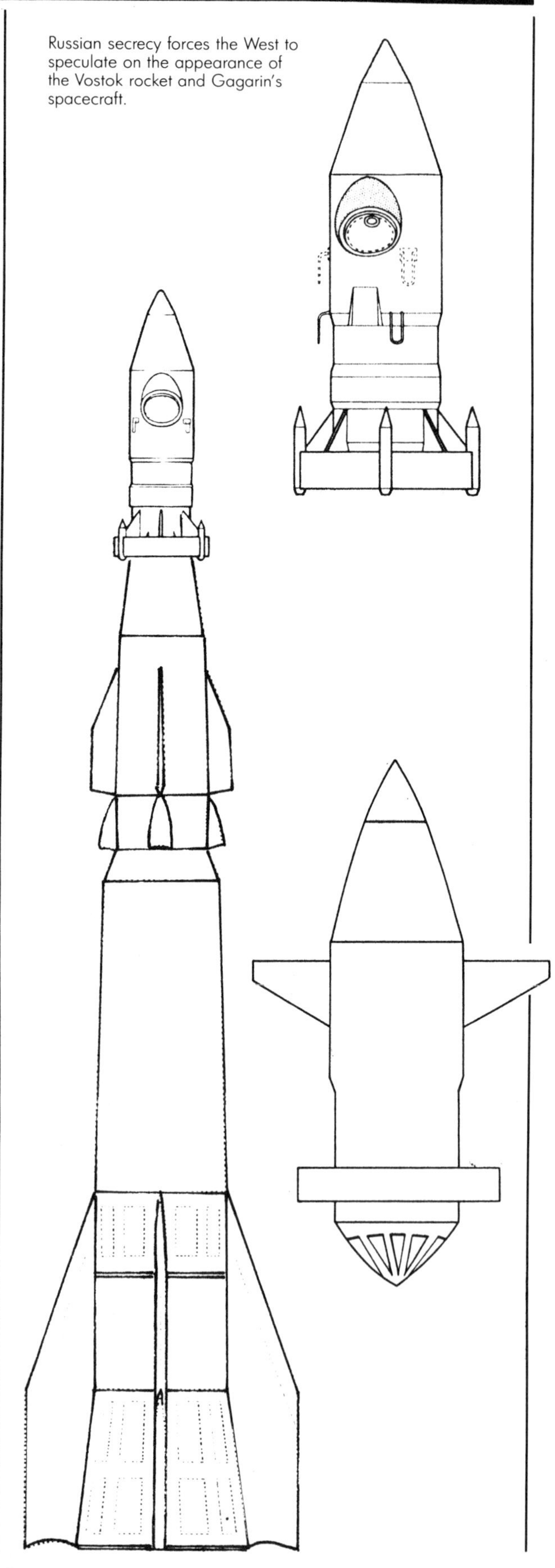

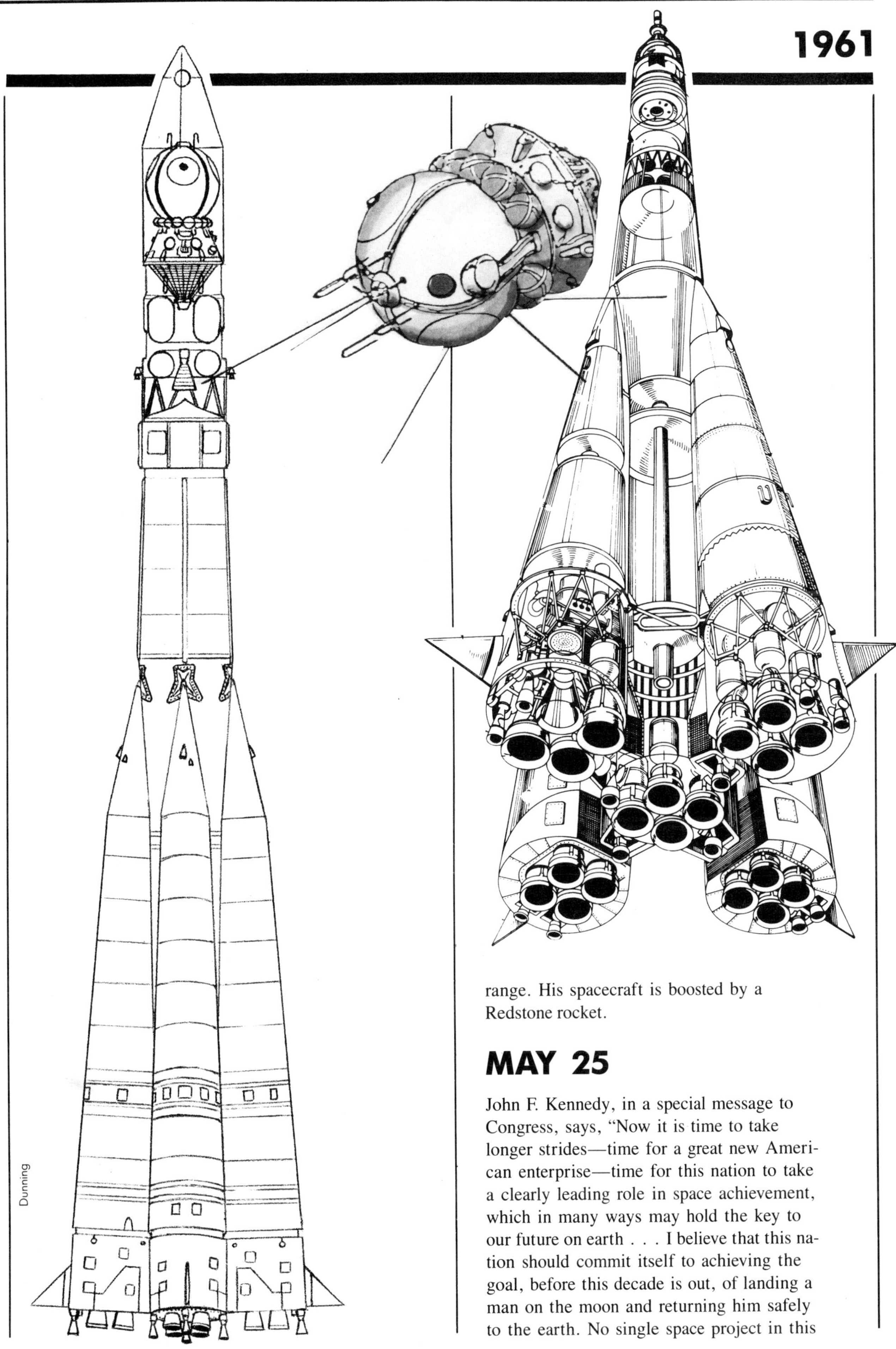

range. His spacecraft is boosted by a Redstone rocket.

MAY 25

John F. Kennedy, in a special message to Congress, says, "Now it is time to take longer strides—time for a great new American enterprise—time for this nation to take a clearly leading role in space achievement, which in many ways may hold the key to our future on earth . . . I believe that this nation should commit itself to achieving the goal, before this decade is out, of landing a man on the moon and returning him safely to the earth. No single space project in this

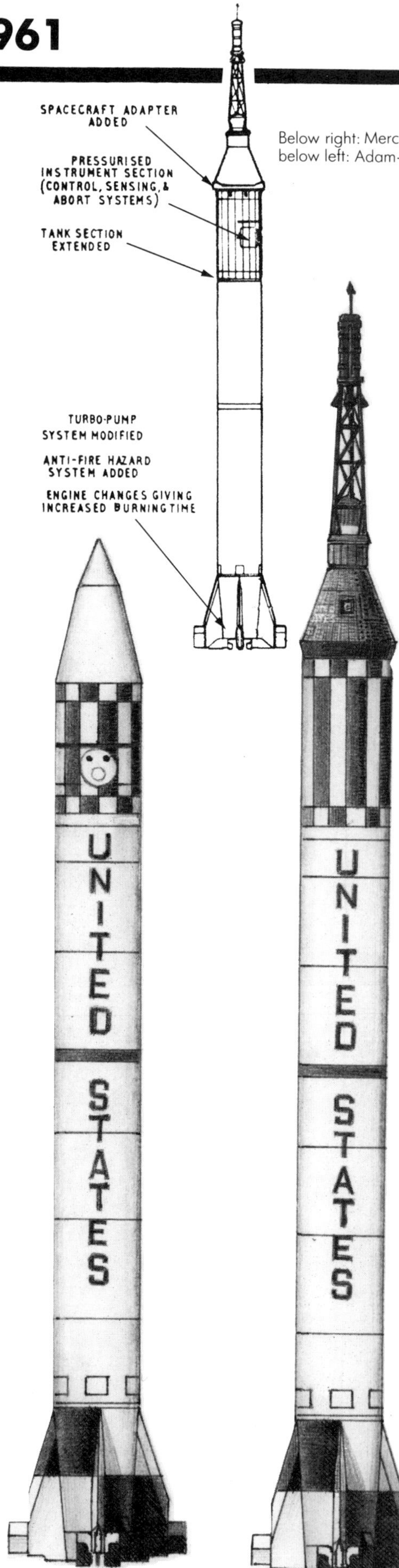

SPACECRAFT ADAPTER ADDED

PRESSURISED INSTRUMENT SECTION (CONTROL, SENSING, & ABORT SYSTEMS)

TANK SECTION EXTENDED

TURBO-PUMP SYSTEM MODIFIED

ANTI-FIRE HAZARD SYSTEM ADDED

ENGINE CHANGES GIVING INCREASED BURNING TIME

UNITED STATES

UNITED STATES

Below right: Mercury-Redstone; below left: Adam-Redstone.

Dunning

period is more impressive to mankind, or more important for the long-range exploration of space; and none is so difficult or expensive to accomplish . . . in a very real sense, it will not be one man going to the moon—if we make this judgement affirmatively, it is an entire nation, For all of us must work to put him there."

JULY 1

Eugen Sänger begins working for Junkers, and eventually ERNO and Dornier, on a space transport system, a study completed in 1964. The RT-8-01 *Raumstransporter* is initially a one-man spacecraft intended for antipodal flights or transport missions to a 186-mile earth orbit. Its launch weight is 187 tons and it can place a 2.75 ton payload in orbit. It is carried aloft piggyback fashion; both craft are fueled with liquid hydrogen and liquid oxygen. The booster stage is manned and recoverable. Both are given their initial impetus by a horizontal catapult using a captive steam-rocket booster (which does not leave the 3.2 km track). This gives the paired rockets a takeoff speed of 900 km/h (560 mph). They have a combined thrust of over 440,000 lbs and a maximum acceleration of 3 g. First-stage cutoff occurs after 150 seconds at about 37 miles altitude. The second stage continues into orbit at 186 miles. After reentry, the orbiter makes a glider-type landing on skids; conventional jet engines allow the pilot maneuverability. Launch weight of the final design is 150 to 200 tons, with a payload of 2 to 3 tons. It is expected that the project can be realized within 15 years. This proposal leads directly

H. Sänger

to the German "Study Project 623" [also see 1963].

In 1964 Sänger writes: "Considering that it will take at least 10 years to realize such an ambitious project as the Aerospace Transporter, work on it must be started without delay. If development in this field has not yet gathered momentum in the United States, nor possibly in the Soviet Union, this is due to the fact that the entire intellectual and material resources available to space research in the countries are at present being concentrated on pioneer projects, in particular, the race to the Moon.

"As soon as this effort is relaxed, these countries will direct their full capacity towards the Aerospace Transporter phase of spaceflight, as can clearly be seen from the preliminary work of the American aerospace industry in this area.

"There is therefore at the moment a unique, but only a short-lived opportunity for Europe, with its great intellectual and material resources, to become active in a sector of spaceflight in which the major space powers have not yet achieved an insuperable lead . . . "

JULY 21

The second and last suborbital Mercury flight is made by Virgil I. Grissom in Liberty Bell 7. His capsule is launched by the Army Redstone rocket, which has 78,000 pounds of thrust.

AUGUST 11–15

The crews of Vostok 3 and Vostok 4 make the first dual spaceflight and the first television broadcast from space.

SEPTEMBER 21

Eurospace is founded by eighty-six European business firms for the common industrial development of space.

OCTOBER 27

The first test flight is made of the Saturn SA-1 (booster only).

ca. 1961

Nova is developed. It is a giant booster which will never get beyond NASA's planning stage. It is patterned after the Saturn C-3 and employs the powerful F-1 engines on at least three of its four or five stages. The first stage is powered by six to eight engines, each of which is rated at 1,500,000 pounds thrust, for a total thrust of from 9,000,000 to 12,000,000 pounds. The second stage has two F-1 engines, and the third stage four J-2 engines. The engines used on the upper stages depend upon the mission. The mammoth rocket stands at least 350 feet tall, with a first-stage diameter of 44 feet. It is capable of putting a 300,000-pound payload into earth orbit, a 100,000-pound probe into lunar orbit, or launching a manned lunar return vehicle weighing 50,000 pounds [see below]. A chemical-nuclear Super Nova for manned lunar missions in the 1970s is also proposed.

The Nova moon mission, a direct-flight concept, is developed because it is timesaving. For a period it is the preferred approach to the lunar landing program. The first stage of the Nova-class booster burns for 145 seconds, reaching an altitude of about 35 miles. At this point, the first stage shuts down and separates, parachuting back to earth for recovery. The second stage takes over, burning for 177 seconds after which time it cuts off and the third stage fires. The rocket is now in a path almost parallel to the surface of the earth at an altitude of 150 miles. For the next 60 hours, after the burnout of the third stage, the cone-shaped lunar lander coasts toward the moon's orbit. Rotating so its retro-rockets face the approaching moon, the four engines fire and the final descent begins. The astronauts are guided down to the lunar surface by a small automatic beacon which has already made a soft landing. Four 40-foot landing legs are extended and the ship settles down on them. Meanwhile, a second, unmanned lander—a duplicate of the manned spacecraft—lands nearby. This will act as an emergency backup should anything go wrong with the main ship. After 12 days of exploration, the astronauts fire the single fifth-stage motor, using the now-empty fourth stage as a launching platform. After 220 seconds of thrust, the fifth-stage booster is discarded. The capsule begins its 60-hour return flight back to the earth. It makes a ballistic reentry and at 30,000 feet releases its parachute. It is recovered by ship after making a landing at sea.

The Apollo Command Module is first envisioned as a lifting body. This gumdrop-shaped vehicle is to be contained within the top stage of a Saturn C-3 or Nova booster. The entire vehicle makes the descent to the lunar surface. After exploration is complete, a booster launches the return module away from the moon. The booster is jettisoned and the aerodynamic Command Module returns to the earth, entering the atmosphere under control.

The atomic pulse rocket is in a pilot study stage at the U.S. Defense Department [also see Orion, 1960]. In the envisioned future version, a thousand 1-kiloton atomic blasts, firing at the rate of one bomb per second, lift the Empire State Building-sized rocket. A wheel-like structure at the top of the rocket provides artificial gravity for its passengers. The rocket is expected to carry such large amounts of payload that transportation charges for freight to the moon may drop to as little as $6.74 per pound.

Kraft Ehricke designs a moon rocket for Convair. An Atlas ICBM will be used to boost the moon lander away from the earth. The lander itself will consist of a cluster of six cylindrical fuel tanks, a personnel module and, atop that, a lifting-body shuttle craft for the return to earth

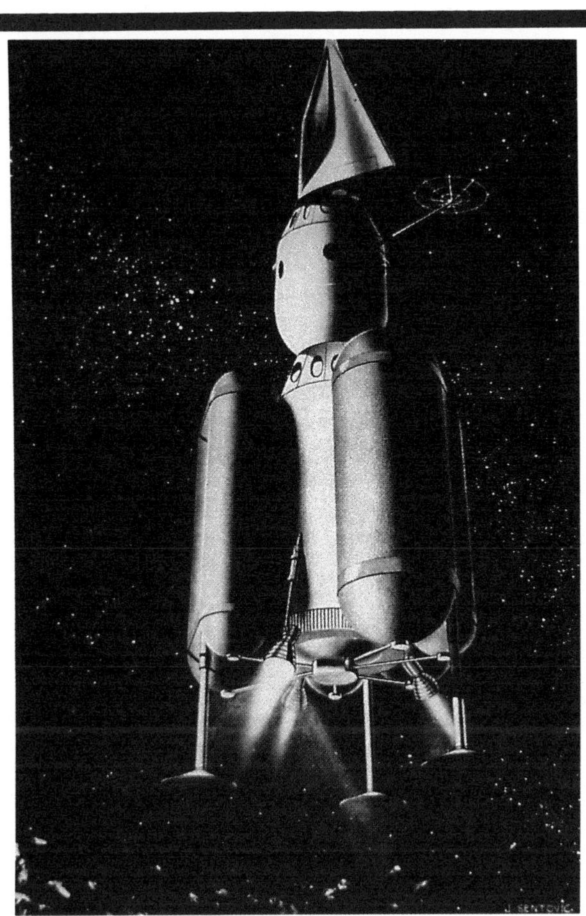

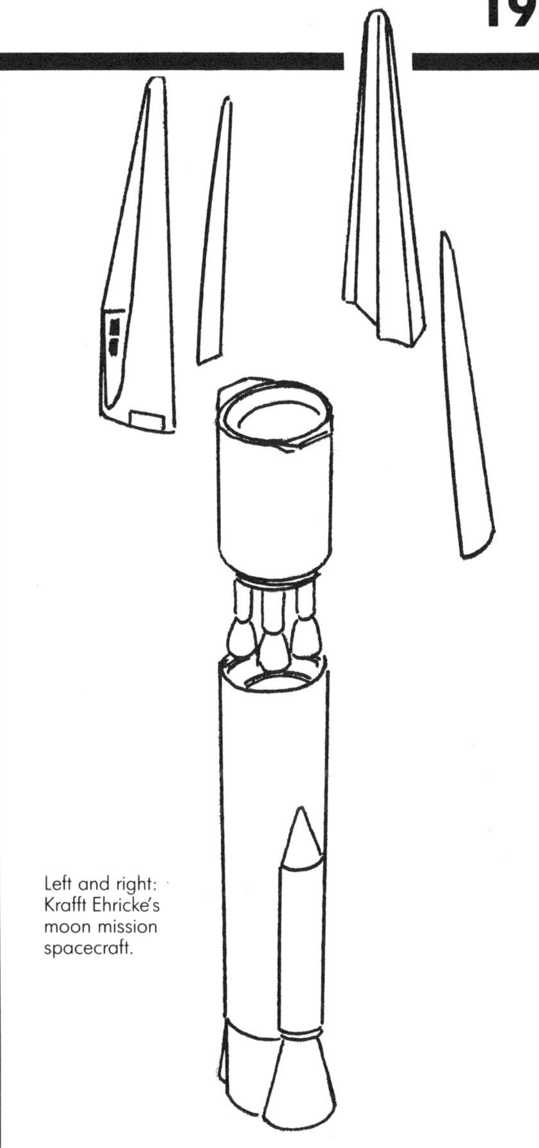

Left and right: Krafft Ehricke's moon mission spacecraft.

1961–1974

Inspired by President Kennedy's resolve to place an American on the moon before the end of the decade, the Soviets begin their own lunar landing program in earnest.

A newly developed booster, the N-1, will be the main booster for the giant rocket. Designed by Sergei Korolyov and originally intended for launching heavy satellites, the N-1 is a three-stage rocket fueled by liquid oxygen and kerosene. The three stages are referred to as "blocks," and are designated, starting from the base, as Block A, Block B and Block V. Block A has thirty engines, with a combined thrust of 10.1 million pounds (which rather than being gimballed for steering are to be selectively throttled). The second stage, Block B, has eight engines (with a total of 3.1 million pounds of thrust) and the third stage has four engines (which are gimballed) producing 360,800 pounds of thrust.

The N-1 will boost the lunar vehicle combination, designated collectively the L3. This is composed of Block G, a single-engined fourth stage for the N-1; Block D, a fifth stage (also with a single engine); the "Lunar Cabin," or LK, which is to be the lunar landing vehicle; and the Lunar Orbital Cabin (LOK), which is a modified Soyuz. Atop the rocket is an escape rocket, similar to those used by the Americans, that pulls the manned portion away from the main body in case of accident during launch.

The LK has a single engine, fueled by UDMH and nitrogen tetroxide, which is used for both landing and ascent. The manned cabin is sphere-shaped like most other Soviet spacecraft. A cockpit on one side gives the lone cosmonaut a view of the approaching lunar surface.

The lunar mission will begin with a takeoff from the Baikonur Cosmodrome with two cosmonauts aboard. After achieving earth orbit, the Block G engine fires, inserting the spacecraft into lunar trajectory. Once it reaches the moon, the Block D engine fires to place the spacecraft in lunar orbit.

One of the two cosmonauts then performs an EVA to enter the LK, which is still inside its shroud beneath the LOK. The spacecraft then separate, the shroud is jettisoned, and the LK uses Block D as the initial descent engine, beginning the drop toward the lunar surface. After slowing the lander for descent, Block D separates, ultimately crashing onto the moon. Small auxiliary retrorockets help cushion the final touchdown.

Once the cosmonaut's exploration and experiments are completed, the lower part of the lander and its legs act as a launch platform for the spherical cabin, which then ascends to meet the orbiting LOK. Another EVA allows the cosmonaut to rejoin his waiting comrade. The LOK engine then allows the spacecraft to leave lunar orbit and return to the earth, where it reenters the atmosphere and is recovered in a manner not unlike the unmanned Zond spacecraft.

The mission is to take place during the third quarter of 1968. Cosmonauts are chosen and are involved in a year and a half of training before the program is cancelled.

Lack of adequate funding ($0.5 billion compared to the $6 billion the United States is spending) and repeated failures of the N-1 booster (in three disastrous explosions) ultimately kill the Soviet moon landing program. The N-1 is formally cancelled in 1974. The two remaining N-1's as well as the completed LK and LOK are dismantled and put into storage.

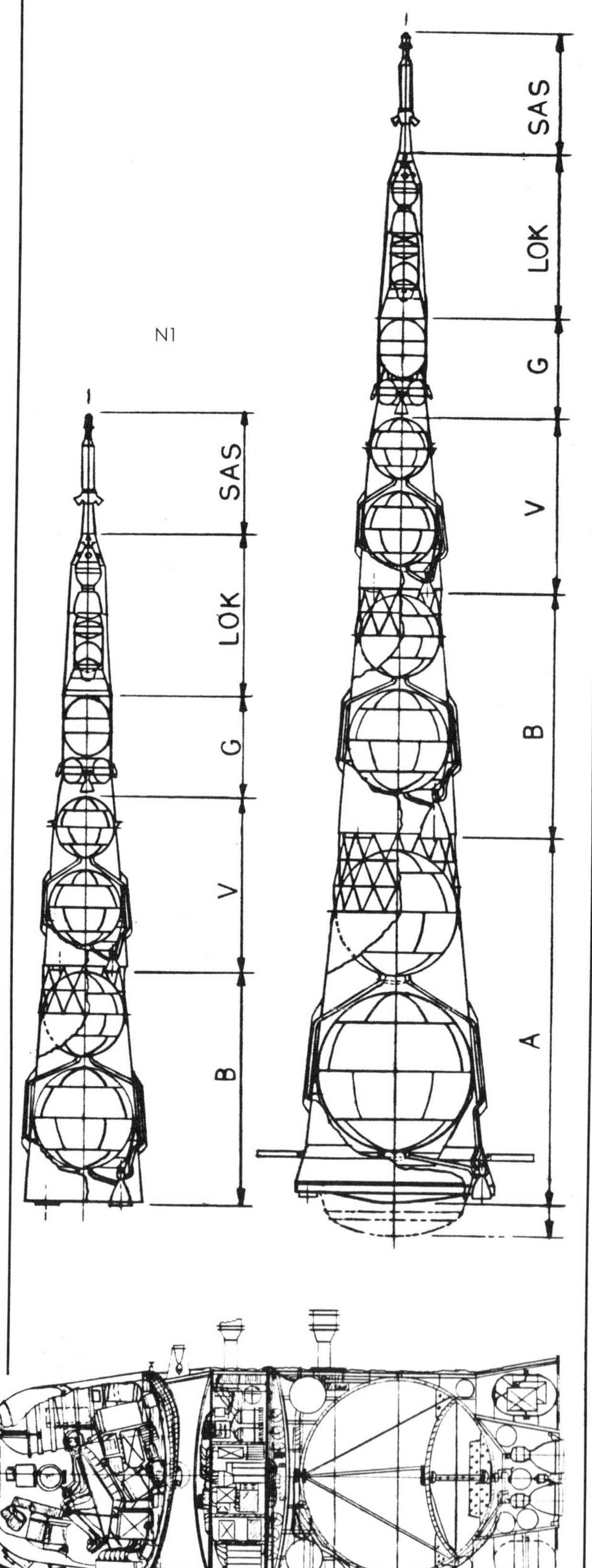

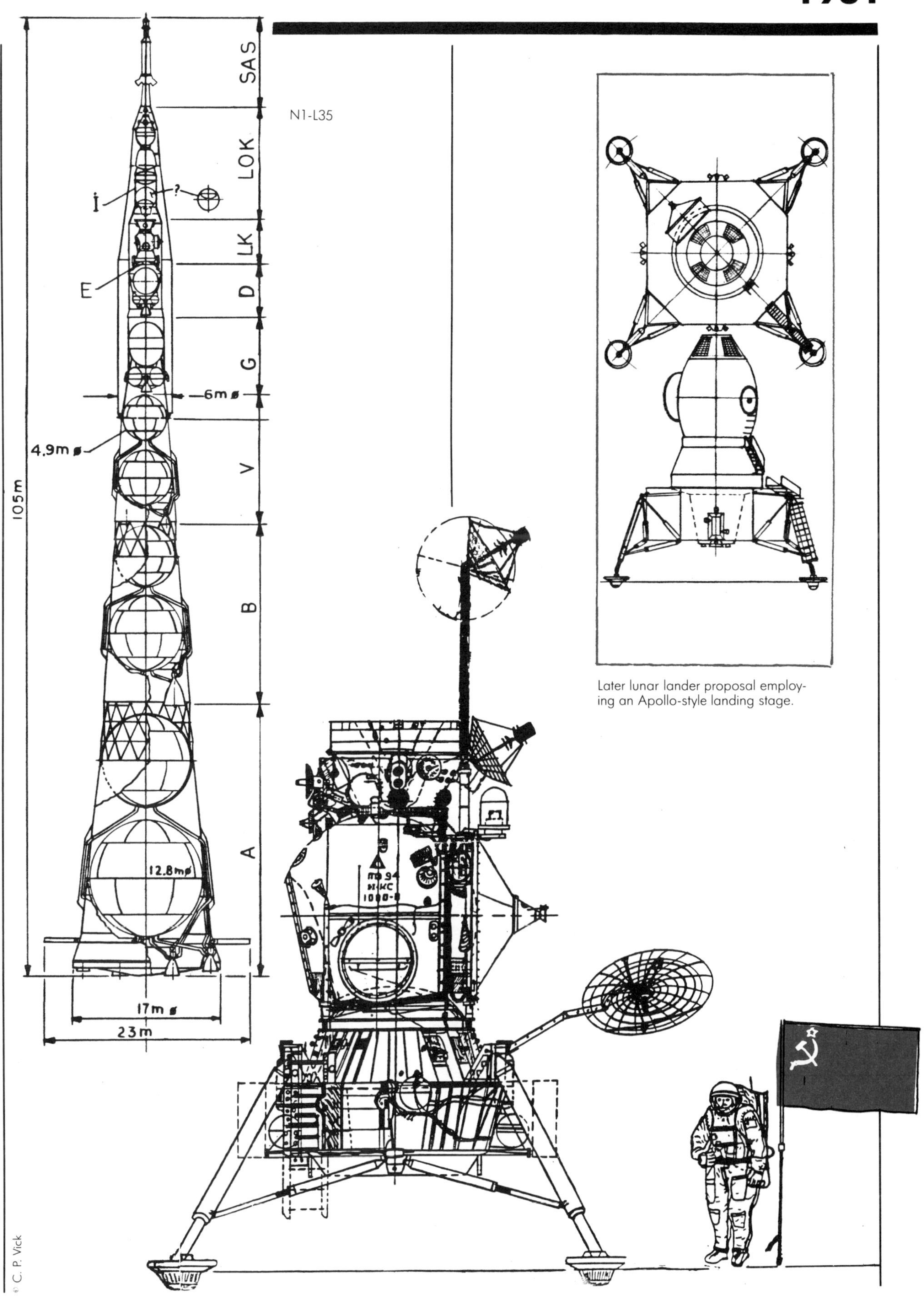

Later lunar lander proposal employing an Apollo-style landing stage.

PART VI

THE DAWN OF THE SPACE AGE

For all our vaunted accomplishments in space travel in the last 30 years we are still at the very beginning of the space age. Our wonderful machines—Vostoks and Soyuzes, Mercuries, Geminis and Apollos, space planes and space shuttles—are but the Conestoga wagons and Model Ts of our initial, tentative voyages off the surface of our planet.

At the time of this writing human beings still have not established any permanent presence in space and no human being has travelled farther from the home planet than the moon. Worse, the space age, so far as the present generation is concerned, may be over. To the generation that grew up with the promises made about spaceflight in the 1950s and 1960s, space exploration, such as it is, has seemed to be a tremendous cheat, a vast anticlimax. The United States abandoned its ambitious moon landing program halfway through, it allowed its only space station to fall from the sky while still scarcely used, it scaled back its space shuttle until the inevitable disaster struck and now what had once seemed to be a real commitment to space exploration is argued in Congress as an unneccessary and even frivolous expense. The future of space exploration is in far greater question, even jeopardy, than it has ever been in past decades, such as the 1950s or 1960s, when the eventual commonplaceness of space travel was taken for granted.

Yet there have been more seriously proposed projects for spacecraft and manned space exploration during these last three decades than in all the centuries before combined; indeed, most of them have been developed during the last 10 years. Designs for spacecraft have proliferated to the extent that the following section of this book cannot even pretend to comprehensively catalog all of them. Yet, one cannot but feel that this enthusiasm is mixed with a real sense of desperation.

Marquard

1962

The Soviets may be developing a system using manned rocket-planes to launch manned spaceships and satellites into orbit. Two manned, winged rocket-planes are attached to opposite sides of a large booster rocket. Together they act as the first stage. They eventually detach and fly back to a landing as conventional aircraft. The expendable second-stage booster continues on, inserting the third stage into orbit. The orbiter is intended to be recoverable by a system of parachutes, retrorockets and folding wings. These latter may allow the orbiter to "skip-glide" back into the atmosphere.

The manned planes have broad, sweptback wings and only a vertical tail surface. Each wing contains two or three air-breathing en-

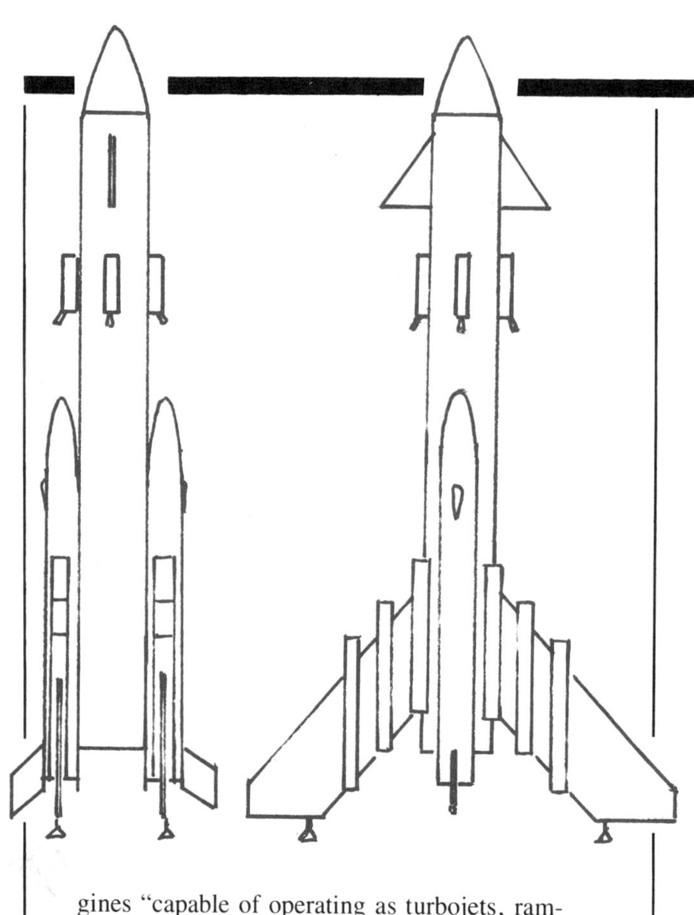

gines "capable of operating as turbojets, ramjets or rockets."

The Mikoyan Design Bureau begins studies for an aerospace vehicle consisting of a hypersonic carrier aircraft which will boost a small one-man spaceplane into orbit (similar to the Sänger concepts which have long seemed to fascinate the Soviets—see 1947). The project is named "50-50." The first stage will reach a speed of Mach 5.5 to 6. The orbital stage weighs 10.3 tons and the total weight of both vehicles is 140 tons. The bureau builds a prototype of the spaceplane powered by a turbojet. This model is 8 m long, 7.4 m wide and 3.5 m high. The test version is flown in 1965 by Igor Volk. It is airlaunched by a Tu-95 bomber at an altitude of 8 to 10 km. Three subsonic tests are eventually completed. The project is abandoned in 1968 with these words from Minister of Defense Gretchko: "We will not engage in nonsense."

John Small and Walter J. Downhower of JPL describe a manned lunar mission utilizing a lunar surface rendezvous technique.

The three-man mission would require five different types of spacecraft: a Surveyor, a modified Surveyor, a lander, a return spacecraft and a refueling spacecraft. The two first spacecraft would be for unmanned reconnaissance of the landing site. Then three 10,000-pound unmanned refueling spacecraft would be set down within 45 feet of a fourth unmanned lander. This latter would consist of a bus and a command module. The manned spacecraft would be evolved from existing Apollo command modules. Solid fuel units would be transfered from the first three spacecraft to the fourth using specially designed transfer tractors operated by remote control. Only when this operation was satisfactorily completed would a manned spacecraft, another bus and command module, be sent from earth. After exploration, the crew would transfer to the first bus/command module for takeoff and return to the earth.

Step Spacecraft Booster Mission 1 Ranger and/or Surveyor Atlas-Agena Survey landing site 2 Surveyor Atlas-Agena Establish homing beacon at site 3 Buspropulsion payload Saturn (3) Cache return propulsion units 4 Buscommand module (unmanned)evironmental support capsule Saturn Cache return spacecraft 5 Automatic refueling of return

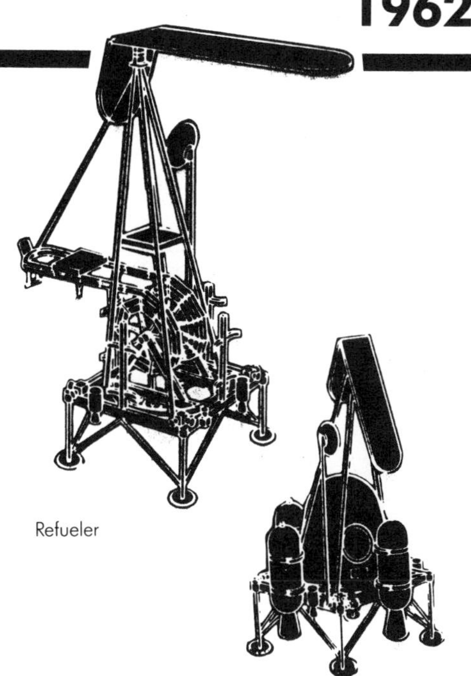

spacecraft 6 Buscommand module (manned) Saturn Carries crew of three to lunar surface 7 Checkout of return spacecraft 8 Buscommand module (manned)evironmental support capsule Lunar-assembled return system Lunar launch, return transit and reentry.

Baron Münchausen is a film by Czechoslovakian director Karl Zeman. A modern astronaut landing on the moon discovers there the spirits of Jules Verne's three space travelers, Cyrano de Bergerac and, of course, the Baron himself.

Philip Bono develops a resusable VTOL/ SSTO vehicle for Douglas. It is evolved from an expendable design called the One-stage Orbital Space Truck (OOST). The new version is dubbed ROOST, the R standing for Reusability. A balloon, inflated by residual hydrogen from the fuel tanks, allow the spacecraft to be recovered by helicopter and towed back to its launch site. It would be capable of orbiting a payload of 1 million pounds.

ROOST ultimately became the basis for Bono's Rombus vehicle [see below].

NASA begins a study for a "Post-Saturn Launch Vehicle Study." A $500,000 contract is awarded the McDonnell-Douglas aerospace firm. The 2-year technical study is directed by Philip Bono. In 1965, NASA allows patents on the resulting reusable booster design, the ROMBUS (Reusable Orbital Module-Booster and Utility Shuttle).

The enormous plug-nozzle rocket (14 million pounds including payload and external tanks) can develop 18 million lbs of thrust. It can place a payload of 400 to 500 tons into orbit and then return to the earth for up to 100 more flights. Average turnaround time is only 1 1/2 weeks. ROMBUS has the advantages of compactness, reusability and reliability. It is expected to be developed by the mid-1980's.

The plug-nozzle engine is the key to its success. Conventional rocket engines cannot withstand tail-first reentry. With a plug-nozzle the same propulsion system used for ascent is also used for orbit insertion and retrothrust for reentry.

Without its conical aerodynamic nose fairing, ROMBUS stands 95 feet tall. Its base diameter is 80 feet. Strapped around it at takeoff are eight external hydrogen tanks, jettisoned after initial boost. The basic structure, sans payload, weighs only 500,000 pounds.

ROMBUS returns to earth very much in the same fashion as an Apollo capsule. It descends base-first, using its large, flat plug-nozzle both as a heat shield and to create drag to reduce speed. Landing is accomplished by using the main engine, hovering above the landing site until its four landing legs are extended.

ROMBUS is expected to bring costs of launching material into earth orbit down to $25.00 per pound after only twenty reuses. Toward the end of its projected 10-year service life and 100 launches, costs could be reduced to as little as $10.00 per pound. [Also see 1969.]

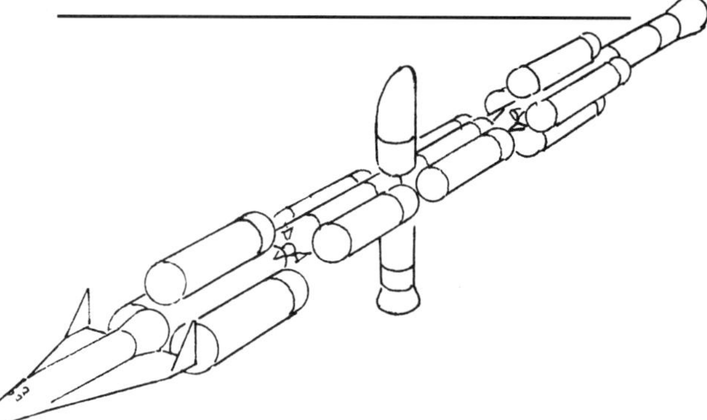

Northrop develops an Orbital Rendezvous Base System (ORBS), which will act as a kind of orbiting launch facility 300 miles above the earth. The technique of going to the moon in two steps—earth orbit to moon orbit—is expected to save at least 2 years from previous timetables. The ORBS system calls for using two or more launches of separate payloads, depositing sections of a moonship into earth orbit where they are assembled at a permanent station. From there a crew of three astronauts continues on to the moon.

The advantage of this system is that there is no need to wait for the development of the giant Nova booster—which will not be ready before 1968—for a direct flight from the earth to the moon. The ORBS system can use the Saturn launch vehicle.

At least five Saturn C-1s are needed to supply the materials for the spacecraft of the Northrop scenario, and five more to fuel it. With the future Saturn C-3, fewer launches are required: one C-3 and one C-1 to set up the orbital launch site, and two more C-3s to bring up the moonship and its crew (where only two Saturn C-4s are needed).

The ORBS station requires seven modules, all of approximately the same size (each can be segmented to stay within the payload limits of the Saturn C-1), launched about a week apart. They occupy an orbit 300 miles above the earth, inclined 32 degrees to the equator. One life-support module and the base unit remain permanently in orbit, while the four propellant tanks and the other life-support command module are assembled into the actual moonship. The first life-support module is launched, unmanned, followed by the base unit which contains communications and other equipment. These two modules form the permanent and reusable launch station. Crew arrival follows the launch of the four fuel tanks.

The assembly is expected to take 6 to 10 days, followed by the 14-day three-man lunar landing mission. Northrop thinks that one crew could handle both projects, while NASA prefers that two different sets of astronauts be used. Shuttle flights between the launch station and the earth are accomplished by a lifting-body spacecraft.

Use of an ORBS-type launch station, Northrop feels, has many distinct advantages: a lunar launch window is available on every 94-minute orbit; use of Saturn boosters eliminates having to wait 2 or 3 years for man-rated Nova rockets; ORBS can act as an emergency way station between the earth and the moon; it can aid in perfecting orbital assembly techniques; and it can help develop a regular space shuttle service.

The first mockups of the Apollo command and service modules are ordered. They are put on display April 4 and officially inspected on July 10.

The Russians consider the earth orbital rendezvous method the most reliable for a manned lunar mission, although some consideration is being given to a direct ascent using the Mastodon booster.

Ames Research Center proposes a variable geometry winged reentry vehicle. It will enter the atmosphere in a semi-ballistic configuration using folding wings for final point landing control.

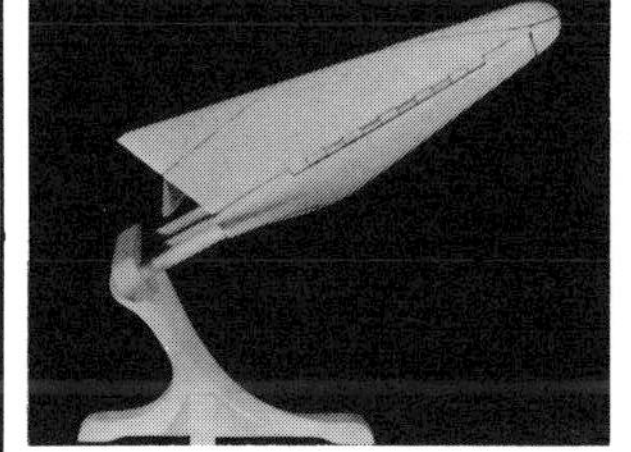

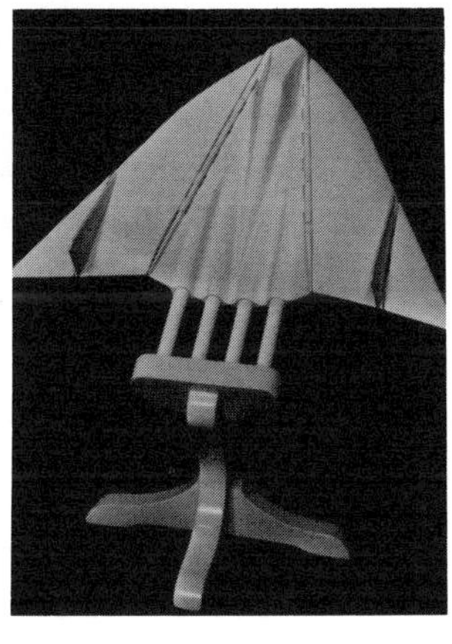

NASA

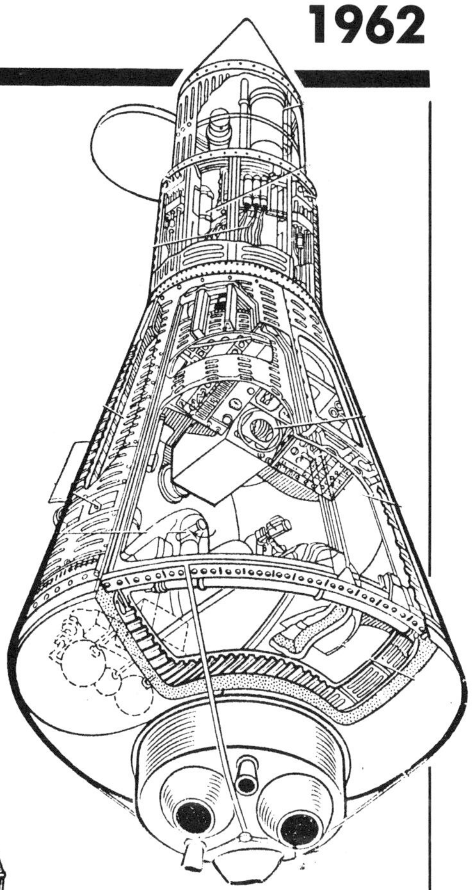

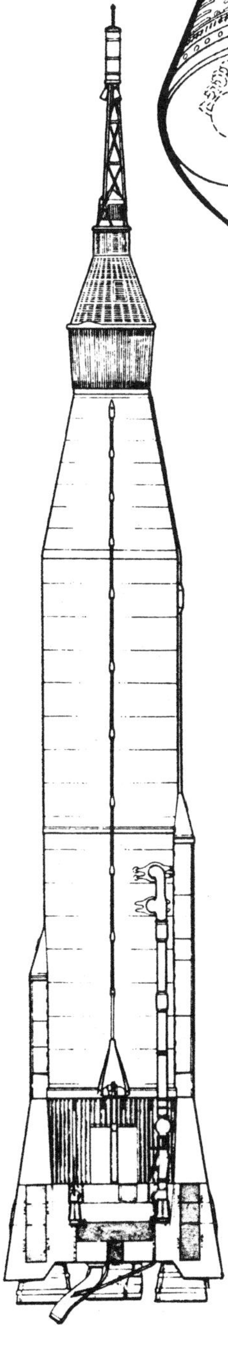

FEBRUARY 20

John H. Glenn, Jr. becomes the first American to orbit the earth. Launched in the Mercury 6 (Friendship 7) capsule atop a 95-foot Mercury Atlas rocket, Glenn achieves orbital velocity in just 5 minutes. *Friendship 7* makes three orbits, from 100 to 160 miles high, in 4 hours and 55 minutes. Glenn splashes down in the Atlantic Ocean, missing the projected target site by just 40 miles.

The Mercury spacecraft are tiny, just large enough to contain their human cargo, life support and instrumentation. The small size is dictated by the relatively low-thrust launch vehicles available. The 3,200-pound capsules consist of a conical pressurized cabin with a cylindrical parachute housing on top and a strapped-on retrorocket package on the bottom. This consists of three 1,000-pound-thrust solid-propellant rockets, and is jettisoned once it has been used to slow the capsule down for reentry. The slightly curved bottom of the capsule is the ablative

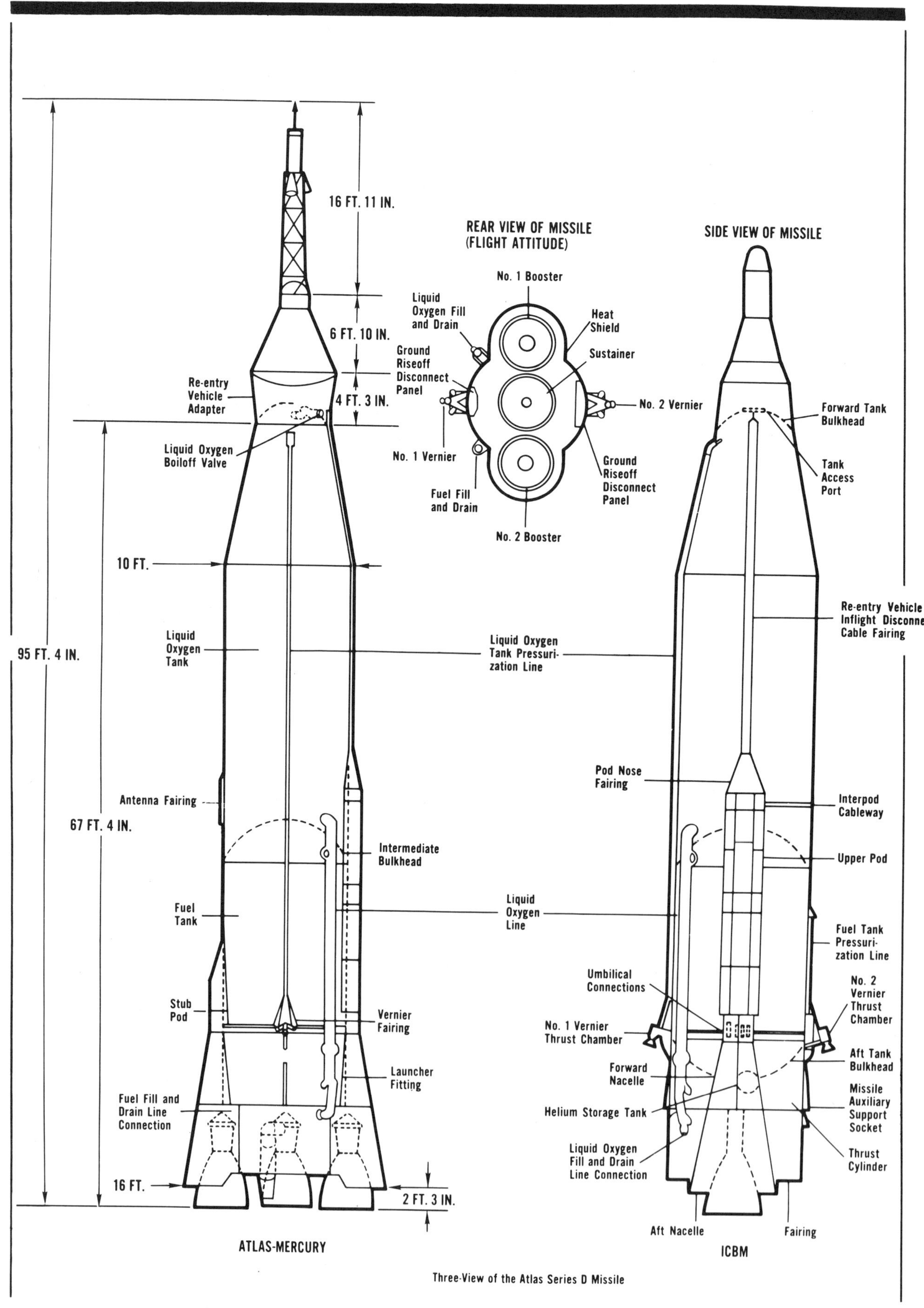

Three-View of the Atlas Series D Missile

heat shield, which has to protect the astronaut from reentry temperatures of up to 3,000 degrees F. Atop the parachute container is a tower carrying the escape rocket, jettisoned immediately after takeoff. The Mercury capsule is 9 feet 6 inches tall (25 feet 6 inches including the 16-foot escape tower), and 6 feet 2 inches wide across the heat shield. It weighs 4,265 pounds at liftoff (including the tower) and weighs 2,493 pounds after splashdown. The Air Force Atlas rocket as the launch vehicle provides 360,000 pounds of thrust for the orbital missions.

MAY

NASA's Future Projects Office at the George C. Marshall Space Flight Center selects three contractors to undertake a 6-month study of manned missions to Mars and Venus that will become known as Project EMPIRE (Early Manned Planetary-Interplanetary Roundtrip Expeditions). The project ultimately consumes some 6,000 man-hours. The contractors are General Dynamics/Astronautics, Lockheed Missiles and Space, and the Aeronutronic Division of Ford. It is eventually decided that manned missions to Venus and Mars will be feasible during the early 1970s. The final report is published in December and followup studies continue on into 1963.

The Aeronutronic spacecraft is to be launched from a 300-km earth orbit between 19 July and 16 August 1970 for the Venus-Mars flyby mission. The spacecraft has 750 ft^3 of habitable volume per man (plus a 300 ft^3 "storm cellar" in case of solar flares). Power is provided by an on-board SNAP-8 nuclear power system. Nuclear propulsion inserts the spacecraft into its interplanetary trajectory; course corrections and attitude control are accomplished with a conventional system of chemical rockets.

Once the first-stage injection is completed, the spacecraft takes on its flight configuration. Two living modules are deployed at the ends of two arms of the Y-shaped spacecraft; the SNAP-8 reactor is at the end of the third arm. (A solar-powered version

Solar powered spaceship by Constantin van Lent (see Ehricke above).

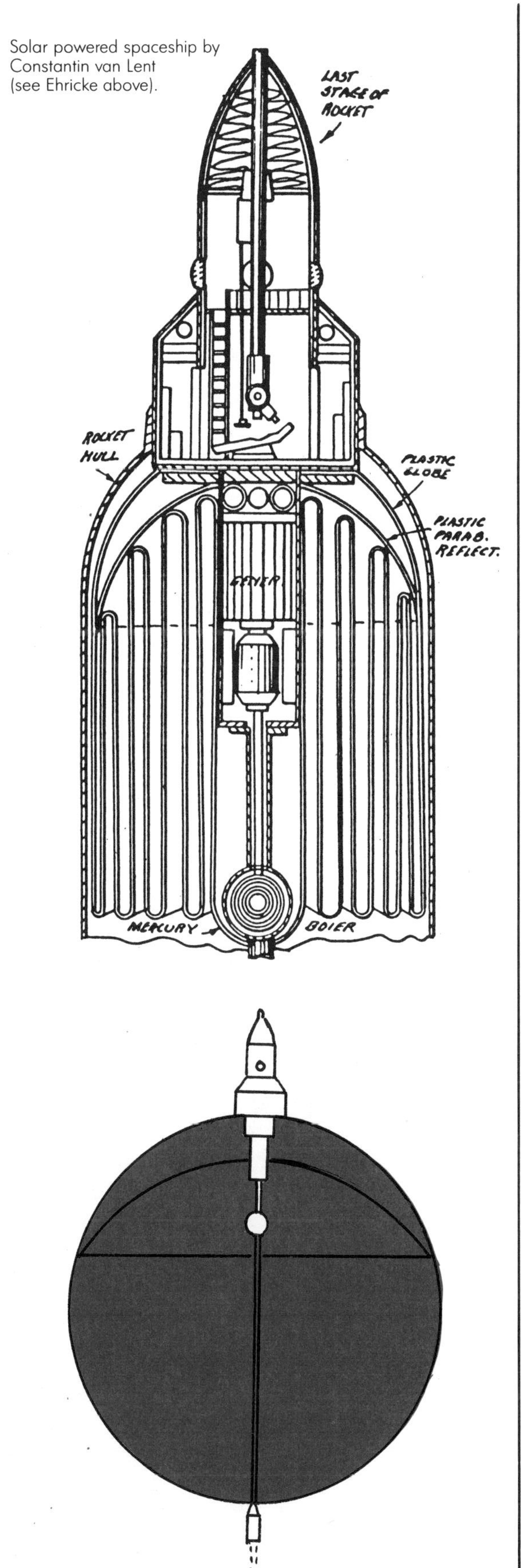

THE EVOLUTION OF MERCURY

Summer 1958, first approximation of Mercury capsule; below, Martin reentry vehicle to be boosted by Titan.

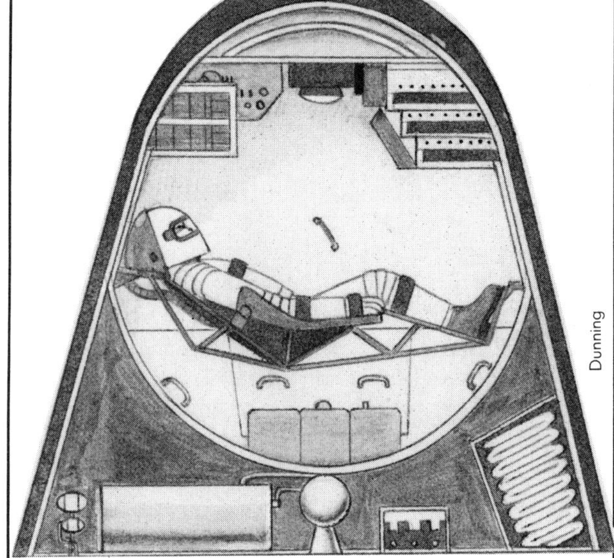

Dunning

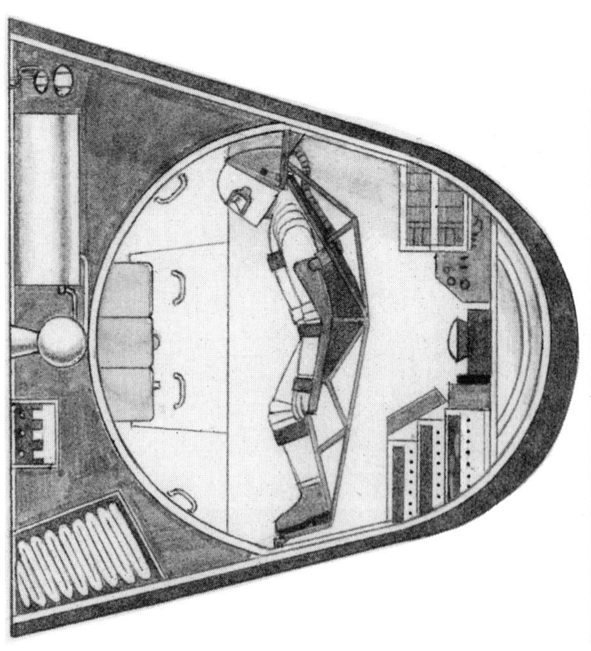

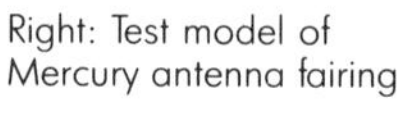

Right: Test model of Mercury antenna fairing.

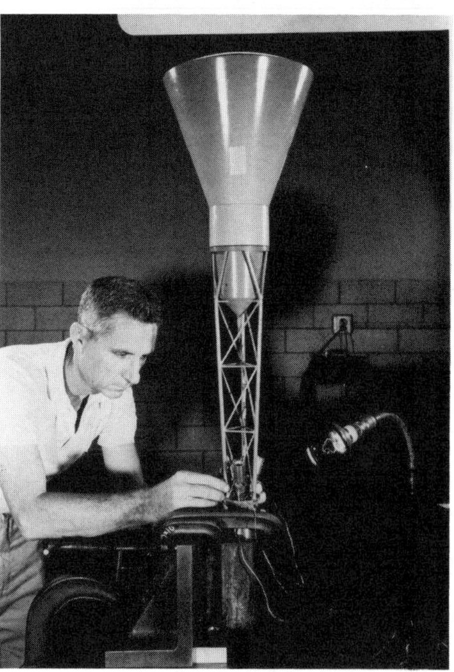

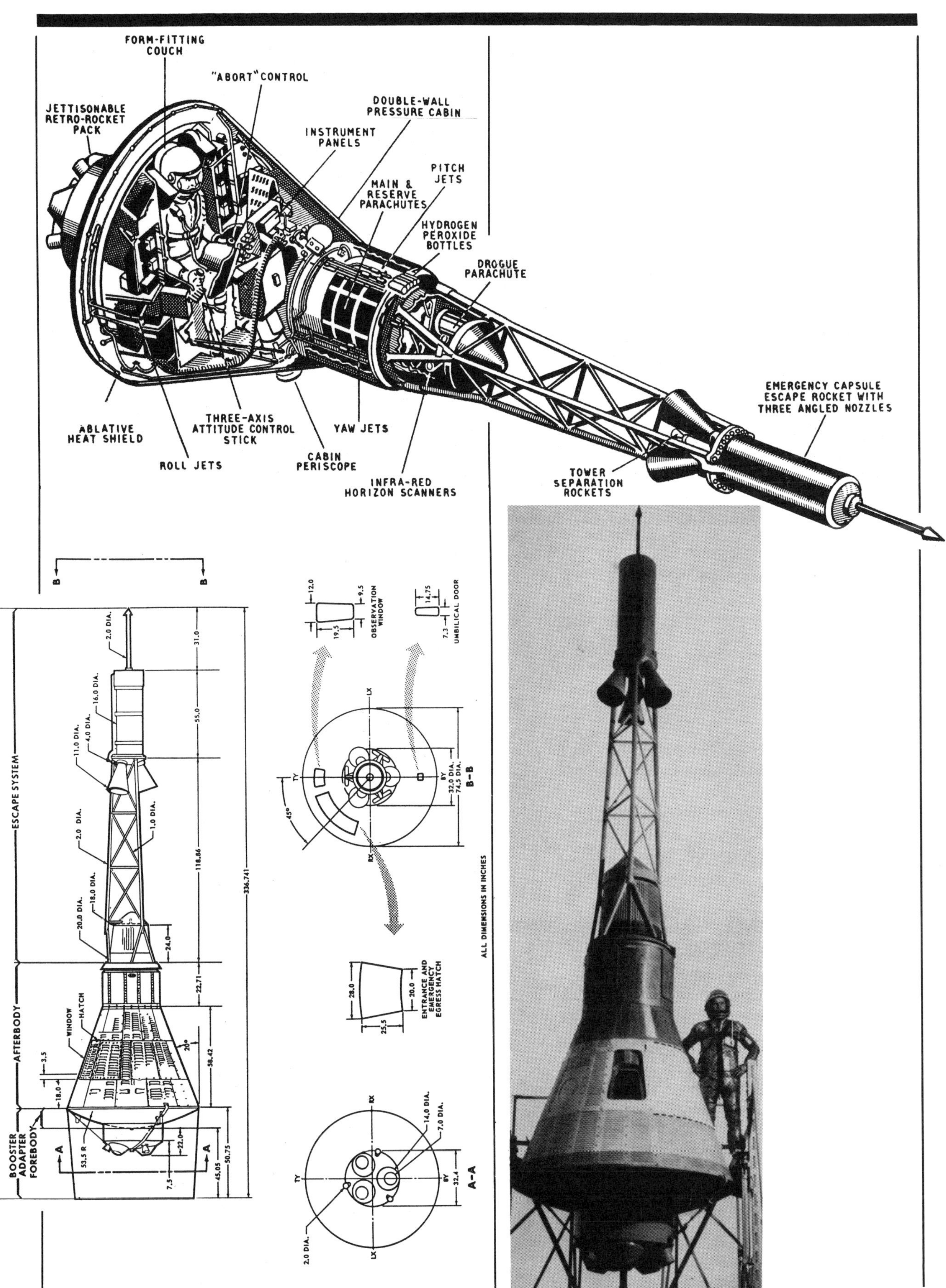
FORM-FITTING COUCH
"ABORT" CONTROL
JETTISONABLE RETRO-ROCKET PACK
DOUBLE-WALL PRESSURE CABIN
INSTRUMENT PANELS
PITCH JETS
MAIN & RESERVE PARACHUTES
HYDROGEN PEROXIDE BOTTLES
DROGUE PARACHUTE
EMERGENCY CAPSULE ESCAPE ROCKET WITH THREE ANGLED NOZZLES
ABLATIVE HEAT SHIELD
THREE-AXIS ATTITUDE CONTROL STICK
YAW JETS
ROLL JETS
CABIN PERISCOPE
INFRA-RED HORIZON SCANNERS
TOWER SEPARATION ROCKETS
ALL DIMENSIONS IN INCHES

eliminates the third arm, with the two living modules now opposite one another.) By rotating the vehicle at 3 rpm, a semblance of gravity (at 0.3 g) relieves the tedium of the 21 months of weightlessness. There is crew of at least six. Upon return to earth orbit, the crew transfers to a reentry vehicle. For the 1970 mission the nuclear engine is expected to produce a thrust of 200,000 pounds for 800 seconds. For the 1972 dual-planet mission a thrust lasting 1,000 seconds is necessary.

The General Dynamics study is headed by Krafft Ehricke (the Pininfarina of spacecraft design—also see 1958 and 1960). The result is a nuclear-powered Mars capture mission for eight astronauts that will occur during the 1973–1975 period. The round-trip mission requires between 400 and 450 days with 30 to 50 days in Mars orbit. The mission requires two spacecraft: a crew ship and a companion service ship. The manned spacecraft includes a crew compartment that is transferrable to the service ship in case of emergency. The crew ship also contains an earth-reentry module and a space taxi, which serves as a commuter vehicle between the two ships.

One of the configurations considered, C-22, relies on the NERVA and Phoebus nuclear systems. Earth departure requires four Phoebus engines and Mars arrival requires one. Escape from Mars orbit requires the NERVA system. The complete vehicle is 347 feet long and 75 feet wide at the time of its departure from the earth. The crew space is divided into three sections: a command module, mission modules and an earth reentry module. The eight-man command modules each contain a control room seating three and a lower compartment with sleeping quarters for five. The internal mission modules house the life support systems, food storage and so on. The external modules contain the space taxis. The reentry vehicle is an Apollo-derived spacecraft.

The service ship carries a manned Mars lander/return vehicle, in addition to a number of unmanned probes and sample returners. The Mars lander (the Mars Excursion Vehicle) will have its descent slowed by annular drag bodies. The lander is between 36 and 40 feet tall, spanning 24 to 30 feet between its landing legs.

Altogether, four different vehicles are considered in General Dynamics' final report, although all fly similar missions. Spacecraft C-22 is described above. Spacecraft C-23 is similar to the C-22, with advanced nuclear engines. It is 312 feet long and 70 feet in diameter. C-26 is 530 feet long with a cylindrical, rotating crew module 79 feet long. C-28 is 570 feet long. The great lengths of the C-26 and C-28 are dictated by the 33 foot diameter of the Saturn V booster. Eight Saturn V launches and seven orbital rendezvous maneuvers are required to assemble C-22. Later, Ehricke expands on trips to Mercury through Saturn, emphasizing flights to Mars and Venus from 1973 to 1982.

Lockheed's proposed Venus flyby mission departs on 11 November 1973 in order to arrive during Venus's conjunction during 1974. Its Mars flyby mission departs on 24 September 1974 for Mars's opposition during 1975. The Venus expedition lasts 370 days and the Martian expedition requires 670 days. Both pass within 500 miles of their target planets.

The spacecraft is a slowly rotating vehicle that creates an artificial gravity of 0.4 g for the astronauts. The command module is a modified Apollo with an internal volume of about 8.5 m^3. It serves as the crew's launch vehicle, emergency escape vehicle during launch, earth reentry vehicle and command center during the interplanetary cruise. A mission module houses the crew's living quarters, as well as the life support systems. The internal volume of the mission module is 113 m^3.

The solar-powered version of the spacecraft connects the mission and command modules by rigid, 25-meter spokes. The mission module is a cylinder 3.66 m in diameter and 12.1 m long. There is a radiation shelter as well. The nuclear-powered version is almost identical except for the addition of a third spoke to support the reactor.

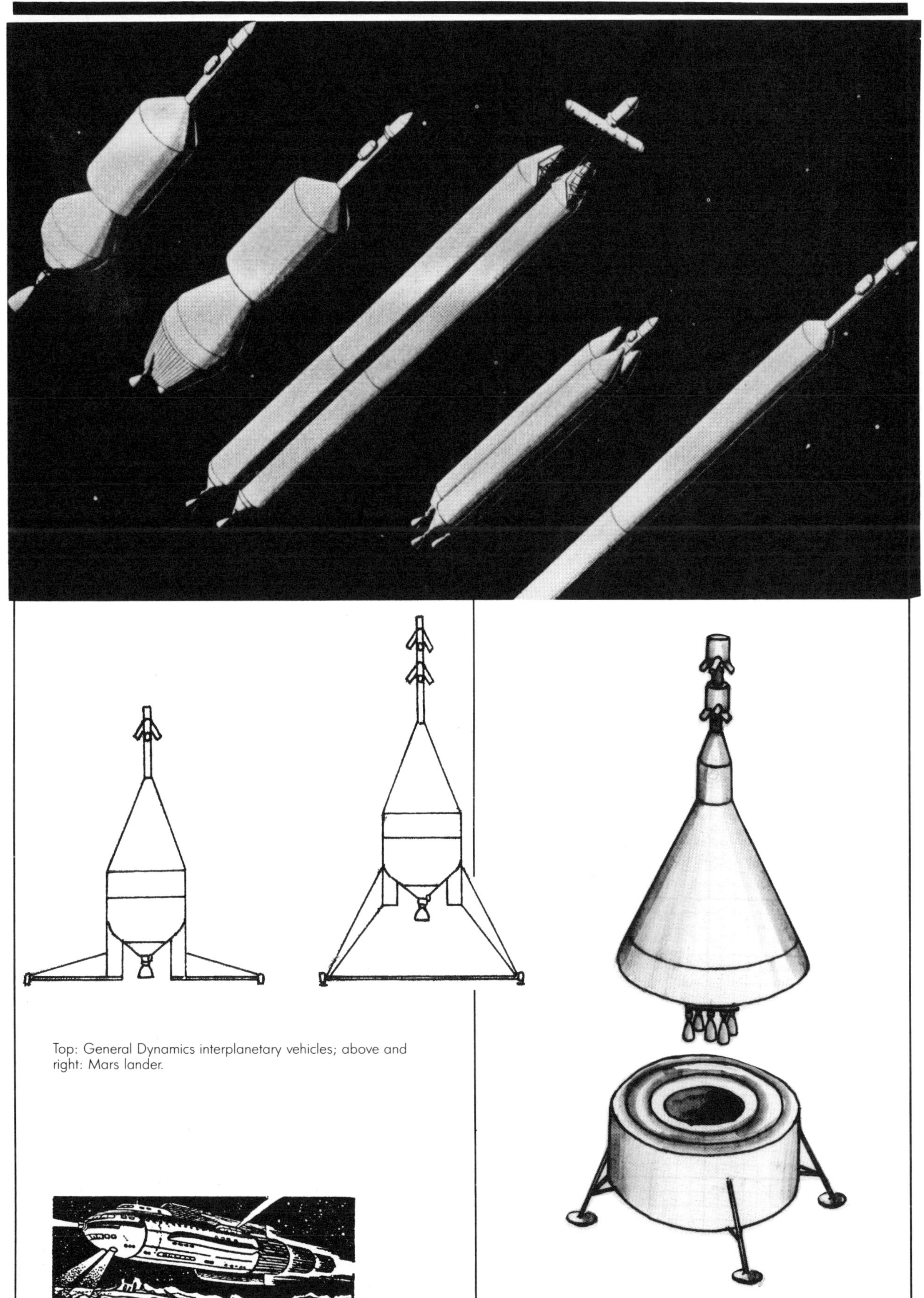

Top: General Dynamics interplanetary vehicles; above and right: Mars lander.

1962

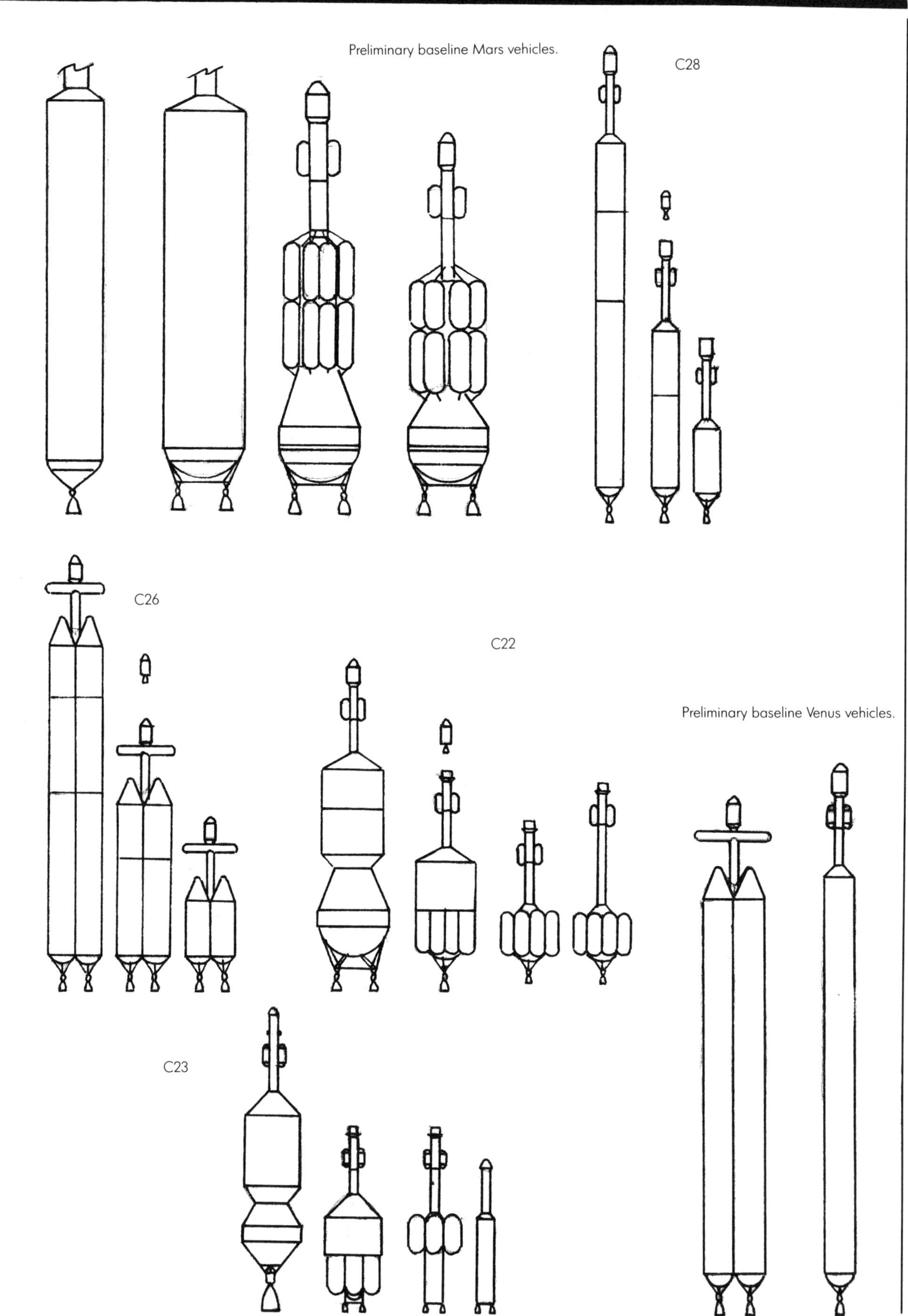

Preliminary baseline Mars vehicles.

Preliminary baseline Venus vehicles.

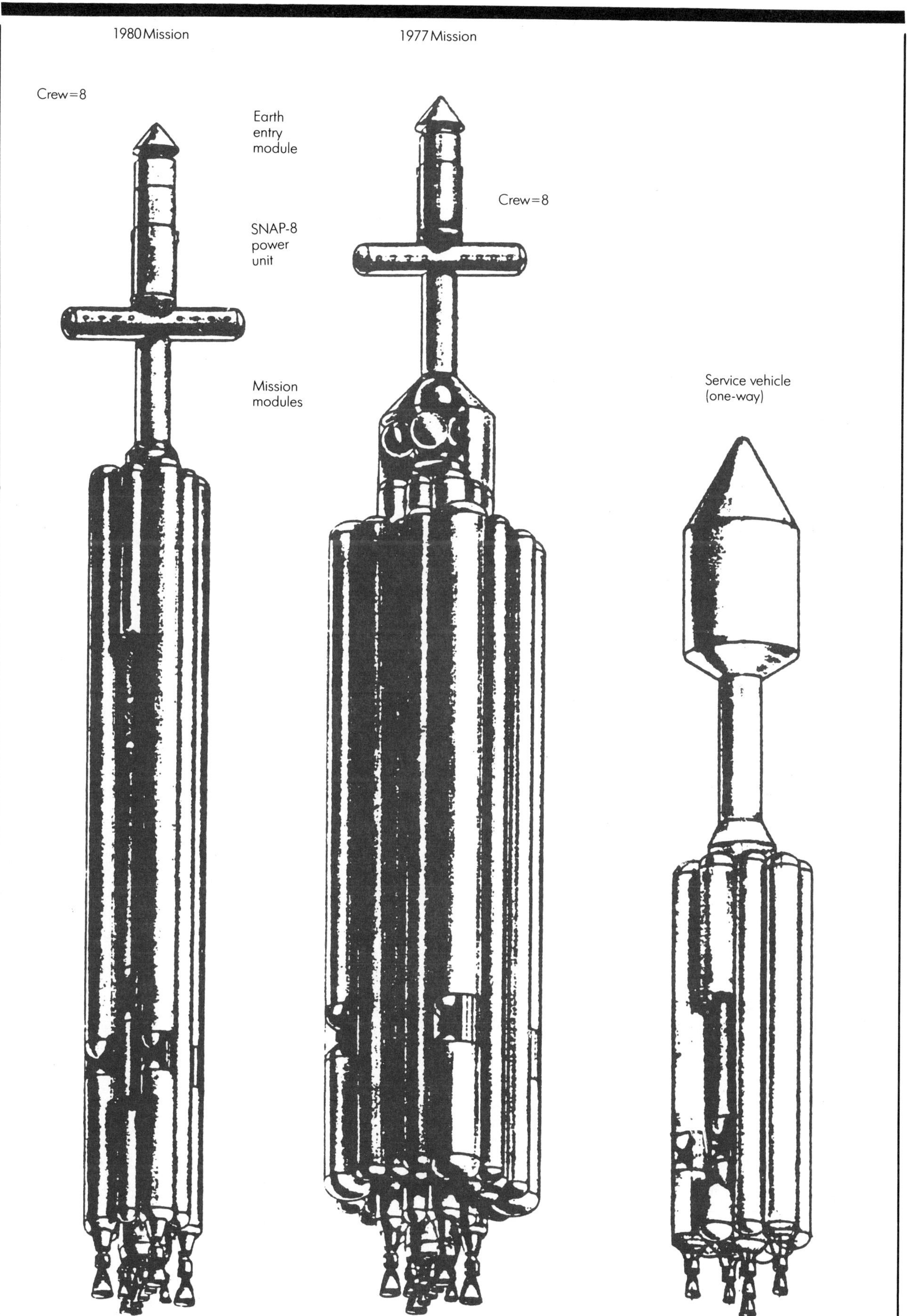
1980 Mission
Crew=8
Earth
entry
module
SNAP-8
power
unit
Mission
modules
1977 Mission
Crew=8
Service vehicle
(one-way)

1962

Mars or Venus atmosphere drag brake (General Dynamics/Convair)

Interchangeable components
Mars convoy Venus convoy

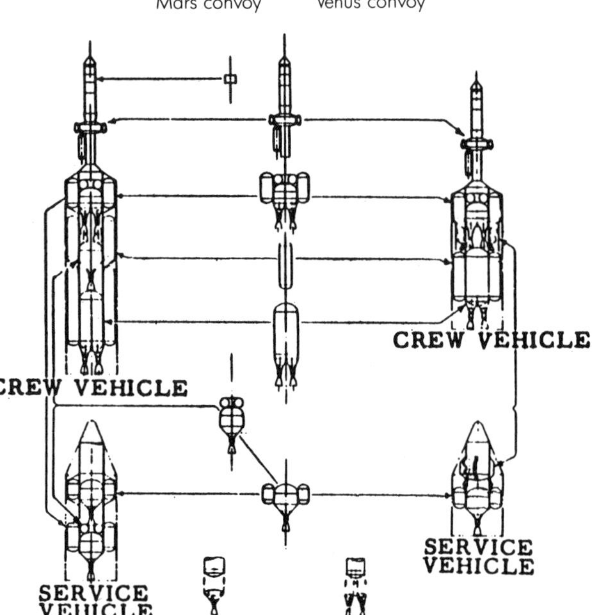

General Dynamics/Convair

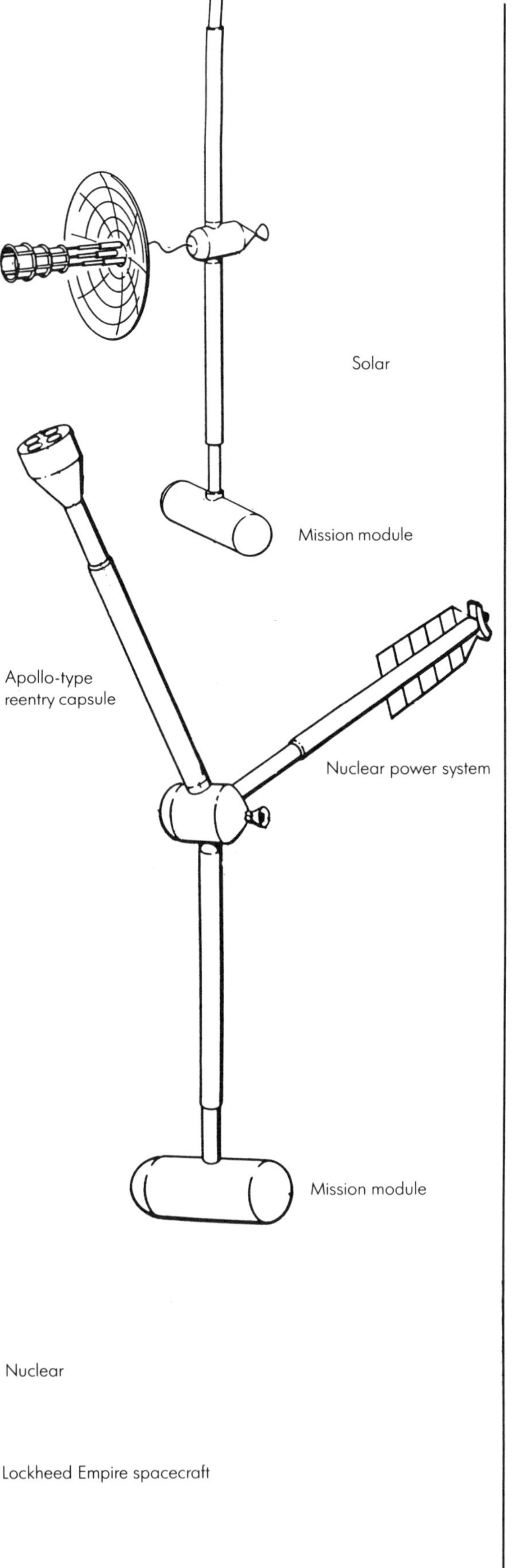

Lockheed Empire spacecraft

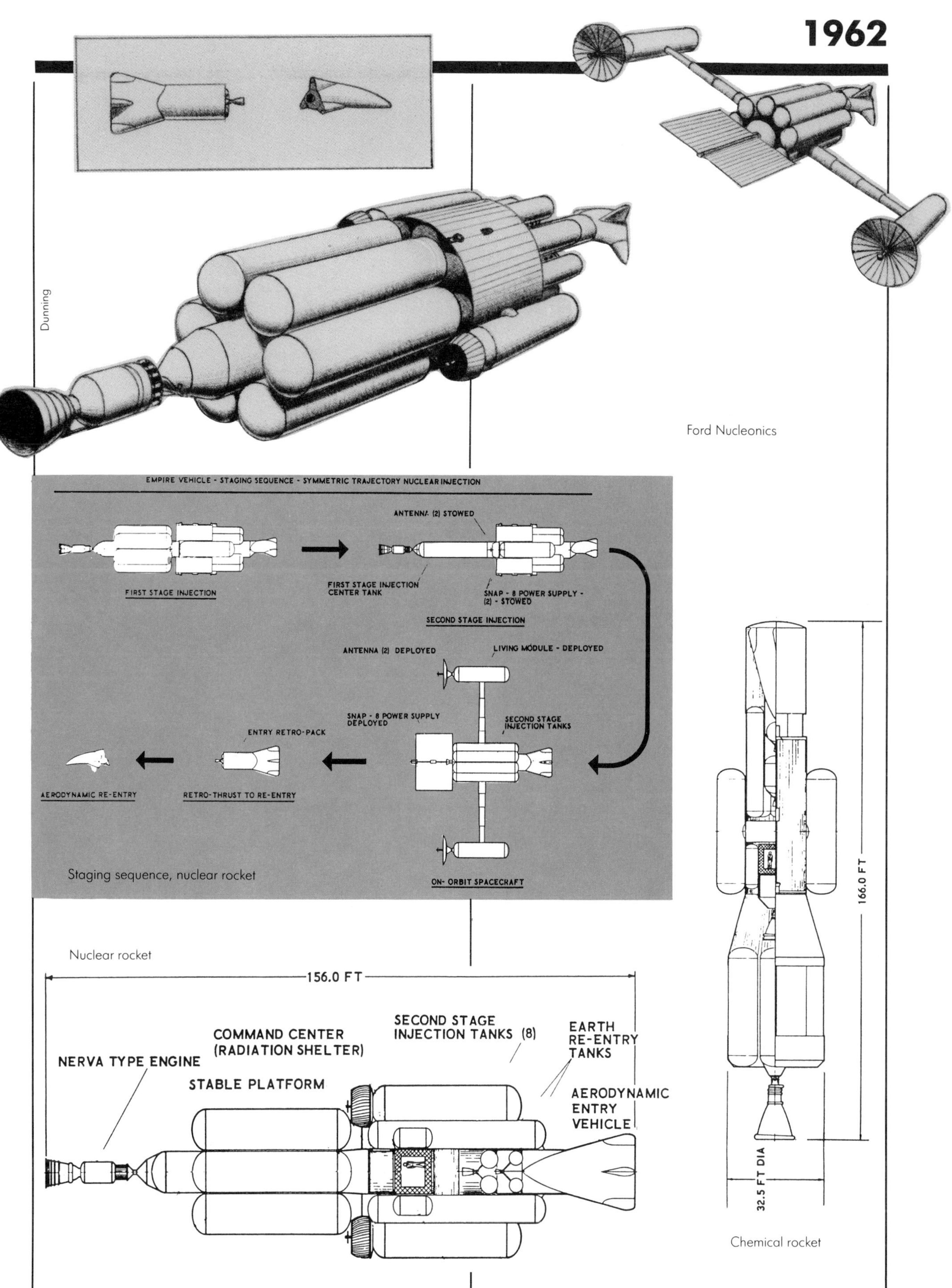

Dunning

Ford Nucleonics

Staging sequence, nuclear rocket

Nuclear rocket

Chemical rocket

1962

Robert L. Forward designs a kind of bathyscaphe for the manned exploration of Uranus. The flotation chamber contains pure hydrogen and is ten to thirty times larger than the manned module it carries. It is quite similar to the deep-sea bathyscaphes used by Jacques Piccard. Motive power is provided by combining an oxydizer with the hydrocarbons in the Uranian atmosphere where propellors could be used.

J. M. Cord and L. M. Seale, engineers with Bell Aerosystems, suggest that a U.S. moon landing could be advanced by 1 1/2 to 2 years by sending an astronaut on a one-man one-way trip in an Apollo-derived capsule. In their proposal the lone astronaut remains on the moon for up to 3 years, supplied by unmanned cargo rockets, until another Apollo-type spacecraft can land and pick him up. A capsule of 2,190 pounds gross weight is used for the landing (far less than the weight of a round-trip spacecraft) which would require a booster of only 450,000 to 1.1 million pounds thrust. About twenty-two cargo rockets are required to sustain the one-man base for a year. The cargo vehicles are 10 feet in diameter and 10 feet in length and carry 910 pounds of cargo. The empty shells can be used for additional living space.

Cord and Seale also suggest that the one-way techniquc bc uscd for Venus and Mars landings.

In 1964 Hank Searles publishes a novel titled *The Pilgrim Project* that utilizes the theme of the "one-way" lunar mission. It is later made into a film (*Countdown*, 1968) by Robert Altman.

JUNE 27

The X-15 sets a speed record of 4,105 mph.

JULY 17

Joseph F. Shea, NASA's deputy director of Manned Space Flight, tells the American Rocket Society that the first manned lunar landing will take place within an area 10 degrees on either side of the lunar equator and between longitudes 270 degrees and 260 degrees.

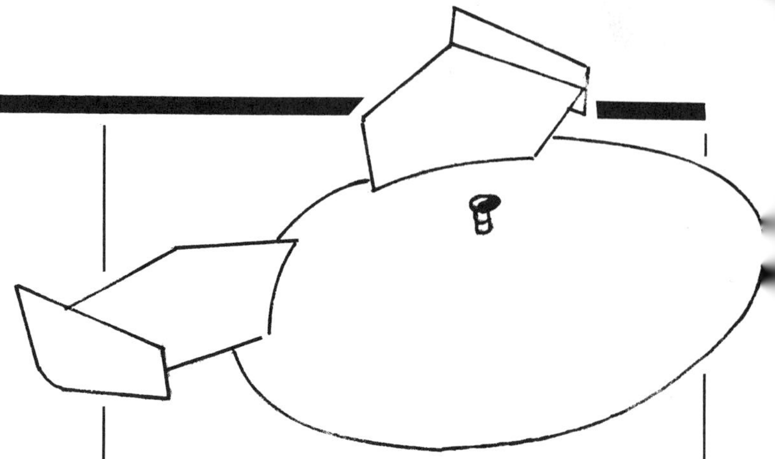

Design for post-Dyna-Soar reentry vehicle.

AUGUST

Ten U.S. Air Force pilots take a month-long simulated spaceflight while sealed in a simulated spaceship cabin. The results are so encouraging that it is declared that they could have remained in the cabin for 40 days with no difficulty. This ignores any consequences of prolonged weightlessness.

AUGUST 24

Life magazine speculates that the "[l]ogical next step in Russia's headlong space program may be a fantastic maneuver . . . the rendezvous and joining of two spaceships in orbit and the transfer from one craft to the other." The article is inspired by the dual launch of Vostoks 3 and 4 the previous week. The illustration accompanying the article is based upon the somewhat misleading photographs of the spacecraft that have been released, showing them with their protective fairings still in place. The retractable ejection seats, with their circular hatch covers, are also based upon a similar misinterpretation of photographs of the Russian collapsible airlock.

Life speculates that the spacecraft will dock rear to rear and that one of the cosmonauts will make an EVA transfer from one ship to the other via the retractable ejection seats.

SEPTEMBER 12

The Apollo Lunar Module, in an early form, is inspected by President John F. Kennedy. It is a full-scale model on display at the Manned Spacecraft Center in Houston, Texas. It has an overall cylindrical shape with three landing legs. The crew compart-

ment is a domed cylinder with large side windows and a clear roof. Spherical fuel tanks surround the base of the crew compartment.

SEPTEMBER 20

Astronauts for the Dyna-Soar project [also see 1954, 1957 and 1963] are chosen: Capt. Albert H. Crews, Maj. Henry C. Gordon, Capt. William J. Knight, Maj. Russell L. Rogers, Maj. James W. Wood, all of the U.S. Air Force, and Milton O. Thompson of NASA.

A full-scale mockup of the X-20 was viewed in September 1961; in December the U.S.A.F. announced the development of the Titan III booster. It is believed that the Titan III will accelerate the X-20 program by 2 years, permitting scheduling of unmanned flights in 1964 and manned flights in 1965.

The Boeing X-20 Dyna-Soar is intended to be launched by the Titan 3C. It is a delta-winged spaceplane 35 feet long with a wingspan of 20 feet and a height of 8 feet, described by one writer as looking "like a cross between a porpoise and a manta ray. "It is an aerospace glider with multi-orbit capabilities. The name is a hybrid of "dynamic soaring."

"The choice of flight paths available to the Dyna-Soar pilot is almost infinite," writes Boeing program manager G. H. Stoner, "By combining the high speed and extreme altitude of his craft with his ability to maneuver, he is able to pick any air field between

NASA Lewis Research Center Mars spacecraft (1961)

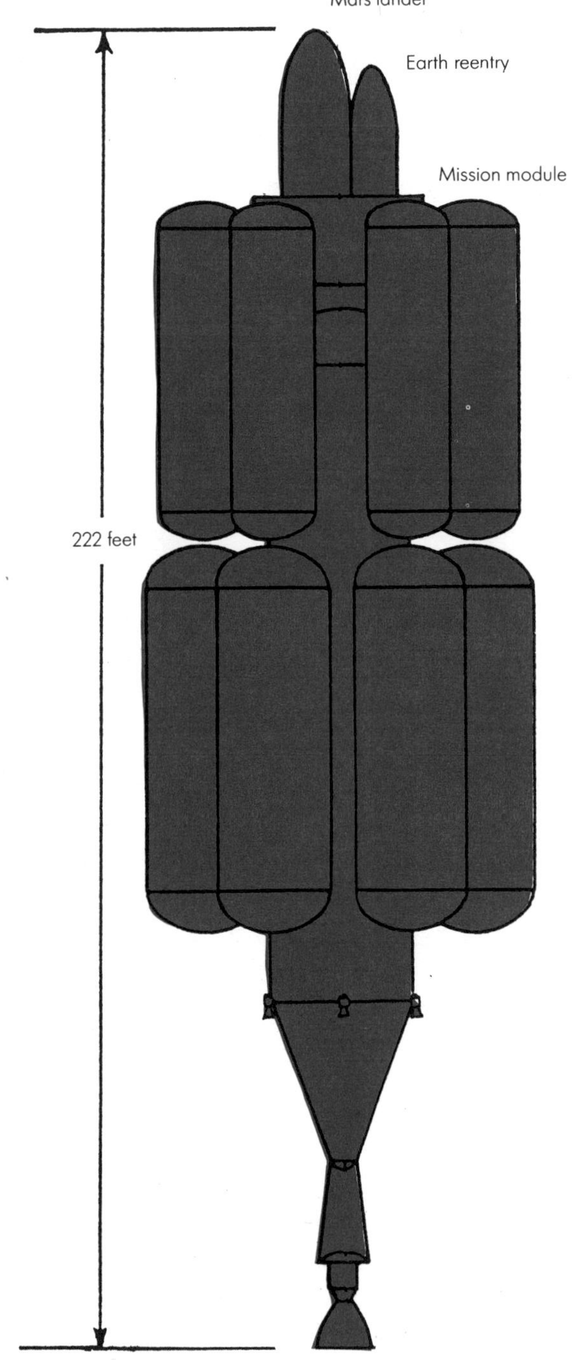

Point Barrow, Alaska and San Diego, California with equal ease."

NOVEMBER 7

NASA awards Grumman Aircraft Company the contract for the construction of the Apollo Lunar Excursion Module.

1962–1964

Sergei Korolev works at designing a spacecraft which can be assembled in orbit, using post-Vostok technology. This eventually becomes the Soyuz spacecraft [see 1966]. In Korolev's original plan, the Soyuz is comprised of three units: the Soyuz-A, which is the manned module, in itself composed of both a cylindrical orbital module attached to a domeshaped descent module, and an instrument module that also contains the retrorockets; Soyuz-B (which is to be launched first), an unmanned rocket booster about 7.8 m long, with a rendezvous module attached; and Soyuz-V, an unmanned 4.2 m-long tanker loaded with fuel for the Soyuz-B rocket. The overall Soyuz-A is about 7.7 m long.

Ultimately canceled as being too complicated, the Soyuz complex, fully fueled, would have been capable of a circumlunar flight [also see General Electric, 1960].

1962–1969

The Soviets study the "50/50" spaceplane concept in which a small (10,300 kg) lifting-body spacecraft, nicknamed "Lapot," and its booster rocket are launched from the back of a highly modified SST. The scheme is similar to one once proposed for the U.S.A.F. Dyna-Soar. The booster rocket is similar to the SL-8, and is about 21 m long.

Atmospheric drop tests are made beginning in 1965 and self-powered tests follow soon after, when the spaceplane is equipped with its own jet engine.

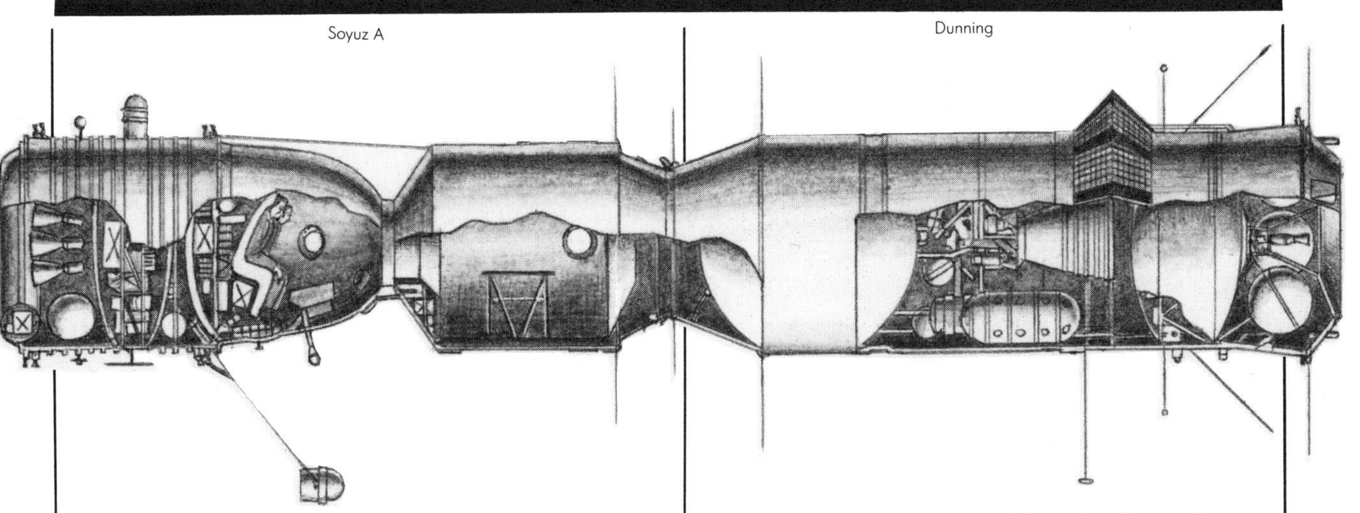

Soyuz A

Dunning

During this period the Federal Republic of Germany spends DM 16.5 million on space-plane and shuttle research.

ca. 1962

Ernst Stuhlinger and J. C. King propose a flight to Mars using five lozenge-shaped electrically propelled spacecraft, each with a crew of three. The 1 1/2-year mission allows 29 days on the surface of the planet.

1963

Bruce H. Neuffer of General Electric proposes a single stage to orbit spaceship. It is intended to place in orbit an integral, fully assembled space station. This payload would have been the empty propellant tanks configured into a kind beaded torus. The SSTO was propelled by a pressure-fed plug nozzle engine.

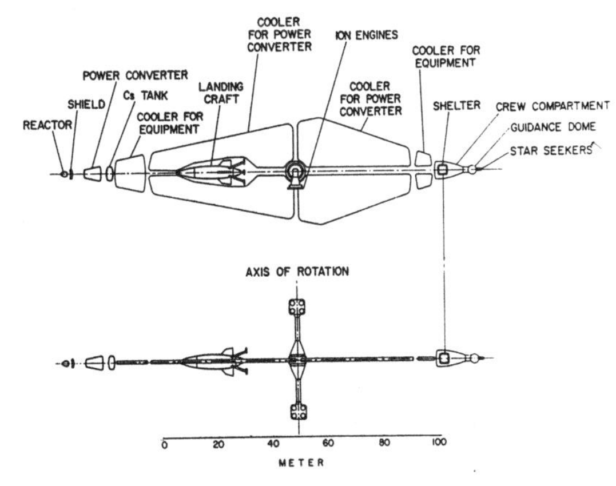

1962

Eugen Sänger, in an address given on the occasion of his appointment to the chair for Spaceflight Technique at the Berlin Technical University, outlines these next steps toward the conquest of space. After a manned landing on the moon, a project which he expects to occur within the next few years, the next phase of practical spaceflight will begin: regular hypersonic transport across the earth, followed by manned space stations and a permanent base on the moon. The first will connect points on the earth from 500 to 12,400 miles apart, with flight times of no more than 2 hours. Following this will be scientific and commercial satellites that will return to earth once their missions have concluded; then large, manned space stations will be built for scientific and economic purposes, as well as transit points for continuing and escalating earth-moon transport.

"This increase in missions," he writes, "and in transport-volume which requires a regular transportation system within the near space, introduces a whole range of new and greater demands on the spacecraft. In the first place, the average of only 50 percent reliability of the ballistic space vehicles of today is much too low for these tasks, especially in view of the fact that the spacecraft in this second phase of practical spaceflight have to carry not only crews, but also passengers."

Study Project 623, initiated by the German Federal Republic's Commission for Spaceflight Technique, is a research program for German industry to determine the design parameters for a space transporter. A yearly budget of 6.6 million DM is recommended as well as cooperation with EUROSPACE. [Also see Sänger, 1943.]

The M2-F1 is a lifting body reentry vehicle developed for NASA's Ames Research Center by Northrop. It is designed to give an astronaut 1,000 miles of lateral maneuverability after reentering the earth's atmosphere, allowing a horizontal landing at almost any location in the United States. The spacecraft is a boat-shaped half-cone with a blunt nose equipped with elevons, vertical fins, cockpit blister and landing gear. The first piloted tests are made at Edwards Air Force Base.

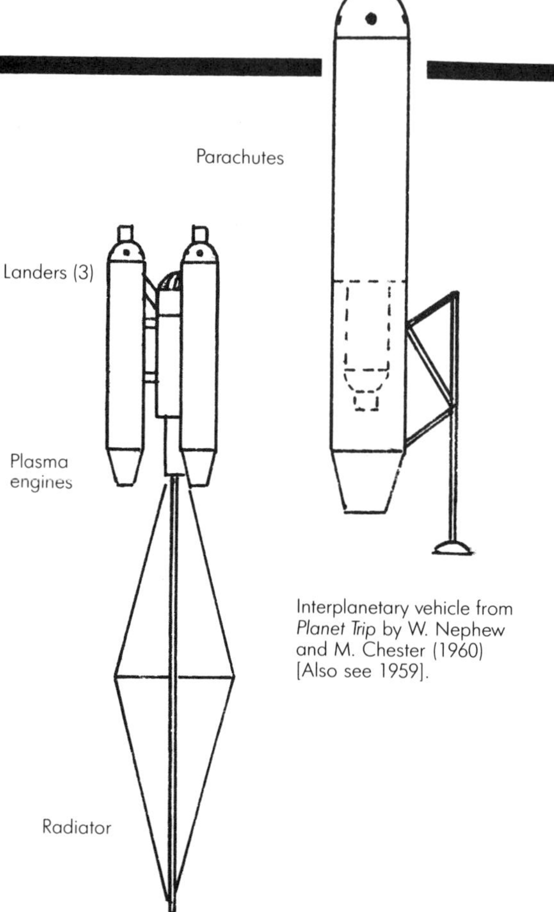

Interplanetary vehicle from *Planet Trip* by W. Nephew and M. Chester (1960) [Also see 1959].

The main difference between the test version and the orbital M2 is an enlarged canopy and an added dorsal fin.

Ames has been studying the M2 and similar configurations for 3 years and has discerned that a lifting-body spacecraft "properly designed and equipped with controls, [could] have a horizontal landing capability equal to or better than the X-15 . . . This means that an astronaut could accurately choose his landing place following a space mission and land horizontally." The astronaut additionally has reentry stresses reduced from about 8 g's to less than 2 g's.

Dandridge Cole designs an asteroid habitat. It is ellipsoidal, 18.6 miles long and rotates on its major axis. Mirrors reflect sunlight into the hollow, landscaped interior. A similar concept could be used as a generation starship.

Yuri Gagarin, in a speech before the International Astronautical Federation in Paris, says that a flight to the moon will require a spaceship weighing several tens of tons. As no rockets that large now exist, the problem

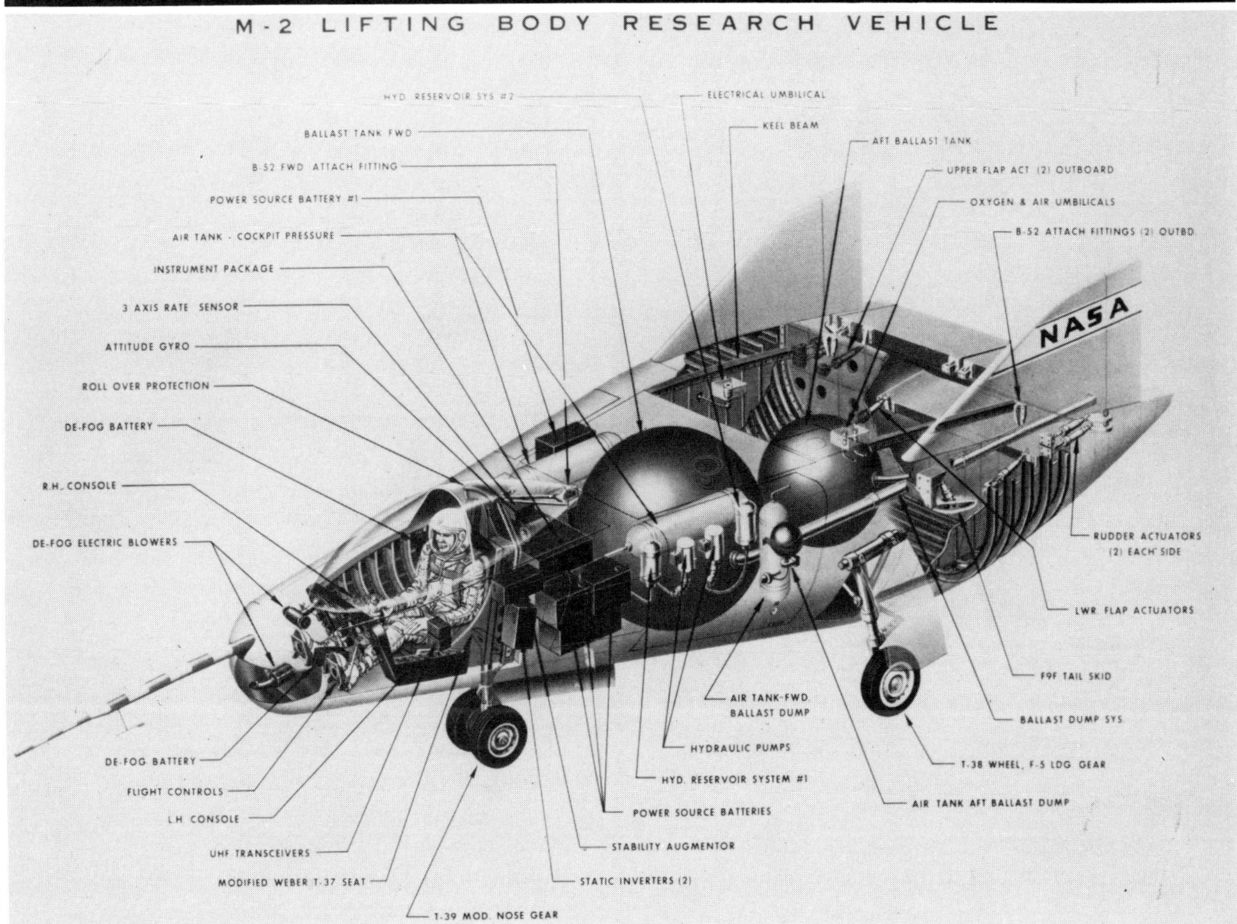

NASA

M2

can be solved in a number of ways. One technique involves the assembly of a spaceship in orbit. A large spacecraft can be made of smaller components. It is with this technique in mind that the Russians have carried out group spaceflights, such as those of Nikolayaev and Popovich, and Tereshkova and Bykovsky. In these, two Vostok spacecraft are brought into close proximity (2.5 to 4 miles).

Philip Bono envisions a manned Mars mission using his ROMBUS [see above] as the launch vehicle. It can boost an atomic-powered spacecraft to an altitude of 100,000 feet, and then return to the earth. Once at this altitude, the nuclear engines of the Mars ship can be safely ignited. Five propulsion modules, each incorporating one Nerva 2 atomic engine, are assembled in earth orbit. This requires 6 months. Each engine produces 200,000 pounds of thrust.

The Mars rocket is constructed in three stages. The first, the stage required for escaping from earth-orbit, is a cluster of three 33-foot Nerva engine modules. The other two stages have a single Nerva engine apiece. Once in earth orbit, the engines of the 2 million pound spacecraft are started for a second time for the insertion into the transmartian trajectory. The trip to Mars takes 170 days. Five of the crew members then explore the surface of the red planet for 30 days. The return to the earth requires an additional 250 days.

An alternate plan is to use the ROMBUS rockets themselves in a minimum-cost Mars mission called Project Deimos. A toroidal crew compartment is added atop the spherical oxygen tank near the center of the ROMBUS body. It is protected during ascent by a jettisonable fairing. Nearly 14,000 pounds of supplies are required for the trip. A 55,000-pound Mars landing module is attached to the top of the ROMBUS, at the center of the crew torus. Included in the payload are a Mars rover and collapsible surface quarters. The complete ship weighs 14,000,000 pounds.

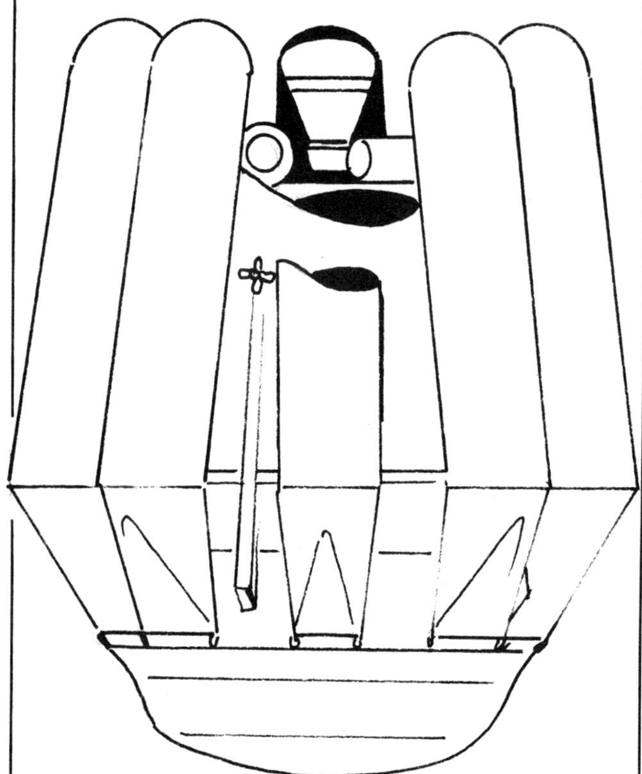

While in earth orbit, the ship has its depleted fuel tanks refilled by a fleet of reusable ROMBUS tankers, each carrying a payload of 400 tons of fuel, mainly in the form of detachable external tanks. Only 1 month is required for the refueling operation.

After launch from earth orbit, four of the external hydrogen tanks are jettisoned, another pair upon arrival at Mars orbit. The final two are discarded after escaping from Mars. The landing module contains its own propellant supply. It makes the round trip to the Martian surface while the main ship remains in orbit. The 24-foot-tall conical module also contains the earth-return capsule.

Although the X-20 Dyna-Soar is cancelled this year due to a lack of definition of the program and controversy over its merits, it still provides valuable information toward developing aerodynamic and structural techniques for future aerospace vehicles.

McGill University uses a space gun for upper atmosphere research, and studies on reentry physics, high-speed ramjets, and so forth. The High Altitude Research Project, the inspiration of legendary big gun designer Gerald V. Bull, employs a 45-year-old naval cannon erected on the south coast of Barbados. It is a 16-inch naval rifle, 70 feet long originally and now extended to 118 feet. The first eighty-four shots are made at a cost of only $1 million. Each of the "Martlett" vehicles costs $2,000. The hope is that Canada will be able to put a rocket-assisted Martlett IV into orbit in 1967.

General Electric develops a manned Mars mission. The complete spaceship is to be assembled in earth orbit. A payload and propulsion module are attached to the rear of a winged Mars excursion module. The propulsion module contains a nuclear rocket engine, and the spaceship's systems are powered by a SNAP-8 or a SNAP-50 nuclear generator. The crew compartment and fuel tank sections are 22 feet in diameter. The two-deck crew quarters are located immediately behind the landing craft. Besides the area set aside for life support there is a "storm cellar" for protection during solar flares. Three tankers are launched each carrying four nearly spherical tanks, each containing more than 45,000 pounds of liquid hydrogen. Six of these auxiliary fuel tanks, containing 140 tons of hydrogen, will be expended during the orbital launch maneuver.

When the ship arrives at Mars it will establish itself in a highly eccentric orbit: 200 miles at perigee and several thousand miles at apogee. Four of the remaining six tanks are used during the capture maneuver. After transferring to the Mars excursion vehicle, two of the four crewmembers fire the braking rockets to start their descent into the Martian atmosphere. The major portion of its deceleration is accomplished via atmospheric drag. After descending to a few thousand feet, the excursion vehicle is eventually maneuvered for a vertical, tail-first landing. The 35,000-pound vehicle has 1 minute of hovering capability. It then lands on three shock-absorbing struts.

In addition to the two-man crew, the module carries 5,000 pounds of equipment. The explorers are limited to a stay of 5 days on the surface of Mars. At the end of this period, all equipment except for the minimum

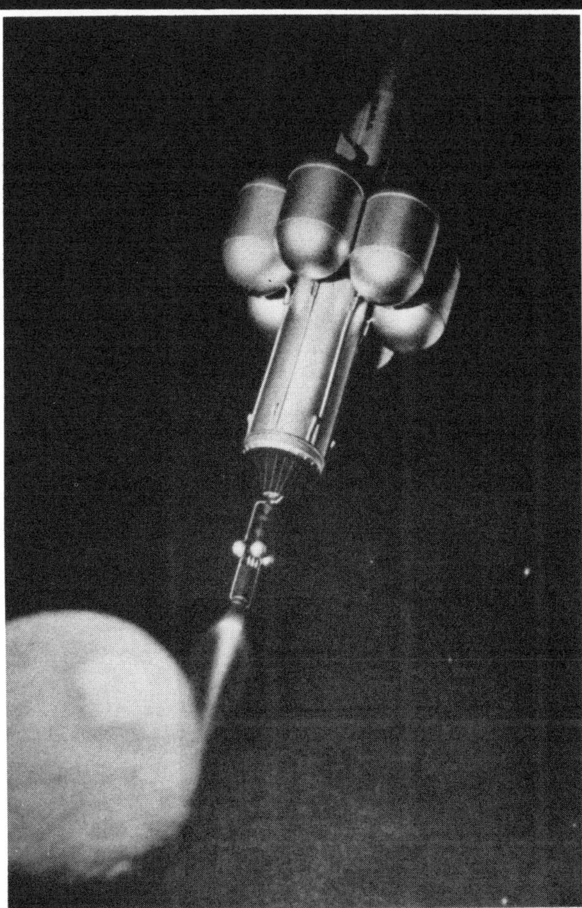

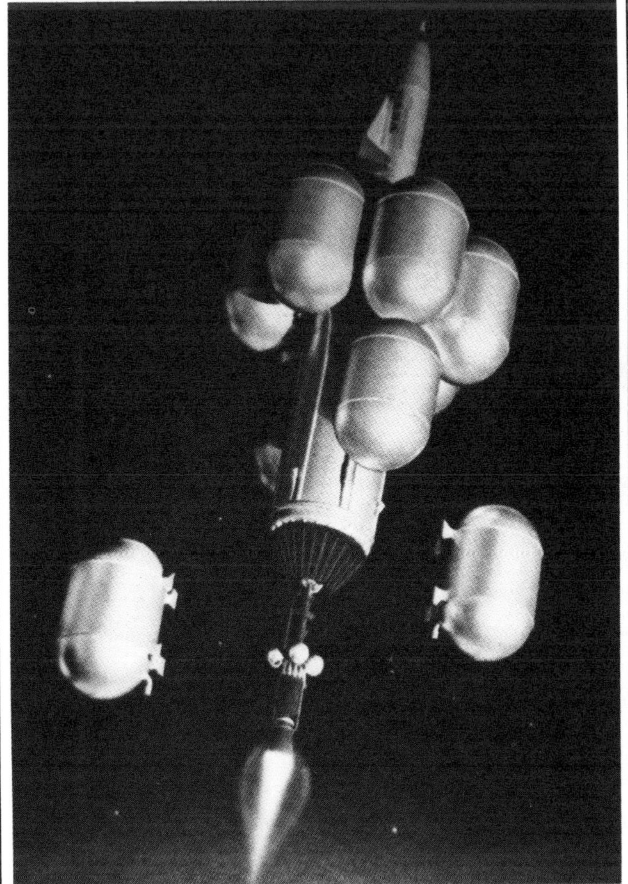

Top, left to right: Earth departure; jettisoning earth departure fuel tanks. Bottom, left to right: Mars arrival; Mars landing.

needed for life support and a few hundred pounds of samples is abandoned. The second stage of the lander uses the first stage as a launch platform. This then makes a rendezvous with the orbiting nuclear spacecraft, after which the excursion module is abandoned.

The last two hydrogen tanks are emptied, as is the aft compartment of the main tank, to escape Mars orbit. Upon arrival back at the earth, the nuclear engine is used to brake the vehicle into a high-altitude parking orbit.

The Space Technology Laboratories of TRW develops a study of manned Mars missions for a post-Apollo space exploration program.

The early missions would include a 400 to 450-day manned flyby of Mars and Venus. It would be placed in earth orbit by a Saturn V launch and assembled by rendezvous docking. Its crew of four would use an Apollo-derived command module with a new mission module developed from the earth orbit station. A launch date sometime in the 1970s was projected.

Next would be a six-man Mars/Venus orbiter. The crew modules would be similar to those of the flyby vehicle and a launch date in the early 1980s was expected. Both the flyby and orbiter spacecraft would carry unmanned landers.

The manned Mars landing mission would be more complex. This much larger spacecraft would carry a crew of six to eight members. The earth departure tank and the three-part Mars departure stage will be placed in earth orbit in four separate launchings. A fifth booster places the manned module into space and the components move to a rendezvous for assembly. Four tankers then are launched to fuel the completed 142-foot-long spacecraft. After insertion into the interplanetary trajectory, the spacecraft splits into two units—the spaceship proper and its empty booster—connected by a cable and spun to create an artificial gravity. (A variation on the mission sequence allows for a zero-g mission plan.) The configuration is despun upon arrival at Mars and the booster counterweight is jettisoned. Aerodynamic braking slows the ship, which goes into an elliptical orbit around Mars. It then shifts into a parking orbit well above the atmosphere. The Mars excursion module (MEM) is ejected and descends toward the surface, slowing by means of aerodynamic braking and making the final descent by parachute. Its engines will allow it to hover while choosing a landing site.

The exploration of Mars will last 10 days. The base of the lander will act as the launcher for the second stage, which will take off and meet with the waiting orbiter. Before departure for the earth, the MEM propulsion section is jettisoned, as well as the orbiter's Mars entry heat shield. Once again, the ship is divided into two components so that it can be spun. This time the counterweight is the Mars departure unit/solar array.

Upon arrival back at the earth, the ship is again despun, the counterweight jettisoned and the earth reentry module ejected. Its strap-on retrorockets slow it for its reentry. It is a lifting body, although the final descent is made by parachute.

The spacecraft—which is 58 feet 4 inches long and about 20 feet wide—consists of three main components: the Mars excursion module; the mission module—which contains the life support systems, medical and hygiene station, command station, recreation, and so forth; and the earth return module, which is also used as the sleeping quarters. The mission module also contains a solar flare radiation shelter (there are half a dozen different possible configurations for this, including individual, water-shielded suits). Power is provided by a solar array consisting of two circular panels, each with an area of 750 ft^2.

The earth return vehicle is a lifting body 21 feet 3 inches long, 6 feet 5 1/2 inches high and 12 feet 7 inches wide.

R. J. Beale and E. W. Speiser of the Jet Propulsion Laboratory propose a 16,000-lb electrically propelled "spacebus" for a trip to Mars, Mercury or Venus. For Mars or Venus missions, the spacebus carries an aerobraking capsule for the manned landing.

The bus's ion engines are powered by a 300-kWe nuclear reactor. The nuclear reactor with its radiation shield is carried at the forward end of the vehicle. Directly behind this is the electrical generation equipment and machinery. In the central area is the momentum transfer wheel, used to the counterbalance the angular momentum of the single power-conversion-system turbo-alternator. Flanking the spacecraft are a pair of radiator panels. The aerobraking capsule is carried at the aft end of the bus. At the extreme aft end of the spacebus and in the plane of the radiators are a pair of gimbal-mounted ion engines along with their cesium fuel tanks.

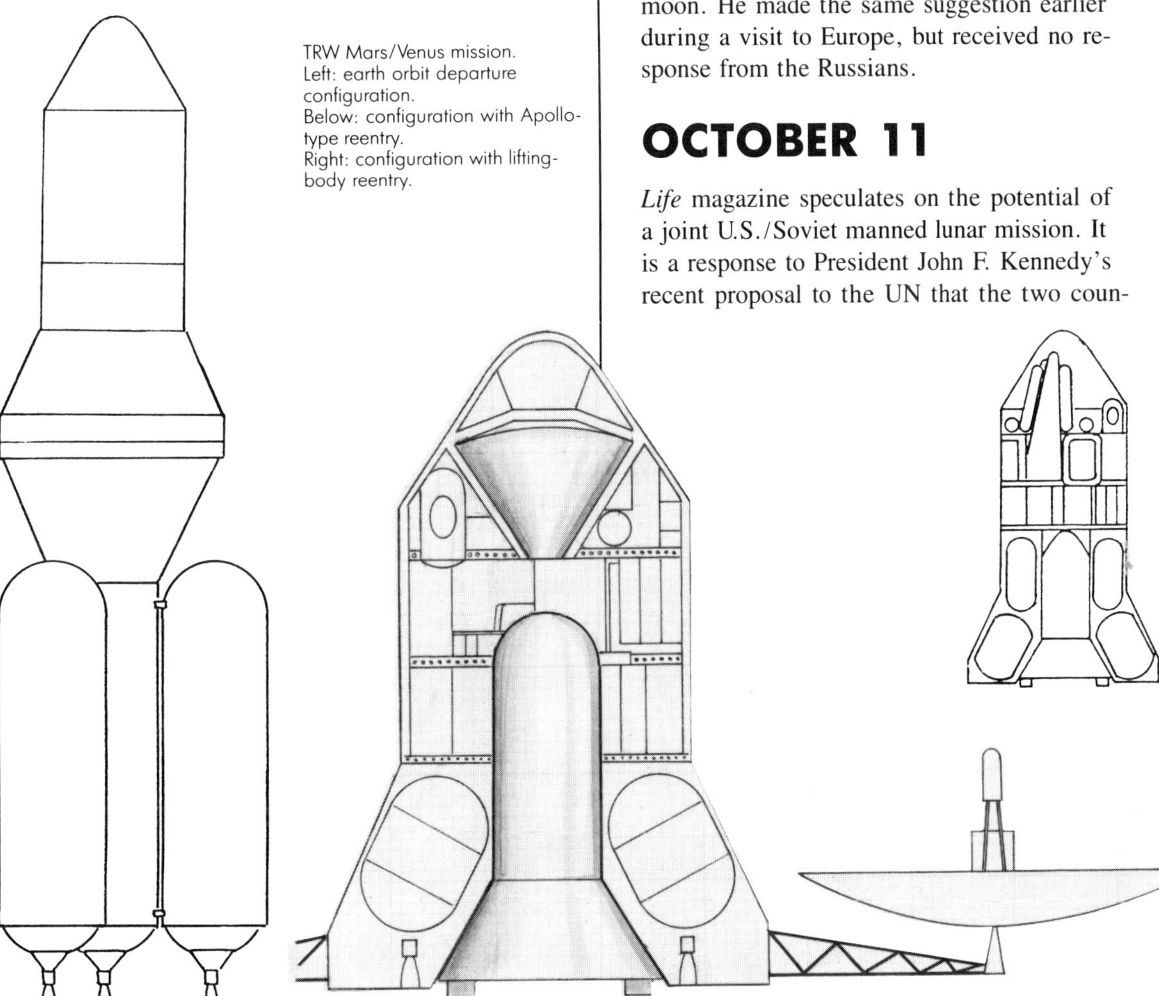

TRW Mars/Venus mission.
Left: earth orbit departure configuration.
Below: configuration with Apollo-type reentry.
Right: configuration with lifting-body reentry.

JUNE 16–19

Valentina Tereshkova, a junior lieutenant, becomes the first woman to fly in space. She makes forty-eight orbits of the earth in her Vostok 6 spacecraft in a flight lasting nearly 71 hours (2 days 22 hours 50 minutes).

AUGUST 22

The X-15, piloted by Joseph A. Walker, reaches an altitude of 67 miles, qualifying Walker for astronaut status.

SEPTEMBER 20

President John F. Kennedy suggests in a speech to the United Nations General Assembly that the United States and the Soviet Union unite in a joint manned flight to the moon. He made the same suggestion earlier during a visit to Europe, but received no response from the Russians.

OCTOBER 11

Life magazine speculates on the potential of a joint U.S./Soviet manned lunar mission. It is a response to President John F. Kennedy's recent proposal to the UN that the two coun-

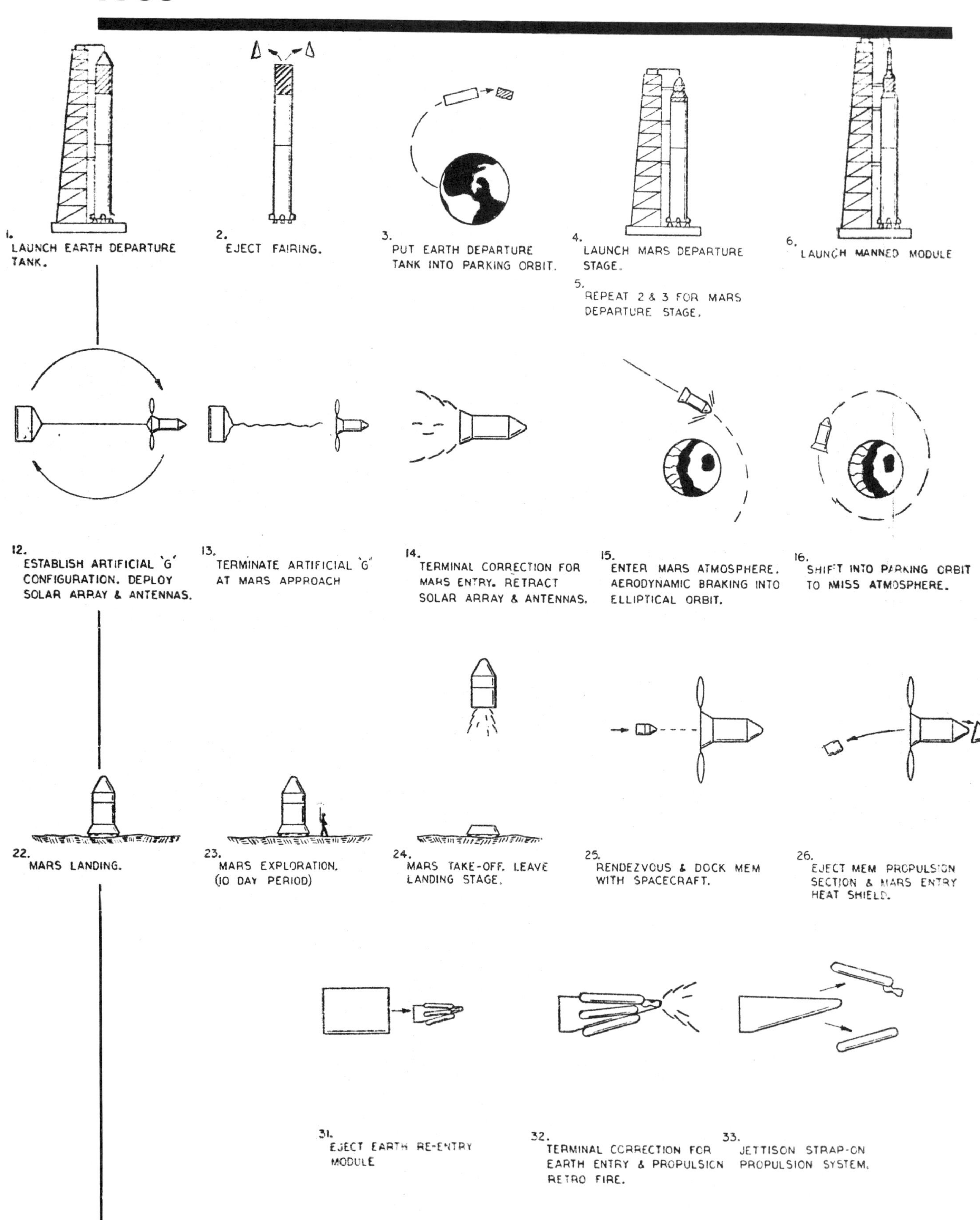
1.
LAUNCH EARTH DEPARTURE TANK.
2.
EJECT FAIRING.
3.
PUT EARTH DEPARTURE TANK INTO PARKING ORBIT.
4.
LAUNCH MARS DEPARTURE STAGE.
5.
REPEAT 2 & 3 FOR MARS DEPARTURE STAGE.
6.
LAUNCH MANNED MODULE
12.
ESTABLISH ARTIFICIAL 'G' CONFIGURATION. DEPLOY SOLAR ARRAY & ANTENNAS.
13.
TERMINATE ARTIFICIAL 'G' AT MARS APPROACH
14.
TERMINAL CORRECTION FOR MARS ENTRY. RETRACT SOLAR ARRAY & ANTENNAS.
15.
ENTER MARS ATMOSPHERE. AERODYNAMIC BRAKING INTO ELLIPTICAL ORBIT.
16.
SHIFT INTO PARKING ORBIT TO MISS ATMOSPHERE.
22.
MARS LANDING.
23.
MARS EXPLORATION. (10 DAY PERIOD)
24.
MARS TAKE-OFF. LEAVE LANDING STAGE.
25.
RENDEZVOUS & DOCK MEM WITH SPACECRAFT.
26.
EJECT MEM PROPULSION SECTION & MARS ENTRY HEAT SHIELD.
31.
EJECT EARTH RE-ENTRY MODULE
32.
TERMINAL CORRECTION FOR EARTH ENTRY & PROPULSION RETRO FIRE.
33.
JETTISON STRAP-ON PROPULSION SYSTEM.

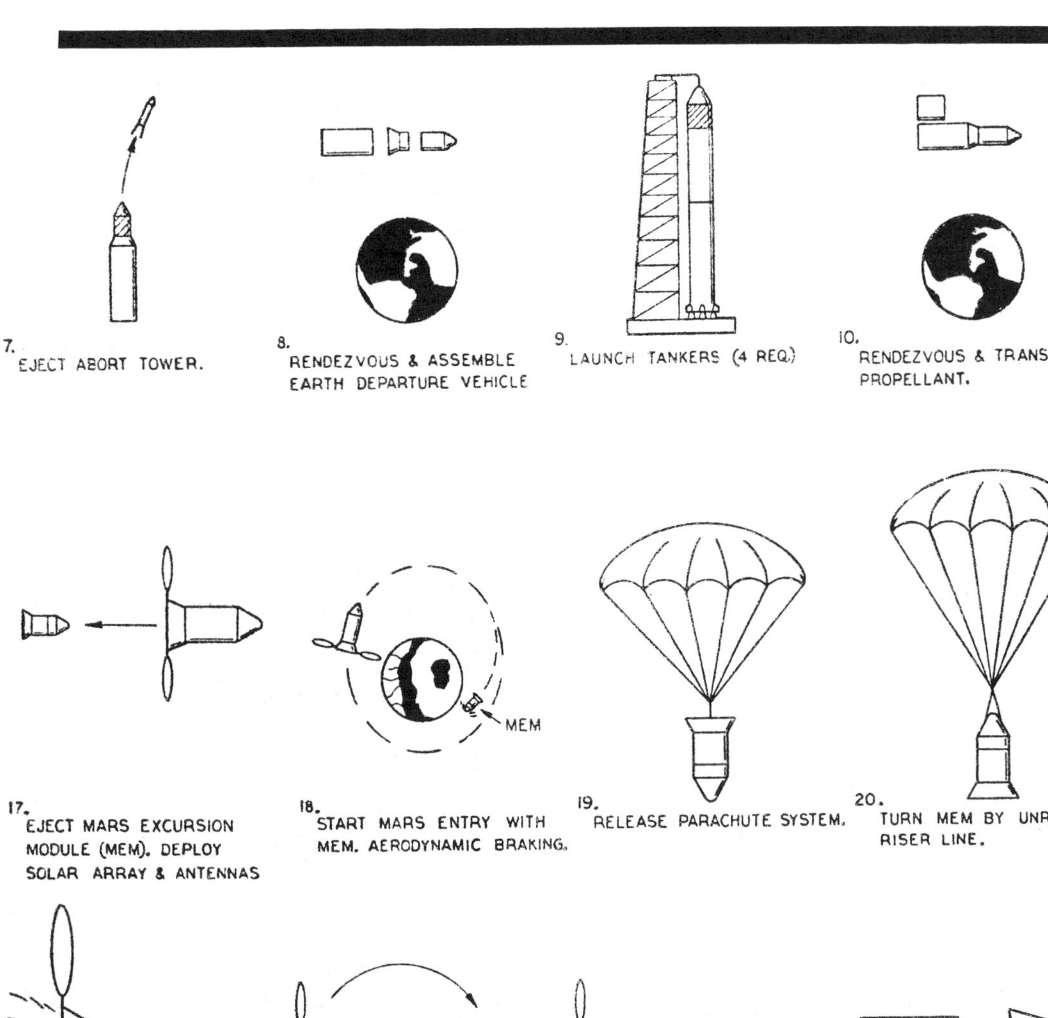

7. EJECT ABORT TOWER.

8. RENDEZVOUS & ASSEMBLE EARTH DEPARTURE VEHICLE

9. LAUNCH TANKERS (4 REQ.)

10. RENDEZVOUS & TRANSFER PROPELLANT.

11. EARTH DEPARTURE & INITIAL CORRECTION

17. EJECT MARS EXCURSION MODULE (MEM). DEPLOY SOLAR ARRAY & ANTENNAS

18. START MARS ENTRY WITH MEM. AERODYNAMIC BRAKING.

19. RELEASE PARACHUTE SYSTEM.

20. TURN MEM BY UNREEFING RISER LINE.

21. HOVER & TRANSLATE TO CHOOSE LANDING AREA.

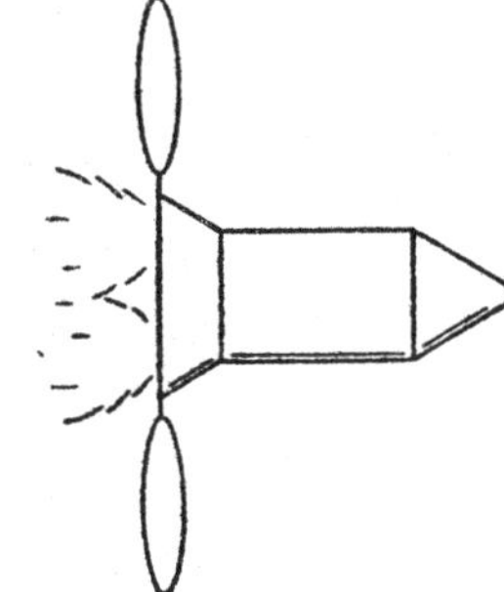

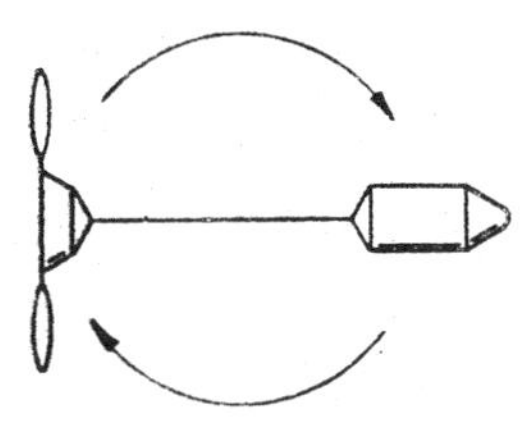

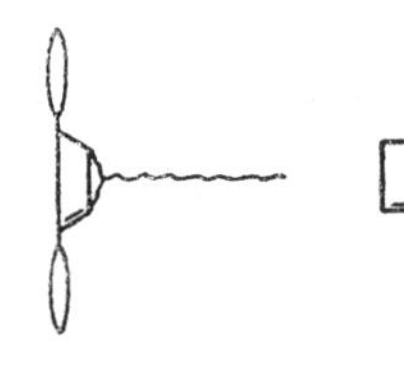

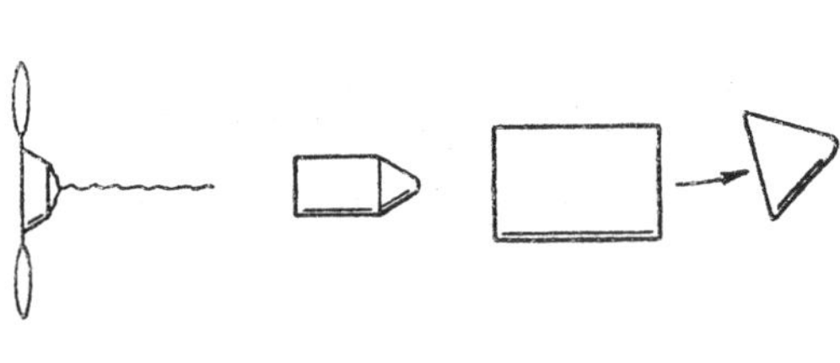

27. MARS DEPARTURE & INITIAL CORRECTION.

28. ESTABLISH ARTIFICIAL 'G' CONFIGURATION.

29. TERMINATE ARTIFICIAL 'G' AT EARTH APPROACH.

30. JETTISON NOSE COVER.

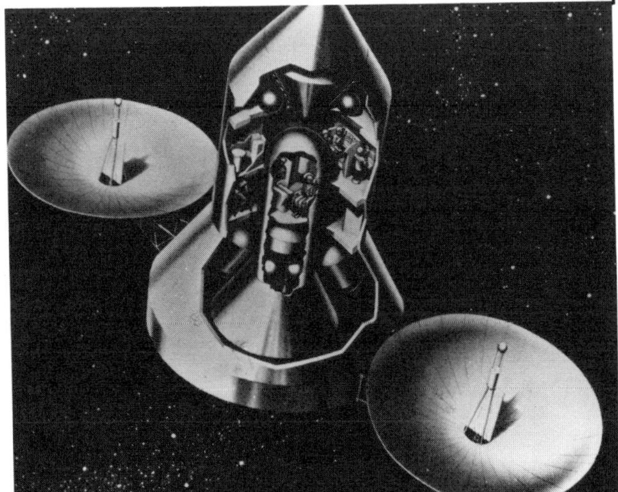

34. EARTH ENTRY, AERODYNAMIC BRAKING.

35. RELEASE PARACHUTE SYSTEM & LAND.

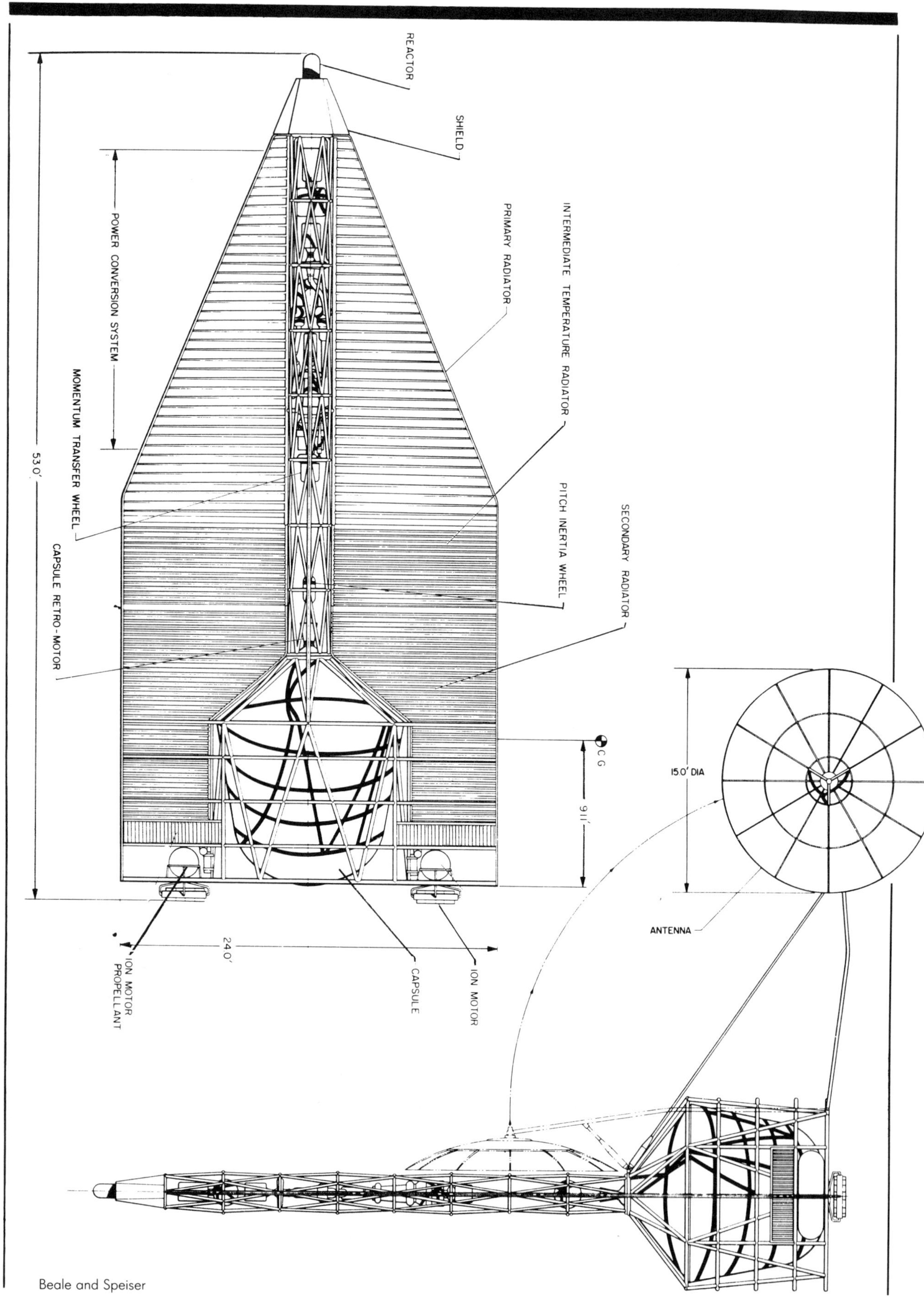

Beale and Speiser

tries cooperate in going to the moon. Although the proposal draws dissent from many space experts—who cite the difficulties that incompatible equipment and technology produce, to say nothing of the possible disclosure of secret information—an engineer at Republic Aviation, Thomas Turner, devises a scheme that he thinks might work.

The joint flight plan will entail the Soviet launch of one of their three-man Voskhod spacecraft, carrying only a two-man crew. Once this has gone into orbit, the United States will launch one of its lunar excursion modules (the LEM, later LM, or "bug" as *Life* prefers to call it). The two spacecraft will rendezvous and join, continuing on to the moon together, the single American astronaut that accompanied the lunar module now occupying the empty third seat in the Russian spacecraft. Once in lunar orbit, the lander, carrying one Russian cosmonaut and the American astronaut, will descend to the surface. After the exploration is complete, the ascent module will return to the waiting Soviet spacecraft for the return to the earth.

Earlier joint mission schemes have envisioned mating the necessary vehicles atop a single large booster, producing most of the worrisome security problems. Turner's scenario, however, alleviates this difficulty. The only hardware that have to be shared are the docking collar, which could easily be developed in its entirety by one nation. In a very real sense, Turner's plan will be realized in the 1975 Apollo-Soyuz Test Mission [see below].

Bottom, NASA post-X-15 hypersonic research aircraft. Two upper models capable of launching secondary stages to orbit.

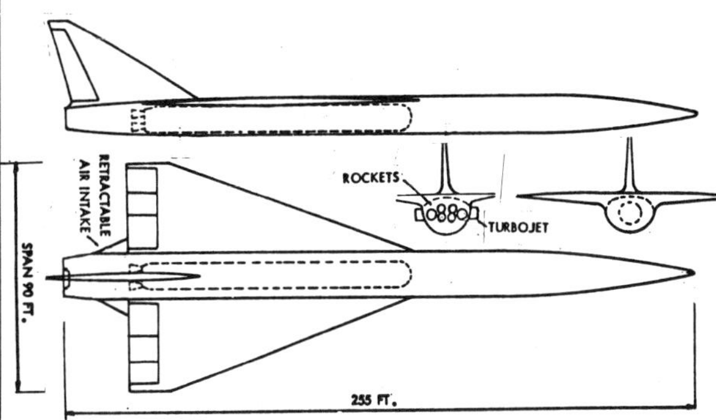

ca. 1963

The French aeronautical firm of Dassault Aviation begins development studies on reusable horizontal takeoff, two-stage spacecraft designs. The booster eventually developed is derived from a large delta-wing Mach 4 bomber powered by six turboramjet engines. This launches a small "space taxi" carrying a 2,000 pound payload.

The taxi—similar to that developed by the Centre National d'Etudes Spatiales (CNES) with a payload capacity of 3 to 4 tons—deploys variable-aspect wings and tail surfaces for its return flight to the earth. Its own turbofan engines allow it to maneuver for landing.

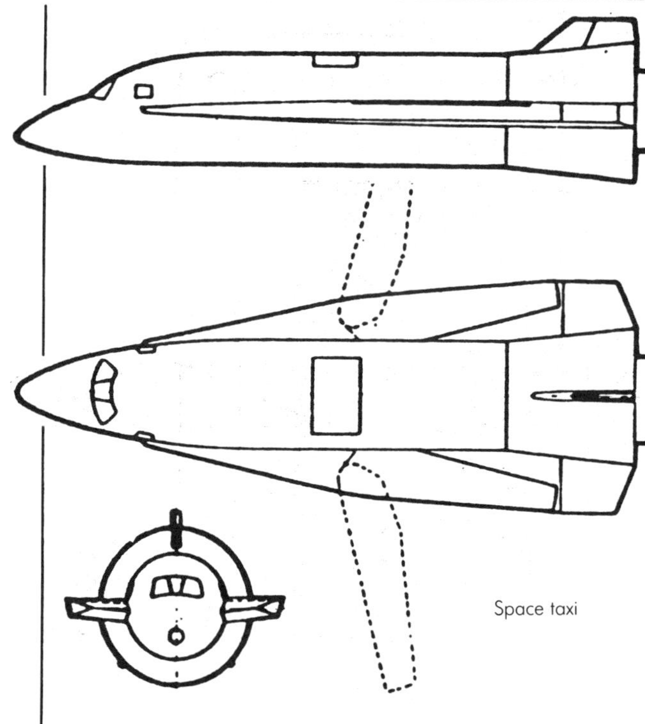

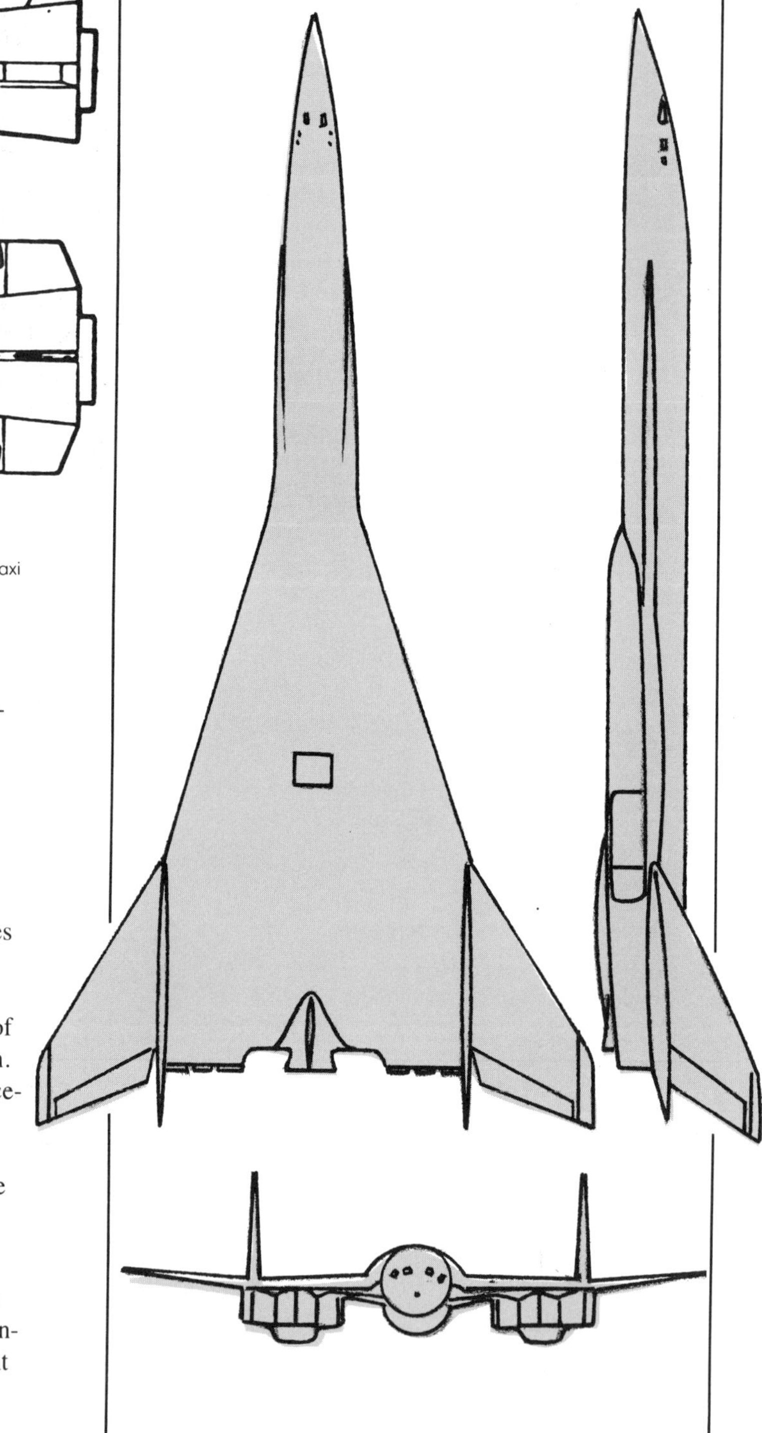

Meanwhile, Nord Aviation proposes its *Mistral* aerospace transporter project, which eventually will become a joint French-German effort involving Nord Aviation, SNECMA, and ERNO (Entwicklungsring Nord). *Mistral* is a 170-foot two-stage, winged space transport. The first stage is powered by a combination of ramjet engines with 72 tons of thrust and the 85-foot second stage, carried beneath the first, which has four rocket engines providing 35 tons of thrust each, and two of 0.7 tons thrust each. With a launch weight of 300 tons, the spacecraft can place 3 tons of payload into an orbit 300 km above the earth. The stages separate at a speed of Mach 7 at an altitude of 35 km, the second stage achieving an orbit of 186 miles.

Sud Aviation proposes similar projects with winged rocket-turbojet-powered and turbofan-ramjet-powered boosters (the latter is a joint study with the French Ministry of Defense and is not intended to launch manned vehicles).

ERNO carries out tests of all of these designs, and others, by performing drop tests of models from Luftwaffe Transall C-160 transport aircraft over the Baltic and Mediterranean.

In the search for more economical methods of boosting payloads into orbit, it is proposed that the stages of the Saturn V be converted for recovery and reuse. A design for a reusable Saturn V first stage requires attaching wings with a 150-foot span to the S-IC stage with six jet engines that allow it to fly back and land (automatically) like a conventional aircraft.

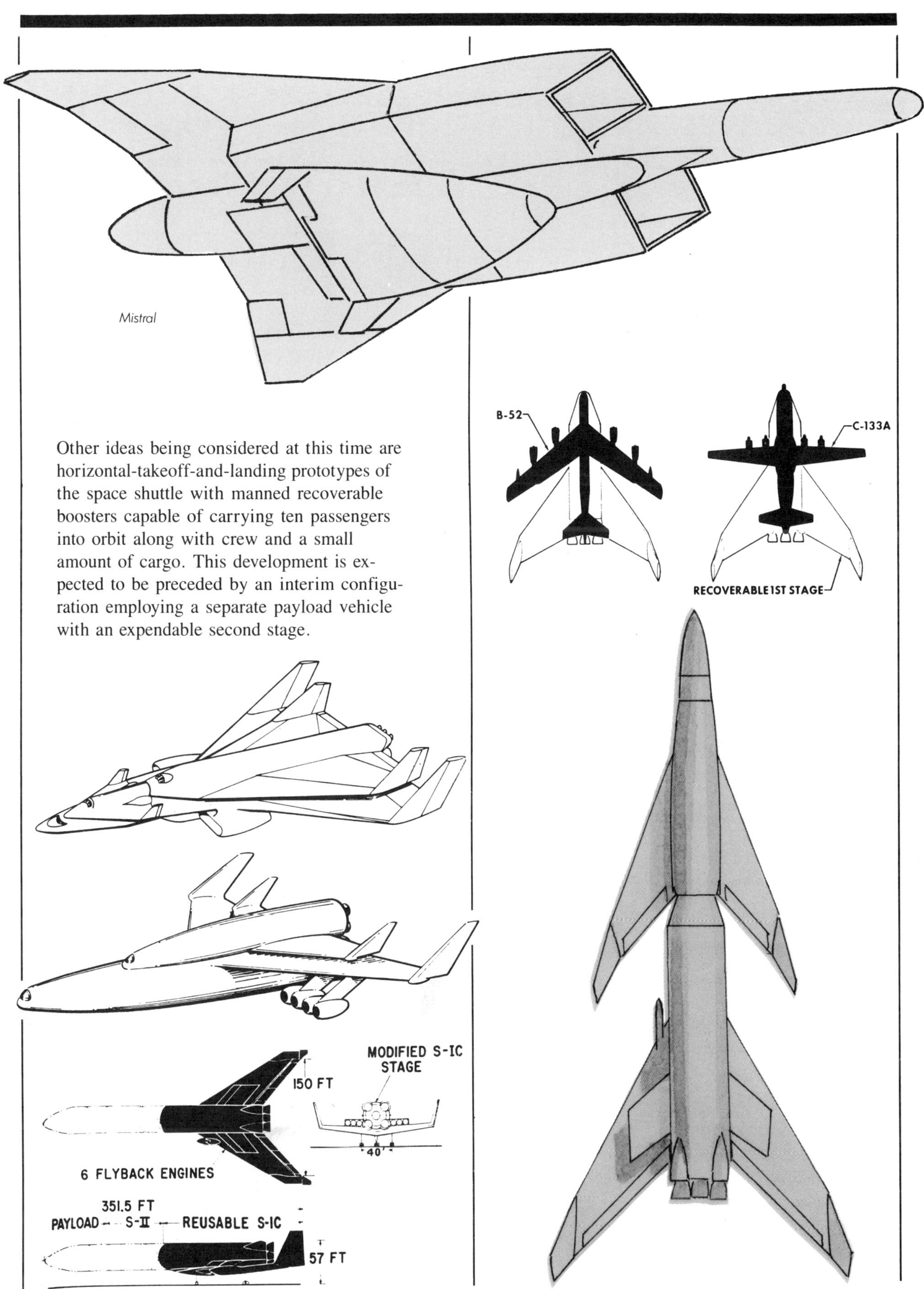

Mistral

Other ideas being considered at this time are horizontal-takeoff-and-landing prototypes of the space shuttle with manned recoverable boosters capable of carrying ten passengers into orbit along with crew and a small amount of cargo. This development is expected to be preceded by an interim configuration employing a separate payload vehicle with an expendable second stage.

1963

Other money and material-saving proposals include ground-based accelerators. Liquid-fuel rocket sleds are one consideration. The spacecraft's own engines are used with the sled itself only supplying the propellants [see Lent, above]. Another approach uses a "linear steam turbine," which employs turbine blades spaced along a track.

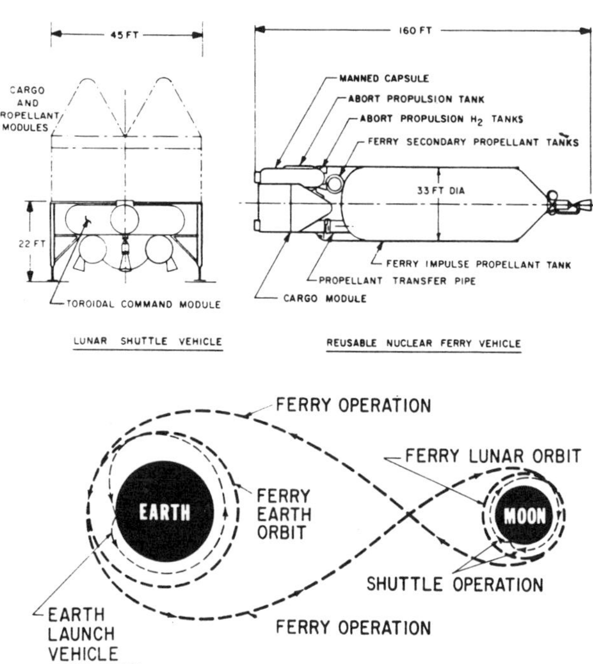

A reusable nuclear ferry rocket is proposed by NASA to cycle passengers and cargo between earth orbit and lunar orbit. Ordinary chemical rockets will then shuttle passengers and cargo to and from the surface of the earth and moon. Such a vehicle will weigh some 550,000 pounds, including 300,000 pounds of propellant.

1963–1965

Dwain F. Spencer and L. D. Jaffe at JPL propose staged reaction rockets for starflight.

The Prime Project, involving lifting body research using Martin SV-5D vehicles, shows the feasibility of maneuvering lifting body spacecraft during the hypersonic speeds of reentry, down to about Mach 2. The program is complementary with Asset [see above]. The first test flight is performed by the SV-5D. The second flight is that of the Air Force X-24A supersonic manned lifting body, with a configuration based on the SV-5 shape. At Edwards Air Force Base it explores reentry from Mach 2 to landing.

"Asset" and "Start" are programs that demonstrate that winged and lifting body spacecraft can successfully make reentry.

Asset (an acronym for Aerothermodynamic/elastic Structural Systems Environmental Test), a McDonnell Douglas program, produces six test vehicles launched by Thor boosters. The flight of Asset 1 on 18 September 1963 is the West's first hypersonic maneuverable boost glide flight through the earth's atmosphere. The last flight takes place on 23 February 1965. Asset spacecraft are launched to altitudes of up to 202,000 feet, from which they descend at speeds up to Mach 18.8. Asset is the first successful reusable non-ablative flightworthy airframe.

The SV-5 Start (Spacecraft Technology and Advanced Re-entry Test) is initiated by Martin in 1965 as the followup program to Asset. Start is a wingless, V-shaped lifting body, equipped with attitude thrusters and flaps for reentry. Launched by an Atlas booster, it is covered with an ablative coating to resist the heat of reentry.

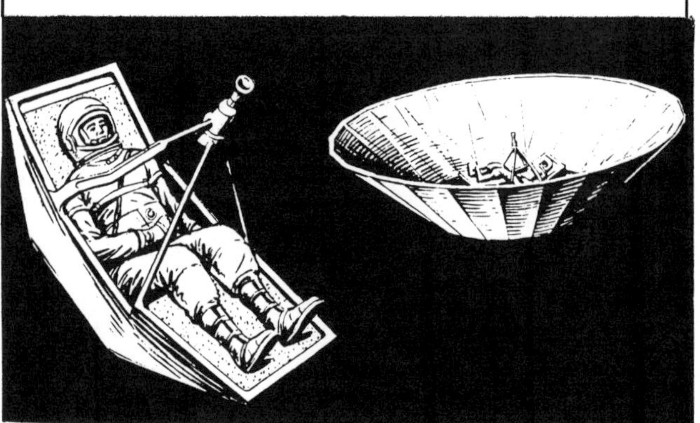

Douglas "space parachute" rescue system.

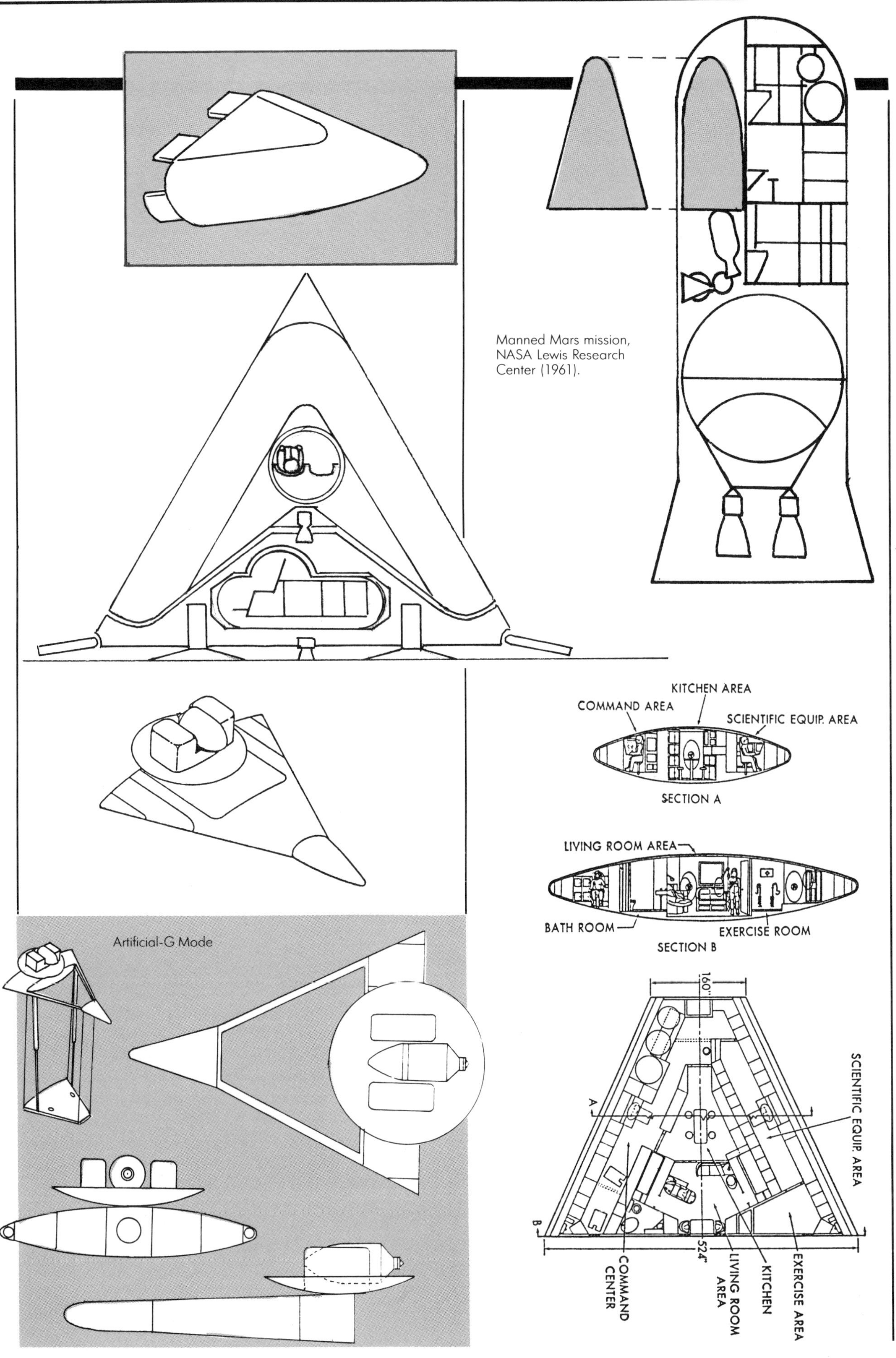

Manned Mars mission, NASA Lewis Research Center (1961).

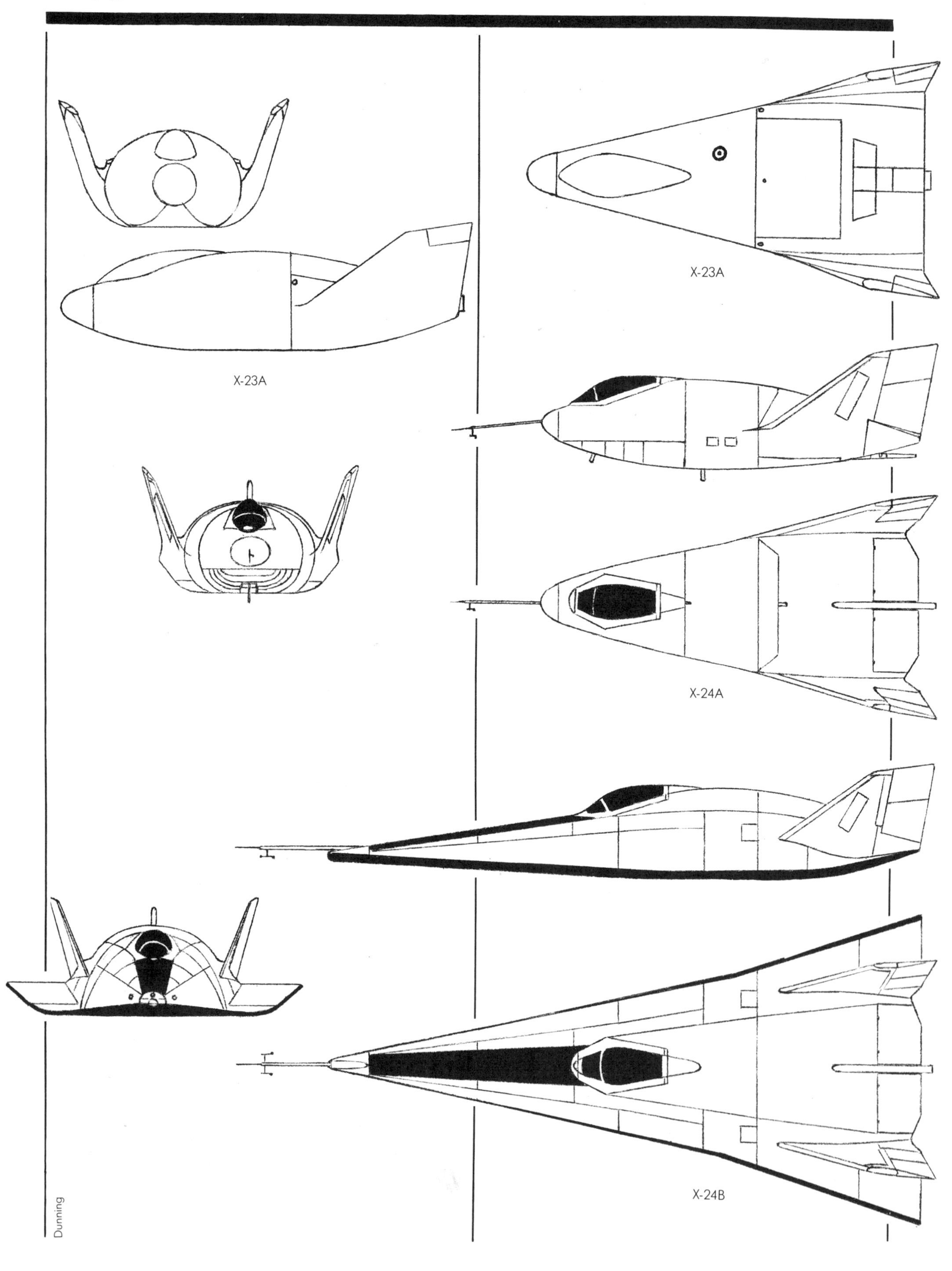
X-23A
X-23A
X-24A
X-24B
Dunning

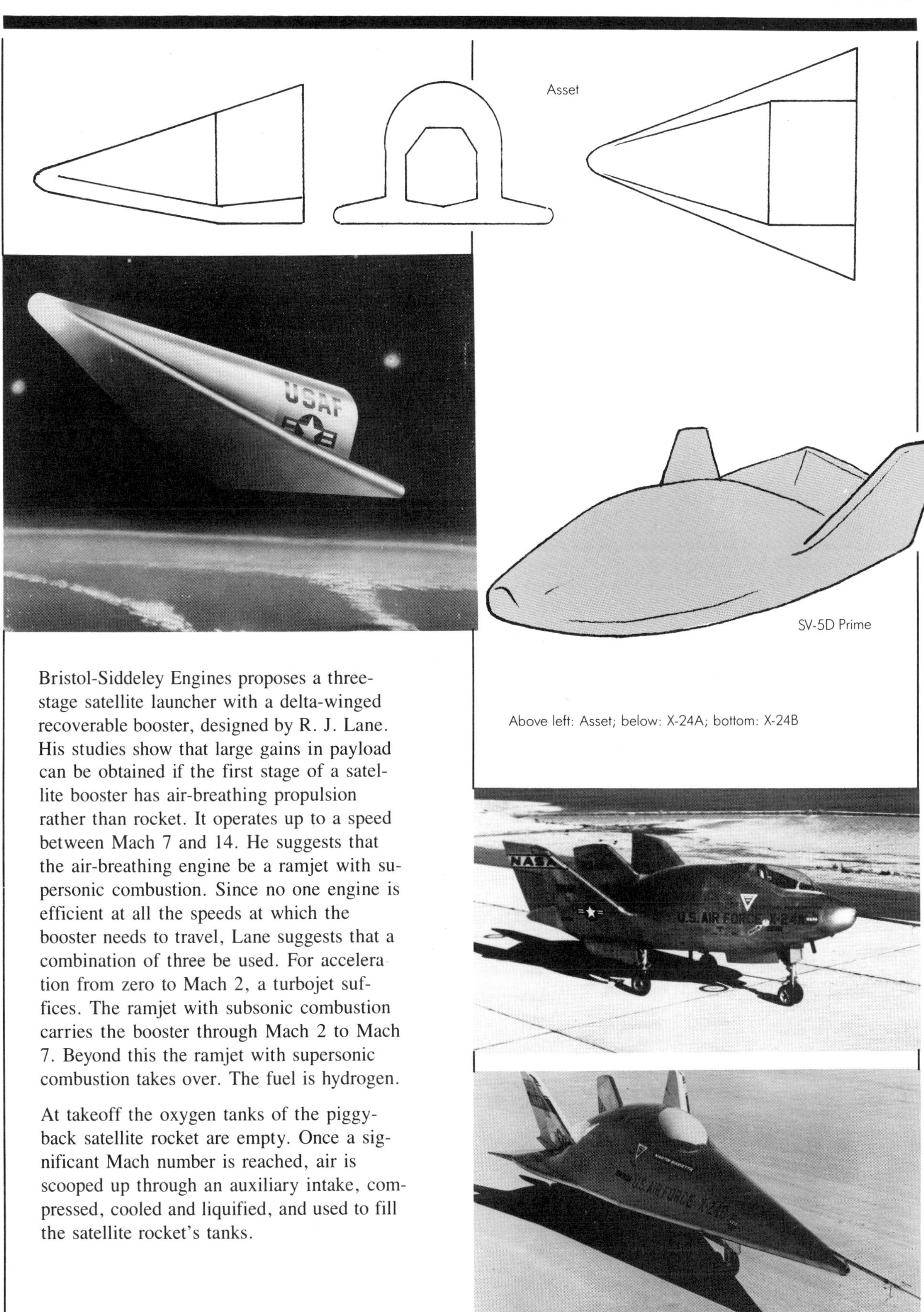

Above left: Asset; below: X-24A; bottom: X-24B

Bristol-Siddeley Engines proposes a three-stage satellite launcher with a delta-winged recoverable booster, designed by R. J. Lane. His studies show that large gains in payload can be obtained if the first stage of a satellite booster has air-breathing propulsion rather than rocket. It operates up to a speed between Mach 7 and 14. He suggests that the air-breathing engine be a ramjet with supersonic combustion. Since no one engine is efficient at all the speeds at which the booster needs to travel, Lane suggests that a combination of three be used. For acceleration from zero to Mach 2, a turbojet suffices. The ramjet with subsonic combustion carries the booster through Mach 2 to Mach 7. Beyond this the ramjet with supersonic combustion takes over. The fuel is hydrogen.

At takeoff the oxygen tanks of the piggyback satellite rocket are empty. Once a significant Mach number is reached, air is scooped up through an auxiliary intake, compressed, cooled and liquified, and used to fill the satellite rocket's tanks.

1964

Douglas Aircraft, under a NASA contract from the Marshall Space Flight Center, produces a report on "Manned Mars Explorations in the Unfavorable (1975–1985) Time Period. "The 96-meter nuclear-powered spacecraft would carry its crew on a 460 day multiple mission to Mars. While the six-person crew would live and work in a weightless environment, they would have individual "staterooms" as well as a centrifuge for exercise. This latter would be a torus equal to the outside diameter of the spaceship (approximately 18 meters) and would resemble, in function if not scale and amenities, the centrifuge in the film *2001* [see 1968].

C. W. Benfield of Minneapolis-Honeywell designs an unusual aerospace plane. It is disk-shaped, 2,000 feet in diameter, 100 feet thick at the center, tapering to sharp edges at the rim. A large circular indentation of 450,000 square feet in the center of the underside provides a ground effect when the twenty-four turbojet engines that surround it are operating. These can be converted to fission power, providing 28,000 pounds of thrust apiece.

The Soviet Union, after a series of unmanned lunar probes, begins serious studies for a manned lunar mission.

The Soviet agenda is composed of two independent projects: a flight around the moon by a manned spacecraft (a modification of the Soyuz) launched by a Proton rocket, and the landing of a lunar module with one cosmonaut aboard. The second part is to be a two-man mission, with one cosmonaut remaining in lunar orbit while the other makes the landing. The first part of the program is carried out in an unmanned mode with four probes successfully circling the moon. The Zond 5 is the first spacecraft to circle the moon and safely return to the earth (September 1968). According to Vasily Mishin, Sergei Korolyov's deputy and successor as chief designer, these probes are in fact test flights to the moon and back to the earth of

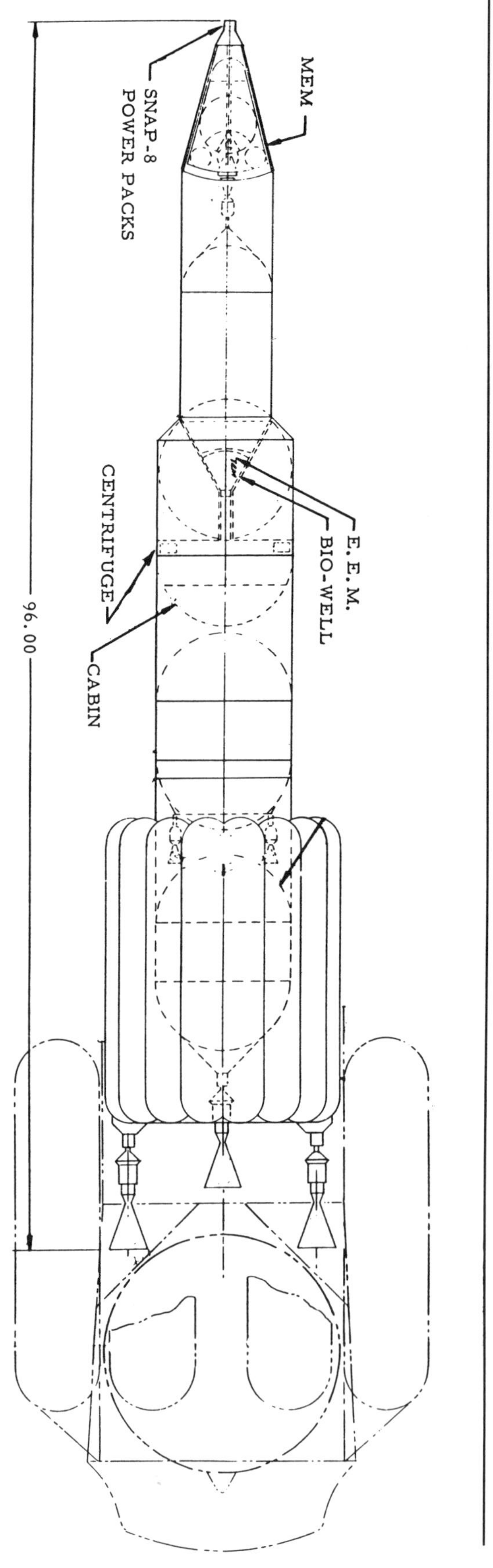

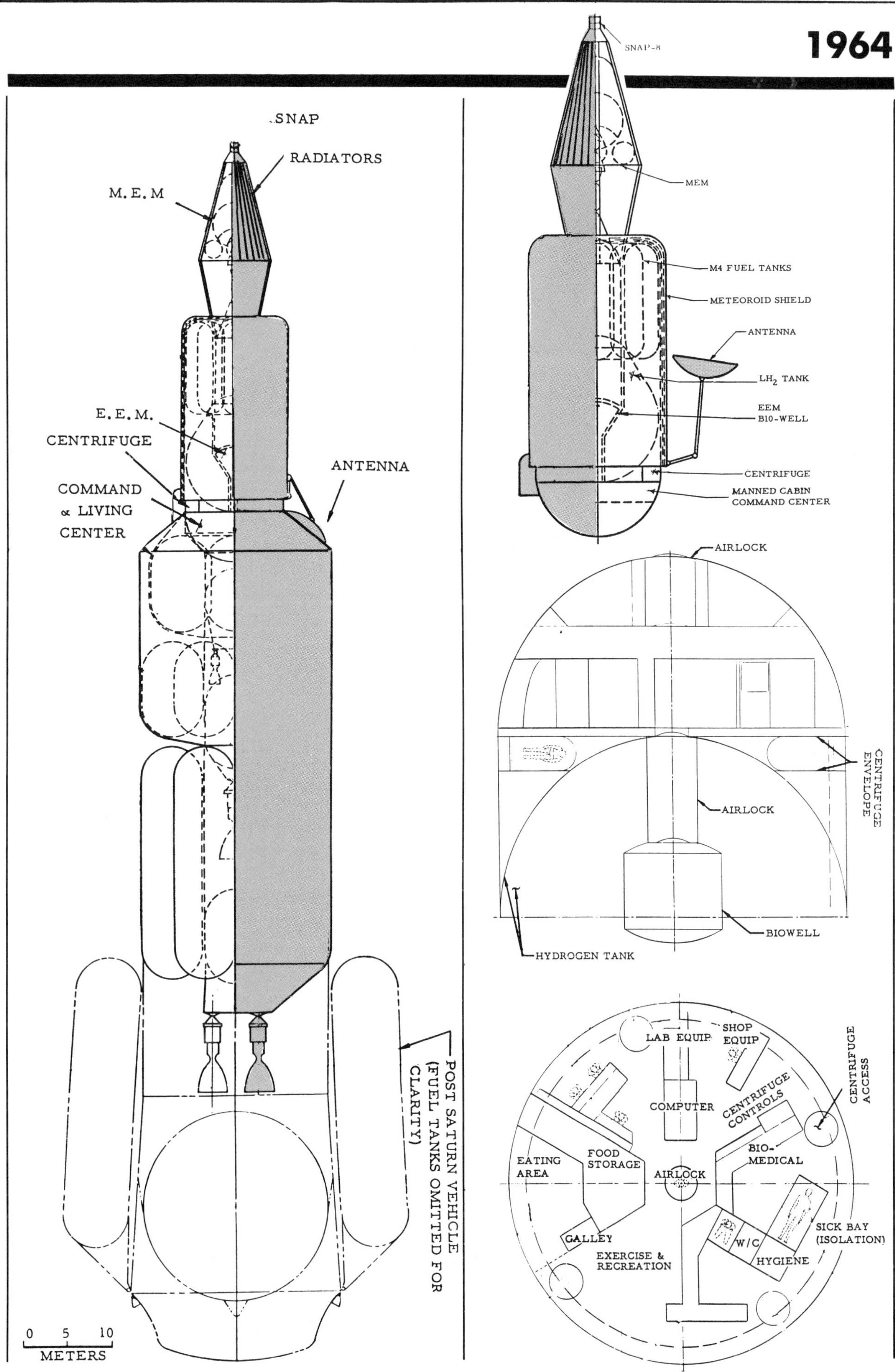
SNAP
RADIATORS
M. E. M
E. E. M.
CENTRIFUGE
COMMAND & LIVING CENTER
ANTENNA
POST SATURN VEHICLE (FUEL TANKS OMITTED FOR CLARITY)
0 5 10
METERS
SNAP-8
MEM
M4 FUEL TANKS
METEOROID SHIELD
ANTENNA
LH_2 TANK
EEM
B10-WELL
CENTRIFUGE
MANNED CABIN
COMMAND CENTER
AIRLOCK
CENTRIFUGE ENVELOPE
AIRLOCK
BIOWELL
HYDROGEN TANK
LAB EQUIP
SHOP EQUIP
COMPUTER
CENTRIFUGE CONTROLS
CENTRIFUGE ACCESS
EATING AREA
FOOD STORAGE
AIRLOCK
BIO-MEDICAL
GALLEY
W/C
SICK BAY (ISOLATION)
EXERCISE & RECREATION
HYGIENE

a two-man descent vehicle. They are designed by Yuri Semenov.

In the flyby mission a Proton booster is launched from Baikonur. The Zond spacecraft makes a half-orbit of the earth before lunar injection by the Proton fourth stage. [See 1961–1974.]

The Soviet lunar flyby mission is canceled after Apollo 8 in December 1968 and the manned landing is abandoned after the Apollo 11 success robs it of its propaganda value. According to Mishin, the project is also bogged down in inefficiency, bureaucracy and low technological expertise. Five hundred separate organizations representing twenty-six different government ministries and departments are involved.

Ling-Temco-Vought proposes a twenty-eight-passenger earth-moon shuttle to NASA. A lifting body carrying passengers is to be boosted into earth orbit. There it makes a rendezvous with a 125-foot-long nuclear-powered ferry. This boosts the combined vehicle to the moon and into 100-mile orbit around it. The lifting body then is swung down 17 feet on tubular struts and a smaller, spherical transfer vehicle attaches itself to the chemical stage of the nuclear booster and an airlock at the rear of the lifting body, through which the crew transfers. The spherical spacecraft then shuttles the passengers down to the lunar surface. The passenger section can make up to fifty round trips.

Franklin P. Dixon reports on a Mars excursion module that would support a 10-day stay on the Martian surface by two astronauts. The lifting body spacecraft would allow considerable lateral range in choice of landing sites. A crew of eight would remain in orbit, in the Mars mission module.

Constantin P. Lent publishes his own variation on the Saturn V Apollo moon rocket, a 150-foot-tall spacecraft with four 80-foot solid-fuel boosters surrounding the first stage. The lander is equipped with his standard "landing pick" as well as a

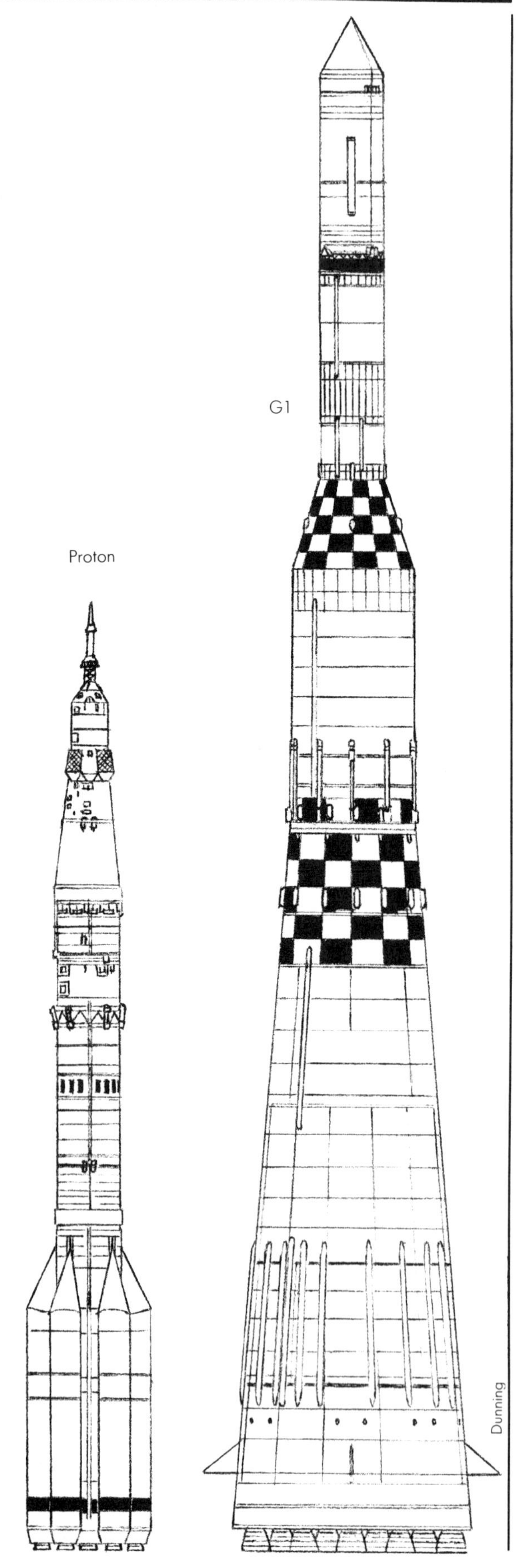

thick, pancake-shaped cushion inflated by mercury vapor that is attached to the base. This is intended to keep the lander from sinking in the deep seas of dust that Lent expects to find on the moon.

JANUARY 5

Eugen Sänger, in a speech to the delegates of the European aerospace industries, says: ."‥ An efficient, economical system of space transportation will become topical at a time when the problems of rendezvous technique have been solved in Earth orbit; furthermore, when the first successful manned landings on the moon have taken place, the construction of large manned permanent lunar bases has to be started. For undertakings like these, which may be realized, presumably during the next decade—an economic system of spaceflight becomes an absolute prerequisite . . . "

Sänger, a pioneer in the story of the space shuttle, dies 18 days later.

APRIL

Winston P. Sanders publishes the story "Sunjammer" in *Analog*. He describes a solar sail spaceship as a disk 4 1/2 miles in diameter. It is constructed of micron-thick aluminized plastic stiffened by foam-filled struts. It weighs nearly 100 tons and slowly rotates to keep from collapsing. The disk is unmanned, with the instrumentation and controls contained within a pod at its center. The purpose of the sails is to haul 30foot-diameter containers of liquified gas from Jupiter, where it has been collected by manned "scoopships," to the earth. They are accompanied by manned "herdships" [also see May 1952].

MAY 28

The Saturn SA-6 carries the first Saturn payload, a "boilerplate" version of the Apollo capsule.

JUNE

NASA's Ad Hoc Committee on Hypersonic Lifting Vehicles endorses development of two-stage-to-orbit shuttlecraft.

OCTOBER 12

The first multi-man spaceflight is made when the normally two-man Voskhod 1 takes a third crewmember into orbit (occupying the space usually alloted to an airlock). The cosmonauts are K. Feokstitov, V. Komorov and B. Yegorov. The cosmonauts have neither pressure suits nor ejection seats—there is no room in the crowded capsule. The vehicle is launched from Tyuratam by

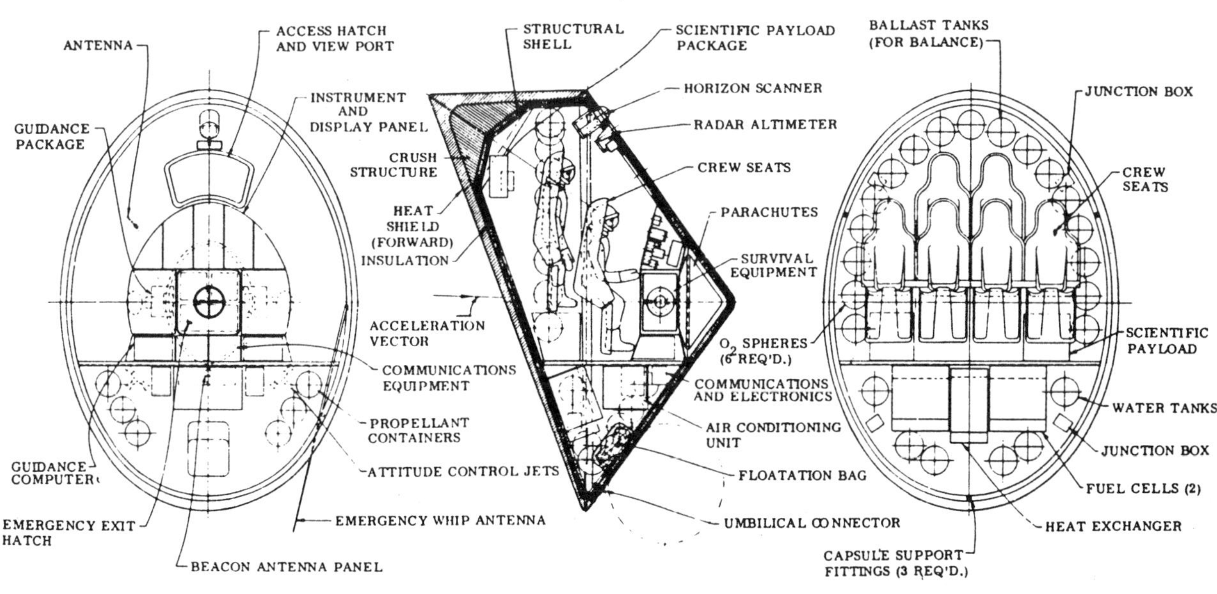

Temco reentry vehicle

1964

NASA

Delta-winged X-15

an A-2 rocket.

This is the first flight of the Voskhod spacecraft, which is little more than an upgraded Soyuz spacecraft weighing 5,320 kg. Like Soyuz, Voskhod has a spherical reentry capsule. On top is a reserve solid-fuel retrorocket package.

OCTOBER 30

Joseph Walker makes the first "flight" of the Bell Lunar Landing Research Vehicle (LLRV). Hundreds of flights are eventually made, lasting up to 10 minutes and reaching altitudes of 800 feet. The purpose of the vehicle is to train astronauts in lunar landing techniques.

The ungainly looking craft is powered by a single, gimballed, vertically mounted General Electric fan jet engine with 4,200 pounds of thrust. Additional control is provided by two 500-pound throttleable hydrogen peroxide motors and sixteen attitude-control jets.

With a gross takeoff weight of 3,710 pounds it can climb at a rate of 100 fpm and stay aloft for 14 minutes. Its maximum altitude is 4,000 feet. Controlled by a single pilot the LLRV is 10 feet 6 inches tall and spans 13 feet 4 inches across its legs.

In case of emergency, the pilot is equipped with an ejection seat (which Neil Armstrong is once forced to use).

To simulate a lunar landing, the pilot takes the LLRV to an altitude of about 700 feet on the power of the main engine. As he approaches this altitude, he tilts the vehicle forward about 5 degrees in order to develop a forward velocity. When this is achieved, he returns the LLRV to a vertical position and the jet thrust is reduced. The two 500-pound lift rockets are fired to maintain the descent rate. The vehicle attitude is varied slightly to keep to the planned flight path. At about 100 feet the rocket thrust and pitch attitude are increased to reduce the rate of descent and forward velocity. As the latter stops, the LLRV's attitude is reduced to vertical and the final touchdown is made with minor corrections made by the thrusters.

1964–1965

MUSTARD is a reusable space shuttle developed by the British Aircraft Corporation, after investigating a number of horizontal and vertical takeoff configurations (after the Royal Aircraft Establishment has been studying various two-stage horizontal takeoff designs). The thirty members of the design team, led by Tom Smith, are under contract to study hypersonics. Eventually their ideas evolve from "Concordish" aircraft to cargo-carrying spacecraft. They eliminate the concept that booster stages need to be stacked because two different rocket designs increase expense. Instead, three identical units are placed side by side. The three will take off vertically; two acting as boosters while the

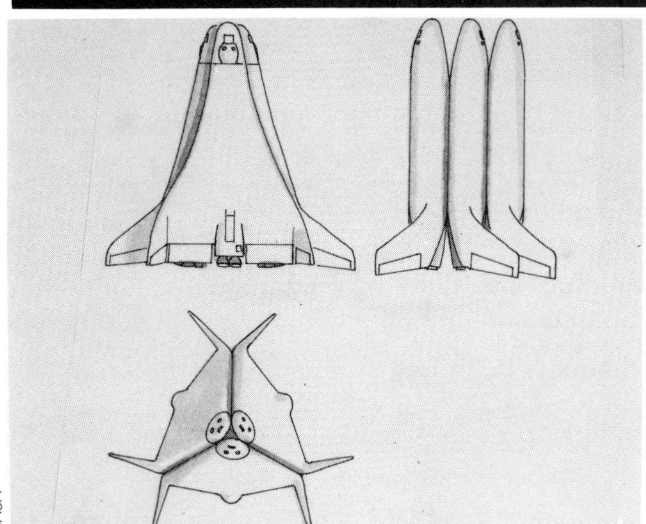
NASA

LLRV

third goes into orbit. An advantage is that all three motors are firing simultaneously and fuel can be pumped into the orbiter so that at separation it still has full tanks.

Each MUSTARD (**M**ulti-**U**nit **S**pace **T**rans-port and **R**ecovery **D**evice) ship is a "braced balloon," like the innovative Atlas ICBM. They can be flown back to base by a pilot or by remote control and are totally reusable. Each is a delta-shaped lifting body, considerably larger than today's shuttle. An additional booster unit can be added for heavier payloads, or even for flights to the moon or beyond, since the orbiter reaches space with a full load of fuel. Its designers calculate a cost break-even point after 50 to 100 missions, with major overhauls for the orbiter after 25 to 50 missions and after 200 missions for the boosters.

The MUSTARD concept is shelved by the British government. In 1970, its designers are invited to join a North American Rockwell study team, where they contribute their knowledge and experience of large airframe design gained from the Concorde. The only scale model ever made of MUSTARD is now in the Smithsonian Institution.

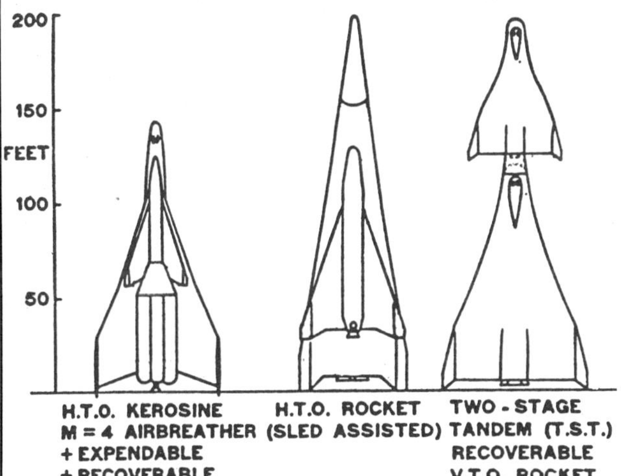

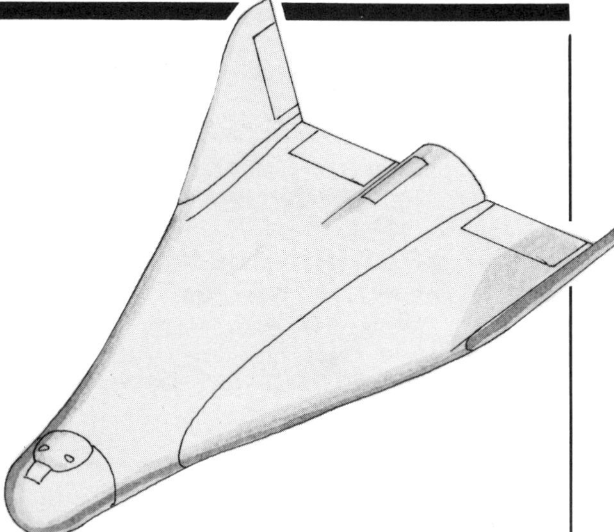

Working with the U.S. Air Force Flight Dynamics Laboratory leads Lockheed into developing a lifting-body derived space shuttle. The result is a stage-and-a-half design that includes a slim delta-shaped lifting body orbiter flanked by a pair of large cylindrical fuel tanks. The idea combines the idea of a high lift/drag ratio reentry shape with the parallel strap-on tank configuration developed by Alfred Draper and Charles Cosenza. The result is the Lockheed 8MX.

The orbiter's fuselage is 164 feet long

(186.5 to the tips of its tailfins), 44 feet high (including fins) and spans 106 feet. A pair of small swing wings extend from the sides of the orbiter for its gliding landing. Two 23.67-foot-diameter fuel tanks, in an inverted V, flank the orbiter. The completely assembled rocket is 239.5 feet tall. The orbiter has five engines producing 5.13 million pounds of thrust. It can carry 50,000 pounds of payload into orbit.

A partially expendable system such as this, as opposed to a fully reusable system, such at the aborted Aerospaceplane [see 1957–1964], is far more economical in terms of both research and development. And since propellants are carried externally, this frees more volume for payload. Lockheed submits a similar proposal to NASA, which it calls "Starclipper." This orbiter is 166 feet long, 91 feet wide, and 38 feet tall.

At about this same time McDonnell-Douglas proposes its own stage-and-a-half concept, using parallel tanks, which is submitted to the NASA Space Shuttle Task Group. The McDonnell-Douglas design has the 130-foot shuttle surrounded by four 24-foot-diameter fuel tanks, two of them 150 feet long and the other two 273 feet long.

Lunar lander proposed by Atomic Energy Commission

ca. 1964

In Europe the Orbital Fighter (manned two-stage transport, ramjet first stage) is proposed by the Royal Aircraft Establishment. The EUROSPACE glider is a smaller European version of the X-20 Dyna-Soar.

Ernst Stuhlinger proposes an electrically propelled ferry that will go from earth orbit to moon orbit.

The Astro A2 is a two-stage transport designed by Douglas, employing a vertical launch.

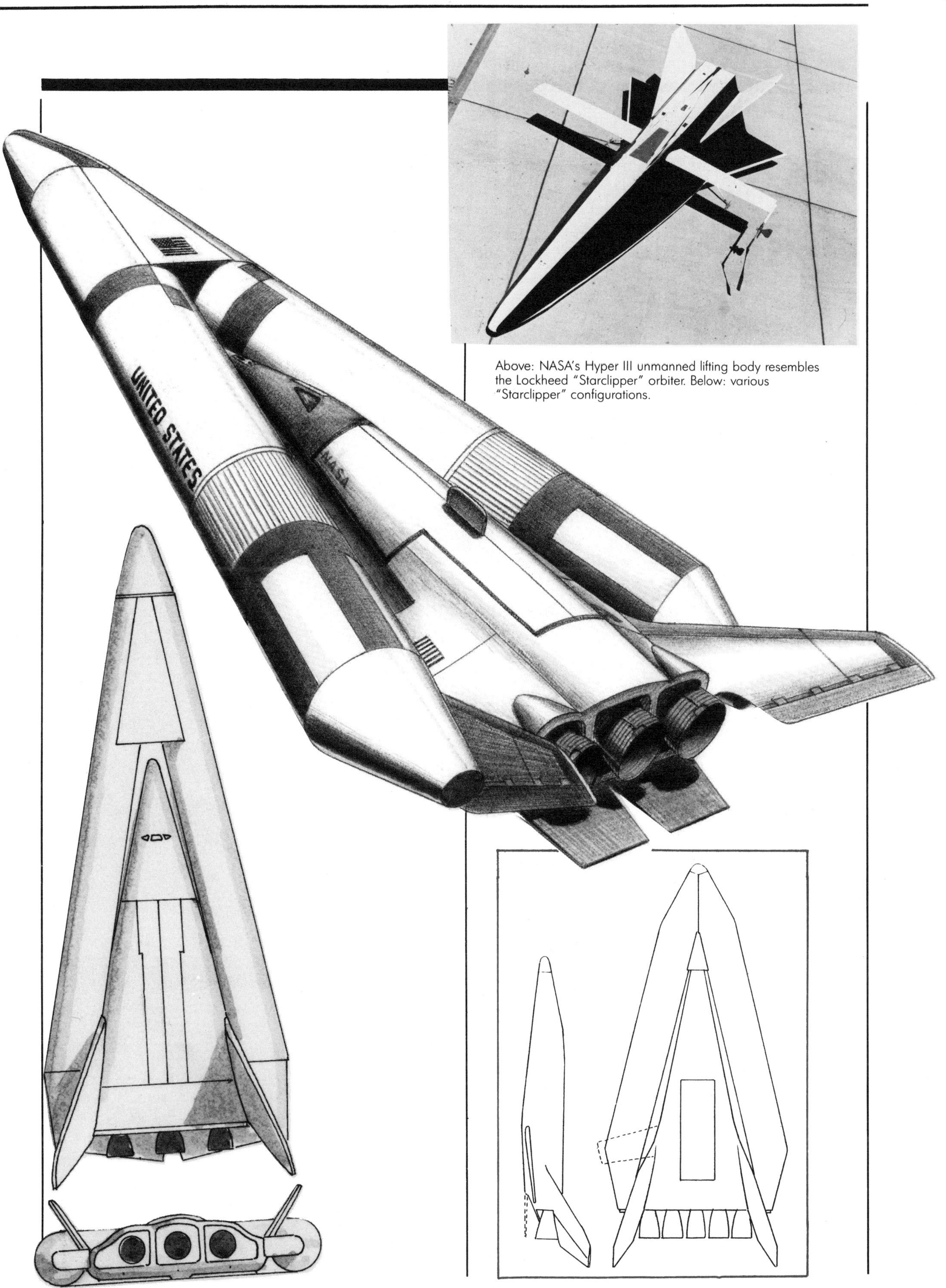

Above: NASA's Hyper III unmanned lifting body resembles the Lockheed "Starclipper" orbiter. Below: various "Starclipper" configurations.

1965

Douglas Missile and Space Systems proposes a manned flyby of Mars. The 22-month-long flight could be launched within 8 years. The mission, developed by M. W. Root under the direction of T. J. Gordon, calls for the assembly of a 250-foot-long spacecraft in earth orbit. This consists of three Saturn S-IVB stages linked in tandem and propelling a 100-ton spacecraft made up by joining a six-man Apollo command module to a 21.5-foot-diameter, 42-foot-long space laboratory (similar to the already planned Manned Orbiting Research Laboratory). The S-IVB stage, which is built by Douglas, is powered by a 200,000-pound-thrust Rocketdyne J-2 engine, using hydrogen and oxygen as propellants. Since virtually all of the hardware is either currently in production or in an advanced state of engineering study, the mission could be accomplished within the 1973–1977 timeframe without the expense or time required to develop new vehicles.

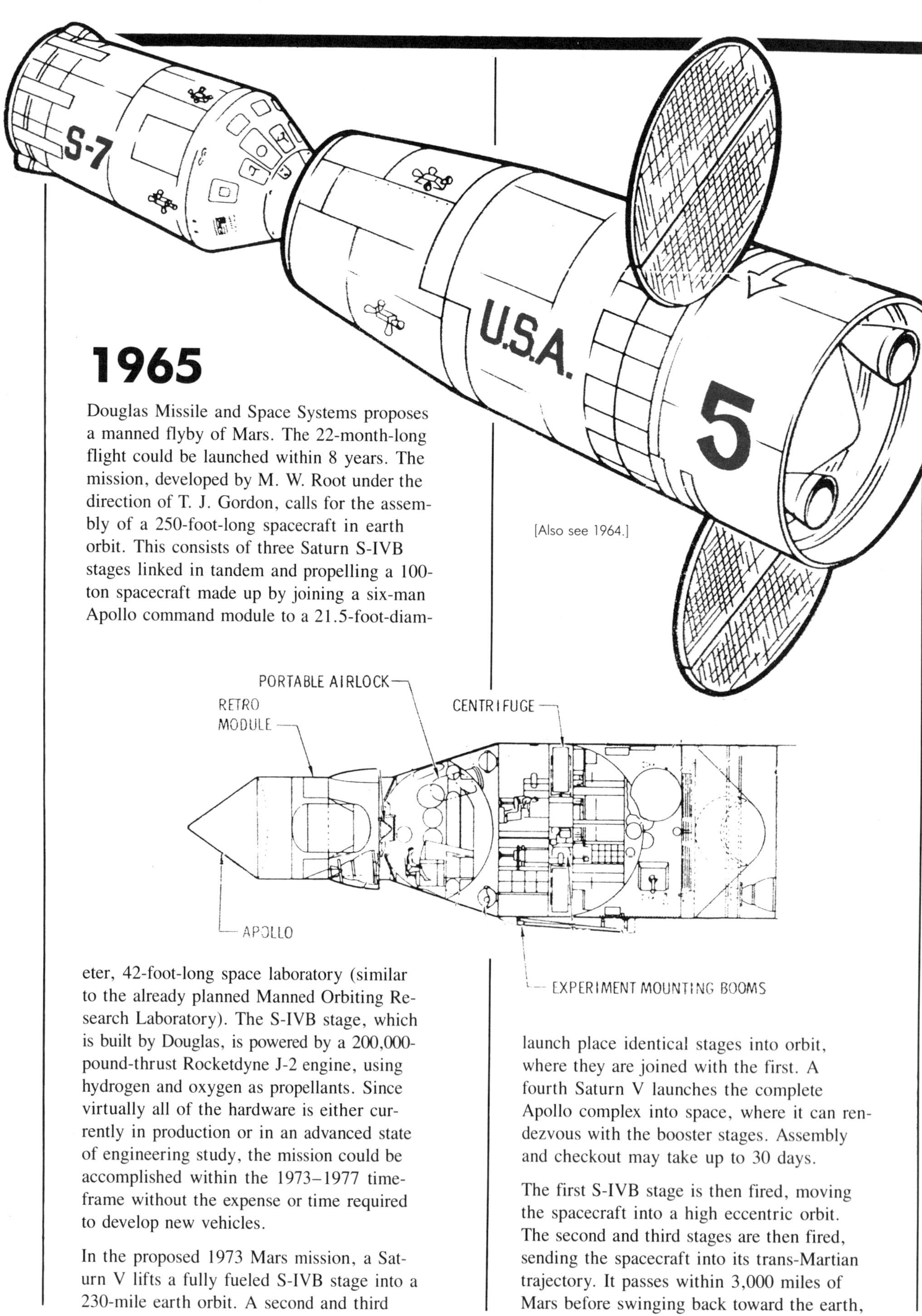

[Also see 1964.]

In the proposed 1973 Mars mission, a Saturn V lifts a fully fueled S-IVB stage into a 230-mile earth orbit. A second and third launch place identical stages into orbit, where they are joined with the first. A fourth Saturn V launches the complete Apollo complex into space, where it can rendezvous with the booster stages. Assembly and checkout may take up to 30 days.

The first S-IVB stage is then fired, moving the spacecraft into a high eccentric orbit. The second and third stages are then fired, sending the spacecraft into its trans-Martian trajectory. It passes within 3,000 miles of Mars before swinging back toward the earth,

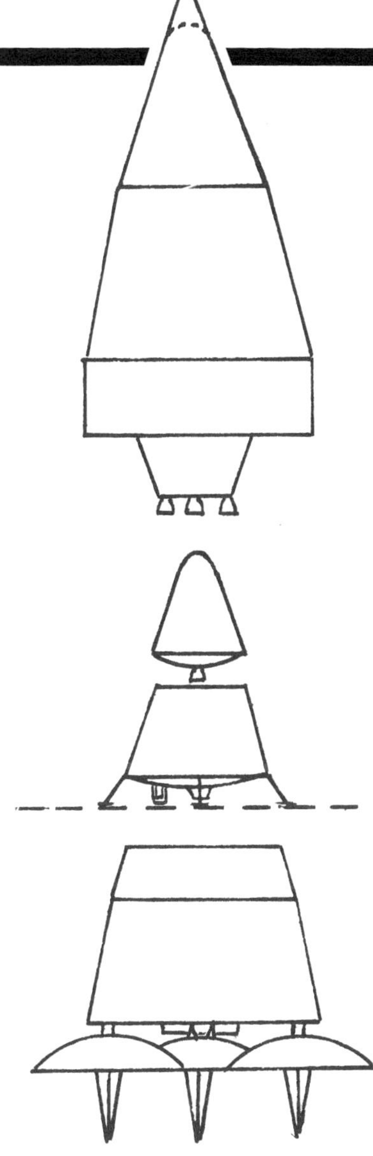

where it completes its mission after 655 days. During the Martian flyby, the crew launches 5-ton unmanned probes that orbit Mars and/or land on its surface.

NASA studies a horizontal takeoff two-stage spaceplane concept in which the 208.8-foot vehicle gets its initial impetus from a rocket-propelled ground accelerator. The spaceplane sits in a 145-foot-long F-1 engine-powered cradle that tilts the vehicle up at an angle of 20 degrees at the moment of takeoff. The cradle is equipped with waterbrakes that extend 26 feet beneath the trestles that support it.

The first stage of the spaceplane has a wingspan of 109 feet and carries the lifting body orbiter within a well on its upper surface.

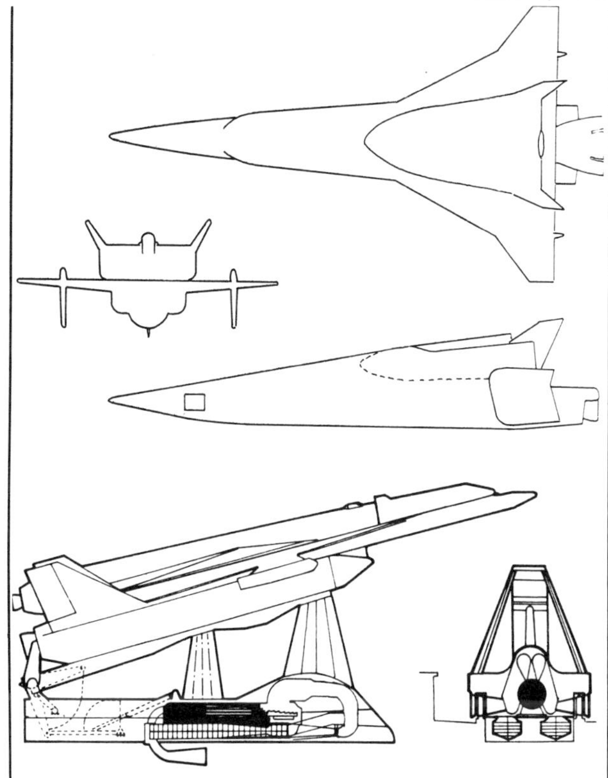

Robert Salkeld proposes using orbiters air-launched from specially modified C-5 cargo aircraft. A small single stage to orbit (SSTO) could provide economical delivery of large payloads into earth orbit. This could eventually lead to the development of a slightly larger version of the SSTO that could take off from a runway using its own integral rolling gear.

The small air-launched vehicle could provide a quick response to an orbital emergency since it would have a fast warning-to-rendezvous time.

Salkeld suggests a way of combining the advantages of the smaller and larger spaceplanes so that the heavy rolling gear of the latter is replaced by the lighter gear required for landing only. The larger spaceplane could then be air-launched from a specially-developed subssonic aircraft.

The small spaceplane would carry a crew of two and could fit in the cargo bay of a C-5. It would possess variable geometry wings for reentry and could carry three passengers in addition to the equipment needed for an extravehicular rescue. The spaceplane would be ejected from the rear doors of the C-5 by sliding along a ramp, assisted by a drag parachute. Two configurations include the addition of Minuteman II solid fuel boosters.

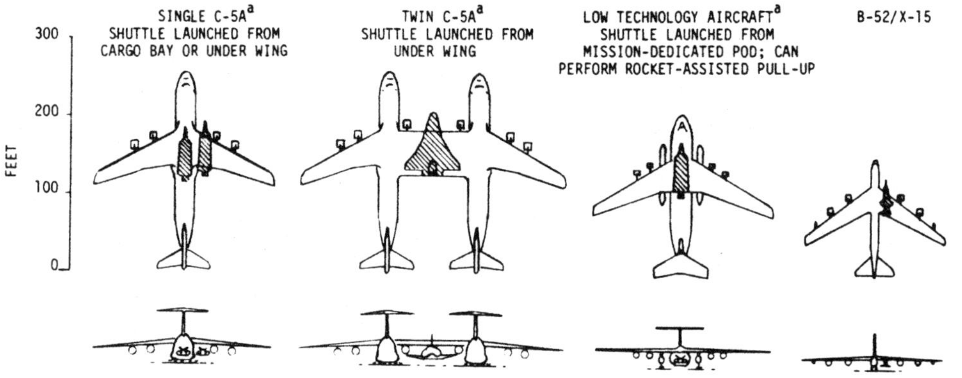

	SINGLE C-5A[a] SHUTTLE LAUNCHED FROM CARGO BAY OR UNDER WING		TWIN C-5A[a] SHUTTLE LAUNCHED FROM UNDER WING	LOW TECHNOLOGY AIRCRAFT[a] SHUTTLE LAUNCHED FROM MISSION-DEDICATED POD; CAN PERFORM ROCKET-ASSISTED PULL-UP		B-52/X-15
TAKE-OFF WT.LB.	764,500		1,600,000	725,000		400,000
LAUNCH WT.LB.	200,000		520,000	200,000		31,300
OP'N'L RADIUS.NM	1,500		1,500	2,000		
PAYLOAD, LB.[b]						
100 NM:POLAR	1,000[c]	2,800[d]	5,000	2,800[c]	4,800[d]	
100 NM/28.5°	3,500[c]	5,500[d]	12,000	6,500[c]	7,500[d]	(SUBORBITAL)

[a]DESIGNS & PERFORMANCE NOT OPTIMIZED; [b]3-g LIMIT; [c]EXTERNAL TANKS RETAINED TO ORBIT; [d]TANKS DROPPED DURING ASCENT; [e]ROCKET ASSISTED PULL-UP (LAUNCH AT 60,000 FT. 45° FROM LOCAL VERTICAL)

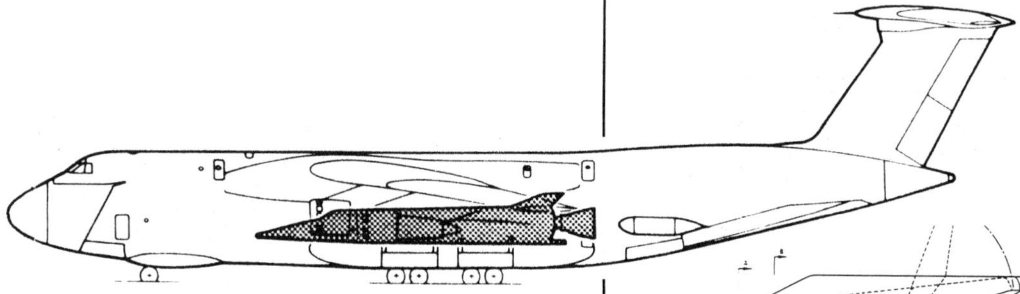

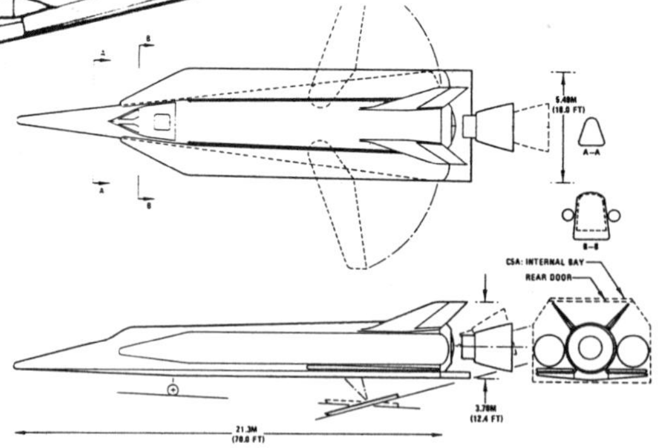

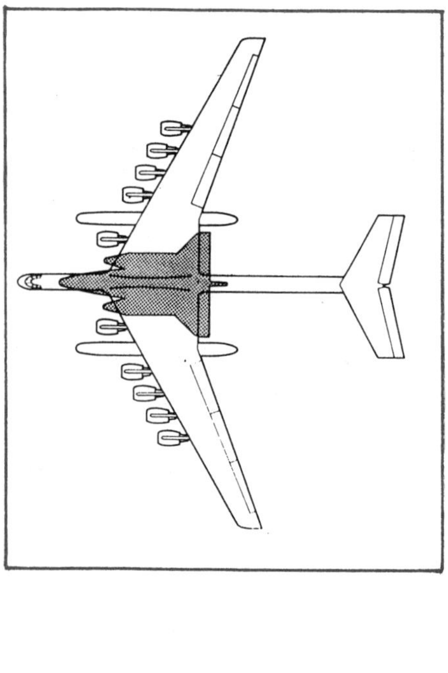

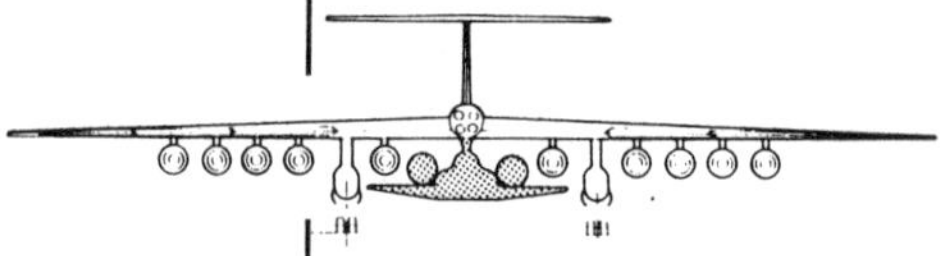

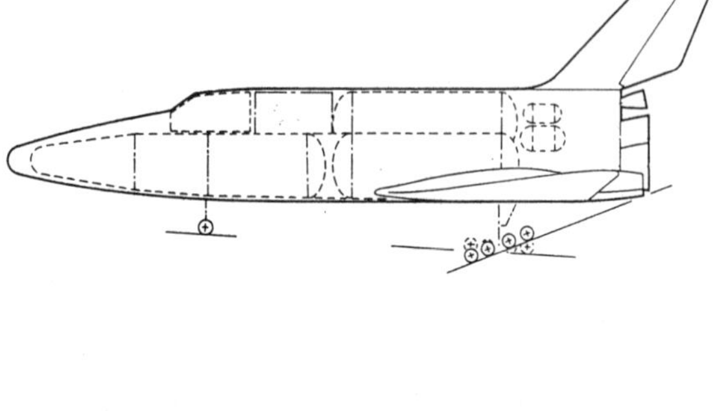

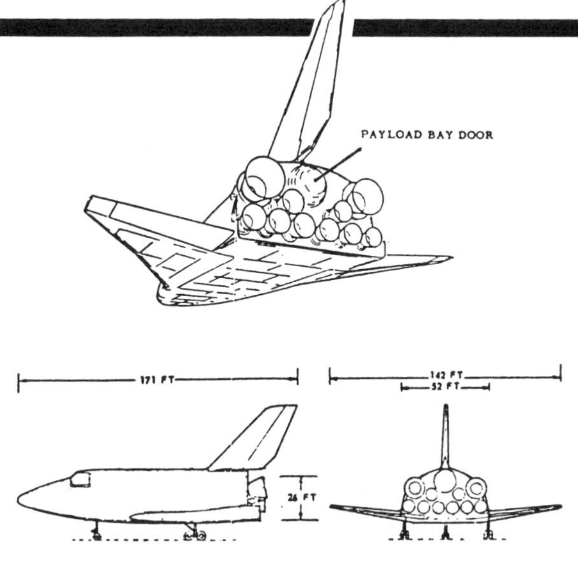

Salkeld's SSTO for runway takeoff.

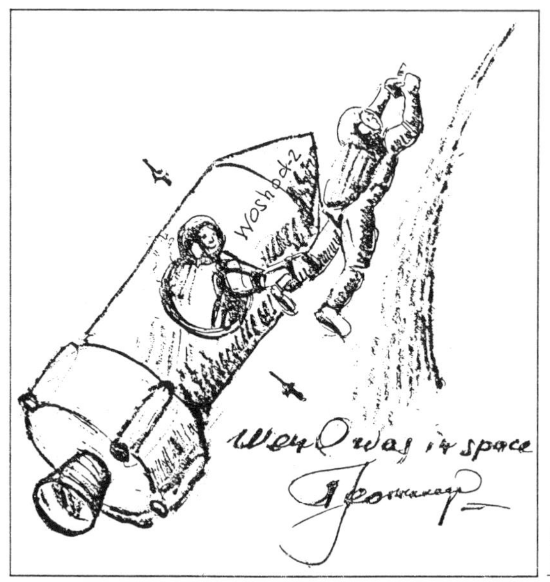

MARCH 18–19

Alexei Leonov performs the first spacewalk during the seventeen-orbit flight of Voskhod 2. The walk almost ends in disaster when Leonov discovers that his inflated spacesuit is now too large to fit through the airlock door. After several minutes of exhausting struggle he manages to release enough air from his suit to squeeze through the opening.

MARCH 23

Gemini 3 (*Molly Brown*) is the first manned flight of the Gemini two-man spacecraft series. Astronauts Virgil I. Grissom and John W. Young make three orbits of the earth during the 4 hour 53 minute flight. They are launched by a modified version of the military Titan II rocket. The 89-foot-tall, 10-foot-wide booster has a first stage thrust of 430,000 pounds. The combined Titan-Gemini spacecraft is 108 feet tall.

There are twelve manned and unmanned Gemini flights in all, including Gemini 3, between April 1964 (with the lauch of the unmanned Gemini 1 and the simultaneous selection of the two men who would pilot Gemini 3) and 11 November 1966. The goal of the series is to test the effects of long-duration spaceflight, EVA activities, orbital maneuvering, and rendezvous and docking techniques . . . mostly with the future Apollo moon flight in mind.

The Gemini spacecraft resembles outwardly its immediate predecessor, the Mercury cap-

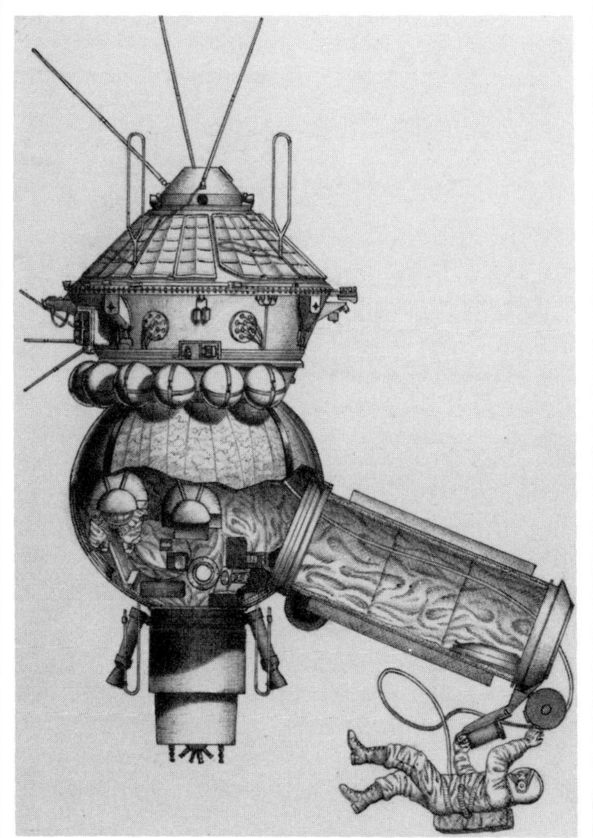

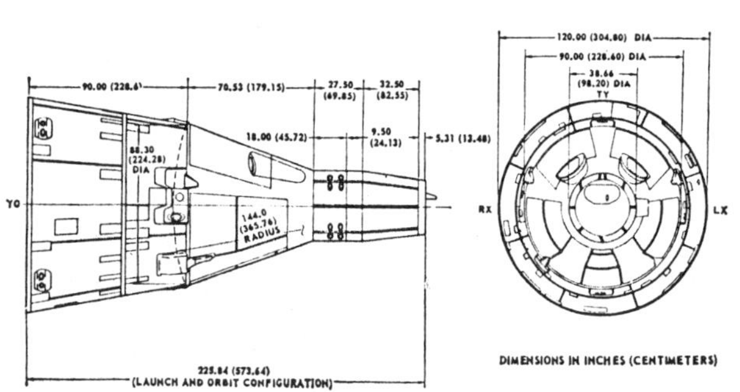

sule. The conical spacecraft is comprised of a reentry module and an adaptor module. The former is 11 feet tall and tapers from 7 feet 6 inches at its base to 3 feet 3 inches at the top. It is 50 percent larger in cabin volume than Mercury. The cylindrical "nose" contains docking facilities and parachutes. There is a redundant system of eight 25-pound attitude-control thrusters. The cabin is equipped with side-by-side ejection seats with separate hatches (rather than the escape tower that both Mercury and the later Apollo featured). A skirtlike addition at the rear of the capsule contains eight 25-pound control thrusters, propellant tanks and hydrogen and oxygen for the fuel cells (introduced with Gemini 5). This section is jettisoned before reentry.

JUNE 19

The first launch of the Titan III-C takes place. The purpose of this booster is to put the Manned Orbiting Laboratory into orbit.

ca. 1965

TRW develops the Janus lifting-body design for a manned orbital spacecraft. Janus is a blunt vehicle, shaped somewhat like a half gumdrop, split along its longitudinal axis. Sunk flush within its flat upper surface is a delta-winged aircraft. The lifting body furnishes a large payload volume as well as acting as the heat shield during reentry. Its configuration adds a substantial amount of aerodynamic maneuverability as well. After reentry, the aircraft separates from the main body and proceeds to its landing site.

The estimated size of the spacecraft, based on a 2-week mission at a low altitude with a crew of three, yields an overall length of

NASA

Proposed Rogallo wing recovery.

26.8 feet, a span of 16 feet and a thickness of 10 feet. The useful payload volume is 860 ft^3. The gross weight of the spacecraft is 16,000 pounds, including 4,000 pounds for the aircraft, its equipment and crew.

The aircraft is delta-winged, with a 62 degree sweep-back angle and a span of 13.3 feet. Vertical stabilizers extend down from each wingtip. It lands on skids on the ends of the stabilizers and a single nosewheel. The plane is 21 feet long and powered by a single, small turbojet engine. It will separate from the lifting body at 30,000 feet at a speed of Mach 0.6. The abandoned pod continues to descend by parachute for eventual recovery.

1965–1967

The M2-F3, a NASA lifting body, is originally built as the M2-F2 by Northrop. It makes its initial flight air-dropped from beneath a B-52 bomber. It crashes in 1967, se-

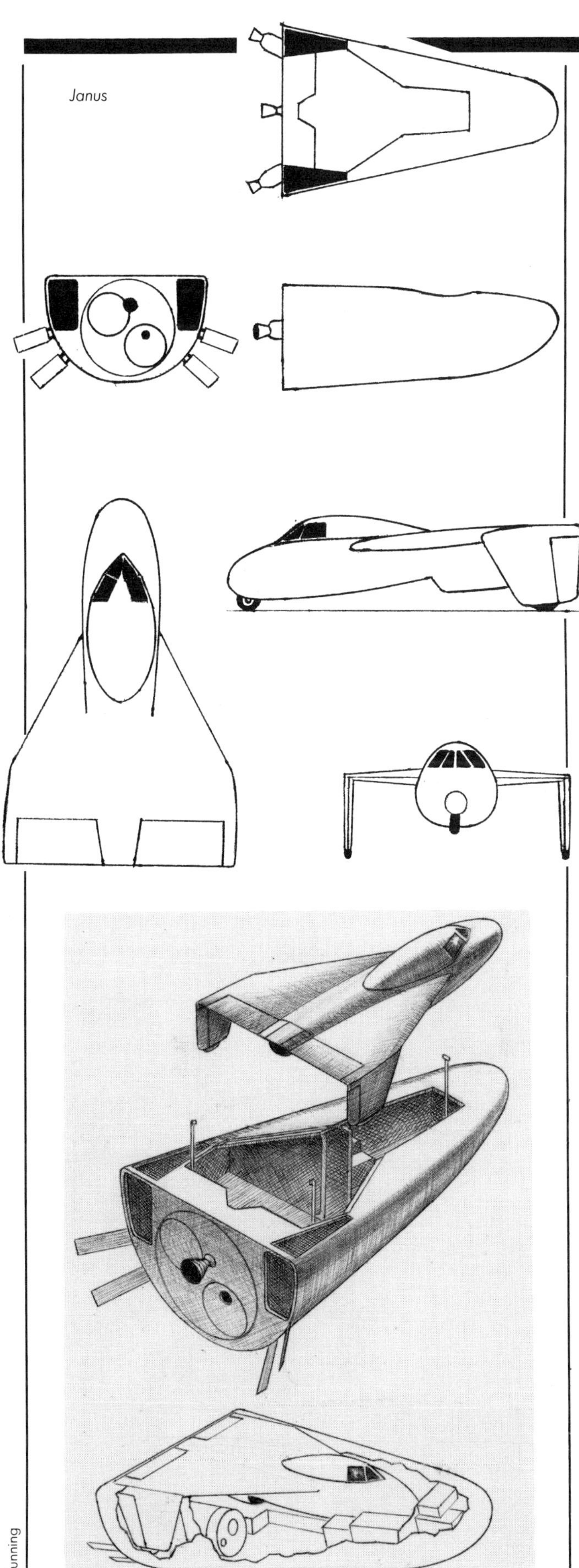

North American toroidal manned Mars spacecraft (1962).

LAUNCH OR INJECTION
SELF-DEPLOYING
SPIN-UP
• SELF-DEPLOYING
• G FIELD ESTABLISHMENT
• AMPLE CREW, EQUIPMENT & STORAGE SPACE
• CONTROL STABILITY BENEFIT

verely injuring its pilot, Bruce Peterson. It is then rebuilt as the M2-F3, with the addition of an extra vertical fin. With twenty-seven research flights, with speeds up to 1,064 mph, before retirement, it is now on display at the National Air and Space Museum.

1966

A NASA manned Mars/Venus flyby-landing mission study is completed. A flyby spacecraft consisting of an Apollo-derived mission module is to be launched from earth orbit by a Saturn IVB stage. After injection into the interplanetary trajectory, the mission module is docked with the command module/service module. The two units are then extended on cables—the mission module at one end and the command/service module at the other—and spun around their common center to provide an artificial gravity during transit. The spacecraft is despun for midcourse corrections and flyby surveillance.

The mission module is 21 feet 8 inches at its base and is divided into two "floors." The lower one contains sanitation and medical facilities, sleeping quarters, an off-duty area (and "corpse provisions!"). The upper floor contains the laboratories, food preparation area, and mission controls. Attached to the upper end of this floor by an intramodule crew transfer chamber is a six-man Apollo-type earth reentry vehicle. This also acts as a radiation shelter and emergency quarters. To the base of this is attached a 12 foot 10 inch propulsion module. The spacecraft car-

1966

ries several hard and soft landers for deployment during the planetary flybys. A large number of alternative spacecraft configurations are also studied.

The eight-man Mars landing mission requires a similar, though much larger, spacecraft, with the addition of a Mars excursion module and a larger booster for leaving earth orbit. Four astronauts are to make the descent onto Mars. The lander includes a four-man Apollo-derived module for landing and return to the orbiter, a surface rover and living quarters.

NASA

M2-F2

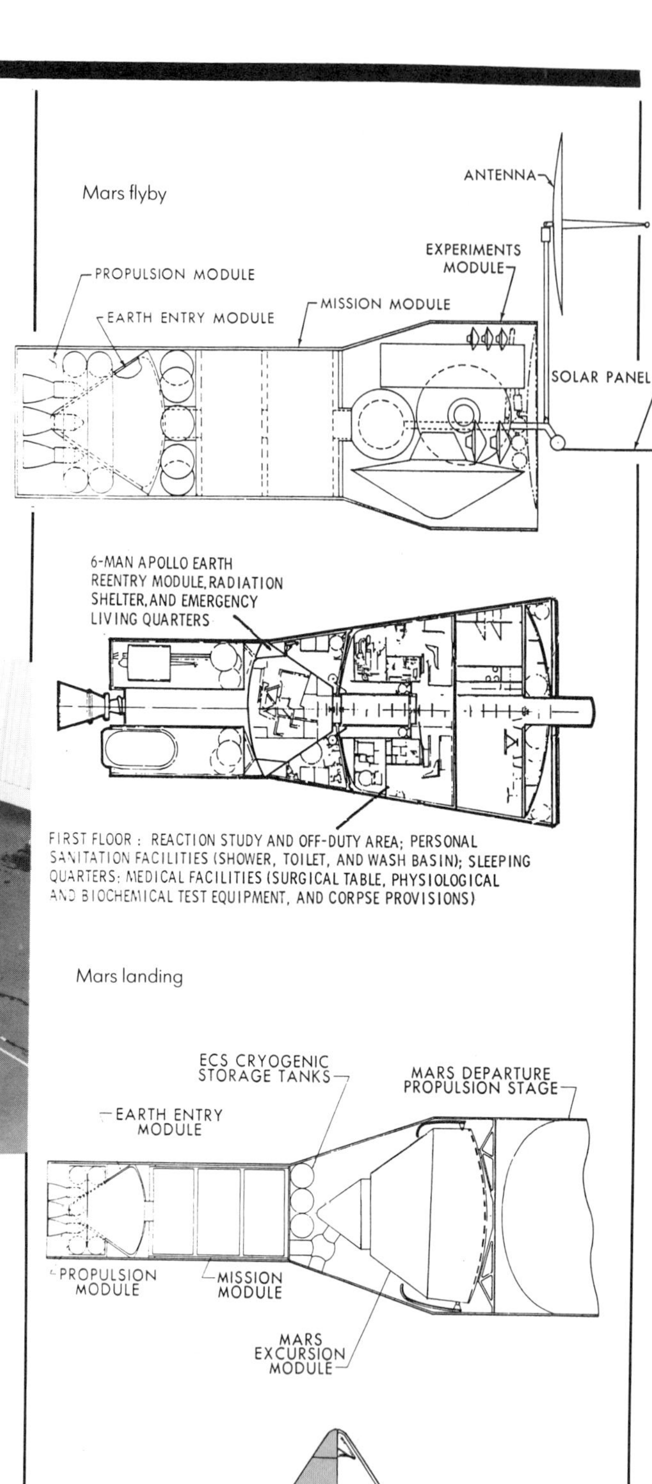

EUROSPACE proposes using its ELDO-A launch vehicle to test configurations for a future European manned shuttle vehicle. Two potential reentry vehicles are considered: a winged spacecraft and a lifting body.

Eventually a full-scale vehicle would be developed using a recoverable booster based on the ELDO-C that would place the spaceplane into an orbit at an altitude of 300 nautical miles. A possible manned configuration would consist of two modules. The crew, life support and other equipment would be carried in the reentry vehicle, while other equipment, including manuvering engines, would be carried in an expendable module. The EUROSPACE proposal bears many resemblances to the later French *Hermes* spaceplane [see below].

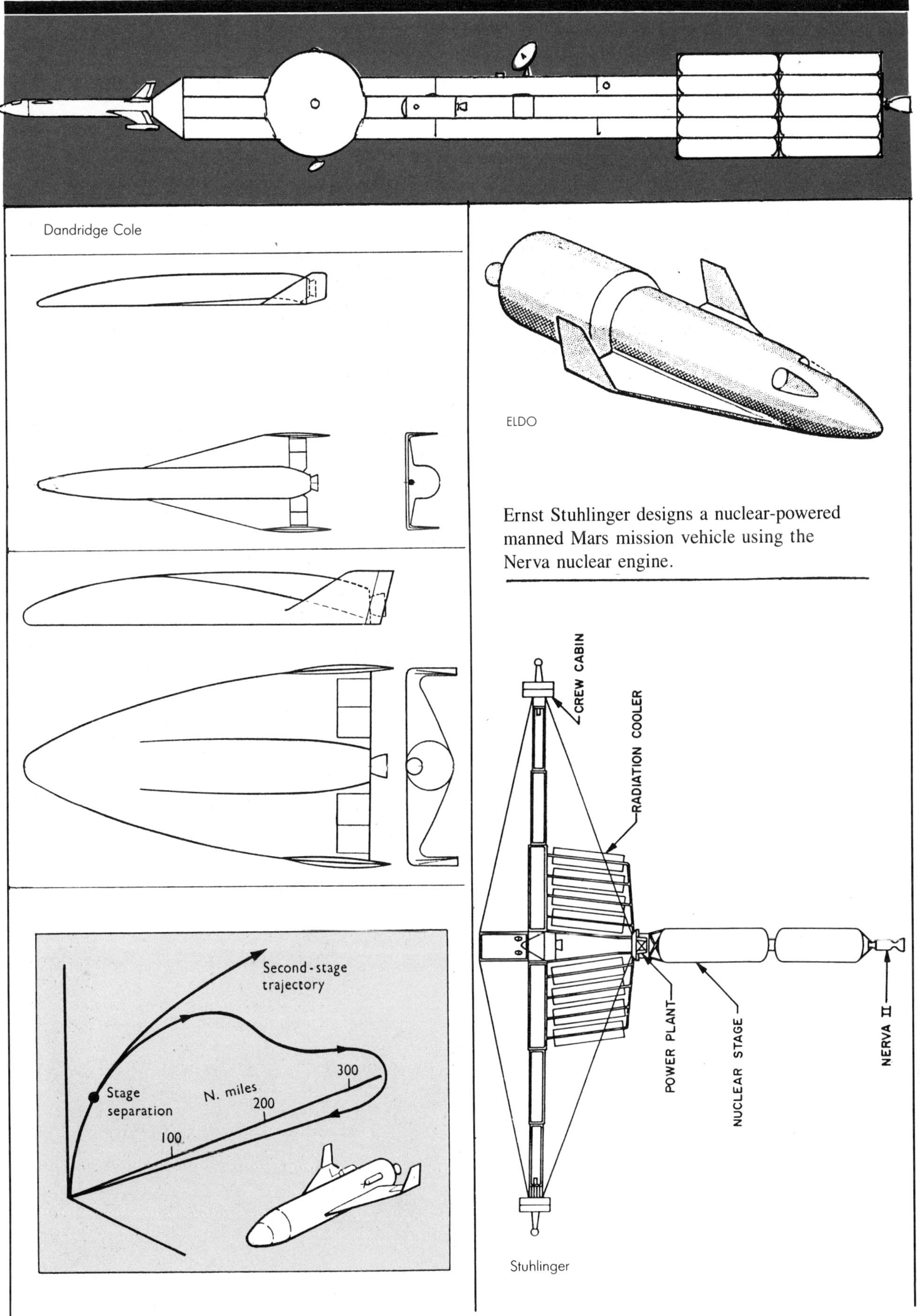

Dandridge Cole

ELDO

Ernst Stuhlinger designs a nuclear-powered manned Mars mission vehicle using the Nerva nuclear engine.

Stuhlinger

1966

North American Aviation proposes a six-man Mars mission to depart earth around Easter of 1985 to spend 3 weeks exploring the red planet before returning to the earth by Hallowe'en of the following year. NAA's M. W. Bell describes the mission at the "Stepping Stones to Mars" conference sponsored by the AIAA. The round trip is to take 562 days, including 20 days in orbit around Mars. The takeoff is on 5 April 1985 and the spacecraft will arrive at Mars on 22 April 1986, after making a Venus flyby en route. Three of the astronauts remain in orbit above Mars while three others descend to the surface. After about 3 weeks of exploration, they rendezvous with the orbiter for the return to earth, where they will arrive on 19 October 1986.

Most of the hardware can be derived from the Apollo program. Specially designed modules include a mission module with living quarters and laboratory, an excursion module, and a heat shield for aerodynamic braking in the Martian atmosphere. Bell thinks that it would be nice to have freeze-dried birthday cakes on board for the crew.

French designer François d'Allegret develops an Empire State Building-sized spaceship to transport 7,000 passengers to Mars.

The design, half-serious and half-satirical, is called the *Astronef 732*. The uppermost of its seven stages is a chapel, with the sixth stage, directly below, an "ivory tower." From the first stage up, some of the *Astronef 732's* features include: First Stage—nuclear engines and 16 electrically powered emergency rockets; Second Stage—30 cabins for the pilots and 72 gymnasiums; Third Stage—5,000 second-class cabins and 6 atomic rockets; Fourth Stage—3,000 first-class suites, swimming pool, and theaters; Fifth Stage—an "ion-powered satellite for adventure and exploration," with a capacity of 500 adults and 2 children.

The total cost of the project is $7 billion.

FEBRUARY 26

The Saturn 1B makes its first flight.

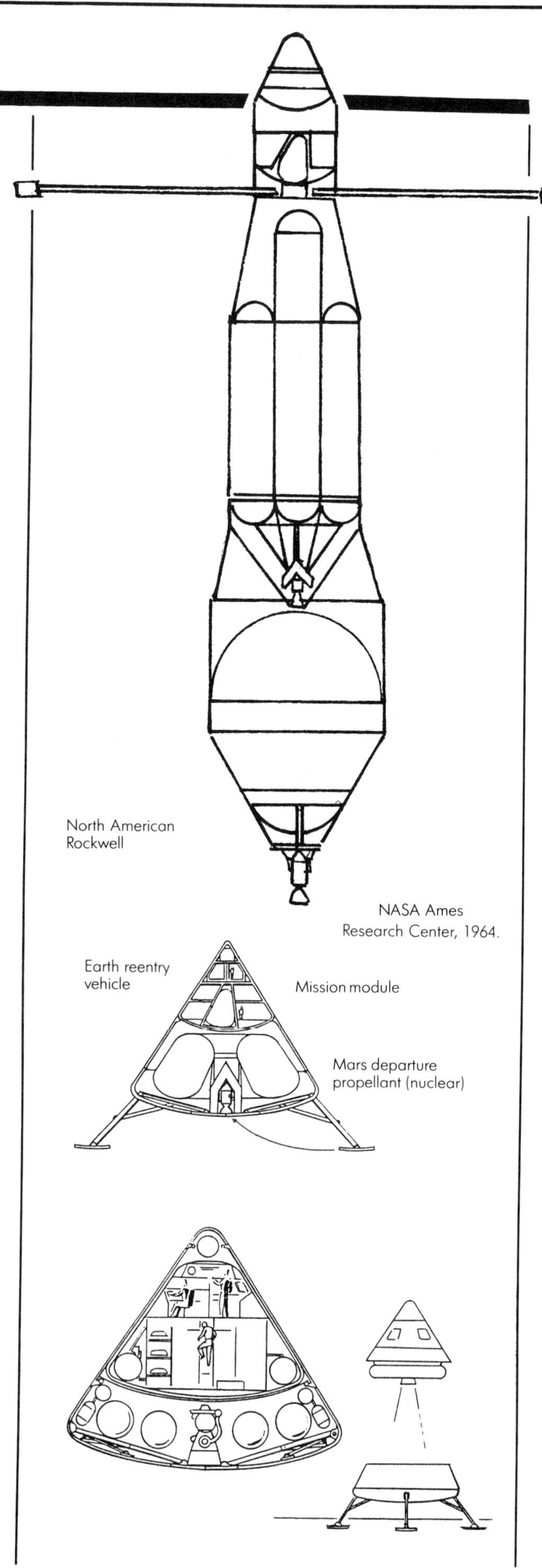

North American
Rockwell

NASA Ames
Research Center, 1964.

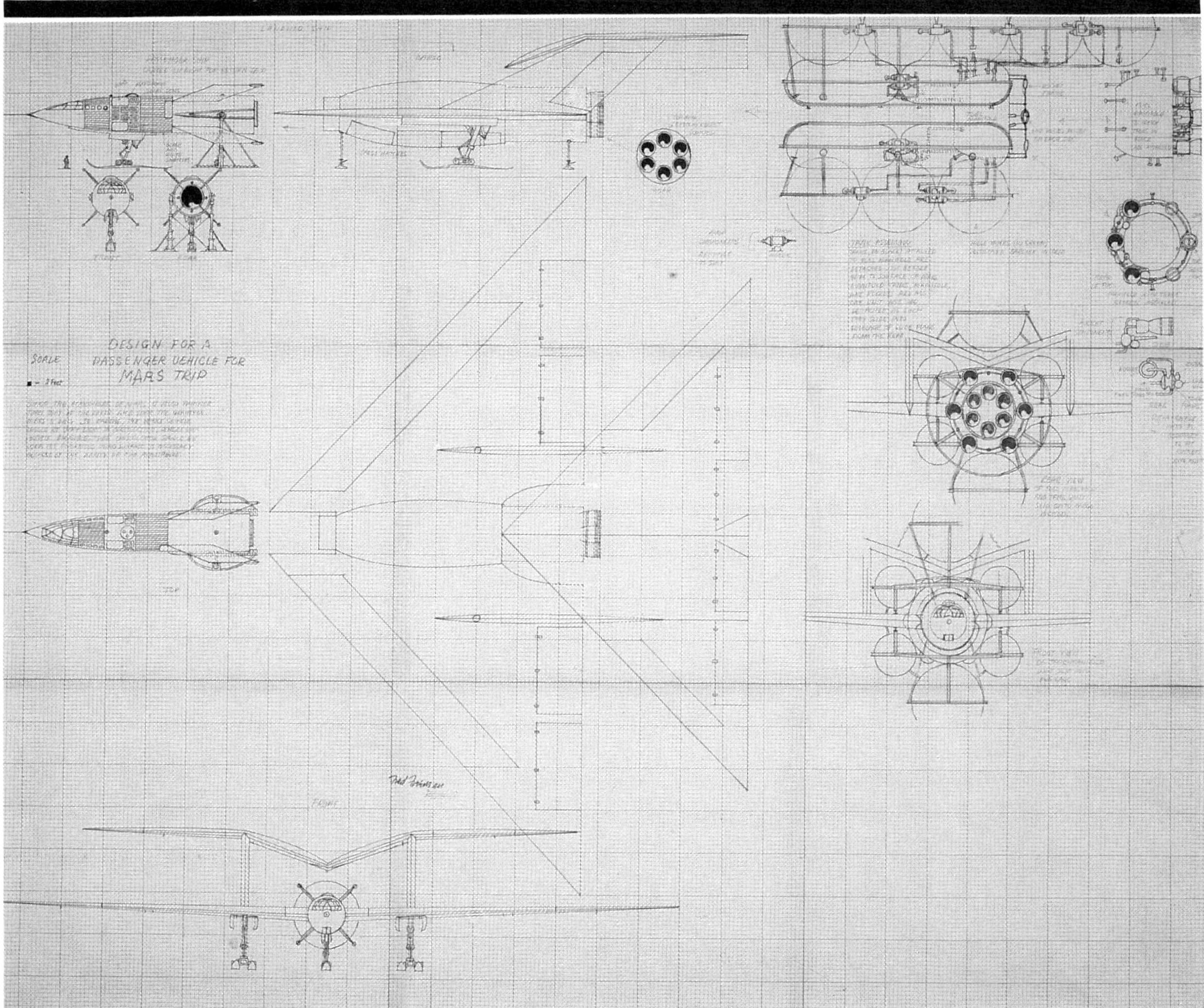

USRC

Manned Mars spacecraft, similar to von Braun's *Collier's* gliders [see 1952–1954], designed by Fred Freeman.

MARCH 3

The first test is made of the NERVA atomic rocket engine.

MID-SEPTEMBER

The Ad Hoc Subpanel on Reusable Launch Vehicle Technology meets to determine post-Apollo successor programs. This panel has been established by the joint Department of Defense-NASA Aeronautics and Astronautics Coordinating Board (AACB) and consists of NASA and DoD joint chairmen and ten NASA and eight DoD representatives.

They determine that no one system is capable of satisfying the future needs of both NASA and the DoD. Nevertheless, the panel feels that second generation reusable spacecraft will be operational by 1974, including the use of vertical takeoff and horizontal landing boosters; partially reusable vehicles, including horizontal takeoff and landing boosters, by 1978; and fully reusable scramjet air-breathing boosters by 1981. These three time periods are distinguished by their spacecraft and termed Class I, II and III.

Class I includes such spacecraft as these:

1. A Titan IIIM booster launching a second generation lifting-body, horizontal-landing spacecraft (though a near-ballistic, land-landing spacecraft will also be viable in all Class I designs).
2. A Saturn IB carrying a 40,000-pound (including payload) lifting-body spacecraft.

3. An advanced expendable booster launching a 95,000-pound spacecraft.
4. An advanced expendable booster that can launch a 35,000-pound spacecraft carrying a payload of 15,000 pounds.
5. A vertical takeoff, horizontal landing (VTOHL), reusable booster stage carrying a lifting-body spacecraft and its expendable booster. The winged launcher is 208 feet long with a span of 97 feet, powered by rockets with a thrust of 1.7 million pounds. The spacecraft, including payload, will have weighed 25,000 pounds.

Class II includes:

1. A VTOHL fully reusable rocket booster with a stage II spacecraft weighing 253,500 pounds. First stage propulsion includes rockets with a thrust of 2.4 million pounds and four turbofans for the return flight and landing.
2. A VTOHL fully-reusable rocket booster carrying a 275,000 pound (295,000 pounds with payload) lifting-body in a parallel-staging arrangement.
3. A fully reusable horizontal takeoff and landing (HTOHL) booster carrying a 290,000-pound second-stage spacecraft. The launch vehicle is 330 feet long with a wingspan of 131 feet, propelled by subsonic hydrogen-fueled turbofan-ramjets. The lifting body is capable of handling 35,000 pounds of payload.

Class III is represented by an advanced edition of the third Class II vehicle, propelled by Mach 0–6 hydrogen-fueled turbofan-ramjets, and a Mach 6–12 scramjet.

1966–1969

The *Star Trek* television series introduces the starship *Enterprise*. The anti-matter-propelled vehicle has a crew of 430 men and women as well as several 24-foot shuttlecraft for planetary landings (to supplement its matter-transmitting "transporters").

The starship is comprised of two units: a

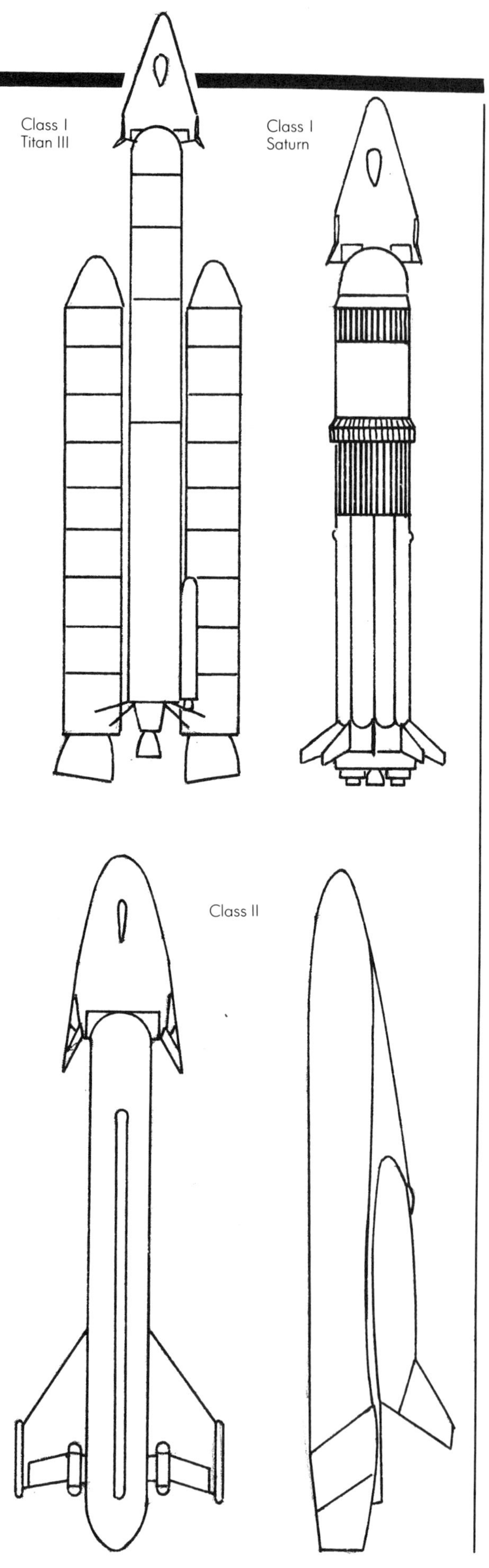

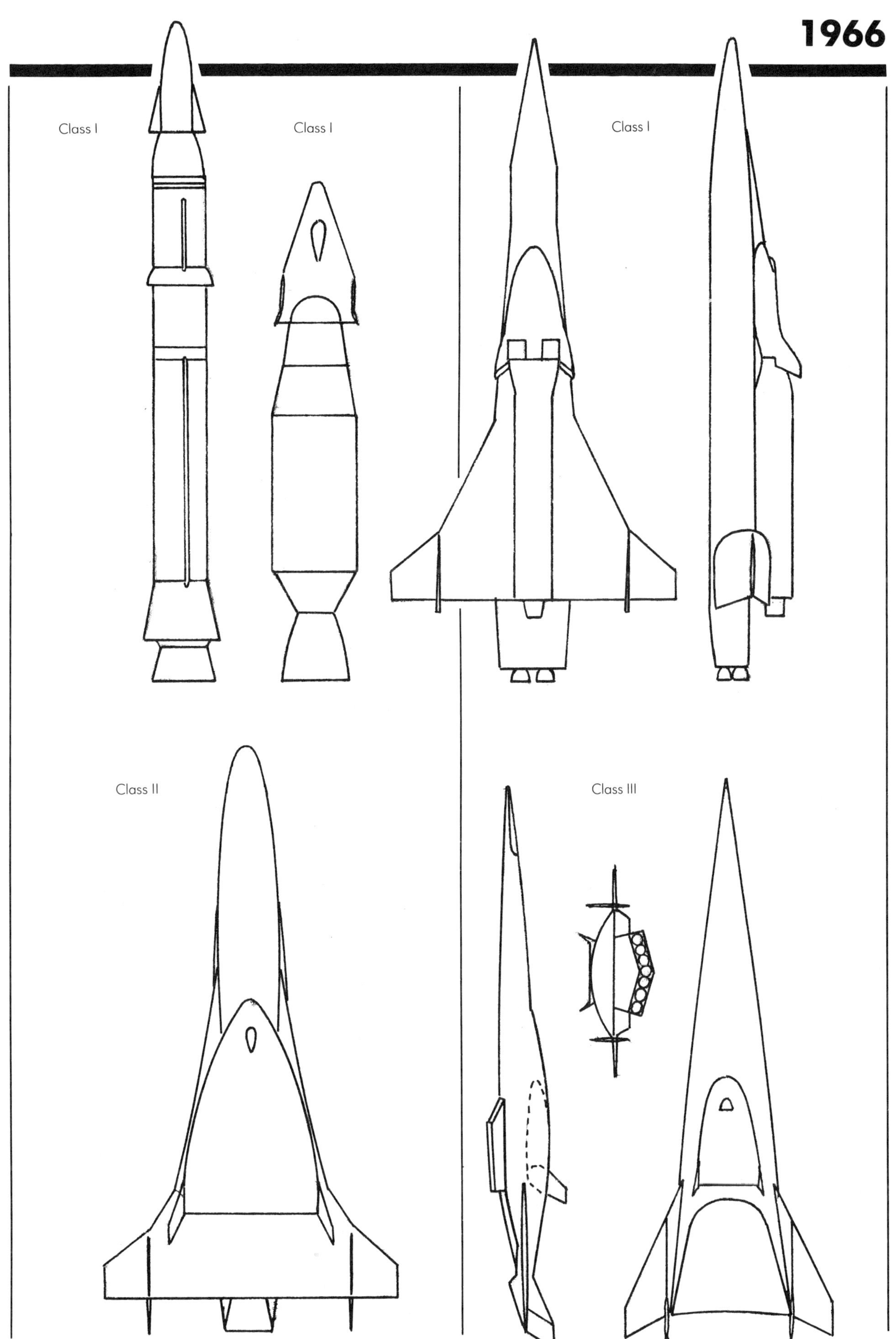
Class I
Class I
Class I
Class II
Class III

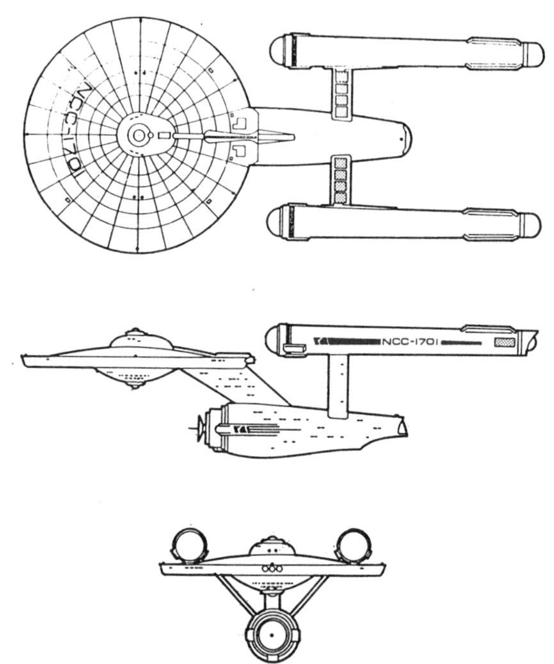

saucer-shaped primary hull 127.1 meters in diameter (which can be detached in an emergency) and a secondary hull 103.6 meters long and 34.1 meters in diameter. Two 153 m × 17.3 m cylindrical spacewarp engines are mounted on outriggers on the secondary hull. These give the *Enterprise* a cruising speed of "warp factor" 6 (i.e., six times the speed of light) and a top speed of warp factor 8.

The *Enterprise* is severely revised twice: once for *Star Trek: The Motion Picture* and again, even more drastically, for a new television series, *Star Trek: The Next Generation*.

1967

Philip Bono evolves an elaborate scenario for developing large, economical, reusable spacecraft and boosters. They are all based upon the projected use of the plug-nozzle engine.

Perseus is a shuttle launcher designed by Philip Bono and evolved from his ROMBUS family of plug-nozzle boosters [see above]. A standard shuttle orbiter with its belly tank will be mounted atop the Perseus booster. After launching the shuttle, the Perseus will reenter the atmosphere, slowing by retro-firing its rockets, and finally make a dry land recovery at the original launch site. The Perseus Mark II will have its internal tanks full (where the Mark I's are only partly filled). This will enable a single-stage-to-orbit mission. The shuttle will have been modified by removing its main engines, becoming a pure glider. The entire shuttle system will now be reusable, with nothing thrown away.

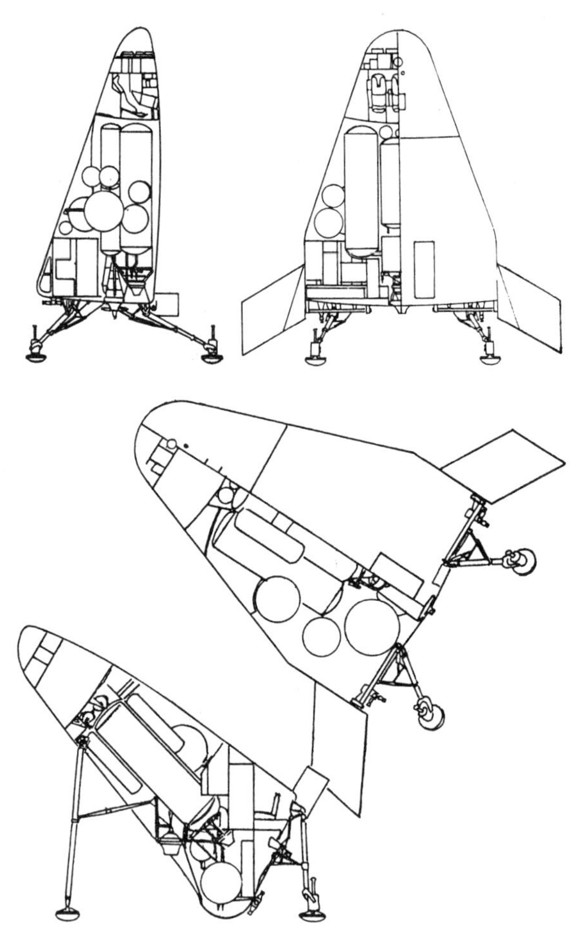

Mars lander proposed by Philco.

SASSTO/SARRA

The first of these is SASSTO; the acronym stands for Saturn Application Single-Stage-to-Orbit. It has evolved from studies toward recovering the Saturn IV-B booster. There are three steps in its evolution. Initially, a sea-recovery technique is developed. Second, the project progresses to land recovery of the booster. The stage is retrieved from a 150-mile orbit. It begins reentry by firing four 6,500-pound thrusters. Just before entering the atmosphere, three "ballutes," each 28 feet in diameter, are released for drag brak-

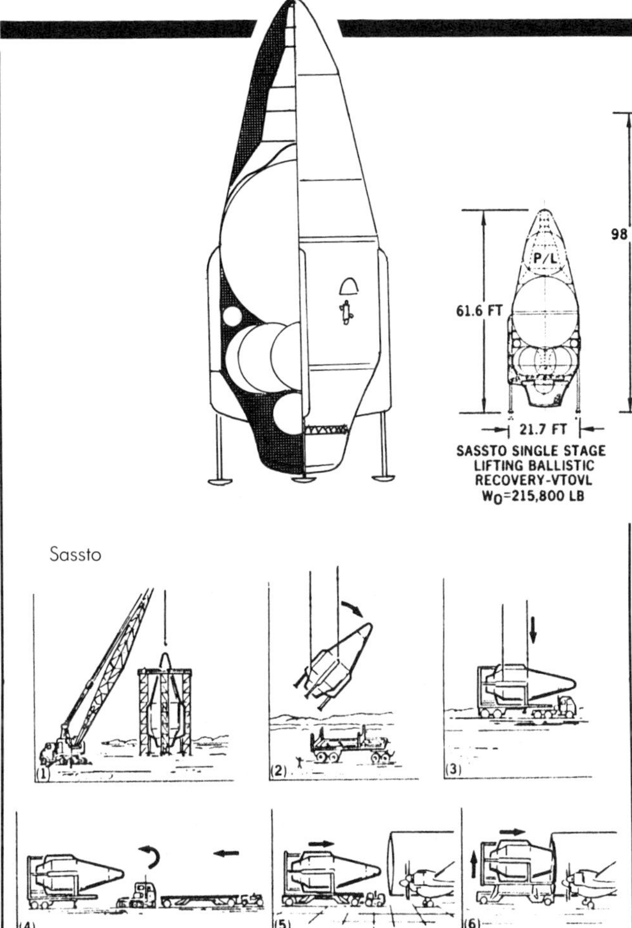

Sassto

ing and stability. These are woven from steel mesh and can withstand up to 2,000 degrees for 2 minutes. The stage itself is protected by an ablative heat shield. At 30,000 feet three 124-foot-diameter ring-sail parachutes are deployed. The shock of landing at 40 fps is absorbed by a crushable aluminum honeycomb bumper. The booster is kept from tipping over by four legs.

The third step in the evolution of the SASSTO requires major redesigning, in which it becomes a squat, 45-foot-tall rocket with a plug-nozzle engine and four landing legs. It is then capable of lifting a two-man Gemini capsule (protected during launch by an aerodynamic fairing) into earth orbit. The entire single-stage booster is recoverable. Liftoff weight is 216,000 pounds, the engine providing a thrust of 270,000 pounds. A payload of 8,100 pounds can be delivered into orbit.

The SASSTO could be converted into a Saturn Application Retrieval and Rescue Apparatus, or SARRA. Its job is the rescue of stranded or incapacitated astronauts, and the retrieval from orbit of satellites that are either no longer operating or in need of repair.

Hyperion

Hyperion is the next step in the space program developed by Philip Bono. The Hyperion passenger transport is a 100-foot-tall plug-nozzle rocket, similar in outward form to the SASSTO. The upper portion of the rocket has three levels: an upper cargo space and two levels for passengers. Each contains fifty-five acceleration couches. To circumvent the waste of "throwaway" boosters, or the trouble and expense of recovering them, Bono proposes using a "captive" booster, one that does not leave the earth's surface. This is a rocket sled, operating on a track. This concept has already been suggested by a number of scientists, notably Eugen Sänger [see 1943].

The spacecraft, which is developed from the SASSTO/SARRA [see above], is placed horizontally onto the cradle of the Hyperion sled, which is about 40 feet long and 30 feet wide. The sled runs almost frictionlessly on an air cushion while travelling along dual tracks at 680 mph. The sled contains only propellants; all of the thrust is provided by the spacecraft's own engines. The rocket travels 2 miles across the level floor of a valley, then up the side of a lofty mountain for another mile. Once the rocket is released, retrorockets, brakes and gravity combine to bring the sled to a halt. By using this system, the payload of the SASSTO/SARRA could be increased by 20 percent.

A drawback to this scheme is that it is lim-

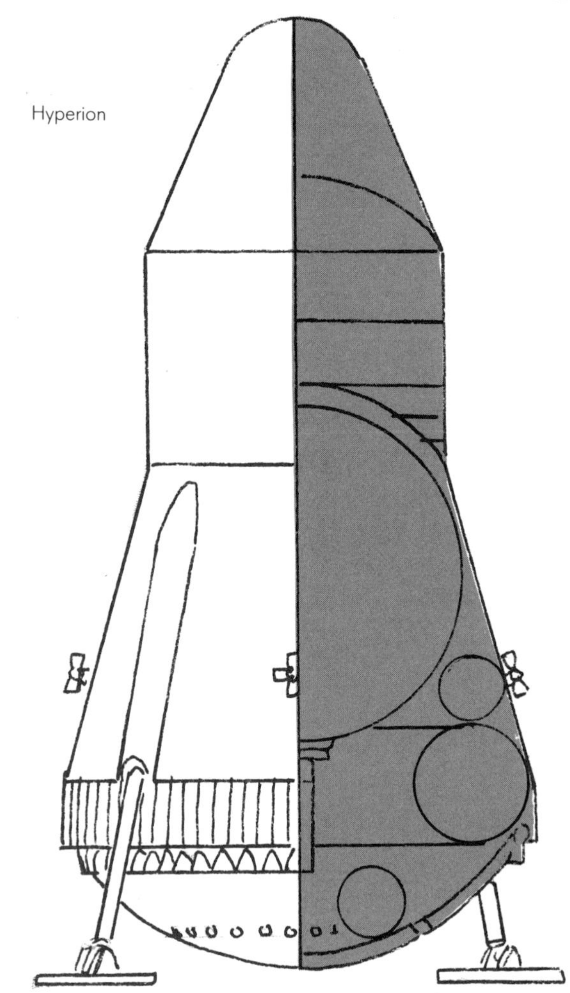
Hyperion

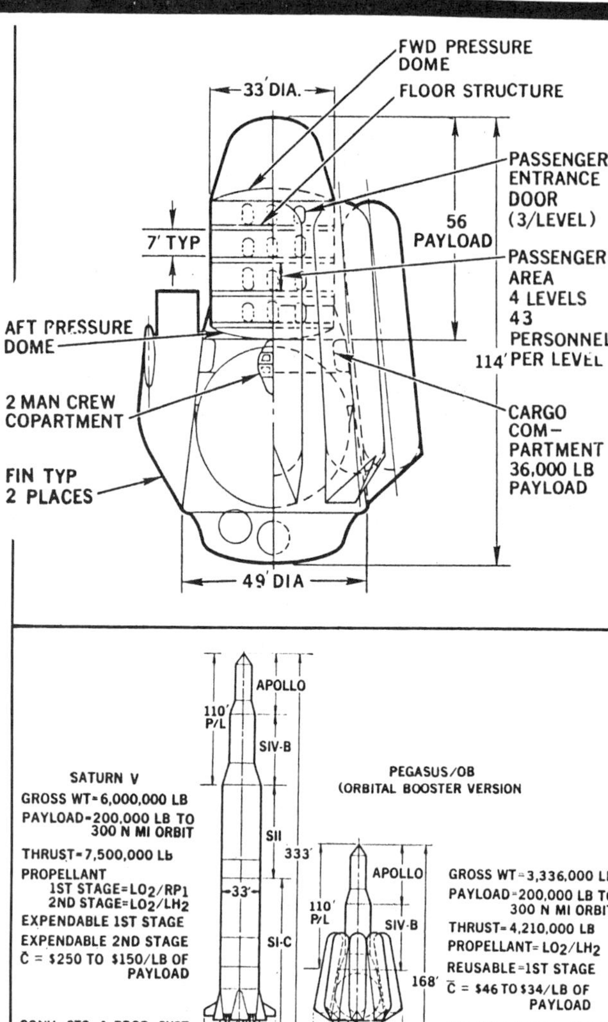

ited to areas with mountains 1 mile in elevation (which makes the New York to London flight difficult).

Used as a suborbital passenger transport, the rocket could make the trip from London to New York in only 26 minutes. The travelers on board would not be subjected to more than 3 g's during the boost phase. Freigher versions could haul up to 5 tons of cargo.

Pegasus

Pegasus is virtually the same size and of the same general outward appearance as the Hyperion passenger transport. However, it is launched vertically, using jettisonable external hydrogen tanks as in the ROMBUS vehicle [see 1962]. It is 114 feet tall, the diameter of its payload section 33 feet, the diameter of its base 50 feet. Four decks hold acceleration couches for 172 passengers. Its gross weight is 3,340,000 pounds, its plug-nozzle engine producing a thrust of 4,200,000 pounds. With the exception of the external tanks, the rocket is completely reusable.

Ithacus

Ithacus is a mammoth troop transport. Superficially resembling the Pegasus [see above], it is more than three-fifths larger, towering 200 feet above its launch site. In its huge passenger-compartment, six decks hold 200 acceleration couches each, for a total of 1,200 passengers.

Project Selena

Project Selena employs Bono's ROMBUS-type launch vehicle for a manned mission to the moon and the construction of a temporary lunar base. Once the ROMBUS is in

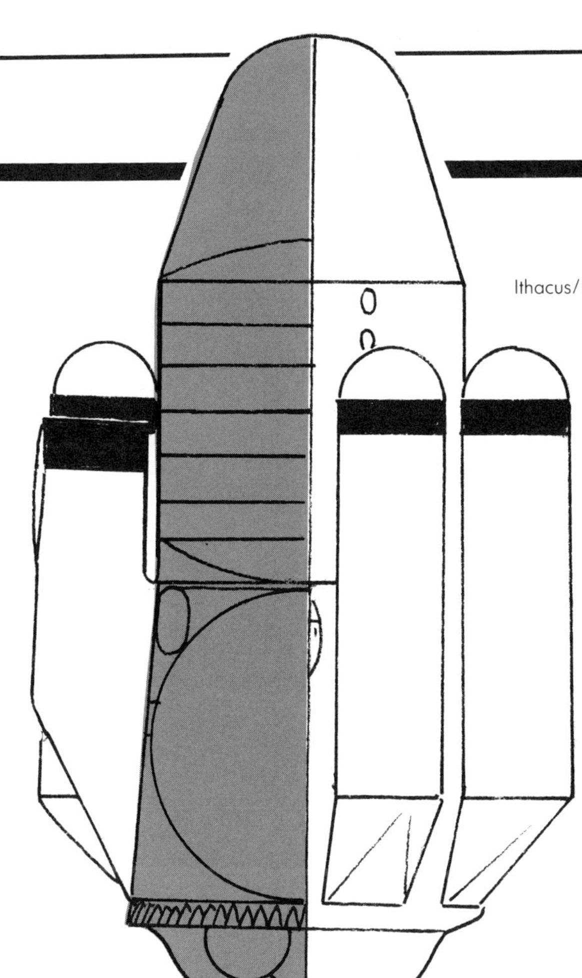

Ithacus/Pegasus

orbit, additional launches carry full external tanks to replace the ones used in reaching earth orbit. It uses all but two of these in leaving the earth. These remaining tanks of fuel supply the engine for the descent to the lunar surface. The empty tanks are then lowered to the ground and adapted for living quarters. Additional tanks from other manned and unmanned landers can be combined to make a base of any desired size. [Also see 1991.]

Robert Salkeld of the System Development Corporation proposes an improvement on planned space shuttle performance: mixed-mode propulsion. Two or more propulsion systems can be combined, to be operated either sequentially or in combination to increase performance and lower vehicle weights. The first mode must have higher propellant density and density-impulse and the second mode must have higher specific impulse. Salkeld shows how the mixed-mode concept could be employed in a vertical takeoff, horizontal-landing shuttle, using liquid oxygen, kerosene (RJ-5) and liquid hydrogen. One version of the Salkeld shuttle uses an aerodynamic shape developed by NASA's Langley Research Center. It has a 90-foot by 15-foot cargo bay situated above its fuel tanks. Fully loaded, the 366,100-pound vehicle weighs 4,990,000 pounds. The shuttle is 171 feet long.

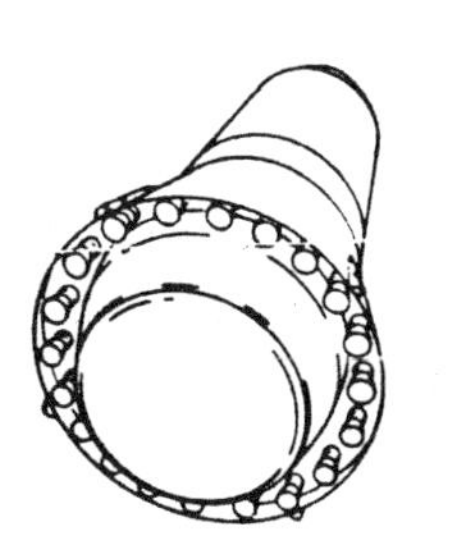

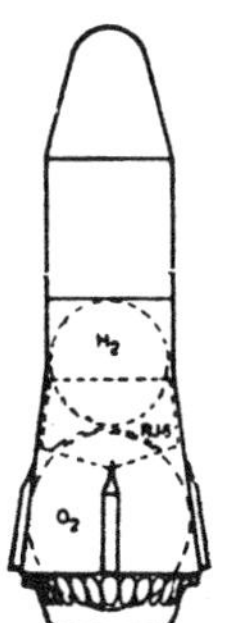

Takeoff is by its first-mode propulsion: ten liquid oxygen/RJ-5-burning engines. At about 260,000 feet, two of these switch to burning hydrogen.

JANUARY 27

A fire on board the Apollo 1 spacecraft claims the lives of astronauts Grissom, White and Chaffee. One result of the disaster is the abandonment of a 100 percent oxygen atmosphere inside future Apollo spacecraft.

APRIL 23

Soyuz 1 is the the first flight of the Soyuz spacecraft. Cosmonaut Vladimir Komarov is killed when his parachute fails to open at the end of his 1 day 2 hour 47 minute flight.

Nevertheless, the Soyuz spacecraft is an enormously successful series, making some seventy-three manned and unmanned (under the Kosmos designation) flights. During this time the spacecraft undergoes a great many design changes and variations, some minor and some major. In 1980 the Soyuz will be revised again under the new Soyuz-T designation.

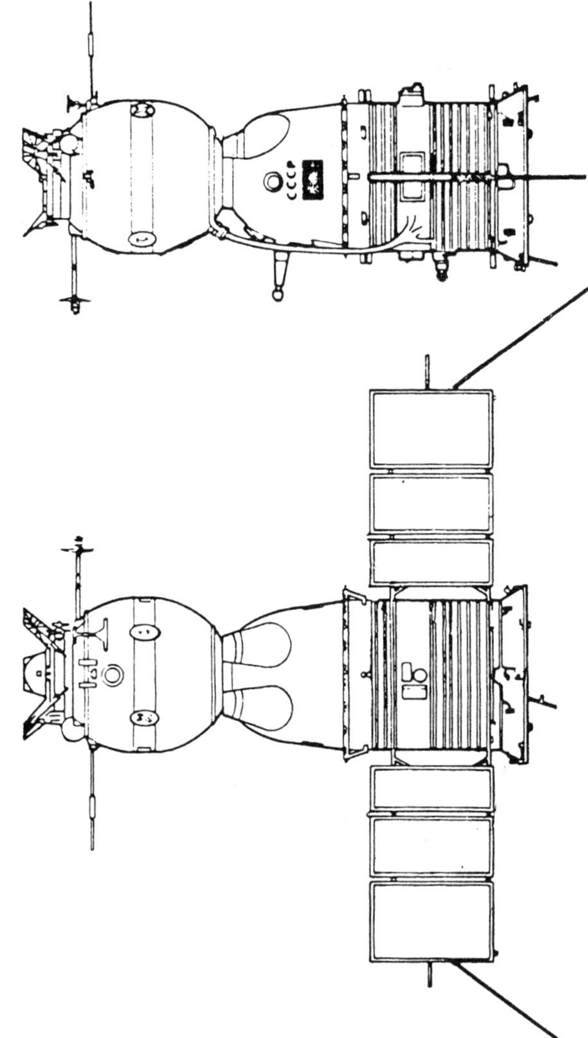

All Soyuz spacecraft are launched from Tyuratam by the A-2 booster. The vehicle consists of three components: the orbital module, the descent module and the instrument module. The former is an ovoid weighing about 1 ton. It contains the life support and controls as well as the descent module. The forward end is equipped with docking or other specialized equipment. The other end is connected to the dome-shaped descent/command module, which is capable of holding up to three cosmonauts. The rear instrument module contains the retrorockets, maneuvering, attitude and thermal control. All but two of the Soyuz spacecraft are powered by twin solar arrays attached to the instrument module.

NOVEMBER 9

The first flight of Saturn V takes place. The Apollo 4 it carries makes the first flight (unmanned) of the complete Apollo spacecraft: a Command/Service Module and a mockup of the Lunar Module.

1968

2001: A Space Odyssey is a landmark film in its realistic portrayal of spacecraft and spaceflight. Neither before nor since has spaceflight been depicted with such a sense of scale and realism. It is originally released in a 70-mm Cinerama format, to be shown in specially designed theaters. The screen size and film format have only been exceeded by the relatively new IMAX system.

The spacecraft designed by Tony Masters and Harry H.-K. Lange include a single-stage-to-orbit spaceplane, the PanAm *Orion*; an earth orbit-to-moon shuttle, the *Aries 1B*; a deep-space vehicle, the *Discovery*; waldo-equipped manned "pods" for extravehicular activity; and an advanced space station that boasts both a Howard Johnson's restaurant and a Hilton hotel.

The spherical *Aries 1B* is an earth orbit to lunar surface passenger shuttle.

The nuclear-powered *Discovery* is 700 feet long and propelled by enormous nuclear gas-core "Cavradyne" engines. These are separated from the living quarters by hundreds of feet of tankage, storage and structure. The living quarters are contained in a sphere that also contains a 38-foot-diameter centrifuge habitat, a zero-g command center, an airlock and a pod bay for three one-man repair and inspection vehicles.

The *Orion* is a relatively small boost-glide passenger shuttle craft that takes off horizontally with the aid of a jettisonable booster. It bears a striking and not coincidental resemblance to the Sänger antipodal bomber [see 1944].

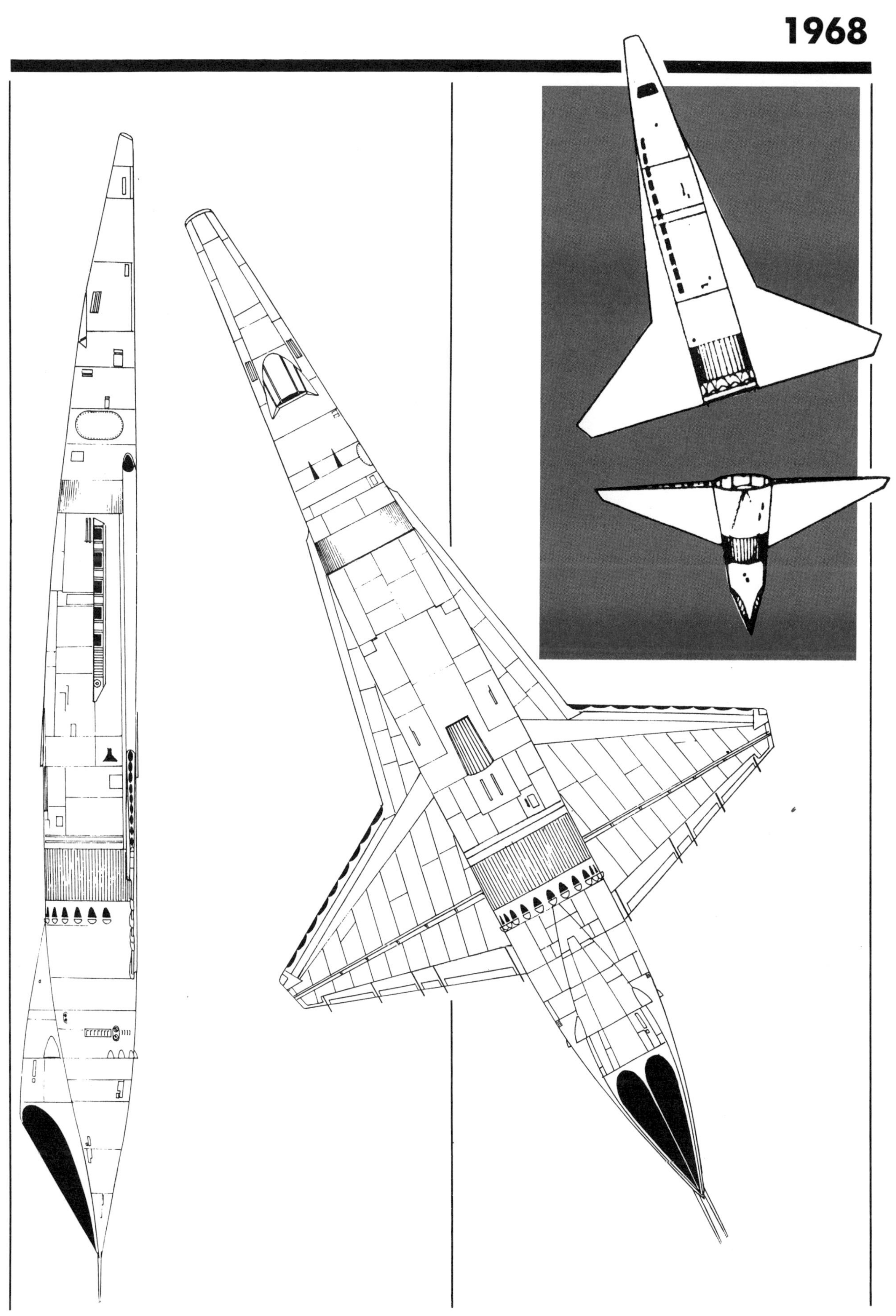

1968

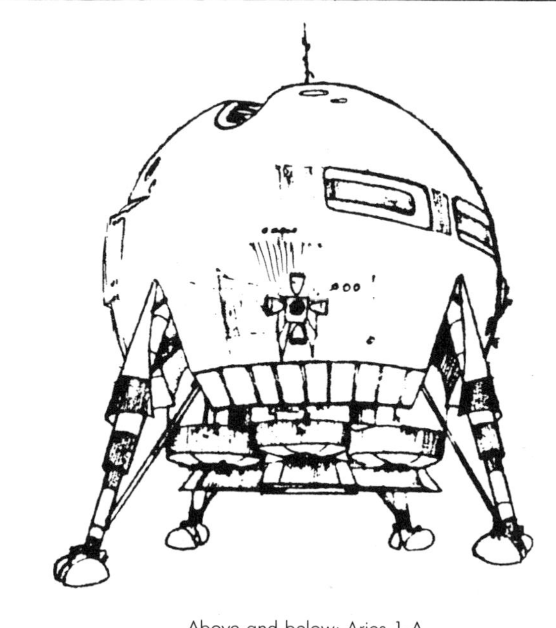

Above and below: Aries 1-A
Right: Aries 1-B

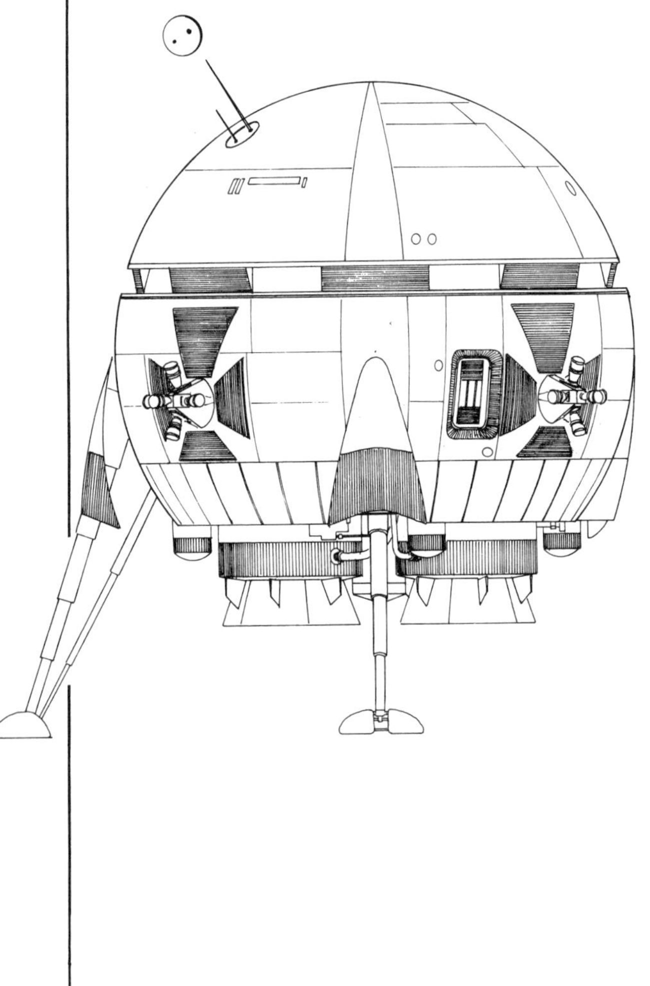

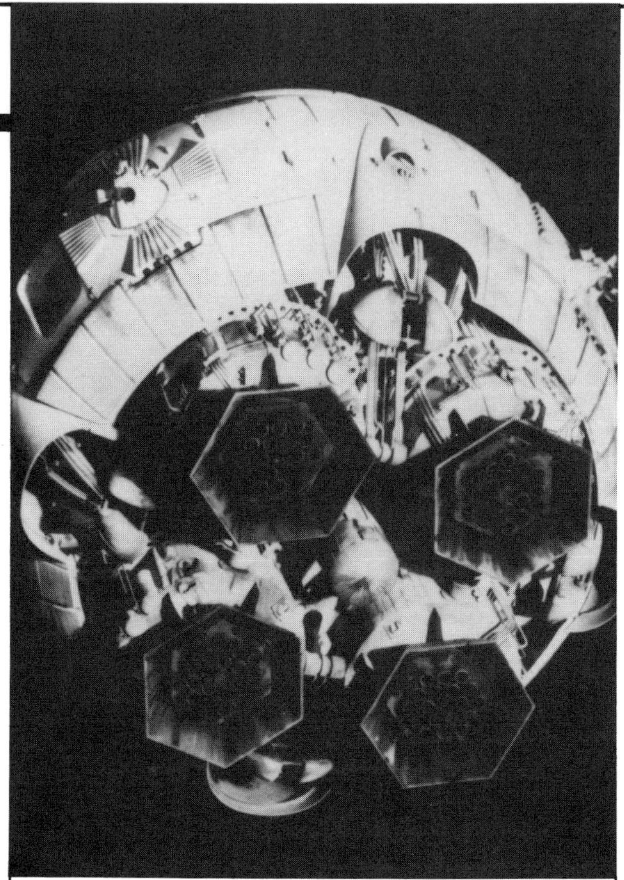

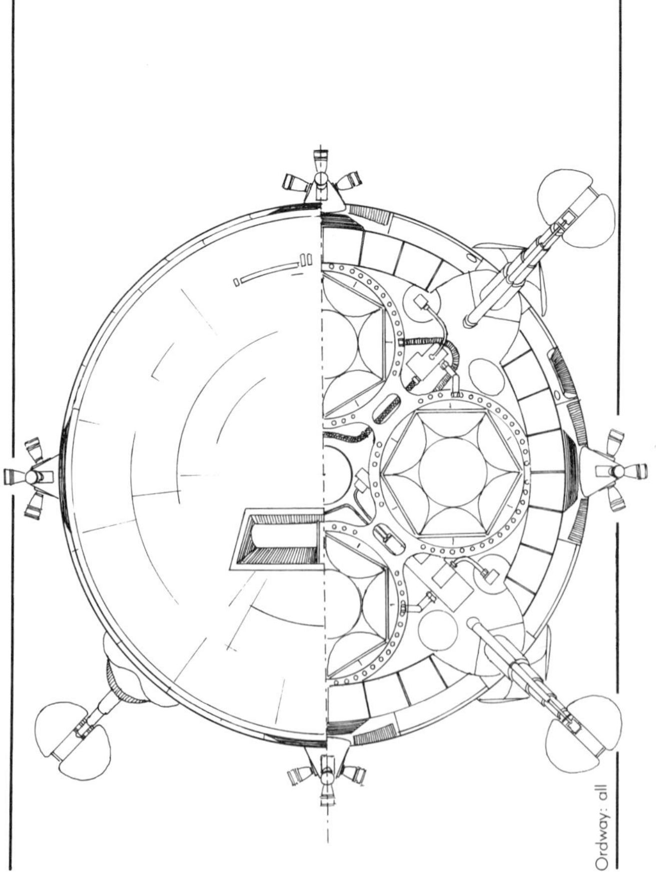

Ordway: all

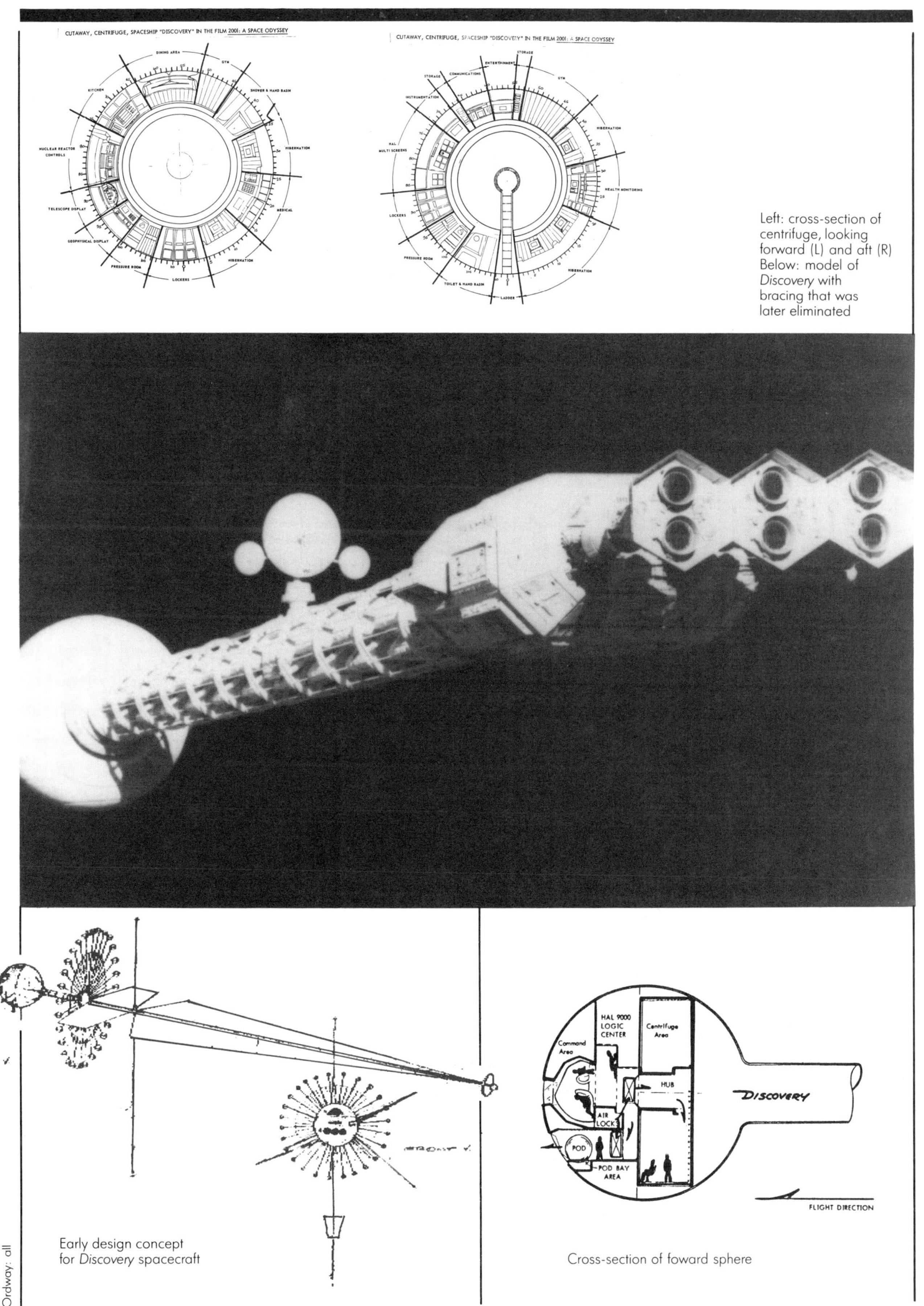

Left: cross-section of centrifuge, looking forward (L) and aft (R)
Below: model of *Discovery* with bracing that was later eliminated

Early design concept for *Discovery* spacecraft

Cross-section of foward sphere

Ordway: all

JANUARY 12

The Space Division of North American Rockwell presents the final report on its manned Mars mission study. Begun in 1967, it is the most comprehensive study done to date.

A large number of variations on the basic spacecraft are proposed. Fundamentally, the mission consists of a large spacecraft with spin-created artificial gravity and a Mars excursion module (MEM) that either makes its entry into the Martian atmosphere as a retro-braking low lift/drag Apollo-derived vehicle or as an aerobraking lifting-body glider. A low lift/drag shape is finally settled upon. A lifting-body reentry vehicle is provided for the eventual return to the earth. The ship gets its power from a nuclear generator, though emergency solar panels are provided.

Manned flyby, orbital and landing missions are all considered, all to be launched from earth orbit after orbital assembly of the spacecraft. The landing missions consider either aerobraking or retrobraking modes. The former requires a much smaller interplanetary vehicle, because of its smaller planetary excursion vehicle.

The spacecraft is highly evolved from existing Apollo hardware and technology.

In 1985, the Marshall Space Flight Center will update the Rockwell designs, primarily making changes in engines and propellant. MSFC considers seven different vehicle variations: a minimum Mars excursion module (a 4-day stay for a crew of two); stays on the surface of 30 days, 60 days, and 300 days; an unmanned cargo lander; a MEM using propellant produced from Martian raw materials; and a reusable single-stage MEM.

OCTOBER 11

Apollo 7 is the first manned mission of the Apollo program. Astronauts Walter M. Schirra, Jr., Donn F. Eisele and R. Walter Cunningham complete 163 orbits of the earth in just over 260 hours. The flight includes a live television broadcast from space.

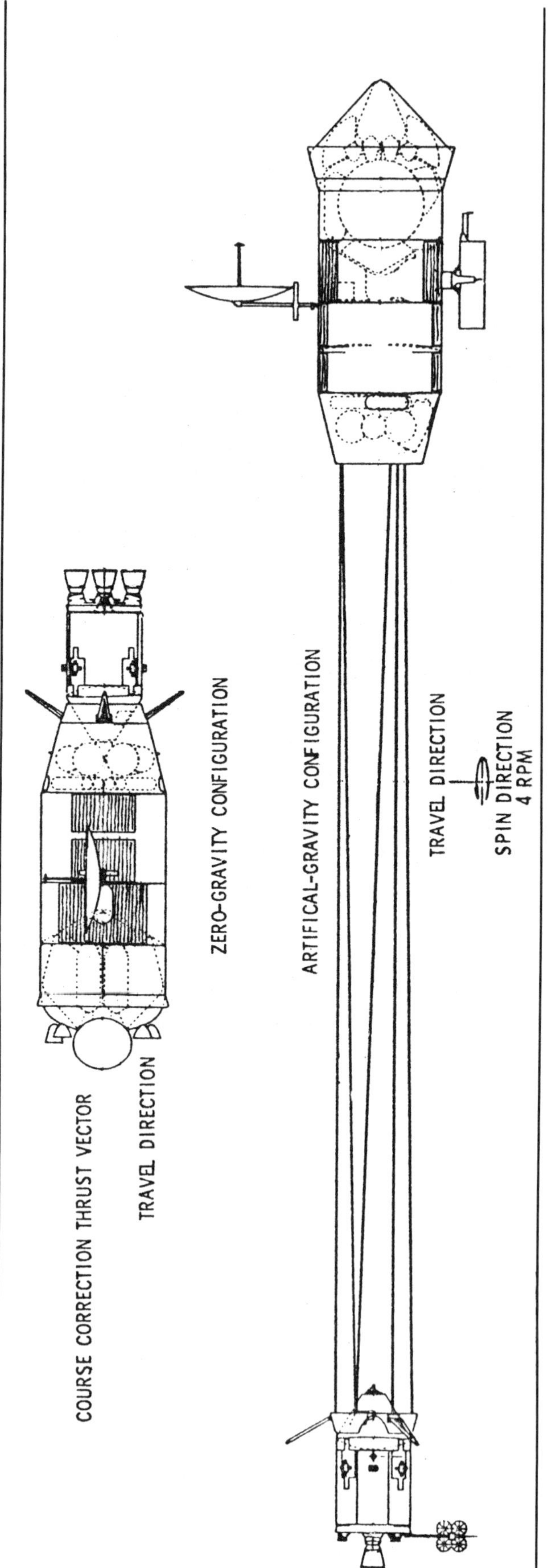

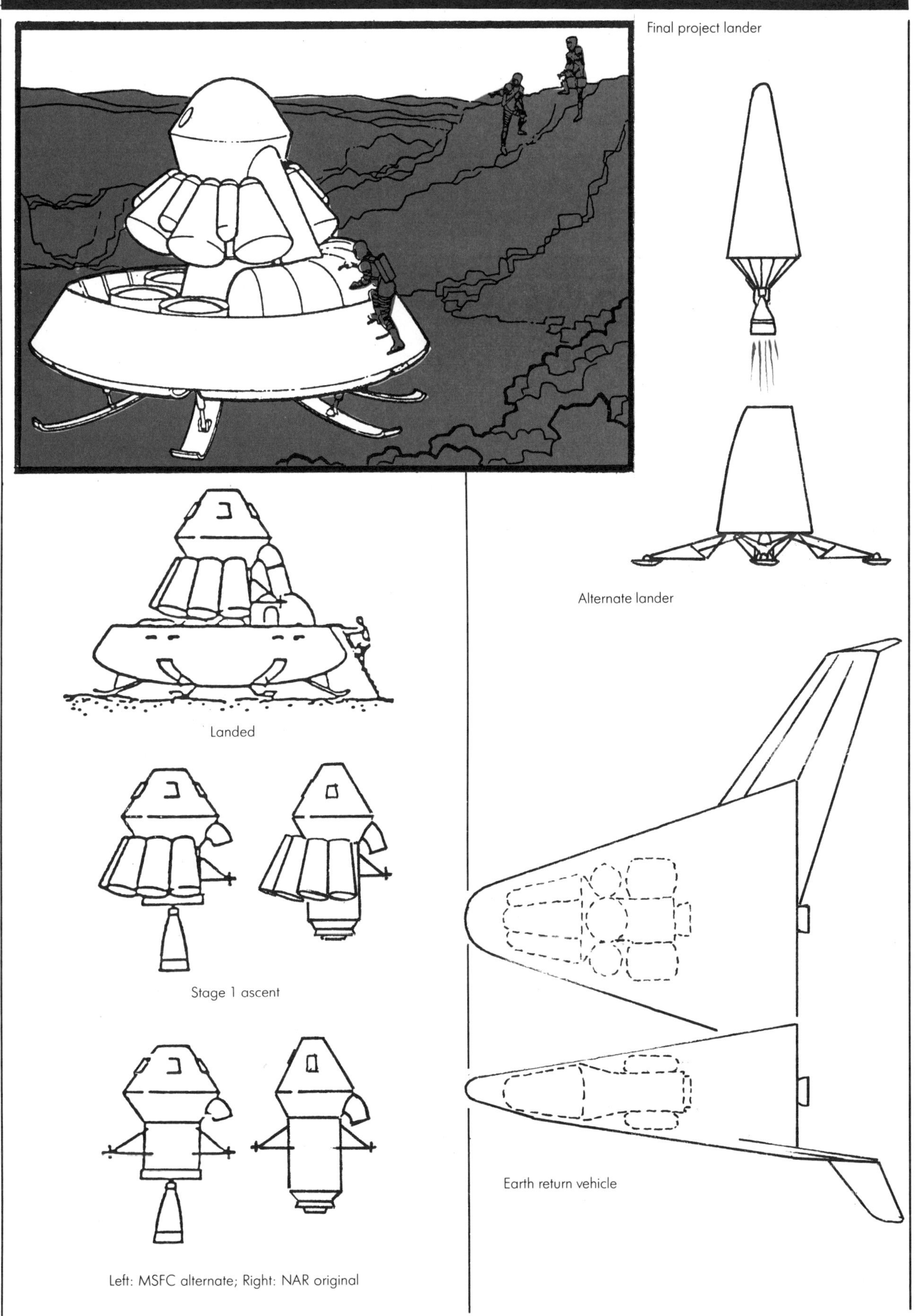

Final project lander

Alternate lander

Landed

Stage 1 ascent

Left: MSFC alternate; Right: NAR original

Earth return vehicle

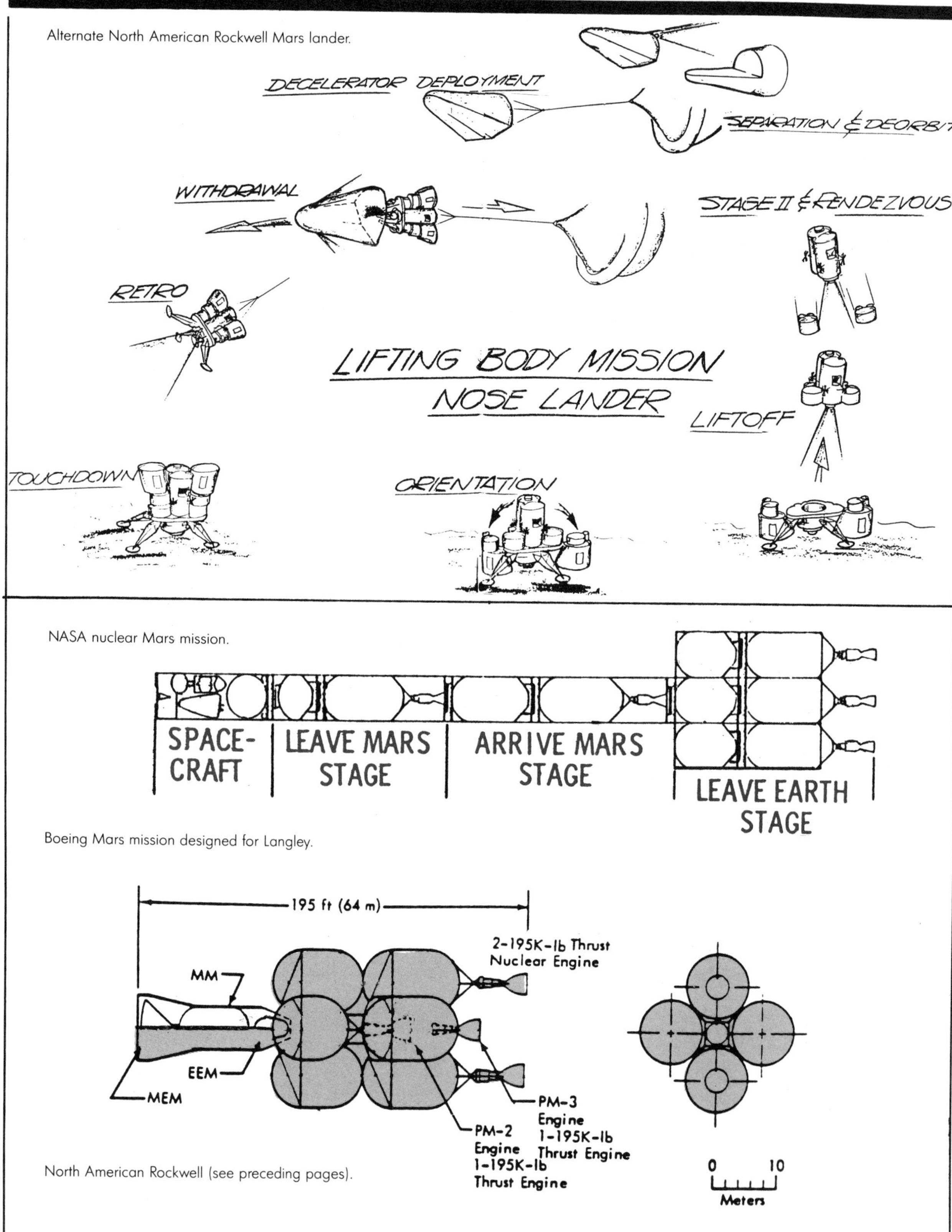

Alternate North American Rockwell Mars lander.

NASA nuclear Mars mission.

Boeing Mars mission designed for Langley.

North American Rockwell (see preceding pages).

OCTOBER 30

The Manned Spacecraft Center and the Marshall Spaceflight Center issue a joint Request for Proposal for an 8-month study of an "Integral Launch and Reentry Vehicle System." The goal is the requirement to place from 5,000 to 50,000 pounds of payload into earth orbit. This marks "Phase A" of a four phase NASA space shuttle study program initiated in 1965. The next three phases will be: Phase B—project definition; Phase C—design; and Phase D—development and operations. Needless to say, this original outline will be revised as necessary. The idea is simply to reduce the number of firms who will be carrying the shuttle development process through Phase D, while at the same time encouraging competition at the outset.

DECEMBER 21–27

Apollo 8 makes first manned flight around the moon. Astronauts Frank Borman, James A. Lovell, Jr. and William Anders orbit the moon ten times, coming as close as 70 miles to the surface. Live television coverage is broadcast to the earth.

On Christmas Eve, astronaut Borman reads the first ten verses of Genesis, the story of creation. "God bless all of you," he concluded, "All of you on the *good* earth."

ca. 1968

Boeing proposes recovering the Saturn V first stage at sea. The plan requires fixing V2-like fins onto the booster. Airbrakes on these would slow the booster's descent until a parachute could be deployed.

Lockheed proposes a two-stage spaceplane concept. The delta-winged first-stage aircraft, although capable of sustained flight with its jet engines, for takeoff will have to be itself boosted along a rail by rockets until it reaches a speed of 450 mph. The on-board rocket engines are then fired, accelerating the combined aircraft to a speed of about 4,000 mph at an altitude of 150,000 feet. The second-stage vehicle then separates, and the first-stage aircraft returns to its starting point, using conventional jet engines. The ten-passenger, two-pilot lifting body then continues accelerating under its own power either into orbit or along a ballistic path to some destination.

Three different second-stage shuttles can be available, depending upon the mission requirements.

1969

The G-I-e (SL-X) Soviet "superbooster" is developed at least partly with a manned moon flight in mind.

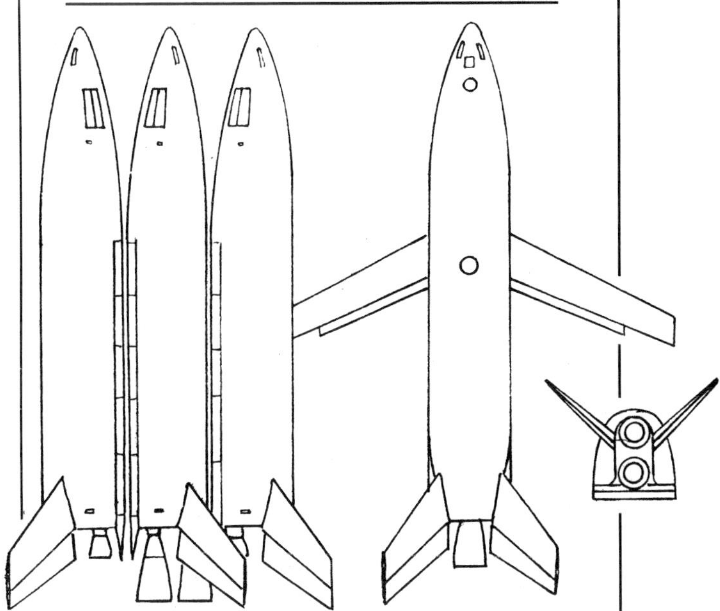

The Convair Triamese is a fully reusable NASA space shuttle concept, developed from the company's studies of variable-geometry lifting reentry vehicles beginning in 1965. It requires three nearly identical winged boosters, two of them 221 feet tall and one 190 feet tall. The two slightly larger rockets together form the two-stage booster for the third. The wings have a 130.3-foot-span when unfolded. They are to be launched in a parallel configuration, similar to the British MUSTARD concept [see above] which helped to inspire it. Two return to earth while the third goes into orbit. The wings on all three units are folded into the body during takeoff, as are the Rolls-Royce jet engines used for landing.

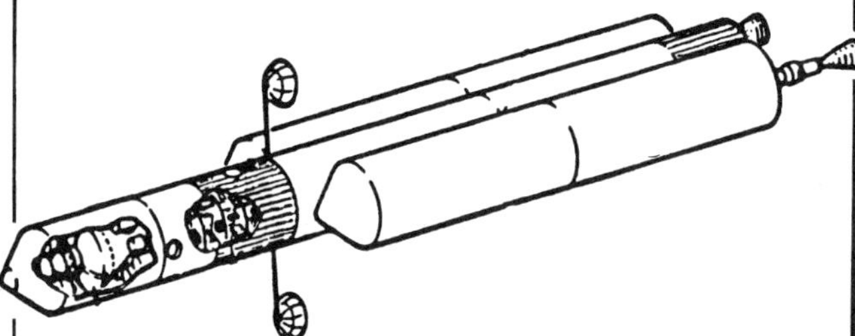

Wernher von Braun expects a manned expedition to land on Mars by the mid-1980s. "All elements, propellants and crews are carried into an earth orbit by a series of flights of the reusable shuttle vehicle plus two Saturn V flights for extra-large payload units. The interplanetary expedition consists of 12 men travelling in two ships which fly in formation." The two ships are boosted by atomic rockets into an circumsolar orbit. The boosters eventually return to the earth for reuse. The manned ships enter Mars orbit 270 days later, braking with their own nuclear engines. A two-stage chemical rocket (similar to the Apollo LM) carries the explorers to the Martian surface for a stay of 80 days. The return trip carries the expedition close enough to Venus for unmanned probes to be dropped into its atmosphere. The nuclear engines are used for a final time to bring the two ships back into an earth orbit where they are refitted for another flight. The total time for the round trip is 640 days.

Rudolf Nebel and W. H. Kurpanek, under a NASA contract, rather blatantly reinvent E. F. Northrup's space gun [see 1937]. They develop a "method of placing a space vehicle into orbit by electromagnetically accelerating it in an evacuated 10-mile-long tube." Like Northrup, they envision the gun running up the slope of a mountain, so that its initial run is horizontal. The projectiles they have designed are cylindrical, with folding, curved wings.

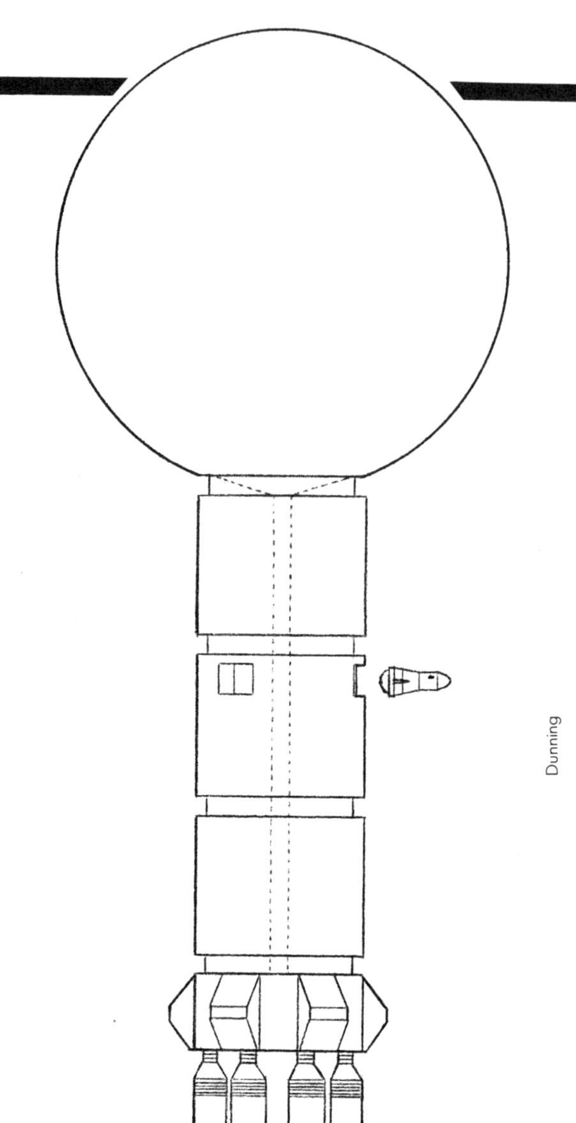

Dunning

Robert Enzmann designs a fusion-powered starship capable of making the trip to the nearest stars. It is fueled by nearly 3 million tons of supercold deuterium, contained in a metal sphere nearly 1,000 feet in diameter. The sphere will also serve as a radiation shield for the three habitat cylinders carried behind it. Each of these, 300 feet long and 300 feet in diameter, are divided into twenty decks, each with more than 100 rooms. The habitats, or portions of them, can be rotated to provide artificial gravity. At the rear of the 2,264-foot-long ship are the twenty-four fusion engines that can boost the ship to within 9 percent of the speed of light. At this speed, the starship can reach Alpha Centauri in 60 years, earth time.

A NASA manned Mars mission and base concept requiring a nuclear-powered spaceship is developed for the Manned Spaceflight

Center. A series of Saturn-derived launch vehicles place the components of the Mars spacecraft into earth orbit for assembly. The Mars spacecraft itself and its nuclear booster stage are launched by the first two stages of a pair of Saturn Vs; fuel, crew and expendables are carried into orbit by a winged shuttle. Already in orbit are a pair of nuclear boosters for each spacecraft.

There are two manned spacecraft, each with a crew of six. The spacecraft each consist of three nuclear boosters; a large lifesupport module, 22 feet in diameter and 75 feet long, divided into four compartments; and a Mars lander occupying a bay 33 feet in diameter. The life support module and lander bay together are 110 feet long. The complete ships are 270 feet long. The nuclear stages are 33 feet in diameter (the same diameter as the Saturn V boosters that carried them into orbit) and 160 feet long. Two of each ship's nuclear boosters are used for earth departure. (Retrorockets on the jettisoned boosters enable them to return to earth orbit for reuse.) The remaining booster is used for braking into Mars orbit and for departure from Mars.

If it develops that long-term spaceflight requires artificial gravity, then two of these spacecraft can be placed end to end, boosters facing outward, and spun around the transverse axis to provide artificial gravity. The giant cone-shaped lander also acts as the Mars base itself with its three floors of living space.

The Apollo-derived landers use a combination of aerodynamic braking and retrorockets for atmospheric entry and landing. The base of the lander acts as a launch pad for the ascent vehicle, which is discarded in Mars orbit at the end of the mission.

The mission is to be launched from earth orbit on November 12, 1981. It will arrive at Mars August 9, 1982 and will depart on October 28. It will return to the earth August 14, 1983 after making a flyby of Venus on February 28, where probes will be dropped.

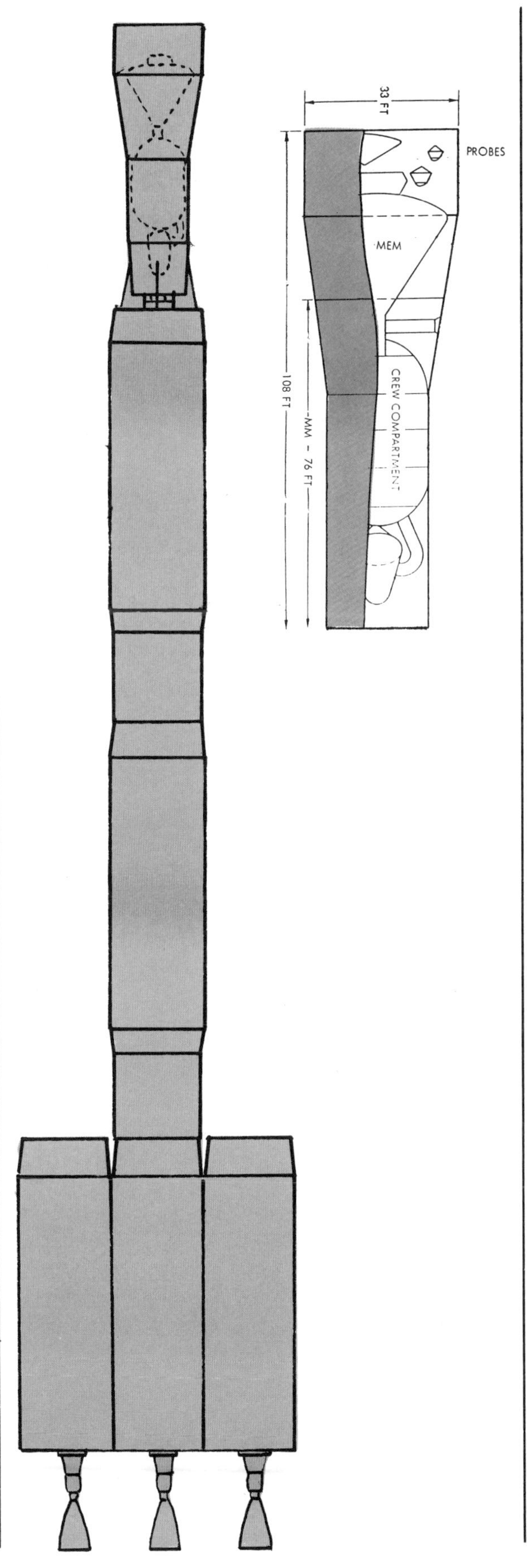

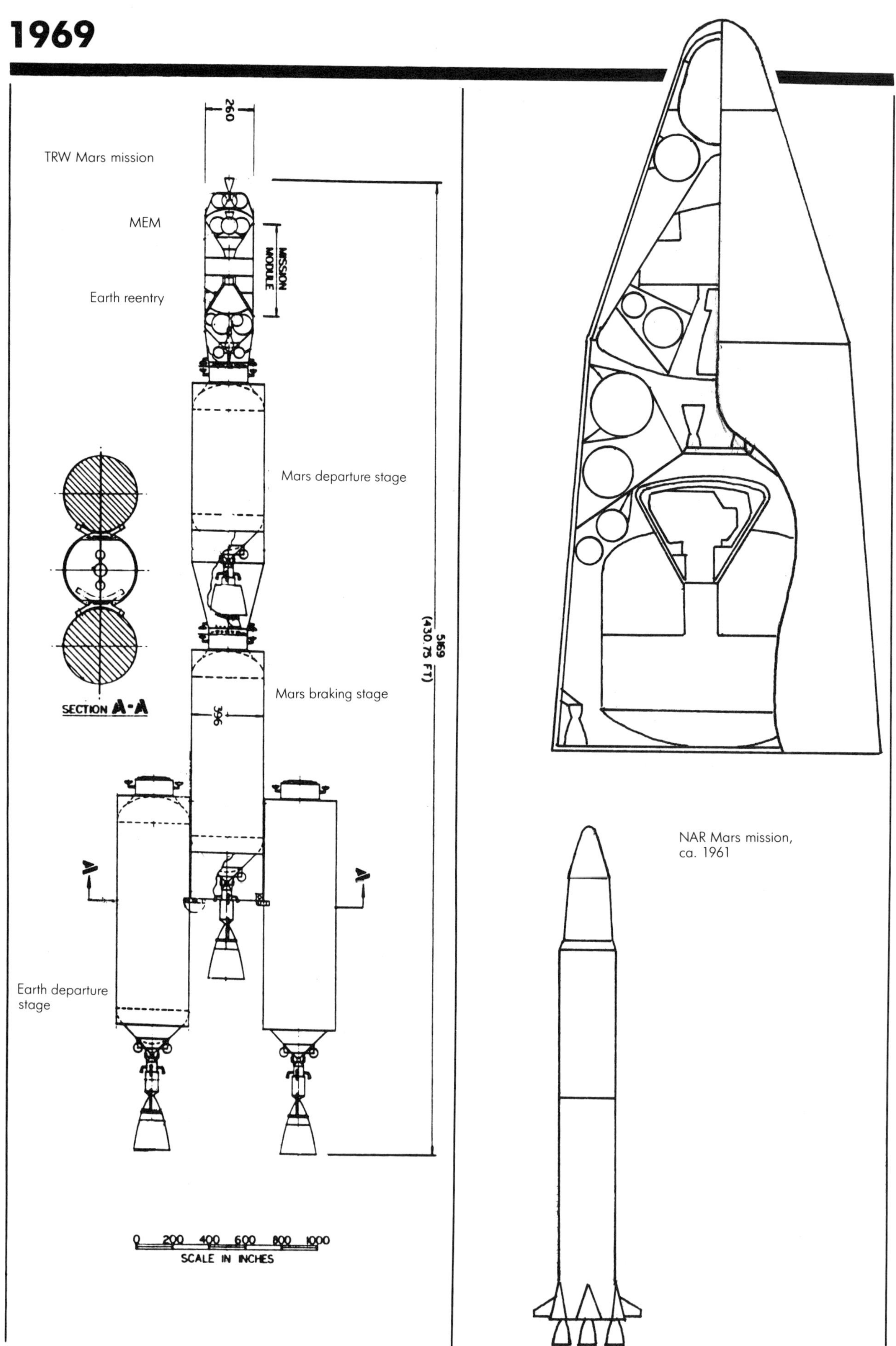
TRW Mars mission
260
MEM
MISSION MODULE
Earth reentry
Mars departure stage
5169 (430.75 FT)
Mars braking stage
396
SECTION A-A
A
A
Earth departure stage
0 200 400 600 800 1000
SCALE IN INCHES
NAR Mars mission, ca. 1961

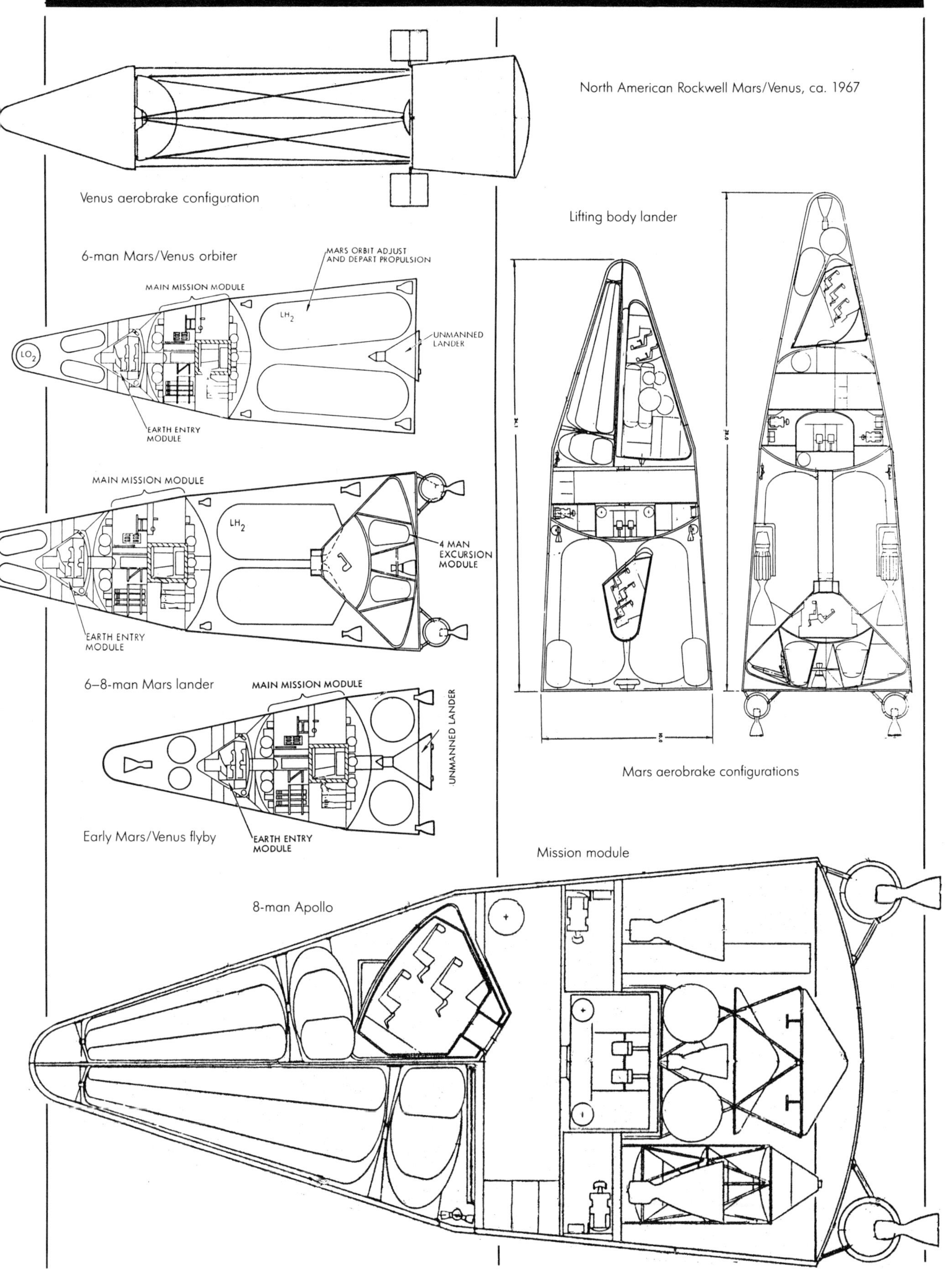
North American Rockwell Mars/Venus, ca. 1967
Venus aerobrake configuration
6-man Mars/Venus orbiter
MARS ORBIT ADJUST AND DEPART PROPULSION
MAIN MISSION MODULE
LH_2
LO_2
UNMANNED LANDER
EARTH ENTRY MODULE
MAIN MISSION MODULE
LH_2
4 MAN EXCURSION MODULE
EARTH ENTRY MODULE
6–8-man Mars lander
MAIN MISSION MODULE
UNMANNED LANDER
Early Mars/Venus flyby
EARTH ENTRY MODULE
Lifting body lander
Mars aerobrake configurations
Mission module
8-man Apollo

1969

Gerard K. O'Neill begins his studies of space colonies at Princeton University.

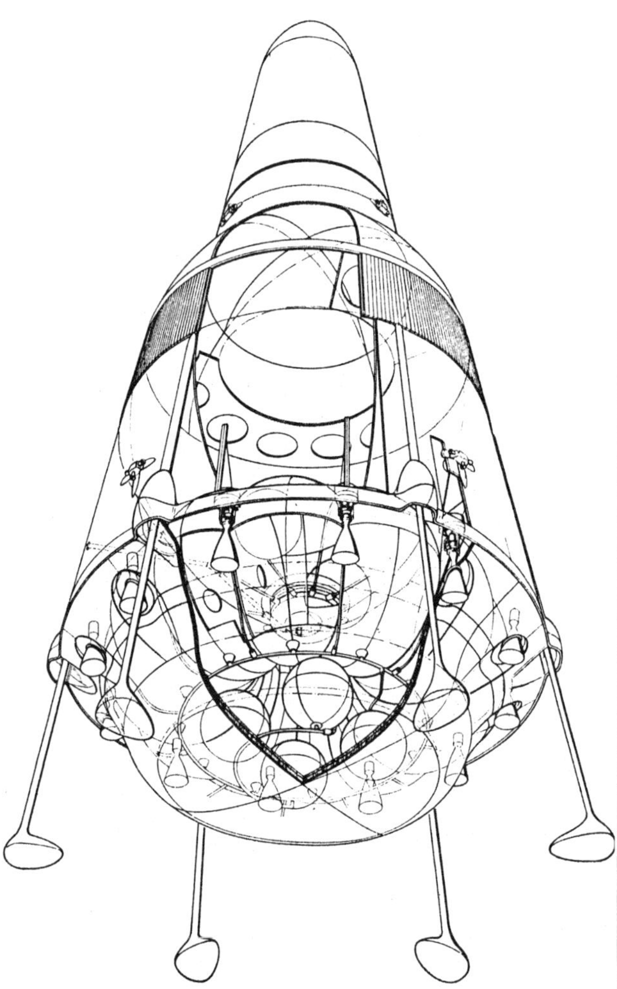

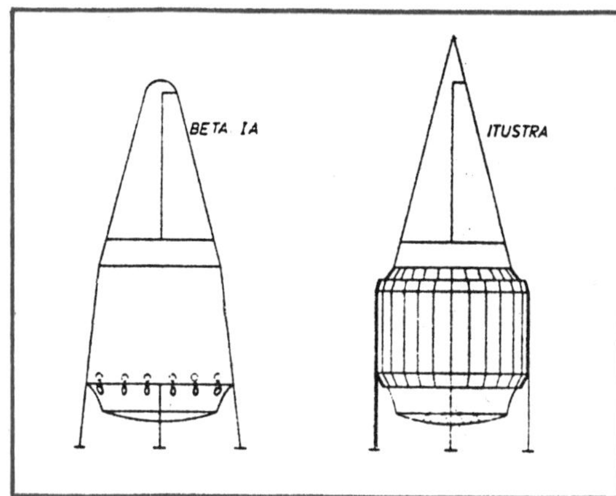

Designer D. Koelle of Messerschmidt-Bolkow-Blohm (MBB) works on a VTOL/SSTO concept called BETA (Ballistisches Enistufiges Trr-Aggregat). It was inspired by the Saturn Application SSTO work done by Philip Bono [see above and also see 1987].

BETA would have a launch weight of 130 tons, with a propellant mass of 115 tons, and a very small ratio of length to diameter. Its 7.8 meter reentry heat shield also acts as a plug nozzle. It takes off and lands on a set of six legs.

BETA could carry a payload of 2.2 to 3.1 tons with this being perhaps doubled by the 1980s as technology develops. A five ton payload could be accomplished today by increasing the size of the vehicle's propellant mass to 145 tons. With the addition of a second "kick" stage a payload of about 250 kg could be launched to the vicinity of the planet Mercury or 700 kg into a geostationary orbit.

JANUARY 14–17

Soyuz 4 and Soyuz 5 achieve the first docking of two manned spacecraft. They remain attached for 35 minutes while the crews of the two spacecraft are exchanged. In the past, the Russians have accomplished the docking of unmanned spacecraft.

FEBRUARY

NASA awards Space Shuttle Phase A study contracts to Lockheed, General Dynamics, McDonnell-Douglas and North American Rockwell.

NASA's Space Shuttle Task Group has meanwhile divided the concept studies into three categories: Class I: recoverable orbiters with expendable boosters; Class II: stage-and-a-half vehicles; and Class III: fully reusable two-stage-to-orbit shuttles. Payload capacities range up to 50,000 pounds (weight and size, up to 15 feet by 60 feet, are determined by anticipated space station component requirements). From these, six configurations are developed that receive detailed review.

First is a Class I McDonnell lifting body dubbed MURP (Manned Upper Reusable Payload) which has room for a crew of ten. It is equipped with engines for landing and re-

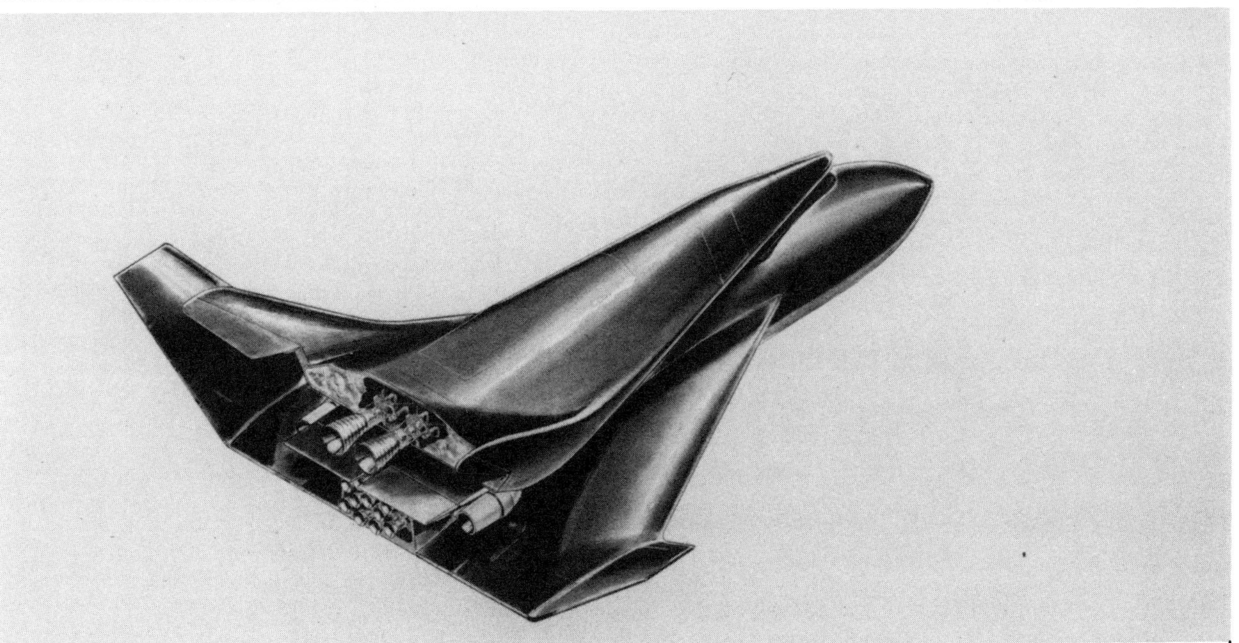

tractable swing-wings. Second and third are the Class II Lockheed Star Clipper [see 1969] and a McDonnell parallel tank stage-and-a-half design [see 1969]. The remaining three proposals are all Class III. One is a NASA-Langley Research Center 130-foot orbiter based on the HL-10 lifting body, launched by a booster also based on the HL-10. Both are equipped with air-breathing engines for landing. Second is a development of Convair's Triamese [see 1969]. The new version takes two different forms: the FR-3A "Biamese" and the FR-4, a revised Triamese. The former is a two-element booster-orbiter combination. The booster is 240 feet long and carries a parallel orbiter that is 158 feet in length. The orbiter and booster are both equipped with swing wings for maneuvering and landing. The new Triamese maintains the three identical units that characterized its original form, but eliminates the propellant cross-feed arrangements. The sixth and last concept being considered is a controversial and influential straight-wing orbiter designed by Max Faget [also see 1970] of the Manned Spacecraft Center.

The Faget orbiter bears a strong resemblance to the final Space Shuttle design of the late 1970s. It is intended to reenter the atmosphere at a steep angle of attack, at 40,000 feet leveling out to a lower angle after which it will ignite its landing engines. This design forms the basis for most NASA shuttle studies from 1968—dozens of variations

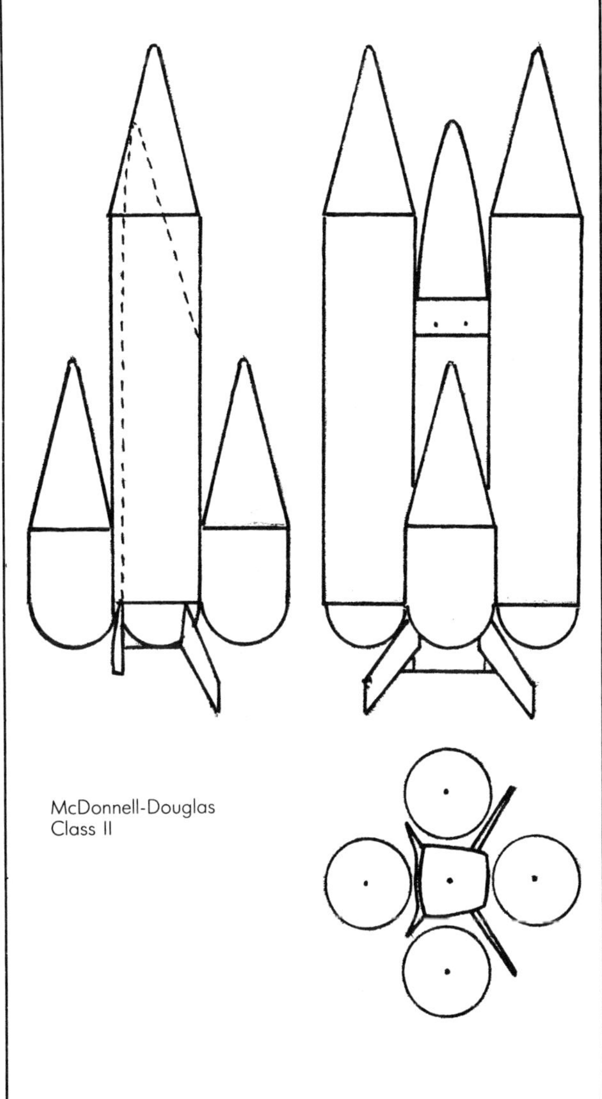

McDonnell-Douglas
Class II

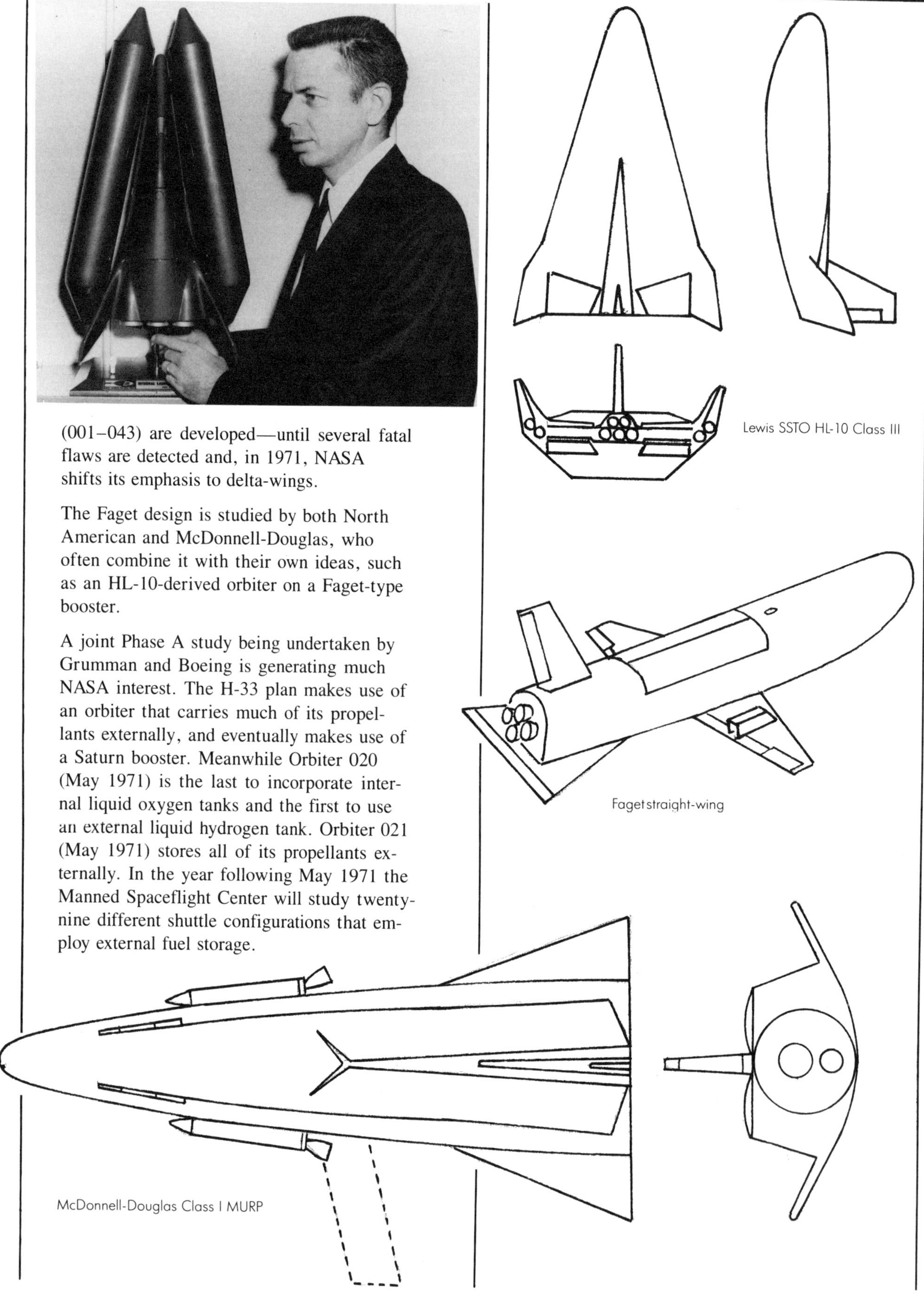

(001–043) are developed—until several fatal flaws are detected and, in 1971, NASA shifts its emphasis to delta-wings.

The Faget design is studied by both North American and McDonnell-Douglas, who often combine it with their own ideas, such as an HL-10-derived orbiter on a Faget-type booster.

A joint Phase A study being undertaken by Grumman and Boeing is generating much NASA interest. The H-33 plan makes use of an orbiter that carries much of its propellants externally, and eventually makes use of a Saturn booster. Meanwhile Orbiter 020 (May 1971) is the last to incorporate internal liquid oxygen tanks and the first to use an external liquid hydrogen tank. Orbiter 021 (May 1971) stores all of its propellants externally. In the year following May 1971 the Manned Spaceflight Center will study twenty-nine different shuttle configurations that employ external fuel storage.

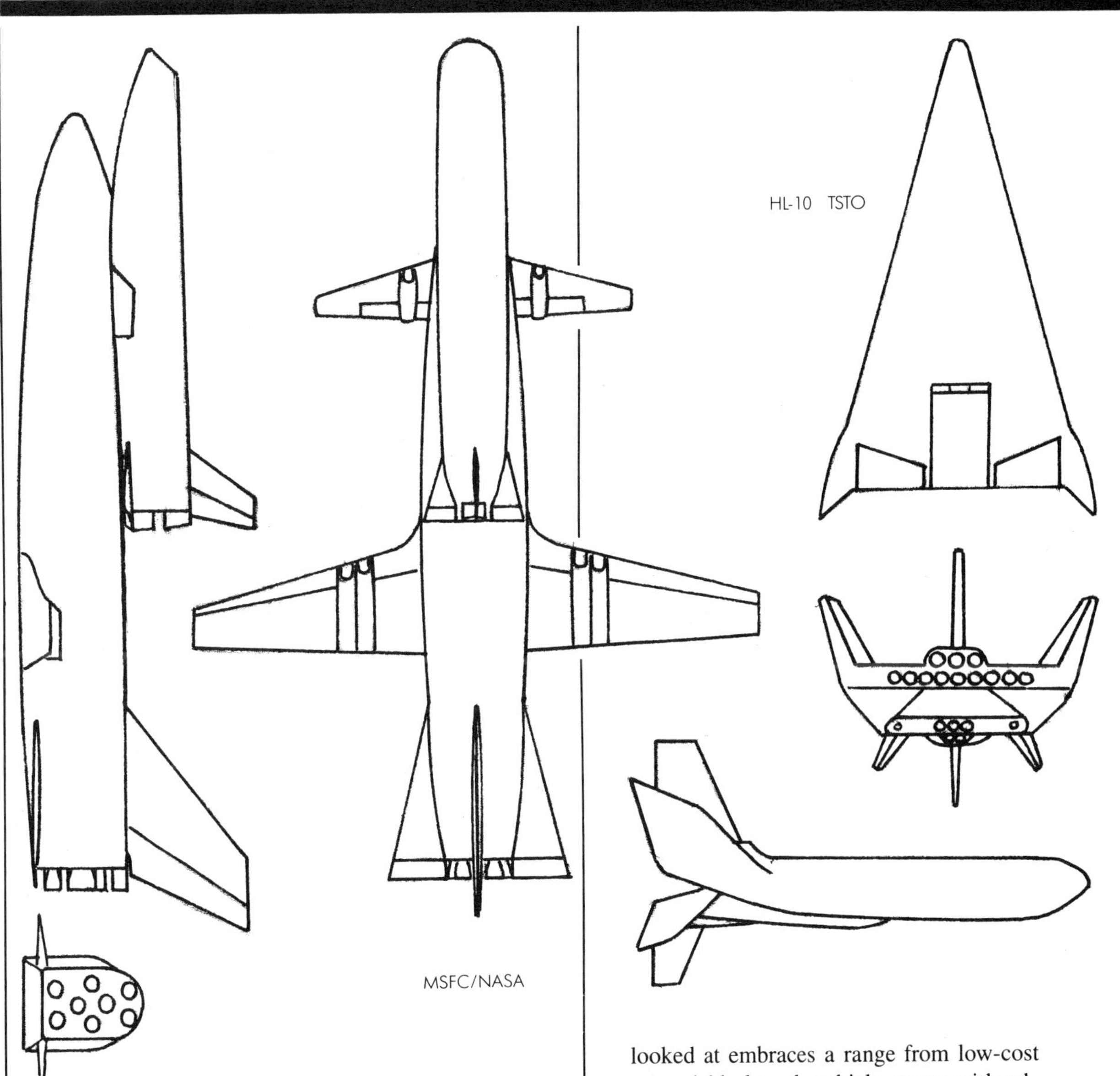

APRIL 29

NASA publishes its Manned Spaceflight Program. Four studies of low cost space transportation systems have been undertaken, at a cost of $300,000 each. The contractors were McDonnell Douglas, North American Rockwell, General Dynamics and Lockheed. NASA hoped to discover a conceptual design that would be an order of magnitude reduction in costs, achieve a significant increase in safety and be flexible and versatile enough to respond to a variety of missions. Low-g forces and shirtsleeve environments are looked for, with large cargo capacities and cheap expendable elements.

The spectrum of concepts that are being looked at embraces a range from low-cost expendable launch vehicle stages with advanced reusable spacecraft, to completely reusable systems having launch, orbit and reentry functions completely integrated.

McDonnell Douglas's share of the project will not concern itself with low lift-to-drag ratio spacecraft, flyback reusable stages or the fully reusable Triamese, instead concentrating on expendable, low cost liquid or solid fuel first and second stages with reusable medium lift-to-drag ratio spacecraft, and stage and a half concepts. North American Rockwell will concentrate on low cost expendable launch vehicles with reusable spacecraft. Lockheed will study stage and a half designs and the fully reusable Triamese concept. General Dynamics will concern itself with the Triamese, flyback first stages with expendable upper stages and low cost expend-

1969

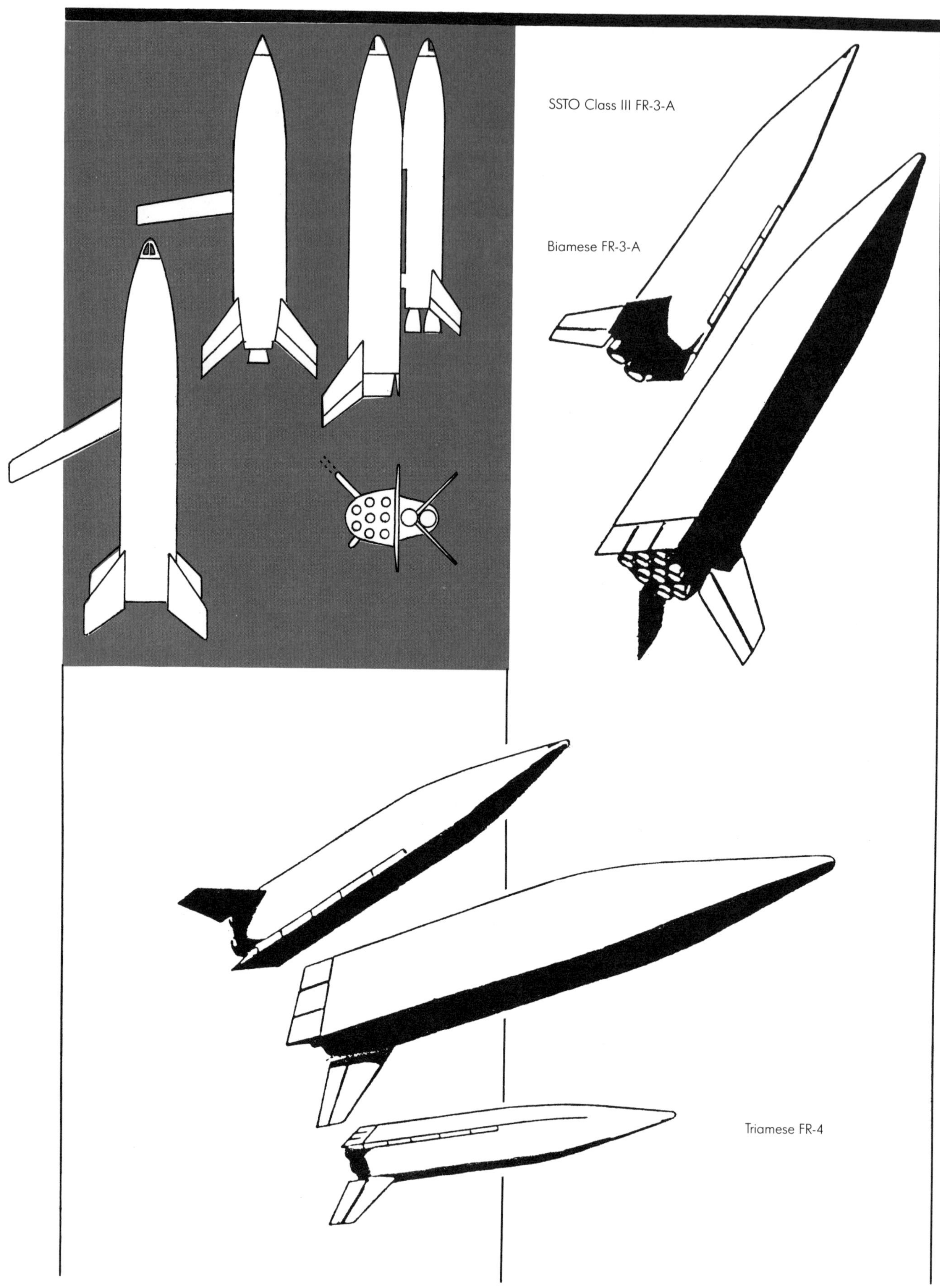

able launch vehicles with reusable medium lift-to-drag ration spacecraft.

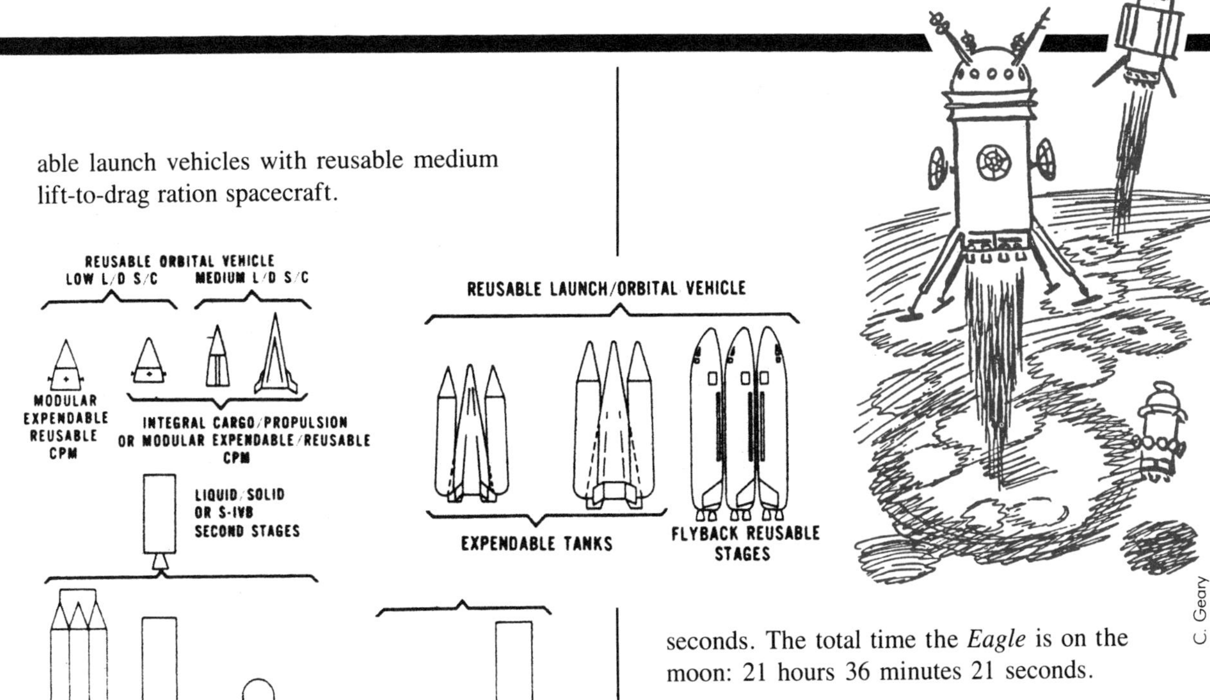

MARCH 20

First test-firing of the Nerva-XE atomic engine takes place.

JULY 16–24

Apollo 11 makes the first manned landing on the moon. Astronauts Neil Armstrong, Michael Collins and Edwin Aldrin, Jr. make the journey in the Command Service Module *Columbia*. While Collins remains in the orbiting command module, Armstrong and Aldrin make the descent in the Lunar Module *Eagle* to the lunar surface where they land in the Mare Tranquilatatis at 4:17:43 EDT on July 20. Armstrong puts the first human footprint on the surface of the moon at 8:56 EDT, Aldrin following about 15 minutes later. As Armstrong takes his first step he utters the now famous—and infamous—words, "That was one small step for man, one giant leap for mankind." Together they erect an American flag and collect 48.5 pounds of samples. The total time spent outside on the moon is 2 hours 31 minutes 40 seconds. The total time the *Eagle* is on the moon: 21 hours 36 minutes 21 seconds.

Lunar Module

The two-stage spacecraft is made primarily of aluminum. The first stage is the descent stage for the lunar landing, and the second stage is the ascent stage for the takeoff from the moon. The descent stage is an octagonal structure containing hypergolic fuel and a large gimballed engine whose thrust is throttleable between 1,050 and 9,850 pounds. The engine compartment is surrounded by four square bays containing propellant and lunar exploration equipment. Between these are triangular bays containing engine controls, and tanks for water, helium and oxygen. The folding landing legs are equipped with 37-inch circular feet.

The ascent stage has a pressurized cabin measuring 92″ × 42″ in area with 160 ft^3 of habitable volume. A vertical tunnel leads to the docking and transfer hatch. A lunar EVA hatch is in front between the windows. Surrounding the control cabin are tanks for the hypergolic fuel, equipment and batteries. An ascent engine has a fixed thrust of 3,500 pounds. Attitude control is provided by four sets of four 100-pound-thrust rockets.
the hypergolic fuel, equipment and batteries. An ascent engine has a fixed thrust of 3,500 pounds. Attitude control is provided by four sets of four 100-pound-thrust rockets.

1969

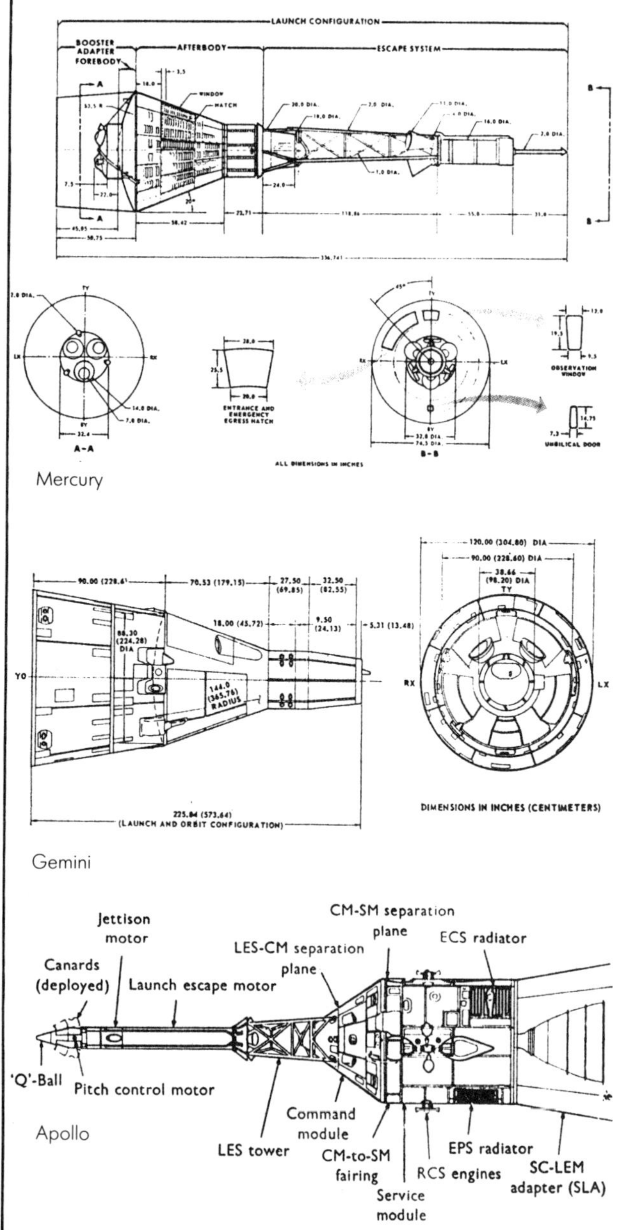

Mercury

Gemini

Apollo

Command/Service Module

The cone-shaped command module is a stainless-steel honeycomb structure 12′ 5″ tall and 12′ 10″ inches in diameter. It is fitted to one end of the cylindrical service module, which is 24′ 7″ long. At the top of the command module are the parachutes for the descent to earth and the docking probe, by which access to the lunar module is obtained. The cabin is pressurized and has a habitable volume of 210 ft^3. This contains three couches for the crew. There are ten attitude-control jets of 93 pounds thrust each. The blunt rear end of the command module is an ablative heat shield.

The service module is made of an aluminum honeycomb, and is divided longitudinally into six compartments containing fuel cells and their hydrogen and oxygen fuel, an attitude control system consisting of sixteen 100-pound jets and the service propulsion system engine of 20,500 pounds thrust along with its hypergolic fuel supply. The service module powers all maneuvers, corrections, lunar orbit departure and retroburn.

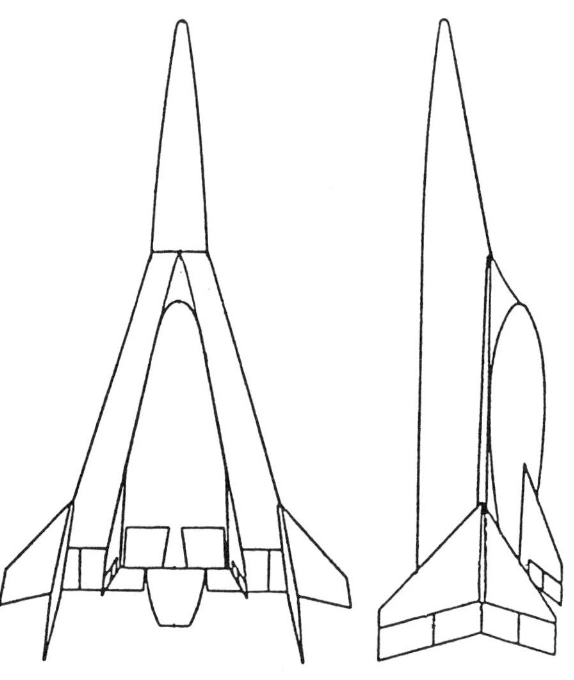

LATE 1960s

Messerschmidt-Bölkow-Blohm (MBB), after studying a number of possible space transport systems, settles upon a fully reusable two-stage shuttle employing a winged booster and a lifting-body orbiter.

ca. 1969

Martin submits an unsolicited space shuttle design to NASA called the "Spacemaster." It is a lifting body orbiter held between a pair of parallel boosters, all joined by the stubby wing and tail surfaces. The joined boosters (compared to an F-82 Twin Mustang) act as a single piloted launch vehicle which can return to earth under its own power.

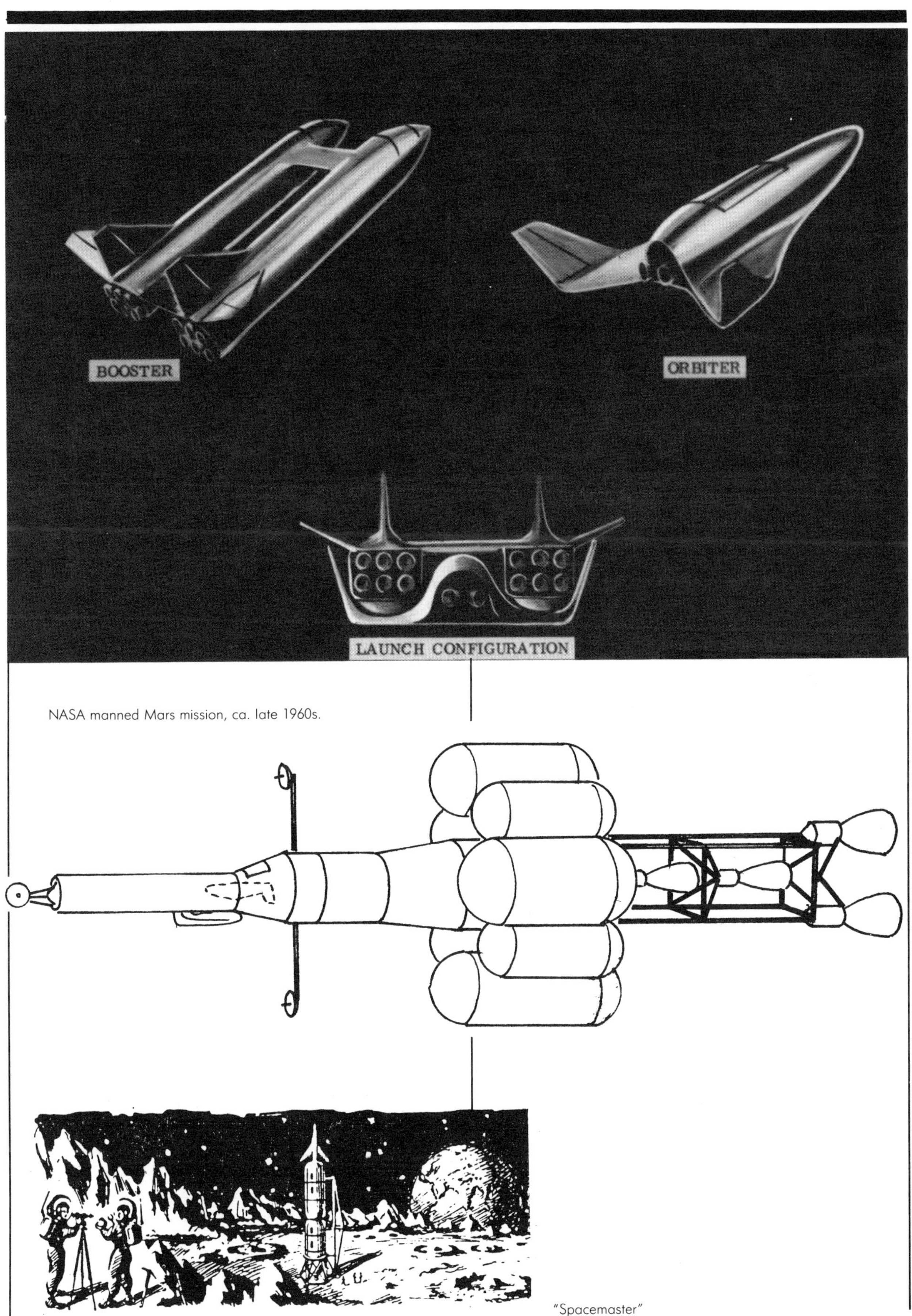

NASA manned Mars mission, ca. late 1960s.

"Spacemaster"

1969

GENERAL DYNAMICS/CONVAIR MANNED MARS/VENUS MISSIONS OF THE 1960s

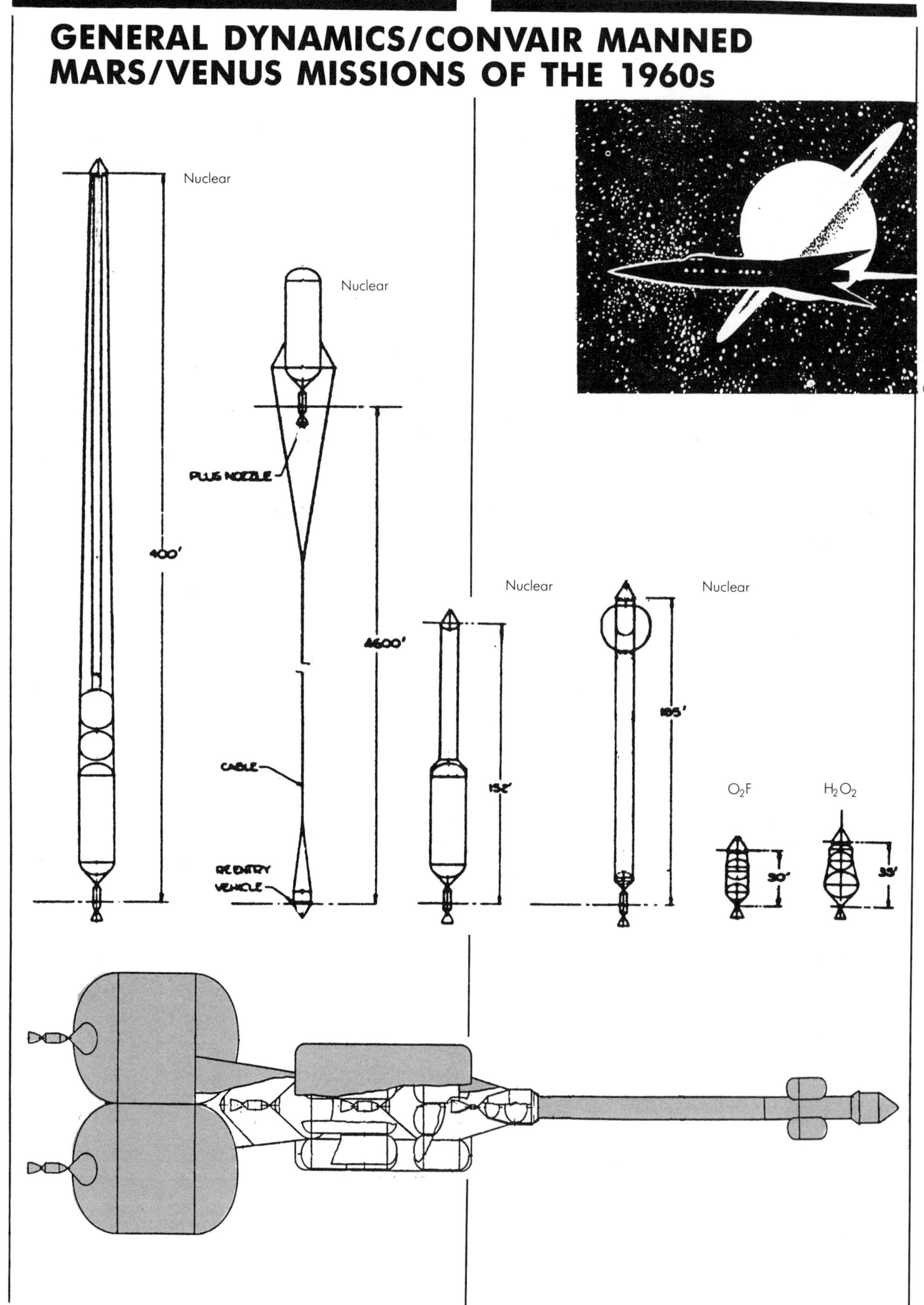

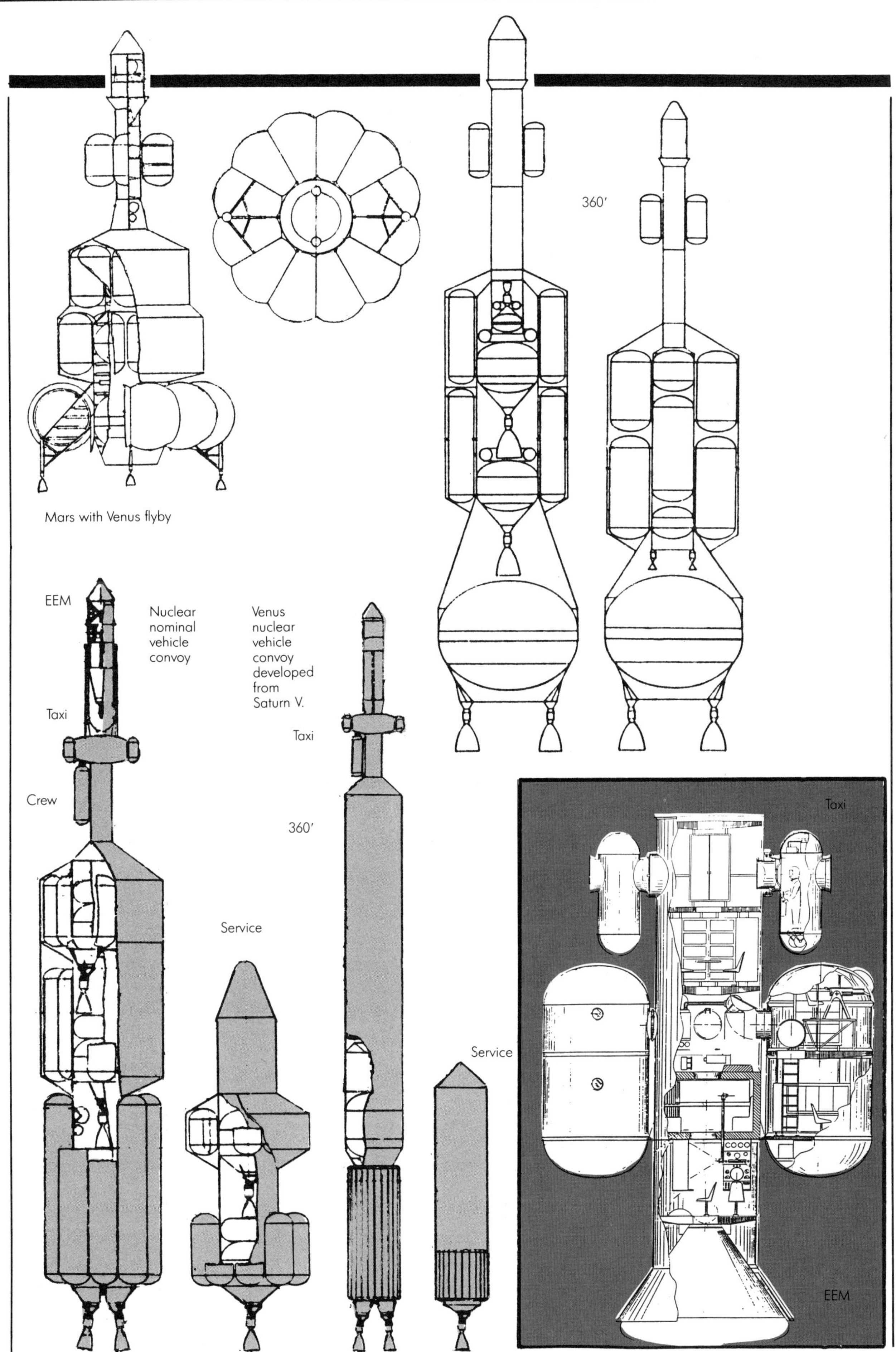
Mars with Venus flyby
360′
EEM
Nuclear nominal vehicle convoy
Venus nuclear vehicle convoy developed from Saturn V.
Taxi
Taxi
Crew
360′
Service
Service
Taxi
EEM

EVOLUTION OF APOLLO I

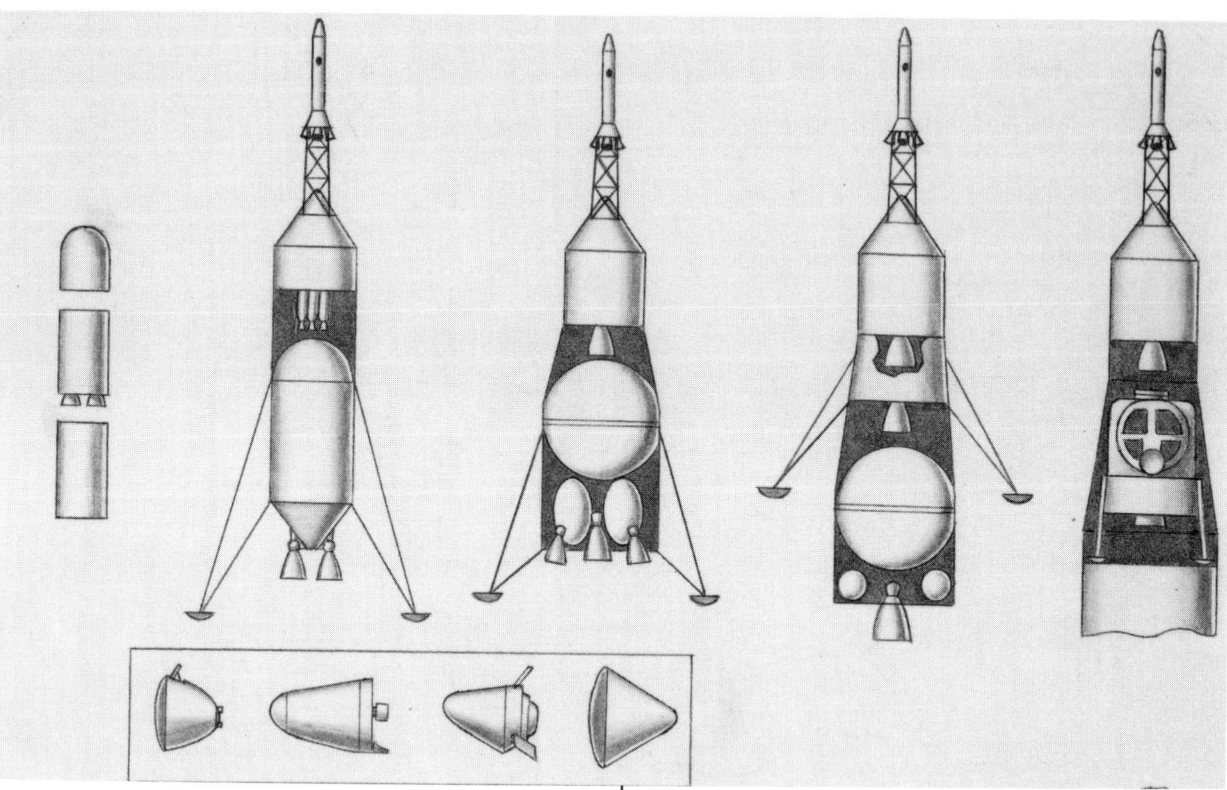

Early evolution of Apollo. Left to right: 7/61, 12/61, 4/62, 7/62. Inset: alternative reentry vehicles.

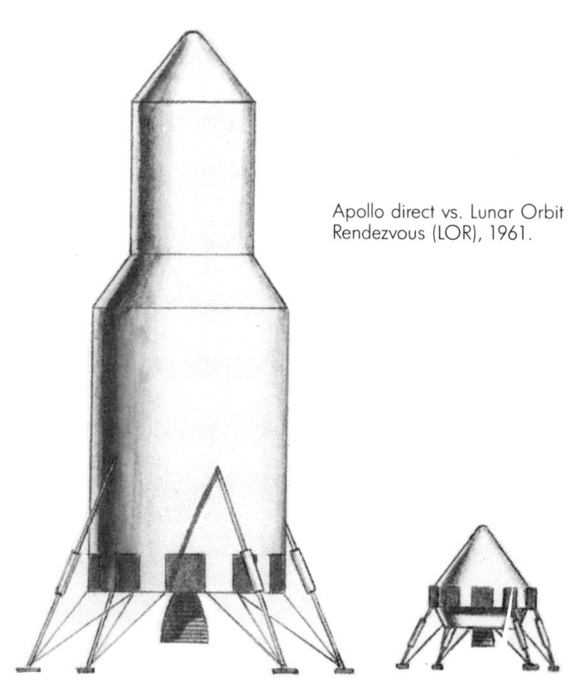

Apollo direct vs. Lunar Orbit Rendezvous (LOR), 1961.

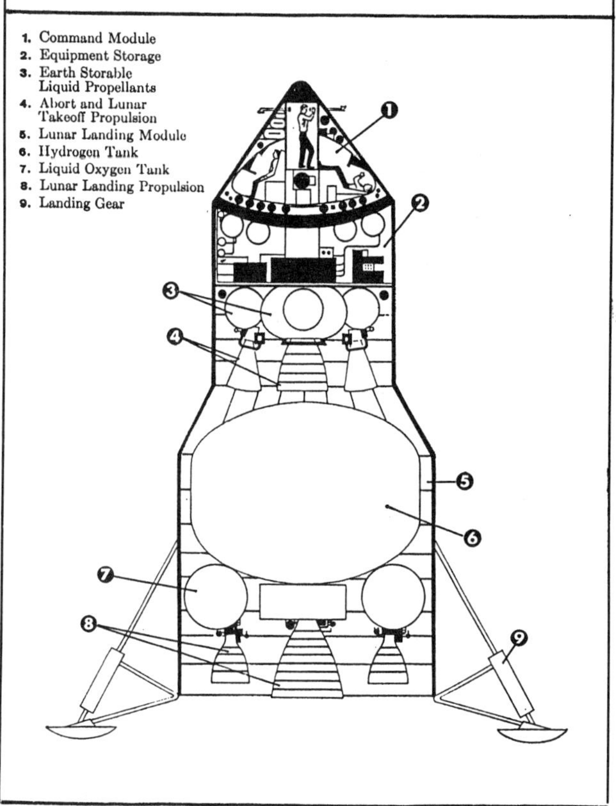

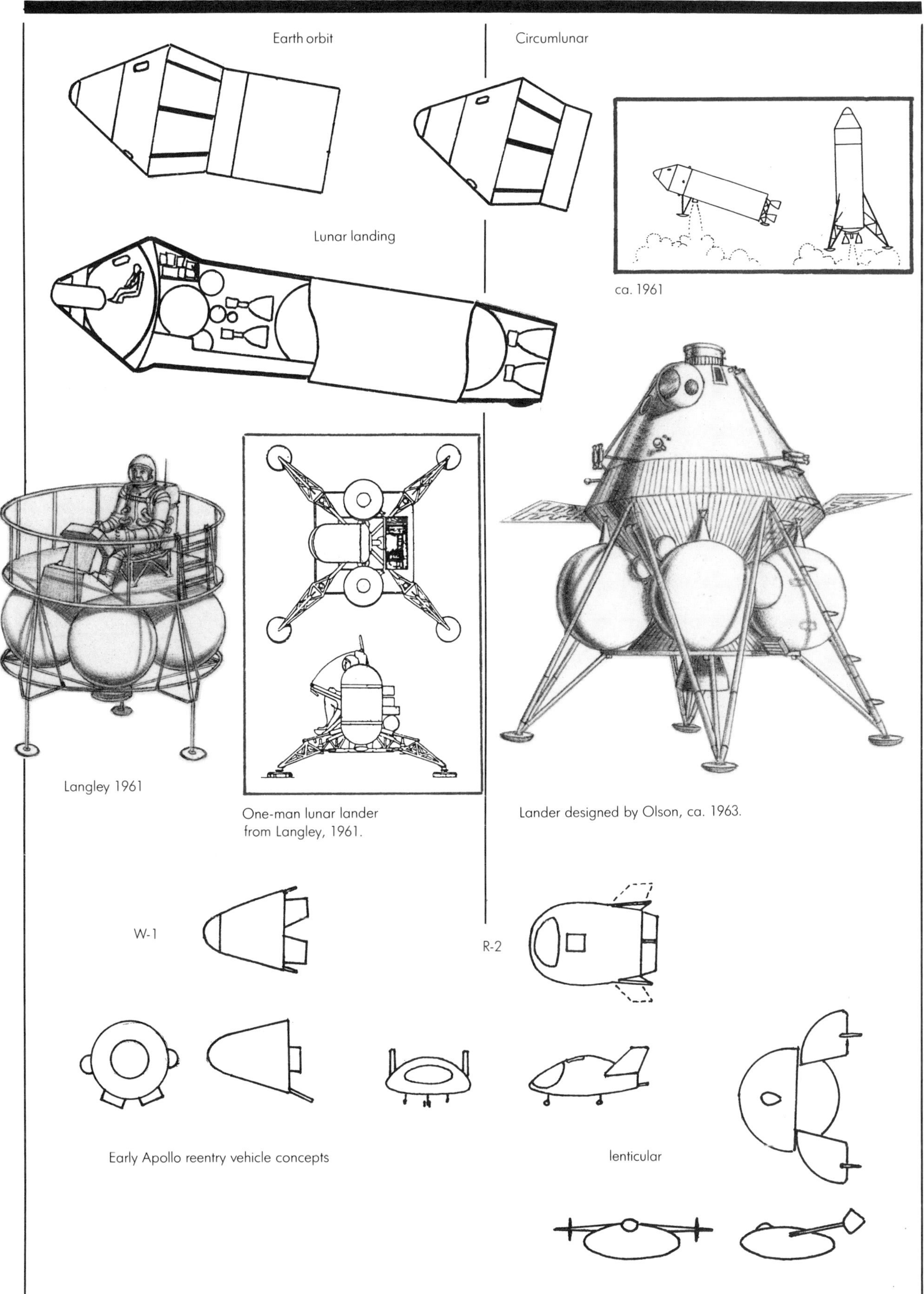

One-man lunar lander from Langley, 1961.

Lander designed by Olson, ca. 1963.

Early Apollo reentry vehicle concepts

APOLLO II

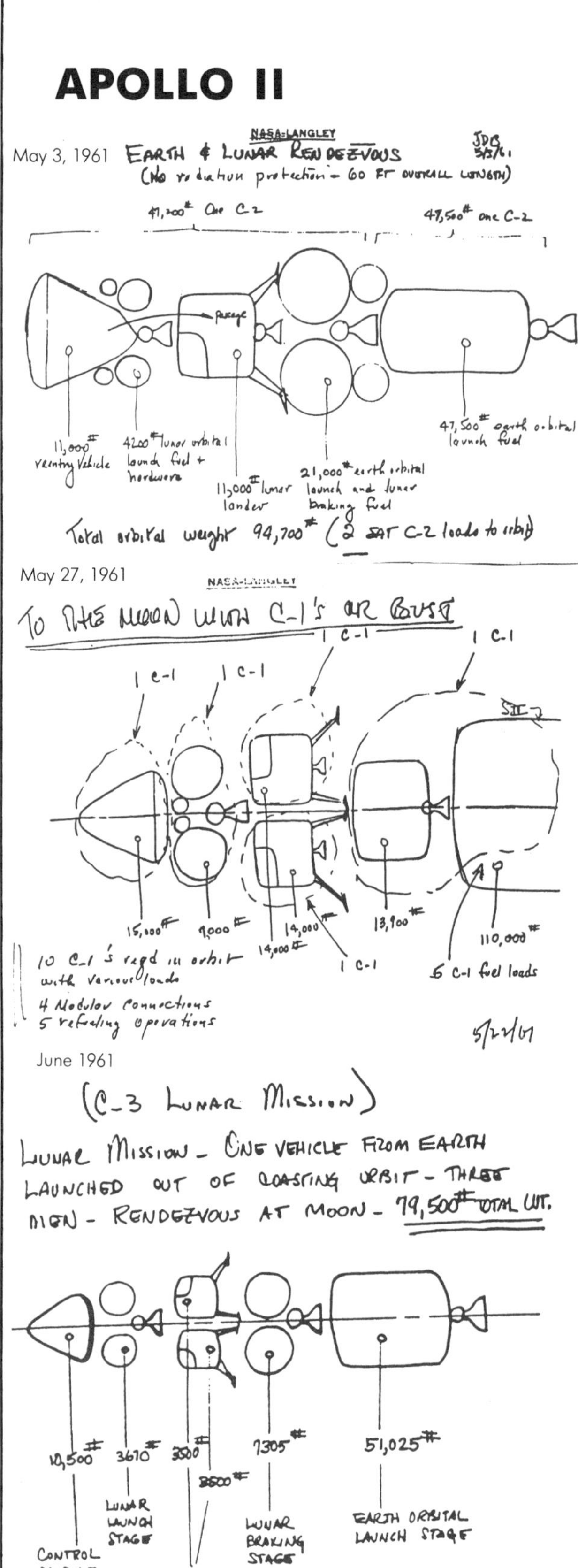

Left to right: Command Service Module with Descent Stage, landing configuration, liftoff from moon (1962).

Dunning

Above: Mock-up of Apollo spacecraft, 4/62. Right: Tranlunar flight configuration, 1962.

NASA (3)

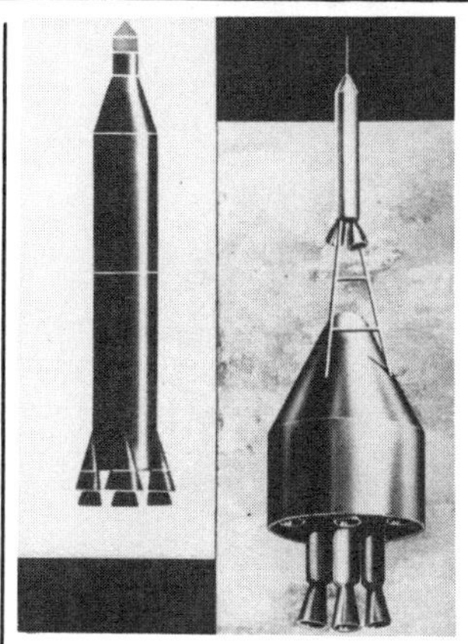

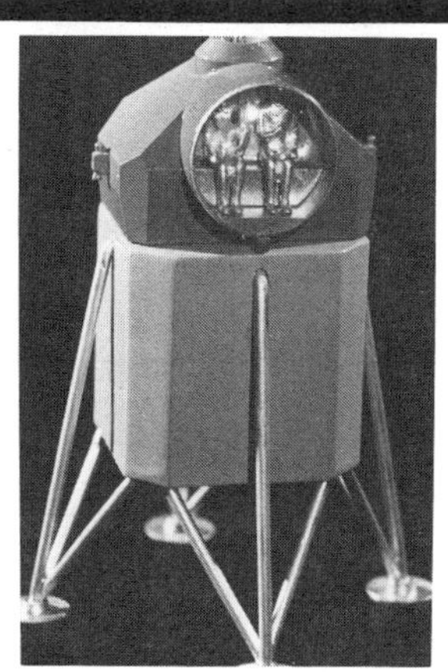

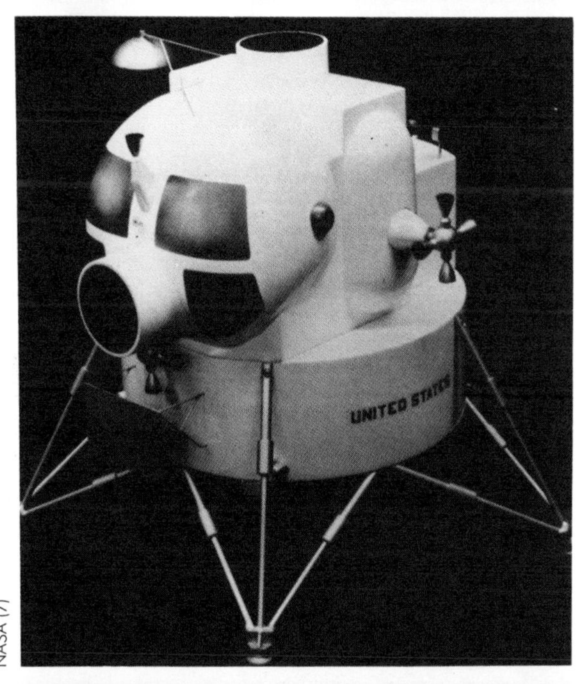

NASA (7)

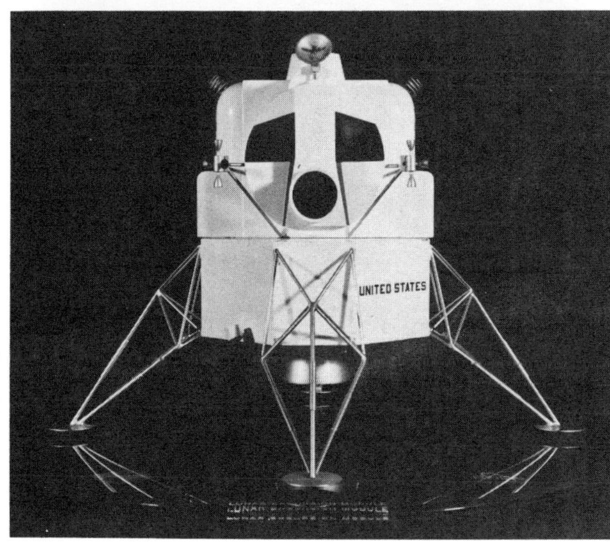

Left to right, top to bottom: Dr. J. F. Shea with Apollo models for direct ascent (L) and LOR (R), 1962; Saturn C-5 Apollo, 1962; Lunar Excursion Module, 1963; Grumman LEM, 1962; LEM, 1963; LEM, 1963; copper model of Grumman LEM, 1964.

APOLLO III

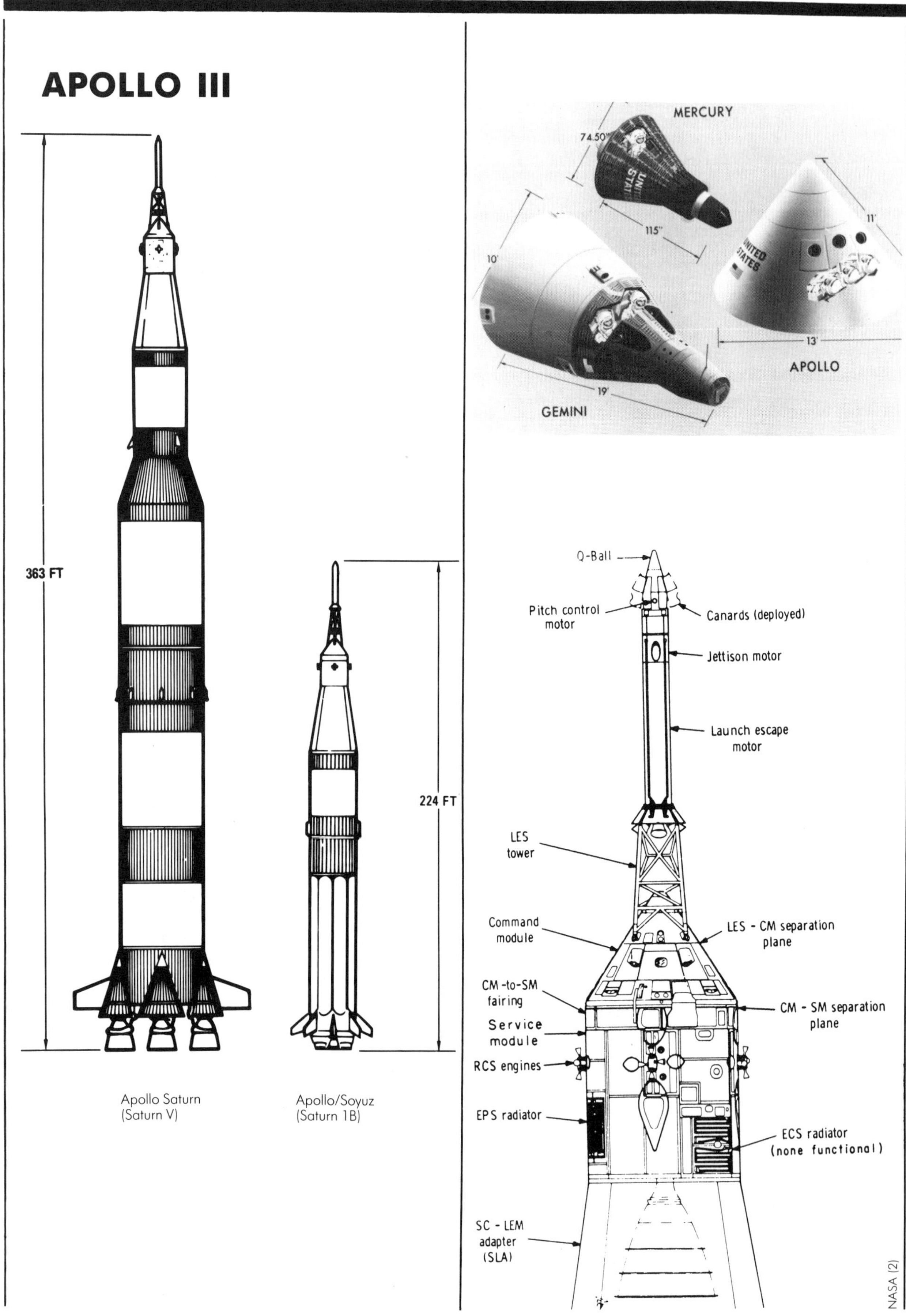

363 FT
224 FT
Apollo Saturn
(Saturn V)
Apollo/Soyuz
(Saturn 1B)
MERCURY
74.50"
115"
10'
11'
13'
19'
APOLLO
GEMINI
Q-Ball
Pitch control
motor
Canards (deployed)
Jettison motor
Launch escape
motor
LES
tower
Command
module
LES - CM separation
plane
CM-to-SM
fairing
CM - SM separation
plane
Service
module
RCS engines
EPS radiator
ECS radiator
(none functional)
SC - LEM
adapter
(SLA)
NASA (2)

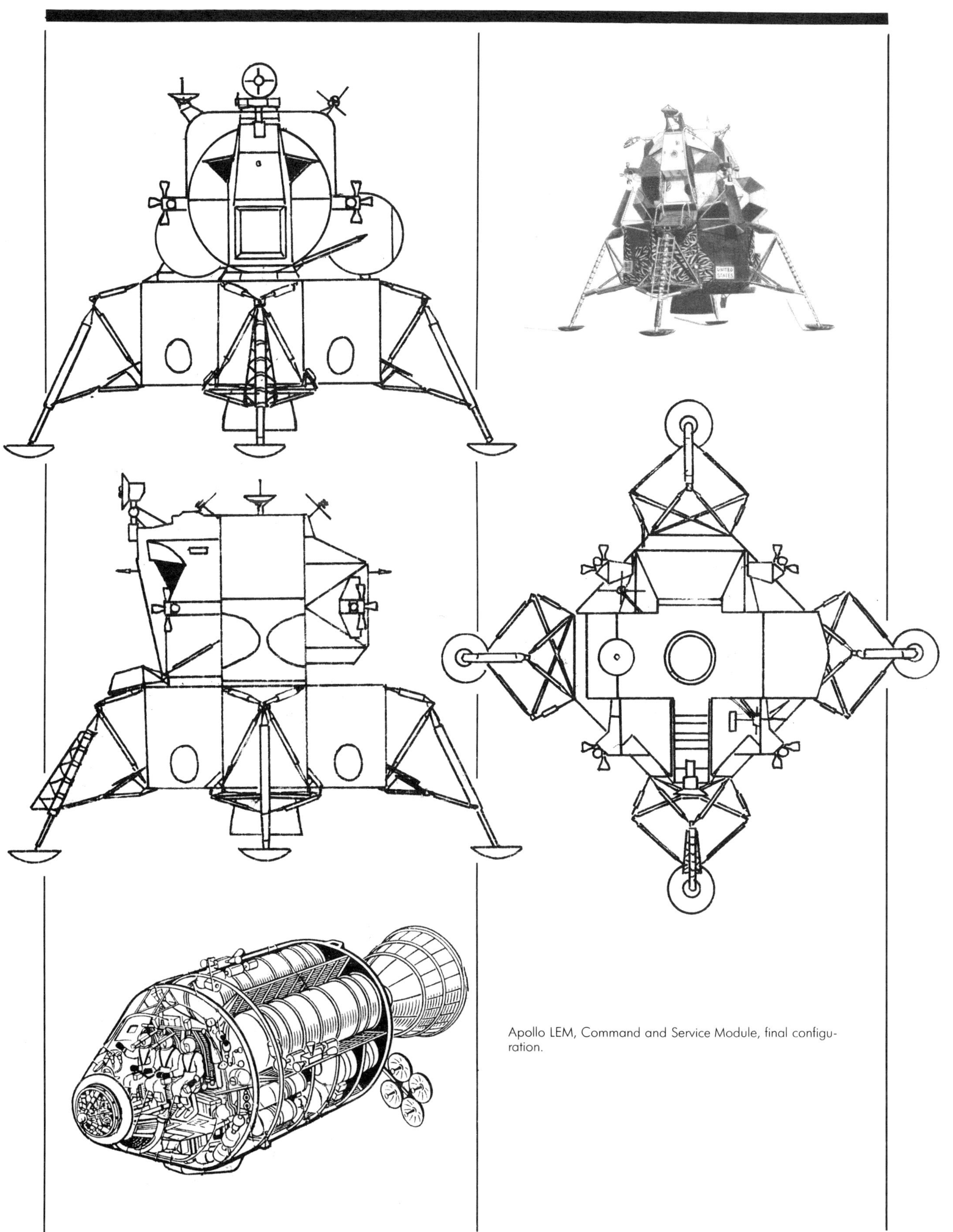

Apollo LEM, Command and Service Module, final configuration.

Ion-propelled spaceship for deep space mission designed by Chesley Bonestell, 1970.

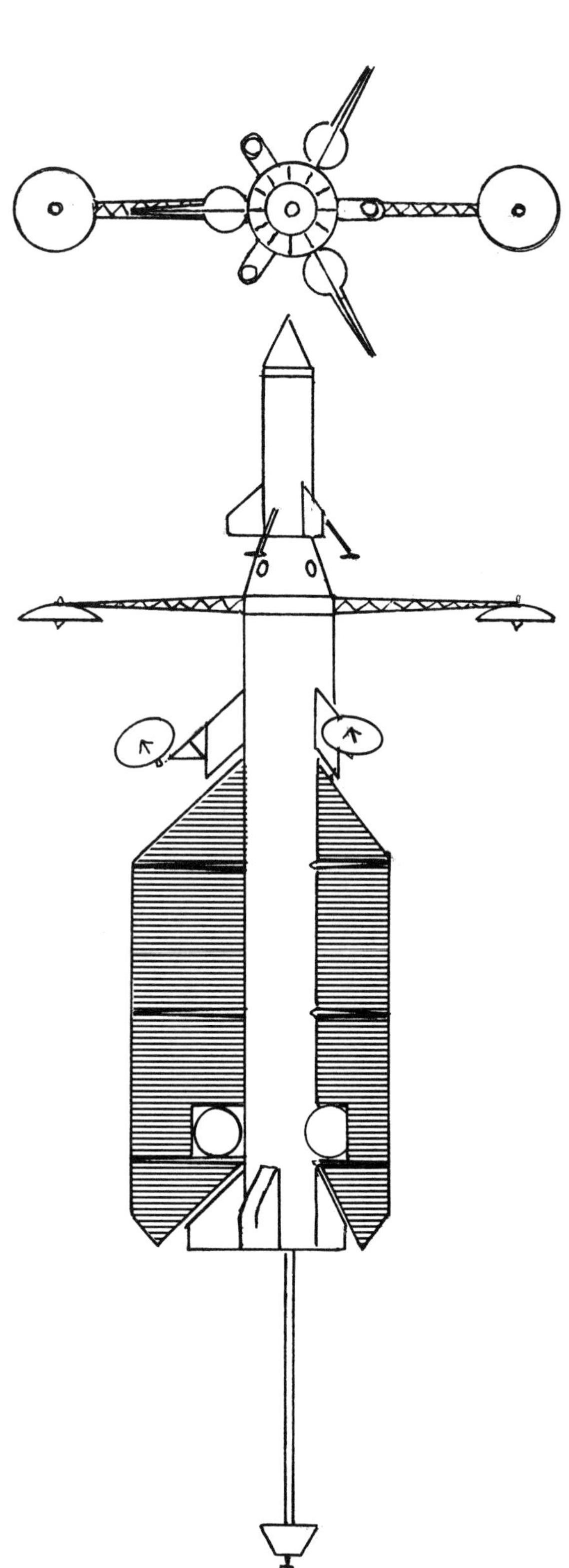

1970

Engineer Max Faget [also see 1969] designs a two-stage space shuttle for NASA that he believes could begin flying as early as 1975. It consists of two winged components, both manned and recoverable, landing like conventional aircraft. The first stage is 203 feet long, about the same size as a DC-3. It has, in addition to its rocket engines, four turbojet engines, two over each wing, and can carry a 12,500-pound payload. Riding piggyback is the orbiter, a 122-foot vehicle with six jet engines, three over each of its small, narrow wings.

At this same time, NASA is considering another fully reusable shuttle design. The orbiter, about the size of a DC-9 airliner, is launched atop a booster that can be flown back for reuse like a conventional aircraft. A new engine, capable of producing from 400,000 to 550,000 pounds of thrust, is to be developed for use in both components. It is expected that the orbiter will have a life of 100 flights and be capable of handling payloads of 65,000 pounds which measure up to 15 feet in diameter and 60 feet in length. Turnaround time is to be only 2 weeks between missions.

The cost would be $10 to $14 billion for a vehicle ready for flight in the mid-1970s.

Boeing and Lockheed, in conjunction with TWA, have a team working on a chemically fueled reusable space vehicle that will take its passengers to a low earth orbit and home again. The piloted lower stage of the projected spacecraft is about the size of a 747 Jumbo Jet. It is launched as a rocket-powered booster for the passenger- or cargo-carrying second stage. The orbiter is capable of carrying up to fifty passengers.

George H. Stoner of Boeing expects that the spacecraft will require a shorter development period than that needed for the Saturn V rocket. He thinks it will be possible that within 15 years normally healthy men and women will make regular trips into space and back. One configuration considered has the manned shuttle carried at the nose of the winged booster. Solid fuel strap-on rockets help to thrust the shuttle into orbit after its release from the first stage. The booster then returns to earth to land as a conventional aircraft, using folded and swinging wings and tail surfaces. The shuttle returns to earth on wings that are folded against its sides during takeoff.

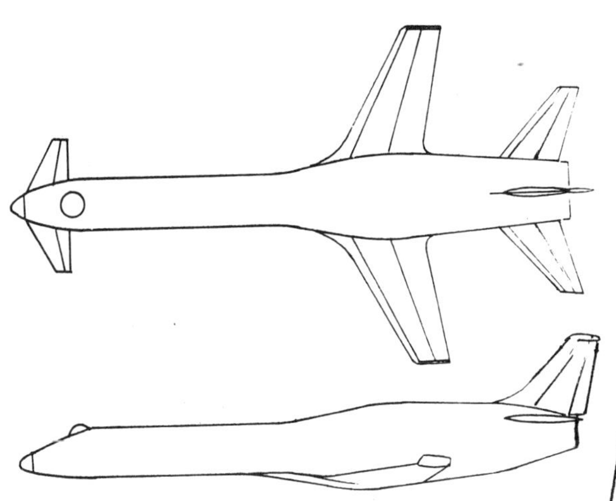

NASA develops the "Blue Goose" space shuttle design. It is probably one of the least attractive spacecraft ever proposed, at least on aesthetic grounds (and one of the last flings of Faget's straight wing influence [see 1969]). Two versions are produced, the main difference being the configuration of the tail surfaces: one has a "T-tail" and the other a conventional tail.

The "Blue Goose" has two engines in the rear as well as a "de-orbit" engine behind a swinging nose cap. Initially designed with variable-geometry canards, NASA engineers

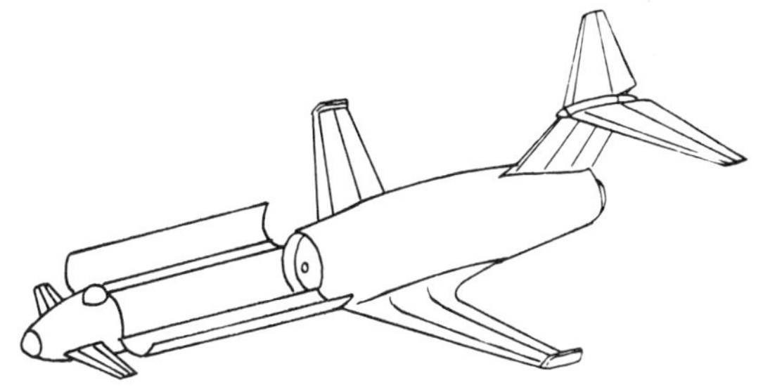

later opt for a bizarre system where the main wings are able to slide 12 feet fore and aft along their roots.

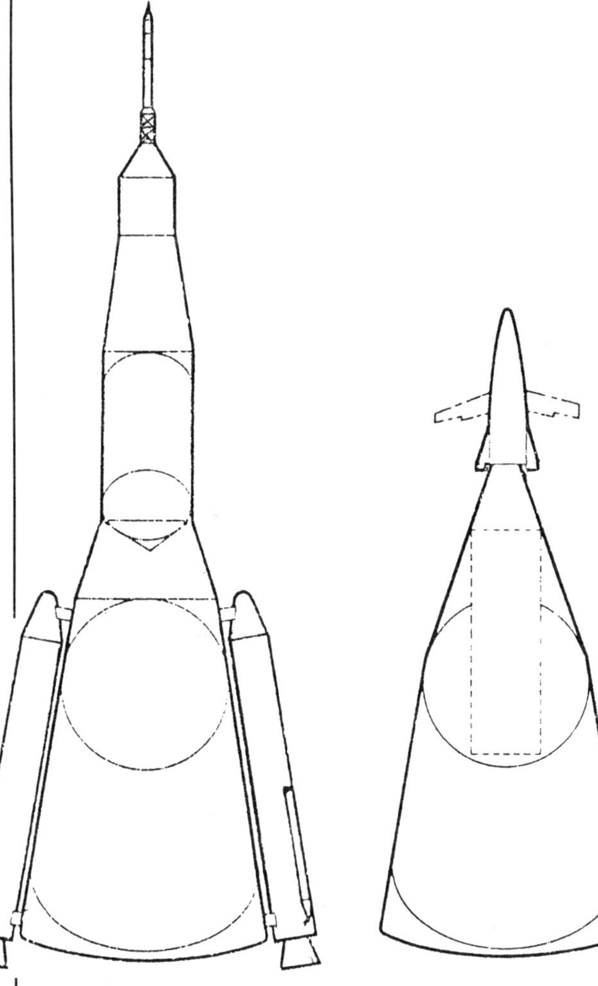

Edward Gomersall authors an unpublished internal NASA Ames Research Center working paper in which he describes an SSTO spacecraft. Promoted as a space shuttle, the vehicle would carry a winged staged for passenger use.

Gomersall's spacecraft is based on available Saturn J-2[S] propulsion technology, and could be uprated by the use of strap-on solid propellant boosters. The design is suppressed by strong opposition from within NASA headquarters and the Marshall and Johnson spaceflight centers.

MARCH 19

The X-24A lifting body research aircraft makes its first powered flight. Pilot Jerald Gentry makes a successful 7-minute flight. The vehicle is powered by a single Thiokol XLR-II rocket engine with a thrust of 8,000 pounds. An attempt at a supersonic flight on August 26 fails and the aircraft is damaged. It cannot be flown again until October 14 when John Manke takes the lifting body to Mach 1.18 and an altitude of 67,900 feet. The X-24A eventually achieves its maximum speed and altitude on October 27 when it reaches Mach 1.35 and 71,400 feet. [Also see 1971.]

JULY

NASA awards advanced Phase B space shuttle study contracts to McDonnell Douglas and Rockwell. Both companies study fully reusable configurations, with piloted boosters and second-stage orbiters.

Rockwell's two-stage-to-orbit design utilizes a modified Faget booster with a V tail. The manned booster has twelve main engines as well as a cluster of four turbojets around its nose for landing. The orbiter is an exceedingly elegant design that closely recalls the image of the classic Bonestellian "spaceship." In 1971, Rockwell presents a refined version with a delta-winged orbiter attached to a delta-winged canard booster.

McDonnell Douglas's design (presented in 1971) has a delta-winged orbiter carried by booster with swept wings and canards.

ORBITER CONFIGURATION
(1500 N MI CROSS RANGE)

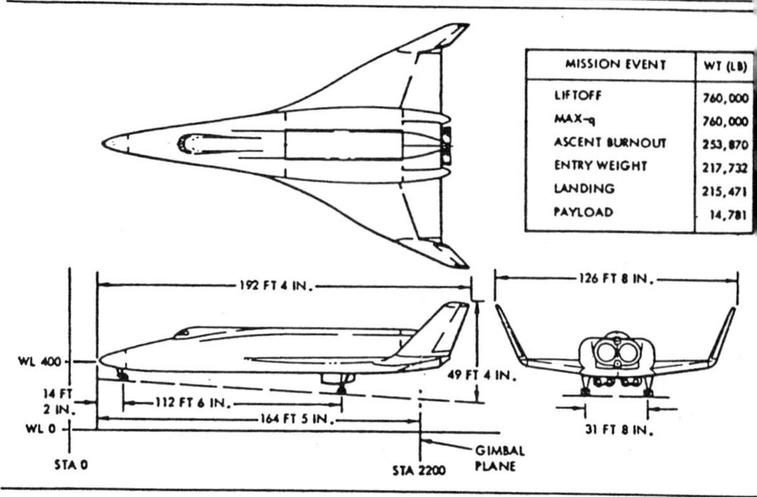

MISSION EVENT	WT (LB)
LIFTOFF	760,000
MAX-q	760,000
ASCENT BURNOUT	253,870
ENTRY WEIGHT	217,732
LANDING	215,471
PAYLOAD	14,781

BOOSTER CONFIGURATION

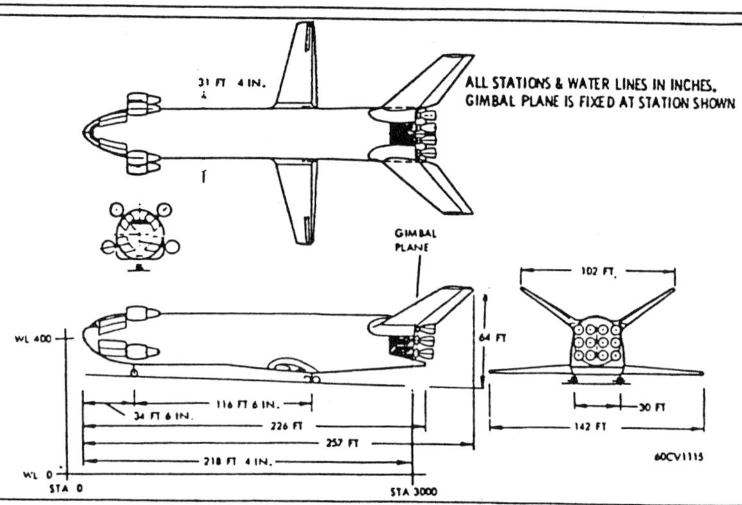

ca. 1970

The *Pilgrim Observer* is a plastic model kit designed by G. Harry Stine and issued by Model Products Corp. (MPC). It is a manned atomic spacecraft fictionally worked into NASA's 1970–1980 space program. It is intended to provide a manned expedition to Mars, Venus and return, using a realistic mission profile.

The *Pilgrim* is a ten-man vehicle intended for long-duration space voyages of up to 5 years. It is capable of travelling to any part of the solar system out to the orbit of Saturn. Strictly a deep-space ship, it is not designed to land on any other world. It is 100 feet long and 33 feet in diameter in its launch configuration, when it is launched folded within the third stage of a Saturn booster. Once in orbit, three modular arms are deployed, giving the *Pilgrim* three spokes with a tip-to-tip diameter of about 150 feet. One of these arms contains the crew living

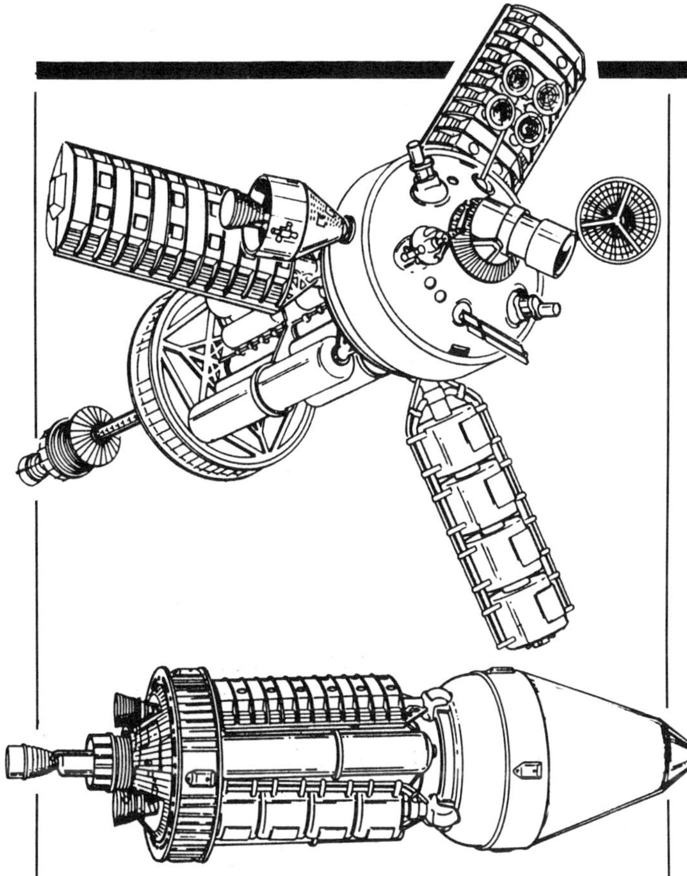

quarters, the second a hydroponic farm (using pumpkin plants) that allows the *Pilgrim* to operate on a closed ecological system, and the third arm contains the nuclear power plant. It is an advanced version of the SNAP (Space Nuclear Auxiliary Power) devices that have been successfully tested on board Nimbus-III satellites and Apollo 12. The *Pilgrim* reactor is a stack of Brayton cycle power units.

The propulsion system consists of three uprated J-2 250,000-pound thrust engines burning liquid oxygen and hydrogen. It is also equipped with a Nerva 28 nuclear rocket engine using liquid hydrogen for its working fluid. Its maximum thrust is 250,000 pounds. A dome-shaped "shadow shield" on the front of the reactor core protects the crew in the main body of the ship. The Nerva engine is also at the end of an extendable boom, removing it from the immediate proximity of the ship and increasing the size of the shadow cone.

The ship rotates about its long axis at about 2 rpm to produce an artificial gravity of 1/10 g.

The main, non-rotating body of the *Pilgrim* contains the Main Control Center and the Service Section. The former is in a zero-g condition and provides a stable platform for astronomical observation, scientific experiments and communications. Mounted here are the various antennas and several types of optical and radio telescopes. The Service Section consists of the propulsion system and storage bays.

Two auxiliary spacecraft are carried aboard: a modified Apollo command and service module (Apollo-M) and a one-man EVA vehicle. The Apollo-M can be used during the construction of *Pilgrim* as a shuttle between it and the nearby space station. During flight, it can be used to separate scientific experiments from the ship and for inspection of astronomical bodies such as asteroids. The one-man EVA craft (OMEVAC) is a small vehicle with a pair of remote handling arms. It can be used for outside inspections and repairs.

The mission scenario is: Two months are needed to prepare the ship after it has been inserted into earth orbit. Launch from earth orbit is expected to take place in June 1979. The mission is to be an earth-Mars-Venus-earth tour, with short periods spent orbiting Mars and Venus. On the Mars-Venus leg, a flyby of the asteroid Eros will be possible. The total length of the trip is 710 days. The crew consists of four astronauts spec ializing in spacecraft operation, and six scientist-explorers.

From its 200-mile earth orbit, the *Pilgrim* begins its transaerean orbit insertion by burning its three J-2 engines. This leg of the journey takes 227 days, ending in a braking burn of the Nerva 28 in order to achieve Martian orbit. It orbits Mars for 48 days at altitudes ranging from 500 to 5,800 miles. Then the Nerva is again fired, thrusting the spaceship into its transvenerian trajectory. The flight to Venus requires 246 days, including a close approach to the asteroid Eros 145 days after leaving Mars. The Nerva engine is again fired, slowing the ship down so that it enters an orbit around Venus. It circles the planet at an altitude of 500 miles for 55 days. The flight from Venus to earth needs 140 days, ending in an orbit 200 miles above the earth.

When orbiting Venus and Mars, small unmanned instrument packages are to be dropped, soft-landing on the planets. Automatic orbiters will be left circling the two planets, as well as Mars' moon Phobos. Another orbiter will be placed around Eros. In all, more than 150 experiments will be performed during the voyage.

A lunar-shuttle mission is developed to take advantage of the NERVA nuclear-powered rocket system.

A 75,000-pound thrust NERVA rocket will be placed into a 300-mile earth orbit by a conventional Saturn V. The shuttle (propellant tanks, service module and command module) will be assembled from components delivered into orbit by subsequent launches. After the fuel is delivered, the shuttle will depart from its parking orbit and eventually arrive in lunar orbit, where it will rendezvous with a lunar-orbiting space station and transfer its payload of crew or supplies. The shuttle then will return to its original earth orbit.

1971

The space tug is a major component of the lunar base as envisioned in the late 1970s and early 1980s. Even the lunar base envisioned by the National Commission on Space in 1986 [see below], as radically different as most of its spacecraft are, relies on its own version of the space tug.

This tug is a modification of one proposed by the European Launcher Development Corporation (ELDO). It is a cylindrical rocket, the top third containing a control cabin, the lower portion propellant tanks (liquid oxygen and liquid hydrogen). A docking unit on the control cabin allows it to link up with other spacecraft. It is also equipped with four retractable landing legs.

The basic tug can be used alone or "stacked" in a multistage configuration. Its maximum fully loaded mass is about 12.5 tons. The engine is Rocketdyne's Advanced Space Engine.

Two tugs can be carried into earth orbit in a shuttle's cargo bay. They are stacked and one acts as a booster to launch the second tug toward the moon. The first tug returns to the shuttle orbit; the second carries some 7 tons of cargo into lunar orbit. It dumps the cargo in orbit, returning to earth orbit with its remaining fuel. A third tug, one equipped with landing legs, rises from the lunar surface, picks up the cargo and returns with it to the moon base. Instead of cargo, up to six astronauts can be carried in each lander mission.

Two Russian spacecraft—Soyuz II and Salyut I—dock for about 22 days to become what the Soviet press term "the first manned orbiting scientific laboratory." The Soyuz II crew of three cosmonauts die while returning to the earth when their spacecraft accidentally depressurizes.

The H-33 space shuttle concept is considered by NASA. It has a reusable booster, 245 feet long, but the orbiter's fuel tanks are external and expendable. The delta-winged orbiter can carry a crew of two with two passengers. The two vehicles are designed to separate at an altitude of 190,000 feet, after which the external tanks are jettisoned.

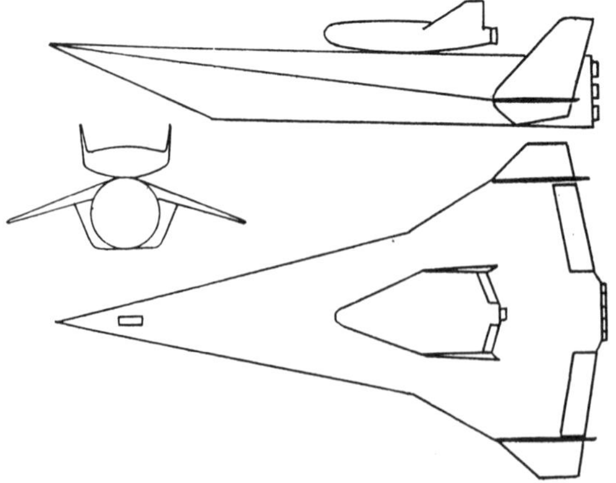

The British firm of Hawker-Siddeley proposes a two-stage fully reusable shuttle whose first stage is derived from Terence Nonweiler's "wave rider" caret-wing configuration. The 200-foot-long, 121-

foot wingspan booster carries a lifting-body orbiter. Its payload capacity is 8,000 pounds.

The redesigned X-24A lifting body [see 1970] is redesignated as the X-24B and as such becomes an immediate predecessor to the space shuttle. Among the changes that builder Martin Marietta makes include extending the length from 24 feet to 39 feet and increasing the width from 14 feet to 19 feet. The rounded underside is replaced by a flat surface. Perhaps the most striking change is the replacement of the original blunt nose with a long, pointed one. The new design doubles the lift-generating surface and increases cross-range maneuverability over three times, from 500 to 1,500 miles.

Although designed to research Mach 5 + hypersonic flight, the X-24B in reality cannot approach anything like that speed since its propulsion system has been left basically unchanged.

The redesigned aircraft is finally delivered in October 1972 but will not make its first powered flight until November 15, 1973. The most successful of the six lifting bodies studied at the time, the X-24B provides invaluable information toward the development of the space shuttle.

A proposed X-24C hypersonic lifting body, a combination of the X-15 and the X-24B, is considered but not built. It would have flown at speeds of Mach 6+.

The Phase B space shuttle study carried out by North American Rockwell concludes with a vertical takeoff horizontal landing booster and orbiter. The manned booster will have twelve engines. After separation at 200,000 feet, the 267-foot booster will fly back and make a conventional landing. The 210-foot orbiter will be able to make at least 100 round trips into space.

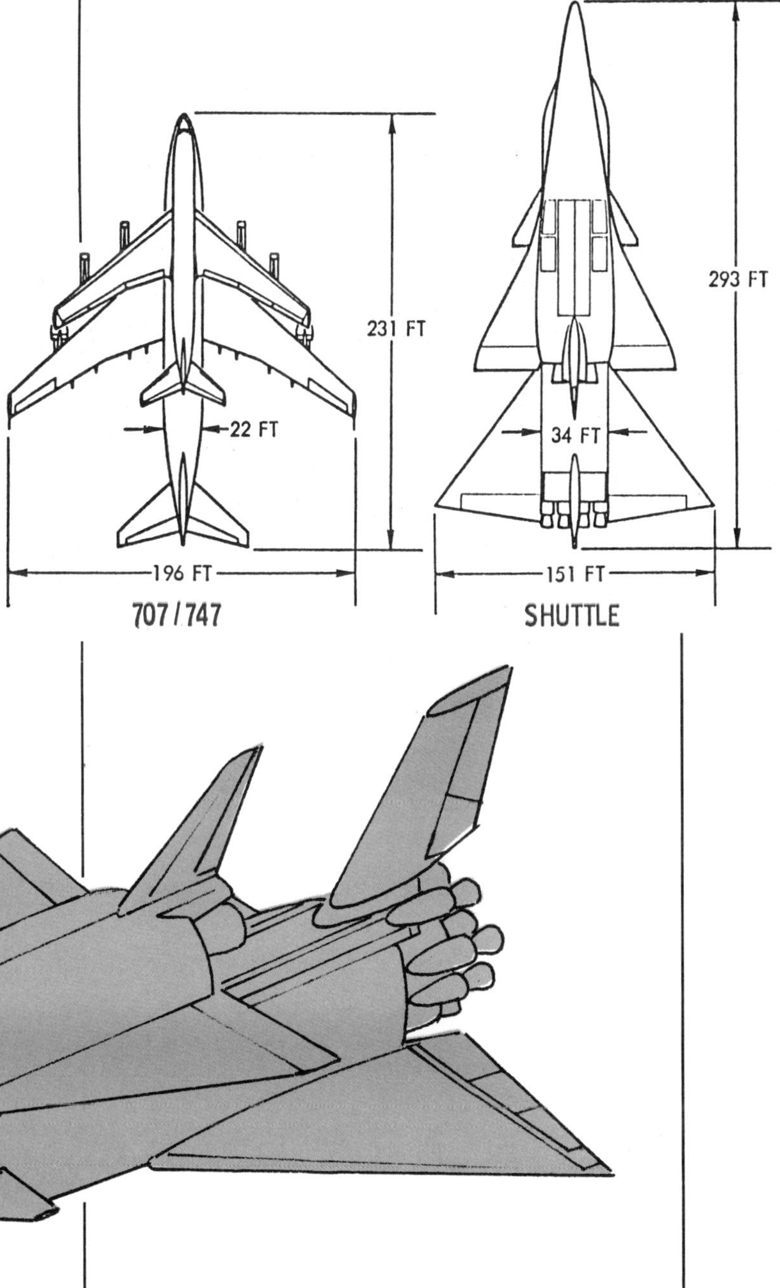

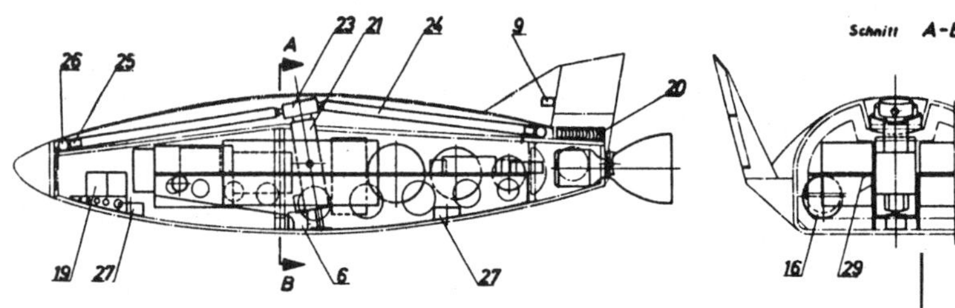

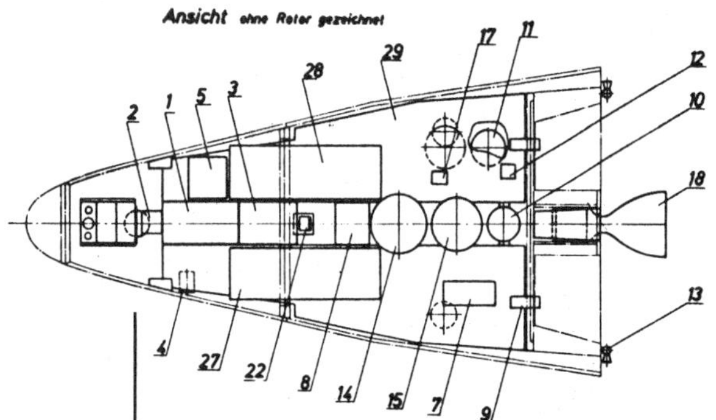

There are plans to test a 3-meter subscale model of an Austrian lifting body design at the Australian Woomera rocket range. The lifting body shuttle, designed by MBB, would be part of an international post-Apollo program.

Chrysler's Space Division completes its study on Project SERV (Single-stage Earth-orbital Reusable Vehicle).

It is a VTOL spacecraft with an aerospike-propelled core stage 67 feet tall and 90 feet in diameter at the base. It carries either a cargo module or a winged passenger vehicle (with a two-man flight crew). Like most other aerospike SSTOs, SERV uses its broad base as a reentry heatshield and eventually makes a base-first touchdown on land.

The Manned Upper-stage Reusable Payload (or MURP) is a high lift/drag ration swing-winged spacecraft. It can carry a crew of two and ten passengers along with a payload of 2,500 pounds. It is equipped with a jet engine for maneuverability when landing.

The SERV can also carry a personnel module into orbit, for the simple ferrying of humans and cargo to a space station and back. This module a large, flat cone similar to the Apollo command module and is designed for both sea and land recovery, though it normally returns to earth with SERV.

MAY

Beginning with Orbiter 022B and continuing through Orbiter 054 a year later, NASA's Spacecraft Design Division (SSD) begins studying delta-winged shuttles—finally abandoning the straight-wing Faget configurations [see 1969]. The 036 series (036–036C) features the same three-engine (Saturn J-2S) cluster as the final design [see]. The payload bays are only 15 feet by 40 feet. The 036 family is scaled up to handle a 15-foot by 60-foot payload bay, with four J-2S engines. This is the 040. The 040C, with increased cargo bay and three new high-pressure engines, evolves into the present-day shuttle. It is Rockwell's basic design at the beginning of Phases C/D [see 1969]. Although NASA continues to work on shuttle variations—up to Orbiter 054—it is now a foregone conclusion that the final version will be some derivative of 040.

The fly-back booster is finally abandoned this month with Orbiter 026, with the engineers of SSD concentrating on parallel-burn approaches, in which the shuttle's main engines will be burning at the same time as the booster's. These most often take the form of sequential, piggyback, staging or stage-and-a-half designs. [Also see February 18, 1972.]

ca. 1971–1972

Numerous aerospace firms have developed independent space shuttle proposals. While the craft itself is more or less the same in all the designs, the method of boosting it into orbit varies considerably.

McDonnell Douglas has proposed three concepts: one with a recoverable liquid-fuel booster, another with an expendable liquid-fuel booster, and third with a recoverable, two-stage liquid fuel booster. North American Rockwell and General Dynamics have also developed three: an expendable solid-fuel booster, a recoverable liquid fuel booster, and a recoverable two-stage booster. Grumman Aerospace and Boeing propose three: a recoverable liquid fuel booster, ex-

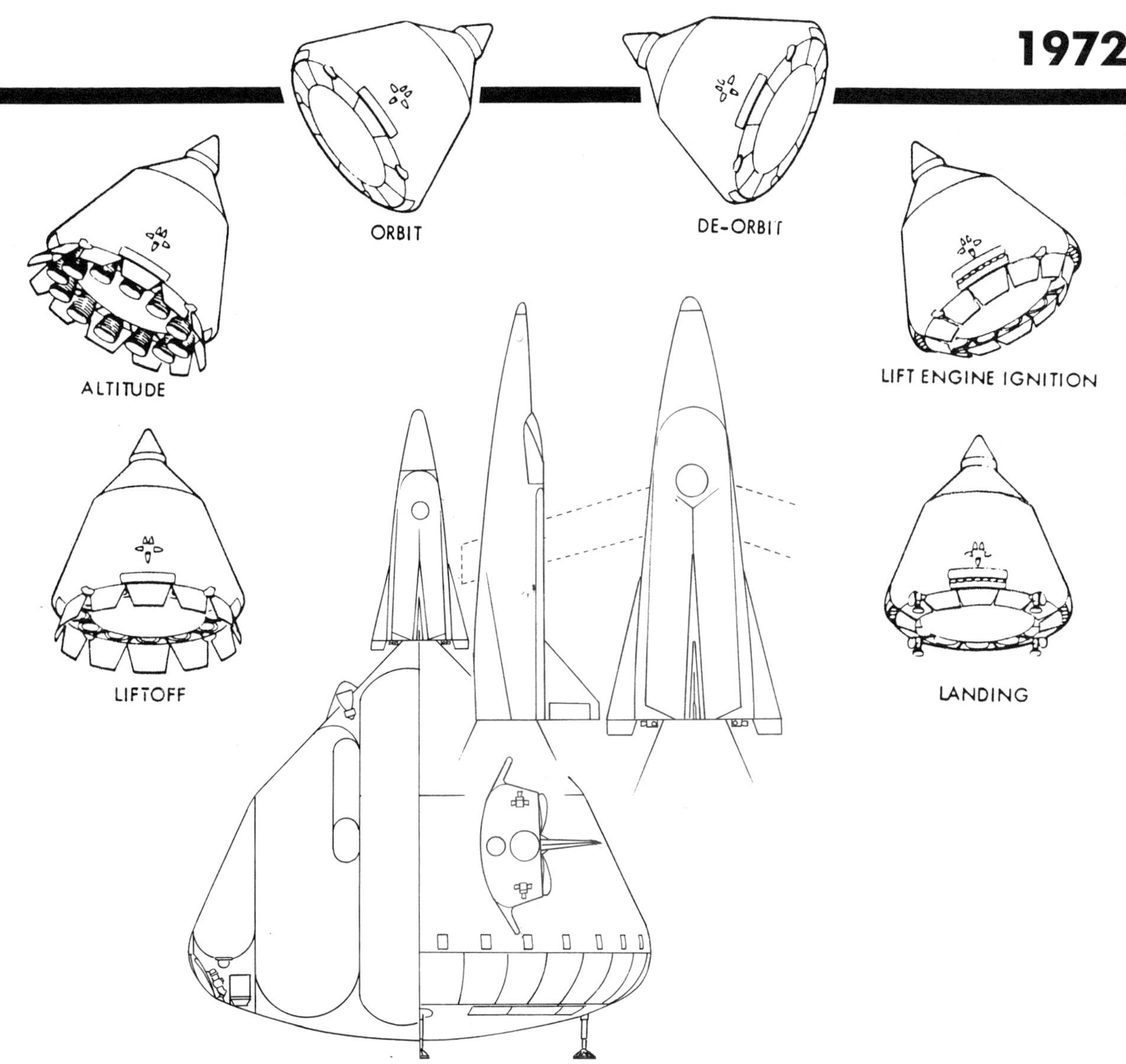

pendable solid fuel boosters with an external tank for the shuttle (very similar to the final, accepted Rockwell design), and a recoverable unmanned booster.

1972

Engineer George Detko, at the NASA Marshall Space Flight Center, proposes a small VTOL SSTO spaceship. It would have a gross liftoff weight of only 50,000 pounds, carrying a crew of two either to orbit or to an antipodal delivery point. This design is a major influence on Gary Hudson's *Phoenix* [see 1982–1991].

JANUARY

President Richard M. Nixon endorses development of the space shuttle, now termed the Space Transportation System (STS).

FEBRUARY 18

Space shuttle orbiter configuration 040C [also see May 1971] inaugurates the now-familiar external fuel tank flanked by a pair of solid-fuel boosters. The main differences between 040C and the present-day shuttle are that the solid fuel rockets are attached to points close to the shuttle, between the external tank and the shuttle's wings, and that the external tank is outfitted with a pair of large fins.

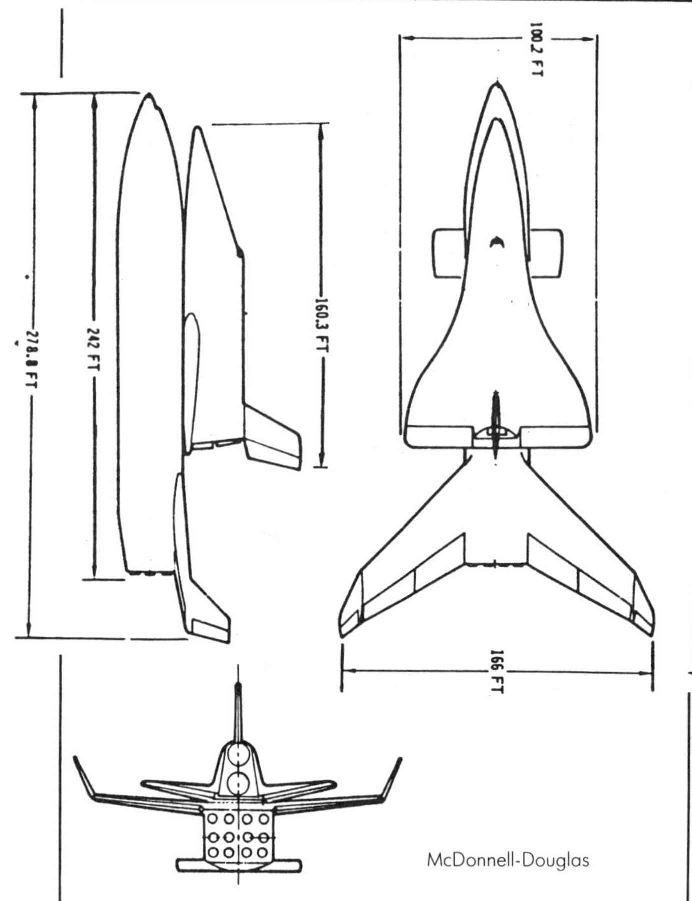

McDonnell-Douglas

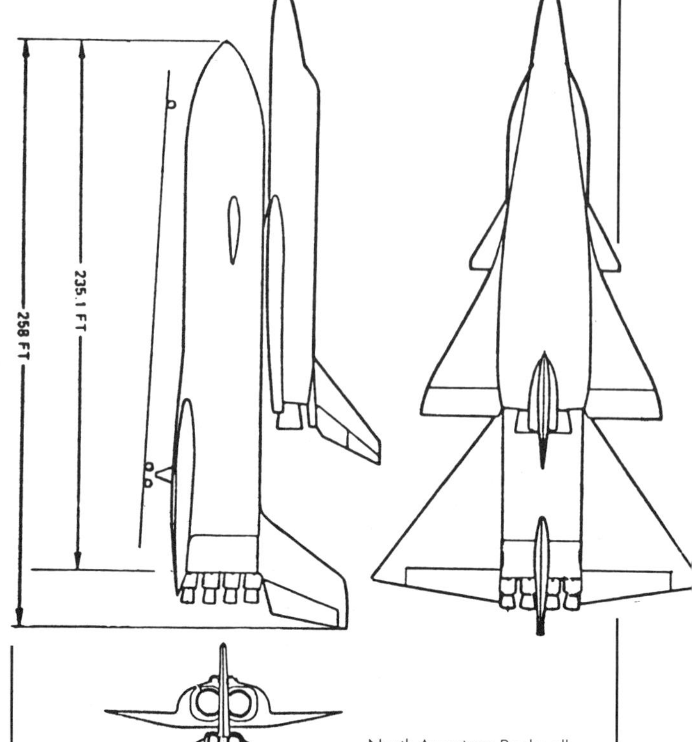

North American Rockwell

MARCH

NASA decides to go with a parallel-burn launch system with the Space Shuttle, as opposed to sequential staging.

JULY 25

Rockwell (formally North American Rockwell) is given the final go-ahead by a joint NASA-Air Force Source Evaluation Board to proceed with shuttle development. Between March 1972 and mid-1974 the shuttle design will be refined considerably, from vehicle ATP (Authority to Proceed—the initial concept) through vehicles PRR, 2A, 3,4, and 5,6.

Rockwell, Lockheed, McDonnell Douglas and Grumman all submit competitive designs, all based on the 040C prototype.

1973

The X-24-B is a lifting body research aircraft developed for NASA by Martin and the only rocket-powered aircraft in operation at the time of its retirement in 1975. An unmanned, remotely piloted version with "switchblade" folding wings, the "Hyper III," is developed to test hypersonic reentry from space, with landing as a conventional airplane.

Coggins

Romick

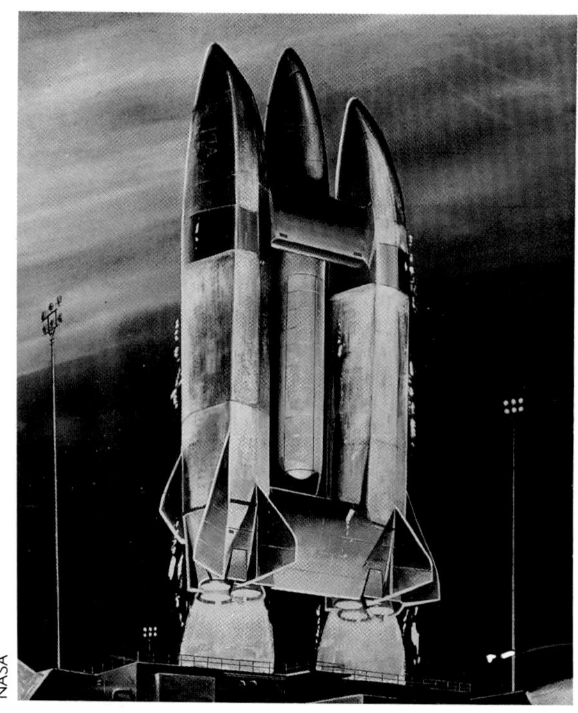
NASA

Above: A 1969 fully reusable, two-stage space shuttle proposal.
Below: The *Phoenix* family of single-stage-to-orbit spacecraft proposed by Gary Hudson. Left to right: The E excursion vehicle, the manned and unmanned LP, and the C cargo ship. Art by Ron Finger.

Above: The *Orion* spaceplane from the 1968 motion picture *2001: A Space Odyssey*. Below: Thermonuculear spaceship illustrated by Pat Rawlings.

Hudson

NASA

NASA

NASA

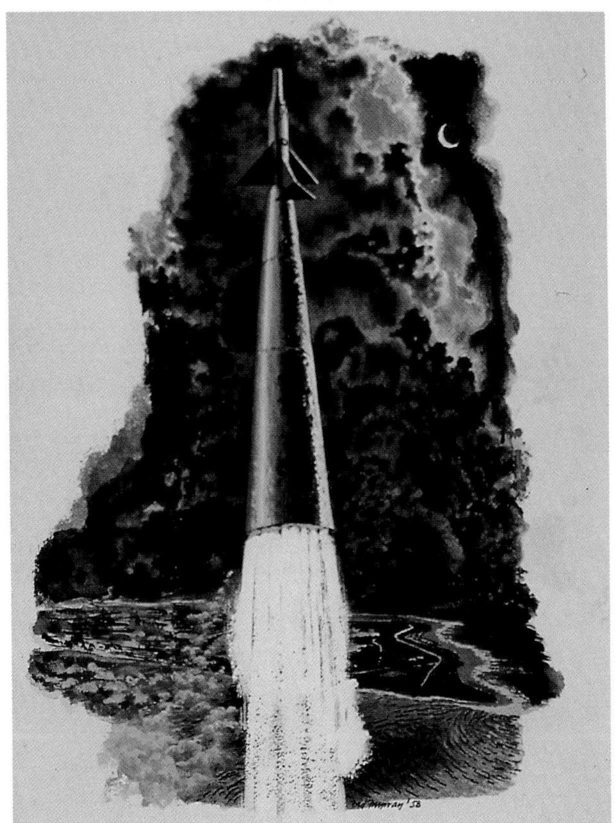

Freeman/Liebermann

Above: The takeoff of the moon-bound spaceship from the 1958 Wernher von Braun/Fred Freeman book *First Men to the Moon*.

Left: Top to bottom: Nine NASA space shuttle proposals from the early 1970s. Top: Martin Marietta/McDonnell Douglas concepts: 1. Expendable solid fuel booster; 2 & 3. Recoverable liquid fuel boosters. Center: North American Rockwell/General Dynamics concepts: 4. Recoverable liquid fuel booster; 5. Expendable solid fuel booster; 6. Recoverable booster and manned reusable orbiter. Grumman Aerospace/Boeing concepts: 7. Recoverable liquid fuel booster; 8. Expendable solid fuel booster; 9. Recoverable unmanned booster and manned orbiter.

Deutsche Aerospace

Above: A 1:8.5 scale model of the *Sänger II* spaceplane.

Right: Spaceship *Mama Spank* from the graphic novel *Starstruck* (1985), designed by Michael Kaluta.

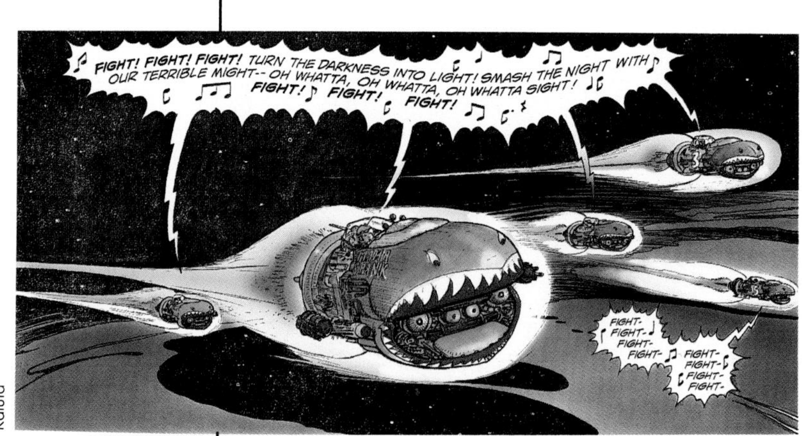

Kaluta

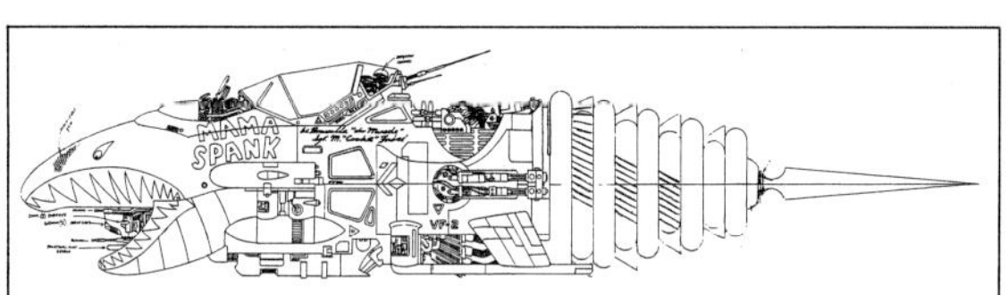

Below left: The Japanese HOPE II spaceplane.

NASDA

Pearson

Above: A small scale manned Soviet spaceplane used for drop tests.

British Aerospace

Above: The interim HOTOL spaceplane proposed as a joint British-Russian venture.

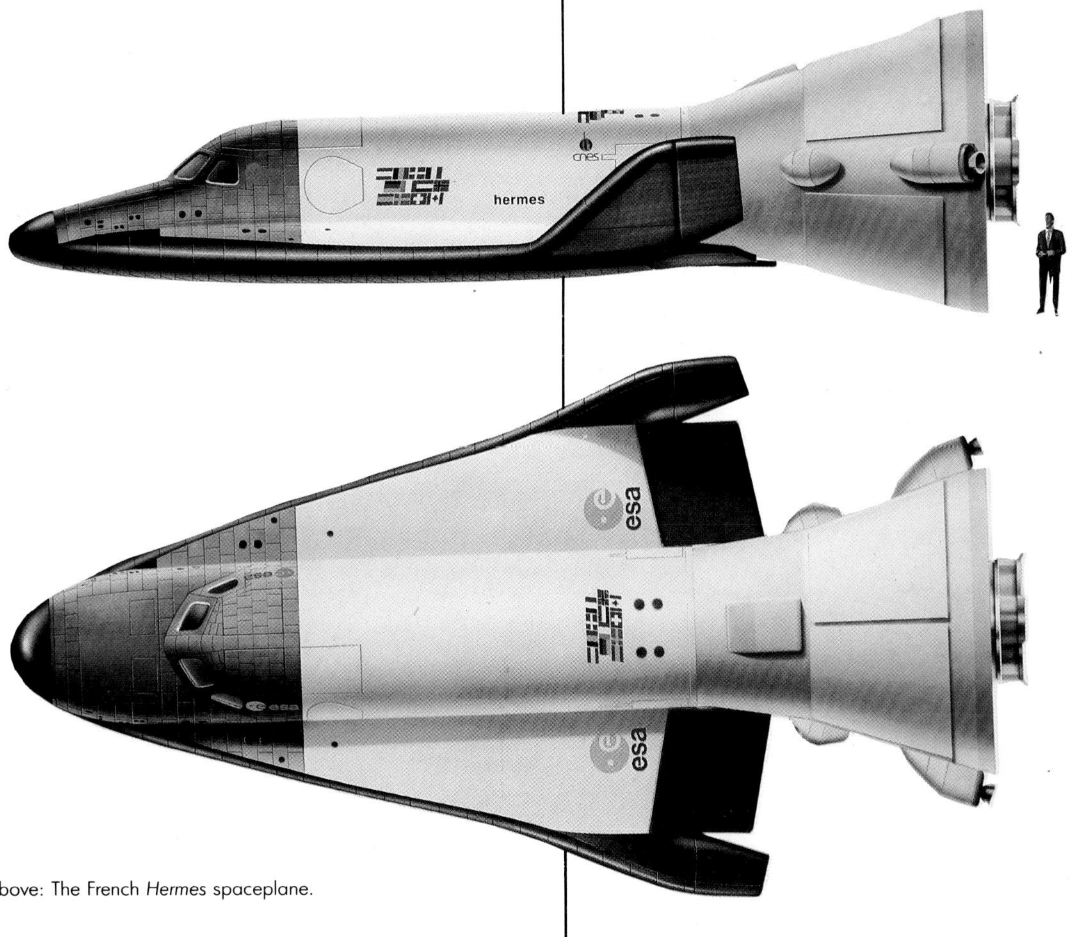

Above: The French *Hermes* spaceplane.

Ducros

Above left: The single-stage-to-orbit airlaunched *Skyrocket*, a private venture proposed by Gary Hudson.
Above right: Ion-propelled spaceship illustrated by Pat Rawlings.
Bottom: A Boeing proposal for a manned Mars lander, Boeing.

Facing page: McDonnell Douglas's *Delta Clipper*, a single-stage-to-orbit launch vehicle.
Right: A second generation space shuttle designed by John Frassanito and Associates.
Below: A fusion spaceship designed at JPL by Humphrey Price.

Hudson

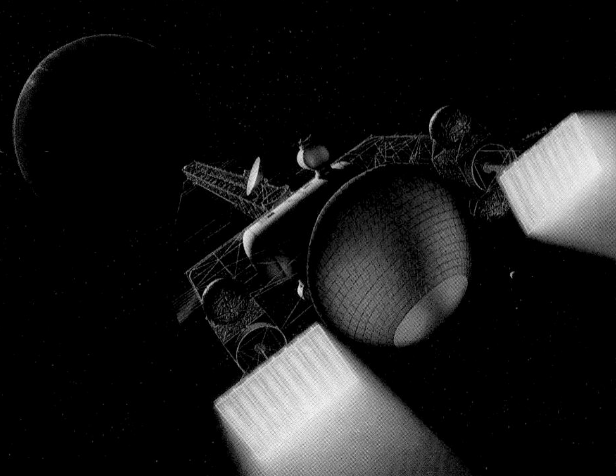

Rawlings

Boeing

McDonnell-Douglas

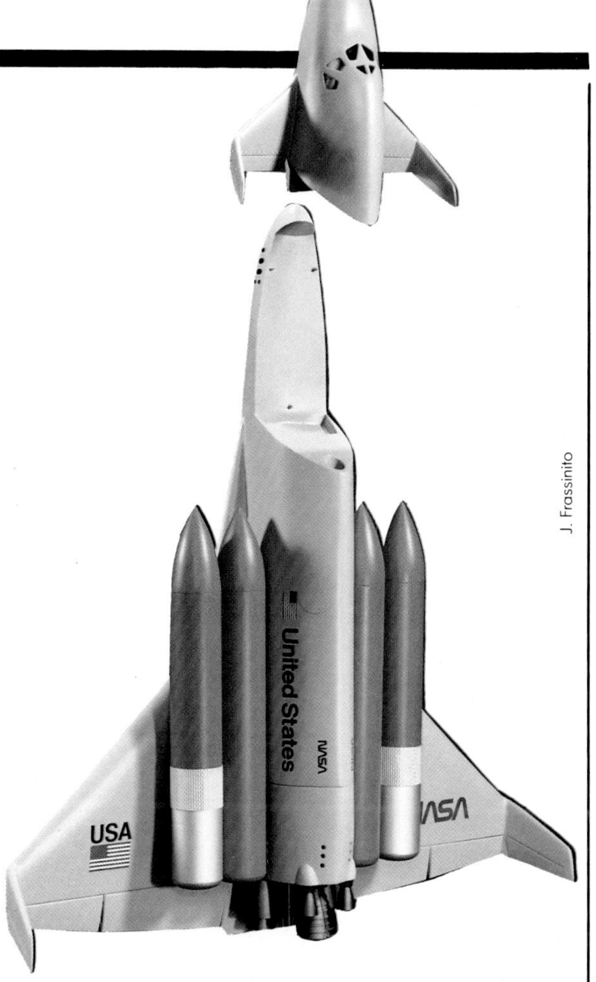

J. Frassinito

JPL

Boeing

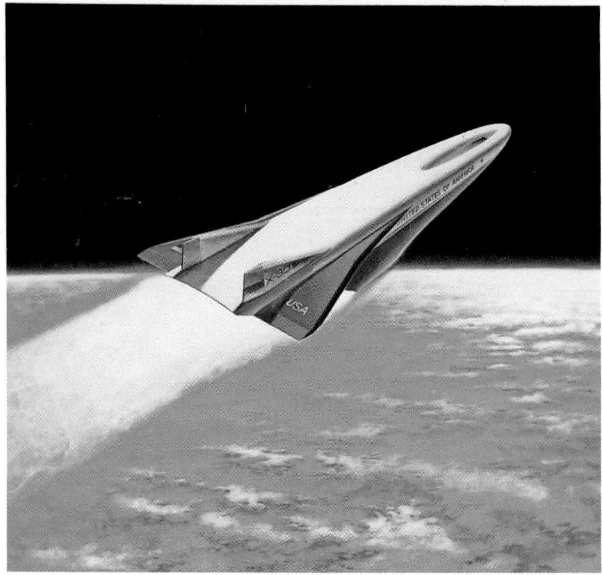

NASA

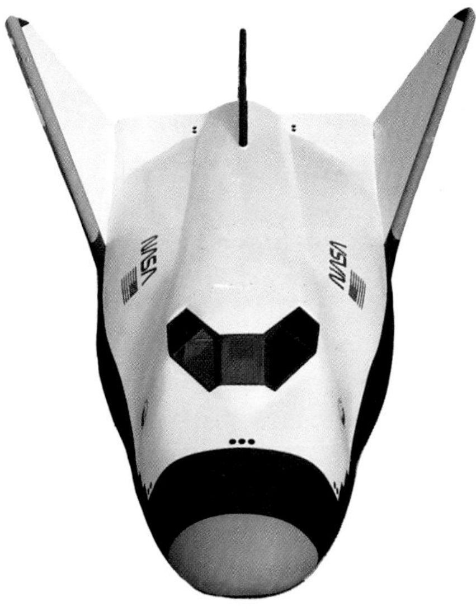

NASA/Langley

Top: A nuclear spaceship for a manned mission to Mars, designed by Benjamin Donahue.
Above: The U.S. X-30 National Aero-Space Plane.
Right: The HL-20 Personnel Launch System proposed by Langley.

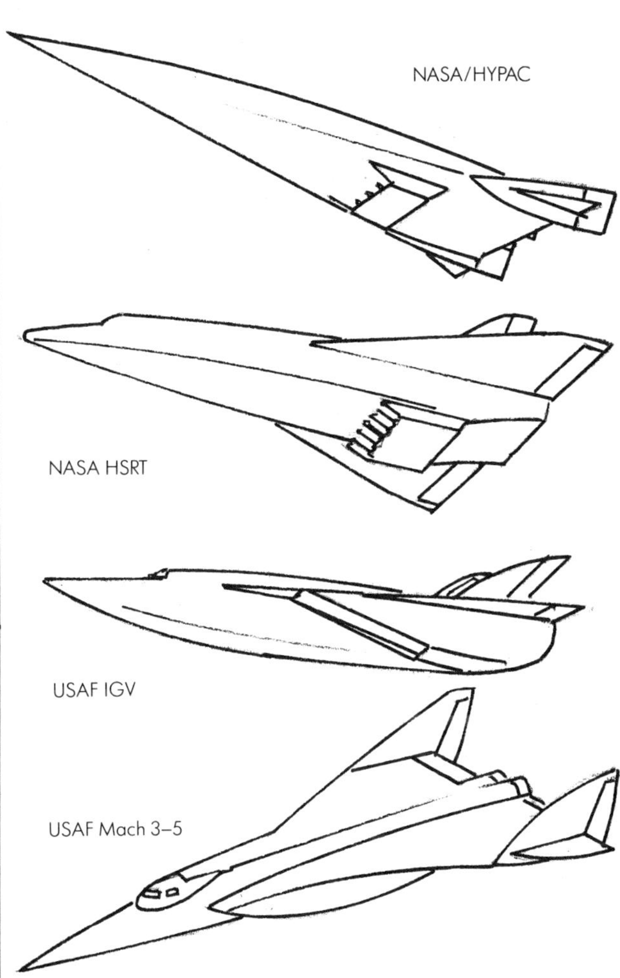

NASA/USAF 1972–1974

1973–1977

The British Interplanetary Society completes its *Daedalus* starship study, done under the leadership of Alan Bond. Like the BIS moonship program of the late 1930s [see 1939], the *Daedalus* project is one of the first detailed approaches to solving the problem of flight to the stars. The designed goal for the project is Barnard's Star, 6 light-years distant. It is to be a one-way voyage and the only passenger is a small unmanned probe. Unlike the moonship study, *Daedalus*' designers have to rely on plausable but as yet unrealized technology.

It is a monster spacecraft, more than twice as tall as a Saturn V and weighing twenty times as much, or 54,000 tons. It can carry some 450 tons of payload, incuding a semi-intelligent computer to operate the ship. The latter is necessary because of the great time-lag in communications at interstellar distances—controlling the ship from the earth is not possible; an emergency would be over long before we even knew it had happened. At the base of the starship is the vast bell of its main engine, 400 feet in diameter. "Microexplosions" of deuterium at the rate of 250 per second are detonated in the engine, accelerating the ship eventually to 10 or 13 percent of the speed of light. The engine operates by igniting pellets of deuterium and helium 3 by electron beams. The pellets are injected by a gun into the center of the engine bell where they are held in place by a cusp-shaped magnetic field. When they are at the target point, they are hit by a large number of electron beams fired simultaneously. The magnetic field channels the plasma created by the resulting fusion explosion out of the end of the rocket. Fuel pellets, 46,000 tons of them, are contained in spherical tanks just ahead of the engine. This is the first stage of the starship.

The second stage, mounted ahead of the first in conventional fashion, is a virtual duplicate of the larger first stage, except in size. Its spherical tanks carry 4,000 tons of fuel. The two stages accelerate the payload to a speed of 24,000 mps, so it will reach Barnard's Star within 50 years after launch. The payload is a squat cylinder protected from space dust by a disk-shaped beryllium erosion shield. It carries eighteen expendable probes, telescopes, and sensing devices, as well as the master computer. The probes, each with its own rocket propulsion system, will make the close planetary approaches as *Daedalus* flies through the target system. The beryllium shield will be of no use in the debris-rich regions of a solar system, so an auxiliary shield will be created ahead of the starship. A dense cloud of cigarette-smoke-sized particles will be flown about 125 miles ahead of the ship. This will effectively destroy obstacles of up to half a ton in mass. Four powerful nuclear generators will beam the information gathered by *Daedalus* back to the earth, taking 3 years to send, repeating its message over and over.

The designers of *Daedalus* expect that, once the technology becomes available, designing, building, testing and fuelling the starship will require 15 to 20 years.

1974

A. Bond proposes a spacecraft capable of interstellar flight, the Ram Augmenter Interstellar Rocket (RAIR). It could accelerate a large starship to 20 to 30 percent of the speed of light. Interstellar protons are used for most of the reaction mass, a small fraction being used to react with on-board lithium in a lithium-proton reactor. [Also see 1960.]

1975

APRIL-JULY, 1975

The U.S.S.R. makes twenty manned launches during this period.

JULY 15–24

In the Apollo-Soyuz Test Project Alexei Leonov and Valery Kubasov dock their Soyuz 19 with the Apollo 18 spacecraft manned by Thomas P. Stafford, Vance D. Brand and Donald K. Slayton (at 51 the oldest man to fly in space). The two spacecraft remain attached for 2 days as part of the first international space mission.

"Together," says Alexei Leonov later, "we have done an irreversible thing. The machine of Apollo Soyuz is operating now and no one can stop it."

1976

P. A. Kramer and R. D. B%er present their concept for Project ITUSTRA (Integrated Turbo-Ramjet/Rocket Performance Potential), a reusable airbreathing VTOL space transport. The spacecraft would resemble D. E. Koelle's BETA 1A [see 1971]. The 1 1/2 stage ITUSTRA would weigh 155 tons, its conical body surrounded by a cluster of airbreathing engines while retaining the plug nozzle engines at the base. The coneshaped nose acts as a common supersonic inlet. At

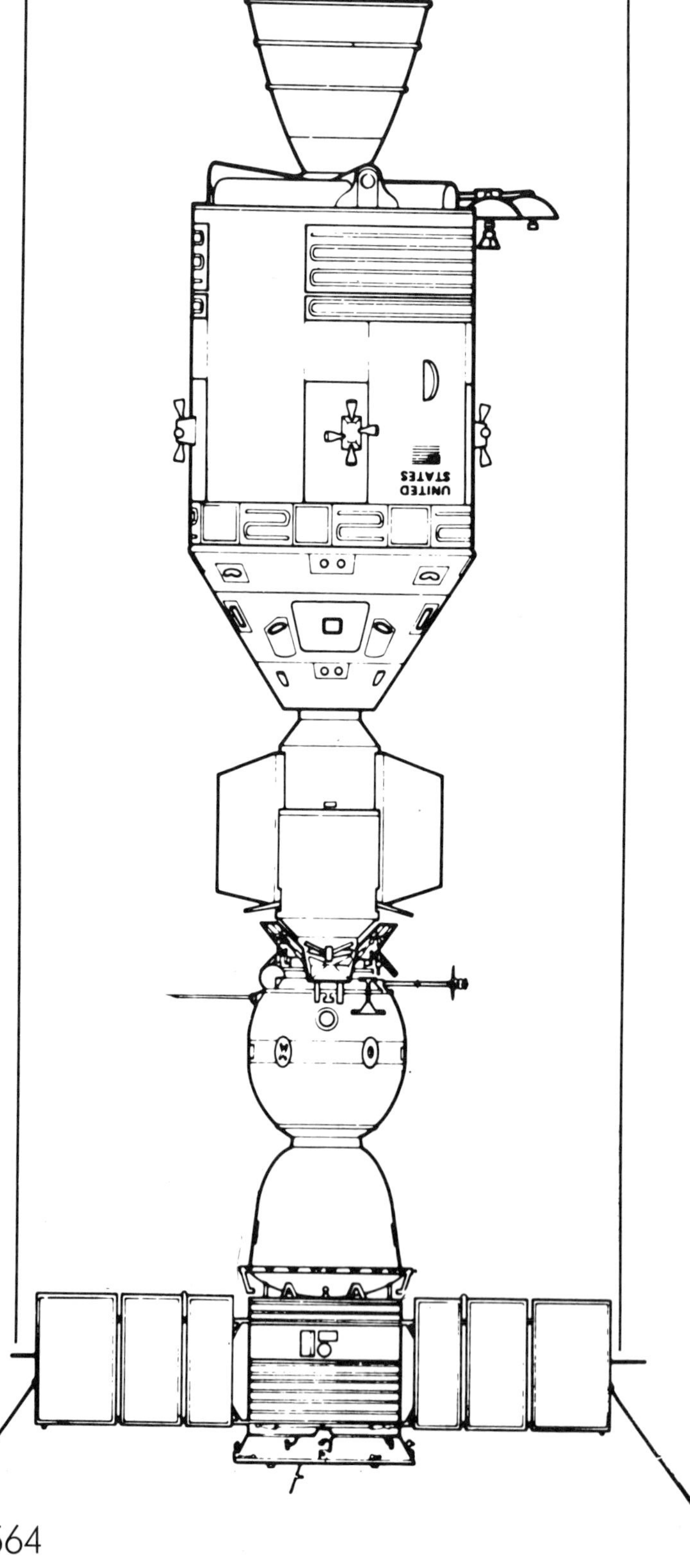

stage seperation, the ring of airbreathing engines is jettisoned for later recovery.

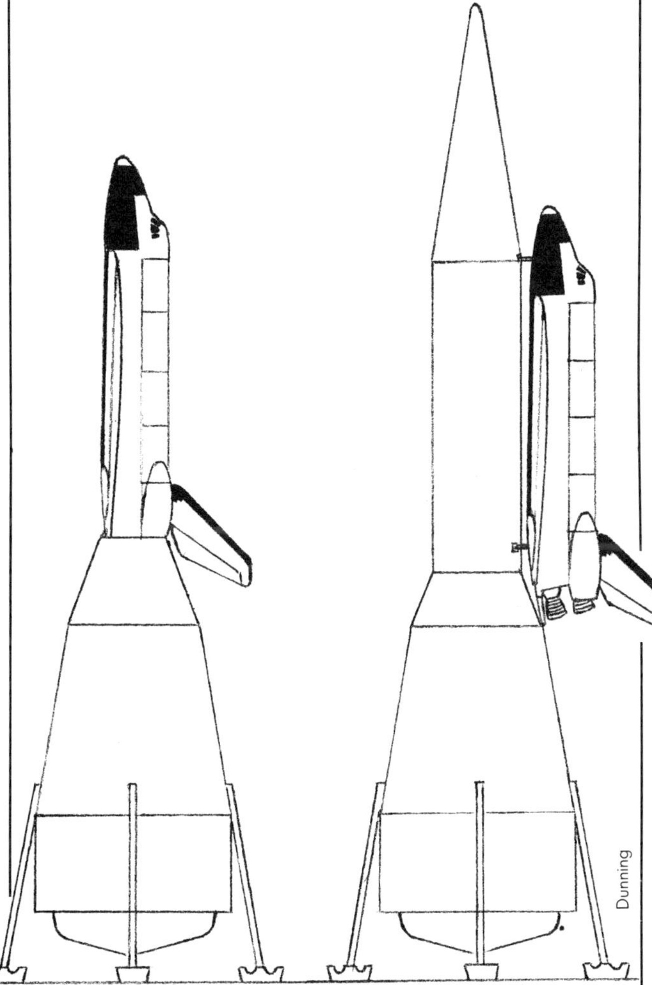

Philip Bono proposes an economical improvement on the existing NASA space shuttle system: replace the solid-fuel boosters with a new liquid-fuel booster. This is an adaptation of his ROMBUS [see 1967]. The main difference between the old ROMBUS and the new booster, called *Perseus*, is the elimination of the external hydrogen tanks. In the *Perseus Mk.I*, the standard shuttle orbiter along with its external hydrogen tank is mounted atop the revised ROMBUS, the aft end of the hydrogen tank attached to the forward end of the booster. Once the shuttle has been given its initial boost, the *Perseus* will return for a land recovery at the original launch site. The complete shuttle system will thus be reusable.

The *Perseus Mk.II* is a slightly upgraded version. The shuttle's hydrogen tank is dispensed with, and the orbiter will ride directly atop the booster. In this case, the *Perseus* will follow the shuttle into orbit, making a reentry and land recovery at its original launch site after making one orbit of the earth.

Another advantage to the *Perseus* system is that each of the boosters will use up to twenty-four or more of the space shuttle main engines, eliminating the need to develop costly new propulsion systems.

The Soviets decide to develop the #C (Energia, or Energy) booster as a launcher for a reusable spacecraft [also see 1980–1988].

Robert Bussard publishes "A Program for Interstellar Exploration" in the *Journal of the British Interplanetary Society*

A NASA task group publishes *A Forecast of Space Technology 1980–2000* in which a large number of near future manned spacecraft are described (among an equally large number of unmanned spacecraft). The task group considers only those spacecraft intended for carrying specific payloads from the earth's surface into earth orbit, and the vehicles forecast are divided into four groups based on an ascending scale of payload capacity. The forecasts are based upon the ideas of twenty-six industry consultants.

Level-I considers six manned spacecraft capable of lifting up to 30,000 pounds into orbit, of which three are single-stage-to-orbit spaceplanes. They are all designed for vertical takeoff and horizontal landing (VTOHL). The first of these is a relatively small 127-foot vehicle with a 106-foot wingspan and a gross liftoff weight of 2.5 million pounds. It has two main engines and eight maneuvering engines. The second is 158 feet long with a wingspan of 131 feet. Its liftoff weight is 2.6 million pounds, including 30,000 pounds of payload to or from earth orbit. The third spaceplane is 203 feet long, with a wingspan of 140 feet, three main en-

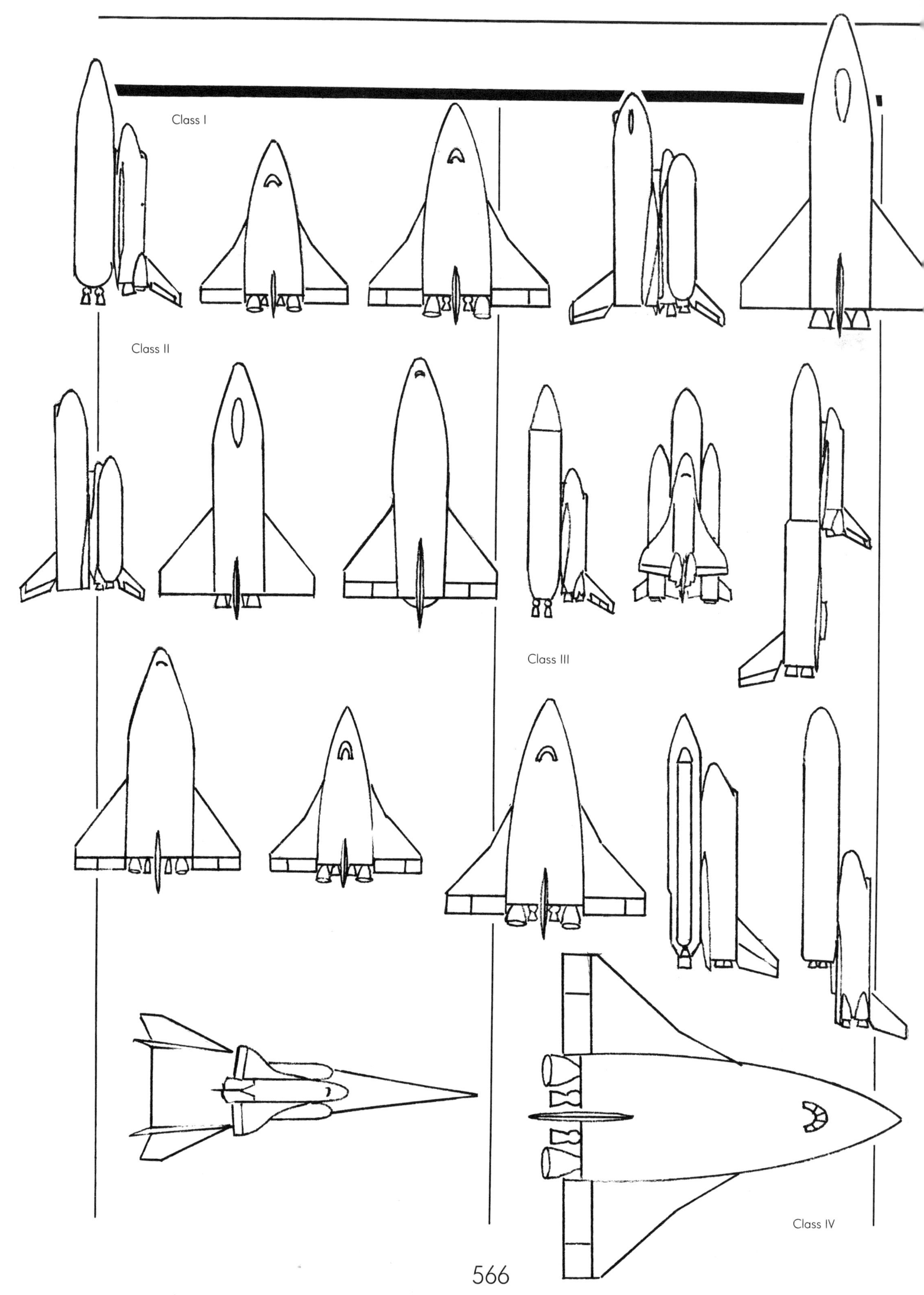
Class I
Class II
Class III
Class IV

gines, and a liftoff weight of 3 million pounds.

Level-I also proposes three manned VTOHL shuttles, not dissimilar to the present-day shuttle. The second pair are 1 1/2-stage vehicles. One is a standard shuttle with a liquid-fuel booster with four main engines in place of the familiar external tank and solid-fuel strap-ons. It can carry 30,000 pounds to and from earth orbit. The other involves a manned, recoverable, glide-back booster 156 feet long, with eight main engines. The shuttle itself has three main engines and two strap-on expendable fuel tanks.

Level-II contains the class of vehicles capable of lifting up to 60,000 pounds into orbit. Ten are manned spacecraft. Four of these are VTOHL single-stage-to-orbit spaceplanes. The first is a 203-foot, 140-foot wingspan rocket weighing 3.3 million pounds at takeoff. It is powered by seven main engines. The second shuttle is 184 feet long, with a wingspan of 148 feet and a liftoff weight of 4.2 million pounds. It is powered by six main engines and two F-1 Saturn engines. It can carry 65,000 pounds of payload. The third is the smallest at 165 feet and with a 137-foot wingspan, carrying 60,000 pounds to orbit.

Four of the designs for Level-II are upgrades of the present-day shuttle. The first employs a 200-foot liquid-fuel booster with four main engines. The second adds a pair of recoverable 140 feet liquid-fuel strap-on boosters to the shuttle's external tank. This version can carry 65,000 pounds to earth orbit. The third shuttle is boosted by a 177-foot manned, recoverable rocket. The shuttle itself is flanked by a pair of 122-foot recoverable fuel tanks. The fourth shuttle's external tank is the upper stage of a two-stage booster. The combined spacecraft is 275 feet tall. The first stage is a manned spacecraft powered by five F-1 engines. This shuttle can carry 83,000 pounds into space.

The ninth Level-II spacecraft is a horizontal takeoff and landing (HTOHL) ram jet-scram jet booster carrying a space shuttle with two external fuel tanks on its back. The wedge-shaped spaceplane carrier is 306 feet long with a wingspan of 130 feet, weighing 2.1 million pounds at takeoff. The booster by itself can be adapted as a high-speed point-to-point transport as a competitor to the SST. It is expected to be available about 2000.

Level-III includes those spacecraft intended to carry up to 400,000 pounds of payload into space. Only three are to be manned. Two of these are shuttle-derived spacecraft. One is 173 feet long and is boosted by a 312-foot liquid-fuel rocket. The combined spacecraft weighs 7 million pounds at takeoff and can carry 400,000 pounds into orbit. The second is 196 feet long and is boosted by a 245-foot external tank with strap-on boosters. The third Level-III manned rocket is a VTOHL spaceplane 244 feet long, with a wingspan of 201 feet. It is powered by twelve engines, and can carry 400,000 pounds into space and 50,000 pounds back to the earth.

Level-IV includes those spacecraft capable of lifting 2 million pounds of payload into space. Only one of these monster spacecraft is to be manned: a 384-foot VTOHL spaceplane with a 320-foot wingspan. Its ten engines can carry 1,000 tons of payload into orbit.

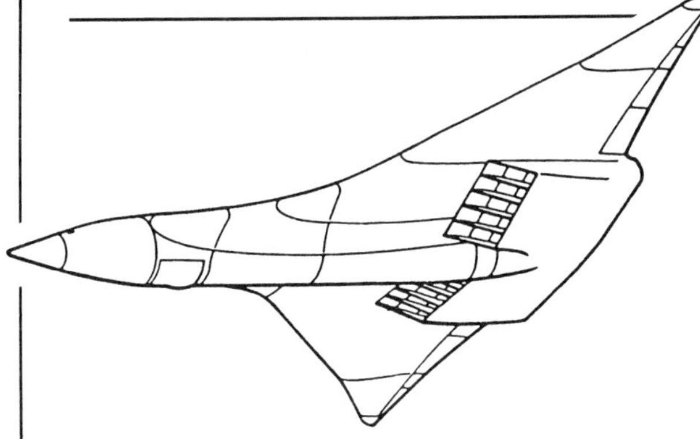

Rockwell investigates a fully reusable single-stage-to-orbit scramjet aerospaceplane design. It would provide support for the construction of large orbiting power satellites.

P. Seigler of Earth/Space, Inc. proposes the privately built SSOAR (Single Stage to Orbit And Return), a VTOL SSTO spacecraft. It has a lox/hydrogen-fueled aerospike engine.

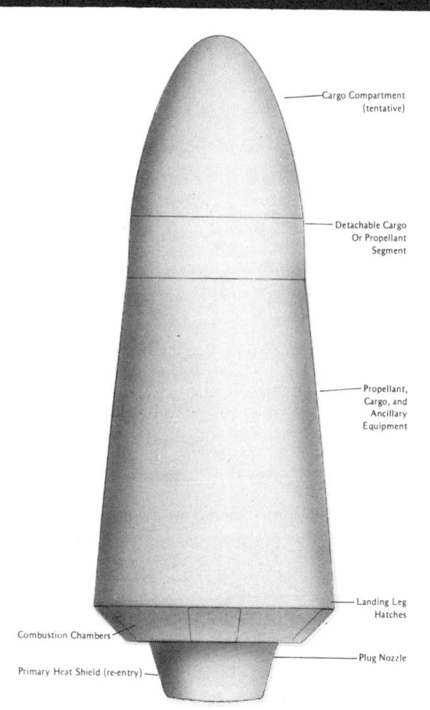

SEPTEMBER

Fabrication of the first Shuttle, OV-101, the *Enterprise* (begun on June 4, 1974) is completed. This shuttle does not possess all of the features necessary for an operational orbiter and is not intended to fly in space. The *Enterprise* is used exclusively for atmospheric flight tests.

1976–1979

The upper stage of the Soviet "50–50" spaceplane [see 1966–1969], code-named "Spiral," is dropped from a TU-95 bomber. It makes a subsonic manned flight. [Also see 1982–1984.]

1977

Boeing designs a VTOL SSTO vehicle capable of carrying a 50,000 pound payload. It would be launched from and recovered in an artificial lagoon.

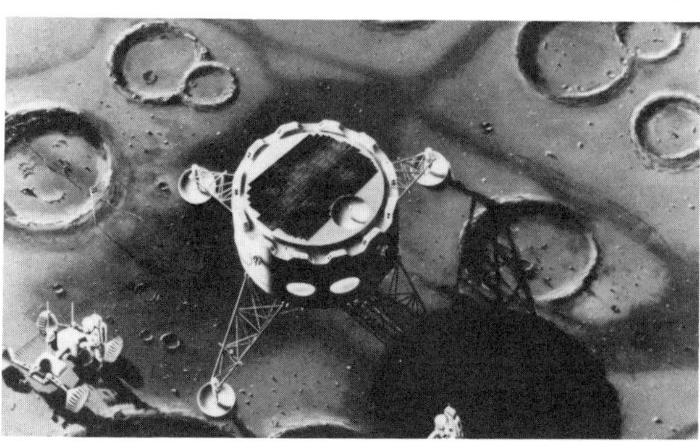

Right, middle: NASA LM-B lunar lander; bottom: propulsion module has returned to orbit, leaving crew cabin and ascent stage.

Boeing space freighter

1978

JANUARY 10

The manned spacecraft Soyuz 27 docks with the orbiting Salyut 6 and Soyuz 26, completing the first three-spacecraft complex.

OCTOBER

Radio Moscow reports that a spaceplane is being developed; its description is almost identical with the "Lapot" spaceplane of 1962–1969 [see above]. Meanwhile, a 4.5 km runway has been constructed at Baikonur and landing tests are recommenced. A group of cosmonauts are named as eventual spaceplane pilots. A conventional rocket booster is to be used instead of the SST that was planned earlier. Test flights of two scale models at a time are made from 1976 to 1979, using the Proton rocket as the booster. Photographs taken by the Australians of the Indian Ocean recovery of the models create a sensation. Until now, no one has seriously suspected that the Russians are developing spaceplanes or space shuttles.

1979

Salvage-1 is a made-for-television movie in which a junkyard owner realizes that the Apollo equipment left on the moon by the astronauts has been legally abandoned. He decides to build his own spaceship, travel to the moon, salvage the NASA equipment and sell it back on earth for a fabulous profit.

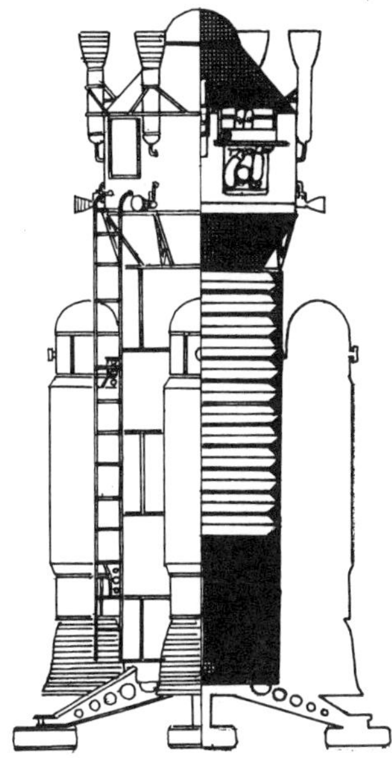

The enterprising junkdealer, played by Andy Griffith, builds his 33-foot spaceship, *The Vulture*, from the scrap metal and spare parts he finds in his junkyard. The main body is made from the cargo tank of a Texaco gasoline truck, the manned control module from the mixer of a cement truck (the truck's rear-view mirror gives a look back toward the earth), its landing legs are cushioned with auto tires, the astronaut's seats are recycled car seats and the throttle is a converted T-bar transmission from a 1971 Javelin. The main engines and the retrorockets are NASA surplus. The fuel tank is accordian-pleated; as the fuel is consumed, it collapses, creating cargo space for the return trip. *The Vulture* is fueled by "monohydrazine," a new fuel that allows almost unlimited thrust: *The Vulture* and its crew never has to worry about escape velocity; they can travel to the moon at a constant, comfortable 50 mph, if they wish. This is one of the few instances where a point is made of the fact that achieving escape velocity is not required for a spacecraft to leave the earth if sufficient energy is available.

M. Villain of Aerospatiale presents the results of a study initiated in 1977 by CNES (France's Centre National d'Etudes Spatiales). The mandate to create "a manned space vehicle which could be launched by an improved Ariane" booster results in the *Hermes* spaceplane [also see 1985]. It is to weigh no more than 6,500 kg, later uprated to 10,000 kg with a six-man crew (or a two-

1979

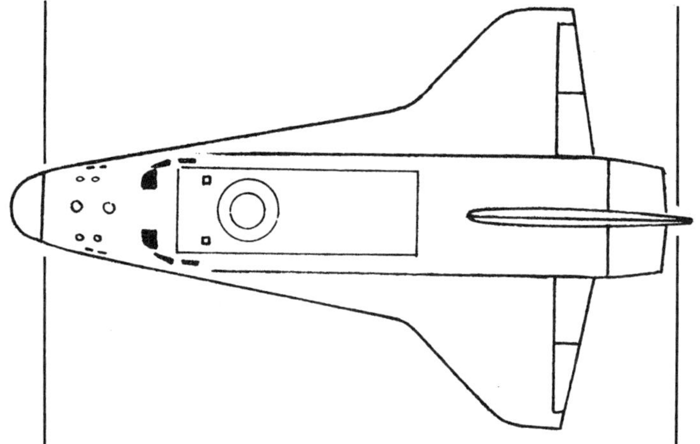

man crew and a payload of 1,500 pounds) and a mission duration of up to 7 days. *Hermes* is to be launched from the French space center in Guyana.

The initial *Hermes* proposal bears a strong, and intentional, resemblance to a scaled-down U.S. space shuttle. It has a length of 12 m and a wingspan of 7.4 m.

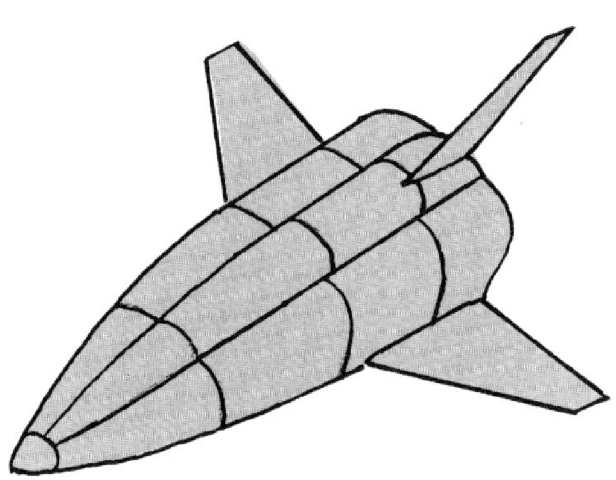

NASA's Langley Research Center develops a concept for a rocket-powered single-stage-to-orbit spaceplane. It could carry a payload of 5 tons into a polar orbit or 15 tons into an east-launched orbit. The gross weight of the spaceplane is 1,105 tons.

JANUARY 25

NASA announces the names of the first four space shuttles, named in honor of the ships of American explorers. The first, scheduled for launch in 1979, is named *Columbia* after Robert Gray's sloop. Gray discovered the Columbia River, which he named for his ship, in 1792. *Challenger* is named for the ship that gathered fifty volumes of oceanographic data from December 1872 to May 1876.

The *Discovery* is named after a ship that was involved in the search for the Northwest Passage in 1610–1611. *Atlantis* is named in honor of the first American oceanographic vessel, a two-masted ketch that logged half a million miles between 1930 and 1966. The *Enterprise*, the first orbiter built, is named both for a ship that explored the arctic from 1851 to 1854 and for the spaceship featured in *Star Trek* [see above].

Irving

ca. LATE '70s

Boeing proposes a large Reusable Aerodynamic Space Vehicle (RASV) and a smaller Air-Launched Sortie Vehicle (ALSV), which can be carried on the back of a 747. The latter spacecraft is a small, wedge-shaped shuttle with an attached external fuel tank, designed to be carried piggyback on a specially adapted 747. At an altitude of 50,000 feet the ALSV detaches and, with its own engines fueled by the external tank, continues on into orbit.

ca. 1979

R. Biechel and Robert Salkeld, and others, have been doing work on single stage to orbit advanced space transportation systems (ASTS). The ASTS could take any of several forms, depending upon whether passengers or cargo are being carried. Passengers and smaller payloads can be lifted by a winged SSTO, while heavier cargoes can be carried by single stage heavy lift launch vehicles (HLLV), such as the VTOL plug-nozzle vehicles being promoted by a number of other engineers.

One small SSTO spacecraft suggested by

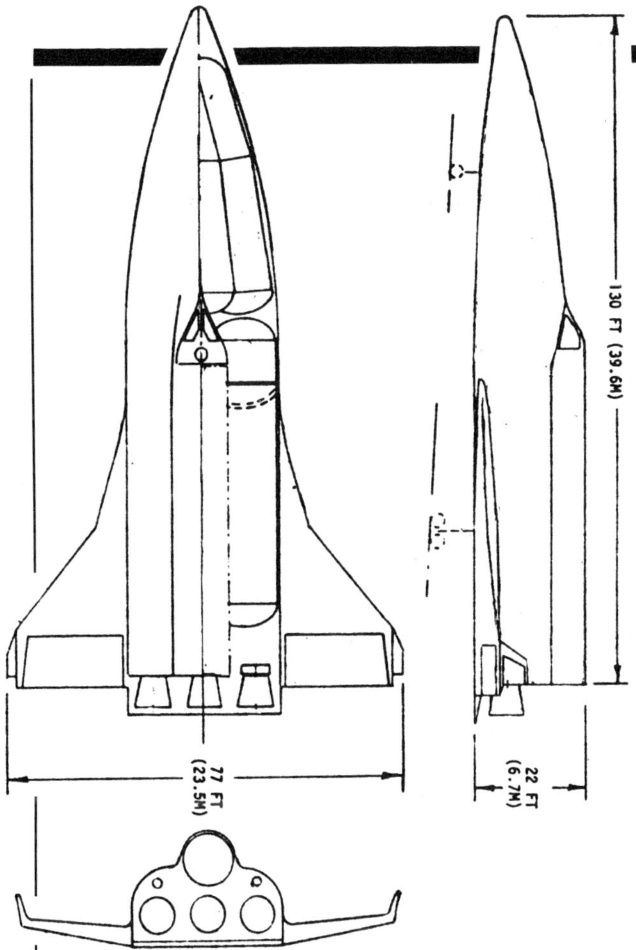

Salkeld would be a horizontal takeoff, horizontal landing shuttle 171 feet long, with a wingspan of 142 feet. It would have mixed-mode propulsion. The payload bay door would be in the tail, amidst the ten engine nozzles.

NASA Langley has developed a design for a state-of-the-art SSTO vehicle 130 feet long, with a wingspan of 77 feet and three engines. It could carry a payload ranging from 21,100 to 33,000 pounds.

The ASTS may require air launching, using any one of several different carriers. This might be a C-5A launching the shuttle from under its wing or from its cargo bay; a twin C-5A launching the shuttle from under from under its wing or from its cargo bay; a twin C-5A launching the shuttle from under the central common wing; or a yet-to-be-developed low technology aircraft (with rocket-assisted pull-up) launching the shuttle from a mission-dedicated pod.

1980

C. E. Singer, in his paper "Interstellar Propulsion Using a Pellet Stream for Momentum Transfer," describes a method of spacecraft propulsion in which high velocity pellets will impact an Orion-type shock absorber mounted on the spacecraft. The pellets could be ejected by an orbiting, solar-powered mass driver.

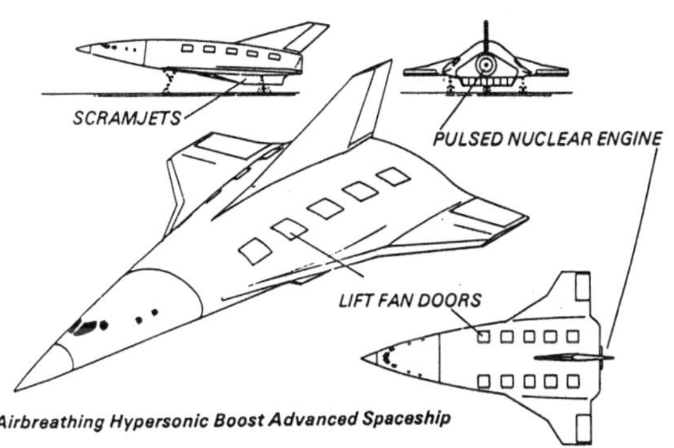

Airbreathing Hypersonic Boost Advanced Spaceship

Gary Hudson discusses the needs, requirements and design for an advanced solar system spaceship. Ideally it would be a compact, single-stage reusable vehicle capable of high-speed, high-performance operation throughout the solar system, including the atmosphere of the earth. Operated like a commercial aircraft, it would be safe, reliable and economical (carrying a payload of 20 percent of its gross weight).

Hudson believes that such a spacecraft, normally only to be found in the wishful thinking of science fiction, would be possible and practical using some version of an advanced fission/fusion pulsed engine. A composite structure would be used for the pusher plate, perhaps some technology evolved from the carbon-carbon compounds used in heat shields. Propellant would be ordinary water (which has the advantage of being more or less readily available throughout the solar system) heated by the detonation of the fuel pellet. This pellet (composed of a $^{238}U/^{235}U$ hybrid, a $^{238}U/LiD/^{239}Pu$ layered capsule or antihydrogen) in turn is detonated by an electron or heavy ion beam.

EVOLUTION OF THE SPACE SHUTTLE I

North American Rockwell/Convair 1970

1970

General Dynamics/ Convair 1970

1971

1961

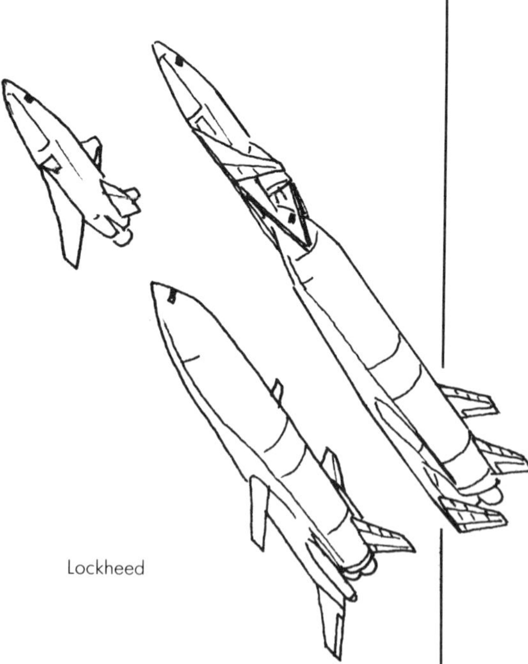

Lockheed

Boeing *Star-Raker*

Grumman 1970

Grumman 1970

McDonnell-Douglas 1971

McDonnell-Douglas 1969

EVOLUTION OF THE SPACE SHUTTLE II

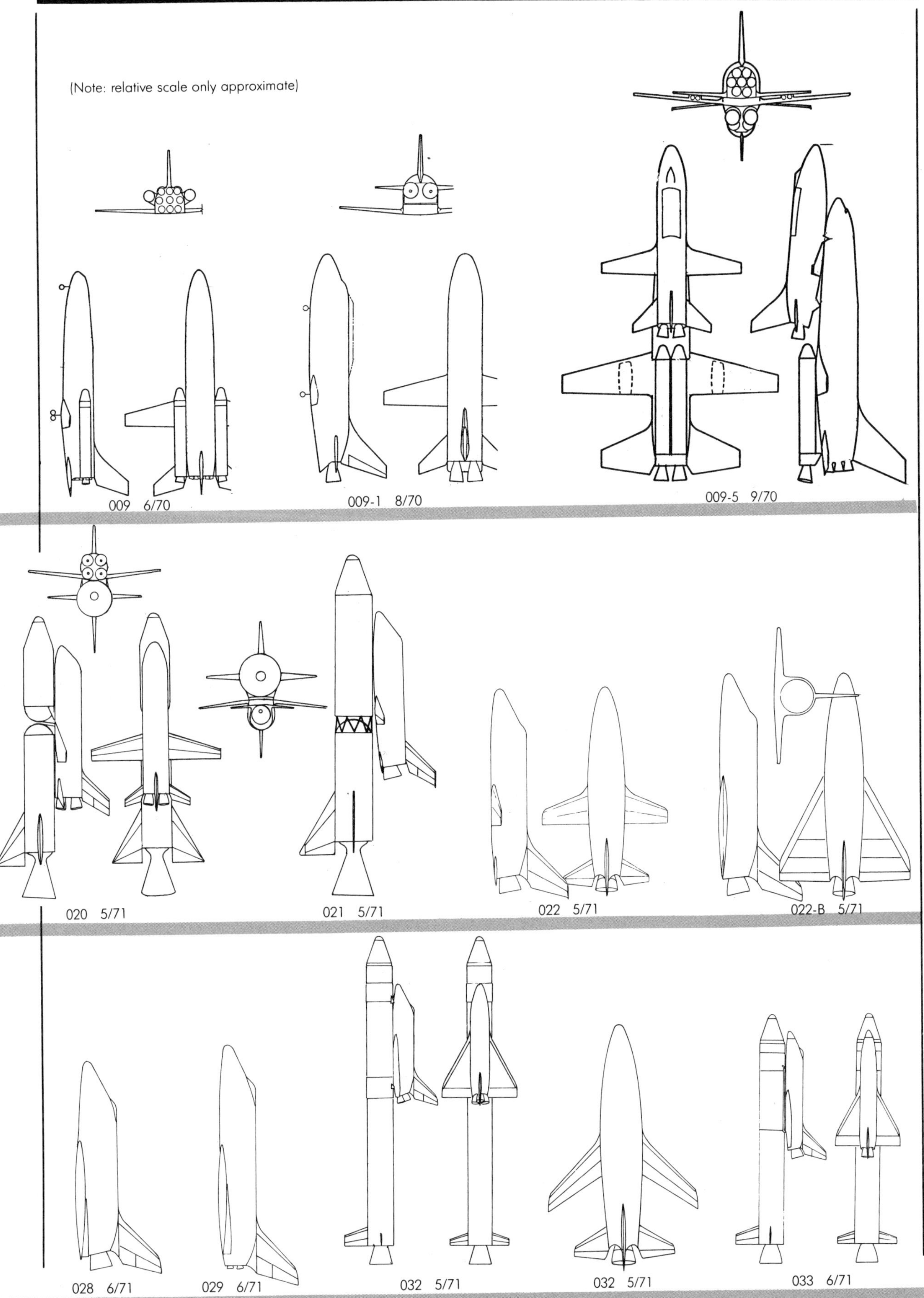
(Note: relative scale only approximate)
009 6/70
009-1 8/70
009-5 9/70
020 5/71
021 5/71
022 5/71
022-B 5/71
028 6/71
029 6/71
032 5/71
032 5/71
033 6/71

EVOLUTION OF THE SPACE SHUTTLE III

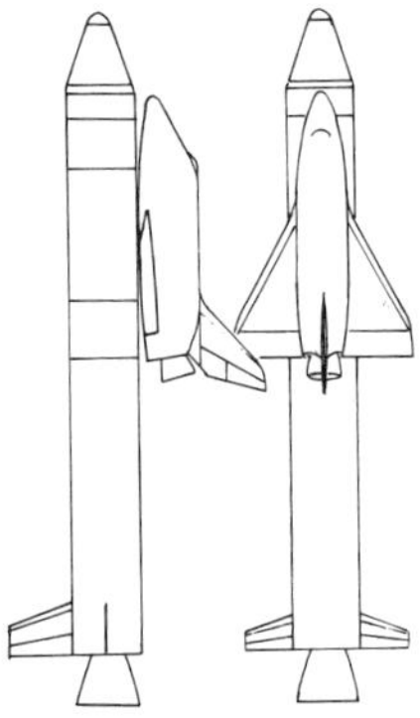

(Shaded areas added to 035 configuration)

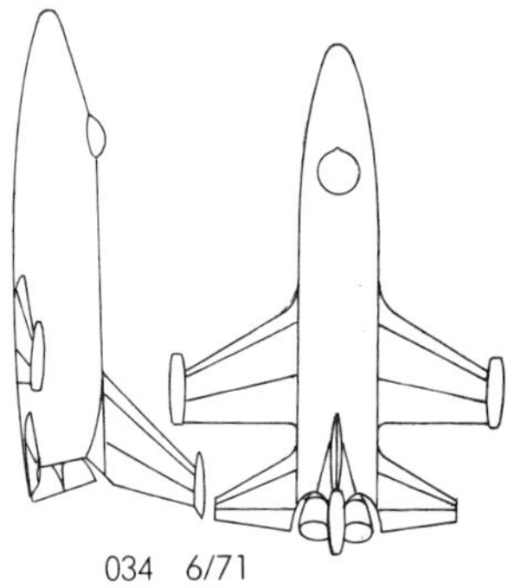

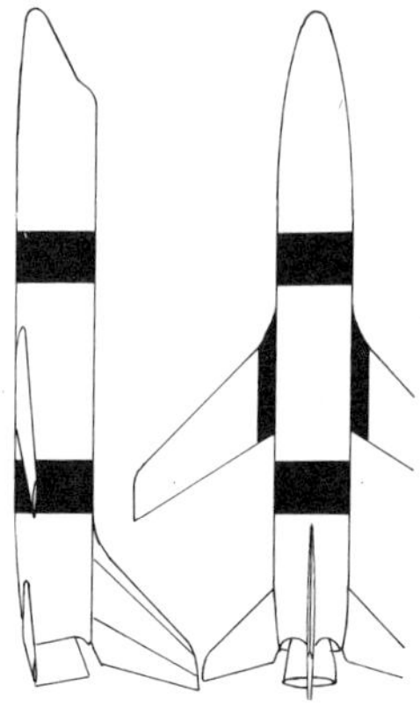

034 6/71

034 6/71

035A 6/71

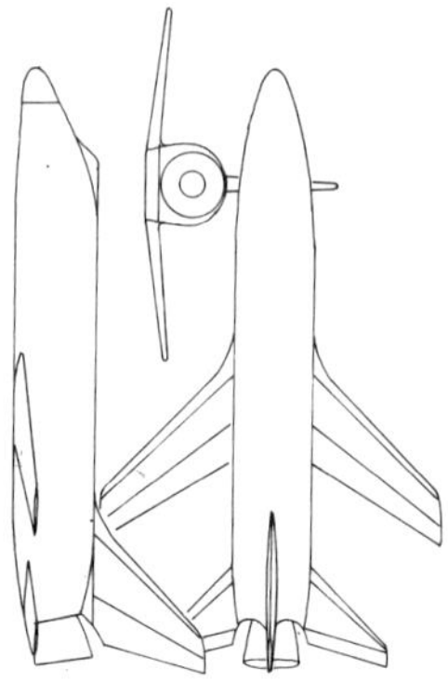

035 6/71

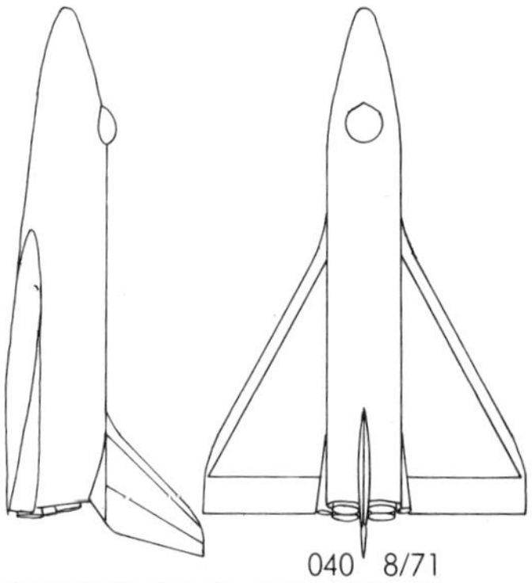

040 8/71

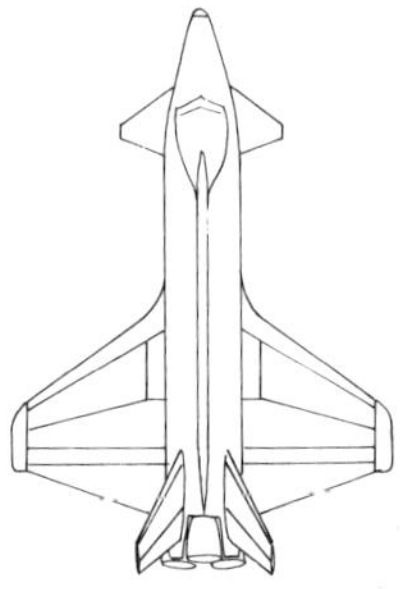

040C 1/72

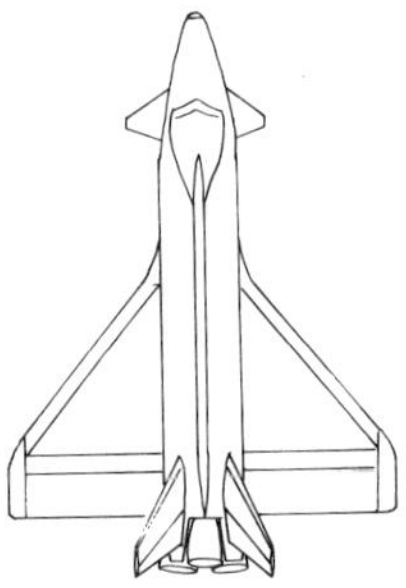

040-C-5 2/72

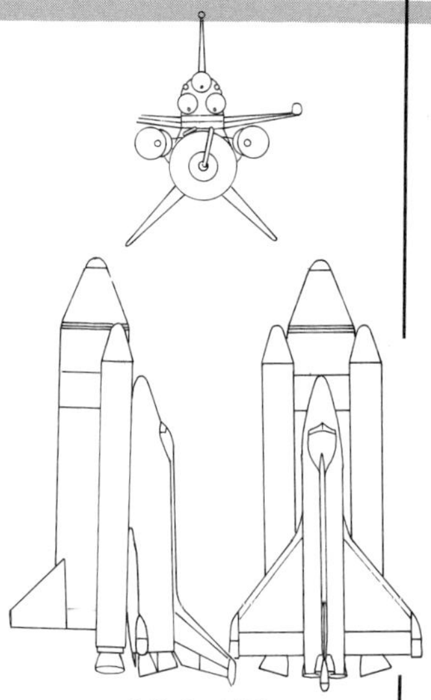

040-C 1/72

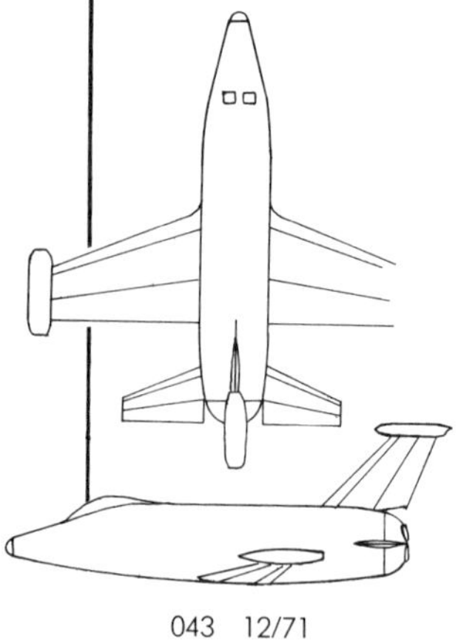

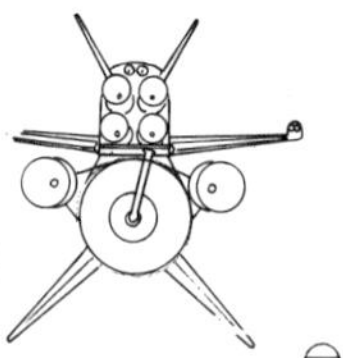

048A 3/72

043 12/71

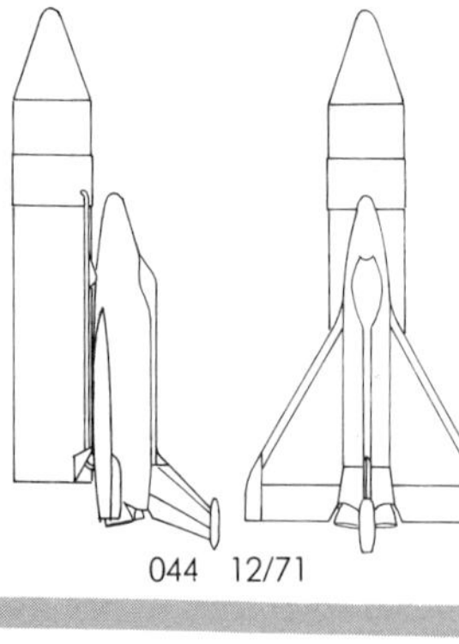

044 12/71

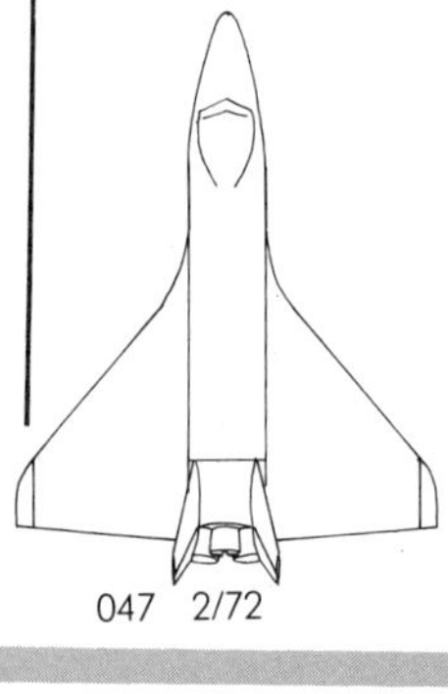

047 2/72

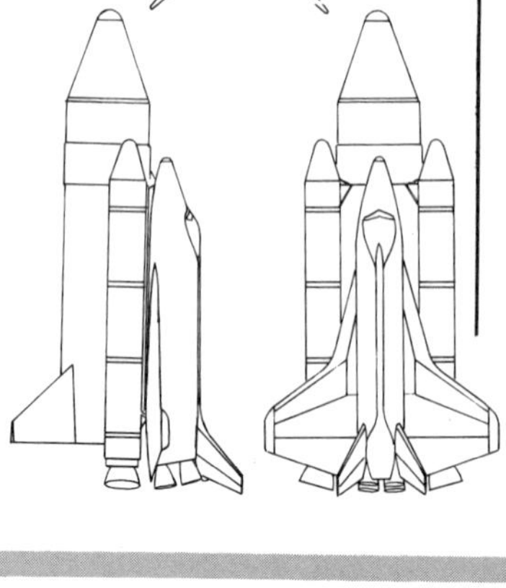

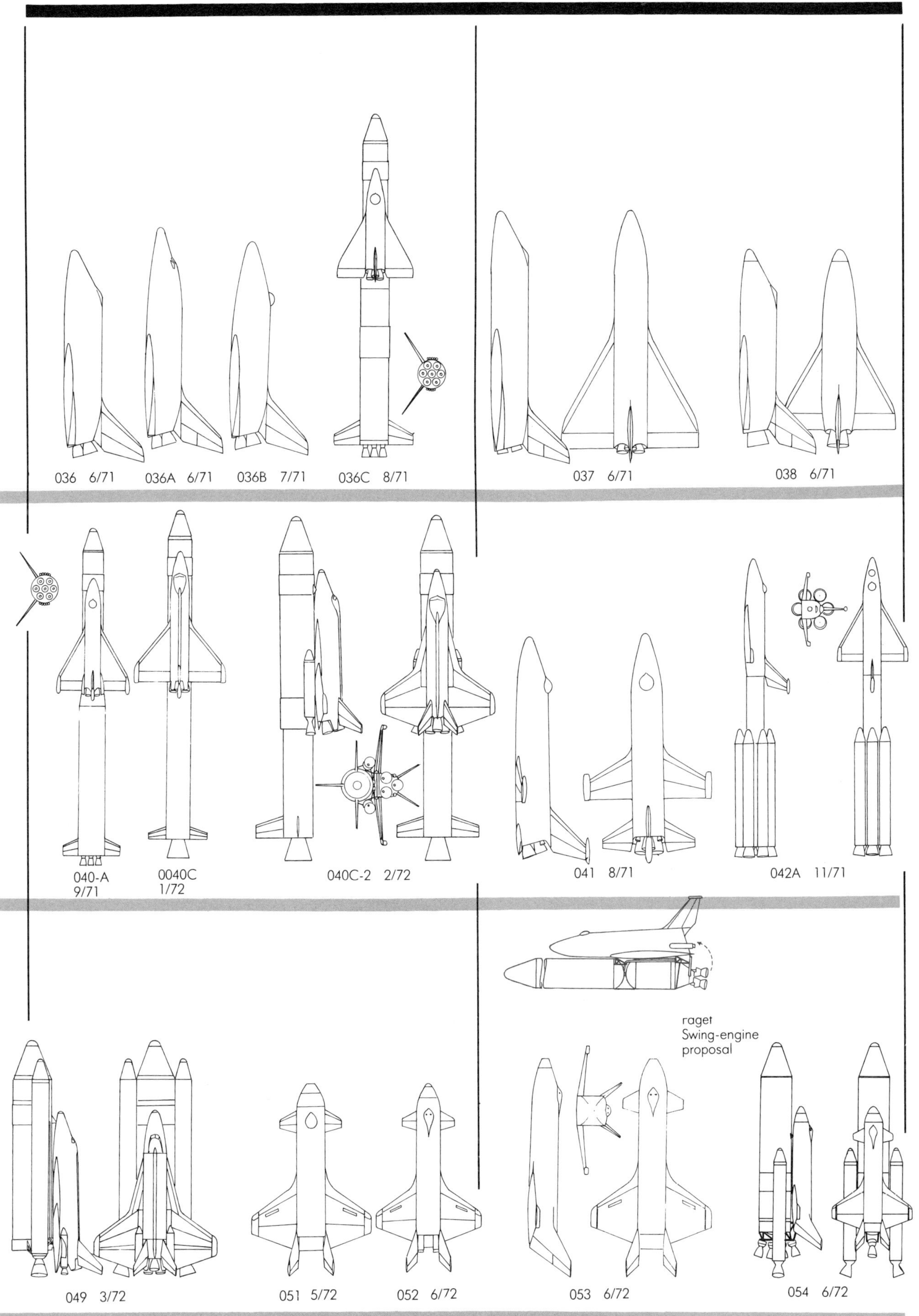
036 6/71
036A 6/71
036B 7/71
036C 8/71
037 6/71
038 6/71
040-A
9/71
0040C
1/72
040C-2 2/72
041 8/71
042A 11/71
raget
Swing-engine
proposal
049 3/72
051 5/72
052 6/72
053 6/72
054 6/72

EVOLUTION OF THE SPACE SHUTTLE IV

Competitive designs submitted to NASA, based on 040C design

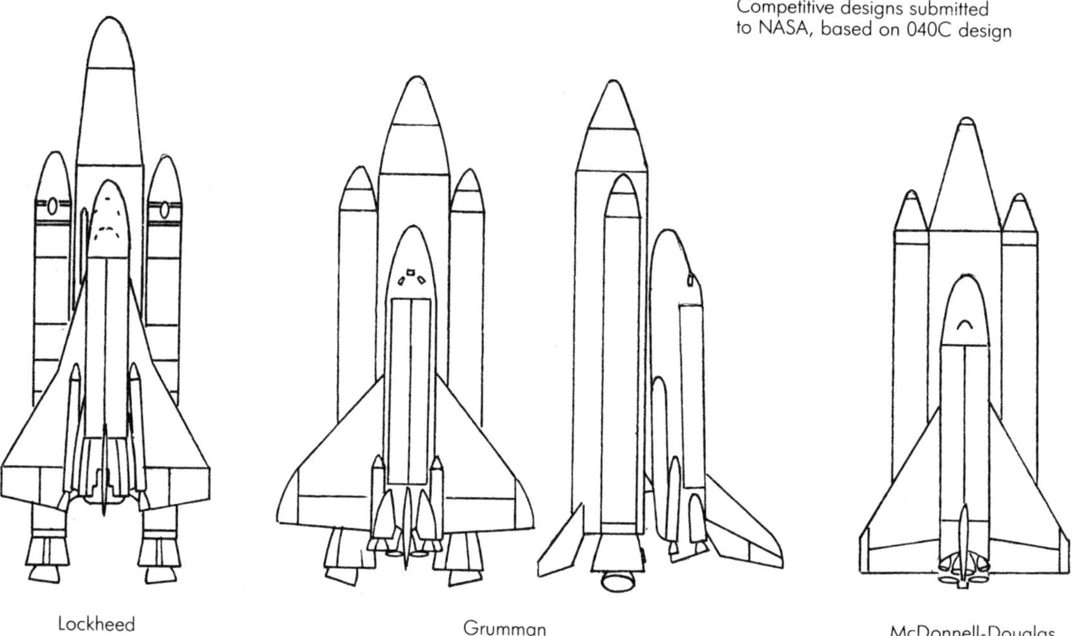

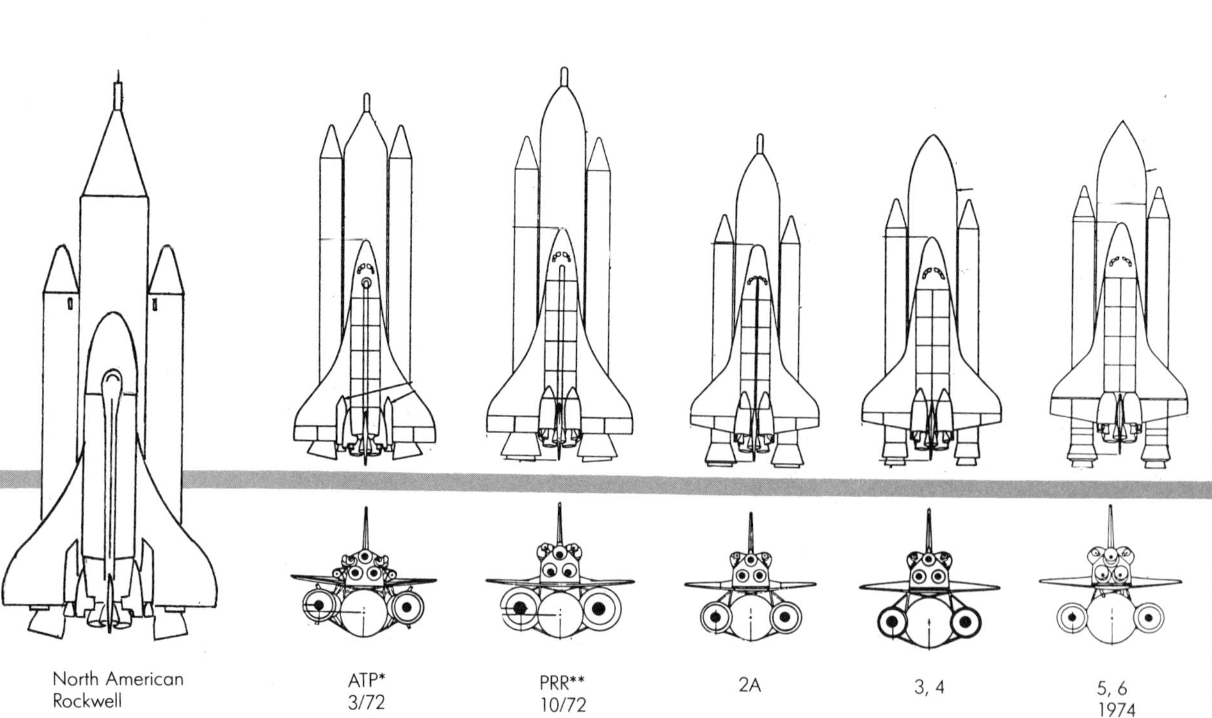

*Authority to Proceed
**Program readiness review

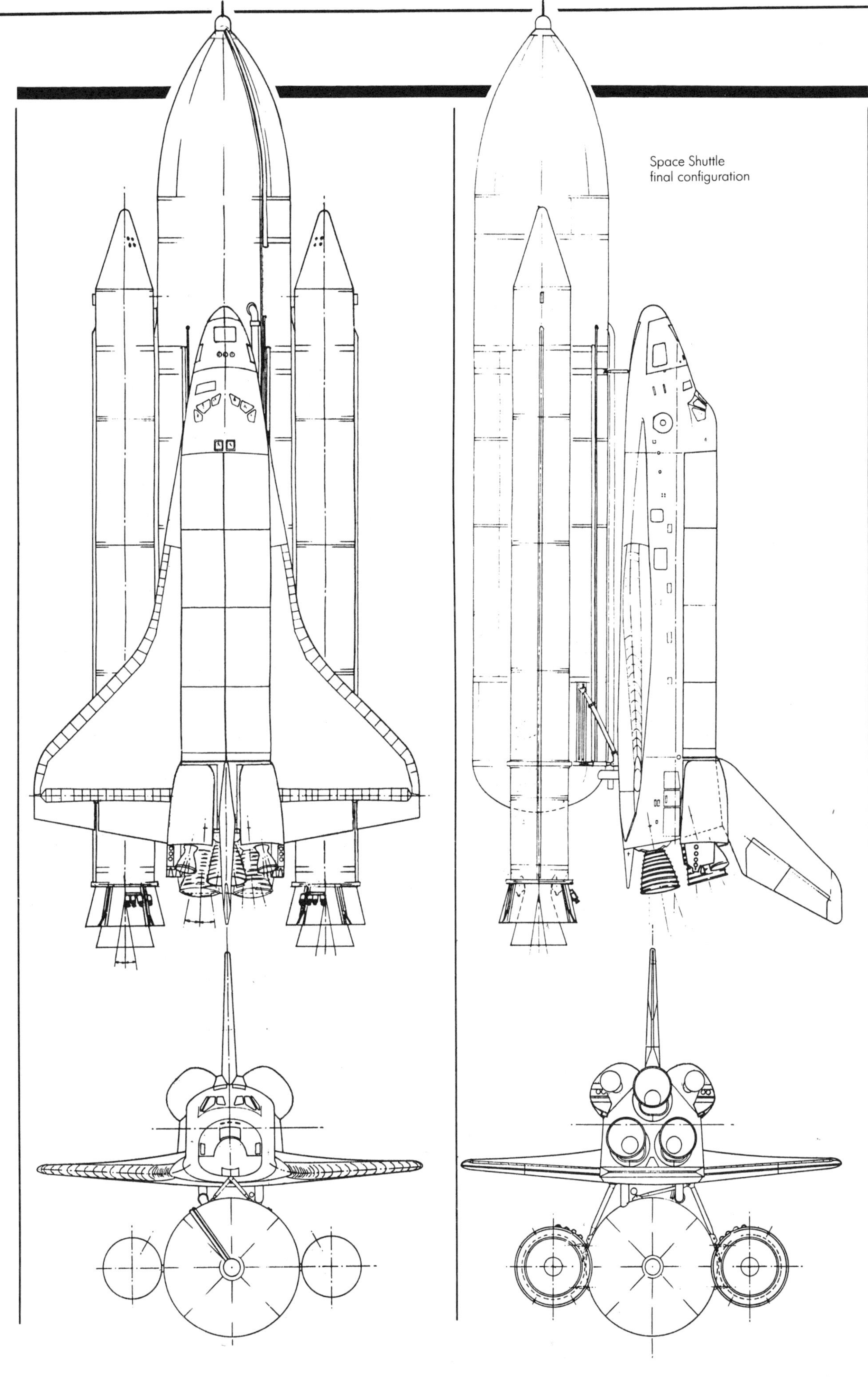
Space Shuttle
final configuration

For transit within the earth's atmosphere, the solar system spaceship would lift off in its VTOL mode using tip driven lift fans, ascending to about 30,000 feet. At this point the vehicle would pitch hard over and ducts would divert the lift fan air rearward. As the spaceship falls, its scramjets would ignite and at Mach 1.2 it would begin to climb. The scramjets would shut down at 11,000 ft/sec and the spaceship would coast to orbit where it would fire its nuclear pulse engine.

JUNE 5

In the first manned flight of the Soyuz T series, the Soyuz T-2 docks with the Salyut 6.

Coggins

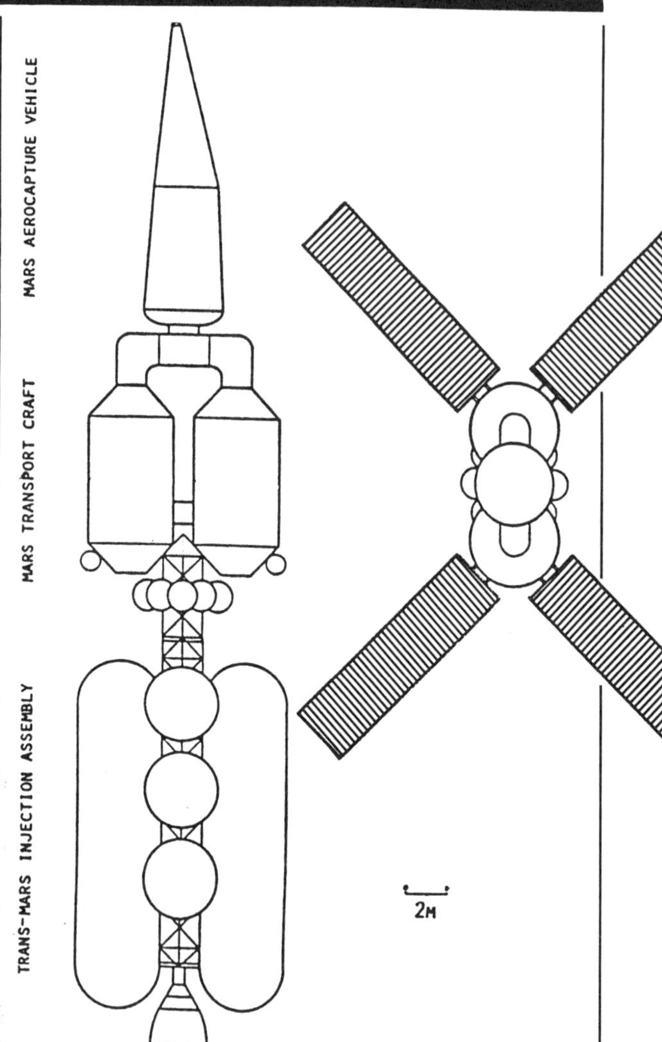

1981

Among the manned Mars mission spacecraft proposed at the first Case for Mars conference is one developed by Robert L. Staehle of the World Space Foundation. He suggests a combination of space shuttle-derived systems, solar sails and aerocapture landers. All of the spacecraft are assembled in earth orbit. The mission requires eight astronauts, four of whom will descend to the surface of Mars for a period of exploration lasting several weeks. Unmanned solar sails act as cargo carriers for delivering equipment, consumables, a pair of aerocapture landers and propellant to Mars orbit to await the later arrival of the manned spacecraft. The manned Mars transfer craft (MTC) is equipped only for a one-way trip. Upon arrival, the crew of the MTC transfers to an aerocapture vehicle for the descent to the Martian surface. Other spacecraft, similar to the MTC and delivered by solar sail, provide an artificial gravity environment in Mars orbit. These spacecraft are reprovisioned and refueled from the solar sail cargo craft for the return to earth. Once back home, the crews transfer to Apollo-derived entry vehicles.

The solar sail cargo vehicle (SSCV) has a 2 km square solar sail of aluminized Kapton, 2.5 μm thick.

The trans Mars injection stage uses a pair of modified space shuttle main engines fueled by liquid hydrogen contained in two cylindrical tanks and liquid oxygen in six spherical tanks. Ahead of the booster is the Mars transport craft (MTC) which consists of two Spacelab modules, external storage containers and connecting tunnels. The crew inhabits the MTC during the voyage to Mars. Attached to the forepart of the MTC is the Mars aerocapture vehicle (MAV). Four solar panels generate power during the flight. The MAV is an aerodynamic shell and an orbital propulsion system surrounding an Apollo-derived command service module and crew accommodations. No specific lander design is suggested.

The four astronauts who remain in orbit around Mars occupy a rotating station. This is created from a pair of MTC modules attached by cables and inflatable tunnels and rotating around a common Spacelab-derived hub.

Bill Kaysing self-publishes a pseudoscientific classic, *We Never Went to the Moon* (subtitled "America's 30 Billion Dollar Swindle!"), in which he attempts to prove that the Apollo landings were a hoax. Mentioning two of his primary arguments will illustrate the level of Kaysing's logic and knowledge.

One of the most important pieces of evidence he dwells upon is that *before* the Apollo landings almost all artists showed the engine of the lunar module creating a crater in the surface as it landed. There was, however, no visible crater beneath the lander in the Apollo photographs. Therefore, Kaysing concludes, the photos are fakes because they do not match the artists' preconceptions!

Kaysing's second major argument centers around the fill lighting seen in the lunar surface photos. That is, detail is visible on surfaces turned away from the sun. According to Kaysing, this proves that the photos were really taken in a studio. The photographers he claims to have consulted apparently never told him about reflected or bounce lighting in which the light on the shaded sides of the lander and the astronauts is simply being reflected from the brilliantly illuminated surface around them.

Kaysing goes on in this fashion for some 200 pages.

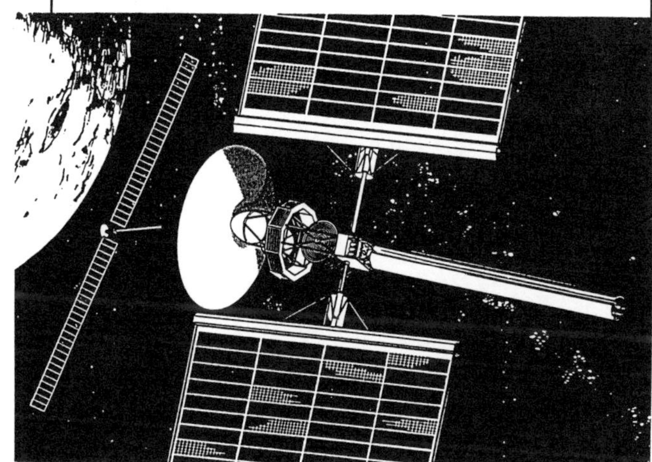

David Bauer, C. Julian Vahlberg and John Barber prepare a study for NASA on the feasibility of using a pellet-firing railgun for spacecraft propulsion.

The railgun would be powered by a huge solar cell array and would fire pellets made of any one of several suggested materials. The propellant would not be stored in pellet form, but would rather be manufactured from bulk. For example, if the propellant mass is to be water, then the water could be stored in tanks to be frozen and formed into pellets as needed.

The authors took some care to consider the threat to other spacecraft from their machine gun-like spaceship. One solution would be to use pellets that would decompose. Ice pellets, for example, would simply sublime to harmless water vapor.

APRIL 12

The space shuttle *Columbia* makes its first orbital test flight.

JUNE 11

According to cosmonaut and professor of technical sciences Konstantin P. Feoktistov the Soviet Union must develop a low-cost space transportation system similar to the American space shuttle.

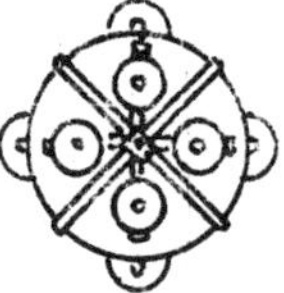

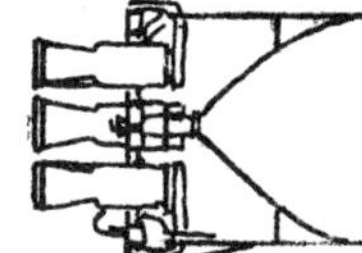

ca. 1981

Truax Engineering develops the X-3 "Volksrocket," or *Arriba One*, a manned rocket 24.2 feet long and 2.08 feet in diameter, with a maximum weight of 3,100 pounds. It is powered by four Rocketdyne LR101-NA3 engines with a combined 3,320 to 4,000 pounds thrust (fueled by LO_2 and RP-1). Designed to carry a human passenger to 286,000 feet on a suborbital flight, it is a direct spinoff of the rocket that Truax designed and built for daredevil Evel Kneivel's jump across the Snake River Canyon, although that one was powered by steam. In the summer of 1980 Truax completes successful static firing test of the X-3 prototype (the "Space Cycle"), with the engines producing a peak thrust of 1.8 metric tons.

The Volksrocket is planned to reach an altitude of over 50 miles, with both rocket and passenger returning unharmed to the earth. A suborbital flight to this altitude qualifies the rocket's passengers for astronaut status. The rocket is to make a vertical takeoff from Fremont Airport (near San Francisco) and reach brennschluss at 113,000 feet, 100 seconds later. The rocket then coasts to its maximum altitude of 50 miles. The entire rocket is recovered by parachute. At an altitude of between 100,000 and 150,000 feet a small drogue chute opens, stabilizing the rocket in a tail-first attitude. At 20,000 feet the main chute opens. The rocket safely splashes down into the Pacific Ocean 6 minutes after takeoff.

Truax anticipates that tickets for flights on the Volksrocket will cost approximately $10,000 each.

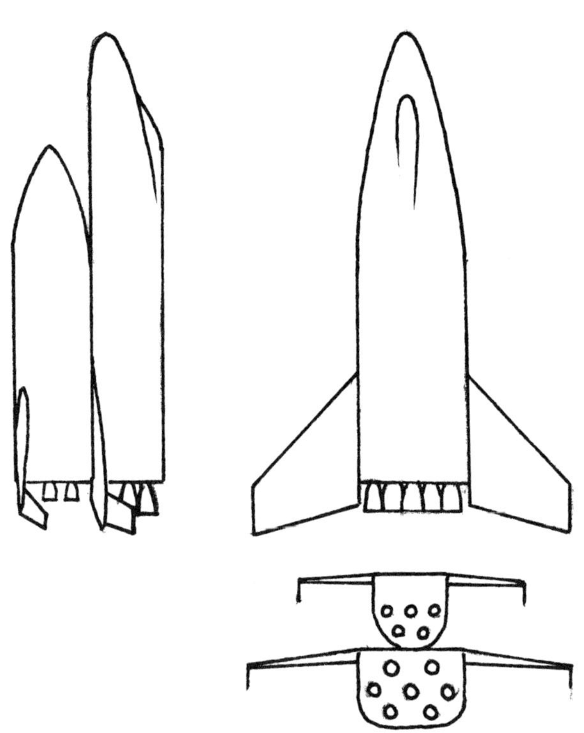

1982

NASA's Langley Research Center begins studying Future Space Transportation Systems (FSTS) in order to begin developing the second-generation of space shuttles. By 1984 Langley has developed a two-stage vehicle with a cargo bay measuring 20 feet × 90 feet in a spacecraft weighing 2,747,000 pounds. The cargo bay is redesigned as an add-on pod to be carried piggyback on top of the orbiter's fuselage. The liquid fuel booster is powered by five en-

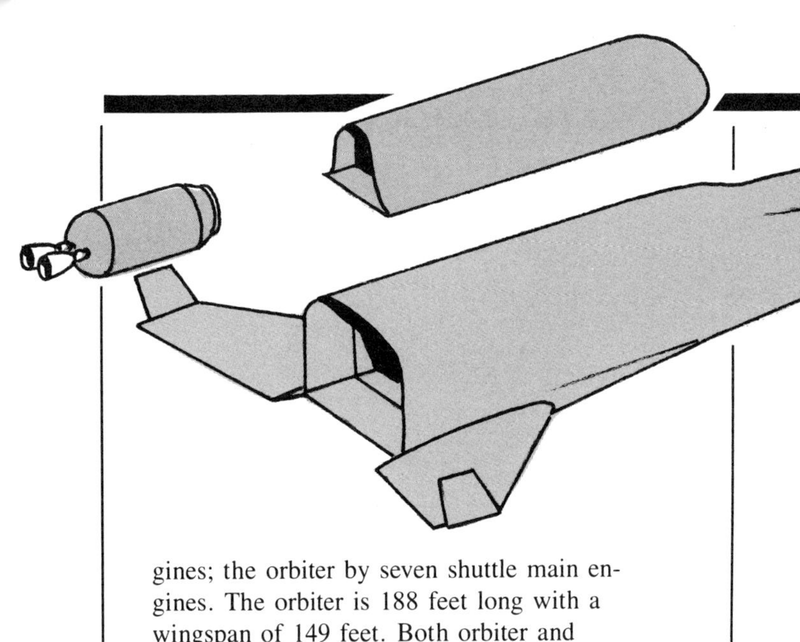

gines; the orbiter by seven shuttle main engines. The orbiter is 188 feet long with a wingspan of 149 feet. Both orbiter and booster are delta-winged with vertical winglets at the tips. The gross weight at takeoff is 4,897,000 pounds.

Assembly of the elements is simplified. The operation takes place while the spacecraft is horizontal. The orbiter is raised on jacks while the booster is slid beneath it. Afterwards, the payload is loaded on pallets into the rear of the payload pod. The complete vehicle is then towed to its launch pad and erected vertically for takeoff. After launch the unmanned booster will land automatically on a conventional runway.

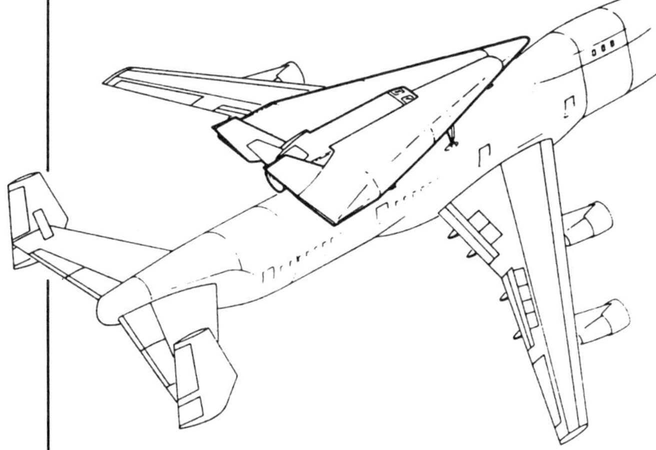

General Dynamics/Convair studies concepts for a second-generation space shuttle intended for the U.S. Air Force's Advanced Military Space Capability. Some of the variations considered include the piggyback launching from a modified 747 of a delta-shaped lifting body, the vertical launch of a single-stage-to-orbit lifting body, and a ground sled booster launching.

Japan, which has been studying winged spacecraft concepts since 1978, announces its preliminary shuttle design. The 22,000 pound NASDA minishuttle would be 46.6 feet long with a wingspan of 24.6 feet. It would carry a crew of four. The shuttle would not possess a main propulsion system of its own, but would be equipped with two small on-board jet engines that would allow it to fly over 600 miles after reentry to any runway of over 8,000 feet in length. The Japanese shuttle would be launched by an H-2 booster. [Also see HOPE, 1990.]

AUGUST 19

The first coed spacecraft crew is flown aboard the Soyuz T-7 flight. Cosmonauts L. Popov, A. Serebov and S. Savitskaya dock with the Salyut 7.

OCTOBER

An informal meeting of the Engine and Vehicle engineers of British Aerospace leads them to establish the potential for the HOTOL spaceplane. The name is an acro-

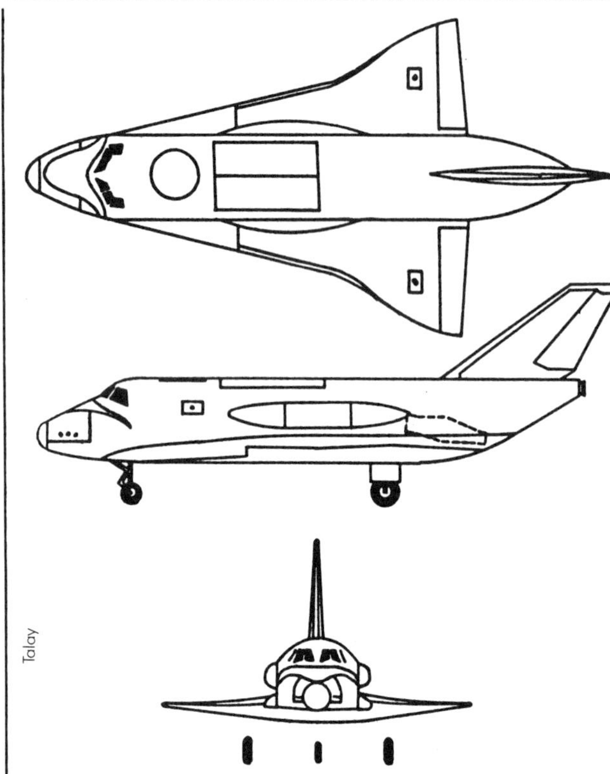

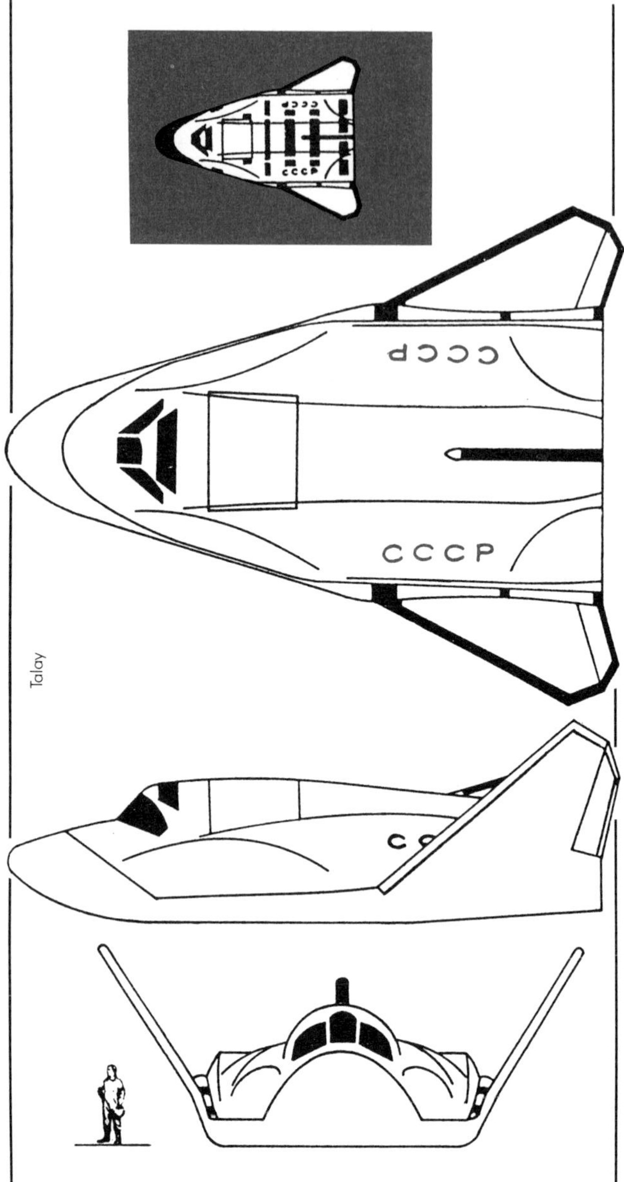

nym for HOrizontal Take-Off and Landing. In April 1983 a working group is established between BAe Stevenage and Rolls-Royce to explore the concept and in April the Concorde team is introduced to the project by the BAe Advanced Projects Group. In September a contract from DTI to study future European launch vehicle concepts establishes the economic basis for pursuing the HOTOL; at the same time Rolls-Royce says that a "first look" shows that the engine is viable. One version of HOTOL would carry sixty passengers halfway around the world in slightly over an hour. Tests are planned to begin in 1996, hopefully with the cooperation of the European Space Agency. [Also see 1984, 1985, 1991.]

NOVEMBER 11

The U.S. space shuttle *Columbia* makes its first operational flight.

1982–1984

Flight tests are made in the U.S.S.R. of a scale model spaceplane. These are the BOR-4 (Cosmos 1445) flights, in which a spaceplane almost identical with ones flown in 1976–1979 is flown in suborbital tests of the "50–50" "Spiral" model. This is followed in 1983 by suborbital flights of the B-5 model of the *Buran* space shuttle. [Also see 1966–1969 and 1982–1984.]

A 1/8th scale model is launched into space from Kasputin Yar as Cosmos 1374 in 1982. It is recovered in the Indian Ocean after a single-orbit flight. The recovery of the second test vehicle, Cosmos 1445, is photographed by the Royal Australian Air Force and creates a great deal of discussion and controversy.

Two more B-4 test flights are made. Another scale model, the B-5, the *Buran* (Snowstorm), is launched to test the overall configuration. The B-5 weighs 1.4 tons and makes several suborbital flights between 1986 and 1988, providing data on *Buran* performance from Mach 16 to Mach 2.

1982–1985

The U.S. Department of Defense/DARPA and NASA define the aero-space plane concept as a hydrogen-powered horizontal-take-off-and-landing aircraft capable of operating between Mach 12 and 25. Participants in the proposed program are the Air Force, Navy, DARPA, SDIO and NASA. An experimental aircraft (eventually designated the X-30) is proposed to test the new technologies required.

1982–1991

Pacific American Launch Systems is created [also see above] to develop a commercial VTOL SSTO space transport. Its spacecraft are developed from founder Gary Hudson's earlier *Phoenix* designs [see 1969–1980].

The company's new family of spaceships is comprised of two large (400,000 pound gross liftoff weight) vehicles, *Phoenix C* (for Cargo) and *Phoenix E* (for Excursion), and two small (under 70,000 pound GLOW) vehicles (the *Phoenix LP*, which also came in manned and unmanned versions). The manned *Phoenix E* could be used for missions to the moon or Mars. It would employ a twenty-four-nozzle aeroplug engine. A number of smaller configurations are considered, grouped under the designation *Phoenix M* (for Medium). These use conventional bell nozzles rather than plug nozzle engines.

1983

MAY

Phase I studies for a Transatmospheric Vehicle (TAV)—precursor of the National Aero-Space Plane—begin with a contract to Batelle Laboratories, which in turn works

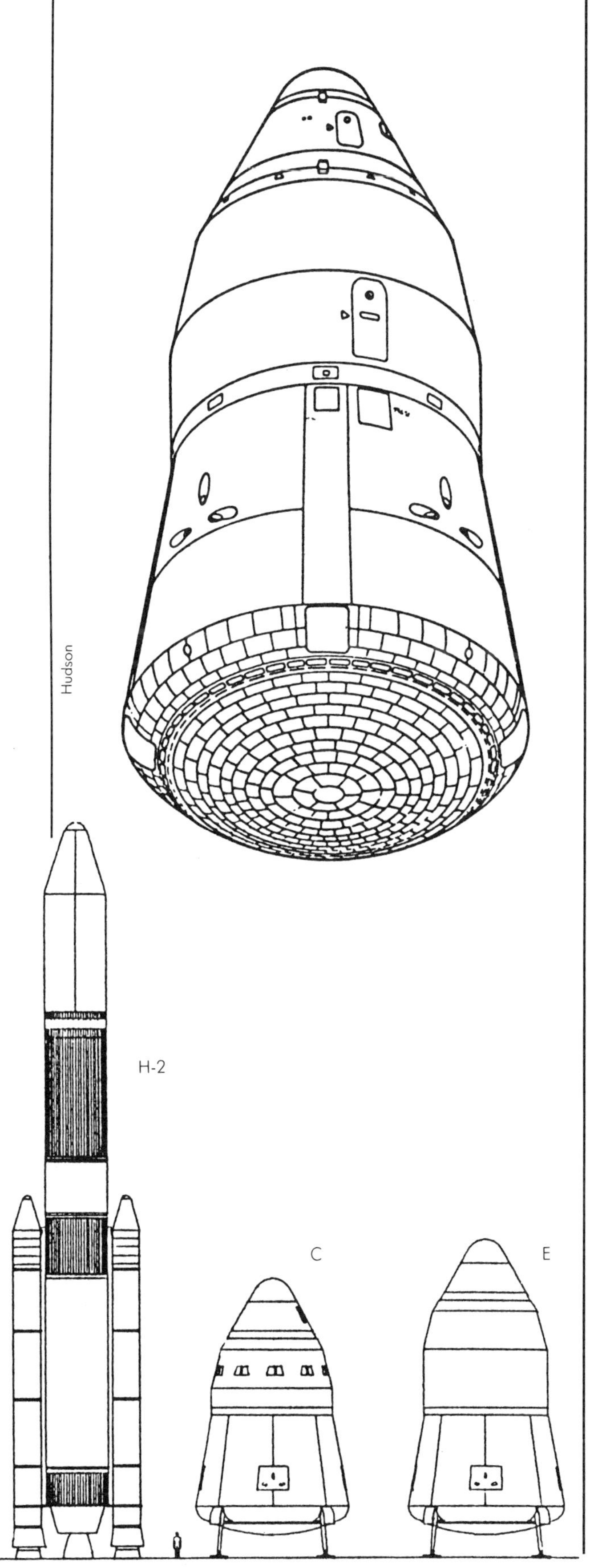

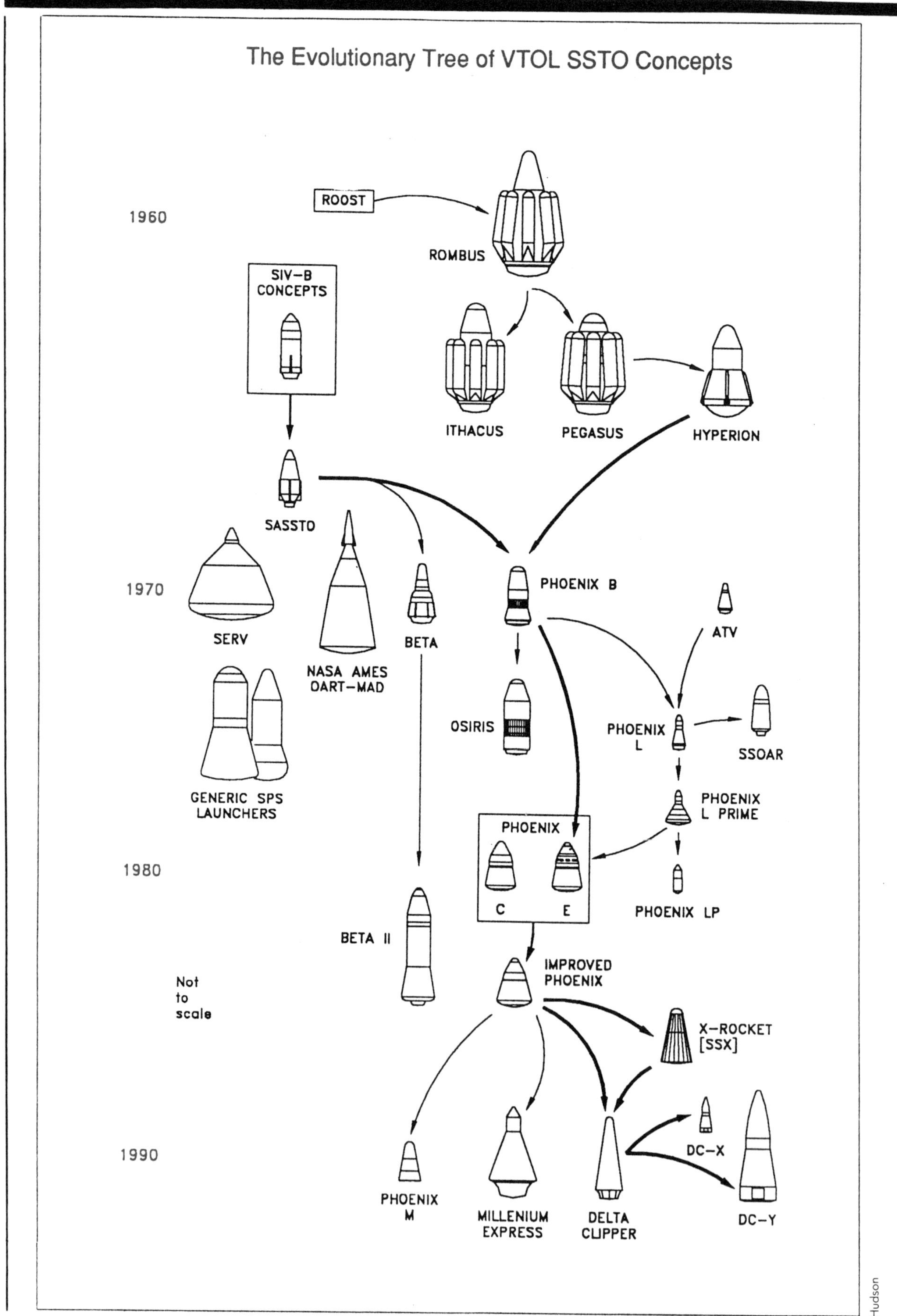
The Evolutionary Tree of VTOL SSTO Concepts
1960
ROOST
ROMBUS
SIV-B CONCEPTS
ITHACUS
PEGASUS
HYPERION
SASSTO
1970
SERV
NASA AMES OART-MAD
BETA
PHOENIX B
ATV
GENERIC SPS LAUNCHERS
OSIRIS
PHOENIX L
SSOAR
PHOENIX L PRIME
PHOENIX
C
E
PHOENIX LP
1980
BETA II
IMPROVED PHOENIX
Not to scale
X-ROCKET [SSX]
DC-X
1990
PHOENIX M
MILLENIUM EXPRESS
DELTA CLIPPER
DC-Y
Hudson

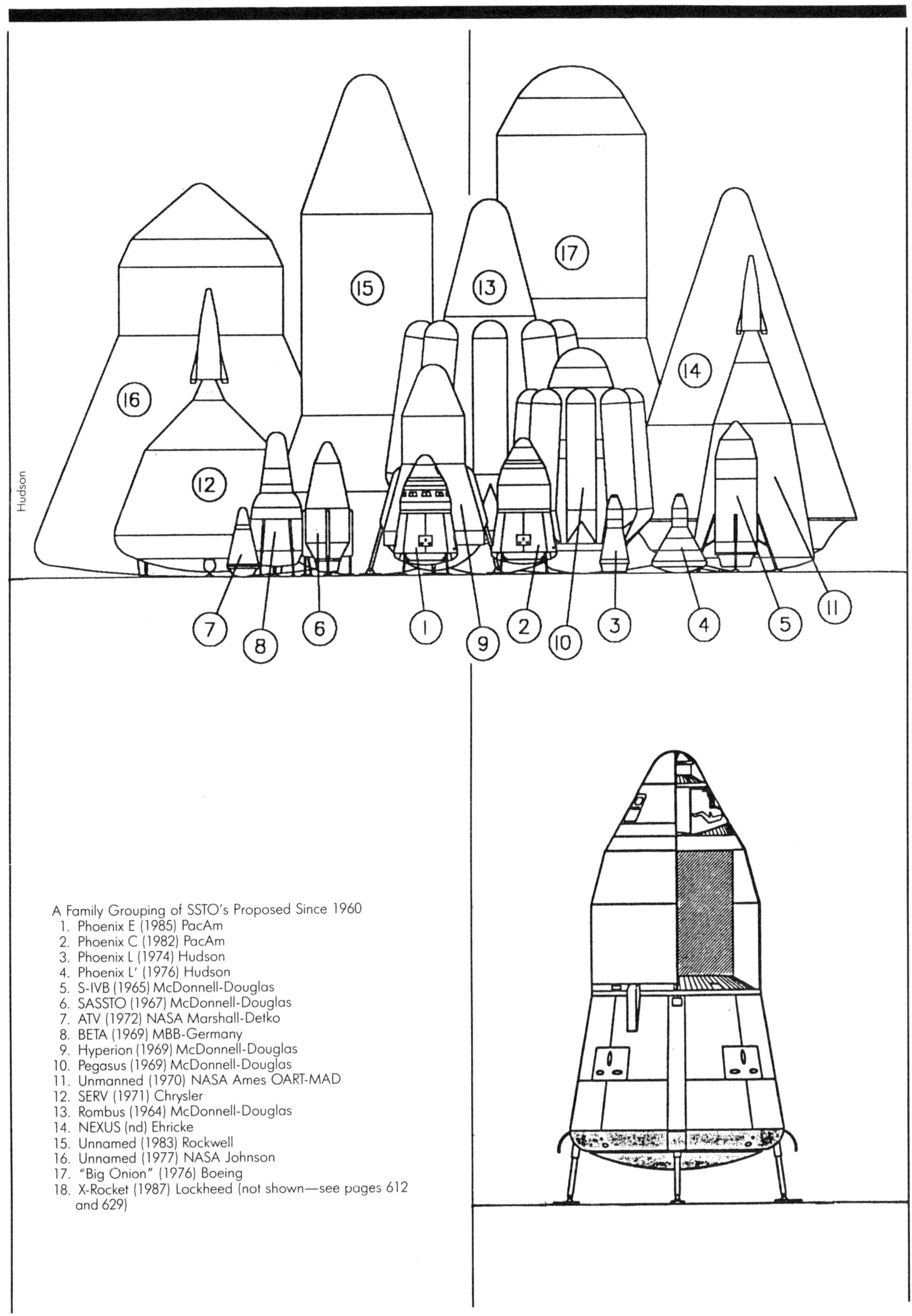

A Family Grouping of SSTO's Proposed Since 1960

1. Phoenix E (1985) PacAm
2. Phoenix C (1982) PacAm
3. Phoenix L (1974) Hudson
4. Phoenix L' (1976) Hudson
5. S-IVB (1965) McDonnell-Douglas
6. SASSTO (1967) McDonnell-Douglas
7. ATV (1972) NASA Marshall-Detko
8. BETA (1969) MBB-Germany
9. Hyperion (1969) McDonnell-Douglas
10. Pegasus (1969) McDonnell-Douglas
11. Unmanned (1970) NASA Ames OART-MAD
12. SERV (1971) Chrysler
13. Rombus (1964) McDonnell-Douglas
14. NEXUS (nd) Ehricke
15. Unnamed (1983) Rockwell
16. Unnamed (1977) NASA Johnson
17. "Big Onion" (1976) Boeing
18. X-Rocket (1987) Lockheed (not shown—see pages 612 and 629)

1984

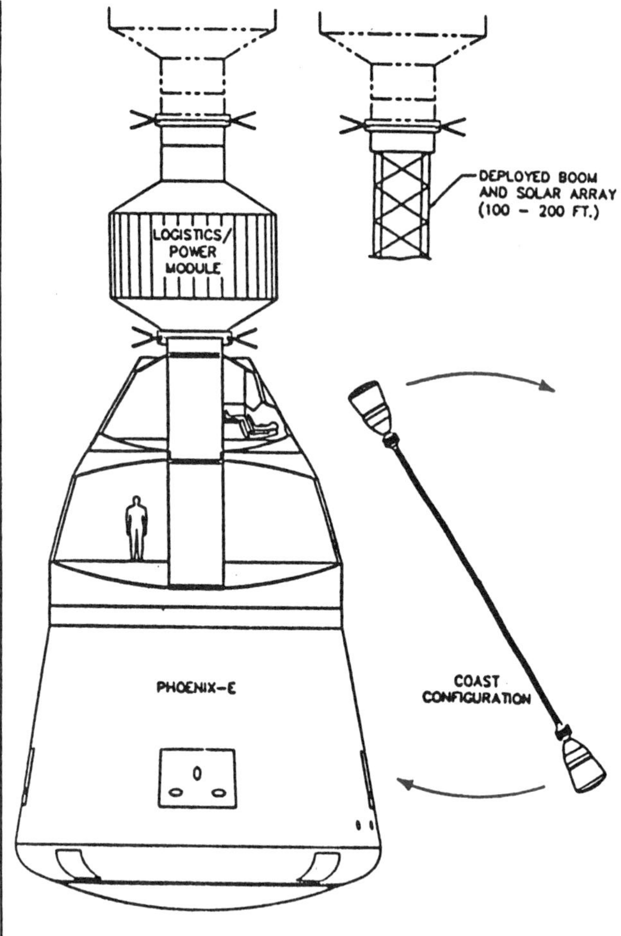

Phoenix-E in Mars mission configuration

with Boeing, General Dynamics, Lockheed, and Rockwell. McDonnell Douglas independently submits its own TAV design.

McDonnell-Douglas TAV

The Phase I study, which ends in December, results in fourteen different TAV concepts. Phase II is initiated in August 1984.

1984

Historian Harry A. Butowsky of NASA and astronaut Wally Schirra survey 350 sites of potential historic interest and recommend twenty-five as possible National Historic Landmarks. In January 1986 NASA and the Department of Defense agree to designate twenty-two "Man in Space" landmarks.

The Soviet space shuttle *Buran* (Snowstorm) is rolled out for the first time [see 1989].

Avions Marcel Dassault-Breguet Aviations and Aerospatiale design the Star-H two-stage hypersonic spaceplane for CNES. It is an air-breathing first stage up to Mach 6 and an altitude of 100 to 120 kilometers. A cryogenic second stage continues on into orbit, carrying a 20-ton payload. The Star-H will eventually replace the Ariane 5.

In March the initial presentation for the HOTOL spaceplane is made to DTI and RAE Farnsborough and in September HOTOL is unveiled at the air show at Farnsborough [also see 1982 and 1986]. In December the spaceplane is presented to the European Space Agency (ESA).

Third Millenium, Inc. proposes its "Space Van" system, an alternative to the space shuttle. It is a "minishuttle" that could have a turnaround of only 7 days (as compared with the 35 days required for the shuttle).

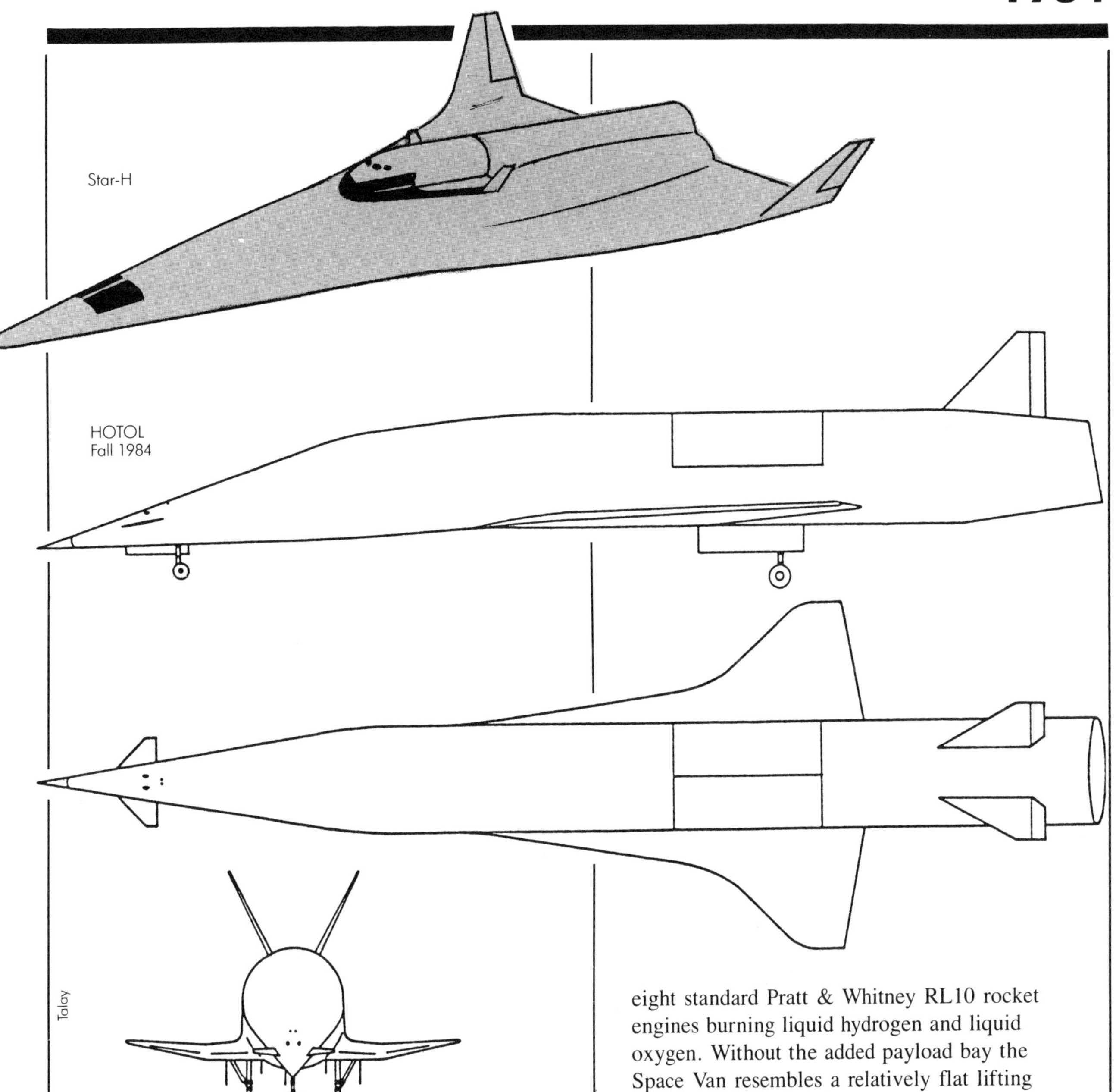

The complete proposed system includes: a specially modified Boeing 747 which acts as the Space Van's first stage; the Space Van orbiter; a reusable space stage; and an orbiting space depot. The otherwise lightly loaded 747 allows separation at the highest possible altitude.

An earlier, smaller, version of the orbiter has a wingspan of 42 feet and a length of about 66.5 feet. The nearly identical, final version is 70.54 feet long with a wingspan of 49.21 feet and has a gross weight of 141,120 pounds (64 metric tons). It is propelled by eight standard Pratt & Whitney RL10 rocket engines burning liquid hydrogen and liquid oxygen. Without the added payload bay the Space Van resembles a relatively flat lifting body. The bay forms a nearly cylindrical body sitting atop the main structure. Payloads up to 7.2 feet in diameter and 39.4 feet long can be accommodated.

In its orbiter mode, the Space Van can carry up to 1,764 pounds into space. Its full-lift capability, however, requires the use of the space stage. This consists of a single RL10 engine and its fuel, which will boost a 6,600-pound payload to a 280-mile orbit. The space stage can accommodate payloads up to 7.2 feet in diameter and 16.4 feet long. The space stage is contained within the long cylindrical fairing above the main fuselage. At the staging point, all eight main engines shut down and the aft section of the payload bay opens, allowing the space stage and its

1984

payload to roll out and separate from the orbiter. The space stages are easily assembled into orbital transfer vehicles by clustering seven of the propellant tanks with a pair of RL10 engines.

The Space Van launch site is to be located at Pago Pago, American Samoa, where the 747 carrying the Space Van lifts off and climbs toward the separation point at an altitude of 39,400 feet. There it pulls up into a gentle climb of 10 to 20 degrees at Mach 0.8. After separation, the 747 returns to its island base. The Space Van fires its eight engines and continues on into space. When carrying a space stage, it only needs to accelerate to a speed of Mach 15. After separation of the space stage, it immediately reenters the earth's atmosphere. When reentering after either orbiter or booster mode, it returns to the Pago Pago base. [Also see 1986.]

Early version

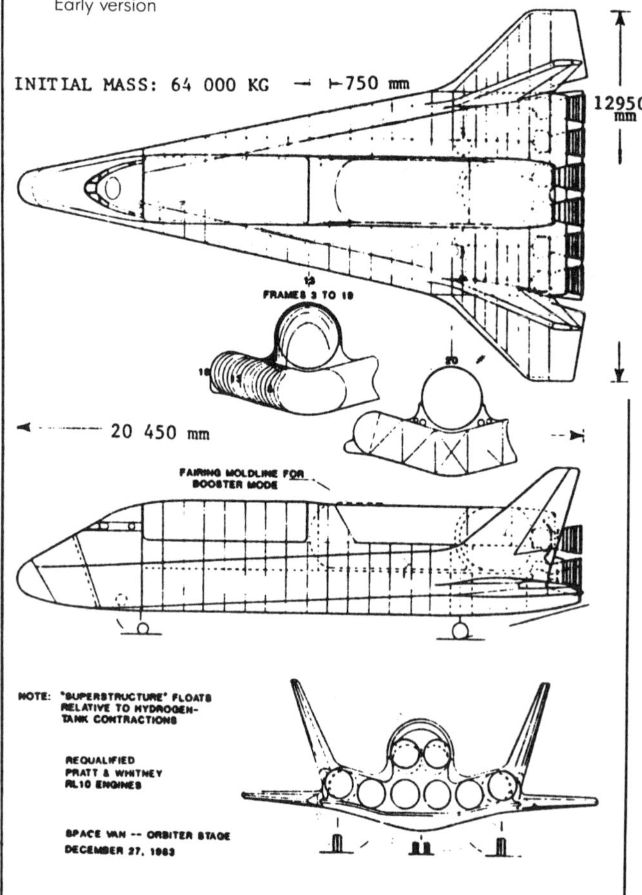

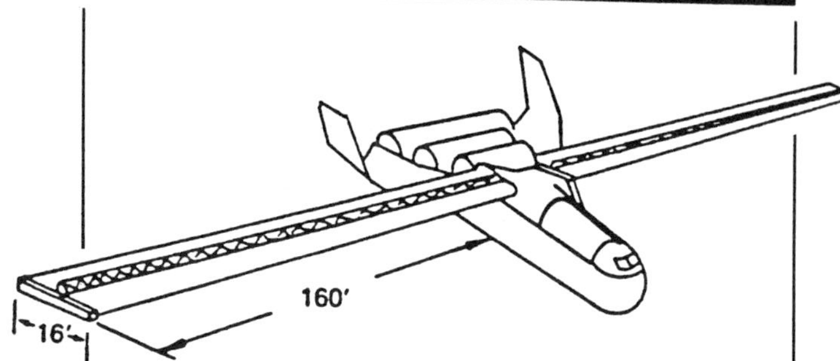

Gordon Woodcock and Timothy Vinopal, both of the Boeing Space Station Program, propose spacecraft for a manned Mars mission. It requires a fleet of seven virtually identical spacecraft: three lander/ascent vehicles, one unmanned lander carrying exploration equipment, and three earth return vehicles that will wait in orbit around Mars. Each of these is an aerodynamic spacecraft that resembles some of the alternate space shuttle designs, though larger (weighing about the same as a 747 airliner). Each has a single space shuttle main engine. All are boosted from the earth by a 88-foot by 26-foot earth departure stage using a single space shuttle main engine. En route, deployable solar cell arrays and fuel cells are used for power.

The mission will land eighteen people on the surface of Mars, where they will stay in the habitats that each lander carries. In an emergency, any two of the landers could carry all of the astronauts. For return to Mars orbit and the waiting return vehicles, the forward third of each manned lander constitutes an ascent stage that uses the main body of the lander as its launch pad.

Steven Welch and the Mission Strategy/Vehicle Design Workshop propose a manned Mars mission at the Case for Mars II conference. In this scenario, the amount of mass leaving the earth is minimized by having a powered flyby of Mars made by an Interplanetary Habitat vehicle. Aerobraking Mars Crew Shuttles, attached to the IHV, make the descent to the planet. Each flyby of Mars delivers fifteen astronauts in three Mars shuttles. It also picks up the fifteen astronauts delivered by the earlier IHV and returns them to earth. In the earlier parts of the program fewer astronauts are picked up (nine) than delivered in order to quickly build up a large base population.

Space Van

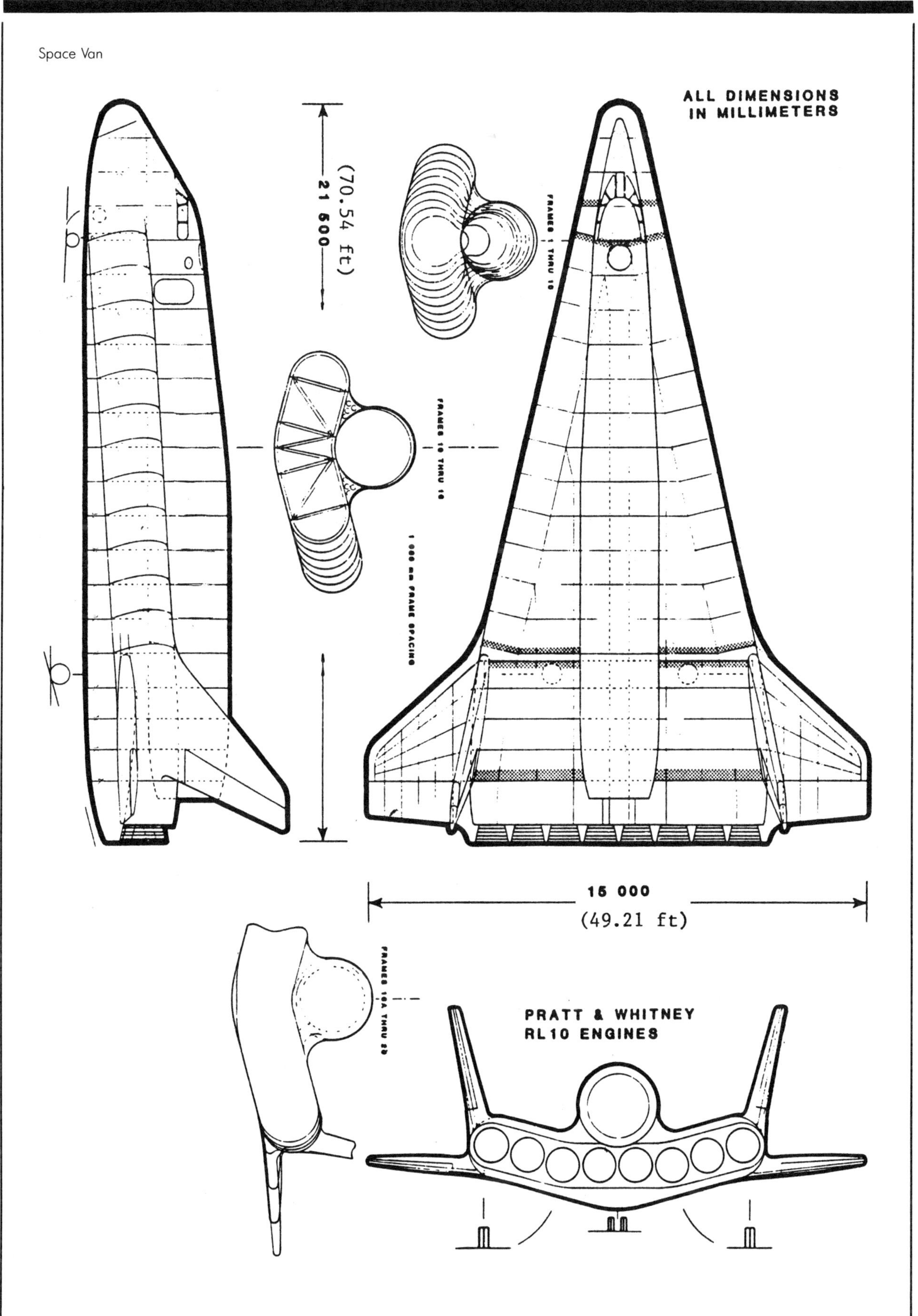

ALL DIMENSIONS IN MILLIMETERS
21 500
(70.54 ft)
FRAMES 1 THRU 10
FRAMES 10 THRU 18
1 000 mm FRAME SPACING
15 000
(49.21 ft)
PRATT & WHITNEY RL10 ENGINES

1984

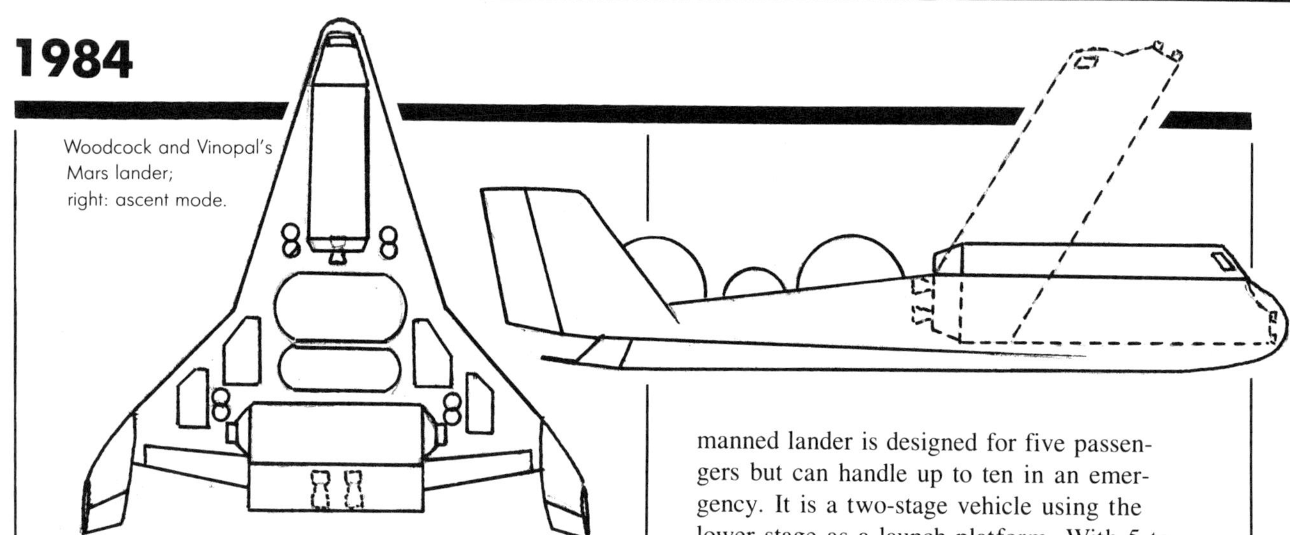
Woodcock and Vinopal's Mars lander; right: ascent mode.

The spacecraft is made up of these components: a Trans Mars Injection Stage (TMIS); the Interplanetary Spacecraft (IS), attached to the TMIS, one-third of the IH; the Interplanetary Habitat (IH), three IS's joined together; the Mars Crew Shuttle, an aerocapture vehicle carrying a crew of five to ten; and a Cargo Lander (a variation of the crew shuttle), which can deliver 24 tons of cargo to the surface of Mars.

The IS is composed of two space station modules. Each IS carries five to ten crewmembers. Three IS's are joined in a triangular or Y-shape at a central hub and docking port to make up one IH. The 2-meter-square box-beam connector contains a tunnel allowing shirt-sleeve access between the three arms. The complete IH rotates to provide 1/3 g.

The aerocapture vehicle is a 23-meter by 5.75-meter biconic shape. It descends by aerocapture, parachute deceleration and rockets for final touchdown. The manned version uses five first-stage engines, while the one-way cargo lander has only two engines. The manned lander is designed for five passengers but can handle up to ten in an emergency. It is a two-stage vehicle using the lower stage as a launch platform. With 5 to 6 tons of cargo, the 30-ton lander's capacity is about 8 tons. The use of methane-O_2 propellant allows the manufacture of fuel on Mars.

JULY 16

The National Commission on Space is approved by Congress under Public Law 98–361 to assist the United States in developing a plan which, "(1) defines the long-range needs of the Nation that may be fulfilled through the peaceful uses of outer space; (2) maintains the Nation's preeminence in space science, technology, and applications; (3) promotes the peaceful exploration and utilization of the space environment; and (4) articulates goals and develops options for the future direction of the Nation's civilian space program."

On March 29, 1985, President Ronald Reagan publicly announces the appointment of fourteen members to the commission: Dr. Thomas O. Paine (chairman), Dr. Laurel L. Wilkening (vice chairman), Dr. Luis W. Alvarez, Neil Armstrong, Dr. Paul J. Coleman, Dr. George B. Field, Lt. Gen. William W. Fitch, Dr. Charles M. Herzfeld, Dr. Jack L.

NASA Mars mission, 1983

Lander

MM (crew=4)

Nuclear engine

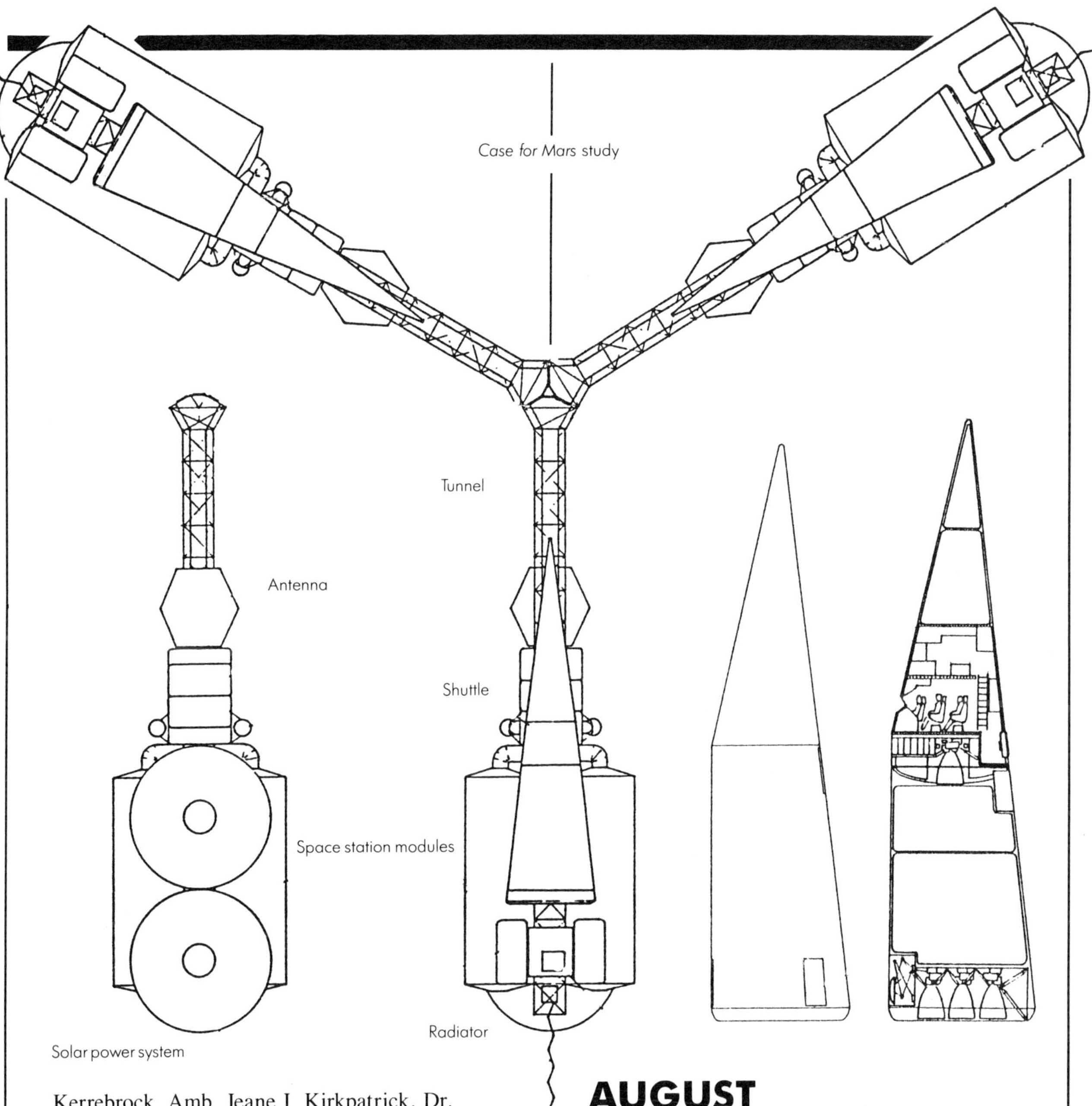

Kerrebrock, Amb. Jeane J. Kirkpatrick, Dr. Gerard K. O'Neill, Dr. Kathryn D. Sullivan, Dr. David C. Webb, and Brig. Gen. Charles E. Yeager. A fifteenth member, Gen. Bernard A. Schriever, is added later. Joining the fifteen voting members are four Congressional advisors: Representatives Don Fuqua and Manuel Lujan, Jr., and Senators Slade Gordon and John Glenn, Jr.

The goal of the commission is to submit a report to the president and congress on April 11, 1986, identifying long-range goals, opportunities and policy options for the U.S. civilian space program to the year 2035. [Also see 1986.]

AUGUST

Phase II of the Battelle TAV study begins [see May 1983]. This emphasizes the TAV's potential military effectiveness. The U.S. Air Force Space Command is hoping for a single-stage-to-orbit configuration, though it will settle for a two-stage system.

The TAV as now envisioned is a relatively large vehicle with a takeoff weight on the order of 1 to 1.5 million pounds. It has a 1,000 to 2,000 ft^3 payload bay capable of carrying 10,000 to 20,000 pounds of cargo. Its propulsion system is capable of ten to twenty flights between major overhauls (the first TAVs will use uprated space shuttle main engines until higher technology engines are available). Space Command hopes

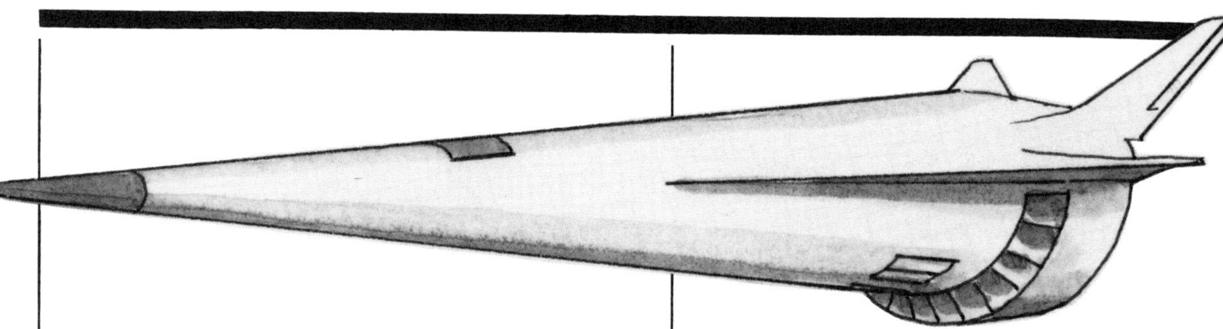

that the TAV will be capable of flights every other day, if need be.

The TAV, as such, will be gone by early 1989, replaced by the National Aero-Space Plane (the name National Aero-Space Plane will replace earlier designations as of December 1, 1985).

1985

A 75-minute simulated space voyage is offered by Rediffusion Simulation, Ltd., a British manufacturer of flight simulators. The space simulator, the first of which has been installed at Toronto's CN tower, consists of a forty-seat module mounted on telescoping, hydraulic legs that allow it to roll and pitch as though in actual flight. Realistic scenes of planets and spacecraft are projected onto a large screen.

NASA and the U.S. Department of Defense approach Congress with the suggestion that a wholly reusable spaceplane be developed as an alternative and replacement for the space shuttle.

The spaceplane is considered a viable alternative to the space shuttle. It is a horizontal takeoff and landing aircraft capable of attaining earth orbit in a single stage. It can take off and land from most large civil or military airports. Attaining a speed of Mach 25 (about 18,000 mph) it can climb to orbit in less than half an hour. It can ascend to the fringe of space, either continuing on into orbit to perform tasks now undertaken by the shuttle, or arcing back down to land half the planet away. The spaceplane can make the trip from Dulles to Tokyo in 2 hours and 18 minutes, hence its nickname: the Orient Express.

As initially conceived, the government "baseline" model, the spaceplane is a low profile, wedge-shaped wing-body rocket 150 feet long—28 feet longer than the space shuttle—and constructed with a skin of titanium alloy and carbon. It is originally expected that two prototypes can be built for $3 billion—about the cost of one new shuttle.

The spaceplane requires three different propulsion systems, one each for takeoff, supersonic flight and insertion into orbit. In the first leg of its journey into space, the spaceplane takes off under the power of turbojet engines. Once these have accelerated the spaceplane to about Mach 3 they convert into ramjets. At slower speeds, the turbojet engine uses the power of its fan-like compressors to force air through it. At extremely high speeds, the air is naturally compressed as it enters the engine—there is no longer any need for the compressor or turbine. Yet even the ramjet has an upper speed limit: about Mach 5 or 6. Beyond this speed, the spaceplane needs to switch to another set of engines: the supersonic combustion ramjets (or scramjets). Liquid hydrogen replaces the kerosene-like jet fuel previously used. Still, the high-powered scramjet engines will not be able to take the spaceplane on into orbit because they are air-breathing. Some sort of rocket is required for this final step. A number of rocket engines exist that the spaceplane can employ to boost it past Mach 20.

The spaceplane itself is a flat, needle-nosed delta with either a single rudder or twin rudders at the wingtips. Research on the spaceplane is being undertaken by the Department of Defense and NASA, under the direction of the Defense Advanced Research Projects Agency. The project has been named the National Aero-Space Plane (NASP) and an experimental aircraft has been designated to test the concept—the X-30A.

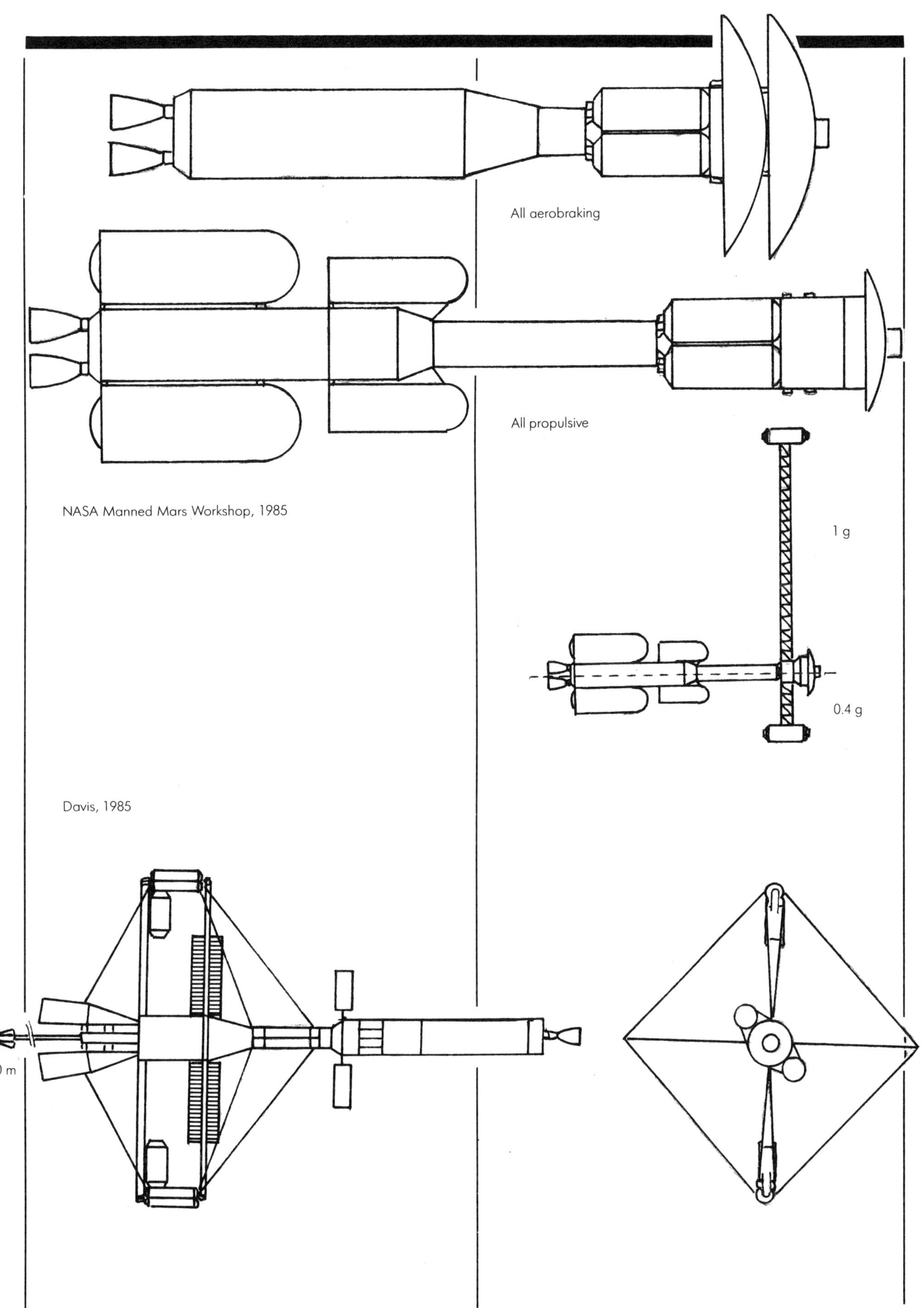
All aerobraking
All propulsive
NASA Manned Mars Workshop, 1985
1 g
0.4 g
Davis, 1985
200 m

1985

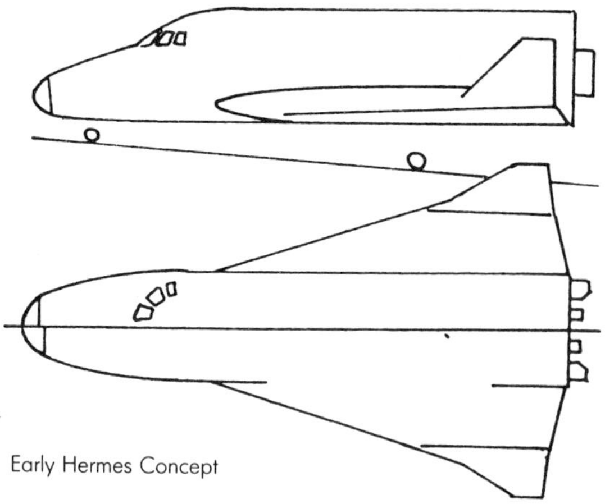

Early Hermes Concept

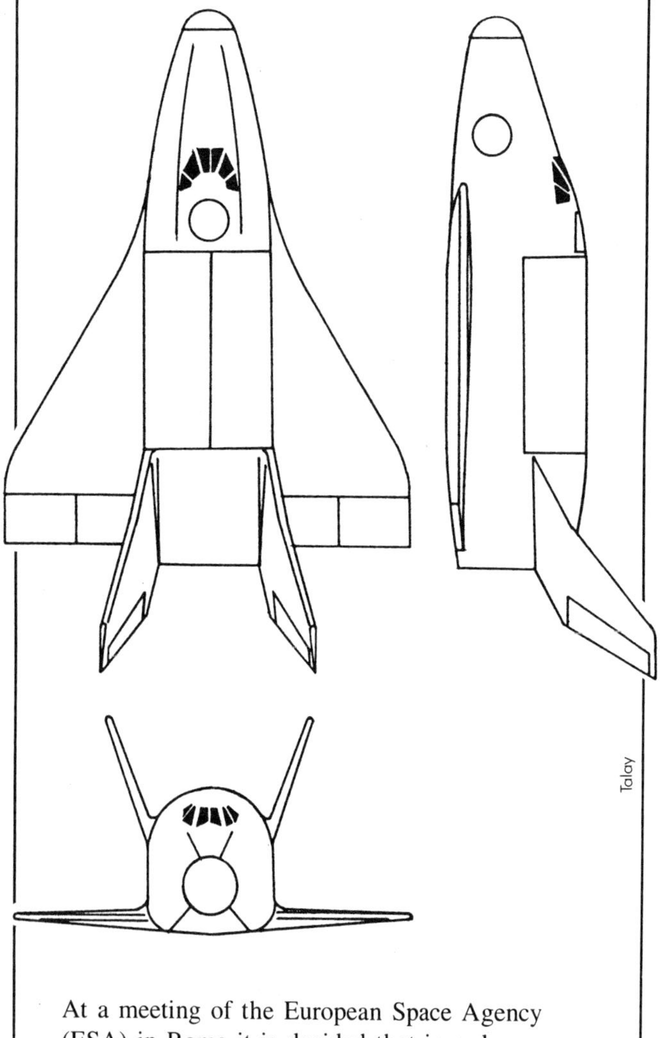

Talay

At a meeting of the European Space Agency (ESA) in Rome it is decided that in order for Europe to acquire autonomy in manned spaceflight it should have the capability to transport personnel into space and return them to the earth. In The Hague, November 1987, the *Hermes* spaceplane, now considerably upgraded since its 1979 debut [see above], is endorsed.

Basic specifications set by the Centre National d'Etudes Spatiales (CNES), the French space agency, now calls for a delta-winged, reusable spaceplane, 15 to 18 meters long with a wingspan of 9 to 11 meters. The weight is limited to 13 to 17 tons (imposed by the choice of the Ariane 5 as the launch vehicle), including 2.5 tons of propellant. A payload of 4 to 5 tons, as well as a crew of up to six astronauts (two crew-members and four passengers), is carried in a cargo bay measuring about 3 m wide with a volume of 35 m^3. A 7-meter remote arm allows payload manipulation from within the cockpit. *Hermes* will be ready for its first flight in 1995 or 1996.

A typical mission will take the craft to a circular orbit 400 km above the earth. After up to a month in orbit (or 90 days if attached to a space station by its airlock) Hermes will reenter the atmosphere at about Mach 25, finally gliding to a landing at

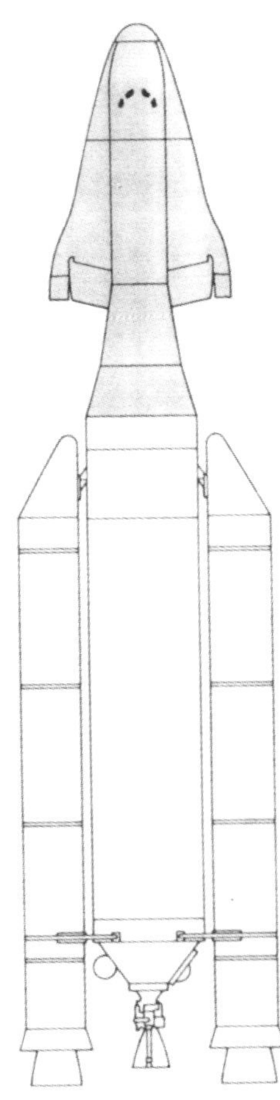

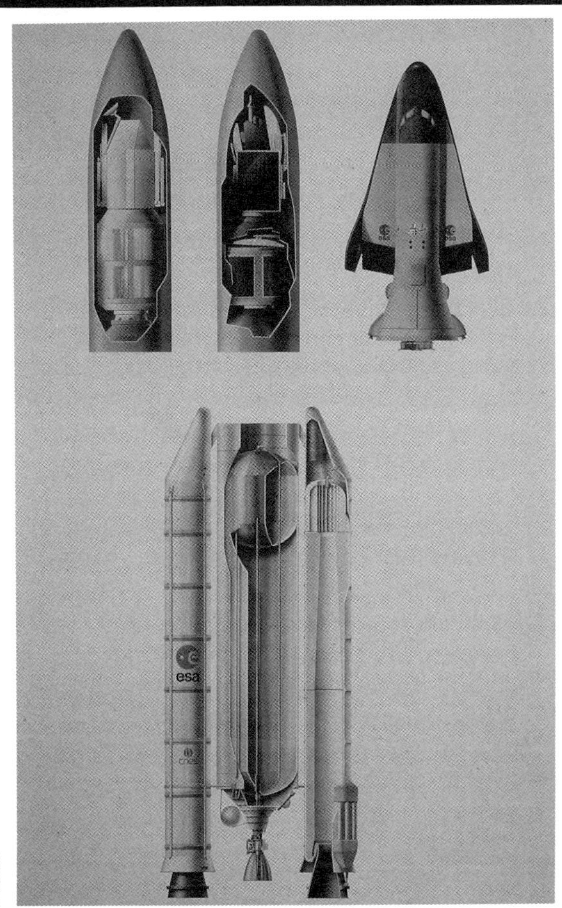
Ducros

Kourou, French Guiana, or at Istres, France (near Marseilles). Hermes is expected to make up to six flights a year.

Two proposals are offered, one from Aerospatiale and another from Dassault-Breguet. The former proposes a delta-winged minishuttle 15.5 m long with small vertical stabilizers on the wingtips. The wingspan is 11 m. Another larger vertical stabilizer is located behind the payload bay door. This design is basically a refinement of their original 1979 proposal. The pilot's seats are placed on the forward floor section of the cockpit and can be raised for maximum visibility during landing; seats for the passengers are located behind the pilots. All of the seats can be folded flat for sleeping. The cockpit volume is about 26 m^3.

The Dassault-Breguet design also incorporates delta wings with modified winglets as stabilizers. It is slimmer than the Aerospatiale design and has no large central vertical stabilizer. It has ejection seats for crew safety.

The crewmembers would dine in fine French fashion, an abbreviated version of their menu looking something like this:

Entrées: Pâté de poivre vert, Crème de crabe, Boeuf assaisonné

Plats cuisinés: Sauté Veau Marengo (lyophilisé), Boeuf bourguigon, Fondue de queue de boeuf à la tomate confite et aux cornichons, Compote de Pigeon aux dattes, Canard à la cuillère et aux artichaux

Fromages: Fromage de gruyère, Fromage de cantal

Desserts: Crème au chocolat, Gâteau de riz, Palets de pâtes de fruit

Pain: Pain de seigle, Pain de Mie, Boule de Pain

Another variation of the *Hermes* bears a resemblance to the MSC space shuttle proposals of February-March 1977, with delta wings and twin V-tailfins.

Near the end of this year CNES, the French Space Agency, appoints Aerospatiale and Dassault-Breguet as the prime contractors for the development of the manned mini-shuttle, *Hermes*. *Hermes* is to be atop the yet-to-be-developed Ariane 5 heavy lift launch vehicle. It is a service vehicle for space stations, transporting cargo and crew. The cargo bay has a capacity of 35 m^3. After separation from the Ariane 5 booster, propulsion is provided by two 20-kN engines.

It is hoped that the first manned, orbital flight will take place just over 11 years from concept definition, nearly a full year earlier than its rival, the British HOTOL. Although France is anxious for ESA to adopt the project, it is given only a lukewarm reaction at a January 1985 meeting of ESA's Ministerial Council.

By 1989, the *Hermes* design has been severely trimmed. The payload bay has been deleted, as well as the payload bay doors, which would have carried radiators. The payload capacity has been reduced to just 3 tons. Its reentry weight is set at only 15 tons which dictates that it can carry only 1.5 tons back from orbit. Even at that, it must jettison its propulsion system, airlock and resource module. The original 17-ton vehicle grows heavier and heavier. The aftermath of the Challenger disaster [see 1986] has meant that an ejectable crew cabin has to be added to the design—an additional 1.5 tons. At one point the *Hermes* launch weight reaches 24.4 tons. These weight increases mean upgrading the Ariane 5, consequently making it less useful for commercial satellite launches. The increased weight also requires a larger wing area for the *Hermes*, which is not possible. Therefore a 15-ton limit is set for its weight.

In his book *Mars One Crew Manual*, Kerry Mark Joëls describes an international manned Mars mission. A crew of eleven uses hardware provided by the United States, the Soviet Union, and ESA. Two vehicles are assembled in earth orbit: the manned mother ship and an unmanned Mars Rover carrier.

The mother ship is boosted from earth orbit by a pair of Space Shuttle Uprated Engines (SSUE) fueled by a cluster of four tanks containing liquid oxygen and liquid hydrogen. The rover carrier is boosted by a single SSUE. The mother ship additionally has a return booster consisting of a cluster of eight tanks and one SSUE.

At the forward end of the return booster is an access tunnel surrounded by two clusters of four storage tanks, for life support consumables. At right angles to the tunnel are four modules: habitat, laboratory and two for storage. On the outboard end of each module is an octagonal lattice pallet holding science packages, probes, comsats, consumables tanks and two lenticular aeroshells. The latter, after entering the Martian atmosphere, release an unfolding unmanned airplane probe powered by a catalytic hydrazine engine.

At the forward end of the mother ship is mounted a Mars Excursion Module (MEM). It is 9.1 m in diameter at its base and 7.2 m high. A 2.6 m retropack is fitted to its heat shield and is used to brake from orbit and enter the Martian atmosphere. After entry, the MEM jettisons its aeroshields and descends via a triad of parachutes. At a relatively low altitude, the parachutes are cut free and the final descent is made using the descent stage engine, the MEM finally landing on six extendable legs.

At the same time the 10.65 m lifting body rover carrier sheds its booster, retrofires, deorbits, enters the atmosphere and releases the rover in midair to make its own parachute-and-rocket-braked landing. The four-man MEM crew will later rendezvous with it to begin their 25-day exploration of Mars. Meanwhile the seven remaining crew members still in orbit are performing experiments and photoreconnaisance. At the end of the 25 days, the crew of the MEM will fire its ascent stage, separating from the spent descent stage, which will now serve as a launch pad. As the ascent stage's strap-on propellant tanks empty, they will be jettisoned. The MEM will eventually rendezvous with the waiting mother ship, and the crew will transship with their samples. The expedition will return to earth 1 year and 9 months after departure.

The motion picture *Lifeforce* features second-generation space shuttles. One of them, the *Churchill*, takes part in a manned mission to Halley's Comet. It is equipped with a NERVA nuclear engine mounted at an angle beneath the fuselage and an extendable solar panel array (stored folded in the payload bay when not in use).

The Soviet Union appears to be actively developing a winged, reusable replacement for their aging Soyuz and Progress vehicles. A subscale (approximately 1:3) model of a lifting body spaceplane has been test flown four times since 1982. The models, which weigh about 2,200 pounds, are launched by a C-1 booster. A new booster capable of lifting over 30,000 pounds into orbit would be able to launch the full-size spaceplane.

Engineer Theodore Talay has estimated that the full scale vehicle would be about 48 feet long with a 39-foot wingspan, carrying a crew of three to five cosmonauts.

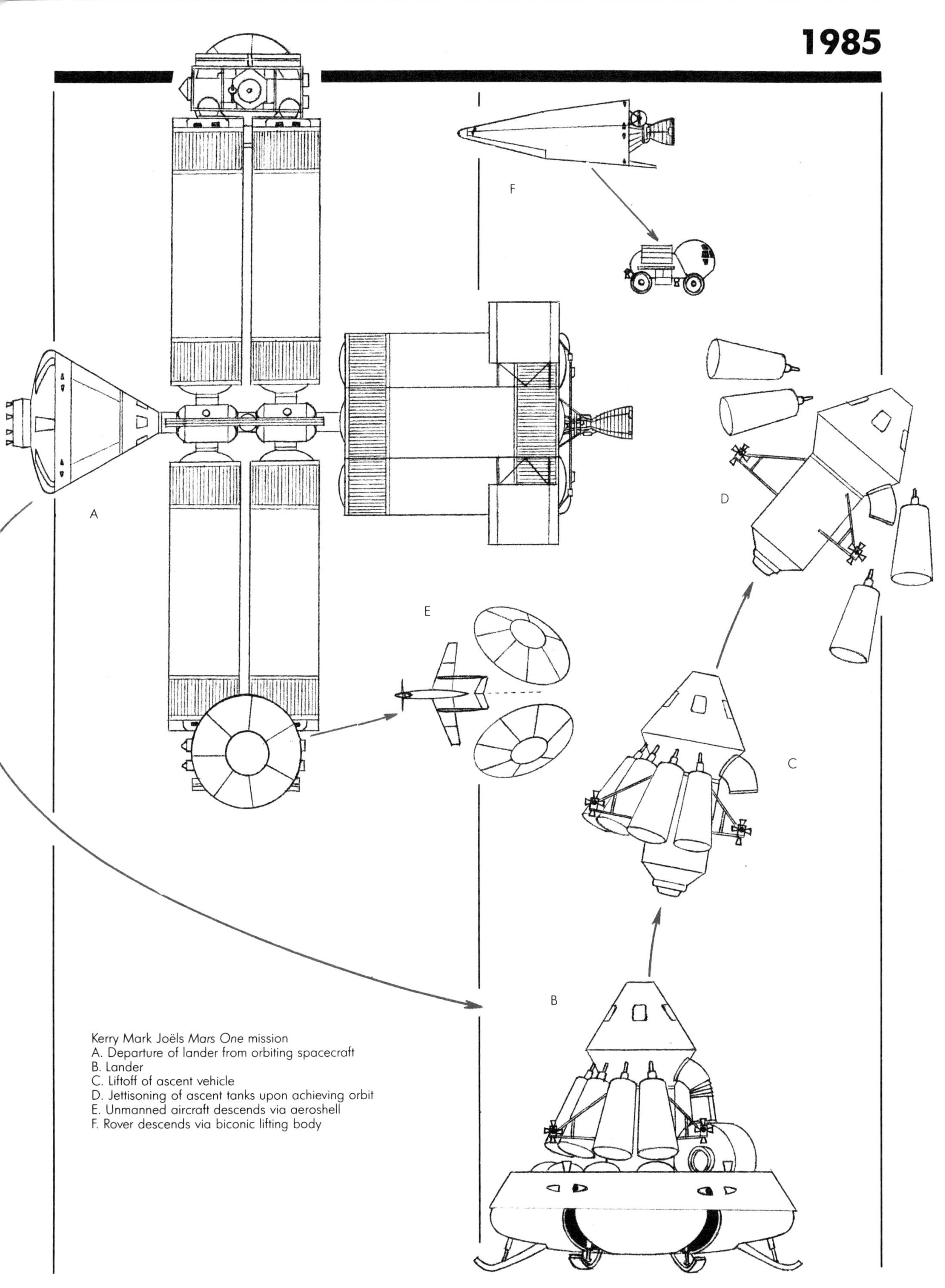

Kerry Mark Joëls *Mars One* mission
A. Departure of lander from orbiting spacecraft
B. Lander
C. Liftoff of ascent vehicle
D. Jettisoning of ascent tanks upon achieving orbit
E. Unmanned aircraft descends via aeroshell
F. Rover descends via biconic lifting body

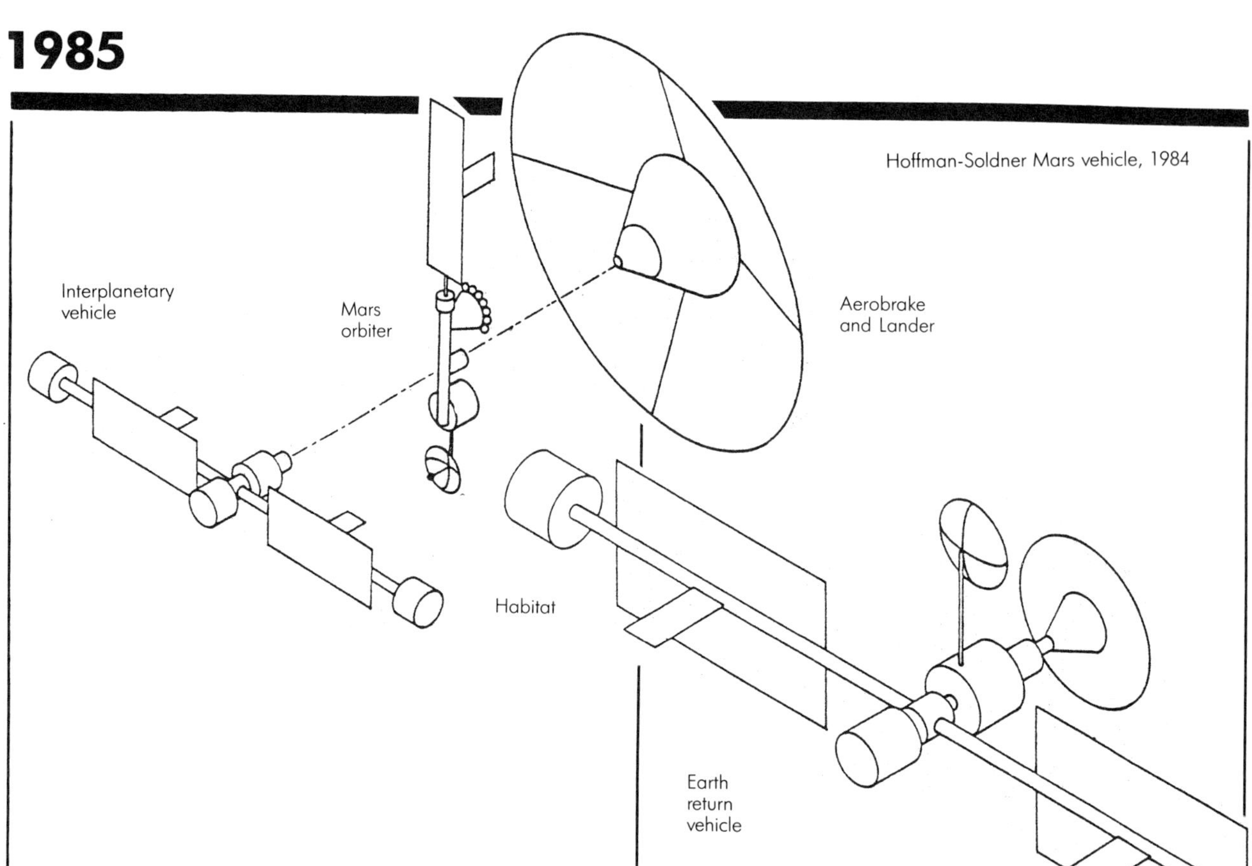

Hoffman-Soldner Mars vehicle, 1984

FEBRUARY

The U.S. Department of Transportation approves a proposal for the first three launches by the private sector—the result of agreement between a group of underwriters and D. K. Slayton's Space Services, Inc., of Houston. The plan is to launch at least 15,000 corpses into orbit beginning in 1987. [Also see Neil R. Jones, 1931.]

SPRING

Students at the University of Texas at Austin, sponsored by the Universities Space Research Association, design a manned Mars mission.

In 1985 the class designs a single stage ascent/descent vehicle. It would carry 0.5 metric ton of payload in addition to a rover and four astronauts. The biconic vehicle also serves as a habitat after landing. The spacecraft is refueled using propellants produced from native Martian materials. Equipment and supplies for doing this are landed earlier in an unmanned cargo ship.

The manned descent/ascent vehicle designed in 1986 is a two-stage 165.6 metric ton vehicle. This carries the ascent vehicle, the permanent habitat, 0.5 metric ton of other payload and four astronauts to the surface of Mars. It is a flattened Apollo-style ballistic lifting body 36 meters in diameter and 6.4 meters tall. The 2.8 meter-diamter ascent vehicle is located in the center.

Also designed is a vehicle for the exploration of the Martian moons. It is an "open truss flatbed bus" massing 9.8 metric tons.

APRIL

Society Expeditions announces "Project Space Voyage," which will offer tourists up to eight orbits of the earth for $50,000. Flights will begin in 1992, using a newly developed reusable spacecraft. In the interim period, Society Expeditions president T. C. Swartz proposes an arrangement with NASA that would allow the space shuttle to carry a tourist module in its payload bay. This will hold twenty-two passengers paying $1 million per seat. These trips will last 3 days (forty-eight orbits) and there will be three to five flights per year. Eventually up to 300 passengers per year will be flown by the mid-1990s. However, it is planned that Society Expeditions will ultimately develop its own private spacecraft.

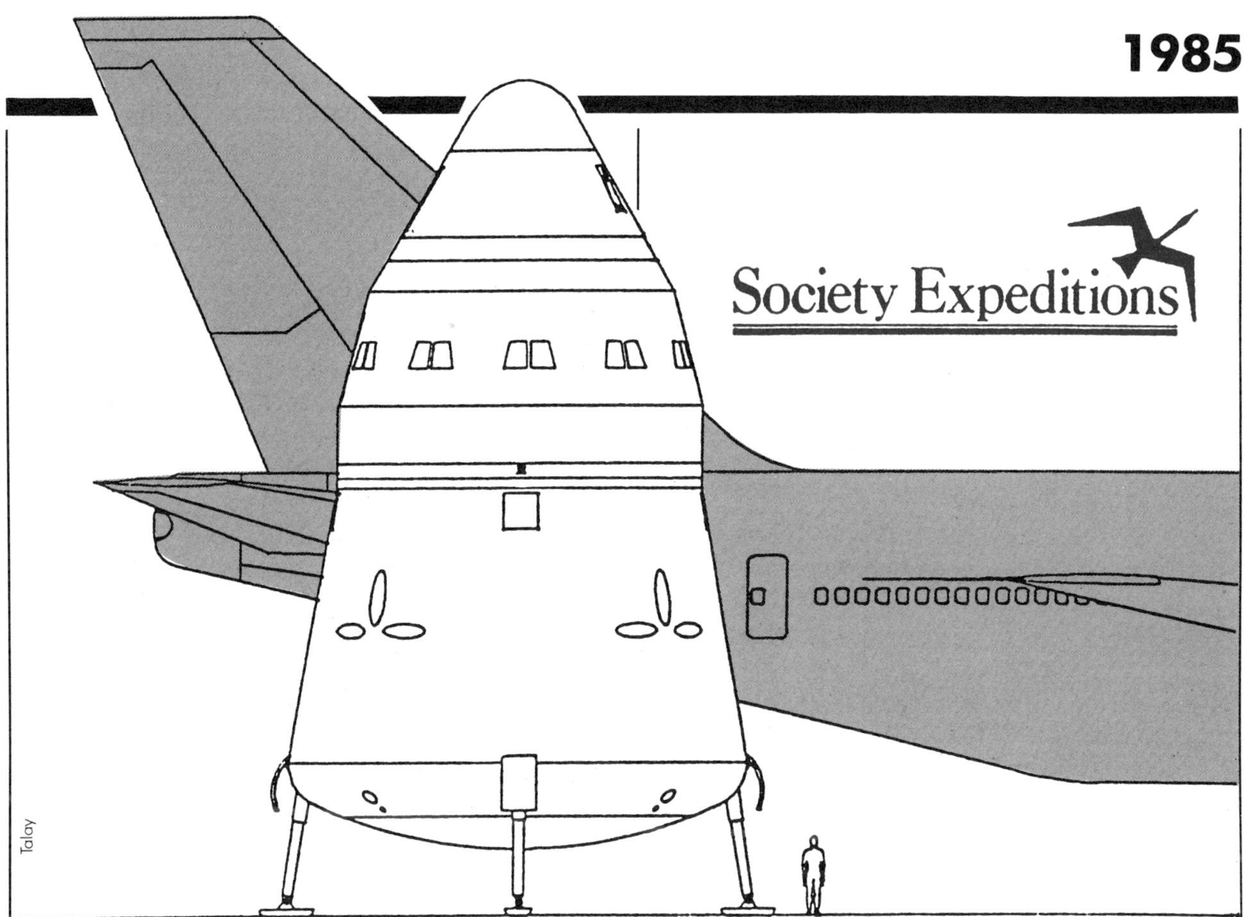

Plans for designing a suitable tourist module are halted by NASA's announcement that not only are the shuttles not to carry tourist passengers, but that its launch schedule is already filled by commercial and scientific missions.

Meanwhile, Society Expeditions has joined with Pacific American Launch Services (later Pacific American Spaceship Company) of Redwood, California, and announces that it will be able to offer flights no later than 1991 on a single-stage-to-orbit rocket of its own design, the *Phoenix*. Ticket prices are $50,000 per seat.

The *Phoenix* spacecraft has been under development by engineer Gary Hudson since 1982 as a low-cost alternative to the shuttle. A true spaceship that can be refueled in orbit for orbit-to-orbit capability or for trips to the moon and beyond, it is a squat, cone-shaped vehicle 57 feet tall and 31 feet wide at its base. It somewhat resembles the designs of Philip Bono [see 1967], upon which it is based, though on a much smaller scale. Propelled by a liquid oxygen and liquid hydrogen-fueled aerospike engine with forty-eight combustors, it can be launched by five people in less than 2 hours from hangar to orbit. Cost per pound in low earth orbit can be as low as $100, which could be reduced to $20 if there are a large number of flights. There are four versions of the basic design: an all-cargo, unmanned vehicle; a cargo rocket with a crew of two to four; a tanker; and the twenty-passenger, two-crew Society Expeditions version, the *Phoenix E*. Common to all versions is a main stage that comprises the lower half of the spacecraft, 25 feet in diameter at its top and 32 feet in diameter at the base. The bottom is a spherical segment with a 36-foot radius that acts as a heat shield during reentry. A single, circular aerospike engine comprised of forty-eight individual combustors surrounds the base. Maximum sea-level thrust is 640,000 pounds. Propellants are triple-point liquid oxygen and 50 percent slush liquid hydrogen. Available to be added to the main stage are a payload shroud, pilot module, propellant extension kit (allowing for an additional 250,000 pounds of propellant), and an excursion module capable of carrying up to twenty passengers and two crewmembers.

The *Phoenix E* is to be 57 feet tall (10 feet shorter than the purely cargo versions). All of the *Phoenix* spacecraft land on four extendable legs. Twenty adjustable seats are arranged around the cabin's circumference, with a window for each passenger. In the center are seats for the two flight attendants, lavatories, etc. A central core airlock allows access to the pilot module above. Peak acceleration and deceleration are to be no more than 3 g. Each vehicle will have a life of 1,000 flights.

Society Expeditions begins accepting reservations for a $200 registration fee and a $5,000 deposit. The first five flights are immediately booked.

Hudson's *Phoenix* would also provide an inexpensive and reliable means of transporting cargo and human beings to and from lunar settlements (it could carry a 20,000-pound payload to low earth orbit or to the lunar surface after refueling). The *Phoenix* could place a twelve-person settlement on the moon in 2 years from the start of launches for an estimated median cost of $54 million (about $214 per pound to the lunar surface). Hudson concludes that by using the *Phoenix* a lunar settlement could be established and operated for a year for certainly under $1 billion and perhaps for as little as $100 million. Given an effective reusable SSTO vehicle such as the *Phoenix*, a lunar settlement could be privately built.

JUNE

Models of HOTOL [see below] and the Aerospatiale *Hermes* [see 1985] are unveiled at the Paris Air Show.

NOVEMBER

A proof of concept study agreement on the HOTOL spaceplane is reached between British Aerospace (BAe), Rolls-Royce, DTI and MOD. [Also see 1982, 1983, 1984 and 1991.]

HOTOL is expected to deliver up to 7,000 kg of payload to low earth orbit in its 9 m x 4.5 m cargo bay. The cylindrical fuselage is over 54 m long with a wingspan of 17 m. It is slightly larger than the Concorde and about twice as long as the American shuttle. HOTOL employs air-breathing engines at

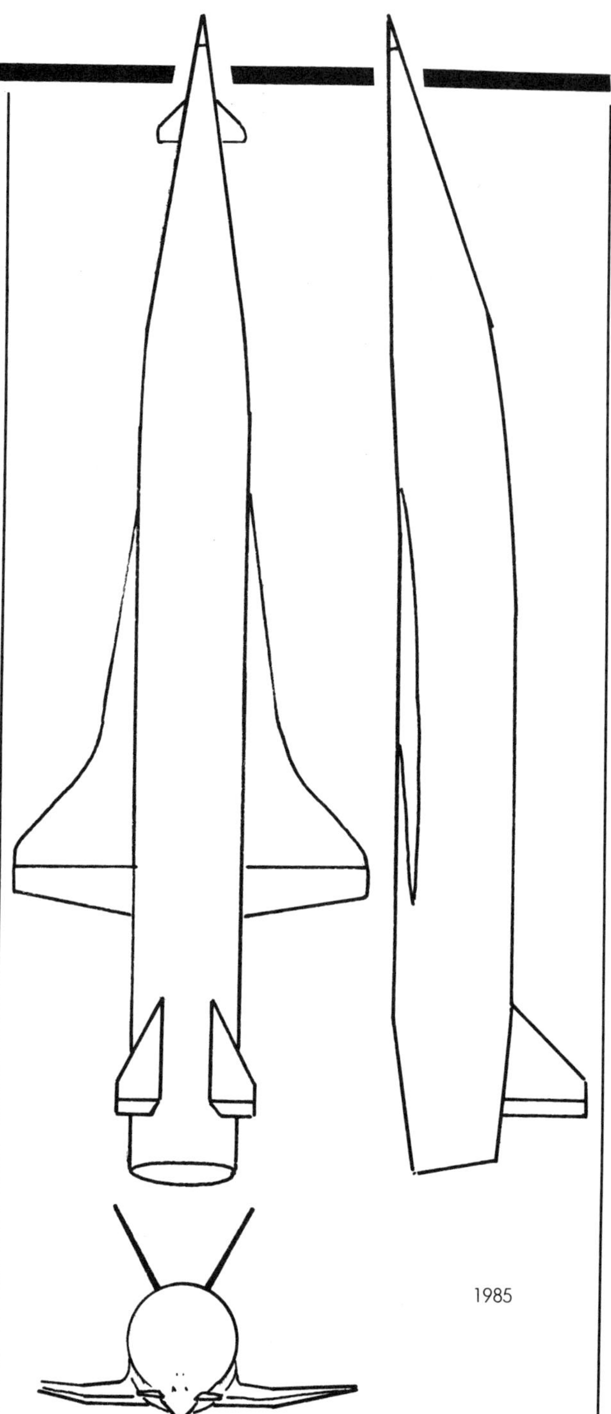

1985

lower altitudes, switching to a hydrogen/oxygen rocket. Most of the forward fuselage is occupied by a large pressurized liquid hydrogen tank; a smaller liquid oxygen tank is placed behind the payload bay. HOTOL, unlike many other spaceplane concepts, is unmanned. It is envisioned, however, that eventually one vehicle out of a fleet of five or six will be dedicated to manned flight, though the passengers will have nothing to do with takeoff or landing, which will still be automatic. Even the commercial passenger version, which is advertised as being ca-

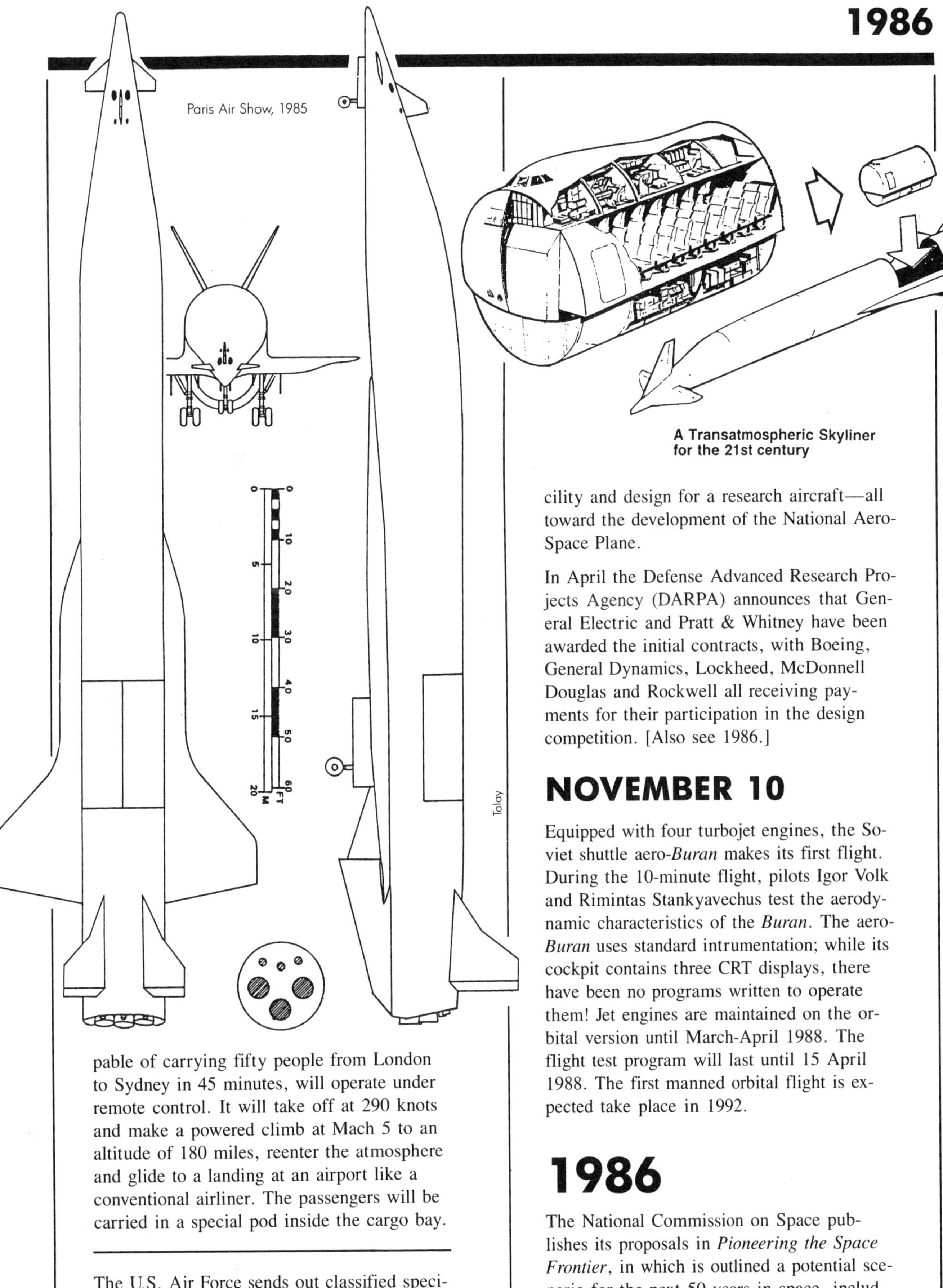

A Transatmospheric Skyliner for the 21st century

pable of carrying fifty people from London to Sydney in 45 minutes, will operate under remote control. It will take off at 290 knots and make a powered climb at Mach 5 to an altitude of 180 miles, reenter the atmosphere and glide to a landing at an airport like a conventional airliner. The passengers will be carried in a special pod inside the cargo bay.

The U.S. Air Force sends out classified specifications to some fifteen aerospace firms, requesting bids for work on engines, test facility and design for a research aircraft—all toward the development of the National Aero-Space Plane.

In April the Defense Advanced Research Projects Agency (DARPA) announces that General Electric and Pratt & Whitney have been awarded the initial contracts, with Boeing, General Dynamics, Lockheed, McDonnell Douglas and Rockwell all receiving payments for their participation in the design competition. [Also see 1986.]

NOVEMBER 10

Equipped with four turbojet engines, the Soviet shuttle aero-*Buran* makes its first flight. During the 10-minute flight, pilots Igor Volk and Rimintas Stankyavechus test the aerodynamic characteristics of the *Buran*. The aero-*Buran* uses standard intrumentation; while its cockpit contains three CRT displays, there have been no programs written to operate them! Jet engines are maintained on the orbital version until March-April 1988. The flight test program will last until 15 April 1988. The first manned orbital flight is expected take place in 1992.

1986

The National Commission on Space publishes its proposals in *Pioneering the Space Frontier*, in which is outlined a potential scenario for the next 50 years in space, including the establishment of a permanent base on Mars by the end of that period.

The first thing that the Commission believes is neccessary is the creation of a more reliable means than the space shuttle for transporting humans and cargo to and from orbit. It believes that costs could be reduced below $200 a pound by the year 2000. The commission emphasizes that "Above all, it is imperative that the United States maintain a continous capability to put both humans and cargo into orbit; never again should the country experience the hiatus we endured from 1975 to 1981, when we are unable to launch astronauts into space." One way of accomplishing this is separating the functions of one-way cargo transport from the round trip transport of people and high value cargo. This is accomplished by the development of heavy-lift shuttle-derived launch vehicles and the National Aero-Space Plane. The Commission hopes that the commercial sector is able to assume the design, development, fabrication and operation of space vehicles and launch and landing facilities.

In order to effectively and cheaply move large payloads beyond low earth orbit, a new class of vehicles needs to be developed. Piloted and unpiloted chemically propelled transfer vehicles are to be based on combinations of standardized units, all—except the lunar lander—employing aerobraking shells. These vehicles include cargo and passenger earth orbit vehicles, cargo and passenger Mars orbit vehicles, cargo and passenger earth orbit-moon orbit vehicles, Mars landers and lunar landers. The Commission has also studied both solar and nuclear-powered small unmanned cargo vehicles. Another addition to the National Commission on Space fleet is a large nuclear electric propulsion cargo vehicle capable of delivering the spaceports to Mars, needed for the mission described below.

The Mars base is established and maintained by the use of earth, lunar and Martian spaceports and cycling spacecraft. The cycling spaceships (dubbed "space castles" by the artists working on the report) have elongated elliptical orbits that enable them to shuttle back and forth between the orbits of earth and Mars. They contain all of the neccessities for an extended journey in space and are sufficiently spacious to provide comfortable quarters for passengers as well as room for research facilities, food production and recyling, and closed-ecology life support. Up to seventeen passengers could be accommodated. The cycling spaceship consists of eight or more standard space station modules, some of which could be attached to the spin axis to provide a microgravity environment for research. Propellants, nuclear power supply and the transfer vehicle hangar are also on the spin axis. The habitats are attached to the opposite ends of a pair of tethers, so that artificial gravity can be provided during the journey. Two cycling spacecraft are required to support the initial traffic to and from Mars.

A typical trip to Mars will require the following steps. The crewmembers first travel from the earth to the orbiting earth spaceport in a passenger transport vehicle (perhaps NASP, for example). There, they board another transfer vehicle that takes them to the lunar libration point spaceport where the transfer vehicle is refueled. Leaving the libration point spaceport, the transfer vehicle makes a rendezvous with one of the cycling spacecraft bound for Mars. The crew transfers to the cycling spacecraft. The transfer vehicle is hangared and refueled during the 5 to 7 month trip. As the cycling spacecraft passes by Mars, the crew reboards the transfer vehicle and leaves the cycling spaceship. After an aerobraking maneuver in the Martian atmosphere, the crew docks at the orbiting Mars spaceport. The spaceport has been delivered by a nuclear electric propulsion freighter and assembled in Mars orbit. It is a virtual twin of the lunar libration point spaceport. At the spaceport, the crew boards a Mars lander for the journey to the surface of Mars.

In his book *The Mars Project* Senator Spark M. Matsunaga advocates a joint manned mission to Mars (the Senator has introduced seven Congressional resolutions since 1982 that deal with international, and particularly U.S./Soviet, cooperation in space).

Right: cycling space station and detail of habitiation module

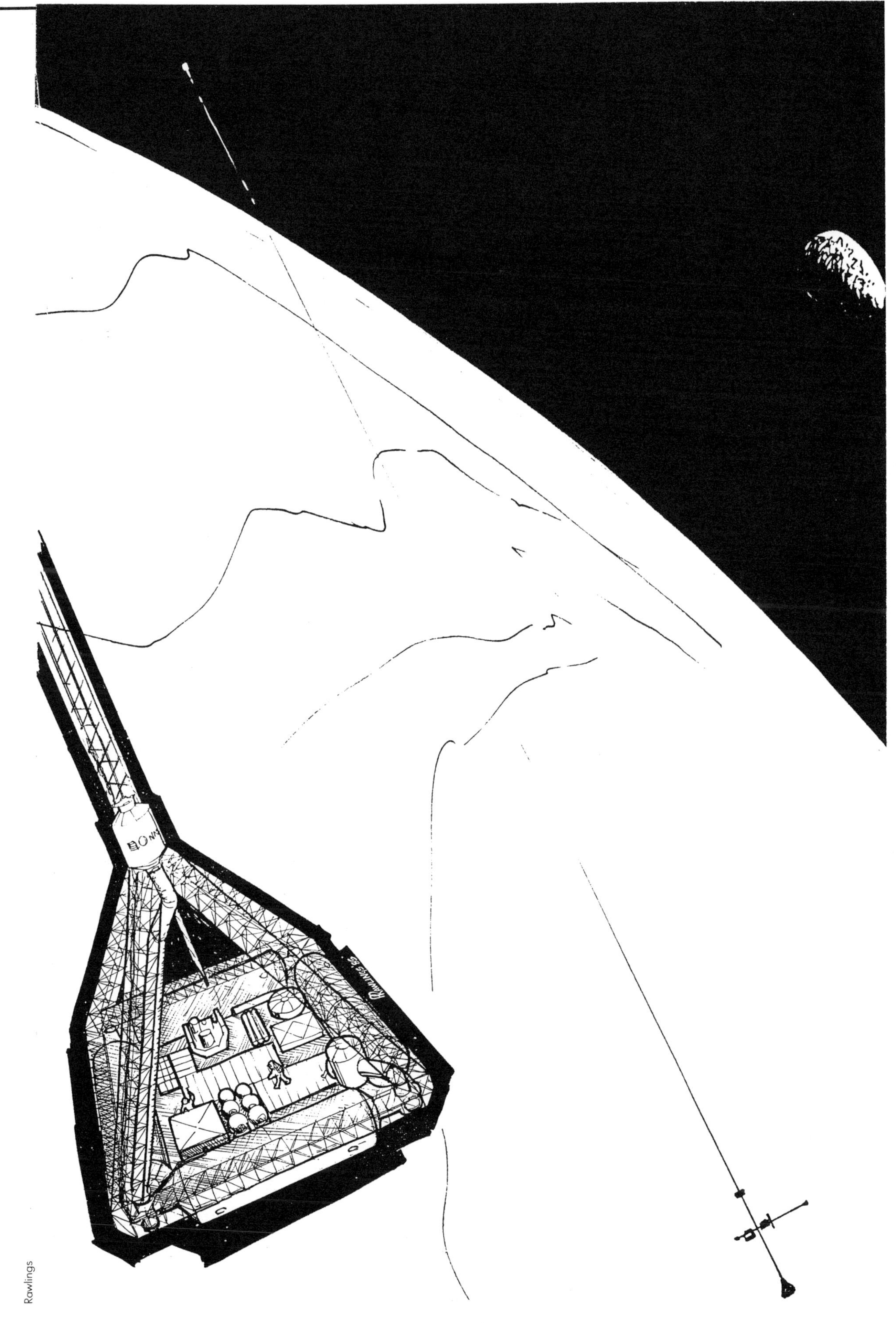

Rawlings

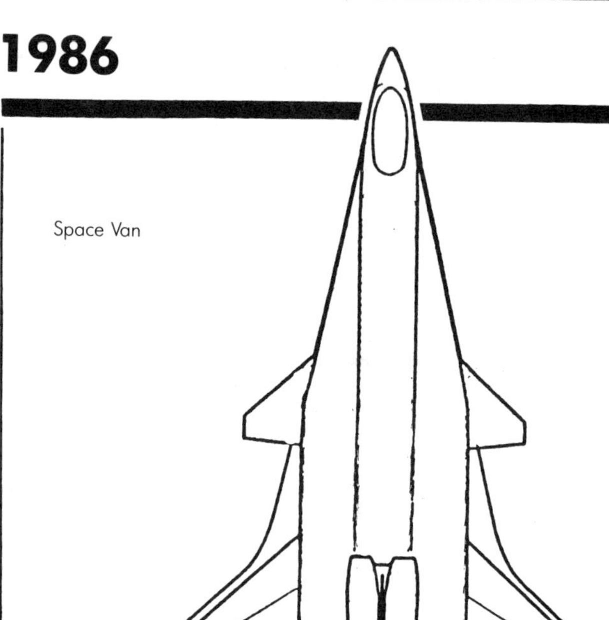

Space Van

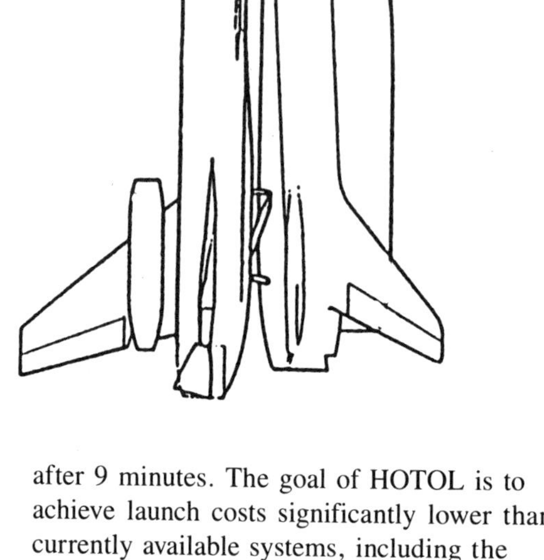

The British HOTOL spaceplane [see 1984] has gone through several stages of modification. HOTOL originally possessed a V-shaped pair of vertical tail fins, in addition to a pair of small canards. Later, a third, vertical canard was added to the nose. This last alteration is revised slightly this year by spacing them 120 apart; additionally, the two vertical tailfins are eliminated. In August, at the Farnsborough Air Show, the model of HOTOL exhibited is even further modified by the deletion of all but the vertical canard. Testing in the 5.5 m wind tunnel at BAe Warton has shown that sufficient control is provided by the single fin. Other modifications from the original concept include moving the under-fuselage air intake further forward.

It is expected that the intitial studies will be completed by the middle of 1987 at a cost of £3 milllion, half of which will be provided by the British National Space Centre (BNSC). The complete development program will include twelve test flights and seven orbital test flights, beginning in early 1996 and concluding with the first commercial launches between 1998 and 2000.

HOTOL's revolutionary propulsion system, Swallow, will allow the spaceplane to take off from a conventional Concorde-length runway, employing a reusable trolley rather than a standard landing gear. It will clear commercial airspace in 4 to 5 minutes, reaching Mach 5 and an altitude of 26 km after 9 minutes. The goal of HOTOL is to achieve launch costs significantly lower than currently available systems, including the U.S. space shuttle. More than thirty different configurations will be studied before the present design is settled upon.

C. E. Singer describes an antimatter manned rocket in his paper "Faster Non-Nuclear Worldships." Miniscule pellets of antimatter are used to raise larger fusion "bomblets" to ignition temperatures.

The Space Van system of Third Millenium, Inc. [also see 1985] is redesigned. Where the orbiter was originally to be carried piggy-back atop a specially modified 747, it is now to have its own Mach 2 booster. The decision to make the change is based on the relatively small payloads that can be delivered, to say nothing of the development costs of the major redesigning required of the 747. In the revised version, the Space Van itself is unchanged. However, it is now a two-stage vertical takeoff vehicle. The booster is a manned winged rocket-plane with three Rocketdyne H-1 engines for takeoff and two air-breathing turbojets for landing. The latter are mounted on either side of the vertical tail.

Instead of the Pago Pago island launch site, the Space Van system is to be launched from a mobile, floating platform. [Also see Hanstein, 1930.] This is a semisubmerged hull with two stabilizing outriggers, 330 feet long and displacing 22,050 tons. The launch platform is on a superstructure raised on pylons above the trimaran hull.

TMI also suggests a scaled-down version of the Space Van called the Minivan, which uses only three RL10 engines. Its booster uses only one H-1 engine.

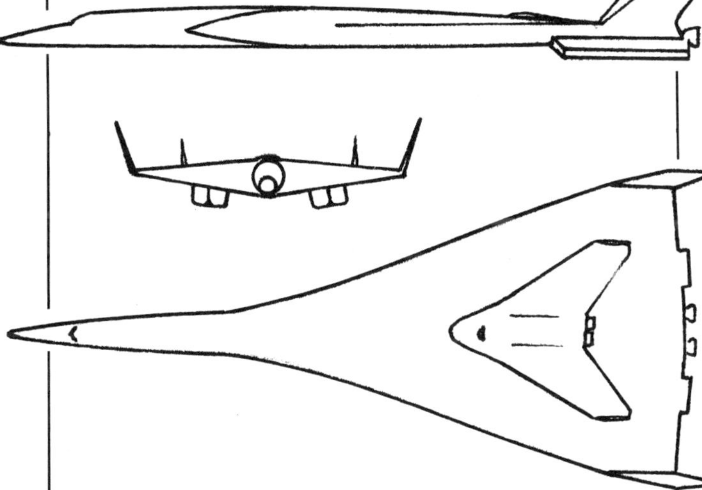

D. M. Ashford of British Aerospace develops the Spacecab II spaceplane concept in which a Mach 4 supersonic piloted booster launches a smaller orbiter. Ashford has designed his project to employ as much "off the shelf" technology as possible, with the goal of reducing costs per flight to 1/2000th that of the U.S. space shuttle and 1/250th that of the French *Hermes*.

The booster stage is 64.8 m long, with a wingspan of 28.3 m. The moth-like orbiter, riding piggyback within a well atop the booster, has a span of 16.3 m. To reach top speed, the booster first reaches Mach 2 via four turbojets (identical to the Concorde's), then a pair of Viking IV rocket engines take over, increasing the speed to the desired Mach 4. The orbiter carries a two-man crew, and its payload bay has a capacity of 6 m^3. It is powered by five HM 7 engines (the third stage engines for the Ariane), which can take it from Mach 4 to orbital velocity.

Ashford hopes that the Spacecab II system will cost no more than $75,000 per flight for the booster—the same as for the Concorde—and about the same again for the orbiter, making the total cost per mission only $150,000.

The Sänger II spaceplane is a West German design created by MBB with cargo, transport and global passenger versions. It is submitted to ESA by Germany's Ministry of Research and Technology. MBB is cooperating with Rolls Royce, MTU and MBW in studying future air-breathing propulsion systems, and with Saab in studying flight dynamics during reentry. An important feature of the Sänger is the ability to reach any orbit desired after takeoff from conventional European airports. Both stages of the piggyback design are capable of landing at the same conventional airport. The Sänger II combines two different lines of development: a hypersonic aircraft about the same size as the present-day Boeing 747 and a shuttle-like vehicle. The booster vehicle by itself is referred to as the Sänger, while the shuttle it carries is called either the Horus or the Cargus, depending upon whether it carries passengers or cargo. The Sänger alone, as a hypersonic transport similar to the U.S. Aero-Space Plane, is capable of transporting 130 passengers a distance of 13,000 km. It has a wingspan of 33 feet and is powered by six air-breathing turbo ramjets, giving it a cruising speed of Mach 4.5. At a speed of Mach 3.5 and an altitude of 20 km it ignites its ramjet engines, propelling the booster to Mach 6.8 and an altitude of 31 km. A short pull-up maneuver then occurs, slowing the plane to Mach 6.6 at 37 km before releasing the upper stage.

The manned upper stage, Horus (Horizontal Upper Stage), is a derivative of the French *Hermes*. With a crew of two it could be used for space station support. An expendable cargo version of the upper stage (Cargus, or Cargo Upper Stage) can carry payloads of up to 15 tons into low earth orbit. The combined vehicle has a mass of about 380 tons (with the first stage accounting for 295 tons). It is expected to deliver the same payload as the Ariane 5, but at

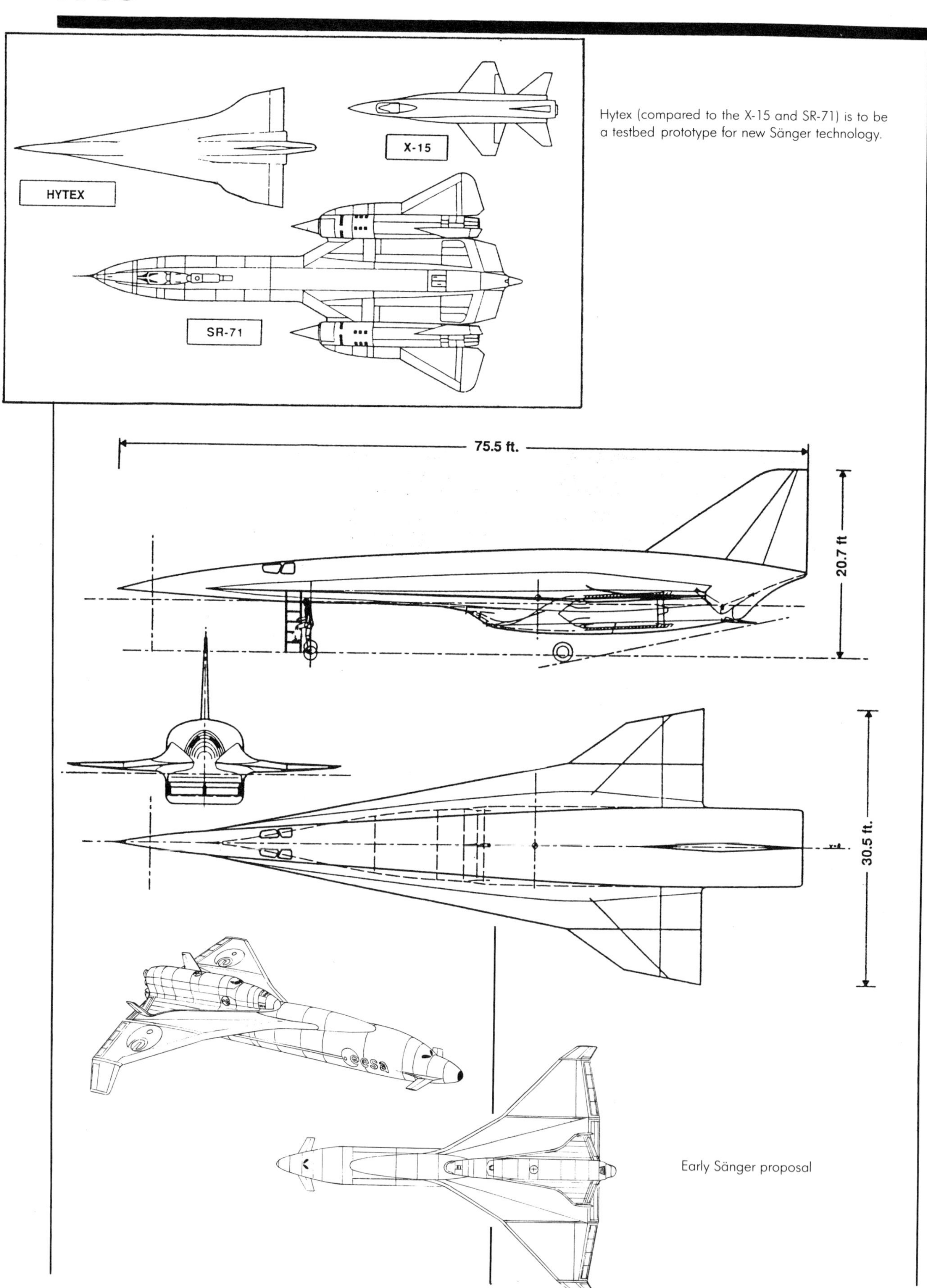

Hytex (compared to the X-15 and SR-71) is to be a testbed prototype for new Sänger technology.

Early Sänger proposal

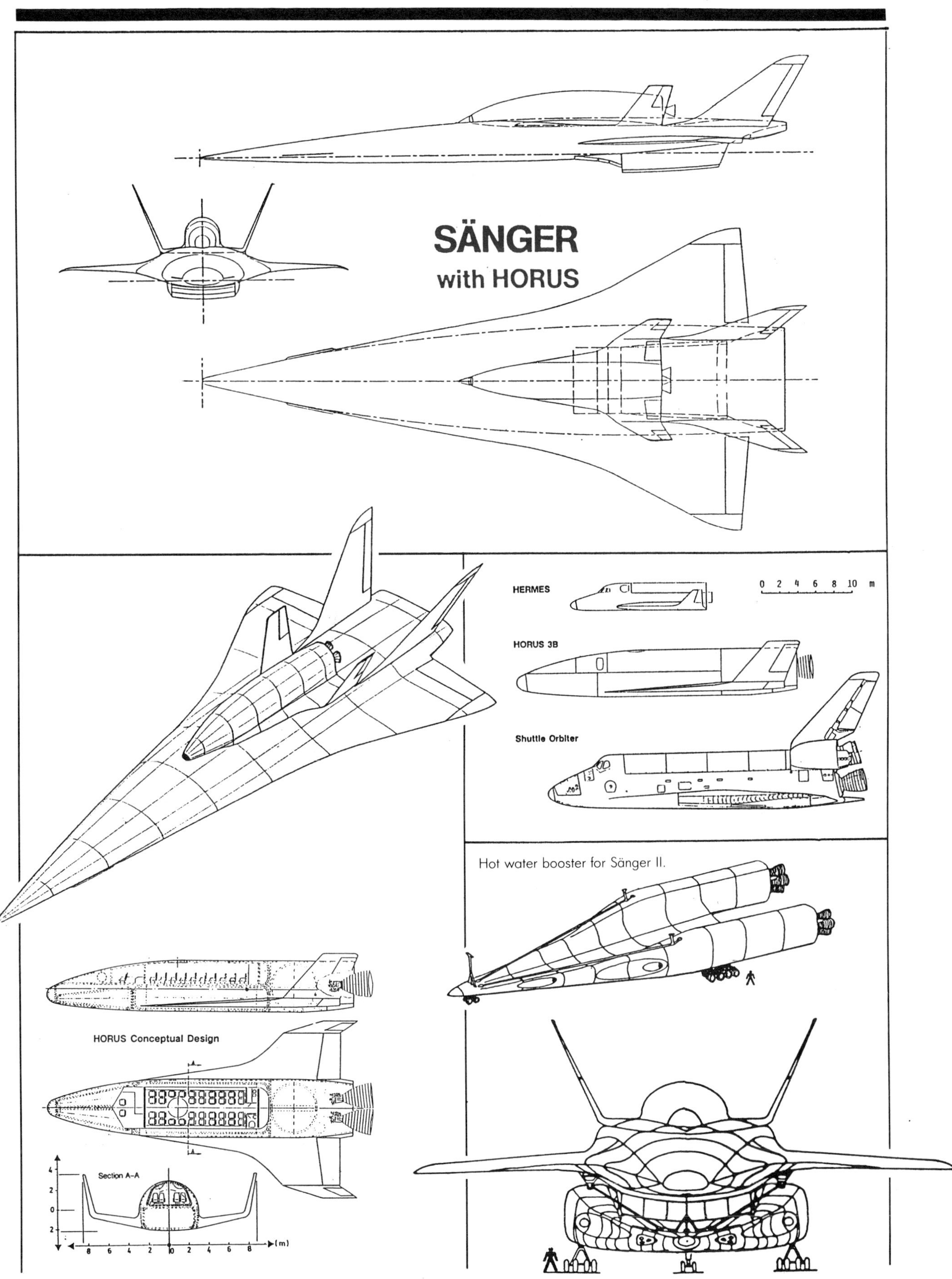
SÄNGER
with HORUS
HERMES
0 2 4 6 8 10 m
HORUS 3B
Shuttle Orbiter
Hot water booster for Sänger II.
HORUS Conceptual Design
Section A-A
(m)

about one-third the cost. (The 1991 design of the Sänger II Horus will allow it to carry three astronauts and up to 6,600 pounds of cargo into low earth orbit. The unpiloted Cargus will deliver up to 15,400 pounds.)

Boeing proposes a horizontal takeoff spaceplane that employs a reusable launch carriage with its own jet engines.

William R. Snoddy of NASA's Marshall Space Flight Center describes a chemically fuelled spacecraft for a manned mission to Mars, one version for a 2001 opposition mission and another version for a 2003 conjunction mission. The former is 70 meters long overall, consisting of an earth departure stage, Mars departure stage, three laboratory and habitat modules, landers for Mars and Deimos (or Phobos), and an aerobrake for Mars and earth. Power is supplied by a pair of solar panel wings 63 meters across. The latter is 67 meters long and requires slightly smaller earth and Mars departure stages. Otherwise it is identical to the first.

If necessary artificial gravity could be produced by extending a pair of habitat/laboratory modules on retractable booms and rotating the entire vehicle, though Snoddy has reservations about the engineering practicality of doing this.

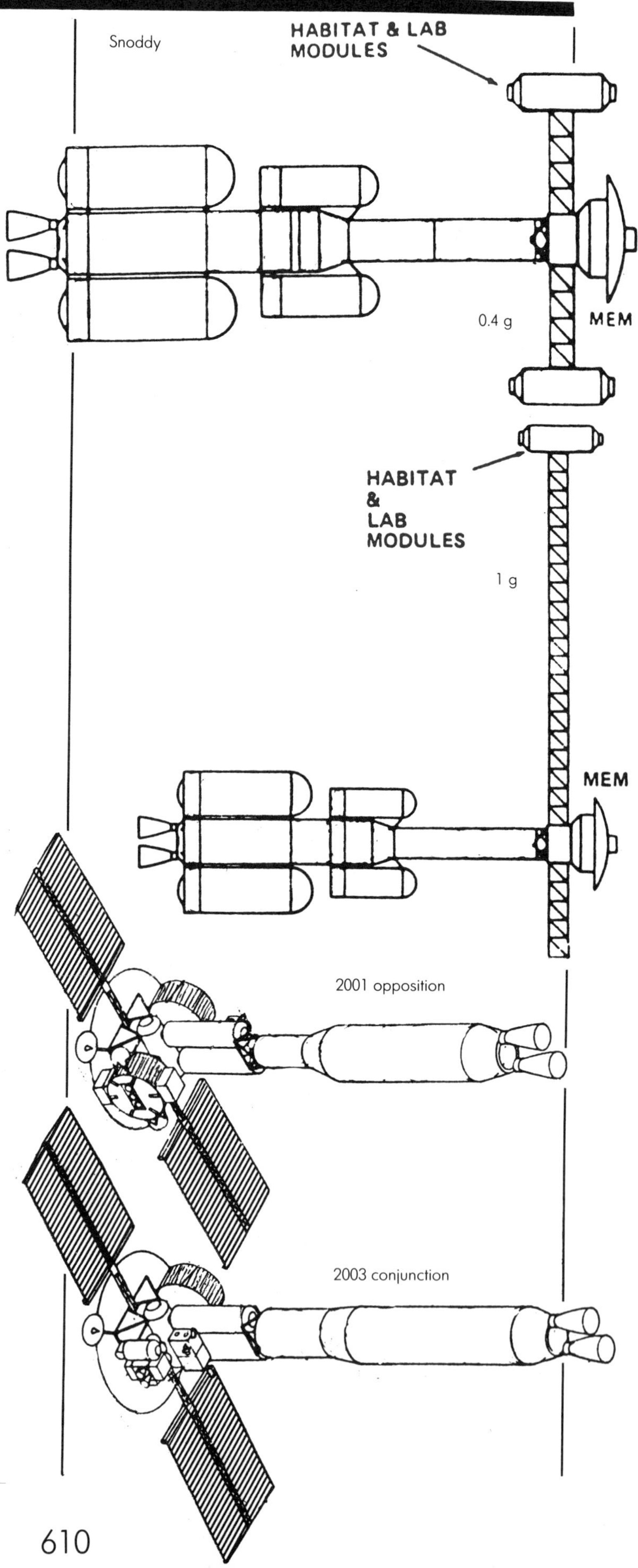

Snoddy

2001 opposition

2003 conjunction

JANUARY

The National Aero-Space Plane (NASP) project is publicly announced. The U.S. Congress establishes the National Aero-Space Plane Headquarters at Wright-Patterson Air Force Base and releases $200 million for research and development (in the following 3 years a total of $2.4 billion will be spent or allocated—about a third coming from private industry).

JANUARY 28

STS 51-L, *Challenger*, explodes, killing the seven astronauts aboard: Commander F. R. "Dick" Scobee, Cdr. Michael J. Smith, Judith A. Resnick, Ronald E. McNair, Col. Ellison S. Onizuka, Gregory B. Jarvis, and

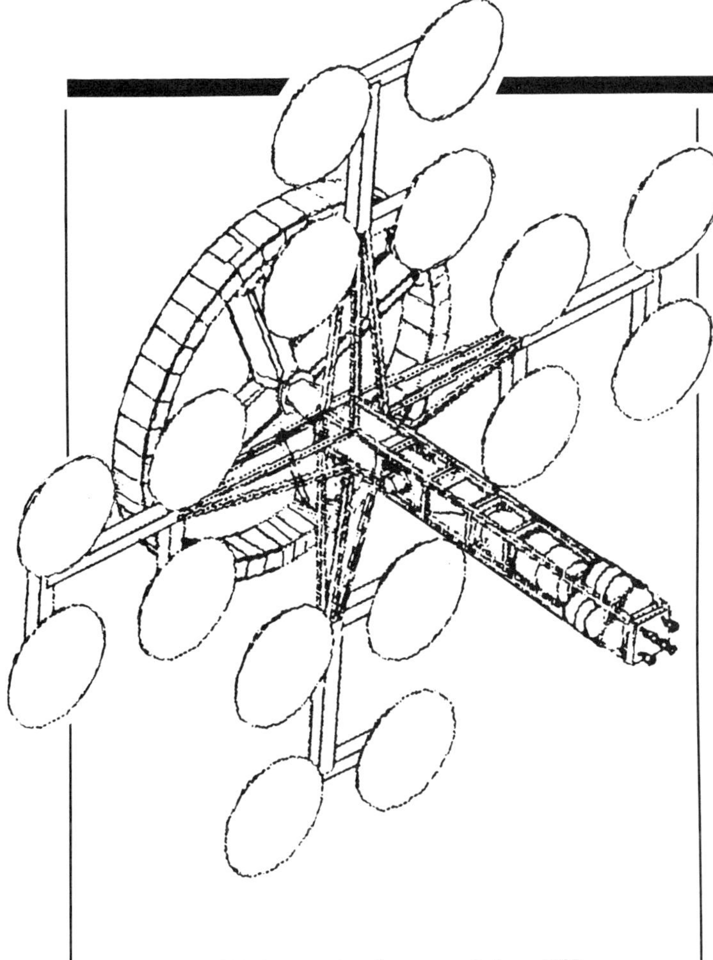

University of Michigan cycling "space castle," ca. 1985–1990.

Sharon C. (Christa) McAuliffe, the highly advertised "first teacher in space." This disaster is a major setback for the American space program and another shuttle flight will not take place until 1988.

The physical cause of the explosion is the failure of one of the O-rings that seal the segments of the solid rocket boosters (SRBs). This allowed white-hot gasses to escape, igniting the huge hydrogen-filled external tank. The disaster is the tragic result of serious flaws both in engineering and management. There have been doubts about the integrity of the O-rings since 1984 and NASA was warned repeatedly, up until minutes before the launch, that the ongoing cold weather was seriously jeopardizing the rings' integrity. Unfortunately, political and commercial pressures overrode NASA's up-until-then legendary caution.

APRIL 7

The Defense Advanced Research Projects Agency (DARPA) and NASA announce contracts for the National Aero-Space Plane. Five are for airframe studies to Boeing, General Dynamics, Lockheed, McDonnell-Doug-

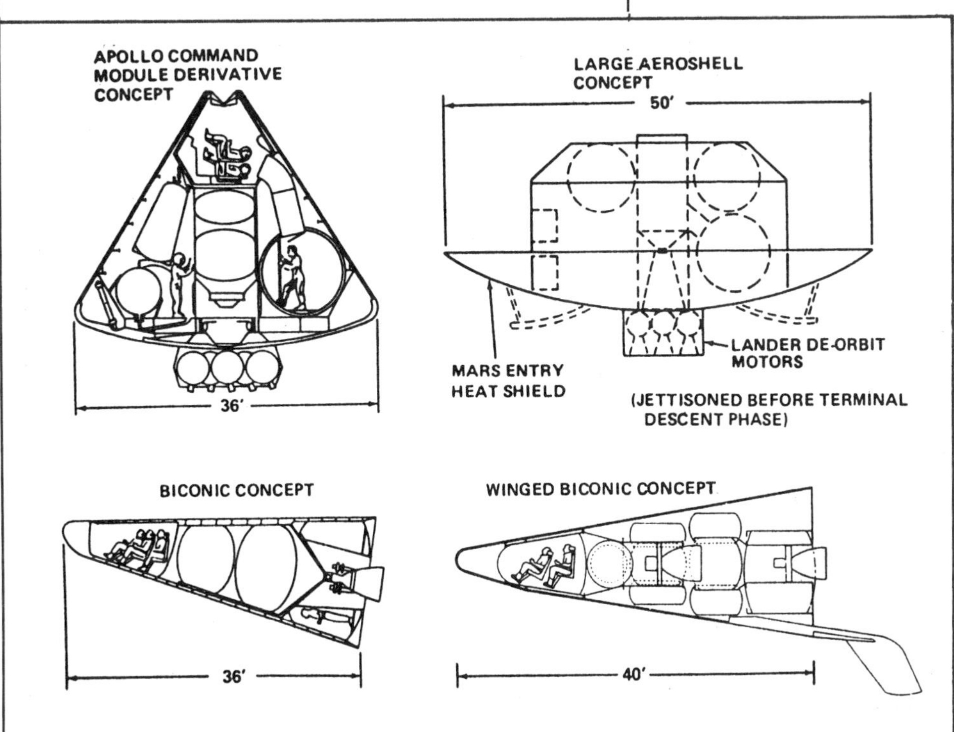

2.5 million kg mass in orbit, orbital period 2.135 years; 3 crew and 17 passengers carried in 7.5 × 4 m torus rotating at 3.22 rpm to create 0.4 g.

las and Rockwell. After one year, NASA and the DoD will narrow the field down to two or three main contractors.

In addition to the government "baseline" conception [see 1985], the contractors consider at least three other configurations: a blended wing-body with bottom-mounted propulsion; a cone-shaped body with an annular propulsion module; and a combination body with bottom-mounted propulsion.

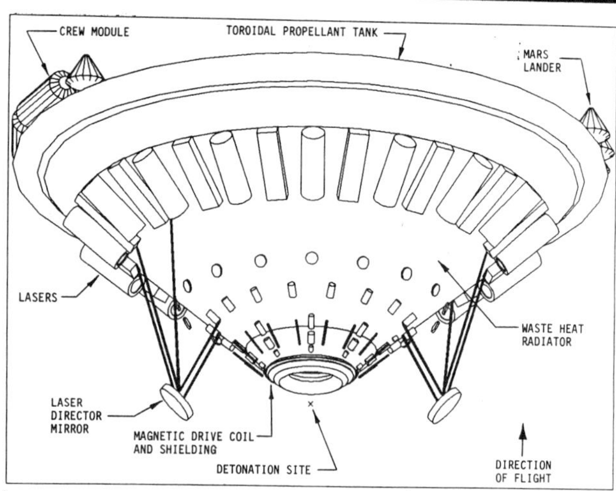

1986–1987

A study led by Charles Orth at Lawrence Livermore National Laboratories develops a conceptual design for an Inertial Confinement Fusion rocket called VISTA (Vehicle for Interplanetary Space Transport Applications). The vehicle uses many high energy lasers mounted around the surface of a cone with mirrors directing the beams to a detonation site at the apex. There, pellets of deuterium, tritium and hydrogen expellant mass are ejected (at a rate of 5 to 30 per second) into the focus where they are detonated, creating a high energy plasma that is directed by a drive coil's magnetic field. This can achieve an acceleration of 0.02 g or greater.

The VISTA manned Mars mission vehicle can make a one-way trip to Mars in about 45 days. The weight of this vehicle is 1,600,000 kg plus 250,000 kg of accommodations, landing craft and shielding. To this is added 4,150,000 kg.

An antimatter variation is proposed by John Callas of JPL. Instead of imploding deuterium pellets, Callas's kilometer-wide cone-shaped vehicle employs protons and antiprotons. It will take only 100 years to travel to Epsilon Eridani (10.7 light-years distant). The upper radiator area and larger propellant tank can be jettisoned halfway through such a mission by placing a second toroidal propellant tank partway up the cone.

Maxwell Hunter of Lockheed's Advanced Development Division conceives the VTOL SSTO X-Rocket, a fully reusable conical vehicle of 500,000 pounds gross weight. It is propelled by a cluster of uprated RL10 engines. The 20,000-pound payload bay is 20 feet high, 15 feet in diameter at the forward end and 20 feet at the lower. The entire vehicle is 60 feet tall, with a 14-foot aerospike, and 34 feet in diameter at the base.

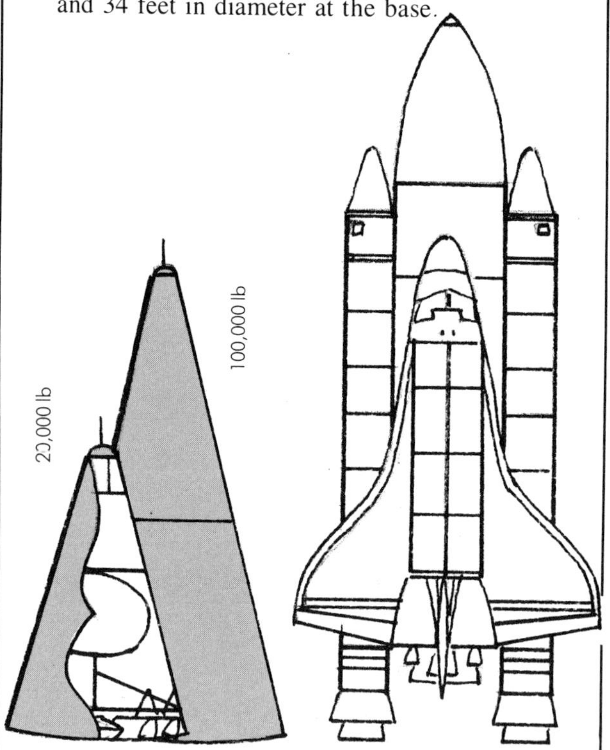

Hunter argues forcibly for the merits of the single-stage-to-orbit concept, concluding: "The desireability of single stage vehicles, far from being a new idea, is obvious. Single stage vehicles have been examined since the rocket equation was invented." He emphasizes that the blunt-cone SSTOs that he and others advocate are inherently safer and more reliable than current multistage and shuttle plans—with the safety and recoverabilty of both the crew *and* the rocket considered. This is because " . . . all space rockets spawn from the ammunition paradigm, since only rockets could make it to orbit when Sputnik called. The NASA 'man-rating' procedures are really a very expensive, understandably neurotic, method of creating *man-rated ammunition*."

Work on the design is discontinued by the defense-oriented company. [Also see 1989].

1986–1988

NASA's Advanced Programs Office develops the Lunar Base Systems Study in the Engineering Directorate of the Johnson Space Center.

1987

H. D. Froning, Jr. proposes a space transportation system propelled by antimatter air-breathing propulsion. The proposed spacecraft is a single-stage-to-orbit spaceplane not dissimilar to the National Aero-Space Plane.

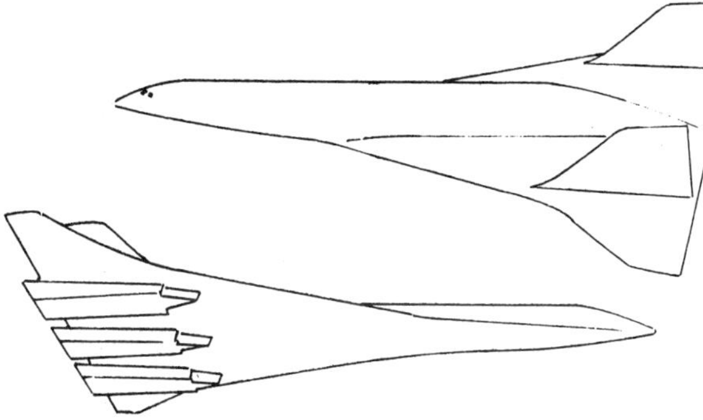

SEP displays a model of a single-stage-to-orbit aerospace plane at the 38th IAF Congress in Brighton.

The 1987 Paris Air Show features a full-scale model of the *Hermes* spaceplane and a 1/5th scale model of the HOTOL. Also shown are models of the Teledyne Brown air-launched spaceplane [see below] and the X-30 Aero-Space Plane.

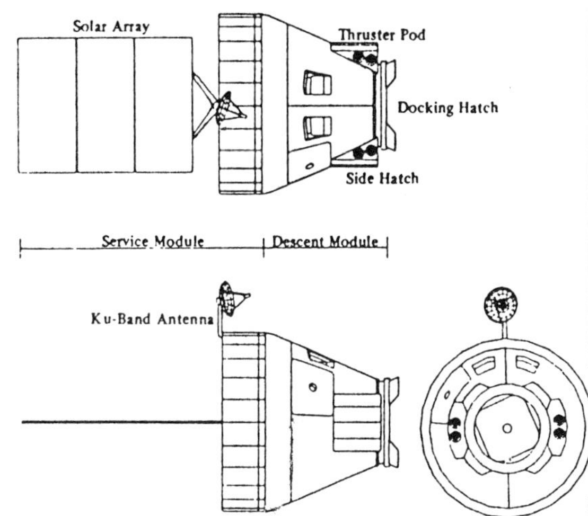

British Aerospace unveils its alternative to the *Hermes* spaceplane. The Multi-Role Capsule (MRC) can carry a crew of four to a space station. However, the MRC will have a cargo capacity of only 250 to 500 kg as opposed to the *Hermes*'s 3 tons. An unmanned cargo carrier, for launch by the Ariane 5, is also suggested as an alternative to *Hermes* and to support an MRC.

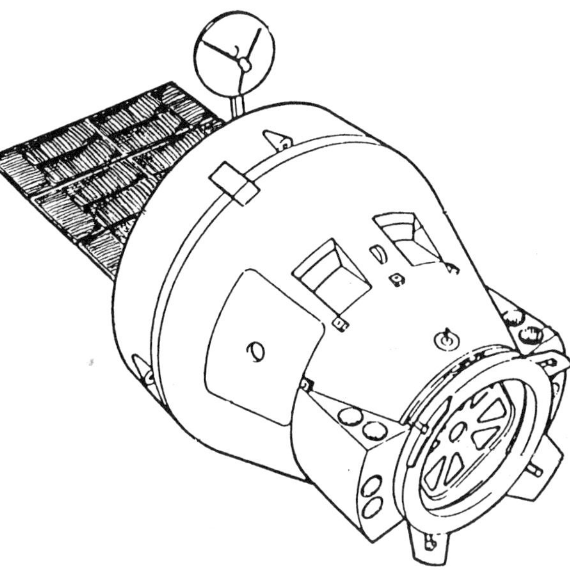

The MRC measures 4 m across and weighs about 7 tons (with a return weight of about 5 tons). As a rescue vehicle, it can carry six passengers ("albeit in rather cramped conditions"). It will return to the earth by means of parachute and splash down into an ocean.

The MRC consists of a descent module and a service module. The latter is a short, cylindrical structure that is jettisoned before reen-

	VOSTOK/ VOSKHOD	MERCURY	GEMNI	SOYUZ	APOLLO	MRC
CREW						
NORMAL CONTINGENCY	1-3 -	1 -	2 -	1-3 -	3 5	4 6
RECOVERY	BALLISTIC LAND	BALLISTIC WATER	SEMI BALLISTIC WATER	SEMI BALLISTIC LAND	SEMI BALLISTIC WATER	SEMI BALLISTIC WATER
CONFIGURATION	RE-ENTRY SPHERE + INSTRUMENT SECTION	CAPSULE + RETRO PACK	RE-ENTRY MODULE + RETRO MODULE + EQUIPMENT MODULE	ORBIT MODULE + DESCENT MODULE + SERVICE MODULE	COMMAND MODULE + SERVICE MODULE	DESCENT MODULE + SERVICE MODULE

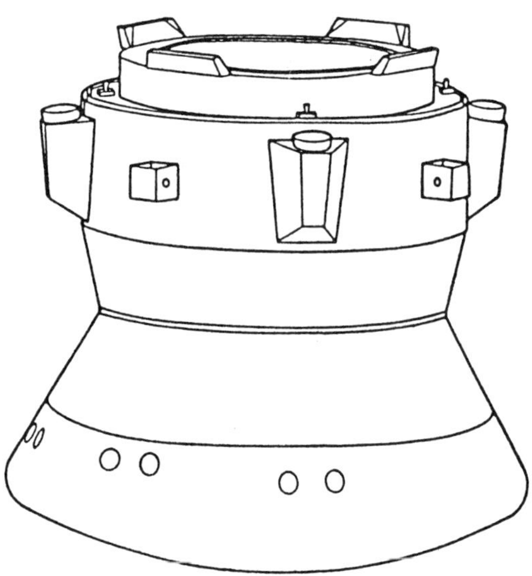

Aerospatiale escape capsule

try. It houses a solar array and the various communications antennas. The descent module is divided into three sections: a forward cabin with docking port, thrusters, galley and hygiene facilities; a midcabin with crew couches and controls; and a rear cabin housing batteries, propellant, the air supply and a payload bay. The launch vehicle is to be an Ariane 4 rocket (after the class has been man-rated and an escape system has been developed). The capsule can also be delivered into orbit aboard the U.S. space shuttle.

Bernard Carr proposes an air launch for the HOTOL spaceplane. His suggestion is that it can be carried aloft by a modified Boeing 747 Jumbo jet. The takeoff weight of the 1987 version of HOTOL is 230 tons, while the maximum takeoff weight, including fuel, of the 747 is only 233 tons. Carr believes, however, that the next generation of the 747, the series 500, will be able to handle the weight easily. Modifications to the aircraft will include removal of the vertical tailfin, replacing it with two fins at the tips of the elevators. [Also see 1991.]

Aerospatiale develops two variations on a spaceplane design. One is a single-stage-to-orbit system using rocket propulsion. The other is a two-stage air-breathing/rocket engine combination.

Teledyne Brown develops a low-cost spaceplane design by Ernst Stuhlinger. It is an unmanned vehicle air-launched from the back of a modified Boeing 747 [see above]. It is capable of carrying 6,300 kg to a space station orbit, or 4,000 kg to a low polar orbit.

D. M. Ashford and P. Q. Collins propose the "Spacebus" orbital and suborbital passenger transport.

Spacebus is a two-stage spaceplane with fully recoverable manned stages. The orbiter is capable of carrying up to seventy-four passengers on a suborbital journey and up to fifty into orbit. The cost per passenger ranges from $2,000 to $5,000 (with a fleet size of thirty-five to seventy spacecraft and

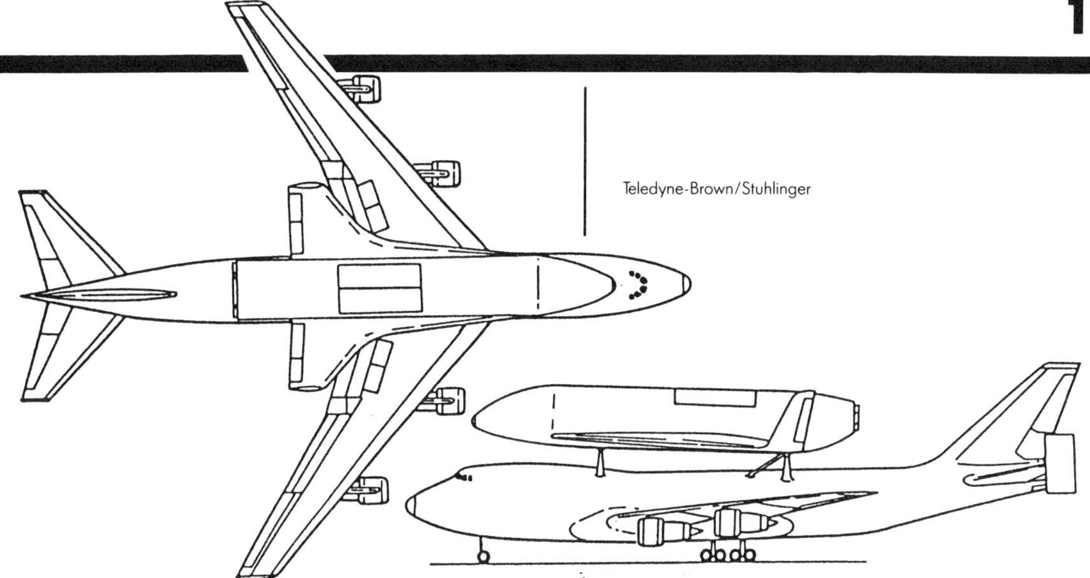
Teledyne-Brown/Stuhlinger

depending upon whether the flight is suborbital or orbital). Up to two launches per month are envisioned for each spacecraft. The horizontal-takeoff-and-landing spacecraft is able to use existing commercial airports.

The booster stage is a large supersonic aircraft similar in outline to the Concorde, but twice the size. Four turboramjet engines provide thrust for takeoff, acceleration to Mach 4, and return to its airport. Four HM 60 or J2S rocket engines take the booster from Mach 4 to Mach 6, where the second stage orbiter separates. The orbiter has two rocket engines fueled by hydrogen and oxygen. During takeoff it remains partially buried in the top surface of the booster.

Until the Spacebus achieves passenger rating—which might take as many as 1,000 flights—many of these test flights could be combined with space station servicing. Tourist flights could begin as early as 4 years after the prototype enters service. Spacebus could begin operating within 7 years, carrying passengers in 11 and "approaching maturity" within 16 to 21 years.

The authors also propose a scaled-down version of their project, called "spacecab," that serves the same purpose as the *Hermes* spaceplane. Spacecab carries a crew of two and six passengers, along with a payload of cargo. Using existing engines means it could be developed for about $2 billion. It is an updated version of the Dassault Aerospace Transporter of 1964. Development of Spacecab would also lay the groundwork for Spacebus.

If the authors' timetable is even remotely correct, space tourism could begin as early as 1999.

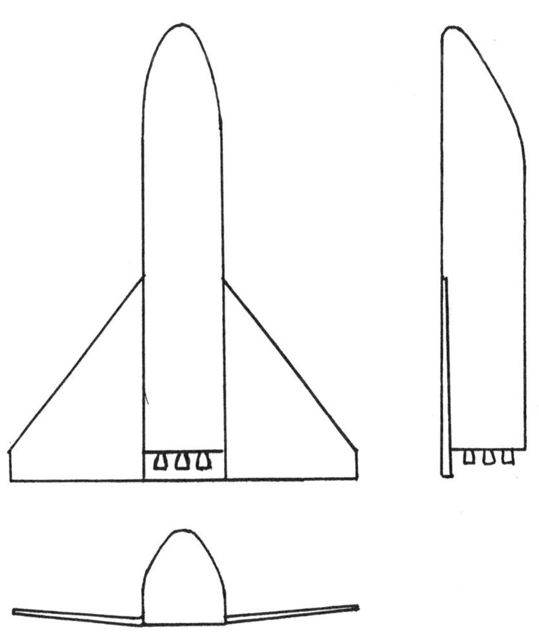

J. R. Olds, a student at the North Carolina State University Department of Aerospace Engineering, designs a single-stage-to-orbit space station service vehicle.

The final design is a delta-winged vehicle with a fuselage 252 feet long (with an additional 13 feet of length added by the main engines) and a wingspan of 208 feet. A payload bay 60 feet long and 15 feet in diameter (the same size as the space shuttle's) accommodates a space station crew module. The shuttle is equipped with either a vertical tail or winglets.

Olds's spacecraft can carry 20,000 pounds to orbit (as opposed to the space shuttle's 41,400) and return 40,000 pounds to the earth. It launches vertically and lands on a conventional 10,000-foot runway.

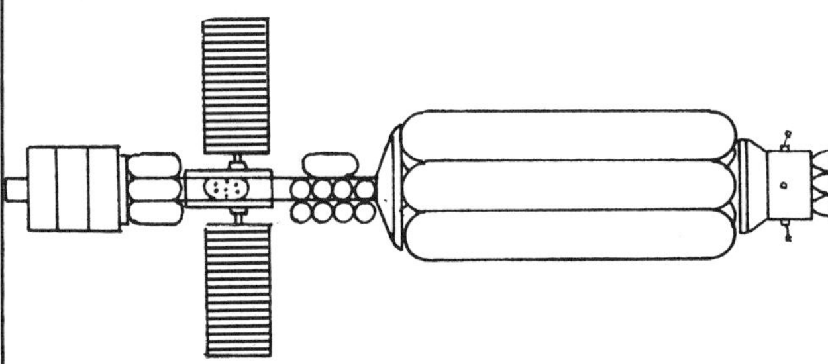

In response to a contest sponsored by the University of Wisconsin, a group of Wausau West High School students develop a manned Mars mission. Their Mars Transit Vehicle, christened the Barsoom, is a modular craft assembled in earth orbit. All of the modules are attached to a central "spine" 3 meters wide and 124 meters long. The first three modules are squat cylinders each measuring 9.8 meters in diameter and 3.1 meters thick. Two are intended for habitation and the third holds the life support machinery. The spacecraft is powered by a pair of solar arrays, 9 meters by 18 meters. Half of the ship's length is taken up by the eight 6.5 × 46.3 meter propellent tanks which fuel the five rocket engines. Four of the tanks are used to escape the earth, while the second cluster of four is used to leave Mars.

Langley Research Center proposes a phased approach to the development of its 2005 Shuttle II concept. The evolution would begin with a liquid oxygen/liquid hydrogen core stage with solid fuel strap-on boosters that would be operational in 1995. This would form the basis for the first three vehicles. The second would combine the core stage with an unmanned Shuttle II flyback booster, replacing the solid fuel strap-ons and increasing the payload capacity from 100,000 kg to 150,000 kg. This would be operational by 1998. In that same year a small, manned interim vehicle called STAR (Space Taxi and Recovery) would be flown atop the core stage alone. The STAR vehicle would be a small lifting body 24.6 feet long and 20.9 feet wide. Finally, the full-scale Shuttle II would begin flying in 2005. For delivery of payloads to orbit and for space station and satellite servicing, the manned orbiter would use a piggyback cargo module. For personnel transport the cargo module would be replaced by a STAR vehicle, also carried piggyback.

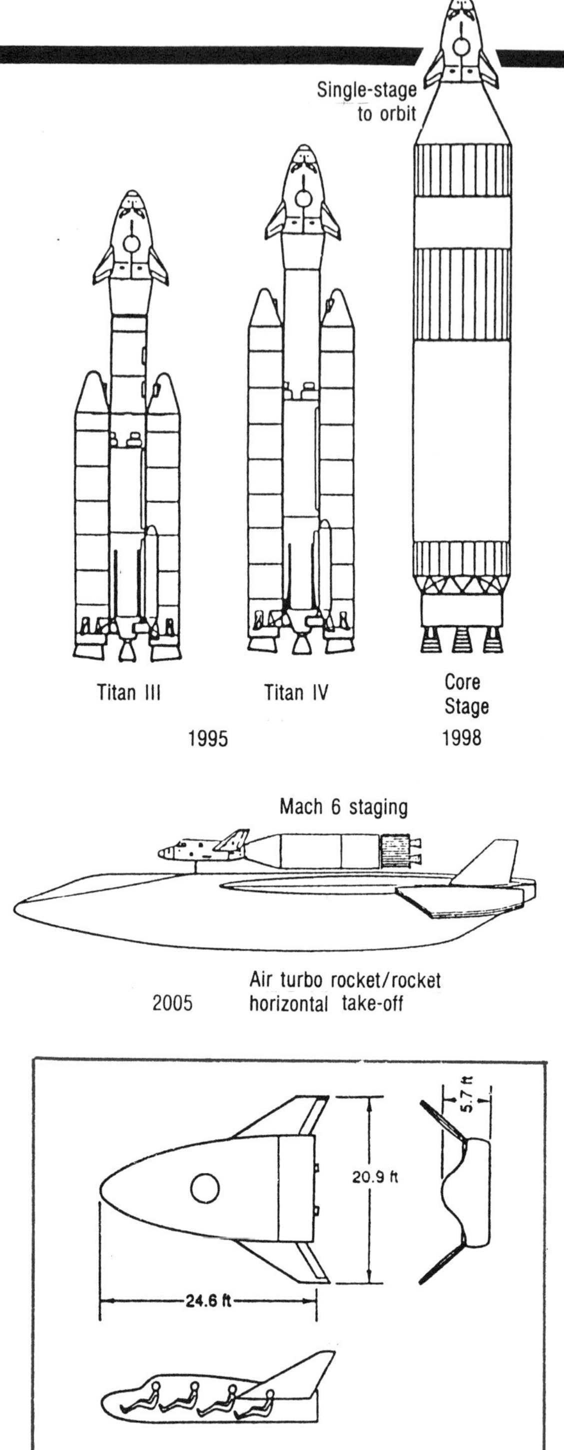

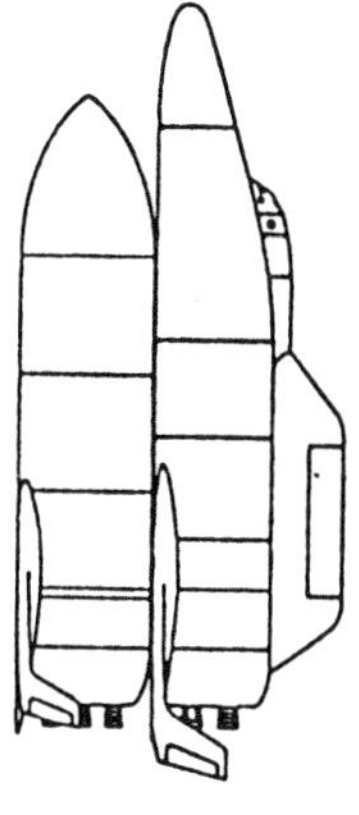

SHUTTLE II
2005
Delivery & Servicing

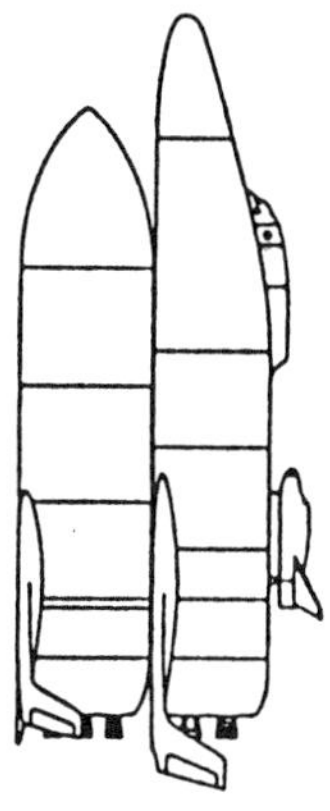

SHUTTLE II
+ STAR VEHICLE
2005
Personnel Transport

MBB designer D. Koelle revives his BETA SSTO concept [see 1969] as the larger, more efficient BETA II.

OCTOBER 7

National Aero-Space Plane contracts for the airframe are awarded to three of the five original contenders (eliminating Boeing and Lockheed): General Dynamics, McDonnell Douglas and Rockwell International. Each receives $25 million to move into the next phase of development. Pratt & Whitney and Rocketdyne (a division of Rockwell) are to design the critical propulsion system. The first flight of the X-30 test-bed is tentatively scheduled for 1993.

1988

NASA Administrator James C. Fletcher believes that the moon, rather than Mars, ought to be the goal for a potential U.S.-Soviet joint manned planetary mission.

"Going to the moon together," he says, "will give the two leading spacefaring nations in the world an opportunity to build a stable base for further cooperation, which could, one day, lead to a cooperative mission to Mars." Fletcher stresses that a foundation of cooperative unmanned activities will be necessary first.

Factors favoring the moon over Mars are (1) the relatively short timetable required for a return to the moon, as opposed to the time demanded for a Mars mission, which will "probably encompass 4 or 5 presidential administrations." (Fletcher does not believe that the United States and the U.S.S.R. have yet demonstrated a relationship that promises that much stability); (2) the experience of mutual cooperation; and (3) the immediate availability of known and proven technology as opposed to the many unknowns involved in a Mars mission.

The *Hermes* escape system is changed from a jettisonable crew cabin to individual encapsulated seats, or Crew Escape Modules (CEM). Each astronaut is encased within a three-finned cylindrical capsule equipped with a solid-fuel rocket that can remove the CEM from an endangered *Hermes* in 3 seconds. The total system weighs only 1 ton as opposed to the 3 tons required by the current system.

The initial Proof of Concept Study on the HOTOL spaceplane, which was completed at the end of 1987, at a cost of million (half of which was provided by the British government), encourages both Rolls Royce and British Aerospace to continue with the project. The target of achieving the launch of 7 tons of payload for $5 million remains and most of the original features of the spaceplane have survived the 2-year study intact. The Rolls Royce RB-545 engine has been relatively unchanged since the inception of the program, but changes in the vehicle's size and mass have led to modifications in the relationship of the engine and the air intake, as well as revisions in performance and trajectory.

A setback occurs, however, on July 25 when Trade and Industry Minister Kenneth Clarke announces that there will be no further government funds available for the HOTOL project. Clarke expects that Britain's role in ESA will satisfy its requirements for access to space. Rolls Royce and BAe were prepared to assume half of the £4 to 6 million cost of the Definition and Initial Development Phase. They now have to look for foreign partners in the project. ESA will not be

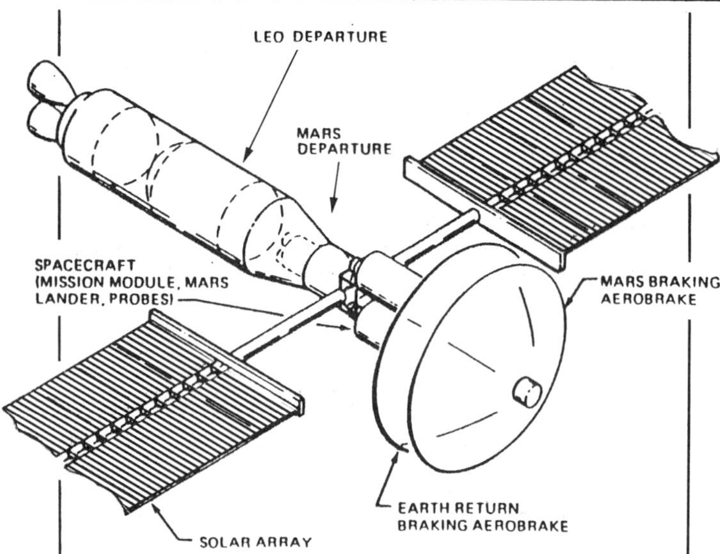

Above: MSFC All-aerobraking 1999, 2001 Mars mission, 1985.
Right: MSFC All-propulsive Mars mission.

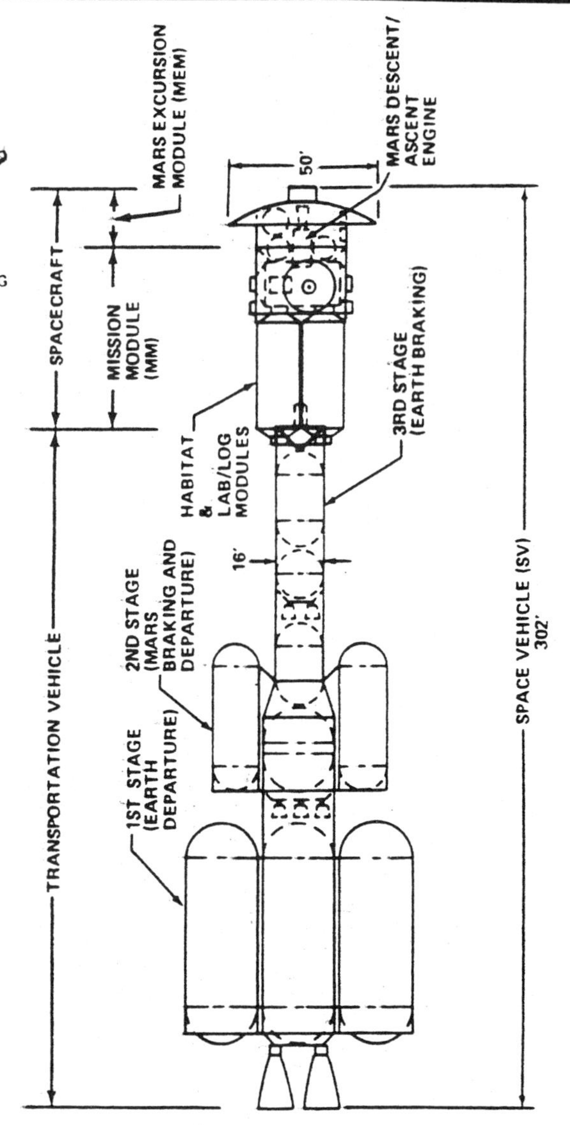

willing, since it is backing its own spaceplane, *Hermes*. West Germany is one possibility, since it might be able to combine HOTOL technology with its Sänger project to produce a hybrid vehicle.

NASA unveils the new Discovery space shuttle. It has been redesigned with a great many new safety features, reflecting the concern following the *Challenger* accident. Although most of the attention has focussed on the solid rocket booster (SRB) joints, there have been some 220 changes to the orbiter itself as well as about 30 modifications of its main engines. Among these features is a new crew escape system.

Gregory L. Matloff discusses the possibilities of interstellar travel by "hitchhiking" on comets. " . . . comets," he writes, "may have the best resource base for ultimate human colonists. Rich in water and oxygen, and with rocky and metallic cores, these objects are small enough to produce no significant gravity wells to hinder ultimate landing missions and yet are massive enough to provide the materials for a vast number of space habitats."

Such a hitchhiking trip from star to star may require up to 50,000 years or more, perhaps making the crossing by "island-hopping" from comet to comet. Although a seemingly enormous number of years, it is on the same order as the 40,000 years the Polynesians required to colonize the Pacific Basin.

A human colony may first be established on a comet orbiting the sun near the plane of the ecliptic. Colonies may begin with as few as ten to twenty human beings, with no genetic disadvantages. Using a solar-powered mass driver or a large hydrogen-oxygen rocket, the comet is gradually nudged from its original orbit into a highly elliptical one that passes close to the sun. Reflectors concentrate sunlight on water from the nucleus, creating hydrogen and oxygen both for fuel and for use by the colonists. During the

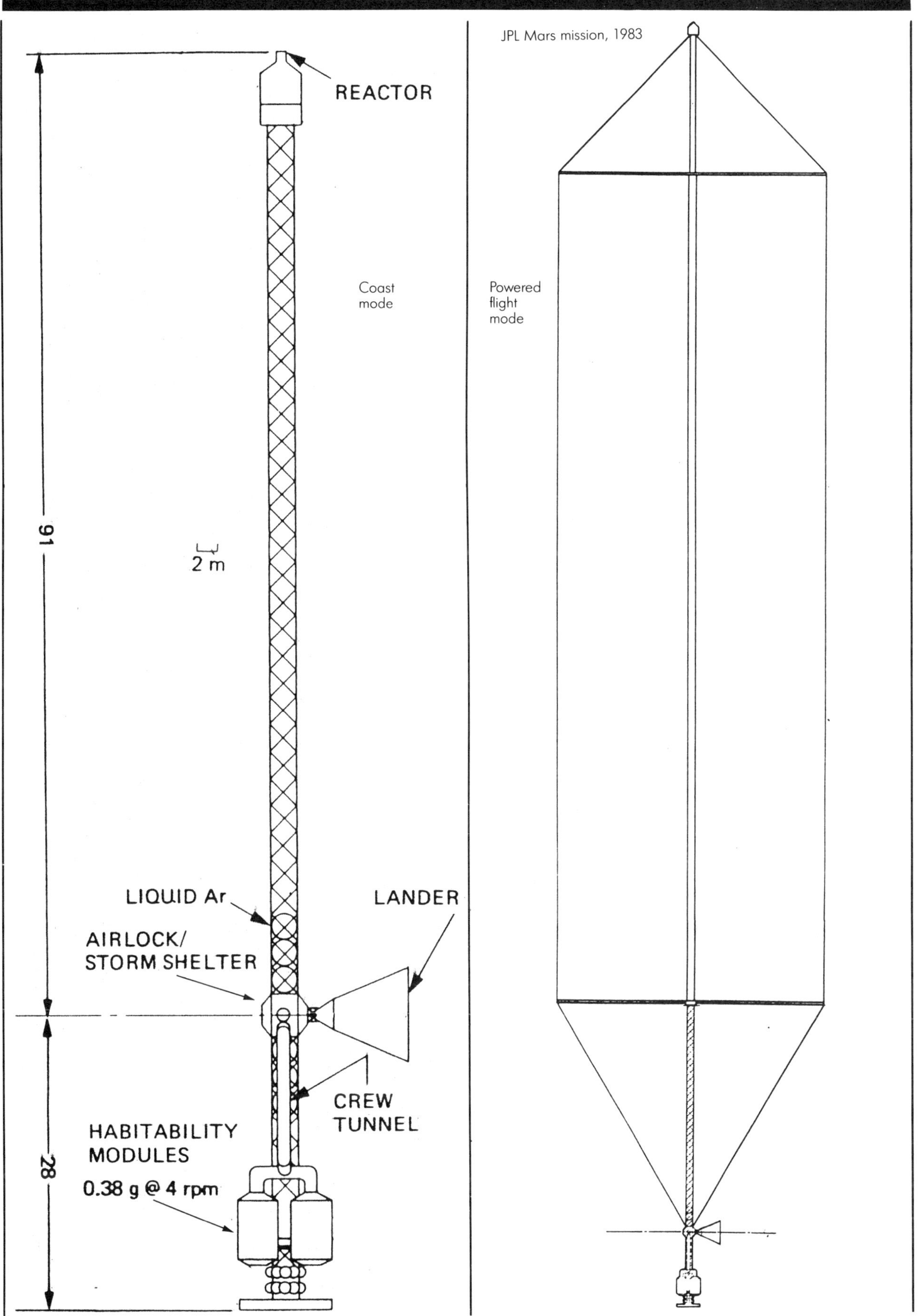

JPL Mars mission, 1983

long fall sunward, the comet's speed is augmented by rocket or mass-driver. Depending upon the comet's original distance from the sun, it could take up to 500 years to reach perihelion. A powered maneuver at perihelion will increase the comet's solar system departure velocity. It will then require some 5,000 years to reach Alpha Centauri. By then the population will have increased to several thousand. They will (supposedly, and presuming they will wish to leave their home world of 5 millenia) devote their energy to the construction of "worldships" for entering the new stellar system. The remaining population will perform a grazing flyby of Alpha Centauri and continue on its way.

Crossing the galaxy in this manner, concludes Matloff, humans could occupy a significant portion of the Milky Way within 10 million years.

Eagle engineering, under a NASA contract, designs a second-generation lunar lander. The reusable vehicle, which bears an outward generic resemblance to its Apollo ancestry, is about 12 meters tall with a six-person manned module 4.5 meters in diameter. The crew module is removable and the lander can be flown without it. Also described is a slightly smaller multipurpose lander. [Also see 1989.]

General Dynamics publishes the *Planetary Explorer*, volume 49, number 4, December 2038; it is a special edition in which the first colonizing expedition to Mars is covered. Mimicking the general appearance of *National Geographic*, the *Planetary Explorer* details the flight of the Marsbound spaceship *Spirit of Galileo* from earth orbit and lunar base to its final arrival at the red planet.

A large number of spacecraft are described, in addition to the *Galileo*. By the year 2038 several nations are operating spaceplanes, including the American CTS built by (of course) General Dynamics; the Sänger C-23; the Chinese Long March-12D (built by Great Wall Industries of Beijing); the Japanese H-10 (built by Boeing Chrysler); the Convair 1100; and the Russian-built Ilyushin RC-8. Lunar transfer vehicles and lunar descent buses carry passengers and cargo from earth orbit to lunar orbit and to the surface of the moon.

The *Galileo* (designated MMSC-103) is over 1,000 feet long and masses 6,000 metric tons (9,400 metric tons fully loaded). It was built by NASA in 2031 as a Kepler class Solar Clipper intended to augment the current fleet of slower, ion drive freighters then operating between earth and Mars. At flank speed it is capable of crossing the solar system in just under 2 years. The spaceship is vaguely axe-shaped, the "blade" being an array of thirty-six solar panels that provide auxillary power for the eight boom-mounted SN-800 nuclear electric generators.

Propulsion is provided by a pair of electric laser-fusion drive systems (ELFD) that provide up to 50 pulses of energy per second. They can accelerate the ship continuously, running as long as there is a supply of fuel pellets. Ion engines steer the ship.

On either side of the long central core are a

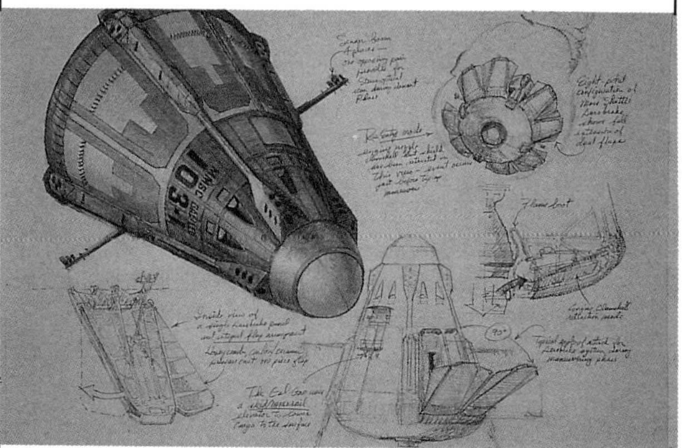

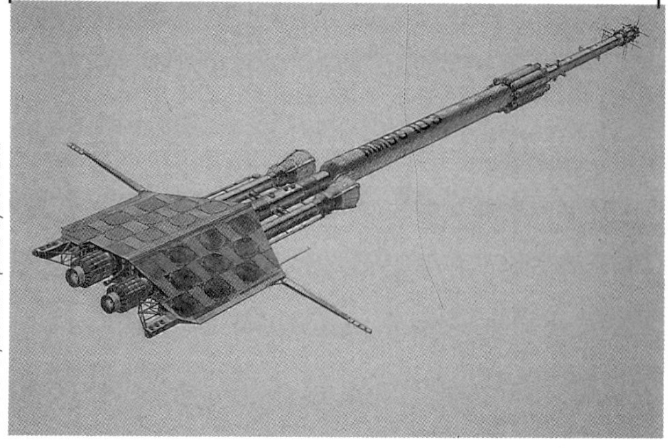

General Dynamics Space System Div.

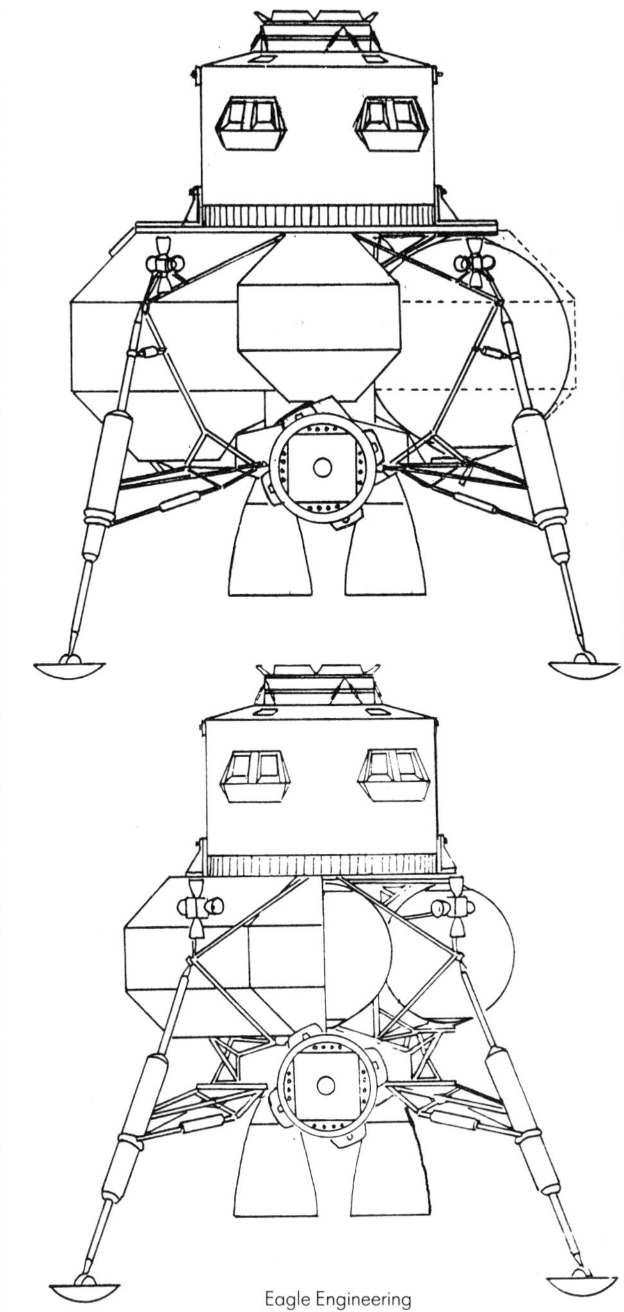
Eagle Engineering

pair of Mars shuttles, for the descent to the Martian surface. Crew quarters are amidship, sheltered beneath 10 feet of lunar regolith as solar radiation shielding.

Gary Hudson proposes a VTOL SSTO spacecraft for a high velocity round trip to Mars: a single-stage chemical rocket, refueled in earth orbit, that can make the one-way trip in less than 130 days and potentially in as few as 90. Aerobraking is necessary at Mars as is refueling at Phobos (perhaps by extracting hydrogen and oxygen from native materials). Two Phoenix-class SSTOs would be required, as well as two Propellant Precursor Tankers (PPTs) and a logistics/power module for each set of Mars-bound SSTOs.

Unfueled PPTs are launched into low earth orbit by Phoenix C cargo rockets. They are fueled and launched into a Mars transfer orbit. There they aerobrake and circularize at Phobos. Two manned Phoenix E SSTOs are launched and refueled in earth orbit. They are then launched into a high-velocity transit to Mars, arriving about 128 days later. During the transit, the two vehicles are joined by an extendable truss (with a foldout solar array). This allows the combined spacecraft to rotate about their common center to create an artifical gravitational field. After arriving at Mars and circularizing their orbit, the spacecraft refuel from the waiting tankers. One of the SSTOs makes a landing on Mars where it remains for a week to 10 days of exploration. It then returns to orbit, refuels once more and the recombined spacecraft return to the earth.

JANUARY

The European Space Agency (ESA) initiates its multilevel study of winged reusable launchers. This work, called the ESA Winged Launch Configuration Study (WLC), is performed by a group of European companies with MBB being the prime contractor. Others involved include British Aerospace, Rolls-Royce, MTU, Fiat Avio and VDK System, with specific contributions by Dornier and MBB/ERNO.

MBB has the responsibility for the development of a two stage to orbit vehicle, drawing on the parallel work being done on the Sänger spaceplane, while British Aeropace will oversee the development of a single stage to orbit rocket, drawing on their experience with the HOTOL spaceplane. [Also see 1989 and 1991.]

1988

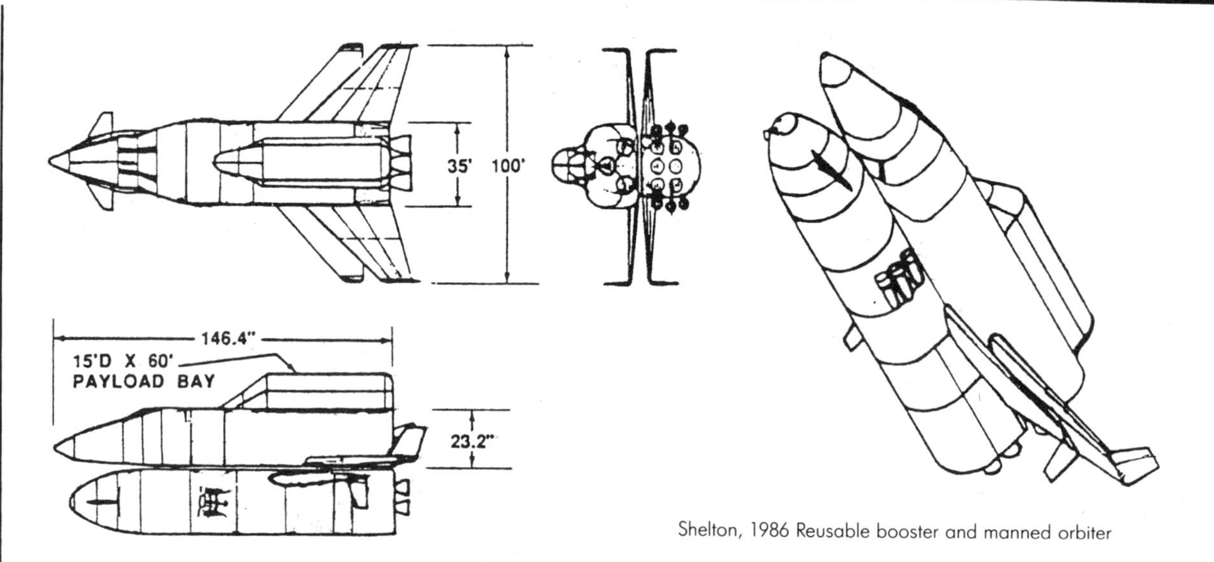

Shelton, 1986 Reusable booster and manned orbiter

SUMMER

The International Space University holds its inaugural session at the Massachusetts Institute of Technology in Boston. While attending ISU, architecture student Madhu Thangavelu designs a unique lunar base/spacecraft that combines cheapness, safety and ease of assembly: MALEO (Module Assembly at Low Earth Orbit). In Thangavelu's scheme, the lunar base is assembled in earth orbit rather than on the moon. The units of the base are triangular assemblages of three cylindrical manned modules within a truss framework: a habitation module for a crew of four to eight; a laboratory module and a power and logistics module (in the original ISU version, the MALEO consisted of four modules arranged in a square.) In the center of the triangle is the lunar descent and landing system. The unit settles down on large landing gear located at each corner.

Other ISU spacecraft associated with the lunar base are a small lunar lander and an Orbital Transfer Vehicle (OTV). These and the modular lunar base could be used separately or in any one of a number of combinations.

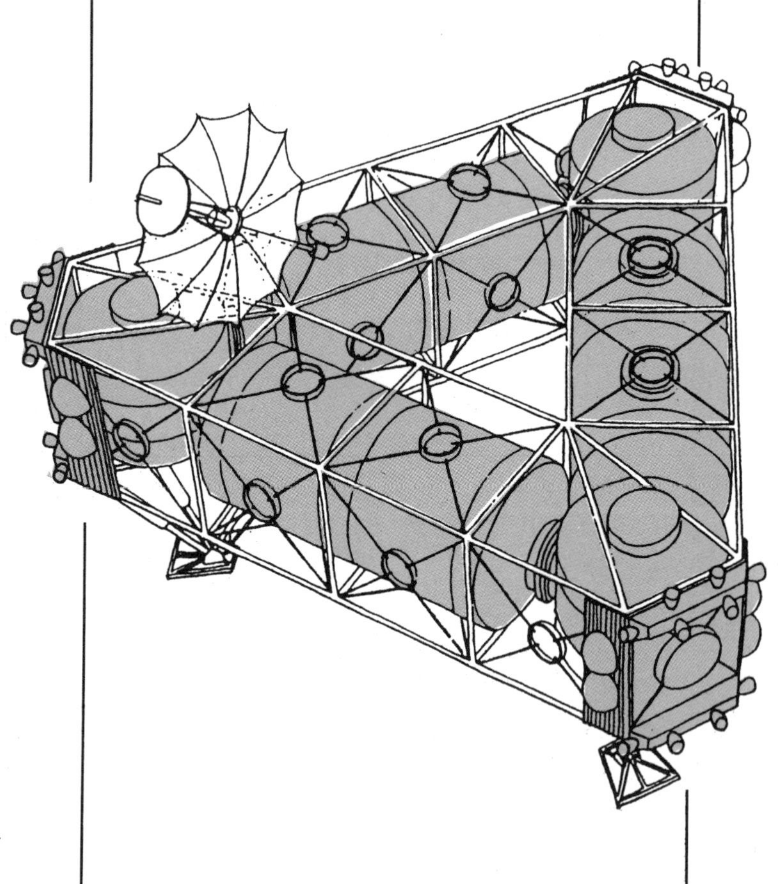

NOVEMBER

NASA initiates its Office of Space Flight Next Manned Transportation Systems Study (NMTS) to improve and evolve the existing space shuttle transportation system, develop an advanced manned launch system, and to replace the current space shuttle system or develop a less complex personnel launch system (PLS). Langley Research Center is given the lead responsibility for the AMLS study [see 1991].

ca. 1988

The Soviets are considering the possibility of a manned mission to Mars in the 1990s, perhaps using nuclear powered vehicles derived from the Mir space station. One observer of Soviet spaceflight, Phillip Clark, speculates on how the Russians might accomplish such a mission.

Four Energia launches would be required to place the necessary components of the spacecraft into earth orbit, while a fifth launch, this time of a Voskhod, would carry the three-person crew. The two Mars spacecraft would each have a 220 metric ton booster with a Proton second stage. One would carry a Salyut/Mir habitat and the other a Mars lander. Once reaching Mars, using their Proton stages for braking, the two spacecraft would join "nose to nose" and the two of the crewmembers would transfer to the lander. The lander would then descend to the surface using a combination of aerobraking and parachutes, making the landing on its retrorockets.

After the exploration period was completed, the cosmonauts would rejoin the orbiting Salyut/Mir and transfer to it, discarding the lander. The remaining fuel in the Proton would be used to leave Mars orbit. The return to the surface of the earth would be made in the Soyuz descent module.

1989

An 80-foot mockup of the National Aero-Space Plane is exhibited by the United States at the Paris Air Show. Constructed by the mechanical and aerospace engineering students of Virginia Tech, the model is made of a shell of expanded polystyrene foam over an aluminum truss frame. Construction began in January and eventually moved to the Virginia Tech Airport in March. An official rollout ceremony was held on May 5.

The Soviet Union may be developing a small spaceplane. It is expected to be operational by 1990, using a cluster of five of the new SL-16 rockets as its booster. The first

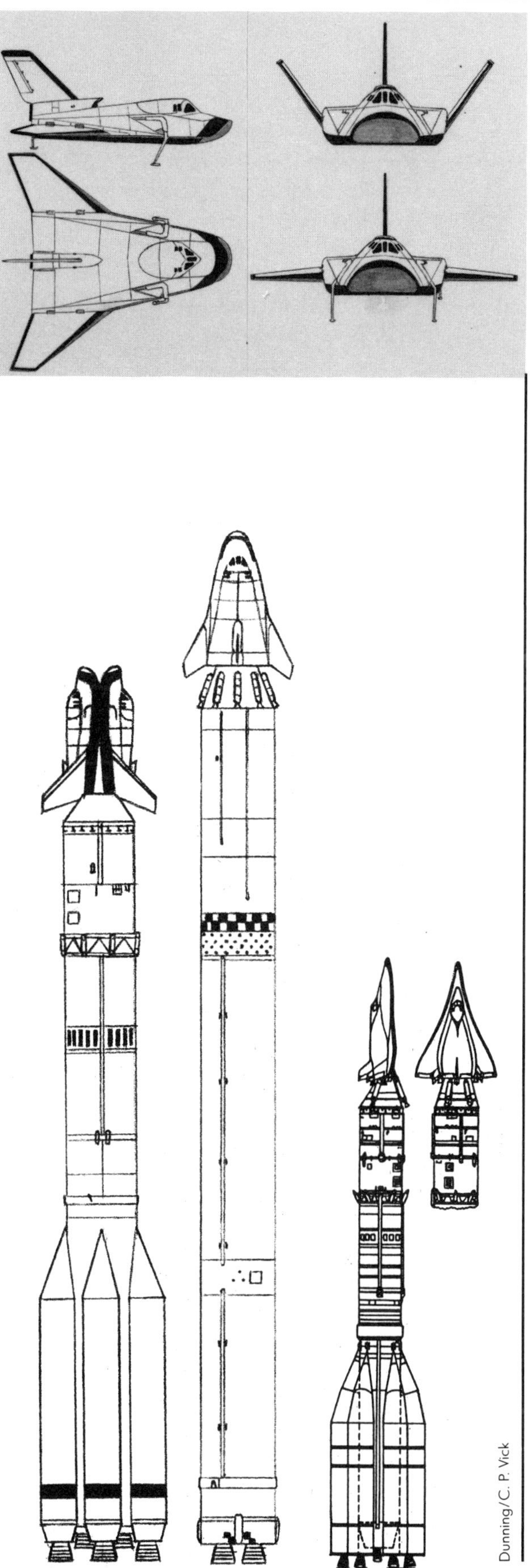
Dunning/C. P. Vick

orbital test flight of a model, Kosmolyot I ("spaceplane I") is made as Cosmos 1767. An earlier version, not dissimilar to the old U.S. Air Force Dyna-Soar [see 1963], was launched by the Proton SL-13, a booster similar in performance to the Titan that launched Dyna-Soar. This program is abandoned when the SL-13 fails to be man-rated. The SL-16 is inherently safer.

One observer of the development of the Kosmolyot suggests these specifications: that it has a wingspan of 13 m, a total length of 18.5 m and that it can carry up to ten passengers and a payload of 5 tons (in a 3 m × 5 m cargo bay). The total weight is 40 tons. Another viewer thinks that the weight might be closer to 20 to 23 tons—similar to ESA's *Hermes*. It is suggested that the *Hermes* and Kosmolyot may even be interchangable.

Kosmolyot II has wings with a sharp 45 degree dihedral, a vertical fin and a fatter fuselage than its predecessor. It resembles the Martin X-24B lifting body [see 1973].

It is believed that the test models represent a full-size vehicle of about 15 tons, with a wingspan of 9.4 m, a length of 16.25 m, and a height of about 3 m.

With its single large maneuvering engine, Kosmolyot will skim the upper atmosphere doing high-resolution reconnaisance at the relatively low altitude of 80 km (50 miles). It will then boost into a higher orbit to rendezvous with a space station.

NASA's Office of Exploration plans manned lunar and Mars missions in response to President Bush's initiative [see below].

As described by Ivan Bekey, the basic vehicle for establishing the lunar base is a reusable lunar transfer vehicle, which consists of a cluster of four boosters and a 60-foot conical aeroshell. It can carry an descent/ascent vehicle capable of transporting a payload of 44,000 pounds to the lunar surface. The lander somewhat resembles the Apollo excursion module. The transfer vehicle will cycle continously between earth orbit and lunar orbit, aerobraking on the earth-return leg sav-

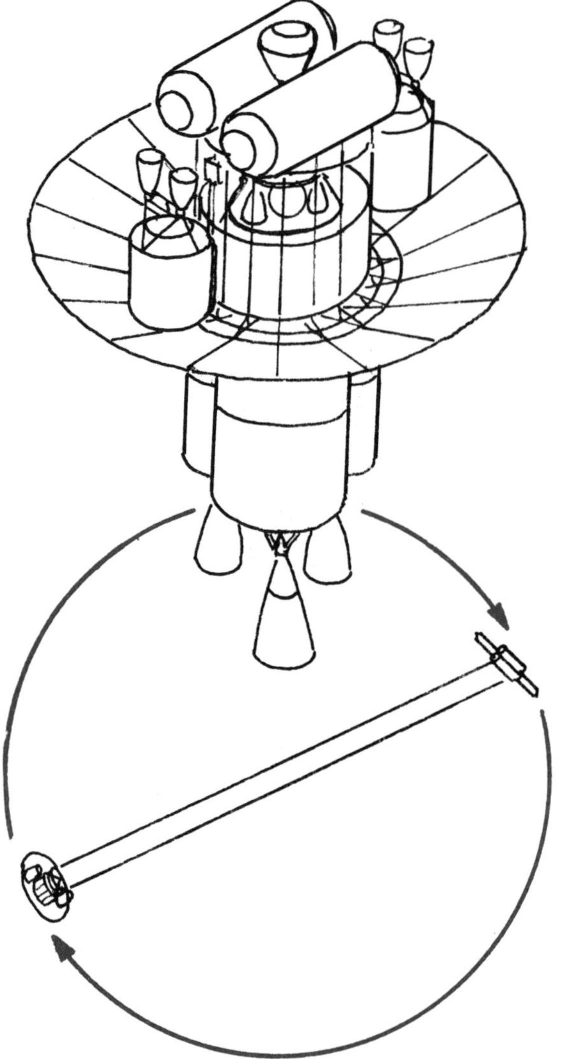

ing up to 50 percent of the fuel required. The schedule proposed for the settlement of the moon includes a period of preparation and development until 2003, after which a program of two missions per year will establish a tended lunar outpost between 2005 and 2010 and a permanent residence after that.

The Mars mission is planned to take advantage of the heavy-lift Shuttle Z (which can carry from 275,000 to 360,000 pounds of payload into space). The Mars spacecraft will be derived from the lunar transfer vehicle. Two launches will suffice to place the components of the Mars ship into orbit, where it will be assembled using the nearby facilities of the Freedom space station. Three more Shuttle Z launches will deliver the three stages of the earth orbit departure booster. A single conventional shuttle flight transfers the crew to the Mars ship.

The Mars ship consists of a 100-foot-diameter aerobrake shell for both earth and Mars braking. A smaller aerobrake for the descent module remains folded until needed. In the "dish" of the aerobrake, opposite the trans Mars injection stage, are the two cylindrical habitation modules—derived from those used for the space station—an Apollo-type crew return capsule, the descent/ascent vehicle and its aerobrake, and the two boosters for the return to earth. Artificial gravity is provided by tethering the vehicle to a counterweight and rotating the whole assembly around its center.

The first Mars mission will only go as far as Phobos, carrying three astronauts in the normally five-man spacecraft. Part of the mission's function is to land an unmanned ascent vehicle on the Martian surface, as a backup for the later manned lander. Other objectives will be to explore Phobos (and Mars via unmanned probes) and set up a pilot propellant plant. The spacecraft will return to earth after about 1 month.

A year later the second mission will take a crew of five to the Martian surface, where they will remain for about 1 year. A third mission will go to both Mars and Phobos, where it will erect the first half of a full-scale propellant plant. A fourth mission will complete the facility. The basis is then established for a permanent Mars infrastructure.

Under a NASA contract from Johnson Space Center, Eagle Engineering and Lockheed develop spacecraft to support a manned lunar outpost. The spacecraft are all dependent upon some configuration of the Freedom space station for support.

The lunar vehicles generally consist of an orbital transfer vehicle (OTV) with an aerobrake shell and a lander. The OTV and lander have four engines each, using liquid oxygen and liquid hydrogen. They can operate with or without a crew and carry a variety of payloads. When manned, a crew module is attached to the OTV or the lander, or both. If the vehicle is intended for reuse, its cargo capacity is 6 metric tons; if the vehicle is expendable, its one-way cargo capacity is 25 metric tons.

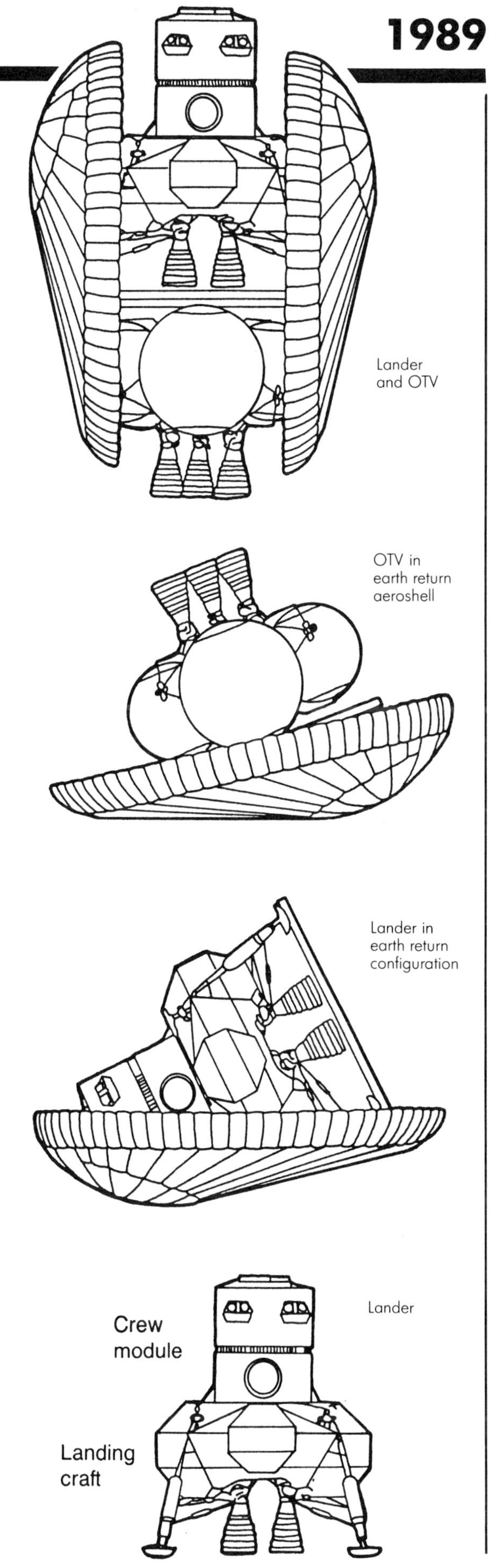

Lander and OTV

OTV in earth return aeroshell

Lander in earth return configuration

Lander

Four different configurations of the spacecraft are evolved, depending upon the mission requuirements. For exploration, the elements assembled include an OTV, a lander with crew module and an aeroshell. For unmanned cargo delivery delivery there is an expendable OTV carrying a lander with a large cargo module. The OTV has no aeroshell and is left in lunar orbit. For crew rotation an OTV carries one crew module to lunar orbit where it meets another landing craft with its own crew module. The two spacecraft crews will simply trade places. As a propellant tanker, an OTV with extra fuel tanks can deliver 40 metric tons of propellant to lunar orbit, enough to fuel two landing craft.

An alternative design makes the OTV an integral part of the aeroshell with the landing craft returning to earth orbit at the end of every mission. The lander's engines are used to propel the entire vehicle using the OTV's propellants.

The standardized crew module is a cylinder 4.5 meters in diameter and 6 meters high, divided into two floors. The upper section is the flight deck/habitation area and the lower compartment is the airlock chamber. There is a hatch at either end of the cylinder and one in the side of the airlock chamber. The upper hatch is for docking with other modules and the lower for entry and exit on the lunar surface. [Also see 1988.]

Dassault is working under a CNES contract to develop a two-stage spaceplane that closely resembles the German Sänger II. The French spaceplane, the Star-H, uses an air-breathing first stage to boost to Mach 6 and an altitude of 100,000 to 120,000 feet. A cryogenic second stage continues on into orbit. Star-H is able to carry up to 20 tons into orbit, perhaps eventually replacing the Ariane 5 as the launcher for *Hermes*, although this is a possibility denied by the minister of research and technology, Hubert Curien.

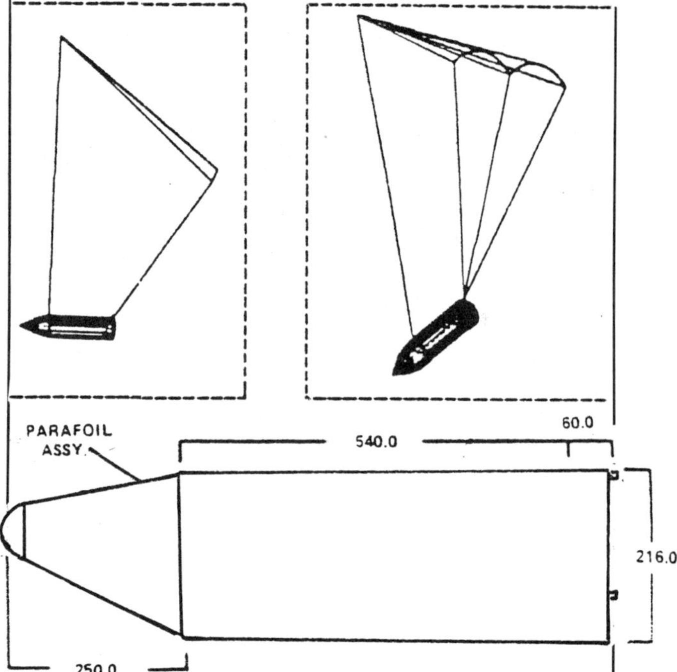

MSFC biconic reentry vehicle, 1987.

By late 1989, the French *Hermes* spaceplane [see 1985] consists of four major components: the crew compartment, the cargo/living area, the Hermes Resource Module (MRH), and a propulsion module. The latter consists of the two engines needed to boost the spacecraft into its correct orbit after separating from the Ariane 5 core stage. After completing its burn, the propulsion module is jettisoned. The MRH is a 28 m^3 truncated cone attached to the rear of the spaceplane. It acts as an airlock as well as accommodating scientific apparatus and cargo and carrying the radiators from the deleted cargo bay doors. The MRH will also be jettisoned before reentry. A tunnel links the MRH with the 25 m^3 pressurized payload/living area. A second tunnel links this with the cabin.

The large number of disposable parts is severely criticized because of the great cost they will incur. Its reduced cargo capacity will not be able to adequately support the Columbus space station—one of the prime duties of *Hermes*. So much of the original design has been compromised that manned capsules, similar to Apollo, are suggested as a viable alternative to the spaceplane [see 1987].

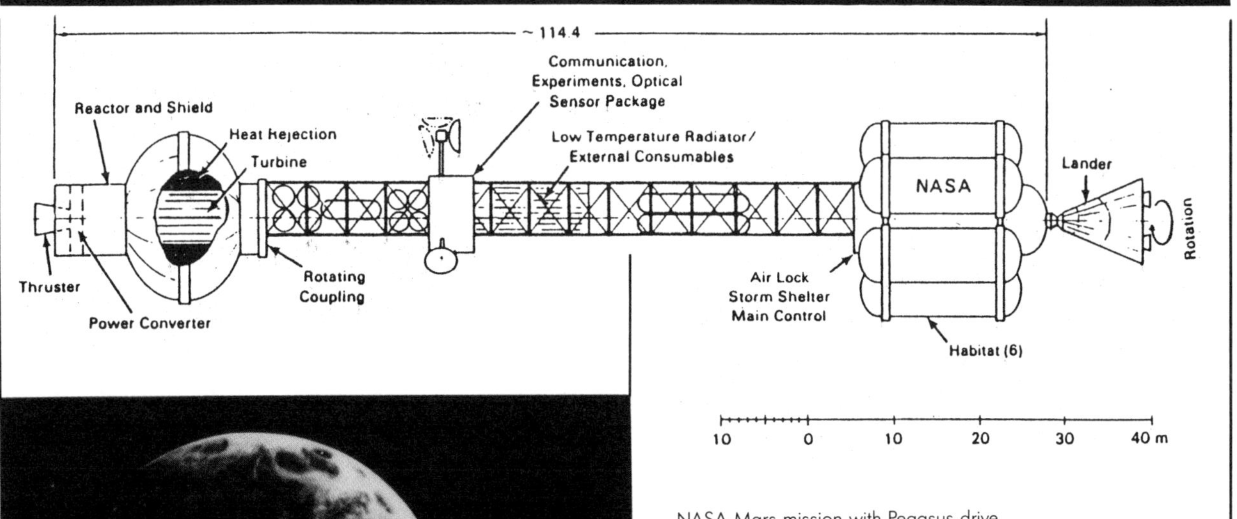

NASA Mars mission with Pegasus drive.

NASA

Ames Research Center's Space Science Division Mars Study Project publishes a scenario for the future of Mars exploration:

Phase	Date	Activity
Precursor	1965–2016	Obtain environmental knowledge and identify sites
Emplacement	2016–2024	First human missions with limited mobility on surface; small crews and short stays
Consolidation	2024–2034	Crew overlapping; resource utilization; base construction
Utilization	2030	Permanent, self-sufficient base

The European Space Agency (ESA) reports on the first two phases of its winged launcher configuration study, begun in 1988. Progress on the next two phases are reported in 1991. [See 1988 and 1991.]

A conclusion reached by NASA's Langley Research Center concerning the second generation Shuttle II is that single-stage rocket systems are not competitive in size and weight with two-stage vehicles, considering an assumed 1992 technology readiness date. Therefore, for operations in 2005, only two-stage concepts are being considered in Langley's Advanced Manned Launch Systems study. Both vertical and horizontal take-off vehicles are included. All are sized for the same polar servicing mission and technology levels. The five concepts studied include:

1. A two-stage fully reusable AMLS with an unmanned glide-back booster. At lift-off all engines are firing. The orbiter carries its payload in a piggyback container.
2. A partially reusable version of the first AMLS, with the addition of a pair of drop tanks that carry most of the hydrogen propellant, reducing the overall size and weight of the spacecraft as well as tank inspection.
3. A booster-core-glider concept that places all of the propellants in an expendable booster stage. This eliminates all tank inspection problems associated with the orbiter. A winged, unmanned glideback booster is attached to the core stage.

1989

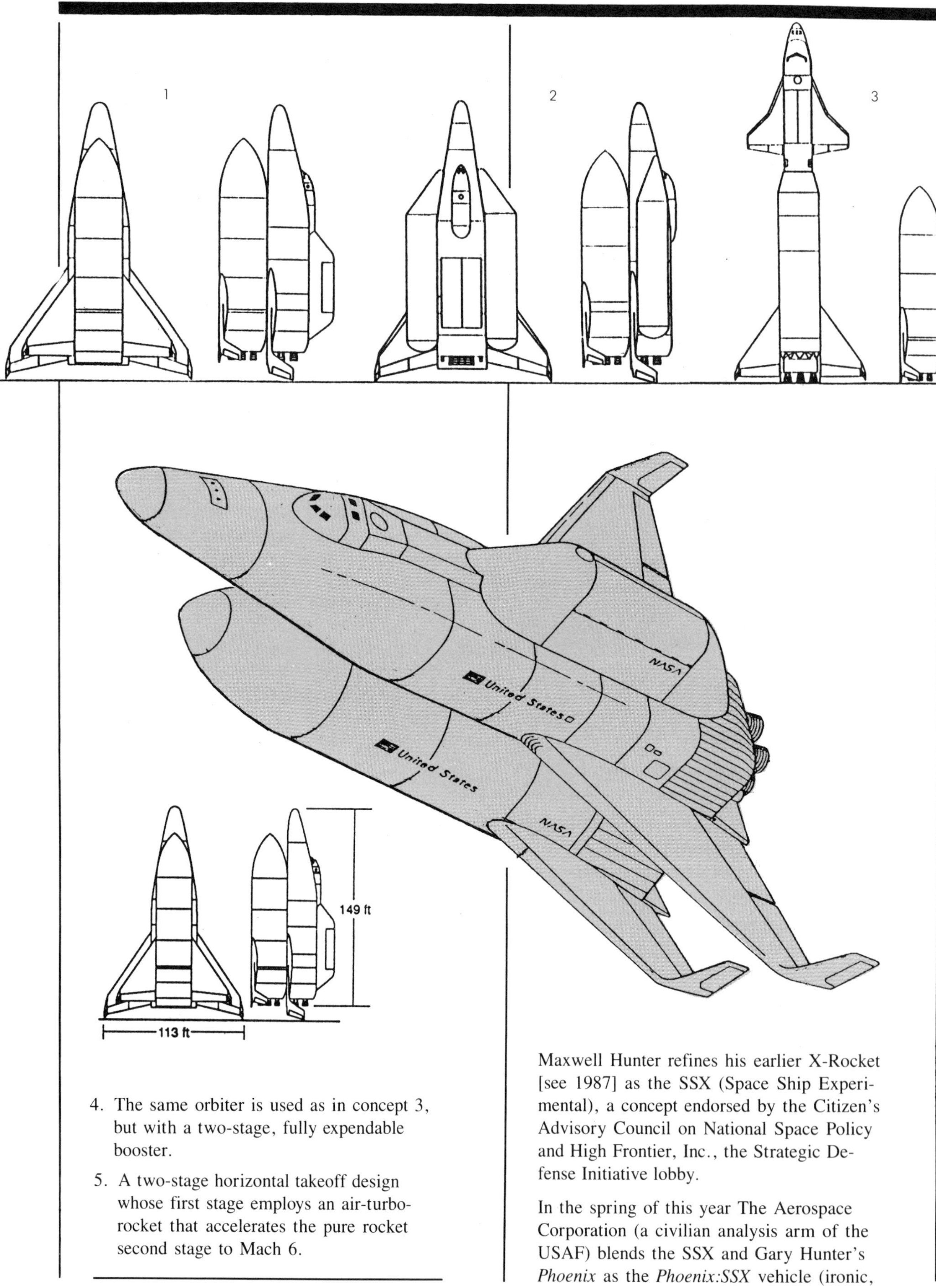

4. The same orbiter is used as in concept 3, but with a two-stage, fully expendable booster.
5. A two-stage horizontal takeoff design whose first stage employs an air-turbo-rocket that accelerates the pure rocket second stage to Mach 6.

Maxwell Hunter refines his earlier X-Rocket [see 1987] as the SSX (Space Ship Experimental), a concept endorsed by the Citizen's Advisory Council on National Space Policy and High Frontier, Inc., the Strategic Defense Initiative lobby.

In the spring of this year The Aerospace Corporation (a civilian analysis arm of the USAF) blends the SSX and Gary Hunter's *Phoenix* as the *Phoenix:SSX* vehicle (ironic,

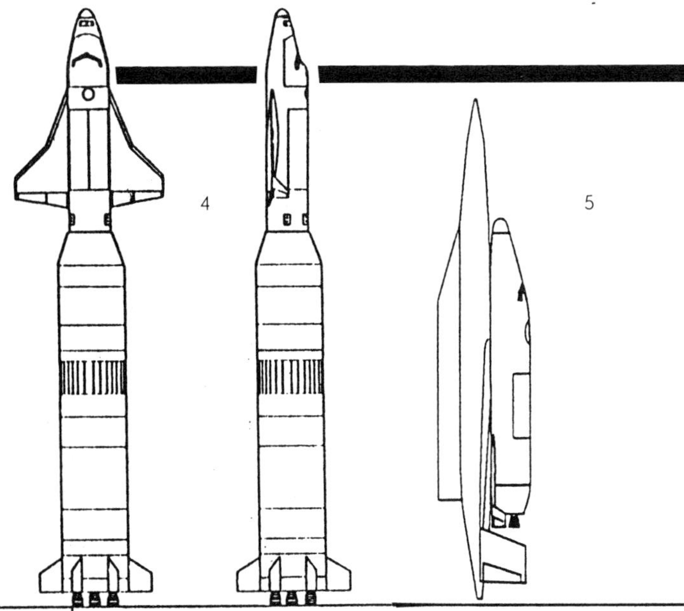

since a senior Aerospace Corporation executive had earlier derided both the *Phoenix* and the X-Rocket).

JULY 20

On the 20th anniversary of the Apollo 11 moon landing, President George Bush remarks "We must commit ourselves anew to a sustained program of [human] exploration of the Solar System . . . and yes, the permanent settlement of space . . . And for the next century—back to the Moon. Back to the future. And this time, back to stay. And then—a journey into tomorrow—a journey to another planet—a [human] mission to Mars."

Early in 1990 President Bush adds, "I believe that before Apollo celebrates the 50th anniversary of its landing on the Moon the American flag should be planted on Mars."

Bush calls upon Vice President Dan Quayle to lead the National Space Council in determining what is required to transform these goals into reality. One result of these studies is that on May 11, 1990, Bush sets a timetable for an expedition to Mars. The president declares that the United States is able to land astronauts on Mars within 30 years. Commenting on what will become the Space Exploration Initiative (issued earlier in the year as a series of policy directives, the studies collectively form the Space Exploration Initiative), Office of Management and Budget Director Richard Darman says that SEI is " . . . at the heart of the American Romance . . . Man is meant to pioneer, to explore, to expand, to advance, to reach and exceed new frontiers . . . in this land of the Wright brothers and the Right Stuff, we must get beyond the policy equivalent of the fear of flying."

However, the 18 October 1990 House/Senate conference report, in considering NASA's 1991 budget, states: "The conferees have deferred consideration of the proposed Space Exploration Initiative due to severe budgetary constraints which limit the agency's ability to maintain previously authorized projects and activities. It is implicit in the conduct of the nation's civil space program that such human exploration of our Solar System is inevitable. Although initiation of a focused program is not possible at this time, the conferees recognize that relevant research and technology development activities should be continued by NASA and that preliminary conceptual design studies are considered if appropriate resources are identified or made available." The mission studies, a $37 million budget item, remain unfunded. The full cost of SEI is estimated in the neighborhood of $400 billion.

NOVEMBER 15

The Soviet VKK (Vozdushno-Kosmicheskiy Korabl, or Air-Spacecraft) *Buran* is successfully launched from Baikonur, making a perfect landing 3 hours and 25 minutes later on a runway only 12 km from its launch site.

The flight is unmanned, the landing being handled by an onboard computer (that James Oberg, an expert observer of Soviet spaceflight, remarks is *only* capable of landing at Baikonur).

The second orbiter, *Ptichka* (Birdie), is being prepared, along with at least one other.

Buran is covered with 38,000 fragile ceramic tiles. It is slightly larger than its U.S. counterpart, with a wingspan of 24 m, a fuselage diameter of 5.6 m and a length of 36 m. It has a split-level crew compartment with a volume of nearly 70 m^3 and can accommodate two to four cosmonauts and six passengers. Its maximum launch weight is 105 tons and landing weight is 82 tons. It can carry 30 tons into orbit and return to

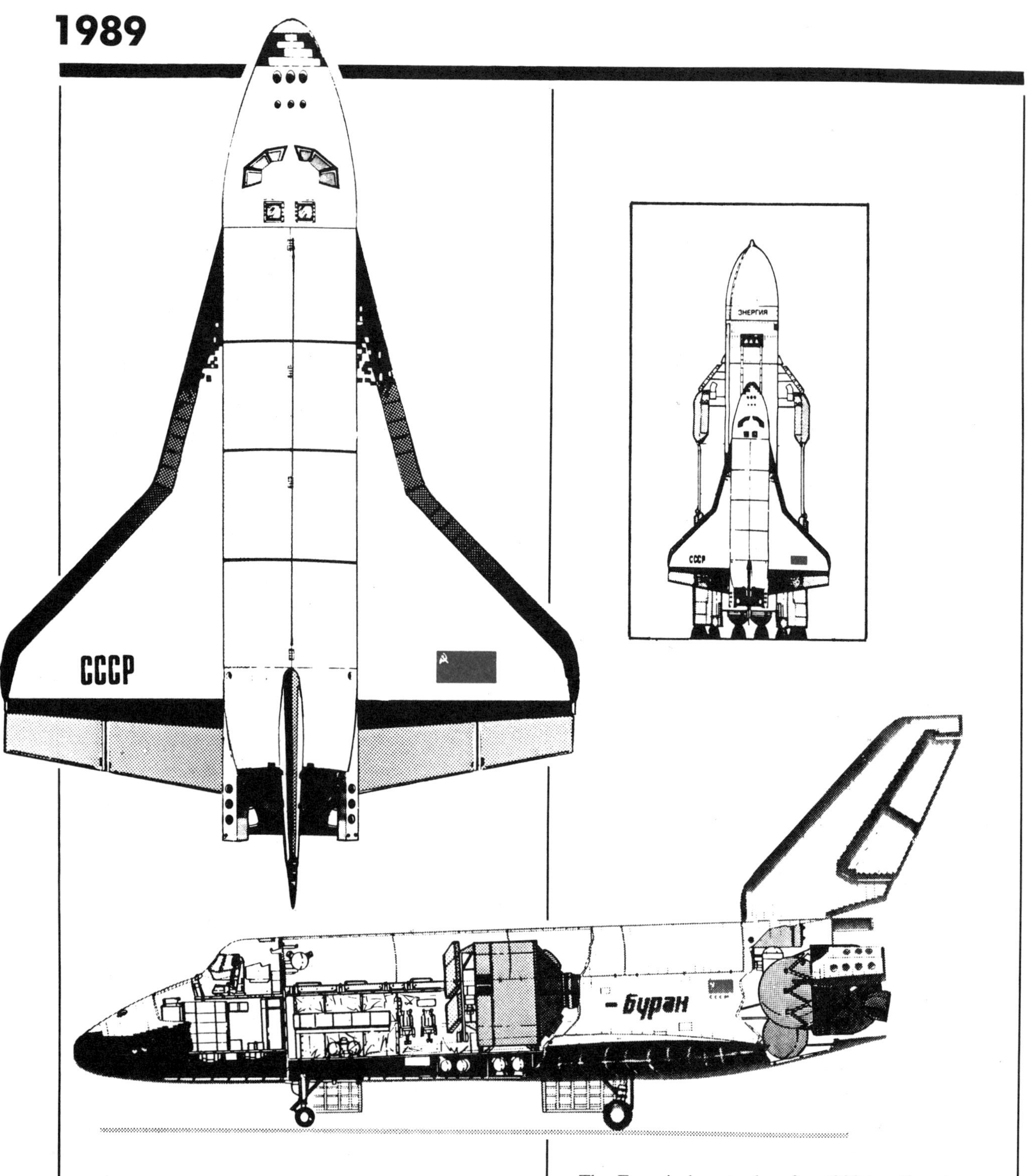

the earth with 20. Instead of the strap-on solid fuel boosters of the American shuttle, *Buran* is lifted by the huge 3,500-ton-thrust Energia rocket. The total weight of the combined system at launch is 2,400 tons, of which 90 percent is fuel.

The outward resemblance between the Soviet and American shuttles is explained by the Russians as a coincidence "prompted by laws of aerodynamics."

The Energia booster has four 200-ton-thrust engines fueled by hydrogen and oxygen. In addition, it has four strap-on liquid fuel boosters, each with a four-chamber RD-170 engine with 800 tons of thrust, fueled by kerosene and oxygen. The boosters are equipped with parachutes and are recoverable. Since the Energia carries its own propulsion system it can be used independently of the shuttle to launch unmanned cargo containers, something that the U.S. system is presently incapable of doing.

In place of the main engines of the U.S. shuttle, the *Buran* has smaller orbital maneuvering engines. Like the U.S. shuttle, however, *Buran* is equipped with fuel cells for electrical power, a first for the Soviet space program.

Buran is launched after 10 years of development and a final cost of $10 billion.

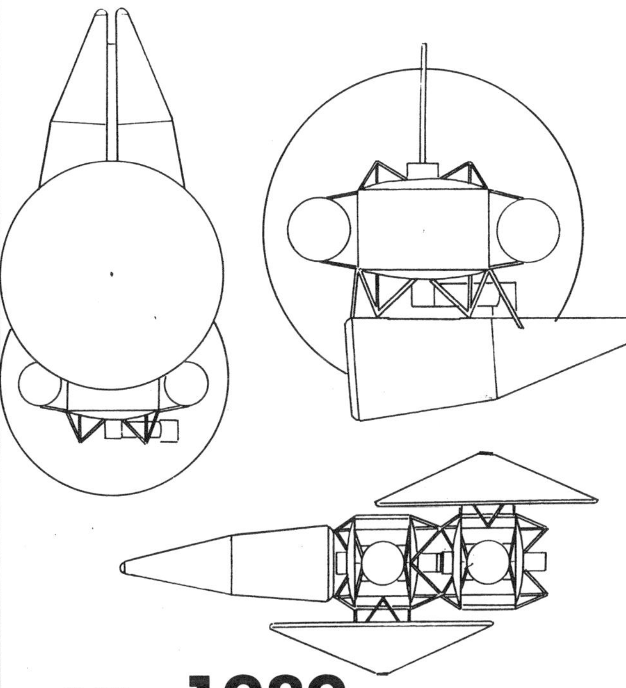

ca. 1989

L. G. Lemke and M. A. Smith of NASA's Ames Research Center prepare an adaptation of the Case for Mars manned Mars mission scenario [see 1984]. The authors do not consider the Case for Mars spacecraft a workable design. They replace the original fifteen astronauts with ten and use a two-unit rotating spacecraft connected by cables rather than a three-unit structure connected by rigid trusses. The cables connecting the two habitats are about 168 m long. A rotation period of 2 rpm gives 0.375 g. An electric "elevator" is connected to the cable and allows single astronauts to travel back and forth between the modules.

A heavy lift launch vehicle is used to place the components of the spacecraft into orbit. The habitation module is a single unit rather than a pair of space station modules (as in the Case for Mars spaceship). This is a barrel-shaped structure 8.4 m in diameter, divided lengthwise into two floors. A central tunnel surrounded by fuel tanks serves as plenum chambers for the airlocks and emergency radiation storm shelters.

The Mars landing shuttles (similar to those in the Case for Mars study) are connected to the main habitat by pressurized tunnels. The two-stage, 20-meter shuttles are carried fully fueled.

1990

NASA creates an internal report: Report of the 90-Day Study on Human Exploration of the Moon and Mars. Drafted by NASA-industry teams led by the Johnson Space Center, it is criticized as being "a hodge podge of wish lists thrown together to satisfy each NASA center." A panel that includes Thomas Paine [see 1986], John Logsdon and Tom Clancy gives the report a thumbs down.

Buzz Aldrin proposes a manned Mars mission. His scenario favors the cycling spacecraft suggested by the National Commission on Space [see 1986].

Aldrin's plan will work according to this schedule:

1997: First space taxi is launched.

1998: Construction begins on space station *Freedom* and its "starport."

1999: Starport is complete; space taxi makes first lunar landing.

2002: Low lunar orbit spaceport is established.

2004: Advanced taxis and landers are developed.

2005: Lunar starport is transferred to L-2; lunar propellants are developed.

2006: Mars starport is tested.

2007: Mars starport is transferred to Mars orbit.

2009: First manned mission to Mars via noncycling taxi takes place; this taxi

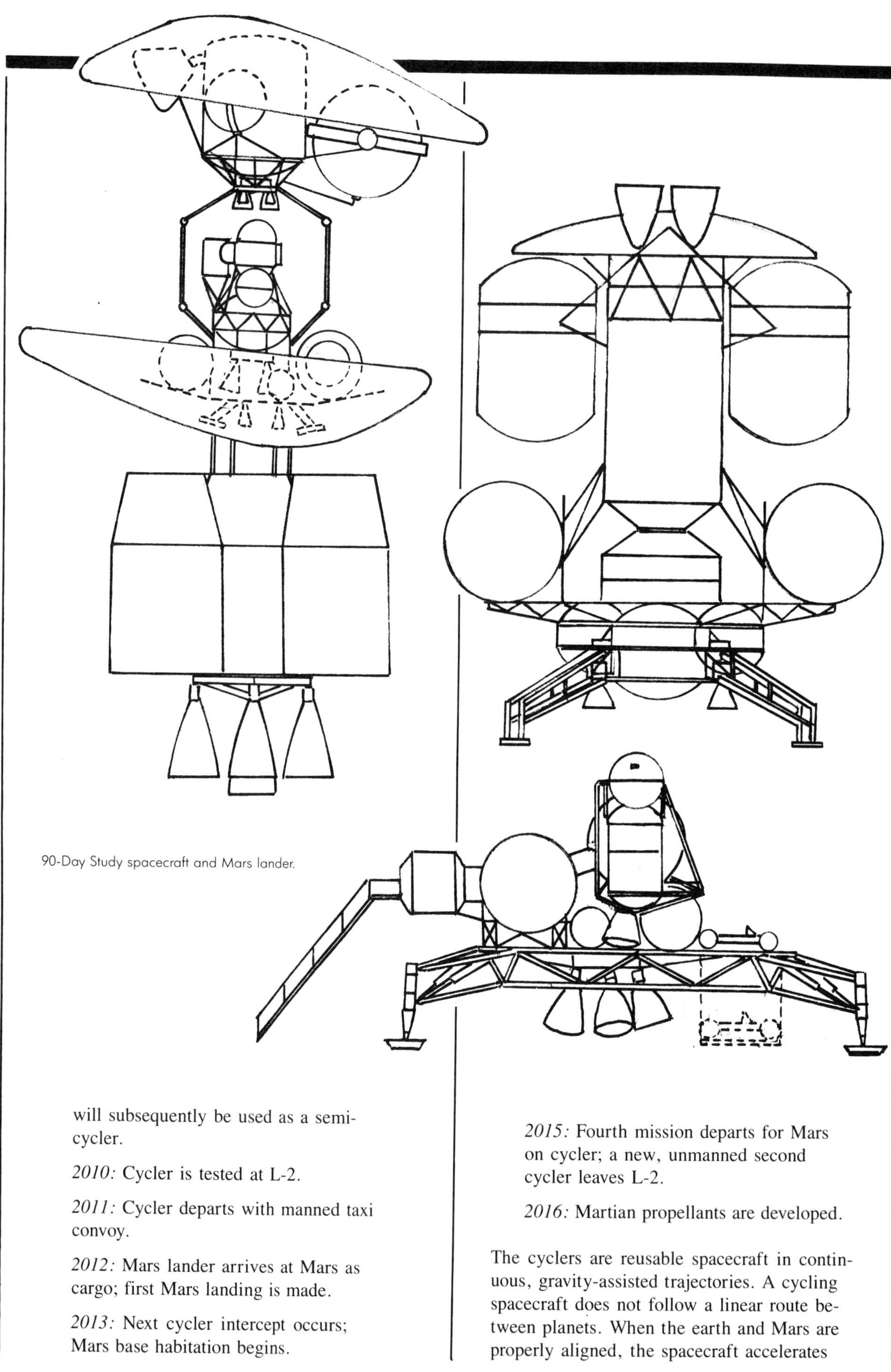
90-Day Study spacecraft and Mars lander.

will subsequently be used as a semi-cycler.

2010: Cycler is tested at L-2.

2011: Cycler departs with manned taxi convoy.

2012: Mars lander arrives at Mars as cargo; first Mars landing is made.

2013: Next cycler intercept occurs; Mars base habitation begins.

2015: Fourth mission departs for Mars on cycler; a new, unmanned second cycler leaves L-2.

2016: Martian propellants are developed.

The cyclers are reusable spacecraft in continuous, gravity-assisted trajectories. A cycling spacecraft does not follow a linear route between planets. When the earth and Mars are properly aligned, the spacecraft accelerates away from the earth, loops outward, and swings close to Mars about 5 months later. Instead of braking into a Martian orbit or landing, the cycler drops off smaller manned vehicles or cargo pods. The cycler then continues on, using the boost it receives from Mars' gravity to curve outward for 8 more months before swinging back toward the earth. This unmanned return trip takes 21 months.

A typical journey to Mars and back might follow this scenario:

1. Two or three space taxis carry the crew from the low earth orbit starport. Entering a staging orbit, they await their rendezvous with the cycler. The taxis mate with refueled propulsion stages from L-2.
2. After leaving the staging orbit, the convoy intercepts the cycler, beginning the 5-month trip to Mars.
3. Before encountering Mars, the propulsion stages for the return trip depart for the Mars starport, arriving ahead of the crew.
4. The crewmembers depart the cycler in their taxis, arriving at Mars 2 or 3 days later, aerobraking in the Martian atmosphere before rendezvousing with the starport.
5. The now unmanned cycler continues on its 21-month-long loop back to the earth.
6. The semi-cycler (so-called because it does not continuously cycle between the two planets), returning on the earth-Mars leg of its trip, aerobrakes into Mars orbit about 16 months after the crewmembers' arrival and 4 months before their departure.
7. The crewmembers mate their taxis and the propulsion stages with the semi-cycler and depart from Mars starport.
8. Just before earth encounter, the taxis and propulsion stages separate from the semi-cycler and head for the low earth orbit starport.
9. The now unmanned semi-cycler continues on its 14-month return to Mars where it aerobrakes for recovery by the next mission crew.

The starports are octotetrahedrons built of lightweight beams. The LEO starport can be attached to the keel of the *Freedom* space station. The facility consists of six concave ports, separated by six tetrahedrons, all connected at the apex to a central pressurized command module with multiple ports. The starport provides facilities for repair, fueling, storage, maintainance, etc. The lunar and Martian starports are similar, with perhaps the addition of rotation to provide artificial gravity.

The core of the cycler is based on the same octotetrahedron as the starports. The three-component cycler will consist of this central starport hub, a habitation module and a nuclear power plant. The three components are attached by tethers and spun to create an artificial gravity, the length of the tethers determining the g-force. The three components are deployed after the booster stages launched the cycler into its trajectory.

The semi-cycler consists of a habitat module, the propulsion stages and transfer taxis.

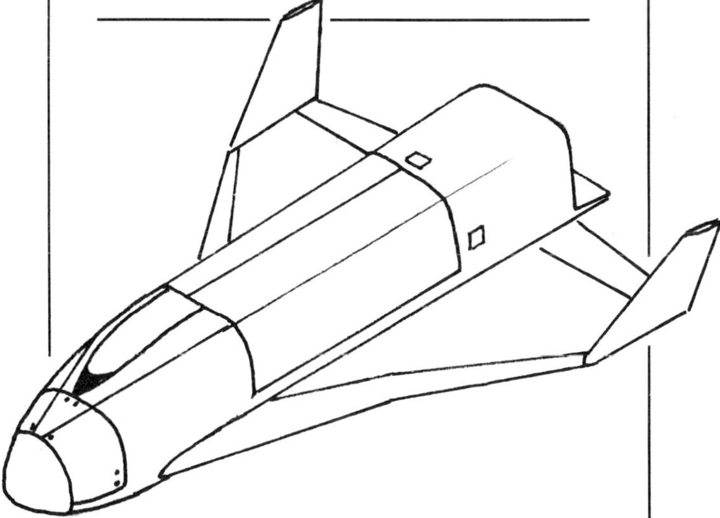

The Japanese National Space Development Agency (NASDA) develops its H-II Orbiting Plane, or HOPE spaceplane. Similar to the French *Hermes*, it is to be launched atop the Japanese H-II booster from the Tanegashima

1990

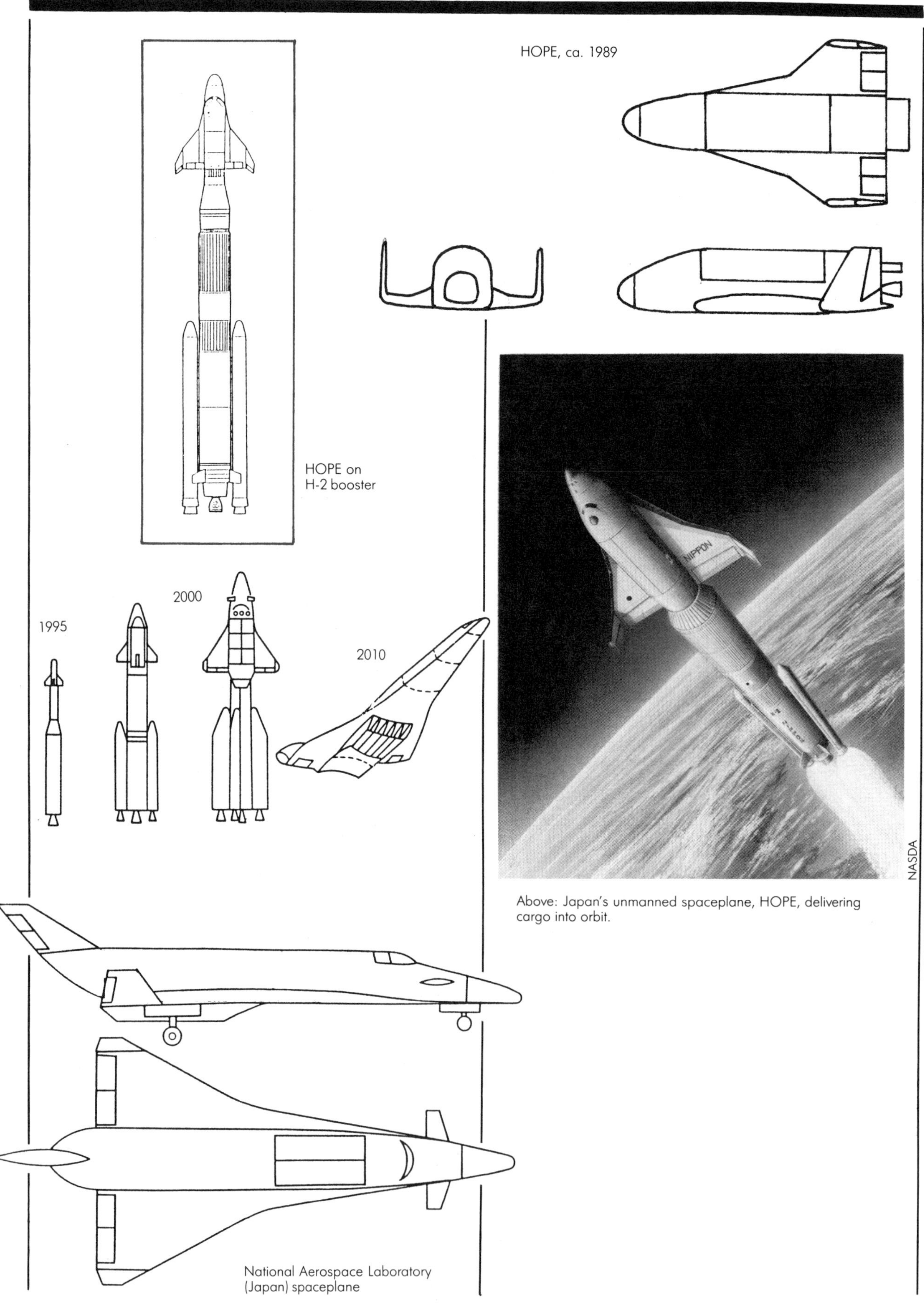

Above: Japan's unmanned spaceplane, HOPE, delivering cargo into orbit.

Mitsubishi aerospace plane

Future HOPE

Space Center. It will achieve a parking orbit 250 km above the earth approximately 14 minutes after takeoff. The unmanned H-II is expected to be operational in 1992, with the spaceplane's first flight occurring in 1996. HOPE is 11.3 m long, 2.3 m wide and 2.5 m high. The wingspan is 6.24 m with two 12.3 m vertical tip fins. The 8.8 ton spacecraft has a payload capacity of 1.2 tons and will be able to remain in orbit for up to 100 hours.

There have been five variations on the HOPE concept: (1) a 10-ton unmanned spaceplane (U1) to be developed first, capable of carrying a 3-ton payload; (2) a 10-ton manned spaceplane (M1) with a crew of two and a 1-ton payload; (3) a 20-ton spaceplane (M2) with a crew of four and 4 tons of payload; (4) a 29-ton vehicle (M4) with a crew of two and a payload of 1 ton, plus internal propulsion; and (5) a 10-ton manned spaceplane (M1J) with a jet engine, two crewmembers and a 1-ton payload.

In addition, Japan is also working on a larger spaceplane more akin to a smaller HOTOL or the Sänger, though it is not expected to fly before 2006. This will carry four astronauts to orbits ranging up to 310 miles. A small test vehicle will be launched atop an H-I booster in the early 1990s. Larger manned versions could be launched by the H-II in the late 1990s. The large spaceplane will have scramjet propulsion mounted beneath the fuselage and will eventually lead to full-size spaceplanes similar to the U.S. NASP.

Martin Marietta Astronautics Group considers building a three to four-person habitat in one of the dry valleys of Antarctica to test resource recycling and other hardware in a Mars-analogous environment. Engineer Benton Clark proposes a Mars Transfer Vehicle that can house up to nine astronauts in habitats derived from those of the Freedom space station. The propulsion system employs existing chemical engines (though the craft could be modified to use a nuclear engine if a practical one is developed by that time). The MTV carries a descent module with the two habitat modules at the ends of arms, allowing artificial gravity to be produced via rotation.

Senior engineers Robert Zubrin and David Baker describe Martin Marietta's Mars Direct scheme, which uses a shuttle-derived heavy-lift vehicle known as Ares (with two Advanced Solid Rocket Boosters, four space shuttle main engines, a modified external tank and a hydrogen/oxygen upper stage with 250,000 pounds of thrust). In 1997 Ares will launch a 40-ton automated payload to Mars which will aerobrake into the Martian atmosphere and then parachute to the surface. The payload is to consist of a two-stage ascent and earth-return vehicle (ERV, without fuel), 6 tons of liquid hydrogen, a 10-kW nuclear reactor, a small compressor and an automated chemical processing plant. After landing, the reactor is moved several hundred meters away where it will begin to operate the compressors and processor. The

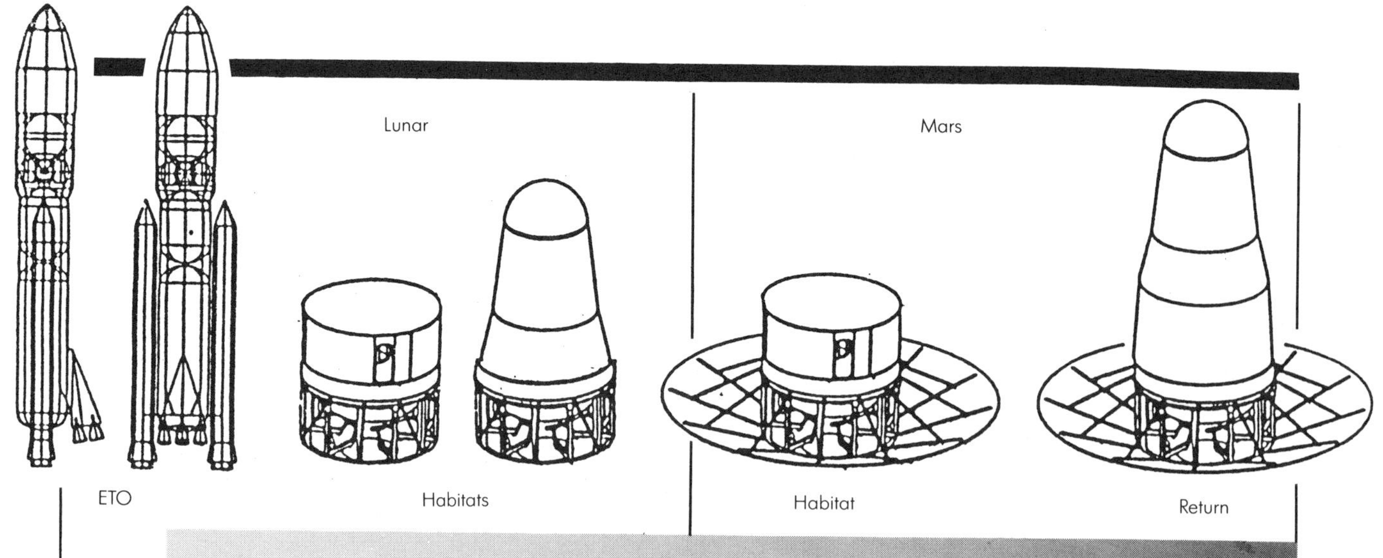

Martin Marietta

hydrogen is combined with the Martian atmospheric CO_2 to produce methane and water. The former is liquified and stored. The water is broken down to make oxygen, which is stored, and more hydrogen, which is recycled. This process will ultimately create 24 tons of methane and 48 tons of oxygen. In order to efficiently burn the methane, an additional 36 tons of oxygen is produced by breaking down the Martian CO_2. This entire process takes 10 months at the end of which there will be enough fuel for the ERV and 11 tons left over for the long-range ground vehicles. Once the earth controllers have confirmed this operation, two more Ares boosters are to be launched. One payload is identical to the first, the other is a habitation module containing a crew of four with enough provisions for 3 years, a methane/oxygen-powered rover, and an aerobrake/landing engine assembly. The manned vehicle is to land in 1999 at the 1997 landing site,

Martin Marietta

25.5 m
5.6 m
3.0 m
4.9 m
2.8 m
3.1 m
8.4 m

Earth return vehicle

where a fully fueled ERV awaits. The second ERV lands nearby and begins making propellant for the next human landing. The crew remains on Mars for a year and a half. With an 11-ton supply of fuel, they have enough to roam over 16,000 km of Mars.

A variation of the Mars Direct plan allows similar lunar missions. The Ares launches a 59-ton payload to the moon (one or more unmanned habitation modules are landed first). The crew then lands in an ERV without a lower stage. After conducting their mission, the crew returns to earth orbit in the ERV. A crew of twelve can be accommodated using the same hardware as the four-man Mars mission. A single Ares is required to take the necessary oxygen for the return trips; for safety enough for four missions can be carried by one Ares. By then end of the fourth mission a fully operational oxygen processing plant is running.

Mars Direct is a 10-year program at one-fourth the cost of the $0.5 trillion, 30-year SEI program.

The General Dynamics Space Systems Division proposes a diverse number of spacecraft designs intended for the manned exploration of Mars. Engineers there believe that President Bush's goal of landing on Mars by 2019 could be achieved at least a decade earlier.

Boeing Aerospace and Electronics Division completes a NASA study for earth-moon transfer vehicles and Mars ascent and descent vehicles, as well as surface rovers.

Ed Repic, Chief Engineer for Rockwell International's Space Division SEI program believes that the moon should be reached again before any manned landings are made on Mars. For the Martian journey Rockwell engineers are considering dual mode, nuclear thermal and nuclear electric propulsion. An inflatable Mars transfer vehicle would save valuable space.

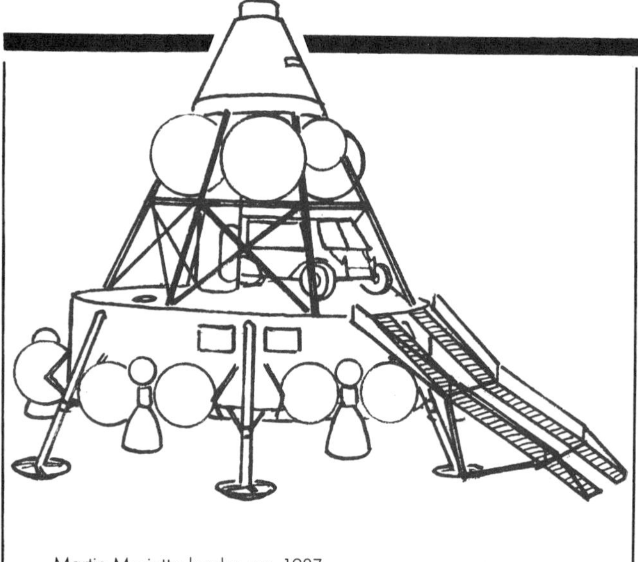

Martin Marietta lander, ca. 1987.

The Soviet Union looks for international partners for an air-launched spaceplane project. The small spacecraft, which can be flown either manned or unmanned, is to be carried, along with its large external tank, on the back of a giant Antonov An-225 aircraft. The manned version can carry 7 tons of payload into an orbit 200 km above the earth. The unmanned version has a payload capacity of 8 tons. The spaceplane with its tank weighs 250 tons. [Also see the Interim HOTOL project, 1991.]

Martin Marietta

The civilian cosmonaut training manager, Aleksandr Aleksandrov, reveals that at one time the Soviet Union considered recruiting a husband and wife team of cosmonauts for a long duration spaceflight. The idea of flying mixed couples on such flights raised "moral and ethical problems"; however, flying mixed couples on short duration flights, such as those of the American space shuttle, is not ruled out.

Aerospatiale announces that it has nearly finalized the design of the *Hermes* spaceplane. *Hermes* will now be inserted directly into orbit by the Ariane 5 and will not require the expendable propulsion module that was an earlier feature. *Hermes* will accommodate three astronauts occupying ejection seats. Four windows will be provided in spite of the weight and technical problems they present. Unlike HOTOL, whose reentry does not require it to have any special thermal protec-

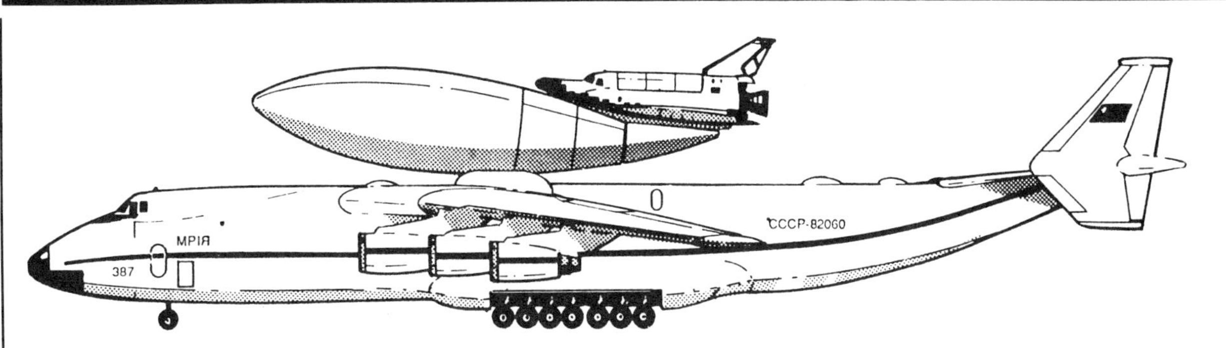

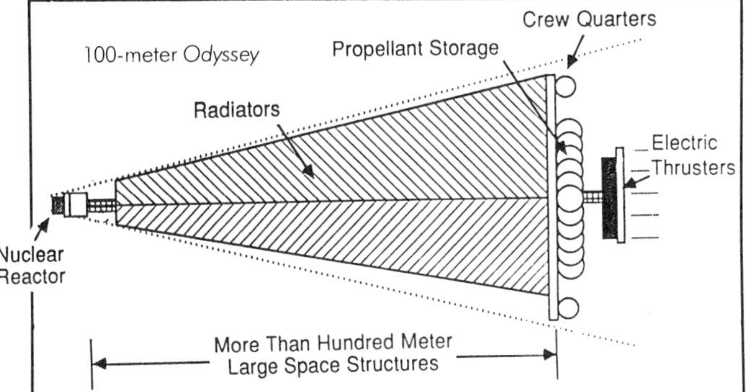

tion, *Hermes* will have flexible quartz-fiber blankets on its upper surfaces and carbon/silica tiles on the lower. Nose and leading edges will be protected by carbon-carbon.

Subsonic flight tests could begin in 1996 with an unmanned orbital flight in 1998 and the first manned flight the next year. *Hermes* could be fully operational by 2015.

In spite of Aerospatiale's optimism, there are still doubts within the industry about this spaceplane's viability. Increasing costs, questions about its usefulness and concern about its weight are all factors. It is intended to service the Columbus Free-Flying Laboratory. However, with a maximum of only four flights a year, each of 12 days duration, Columbus will be unmanned 10 months out of each year. Patrick Baudry, sole test pilot for *Hermes*, maintains his belief in the necessity of developing the spaceplane, especially if a European single-stage-to-orbit vehicle is contemplated for the future.

Meanwhile, Aerospatiale is studying an advanced version of the Ariane 5 with a recoverable liquid-fueled booster stage. This could land like a conventional aircraft back at the Guiana Space Center at Kourou.

The White House/NASA Advisory Committee on the Future of the U.S. Space Program presents its report. The eleven-member Committee is headed by Norman R. Augustine, chief executive officer of Martin Marietta.

Space science, the committee decides, warrents the highest priority, above the space station, aero-space plane, and manned missions to the planets. Nevertheless, the United States will honor its commitments to its foreign space station partners. Instead of going ahead with the construction of a fifth shuttle orbiter, the report recommends development of a new heavy-lift vehicle. Shuttle operations would be reduced to the bare minimum by the early twenty-first century.

NASA Langley develops a concept for a near future two-stage Advanced Manned Launch System for space station support. The vertically launched shuttle vehicle could deliver 10 people or 40,000 pounds to an orbiting space station. The parallel liquid oxygen / liquid hydrogen-fueled booster is equipped with wings for an unmanned return to earth. The 146-foot manned orbiter is mostly a huge fuel tank carrying a jettisonable crew module and 15-foot by 30-foot cargo bay piggyback. The crew compartment and cargo bay are connected by a tunnel.

Also proposed is a single-stage-to-orbit, vertical takeoff/horizontal landing version that is 179 feet long, fueled by triple-point oxygen and slush hydrogen.

An engineering model of the proposed HL-20 PLS (Personnel Launch System), or "space taxi," is studied at NASA's Langley Research Center. This mini-shuttle, not dissimilar to the *Hermes*, is to be launched by a Titan IV. It can service satellites, undertake observational missions of up to 3 days in duration, or deliver payloads or up to 10

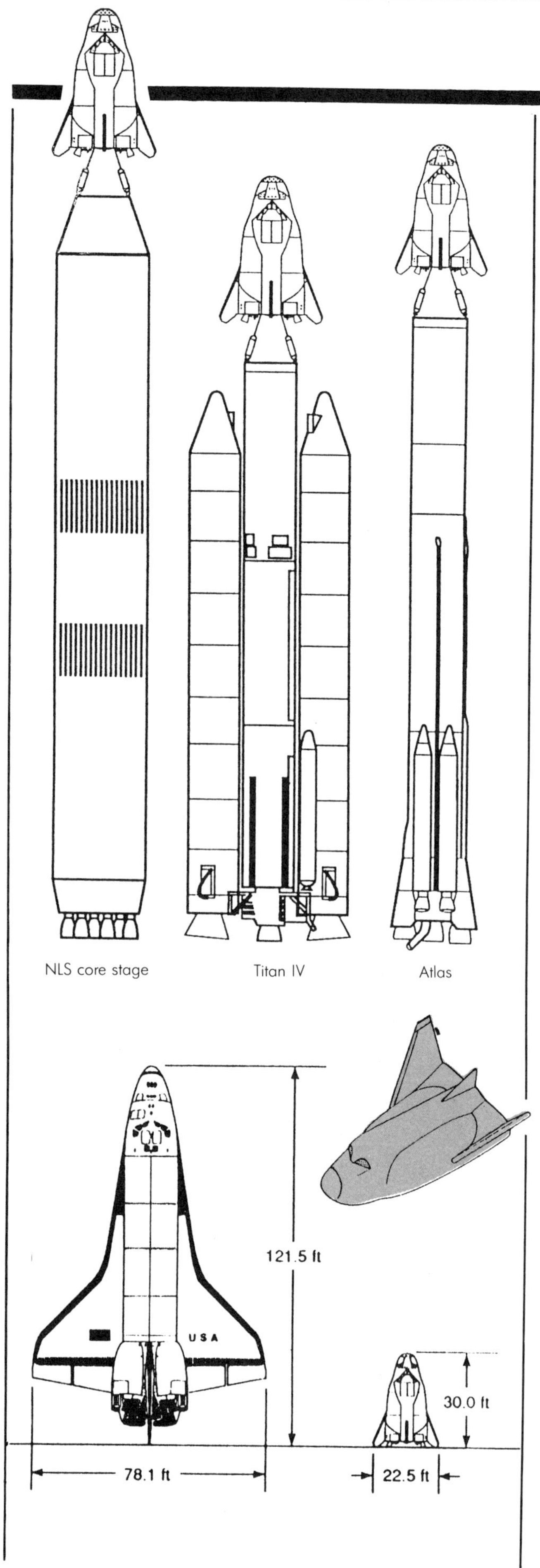

people (two crewmembers and eight passengers) to a space station. The HL-10 is similar to several lifting-body vehicles flown in the United States in the 1960s and early 1970s, such as the HL-10, M2-F2, X-24A and X-24B.

The HL-10 is approximately 30 feet long (the actual length depends upon the number of crewmembers in the final version chosen —dimensions vary from a 30-foot length and 23.5-foot wingspan to lengths of 28.2 feet and 24.6 feet). It weighs 1/10th as much as the space shuttle. The vehicle could be launched by a NLS Core Stage, a Titan IV or an Atlas IIAS (in a four-man version). It can land anywhere within 700 miles to either side of its orbital path and could be developed in 8 to 10 years.

Langley, on a $1 million-a-year budget, has other space shuttle alternatives planned, including one that resembles an Apollo capsule and another that resembles a bullet. Both would land by parachute.

Some Soviet space observers have noted the striking resemblance between the HL-20 and the small Soviet spaceplane tested a few years earlier [see above]. Their belief is that, despite the protests of NASA Langley, the American spaceplane was actually developed from studies of its Russian counterpart.

A plan for a permanent Mars base is created by J. R. French, R. L. Staehle, C. R. Stoker, C. Emmart, and S. B. Welch. The vehicles required by the scheme are assembled and checked out an earth-orbiting assembly dock. Each spacecraft consists of three arms, each with a crew of five. These are launched separately by trans-Mars injection stages. Once on the way to Mars, they link up into a Y shape, rotating around the central axis at 3 rpm to provide 0.38 g of gravity artificially. The complete Interplanetary Vehicle makes a flyby of Mars, utilizing a brief powered maneuver to enter an earth-return trajectory. During the flyby, each crew of five boards a Mars Crew Shuttle and aerobrakes to a landing on Mars. Once there, they direct the landing of the cargo vehicles, which have been launched separately. These

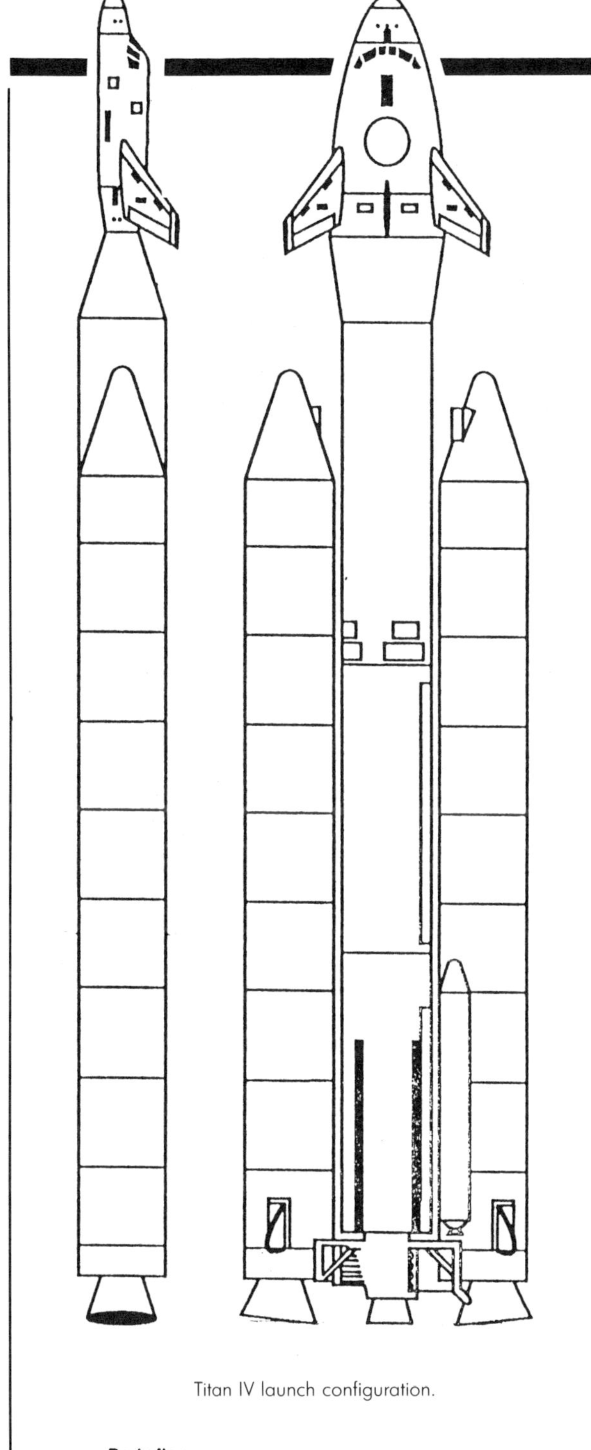

Titan IV launch configuration.

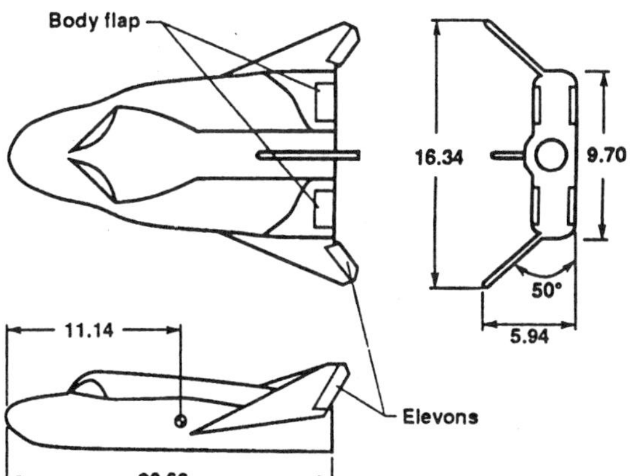

contain the materials necessary for setting up the permanent base.

At the next launch opportunity, another Interplanetary Vehicle is sent to Mars carrying a second crew. Upon arrival, the first crew transfers to it (using fuel they have manufactured from native materials) and the new crew takes their place. The original crew returns to the earth, transfering to the space station while the IV enters a parking orbit until next needed.

The Interplanetary Vehicle is about 60 m long, including its radiator panels. Each of the three arms is an independent spacecraft with about 50 m^3 per crew member of habitable space. In an emergency, any two of the arms could take on the crew of the one that is disabled. (On missions with smaller crews only two arms will make up an Interplanetary Vehicle.) Space is also available in the shuttle craft. The crew module is carried within the cluster of propellant tanks to provide solar radiation shelter. The engines used for midcourse maneuvers and earth-orbit insertion on the return leg are Space Shuttle Orbital Maneuvering System engines. The rotation axis points toward the sun and the sunward side of the spacecraft carries solar power collectors.

The Mars crew shuttles are 20-meter-long aerocapture vehicles which land on Mars in three steps: aerocapture, parachute and retro landing rockets. They can carry a total payload of 8 tons. The passenger shuttles each carry a crew of five (ten in an emergency) and a small amount of cargo.

Richard Reinert and Mark Crouch, of Ball Space Systems Division, propose a manned Mars Mission for the year 2000, which employs two 54-meter-long spacecraft consisting of a propulsion module, a mission module, an EVA support module, a command and service module, a logistics module and a Phobos excursion module. The two spacecraft will depart earth on December 17, 1998, arriving at Mars on September 28, 1999. The eight-member crew will leave Mars on January 25, 2001, and arrive home on September 2, 2001.

manned mission for 1999. Their approach does not require any in-orbit assembly, refueling or space station support.

In the scenario a single shuttle-derived Ares heavy lift booster lands a 40-ton payload on Mars in 1997. This includes an unfueled earth return vehicle (ERV) and a nuclear-powered processing plant that will create fuel for the ERV from native materials. The ERV automatically fills itself with methane/oxygen propellant. Once this is completed, the mission continues.

In 1998 two more Ares boosters lift off. One is a manned spacecraft with a four person crew. The squat cylindrical habitat is 8.4 meters in diameter and 4.9 meters tall and is divided into two decks. Using the expended booster as a counterweight connected to the habitat by a tether 1,500 meters long, the spacecraft can be rotated at 1 rpm to provide 3/8 g of artificial gravity—equivalent to that of Mars. Upon arrival at Mars, the booster and tether are jettisoned and the ship lands via aerobrake, parachutes and retrorockets near the processing plant and ERV. Exploration continues for a year and a half. The crew then returns in the waiting ERV.

At the same time that the manned spacecraft is launched, a third Ares is also launched, carrying a second ERV and fuel manufacturing plant. This lands about 800 km from the first mission's location and begins manufacturing fuel for the second manned mission, which is sent to Mars in 2001. These overlapping missions can be continued indefinitely, establishing a new base every 2 years.

In the second phase of the program, nuclear thermal propulsion is used to cut crew transit times in half. This also allows for greater cargo capacity and the use of CO_2-propelled ballistic "hoppers."

The twenty-seven-member Synthesis Group, headed by Thomas Stafford, reports on its recommendations for future manned spaceflight. Seven goals are established, five for the moon and two for Mars. The group dismisses Phobos missions in favor of direct landings by six astronauts on the planet itself. The two alternatives for the Mars missions are chemical versus nuclear propulsion.

The report of the Synthesis Group is published in 1991; *America at the Threshold* describes the goals of the Space Exploration Initiative inaugurated by President George Bush [see 1989].

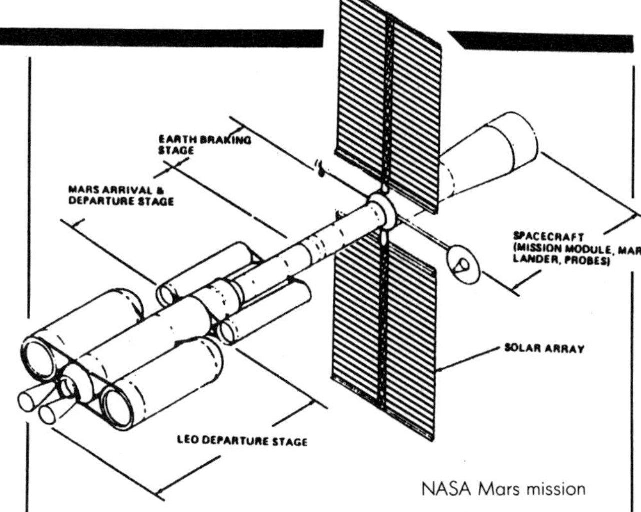

NASA Mars mission

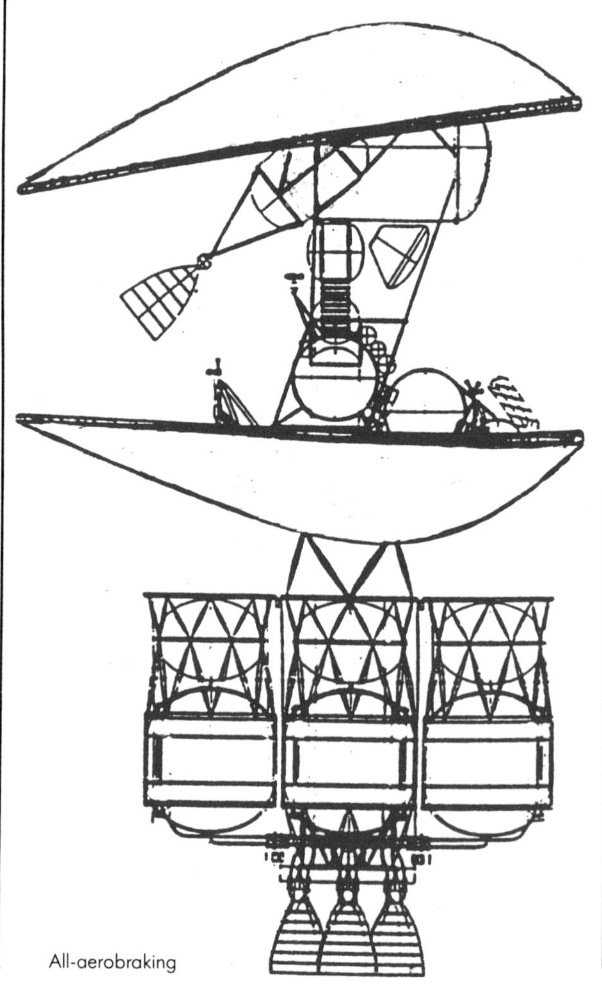
All-aerobraking

MARS TRANSPORTATION CONCEPTS, ca. 1991

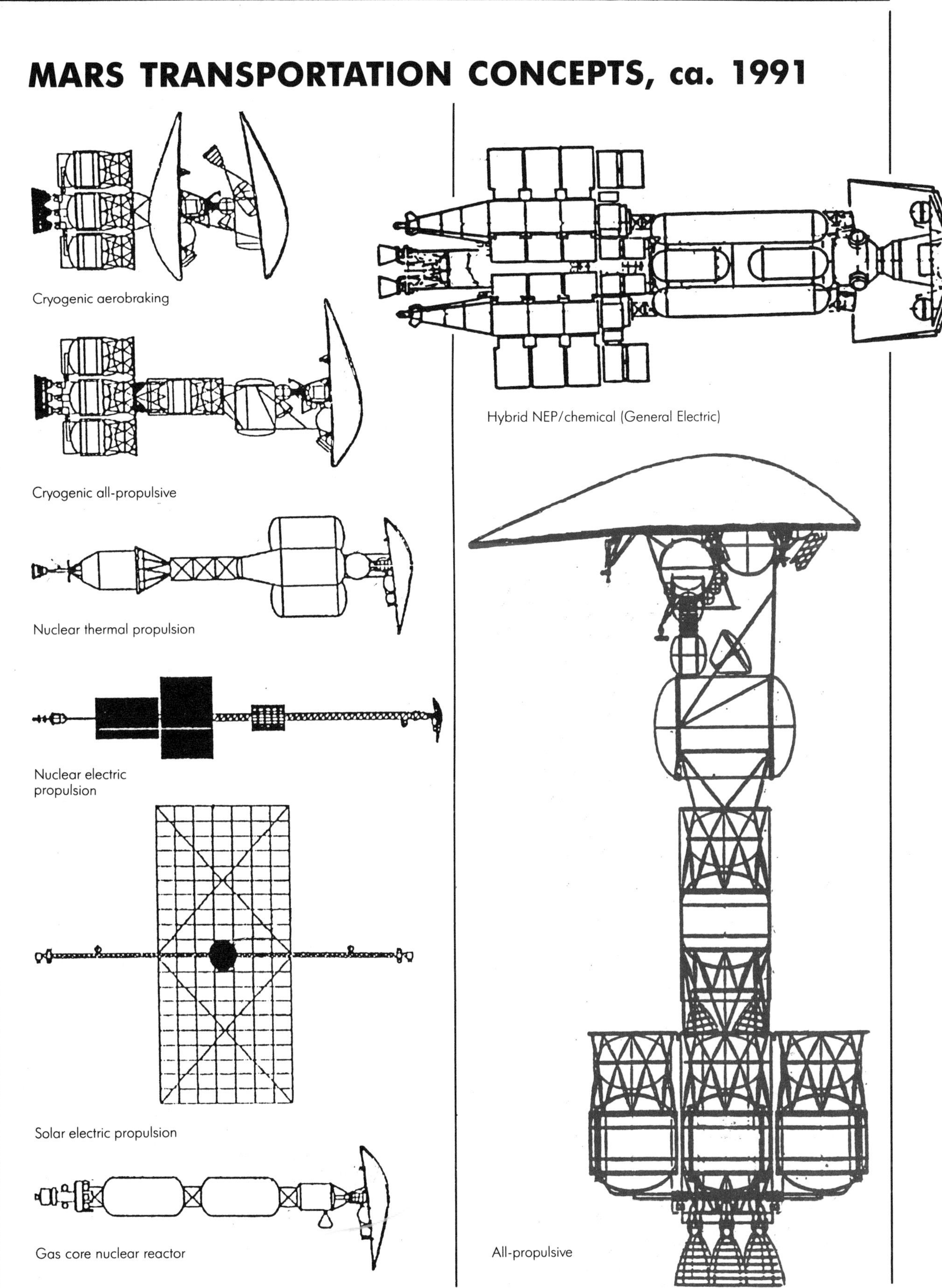

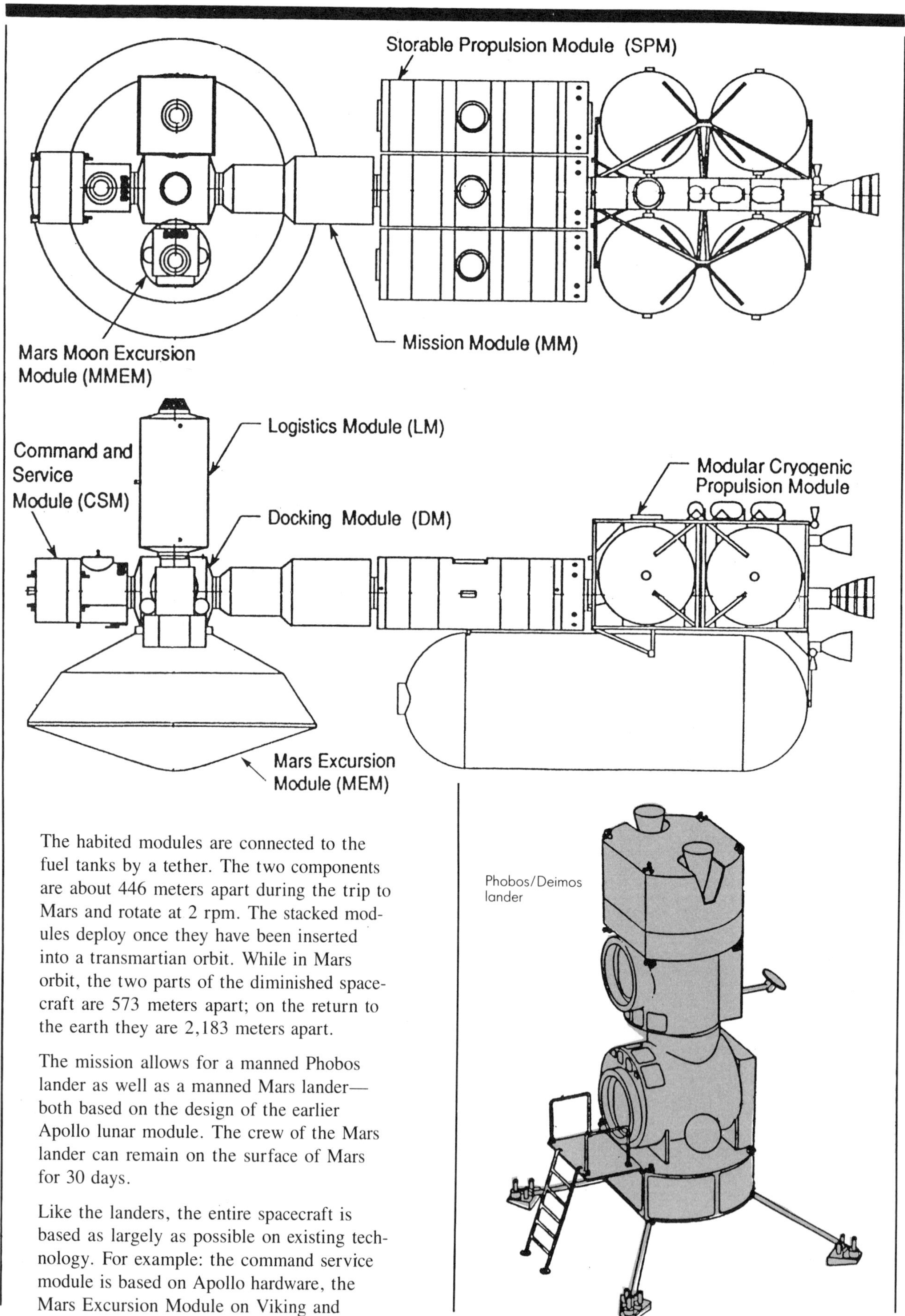

The habited modules are connected to the fuel tanks by a tether. The two components are about 446 meters apart during the trip to Mars and rotate at 2 rpm. The stacked modules deploy once they have been inserted into a transmartian orbit. While in Mars orbit, the two parts of the diminished spacecraft are 573 meters apart; on the return to the earth they are 2,183 meters apart.

The mission allows for a manned Phobos lander as well as a manned Mars lander—both based on the design of the earlier Apollo lunar module. The crew of the Mars lander can remain on the surface of Mars for 30 days.

Like the landers, the entire spacecraft is based as largely as possible on existing technology. For example: the command service module is based on Apollo hardware, the Mars Excursion Module on Viking and

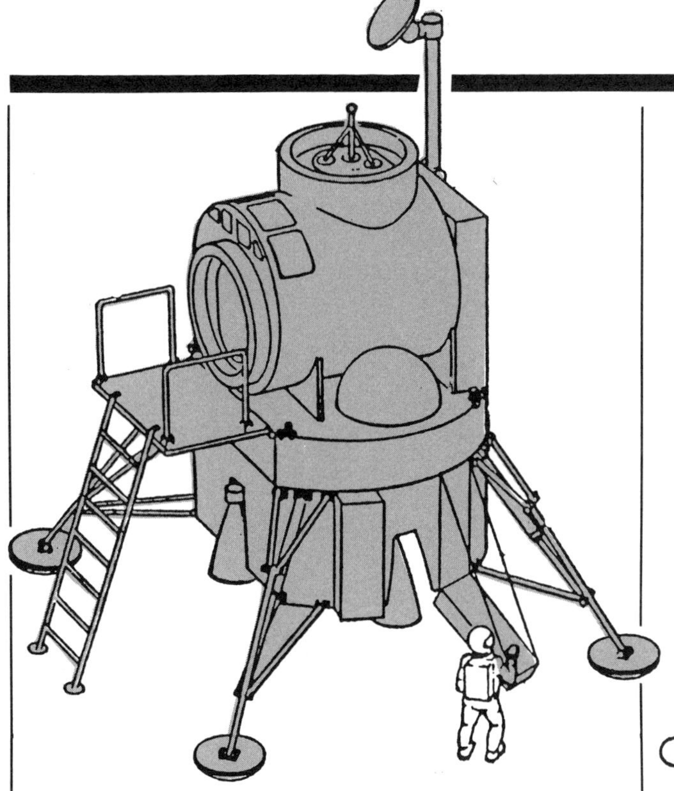

Apollo technology, and the propulsion and fuel storage on Titan, Ariane, Mir and Soyuz designs. The spacecraft is assembled in earth orbit, requiring fifteen launches: nine shuttle-derived cargo boosters, one manned shuttle, and five heavy lift launch vehicles.

The design project for the summer session of the International Space University is an International Asteroid Mission. [See p. 657.] The plan calls for a manned mission that will cycle between the earth, the moon and the asteroid:

1. The first and second stages of the vehicle leave low earth orbit, where the spacecraft has been assembled. The first stage consists of four Ariane boosters. The second stage consists of two Ariane fuel tanks and the habitation module. The tanks are 23 meters long and 5.4 meters in diameter. The crew habitation module weighs 100 tons; it is 15 meters long and 6 meters in diameter. In addition there are 8.23 × 5.52 meter solar panels and two 6 × 6 meter radiator panels.
2. The first stage separates.
3. After 516 days the manned spacecraft contacts the asteroid. Landing is made using a penetrator system and tripod landing gear.
4. The crew refuels from a processing plant that manufactures propellant from available materials.
5. After a stay of 275 days the vehicle and its crew depart the asteroid and head for low lunar orbit.
6. The crew and cargo transfer in low lunar orbit.
7. The crew transfers to the earth in a aerobraking vehicle.

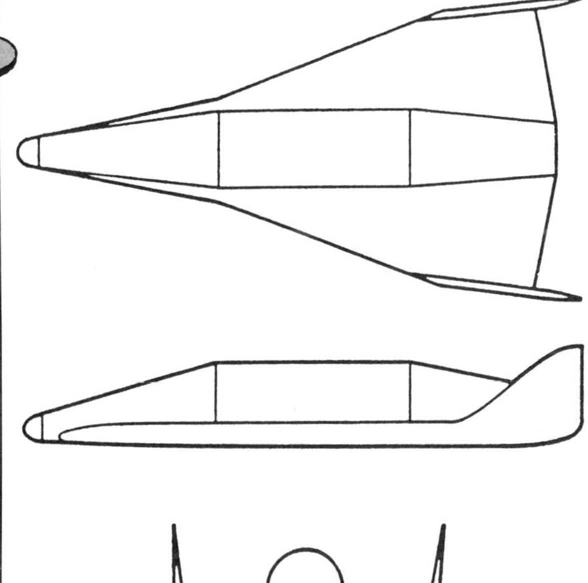

M. E. Tauber, J. V. Bowles and L. Yang propose the use of atmospheric braking upon arrival at Mars of a five-person mission. The 14 to 16 month mission will take advantage of the 2003 conjunction of earth and Mars. The Mars entry vehicle will strike the Martian atmosphere at a speed of 7.8 km/s, skip off the atmosphere and enter into a low orbit. This proposed high-lift entry vehicle strikingly resembles the Dyna-Soar spaceplane project. The mission will include, in addition to the 15,000 kg winged lander, an earth reentry capsule, a Mars orbital ascent vehicle and interplanetary living modules.

The Synthesis Group's schedule for the exploration of the moon and Mars includes four different scenarios:

1. Emphasis on Mars exploration
   2003—Surface rover on Mars.
   2005—Humans return to the moon.
   2012—Cargo rockets to Mars.
   2014—First humans on Mars (remaining on surface 30–100 days).
2. Emphasis on moon and Mars science
   1999—Lunar orbiter scouts for landing sites.
   2003—Humans return to the moon; first Mars surface rover.
   2014—Humans land on Mars.
3. Emphasis on permanent lunar base, with Mars exploration
   2000—Orbiters scout lunar landing sites.
   2004—Humans return to the moon.
   2014—Humans land on Mars.
4. Emphasis on returning resources and energy to the earth
   1999—Orbiters scout lunar landing sites.
   2004—Humans return to the moon.
   2016—Humans land on Mars.

The group's plans depend upon standard mission plans and conservative technology using few shortcuts. Its conservative approach recalls the heritage of the Apollo program (including the revival of the Saturn V's F-1 engine). Required is the development of new heavy lift boosters and nuclear propulsion.

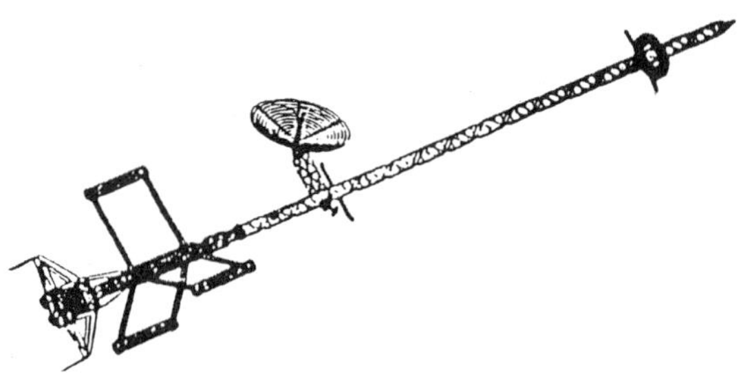

Project Kepler interplanetary transfer vehicle, University of Michigan, 1986.

NASA and the U.S. Department of Energy believe that if nuclear-propelled spacecraft are used, Mars could be reached by 2005. Either nuclear electric or nuclear thermal propulsion is needed. The nuclear electric system is basically that proposed more than three decades earlier by Ernst Stuhlinger [see above]. This requires long transit times to Mars, however. The nuclear thermal engine dramatically reduces the time needed to travel to Mars. One model of a nuclear thermal engine—developed from the NERVA program of the 1960s and 1970s—has its uranium-graphite core superheat gaseous hydrogen which is then exhausted at high velocities.

British Aerospace and the Soviet Ministry of Aviation study the possibility of launching a HOTOL-type spaceplane from the back of an Antonov An-225 heavy transport. This concept has been dubbed "Interim HOTOL," while the prime program remains on hold. The An-225, the world's largest production aircraft, already is used to transport the *Buran* shuttle. It can be modified to handle the 250-ton British spacecraft. In effect, the An-225 would replace HOTOL's classified air-breathing first-stage engine. The HOTOL would be released at an altitude of 9 kilometers. Wind tunnel tests of the combined air-and-spacecraft begin in the Soviet Union in early 1991. After 7 months of study, engineers conclude that the scheme is viable.

M. Q. Hassan suggested the possibility of using the An-225 for this purpose in 1989, and B. Carr first suggested air-launching the HOTOL spaceplane in 1987 [see above and 1991].

The Case for Mars Conference proposes an earlier date than that of President Bush's 2019 for placing the first explorers on Mars. The new mission could be flown both faster and cheaper as well and without the need for either a lunar base or a space station. The spacecraft is simpler than those in NASA's "90-Day" report [see 1989].

One mission suggested by Martin Marietta has its spacecraft carried into orbit by a shut-

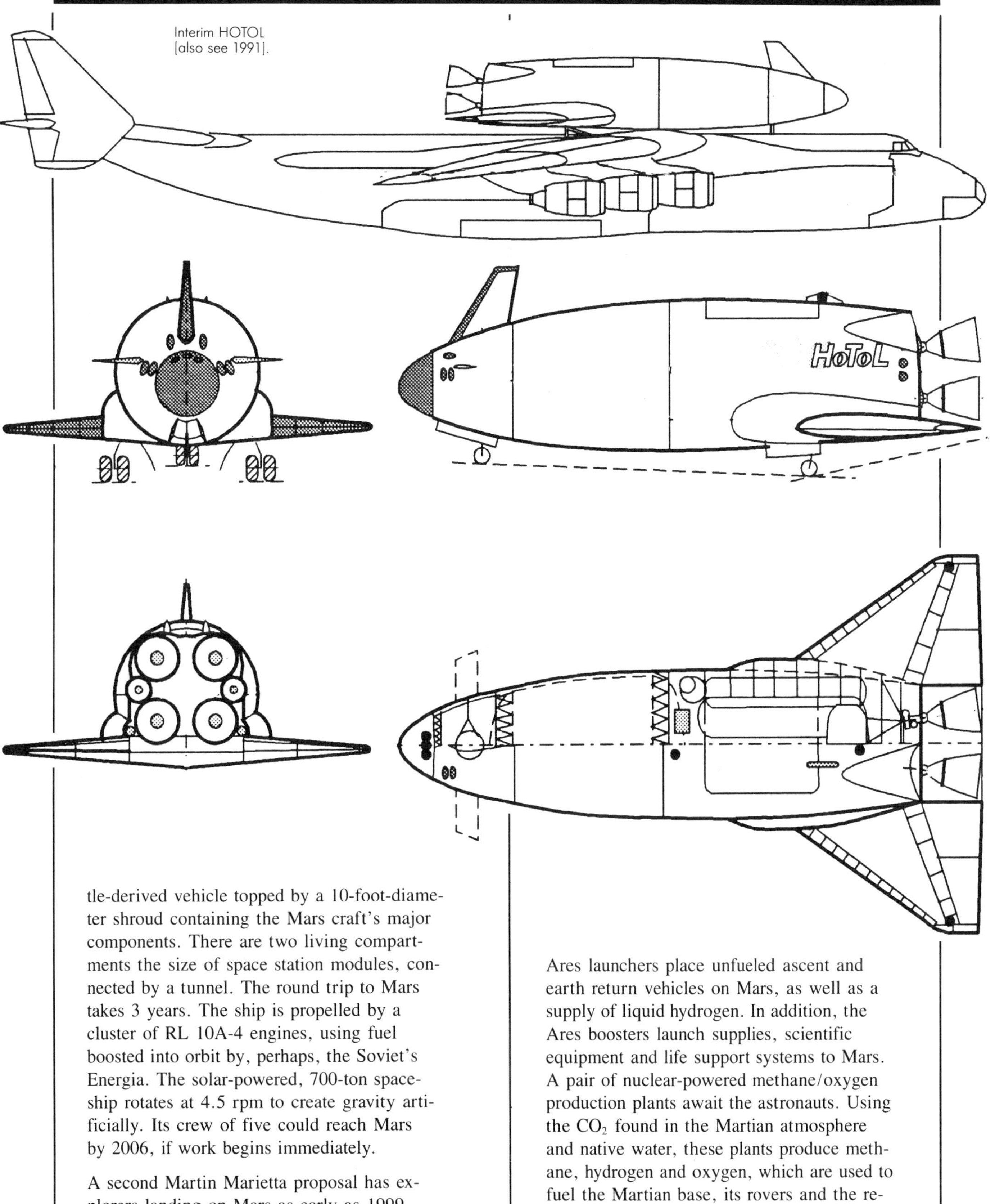

Interim HOTOL [also see 1991].

tle-derived vehicle topped by a 10-foot-diameter shroud containing the Mars craft's major components. There are two living compartments the size of space station modules, connected by a tunnel. The round trip to Mars takes 3 years. The ship is propelled by a cluster of RL 10A-4 engines, using fuel boosted into orbit by, perhaps, the Soviet's Energia. The solar-powered, 700-ton spaceship rotates at 4.5 rpm to create gravity artificially. Its crew of five could reach Mars by 2006, if work begins immediately.

A second Martin Marietta proposal has explorers landing on Mars as early as 1999. Three shuttle-derived Ares boosters each launch 40 metric tons on a direct flight to Mars. Over a period of 3 years the three Ares launchers place unfueled ascent and earth return vehicles on Mars, as well as a supply of liquid hydrogen. In addition, the Ares boosters launch supplies, scientific equipment and life support systems to Mars. A pair of nuclear-powered methane/oxygen production plants await the astronauts. Using the CO_2 found in the Martian atmosphere and native water, these plants produce methane, hydrogen and oxygen, which are used to fuel the Martian base, its rovers and the return launch vehicle.

The explorers can remain on Mars for 500 days, with an additional 12 to 16 months for

1990

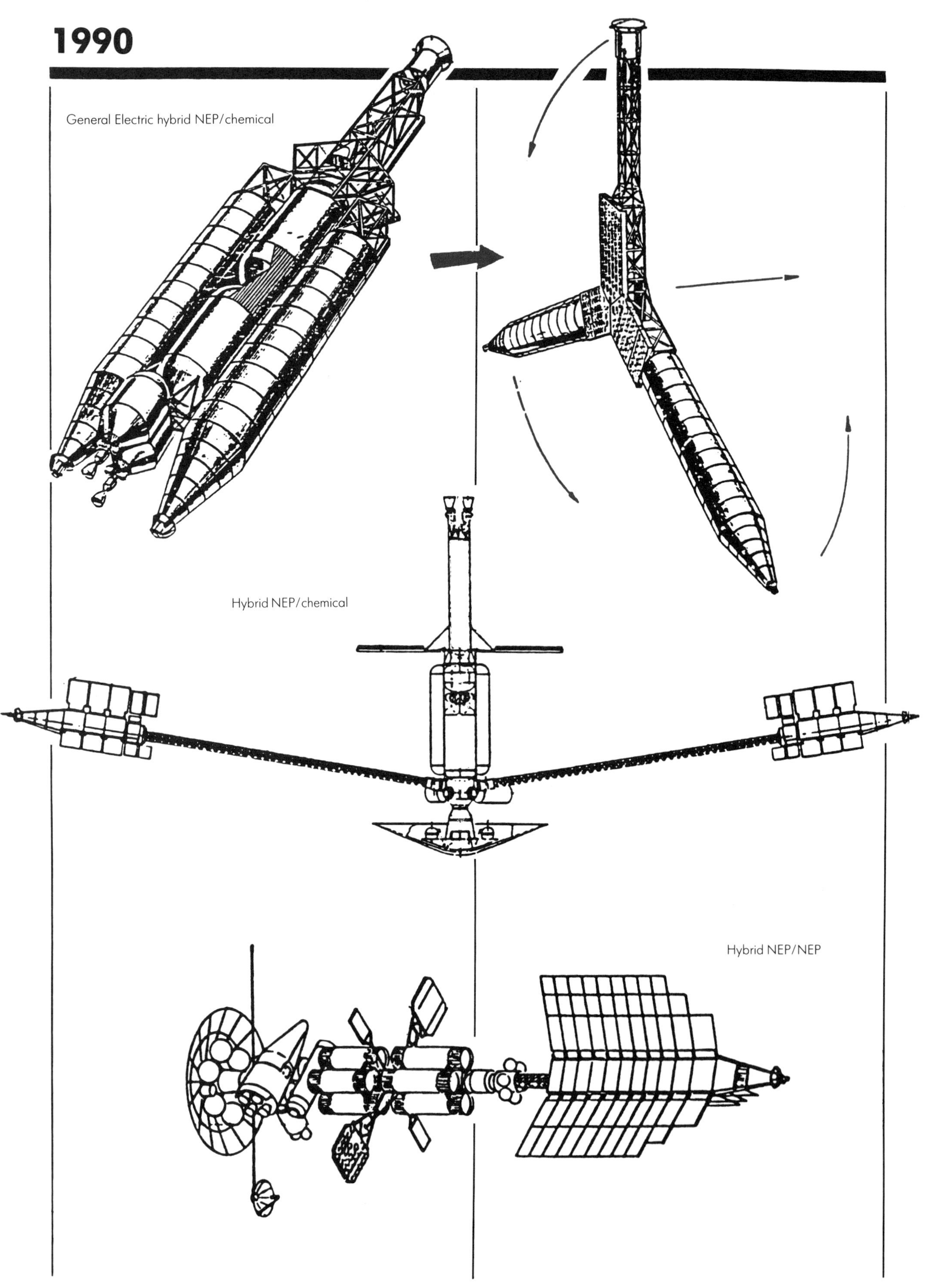

General Electric hybrid NEP/chemical
Hybrid NEP/chemical
Hybrid NEP/NEP

the round trip. Later missions will arrive at 2-year intervals. A population of 100 people might be on Mars by the year 2020.

Messerschmidt-Bölkow-Blohm completes initial feasibility studies for an experimental Mach 5.5 aircraft. It would be used to demonstrate technologies for the first stage of the Sänger II two-stage orbiter system. The project, Hytex (**hy**personic **t**echnology **ex**perimental vehicle), is some 75.5 feet long, 20.7 feet high and has a wingspan of 30.5 feet. It is powered by turbo-ramjet engines.

The Lunar and Mars Exploration Program Office at NASA's Johnson Space Center advocates an "aggressive Mars architecture" approach to a manned Mars mission. A landing on Mars will precede a lunar outpost by about 5 years. The first Mars landing, by eight astronauts, will take place in 2004 using nuclear propulsion for the transit.

Martin Marietta proposes a "straight arrow" approach for the near-term exploration of Mars. Intended for a landing in 2006, the five-man spacecraft consists of two (or possibly three) cylindrical habitats connected by a rigid arm. The aeroshell lander and propulsion systems are located on the rotation axis. The habitat modules are 14.5 feet in diameter and 42 feet long. Power is provided by solar arrays.

H. Lacaze and J. P. Bombled of Aerospatiale describe a recoverable manned shuttle and booster system. It is a two-stage vertical takeoff vehicle. The delta-winged first stage is unmanned with six or seven uprated Vulcain engines fueled by liquid oxygen and liquid hydrogen. The manned shuttle has a single Vulcain engine and is also fueled by liquid oxygen and liquid hydrogen, with the latter carried in a pair of jettisonable overwing external tanks. It has a double delta wing and a single vertical stabilizer.

The designers consider four different configurations of the combined vehicle: 1. the rockets are stacked into a 77-meter-tall vehicle;

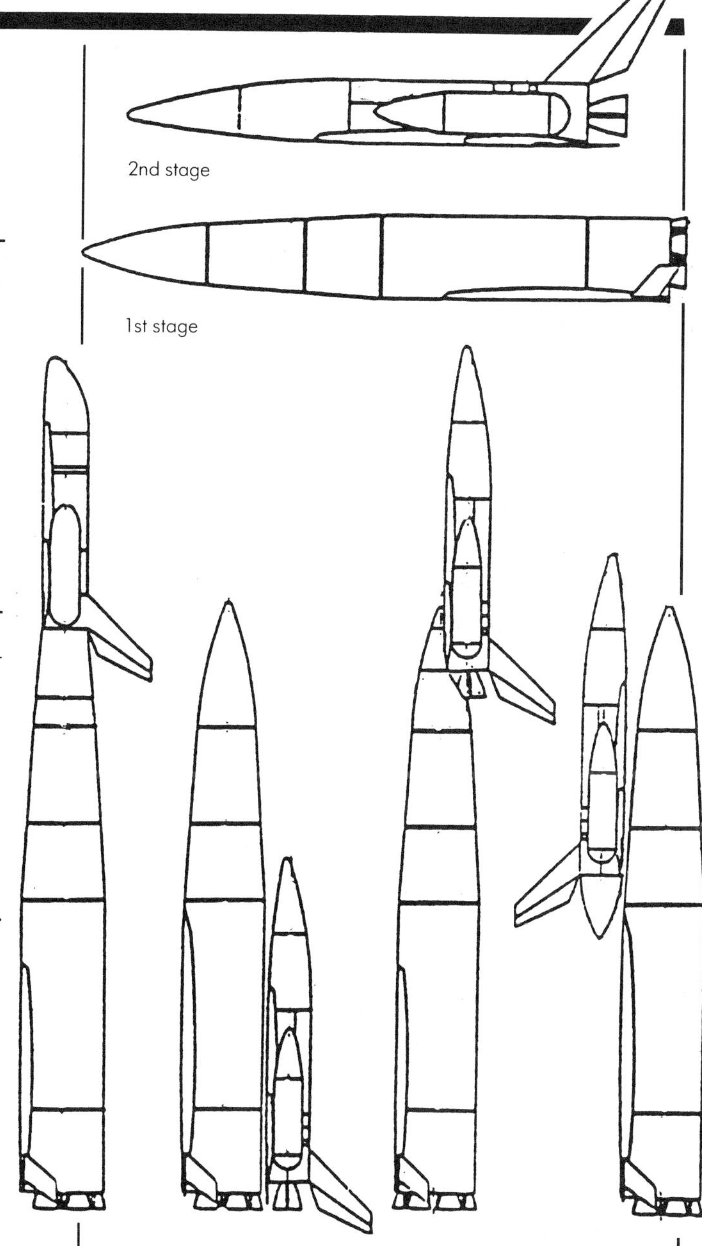

2. a belly-to-belly or belly-to-back parallel configuration; 3. and 4. two different staggered arrangements. The authors settle on the parallel belly-to-back arrangement with all engines burning at takeoff.

Benjamin Donahue, of the Advanced Civil Space Systems division of the Boeing Defense and Space Group, designs a Mars Nuclear Thermal Propulsion (NTP) Vehicle for a manned mission to Mars. Donahue's NTP is the current NASA "preferred concept" for a spacecraft intended for manned Mars missions during the 2014–2020 period.

FEB. 1, 09:40 GMT

Cosmonaut Aleksander Serebrov makes the first EVA on the new YMK maneuvering unit, which the cosmonauts have dubbed the "space motorcycle." The device is similar to the American MMU (Manned Maneuvering Unit) The total time spent on the test flight is 4 hours and 59 minutes and Serebrov maneuvers to a distance of 33 m from the Kvant-2 hatch on the MIR space station. Serebrov is enthusiastic about the performance: "there is nothing in [the YMK's] work which we did not like. The new vehicle will permit cosmonauts to increase their maneuverability in space, to carry up to 100 kilograms of tools, parts and instruments. In case of an a ccident with some cosmonaut he can be transported with the help of the vehicle." For safety reasons, Serebrov remains tethered during the YMK test flight. The unit possesses thirty-two individual jets and is capable of traveling up to 100 m from its parent spacecraft.

MAY

The team of contractors developing the National Aero-Space Plane narrow the X-30 research design to a single configuration featuring small wings, twin vertical stabilizers and a two-person crew. Propulsion will be provided by three to five scramjets and one rocket that will create the additional thrust necessary to achieve orbit. The spaceplane will measure between 50 to 65 m in length. It is expected that test flights will begin in 1997 and the first orbital flights in 1999.

AUGUST

The Strategic Defense Initiative Organization (SDIO) funds four of six aerospace industry respondents to its request for reusable spacecraft designs (not accepted were the Horizontal Takeoff/Horizontal Landing air-launched concepts of Grumman and Third Millenium). The four companies receiving funding are McDonnell Douglas, Rockwell, General Dynamics and Boeing.

McDonnell Douglas and General Dynamics begin work on VTOL spacecraft while Boeing proposes an improvement on its sled or rail-launched HTHL. Rockwell meanwhile develops a Vertical Takeoff/Horizonal Landing vehicle.

General Dynamics Space Systems Division develops its VTOL SSTO *Millenium Express*. Propelled by a Rocketdyne aerospike engine, it would carry 10,000 pounds and a crew of two into a polar low earth orbit.

McDonnell Douglas is eventually awarded a contract to develop a demonstrator vehicle, called the DC-X, scheduled for flight in the spring or early summer of 1993. The DC-Y prototype of the orbital operational vehicle is scheduled for its first suborbital flight in 1995 and its first orbital mission in 1997. [Also see 1991.]

OCTOBER 17–19

The First International Hypersonic Waverider Symposium is held, promoting and studying the use of a concept originally developed by Terence Nonweiler in the 1960s [see above]. The unusual wing design of the waverider creates a shockwave that remains attached to the wing's leading edge. The high pressure trapped beneath creates lift. Some work was done in the 1960s using the concept for a Mach 6 airliner. To date, most work on the waverider has been done by the Association in Scotland to Research into Astronautics (Glasgow), whose members achieved the first rocket-launched waverider in 1985.

ca. 1990

The Japanese proceed with an aero-space-plane project. Initiated by the cabinet level Science and Technology Ministry, via the National Aerospace Laboratory, the design studies are being undertaken by Mitsubishi's Heavy Industry facilities. An experimental prototype could lead to an operational spaceplane by 2006.

Boeing develops the Beta Booster for the Air Force Systems Command. A winged, horizontal takeoff and landing manned

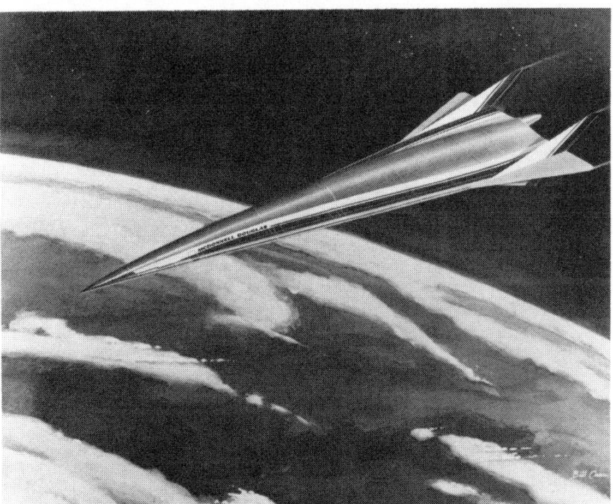

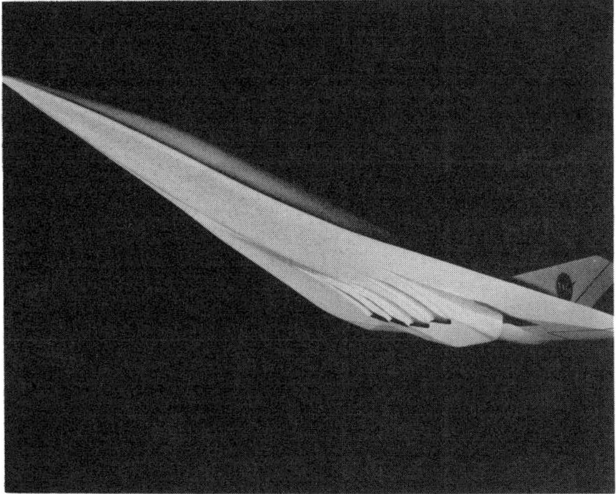

National Aero-Space Plane proposals, left to right, top to bottom: North American Rockwell, North American Rockwell, General Dynamics, McDonnell-Douglas, NASA.

spaceplane booster is used to loft a manned orbiter into space. The manned rocket-powered orbiter is carried internally beneath the fuselage of the air-breathing booster. The concept is further developed by Boeing engineers V. Weldon and L. Fink, under the sponsorship of the NASA Langley Research Center.

The final vehicle can carry 10,000 pounds of payload, in addition to a crew of two, into a low polar orbit. The orbiter's payload bay can accommodate cargos measuring 20 feet long and 14 feet in diameter. The gross takeoff weight is less than 1.25 million pounds (in the same class as an advanced 747). The first stage is propelled by hydrogen-fueled ramjets, the second stage by rockets, also fueled by hydrogen.

Two versions of the Mach 6.5 booster stage are studied: one with six turbofans and one with eight. The final configuration has ten JP-7-fueled turbojets and two hydrogen-fueled ramjets. The booster is 248 feet long (281 feet including the swept-back vertical stabilizer) and a wingspan of 117 feet.

The orbiter, which resembles a stretch version of the present-day space shuttle, is 131′ 8″ long with a wingspan of 46′ 8″. It can carry a 10,000-pound payload.

1991

The Pentagon's Strategic Defense Initiative Organization (SDIO) requires a single-stage reusable shuttle vehicle that can carry large payloads at a fraction of current space shuttle costs. Boeing, General Dynamics, McDonnell Douglas (teamed with Martin Marietta) and Rockwell are asked to submit proposals. Two of the companies will be selected to produce a prototype with suborbital tests beginning as early as 1995.

One of the major advantages to single-stage operations is the elimination of costly and time-consuming stacking and assembly of the spacecraft's components. Weekly flights might then be possible, with cargo costs cut to $100 to $1,000 a pound.

Boeing's initial concept takes off and lands horizontally, much like the National Aero-Space Plane. Rockwell proposes a vehicle that takes off vertically but lands horizontally. The McDonnell Douglas and General Dynamics designs take off and land vertically. These latter are cone-shaped rockets with a wide, rounded base that acts as a heat shield during reentry. The rocket then restarts its engines and lands like a lunar module or Harrier jet, settling down onto four extendable legs. This vehicle can lift a

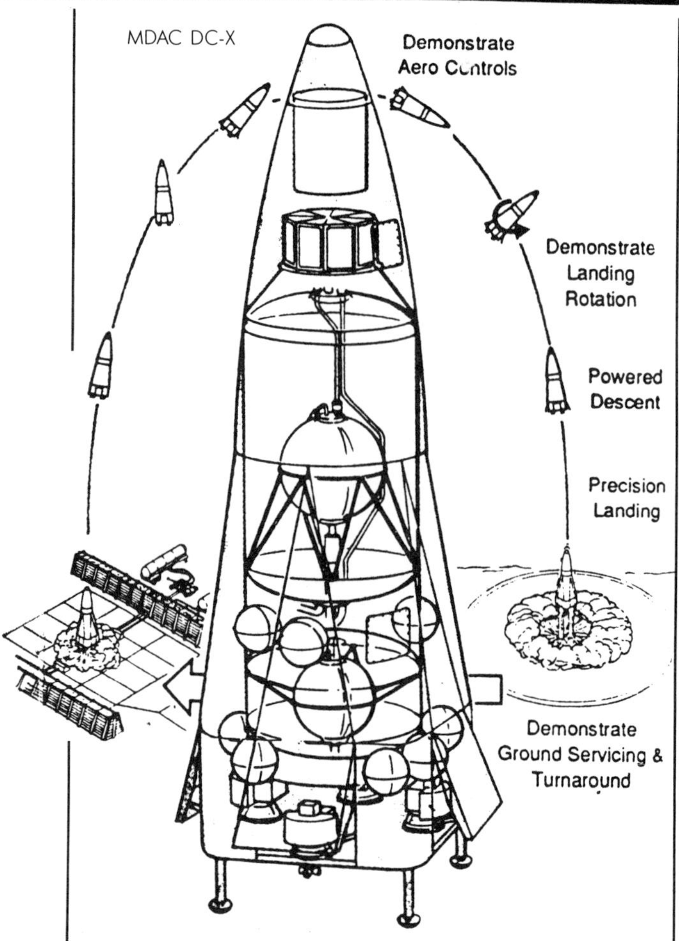

10,000 pound payload and a crew into a polar orbit, and twice as much into an equatorial orbit. This concept, though on a larger scale, reflects the work done by Max Hunter, developer of the SSX Space Ship Experimental project.

A derivative of the Saturn V booster is recommended for manned lunar and Mars missions. The recommendation is made by the White House Synthesis Group (headed by former astronaut Thomas P. Stafford) defining lunar base/Mars mission options. The proposed vehicle could place up to 600,000 pounds into orbit, compared to the 250,000-pound capacity of the original Saturn V. The new rocket would use six space shuttle-derived hydrogen-oxygen strap-on boosters. The core would use five Rocketdyne F-1 engines, which would be placed back into production (all remaining engines from the Apollo era are in museums).

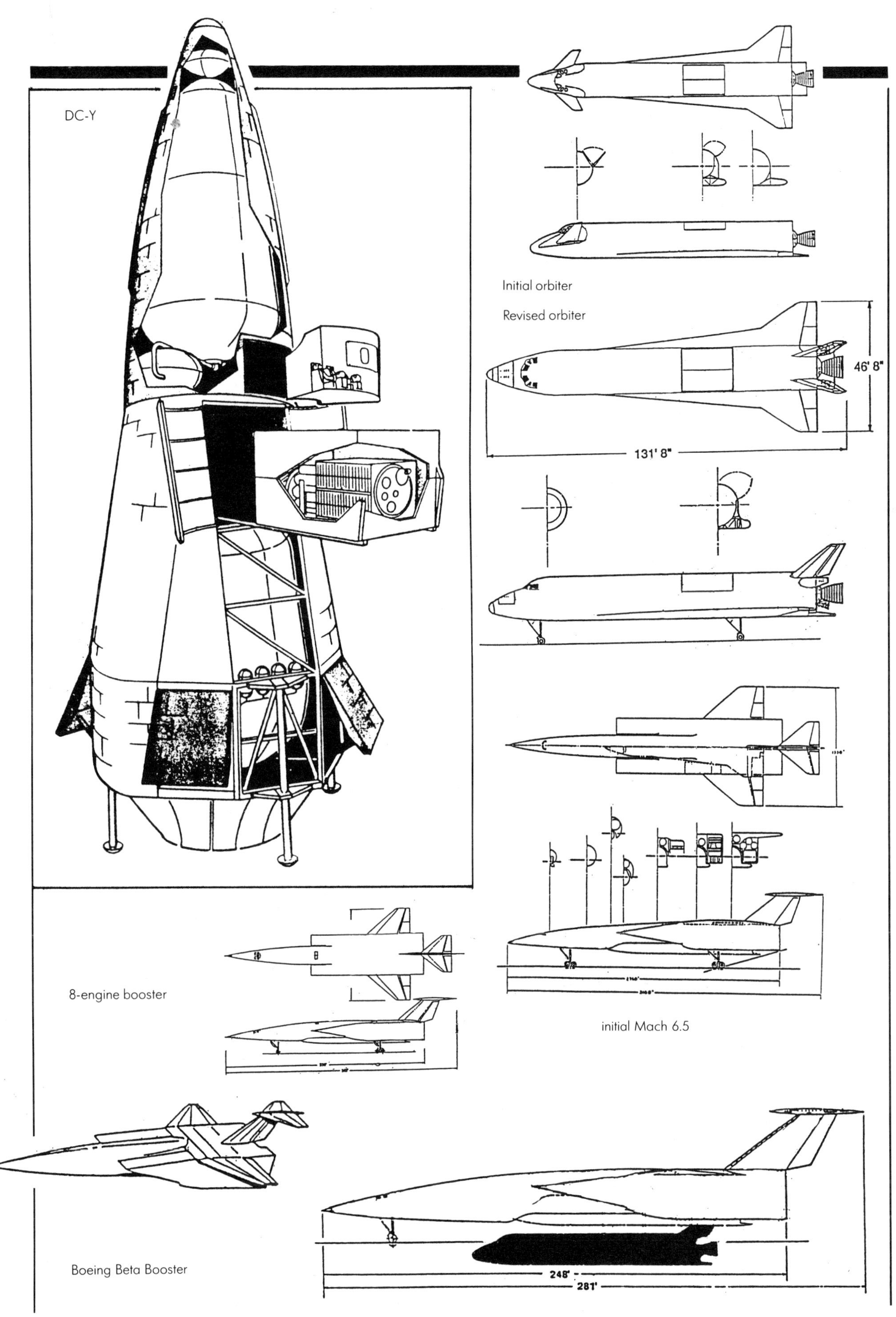

DC-Y
Initial orbiter
Revised orbiter
46' 8"
131' 8"
8-engine booster
initial Mach 6.5
Boeing Beta Booster
248'
281'

1991

A Stanford University study group proposes a low-cost manned Mars mission that is "safer, cheaper and quicker" than current NASA plans. This mission will only cost $60 billion compared to the $500 billion NASA requires and could place a team of six (three men and three women) on Mars in 2012, 7 years earlier than NASA's plan.

The key to the low-cost mission is the Soviet Energia booster. The United States will not be required to develop either a space station or a new rocket of its own. The landing will be made in a vehicle similar to the Apollo lunar module. The transit to Mars will take 9 months, with an artificial 1/3 gravity created by rotating the 60-foot transit vehicle at four revolutions per minute with a 5-ton counterweight.

Before the astronauts leave earth, several Energia rockets will transport prefabricated "apartments" and a "garage" containing two tracked ground vehicles onto Mars, as well as the return rocket to be used by the astronauts. The landing site may be in the region of Candor Chasma, in the Vallis Marineris rift valley.

There would be two missions to Mars: a crew of six who will remain on Mars for a year, and a second crew who will land approximately 3 or 4 months after the first crew has departed.

R. Amekrane of the Technical University of Berlin describes the use of a hot water rocket to propel a booster sled (a "zero stage" acceleration trolley) for the Sänger II spaceplane.

Rockwell and Lockheed are the prime contractors selected by NASA to develop an assured crew return vehicle (ASRV). This will be used to guarantee the international crew of the Freedom space station the ability to return safely to the earth. The ASRV will provide the space station crew with these advantages: the ability to evacuate an injured or seriously ill crewmember; safe and speedy evacuation of the space station; and a back-up in case space shuttle flights are interrupted.

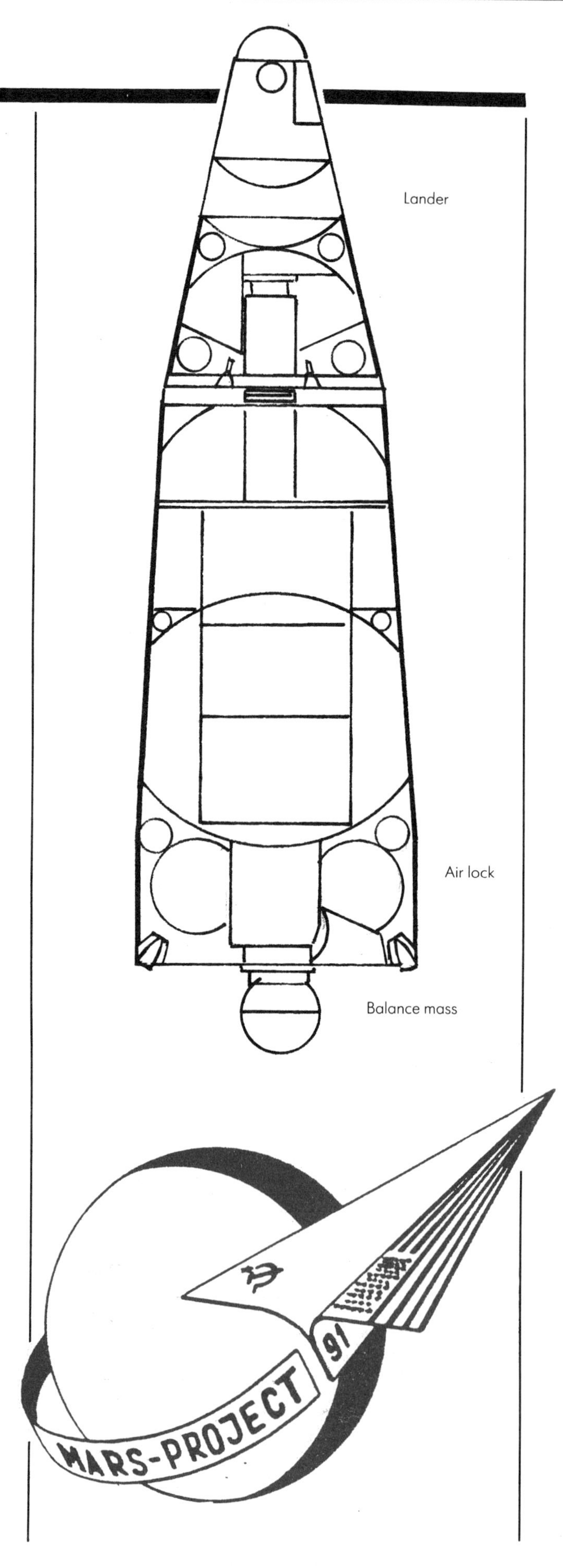

1991

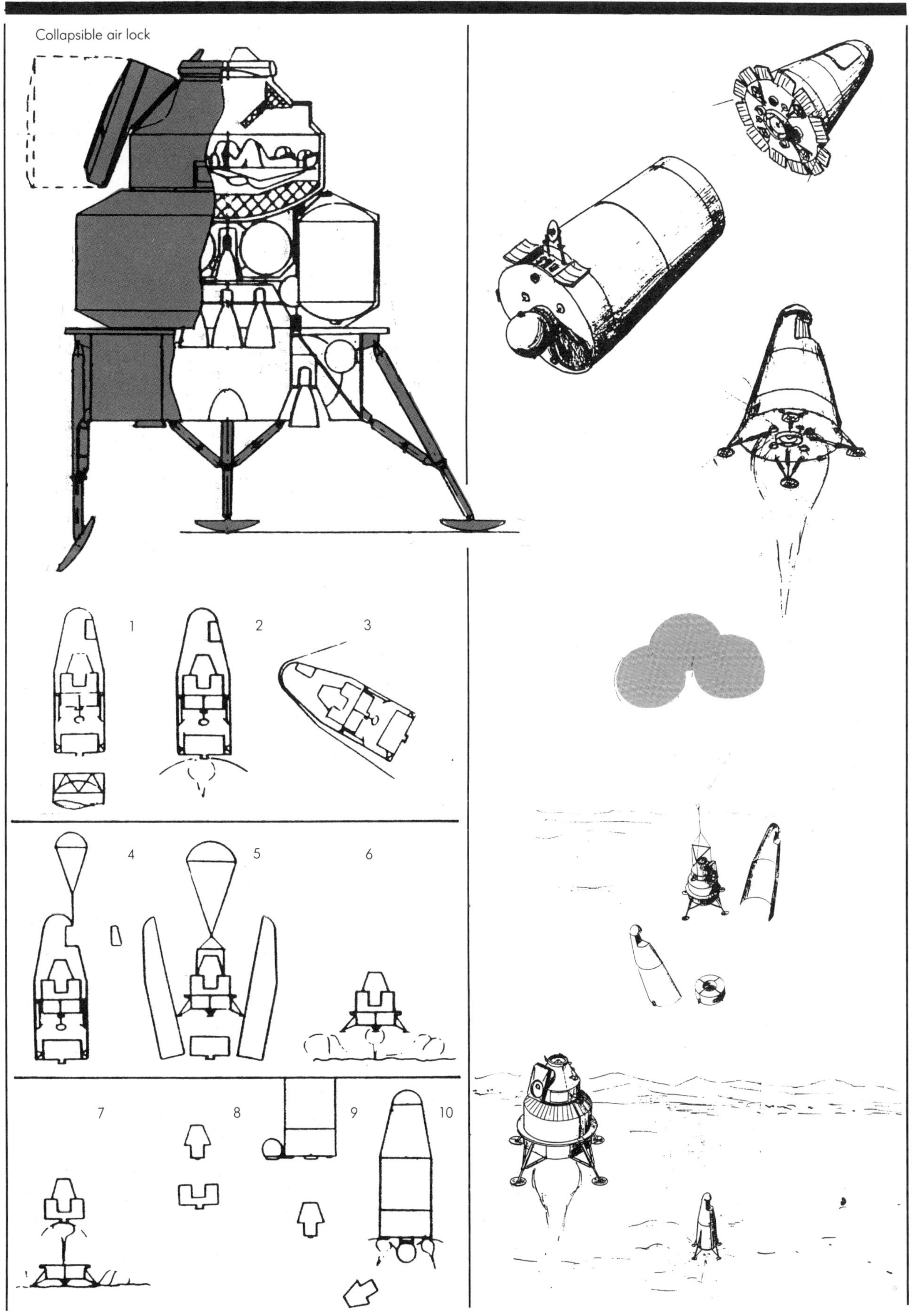

1991

Simplicity is the project's philosophy, as well as ease of operation, reliability and cheapness. Several configurations are being studied, including lifting body, Apollo and Discoverer shapes, as well as a new low lift-to-drag ratio capsule with a separable deorbit propulsion module.

U. N. Zakirov, of the Intercosmos Council of the U.S.S.R. Academy of Sciences, proposes a possible manned interstellar mission to Proxima Centauri. The expected round trip would require a total of 50 years. The spacecraft that Zakirov describes would employ "a new scientific solution—multiple-step apparatus with different types of engine on the basis of synthesis and annihilation reactions, which can reach a star within 25 years . . . " The returning vehicle would employ atmospheric braking in the atmospheres of the solar system's planets.

The Orion system is revived by Bruno Augenstein of the RAND Corporation. He suggests its use for expeditions of less than 5 years' duration (one way) to the inner Oort cloud (a "deep freeze" of comets that surrounds our solar system). The Orion concept has been improved by Johndale Solem of the Los Alamos National Laboratory. The spacecraft would have a large canopy attached to it by tethers. The nuclear explosions would occur between the ship and the canopy.

Gilles Primeau, president of Aerocorp Technologies, proposes the Heracles Project, a spacecraft propelled by muon catalyzed nuclear fusion that would be capable of reaching Mars in 3 days. Primeau's unique propulsion system utilizes the negative muon (μ), a particle with the electrical charge of an electron but which is 207 times heavier, and which can act as a catalyst in fusion reactions. It allows fusion to take place at temperatures and pressures orders of magnitude below those encountered in the Tokamak or laser confinement fusion reactors.

If Primeau is correct in his assumptions, his propulsion system would yield a maximum

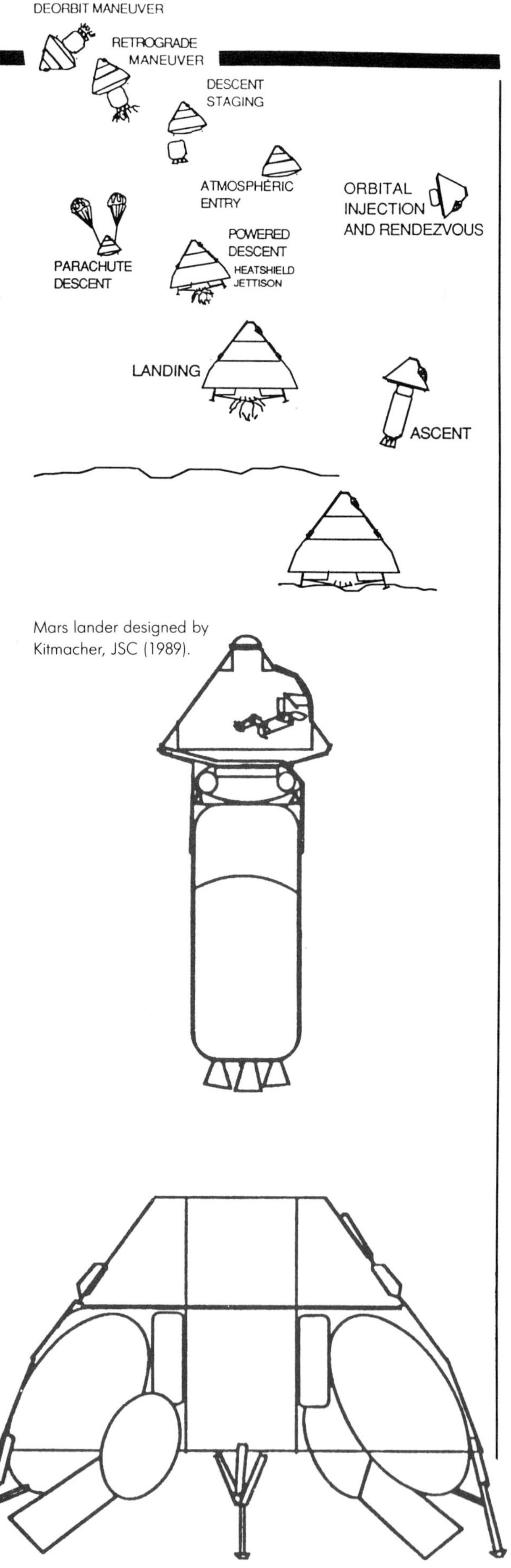

Mars lander designed by Kitmacher, JSC (1989).

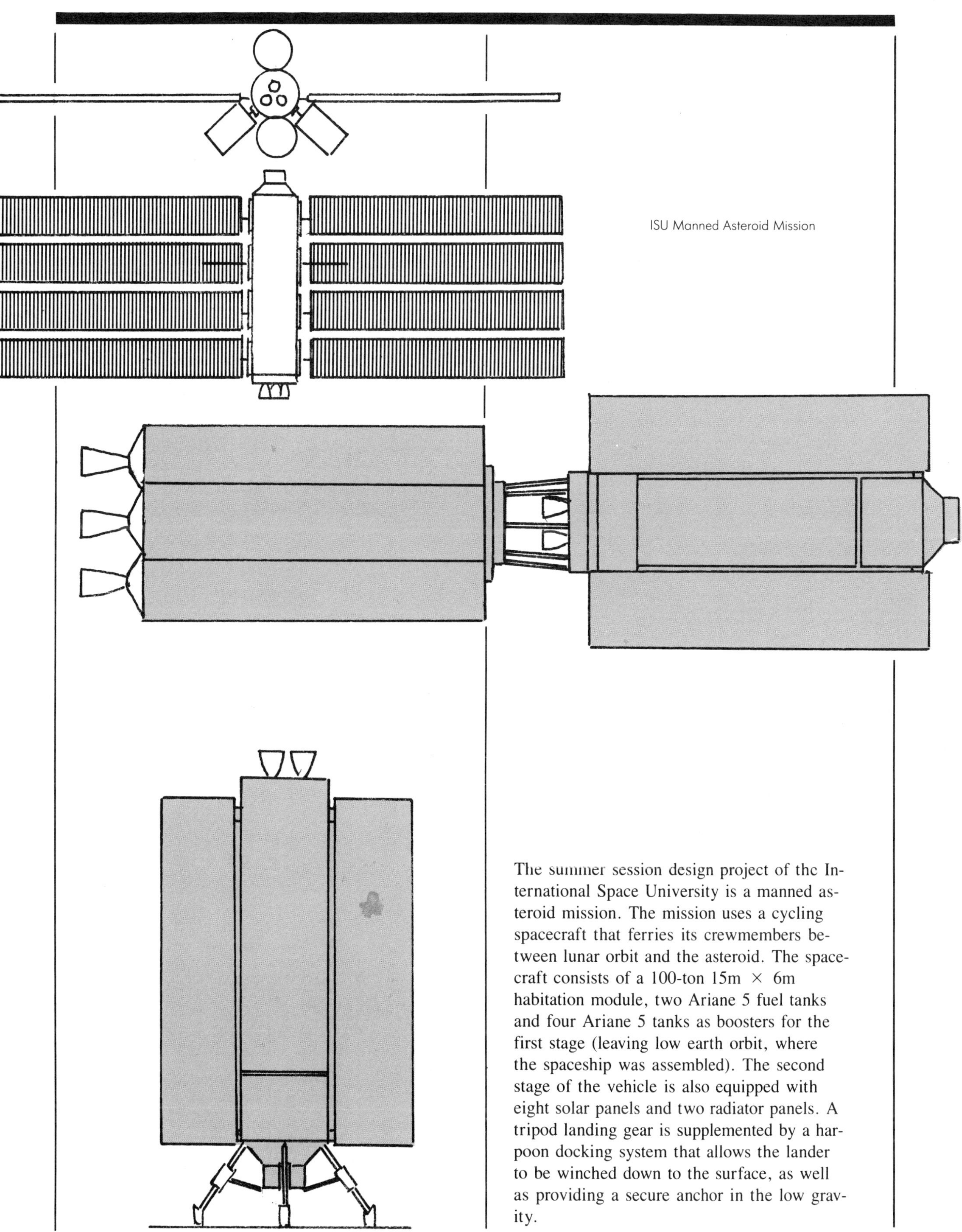

ISU Manned Asteroid Mission

The summer session design project of the International Space University is a manned asteroid mission. The mission uses a cycling spacecraft that ferries its crewmembers between lunar orbit and the asteroid. The spacecraft consists of a 100-ton 15m × 6m habitation module, two Ariane 5 fuel tanks and four Ariane 5 tanks as boosters for the first stage (leaving low earth orbit, where the spaceship was assembled). The second stage of the vehicle is also equipped with eight solar panels and two radiator panels. A tripod landing gear is supplemented by a harpoon docking system that allows the lander to be winched down to the surface, as well as providing a secure anchor in the low gravity.

theoretical specific impulse of over a million seconds, as compared to 500 seconds for chemically fueled rockets. The muon-catalyzed system could also operate for extended periods of time. With nucleons expelled at an average energy of 10 MeV, a spaceship of one million kg could ultimately maintain an acceleration of 1 g, using less than a quarter of a kg of deuterium fuel per second. This acceleration would provide an earth-normal artificial gravity. An acceleration of 1 g also translates into a one-way travel time to Mars of under three days!

The 73 meter-long SSTO has a takeoff mass of 338 metric tons. The fuselage is 10.8 meters wide with a wingspan of 28.665 meters. Its propulsion system would take it from a standing start to Mach 25.

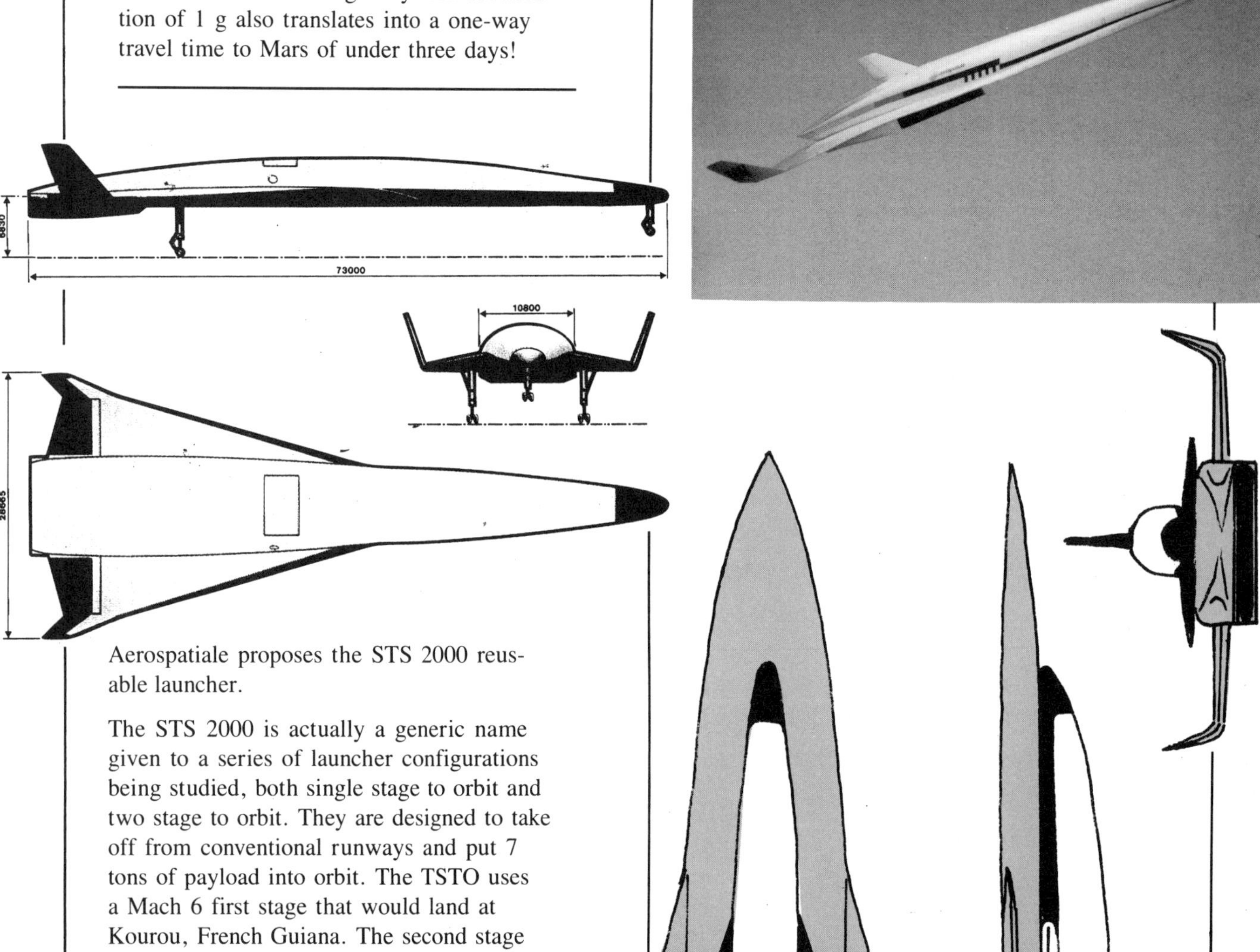

Aerospatiale proposes the STS 2000 reusable launcher.

The STS 2000 is actually a generic name given to a series of launcher configurations being studied, both single stage to orbit and two stage to orbit. They are designed to take off from conventional runways and put 7 tons of payload into orbit. The TSTO uses a Mach 6 first stage that would land at Kourou, French Guiana. The second stage continues from Mach 6 to orbit. The first stage of the TSTO is 60 meters long and weighs 83 metric tons at takeoff (the two vehicles have a combined mass of 204 metric tons). It has a wingspan of 31.42 meters. The second stage is 38 meters long with a wingspan of 16 meters and a separation weight of 121 metric tons. Both orbiters could glide back to a landing at virtually any orbit.

J. G. Pearsall proposes using an airship for an air launch of the HOTOL spaceplane. Similar in size to one of the large prewar rigid airships—such as the R-100 or *Graf Zeppelin*—the 64-meter-diameter dirigible would weigh 262 metric tonnes. It would be cylindrical, about 320 meters long and made of Kevlar-29. The airship would carry the HOTOL to an altitude of around 7 km at which point the spaceplane would be dropped vertically in order for it to reach the speed necessary to start its engines.

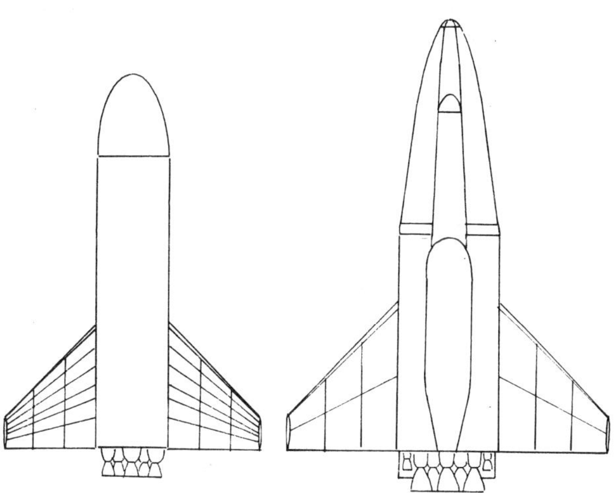

The NASA Langley Research Center develops an advanced manned launch system (AMLS). The near-term AMLS is a two-stage reusable rocket system while the future, advanced technology AMLS would probably be a single stage to orbit (SSTO) vehicle. An initial operating date of 2005 will allow the AMLS to gradually replace the aging shuttle fleet.

The near-term AMLS consists of a 149-foot 113-foot-wingspan manned orbiter with a parallel 129-foot unmanned winged booster. The orbiter's payload bay is carried piggyback just behind the jettisonable crew module. The stages separate at Mach 3. The gross weight at liftoff of the complete system is to be 2.6 million pounds (compared to the present shuttle's 4.5 million pounds).

An advanced two-stage AMLS will take advantage of improved space shuttle main engines and the use of slush hydrogen and triple-point oxygen propellants. The gross weight of this vehicle will be only 1.5 million pounds, a reduction of 47 percent from the near-term AMLS.

The SSTO will have a gross weight of 2.7 million pounds. [Also see November 1988.]

Engineers at NASA's Lewis Research Center (in particular S. K. Borowski) are pursuing the idea of reusable vehicles propelled by nuclear thermal rockets (NTR) to deliver astronauts and cargo to and from the moon. With the addition of modular elements, the lunar transit vehicle could be transformed into a spacecraft that could land humans on Mars early in the twenty-first century. The lunar transit vehicle (LTV) could be ready by 2000–2005 at a cost of $3–5 billion. The manned Mars mission could be ready between 2014 and 2019.

The 63.6 meter LTV masses 218 tons in low earth orbit and requires that its nuclear engine burn for a total of 53 minutes. It carries the same lunar excursion module as a chemically fueled aerobraking LTV. Propulsion is provided by a reusable lunar vehicle core that consists primarily of a main propellant tank and a Nerva engine. The NTR is fueled and refueled in earth orbit by tankers. With a space station-serviced lunar excursion module attached it boosts out of earth orbit. Arriving in lunar orbit, the lander detaches and makes the descent to the surface. Later, the same lander or a different one from an earlier mission makes a rendezvous with the NTR for the return boost to the earth.

The 638–668 ton manned Mars spacecraft is 105.6 meters long and is built up from four basic components: a 47.6 meter reusable lunar vehicle core, 30 meters of propellant tanks, and a Mars excursion module and mission module that together are 28 meters long.

Boeing Advanced Civil Space Systems, after a 3-year-long study, proposes a Mars transfer vehicle.

BOEING MARS MISSION PROPOSALS, ca. 1990–1991

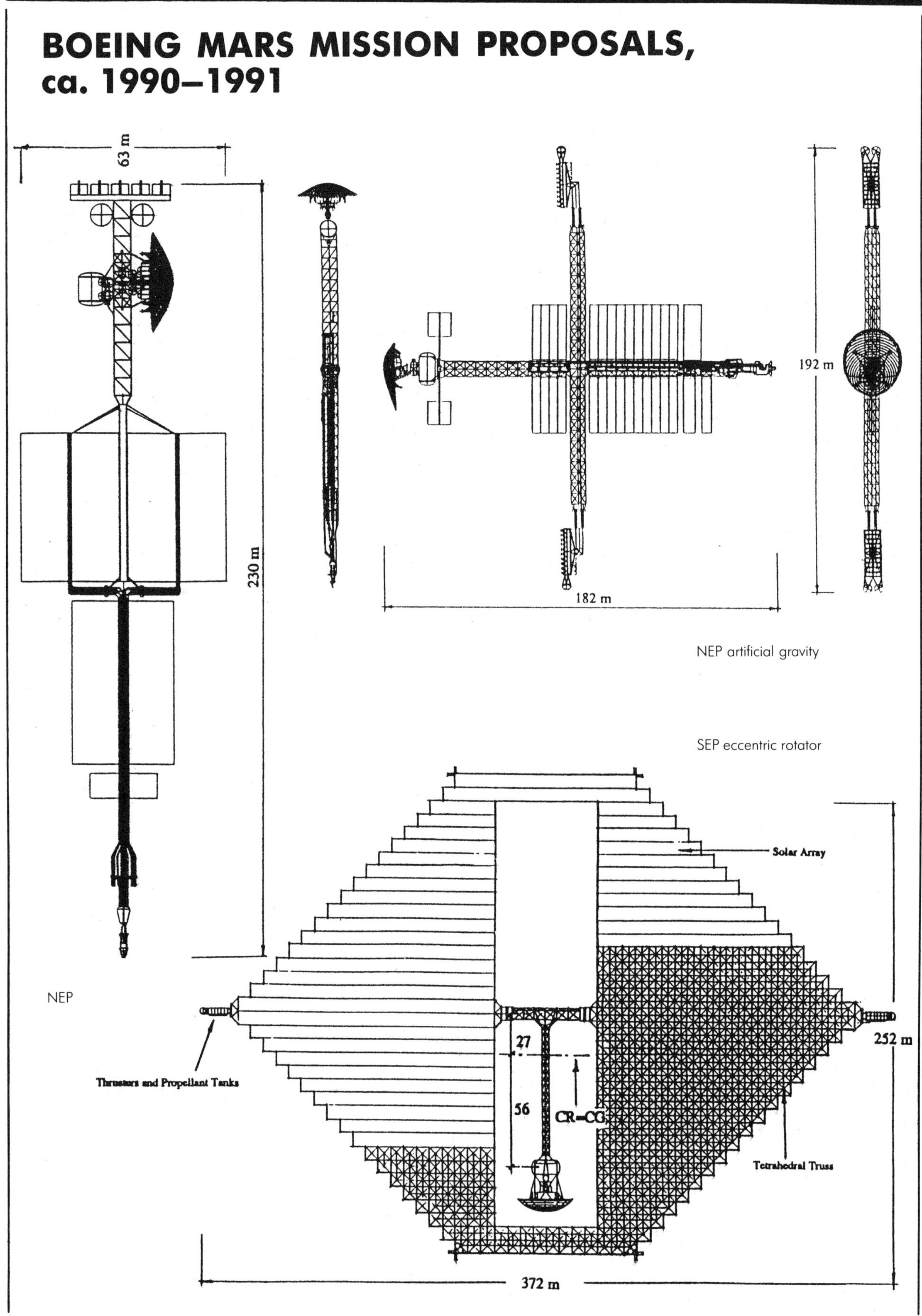

NEP artificial gravity

SEP eccentric rotator

NEP

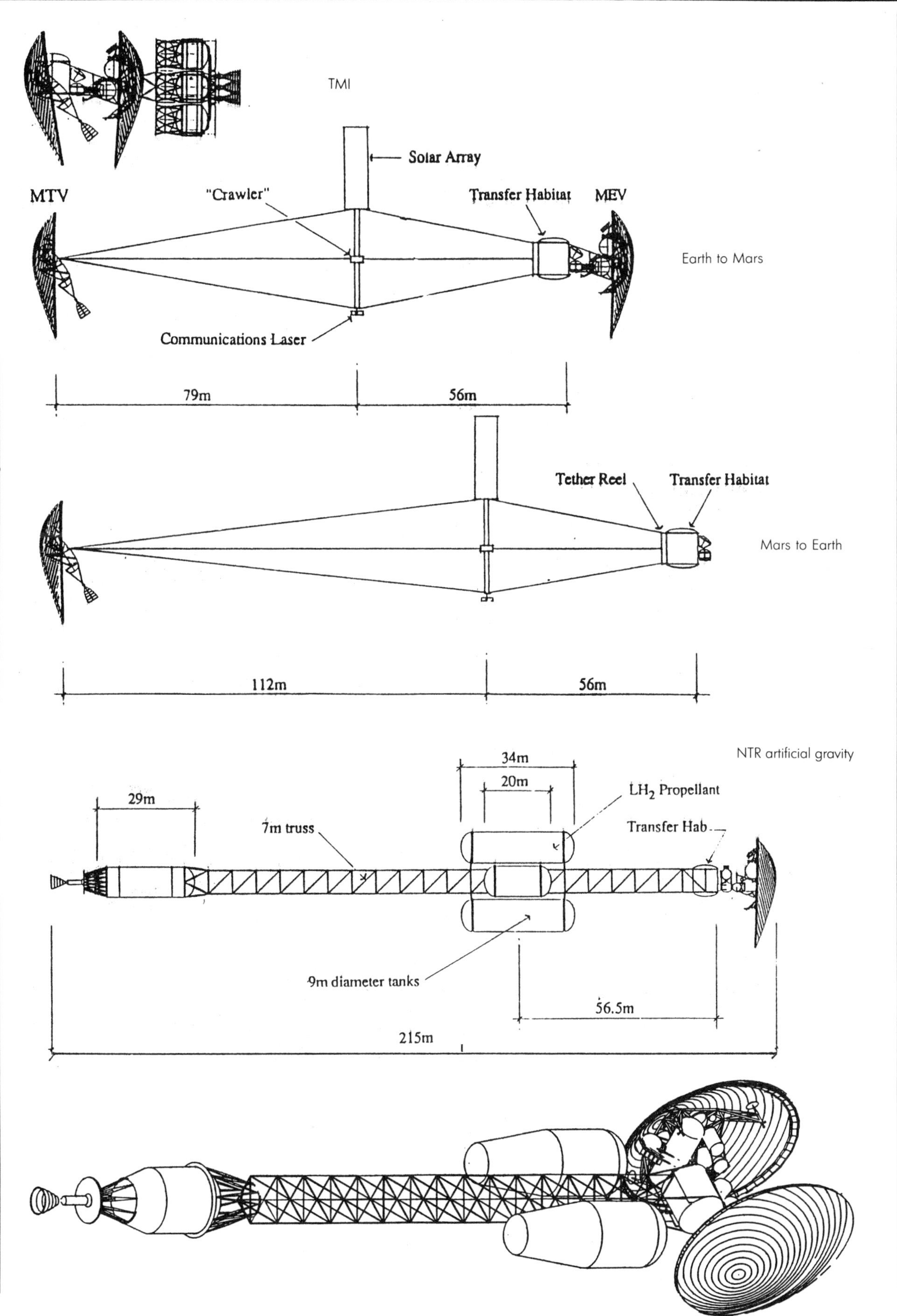
TMI
Solar Array
MTV
"Crawler"
Transfer Habitat
MEV
Earth to Mars
Communications Laser
79m
56m
Tether Reel
Transfer Habitat
Mars to Earth
112m
56m
NTR artificial gravity
34m
20m
29m
LH_2 Propellant
7m truss
Transfer Hab
9m diameter tanks
56.5m
215m

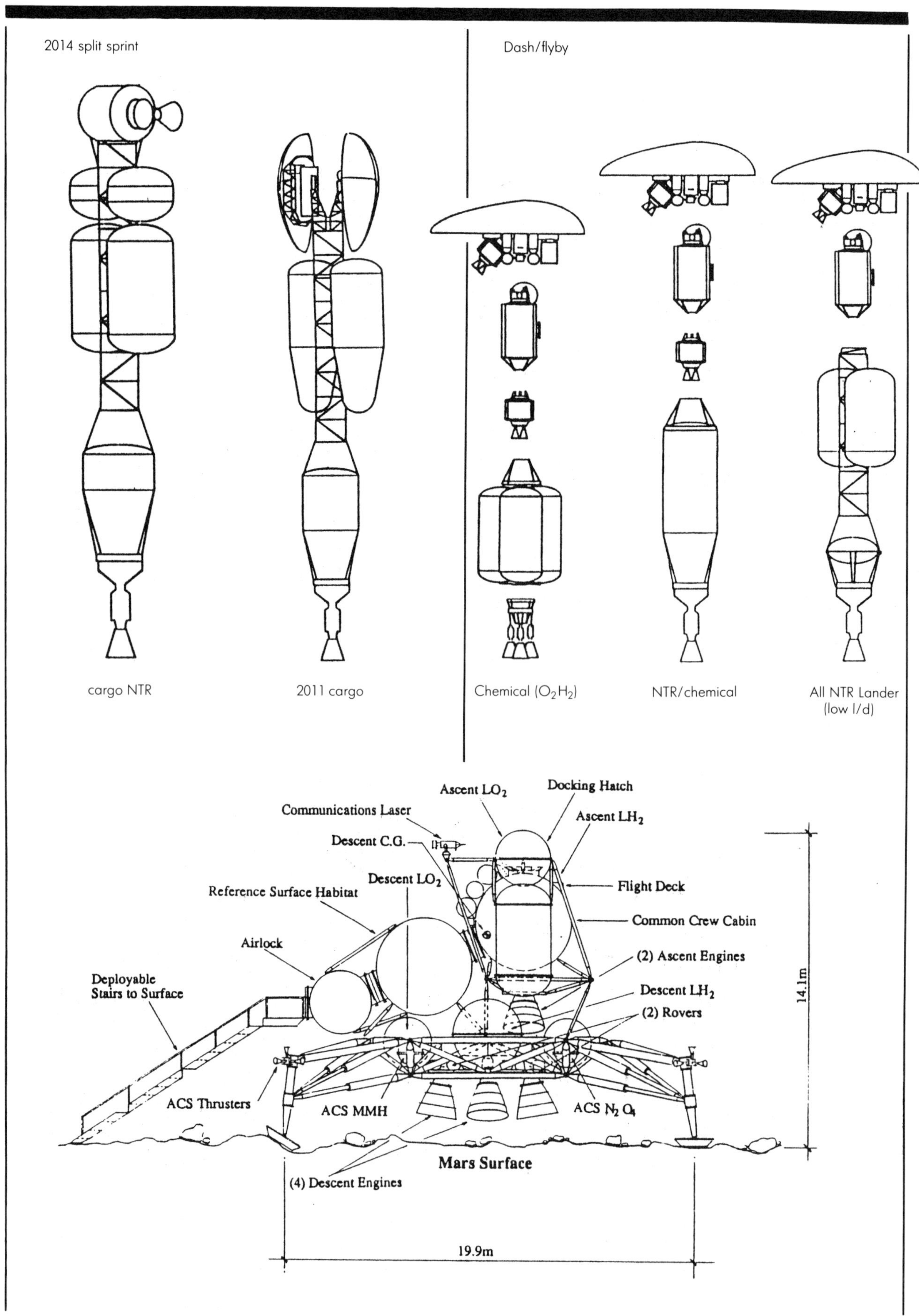
2014 split sprint
Dash/flyby
cargo NTR
2011 cargo
Chemical (O_2H_2)
NTR/chemical
All NTR Lander (low l/d)
Ascent LO_2
Docking Hatch
Communications Laser
Ascent LH_2
Descent C.G.
Descent LO_2
Flight Deck
Reference Surface Habitat
Common Crew Cabin
Airlock
(2) Ascent Engines
Deployable Stairs to Surface
Descent LH_2
(2) Rovers
14.1m
ACS Thrusters
ACS MMH
ACS N_2O_4
Mars Surface
(4) Descent Engines
19.9m

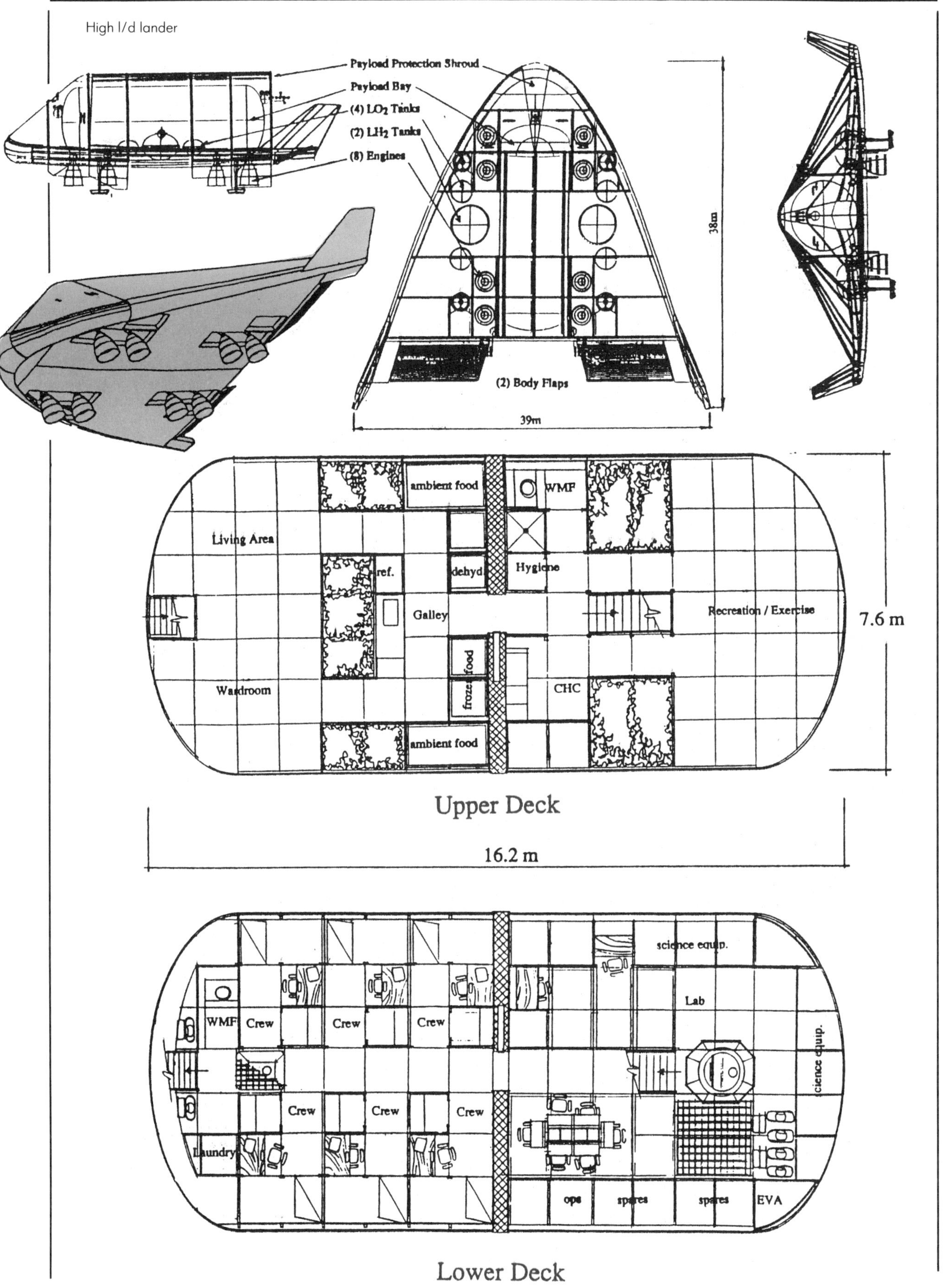
High l/d lander
Payload Protection Shroud
Payload Bay
(4) LO2 Tanks
(2) LH2 Tanks
(8) Engines
38m
(2) Body Flaps
39m
ambient food
WMF
Living Area
ref.
dehyd
Hygiene
Galley
Recreation / Exercise
7.6 m
frozen food
Wardroom
CHC
ambient food
Upper Deck
16.2 m
science equip.
Lab
WMF
Crew
Crew
Crew
science equip.
Crew
Crew
Crew
Laundry
ops
spares
spares
EVA
Lower Deck

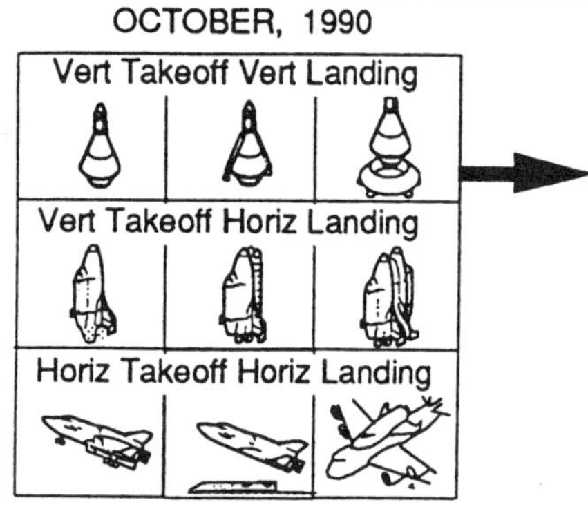

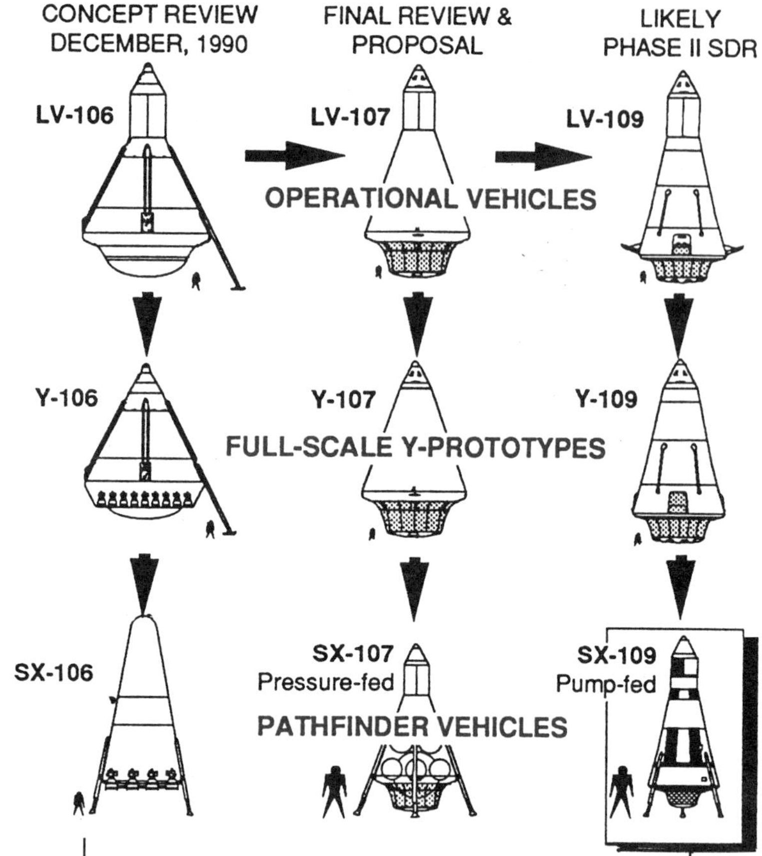

General Dynamics develops the *Millenium Express*, a vertical takeoff and landing (VTOL) concept for a single stage to orbit (SSTO) vehicle. The study is done under a contract from the Strategic Defense Initiative Organization.

The General Dynamics Model 107 VTOL SSTO (a slightly longer, slimmer version of the company's baseline VTOL, Model 106) employs a twelve module plug nozzle engine, resembling the concepts advocated in 1967 and 1976 by Phillip Bono. The 105-foot-tall rocket takes off vertically and reenters base-first, like an Apollo command module. The booster segment is a 76.2-foot cone 53.5 feet in diameter at its base. The remainder of the vehicle's height is taken by a variety of payload modules, including a two-person capsule, a six-person crew ferry and an expendable cargo carrier. The 1.27 million pound spacecraft will have a life of some 500 flights, with a turnaround time of only 104 hours, requiring only fifty people.

General Dynamics considered numerous other SSTO options, divided into three categories: vertical takeoff and landing (VTOL), vertical takeoff and horizontal landing (VTHL) and horizontal takeoff and landing (HTOL). Each of these was subdivided into pure SSTO concepts versus augmented SSTOs (those requiring staging or other added booster systems). These two categories were further subdivided. The pure SSTOs went into two groups: those requiring only two or three space shuttle main engines and those employing either plug/aerospike or scramjet engines. The augmented SSTOs were subdivided into three groups: assisted takeoff (employing strap-on boosters or electromagnetic launchers), drop tanks and "other" (including "zero stages," biamese configurations and piggyback carriers).

Once the basic Model 106 had been selected, three design iterations were evaluated, beginning with Model SX-106, powered by four RL10 engines. This design was soon abandoned in spite of the long experience General Dynamics had with the RL10 Centaur engines. Model 107's plug nozzle engine was developed in cooperation with Aerojet and Rocketdyne. Together they evolved the SX-107 (pressure-fed) and SX-109 (pump-fed) Pathfinder vehicles. From the operational Y-107 and Y-109 prototypes will be developed the final LV-109 *Millenium Express* SSTO. This spacecraft will stand about 113 feet tall, about 80 feet of which will be its booster stage.

Before the operational vehicle is built, General Dynamics will test a Pathfinder, an inexpensive small-scale version of the final spacecraft. Its first flight will take place in mid-1993. The full-scale Y-Prototype test vehicle will make its suborbital flights from late 1995 to mid-1998. The operational vehicle will begin its flights in 1999.

Eventually, an entire family of space vehicles will be evolved from the basic Path-

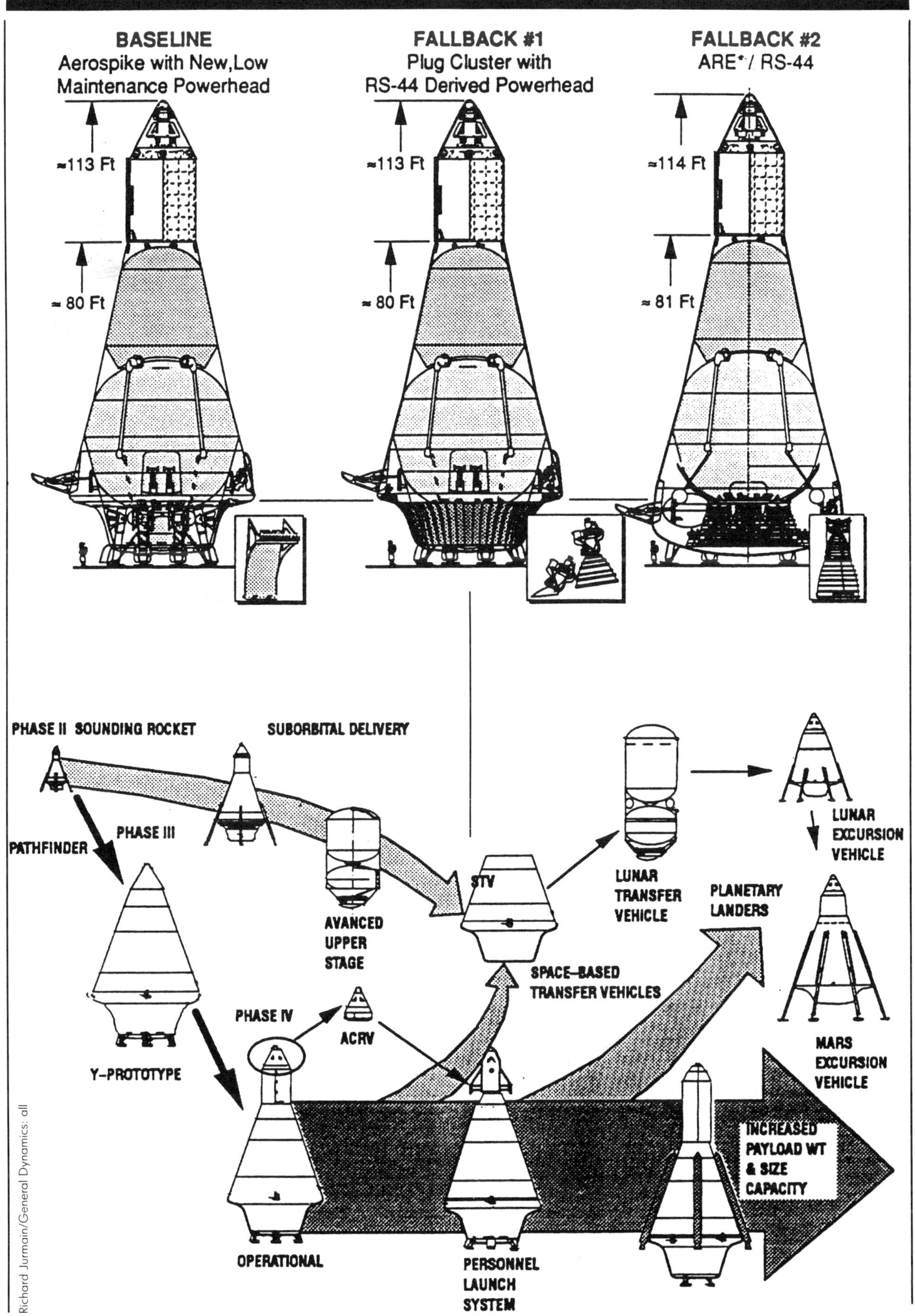

Richard Jurmain/General Dynamics: all

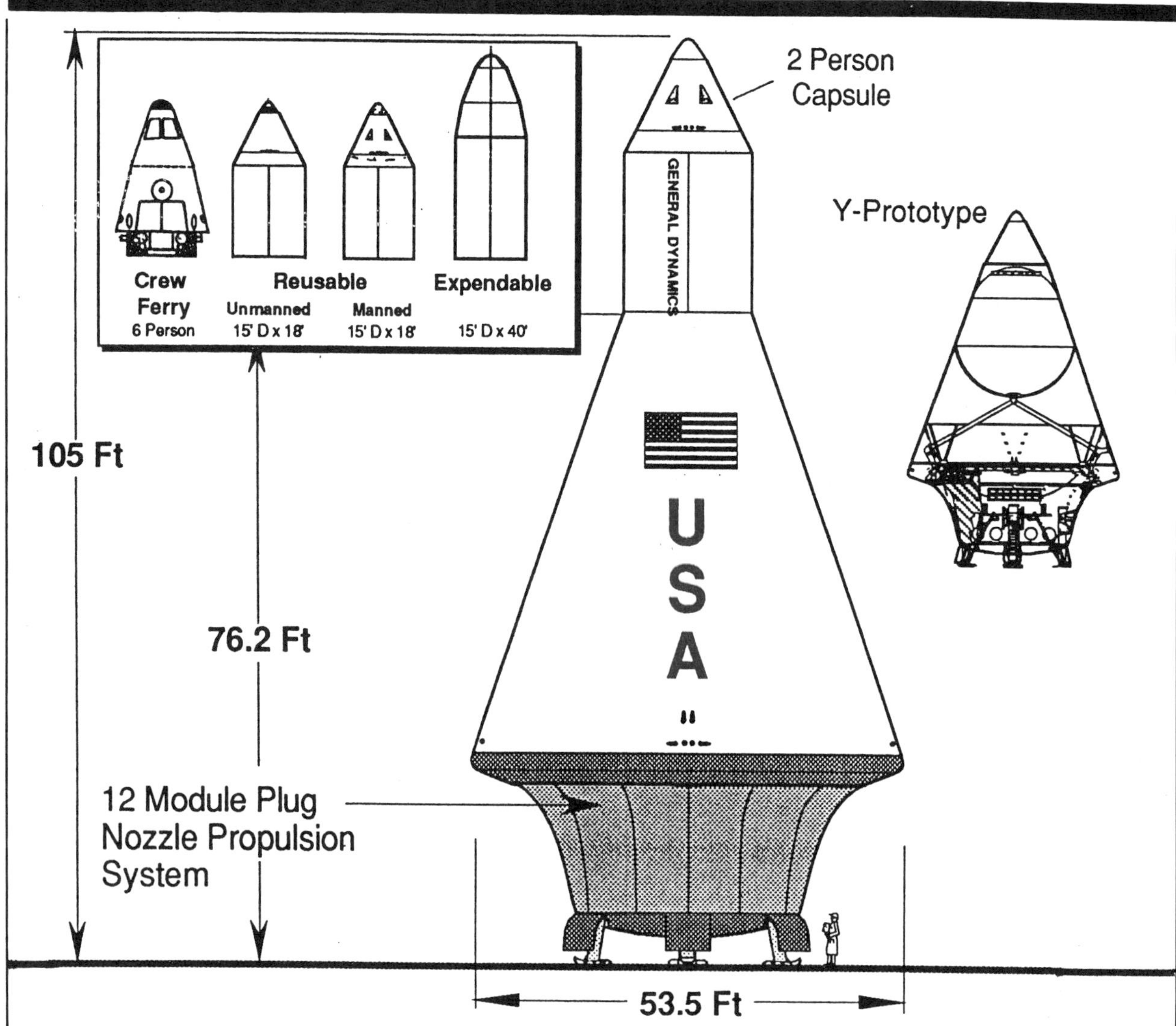

finder and LV-109, including vehicles for suborbital delivery, space-based transfer, lunar transfer and lunar and Mars landers. The *Millenium Express* booster could also carry a winged orbiter similar to Langley's Personnel Launch Vehicle.

The European Space Agency (ESA) completes its studies on fully reusable single stage and two stage to orbit winged launchers, a project begun in 1988. [Also see 1989.] The two stage to orbit (TSTO) vehicle developed is propelled by hydrogen-fuelled turbofans and ramjets in the first stage and hydrogen/oxygen rockets in the second. Its launch and landing site will be in southern France. It can deliver 7,000 kg into orbit.

The final TSTO resembles the German Sänger spaceplane. The first stage is 81.8 m long with a five engine turboramjet installation (the baseline design uses plug nozzles). The second stage is 34.2 m long with a wingspan of 18 m. Work on this vehicle will continue with a new phase beginning in 1992.

Options developed by the space exploration initiative (SEI) of President Bush include lunar landers capable of transporting cargos of 10 to 30 metric tons on missions lasting from a few days to months. Lander concepts include single-stage, two-stage and single-propulsion/avionics modules.

The lunar transportation systems include ground-based, direct launch; ground-based, earth-orbit rendezvous/dock; and space-based, on-orbit assembly. The ground-based, direct launch system is composed of three

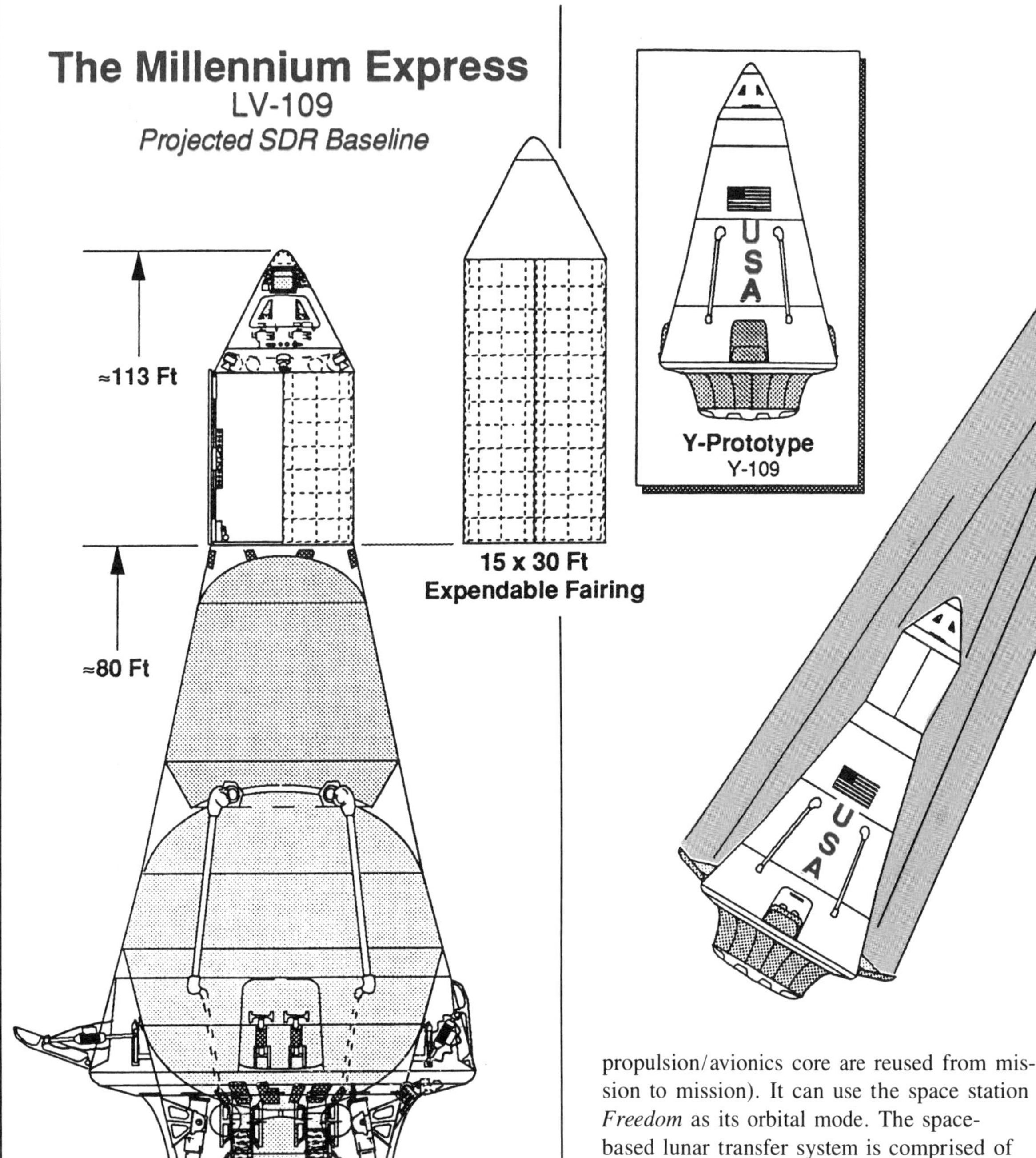

Richard Jurmain/General Dynamics: all

main elements: the translunar injection stage, the lunar orbit insertion/trans-earth injection stage, and the lunar lander. This configuration utilizes two crew modules: a lunar transfer module and a lunar excursion module for the lander. The ground-based, earth-orbit rendezvous/dock concept uses the same three main elements as above.

The space-based configuration is a more advanced concept that permits a greater amount of reusability (the crew module and propulsion/avionics core are reused from mission to mission). It can use the space station *Freedom* as its orbital mode. The space-based lunar transfer system is comprised of three main elements: the single propulsion/avionics (P/A) module core, expendable drop tanks, and a single crew transfer/excursion module. The vehicle is assembled in orbit. The P/A module contains the aerobrake, engines, major subsystems and propellant tanks for the lunar descent/ascent. The single crew module functions for both transfer and excursion.

For the mission to Mars three options are considered: (1) a cryogenic, all-propulsive vehicle, (2) a cryogenic aerobrake vehicle, or (3) a nuclear thermal propulsion (NTP) vehicle. Mars transfer system propulsion in-

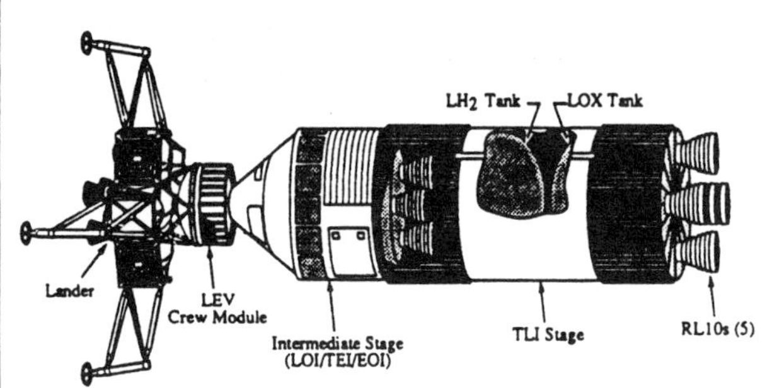

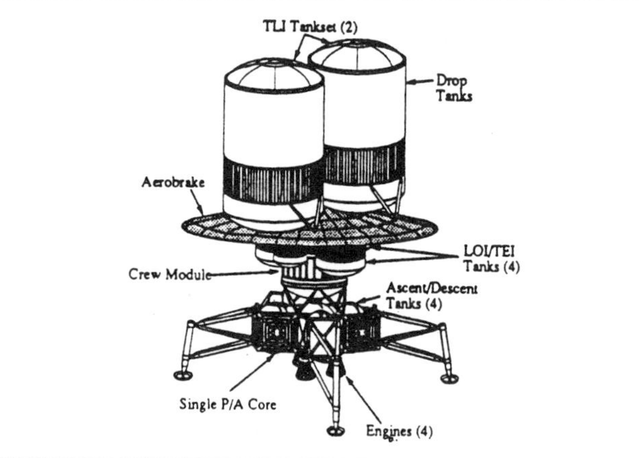

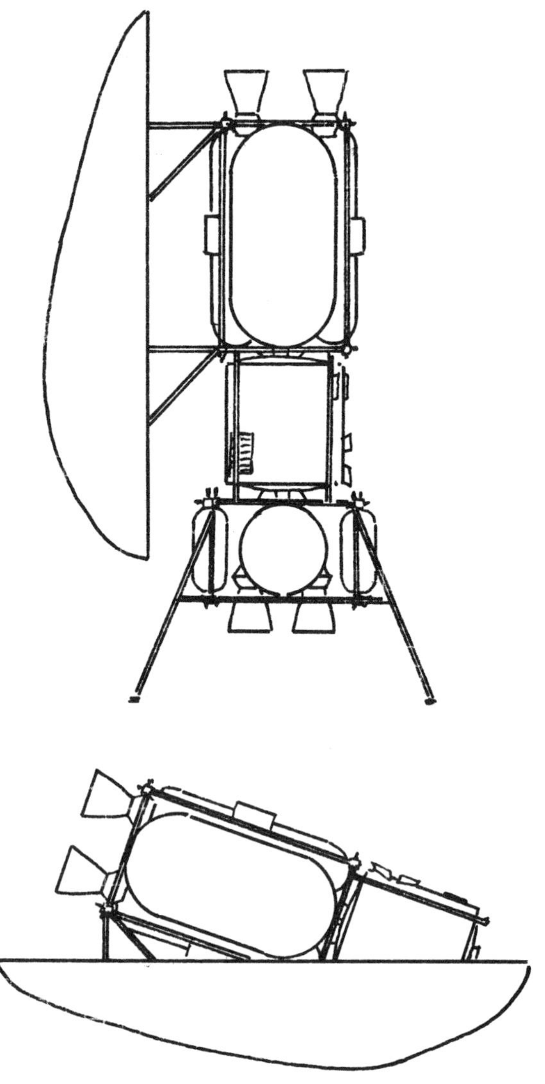

cludes chemical, chemical aerobrake, nuclear thermal, nuclear electric and solar electric. All vehicles are designed to support crew sizes of from four to eight astronauts, payloads of 10 to 25 metric tons and mission durations of from 400 to 1,000 days. All of the concepts employ cryogenic propulsion/aerobraking for the lander with the return vehicle remaining in orbit around Mars.

The first vehicle is 64 meters long with a 30-meter aerobrake (common to all three concepts). A trans-Mars insertion (TMI) stage propels the spaceship into its trajectory for Mars. The second stage acts for both arrival and departure from Mars orbit. There is a crew module for four astronauts and an aerobrake with a lander stowed within it.

The second vehicle is the one defined by NASA's 90-day study. It uses a TMI stage similar to the first vehicle's. This spaceship has two aerobrakes, however. One is for the Mars lander and the other is for the Mars transfer vehicle. After TMI the vehicles separate and are independently aerocaptured at Mars. Once in orbit they rendezvous and dock. The crew transfers from the transfer vehicle to the lander and then descends to the surface. A small Apollo-style capsule is used for the final return to the earth.

The NTP vehicle can be a reusable system if nuclear propulsion is used for capture upon its return to the earth.

NASA's Langley Research Center studies several design options for advanced manned launch systems [also see above]. Among those considered are:

• *Near-term two-stage.* This is a fully reusable two-stage AMLS vehicle that is described above [1991].

• *Near-term SSTO.* This is an adaptation of the orbiter stage of the two-stage AMLS. The 391-foot spacecraft has a gross weight of 25.1 million pounds.

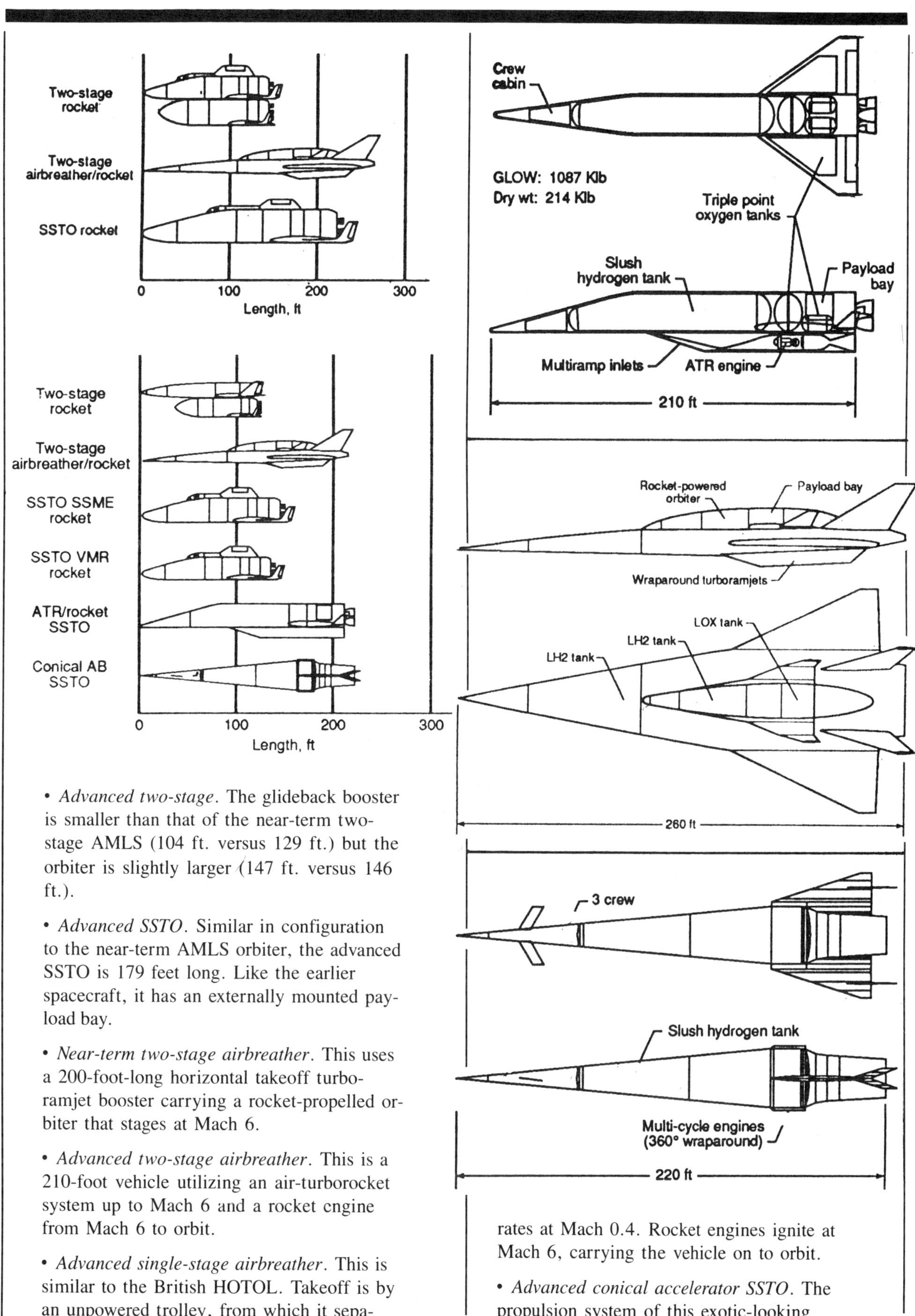

• *Advanced two-stage*. The glideback booster is smaller than that of the near-term two-stage AMLS (104 ft. versus 129 ft.) but the orbiter is slightly larger (147 ft. versus 146 ft.).

• *Advanced SSTO*. Similar in configuration to the near-term AMLS orbiter, the advanced SSTO is 179 feet long. Like the earlier spacecraft, it has an externally mounted payload bay.

• *Near-term two-stage airbreather*. This uses a 200-foot-long horizontal takeoff turboramjet booster carrying a rocket-propelled orbiter that stages at Mach 6.

• *Advanced two-stage airbreather*. This is a 210-foot vehicle utilizing an air-turborocket system up to Mach 6 and a rocket engine from Mach 6 to orbit.

• *Advanced single-stage airbreather*. This is similar to the British HOTOL. Takeoff is by an unpowered trolley, from which it separates at Mach 0.4. Rocket engines ignite at Mach 6, carrying the vehicle on to orbit.

• *Advanced conical accelerator SSTO*. The propulsion system of this exotic-looking spacecraft uses slush-mode hydrogen with low speed, ramjet, scramjet and rocket cycles. The propulsion package wraps completely around the rear of the 220-foot conical vehicle.

SUMMER

The International Space University, during its Toulouse, France, session, develops what may inarguably be the most thoroughly explored study undertaken to date for a manned Mars mission. In a final report consisting of well over 600 pages the students consider quite literally every conceivable aspect of an international mission, from crew sex to merchandizing, from international treaties to hourly exploration schedules. Unlike most previous Mars mission studies where most consideration is given to the engineering and technology, ISU's organization is such that all facets are considered as part of a Mars mission gestalt. To do this, groups of students work in various specialties, preparing their particular contributions to the whole mission. These include the life sciences, humanities, policy and law, architecture, engineering and so forth. It is unfortunate that the only aspect of this remarkable achievement that can be considered in any detail here is the spacecraft.

The vehicle of choice is a 200-meter nuclear electric spacecraft that provides its crew with spin-created (2 rpm) artificial gravity. Onboard power is provided by the same reactor that provides the energy for the propulsion system. Two cylindrical habitats are at the end opposite the reactor. One cylinder provides living space for the crew of eight; the other is a work area. Separating the habitats from the reactor is a square aluminum truss which also supports a pair of narrow triangular radiators. The 5 degree angle they make with the reactor keeps them in the shadow of its radiation shielding. Six hundred metric tons of propellant (liquid xenon) are stored in eight cylindrical tanks, four at either end of the truss. (The ISU students also study a nuclear thermal rocket for the manned mission. This configuration is based on a 160-meter truss. The habitats are

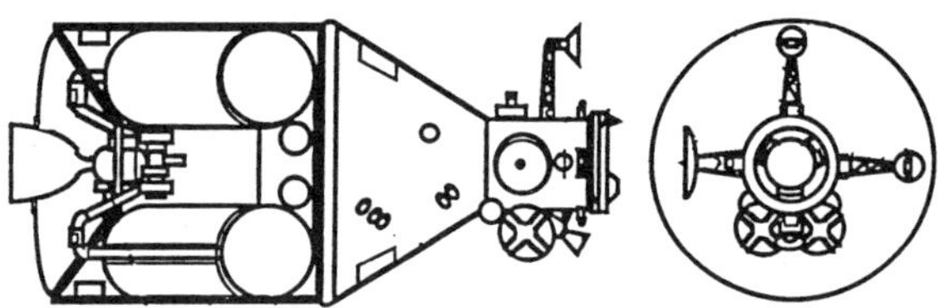

Vehicle configuration

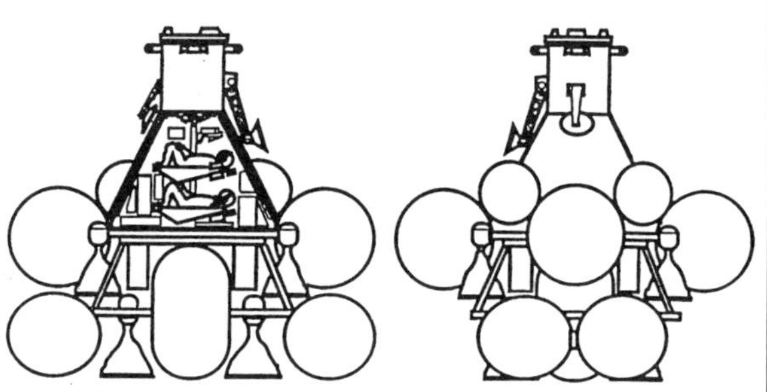

Ascent module

the same. It lacks the radiators of the other spacecraft but must carry a great deal more fuel, which is carried in tanks covering most of the length of the truss.)

Cargo is sent in its own unmanned vehicle ahead of the manned spacecraft. Among the material carried are the biconic landers that the astronauts will use to descend to the Martian surface and the earth transfer capsules which will transport the crewmembers between the earth and the Mars ship.

Once the spacecraft has arrived in orbit around Mars, a number of Navigation Land Capsule landers are launched to the surface to create a navigation network that will increase the accuracy of the manned landings.

Both crew and cargo then use the biconic landers to land on Mars. These are 23 m long, 5.5 m in diameter and have an initial mass of 70,000 kg. They descend aerodynamically, making the final landing on retrorockets. The manned landers each carry a crew of four. Contained within each manned lander is an ascent vehicle. The cargo land-

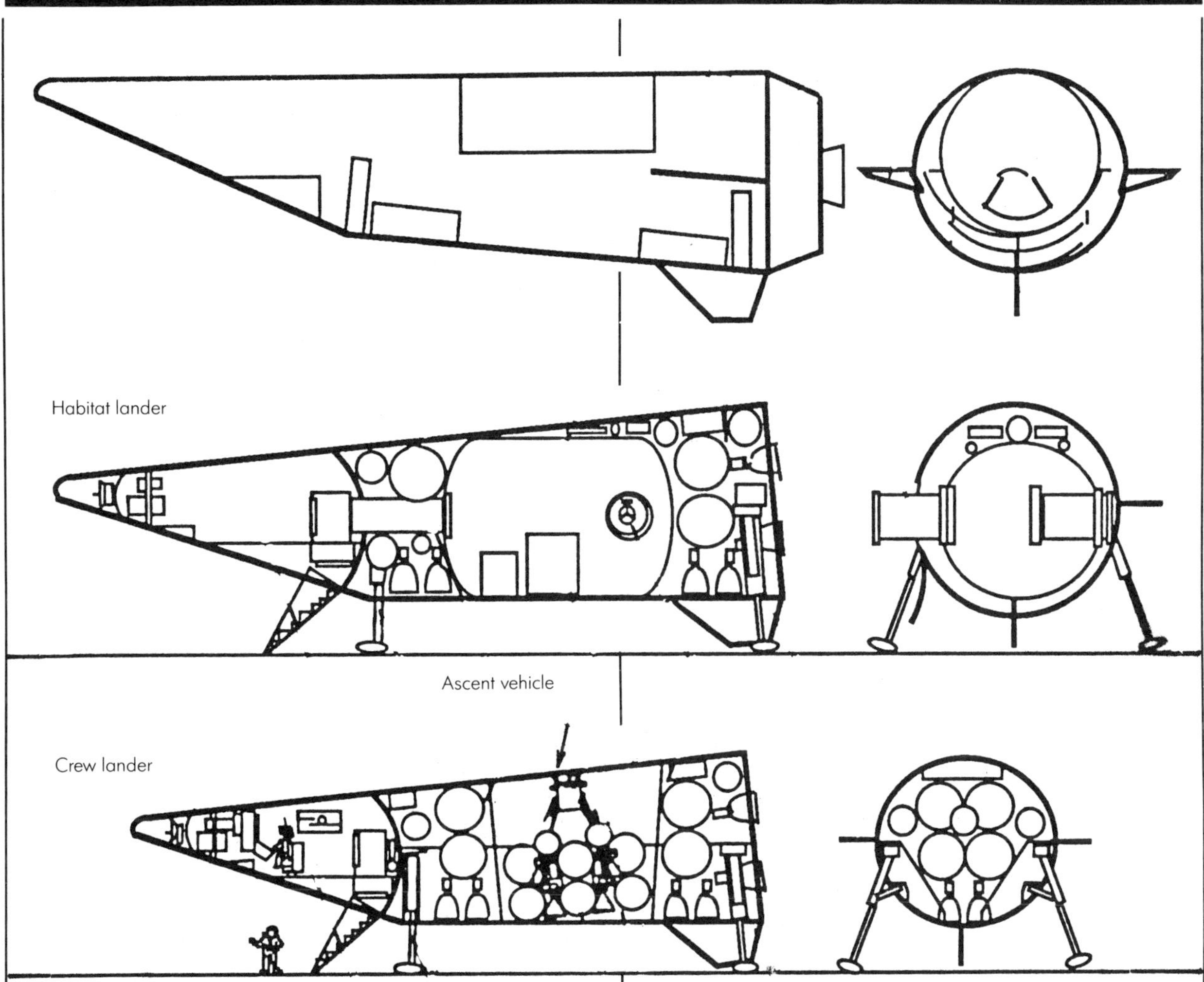

ers, which are of the same dimensions and landing weight as the manned landers, carry surface rovers rather than ascent vehicles. One of the unmanned landers is a habitat module that provides living and working space for the astronauts.

The 8.4 m × 4.1 m earth transfer capsules are similar to the Apollo Service and Command Modules. Each carries a crew of four and can also act as an emergency vehicle.

JUNE

White House National Space Council Executive Secretary Mark Albrecht confers with Soviet officials on a long-term strategy for a joint mission to Mars that will use the Soviet Union's Energia rocket. In addition to U.S. and Russian involvement the international mission will also include Japan and Europe. The plan calls for landing prefabricated housing before the crew of three men and three women leave on the 9-month journey. They will stay on the Martian surface for one year.

JUNE 21

British Aerospace and the Soviet Ministry of Aviation Industry complete the first phase of the An-225/Interim HOTOL spaceplane concept. This would make use of the giant heavy lift (*Dream*) aircraft as a first stage.

The An-225 was designed in 1985–1988 as transport for nonstandard oversize cargo either internally or piggyback. In the latter case it has provided transport for the *Buran* space shuttle, logging 32 hours with it aboard. The An-225 can carry external cargo with the maximum dimensions of 7 to 11 meters in diameter and a length of up to 70 meters. The giant aircraft has a wingspan of 88.4 m and a length of 84 m. Modifications for use as a HOTOL first stage would include the installation of two extra D-18 engines in addition to the existing eight (or a

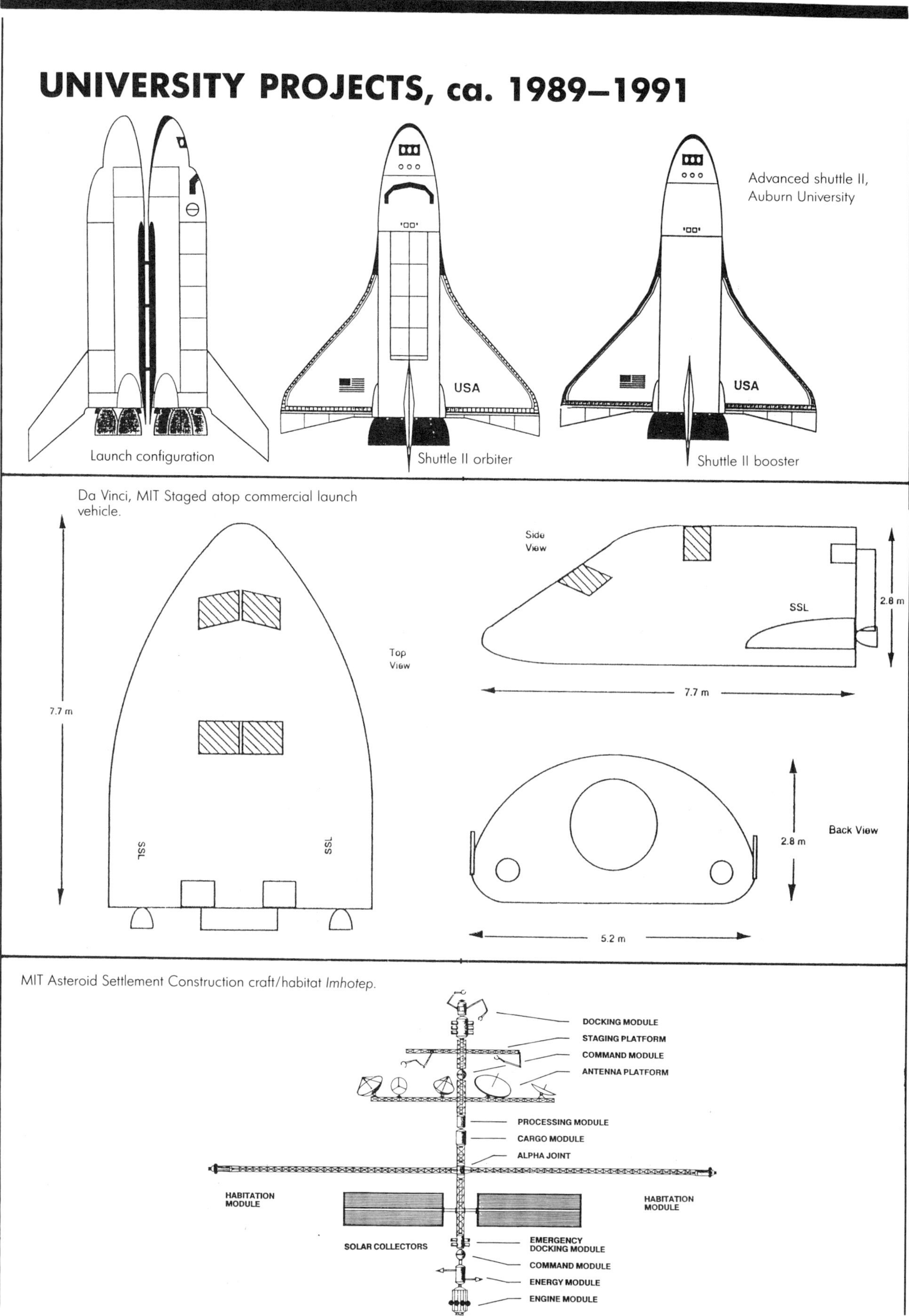
UNIVERSITY PROJECTS, ca. 1989–1991
Advanced shuttle II, Auburn University
USA
USA
Launch configuration
Shuttle II orbiter
Shuttle II booster
Da Vinci, MIT Staged atop commercial launch vehicle.
Side View
SSL
2.8 m
7.7 m
Top View
7.7 m
SSL
SSL
Back View
2.8 m
5.2 m
MIT Asteroid Settlement Construction craft/habitat Imhotep.
DOCKING MODULE
STAGING PLATFORM
COMMAND MODULE
ANTENNA PLATFORM
PROCESSING MODULE
CARGO MODULE
ALPHA JOINT
HABITATION MODULE
HABITATION MODULE
SOLAR COLLECTORS
EMERGENCY DOCKING MODULE
COMMAND MODULE
ENERGY MODULE
ENGINE MODULE

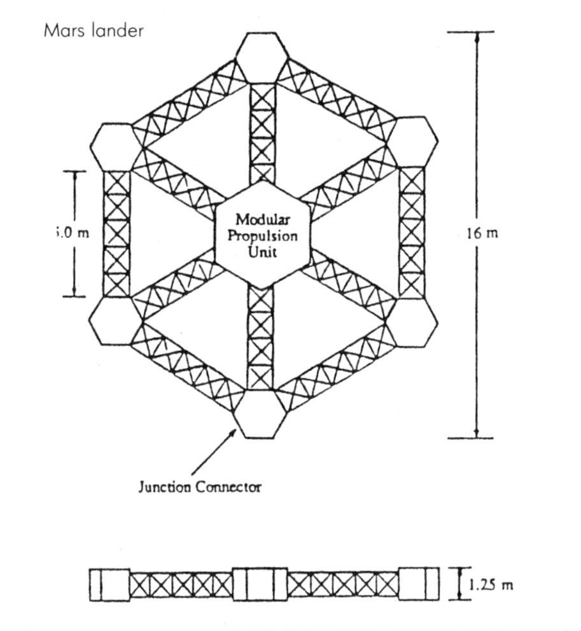

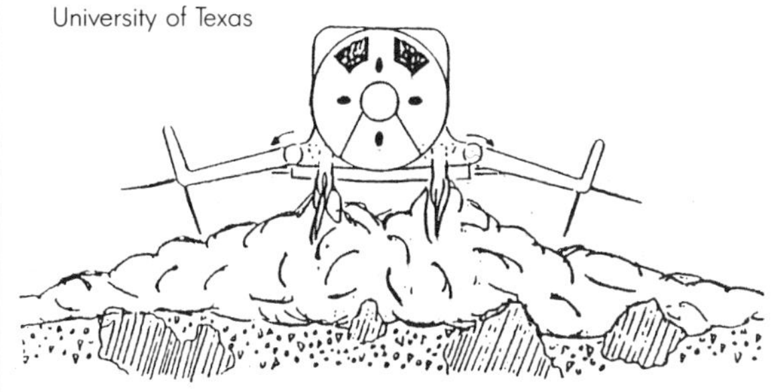

Hopper Hovering Prior to Landing

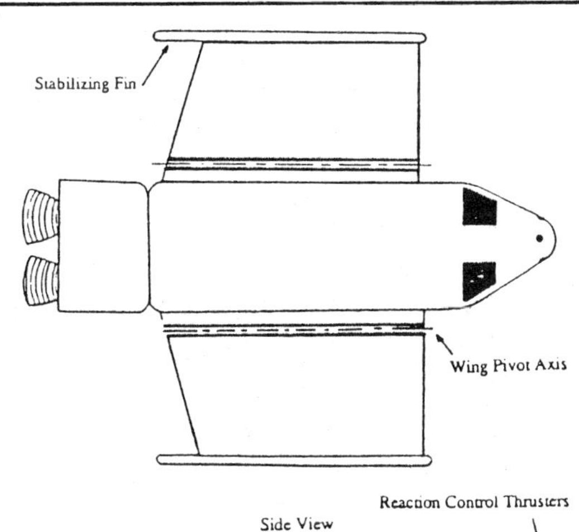

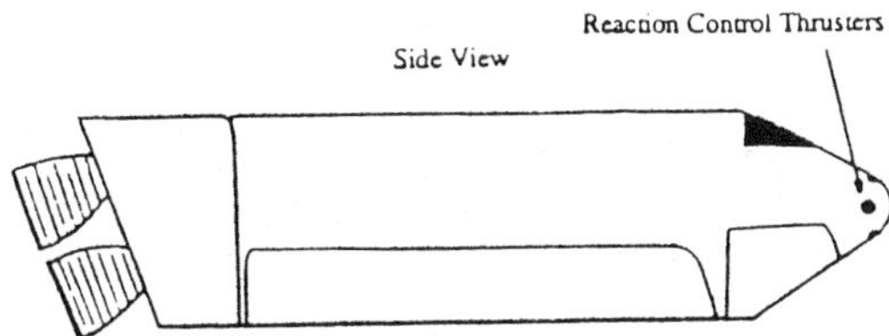

Top and Side View of the Mars Ballistic Hopper Using a Modified Low Profile Design. Wings are added to provide lift, drag, and stability at the high Mach numbers existing throughout the aerobrake phase of the trajectory. The wings also pivot down to serve as a component of the landing gear.

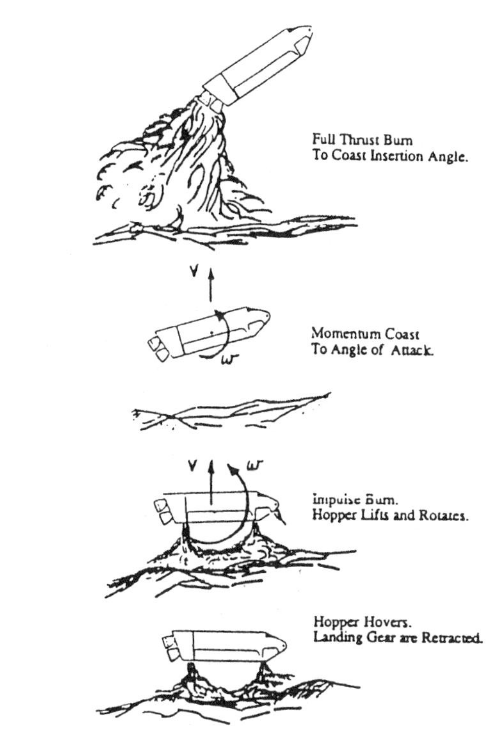

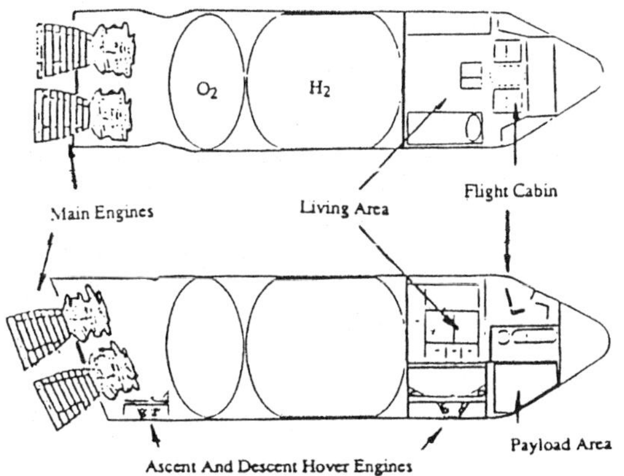

change to six Rolls-Royce Trent engines), strengthening of the fuselage, addition of auxiliary equipment and operators and the fitting of the HOTOL's support assembly.

The HOTOL orbiter would undergo changes itself. It would have the same length and wingspan as the U.S. space shuttle or the Soviet *Buran*, but with a considerably fatter body. Deployable foreplanes would assist control during the final stages of recovery. The HOTOL vehicle would be 36.45 m long with a wingspan of 21.6 m. The fully loaded orbiter would weigh 250 metric tons, including a payload of 5.5 to 8 metric tons.

At an altitude of 9,400 m the An-225 would begin a powered dive, reaching a speed of Mach 0.8. At an altitude of 8,800 m it would pull up for separation. At 9,200 m and the point of negative g-loading the An-225-orbiter connection would be broken. The An-225 would drop to 9,100 m as the orbiter's main engine ignites, 4 seconds after separation.

AUGUST 16

McDonnell Douglas Space Systems is selected by the Strategic Defense Initiative Organization (SDIO) for the 24-month Phase II SSTO study [see above]. McDonnell Douglas's design, developed by engineer Max Hunter (from a program he masterminded from 1959–1960, the Reusable Interplanetary Transport Approach, or RITA), is designated the SSX, or Delta Clipper. Resembling an oversize Mercury capsule, the Delta Clipper could be built and operational by 1996 for a total price of $2 billion. Once it is flying the cost of ferrying payloads to orbit could be as low as $100 per kg. McDonnell Douglas will develop a one-third size prototype SSTO called the DC-X. This will be flown in suborbital takeoff and landing tests beginning in the spring of 1993 at White Sands Missile Range. The results will be incorporated in the operational prototype, the DC-Y.

DECEMBER

Benjamin Donahue of the Boeing Defense and Space Group [also see 1990] completes his design for a nuclear thermal propulsion (NTP) spacecraft intended to satisfy the requirements of the Synthesis Report [see above] for a manned Mars spacecraft. Donahue presents his design to the Marshall Space Flight Center. This nuclear thermal propulsion (NTP) spacecraft has become NASA's current preferred concept for the first manned Mars missions of the 2014–2020 period.

The design illustrated represents a lunar "dress rehearal" version which will be used to validate the unique Mars mission hardware and operations in a lunar landing mission. The ultimate Mars mission spacecraft will have more propellant tanks and a lander equipped with an aeroshell. The dress rehearsal NTP is approximately 80 meters long and 14 meters in diameter, with its propellant tanks, crew module and lander "suspended" from an overhead truss. It has two NTP engines providing 75,000 tons of thrust each. The crew module provides full-service systems with private quarters, galley/wardroom, command and control area, health maintenance, and exercise and recreational equipment and space.

The spacecraft is divided into three more or less equal sections that can be launched into earth orbit preassembled. The only orbital assembly required is at the two connecting points, with only a single interconnect at each point for all lines. The complete vehicle requires three launches of about 150 metric tons each.

The lunar lander is designed to allow it to offload its cargo container without requiring cumbersome and complicated cargo unloading mechanisms. The container can be lowered directly onto a "flatbed" wheeled transporter. The lander can deliver up to 30 tons of surface habitation hardware or other cargo to the lunar surface. For the option of delivering a Mars excursion vehicle ascent stage to the moon for flight testing, a two-stage spacecraft is required.

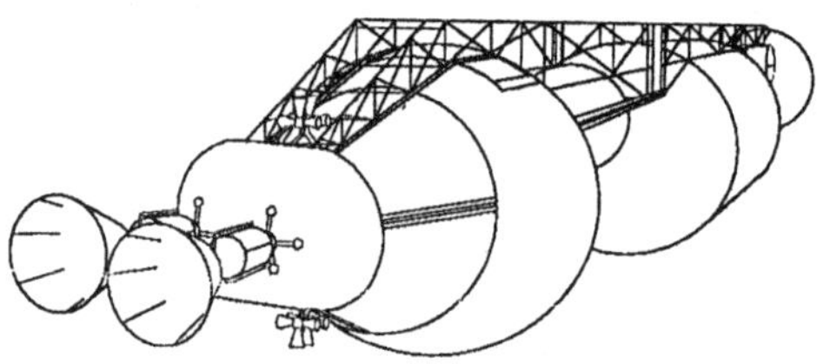

ca. 1991

The Department of Defense program codenamed "Timberwind" is a nuclear rocket designed to be launched from the surface of the earth, or started once in orbit as a kind of "superbooster. "The Strategic Defense Initiative Office decries the project, which would have been tested in Antarctica, as displaying "remarkable contempt for the safety issues."

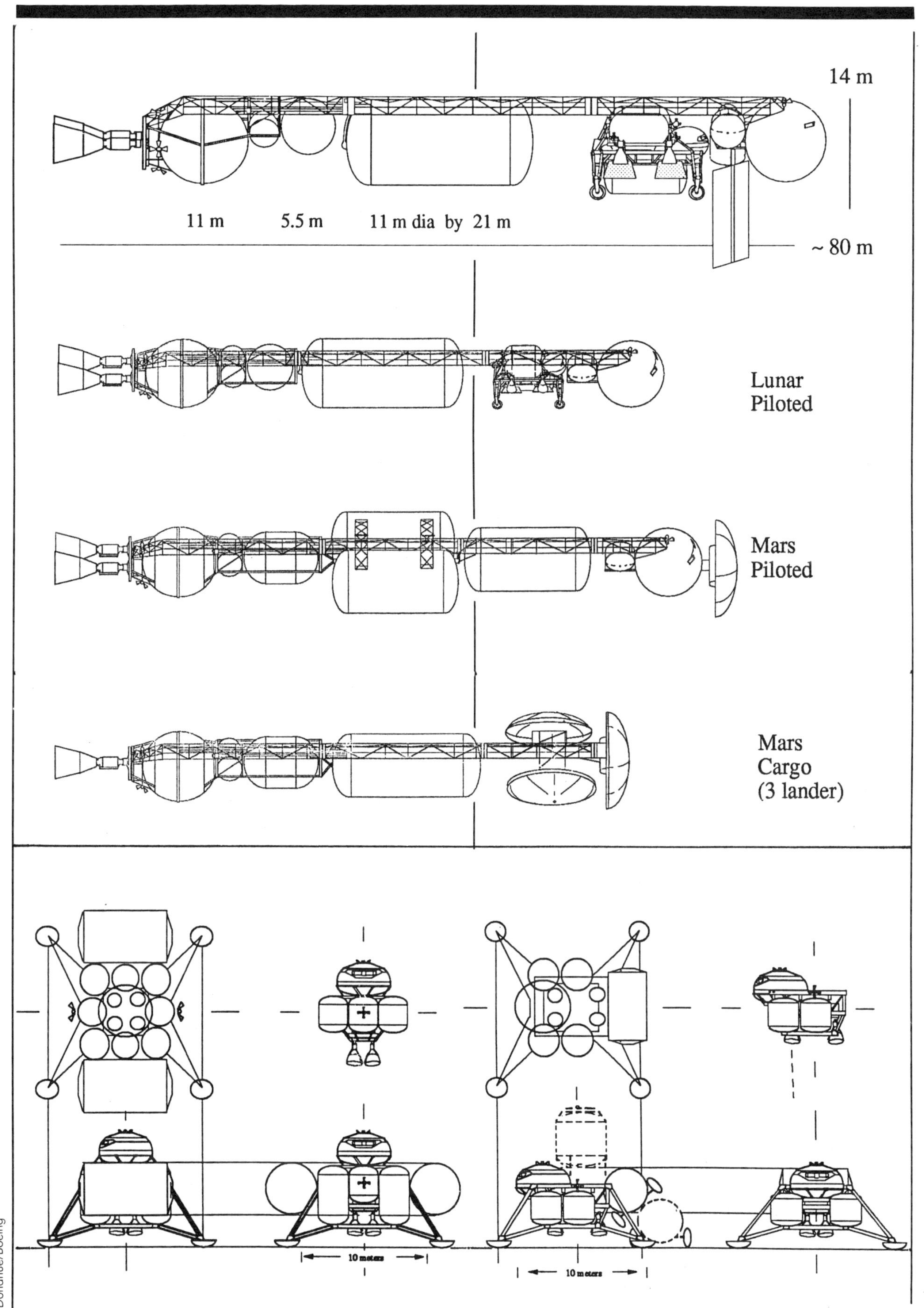
14 m
11 m
5.5 m
11 m dia by 21 m
~ 80 m
Lunar
Piloted
Mars
Piloted
Mars
Cargo
(3 lander)
10 meters
10 meters
Donahue/Boeing

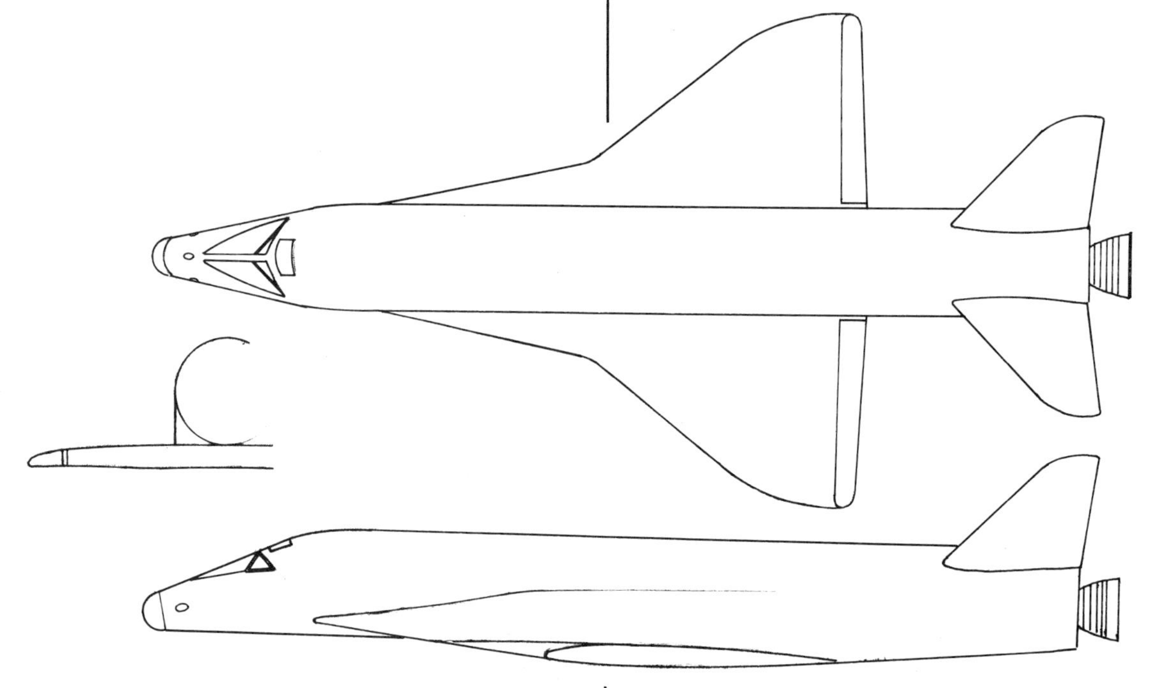

1992

Gary Hudson begins construction of a 60 percent scale model of his SkyRocket spaceplane. The goal of his private space program is to break the speed and altitude records of the X-15 with a relatively inexpensive rocket-propelled aircraft. The full-size SkyRocket would be powered by a LOX-hydrogen Pratt & Whitney RL10 engine. It would have a two-person crew and be capable of a transcontinental flight in under one hour. It would be air launched from the back of a surplus transport aircraft of the 707 or DC8 class.

The model uses foam and fiberglass construction similar to that used in the Voyager around-the-world aircraft. A graphite-epoxy version propelled by a hybrid rocket motor could exceed the X-15 records even while launched from a runway . . . no air launch would be needed.

The full size, or slightly larger, SkyRocket, either runway or air launched, would be capable of achieving orbit.

Ground-based laser-boosted spacecraft proposed by the Rensselaer Polytechnic Institute (1990).

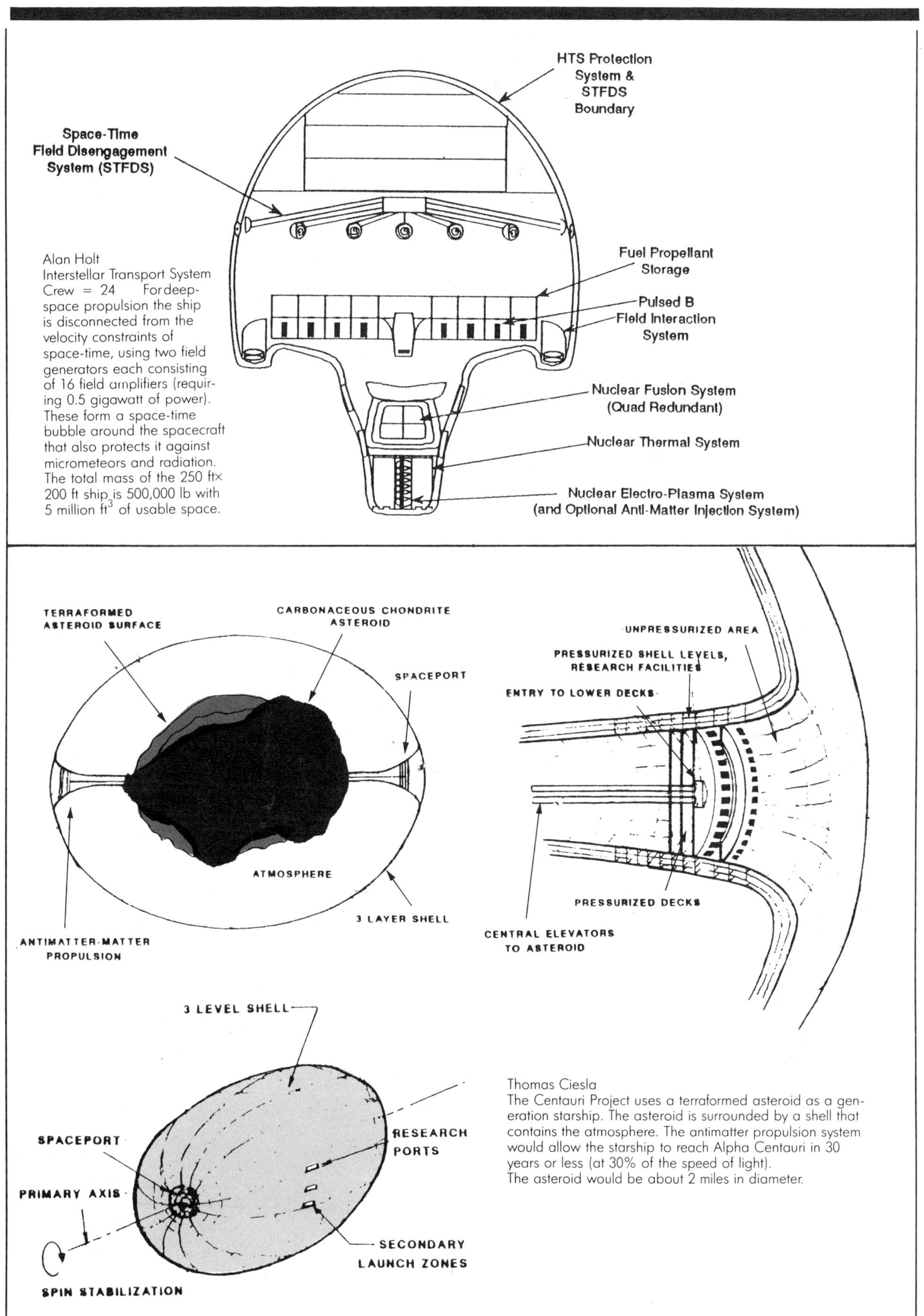

Alan Holt
Interstellar Transport System
Crew = 24 For deep-space propulsion the ship is disconnected from the velocity constraints of space-time, using two field generators each consisting of 16 field amplifiers (requiring 0.5 gigawatt of power). These form a space-time bubble around the spacecraft that also protects it against micrometeors and radiation. The total mass of the 250 ft × 200 ft ship is 500,000 lb with 5 million ft^3 of usable space.

Thomas Ciesla
The Centauri Project uses a terraformed asteroid as a generation starship. The asteroid is surrounded by a shell that contains the atmosphere. The antimatter propulsion system would allow the starship to reach Alpha Centauri in 30 years or less (at 30% of the speed of light).
The asteroid would be about 2 miles in diameter.

1992

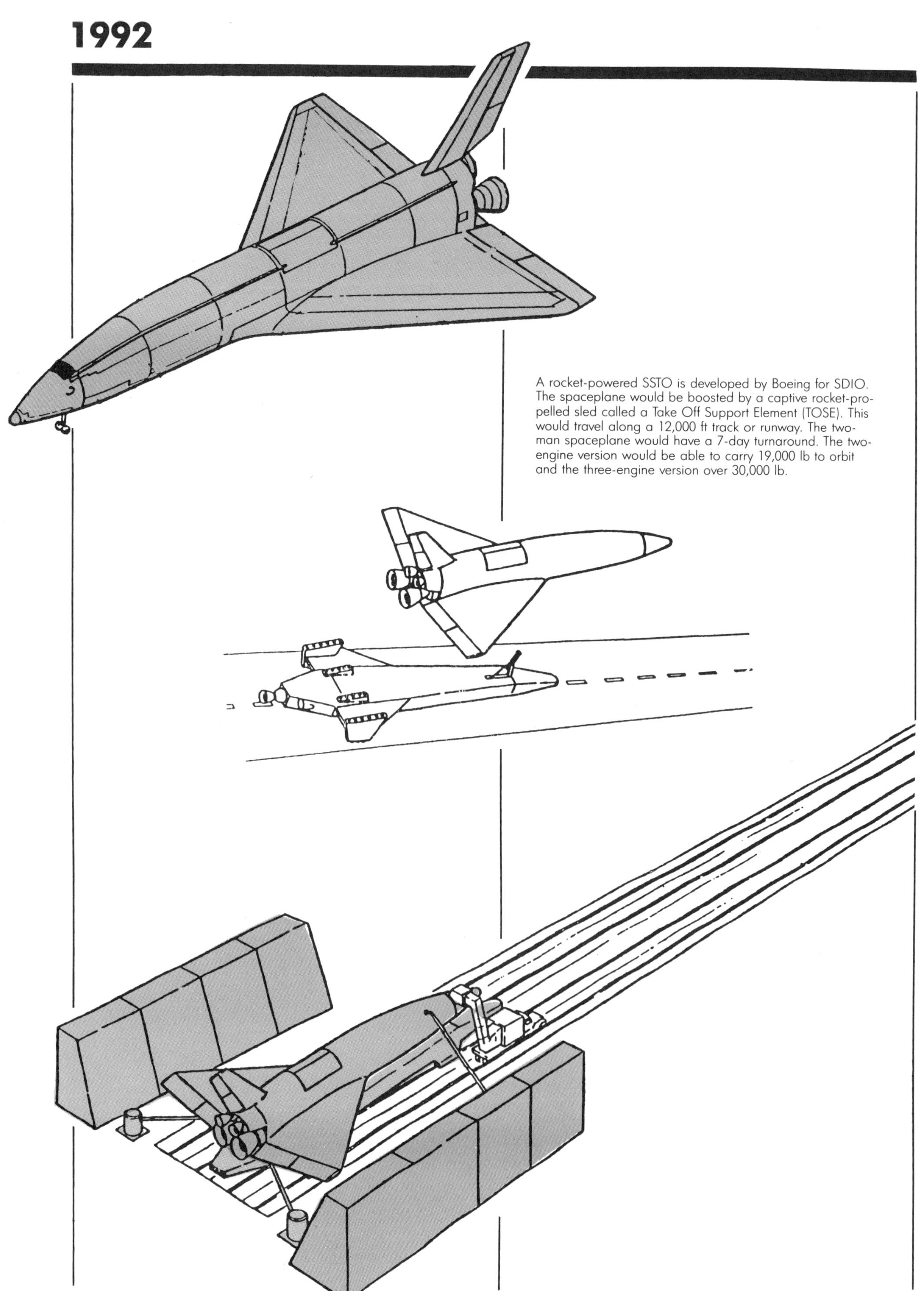

A rocket-powered SSTO is developed by Boeing for SDIO. The spaceplane would be boosted by a captive rocket-propelled sled called a Take Off Support Element (TOSE). This would travel along a 12,000 ft track or runway. The two-man spaceplane would have a 7-day turnaround. The two-engine version would be able to carry 19,000 lb to orbit and the three-engine version over 30,000 lb.

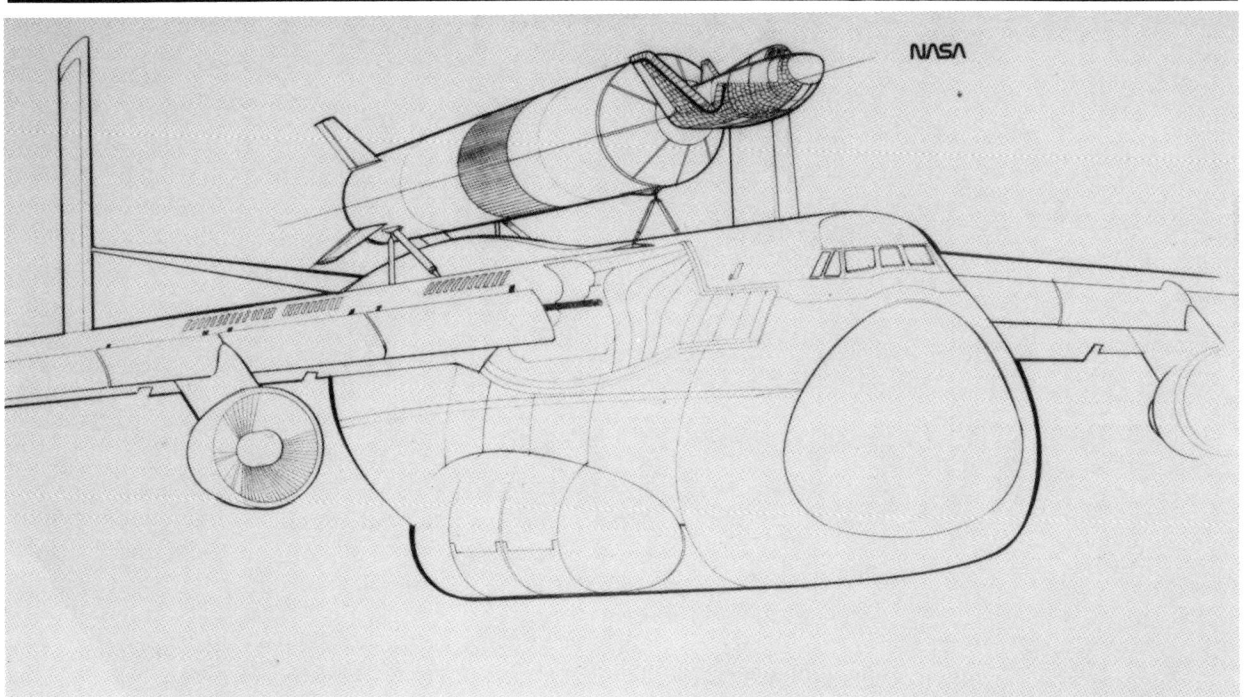

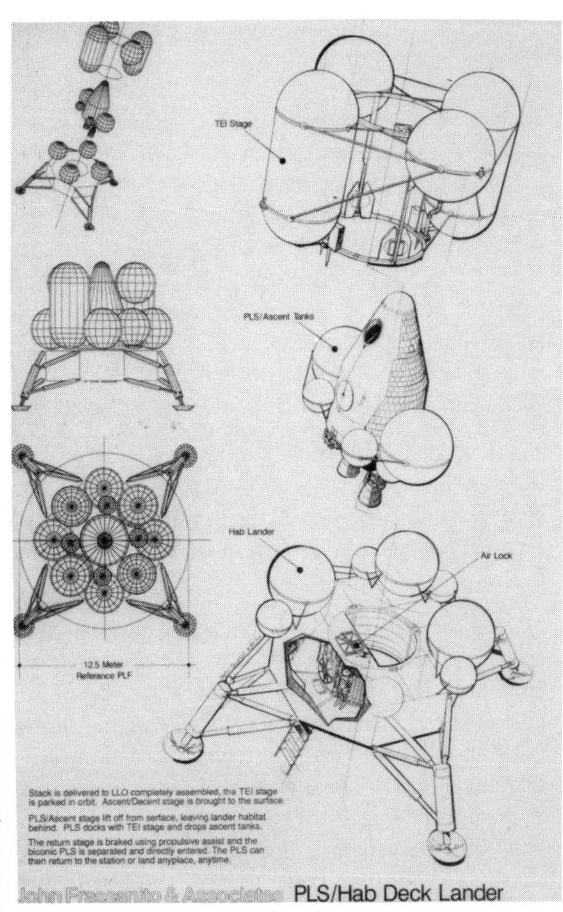

Top: an air-launched PLS developed by John Frassinito and Associates for NASA. Above: second-generation lunar lander combining a PLS with a habitat lander. Frassinito has developed numerous variations of this modular lander. Left: chemical-fueled Mars transfer vehicle by Frassinito. Electrical power would be provided by a free-flying nuclear power source.

EPILOGUE

What does the future hold for spaceflight? *Is* there even a future for spaceflight? A new generation has grown to college age since the last man walked on the moon, an event that is receding ever further into history. As this is being written, the only nation with an active manned spaceflight program is the United States with its irregular Space Shuttle launches. Yet even this half-hearted effort is threatened as the public becomes ever more disenchanted with NASA, a disenchantment that admittedly is as much the result of budget-cutting congressmen looking for an easy target as it is the fault of anything NASA itself—plagued by lack of direction—has done. The American public, entering a new Dark Age of antiscience and antitechnology, looks at NASA's budget and asks itself, "Can't this money be better spent here on the earth?" as if bags of cash were being loaded into the nose cones of rockets and shot into space. They forget three important facts: 1. NASA's money is spent right here, supporting a wide variety of businesses that in turn employ thousands of people; 2. there have been literally billions of dollars in payback from the space program, in spinoff technology and information from weather and earth resource satellites; and 3. NASA's budget is miniscule as compared to the bloated budgets of the Department of Defense. In fact, Americans spend more money on cosmetics and video games than they do on the exploration of space.

There seems, fortunately, to be a renaissance approaching as human space exploration enters a potential new age. Harbingers of this are the new generation of single-stage-to-orbit spacecraft and aerospaceplanes. This is also, for the first time, an international effort, with serious work being done by Japan, France/ESA, Britain and Germany (with the ex-Soviet Union perhaps again a contender in the future).

One of the themes of this book has been the importance of the *romance* of space travel, and that for centuries this romance has been the driving force behind its development. Much of this romance has been tarnished of late: space travel has proved to be expensive, elitist (how few of us will ever be able to travel beyond the earth?), dangerous and, ultimately, boring. That latter might sound surprising, but I believe that it is true. How continually exciting can it be to watch other people have incredible adventures? There is scarcely an adventure on the earth, from climbing Mount Everest to exploring the depths of the seas, that is not ultimately accessible to anyone who cares to take the risk. But spaceflight is not. It has so far been too complex and too expensive to be accomplished by anything other than governments, and large governments at that, who are not about to waste their resources on giving tourists a thrill, even if there were tourists who could afford the tickets.

Spaceflight has also become boring because of the intense effort to *make* it boring. This was certainly not the intention, but it is the inevitable result of a striving for perfection that has been almost entirely successful . . . certainly far more successful than the news media would have us believe. Transatlantic passenger jet flights would still be exciting events if every other plane dropped out of the sky;

but they don't, and few people think twice about the millions of uneventful air miles flown every day. Space enthusiasts have worked hard toward a day when spaceflight will be as commonplace as air travel, but without realizing that this cannot happen while at the same time maintaining *excitement* about space travel. The commonplace and the exciting are mutually exclusive. Just remember that only 60-odd years ago, transatlantic and transcontinental airplane flights were headline news and their pilots heroes. Anyone can fly the Atlantic or from continent to continent now and no one thinks twice about it.

Hopefully, the new hypersonic passenger spaceplanes, such as the U.S. *Orient Express* (the passenger transport edition of the National Aero-Space Plane) or the *Sänger II* passenger transport, will finally allow anyone access to space, or at least to the outer fringes of the earth's atmosphere, where the curvature of the earth will be distinct beneath a jet-black sky. The various single-stage-to-orbit plug-nozzle spacecraft also hold a great deal of promise for relatively cheap passenger spaceflight. Although the latter's promise is greater, it is my opinion that the impact of the former will be important for no other reason than that they *look* terrific.

To date, with the possible exception of the X-15, there have been no spacecraft with an aesthetic appeal equivalent to that possessed by, for example, the DC-3 or Constellation aircraft. In other words, spaceships so far just haven't looked very good. The Lunar Module looked like a windup toy and the Space Shuttle looks like a brick with wings. They are certainly nothing like the sleek spaceships we have been promised for generations (see the first two-thirds of this book, for instance). Who other than an engineer or rabid space enthusiast can get truly excited about the Shuttle? How can it hold its own against the *Luna*, Tintin's spaceship or the spectacular *Collier's* fleet of Wernher von Braun? A citizenry such as the United States possesses, that has for generations purchased automobiles almost strictly on how sexy they look, will probably not wholly embrace spaceflight until spacecraft really look like spaceships.

I suppose that as an artist I am biased about this, but I truly believe that the importance of a spacecraft's design, as distinct from any of its other qualities, has been almost entirely overlooked. Something that has happened to me too many times to be coincidental has emphasized this dramatically. Young people often visit my studio, where there are scores of model spaceships on display. By far the majority of these youngsters are of the generation that has been raised on spaceships of the *Star Wars* and *Battletech* variety, that is, the post-*2001* design sensibility. Nevertheless, they will see a model V2 and immediately recognize it as a "spaceship." Why? Has its distinctive silhouette become a kind of cultural icon that automatically triggers the response "spaceship" in much the same way that the silhouette of a hawk panics baby chickens? Has the V2 shape come to represent "spaceship" simply because it, considered solely as a kind of sculpture, as an example of industrial design, just looks so great? I think that this is exactly what has happened.

For decades before the advent of spaceflight, scientists and engineers pooh-poohed the sleek, needle-nosed, swept-finned spacecraft of science fiction and film, telling us that the idea of a streamlined, single-stage-to-orbit spaceship wa the sheerest nonsense (the same scientists who told us that bubble-helmeted spacesuits would never be practical). But these people were living in the Conestoga wagon era of spaceship design, and their assumption that what was state of the art *then* was as good as anything was going to get was fallacious at best and exceedingly narrow-minded at worst . . . something like assuming that all cars forever were going to look like Model Ts.

Now with the advent of spacecraft that are not only frankly terrific-looking but are exactly what spaceships have been supposed to look like all along, a new enthusiasm for spaceflight may be kindled.

In any event, this book will be the definitive history of manned spacecraft for the forseeable future, or it will prove to be only the briefest prologue. We are teetering on a very narrow fence. It is a very high fence and the fall to either side may well be permanent. If we choose the wrong side, it could be a great many years before the fence is again scaled.

If we choose the right side of the fence, the following are dates that we can look forward to:

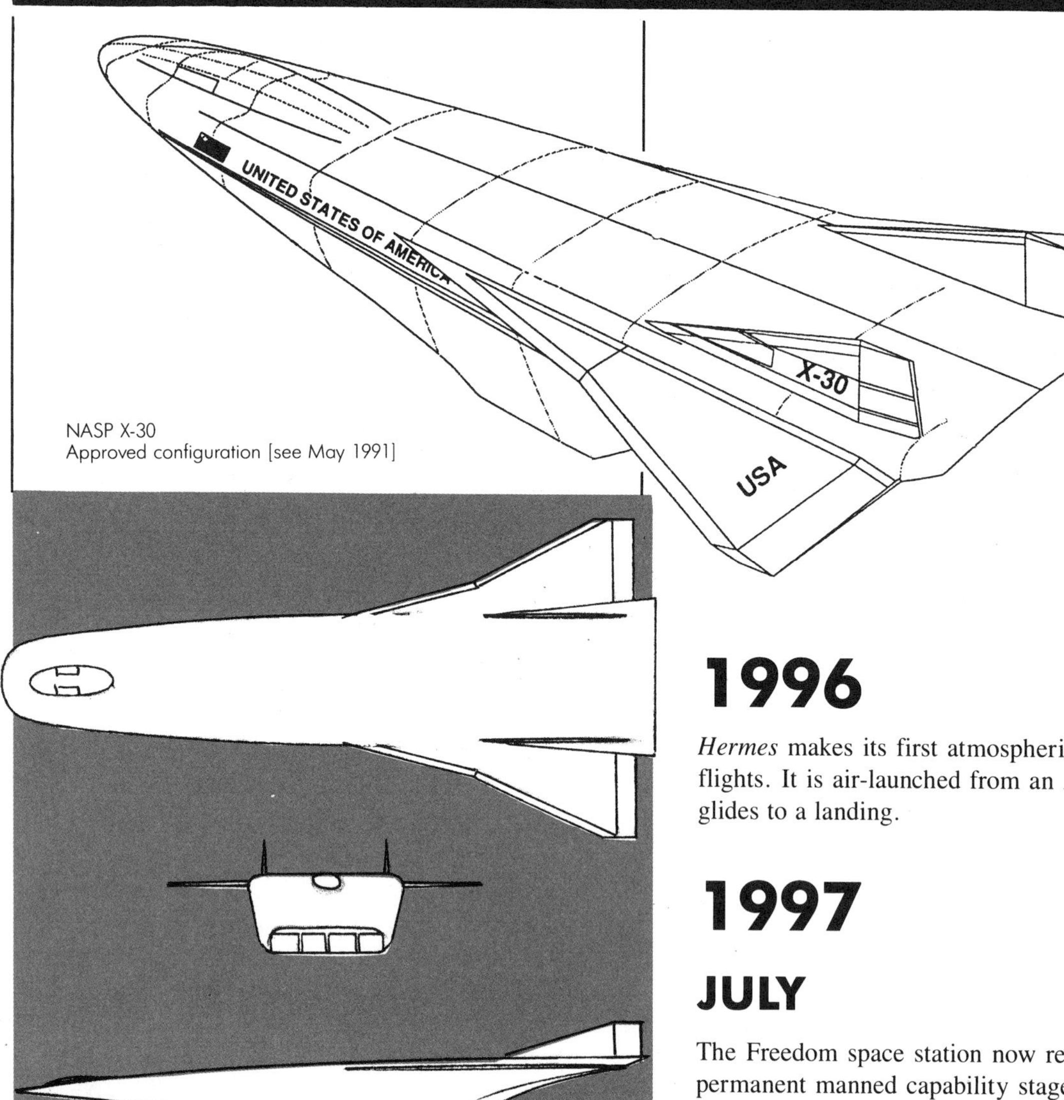

NASP X-30
Approved configuration [see May 1991]

1993

The X-30A National Aero-Space Plane prototype makes its first flight.

1995

MARCH

Assembly of space station *Freedom* begins.

1996

Hermes makes its first atmospheric test flights. It is air-launched from an Airbus and glides to a landing.

1997

JULY

The Freedom space station now reaches its permanent manned capability stage and an initial crew of four is delivered by the Shuttle.

2003

Hermes makes its first unmanned spaceflight, launched by an Ariane 5 booster.

2004

Hermes makes its first two manned spaceflights.

Right: Second generation shuttle proposed by John Frassinito and Associates.

John Frassinito and Assoc.

APPENDIX ONE: U.S. LAUNCH VEHICLES

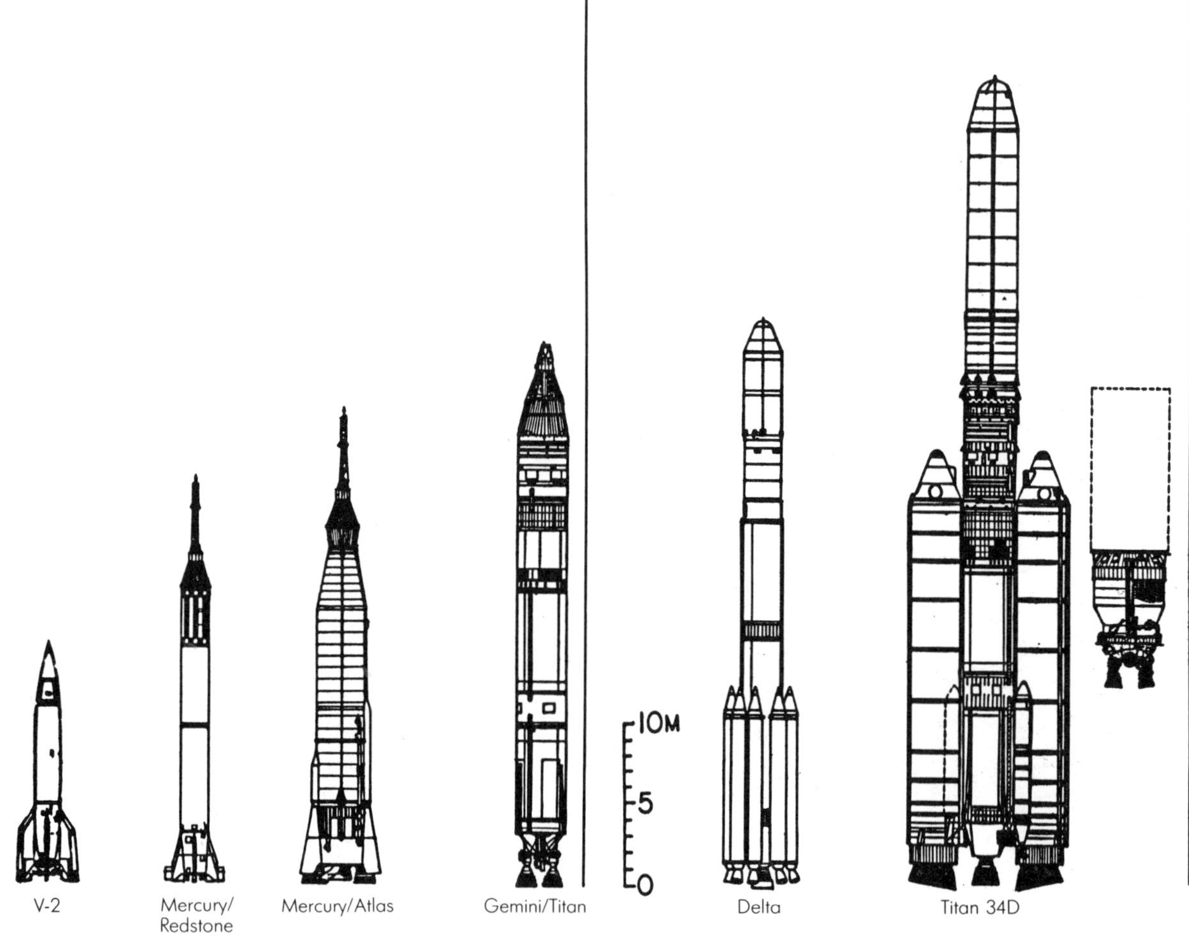

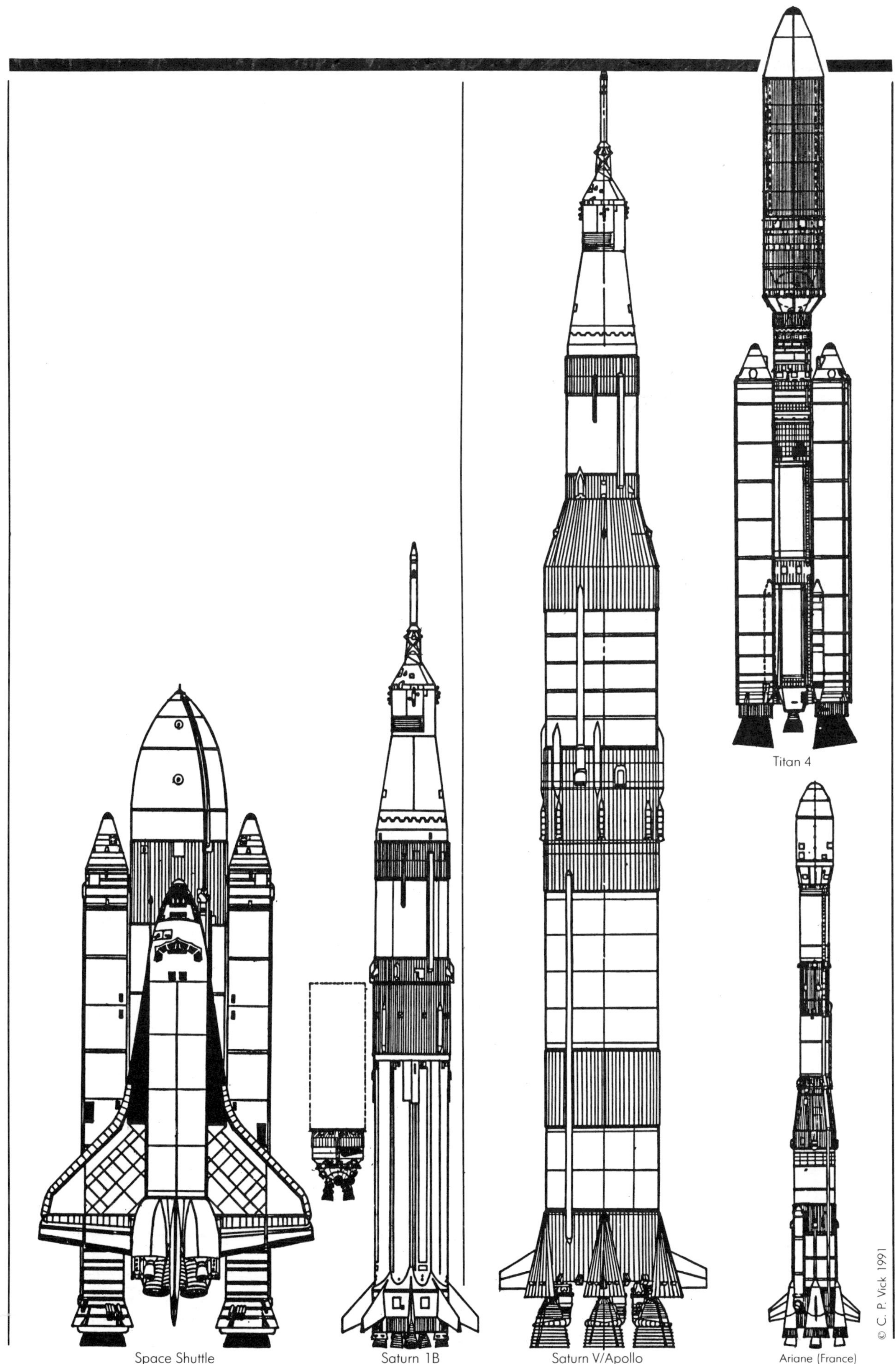
Titan 4
Space Shuttle
Saturn 1B
Saturn V/Apollo
Ariane (France)
© C. P. Vick 1991

APPENDIX TWO: SOVIET LAUNCH VEHICLES

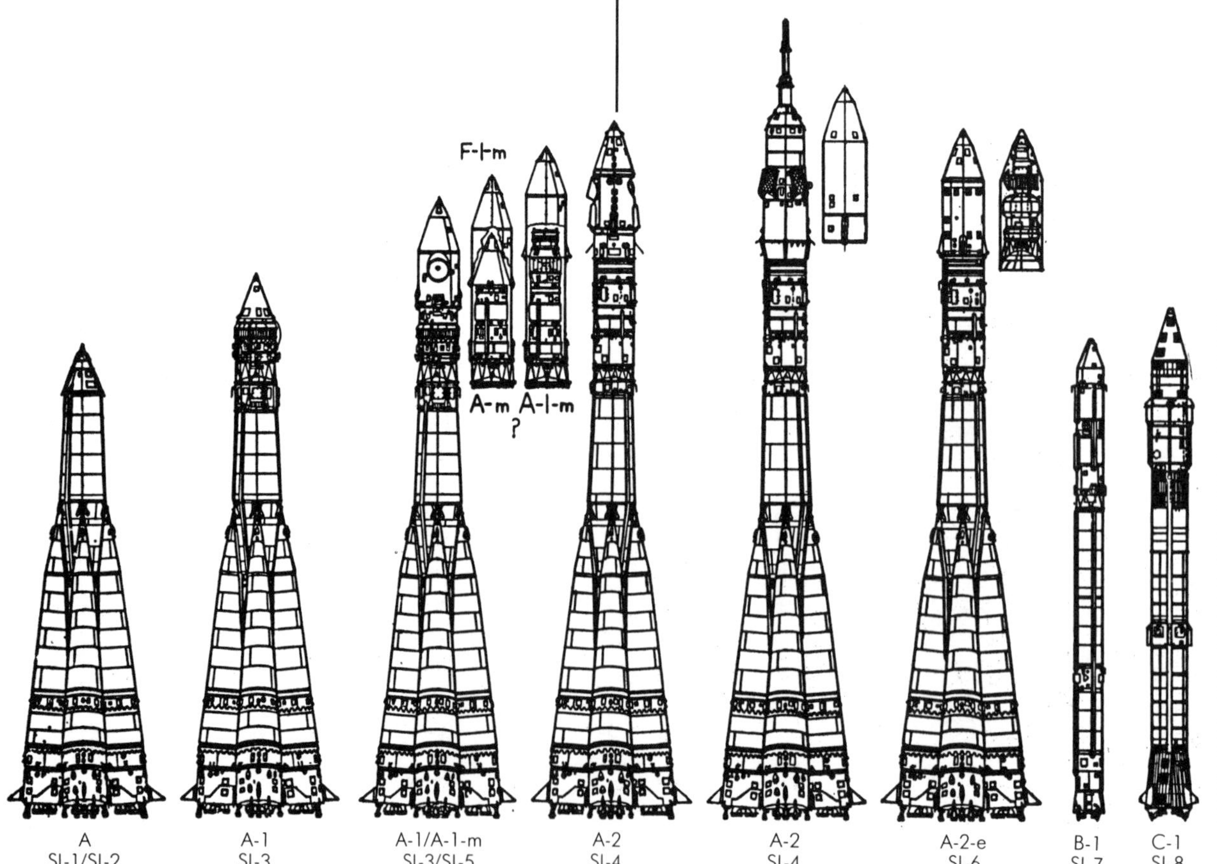

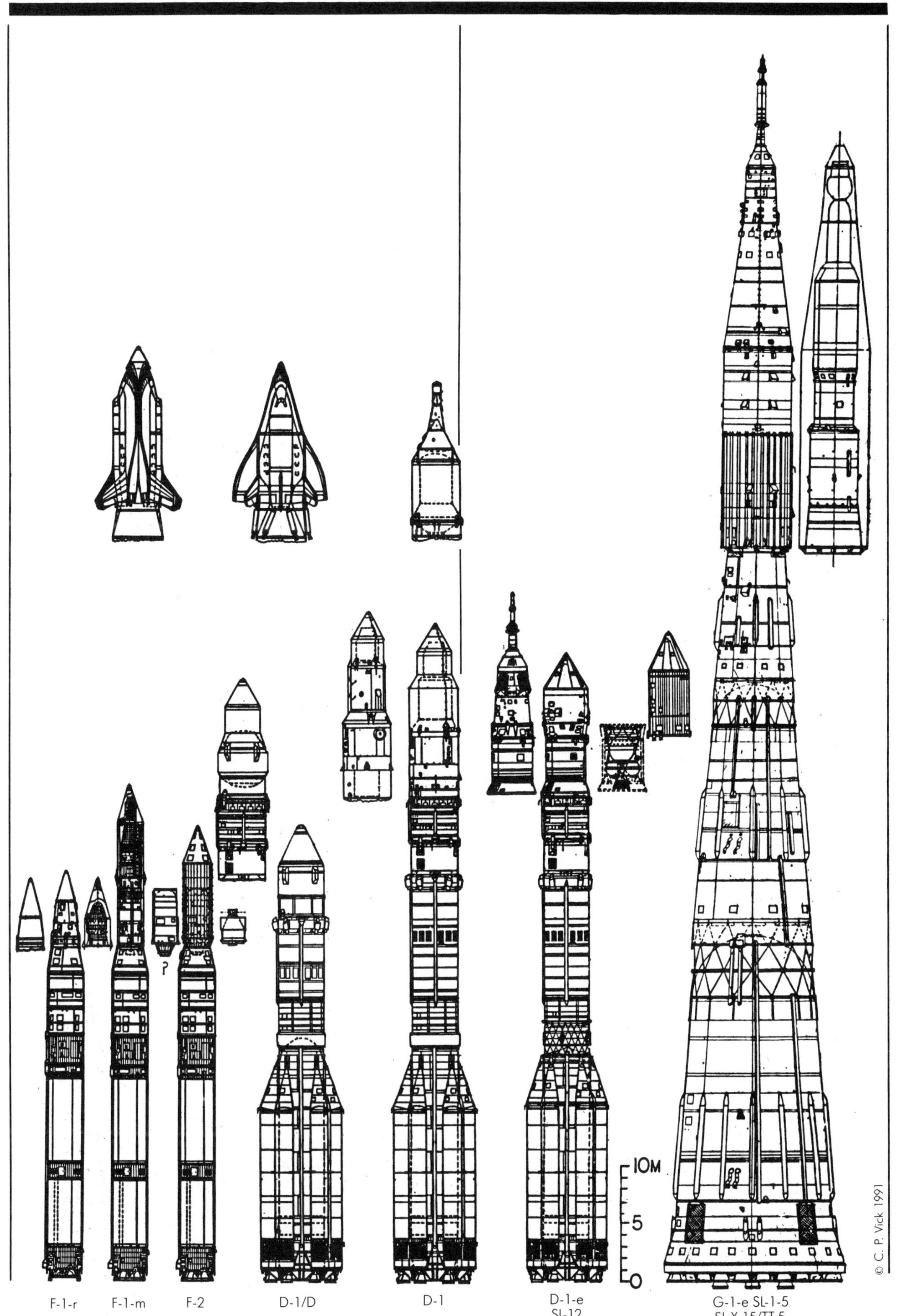
10M
5
0
F-1-r
F-1-m
F-2
D-1/D
D-1
D-1-e
SL-12
G-1-e SL-1-5
SL-X-15/TT-5
?
© C. P. Vick 1991

Energia and Buran

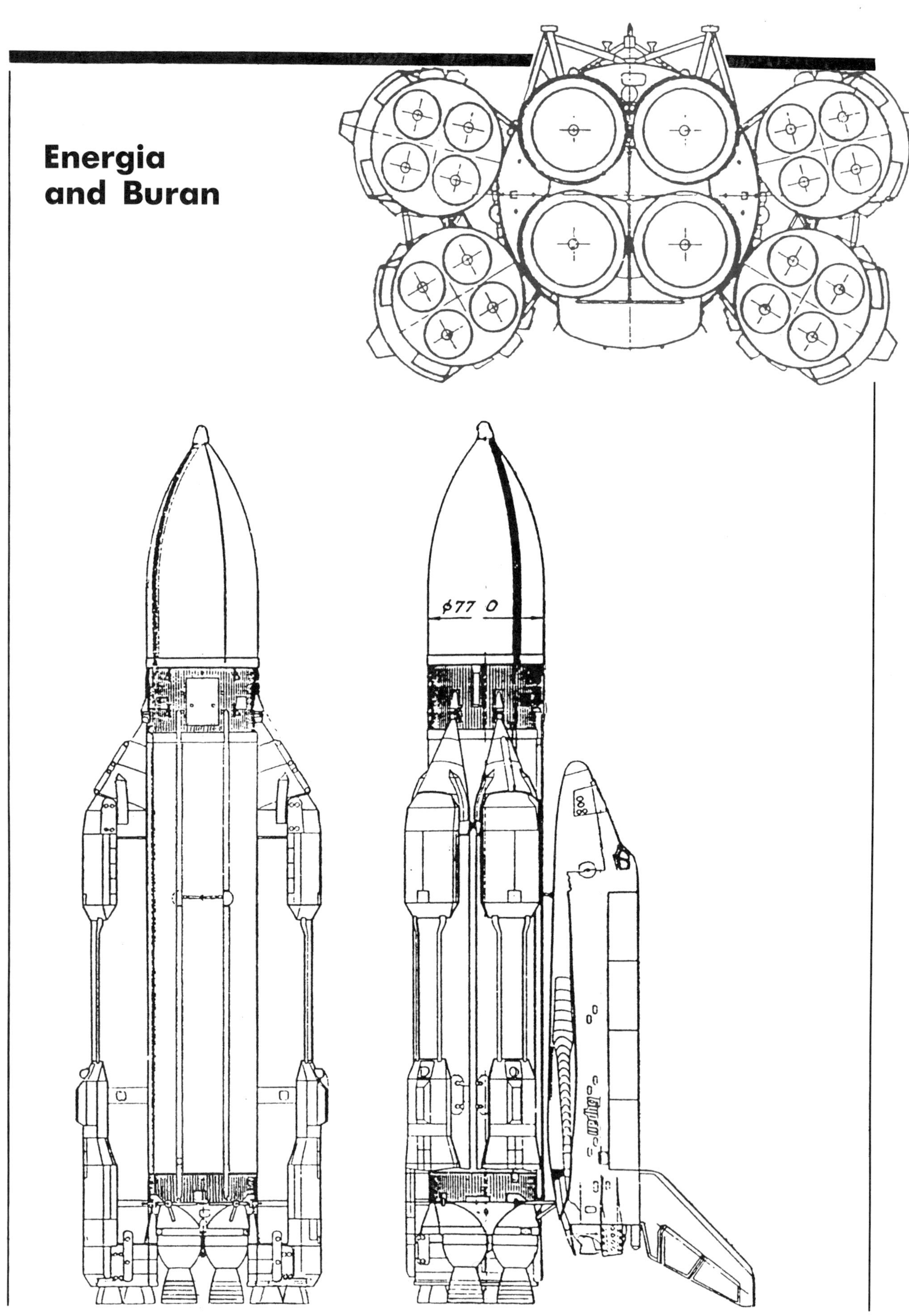

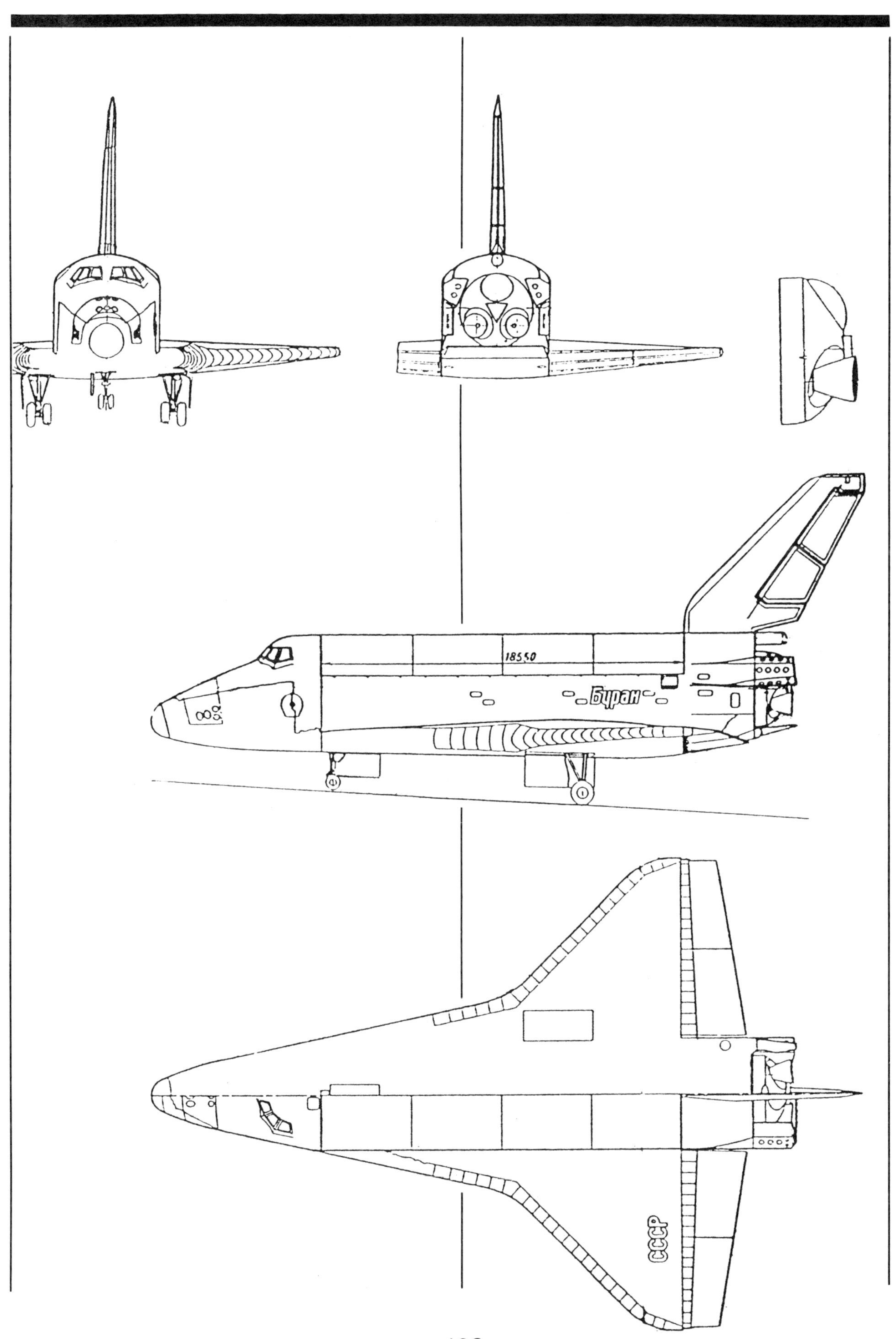
18550
Буран
СССР

APPENDIX THREE

HOPE II

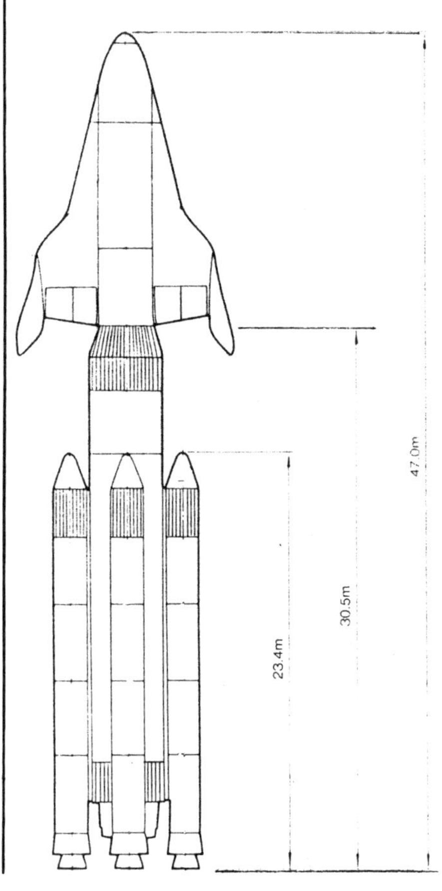

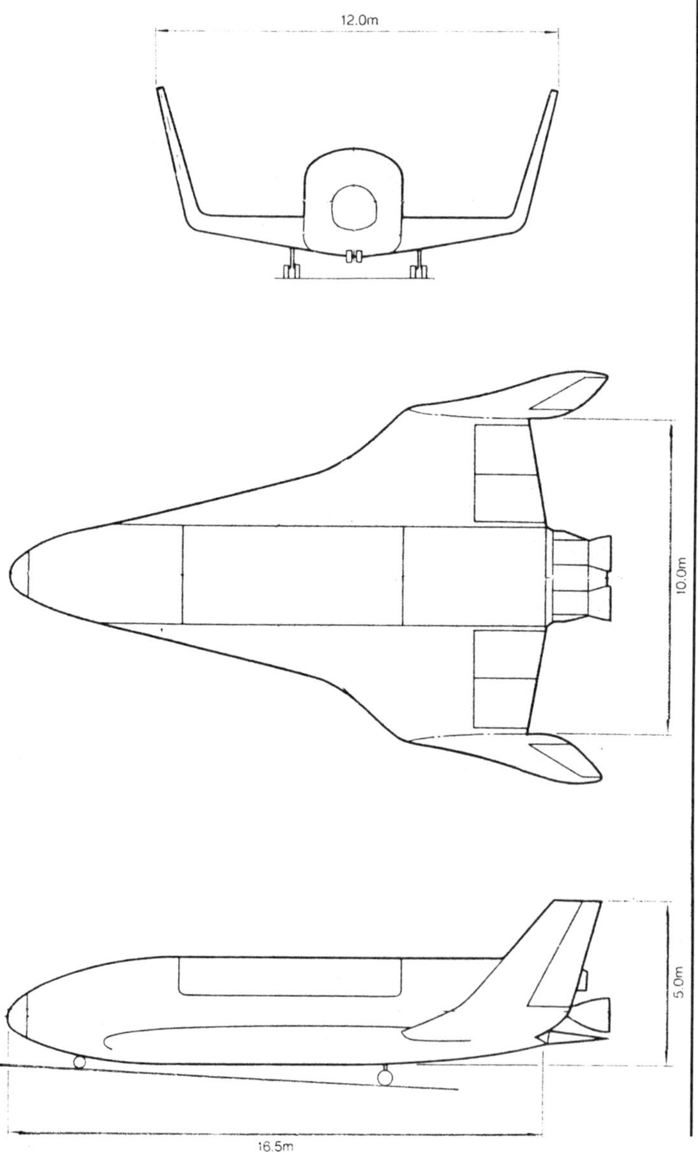

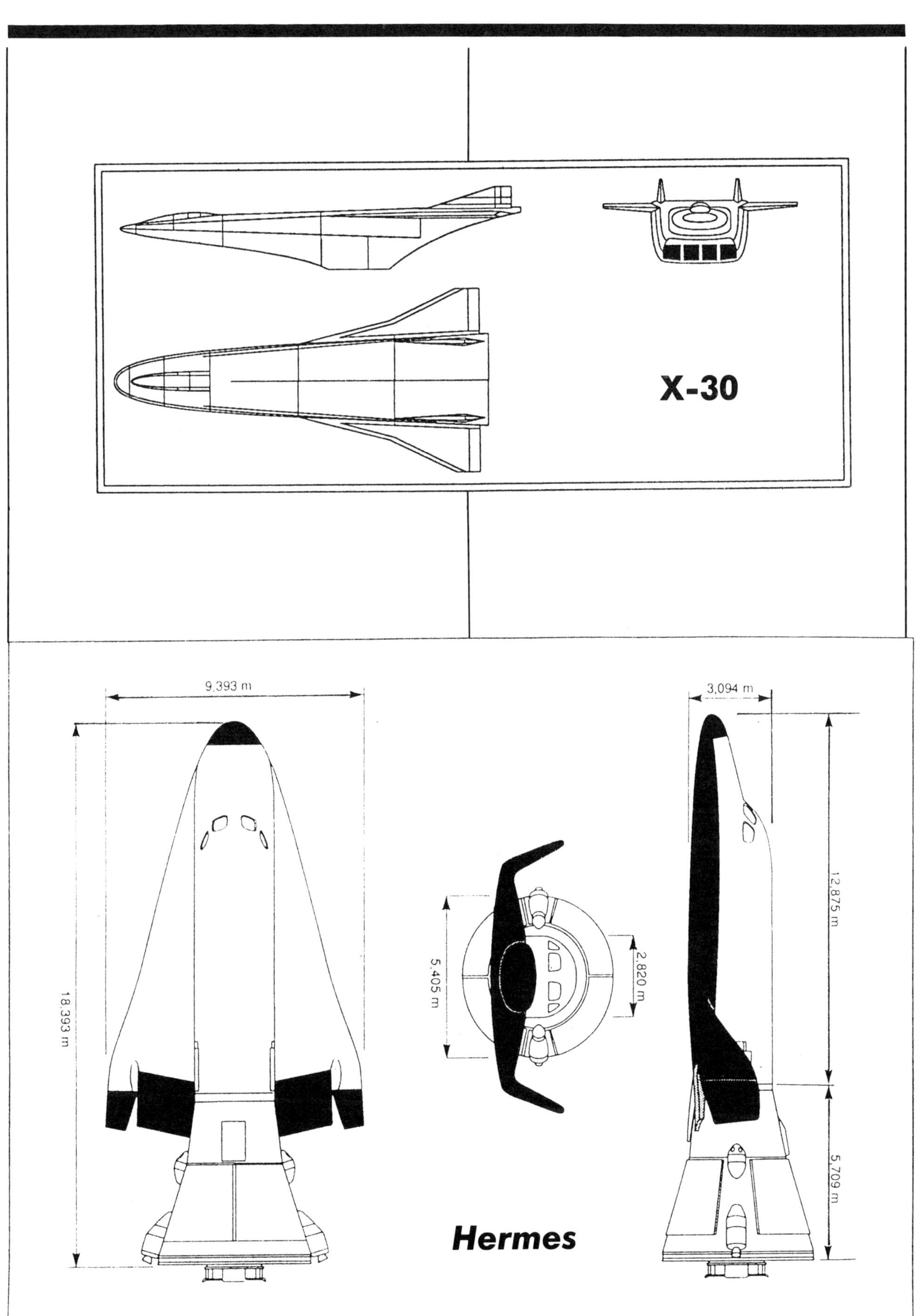
X-30
9,393 m
3,094 m
18,393 m
5,405 m
2,820 m
12,875 m
5,709 m
Hermes

APPENDIX FOUR

German Rocket Aircraft (World War II)

Name	Data	Remarks
Antipodal Bomber	Payload: 672 lbs.; length: 91.8 ft.; thrust: 220,500 lbs.	Project only — begun in 1936, suspended in 1942
Ar-234	48-sec. endurance; 36,000 ft. in 3 min.	Built by Arado with BMW turbojet and rocket
DSF-194		High-speed interceptor investigated by A. Lippisch
DSF-228	109-509A1 engine intended for this design	DSF-Siebel concept; prototype built and flown; was to have been operational to 100,000′
DSF-346	Two liquid fuel rockets for intended Mach 2.4 speed	Similar to above
He-176	Walter liquid fuel rocket on Heinkel airframe	First flew in 1938; not operational
Julia, P-1077	Thrust variable from 440 to 3750 lbs.; range: 40 miles; speed: 500 mph	Also Heinkel development; VTO with four 2600-lb-thrust boosters plus Walter liquid fuel rocket; rate of climb: 40,000 ft/min; not built
Ju-248	26 feet long; range: 100 miles; speed: 620 mph; designed to reach 40,000 ft. in 3 min.	109-509C1 liquid fuel engine; similar to Me-163; only one built
Me-163B	See separate chart	See separate chart
Me-163C	Designed for 590 mph and 52,000 ft. altitude; endurance of 15 min. at 500 mph	Mid-wing monoplane; evolved into Me-262
Me-262	weight: 18,000 lbs.; 2750 lbs. thrust	BMW-718 rocket engine with BMW or Juno jet engines: also designed to use one or two Walter rockets

German Rocket Aircraft (World War II)

Name	**Data**	**Remarks**
Me-263 (also Ju-263)	25.9 ft. long; 31.3 ft. wingspan; max. thrust: 4430 lbs.; weight less than 12,000 lbs.; speed: 620 mph (max.); rate of climb: 13,800 ft/min.	109-509Cl engine; glide-tested only; pure rocket-fighter design; development begun by Junkers, turned over to Messerschmidt
Natter, BP-208	18.75 ft. long: 10.5 ft. wingspan; rate of climb: 35,000 ft/min. (22 G acceleration); cruising speed: 496 mph (620 max.)	Four SR-34 solid boosters by Schmidding and recoverable Walter rocket; primarily wood construction
P-1104	Rate of climb: 39,000 ft/min.	Messerschmidt design; pure rocket interceptor
Ta-183		Focke-Wolfe jet/rocket; twin-boom design
Volsjaeger, He-162	Endurance: 45 min. (design)	Heinkel project
Walli, EF-127, 128	26.4 ft. long; 20.6 ft. wingspan; range: 60 miles; speed: 630 mph	Pure rocket; not built
Zeppelin Ram		Solid propellant; for ramming similar to Japanese Baka

Me-163 Rocket Fighter Program

Model	**Velocity (max. mph)**	**Powered Flight Time (minutes)**	**Rate of Climb (ft./min.)**	**Engine Number**
163A		3.25		R11-203
163BO	555	10	16,000	109-509A
163B1				109-509B1
163C	590	12	12,500	109-509A2

Natter Models

Mark	**Model**	**Engine**	**Powered Flight Time (min.)**	**Rate of Climb (ft./min)**	**Velocity (mph)**
1	BP-208	109-509A1	2–4	37,400	620
2	BA-349A	109-559	2–4		
3	BA-349B	109-5090	2–4	37,300	620

Natter Engine Comparisons

Model	Thrust (lbs)
109-509A1	220–3520
109-509D	880–4400
109-559	330–3750
(109-509A-2)	(possibly up to 4410)

V2 (A-4)

Length	46 ft. 1 in.
Diameter	5 ft. 5 in.
Propellants	Alcohol and liquid oxygen
Thrust	56,000 lbs
Velocity	3,500 mph
Altitude	Peak of operational trajectory: 55 mph

X-1 History

1946	First powered flight; Mach 0.795 at half throttle
1947	Speed of sound exceeded in level flight
1948	Maximum speed attained: 967 mph
1949	Take-off from ground (2300-ft run at full power); reaches 23,000 feet in 100 seconds
1949	Maximum altitude reached: 73,100 feet
1950	Retired to National Air Museum (now National Air & Space Museum) in August

(Note: two other X-1s are built, one of which had a turbopump assembly and exploded in November, 1951)

USAF-NACA-Bell X-1 Series

Name	Remarks
X-1	See chart above
X-1A	Speed record set in December 1953; altitude record in June 1954; air-launched at $ 25,000 ft.; duralumin construction with tempered glass windshield and ejection seat
X-1B	Weight: 18,000 lbs.; carries 1,000 lbs. of instrumentation
X-1C	Cancelled
X-1D	Destroyed in August 1951
X-1E	

X-1

Wingspan	28 ft.
Length	30.9 ft.
Height	10.85 ft.
Weight	Launch: 12,250 lbs Landing: 7,000 lbs
Engine	Engine: Reaction Motors XLR-11-RM-3 (Model A 6000C4) four-chamber engine with 6,000 lbs thrust

Velocity Records of U.S. Rocket Aircraft (ca. 1957)

Airplane	Velocity
X-1	First to reach Mach 1
D-558-1	First to reach Mach 2
X-1A	First to reach Mach 2.5
X-2	First to reach Mach 3
X-15	Designed to reach Mach 5–6

X-15

Wingspan	22.36 ft.
Length	50.75 ft.
Height	13 ft.
Weight	Launch: 38,000 lbs Landing: 12,500 lbs
Engine	Thiokol (Reaction Motors) XLR-99-RM-2 57,000 lbs thrust at sea level

U.S. Rocket Aircraft (ca. 1957)

Name	Velocity (Mach no.)	Altitude (ft.)	Thrust (lbs.)	Length (ft.)	Wingspan (ft.)
X-1	1.4	73,000	6000	31	28
XF-91	$ 1		6000	42′3″	31′3″
D-558-1	2.1	83,000	6000	45	25
X-1A	2.5	90,000	6000	35′7″	28
X-2	3.1	126,000	15,000		
X-15	5–6 (design)	500,000 (design)	60,000		
Future	10	1,000,000	100,000+		

French Rocket Aircraft (ca. 1957)

Airplane	Designation	Manufacturer	Engine thrust (lbs)	Velocity (Mach no.)	Remarks
Durandal	SE.212	Sud	3300 × ·2(?)	$ 1	Delta-wing
Espadon	SO.6025 (6026)	Sud	3300	$ 1	First rocket plane in post-war Europe
Griffon	1500	Nord/SPE-CMAS	4800	1.3	
Harpon		Nord	4800	$ 1	Similar to Griffon
Mirage I	MD-550	Dassault	3300	1	Delta-wing; first flew May 1955
Mirage II		Dassault	2400	$ 1	Similar to I
Mirage III		Dassault	3300	1.7	Similar to I
Mirage IV		Dassault		2	Similar to I
Trident I	SO.9000	Sud	9900	$ 1	SEPR 481 rocket & Viper wing tip jets
Trident II	SO.9050	Sud	6600	1.9	SEPR 631 rocket; reached 59,000′

British Rocket Aircraft (ca. 1957)

Name	Manufacturer	Engine	Remarks
Avro-720	Avro	Liquid fuel rocket & turbojet	Cancelled interceptor design
Fairey VTO	Fairey	Liquid fuel rocket & solid fuel boosters	VTO test vehicle
S-R.53	Saunders-Roe	Viper jet plus Spectre liquid fuel rocket	Recoverable booster for air-to-air missiles
S-R.177	Saunders-Roe	Liquid fuel rocket	Project only

Russian Rocket Aircraft (ca. 1957)

Aircraft	Velocity	Remarks
Yak-21	700 mph	21′ wingspan; 12,000-lb weight; auxiliary liquid fuel rocket
La-17	Mach 1	36′ wingspan; 16,000-lb weight; auxiliary liquid fuel rocket
I-1	1700 mph	Interceptor-type research aircraft; altitude: 100,000′
I-2	< 800 mph	Liquid fuel VTOL interceptor; less than 8 min. flight time.

M2-F3

Length	22 ft. 2 in.
Span	9 ft. 7 in.
Height	8 ft. 10 in.
Weight	Empty: 6,000 lbs Fueled: 10,000 lbs
Speed	1,066 mph (Mach 1.5) max. achieved
Altitude	71,500 ft. max achieved

Mercury

Height (including escape tower)	26 ft.
Width (across heat shield)	74.5 ft.
Weight	Launch: 4,265 lbs Orbital: 2,987 lbs Landing: 2,493 lbs

Freedom 7

Diameter	6 ft. 6,in. max
Length	9 ft. 2 in. at launch
Weight	Launch: 3,650 lbs

D-558-2 #2

Wingspan	25 ft.
Length	42 ft.
Height	12 ft. 8 in.
Weight	Launch: 15,787 lbs Landing: 9,421 lbs
Engine	Reaction Motors XLR-8-RM-6 (Model A 6000C4) four-chamber engine rated at 6,000 lbs thrust

Vostok

Reentry capsule	
Diameter	2.3 m
Weight	2.4 metric tons
Crew	1

Instrument section	
Diameter	2.42 m
Height	2.2 m
Weight	2.3 metric tons

Complete assembly	
Diameter (max	2.42 m
Height	5 m
Weight	5.3 metric tons

Gemini

Length	Orbit: 18 ft. 4 in. Landing: 7 ft. 4 in.
Base diameter	Adapter: 10 ft. Spacecraft: 7 ft. 6 in.

Comparison of Apollo and Dyna-Soar Spacecraft (1961)

	Apollo Orbital	**Apollo Circumlunar**	**Dyna-Soar I Suborbital**	**Dyna-Soar II Orbital**
Crew	3	3	1	1
Weight	20,000 lbs in orbit; 12,500 lbs reëntry	15,000 lbs	About 15,000 lbs	About 10,000 lbs
Booster	Saturn C-1	Saturn C-2	Titan II	Saturn C-1
Shape	Lifting body	Lifting body	Glider	Glider
Lift/drag ratio	0.4 to 0.7	0.4 to 0.7	2 or more	2 or more
Landing method	Parachute	Parachute	Conventional	Conventional
Aerodynamic control	Limited	Limited	Complete	Complete
First scheduled flight	1965–1966	1967–1968	1963–1964*	1964–1965*

*First glider flights from Edwards AFB in 1962; R&D suborbital flight early 1965

Voskhod

Reentry capsule	
Diameter	2.3 m
Weight	3 metric tons
Crew	2 or 3
Instrument section	
Diameter	2.42 m
Height	2.2 m
Weight	2.3 metric tons
Complete assembly	
Diameter (max)	2.42 m
Height	5 m
Weight	5.3 metric tons

Soyuz

Orbital module	
Diameter	2.2 m
Length	2.65 m
Weight	1.2 metric tons
Crew module	
Diameter	2.2 m
Length	2 m
Weight	2.8 metric tons
Crew	2 or 3
Equipment Module	
Diameter	2.2 m (aft skirt 2.7 m)
Length	2.3 m
Weight	2.6 metric tons
Solar panel span	8.37 m
Combined assembly	
Length	7.1 m
Weight	6.6–6.9 metric tons
Propulsion	two 400 kg motors

Apollo-Soyuz Test Project Apollo

Command Module

Base diameter	12.8 ft.
Length	12 ft.
Weight	13,000 lbs

Service Module

Diameter	12.8 ft.
Length	22 ft.
Weight	Launch: 55,000 lbs

Docking Module

Diameter	5 ft.
Length	10 ft.
Weight	4,155 lbs

Soyuz

Orbital Module

Length	8.7 ft.
Weight	2,700 lbs

Descent Module

Diameter	7.5 ft.
Length	7.2 ft.
Weight	6,200 lbs

Instrument Module

Diameter	9.75 ft.
Length	7.5 ft.
Weight	5,850 lbs

Service Module

Length	24.3 ft.
Diameter (max)	12.8 ft.
Weight	54,074 lbs (average: not typical for every mission)

Command Module

Length (less nosecone)	10.6 ft.
Diameter (max)	12.8 ft.
Weight	13,090 lbs (with astronauts)
Habitable volume	210 ft.^{3}

Lunar Module

Height	22 ft. 11 in. with legs extended
Diameter	31 ft. diagonally across landing gear
Weight	Launch: 32,400 lbs LM dry: 8,600 lbs
Volume	Pressurized: 235 ft.^{3} Habitable: 160 ft.^{3}

HERMES

Spaceplane

Total Length	14.584 m
Fuselage Length	12.875 m
Wingspan	9.402 m
Fuselage Height	9.402 m

Resource Module

Length	6.060 m

Mass

Total mass of vehicle in orbit	22,418 kg
Cargo (to orbit)	3,000 kg
Cargo (from orbit)	1,500 kg

SÄNGER II

	First Stage	Passenger Plane (HORUS)
Vehicle Length	92 m	34 m
Wingspan	46 m	18 m
Wing Area	880 m	
Mass	95 Mg+26 Mg (cabin & equipment)	
Payload	35–40 Mg	91 Mg (2 pilots and 4–36 passengers)
Number of engines (max. thrust level)	6 × 400kN	2 × 1050kN
Cruise speed	Mach 4 (Mach 7 stage separation)	Mach 4
Flight range	5,000 km	10,360 km

Proposed reusable space transporters, ca. 1979

Project	Mass (t)	LEO Payload (t)	Type	Stages	Remarks
Konv. SSTO	4.08	29,5	VTHL	1	
Konv. mixed mode shuttle	1.905	29,5	VTHL	1	2 Fl engines+6 SSME
Adv. Langley MM shuttle	1.488	29,5	VTHL	1	
Adv. Langley shuttle	1.633	29,5	VTHL	1	
MM-CCV shuttle	1.488	29,5	VTHL	1	
Winged HLLV	31.162	907	VTHL	1	117 m tall
Boeing: Growth shuttle	2.313	75	VTHL	2	
Konv. HLLV	3.175	181,5	VTHL	2 1/2	2 boosters+ET
Boeing: Winged HLLV	9.566	381	VTHL	2	
Boeing: Winged HLLV (1978)	10.977	424	VTHL	2	
MacConachie/Klich: ETPV SSTO	922	6,8	VTHL	1	
MacConachie/Brien: HLLV SSTO	5.616	227	VTHL	1	
MacConachie/Brien: HLLV, winged tandem	4.082	227	VTHL	2	Turbojet for cruise-back, 1st stage rocket
Chase: SSTO	1.207	29	VTHL	1	
Bill: Adv. shuttle, tandem	2.268	45,4	VTHL	2 1/2	Turbojet for cruise-back; 1st stage rocket
Martin Marietta: Sled launch	1.382	29,5	HTHL	1 1/2	
Langley: Adv. sled launch	1.22	29,5	HTHL	1 1/2	
Langley: Air-breathing booster	1.18	29,5	HTHL	2	
Langley: Adv. air-breathing booster	1.18	62	HTHL	2	2nd stage adv. scramjets
Salkeld: MM-HTHL	644	2,3	HTHL	1	
Salkeld: Airlaunch MM	329	2,9	HTHL	2 1/2	1st stage airplane, 2nd stage MM+ET
Martin: Supersonic TF boosters	1.3	29,5	HTHL	2	
Martin: Subsonic TF booster	1.15	20	HTHL	2	
Martin: Fan-RJ SSTO	1.03	ca. 10	HTHL	1	
Martin: Fan-RJ SSTO	1.03	ca. 19	HTHL	1	

Project	Mass (t)	LEO Payload (t)	Type	Stages	Remarks
Martin: Scramjet SSTO	1.03	ca. 8	HTHL	1	
Martin: Scramjet SSTO	1.03	ca. 22	HTHL	1	
Reed/Ikawa/Sadunas: *Star Raker* SSTO	2.27	89,2	HTHL	1+	Airbreathing/rocket, tridelta flying wing
Mini *Star Raker*	544,3	22,7	HTHL	1+	
Boeing: RASV	657	10	HTHL	1 1/2	
Cornier *L'Enterprise*	500	7,4	HTHL	2+	
MacConachie/Klich: SHIPS SSTO	653	2,3	HTHL	1	
Sled-assisted SSTO	1.261	29	HTHL	1 1/2	
Subsonic TF Booster	1.371	29	HTHL	2	
Hypersonic turbo/ram/scram jet booster	1.049	29	HTHL	2	Mach 10 turbo/ram/scramjet
Hypersonic rocket	1.05	229	HTHL	2	Mach 10 staging
Boeing: Shuttle growth HHLV	2.858	104	VTVL	2	
Boeing: Shuttle derived HHLV	3.22	113	VTVL	2 1/2	4 SRBs, ET
Boeing: SPS-HLLV	10.433	450	VTVL	1	
Koelle *Neptun*	20	400	VTVL	1	
NASA-HLLV	23	400	VTVL	1	
MM-HLLV	4.309	181,5	VTVL	1	
MM-HLLV	19.505	907	VTVL	1	
Langley-HLLV	3.5	181,5	VTVL	1	
Drop tank HLLV	3.175	181,5	VTVL	1 1/2	
Ext. scramjet HLLV	5.035	454	VTVL	1	External combustion ramjet
Boeing: HLLV ballistic	10.472	391	VTVL	2	
Koelle: *Beta 1a*	130	2,7	VTVL	1	
Koelle: *Beta 30*	1.28	72	VTVL	1	
Koelle: *Beta 150*	5.33	330	VTVL	1	
HLLV, SSTO	6.8	227	VTVL	1	MM rocket
Shuttle derived HLLV	2.021	77,1	VTVL	1	2 SRBs, ET, reusable propellant module
Chrysler: SERV, SSTO	2.25	44,03	VTVL	2 1/2	Turbojet lift for landing
Kramer/Bühler: ITUSTRA	155	15,5	VTVL	1 1/2	Integrated turborocket/ramjet+ rocket
Kramer/Bühler: ISTRA	155	15,4	VTVL	1 1/2	Integrated ramjet+rocket

ET = external tank, MM = mixed mode, SSTO = single stage to orbit, VTVL = vertical takeoff and landing, HTHL = horizontal takeoff and landing, VTHL = vertical takeoff and horizontal landing, RJ = ramjet, SRB = solid rocket booster, HLLV = heavy lift launch vehicle

BIBLIOGRAPHY

*Recommended reading

BOOKS

Ahnstrom, D. N. *The Complete Book of Jets and Rockets* (World. Cleveland: 1959)

Akens, A. *A Picture History of Rockets and Rocketry* (Strode. Huntsville, AL: 1964)

Ananoff, A. *L'Astronautique* (Librairie Artheme Fayard. Paris: 1950)

Asimov, I. (ed.). *Before the Golden Age* (Doubleday. New York: 1974)

Astor, J. J. *A Journey in Other Worlds* (D. Appleton. New York: 1894)

*Baker, D. *The History of Manned Space Flight* (Crown. New York: 1981)

Bekey, I. (ed.). *Space Stations and Space Platforms* (AIAA. New York: 1985)

Bell, W. D. *The Moon Colony* (Goldsmith. Chicago: 1937)

Bergaust, E. *The Next 50 Years in Space* (MacMillan. New York: 1964)

Bergaust, E., and W. Beller. *Satellite!* (Hanover House. Garden City, New York: 1956)

Bergaust, E., and S. Hull. *Rocket to the Moon* (Van Nostrand. NJ: 1958)

Bono, P., and K. Gatland. *Frontiers of Space* (MacMillan. London: 1969; 2d ed.: 1976)

Booth, N. *Space: The Next 100 Years* (Mitchell Beazley. London: 1991)

Burgess, E. *Satellites and Spaceflight* (Scientific Book Club. London: 1957)

Caidin, M. *Worlds in Space* (Henry Holt. New York: 1954)

Caidin, M. *War for the Moon* (Dutton. New York: 1959)

Caidin, M. *The Moon: New World for Men* (Bobbs-Merrill. New York: 1963)

Caidin, M. *Countdown for Tomorrow* (Dutton. New York: 1968)

Canby, C. *A History of Rockets and Space* (Hawthorn Books. New York: 1963)

Carassa, F., et al. *Quest for Space* (Crescent. New York: 1986)

Carter, L. (ed.). *Realities of Space Travel* (McGraw-Hill. New York: 1957)

Clarke, A. C. *The Exploration of Space* (Harper. New York: 1951)

Clarke, A. C. *Islands in the Sky* (Signet. New York: 1952)

Clarke, A. C. *Going into Space* (Trend. Los Angeles: 1954)

*Clarke, A. C. *Interplanetary Flight* (2d. ed. Harper. New York: 1960)

Clarke, A. C. *Prelude to Mars* (Harcourt, Brace, Jovanovitch. New York: 1965)

Clarke, A. C. *Rendezvous with Rama* (Harcourt, Brace, Jovanovitch. New York: 1973)

Claudy, C. *The Mystery Men of Mars* (Grosset & Dunlap. New York: 1933)

Cleator, P. E. *Rockets Through Space* (Geo. Allen & Unwin. London: 1936)

Coombs, C. *Skyrocketing into the Unknown* (Wm. Morrow. New York: 1954)

Corliss, W. R. *Nuclear Reactors for Space Power* (USAEC. Washington, DC: 1971)

Cox, D., and M. Stoiko. *Spacepower* (Winston. Philadelphia: 1958)

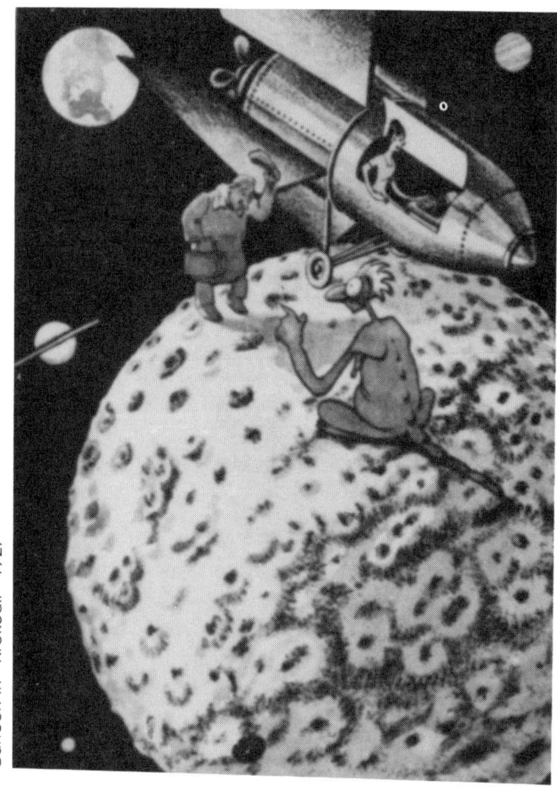

Cartoon in "Krokodil" 1927

Devorkin, D. *Race to the Stratosphere* (Springer-Verlag. New York: 1989)

De Vries, L. *Victorian Inventions* (American Heritage Press. New York: 1971)

*Dornberger, W. *V2* (Hurst & Blackett. London: 1954)

Durant, F. C. *Robert H. Goddard: Accomplishments of the Roswell Years* (Privately published: 1973)

Durant, F. C., and G. James. *First Steps Toward Space* (Smithsonian Institution. Washington, DC: 1974)

Ehricke, K. A., and B. A. Miller. *Exploring the Planets* (Little, Brown. New York: 1969)

Eisner, W. *America's Space Vehicles* (Stirling. New York: 1962)

Emme, E. M. (ed.). *The History of Rocket Technology* (Wayne State University Press. :1964)

Emme, E. M. *Science Fiction and Space Futures* (AAS. San Diego, CA: 1983)

Ertel, I. D., and M. L. Morse. *The Apollo Spacecraft* (3 volumes. NASA. Washington, DC: 1969)

Essers, I. *Max Valier—A Pioneer of Space Travel* (NASA. Washington, DC: 1968–1976)

Essers, I. *Max Valier* (Athesia. Bozen: 1980)

Evans, I. O. *Science Fiction Through the Ages* (vol. 1. Panther. London: 1966)

Faget, M. *Manned Space Flight* (Holt, Rinehart & Winston. New York: 1965)

Farnsworth, R. L. *Rockets New Trail to Empire* (2d ed. Privately published: 1945)

Firsoff, V. A. *Our Neighbor Worlds* (Philosophical Library. New York: 1953)

Frewin, A. *100 Years of Science Fiction Illustration* (Pyramid. New York: 1974)

Furniss, T. *Manned Spaceflight Log* (Jane's. London: 1986)

Gail, O. W. *The Shot into Infinity* (*Science Wonder Quarterly*, vol. 1, no. 1, Fall 1929)

Gail, O. W. *The Stone from the Moon* (*Science Wonder Quarterly*, vol. 1, no. 3, Spring 1930)

Gail, O. W. *By Rocket to the Moon* (Dodd, Mead. New York: 1956)

Galopin, A. *Le Docteur Oméga* (Albin Michel Éditeur, Paris: nd)

Gatland, K. *Development of the Guided Missile* (Icliffe & Sons. London: 1952; 2d ed.: 1954)

Gatland, K. *Project Satellite* (Allan Wingate. London: 1958)

Gatland, K. (ed.). *Spaceflight Today* (Icliffe. London: 1963)

Gatland, K. *Manned Spacecraft* (MacMillan. London: 1967)

*Gatland, K. (ed.). *Space Technology* (Harmony. New York: 1981)

Gatland, K. *Space Diary* (Crescent. New York: 1989)

Gatland, K., and A. M. Kunesch. *Space Travel* (Allan Wingate. London: 1953)

Gaul, A. T. *The Complete Book of Outer Space* (World. New York: 1956)

*General Dynamics. *Planetary Explorer* (General Dynamics. San Diego, CA: 1989)

Goddard, R. H. *Rockets* (ARS. New York: 1939)

Goodwin, H. L. *The Real Book about Space Travel* (Garden City. Garden City, New York: 1952, 1956)

Goodwin, H. L. *The Science Book of Space Travel* (Cardinal. New York: 1955)

Green, R. L. *Into Other Worlds* (Abelard-Schuman. New York: 1958)

Green, W. *Famous Fighters of the Second World War* (Doubleday. New York: 1965)

Green, W. *Rocket Fighter* (Ballantine. New York: 1971)

Greener, L. *Moon Ahead* (Viking Press. New York: 1951)

Griffith, G. *Stories of Other Worlds* [A Honeymoon in Space] (Pearson's Magazine. London: 1900)

Guegan, G. *Ils Ont Marché Sur La Lune* (Casterman. Tournai: 1985)

Hacker, B.C., and J. M. Grimwood. *On the Shoulders of Titans* (NASA. Washington, DC: 1977)

Hale, E. E. *The Brick Moon* (in *His Level Best*, James R. Osgood. Boston: 1873)

Hall, A. (ed.). *Peterson's Book of Man in Space* (5 volumes. Peterson. Los Angeles: 1974)

Hall, C. H. (ed.). *Essays on the History of Rocketry and Astronautics* (2 volumes. NASA. Washington, DC: 1977)

Hallion, R. F. *American Rocket Aircraft* (Privately published: 1974)

*Hallion, R. F. *The Hypersonic Revolution* (2 volumes. Wright-Patterson Air Force Base. Dayton, Ohio: 1987)

Harley, T. *Moon Lore* (Tuttle. Rutland, VT: 1970)

Harper, H. *The Dawn of the Space Age* (Sampson Low. London: 1946)

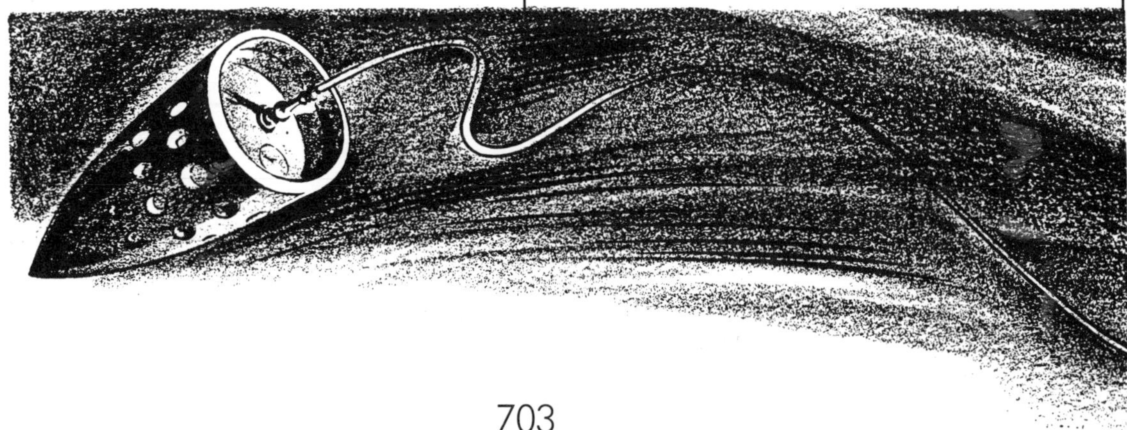

Max Valier, 1930

Harrison, H., and M. Edwards. *Spacecraft in Fact and Fiction* (Exeter. New York: 1979)

Hart, D. *The Encyclopedia of Soviet Spacecraft* (Exeter. New York: 1987)

Healy, R. J., and J. F. McComas (eds.). *Adventures in Time and Space* (Modern Library. New York: 1957)

Hendrickson, W. B. *Winging into Space* (Bobbs-Merrill. New York: 1965)

Hendrickson, W. B. *Manned Spaceflight to Mars and Venus* (G. P. Putnam's Sons. New York: 1975)

Hergé. *Destination Moon* (Golden Press. New York: 1960)

Hergé. *Explorers on the Moon* (Golden Press. New York: 1960)

Heuer, K. *Men of Other Planets* (Pellegrini & Cudahy. New York: 1951)

Heyn, E. V. *A Century of Wonders* (Doubleday. New York: 1972)

Hirsch, D., and H. Zimmerman. *Spaceships* (Starlog. New York: 1980)

Hobbs, M. *Fundamentals of Rockets, Missiles and Spacecraft* (John F. Rider. New York: 1962)

Howard, W. E., and J. Barr. *Spacecraft and Missiles of the World—1966* (Harcourt, Brace & World. New York: 1966)

Humphries, J. *Rockets and Guided Missiles* (Ernest Benn. London: 1957)

International Space University. *International Mars Mission* (ISU. Boston: 1991)

Joëls, K. M. *The Mars One Crew Manual* (Ballantine. New York: 1985)

Kaiser, H. K. *Rockets and Spaceflight* (Pitman. New York: 1962)

Kennedy, G. P. (ed.). *Rockets, Missiles and Spacecraft of the National Air & Space Museum* (Smithsonian Institution. Washington, DC: 1983)

Kennedy, G. P. *Vengeance Weapon 2* (Smithsonian Institution. Washington, DC: 1983)

C. Geary

Klee, E., and O. Merk. *The Birth of the Missile* (E. P. Dutton. New York: 1954)

Knight, D. (ed.). *Science Fiction of the 30's* (Bobbs-Merrill. Indianapolis, IN: 1975)

Kolosimo, P. *Spaceships in Prehistory* (Citadel. Secaucus, New York: 1982)

Kosmodemyansky, A. *Konstantin Tsiolkovsky* (Foreign Languages Publishing House. Moscow: 1956)

Lasser, D. *The Conquest of Space* (Penguin Press. New York: 1931)

Lasswitz, K. *Two Planets* (Popular Library. New York: 1971)

Laurie, A. *The Conquest of the Moon* (Sampson, Low. London: 1889)

Lehman, M. *This High Man* (Pyramid. New York: 1970; rev. ed. Da Capo. New York: 1988)

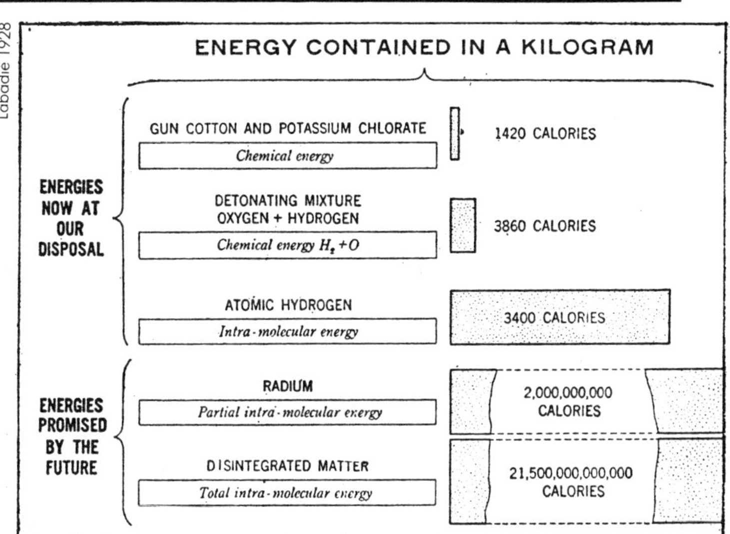

Leighton, P. *Moon Travellers* (Oldbourne. London: 1960)

Leinster, M. (Will F. Jenkins). *Space Platform* (Pocketbooks. New York: 1954)

Leinster, M. (Will F. Jenkins). *Space Tug* (Pocketbooks. New York: 1954)

Lent, C. P. *Rocketry* (Pen-Ink. New York: 1947)

Lent, C. P. *Rockets, Jets and the Atom* (Pen-Ink. New York: 1952)

Leonov, A. *I Walk in Space* (Malysh. Moscow: 1980)

Leonov, A., and V. Lebedev. *Space and Time Perception by the Cosmonaut* (Mir. Moscow: 1971)

Levitt, I. M. *Target for Tomorrow* (Fleet. New York: 1959)

Lewis Research Center. *Spaceflight with Electric Propulsion* (NASA. Washington, DC: 1969)

Ley, W. *Rockets* (Viking Press. New York: 1944)

Ley, W. *Man-made Satellites* (Guild Press. Poughkeepsie, New York: 1957)

Ley, W. *Rockets, Missiles and Space Travel* (Viking Press. New York: 1957)

Ley, W. *Space Pilots* (Guild Press. Poughkeepsie, New York: 1957)

Ley, W. *Satellites, Rockets and Outer Space* (Signet. New York: 1958)

Ley, W. *Space Travel* (Guild Press. Poughkeepsie, New York: 1958)

Ley, W. *Man in Space* (L. W. Singer. Syracuse, New York: 1959)

Ley, W. *Mars and Beyond* (L. W. Singer. Syracuse, New York: 1959)

*Ley, W. *Rockets, Missiles and Men in Space* (Viking Press. New York: 1968)

*Ley, W. (with Chesley Bonestell). *The Conquest of Space* (Viking Press. New York: 1949)

Ley, W., et al. *Complete Book of Satellites and Outer Space* (Maco. New York: 1953)

Locke, G. (ed.). *Worlds Apart* (Cornmarket. London: 1972)

Locke, G. *Voyages in Space* (Ferret Fantasy. London:1975)

Mallan, L. *Mystery of Other Worlds Revealed* (Fawcett. New York: 1953)

Mallan, L. *Secrets of Space Flight* (Fawcett. Greenwich, CT: 1956)

Mallan, L. *Man into Space* (Fawcett. Greenwich, CT: 1960)

Mallan, L. *Space Science* (Arco. New York: 1961)

Maral-Viger. *Les Anneau des Feu* (Hachette. Paris: 1922)

McVey, J. W. *How We Will Reach the Stars* (Collier. Toronto: 1969)

*Miller, J. *The X-Planes* (Specialty Press. Marine on St. Croix, MN: 1983)

Miller, R. *Materials Toward a History of Spacecraft* (unpublished)

Mitchell, J. A. *Drowsy* (Frederick A. Stokes. New York: 1917)

Moore, P. *Earth Satellites* (W. W. Norton. New York: 1956, 1958)

Moskowitz, S. (ed.). *Science Fiction by Gaslight* (World. Cleveland, OH: 1968)

Murphy, L. (ed.). *Rockets, Missiles and Spacecraft of the National Air & Space Museum* (Smithsonian Institution. Washington, DC: 1976)

NASA. *A Forecast of Space Technology 1980–2000* (NASA. Washington, DC: 1979)

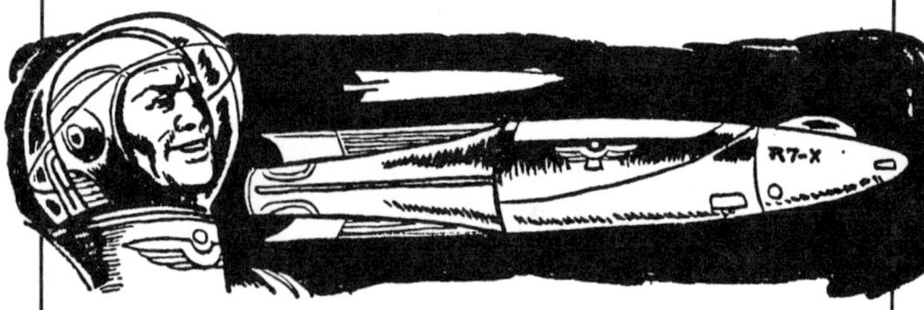

Nephew, W., and M. Chester. *Moon Base* (Scholastic. New York: 1959)

Nicholls, P. (ed.). *The Science in Science Fiction* (Alfred A. Knopf. New York: 1983)

Nicholson, M. H. *Voyages to the Moon* (MacMillan. New York: 1960)

Noordung, H. *Secrets of Space Flying* (*Science Wonder Stories*, vol. 1, no. 2, 3, 4, 1929)

Oberth, H. *Man into Space* (Harper's. New York: 1957)

Oberth, H. *The Moon Car* (Harper's. New York: 1959)

*Oberth, H. *Rockets in Interplanetary Space* (NASA. Washington, DC: 1965)

Oberth, H. *Ways to Spaceflight* (NASA. Washington, DC: 1972)

Office of Technology Assessment. *Round Trip to Orbit* (USGPO. Washington, DC: 1989)

*Ordway, F. I., and R. Liebermann. *Blueprint for Space* (Smithsonian Institution Press. Washington, DC: 1991)

*Parkinson, B., and R. A. Smith. *High Road to the Moon* (BIS. London: 1979)

Pendray, G. E. *The Coming Age of Rocket Power* (Harper. New York: 1944)

Piccard, A. *Entre Terre et Ciel* (Editions D'Ouchy. Lausanne: nd)

Pippin, E. *Space Opera* (Privately published: 1978)

Pizor, F. *The Man in the Moone* (Praeger. New York: 1971)

Poole, L. *Your Trip into Space* (McGraw-Hill. New York: 1953)

Powers, R. M. *Shuttle* (Stackpole. Harrisonburg, PA: 1979)

Pratt, F. *All About Rockets and Jets* (Random House. New York: 1955)

Pratt, F., and Jack Coggins. *Rockets, Jets, Guided Missiles and Spaceships* (Random House. New York: 1951)

Pratt, F., and Jack Coggins. *By Spaceship to the Moon* (Random House. New York: 1952)

Pseudoman, A. (E. F. Northrup). *Zero to Eighty* (Scientific Publishing Co. Princeton, NJ: 1937)

RAND Corporation. *Space Handbook* (USGPO. Washington, DC: 1959)

Reitsch, H. *Flying Is My Life* (Putnam. New York: 1954)

Rockwood, R. *Lost on the Moon* (Cupples & Leon. New York: 1911)

Rockwood, R. *Through Space to Mars* (Whitman. Racine, WI: nd)

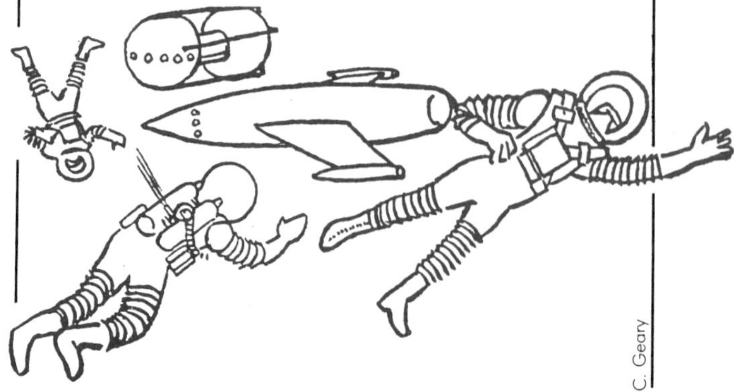

Volier 1929

Romick, D. *Meteor* (Goodyear Aircraft Corp. Akron, OH: 1956)

Ross, F. *Spaceships and Spacetravel* (Lothrop, Lee & Shephard. New York: 1954)

Russell, A. K. *Science Fiction by the Rivals of H. G. Wells* (Castle Books. Seacaucus, NJ: 1979)

Russell, jr., J. L. *Destination Space* (Popular Mechanics Press. Chicago: 1960)

Rycroft, M. (ed.). *The Cambridge Encyclopedia of Space* (Cambridge University Press. London: 1990)

*Rynin, N. A. *Interplanetary Communications* (9 volumes. NASA. Washington, DC: 1971)

Sänger, E. *Rocket Flight Engineering* (NASA. Washington, DC: 1965)

Sänger, E., and I. Bredt. *A Rocket Drive for Long-Range Bombers* (Robert Cornog. Santa Barbara, CA: 1952)

Schroeder, W. *First Stop: The Moon* (Odhams. London: 1959)

Serviss, G. P. *Edison's Conquest of Mars* (Carcosa House. Los Angeles, CA: 1947)

Serviss, G. P. *A Columbus of Space* (Hyperion. Westport, CT: 1974)

Shatalov, V. A., et al. *To the Stars* (Moscow: 1982)

Shelton, W. R. *Man's Conquest of Space* (National Geographic Society. Washington, DC: 1968)

Silverberg, R. *First American into Space* (Monarch. Derby, CT: 1961)

Simmon, S. *Planets and Space Travel* (Doubleday. New York: 1958)

Simons, D. G. *Man High* (Avon. New York: 1960)

*Simpson, T. (ed.). *Pioneering the Space Frontier* (Bantam. New York: 1986)

Siodmak, C. *Riders to the Stars* (Ballantine. New York: 1953)

Smith, E. P. *The Space Shuttle in Perspective* (AIAA. New York: 1975)

Smith, M. *An Illustrated History of the Space Shuttle* (Haynes. Newbury Park, CA: 1985)

Smith, R. A., and A. C. Clarke. *The Exploration of the Moon* (Harper. New York: 1954)

Sparks, J. C. *Winged Rocketry* (Dodd Mead. New York: 1968)

Stanek, B., and L. Pešek. *Space Shuttles* (Hallwag. Bern: 1975)

Sternfeld, A. *Interplanetary Travel* (Central Books. London: nd)

Coggins

H. Lanos, 1921

Stine, G. H. *Earth Satellites* (Ace. New York: 1957)

Stine, G. H. *Man and the Space Frontier* (Alfred A. Knopf. New York: 1962)

Strickland, A. W., and F. J. Ackerman. *A Reference Guide to American Science Fiction Films* (TIS Publications. Bloomington, IN: 1981)

*Synthesis Group. *America at the Threshold* (USGPO. Washington, DC: 1991)

Thomas, S. *Men of Space* (Hillman-McFadden. New York: 1960)

Time-Life (ed.). *To the Moon* (Time-Life. Chicago: 1969)

Tinsley, F. *The Answer to the Space Flight Challenge* (Whitestone. Louisville, KY: 1958)

Tolstoi, A. *Aelita* (Ardis. Ann Arbor, MI: 1985)

Tsander, F. A. *Problems of Flight by Jet Propulsion* (NASA. Washington, DC: 1964)

Tsiolkovsky, K. *The Call of the Cosmos* (Foreign Languages Publishng House. Moscow: nd)

Vaeth, J. G. *200 Miles Up* (Ronald Press. New York: 1951)

Valier, M. *Raketenfahrt* (R. Oldenbourg. Berlin: 1930)

*Verne, J. *From the Earth to the Moon and Round the Moon* (Scribner, Armstrong. New York: 1874

von Braun, W. *The Exploration of Mars* (Viking Press. New York: 1956)

Mr Jex, 1930

von Braun, W. *First Men to the Moon* (Holt, Rinehart & Winston. New York: 1960)

von Braun, W., et al. *Across the Space Frontier* (Viking Press. New York: 1953)

von Braun, W., et al. *The Conquest of the Moon* (Viking Press. New York: 1953

von Braun, W., and F. I. Ordway, III. *History of Rocketry and Space Travel* (Thomas Y. Crowell. New York: 1969)

von Braun, W., and F. I. Ordway, III. *The Rocket's Red Glare* (Anchor Press-Doubleday. New York: 1976)

Weinbaum, S. *A Martian Odyssey* (Hyperion. Westport, CT: 1974)

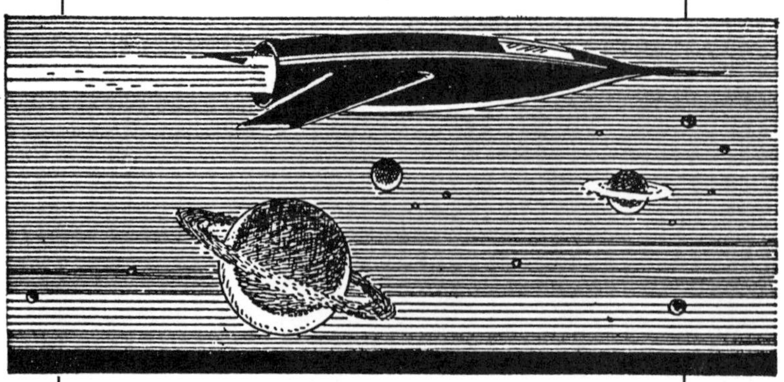

Wells, H. G. *First Men in the Moon* (MacMillan. London: 1920)

Wilding-White, T. M. *Jane's Pocket Book of Space Exploration* (Collier. London: 1976)

Williams, B., and S. Epstein. *The Rocket Pioneers* (Julian Messner. New York: 1955, 1958)

*Winter, F. *Prelude to the Space Age* (Smithsonian Institution. Washington, DC: 1983)

Winter, F. *The First Golden Age of Rocketry* (Smithsonian Institution. Washington, DC: 1990)

Winter, F. *Rockets into Space* (Harvard University Press. Cambridge, MA: 1990)

Woodbury, D. O. *Outward Bound for Space* (Little, Brown. New York: 1961)

Wooldridge, E. T. *Winged Wonders* (Smithsonian Institution. Washington, DC: 1983)

Zim, H. F. *Rockets and Jets* (Editions for the Armed Services. New York: 1945)

SPECIALTY MAGAZINES CONSULTED

Air & Space

Aviation Week & Space Technology

Cinefantastique

Filmfax

Final Frontier

Journal of the British Interplanetary Society

Missiles and Rockets

Rocket-Jet Flying

Space Journal

Space Age

Space World

Spaceflight

Spacemen

C. Schneeman, 1933

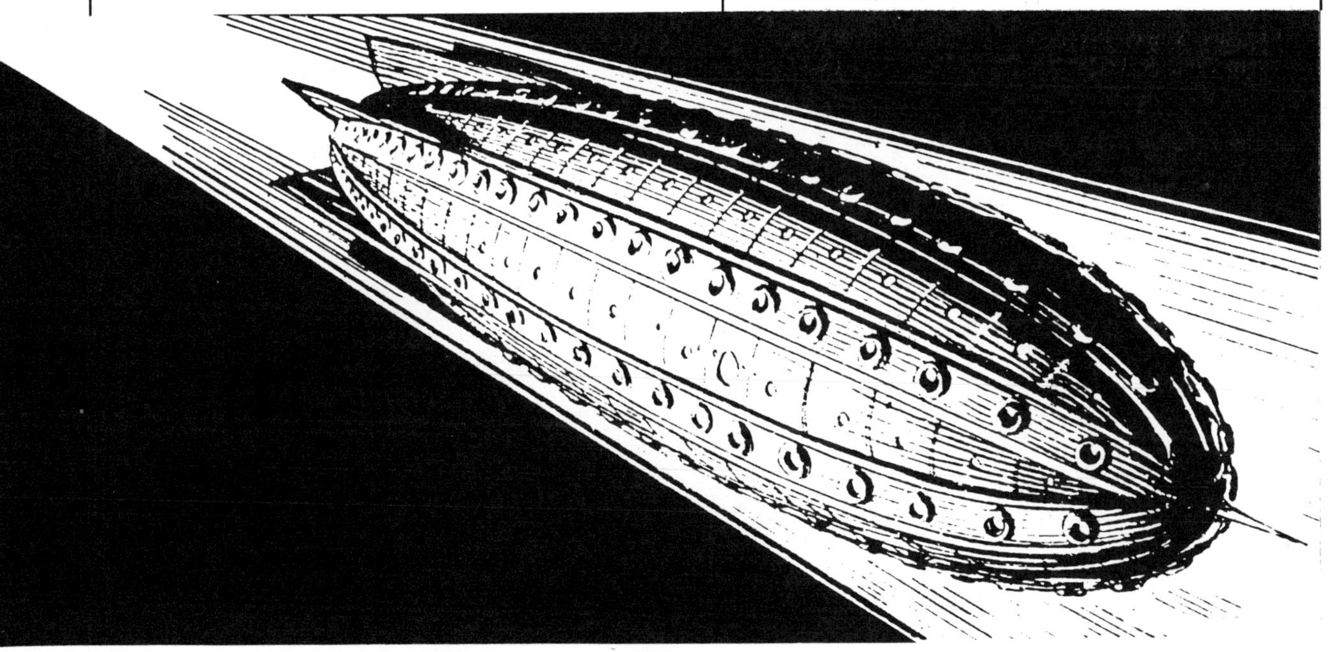

INDEX

PICTURE SOURCES

Forest J. Ackerman
Boeing Defense and Space Group (Boeing)
British Aerospace
Norman Brosterman
Bob Burns
Jack Coggins
Benjamin Donahue
Morris Scott Dollens
David Ducros
General Dynamics Space Systems Division
David A. Hardy
Gary Hudson
International Space University
Mat Irvine
Michael Wm. Kaluta
Saunders Kramer
Randy Liebermann
McDonnell Douglas
Martin Marietta Aerospace
Messerschmitt-Bolkow-Blöhm
NASA
NASDA
Frederick I. Ordway, III
Pat Rawlings
Darrell Romick
Hartmut Sänger
Bob Skotak
Smithsonian Institution/National Air & Space Museum (NASM)
Lee Staton
U.S. Space & Rocket Center (S&RC)
Charles P. Vick

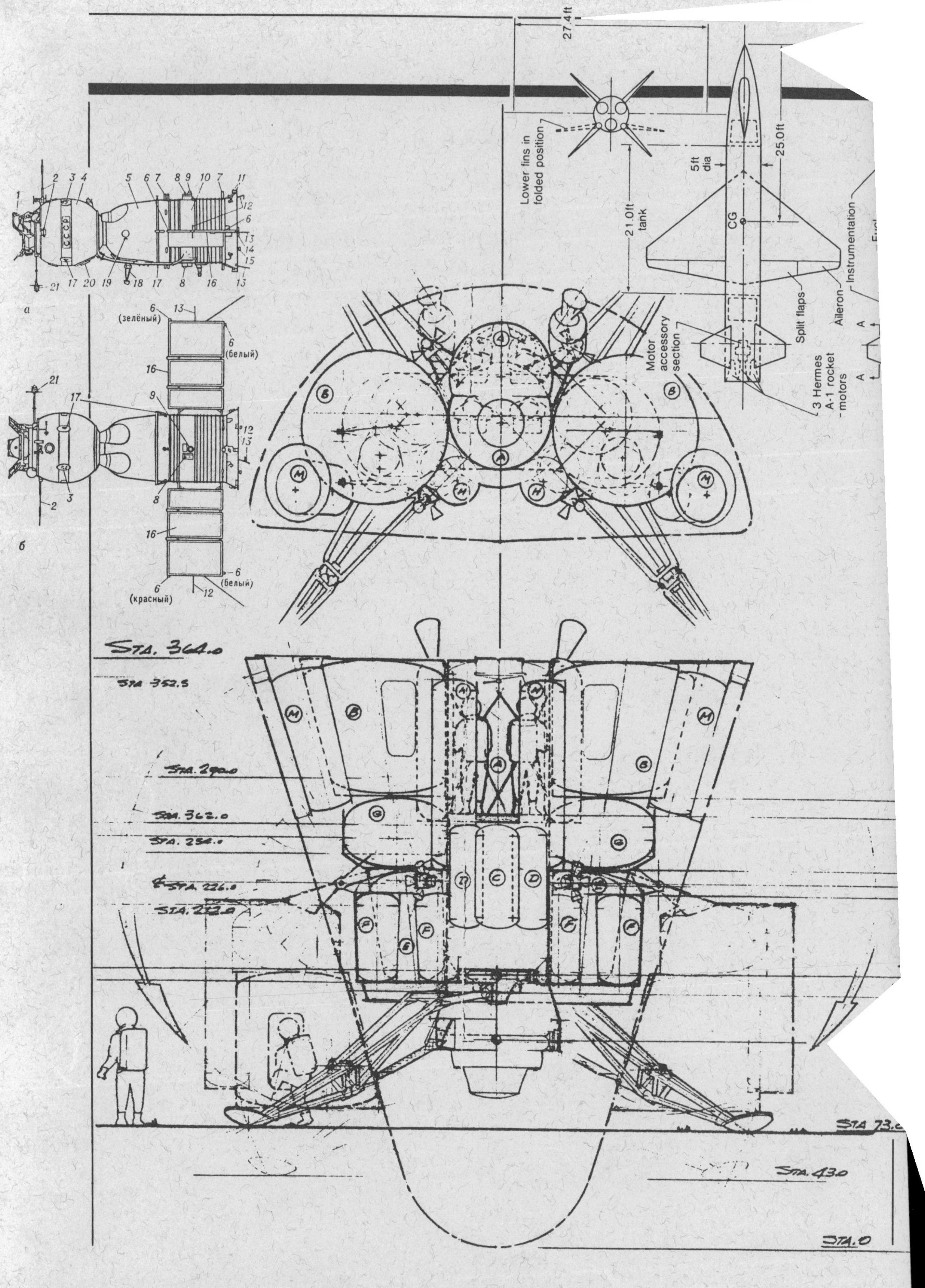